5th ICEG ENVIRONMENTAL GEOTECHNICS

Opportunities, Challenges and Responsibilities for Environmental Geotechnics

Volume II

Edited by Professor H. R. Thomas

Proceedings of the ISSMGE's fifth international congress organized by the Geoenvironmental Research Centre, Cardiff University and held at Cardiff City Hall on 26–30th June 2006

ICEG

BGA

igs

GRC
Geoenvironmental Research Centre
Canolfan Ymchwil Ddaearamgylcheddol

CARDIFF UNIVERSITY
PRIFYSGOL CAERDYDD

thomas telford

Published by Thomas Telford Publishing, Thomas Telford Ltd, 1 Heron Quay, London E14 4JD.
www.t-telford.co.uk

Distributors for Thomas Telford books are
USA: ASCE Press, 1801 Alexander Bell Drive, Reston, VA 20191-4400, USA
Japan: Maruzen Co. Ltd, Book Department, 3–10 Nihonbashi 2-chome, Chuo-ku, Tokyo 103
Australia: DA Books and Journals, 648 Whitehorse Road, Mitcham 3132, Victoria

First published 2006

A catalogue record for this book is available from the British Library

ISBN: 0 7227 3472 5

Printed and bound in Great Britain by MPG Books, Bodmin

Organization

ISSMGE Conference Advisory Committee
Prof. William Van Impe, President, ISSMGE
Prof. Neil Taylor, Secretary General, ISSMGE
Prof. Pedro Seco e Pinto, Vice-President ISSMGE Europe
Prof. Mario Manassero, Co-Chair ISSSMGE TC5
Prof. Luiz Guilherme de Mello, Chair ICEG4
Prof. Marcio Almeida, Secretary ICEG4
Prof. Hywel Thomas, Chair ICEG5
Prof. Steff Jefferis, Vice-Chair ICEG5

Organising Committee
Prof. Hywel R Thomas, Geoenvironmental Research Centre, Cardiff University – 5ICEG Chair
Prof. Jefferis, Environmental Geotechnics Ltd – 5ICEG Vice Chair
Dr David-Huw Owen, Cardiff University – 5ICEG Secretary
Ms Cherrie Summers, Cardiff University – 5ICEG Secretariat
Dr Andy Rees, National Waste Policy Unit, Welsh Assembly Government
Dr John Lewis, TRL Ltd, Wales Office
Mr Martin Lamb, TRL Ltd, Wales Office
Mr Steve L Smith, Welsh Development Agency
Dr Peter Cleall, Geoenvironmental Research Centre, Cardiff University

National Advisory Committee
Mr Jan Hellings, British Geotechnical Association
Prof. Michael Loxham, GeoDelft Environmental Services
Prof. Mark Dyer, Strathclyde University
Dr Kenichi Soga, Cambridge University
Dr Doug Stewart, Leeds University
Dr Murray Reid, Viridis, TRL Ltd
Dr Hugh Potter, Environment Agency
Ms Hilary Skinner, Building Research Establishment

TC5 Core Members
Prof. Marcio Almeida, University of Rio de Janeiro, Brazil
Prof. Akram Alshawabkeh, Northeastern University, Boston, USA
Prof. Ademalek Bouazza, Monash University, Melbourne, Australia
Prof. Evelina Fratalocchi, University of Ancona, Italy
Prof. Takeshi Katsumi, Kyoto University, Japan
Prof. Mario Manassero, Technical University of Torino, Italy
Prof. Charles D. Shackelford, Colorado State University, US
Prof. Peter O. Van Impe, University of Gent, Belgium

Preface

This book is a compilation of peer reviewed papers which were presented at the International Society of Soil Mechanics and Geotechnical Engineering's (ISSMGE) 5th International Congress on Environmental Geotechnics, held at Cardiff, Wales, UK in June 2006. Previous conferences in the series had been held in Edmonton, Canada (1994), Osaka, Japan (1996), Lisbon, Portugal (1998) and Rio de Janeiro, Brazil (2002). Issues associated with Environmental Geotechnics continue to be a major preoccupation for governments, public and private organisations and the general community around the world. This congress brings together people working in industry, academia and the public sector to discuss the latest ideas and developments in Environmental Geotechnics and related fields. The papers in these proceedings reflect the work being undertaken across the discipline.

The theme of the congress was 'Opportunities, Challenges and Responsibilities for Environmental Geotechnics'. The papers in this book are grouped according to the sessions under which they were presented at the congress: Remediation; Testing and Monitoring; Barrier Design including Nuclear Waste Disposal; Mine Sites, Tailing Dams, Sludge Ponds and Underwater Geoenvironmental Issues; Sustainability, Regulation and Risk Management; Fate and Transport Pollutants; and Waste Reuse and Waste Management. The papers have been submitted from authors from across the world and reflect the truly international flavour of the congress.

Additionally, the keynote paper on "Persistent Organic Pollution: Sustainable Solutions" and the three feature papers on "A survey of lethal failures in municipal solid waste dumps and landfills"; "Some Factors affecting Geosynthetics used for Geoenvironmental Applications"; and "The Use of Waste Materials in Construction – Progress Made and Challenges for the Future", are included in these proceedings.

The congress was organised by the Geoenvironmental Research Centre at Cardiff University, supported by the Transport Research Laboratory (TRL) and the Building Research Establishment (BRE). As stated above, the congress series is sponsored by the International Society for Soil Mechanics and Geotechnical Engineering (ISSMGE), supported by the International Geosynthetics Society (IGS) and the British member society of ISSMGE, the British Geotechnical Association (BGA). The congress was organised under the auspices of ISSMGE's Technical Committee on Environmental Geotechnics (TC5).

I should like to express my gratitude to ISSMGE for the honour of organising this congress and to thank the members of the ISSMGE Conference Advisory Committee: Professor William Van Impe; Professor Neil Taylor, Professor Pedro Seco e Pinto; Professor Mario Manassero; Professor Charles D Shackelford; Professor Luiz Guilherme de Mello; Professor Marcio Almeida and Professor Jefferis for their contributions to the overall planning and development of the congress.

I should also like to thank the members of the National Advisory Committee: Professor Mark Dyer; Dr Jan Hellings; Professor Michael Loxham; Dr Hugh Potter; Dr Murray Reid; Ms Hilary Skinner; Dr Kenichi Soga and Dr Doug Stewart: TC5 Core Members: Professor Akram Alshawabkeh; Professor Abdelmalek Bouazza; Professor Evelina Fratalocchi; Professor Takeishi Katsumi and Professor Peter O Van Impe: and the Organising Committee: Dr Martin Lamb; Dr John Lewis; Dr Andy Rees; Mr Steve Smith; Dr Peter Cleall; Dr David-Huw Owen and Ms Cherrie Summers for their contributions to the detailed planning and arrangements for the congress.

The assistance of all members of the committees with regard to the paper review process is also very gratefully acknowledged.

H R Thomas
Congress Chair

Contents

Chapter 4 – Mine Sites, Tailing Dams, Sludge Ponds, and Underwater Geoenvironmental Issues

Chapter 7 – Waste Reuse and Waste Management

MINE SITES, TAILING DAMS, SLUDGE PONDS, AND UNDERWATER GEOENVIRONMENTAL ISSUES

Investigations of Layered Cover Systems Acting as Oxygen Barriers to Limit Acid Mine Drainage

Michel Aubertin, Ecole Polytechnique, Montreal, Canada
John Molson, Ecole Polytechnique, Montreal, Canada
Bruno Bussière, UQAT, Rouyn-Noranda, Canada
Anne-Marie Dagenais, URSTM/UQAT, Rouyn-Noranda, Canada

Abstract
In recent years, a number of closure plans for tailings impoundments having an acid mine drainage (AMD) problem have involved the construction of a cover with capillary barrier effects (CCBE). Under a relatively humid climate, this type of cover system functions primarily as an oxygen barrier, due to a fine-grained material layer that remains nearly saturated. The technique has been applied to a few sites in Quebec, Canada, which have been instrumented and monitored to follow cover behaviour. A comparison of reactive transport simulations with in-situ collected data confirms the efficiency of CCBEs in limiting acid mine drainage. This paper presents a few recent cases where covers have been constructed on reactive tailings, including data gathered at the Manitou and LTA sites. Sample results are shown and discussed.

Introduction
The oxidation of sulphide minerals contained in mining wastes can induce the formation of acid mine drainage (AMD). The prevention and control of AMD remains a major environmental challenge for the industry. To limit acid formation, interactions between the constitutive elements of AMD (water, oxygen and sulphide minerals such as pyrite), must be impeded. Recent studies show that a cover with capillary barrier effects (CCBE) can be used to effectively limit the influx of oxygen under relatively humid climatic conditions. Hence, they have become an integral part of the closure plan for various mine sites in Canada, and elsewhere around the world (Aubertin et al. 2002; MEND 2004).

Designing a CCBE typically requires laboratory and field experiments combined with the application of numerical tools. This paper presents some advances made in this area through the authors' work, referring to recent cases where covers have been constructed on tailings impoundments. First, a brief review of the fundamental principles governing the behaviour of CCBEs is presented. It is followed by a description of field and modelling work that leads to the construction of large scale covers. The application of a full-scale CCBE on actual mine sites is then presented and briefly discussed.

Covers with Capillary Barrier Effects
Layered cover systems made of differently textured soils (and other similar media), which have been labelled "covers with capillary barrier effects – or CCBEs" by the authors (e.g. Aubertin et al. 1997; Bussière et al. 1998a), are raising considerable interest as they can be used instead of more traditional covers that rely on materials having a low saturated hydraulic

conductivity. Capillary barrier effects appear when unsaturated flow occurs in layered soils having different hydraulic properties (Nicholson et al., 1989; Collin and Rasmuson, 1990; Aubertin et al., 1995, 1999, 2006; Bussière et al. 2003a, 2003b). Under unsaturated conditions, the coarse-grained soil located under a fine-grained material tends to drain easily, hence reaching a low volumetric water content θ_w. The drained coarse material (sand or gravel) then has a low unsaturated hydraulic conductivity that limits downward water movement from the fine-grained soil layer above. This fine textured soil behaves as a reservoir where water can be stored by capillary forces. Its high degree of saturation makes it an efficient barrier against oxygen flow, by impeding gas advection and diffusion.

A typical CCBE contains three to five layers made of different materials. Each layer plays specific role(s). In the three-layer system shown in Figure 1, the bottom layer, made of a coarse-grained soil, acts as a mechanical support layer and as a capillary break layer that limits downward movement of water. It prevents desaturation of the moisture-retaining layer above, which is made of a fine-grained material. The top coarse-grained material layer is a drainage layer, which also serves to limit water loss by evaporation from the surface. Other layers may be added for protection against erosion and biointrusion (Aubertin et al., 1995).

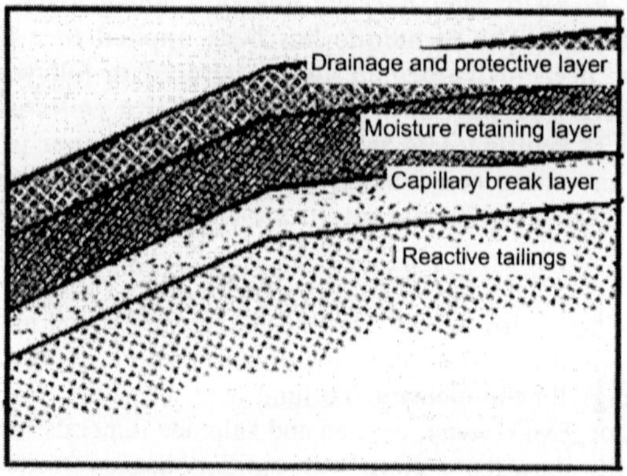

Figure 1. Basic configuration of a CCBE used to limit oxygen ingress to reactive tailings (Aubertin et al. 1997, 1999).

For more than 15 years now, the authors have been conducting a series of studies on the hydrogeological and geochemical behaviour of CCBEs for reducing AMD production. The initial phases of this investigation involved mainly laboratory work, which included characterization of the hydro-geotechnical properties of various soils and tailings and column tests to compare various cover scenarios (e.g. Aubertin et al. 1995, 1997; Bussière et al. 1998b; 2004). Column tests, combined with numerical simulations, were very useful to study the details of the hydrogeological and geochemical behaviour of layered systems, some of which include a water retention layer made of clean (non acid generating) tailings. The efficiency of these systems to limit AMD generation was evaluated by comparing covered columns to reference columns containing only reactive tailings. The results have also been used in conjunction with numerical modelling to better assess various cover scenarios.

Field Tests at the Manitou Site

Once confidence was gained in our ability to properly analyze and design CCBEs, and to predict their behaviour, field work was initiated. To complement the laboratory studies briefly mentioned above, a series of experimental cells were constructed in 1995 to evaluate, on a larger scale and under more realistic conditions, the performance of CCBEs using silty materials (natural silt or mine tailings) as the moisture-retaining layer (Aubertin et al. 1999). Figure 2a presents a view of one of the five covered cells at the end of the construction phase; the test plots were located at the Manitou mine site, near Val-d'Or, Quebec, Canada. The cells were monitored to follow the water content and suction in the cover layers, and also to assess the water quality of the leachate.

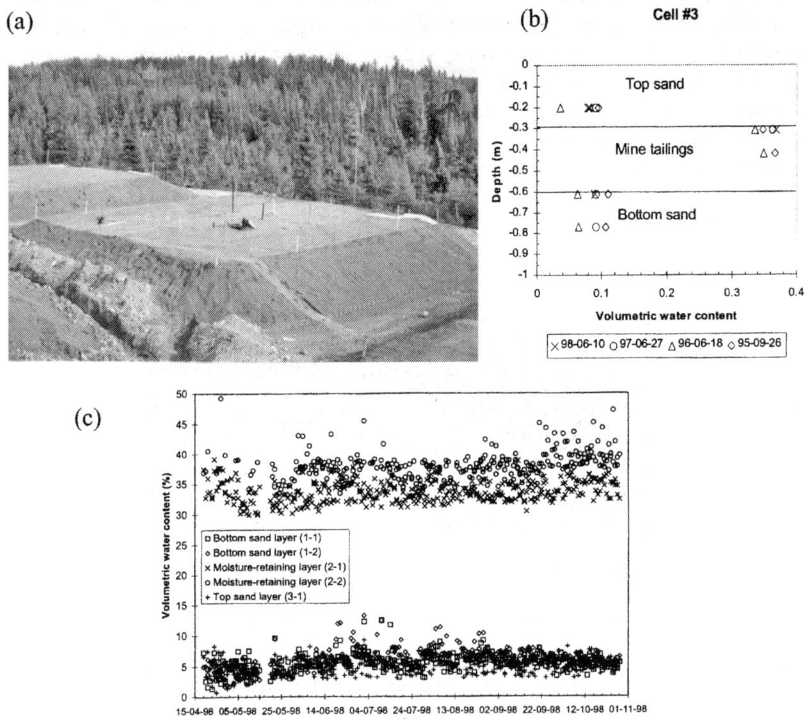

Figure 2. a) Experimental cells constructed at the Manitou mine site; b) typical values of measured water content in the 3-layer system; c) values of θ_w measured in 1998 in one experimental cell (after Aubertin et al. 1999; Bussière et al. 2005).

The field results confirmed the creation of capillary barrier effects in all cells, during four years of monitoring. Figures 2b and 2c present typical results, showing low θ_w in the sand layers creating the capillary effects, and high θ_w in the moisture retaining layer. The water quality of the leachate collected at the bottom of the cells (not shown here) confirmed the efficiency of the CCBE to limit acid mine drainage production: the pH was between 6 and 7 for the covered cells compared to a pH of less than 3 for the control (uncovered) cell. Sulphate concentrations and metal loadings were reduced by more than two orders of magnitude in the covered cells. The information gathered from these field plots over 4 years confirmed the efficiency of CCBEs for controlling AMD production.

Reactive Transport Simulation of a Field Cell

The data obtained from the test plots were compared to results obtained from numerical calculations. The initial simulations focussed mainly on the hydraulic behaviour of the covers (e.g. Aubertin et al. 1999; Bussière et al. 2003a). More recently, the lab and field data have been used in conjunction with a more advanced hydro-geochemical model to better assess the response of CCBEs. The MIN3P model (Mayer et al. 2002) was used to solve the governing equation for Darcy-type fluid flow in a variably-saturated porous medium. Assuming a passive air phase, the mass conservation equation for the water phase is solved by taking into account the aqueous phase (water) saturation, the specific storage coefficient, the hydraulic head, time, and material porosity. In this model, the non-linear functions defining water saturation (S_r) and relative permeability (k_r) are given by van Genuchten (1980), according to:

$$S_r = \frac{\theta_w - \theta_r}{n - \theta_r} = \left[\frac{1}{1 + (\alpha_v \psi)^{n_v}}\right]^m \qquad (1)$$

$$k_r = S_r^{\ell}\left[1 - \left(1 - S_r^{1/m}\right)^m\right]^2 \qquad (2)$$

where n is the porosity, θ_w is the water content, θ_r is the residual water content, ψ is the water pressure (suction); α_v, n_v and m are the van Genuchten (1980) model parameters, determined from the water retention curve ($m=1-1/n_v$), and ℓ is a parameter representing the degree of pore connectivity (ℓ=0.5 is used here).

The MIN3P model uses a general transport equation for advective-dispersive flux of multiple, dissolved phase components, including diffusive gas transport and reactive source-sink terms (from mineral precipitation and dissolution, phase exchange, etc.). Transport of oxygen through partially saturated porous media is governed by Fick's laws, using the effective diffusion coefficient D_e (e.g. Collin and Rasmuson, 1990; Mbonimpa et al. 2003). In this study, the authors have also used the shrinking core model within MIN3P to simulate diffusion-controlled pyrite oxidation assuming spherical grain geometry (Davis and Ritchie, 1986). Further details on the MIN3P model are given in Mayer et al. (2002).

One of the test cells constructed on the Manitou site (corresponding to field Cell 1) is analysed here. It is treated as a vertical 1D column 2.8 m long and discretized uniformly using 101 control volumes. The column is covered with a CCBE (Figure 3), with 1.5m of sulphide-containing tailings at the base. Three porous materials are included in the model: the reactive Manitou mine wastes, the fine-grained capillary barrier made of non-reactive tailings (from the neighbouring Sigma mine), and a non-reactive sand. Material properties affecting the moisture-retention behaviour for each layer are also provided in Figure 3.

The calculations considered 12 aqueous components, 2 redox pairs, 21 secondary species, 2 gases and 10 minerals (Molson et al. 2004). Mineralogical analyses of the Manitou sulphide-bearing tailings by X-ray diffraction showed an abundance of quartz (60%), followed by sericite (15%), chlorite (9%), pyrite (5.5%), feldspars (5%) and gypsum (3%) (Aubertin et al. 1999). Traces of carbonates were also found. Mineralogical data for the cover materials were less precise but included quartz and feldspars. Carbonate content, including calcite and ankerite, was estimated at up to 10%. The primary sulphide mineral in the Manitou tailings is assumed to be pyrite, with the overall oxidation reaction stoichiometry given by:

$$FeS_{2(s)} + 7/2\ O_2 + H_2O \rightarrow Fe^{2+} + 2SO_4^{2-} + 2H^+ \qquad (3)$$

Pyrite is assumed to be uniformly distributed within the Manitou waste at a volume fraction of 0.05. The initial unoxidized grain diameter is taken to be 0.10mm, corresponding to the D_{30} fraction (Aubertin et al. 1999). To simplify the conceptual model, and to focus on the effect of the capillary barrier, the authors assumed, with the exception of the sulphide content, identical mineralogy for the Manitou tailings, Sigma tailings, and sand layers. Where the observed mineral fractions differed, an average was used. Pyrite was only included in the Manitou tailings.

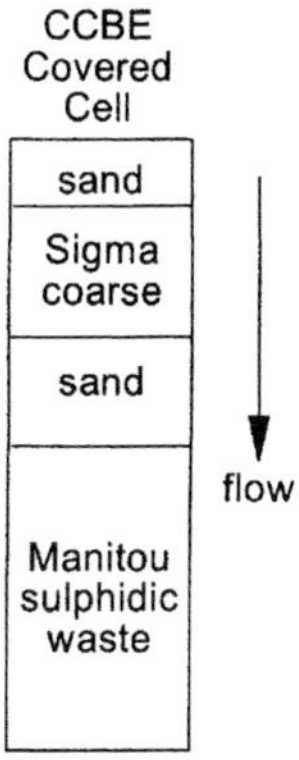

Material	n	θ_r	k_{sat} m/s	α_v, n_v, l (1)	AEV (m)
Sand	0.45	0.1	7.2×10^{-5}	2.9,10.2,0.5	0.28
Sigma	0.41	0.08	1.5×10^{-6}	0.21,2.6,0.5	2.75
Manitou	0.42	0.07	2.7×10^{-5}	0.61,2.6,0.5	2.65

Figure 3. Left: Structure of the covered cell as incorporated in the 1D numerical model, with material properties (after Aubertin et al., 1999). The "Sigma" material was used as the moisture retaining layer, while the Manitou unit formed the reactive tailings; Right: Corresponding material properties: (1) van Genuchten parameters; AEV = Air Entry Value.

The modelled column is open at the top and bottom, and the sides are impermeable. A steady recharge of 0.2 m/yr is applied at the top inflow boundary, and is assigned a typical chemical composition of rainwater which has been equilibrated with atmospheric oxygen (Mayer et al., 2002). At the top surface, oxygen is fixed at its atmospheric concentration (0.2 mg/L_{air}). The water pressure is fixed at the base at 0.0 m, allowing free drainage. The column was assumed initially water saturated and oxygen-free. Effects of temperature are not considered. The dispersivity for each material is assumed 0.5 mm, and the free diffusion coefficients are assumed constant for all components in water (2.38×10^{-9} m^2/s), and for oxygen in air (2.07×10^{-5} m^2/s). The model was run for a total of 1200 days. Execution times were typically on the order of 10-15 minutes. The results for this case are provided in Figure 4, showing vertical profiles of degree of saturation, oxygen concentration, and pH. Other simulated parameters, such as mineral volume fractions and aqueous component concentrations, are not shown here because of space limitations.

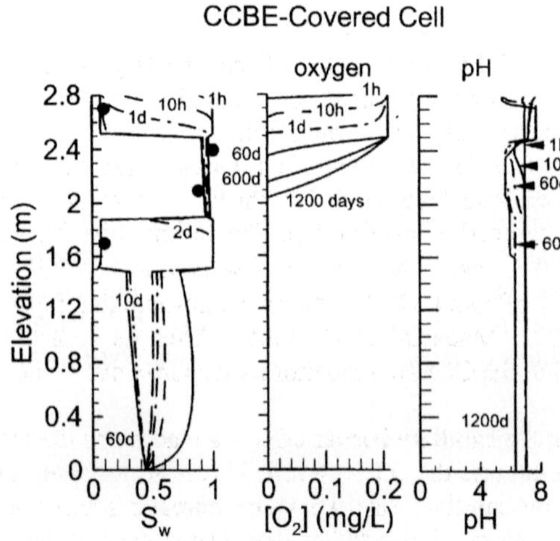

Figure 4. Vertical profiles of the MIN3P simulation results of a covered cell showing changes in saturation (field data also shown), oxygen concentration, and water pH.

The transient response over the first 60 days shows that the water drains rapidly from the two sand layers, while the moisture retention layer (located between 1.9 and 2.5 m elevation) remains close to saturation. The degree of saturation within the reactive Manitou tailings (~0.4-0.5) is above that of the sand (0.2), but remains significantly less than that of the moisture retaining layer in the CCBE. Within 2 days of the start of drainage, the oxygen front has advanced about 0.5 m from the top surface. However, as the front enters the nearly-saturated layer, the diffusive flux is significantly reduced. After 1200 days, for example, the front has only advanced 0.8 m from the top surface. The higher water saturation in the fine-grained material layer has thus significantly reduced the flux of oxygen into the column. In this case, the pH has remained near neutral at about 6.5-7. Other results (not presented here) indicate that there is no pyrite-dissolution front, and there is no significant increase in concentrations of Fe and SO_4 in the seepage. These results compare well with the corresponding field and lab data (Molson et al. 2004).

Additional simulations were also performed for the cell with Manitou tailings without a CCBE. These have shown that the oxygen entering the reactive tailings is rapidly consumed due to pyrite oxidation. As the pyrite is oxidized, a low-pH zone develops. This low-pH water is carried downwards with the natural infiltration, and the low-pH zone gradually widens. The minimum simulated pH (about 1) was somewhat lower than that observed (pH~2-3); the difference was attributed to incomplete background mineralogy in the model, and to suppressed oxidation during the cold winter season (which was not considered in the calculations). The oxidation of pyrite releases aqueous iron and sulphate which then react with other species. These simulated results compared fairly well with those observed in the uncovered field cell (Molson et al. 2004). Overall, the results gathered from the Manitou study have confirmed the efficiency of CCBEs as oxygen barriers. The study has also shown that the available tools can be used to evaluate the field performance of such types of covers, which aim to prevent AMD.

The LTA Cover

The cover constructed at the LTA site was one of the first large scale covers with capillary barrier effects constructed in a humid climate on an acid generating tailings impoundment. The LTA site is located near Malartic, in the Abitibi region (Quebec, Canada). This 60-hectare cover is made of three layers (shown in Figures 1 and 5): a capillary break layer 0.5 m thick, a moisture-retaining layer of 0.8 m thick made of non acid-generating MRN tailings, and a 0.3 m thick protection layer on top. The design objective was to reduce oxygen migration through the CCBE, based on a degree of saturation $S_r \geq 85\%$ (for a porosity n of 0.44, this corresponds to θ_w of about 37%) in the moisture-retaining layer (Ricard et al. 1999; Bussière et al, 2003c, 2005; Maqsoud et al. 2005). After its construction in 1996-1997, instruments were installed in the CCBE to monitor its response over time.

The field data indicate that the capillary barrier effects are active in the cover where the water table is deep enough to desaturate the coarse grained layer at the base. Otherwise, the cover remains nearly saturated by capillary rise when the phreatic surface is close to or in the CCBE. Figure 5 shows values of the volumetric water content obtained at one of the monitoring stations (CS96-1) where the water table is deep. One can see that θ_w of the sand layer is low, while that of the moisture-retaining layer remains high (with a degree of saturation above 85%; Bussière et al. 2003c). This is typical of observations made on the main covered area of the LTA tailings impoundment. Such high water contents have been associated with a very low oxygen flux through the cover (Bussière et al. 2003a, 2003c).

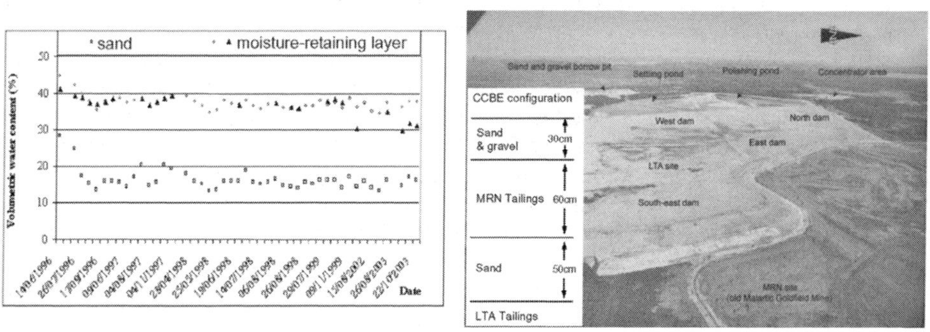

a) b)

Figure 5. a) Volumetric water content measured in the sand layer and the moisture-retaining layer at the LTA site, station CS96-1 (adapted from Bussière et al., 2003c), and b) aerial picture of the LTA site (modified from Golder and Associates 1999).

It is also worth mentioning here that a small aerial portion of the cover constructed on the LTA site is located on the sloping side of the tailings dams. The response of a CCBE is affected by its inclination. It is usually more difficult to maintain the desired θ_w in the moisture-retaining layer along the upper parts of a sloping cover, and this possible desaturation can influence the performance of the cover (Aubertin et al. 1999; Bussière et al. 2003a). A numerical study based on the LTA site was performed to quantify this effect. The results showed that a possible desaturation may sometimes occur near the upper part of the cover along the slope of the LTA site. This effect is, however, temporary, and was not deemed detrimental to the cover performance. Many other lessons have been learned from the

LTA site (Bussière et al. 2003a, 2003c; Maqsoud et al. 2005) and these are now being used to improve the design of new cover systems

Final Remarks

Over the past few years, several additional covers have been constructed in Canada based on the principles described above (e.g. Aubertin et al. 2002; Dagenais et al. 2005a, 2005b; see also the MEND, 2004, report on covers). Field observations on other sites have also confirmed the good performance of well designed CCBEs. The results summarized very briefly in this paper show how covers with capillary barrier effects placed on reactive tailings under relatively wet climatic conditions can be very effective in controlling the oxygen flux, and thus the production of AMD. On the other hand, under arid and semi-arid climates, it would be difficult to maintain a layer of the cover close to saturation at all times. Under these climatic conditions, the cover is subjected to extended dry periods, and the effect of evaporation (and evapotranspiration) would render it ineffective as an oxygen barrier. In this case, a CCBE would aim at controlling the influx of water to the reactive waste rock or tailings, by storage and diversion before its release (e.g. Zhan et al. 2001; Martin et al. 2005). This is another effective way of using CCBEs to control AMD.

Acknowledgements

The authors would like to thank the participants of the NSERC Polytechnique-UQAT Research Chair in Environment and Mine Wastes Management (http://www.polymtl.ca/enviro-geremi/). In particular, the collaboration of Barrick Gold and Golder Associates (for the LTA site) is gratefully acknowledged.

References

Aubertin, M., Aachib, M., Monzon, M., Joanes, A-M., Bussiere, B. and Chapuis, R. P. (1997). Étude de laboratoire sur l'efficacité des barrières de recouvrement construites à partir de résidus miniers, MEND Report 2.22.2b, Ottawa.

Aubertin, M., Bussière, B., Bernier, L. (2002). "Environnement et gestion des rejets miniers". [Manuel sur CD-ROM] Montreal, Qc : Les Presses Internationales de Polytechnique.

Aubertin, M. Bussière, B. Monzon, M. Joanes, A.M., Gagnon, D. Barbera, J.M., Aachib, M. Bédard, C. Chapuis, R.P., and Bernier, L. (1999). Étude sur les barrières sèches construites à partir de résidus miniers : Phase II. MEND Report 2.22.2c, Ottawa.

Aubertin, M., Chapuis, R. P., Aachib, M., Bussière, B., Ricard, J.-F., and Tremblay, L. (1995). Évaluation en laboratoire de barrières sèches construites à partir de résidus miniers, MEND Report 2.22.2a, Ottawa.

Aubertin, M., Cifuentes, E., Martin, V., Apithy, S., Bussière, B., Molson, J., Chapuis, R. P., and Maqsoud A. (2006). An investigation of factors that influence the water diversion capacity of inclined covers with capillary barrier effects, Proc.: Unsaturated Soils Conference 2006, ASCE, Carefree, AZ (to be published).

Bussière, B., Benzaazoua, M., Aubertin, M. and Mbonimpa, M. (2004). A laboratory study of covers with capillary barrier effects made of low sulphide tailings to prevent acid mine drainage. Environmental Geology, 45 : 609-622.

Bussière, B., Aubertin, M. and Chapuis, R.P. (2003a). The behaviour of inclined covers used as oxygen barriers. Canadian Geotechnical Journal, 40 : 512-535.

Bussière, B., Apithy, S., Aubertin, M. and Chapuis, R.P. (2003b). Diversion capacity of sloping covers with capillary barrier effects. Proc. 56th Annual Canadian Geotechnical Conference, Winnipeg, Canada, 9 pages, [CD-ROM].

Bussière, B., Maqsoud, A., Aubertin, M., Martschuk, J., McMullen, J., and Julien, M. (2003c). Results from the monitoring program at the LTA site: Hydraulic behaviour of the cover, Proc. CIM Conf. (CD-ROM), Montréal.

Bussière, B., Aubertin, M., Morel-Seytoux, H.J. and Chapuis, R.P. (1998a). A laboratory investigation of slope influence on the behaviour of capillary barriers. Proc. 51st Canadian Geotechnical Conference, Edmonton, 2 : 831-836.

Bussière, B., Benzaazoua, M., Aubertin, M., Lelièvre, J., Bois, D. and Servant, S. (1998b). Valorisation des résidus miniers : une approche intégrée – Phase II. Final Report submitted to the Ministère des Ressources naturelles du Québec : 198 p. (in French).

Bussière B., Maqsoud A., Aubertin M., Martschuk J., McMullen J., and Julien M. (2005) Cover performance at the LTA site, Malartic, Quebec, Canada. (submitted)

Collin M., and Rasmuson, A. (1990). Mathematical modeling of water and oxygen transport in layered soil covers for deposits of pyritic mine tailings, Acid Mine Drainage : Designing for Closure. GAC-MAC Annual Meeting, 311-333.

Dagenais, A.-M., Aubertin, M., Bussière, B., and Cyr, J. (2005a). Performance of the Lorraine mine site cover to oxygen migration, Proc. SME Annual Meeting, (CD-ROM), Salt Lake City.

Dagenais, A.M., Aubertin, M. Bussière, B., and Martin, V. (2005b). Large scale applications of covers with capillary barrier effects to control the production of acid mine drainage. Post-Mining 2005, Nancy France (to be published).

Davis, G.B., Ritchie, A.I.M., (1986). A model of oxidation in pyritic mine waste. Part 1: equations and approximate solution. *Appl. Math. Model.* 10, 314-322.

Golder Associates (1999). Field performance of the Les Terrains Aurifères. MEND Report 2.22.4.

Martin, V., Aubertin, M., Zhan, J. Bussière, B. and Chapuis, R.P. (2005). An Investigation into the Hydrological Behaviour of Exposed and Covered Waste Rock Dumps. Proc. SME Annual Meeting and Exhibit. Salt Lake City, USA. 10 pages, [CD-ROM]

Mayer, K.U., Frind, E.O., and Blowes, D.W., (2002). Multicomponent reactive transport modeling in variably saturated porous media using a generalized formulation for kinetically controlled reactions. *Water Resour. Res.* 38(9), 13:1-21.

Maqsoud, A., Bussière, B., Mbonimpa, M., Cissokho, F., Aubertin, M. (2005) Suction break to control slope-induced moisture variation in covers. Proc. GeoSask 05 (CD-ROM)

Mbonimpa, M., Aubertin, M., Aachib, M., and Bussière, B. (2003). *Diffusion and Consumption of Oxygen in Unsaturated Cover Materials*, Canadian Geotechnical Journal, 40, 916-932.

MEND (2004). *Design, construction and performance monitoring of cover systems for waste rock and tailings,* Report 2.21.4, Ottawa.

Molson, J.W., Aubertin, M., Bussière, B., and Joanes, A.-M. (2004) Simulating acid mine drainage through mine wastes constructed with capillary barriers, Proceedings: *Quebec 2004: 5th Joint IAH-CGS Conference*, Quebec, October.

Nicholson, R. V., Gilham, R. W., Cherry, J. A., and Reardon, E. J. (1989). *Reduction of acid generation in mine tailings through the use of moisture-retaining cover layers as oxygen barriers*, Canadian Geotechnical Journal, 26, 1-8.

Ricard, J.-F., Aubertin, M., Pelletier, P., and Poirier, P. (1999). *Performance of a dry cover made of tailings for the closure of Les Terrains Aurifères site, Malartic, Québec, Canada*, Proc. Sudbury'99 Conf., Mining and the Environment II, Vol.1, 155-164.

van Genuchten, M. Th., (1980). A closed-form equation for predicting the hydraulic conductivity of unsaturated soils. *Soil Science Society of America Journal*, 44: 892-898.

Zhan, G., Aubertin, M., Mayer, A., Burke, K., McMullen, J. (2001). *Capillary cover design for leach pad closure*, Transaction: SME Annual Meeting, Denver, pp. 104-110.

Determination of Slimes Consolidation Parameters from Instrumented Column Measurements

Bourgeois, F., BRGM, BP 6009, 45060 Orléans Cedex 2 - France
Le Bras, G., Institut de Recherche en Génie Civil et Mécanique - Centrale Nantes, BP 92101, 44321 Nantes Cedex 3 - France
Alexis, A., Institut de Recherche en Génie Civil et Mécanique - Centrale Nantes, BP 92101, 44321 Nantes Cedex 3 - France

Abstract

Prediction of the consolidation behaviour of fine-grained saturated materials is of significant interest in a wide range of industrial and environmental fields. It is classically achieved using the *s'(e)* and *k(e)* constitutive laws that relate effective stress *s'*, permeability *k* and void index ratio *e*. This work compares two approaches for estimation of the consolidation laws from instrumented column measurements: a direct method and a reverse method. The former requires that successive void index ratio *e(h)* and excess pore pressure *u'(h)* time profiles be converted into a set of *(s';e)* and *(k;e)* data points, from which the *s'(e)* and *k(e)* parameters can be determined by direct regression. With the latter, the *s'(e)* and *k(e)* parameters are determined by matching the measured *e(h)* and *u'(h)* profiles with those predicted using a finite element consolidation simulation code. It is found that the reverse method, which is ideally suited for simulation work yields more accurate and less biased parameter estimates.

Introduction

Prediction of the consolidation behaviour of fine-grained saturated materials is of significant interest in a wide range of industrial and environmental fields. It is classically achieved using the *s'(e)* and *k(e)* consolidation laws that relate effective stress *s'*, permeability *k* and void index ratio *e*. There are several ways whereby consolidation laws can be assessed for such materials. A number of geotechnical research and development laboratories throughout the world utilise specially instrumented columns. Such a non-destructive measurement technique, which leaves the sediment unperturbed, is particularly adapted for measuring density and interstitial water pressure on soils with high water saturation levels (Been and Sills, 1981). Typically, such apparatus yield measurements of void index ratio *e(h)* and excess pore pressure *u'(h)* profiles at desired times during consolidation, where *h* is the elevation within the consolidating material. This information represents fundamental information about the mechanical behaviour of the material, from which the consolidation laws can be extracted in a number of ways. This article compares two distinct consolidation parameter estimation approaches, a direct method that is classically used by geotechnicians, and a proposed reverse method that utilises computer simulation of the consolidation process.

Experimental work

Measurement of sediment consolidation using an instrumented column

The measurement technique used in this work relies upon an instrumented column developed at the *Institut de Recherche en Génie Civil et Mécanique* (Alexis et al., 2004; Gallois et al., 1994). Through continuous vertical scans of a column of sediments undergoing consolidation, the apparatus permits simultaneous measurements of void index ratio and excess pore pressure profiles. The experimental rig consists in a 1.5-meter high by 90 mm I.D. transparent PVC column. Interstitial pore water pressure profiles *u'(h)* are acquired using a set of 10 high-precision interstitial water pressure gages that are spaced regularly along the column height *h*. The practical accuracy of the pressure measurement is 3 Pa (Le Bras, 2000). Void index ratio profiles *e(h)* are obtained using a gammadensimeter mounted on a travelling platform that slides up and down along the column at a controlled speed. It has been established that the accuracy of the density measurement is better than 0.005 kg/m^3 (Le Bras, 2000).

Samples characteristics and handling

The test sample is a mixture of water and quarry slimes from an operating quarry in the Loire region in France. The mineral composition of the slimes comprises illite (13wt%), smectite, sepiolite and attapulgite (23wt%), kaolinite (9wt%), calcite (31wt%), quartz (5wt%), feldspars and mica (15wt%). In order to test the consolidation parameter estimation methods for radically different consolidation behaviours, the water-slimes mixture, or slurry, was tested with and without flocculent addition. Where flocculated, a medium molecular weight slightly anionic acrylate-acrylamide co-polymer flocculant was used (Bourgeois et al., 2003). The effect of flocculating the slimes on their physical properties is dramatic. For example, the median particle diameter of the raw slimes is 3 µm, against 150 µm for the flocculated slimes.

Prior to starting the experiment, the 1.5m high instrumented column is filled with a homogeneous water-slimes mixture with known solids concentration (See Table 1). After the mixture has been transferred completely into the column, it is homogenised manually and left to sediment and consolidate under gravity for a period of 32 days. During this period, void index ratio (alt. density) and excess pore water pressure profiles are recorded at a number of pre-set times.

Table 1. Sample characteristics

Sample	Initial solids concentration (g/l)	Initial specific gravity	Initial void index ratio e_0
Unflocculated	100	1.060	27.6
Flocculated	130	1.085	19.0

Experimental measurements

Figure 1 shows the profiles measured with the instrumented column.

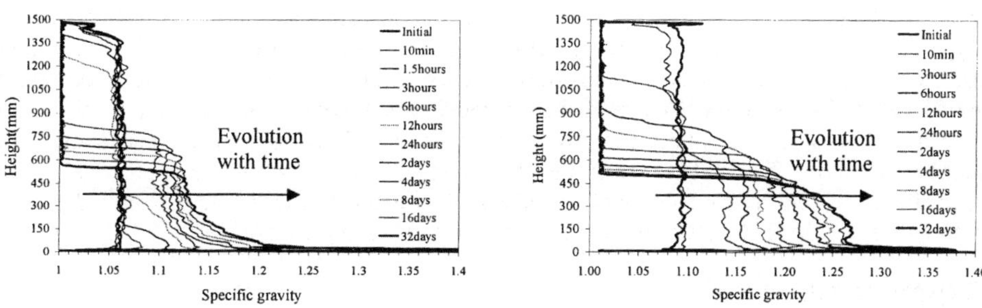

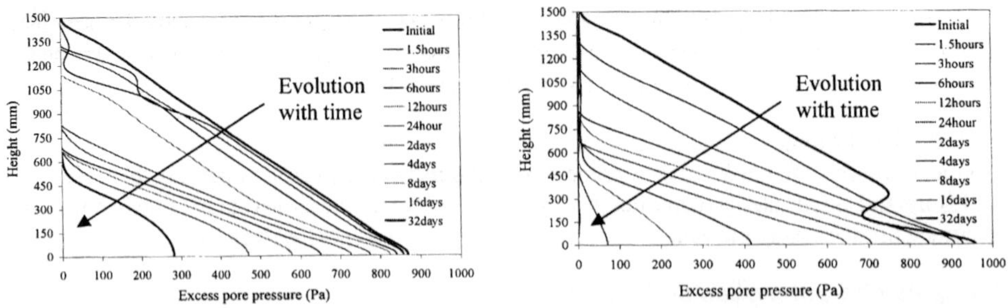

(a) Unflocculated sample (b) Flocculated sample

Figure 1. Specific gravity and excess pore water pressure profiles measured for unflocculated and flocculated samples.

The density profiles show that the flocculated sample settles at a much higher rate than the unflocculated one, which is explained by the significantly faster dissipation rate of excess pore water pressure with the flocculated sample. It is outside the scope of the present article to comment further on the dramatically different behaviour of the two samples. At any rate, these observed behaviours reflect very distinct consolidation properties, which is ideal for testing two consolidation parameters estimation techniques.

Determination of consolidation laws

Several analytical expressions can be found to describe the consolidation laws *s'(e)* and *k(e)* that constitute the basis for predicting the consolidation behaviour of a sediment. Here, the following expressions are used (Abu-Hejleh and Znidarcic, 1995):

$$e = A \cdot (s' + Z)^B \quad (1)$$

$$k = C \cdot e^D \quad (2)$$

The consolidation constants A, B, C, D and Z characterize the mechanical behaviour of the sediment. The extended power law given by Equation (1) is such that $s' = 0$ for all void index ratios greater than a critical value e_{max} given by:

$$e_{max} = A \cdot Z^B \quad (3)$$

Direct estimation of consolidation parameters by regression

Presentation of the method

One classic approach (Gallois et al., 1994; Gallois, 1995) consists in converting the raw void ratio *e(h)* and excess pore water pressure *u'(h)* profiles into a set of *(s';e)* and *(k;e)* data points, and then fitting the consolidation equations (1) and (2) to these data points by regression analysis. The *(s';e)* data points are calculated at elevation *h* using Terzaghi's postulate, i.e. by substracting the hydrostatic pressure and the measured excess pore water pressure from the total stress. It is estimated that the accuracy of the effective stress is 60 Pa. Calculation of the *(k;e)* data points involves using successive *e(h)* time profiles. One starts with calculating the total stress profile by integration of the mass conservation for the solids along the column height. By comparing consecutive time profiles, local values for the time derivative of the total stress are found, from which the solids relative velocity is obtained. Finally, the local permeability is found as the ratio between solids relative velocity and hydraulic gradient, noting that hydraulic gradient is obtained from the derivative of the excess pore water pressure with respect to height in the column (Gallois, 1995).

Results and analysis

Application of the above-described procedure to the measured profiles from Figure 1 yielded the data points *(s';e)* and *(k;e)* plotted in Figure 2. The corresponding values of the estimated parameters are given in Table 2.

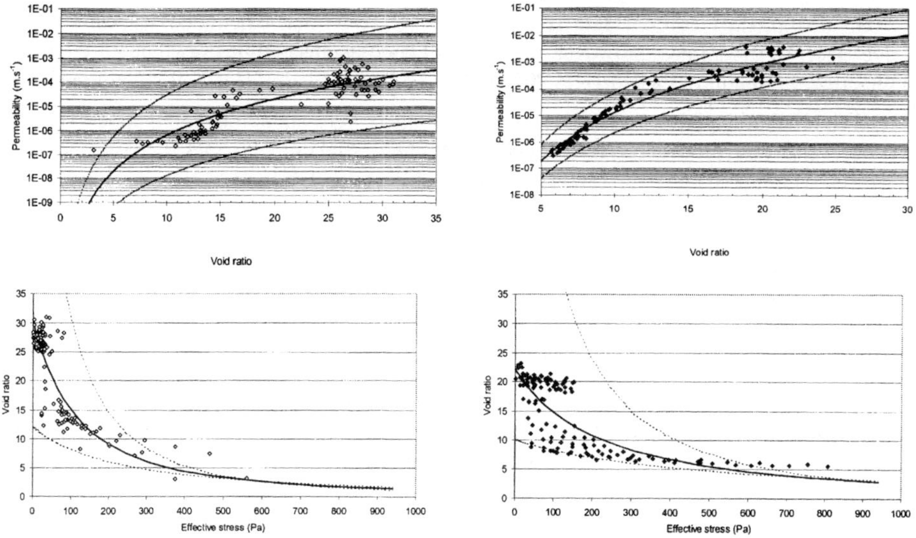

a) Unflocculated sample b) Flocculated sample

Figure 2. Values of *(s';e)* and *(k;e)* data points obtained with the instrumented column. (Symbols = experimental data points; solid line = best fit of equations (1) and (2); Dotted lines = 95% non-linear confidence limits on the estimates of equations (1) and (2)).

Table 2. Best estimates of the consolidation parameters obtained with the direct parameter estimation method (coefficients of equations (1) and (2) are given for *k* in m/s and *s'* in kPa).

Consolidation parameter	Unflocculated sample	Flocculated sample
A	1.80	4.05
B	-1.82	-1.48
C	6.89×10^{-12}	1.03×10^{-11}
D	4.98	6.12
Z	0.212	0.316
e_{max}	30.3	22.2

An interesting feature of these graphs is the clustering of the data points. This comes from the consolidation gradient inside the column of sediments. Since the void index ratio does not vary linearly with height, a vertical profile always consists of fewer data points at low void index ratios, i.e. at high consolidation density. Analysis of all the measured profiles simply amplifies this phenomenon, giving rise to a significant clustering of data points at high void index ratios. It is worth noting that the flocculated sample does not exhibit as strong a clustering as the unflocculated one. This difference is the direct consequence of the relatively greater uniformity of the density profile obtained with the flocculated sample, as seen in Figure 1.

Results and analysis

For estimation of the consolidation parameters with the direct method, the authors chose to use the Law and Bailey (1963) non-linear least-squares parameter estimation method (Moys and King, 1973) for calculating the non-linear confidence intervals on the estimates. The results are shown in Figure 2, and the best estimates are given in Table 2. Observation of the results confirm that the calculation method for determining permeability k and effective stress s' from instrumented column data is reliable in the sense that calculated values have the correct order of magnitude. However, a number of conclusions can be drawn about the limits of the direct estimation method:

- The significant experimental scatter of the *(k;e)* and *(s';e)* data points that is inherent to the data acquisition procedure itself makes parameter estimation by this method unable to extract accurate parameter estimates. Indeed, the non-linear regression analysis yields broad confidence limits across the measured range of void ratios for both permeability and effective stress constitutive equations. Such relationships, when used in consolidation simulation, would yield predictions with correspondingly broad uncertainties.
- Due to the data acquisition technique described earlier, the data are not distributed evenly over the range of void ratios. There are two distinct clusters of data located at high and low void ratio values respectively. Observation of Figure 2 clearly shows that the significant group of data at high void index ratio and low effective stress values forces the regression to significantly underestimate the void ratio at high effective stress values. When used in simulation, such a fitted constitutive law would invariably overestimate the degree of compaction of the sediment, since low values of void ratios equate to high sediment compaction. Moreover, this uncertainty yields difficulties in estimating the void index ratio e_{max} at zero effective stress, which is an important consolidation characteristics.

Reverse estimation of consolidation parameters by simulation

Presentation of the method

The reverse parameter estimation method consists in estimating the set of consolidation constants A, B, C, D and Z from equations (1) and (2) that yield simulated $h_{sim}(t)$ and $u'_{sim}(t)$ profiles that best match the measured profiles $h(t)$ and $u'(t)$. In this work, simulated profiles are obtained using the simulation package CADA (Abu-Hejleh; 1996), which implements a non-linear finite element solution of the 1D consolidation problem.

With this method, the best estimates of the consolidation parameters are obtained by minimising the objective function given by equation (4):

$$\mathrm{Obj} = \left(\frac{u'_{sim}\left(\frac{h_{sim}}{2}\right) - u'_{exp}\left(\frac{h_{exp}}{2}\right)}{u'_{exp}\left(\frac{h_{exp}}{2}\right)} \right)^2 + \left(\frac{\frac{h_{sim}}{2} - \frac{h_{exp}}{2}}{\frac{h_{exp}}{2}} \right)^2 \tag{4}$$

which matches the measured and predicted mid-elevation values at given times during the consolidation process. Such an objective function was chosen due to the shape similarity of the measured height $h(t)$ and excess pore pressure $u'(t)$ profiles (See Figure 1).

The objective function was minimised for 7 discrete times that correspond to the experimental measurement times: 3 hrs, 6 hrs, 1 day, 2 days, 8 days, 16 days and 32 days. Figure 3 gives an example of the consistently good match that was obtained between measured and simulated profiles, thereby confirming that the reverse parameter estimation method is well suited for analysing instrumented column data.

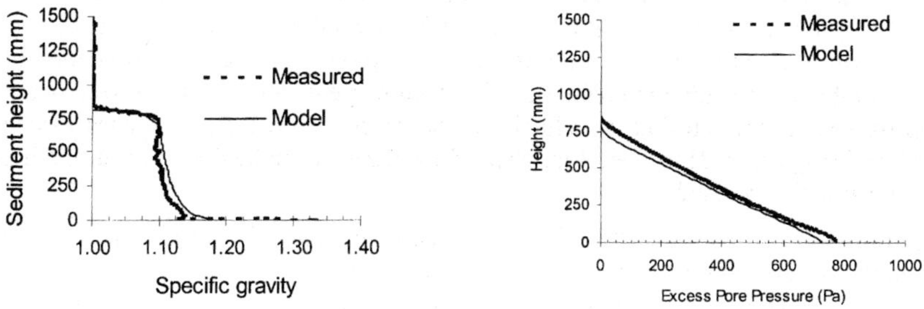

Figure 3. Example of measured and reverse estimated profiles (Unflocculated sample after 1 day of consolidation).

Results and analysis

Figure 4 shows the 7 curves *k(e)* and *s'(e)* that were calculated using the reverse estimated parameters A, B, C, D and Z for the 7 experimental times selected. The bold curves referred to as "Average Fit" correspond to the average *k(e)* and *s'(e)* response calculated by averaging the 7 responses. The consolidation parameters that characterize the "Average Fit" curves are given in Table 3.

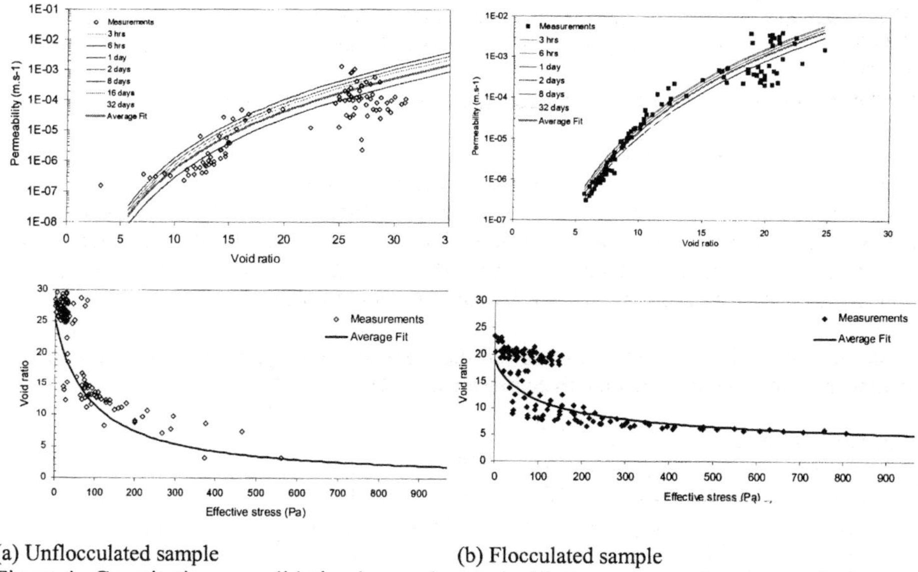

(a) Unflocculated sample (b) Flocculated sample

Figure 4. Constitutive consolidation laws calculated with the reverse estimation method

Table 3. Consolidation parameters estimated with the reverse estimation technique (coefficients of equations (1) and (2) are given for *k* in m/s and *s'* in kPa).

Consolidation parameter	Unflocculated sample	Flocculated sample
A	1.94	5.03
B	-1.10	-0.42
C	2.9×10^{-13}	8.5×10^{-12}
D	6.29	6.23
Z	0.10	0.04
e_{max}	25.2	19.0

Before comparing the reverse estimation scheme with the direct estimation method, it is noted that the estimated constitutive laws are clearly located within the set of *(k;e)* and *(s';e)* data points. Recalling that the estimation method does not make any use of these experimental data points, since it uses directly the *u'(h)* time profiles that are measured by the instrumented column, it is concluded that:

- The concept of estimating the consolidation parameters with the reverse parameter estimation technique is valid.
- The method for determining permeability *k(e)* and effective stress *s'(e)* from the instrumented column *(k;e)* and *(s';e)* data is reliable.

Figures 2 and 4 are directly comparable. From direct observation, it can be readily seen that the proposed parameter estimation scheme yields a significantly reduced scatter of the permeability relationships. Similarly, the benefit of the reverse estimation scheme is self-explanatory for the effective stress law, given that the estimated curves are superimposed with the average fit curve. Moreover, it is evident that the reverse estimated consolidation laws are not biased by the clustering of the experimental data points. This in turn means that the determination of the void index ratio e_{max} at zero effective stress is determined more accurately with this technique than by the direct method. The observation that the calculated values for e_{max} are very close to the experimental values e_0 with the reverse method confirms this conclusion and validates the choice of objective function.

Besides the positive finding that the reverse parameter estimation scheme considerably reduces the uncertainty associated with the constitutive equations (1) and (2) estimated from instrumented column data, the method also ensures that the experimental *e(h)* and *u'(h)* curves are recovered by simulation, as seen earlier in Figure 3. This is of considerable value for simulation purposes of course.

Overall, the simulation-based method is useful for both i) determining the consolidation parameter constants directly from the raw void index ratio and excess pore pressure profiles measured by an instrumented column and ii) for its direct use in consolidation simulation.

Conclusions

Two distinct parameter estimation methods were presented for estimating the consolidation constants of the *s'(e)* and *k(e)* constitutive laws that characterize the consolidation behaviour of a compacting sediment as measured using instrumented columns: a direct estimation method and a reverse estimation method.

It is confirmed that the direct estimation method yields values of *(s';e)* and *(k;e)* of the correct order of magnitude. However, the scattering and clustering of the data points, which are both inherent to the analysis technique, lead to high uncertainties on the estimates of the consolidation law parameters. As a result, the consolidation laws estimated by direct regression are not reliable for consolidation simulation.

The proposed reverse parameter estimation method is found to alleviate some of the weaknesses of the direct estimation method. In particular, it is found to yield narrower confidence intervals on the estimates of the consolidation parameters. Secondly, it is insensitive to the clustering of the experimental data points since it operates directly on the measured *e(h)* and *u'(h)* profiles, thereby providing reliable estimates of the void index ratio at the start of consolidation. Its reliability was demonstrated by applying it to two samples with drastically different consolidation behaviours.

Finally, the method is ideally suited for simulation work. Indeed, since consolidation parameters are estimated using the consolidation code used precisely for simulation, it guarantees that the simulated sediment behaviour matches the measured one in the sense of the void index ratio and excess pore pressure profiles.

References

Abu-Hejleh, A.N., (1996). "Software CADA : Description of Input Data & Output Results" (http://members.aol.com/CADA1234/manual.pdf).

Abu-Hejleh, A.N., and Znidarcic, D. (1995). "Desiccation Theory For Soft Cohesive Soils." *J. Geotech. Engrg. Div.*, American Society of Civil Engineers, **121**(6), 493-502.

Alexis, A., Le Bras G. and Thomas P. (2004). "Experimental bench for study of settling-Consolidation soil formation", *Geotechnical Testing Journal*, **27**(26), 557-567.

Been, K. and Sills, G.C. (1981). Self-weight Consolidation of Soft Soils: an Experimental and Theoretical Study, *Géotechnique*, **31**(4), 519-535.

Bourgeois, F., Baudet, G., Bizi, M. and Gaboriau, H. (2003). Conditioning circuit analysis for slimes management in quarries, *Trans IChemE*, **81**, Part A., 1158-1164.

Gallois, S. (1995). *Modélisation de la sédimentation-consolidation et expérimentions sur les vases estuariennes*, Ph.D. thesis, University of Nantes, France.

Gallois, S., Alexis, A. and Thomas, P. (1994). Settling column parametric tests applied to coastal sediment consolidation, in *Proceedings of the International Conference on Coastal Engineering ICCE'94*, Kobe, Japan.

Law, V.J. and Bailey, R.V. (1963). A Method for the Determination of Approximate System Transfer Functions. *Chem. Eng. Sci.*, **18**, 189-202.

Le Bras, G. (2000). *Etude expérimentale du début de consolidation de sédiments fins naturels*, Ph.D. thesis, University of Nantes, France.

Moys, M. H. and King, R. P. (1973). Computer programme for the estimation of parameters in flotation, National Institute for Metallurgy, Report No. 1568.

Environmental and Geotechnical Problems of Albania's Mine Sites

Prof. Luljeta Bozo , Polytechnic University of Tirana, FIN , Tirana, Albania

Eng. Ylber Muceku , Geological Survey of Albanai, Tirana, Albania

Eng Kujtim Cela, President of ILIRIADA shpk, Tirana, Albania

Abstract

Albania is a country rich in minerals. After the Second World War, the mines and the mining industry were further developed. Often, the underground works are situated near the earth surface; therefore, their exploration has caused many geotechnical phenomena that are very dangerous for the population, as well as for the economy of the country. In this paper, we present some of these phenomena as well as the engineering measures for their stabilization and/or rehabilitation.

Introduction

Albania is a country rich in minerals such as metalliferous, coal sand minerals, etc. As a consequence, after the Second World War, the mining and mine industry were further developed. The extensive exploitation of coal was often accompanied with an impact on the environment. In this paper, we present the issues of this phenomenon in Albania. Environmental impact can be defined as the consequences produced by a development of mining industry that is capable of affecting the characteristics of the environment. The waste ponds of coal in Albania are the following: Basin of Valias (Tirana), basin of Manza (Durres), basin of Memaliaj (Tepelena), basin of Mborje-Drenove (Korca), etc.

Exploitation of coal that is in close proximity with the earth surface is accompanied with dangerous phenomena, such as damage of buildings, sewerage, roads, telecommunication networks, and potential hazardous damages of villages.

The environmental impact statement for mining has to consider not only the actual effects on the environment because of the exploitation, but the development of the potential rehabilitation as well as provide for the rehabilitation costs.

Zones to be studied

The zones studied for the purpose of this report are the coal ponds in Tirana, Durres and Tepelena regions.
General lithology of coal deposits

The coal mineral in Albania is generally found in molasses rocks and less in flysch rocks (Fig. 1). It's situated between claystone and siltstone layers. From the structural point of view this mineral is found in folded structures (syncline and anticline). According to geological construction claystone and siltstone rocks predominate with to sandstone rocks. They represent about 80-90% of the lithological profile. The main coal deposits in Albania are located in following areas (Fig. nr. 1):

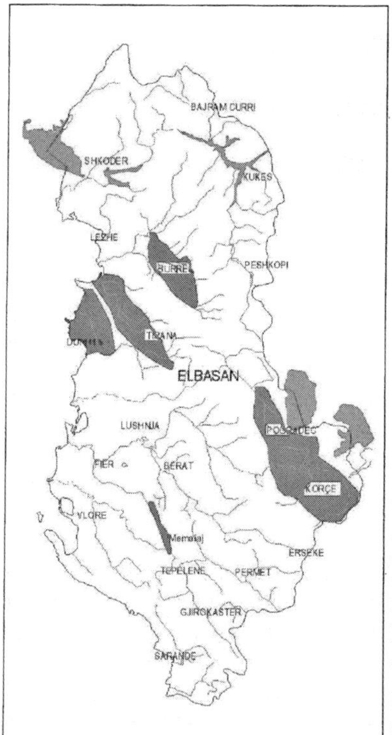

Fig. 1	Schematic map of coal mines distributions in Albania

Coal-bearing flysch formation of Mborja-Drenova and Gore-Moker area.

- Coal-bearing molasses formation of Tirana-Shijak-Fushe Kruja area.
- Coal-bearing molasses formation of Tepelena area.
- Coal-bearing molasses formation of Durresi area.
- Coal-bearing molasses formation of Mati area.

In these areas many coals deposits are found and a lot of mines have operated up to the year 1995. This paper describes the land subsidence of Valiasi, Mezezi and Memaliaj, which are caused by coal mineral exploitation.

- **Valiasi site**

In Valiasi mine, the coal mineral is found in eastern limb of the Tirana syncline, which to west with 2-5° up to 10-14°. Here the claystone rocks are about 44% of the lithological profile. They have greyish and greyish-brownish colour. Concerning mineralogical composition they are ilite and montmorillonite type. In the presence of water, they are subject of swelling phenomenon. So, the claystones increase their volume up to 50%.

The siltstone rocks have greyish and greyish-brownish colour, as 42% of lithological profile. Siltstones in the interaction with water will be swelled, increasing their volume up to 30%. Whereas the sandstones rocks consist of quartz, feldspars, micas and carbonate fine up to medium grained, which are cemented by clay and clay-carbonate material. These rocks represent 6-7% of lithological profile. The coal layers have the different dimension. They range from 0.2-0.5 km up to 1.0-3.0 km in the extension direction, from 80-100 m up to 200-500 m in dip direction and 0.2-0.5 m up to 2-4 m thick. The coal deposits constitute about 8.4% of the whole lithological profile. These rocks-molasses claystone and siltstone layers intercalated with sandstone layers. The above rocks are covered by Quaternary deposits, which make up the first terrace of Tirana River, as well as its bed. They are represented by middle to coarse grain gravels 35-40m up to 60-70m thick, made up mainly of limestone and less sandstone filled by sands, silts and brown clays, which are situated in lower part of soil profile and clay and silts in upper part of soil profile with thickness 5-15m.

As a result of the coal exploitations in this area has occurred the land subsidence furrow shape (Fig. 2, Photo 1, 2) in two parts. The first one, in eastern part of exploitation field has a 0.8 x 3.0 km dimension and other in western part of exploitation field has a 0.5 x 1.0 km dimension.

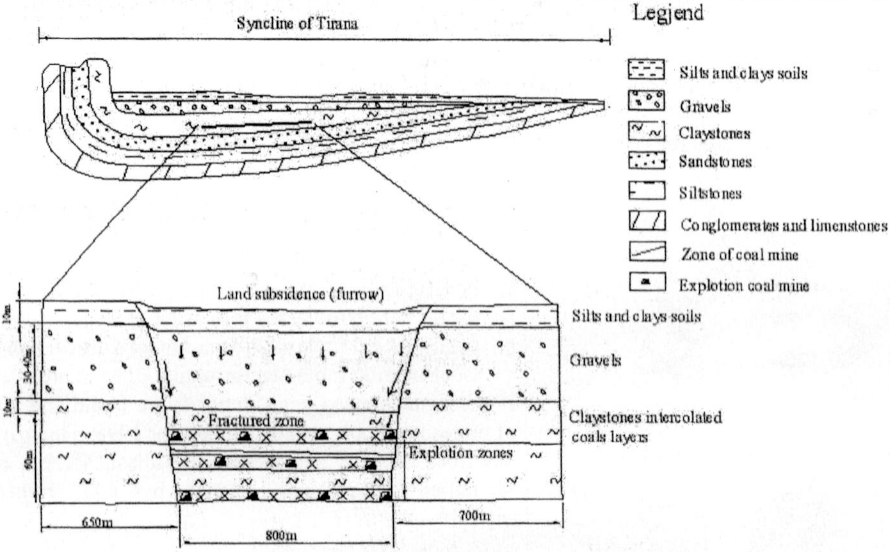

Fig. 2 **Schematic cross section of Valiasi mines (Tirana)**

Photo 1

Photo 2

· **Mezezi site**

Mezezi mine has the same geological construction as the above mentioned mine. But it differs on geological structure (Fig.3), because this mineral deposit is situated on western limb of Tirana syncline, which dip to east direction with 70-75^{0}. Most of this limb, where the coal mineral is found, overlies by Quaternary deposits 10.0-25.0m thick (silts and clay-upper part of soil profile and gravels-lower part). From the coal exploitations on this area many land subsidence have occurred in funnel shape, with a 10.0-20.0 m diameter and 2-5m thick (Fig.3, Photo 3).

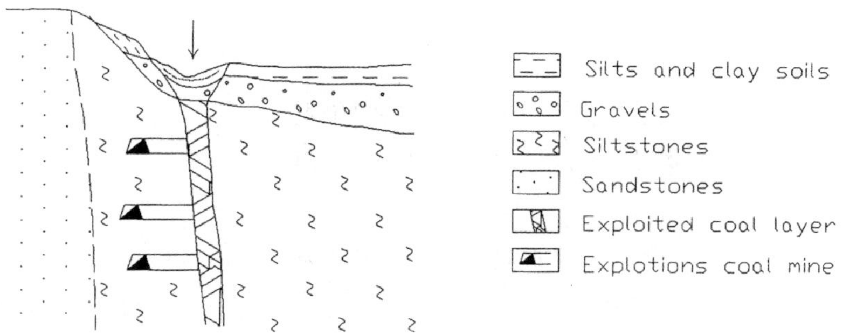

Fig. 3 **Schematic cross section of Mezezi coal mine (Tirana)**

Photo 3

Memaliaj site

Memaliaj Mine is located in coal-bearing molasses formation of Tepelena area (Fig.1). The coal mineral is created by the claystone and siltstone combinations, which make up an inverted syncline structure (Fig.1). The lithological construction of this deposit is almost the same as the above-mentioned sites. But the thickness of the Quaternary deposits is lesser than Mezezi and Valiasi areas. They are 4-10m thick and consist of silt and gravels soils. According to physical and mechanical properties the claystone and siltstone rocks-molasses are soft rocks.

In these regions of Albania quite strong horizontal and lesser vertical movements have been recorded in a zone up to 50-60 m from the margin of open coal's cuts. These movements result in damage to some villages in Memaliaj as Cerik , Mirik , Zhupanik, Licaj etc where some of them (Cerik and Zhupanik) are transferred to other places. Also we did record damage to water supply, sewerage services, drainage network and some rural roads.

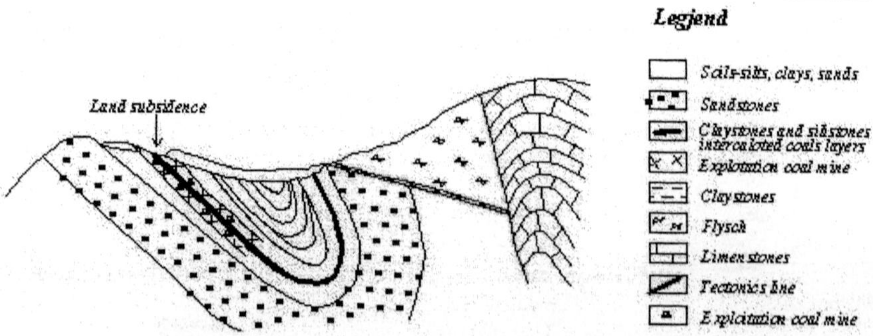

Fig. 4 Schematic cross section of Memaliaj coals mine (Tepelena)

- **Degree of damages**

The houses in the study's zone are 2-5 stores. Some of them have undergone very great settlements, the others dangerous fissures from horizontal movements. The subsidence due to coal extraction and coal mining areas is very important when the thicknesses of soils over the open coal's cuts are 30-50m.

The subsidence zone extent with ray R= (6-10) km from the centre of mining work.

The settlements and the fissures are so enormous they have to be considered very important and dangerous for the habitants or costly for their rehabilitation. In other hand the costs of rehabilitation of sewerage, roads etc are considerable.

3- Analysis of phenomenon

- **Subsidence due to coal extraction**

The earth surface deformation may be explained by covering viscous - elastic to plastic media. Stiffly plastic flow in a rock mass is influenced by underlying workings (fig 5).

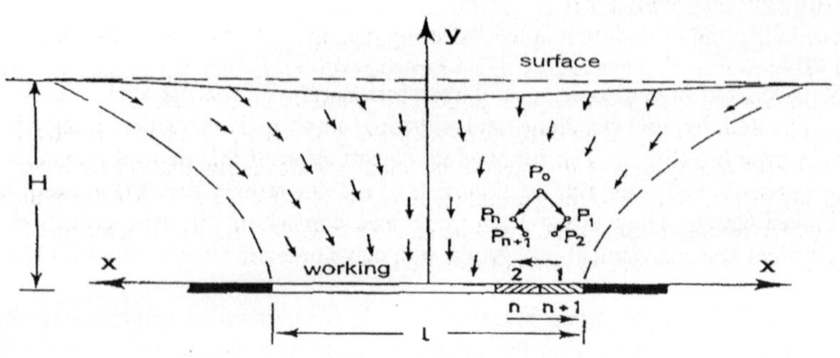

Fig. 5 **Stiffly plastic flow in a rock mass influenced by underlying workings. The final position P_2 or P_{n+1} in the movement of the rock point Po after the mining of strip no 1 depends on whether the direction of mining is from *l* to 2 or from *n* to *n+1***

The downward movements spread very rapidly until it reaches the upper earth surface. In this process six zones of movement can be distinguished (Fig 9) and two different stressed vertical zones (compression and expansion)

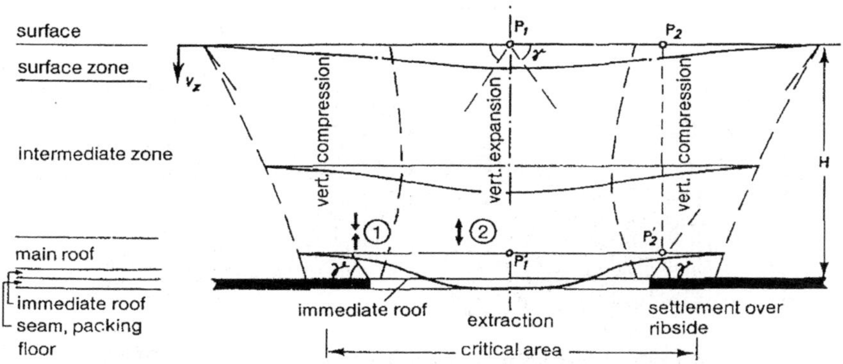

Fig. 6 **Vertical deformation of an undermined rock mass in flat-lying measures (subcritical area of extraction). the rock point Po after the mining of strip no 1 depends on whether the direction of mining is from *l* to 2 or from *n* to *n+1***

- Developed forces

The deformations of earth surface are vertical and horizontal .The first are because of the buildings the serviceable limit state and the second because of the finite limit state for them. Especially for the pipelines, water supply and sewerage the curvature of earth surface is very dangerous. Thus in Memaliaj zone, Mezez zone we have damage of water supply and sewerage. By our calculation, we have determined the vertical deformation which vary from 10cm-150cm (or 1.55-2.45 mm/m relatively) In foundations of buildings seating on native earth, under the influence of a mine working, forces are transmitted to the structure, partially directly as earth pressure and friction force through linear horizontal deformations of the ground and partially as bending forces through curvature of the ground.
As a result of upward Fig 7 or downward Fig 8 of the ground built on, the underside of the foundation becomes detached from the ground in places and stands free of it, thereby losing some of its support. So we have replacements of load and this effect produces bending stresses in the structure and many cracks in them.
The small degree of ground curvature it is not suffering damage. The visible cracks develop in masonry walls given radius of curvature of 1-1.5 km corresponding to a sag f=2cm in a chord 12m length and cracks over 1mm wide appeared in downward from R= 10Km.
Curvature caused by settlements cause damage to a masonry structure of length L where ρz<500L. The value of base pressure stresses on the underside of the foundation we have calculate by the stiffness index Es of the ground (E_s= (0.5-0.7) x104 Kpa….) and from them the bending stress on the foundation which will be smaller the allowing bending stress of material. In many cases this condition it is not satisfied.

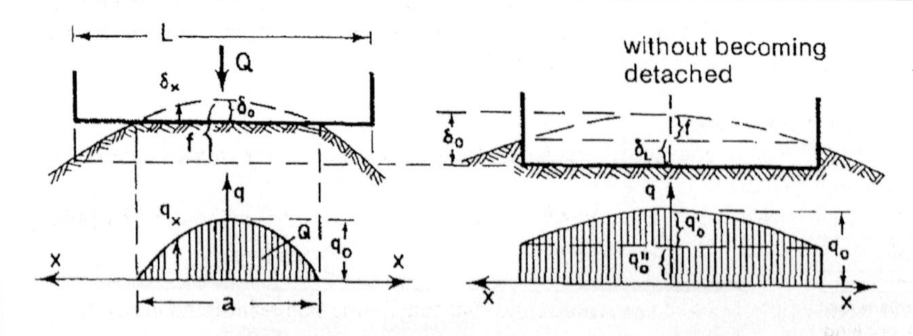

Fig. 7 The base pressure q where the structure rests on an upwarped saddle, by the bedding-index theory. (After Rausch, 1955)

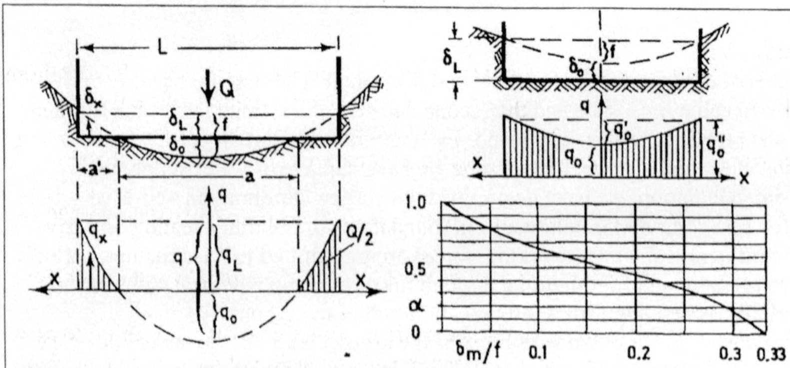

Fig. 8 The base pressure q where the structure rests on a downwarped trough, by the bedding-index theory. (After Rausch, 1955)

From the horizontal relative displacement of the ground which are (35-45) mm/m in a foundation appeared a friction force. They increase up to the middle of the structure (see fig.9) and produce in the ground compression or extension state.

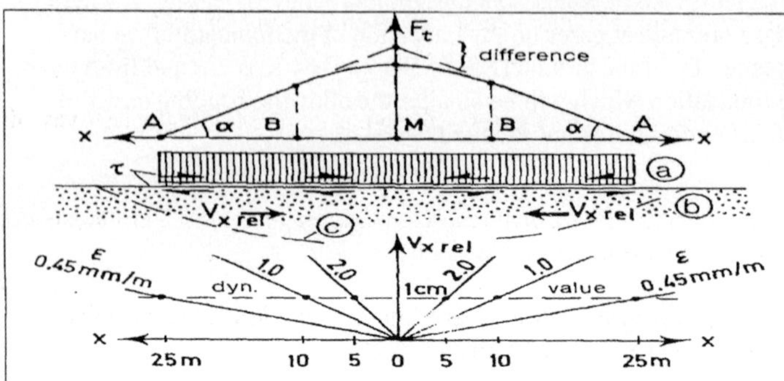

Fig. 9 Friction force F_1 transmitted to the structure (a) as a function of the amount of relative displacement and of linear deformation in the ground (b). (c) Zone of disturbance

Finally the structure warped in such scale that can be destructive for the houses where the length is around 9-10 m. The biggest damage of houses recorded in these zones took place where the horizontal relative displacement is 12-15mm/m.

4- Engineering measures

The engineering measure consists of increasing the rigidity of structure towards the horizontal and bending deformation.
For existing buildings reinforcement of foundation will be made and can be done by putting some (placing) concrete belt in difficult level.

For the new buildings should be provided rigid mat foundation and reinforced concrete structure.
For pipelines sewerage and water supply in the zones where may be strip mining , which is particularly damaging , it is necessary to provide supplementary connections and usage (utilisation) of geotextill or other resistant materials towards the deformation of earth surface.

5- Conclusions

In existing building (2-5) stores, which dominate in mining zones the destructive force is from horizontal displacement at degree 15-35 mm/m and for curvature of earth surface R<= 6km.
The degree of damage of buildings increases with reduction (diminution) of rigidity of structure. It depends from the thickness of rock over coal cuts and from mechanical and physical propriety of soils.
The maximal settlements are at the centre of mining works and they are from 10cm to 100-150cm.
The impact on the land surface started 3-6 months after exploitation of mines and terminates 2-4 years after this process.
The actual development of urban zones should be made in accordance (in conformity) to new regulation laws. These laws enforce the investigation and evaluation of the consequences to the environment of all engineering works, such as dams, pipelines, motorways, hydroelectric schemes and mining projects and urban development.
Geologists and geotechnics are the first in this work and are in a unique position to solve possible issues for the impact study.

6- Literature

Muceku, Y. "The Study of Soil behaviour under static and dynamic load for urban development of Tirana – Vora region" , Ph. D. Thesis p64-p113, 2005

Muceku, Y. "Engineering mapping of Tirana-Durres – Kavaja Region", Geological Survey of Albania, 2003

Dimo, Z. "About stratification conditions and coal reserves perspectives of molasses deposits of Korca Lowland" Geological Science Bulletin, No 4, Tirana

Çela, K. "Building damages in earth surface from Coal Mine Exploitation" , Geotechnical Journal no 3, Albanian Geotechnical Society edition, 2004

Geotechnical assessment of the Mejita tailing dam (Dom. Rep.) in karstic and seismic context

Cazaux D., BRGM, BP 6009, 45060 Orléans Cedex 2, France
Alameda J. G., PROINTEC, Avda. de Burgos 12, 28036 Madrid, Spain
Garcia Tascón, J., PROINTEC, Avda. de Burgos 12, 28036 Madrid, Spain
De León R. O., GEOFITEC SA, Calle Orión 2, Santo Domingo, Dominican Republic
Carbó A. Faculty of Geology. Complutense University of Madrid, Spain.
López O., Mine Department Director, State Secretary of Industry and Trading, Government Offices, Av. de México, Santo Domingo, Dominican Republic

Abstract

The tailing dam geotechnical assessment presented in this paper was carried out beyond the European SYSMIN program in Dominican Republic in 2003. The high seismic and hurricane occurrence risk of the area led to specific studies including: a seismic analysis of the area, an estimation of the liquefaction potential of tailings for dynamic analysis of the stability, an historical survey and the determination of maximum water flows induced by hurricane conditions. The seismicity of the site was studied to determine the magnitude of potential seismic activity and its effect on the existing embankment. A large laboratory and field testing program was defined for obtaining the typical geotechnical data needed for slope stability analysis. The testing program was including more than twenty geotechnical boreholes, soil and rocks classifications and mechanical testing (consolidation, shear and triaxial tests, hydraulic conductivity). The assessment was completed by a very large network of geotechnical monitoring devices installed into and around the dams (inclinometer tubes, vibrating wire piezometers and topographical targets). The stability analysis of the tailing dams focused on static and pseudo-static slope stability and the influence of karstic features within the foundation of the dam. The analyses presented in this paper include the actual static geometrical configuration, the seismic conditions including the liquefaction potential of the tailings, and karstic risk assessment. The results of the stability analysis lead to the necessity of considering in a near future a reinforcement of the dam.

Introduction

In view of developing its mining sector, the Government of the Dominican Republic is in the process of privatising state-owned mining projects. The Pueblo Viejo gold and silver mine located in the central part of the Dominican Republic (Figure 1) is a very important project that includes expansion of the existing property to incorporate additional reserves of precious and base metals. Ore production between 1975 and 1999, date of termination was 5,500,000 troy ounces of gold and 25,000,000 troy ounces of silver and was mainly focused on the oxidised ore.

The Mejita tailing embankment investigation and design were originally carried out in 1972 for two stages construction process. Additional ore reserves had been identified in the late 70's so that upper construction stages were initiated (figure 2). The last stage IV was designed to provide the storage volume required to contain the remaining identified oxide ore tailings and maintain factors of safety. The embankment is 1200 m long for a maximum height of 75m (figure 1). Stage IV was designed to be the ultimate embankment possibility for the site. The abutments have vertical relief remaining to accommodate a higher structure, but the internal drainage system has not been built up since stage II. The top of chimney drain (figure 3) remains 20 meters lower than the actual crest of stage IV. Since 1990, the Mejita impoundment has been filled and replaced by second tailing dam.

The importance of the karstic risk on the tailing dams was revealed by the geological studies, boreholes logging and geophysical profiles carried out in this study and the previous reports. This will allow the definition of a comprehensive emergency response plan which may be the most important asset for management of risk in the case of a tailings failure or release.

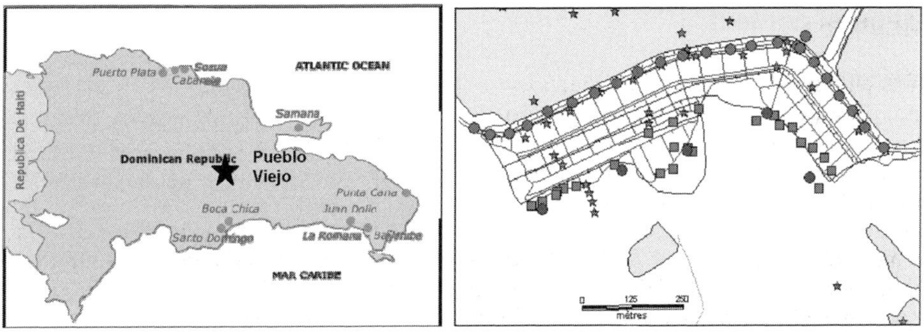

Figure 1: *Left:* Situation of the Pueblo Viejo mine in Dominican Republic. *Right:* map of the last stage of Mejita embankment showing investigations program (points represent drilling and squares and stars represent trial pits)

Construction stage	Year	Max. Height	Full length at crest	Slope	Materials
I	1975	25 m	800 m	3H:1V us 5H:2V ds	Compacted clay and sandy clay with gravels
II ds	1980	40 m	800 m	3H:1V us 5H:2V ds	Compacted clay and sandy clay with gravels
III ds	1984-86	62 m	1100 m	1H:1V us 2H:1V ds	Compacted clay and rockfill
IV us	1987	80 m	1200 m	1H:1V us 2H:1V ds	Compacted clay and rockfill

Figure 2: History of the construction of the Mejita embankment (*ds*: downstream, *us*: upstream)

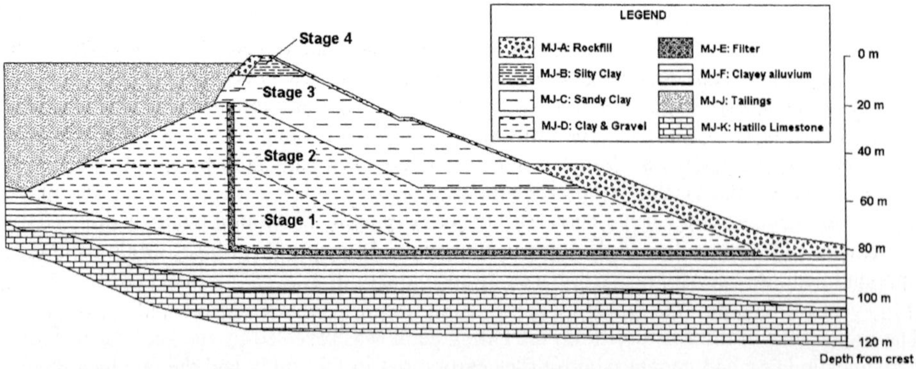

Figure 3: Cross section of the Mejita tailing dams at its highest elevation above Hatillo calcareous foundation

General context

Geological context

The tailings dam and reservoir are situated in the valley of the Rio Mejita. The initial drainage area comprised small creeks which feed substantial amounts of contributing inflow. The embankment crosses the Mejita valley at its narrowest point controlled by three calcareous ridges, two of them being the abutments (Tippets and al., 1974).

The geology of the dam is structurally controlled by various reverse faults. The geological deposits and formations that underlie the embankment can be grouped in five categories at the level of the footprint: alluvium and colluviums of quaternary age, weathered rocks and colluviums of early quaternary age, Hatillo limestone of Cretaceous age and the Los Ranchos volcanic formation. The faults separate the volcanic formation (Los Ranchos) localized at the central part of the dam from the limestone which forms the abutments. The Hatillo limestone was found to be highly karstic with an average open porosity of 10% (CGS, 1998). Cavities vary in general from 2 to 10 cm as well as several caverns up to 5 metres have been observed by drilling. Theses caverns are often filled by clays but sometimes they appear to be voids. At both abutments the limestone is highly fractured and weathered showing frequent clay fillings and pockets. The volcanic formation is deeply weathered along the axis of the dam and presents more than 20 meters of silts and clays which form the foundation of the dam at its highest elevation (Figure 3).

Hydrological context

The watershed above the dam is approximately 5.5 km². The basin is moderately vegetated with steep slopes. A great part of the basin has been disturbed by the mining activity and the dumping of waste rocks. As a matter of fact, the quality of the water drained within the basin is really poor (pH levels as low a 2). The average annual rainfall at mine site is approximately 1700 mm. The highest monthly rainfall record is 730 mm (Hurricane David in 1970) and daily rainfall can reach 500 mm (Hurricane David in 1998). The maximum daily hurricane rainfall for the region is estimated to 1200 mm. An emergency spillway of the dam, located at the east bound, was designed to pass an attenuated maximum probable flood and maintaining a safety level of water behind the embankment.

Seismic activity

The Dominican Republic is situated within the Caribbean Loop of seismic activity. During the twentieth century, more than 30 earthquakes having magnitudes over 6 on the Richter scale have occurred within a radius of 300 km of the mine site. An earthquake of magnitude 8.1, having its epicentre approximately 130 km east of the site (Bay of Samana) was the most severe occurrence (1946). It took place along the "septentrional" fault localized 35 km north of the site at its nearest point (Figure 4). The probability of occurrence of an earthquake of magnitude 8 is 0.983 for a period of 500 years. Estimates of ground accelerations for risk assessment caused by earthquakes have been made considering this highest event. The value of acceleration, according to Joyner and Boore (1993), is 0.41 for magnitude 8 at 35 km (Figure 4). The duration of maximum acceleration is estimated between 0.1 and 0.6 seconds.

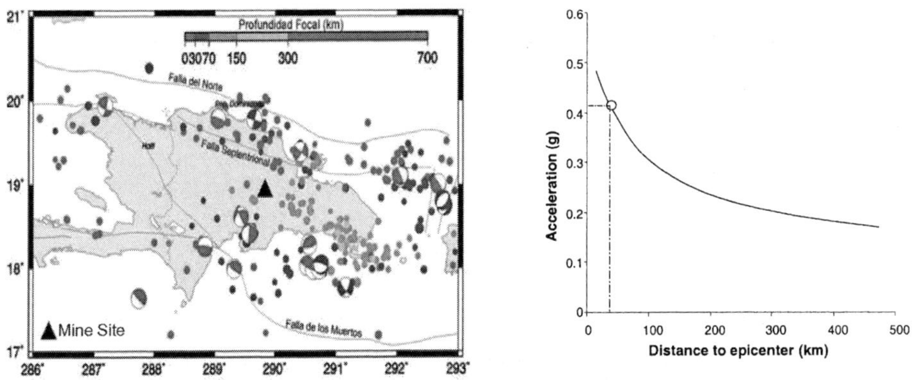

Figure 4: Seismic map of Dominican republic and estimation of acceleration according to Joyner and Boore (1993)

The analysis of the seismicity has been carried out by evaluating the liquefaction potential of the tailings using the Chinese Criterion (Wang, 1979, modified by Finn et al. 1994 and Andrews and Martin, 2000). The basic characteristics are illustrated in figure 5a and reveal that the tailings are potentially liquefiable.

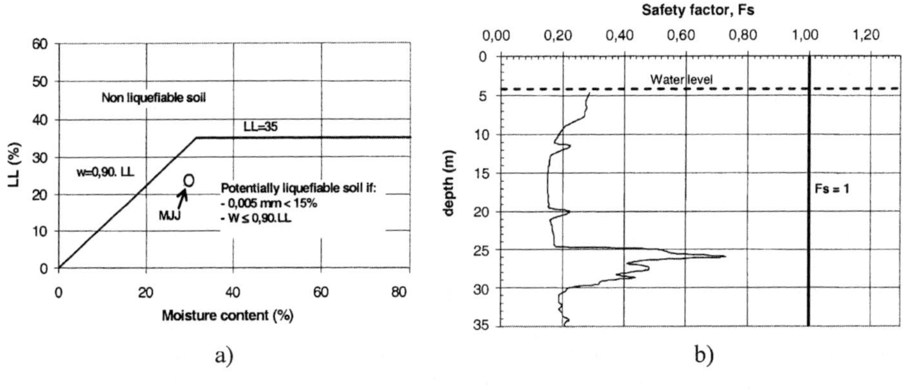

Figure 5: a) Application of the Chinese criterion to the tailings (Geotechnical unit MJJ); b) Variation of the safety factor against liquefaction with depth in the tailings

The potential of liquefaction has been analyzed with the Cyclic Stress Ratio (CSR) and Cyclic Resistance Ratio (CRR) (Eq. 1). The potential of liquefaction is represented as a safety factor Fs (figure 5b) applying a factor of correction due to the magnitude of the predicted earthquake. As shown on figure 5b, the tailings are potentially liquefiable for the seismicity contemplated in the study:

$$Fs = \left(\frac{CRR}{CSR}\right) \times MSF \times K_{\sigma} \qquad \text{Eq. 1}$$

Site investigations and laboratory testing

Site investigation

Site investigations at the site were carried out from September 2002 to April 2003 to address:

- Competency of the bedrock
- Classification and suitability of the material beneath the embankment footprint
- Geotechnical properties of the material used for the construction of the dam

Field investigations comprised geotechnical drilling and testing using continuous wire-line core drilling, thin-walled core sampling and standard penetration testing (SPT). The program was completed by permeability tests in bedrock (water pressure test with packer) and falling-head permeability tests in the compacted clay materials. Geophysical survey was carried out along the dam crest and the footprint by four electrical profiles and eight seismic profiles. As a conclusion of these geophysical studies, is the confirmation of the presence of karstic cavities, advising the accomplishment of a sealing screen. Also, the geophysical profiles show the bad quality of compaction of the last stage of the dam in several areas compared to the consolidated lower levels. Figure 6 resumes the number of investigations carried out on the dam. Boreholes were drilled every 50 meters along the dam crest (Figure 1) and then equipped either by inclinometer tubes or with vibrating wire gauges. The latter were installed at the levels of the various stage of construction, drilling diameter allowing the installation of five sensors in each hole. Figure 6 also summarizes the investigations and monitoring devices carried out on the Mejita Dam.

No of boreholes	No of trial pits	Core drilling	Thin- wall sample	SPT	K test	Resistivity profiles	Seismic profiles
30	20	1760 m	127	271	86	4 1700 m	8 1100 m

Inclinometer tube	VW sensors	Topographical targets
820 m 12 instrumented holes	43 11 instrumented holes	25

Figure 6: summary of field investigation and monitoring programs

Laboratory testing

The laboratory testing program included characterisation and mechanical testing of the various materials encountered in the embankment, the foundation soils and the bed rock:

- Moisture content, grain-size distribution and Atterberg limits
- Pinhole erosion test

- Consolidation tests
- Compression tests, Shear tests, Triaxial tests

Direct shear test and triaxial tests were run to establish design parameters for use in performing stability analyses. According to core drilling observation and the various stage of construction and materials, several homogeneous geotechnical units were defined. They are summarized in Figure 6 with their average characteristics.

Unit	Structure	Material	N_{SPT}	Dry density	w (%)	% < 63μm	USCS Class	Φ´ (º)	C´ (Mpa)
MJ-A	Downstream Rockfill	limestone	-	2.5	-	-	-	40	-
MJ-B	4th step	Silty clay	23	1.6	27.5	78.8	CH	22.51	0.02
MJ-C	3rd step	Sandy clay	37	1.8	16.5	50.1	CL, CH, SC	24.39	0.1
MJ-D	1st and 2nd step	Clay and gravel	41	1.7	19	49.3	CL, CH, ML, SC	25.4	0
MJ-E	filter	Sand and gravel	-	2	25.9	74	GC, CL, CH, ML	31.8	0.09
MJ-F MJ-G	Alluvial tills colluviums	Clay and gravel	46	1.6	24.7	79	GC, CL, CH, MH	24.23	0.28
MJ-H	Bed rock (Los Ranchos)	weathered rock	-	1.6	19.2	74.8	CL, CH	25.23	0.04
MJ-I	Bed rock (Los Ranchos)	Volcanic rock	-	1,6-3,0	22.4	82.9	CH	-	-
MJ-J	tailings	tailings	7	2.4	27.3	52.5	-	27.26	0.09
MJ-K	Bed rock (Hatillo)	limestone	-	2.7	-	-	-	-	-
MJ-K1	Bed rock (Hatillo)	Filling clays (karstic cavities)	-	1.33	32.7	93.8	CH	27.61	0
MJ-L	Bed Rock (Las Lagunas)	Weathered rock and clay	63	1.5	29	92	CH	20.73	0.46

Figure 6: Summary of laboratory results and average values

Stability analysis

Once established the geomechanical model (one for the profile of maxima height and another one with the presence of a limestone promontory), the stability analysis has focused on the possible risks that can affect the dam: slope stability (static and pseudostatic model), possible risk of collapse due to high karstification. In order to complete the stability analysis, an evaluation of the level of alarm of the installed instrumentation has been also made. The safety factors have been referenced to the ICOLD specifications.

Static and pseudostatic analysis

The main conclusion of the general stability analysis is that, in static conditions, the dam is stable and would practically present a safety factor of 1.25, although it does not reach the required safety factor of 1.50 (summary of the most significant cases in figure 7). For pseudo static conditions (modelled for an earthquake with an acceleration of 0,2g), the dam appears to be unstable. In both cases the critical circles take place in the last stages of construction of the dam (figure 8). In order to improve the general stability of the actual state, active actions on the dam have been discussed and considered in a second analysis. Thus, the lowering of the static level within the tailings and the installation of rocky embankment reinforcement on the downstream slope appear to be efficient solutions (figure 9). The pseudo-static safety factor is increased to 1.14 with these active actions which remain technically and economically viable in the local context.

Profile	Description	Pessimistic circle	Static	Pseudos-static	Static with reinforcement	Pseudos static with reinforcement
1 (C-C')	Profile of max. height	upstream	1,65	0,98	NA	1,48
		downstream	1,24	0,81	1,63	1,14
2 (D-D')	Profile on calcareaous promontory	upstream	1,67	1,1	NA	3
		downstream	1,08	0,74	1,96	1,21

Figure 7: results of the stability analyses (circular rupture)

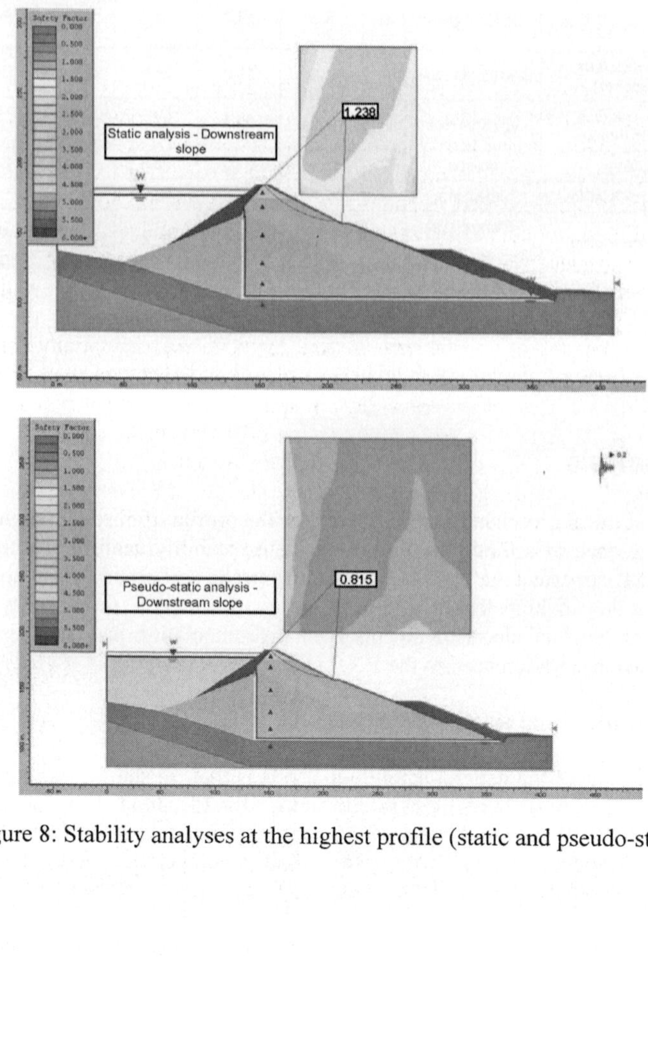

Figure 8: Stability analyses at the highest profile (static and pseudo-static)

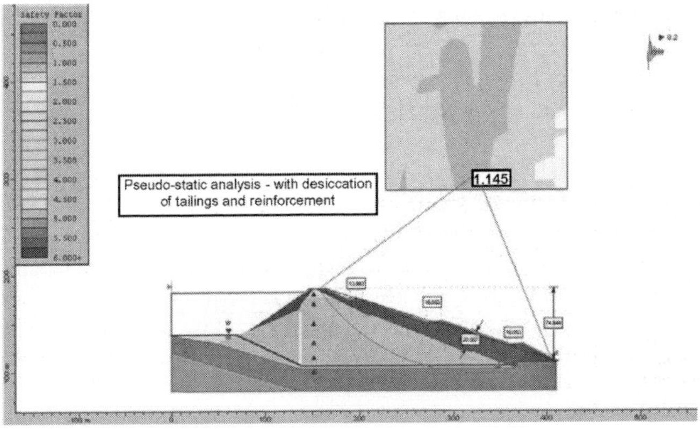

Figure 9: Stability analysis for the maximum height profile, with reinforcement and lowering of water level in tailings

Karstic risk assessment

A finite element modelling (Rocscience PHASE 2) was carried out for evaluating the influence of karstic voids into the calcareous foundation of the dam (figure 11). The risk of collapse due to cavities is real but it has been very difficult to deduce the critical volume and size of a cavity that can affect the foundation. In effect, even if it represents the maximum size observed in the Hatillo formation (according to visual surveys and boreholes data), the size of the cavity used in the model (13 by 9 meters) has a low probability of occurrence in that part of the site. The displacements reach 15 cm within the roof of the cavity, and 10 cm in the slope of the dam. The formation of an instable "chimney" reaching the top of the dam crest is also observed. These results indicate that a systematic consolidation of the ground would greatly improve the stability of the dam. At the moment, even if initially considered, these works are difficult to engage for technical and economical reasons and because the distribution of the cavities is not well-known in spite of the investigations carried out during the program. The problem of a possible extension of the cavities is the future due to acidic filtrations through the foundation does not seem to influence the conditions of stability at present according to the field investigations. Nevertheless, the management of water in the next future by lowering the water in the tailings will limit the infiltration to the cavities. The choice of a large network of monitoring into the dam appears to be the best way of managing the risk regarding technical and economical parameters.

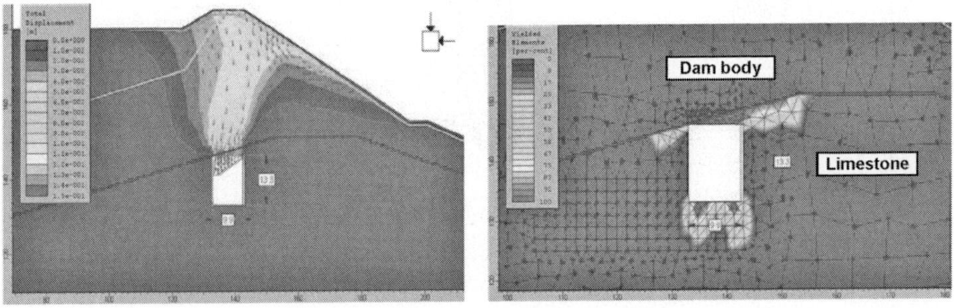

Figure 11: Analysis of the dam displacement above a karstic cavity

Evaluation of alarm levels of instrumentation

In order to fix the alarm levels of the instrumentation devices, a modelling of the section of greater height of the dam has been carried out. For that situation, the inclinometers can reveal displacements that create a new situation, with a smaller factor of safety than the initial. Considering the initial safety factor in the current situation, the displacements have been estimated with a finite elements model, using PLAXIS code, for a new situation that corresponds to a reduction of the safety factor of 5 %. As observed in figures 12 and 13, for a factor of safety 1.15, corresponding to the fixed alarm level, the displacement is about 2 centimetres. Theses displacements are being developed within the highest elevations of the dam (stages III and IV), where the instability is shown with the analysis of the circular failure and by the finite elements method.

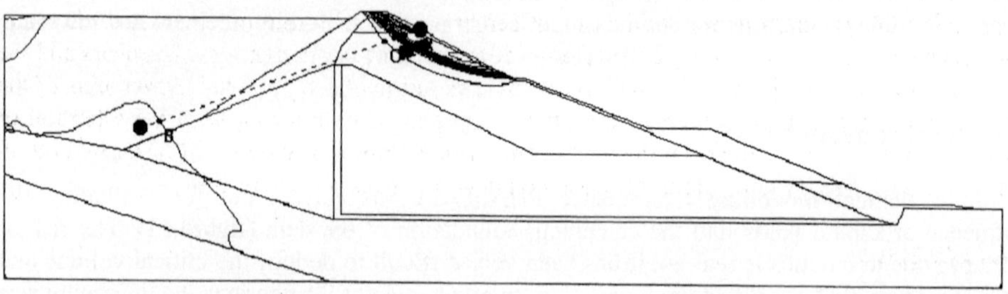

Figure 12: Modelling of the safety factor corresponding to the alarm level

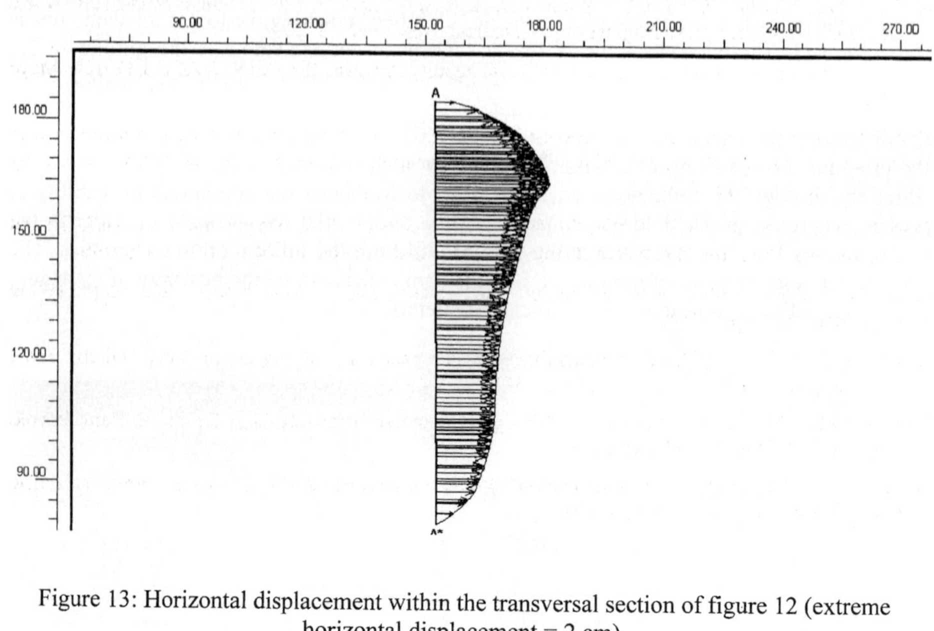

Figure 13: Horizontal displacement within the transversal section of figure 12 (extreme horizontal displacement = 2 cm)

Conclusion and remediation strategy

The result of the general slope stability of the dam has been made for several profiles and several hypotheses, considering the present situation and the accidental situation of an earthquake where the liquefaction of tailings has been considered. From this analysis, it appears necessary to anticipate the installation of reinforcements on the downstream slope, since the general stability is improved in case of an earthquake. It is necessary to recall that the liquefiable character of the tailings acts as a destabilizing force during the earthquake, reason why it is important to study the possibility of "drying" the tailings and designing a general water management system that allows the water collection coming from the run-off. The analysis of the karstic risk has been modelled with a finite elements program, supposing the presence of cavities of different sizes that can cause the possible instability of the dam. It is necessary to indicate that this modelling presents many simplifications since the size of the caverns, within the bedrock of the dam is not known. For that reason, it seems difficult to approach a full program of consolidation with injections, throughout all the axis of the dam. The large network of monitoring of the dam (1500 meters of inclinometers, 84 sensors and the system of topographic landmarks) will allow the evaluation of the potential movements of the embankment, and will constitute in the future a system of monitoring with the intention of anticipating a greater accident. In the next future, new European Projects will be carried out in order to design and build the dam reinforcement and the management of water.

References

Alameda, J., Cazaux, D., Garcia J., "Estudio de la presa de colas de Mejita", Programa SYSMIN 7 ACP DO 024, Mayo 2003, 250 p.

Andrews DC, Martin GR, Criteria for liquefaction of silty sands. In: Proceedings of the 12th World Conference on Earthquake Engineering, Auckland, New Zealand, 2000.

CGS, "Depósitos de caliza de transición", Final report, Program SYSMIN 7 ACP DO 024, Mayo 1998.

Engdahl, E.R., Van der List, R.D. and Buland R., "Global teleseismic eartquake relocation with improved travel times and procedures for depth relocation", Bulletin of the Seismological Society of America, 88, 1998, 722-743.

Finn, W.D.L., Ledbetter, R.H. & Wu, G., “Liquefaction in Silty Soils: Design and Analysis”, Ground Failures under Seismic Conditions, Geotechnical Special Publication 44, ASCE, 51-76, 1994.

Hydro-Triad, LTD., "tailing management and water balance study - Mina Pueblo Viejo, Mejita tailing Impoundment", April, 1986.

Joyner, W.B., and Boore, D.M., Methods for regression analysis of strong-motion data. Bulletin of the Seismological Society of America, 83(2), 1993, 469-487.

Tippets, Abbet, Mc Carthy and Stratton, "Design Recommendations Tailings Retention Dam, Pueblo Viejo project, January 1974, 18 p.

Wang, W., "Some finding in soil liquefaction". Water conservancy and hydroelectric power scientific research institute. Beijing. China, 1979.

Back Analysis of Crown Hole Development Using Numerical Techniques

Owen Davies[1], Mohamed Rouainia[2] and Colin JFP Jones[3]

1. Postgraduate Student, Newcastle University
2. Lecturer of Geotechnical Engineering, Newcastle University
3. Emeritus Professor, Newcastle University

Abstract

This paper represents the back analysis of crown holes which developed on a section of the East Coast Main Line East of Edinburgh. A site investigation identified abandoned coal mines beneath the site, the collapse of which were causing the development of the surface instability. A single crown hole identified by the site investigation was chosen for this study. A representative model of the stratigraphy from borehole records was constructed using a finite difference program. The model was used to carry out a back analysis of the collapse of a single void with the aim of establishing the mechanism of collapse, the size of the underground voids and the reason for the delay of the collapse.

Introduction

This paper describes an analysis of a section of the East Coast Mainline (ECML) 8-9 miles South East of Edinburgh at Dolphingstone. The area was extensively mined during the 18th and 19th Centuries. In 2001 a number of crown holes appeared on the site adjacent to the ECML. A comprehensive geotechnical appraisal of the site revealed extensive old mine workings below the ground. During the site investigation a further 22 new subsidence events occurred, (Donaldson 2002). The risk to the stability of the ECML revealed by the presence of the old workings lead to major remedial works being undertaken including the diversion of the mainline over a distance of approximately 1 mile. As part of the studies initiated by the unexpected subsidence it was decided to investigate the potential collapse mechanisms of the old workings.

This paper describes the back analysis of the development of a representative crown hole using the finite difference method (FDM).

The model properties

For the purpose of the back analysis a single typical crown hole was singled out for investigation. This hole was described as elliptical and having a depth of 0.3m and a maximum surface width of 2m. The detailed geological conditions at this location were investigated by two nearby boreholes.

The geometry of the numerical model was developed to mirror the conditions found on site Figure.1(a). The model was built in a plane strain simulation and used the sequence of strata identified from the borehole records as closely as practically possible. The strata within the model were assumed to lie horizontally. This assumption was made as the dip of the strata was found to be low. In addition it was assumed that the plane modelled lay perpendicular to the strike of the beds. Coal seam depth was determined from mining records and boreholes. The exact depth and thickness of the seam (referred to as P) was refined as the modelling process developed. A phreatic surface at 24m AOD (Above Ordnance Datum) was assumed with a hydrostatic profile beneath. This was assumed as the flow of groundwater was found to be insignificant and the fractured strata behaved as a homogenous hydrological unit.

Material strength parameters

Due to the lack of available specific strength data a Mohr-Coulomb model was used to represent the materials on site. The parameters used for the materials were obtained from the site investigation where available (Donaldson, 2002). Where the parameters were not available they were assumed from the extensive FLAC database (Itasca, 2002) or where not available, published literature (Tomlinson, 2001). All assumptions were based on the actual material descriptions from the site investigation.

As there was no indication of strength or state of the coal from the site investigation all the properties of this material were assumed. The density, bulk modulus and shear modulus were all assumed from literature and the coal was assumed to be elastic. This was assumed as studies of shallow mine workings (below 50m depth) show that mine pillars can show signs of stress and some fracturing at the sides and corners, but stresses are usually too small to cause significant damage, (Bell and De Bruyn, 1999).

Laboratory measured properties for intact material from the site investigation were not used for the full-scale model of the problem. The values were scaled in order to account for the discontinuities and heterogeneities present in the rock mass. Where closely interbedded strata were encountered an assumed homogenous rock mass containing an average of the constituents properties was used.

The mass densities of the superficial material were obtained from the site investigation. For material above the water table drained densities were used. For material below the water table saturated densities were used.

The elastic constants K (bulk modulus) and G (shear modulus) used in the analysis were determined from the expressions:

$$K = E/3(1-2\nu),\ G = E/2(1+\nu)$$

Where (E) is the elastic modulus determined from the site investigation and (ν) is the Poisson ratio. Within coal measures the Poisson ratio has been identified to vary between 0.2 and 0.33 (Riggott, 1986). For this study a Poisson ratio of 0.3 was assumed for the solid geology. The elastic modulus of the rock mass was scaled from that of the intact rock type with respect to the rock mass rating of the material found on site. For hard rocks with a rock mass rating over 55 the expression proposed by Bieniawski (1978) was used:

$$E_{RM} = 2\,(\mathrm{RMR}) - 100$$

Where units are in GPa.

For values of E_{RM} between 1 – 10 GPa, Sarafim and Pereira (1983) suggest a better fit with:

$$E_{RM} = 10^{(RMR - 10)/40}$$

For soft rock masses with horizontal stratification the equations proposed by Ramamurthy (1986) were used as this gave a better approximation for their behaviour:

$$E_{RM} = E_R e^{(0.0217\ RMR - 2.17)}$$

Where:
E_{RM} =Deformation modulus for the rock mass
E_R = Elastic modulus for intact rock type
RMR = Rock mass rating.

RMR was based on values determined by Bieniawski 1984 (cited in Brown and Brady, 1993) which take into account the strength of the intact material, RQD (Rock Quality Designation), joint spacing, joint condition and presence of groundwater.

Cohesion of the rock mass (C_m) was estimated from a simple relationship proposed by Bieniawski (1989):

$$C_m = 5.\ RMR\ kPa$$

The tensile strength of the rock mass (σ_t) was calculated from the Hoek - Brown (1988) equation:

$$\sigma_t = \tfrac{1}{2}\,\sigma_c\,[m - (m^2 + 4s)^{1/2}]$$

Where:
σ_t = Tensile strength
σ_c = uniaxial compressive strength of intact material
m and s are constants depending on the properties of the rock and the extent to which it has been broken before being subjected to failure stresses. These constants can be calculated using the empirical relations presented by (Hoek and Brown, 1988) for interlocking rock masses:

$$m = m_i e^{(RMR - 100)/28},\ s = e^{(RMR - 100)/9}$$

Where m_i is a characteristic of the intact rock. For this analysis m_i was obtained from values for intact rock given by (Hoek, *et al* 1995).

The friction angle of the rock mass (ϕ_m)was estimated from a simple relationship proposed by Bieniawski (1989):

$$\phi_m = 5 + (RMR / 2)$$

The calculated material properties are shown in Table 1.

Table 1.Material properties

	Mass density Kg/m^3	Bulk modulus Pa	Shear modulus Pa	Cohesion Pa	Tension Pa	Friction ^{0}
Topsoil (assumed values as no site data available)	1550	$3.33x10^5$	$2.0x10^5$	7000	0	20
Sand	1530	$2.0x10^7$	$1.6x10^7$	6500	0	37
Gravely,silty clay	1830(drained) 2140(sat.)	$1.4x10^7$	$8.4x10^6$	10000	0	29.5
Moderately strong sandstone	2280	$7.7x10^8$	$3.5x10^8$	$2.5x10^5$	17350	30
Coal	1550	$7.5x10^7$	$3.5x10^7$	-	-	-
Interbedded sandstone siltstone/mudstone	2420	$1.3x10^8$	$6.2x10^7$	$2.8x10^5$	164000	33

Modelling procedure

A model of the strata with the relevant calculated strengths is shown in Figure 1(a). The coal seam depths were assumed from mining records and the water table level related to the site investigation.

Modelling was undertaken using the following procedure

1. Voids were created within the model and their effects analysed. Investigation of the site history indicated the common methods of mining in the area were stoop and room with some limited shortwall. In the middle of the 18th Century pillar size was typically 5.5m by 3m with room sizes varying from 1.8-4.6m (Taylor 1975). These dimensions were used as the limits of room size for the analysis.
2. A pressure was established on the roof , walls and base of the void to simulate hydrostatic pressure at depth.
3. The model was then run to determine if the void was stable. If the void remained stable the cohesive strength of the sandstone roof was reduced to instigate failure. When the void collapsed the model was stopped and monitored to determine which elements on the void roof would fail in tension. This was achieved by reference to a plot of the zone states of the elements within the model. This shows elements which have failed in shear and tension. By referring to this plot throughout a run it becomes clear which elements had failed and needed to be removed. The failed elements were removed from the roof and replaced by a bulked material at the base of the void. The bulked material was assumed to have a volume 1.3 times the volume of the removed elements to reflect a bulking factor of 30%, typical of coal seam stratigraphy (Whittaker and Reddish, 1989). The bulked material was also assumed to have a constant volume under loading.
4. A new hydrostatic pressure was then established on the revised void roof.
5. The models were run until they reached equilibrium. This occurred when the void reached the surface or become arched.

The model grid consisted of uniform square elements (0.25m by 0.25m) throughout. These elements were chosen as it was simpler to sum and multiply uniform elements once collapse had taken place. Preliminary models showed that multiple voids within a seam had no influence on each other. Later models, therefore, only included one void in each seam. Collapse of the voids within the preliminary models was not symmetrical about the void midpoint. The voids in later models were therefore placed within the centre of the model rather than just modelling the collapse of half a void and assuming symmetry. This approach also reduced rounding errors when bulking the collapsed material for placement at the base of the void.

Seam collapse

When a void of 3.5m width was introduced into the P seam (thickness 1.25m) and the sandstone roof weakened the collapse of the void was considered. A study of the zone condition plot showed which elements within the roof had failed in tension. These elements were removed and replaced with a bulked material. The model run was then continued. During this process the collapse propagated towards the superficial geology in the model. At this point the collapse accelerated. Figure 1(b) shows an image of the collapse. The elements within the central region above the void failed in tension whilst the material surrounding the failed region had failed in shear. This suggests that a column of material which had failed in tension would collapse into the void. The collapse was performed manually. The failure of the material was presumed to occur instantaneously. In previous studies of modelling collapse mechanisms, the bulking factor for the superficial geology was assumed to be 75% of the rocks, (Richards *et al*, 2002.). This was the assumed case for this analysis. This particular void reached the surface and created a crown hole 1m wide by 0.5m deep.

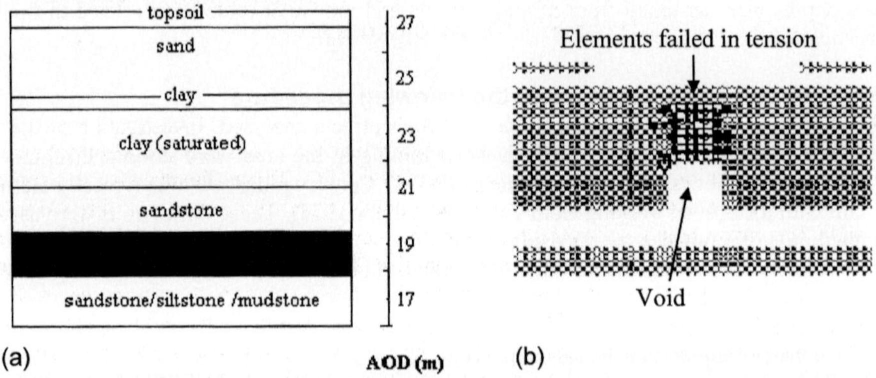

Figure 1 (a) Sequence of strata, (b) Zone state plot of void propagating into the superficial geology.

Various sized voids were considered in the model. A void size 1.5m deep and 3.5m wide was found to create a crown hole very similar in size to that observed on site, Figure 2(a).

It was found during the running of the model that altering the number of timesteps had the effect of altering which elements failed. Although this altered the progress of the collapse the collapse would always show a similar outcome irrespective of when the model was paused. The areas where few elements failed and the collapse slowed and occasionally the void self-arrested were within the top half of the sandstone unit and also the sands. The areas where the collapse accelerated were within the saturated clays.

It was found that when the void dimensions were altered within the assumed boundaries, larger voids were more unstable and tended to generate either a larger volume of bulked material or a larger surface void. This is in keeping with empirical theories of shallow subsidence collapse. The dimension of the void which showed a most profound effect on the surface subsidence was that of the seam thickness. This is also generally accepted and forms the basis of many empirical prediction theories. Voids of widths less than 3m were much more stable. Voids of this size often arched below the surface, Figure. 2(b).

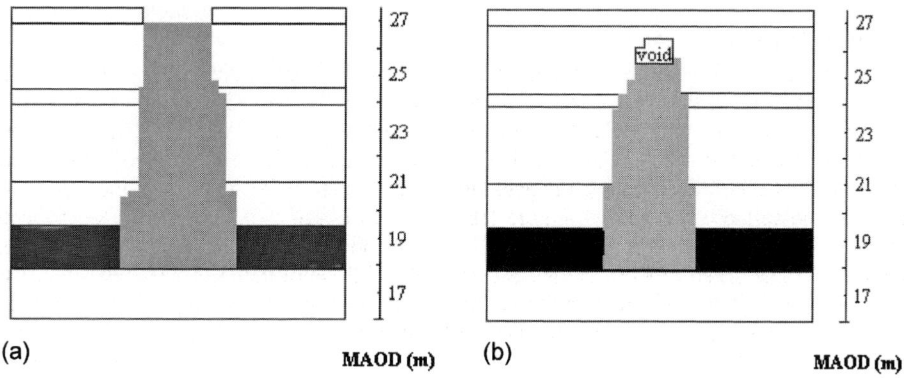

Figure 2(a). Collapse of void 3.5m wide and 1.5m deep forming crown hole 3m wide and 0.5m deep, (b) Arched void formed during collapse of void 3m wide and 1.5m deep. Shaded material represents the bulked, collapsed material.

Results

A void with dimensions of 3.5m width and a seam depth of 1.5m was found to produce the closest approximation of the crown hole observed on site. These dimensions are consistent with both the site investigation findings and known room dimensions for this method of mining. The crown hole that developed within the model was 2m wide and 0.5m deep. This is in very close agreement with the depression that formed on site considering the model resolution. The modelling has also revealed the mechanisms with which the void can propagate towards the surface. The modelling has also shown that it is possible for some considerable sized voids to mitigate towards the surface and become arched as little as 1m below the surface, Figure 2(b). Any ground disturbance nearby could instigate a full collapse. During the site investigation in 2001and the subsequent geotechnical works undertaken in 2002-3 there were a number of recorded instances of crown holes appearing close to drilling and grouting activities.

Discussion and conclusions

The model assumed an elliptical collapse. Variations on the actual shape of the crown hole developed on site cannot be simulated with a 2-dimensional model. The model used therefore has limitations. These limitations were deemed acceptable in this first study of the problem however a 3D model is required to simulate the collapse of non-symmetrical voids.

The collapse modelled was confined to a closed system in which material fell into a hypothetical void within the rock mass. The void was a sealed void and no escape of material was permitted. The actual void would be connected to a tunnel or adjacent room(s) which would permit the escape of some material until these entrances became blocked by fallen debris after which the collapse would become a closed system. This effect will be amplified by steeply dipping seams or mines with flowing water. With this consideration the actual void may be smaller than the model predicts. A 3D model would be able to simulate the loss of material into an adjacent room or tunnel.

The bulking factor of the fallen material was assumed to remain constant throughout the investigation. A larger bulking factor would cause the void to become choked at a deeper level and a smaller factor would allow the void to migrate further towards the surface and the possible formation of a larger crown hole. The assumption of a constant volume under loading contributes to the effects outlined above. The bulking factors used and assumptions made were consistent with the literature for coal seam stratigraphy, however for a more accurate investigation tests would need to be carried out on the relevant materials to determine more representative material properties.

The strengths of the material within the model were assumed using the available data. The materials are also assumed to remain homogenous. Further investigations may vary the material parameters which may affect the profile of the collapse material. Stronger sandstone may arch more before a void reaches the weaker superficial material. This would create a thinner column of bulked material which would have the effect of creating narrower, deeper crown holes on site. Weaker material would arch less therefore creating wider bulked profiles which could increase the probability of the void becoming bulked before reaching the surface.

The model exceeded expectations on correlation with site data. The possible stalling of a collapse of smaller voids has also gone some way to explaining the apparent 200 year delay for the full development of crown holes associated with long dormant mining. This may also explain the existence of holes appearing near drainage paths and areas of high disturbance, such as near the track, which were recorded during the site investigation in 2001 and the remedial works in 2002-3.

Further development

The model should be developed to analyse 3-D conditions and to take into account groundwater flow, seam dip and the effects of material migrating into adjacent voids and roadways. In addition the model can be improved to analyse the effects of strain softening and the presence of interbedded materials. Once these are included the model offers a powerful tool in the analysis of old mine workings and the development of risk scenarios.

Acknowledgements

The support of Railtrack in undertaking the study is gratefully acknowledged. Donaldson Associates Ltd. kindly provided detailed information of the geology and material parameters of the Dolphingstone site.

References

Bell, F.G. and I.A. DeBruyn (1999) "Subsidence Problems Due to Abandoned Pillar Workings in Coal Seams." Bull Eng Geol Env 57, pp 225-237.

Bieniawski, Z.T., (1978), "Determining rock mass deformability. Experience from case histories." Int. J. of Rock Mech and Min. Sci. Vol. 15 pp 237-242

Bieniawski, Z.T. (1989), "Engineering rock mass classifications: a complete manual for engineers and geologists in mining". Wiley New. York

Brady, B.H.G. and E.T. Brown (1993), "Rock Mechanics for Underground Mining". 2nd edition. Chapman and Hall. London.

Donaldson Associates Ltd. (2002). "ECML Dolphingstone Remediation Works Geotechnical Interpretative Report." Prepared for Edmund Nutall Ltd., Unpublished report.

Hoek, E. & Brown, E.T., (1988), "The Hoek- Brown Failure Criterion- a1988 Update". Rock Engineering for Underground Engineering (Proceedings of 15th Canadian Rock Mechanics Symposium, Toronto, October 1988), pp31-38. Toronto: Unoversity of Toronto, 1988.

Hoek, E., Kaiser, P.K. & Bawden, W.F. (1995) "Support of Underground Excavations in Hard Rock". A.A.Balkema. Rotterdam.

Itasca Consulting Group Inc (2002). Flac4.0: Users Guide.

Ramamurhy, T., (1986), "Stability of rock mass: 8th Indian Geotechnical Society annual lecture." Indian Geotechnical Journal 16. pp. 1–73.

Richards, L., Mazengarb, C., Beetham, D., Braithewaite, B. & Smith, W. (2002), "Waihi Underground Mine Workings, Stage II Investigations (Risk Assessment & Mitigation)".

Riggott, N., (1986), "Mine roadways deformation".
Colliery Guardian. Vol. 234, pp. 478+.

Serafim, J.L. & Pereira, J.P.,(1983), "Considerations on the Geomechanical Classification of Bienawski". Int. Symp. Engineering Geology and Underground Construction. Lisbon Theme II. Vol 1. pp II.33-II.42.

Taylor, R.K.(1975) "Characteristics of Shallow Coal Mine Workings and Their Implications in Urban Redevelopment areas" in Bell, F.G. (ed). Site Investigations in Areas of Mining Subsidence. Butterworth and co. (publishers) Ltd. London.

Tomlinson, M.J. (2001) Foundation Design and Construction 7th edition. Pearson Education Ltd. Essex.

Whittaker, B.N. & Reddish, D.J.(1989), "Subsidence: Occurrence, Prediction and Control" Elsevier Science Publishing Company Inc. New York

Treatment of Sewage Sludge Using Electrokinetic Geosynthetics

Glendinning, S, University of Newcastle upon Tyne
Lamont-Black, J, Electrokinetic Ltd.

Abstract
The treatment and disposal of sewage sludge has, historically, included depositing to land, forming lagoons. Once deposited, these materials do not change in material characteristics, even after several decades, leaving behind a legacy of deeply deposited low strength material. These pose a potential danger to the public and a source of contamination for adjacent water courses. Several researchers have shown that electrokinetic dewatering of sludge is an efficient means of driving off water from the sludge, thus improving strengthand reducing volume. This involves the application of a D.C Voltage across the sludge, driving water under an electrical gradient from positive (anode) electrode to negative (cathode) electrode. This paper will describe laboratory and field experimentation using electrokinetic geosynthetics: polymer- based materials containing conducting elements. These have been used to minimise the problem of electrode corrosion problematic with the use of traditional, metallic electrodes. Formed as tubes, or 'wicks', these materials offer a potential means to dewater sewage lagoons *in situ*.

Introcuction
It is well understood that the dewatering of sewage sludge is one of the most challenging technical tasks in the field of wastewater engineering (Metcalfe and Eddy, 1991) and all existing methods have severe limitations. These limitations may be attributed to the different physical states of water associated with the sludge and the manner with which they are bound to the sludge flocs. According to Smollen and Kafaar (1994), water exists in the following physical states:

- Free: water not associated with solid particles
- Interstitial, capillary: mechanically bound water which is trapped in the flocs
- Vicinal: physically bound multiple layers, held tightly to the particle surface by hydrogen bonding
- Chemically bound: water of hydration.

In addition, some sludges such as digested and activated sludges have a substantial intracellular component. Current dewatering methods are based on the use of mechanical pressure to cause the free water in the system to flow. This is performed by feeding a liquid sludge into a machine (centrifuge, belt press, etc.). This water can then be removed by mechanical means. More dewatering is achieved by exerting a greater and greater pressure. However, mechanical pressure alone cannot achieve a sufficiently high potential to drive the interstitial water through the very narrow pore spaces and removal of interstitial water through mechanical means is extremely limited. Vicinal and chemically bound waters are not removed at all by mechanical means. Therefore, once a cake has been formed, however bad the quality, there are few ways of dewatering it further. Currently only potential gradients produced by thermal methods are sufficiently high to remove interstitial and vicinal waters, involving high capital and operating costs. Electrokinetic techniques offer one potential cost-effective solution.

Electrokinetics

Electrokinetic techniques have been developed for treatment of clay soils, since their introduction as a construction technique by Casagrande in 1939 (Casagrande, 1949). Electrokinetics, for these applications, may be defined as the application, or induction, of an electrical potential difference across a soil mass containing fluid, or a high fluid content slurry/suspension, causing or caused by the motion of electricity, charged soil and/or fluid particles.

Electrokinetic phenomena are the result of the coupling between hydraulic and electrical potential gradients in fine grained soils (Acar & Alshawabkeh 1993, Mitchell 1993, Yeung & Mitchell 1993). These phenomena occur due to the presence of the diffuse double layer around the fine grained soil particles and involve the movement of electricity, charged particles and fluids (Mitchell 1993).

Electrokinetic phenomena may be defined in terms of five categories. Of these, the three most relevant are defined below and illustrated in Figure 1:

Electromigration/Ion migration - Applied electrical potential difference induces ion migration within the fluid phase of a charged particle matrix.

Electrophoresis - Applied electrical potential difference induces movement of suspended colloidal particles within a fluid medium.

Electroosmosis (E-O) - Applied electrical potential difference induces fluid flow in a charged particle matrix.

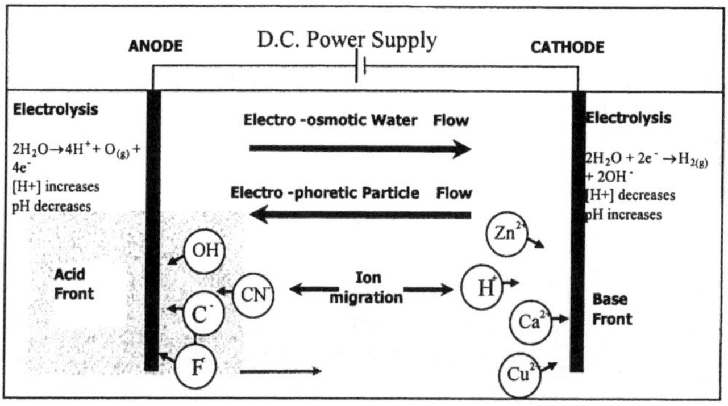

Figure 1 Electrokinetic Effects

Typically electroosmotic dewatering of clay soils is of the order of 1 to 4 orders of magnitude faster than hydraulic dewatering, with a typical value of electro-osmotic permeability (k_e) for a clay soil being 10^{-5} cm^2/Vs, as opposed to hydraulic permeability which ranges from 10^{-9} m/s to 10^{-5} m/s for silts and clays. Actual values of k_e and k_h for a range of soils are shown in Table 1 of the parallel paper on dewatering of tailings (Glendinning et al, 2006). The greatest advantage to be gained from electrokinetic dewatering over hydraulic dewatering is when the ratio of k_e: k_h is high.

Like clay particles, sludge particles are negatively charged. For this reason they too develop a diffuse 'double layer' of water surrounding the particles with the characteristic zeta potential at the boundary between the fixed and mobile portions of this layer. Because the flow of water induced by an electrical potential difference is not limited by pore size, electro-osmosis has the potential to remove interstitial water from the sludge flocs, thus greatly improving dewatering efficiency. Previous research, which has concentrated entirely on improving various dewatering machines, has showed that sludges are shown to vary in electrokinetic performance, but this was not backed up by characterisation of the fundamental electroosmotic parameters e.g. k_e (electroosmotic permeability), with the exception of Yan & Weng (2002) who noted a k_e value about 5 times that of many soils, i.e. $k_e{:}k_h = 5$.

Previous work

The idea of using electrokinetic techniques to dewater sludges is not new and has been investigated by several researchers (Smollen & Kafaar, 1994; Gazbar et al., 1994; Miller et al., 1998). The majority of these have concentrated upon the use of electro-osmosis to enhance the dewatering capabilities of a mechanical process and have incorporated, for example, electrically conducting rollers into a belt press. Other studies have concentrated on laboratory studies using cake materials.

The major drawback with these studies is their practical application. If E-O is to be used only in conjunction with mechanical pressure this severely limits its application within the overall framework of sludge and wastewater treatment. The requirement for mechanical pressure would mean that it could not be applied, for instance, to treat sludge lagoons *in situ*. Researchers who have investigated cake, have not suggested a means of implementation.

Another major deficiency is that all studies have used metallic electrodes. Electrolysis of water at the electrodes always produces acidic conditions at the anode which contributes to rapid corrosion of the metallic electrode. Within a relatively short period of time, the corrosion reduces the electrical contact with the soil and electrical efficiency is reduced. The inability of a metallic electrode to act as an effective drain also means that single polarity treatment, so that the final product is extremely heterogeneous.

EKG Concept

EKG comprises conducting elements coated in a non-corrosive material, incorporated into a geosynthetic material. This design has overcome the problem of electrode corrosion. By encasing the metallic filaments in a relatively inert material, electrode corrosion is eliminated. By forming the electrode as a geosynthetic, EKG overcomes the problem of removing clean water by utilising the drainage and filtration functions of geosynthetics. Their ability to take on a wide variety of shapes and forms to means that they can be manufactured to suit a range of different applications. It is envisaged that EKG could be applied in several ways:

- installed as vertical 'wicks' into sludge lagoons and used to draw water to the surface, removed by pumping, and discharged;
- installed as a combination of basal grid and fabric cover to increase dewatering rates in windrows; and
- formed as the fabric in a belt or filter press to improve dewatering performance.

Lagoon Applications

Electroosmotic consolidation could be applied practically in real situations by adapting the technology knows as prefabricated vertical drains (PVD) or wick drains. These are used routinely to consolidate soft soils *in situ*. PVDs work by creating short pathways for water to

flow out of a low permeability substrate such as fine grained lagoon waste. Using the traditional PVD approach, water flow is caused by creating a pressure difference between the PVD and the material to be consolidated. This pressure difference is achieved by placing a load (usually in the form of a layer of sand and gravel several metres thick) over the surface of the material to be consolidated. This process can take many months or even years to complete and is limited by the hydraulic permeability of the materials and the rate at which the overlying load material can be placed onto a weak substrate (staged loading).
Electroosmotic PVDs or *e*PVDs utilise the higher flow rates that are achievable by E-Owithout any additional load. The advantages of this approach are that it does not require any load and it is much faster. The speed advantage derives from the fact that electroosmotic permeability is higher than hydraulic permeability (for materials such as silts, clays and sludges), and that because no load is required, the full flow rate can be achieved immediately rather than needing to wait for gradual improvements in strength before additional load is applied as is the case with 'staged loading'.

Part of the approach with ePVDs is the reversal of polarity to create homogeneous gains in material strengths between anodes and cathodes. EKG electrodes are designed to operate equally effectively as cathodes or anodes and are thus ideally suited to secure the advantages offered by polarity reversal. This paper will describe laboratory trials and a subsequent field trial undertaken on lagooned sewage sludge.

Laboratory Testing

Tests were performed on a cold digested thickened lagooned sewage sludge with an initial dry solids content of approximately 15%. This sludge had been resident in the lagoon for approximately 3 years. Tests were performed to determine the ability of the materials to support E-O and the degree of consolidation that may be achieved by E-O. The coefficient of electroosmotic permeability, k_e, is used as an indicator of a materials ability to support E-O. The test is non-standard and has been developed at Newcastle University. It comprises measuring the flow of water caused by E-O under constant electrical gradient with a constant supply of water to the anode under zero hydraulic head. E-O consolidation tests were performed in a similar experimental apparatus, with the water supply to the anode switched off, i.e. operating under 'closed anode' conditions. An initial test of 3 days duration was performed, followed by a longer test of 3 weeks duration. All tests were undertaken using an applied back pressure of 50kPa and an applied voltage gradient of 1V/cm. The control test had zero applied voltage.

Both sets of tests were performed using parallel copper disc electrodes. The key features of the experimental set up are illustrated in Figure 2. Three repeat tests (A, B and C) were performed due to the variability of the initial material dry solids.

Results

The sludge demonstrated a k_e value of 1.5×10^{-5} cm^2/Vs, which compares very favourably with the types of clay soils that are most amenable to this treatment (see Table 1).

The results of the 3-day consolidation tests are shown in Figure 3. This illustrates the volume of water extracted from the sample with time and demonstrates the very marked difference between the control (with no voltage applied) and the samples subjected to an electric field.

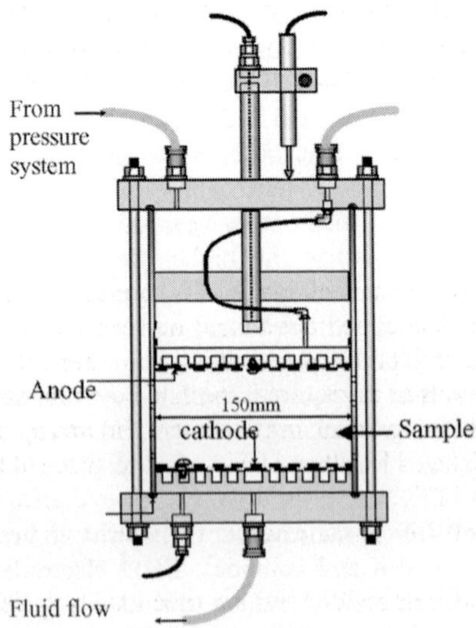

Figure 2 Electrosmotic Cell

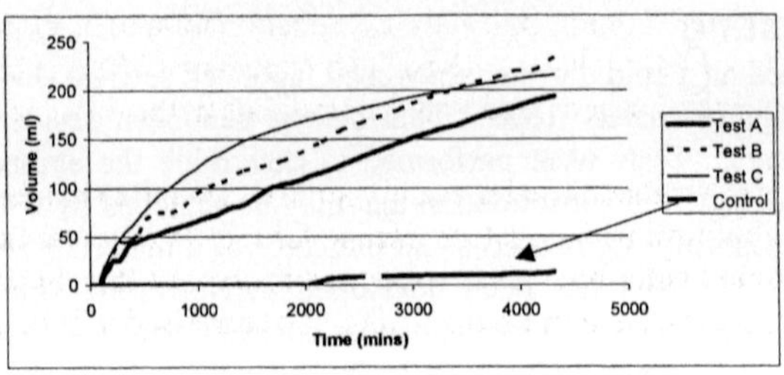

Figure 3. Volume discharge of water during 3 day E-O consolidation test

At the end of the 3-day test the control sample had a dry solids content of 16% with virtually no change in volume of the sample; the average dry solid content of the E-O treated samples was 27%, accompanied by a 10% reduction in volume. Results of the longer, 21-day tests (A and B) are summarised on Table 1.

Table 1 Summary of long term consolidation tests

	Test A	Test B
Initial moisture content	621 %	574 %
% dry solids	13.9% ds	14.84% ds
Test duration	21 days	21 days
Volume of extracted water	837 ml	1083 ml
% dry solids at cathode	31.8%	38.9 %
% dry solids at anode	64.5%	72.5%
Total volume reduction	-40%	-51.7%

The tests concluded that the sludge is a suitable substrate for effective E-O dewatering, which proved considerably more effective than hydraulically driven dewatering alone. Treatment could produce an overall volume reduction of between 40% and 50% over a period of 21 days.

Field Trial

Set up

The effectiveness of electrokinetic consolidation of lagooned sewage sludge cake by means of ePVDs (electrokinetic prefabricated vertical drains) was examined at pilot scale. Two steel containers of 9.7m^3 capacity with tailor-made butyl liners were filled with approximately 8.5m^3 of sludge. The sludge was in a liquid state with a low shear strength (unmeasureable using a hand vane, but estimated to be approximately 1kPa) and a solids content of only 10.6%. The timing of the trial was less than ideal as it took place at the end of the winter during which the lagoons had filled up with rain water. Two electrode arrays were evaluated:

- Rectangular array with Anode-Cathode spacing of 0.9m and Anode- or Cathode-Cathode spacing of 0.4m
- Hexagonal array with spacings of 0.7-0.9m (in practice an equilateral spacing would be chosen; here the hexagons were distorted to suit the containers).

The electrodes were composite EKG Mk5 ePVDs, which comprised six different components including a central perforated drain, filter fabric, primary and secondary conducting elements, and integral and binding knitting yarns. The arrays were run at 30V which is relatively low by electrokinetic standards, but provided voltage gradients in the order of 33V/m.

Water moves towards the cathodes under electroosmosis. Owing to the fact that the trials were performed in containers with no lateral escape for the water it was necessary to remove water. Siphon tubes fitted with foot valves were used to achieve this. In practice these siphons were only partially successful and in future an alternative method will be sought.

Treatment

The electrode arrays were powered up using a step-wise increase in voltage to avoid initial power surges often seen in electrokinetic treatment. After powering up the arrays were left on 24hrs a day with an intermittent current. Active treatment lasted for 92 days and was interrupted on several occasions relating to accidental stoppage of the generator or other factors, resulting in an active treatment period of 63 days or 9 weeks.

Results

Figure 4 shows a plot of sludge volume against time taken from the initial powering up. It can be seen that two separate regression lines have been inserted for each curve, accounting for the initial problems experienced with power supply. After 25th May, these problems with power supply were resolved and this is reflected in a greater rate of volume reduction in the second period, which is especially clear for Skip B. Overall this represents a reductions in volume of 23.0% for skip A and 29.7% for skip B. It is considered that the potential for reduction, however is much greater as the siphons installed to remove water from the cathodes were not particularly efficient and significant quantities of water collected at the surface of the sludge adjacent to the cathode from where it flowed back to the anode and effectively recirculated around the system.

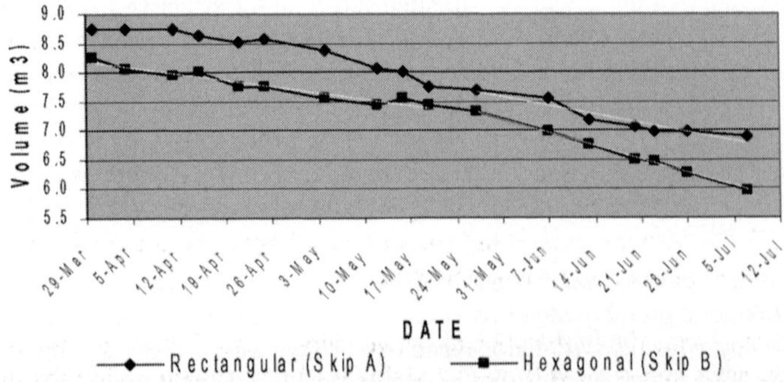

Figure 4 Volume reduction with time for in-situ 'skip' trials

Shear strength was measured at three different depths from the surface (top: 200mm; middle: 520mm; bottom: 825mm) using a hand shear vane in a regular grid pattern across each skip. The average shear strength for both skips at each depth is given in Table 2 and a before and after photograph of the Skip B is shown in Figure 5.

Figure 5 Skip B showing condition of sludge before treatment (left) and after (right).

Table 2 Average shear strengths after treatment in kPa

Skip	Top	Middle	Bottom
A	6.0	7.8	8.2
B	7.7	10.9	13.4

Again, it is felt that higher strengths can be achieved in the same time once the problem of water removal caused by the poor performance of the siphons is corrected.

Power consumption was calculated for the entire treatment period, from the readings made of current and voltage, to be 134 kWh/wet tonne and 105kWh/wet tonne for skips A and B respectively. However, it is felt to severely overestimate the amount of power required to treat this volume of soil if the overall treatment area is scaled up. This is due to the effect of the edges of the container, where there were no electrodes – in a full treatment the equivalent area of sludge would be influenced by a number of electrodes, just as the centre of the skips are being effected in this trial.

Discussion

Dealing with an old sludge lagoon has generally meant removing the sludge altogether and clearing the site. Attempting to remove such material is difficult since it is too thick to pump and will contain tufts of grass, etc., and is too thin to shovel since it will slump completely. Prior to the arrival of the EKG process only two methods for stabilising a lagoon *in situ* were available; either mixing in more and more dry material until the lagoon was effectively solid; or employing a specialist contractor to mix quicklime or cement into the sludge to stiffen it. This second option is very expensive and would be unlikely to be economic unless the site of the lagoon had significant development potential.

It is envisaged that, once the problems with power supply and siphoning have been resolved, the EKG 'wick' drain approach will make a viable in-situ treatment option for sludge lagoon. EKG will be installed into lagoons containing this type of material by firstly creating a working platform over the surface of the lagoon using a stiff geogrid overlain by a granular capping. From this working platform EKG 'wicks' will be lanced vertically through to the base of the lagoon at 1-2m centres. Water will be drawn to the surface from where it will be pumped and discharged. The design of an installation such as this will require initial laboratory characterisation of the sludge, followed by an on-site pilot-scale test, similar to the field trial described above.

References

Acar, Y.B., & Alshawabkeh, A.K. (1993) Principles of electrokinetic remediation. Environmental Science and Technology, 37 (13), pp 2638-2647.

Casagrande, L. (1949) Electro-osmosis in soils. Géotechnique, Vol. 1(3), pp 159-177.

Gazbar, S., Abadie, J. M. and Colin, F., 1994, Combined action of electro-osmotic drainage and mechanical compression on sludge dewatering. Water Science and Technology, 30 (8), pp169-175.

Glendinning, S, Lamont Black, J & Fourie, A., 2006. Dewatering of tailings using electrokinetic geosynthetics (EKG). Proceedings of the 5th International Geoenvironmental Confernce, ISSMGE, Cardiff.

Metcalf, L. and Eddy, H. P., Wastewater Engineering: Treatment, Disposal and Re-Use. Mcgraw-Hill, 1991.

Miller, S., Murphy, A., Veal, C. and Young, M., 1998. Improved filtration of sewage sludges using electrodewatering. CSIRO, Division of Energy Technology, Brisbane.

Mitchell, J. K. 1993. Fundamentals of Soil Behaviour. 2nd Edn., Pub. John Wiley & Sons Inc, New York, USA.

Smollen, M. and Kafaar, A., 1994. Electroosmotically enhanced sludge dewatering: pilot plant study. Water Science and Technology, 30 (8), pp159-168.

Yaun, C. and Weng, C. 2002. Sludge dewatering by electrokinetic technique: effect of processing time and potential gradient, Advances in Environmental Research, 2002, www.elesevier.com/loacte/aer,

Yeung, A.T., & Mitchell, J.K. (1993) Coupled fluid, electrical and chemical flows in soil. Géotechnique, 43 (1), pp121-134.

Dewatering of Tailings Using Electrokinetic Geosynthetics (EKG)

Glendinning, S, University of Newcastle upon Tyne
Lamont- Black, J, Electrokinetic Ltd.
Fourie, A, University of Western Australia

Abstract
Many tailings disposal facilities present stability and environmental contamination risks yet currently the technology available to dewater these materials means that it is too expensive to contemplate for all but the most severe risk scenarios. Electrokinetics represent powerful but relatively under utilised process for ground improvement offering the potential to treat these extremely wet, low strength materials either in-situ or prior to placement. The concept of EKG is the combination of electrokinetic phenomena with the various functions of geosynthetics, greatly reducing the technological barriers to the application of electrokinetics in practice. This paper will present results from a series of laboratory trials and a prototype site trial that conclude that electro-kinetic treatment offers a potential solution to in-situ dewatering of tailings lagoons.

Introduction
Many tailings disposal facilities contain material that undergoes extremely slow self-weight consolidation. Aside from the obvious dangers posed by such poorly consolidated material (eg the potential for flowslides or mudrushes in the event a breach of the impoundment occurs), there are problems of rehabilitation of the facility once it is closed. The extremely low undrained shear strength of such deposits means that earthmoving machinery cannot operate on the surface and placement of cover material and a growing medium is extremely difficult. In order to improve the strength of these materials and to increase the volume of storage space available it would be desirable to dewater the existing impounded tailings *in situ.* However, dewatering of this type is currently too expensive to contemplate for all but the most severe risk scenarios.

An alternative to in-situ treatment, and one which would prevent the same problem arising, would be to deposit the tailings in a thicker, dryer consistency. Much development is currently concentrating on thickening tailings pastes (e.g. Fourie, 2002). However, this thickening is constrained by the rheological necessity to pump the tailings several kilometres in a fluid state. Current technology is not capable of creating a material of sufficient thickness to transport by conveyor. Therefore it would be desirable to dewater the tailings before it is transported to the lagoon to a sufficient degree to allow it to be conveyed and to recycle the water back into the process plant.

Electrokinetic Geosynthetics
Electrokinetic techniques have been developed for treatment of clay soils, since their introduction as a construction technique by Casagrande in 1939 (Casagrande, 1949). Electrokinetics, for these applications, may be defined as the application, or induction, of an

electrical potential difference across a soil mass containing fluid, or a high fluid content slurry/suspension, causing or caused by the motion of electricity, charged soil and/or fluid particles.

Electrokinetic phenomena are the result of the coupling between hydraulic and electrical potential gradients in fine grained soils (Acar & Alshawabkeh 1994, Yeung & Mitchell 1993). These phenomena occur due to the presence of a diffuse double layer around grained soil particles and involve the movement of electricity, charged particles and fluids (Mitchell 1993). Electrokinetic phenomena may be defined in terms of five categories. Of these, the three most relevant are defined below and illustrated in Figure 1:
Electromigration/Ion migration - Applied electrical potential difference induces ion migration within the fluid phase of a charged particle matrix.
Electrophoresis - Applied electrical potential difference induces movement of suspended colloidal particles within a fluid medium.
Electroosmosis (E-O) - Applied electrical potential difference induces fluid flow in a charged particle matrix.

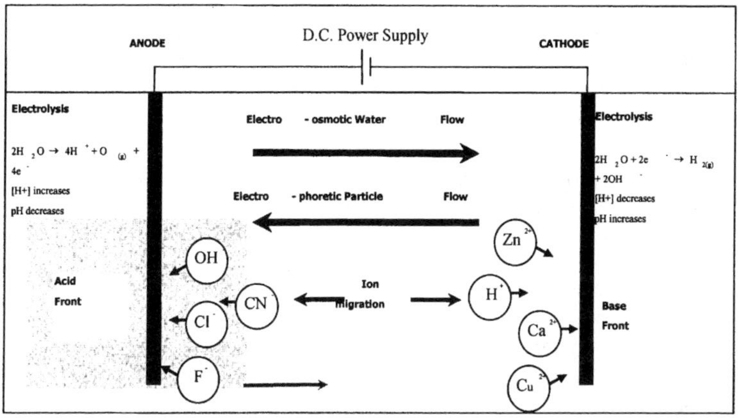

Figure 1 Electrokinetic Effects

Typically electroosmotic dewatering of clay soils is of the order of 1 to 4 orders of magnitude faster than hydraulic dewatering, with a typical value of electro-osmotic permeability (k_e) for a clay soil being 10^{-5} cm^2/Vs, as opposed to hydraulic permeability which ranges from 10^{-9} m/s to 10^{-5} m/s for silts and clays. Actual values of k_e and k_h for a range of soils are shown in Table 1. This table has been reproduced directly from Mitchell (1993); hence the use of non-standard units. The greatest advantage to be gained from electrokinetic dewatering over hydraulic dewatering is when the ratio of k_e: k_h is high.

The concept of EKG is the combination of electrokinetic phenomena with the various functions of geosynthetics. The combination concept of EKG has greatly reduced problems of electrode corrosion, and owing to its structure the electrodes have improved upon the limitations of current density and electrical contact with the soil. By forming the electrode as a geosynthetic, EKG overcomes the problem of removing electroosmotically gathered water by utilising the drainage function of geosynthetics. Newcastle University has demonstrated the use of EKG to in several applications, including the consolidation of super soft clays (Jones et al, 2002: Lamont-Black, 2001) and the construction of reinforced soil using a waste cohesive material (Glendinning et al 2005).

Table 1. Typical values of k_e and k_h (After Mitchell, 1993)

Material	Moisture content %	k_e x 10^{-5} cm^2/Vs	Approx. k_h cm/sec
Fine sand	49.7	4.1	10^{-4}
Clayey silt	31.7	5.0	10^{-5}
Kaolin	67.7	5.7	10^{-7}
London Clay	52.3	5.8	10^{-8}
Na montmorillonite	170	2.0	10^{-9}
Na montmorillonite	2000	12.0	10^{-8}

Applications of EKG

Process dewatering

It is possible to dewater large volumes of liquid slurries using belt or filter presses as is carried out by the water industry on sewage sludge. The process involves the application of pressure to the material, either by direct application (as in a filter press) or by passing two parallel filter belts through a series of rollers of ever decreasing size (as in a belt press). It is envisaged that the filter belts could be formed as an EKG, thus combining pressure driven dewatering with electroosmotic dewatering. This electrokinetically enhanced dewatering machine could be used to dewater large volumes of tailings prior to deposit in a lagoon, thus reducing the volume of the deposited material and providing the opportunity for recycling some of the water.

Lagoon Dewatering

Wick drains are commonly used to consolidate soft soils *in situ*. In some cases they have also been used on tailings lagoons. However, their application is limited by time and by access to the lagoon surface for installing the wicks and for the application of a surcharge (usually several meters of fill). It is envisaged that EKG could be installed into lagoons by firstly creating a working platform over the surface of the lagoon using a stiff geogrid overlain by a granular capping. From this working platform EKG 'wicks' will be lanced vertically through to the base of the lagoon at 1-2m centres. Every second 'wick' would be connected as an anode and the intermediate ones connected as cathodes. Water will be drawn to the surface at the cathodes and pumped to a discharge point. If a uniform final shear strength is desired, the current would then be reversed making the former anodes into cathodes. This process would be repeated several times. The design of an installation such as this will require initial laboratory characterisation of the sludge, followed by an on-site pilot-scale trial.

Laboratory Trials

Laboratory trials were conducted with an initial aim of establishing the susceptibility of 3 different types of tailings to electrokinetic treatment. Therefore tests were performed to determine the electroosmotic permeability (k_e) and a comparison made with values for materials known to be treatable using electrokinetics. These tests were followed by electrokinetic dewatering tests in which samples were allowed to dewater for a set period under the combined action of voltage and pressure.

Tests were also performed to determine electrical resistivity: materials with low resistivity (or high conductivity) will tend to use more power to achieve the same amount of dewatering as materials with high resistivity, therefore making the treatment less economically viable. The pH both before and after treatment was measured to ensure that electrokinetic treatment had no potential to create an environmental hazard due to the generation of extremes in pH by hydrolysis.

Mine Tailings Tested

Diamond mine tailings were tested from three separate sources in South Africa. For the purposes of this paper these will be labelled A, B and C. A and B tailings comprised flocculated thickened tailings. These materials were received in the form of a thick slurry or mud; P tailings were received in an unthickened form and thus had a much thinner initial consistency. An initial characterisation of the materials is shown in Table 2

Table 2 Characteristics of the mine tailings

	Water content (%)	Specific gravity	pH	Bulk density (Mg/m^3)	Dry solids content (%)
A	74.8	2.85	9.5	1.581	57.2
B	166	2.66	7.9	1.308	37.6
C	67.8	2.86	8.6	1.640	59.6

Test Methods

Tests were performed in Electrokinetic 'Rosli' Cells (Figure 2). Electroosmotic tests were run in parallel with control tests in which no voltage was applied to the control samples thus dewatering occurred solely by standard hydraulic means. The k_e was determined with an irrigated anode, allowing water to continuously pass through the sample.

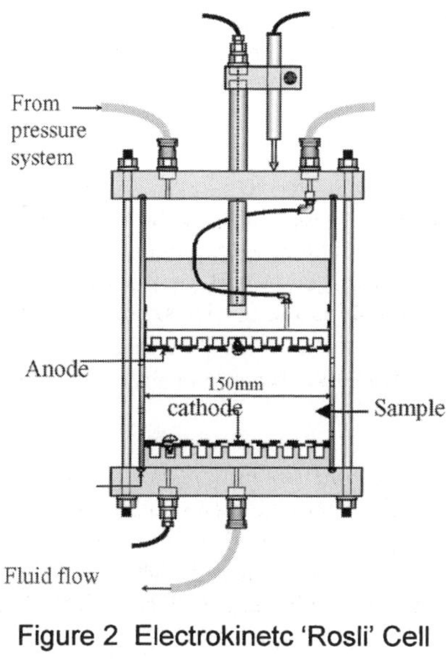

Figure 2 Electrokinetc 'Rosli' Cell

Electroosmotic dewatering tests were carried out with the anode closed (ie no water was allowed to enter the cell) using electrodes formed as an EKG. In this case this meant a piece of polyester woven filter cloth combined with carbon fibre conducting elements spaced at 10mm across its surface. In order to investigate the effect of hydraulic gradient on dewatering performance, applied voltages of 15 and 30V were used in conjunction with electrode separations of 15 and 50mm. All tests were carried out for 35 minutes under a static load of approximately 70kPa. Electrical resistivity was determined according to BS1377 (BSI, 1990).

Results

K_e, pH and electrical resistivity

A summary of the k_e, pH and electrical resistivity values for the three different tailings is shown in Table 3. If these values are compared with those of the reference materials in Table 1, it may be seen that all three tailings samples have relatively high k_e values and therefore are very good candidates for electrokinetic treatment. The electrical resistivity values are all lower than the range considered suitable for electrokinetic treatment by Casagrande (1983). However, this was considered in the context of consolidation of large areas of soil for civil engineering purposes. Other applications in other market sectors will have different suitability criteria, depending on factors such as timescales and treatment alternatives.. The higher value for the C tailings is probably attributable to its unthickened state: the addition of thickeners will increase the polyelectrolyte content, thus reducing resitivity. The pH values are the average of the bulk tailings samples for both 15 and 50mm samples after treatment and show little variation from their initial values.

Table 3 Summary of index values of mine tailings

	Dry solids	k_e cm^2/sV	Resistivity	pH at end of test
	%		Ωm	
A	57.2	$6\text{-}8x10^{-5}$	3.4	9.5
B	38.5	$1\text{-}2x10^{-5}$	3.2	7.3
C	59.7	$7\text{-}9x10^{-5}$	9.3	8.4

An example of the results for the k_e test is shown for A tailings in Figure 3. It may be seen that k_e is not constant with time. This is due to several factors, including electrochemical reactions at the electrodes. The k_e values quoted are taken as over a range, rather than a specific, value.

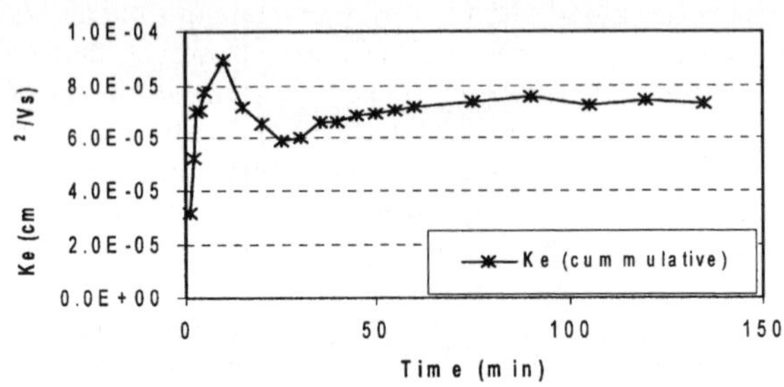

Figure 3 Plot of k_e vs time for K mine tailings

Electrode Configuration and dewatering performance

The % dry solids achieved at the end of the test has been used as an indicator of dewatering performance. A summary of the results for the different cell configurations is given in Table 4.

Table 4 Summary of % dry solids achieved with varying electrokinetic dewatering arrangements

Material (initial % ds)	Thickness	Applied Voltage 0V	15V	30V
A *(57.2)*	15mm	57.6	65.8	69.8
	50mm	58.3	58.1	62.1
B*(38.5)*	15mm	46.0	59.4*	55.7*
	50mm	38.7	42.4	42.0
C *(59.7)*	15mm	61.5	70.5	74.0
	50mm	60.0	62.3	64.6

* 19mm thickness

Discussion and Further work

It was apparent during testing that at the higher voltages, within 7-15 minutes of starting the test, the current dropped to just above zero, effectively terminating the test. This was attributed to the 'corrosion' of the fine carbon fibres caused by electrolysis reactions. Therefore, the 30V dewatering tests are likely to have significantly underestimated the dewatering. It is emphasised that carbon fibre electrodes would not be the chosen material for a working electrode and a much higher quality material has been developed for this application.

All mine tailings tested showed a good electrokinetic dewatering performance. The dewatering was very significant when compared to pressure dewatering. However, the higher resistivity of the C tailings means that less power is consumed in the process. Hence it would appear more economic to dewater unthickened tailings, saving money on the thickening agents and in electrical energy consumption

However, there is further laboratory testing required before the potential for electrokinetics when used in the different applications can be proven. These are summarised below. In each case it is recommended that the next phase of testing would be to choose test configurations more closely mapped to the desired application.

Voltage gradient. The voltage gradients used were in the range 3 to 20V/cm and are realistic when considered in the context of a belt or filter press. However, they are fairly high when compared to previous use of EKG for in-situ consolidation (of the order of 0.5 – 1.0V/cm).

Treatment time. The treatment times used were of the order of 3 times as long as the residence time in a standard belt press, yet very much shorter than times taken to carry our in-situ consolidation: these times were used to enable the feasibility of EK treatment to be assessed rapidly, but it is likely that a higher degree of dewatering is achievable after longer treatment times.

Pressure. The pressure used represents well to the pressures generated in a belt press and to a realistic surcharge of approximately 3.5m of fill in an in-situ application.

Electrode configuration. The parallel plate electrodes are representative of a belt or filter press, producing a one-dimensional electric field. However, electrodes place vertically into a

tailings lagoon produce a two-dimensional field. It is recommended that further tests are carried out in tanks similar to those described by Fourie et al (2004).
Electrode materials. As stated previously, the carbon fibres used in the EKGs were for experimental purposes only and more robust elements have been developed. The way in which these elements are combined with the geosynthetic would be tailored to the application: 2-dimentional electrodes for press applications and tubular, 3-dimentional 'wicks' for in-situ applications.

Changes in pH of the mine tailings as a result of treatment were small with a just discernable reduction in their pH. This appears to indicate that tailings lagoons treated in this way would not change in pH significantly, nor would tailings treated prior to placement. Hence there would appear to be a low risk of a severe environmental impact using electrokinetics to treat mine tailings. However, this would require further investigation using an experimental set up more closely mapped to field conditions to make a full environmental impact assessment. Also, every tailings is unique and would need to be characterised prior to treatment.

Field Trial

A field trial was conducted using vertical EKG 'wicks' inserted into the tailings contained in a 3m diameter by 1m deep plastic tank, as shown in Figure 4. The tailings was material from KwaZulu-Natal in South Africa; a residue from the recovery of ilmenite, zircon, rutile, leucoxene. The material was very fine, with 75% smaller than 2μm. A central cathode was placed at the centre of 6 anodes (placed 0.9m from the cathodes, with Anode-Cathode spacing of 0.9m) from where water was sporadically pumped. Treatment was conducted using a constant 30V during the day and 12V at night, and resulted in the abstraction of 2 cubic metres of water over a 2 month period as illustrated in Figure 5.

Figure 4 Field Trial: in-situ dewatering of mine tailings

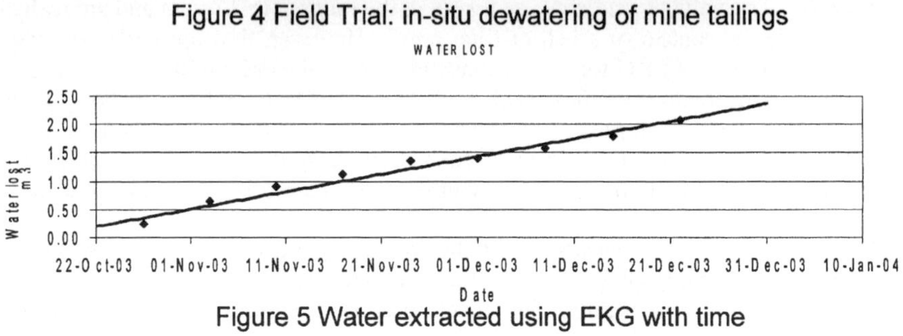

Figure 5 Water extracted using EKG with time

The overall volume of the tailings was reduced from 5.3 m^3 to 3.0 m^3, accompanied by an increase in shear strength from 2kPa to 18kPa and a reduction in water content from 158% to 75%. The overall power consumption was equivalent to 1.1kWh/m^3 water removed or 2.3kWh/wet tonne. This is considered as a fairly low power consumption, which may be attributable to additional hydraulic dewatering caused by daily pumping from the Cathode.

The trial proved that in-situ EK dewatering of tailings could be effected using EKG that the power consumption was not significant when compared to other treatment options, such as in-situ stabilisation using cementitious mixing, or digging out and depositing elsewhere.

Conclusions

Overall, these results show that dewatering tailings using electrokinetic geosynthetics has potential both for the in-situ dewatering of lagoons and for pre-depositional processing. Power consumption for in-situ treatment does not appear prohibitive. Further work is required (and is underway) on more application specific testing in parallel with the development of more robust electrodes.

References

Acar, Y.B., & Alshawabkeh, A.K. (1994) Modelling conduction phenomena in soils under an electric current. In Proc. 13th Int. Conf. on Soil Mech. and Found. Eng., January 5-10, 1994, Vol. 2, New Deli, India. Publ. by Balkema, Rotterdam, The Netherlands. pp 669-672.

British Standard Institution (1990), British Standard Methods of Test for Soils for Civil Engineering Purposes. BS 1377, BSI, 2 Park St. London, WA1 2BS.

Casagrande, L. (1949) Electro-osmosis in soils. Géotechnique, Vol. 1(3), pp 159-177.

Casagrande, L. (1983) Stabilization Of Soils By Means Of Electro-Osmosis. Journal of the Boston Soc. of Civil Engineers section, ASCE, Vol.69, (2), pp 255-302.

Fourie, A B. (2002) In search of the sustainable tailings dam: do high-density thickened tailings provide the solution? 4th Internal Conference on Environmental Geotechnics, Rio de Janeiro, .

Fourie, A B, Johns, D & Jones, C J F P. (2004) In-situ dewatering of mine tailings using electokinetic geosynthetics. Tailings and mine waste. London. pp341-345.

Glendinning, S, Jones, C J F P & Pugh, R C. (2005) Reinforced soil using cohesive fill and EKG. International Journal of Geomechanics, ASCE, Vol. 5 (2), pp138-146

Lamont-Black, J. (2001) EKG: the next generation of geosynthetics. Ground Engineering, Vol. 34 (10), pp. 22-23.

Mitchell, J. K. (1993). Fundamentals of Soil Behaviour. 2nd Edn., Pub. John Wiley & Sons Inc, New York, USA.

Yeung, A.T., & Mitchell, J.K. (1993) Coupled fluid, electrical and chemical flows in soil. Géotechnique, Vol. 43 (1), pp 121-134.

Water Management Simulation Around Tailings Impoundments

Guézennec A.G., BRGM, BP 6009, 45060 Orléans Cedex 2, France
Lachapelle S., BRGM, BP 6009, 45060 Orléans Cedex 2, France
Ménard Y., BRGM, BP 6009, 45060 Orléans Cedex 2, France
Burnol A., BRGM, BP 6009, 45060 Orléans Cedex 2, France
Bourgeois F., BRGM, BP 6009, 45060 Orléans Cedex 2, France

Abstract

The mining and quarrying industry produces very large quantities of fine-grained tailings that are generally disposed of in slurry form. The most common tailings disposal method consists in pumping the tailings into dedicated ponds. The ponds' twofold purpose is to confine the solids within their boundaries and achieve clarification of the process water for re-use. Questions can be raised about the nature of the interactions between a tailings pond and its surroundings. Useful information for tailing impoundment management can be obtained by simulation of both the impoundment filling process with that of water seepage. In this article, the authors present a simulation environment that combines dynamic prediction of the behavior of a tailing pond by sediment transport and consolidation modeling, with simulation of ground water seepage. It is concluded that computer simulation could become a powerful tool for assisting professionals with design, operation and monitoring of tailings ponds.

Introduction

The extractive industry produces very large quantities of fine-grained by-products referred to as tailings that are generally disposed of in slurry form. The most common and economical tailings disposal method consists in pumping the tailings into dedicated impoundments. Their twofold purpose is to confine the solids within their boundaries and achieve clarification of the process water for re-use (Environment Canada, 1987). Whether they are in-pit impoundments or specially-dug pits (Ritcey, 1989), tailing impoundments are becoming increasingly scrutinized with respect to their environmental performance. One area of concern is the control over possible solid and water transfers between the impoundments and hydrogeological structures such as aquifers, particularly with respect to the migration of seepage and pollutants (Ritcey, 1989). Such concerns have led to various monitoring systems devoted to seepage and groundwater monitoring, particularly on mining sites (Hutchinson and Ellison, 1992). As a result, tailing impoundment design is becoming one of the acknowledged pillars of integrated water management strategies used in the extractive sector (De Tomi and Owen, 1995). Another area of concern is the mechanical stability of tailings (Jakubick et al., 2003) once the impoundment is no longer in use. Notwithstanding the problems associated with the nature of the substances contained within the waste materials, whether the settled solids can reach a mechanical state that can carry a load, static or dynamic, is of particular concern in terms of public safety and land reclamation for re-use.

Tailing ponds are used by over 99% of sand and gravel quarries in France. With a total production of 200 Mt of aggregates per annum, sand and gravel wash plants produce about 20 Mt/yr of fine clayey rejects at low solids concentration that rarely exceeds 50 g/L (Gaboriau et al., 2002). These dilute streams are pumped directly into one or more tailing impoundments, where they undergo settling and self-weight consolidation. Increasingly faced with problems of land availability for slimes storage and concerns about environmental performance, quarry operators are seeking solutions for improving dimensioning and operation of tailing impoundments.

Ideally, tailing impoundment design should embed all aspects of solids and water transport within the geometry of the impoundment. Indeed, a satisfactory design protocol should consider sedimentation, mixing and deposition of particles, environmental interactions as well as sediment consolidation over the life of the impoundment as a basis for impoundment sizing. No set of empirical rules could be expected to capture such a complexity.

This article explores the possibility of developing a computer simulation for tailing impoundment design and operation in the quarrying industry. Although researchers have proposed simulation tools for predicting solids and water transport within a tailings impoundment (Consoli, N. and Sills, G., 2000), the originality of the approach comes from the fact that adds another important functionality, that is prediction of water transfers between the impoundment and hydrogeological formations such as subterranean aquifers

Model presentation

Overview of the model

The model takes into account three families of governing processes, that interact with one another:

- ***Water transport:***
  - *In the impoundment:* process water is transported between the impoundment inlet and outlet.
  - *In the surrounding aquifer:* groundwater flow around the impoundment governs water seepage, which is a pivotal point to the overall system water balance..
- ***Solids transport:*** the fine particles suspended in the process water are transported and deposited at preferential locations within the impoundment, depending on local flow field, which is itself modified by particle deposition.
- ***Sediment consolidation:*** after deposition at the bottom of the impoundment, particles form a sediment that undergoes self-weight consolidation, thereby modifying the topography of the impoundment and the flow field inside the impoundment.

The numerical model takes into account the transient nature of the involved phenomena thanks to an appropriate discretisation of the time domain. To reduce the calculation time, a 2-dimensional configuration has been chosen (depth and main direction). The width of the impoundment is supposed to be constant.

Figure 1 presents the discretization scheme for the impoundment and its surroundings. The calculation domain is divided into cells. An impoundment cell *"i"* is characterised by flow parameters (flow rate $v^i(t)$, water seepage $Q_s^i(t)$ and solid concentration $c^i(t)$) and sediment characteristics (sediment height $h_{sed}^i(t)$ and void ratio) whereas an aquifer cell *"k"* is characterised by its hydraulic head ($H_A^k(t)$). The water stage ($h_w(t)$) is the same in all cells. The simulation consists in calculating these parameters at each time step by solving the coupled equations of water transport, solid transport and sediment consolidation. The general algorithm is described in Figure 2.

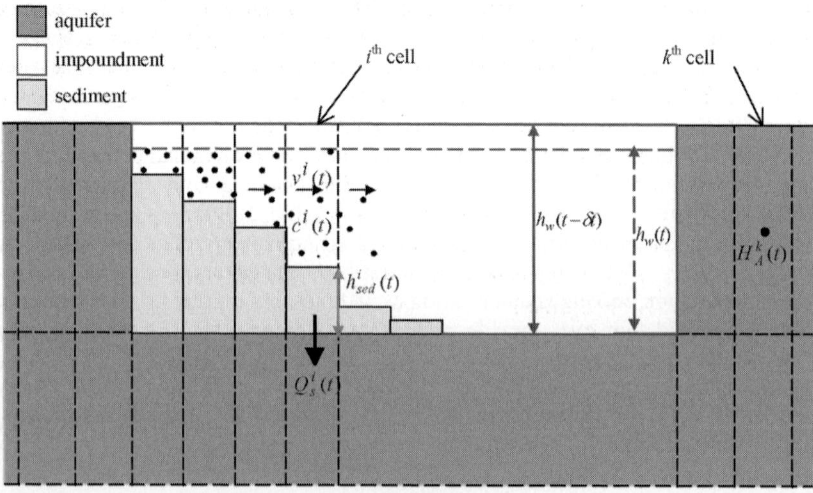

Figure 1: Schematic representation of the discretization of the impoundment and its environment

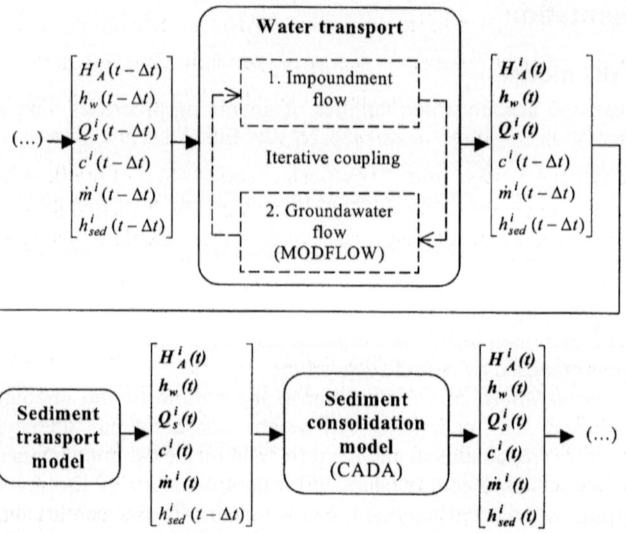

Figure 2: Overall algorithm; description of the simulation calculation over one time step

Hydrodynamic model

The hydrodynamic model is divided into two separate modules that are linked via an iterative coupling, which updates the flow field at each time step. Each module takes the other module's outputs as inputs. Since the two modules calculate the updated water stage value using independent methods, the iterative calculation ends when both modules yield the same value for the impoundment water stage.

- Module 1 solves the water balance in the impoundment by closing a set of simultaneous water conservation equations for each impoundment cell (see figure 1). For each cell, the water balance equation takes into account precipitation, evaporation and seepage at the cell/aquifer interface. For the inlet and outlet impoundment cells, the water balance equation also includes inflow and outflow rates respectively. Water seepage at the cell/aquifer interface is calculated from Darcy's law and the aquifer's head field returned by module 2. The solution of the system of mass balance equations gives the updated impoundment flow field and the corresponding water stage.

- Module 2, which embeds the MODFLOW software (Merritt and Konikow, 2000) solves the groundwater flow field. The calculated aquifer head field, from which the updated cell/aquifer interface flow rate values $Q_s^i(t)$ are inferred, is then input into module 1 for water balance calculation. It is worth pointing out that MODFLOW performs its own global water budget for the impoundment and therefore determines an updated water stage value independently of module 1.

An interesting feature of the coupling between the two modules concerns the boundary conditions at the impoundment/aquifer interface. Indeed, by predicting sedimentation and consolidation of particles fed to the impoundment, updated boundary conditions for each impoundment cell are passed to MODFLOW at each time step, which include sediment height and hydraulic conductivity.

Sediment transport model

At this point in the sequential algorithm of Figure 2, the velocity flow field has been updated. The next calculation step solves the transport equations for the solids between the impoundment inlet and outlet locations. This consists in calculating a deposition function $\dot{m}_{sed}^i(t)$ equal to the mass of solids deposited per unit time and per unit area for each impoundment cell. The deposition model is that of Ariathurai and Krone (1976). Based on the mass conservation of solids in the supernatant fluid, the model describes $\dot{m}_{sed}^i(t)$ as a function of the average concentration of suspended solids $c^i(t)$, of the flow field calculated by the water transport model and of the sedimentation velocity. This one depends on the particle size and on the concentration of suspended solid, and is calculated using a classical law of sedimentation.

Sediment consolidation model

At each time step, self-weight consolidation is implemented for every cell of the impoundment mesh. The consolidation calculation is done through successive calls to the commercial Fortran code CADA (Abu-Hejleh, 1996), which provides an efficient solution to the overall one-dimensional consolidation-desiccation process. The consolidation model relies upon two constitutive laws (Abu-Hejleh, 1996), namely $e = A \cdot (s' + Z)^B$ and $k = C \cdot e^D$, where e is the void ratio, s' is the effective stress, k is the sediment permeability and A, B, C, D are constants that are specific to the sediment. For each cell, the consolidation calculation computes the new consolidated height at the end of the time step and the associated void ratio profile. The latter is used to compute the new equivalent sediment hydraulic conductivity. The new calculated sediment height accounts for the combined consolidation of the sediments deposited during the previous time steps and the current one. Once the consolidation calculation has run through every cell, the impoundment bed topography is updated and the simulation proceeds to the next time step.

By integrating the three above models, the simulation tool is capable of predicting the evolution of the main characteristics of the tailings impoundment over time. Some of the predictive capabilities of the model are illustrated hereafter.

Case study

Presentation of the case study

The impoundment and surroundings are represented in Figure 3. The process water, which contains 30 $mg.L^{-1}$ of suspended solids with a 10 μm mean diameter, is fed to the impoundment at a rate of 300 m^3/h. The outlet rate is set equal to the inlet rate. The consolidation characteristics are described by the following values: A = 5.03, B = -0.42, C = 7.3 10^{-7}, D = 6.23, which were measured experimentally for a clayey aggregate plant tailings.

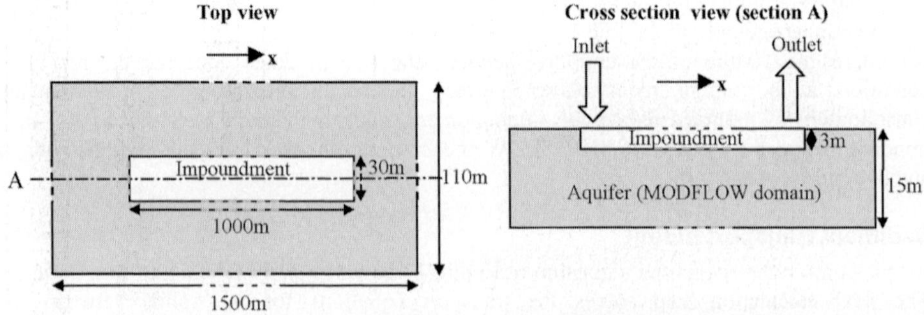

Figure 3: Schematic description of the geometry of the case study

The hydraulic conductivity of the aquifer is set to 10^{-4} m/s. Aquifer head boundary conditions are fixed to 15.25 m and 13.75 m upstream and downstream from the impoundment respectively. The initial head field in the rest of the domain were calculated by MODFLOW prior to other simulation. It corresponds to the equilibrium state of the system before the impoundment comes on stream (no inlet flow nor outlet flow).

Simulation results

Impoundment filling

Figure 4 shows the predicted evolution of the sediment layer with time.

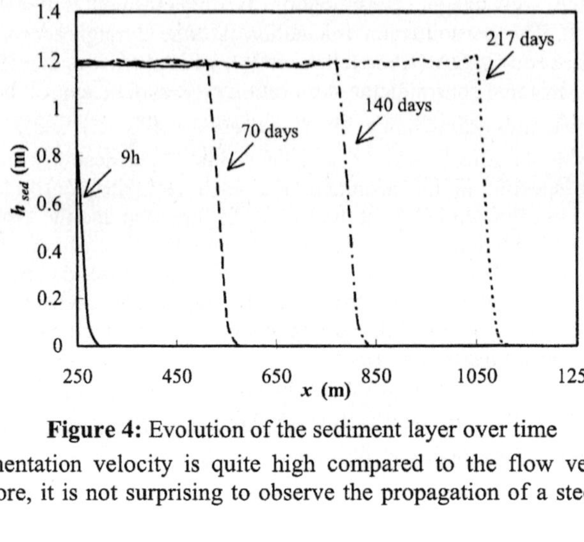

Figure 4: Evolution of the sediment layer over time

Here, the sedimentation velocity is quite high compared to the flow velocity (500 times greater). Therefore, it is not surprising to observe the propagation of a steep sediment front.

More gentle fronts are obtained with solids that exhibit slower settling velocities (with finer particle sizes for example).

As it can be seen in figure 5, the code is able to calculate the evolution of the void ratio profile of the sediment layer. The latter undergoes a compaction process that entails a consolidation of the sediment and, hence, a slight decrease of the sediment height. The consolidation state of the sediment layer is a key-parameter to determine which type of rehabilitation is adapted to the site after the complete filling of the impoundment.

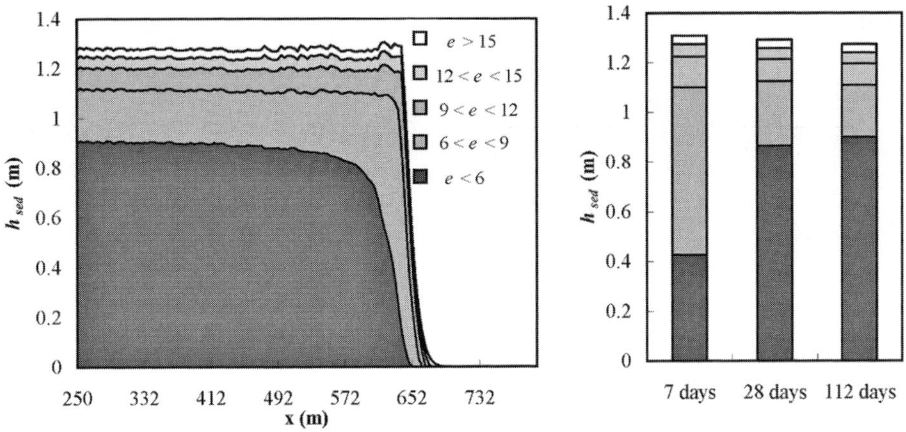

Figure 5: Void ratio profile in the sediment layer after 112 days (left) and evolution of the void ratio profile in the first cell (right)

Knowing the kinetics of the impoundment filling, it is interesting to determine the lifetime of the impoundment, defined here as the critical time at which the solids concentration in the supernatant water exceeds 30 mg/L at the impoundment outlet. In the case studied here, this condition is reached 252 days after the first working day of the impoundment.

Water transfers with the aquifer

Figure 6 shows the evolution of the water flow rates at the interface between the impoundment floor and the aquifer. While the impoundment is fed by the aquifer in the upstream region (positive seepage rates), it is leaking towards the aquifer in the downstream region (negative seepage rates). The variations of seepage rate at both ends of the impoundment come from the fact that the impoundment flow calculation accounts for the water transfer across both the side and bottom of the impoundment end cells. One can directly observe the influence of the deposition of sediments on the impoundment/aquifer interaction. It is readily observed that the formation of the sediment layer yields a progressive "clogging" of the impoundment bottom and sides, hence a gradual decrease of the transfers between the impoundment and the aquifer.

This information, which cannot be derived from empiricism, can only be obtained by simulation and is of direct interest from both production and environmental standpoints. From an environmental viewpoint, this information is particularly interesting for impact evaluation because it permits to predict transfers of pollutants from the impoundment to the aquifer. From a production angle, this information equates to predicting water availability for processing.

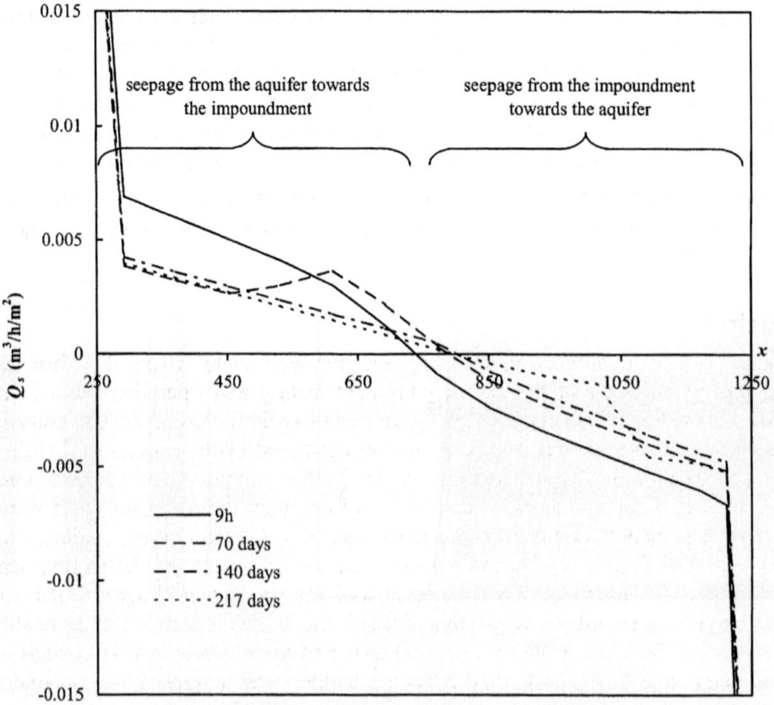

Figure 6: Profiles of the seepage rates along x-axis

As an illustration, the net seepage flow rate (Q_s^{tot}) can be calculated from the data of Figure 6 and compared to the net impoundment feed rate (Q_F) defined as the algebraic difference between the operating inlet and outlet flow rates. Figure 7 shows the evolution of Q_s^{tot} and of h_w with time when the outlet rate is 300 m^3/h.

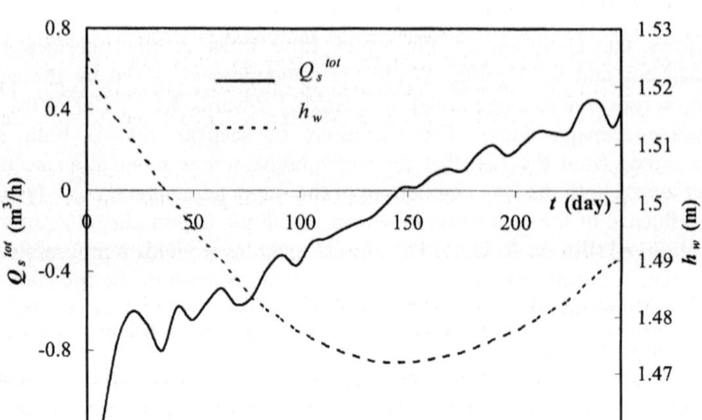

Figure 7: Evolution of the net seepage rate (Q_s^{tot}) and of the water stage (h_w) over time

In this case, Q_F is equal to zero, hence the variations of the water stage are due only to the seepage. At the beginning of the impoundment filling, the net seepage flow rate is negative, which entails a net loss of water for the impoundment and a decrease of the water stage. As time goes by, the impoundment walls are progressively waterproofed by the sediment deposition. This leads to a decrease of the water transfers from the impoundment to the aquifer until Q_s^{tot} becomes positive. At this time (145 days), the seepage process is inverted and the aquifer begins to feed the impoundment. Note that even if Q_s^{tot} is positive, local values of Q_s can be negative, hence allowing for flow from the impoundment towards the aquifer.

Conclusion

This article presents a simulation environment that can yield practical information for quarrying professionals about tailing impoundment design and operation. Based on some simplifying assumptions that permit coupling between transport, deposition and consolidation processes, the simulation model is able to predict the timed evolution of a full-scale tailing impoundment. The wealth of information output by the simulation makes it a potentially valuable design tool for operators. It can yield information about a number of production related issues that include instantaneous slimes storage volume reserves, available quantity and quality of clarified water, consolidation state of the sediment layer, impoundment lifetime. Beyond the benefit of the simulation tool for production purposes, it also holds promises from an environmental standpoint. Indeed, the simulation code can be readily used to predict transfers between a tailing impoundment and hydrogeological formations such as subterranean aquifers or rivers, and their influence on the water budget of the impoundment.

References

Abu-Hejleh, A.N., (1996). "Software CADA : Description of Input Data & Output Results" (http://members.aol.com/CADA1234/manual.pdf).

Ariathurai, R., and Krone, R. B. (1976). "Finite element model for cohesive sediment transport." J. Hydr. Div., ASCE, 102, 323-338.

Consoli, N.C. and Sills, G. C. (2000). "Soil Fomation from tailings: comparison of predictions and field measurements", Géotechnique, 50 (1), 25-33.

De Tomi, G and Owen, D., 1995. "Management of Mine Water Integrated to Mining Systems", in Proceedings of the APCOM XXV 1995 Conference, Brisbane, 9-14 July, 1994, 567-571

Environment Canada, Mining, Mineral and Metallurgical Process Division, 1987. "Mine and Mill Wastewater Treatment". Report EPS 2/MM/3. Minister of Supply and Services Canada Catalog NO. En 49-7/2-3.

Gaboriau, H., Bourgeois, F., Bizi, M., Baudet, G. et Le Berre, P., 2002. "Problématique du recyclage des eaux et de la gestion des rejets fins dans les carrières de granulats", Mines et Carrières, 84, pp. 44-50.

Hutchinson,I. P.G., and Ellison, R. D., 1992. "Mine Waste Management" Lewis Publishers. pp453 to 497

Jakubick, A.T., McKenna, G. and Robertson, A.M., 2003. "Stabilisation of tailings deposits: international experience", in proceedings of the Sudbury 2003 – Mining and the Environment Conference, 25-28 May 2003.

Merritt, L.M. and Konikow, L.F., 2000. "Documentation of a computer program to simulate lake-aquifer interaction using the MODFLOW ground-water flow model and the MOC3D solute-transport model" USGS, IR 00-4167, 146 p.

Ritcey, G.M., 1989. "Tailings Management: Problems and Solutions in the Mining Industry". Elsevier Science Publishers, B.U. Amsterdam, The Netherlands.

Slope Stability Analysis for Plane Failure Under the Influence of a Tension Crack

Haswanto, Department of Mining, University of Trisakti, Jakarta, Indonesia,
Email : haswanto_wa@yahoo.com.
Ghani Rafek, A., Department of Geology, UKM, Bangi, Malaysia.

Abstract
The significance of tension cracks for the stability of slopes has been widely recognized. This study is based on the Plane Failure model for the analytical solution of rock slopes in which the upper slope surface and the tension crack are inclined. In the present paper an attempt has been for made to correct and modify this approach. Application of the analytic solution for rock slope stability analysis for an overburden slope at Air Laya coal mining site PT.Bukit Asam, Tanjung Enim, Sumatera, Indonesia is discussed.
The application of the calculated analytic solution of the safety of factor for dry and saturated condition in the tension crack is carried out. This indicates that the factor of safety increases considerably as the angle of tension crack inclination increases. The factor of safety decreases from dry conditions to saturated conditions.The height of water in the tension crack will have a significant influence on factor of safety. For the influence of earthquake conditions, is was found that the factor of safety increases considerably as angle of tension crack inclination increases.This method is very significant in determining the critical location, depth and angle of the tension crack
Key-words : *Slope stability analysis, plane failure, tension crack.*

Introduction
Plane failures in rock slopes occur when a geological discontinuity strikes parallel or nearly parallel to the slope face and dips at angle greater than the angle of internal friction. Hoek and Bray (1981), Ling and Cheng (1997), Haswanto and Ghani (2004) gave the analytical solution for the plane failure mode in rock slopes. In this analysis, they assumed that the upper slope surface is horizontal and tension crack is vertical. Sharma et.al (1995), Haswanto and Ghani (2004-1;2), gave the analytical solution for the rock slopes in which the upper slope surface and tension crack are inclined. In the present paper an attempt has been made to correct and modify their approach. The objectives of this analysis are : (1) Analytic solution for plane failure model for rock slopes in which the upper slope surface and tension crack are inclined, (2) Application of the analytic solution to rock slope stability analysis for coal mining at PT. Bukit Asam (PT.BA), Tanjung Enim, Sumatera, Indonesia, and (3) to determine the appropriate depth, angle and location of tension crack at the rock slope.

Geometry of the Slope
The geometry of the slope considered in the present analysis is defined in figure.1. The various symbol used in this figure are: ψ_f, slope face angle; ψ_S, upper slope surface angle; ψ_P,

dip of the potensial failure plane; ψ_T, angle of the tension crack; h, height of slope; Z_L, height of tension crack; W, weight of sliding block; U, uplift water force acting on the block; V, water force in the tension crack, acting on the rear face of the block; α, coefficient acceleration of earthquakes.

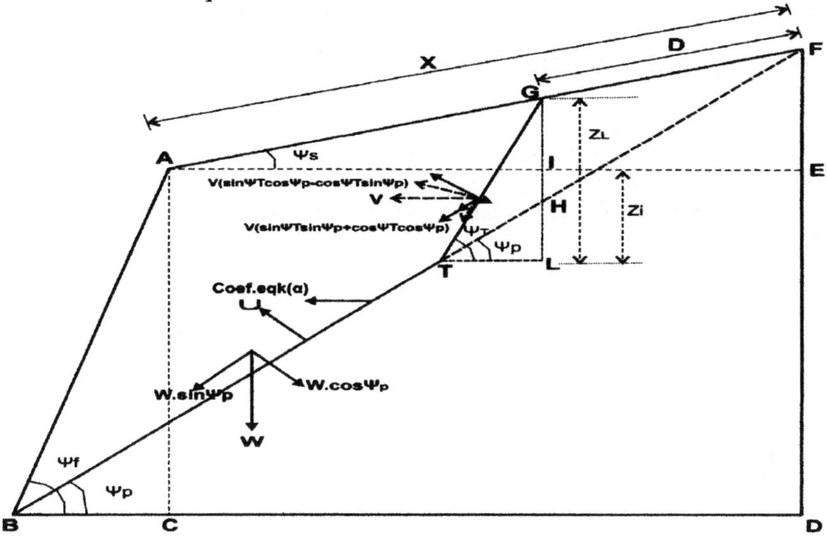

Figure.1. Geometry of the slope.

Plane Failure Analysis for Inclined Upper Slope Surface and Tension Crack

In the present analysis, the general condition as assumed by Hoek and Bray (1981) and Sharma, et.al (1995) and Haswanto and Ghani (2004-1;2) remains the same for plane failure except that the upper slope surface and tension crack are inclined. For the present analysis the following general condition must be satisfied :

1. The failure plane must strike parallel or nearly parallel (approximately $\pm 20^0$) to the slope.
2. The dip of the failure plane must be smaller than the dip of the slope face ($\psi_P < \psi_f$)
3. The angle of internal friction (ϕ) of the failure plane must be smaller than the dip of the failure plane ($\phi < \psi_P$);
4. The upper slope surface and tension crack must be inclined;
5. The dip of the upper slope must be smaller than the dip of the failure plane ($\psi_S < \psi_P$);
6. The tension crack must be present on the upper slope surface.

The following assumptions are made in this analysis :

1. The tension crack is filled with water to a vertical depth of Z_w.The water from the tension crack seeps along the failure surface and escapes out on the slope face where the failure surface daylights.
2. It is presumed that there is no resistance to sliding at the lateral boundaries of the slide.

These equations are equivalent to those given by Hoek and Bray (1981), Sharma, et.al (1995), Ling and Cheng (1997), Haswanto and Ghani (2004-1;2) and the results are :

Area of the Sliding Block

The area of the sliding surface, A, is represented by the length of the surface visible in a cross section drawn through the slope. (see Hoek and Bray, 1981., and Sharma, et.al.,1995, and Haswanto and Ghani, 2004-1;2).

$$A = h\left(1 - \left(\frac{Z}{h}\right)\frac{\sin\psi_T}{(\sin\psi_T - \tan\psi_P.\cos\psi_T)} - \frac{(\sqrt{\text{ctan}\,\psi_F.\text{ctan}\,\psi_P}) - \text{ctan}\,\psi_F}{\text{ctan}\,\psi_S}\right) * \text{cosec}\,\psi_p \quad .. \quad 1$$

Height of the Tension Crack

$$Z = h.\left[\left(1 - \frac{\text{ctan}\,\psi_F}{\text{ctan}\,\psi_S}\right) + \left\{\left(\sqrt{\frac{\text{ctan}\,\psi_F}{\text{ctan}\,\psi_P}}\right) * \left(\frac{c\tan\psi_P}{c\tan\psi_S} - 1\right)\right\}\right] \quad \ldots\ldots\ldots\ldots \quad 2$$

Weight of the Sliding Block :

$$W = \frac{1}{2} * \gamma * h^2 * \left[\begin{array}{l}\left\{1 + \left(\dfrac{\tan\psi_S}{\tan\psi_F} . \dfrac{\tan\psi_P - \tan\psi_F}{\tan\psi_S - \tan\psi_P}\right)\right\} * \left(\dfrac{\text{ctan}\,\psi_F}{\cos\psi_S} . \dfrac{\tan\psi_P - \tan\psi_F}{\tan\psi_S - \tan\psi_P}\right) - \\ \left(\dfrac{Z}{h}\right)^2 * \left(\dfrac{1}{\tan\psi_P.(\cos\psi_S - \sin\psi_S}\right)\end{array}\right] \quad \ldots\ldots \quad 3$$

Horizontal Water Force

The horizontal force, V, due to water pressure in the tension crack, acting on the rear face of the block is derived from :

$$V = \frac{1}{2}.\gamma_W.Z_W^2.\sin^2\psi_T \quad \ldots\ldots\ldots\ldots \quad 4$$

where, γ_W is the unit weight of water and Z_W is the hight of the water column in the tension crack.

Uplift Water Forces

The water seeps through the tension crack into the sliding surface and results in an uplift force, U. This uplift force is calculated from :

$$U = \frac{1}{2}.\gamma_W.Z_W.\sin\psi_T.h.\left(\begin{array}{l}1 - \left(\dfrac{Z}{h}\right)\dfrac{\sin\psi_T}{(\sin\psi_T - \tan\psi_P.\cos\psi_T)} - \\ \dfrac{(\sqrt{\text{ctan}\,\psi_F.\text{ctan}\,\psi_P}) - \text{ctan}\,\psi_F}{\text{ctan}\,\psi_S}\end{array}\right) * \text{cosec}\,\psi_p \quad \ldots\ldots\ldots \quad 5$$

Factor Of Safety

Factor of safety (FOS) can be defined as the ratio of the total forces available to resist sliding to the total forces tending to induce sliding. The FOS of the slope considered in this analysis, can be derived from the equation where c is the cohesive component of strength of the failure plane. These equation are equivalent to those given by Hoek and Bray (1981), Sharma, et.al (1995), Ling and Cheng (1997), and then the equation to develop by Haswanto and Ghani (2004-1;2) and the results are equation 6 and 7.

$$F = \frac{C.A + (W\cos\psi_p - U - V.(\sin\psi_T.\cos\psi_p - \cos\psi_T.\sin\psi_p)).\tan\phi}{W.\sin\psi_p + V.(\sin\psi_T.\sin\psi_p + \cos\psi_T.\cos\psi_p)} \quad \ldots\ldots\ldots \quad 6$$

and Slope stability under the influence of earthquakes is equation 7.

$$F = \frac{C.A + (W(\cos\psi_p - \alpha\sin\psi_P) - U - V.(\sin\psi_T.\cos\psi_p - \cos\psi_T.\sin\psi_p)).\tan\phi}{W.(\sin\psi_p + \cos\psi_p) + V.(\sin\psi_T.\sin\psi_p + \cos\psi_T.\cos\psi_p)} \quad \ldots\ldots\ 7$$

Substituting eq. 1,2,3,4 and 5 in eq.6 for non influence of earthquakes.
Substituting eq. 1,2,3,4 and 5 in eq.7 for influence of earthquakes.

Effect of Inclined Upper Slope Surface and Tension Crack on FOS

To examine the effect of the inclined upper slope and tension crack on the factor of safety (**FOS**), a example was considered. The data for this example is an overburden slope from PT.BA, Tanjung Enim coal mine which is as follows :

Slope face angle (ψ_f)	$= 60^0$
Upper slope surface angle (ψ_S)	$= 0^0$ to 30^0
Dip of potensial failure plane (ψ_P)	$= 40^0$
Angle of tension crack (ψ_T)	$= 70^0$ to 90^0
Height of slope (h)	= 20 m
Cohesion of rock mass (C)	$= 23.56\ kNm^{-2}$
Angle of internal friction (ϕ)	$= 13.53^0$
Unit weight of the rock (γ)	$= 21\ kNm^{-3}$
Unit weight of water (γ_w)	$= 10\ kNm^{-3}$
Height of water column in the tension crack (Z_w)	= 0 and Z m
Coefficient of acceleration for earthquake (α)	= 0.1 to 0.3

The results are presented in tables 1,2 and 3. A plot of the factor of safety against the angle of the upper slope (ψ_S) and angle of tension crack are given in fig.2.and 3. From the calculated result is the overburden slope is not stable because FOS <1 for the slope face of 60^0, and the slope stable for slope face angle of 45^0. In general, this indicates that the factor of safety for non influence of earthquakes decreases considerably as the upper slope inclination increases. For the case where the upper slope is horizontal, slope face angle of 45^0 for dry and saturated conditions, the calculated factor of safety would be 1.46 for dry conditions and 1.32 for saturated conditions. If the upper slope inclination is increased from 0^0 to 30^0, the factor of safety drop to 0.50 for dry conditions and 0.46 for saturated conditions. For the influence earthquake conditions, this factor of safety decreases considerably as upper slope inclination increases. For the case where the upper slope is horizontal and $\psi_T=90^0$, with a slope face angle of 45^0 for dry and saturated conditions and under the influence earthquakes, the calculated factor of safety would be 1.28 for dry conditions and 1.16 for saturated conditions. If the upper slope for inclination is increases from 0^0 to 30^0, the factor of safety changes from 1.28 drop to 0.41 for dry conditions and from 1.16 drop to 0.39 for saturated conditions.

For the case where the angle of tension crack is 70^0, slope face angle 45^0 with dry and saturated conditions, the calculated factor of safety is 1.41 for dry conditions and 1.28 for saturated conditions. If the angle of the tension crack increases from 70^0 to 90^0, the factor of safety increases from 1.41 to 1.46 for dry conditions, and factor of safety increases from 1.28 to 1.32 for saturated conditions. This indicates that the factor of safety increases considerably as angle of tension crack inclination increases.

The impact of the tension crack inclination (ψ_T) on the FOS is shown in fig.3. The effect is appreciable, and the FOS actually increases as angle of the tension crack (ψ_T) increases.

For the influence of earthquakes, it is indicated that the factor of safety increases considerably as angle of tension crack inclination increases. For the case where the angle of tension crack is 70^0, slope face angle is 45^0 for dry and saturated conditions and under the influence of earthquake conditions, the calculated factor of safety would be 1.24 for dry conditions and 1.13 for saturated conditions. If the angle of the tension crack increases from 70^0 to 90^0, the factor of safety changes from 1.24 up to 1.28 for dry conditions and from 1.13 up to 1.16 for saturated conditions.

This indicates that factor of safety decreases considerably as coefficient of acceleration for earthquakes increases. For the case where the coefficient of acceleration for earthquake is 0.1, slope angle is 45^0, and angle of the tension crack of 70^0 for dry and saturated conditions, the calculated factor of safety would be 1.24 for dry conditions and 1.13 for saturated conditions. If the coefficient acceleration of earthquake increases from 0.1 to 0.3, the factor of safety decreases from 1.24 to 0.99 for dry conditions and from 1.13 to 0.90 for saturated conditions.

Table 1. Stability analysis for overburden slope at Air Laya mine site.
(non influence of earthquake conditions)

Case	(α_f)	(α_s)	Dry			Saturated		
			$\alpha_T=70^0$	$\alpha_T=80^0$	$\alpha_T=90^0$	$\alpha_T=70^0$	$\alpha_T=80^0$	$\alpha_T=90^0$
1	60^0	0^0	0.55	0.59	0.61	0.41	0.42	0.43
2		10^0	0.44	0.47	0.49	0.33	0.34	0.35
3		20^0	0.35	0.37	0.38	0.29	0.30	0.30
4		30^0	0.30	0.31	0.31	0.28	0.28	0.28
1	**45^0**	**0^0**	**1.41**	**1.44**	**1.46**	**1.28**	**1.30**	**1.32**
2		**10^0**	**1.10**	**1.13**	**1.15**	0.99	**1.01**	**1.02**
3		20^0	0.78	0.80	0.82	0.71	0.72	0.73
4		30^0	0.48	0.49	0.50	0.45	0.45	0.46

Table 2. Stability analysis for overburden slope at Air Laya mine site.
(dry and under the influence of earthquake conditions)

Case	(α_f)	(α_s)	Coef.earthq. 0.1			Coef.earthq. 0.2			Coef.earthq. 0.3		
			$\alpha_T=70^0$	80^0	90^0	$\alpha_T=70^0$	80^0	90^0	$\alpha_T=70^0$	80^0	90^0
1	60^0	0^0	0.47	0.50	0.52	0.40	0.43	0.45	0.35	0.38	0.40
2		10^0	0.37	0.39	0.41	0.31	0.34	0.35	0.27	0.29	0.30
3		20^0	0.29	0.31	0.32	0.24	0.26	0.27	0.20	0.22	0.23
4		30^0	0.25	0.25	0.26	0.20	0.21	0.21	0.17	0.17	0.17
1	**45^0**	**0^0**	**1.24**	**1.27**	**1.28**	**1.10**	**1.13**	**1.14**	**0.99**	**1.01**	**1.02**
2		10^0	0.96	0.99	**1.00**	0.85	0.87	0.89	0.76	0.78	0.79
3		20^0	0.68	0.70	0.71	0.56	0.61	0.62	0.53	0.54	0.55
4		30^0	0.41	0.42	0.42	0.35	0.36	0.36	0.30	0.31	0.31

Table 3. Stability analysis for overburden slope at Air Laya mine site.
(Saturated and under the influence earthquakes condition)

Case	(α_f)	(α_s)	Coef.earthq. 0.1			Coef.earthq. 0.2			Coef.earthq. 0.3		
			$\alpha_T=70^0$	80^0	90^0	$\alpha_T=70^0$	80^0	90^0	$\alpha_T=70^0$	80^0	90^0
1	60^0	0^0	0.35	0.36	0.37	0.30	0.31	0.32	0.26	0.27	0.28
2		10^0	0.28	0.29	0.30	0.24	0.25	0.25	0.20	0.21	0.22
3		20^0	0.24	0.25	0.25	0.20	0.21	0.21	0.17	0.17	0.18
4		30^0	0.23	0.23	0.23	0.19	0.19	0.19	0.16	0.16	0.16
1	**45^0**	**0^0**	**1.13**	**1.15**	**1.16**	**1.00**	**1.02**	**1.03**	0.90	0.91	0.93
2		10^0	0.87	0.88	0.90	0.77	0.78	0.79	0.68	0.69	0.71
3		20^0	0.61	0.62	0.63	0.53	0.54	0.55	0.47	0.48	0.49
4		30^0	0.38	0.38	0.39	0.32	0.33	0.33	0.28	0.28	0.29

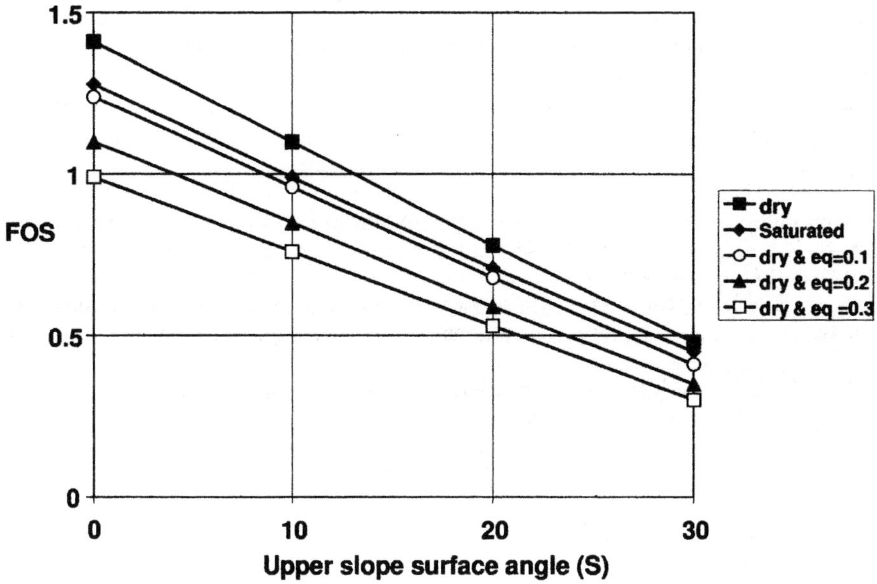

Figure.2. Effect of upper slope angle (ψ_S) on factor of safety (FOS).

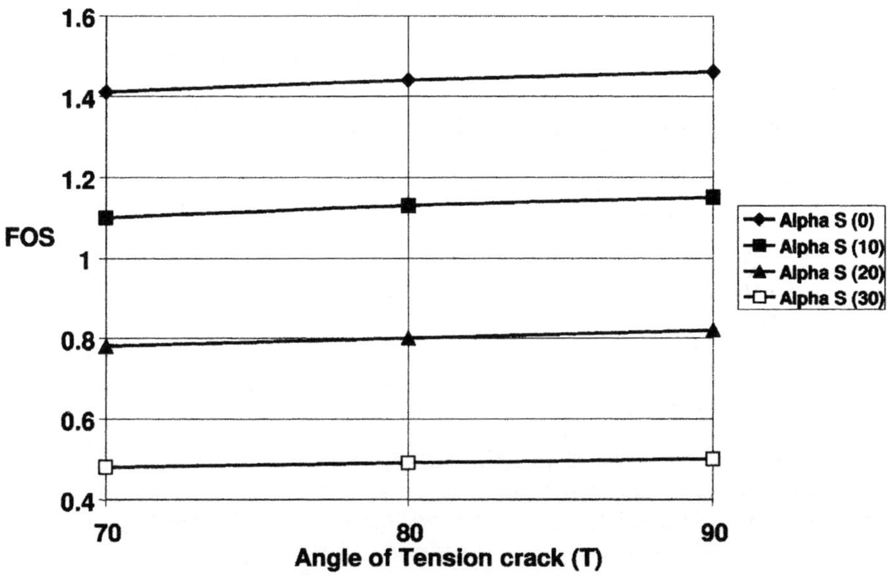

Figure.3. Effect of angle of tension crack (ψ_T) on factor of safety (FOS).

The plane failure analysis for the overburden slope on Air Laya coal mining at PT.BA Tanjung Enim, and the result for the geometry slope is slope face angle (ψ_f) is 45^0, upper slope surface angle (ψ_S) is 0^0 to 10^0, dip of potencial failure plane (ψ_P) is 40^0, angle of tension crack (ψ_T) is 70^0 to 90^0, height of slope (h) is 20 m and for saturated and dry conditions and coefficient of acceleration for earthquake is 0.1 to 0.2.

Conclusion

This study is based on the plane failure model gave the analytical solution for the rock slopes in which the upper slope surface and tension crack are incline.

The application of the calculated analytic solution was to find the influence on the factor of safety for dry and saturated and under the influence earthquakes conditions in the tension crack. The effect is appreciable, and the factor of safety actually increases as angle of the tension crack (ψ_T) increases. This indicates that the factor of safety decreases considerably as upper slope inclination (ψ_S) increases. The factor of safety decreases from dry conditions to saturated conditions. For under the influence coefficient of acceleration for earthquake conditions, this indicates that the factor of safety increases considerably as angle of tension crack inclination increases. The coefficient acceleration of earthquake increases from 0.1 to 0.3, the factor of safety decreases from 1.24 to 0.99 for dry conditions and from 1.13 to 0.90 for saturated conditions.

The height of water in the tension crack will have a significant influence on factor of safety. This method is very significant in determining the critical location, depth and angle of the tension crack.

The plane failure analysis for the overburden on Air Laya coal mining PT.BA Tanjung Enim, and the result for the geometry slope is slope face angle (ψ_f) is 45^0, upper slope surface angle (ψ_s) is 0^0 to 10^0, dip of potencial failure plane (ψ_p) is 40^0, angle of tension crack (ψ_T) is 70^0 to 90^0, height of slope (h) is 20 m and for saturated and dry condition and coefficient acceleration of earthquake is 0.1 to 0.2.

References

Hoek E. & Bray J.W., 1981, "*Rock slope Engineering*", 3rd Edition, The Institution of Mining dan Metallurgy, London, 402 p.

Ling Hoe I., & Cheng Alexdaner H.D., 1997, "*Technical note : Rock sliding induced by seismic force*", International Journal of Rock Mechanics & Mining Sciences; Vol. 34, pp. 1021-1029.

Haswanto, & Ghani Md. Rafek. Abd., 2004(1), "*Permodelan Matematis kelongsoran Bidang Dengan Permukaan Lereng Membentuk Sudut (Bagian 1)* ". Prosiding TPT Perhapi, Palembang, Indonesia.

Haswanto, & Ghani Md. Rafek. Abd., 2004(2), "*Permodelan Matematis kelongsoran Bidang Dengan Permukaan Lereng Membentuk Sudut (Bagian 2)* ". Prosiding TPT Perhapi, Palembang, Indonesia.

Haswanto, & Ghani Md. Rafek. Abd., 2004, "*Analysis of Plane Failure Type Cut Slope Stability Under the Influence of Discontinuities dan Earthquakes : Case Study from Tanjung Enim South Sumatra, Indonesia*", Persatuan Geologi Malaysia, Perlis, Malaysia.

Sharma S., Raghuvanshi T.K., and Anbalagan R., 1995, "*Plane Failure Analysis of Rock Slopes*" Geotechnical and Geological Engineering, Vol 13 , pp 105-111.

Modeling Bioremediation of Acid Mine Drainage in Permeable Reactive Barriers

Paulo S. Hemsi
Department of Civil Engineering, Colorado State University, Fort Collins, Colorado 80523-1372, USA

Charles D. Shackelford
Department of Civil Engineering, Colorado State University, Fort Collins, Colorado 80523-1372, USA

Linda A. Figueroa
Division of Environmental Science and Engineering, Colorado School of Mines, Golden, Colorado 80401, USA

Abstract

Biological, sulfate reducing permeable reactive barriers (PRBs) containing decomposable solid organic substrates can be used for *in-situ* bioremediation of ground water impacted by acid mine drainage (AMD). A new mathematical model for bioremediation of AMD in these PRB systems is presented. The model combines the primary processes of (i) decomposition of polysaccharides in solid (particulate) organic materials, (ii) sulfate reduction based on lactate, (iii) precipitation of insoluble heavy-metal sulfides, (iv) partial volatilization of H_2S to the gas phase, and (v) reversible dissolution-precipitation of calcite ($CaCO_{3(s)}$) and metal carbonates. Based on comparisons of model simulations with experimental data, the performance of these sulfate reducing systems is shown to be affected critically by the availability and decomposition kinetics of polysaccharides within the solid organic materials provided to the system.

Introduction

As shown in Fig. 1, permeable reactive barriers (PRBs) are subsurface emplacements of reactive materials designed to intercept a contaminant plume, provide a flow path through the reactive media, and transform the contaminants into environmentally acceptable forms to achieve remediation goals down gradient of the barrier (US EPA 1998). PRBs are an attractive solution to site remediation primarily because of low operational and maintenance costs (i.e., except for the costs of monitoring). The majority of PRBs to date have been based on the use of zero-valent iron for reductive dehalogenation of volatile organic compounds (VOCs), reduction of mobile and toxic oxyanions (chromium, arsenic, and selenium) into immobile precipitates, and treatment of radionuclide and nitrate bearing ground waters (e.g., US EPA 1998, O'Hannesin and Gilham 1998, US EPA 2002).

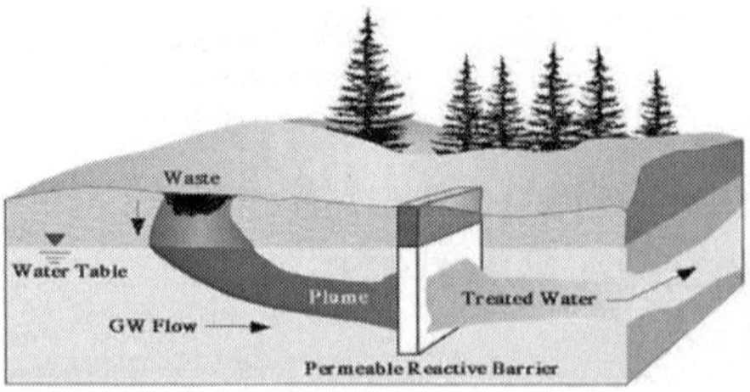

Figure 1: General PRB scenario (US EPA 1998).

Biochemical systems using sulfate reducing bacteria (SRB) for *in-situ* bioremediation of acid mine drainage (AMD) also have been used as PRBs (e.g., Blowes et al. 2000, Benner et al. 2002, US EPA 2002, Amos and Younger 2003). These sulfate reducing PRBs are based on decomposition of emplaced solid organic materials, bacterial sulfate reduction, precipitation of heavy metal sulfides and carbonates, and amelioration of pH. The attractiveness of the use of sulfate reducing PRBs for AMD sites is further increased by (i) the low costs of solid organic substrates, typically detritus (e.g., leaf mulch compost, wood chips, sawdust), and (ii) the typical remoteness of AMD sites, which in some cases precludes using more active remediation technologies. The use of computational modeling to describe the primary biochemical processes responsible for *in-situ* bioremediation of AMD in sulfate reducing PRBs is illustrated in this paper.

Background

A new mathematical model that describes the bioremediation of AMD using sulfate reducing systems recently has been formulated (Hemsi 2005). This model combines the primary processes delineated in Fig. 2, as follows: (i) decomposition of polysaccharides in solid (particulate) organic materials emplaced in the PRB (such as wood chips, sawdust, leaf mulch, etc.), (ii) sulfate reduction based on lactate, (iii) precipitation of insoluble metal sulfides (e.g., $FeS_{(s)}$), (iv) partial volatilization of H_2S to the gas phase, and (v) reversible dissolution-precipitation of calcite ($CaCO_{3(s)}$) and metal carbonates (e.g., siderite, $FeCO_{3(s)}$). These processes are solved using a Runge-Kutta-Fehlberg integrator as a stand-alone algorithm or coupled to flow (MODFLOW 2000, Harbaugh et al. 2000) and multi-species reactive contaminant transport (RT3D v. 2.5, Clement 2003) using reaction operator splitting. The model has been evaluated on the basis of both laboratory and field data involving sulfate reducing systems for AMD remediation (Hemsi 2005).

The experimental data used for testing and calibrating the model pertain to four batch-equilibrium (no-flow) tests (Waybrant et al. 1998), two closed-system column tests (Waybrant et al. 2002 and Hulshof et al. 2003), three open-system bioreactor tests (Bechard et al. 1994), and a full-scale PRB (Benner et al. 2002). In terms of calibration, experiments involving flow-through conditions such as columns and bioreactors are expected to represent field scenarios such as PRBs and wetlands better than are batch-equilibrium tests involving no-flow conditions. The model simulations and evaluations for batch-, column-, and bioreactor-test data are presented and discussed by Hemsi (2005) and Hemsi et al. (2005). An overview of some of the results based on these comparisons is provided herein.

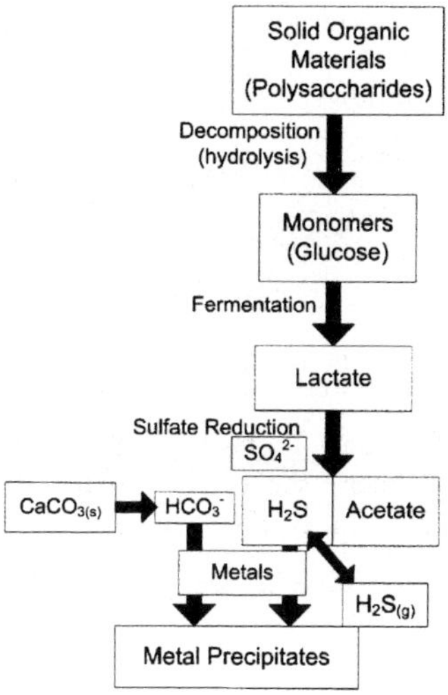

Figure 2: Primary biochemical processes considered in model for metals precipitation in sulfate reducing systems (after Hemsi 2005 and Hemsi et al. 2005).

Results

Batch Test Evaluation

A batch test based on an initial 18.6 g/L of decomposable cellulose, inoculated with microbial populations of decomposers and sulfate reducing bacteria (SRB) was equilibrated with a solution initially containing 4,000 mg/L SO_4^{2-} and 1,110 mg/L Fe^{2+} at pH 5.5 (Waybrant et al. 1998). Model approximations of experimental results are shown in Hemsi et al. (2005). Rates of sulfate reduction were modeled using Monod kinetics (Monod 1949) for SRB (such as *Desulfovibrio vulgaris*) growth coupled to, and limited by, decomposition of cellulose modeled using either first-order (Cases 1 and 2) or Contois kinetics (Cases 3 and 4).

SRB growth using Monod kinetics and SRB decay using first-order kinetics can be written as follows:

$$\frac{d[X_{SRB}]}{dt} = \mu_m [X_{SRB}]\left(\frac{[LA]}{K_{LA} + [LA]}\right)\left(\frac{[SO]}{K_{SO} + [SO]}\right) - d[X_{SRB}] \tag{1}$$

where the first term on the right-hand side is Monod kinetics for growth, $[X_{SRB}]$ is the biomass concentration, μ_m is the maximum specific growth rate coefficient for SRB, [LA] and [SO] are the concentrations of lactate and sulfate, respectively, K_{LA} and K_{SO} are half-saturation constants for LA and SO, respectively, and d is a decay coefficient. Polysaccharide decomposition using first-order kinetics can be written as follows:

$$\frac{d[CE_i]}{dt} = -k_{f,i}\,[CE_i] \tag{2}$$

where $[CE_i]$ is the equivalent concentration of polysaccharide 'i', and $k_{f,i}$ is a first-order decomposition rate coefficient. Also, polysaccharide decomposition using Contois kinetics (Contois 1959) can be written as follows:

$$\frac{d[CE_i]}{dt} = -k_{c,i}\,[X_{d,i}]\,\frac{\left([CE_i]/[X_{d,i}]\right)}{K_{c,i} + \left([CE_i]/[X_{d,i}]\right)} \tag{3}$$

where $k_{c,i}$ is a Contois decomposition rate coefficient, $[X_{d,i}]$ is the biomass concentration of decomposer bacteria 'i', and $K_{c,i}$ is a half-saturation constant. This expression transitions between the two limiting cases of first-order kinetics with respect to [CE] at low $[CE]/[X_d]$ ratios, and first-order kinetics with respect to $[X_d]$ at high ratios (i.e., exponential X_d growth). Decomposition parameter values for 35 °C from the literature (Vavilin et al. 1996) were used initially and then adjusted for temperature in subsequent simulations.

As shown in Fig. 3, Case 4 with Contois kinetics for cellulose decomposition provides a better approximation to experimental data than do Cases 1 to 3. Also, Case 4 provides a good approximation to measured Fe^{2+} versus time, as shown in Fig. 4a.

For Case 4, the patterns in SRB dynamics with time (e.g., Postgate 1984) are illustrated in Fig. 4b, where the end of the SRB stationary phase at 120 days is marked by the depletion of SO_4^{2-} from the batch flask which also occurs at 120 days. The approximation of experimental data with the new model was affected critically by the type of kinetics employed for polysaccharide decomposition, with Contois kinetics appearing to provide better results than first-order kinetics (compare Figs. 3a and b).

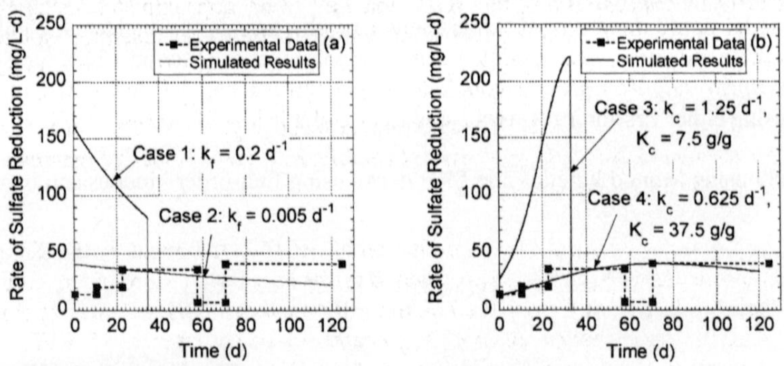

Figure 3: Model approximation of rates of sulfate reduction for a batch test using cellulose: (a) first-order decomposition kinetics; (b) Contois decomposition kinetics (Hemsi et al. 2005).

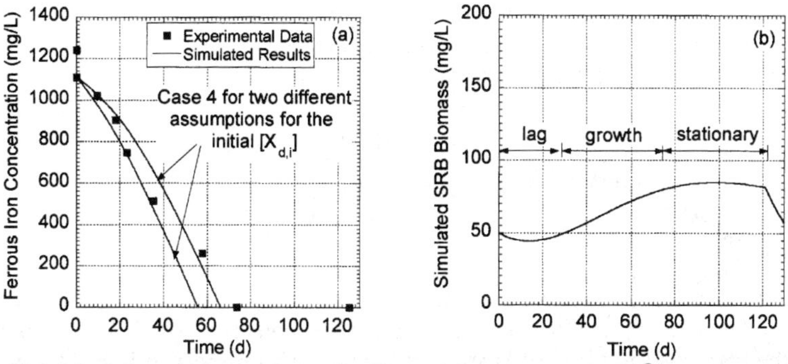

Figure 4: (a) Model approximation of effluent ferrous iron (Fe^{2+}); (b) simulation of sulfate reducing bacteria (SRB) dynamics for a batch test using cellulose (Hemsi et al. 2005).

Column Test Evaluation

A column test with specific discharge of 14 mm/d and residence time of 15 d based on 22 g of decomposable polysaccharides in leaf mulch, 12 g in sawdust, and 9 g in wood chips was used to neutralize an influent solution containing from 1,400 to 3,900 mg/L SO_4^{2-}, and from 300 to 1,200 mg/L Fe^{2+}. The column test extended to 425 days (Waybrant et al. 2002).

The first sink for influent ferrous iron (Fe^{2+}) in the column is $FeS_{(s)}$ precipitation due to sulfate reduction coupled to, and limited by, decomposition of polysaccharides. Calcite ($CaCO_{3(s)}$) dissolution in the column which releases HCO_3^- resulted in a second significant sink for Fe^{2+} in the form of precipitation of siderite ($FeCO_{3(s)}$).

Measured effluent concentrations for SO_4^{2-} and Fe^{2+} were approximated using the model coupled in time to flow and contaminant transport (reaction operator splitting) and temperature adjusted parameter values based on the results of the batch test simulations. As shown in Fig. 5, the temporal masses of SO_4^{2-} and Fe^{2+} removed by bacterial sulfate reduction, and $FeS_{(s)}$ and $FeCO_{3(s)}$ precipitation, respectively, based on a cumulative mass approach for the modeled results compared favorably with the cumulative masses calculated based on the measured experimental data.

As shown in Fig. 6, Fe^{2+} removal by precipitation of $FeCO_{3(s)}$ is limited by the dissolution of granular limestone ($CaCO_{3(s)}$), which was added to the reactive mixture in the column. For example, $CaCO_{3(s)}$ is depleted from the column after about 15 pore volumes of flow (PVF) (Fig. 6a), and the precipitation of $FeCO_{3(s)}$ ceases to occur after ~ 15 PVF (Fig. 6b). Therefore, the longer-term removal of Fe^{2+} from the influent solution in this column remains solely attributable to sustained SO_4^{2-} reduction (i.e., from ~ 15 PVF to ~ 28 PVF corresponding to 425 d).

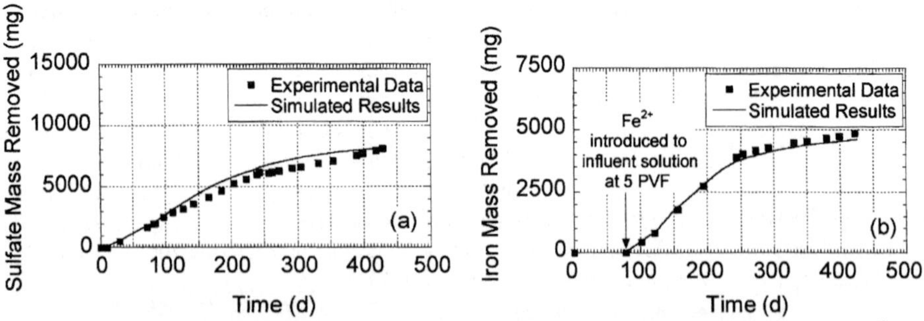

Figure 5: Model approximation of the cumulative masses of (a) sulfate (SO_4^{2-}) and (b) ferrous iron (Fe^{2+}) removed in the column test (Hemsi 2005).

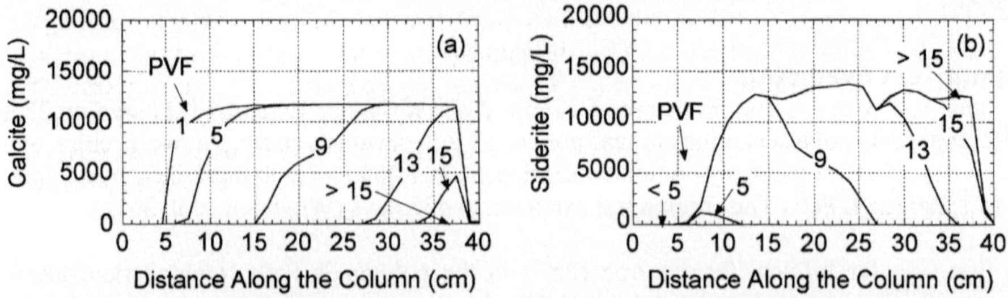

Figure 6: Modeled behavior for (a) calcite ($CaCO_{3(s)}$) and (b) siderite ($FeCO_{3(s)}$) along the column length as a function of pore volumes of flow, PVF (Hemsi 2005). Note: Precipitate concentrations are expressed as mass of precipitate per liter of solution.

Another modeled column test (data not shown) also contained influent Zn^{2+}, ranging from 70 to 120 mg/L (Hulshof et al. 2003). Model approximations to the data for these experimental results are shown in Hemsi (2005).

Extension to the Field Scale

Extending the modeling capability to the field scale, e.g., full-scale PRB, requires the consideration of additional factors, such as (i) temperature and seasonality effects on bacterial activity and the PRB performance, and (ii) the influence of heterogeneity in aquifer hydraulic conductivity (K) on flow and contaminant transport in the vicinity of a PRB.

With respect to temperature and seasonality, lower groundwater temperatures occurring during the winter will likely cause the activity of decomposer and sulfate reducing bacteria to decline during this season, which will result in lower degree of remediation during the winter, i.e., assuming no significant change in groundwater specific discharge. Computational modeling applied to a full-scale PRB located in Ontario (Canada) and with a vertically-averaged (0 to 3.6 m) groundwater temperature of 3 °C during the winter (Benner et al. 2002) indicated peaks in SO_4^{2-} and Fe^{2+} effluent concentrations correlated in time with the winter months (Hemsi 2005).

With respect to the effects of heterogeneity in K, preferential flow and contaminant transport channeling due to heterogeneity will expose the PRB to spatially variable groundwater seepage velocities and contaminant concentrations, with localized values that may be sufficiently high such that the required PRB thickness will be greater than that for a homogenous aquifer. Additionally, aquifer heterogeneity may induce divergent flow patterns that will render the required length of the PRB to be greater than that for a homogenous aquifer, i.e., in order to capture the entire width of the plume. Hemsi and Shackelford (2005) present a quantification of probabilistic factors of safety for scaling the required PRB thickness and length amidst heterogeneity in K. These probabilistic factors of safety were obtained using mathematical modeling of flow and particle tracking (MODPATH 4.2, Pollock 1994) in conjunction with 650 synthetic stochastic heterogeneous aquifers produced with the Turning Bands method. The significant impact of aquifer heterogeneity on both PRB thickness and length emphasizes the need for detailed site characterization for assessing the level of field heterogeneity (Hemsi and Shackelford 2006).

Conclusions

Some results from computational modeling of the primary biochemical processes responsible for AMD remediation in sulfate reducing systems have been presented. The model approximations of batch-test data illustrated the importance of the type of kinetics employed for modeling cellulose (polysaccharide) decomposition, with Contois kinetics providing better results than first-order kinetics. The model approximations of column-test data illustrated the role of iron sulfide ($FeS_{(s)}$) and siderite ($FeCO_{3(s)}$) precipitation in the removal of influent Fe^{2+} for more than 28 pore volumes of flow. Due to the depletion of $CaCO_{3(s)}$, the longer-term removal of influent Fe^{2+} was observed to be primarily attributable to sustained sulfate reduction and $FeS_{(s)}$ precipitation.

References

Amos, P.W. and Younger, P.L. (2003). Substrate characterization for a subsurface reactive barrier to treat colliery spoil leachate. *Water Research*, 37(1): 108-120.

Bechard, G., Yamazaki, H., Gould, W.D., and Bedard, P. (1994). Use of cellulosic substrates for the microbial treatment of acid mine drainage. *Journal of Environmental Quality*, 23(1): 111-116.

Benner, S.G., Blowes, D.W., Ptacek, C.J., and Mayer, K.U. (2002). Rates of sulfate reduction and metal sulfide precipitation in a permeable reactive barrier. *Applied Geochemistry*, 17(3): 301-320.

Blowes, D.W., Ptacek, C.J., Benner, S.G., McRae, C.W.T., Bennett, T.A. and Puls, R.W. (2000). Treatment of inorganic contaminants using permeable reactive barriers, *Journal of Contaminant Hydrology*, 45(1-2): 123-137.

Clement, T.P. (2003). RT3D v.2.5 Update Document, What is New in RT3D version 2.5. Electronic document: http://bioprocess.pnl.gov/rt3d.htm.

Contois, D.E. (1959). Kinetics of bacterial growth, relationship between population density and specific growth rate of continuous cultures. *Journal of General Microbiology*, 21(1): 40-50.

Harbaugh, A.W., Banta, E.R., Hill, M.C., and McDonald, M.G. (2000). *MODFLOW-2000, the U.S. Geological Survey Modular Ground-Water Model – User Guide to Modularization Concepts and the Ground-Water Flow Process*, USGS Open-File Report 00-92, Reston (VA).

Hemsi, P.S. (2005). *Computational Modeling of Biochemical Permeable Reactive Barriers for Acid Mine Drainage*. PhD Dissertation, Colorado State University, Fort Collins, CO, USA.

Hemsi, P.S. and Shackelford, C.D. (2006). An evaluation of the influence of aquifer heterogeneity on permeable reactive barrier design. *Water Resources Research*, in press.

Hemsi, P.S., Shackelford, C.D., and Figueroa, L.A. (2005). Modeling the influence of decomposing organic solids on sulfate reduction rates for iron precipitation. *Environmental Science and Technology*, 39(9): 3215-3225.

Hulshof, A.H., Blowes, D.W., Ptacek, C.J., and Gould, W.D. (2003). Microbial and nutrient investigations into the use of in situ layers for treatment of tailings effluent. *Environmental Science and Technology*, 37(21): 5027-5033.

Monod, J. (1949). The growth of bacterial cultures. *Annual Review of Microbiology*, 3:371-394.

O'Hannesin, S.F., and Gillham, R.W. (1998). Long-term performance of an in situ "iron wall" for remediation of VOCs. *Ground Water*, 36(1): 164-170.

Pollock, D.W. (1994). *User's Guide for MODPATH/MODPATH-PLOT, Version 3: A Particle Tracking Post-Processing Package for MODFLOW, the U.S. Geological Survey Finite-Difference Ground-Water Flow Model,* U.S. Geological Survey, Open-File Report 94-464, Reston (VA).

Postgate, J.R. (1984). *The Sulphate-Reducing Bacteria.* 2nd Edition. Cambridge University Press, Cambridge (UK).

US EPA (1998). *Permeable Reactive Barrier Technologies for Contaminant Remediation,* EPA/600/R-98/125, Washington DC.

US EPA (2002). *Field Applications of In Situ Remediation Technologies: Permeable Reactive Barriers,* Office of Solid Waste and Emergency Response, Washington DC.

Vavilin, V.A., Rytov, S.V., and Lokshina, L. (1996). A description of hydrolysis kinetics in anaerobic degradation of particulate organic matter. *Bioresource Technology*, 56(1): 229-237.

Waybrant, K.R., Blowes, D.W., and Ptacek, C.J. (1998). Selection of reactive mixtures for use in permeable reactive walls for treatment of mine drainage. *Environmental Science and Technology*, 32(13):1972-1979.

Waybrant, K.R., Ptacek, C.J., and Blowes, D.W. (2002). Treatment of mine drainage using permeable reactive barriers: column experiments. *Environmental Science and Technology*, 36(6): 1349-1356.

Ferrite Formation using Precipitate in the Treatment of Acid Mine Drainage for Reducing its Volume

Toshifumi Igarashi Graduate School of Engineering, Hokkaido University, Sapporo, Japan
Tetsuya Yoshida Ataka Construction & Engineering Co., Ltd., Osaka, Japan
Kuniomi Asakura Graduate School of Engineering, Hokkaido University, Sapporo, Japan
Hiroko Miyamae Ataka Construction & Engineering Co., Ltd., Osaka, Japan
Nobuyoshi Iyatomi Nittetsu Mining Co., Ltd., Tokyo, Japan
Koichi Hashimoto Japan Oil, Gas and Metals National Corporation, Kasawaki, Japan

Abstract

Ferrite formation in precipitates by a two-step neutralization process in which magnesium oxide (MgO) was used as the first neutralizer and sodium hydroxide (NaOH) as the second was applied to acid mine drainage (AMD) from the Horobetsu mine, an abandoned sulfur mine, in Hokkaido, Japan, to reduce the volume of precipitates to be disposed. Batch neutralization experiments with MgO and continuous flow experiments by the two-step neutralization were conducted in the laboratory and the actual AMD treatment plant. The results showed that the heavy metals in the AMD were successfully removed by the process, and that the precipitates were magnetized by the neutralization in the continuous flow experiments although there was Al and Si in the AMD that could have prevented the formation of ferrite. This indicates that ferrite can be formed from the precipitates after removing most of the soluble Al and Si by the first neutralization. The change in the chemical forms of the iron precipitates and preventing gypsum and calcite formation in the precipitates by MgO result in a volume reduction of precipitates by two thirds for this AMD when compared with conventional processes.

Introduction

Hokkaido, Japan has a large number of abandoned mines. The effluent from the mines, in particular metal mines, known as acid mine drainage (AMD), must be treated when the acidic and heavy-metal-rich effluent threatens to seriously damage the surrounding environment. Large volumes of precipitates are generated in the neutralization of AMD while a shortage of space for disposal of the precipitates is demanding the development of new AMD treatment systems to reduce the volume of neutralized precipitates. This study investigates a new AMD treatment process to determine its ability to produce lower volumes of precipitates.

Ferritization of precipitates by a two-step neutralization process in which magnesium oxide (MgO) was used as the first neutralizer and sodium hydroxide(NaOH) as the second was applied to AMD from the Horobetsu mine, an abandoned sulfur mine in central Hokkaido, Japan. The AMD contains approximately 400 mg/L of total dissolved iron (Fe) and almost 10 mg/L of arsenic (As).

The two-step neutralization process or the use of MgO as a neutralizing agent have been studied to determine a suitable method of AMD treatment[1)-10)]. However, the formation of ferrite in the precipitates generated by the neutralization has not succeeded in reducing the volume of precipitates when significant amounts of Al or Si are contained in the AMD. The Horobetsu AMD has chemical properties that make ferrite formation difficult because the AMD contains much Al and Si and because precipitates include impurities, such as gypsum and calcite, that interrupt ferritization are easily formed. This study applies a new two-step neutralization process with MgO to this AMD with high ferrous (Fe(II)), sulfate (SO_4^{2-}), and As concentrations to form a more compact and stable precipitate. The optimum conditions of AMD treatment for ferrite formation are also examined.

Experiments

AMD and Neutralizing Agents

Batch and two-step continuous neutralization experiments were conducted. The continuous neutralization experiments were carried out in the laboratory and in an actual AMD treatment plant. Synthesized AMD was used in the laboratory while actual AMD was used in the Horobetsu treatment plant.

A MgO solution was used as the first neutralizing agent. Powdered MgO was added to deionized water and the solution was continuously stirred during the experiments to provide 1.5wt% suspended solution. Three different grades of MgO were used as the neutralizer. The physical and chemical properties of the various agents employed are shown in **Table 1**.

Synthesized AMD was made by dissolving pure reagents in mineral water. The chemical properties of the synthesized AMD as well as the actual AMD are listed in **Table 2**.

Table 1: Physical and chemical properties of the MgO used.

Type	Abbreviation	Physical and chemical properties
Pure Grade by Kanto Chemical Co., Inc.	KK	purity minimum 99% average particle diameter 10μm ignition loss (800°C) maximum 2.0%
First Grade by Sigma Aldrich Japan K. K.	Σ	purity minimum 98% average particle diameter 3.7μm ignition loss (800°C) maximum 10.0%
Second Grade by Combustion of Magnesite	KDB	produced by the combustion of $MgCO_3$ in China the rest of 325 Mesh: 1.60% average particle diameter 19μm ignition loss 4.07% MgO 91.21%, CaO 2.35%, SiO_2 1.67% Fe_2O_3 0.40%, Al_2O_3 0.14%

Table 2: Chemical properties of synthesized and actual acid mine drainage samples.

Elements	Actual AMD (mg/L)*	Synthesized AMD (mg/L)*	Reagents used for synthesized AMD
Si	33	34.4	Mineral water
Al	50	50	$AlCl_3 \cdot 6H_2O$
Fe(II)	300	301	$FeSO_4 \cdot 7H_2O$
Fe(III)	70	62	$FeCl_3 \cdot 6H_2O$
SO_4	2510	2544	H_2SO_4
Mg	13	12	$MgSO_4$
Ca	114	116	$CaCl_2$
As	10	12	As_2O_3 (Arsenic standerd solution for atomic absorption spectrometry)
pH	1.8	1.7	

* The mg/L unit is only for the concentrations of elements.

Batch Experiments

Titration with 1.5% MgO suspended solution was carried out to determine the amount of the neutralizing agent that must be added to the AMD to change the pH of the solution to a specific value. The pH, dissolved oxygen (DO), and oxidation reduction potential (ORP) were simultaneously monitored during the titration.

In the batch experiments, a specified amount of MgO suspension was added to 100 mL of AMD based on the obtained titration curve. The mixing period was set to 30 min in most of the experiments because more than 30 min of mixing caused no further changes in the precipitation/dissolution behaviors of the constituents of the AMD. The conditions of the experiments are listed in **Table 3**. The type of AMD, the grade of neutralizing agent, aeration, and mixing period were changed in the experiments.

Table 3: Batch experiment conditions.

Case	AMD	MgO	Added volume of MgO solution (mL/L)	Aeration	Mixing period (min)
1	Synthesized	KK	45, 58, 62, 66, 70, 80	none	5, 10, 30, 60, 120
2	Synthesized	Σ	70	none	5, 10, 30, 60, 120
3	Synthesized	KK	80	aerated	5, 10, 30, 60, 120
4	Synthesized	KK	45, 58, 62, 66, 70, 80	none	30
5	Synthesized	KDB	45, 58, 62, 66, 70, 80, 90	none	30
6	Synthesized	KK	80	aerated	30
7	Actual	KK	25, 30, 32, 36, 40, 45, 55	none	30
8	Actual	KDB	25, 33, 43, 46, 50, 65, 80	none	30
9	Actual	KK	55	aerated	30
10	Actual	KDB	80	aerated	30

Continuous Flow Experiments

A two-step neutralization process using MgO as the first neutralizing agent and NaOH as the second was adopted. A precipitate recycling system from a ferrite sedimentation tank to a ferrite formation tank was set in the process. A diagram of the process is represented in **Figure 1**. The principle of the two-step neutralization is based on the concept that the removal of Al and Si prior to ferritization is necessary because these elements interrupt ferrite formation. The volumes of the MgO mixing, ferrite formation, and ferrite sedimentation tanks were 0.8, 1.5, and 9.0 L, respectively, by considering a conventional $CaCO_3$-$Ca(OH)_2$ treatment process operated for several decades at the actual treatment plant. The flow rate through the process was 36 L/day, and the volume of sludge, including ferrite from the ferrite sedimentation tank, returned to the ferrite formation tank was 7.2 L/day. A 2M NaOH solution was used in the second neutralization process.

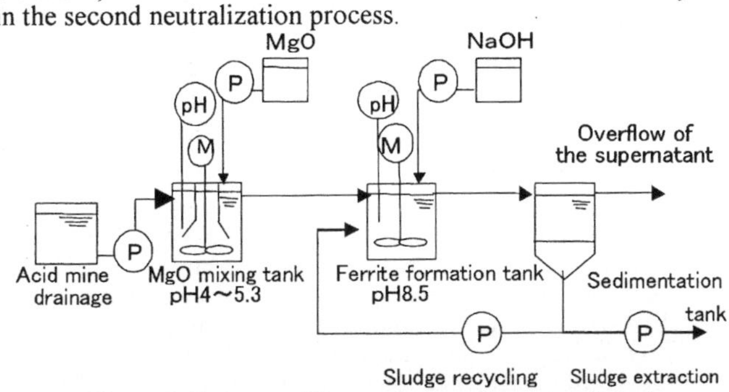

Figure 1: Process of the continuous flow experiments.

Preliminary continuous flow experiments using synthesized AMD were conducted in the laboratory before experiments using the actual AMD in the treatment plant. The pH in the first neutralization tank (MgO tank) was changed from 4 to 5.3 in the laboratory to identify the optimum condition for both As removal and ferrite formation in the following steps while the pH in the second neutralization tank was set at 8.5. The duration of each experiment was from half a day to 30 days.

Continuous flow experiments in an actual plant followed the continuous flow experiments in the laboratory. A pH of 4.3 in the first neutralization tank and the pH of 8.5 in the ferrite formation tank were determined by considering the results of the continuous flow experiments in the laboratory.

Analysis of Water and Precipitates

Water in the tank was sampled periodically and the total dissolved iron (Fe(T)), ferrous iron (Fe(II)), aluminum (Al), silicate (Si), arsenic (As), magnesium (Mg), sulfate (SO_4^{2-}), and calcium (Ca) in the filtrate with a 0.45 μm membrane filter were determined, after measuring pH, ORP, temperature, and DO by probes. The analysis of Fe(II) was based on the 1,10-phenanthroline method, and inductively coupled plasma atomic emission spectrometry (ICP-AES) was used for the analysis of the other constituents. When the As concentration was below 0.1 mg/L, an ICP-AES equipped with a hydride generator was employed for the As analysis. The Fe(II) concentration of the actual AMD was analyzed immediately after the sampling in the treatment plant because Fe(II) is rapidly oxidized to Fe(III) by bacterial activity.

The precipitates formed in the ferrite sedimentation tank were subjected to a characterization of the mineralogical properties by X-ray diffraction analysis (XRD) and of the chemical properties by a wet method. The magnetism of the precipitates after drying in a vacuum was determined to identify magnetite in the precipitates.

Results and discussions

Batch Experiments

The effect of the mixing period in the MgO tank on the precipitation/dissolution behavior of the constituents in the AMD is shown in **Figure 2**. The pH increased rapidly and then approached a constant value in response to the amount of MgO added. A reduction in ORP and Fe(T) concentration was observed with increases in pH although the DO remained unchanged due to intrusion of atmospheric oxygen to the solution by the intense mixing. It is found that the neutralization reaction induced by adding MgO was complete within 30 min, and a mixing period of 30 min was determined as a reference.

The dissolved concentrations of a variety of the constituents in the AMD as a function of pH are presented in **Figures 3 to** 7. The type of AMD, the grade of MgO, and the aeration conditions were changed according to the conditions in Table 3. The quality of the actual AMD when using a second-grade MgO, termed KDB, was somewhat different from that when using a pure MgO, termed KK, due to seasonal fluctuations in the properties of the AMD. According to Figures 3 and 4, the dissolved ferric ion (Fe(III)) corresponding to the difference between Fe(T) and Fe(II) vanished at around pH 3.5 since the dissolved Fe(T) concentration agreed with the dissolved Fe(II) at pH>3.5. A dissolved Fe(II) concentration equal to the dissolved Fe(T) concentration in the pH range vanished at pH 6.5 to 7. Overall, ferric hydroxide ($Fe(OH)_3$) from ferric ions originally in the AMD formed at around pH 3.5, and the oxidation from ferrous to ferric ions and subsequent precipitation of $Fe(OH)_3$ were complete at around pH 7; Al precipitates like $Al(OH)_3$ formed at pH 5 as shown in Figure 5. The dissolved As concentration gradually decreased with pH, as in Figure 6, suggesting the formation of As-bearing precipitates at pH 2 to 6. The dissolved Si concentration also decreased with pH as shown in Figure 7. The decline in Si concentration at pH<6 was smaller than that at pH>6, indicating that Si-bearing precipitates are closely related to the formation of iron hydroxide. The dissolved Mg concentration increased with pH as the amount of Mg added increased with pH while the SO_4^{2-} concentration remained almost constant. The results show a pH envelope of dissolved constituents independent of the type of AMD, the grade of MgO, and aeration conditions. This indicates that the pH of a solution is the primary control of the precipitation/dissolution of constituents.

Continuous Flow Experiments with Synthesized AMD

Changes in the dissolved Fe(II) concentration in the continuous flow experiments with synthesized AMD are presented in **Figure 8**. The pH in the MgO tank was adjusted to 4.4 to 4.8 by changing the amount of MgO added. The dissolved Fe(II) concentration in the ferrite formation tank remained stable and decreased to nearly zero although small variations in the Fe(II) concentration in the MgO tank were observed. A similar pattern of concentration changes was also observed for the other constituents. In addition, the magnetism of the precipitates in the ferrite formation tank was qualitatively determined by observing adhesion between a neodymium magnet and precipitates when placing the magnet on the outer surface of the tank. The above results indicate that the two-step neutralization process was successful in removing the heavy metals, and that ferrite was formed from the precipitates in the ferrite formation tank. Previous peports[11)12)] have found that little ferrite forms from precipitates with high Si or Al concentrations, because of the interruption of ferrite formation by the generation of Si- or Al-containg precipitates among the precipitates. In the process investigated here, the Si and Al did not interrupt the ferrite formation in the ferrite formation tank since the dissolved Si and Al concentration was significantly reduced at the initial neutralization by MgO to form ferrite during the second neutralization by NaOH.

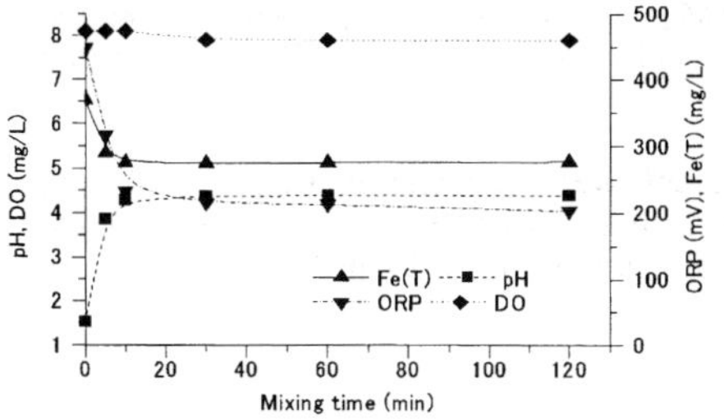

Figure 2: Changes in pH, DO, ORP, and Fe(T) concentration with mixing time in case 1 (62mL/L MgO added).

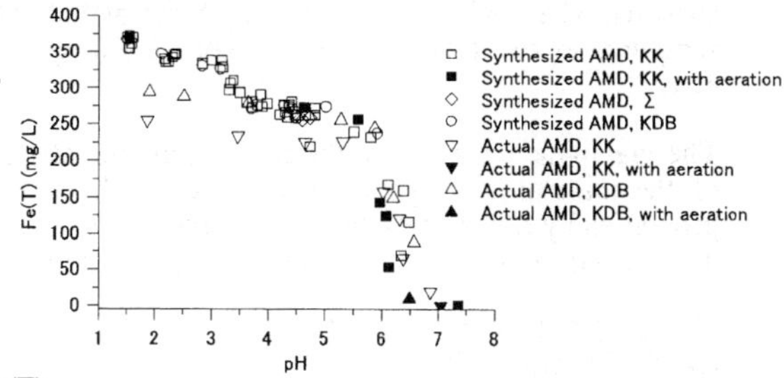

Figure 3: Total dissolved Fe concentration as a function of pH in the batch experiments.

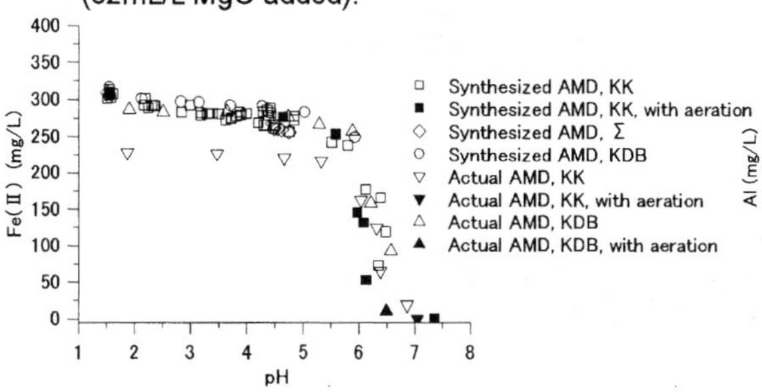

Figure 4: Dissolved Fe(II) concentration as a function of pH in the batch experiments.

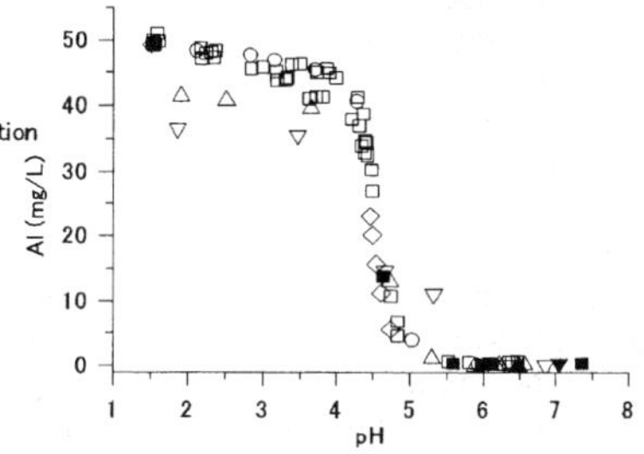

Figure 5: Dissolved Al concentration as a function of pH in the batch experiments.

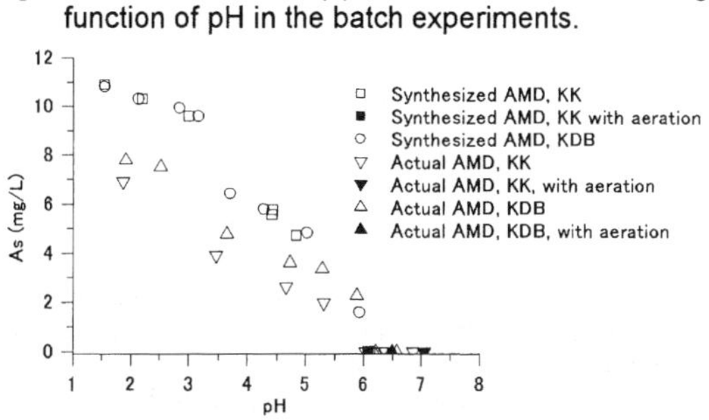

Figure 6: Dissolved As concentration as a function of pH in the batch experiments.

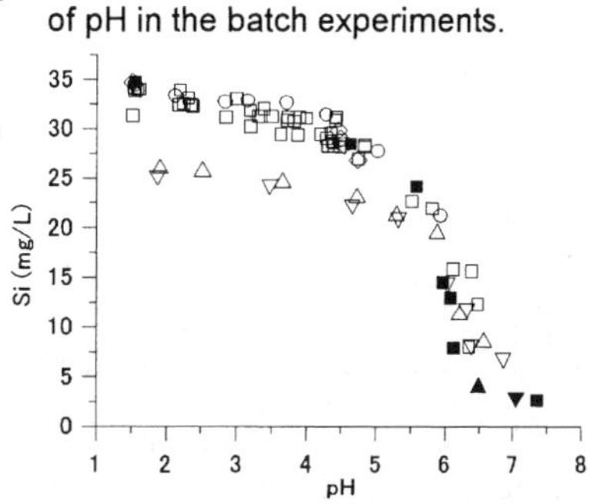

Figure 7: Dissolved Si concentration as a function of pH in the batch experiments.

The pH dependence of neutralization with MgO on the treated water quality of the AMD is presented in **Figure 9**. For short-term experiments the duration of an experiment was half a day while for long-term experiments the duration was longer than one week. Dissolved Fe(II) in the MgO tank decreased with pH from 4 to 5, similar to the results from batch experiments. The only difference between the experiments is the pH region where Fe precipitation occurs. The pH in which Fe hydroxide formed in the continuous flow experiments was lowered by one to two pH values, compared with the pH in batch experiments as shown in Figure 4. The differences in the precipitation/dissolution behavior of the other constituents in the batch and continuous flow experiments agreed with those of Fe(II). This may be due to differences in the spatial distribution of pH in the MgO tank caused by the flow conditions.

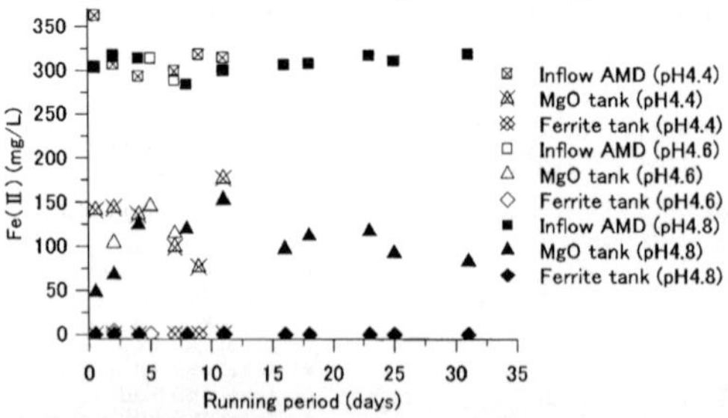

Figure 8: Changes in dissolved Fe(II) concentration of synthesized acid mine drainage in the long-term continuous flow experiments. The pH values in the MgO tank are shown.

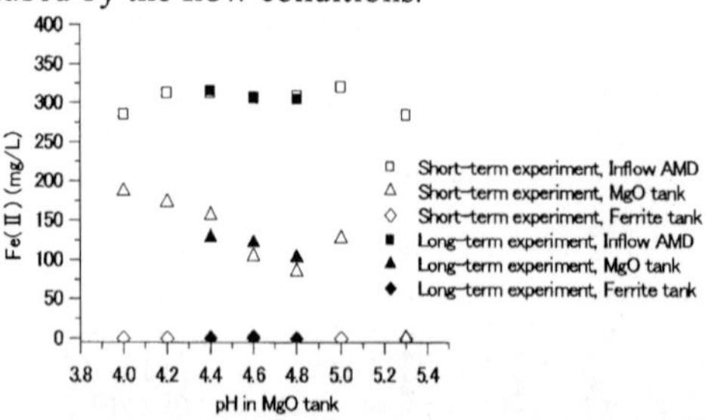

Figure 9: Dissolved Fe(II) concentration as a function of pH in the MgO tank for synthesized acid mine drainage in the continuous flow experiments.

Continuous Flow Experiment with Actual AMD

Continuous flow experiments with actual AMD were carried out by considering the results of the continuous flow experiments with synthesized AMD. The results of the experiments are presented in **Figures 10 to 12**. The dissolved concentrations of Fe(II) and As in the ferrite formation tank were reduced below those in regulations for processed effluents (10 mg/L of Fe(T) and 0.1 mg/L of As, in Japan), although the concentrations in the MgO tank fluctuated somewhat. Concentrations of less than 0.6 mg/L for Al, less than 0.8 mg/L for Fe(II), less than 2 mg/L for Si, and less than 0.06 mg/L for As were achieved in the ferrite formation tank. This shows that appropriate pH control leads to low and stable concentrations of heavy metals in the effluent. The coexistence of 0.6 mg/L Al and 2 mg/L Si in the MgO tank did not interrupt the ferrite formation of the precipitates in the ferrite formation tank as shown by the qualitatively observed magnetic properties of the precipitates.

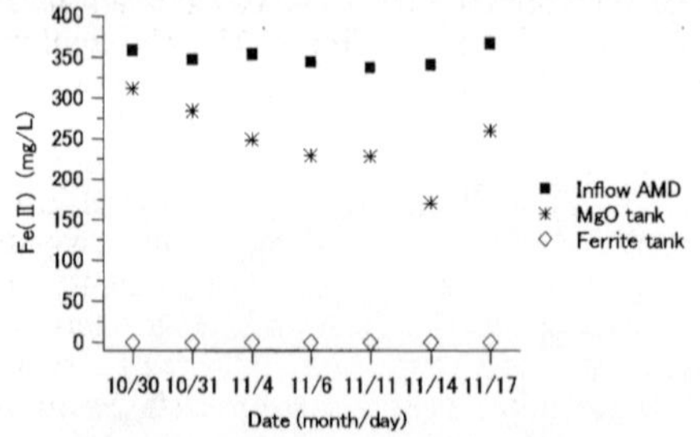

Figure 10: Changes in dissolved Fe(II) concentration of actual acid mine drainage in the continuous flow experiments.

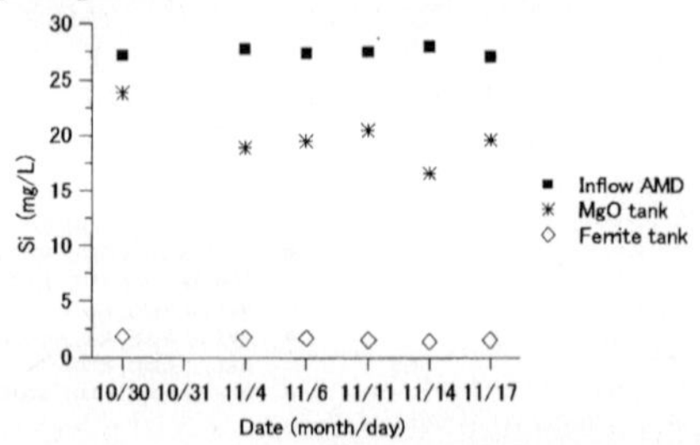

Figure 11: Changes in dissolved Si concentration of actual acid mine drainage in the continuous flow experiments.

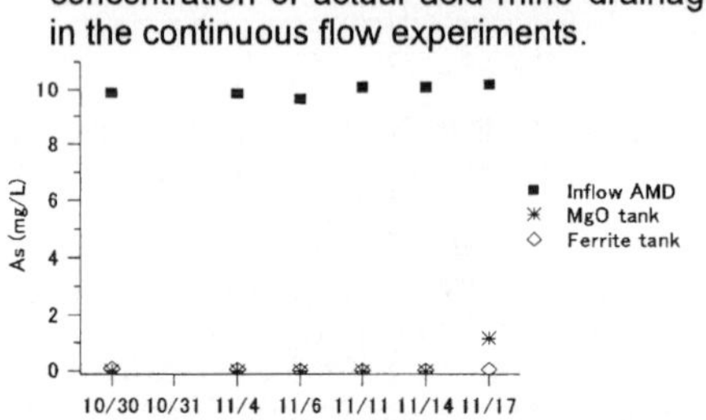

Figure 12: Changes in dissolved As concentration of actual acid mine drainage in the continuous flow experiments.

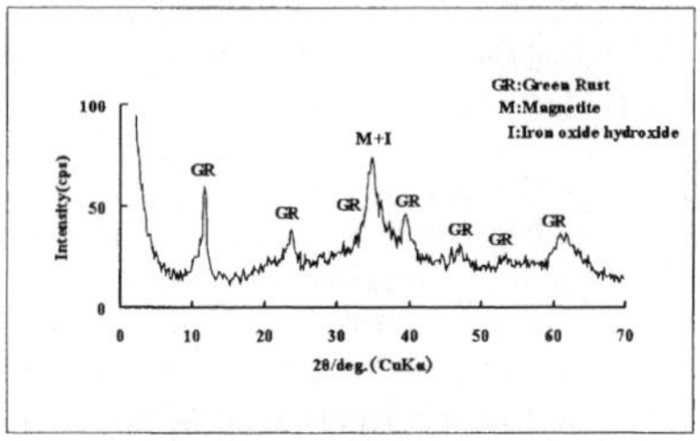

Figure 13: XRD analysis of the precipitates in the sedimentation tank.

The precipitates in the ferrite sedimentation tank were analyzed by XRD after drying out. **Figure 13** shows the results of the X-ray diffractogram. There is a peak of magnetite or iron oxide/hydroxide in the figure, as well as peaks of green rust. Based on the XRD analysis and weak magnetism of the precipitates, the precipitates are likely to contain magnetite.

Ratio of As to Fe in the Precipitates

The molar ratio of As to Fe in the precipitates was calculated based on the differences in dissolved As and Fe concentration caused by the pH increments to better understand the mechanism of As removal. **Figure 14** shows the relation between As/Fe(T) and pH. The As/Fe(T) ratio increased with pH from 2 to 3.5, and remained at approximately 0.1 at pH 3.5 to 5.5. This indicates that the As removal is associated with the Fe removal. The more efficient As removal takes place at pH>3.5 because there are no ferric ions in the solution and because ferrous ions are oxidized to ferric ions, and the formed $Fe(OH)_3$ precipitates adsorb As. The As removal efficiency with respect to Fe removal is lower at pH<3.5 because there are dissolved ferrous ions in the solution. The above results suggest that coprecipitation with Fe is the main mechanism of As removal from AMD.

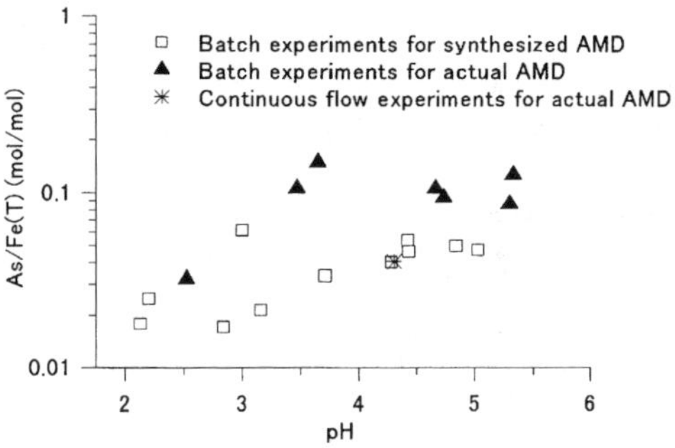

Figure 14: Molar ratio of As/Fe (T) in precipitates versus pH.

Reduction of Precipitates

The mass balance of each element was estimated for the two-step neutralization using MgO and NaOH, and the average concentrations of several constituents in the inflow, the MgO tank, the ferrite formation tank, and the effluent in the continuous flow experiments with actual AMD were calculated. It is assumed that the differences between inflow concentrations and outflow concentrations for elements in a tank are transferred to precipitates formed in the tank. The chemical composition of the precipitates obtained from the mass balance calculations and the analyzed chemical composition of precipitates are shown in **Figure 15**. The analyzed compositions agree well with the calculated ones.

The fate of the elements in the process is calculated as shown in **Figure 16** after confirming the mass balance of each element. The precipitates generated by the process adopted in this study amount to 41.91 g/day, corresponding to one third of the conventional two-step neutralization using $CaCO_3$ and $Ca(OH)_2$, assuming that the water content of precipitates and the scale factor of the treatment plant are ignored. Most of the Fe, Si, Al, and As are removed from the AMD and included in the precipitates whereas most of the Mg, SO_4, and Ca remain in the effluent. The reduction in the volume of precipitates results from changes in the chemical form of Fe and the absence of precipitation of calcite and gypsum.

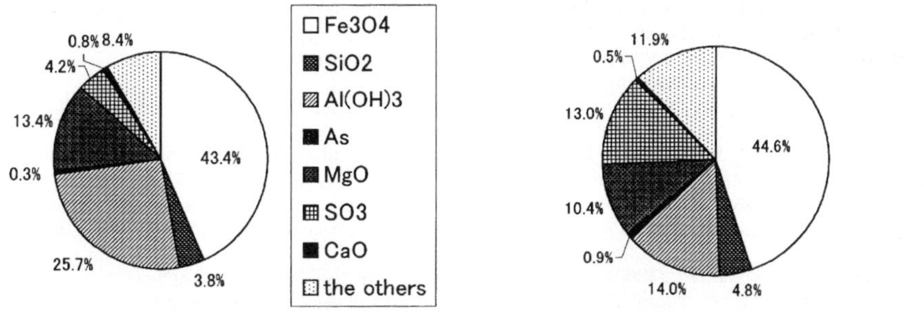

(a) Calculated from the water quality (b) Analysis of precipitates

Figure 15: Chemical composition of precipitates by the two different methods.

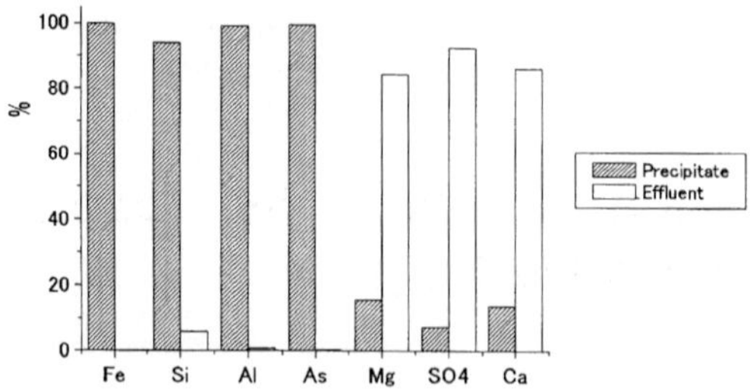

Figure 16: Fate of constituents contained in AMD.

Conclusions

The conventional AMD treatment process was revised to consider the properties of the precipitates generated. Two-step neutralization with MgO and NaOH was applied to Horobetsu AMD. The results obtained are as follows:

(1) The solid-liquid partitioning of constituents in AMD was elucidated by changing the pH in the first neutralization. The optimum conditions for continuous flow experiments with actual AMD were determined based on batch and continuous flow experiments with synthesized AMD.
(2) Ferrite formation in the ferrite formation tank was confirmed with XRD analysis and qualitative magnetism measurements.
(3) The As was removed from AMD in the MgO mixing tank by coprecipitation with Fe hydroxide.
(4) The amount of precipitates generated is estimated to be one third of those with the conventional process.

Acknowledgements

We wish to thank Mitsubishi Materials Natural Resources Development Corp. and Dowa Engineering Co., Ltd. for generous support with the experiments in the actual plant.

References

1) Schiller, J.E. and Khalafalla, S.E. (1984) Magnesium oxide for improved heavy metals removal, *Mining Engineering*, 36, 171-173.
2) Mehta, R.K. and Zhang, L. (1992) Chemoremediation of acid mine drainage by a two-step ferrite coprecipitation followed by magnetic separation, *Proceedings of the Sixth International Conference on Ferrites* (ICF 6), Tokyo and Kyoto, pp.253-257.
3) Oh, J. (1992) New ferrite process and its chemistry for removal of heavy metal ions from waste water, *Proceedings of the Sixth International Conference on Ferrites* (ICF 6), Tokyo and Kyoto, pp.211-217.
4) Toshida, T. and Tamaura, Y. (1992) Ferrite process; Removal of heavy metal ions from waste waters and stabilization into ferrites, *Proceedings of the Sixth International Conference on Ferrites* (ICF 6), Tokyo and Kyoto, pp.258-252.
5) Matsuoka, I. (1996) Mine drainage treatment, *Shigen-to-Sozai*, 112, 273-281 (in Japanese).
6) Wang, W., Xu, Z., and Finch, J. (1996) Fundamental study of an ambient temperature ferrite process in the treatment of acid mine drainage, *Environ. Sci. Technol.*, 30, 2604-2608.
7) Bekin, A. and Matsuoka, I. (1998) Applicability of magnesium oxide as a neutralizing agent in mine water disposal, *Shigen-to-Sozai*, 114, 553-559 (in Japanese).
8) Perez, O.P., Umetsu, Y., and Sasaki, H. (1998) Precipitation and densification of magnetic iron compounds from aqueous solutions at room temperature, *Hydrometallugy*, 50, 223-242.
9) Perez, O.P. and Umetsu, Y. (2000) ORP-monitored magnetite formation from aqueous solutions at low temperature, *Hydrometallugy*, 55, 35-56.
10) Lahav, O., Morgan, B.E., Hearne, G, and Loewenthal, R.E. (2003) One-step ambient temperature ferrite process for treatment of acid mine drainage waters, *Jour. Environ. Eng. ASCE*, 192, 155-161.
11) Sasaki, H. and Takada, T. (1999) Possibility of magnetite formation in the treatment of acid mine drainage, *Proceedings of Annual Conference of the Mining and Materials Processing Institute of Japan*, C1-3, pp.9-12 (in Japanese).
12) Takao, H. and Sasaki, H. (2000) Formation of magnetite in the treatment of iron-containing acid mine drainage, *Proceedings of Annual Conference of the Mining and Materials Processing Institute of Japan*, C2-4, pp.69-72 (in Japanese).

Basic requirements on the capping system of the uranium mine tailings

Eva Kolarova, Geotechnical Department of the Czech Technical University in Prague, The Czech Republic
Ivan Vanicek, Geotechnical Department of the Czech Technical University in Prague, The Czech Republic

Abstract

Extensive investigation and mining activities for the uranium ore started in the Czech Republic only at the end of Second World War firstly for military and later for energy purposes. In the new political and economic situation of the eighties and nineties the curtailment of these activities began and most of the mines and concentration plants (mills) were closed. The paper concentrates on the uranium mine tailings from the uranium mill MAPE Mydlovary, (the other ones are Straz pod Ralskem and Dolni Rozinka), especially on the capping system, which have to fulfill such different demands as low permeability, surface drainage, shielding gamma radiation and aerial emission of radon.

Introduction

The construction of the uranium mill in Mydlovary, situated in the South of Bohemia, Figure 1, started in 1959 and this uranium mill operated from 1962 up to 1991. The capacity was increased step by step from the 300 000 t per year up to 600 000 t per year. Between the years 1962 and 1991 nearly 17 millions of tons of the uranium ore with average uranium content 0.184% was milled and processed there and the tailings after that were deposited by the hydraulic method to the tailing ponds. Roughly 36 mil tons of the tailings are stored there still with some content of uranium (2 320 tons) with radioactivity E14 Bq Ra 226. Two different methods of processing were used there – by acid leaching (12.8 Mt), or by alkaline method (3.9 Mt). Tailings were stored into eight different tailing ponds, the first one is the classical tailing dam (KI), constructed on the surface, the rest were deposited into depressions after lignite surface mining. Total inner area is roughly 23 km^2, with the infiltration area about 28 km^2, Kolarova (2002).

Remediation activities started at the end of the eighties. From the beginning some theoretical demands from the environment protection point of view were defined and after that the practical application started, which had mostly an experimental character. The first tailing pond is partly remediated, but during the application of the protection methods some problems appeared such as

- slope instability, especially on places, where the periphery of the tailing dam is very close to another pit after surface mining,
- leachate spreading was recorded as a result of not so good a sealing system based mostly on the natural geological barrier,
- high settlement of the capping system.

For other tailing ponds the two initial points are not so important, due to the fact that the deposited material is situated in old excavated pits of lignite mines with relatively good clayey subsoil. Therefore the main focus is now on the capping system. The design should eliminate as much as possible the differential settlement, to prevent tensile cracks of the sealing clay liner and to guarantee surface drainage. The remedial works are moving so far relatively slowly. In principle the reasons are two. Firstly, water, which now covers the surface of tailing ponds, creates a sufficient protection barrier. Secondly, the total remedy effort requires a huge amount of materials. If the capping system is composed on average from 1.5 m of material, then the total volume of the material for the remediation is close to 40 millions m^3. Technically feasible is the transfer of the deposited tailings from the smaller, and initially, from shallower tailing ponds (as e.g. K IV/D) to the bigger ones. In this way it is possible to reduce the amount of the material for the capping system and to remedy excavated places very promptly. Against this solution there are two concerns. Firstly, the environmental one, because re-excavation and transport can bring new problems with the evolution of radio nuclides into the environment and can also lead to the contamination of the traffic way. The second concerns is a financial one because the price of this solution is higher.

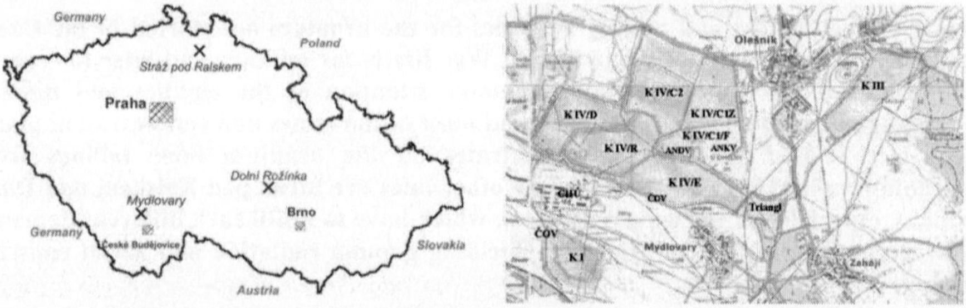

Figure 1: Location of the concentration plant MAPE Mydlovary on the map of the Czech Republic and detailed placing of uranium tailing dams (KI, KIII and different KIV), concentration plant (CHÚ), sewage plant (ČOV), mine water plant (ČDV), and tailing dam for ash from the power plant Mydlovary (Triangl).

Basic requirements for the capping system

The main aim of this area remedy effort is to find an optimal solution for the capping system which is able to fulfill all demanded technical criteria with minimum claims on the protection material. The demands on the capping system can be as follows.

- Shielding of the gamma radiation.
- Restriction of the aerial emission of radon – Rn.
- Prevention of dust spreading with different radio nuclides as the result of wind surface erosion.
- Limitation of rain water infiltration.
- Securing of the surface drainage of rain water.

<u>Shielding of gamma radiation</u>

One of the basic conditions is the reduction of gamma radiation on the surface of the remediated area. According to the Czech codes the recommended maximum value is 0.1μSv per hour plus natural background, altogether maximally 0.4 μSv per hour. Another possibility is to restrict access to this area. The estimation of the thickness of the shielding layer for deposited material after uranium ore mining or milling and processing is very often close to 1 m, with a maximum up to 2.5 m. Proposed thickness has to be checked by equation for the radiation decay during passage through the capping system.

Restriction of the aerial emission of radon – Rn

The aerial emission of radon from remediated areas is directed by the content of Ra^{226} in deposited material. Only a limited part of radon can reach the atmosphere. How big it will be depends on the values of the emanation coefficient and permeability for gases. Therefore the capping system has to be composed from material with low permeability for gases. According to international standards the accepted level of aerial emission should be lower than 0.8 $Bq.m^{-2}.s^{-1}$. From this point of view the most problematic tailing pond is K IV/R where the aerial emission of radon directly on the surface of deposited tailings is between 20 – 24 $Bq.m^{-2}.s^{-1}$ above the natural background. For the rest of the tailing ponds it is much lower, in the range of 5 – 9 $Bq.m^{-2}.s^{-1}$. The basic equation for the radon emission on the surface of the deposited material is (1):

$$F_t = A * E * \rho * (\lambda * D_t)^{1/2} \ Bq.m^{-2}.s^{-1} \qquad (1)$$

Where:
A – activity of Ra^{226} ($Bq.kg^{-1}$)
E - emanation coefficient
ρ – volume density of the deposited material ($kg.m^{-3}$)
λ – decay constant Rn ($2.1 \times 10^{-6}.s^{-1}$)
D_t – diffusive coefficient of the deposited material Rn ($m^2.s^{-1}$)

For a multilayer-capping system the equation is more complicated, as it is necessary to know the diffusive coefficient of each layer.

Prevention of dust spreading with different radio nuclides as the result of wind surface erosion

This condition can be fulfillled relatively easily by creation of a re-vegetation surface, composed for example from 0.1-0.2 m thick biologically active layer.

Limitation of rain water infiltration. Securing of surface drainage for rain water.

Both these conditions are very similar to the conditions for classical municipal waste landfills. On the one hand it is necessary to construct a special sealing layer, usually under the condition that the coefficient of permeability is equal or lower than 1.10^{-9} $m.s^{-1}$ and on the other to guarantee that the dominant amount of rain water will be drained from the surface or via a drainage layer or will be evaporated from the remediated surface. To guarantee this condition the inclination of the remediated surface should be at least 1%. So it means that for the tailing ponds with relatively large areas the thickness will be strongly affected by this condition especially in the middle of the tailing pond. A minimum condition for the thickness along the periphery of the tailing pond will be in most cases directed by aerial emission of Rn.

Basic composition of the capping system

For fulfillment of the above mentioned conditions the basic composition of the capping system should consist of, Vanicek, I (2002), Gens, A (2003), Figure 2:

- deposited tailings,
- sub base layer, minimum thickness of which around the tailing pond should fulfilll the requirements for protection against aerial emission of Rn, and increasing in such a way that even after the settlement (with respect to the different thickness of deposited tailings and with the different thickness of the capping system eliciting surcharge loading above layer of tailings) will guarantee the required inclination of the surface

of the capping system. Collected water will not be affected by deposited tailings and can be drained into the water course to restore the original hydro geological system,
- sealing layer,
- drainage layer,
- protection filter,
- mantle, partly biologically active, appropriate for plant development,
- final biologically active layer – humus, top soil.

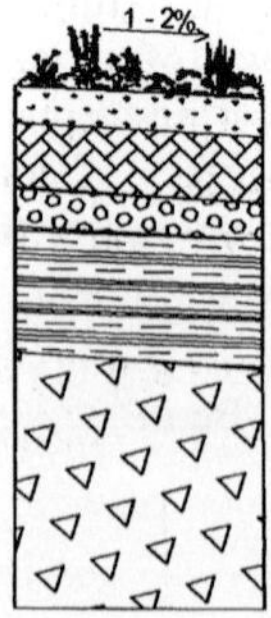

Figure 2: Basic composition of the capping system.

The sub base layer, which can consist of unsorted soil, probably will reach the largest volume. Total volume will depend on the recommended arrangement of the surface. Here we can speak about roof-shaped dewatering or about valley dewatering – Figure 3. Preferably this layer should be permeable to fulfill the condition that the deposited and saturated tailings should be drained during consolidation even if from the upper part. Around the tailing pond the collection of water from the process of consolidation has to be arranged and collected water treated in the sewage disposal plant. In our case we suppose that the sub base layer will be deposited directly on the surface of the uranium tailings. In Germany, the strengthening of this contact by geosynthetics is proposed to eliminate low bearing capacity of the deposited uranium tailings, Leupold et al (2003).

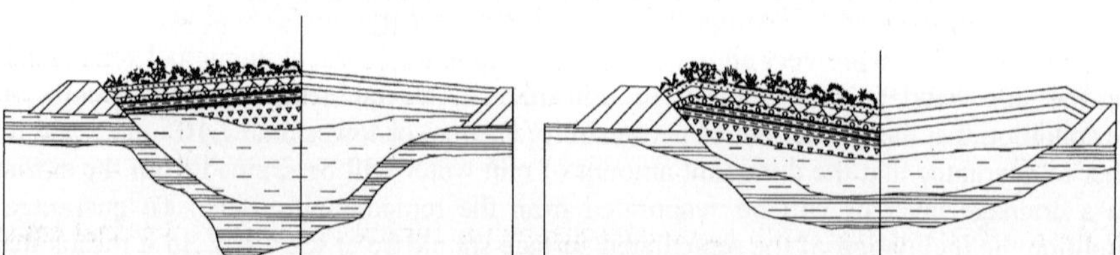

Figure 3: Recommended arrangement of the surface, roof - shaped dewatering and valley dewatering, not in scale.

Under the effort to optimize the total thickness of the capping system not only from the point of view of protection against aerial emission of radon, but also to minimize the surcharge loading to decrease the total settlement the calculation of radon emission was done for two cross sections. The first cross section is using the compacted clay liner and the second one is using the geosynthetic clay liner as a main component to fulfill the condition of the permeability of the sealing layer.

<u>*Cross section using compacted clay liner and grain filter*</u>

The first case represents the classical case of using natural materials for both the sealing and the drainage layers. The thickness of clay liner is proposed 0.6 m, compacted in 3 layers each having 0.2 m. The thickness of the other layers is as follow – sub base 0.3 m, drainage layer 0.3 m, mantle 0.5 m and top soil 0.15 m – see Figure 4,b and table 1. In this table the radon emission is calculated for the extreme case that the radon emission on the top of the deposited uranium tailings is 24 $Bq.m^{-2}.s^{-1}$. Calculated radon emission on the remediated surface is 0.11 $Bq.m^{-2}.s^{-1}$ fulfillling demanded value.

Cross section using geosynthetic clay liner and geodrain

The second case (Table 2 and Figure 4a) represents utilisation of geosynthetics materials. For sealing layer geosynthetic clay liner GCL – known also as bentonite mattresses - is proposed and for drainage layer geodrain – e.g. type Colbonddrain. The thickness of each layer is roughly 10 mm. The thickness of the sub base layer (0.7 m) and mantle (0.65 m) is bigger than in the previous case. The final value of radon emission is also lower than the required value.

Table 1: Calculation of radon emission on the top of capping system with CCL.

layer	depth of layer [m]	radon emission on the surface of layer [$Bq.m^{-2}.s^{-1}$]
deposited tailings – initial level		24
sub base layer	0.3	5.17
compacted clay liner CCL	0.6	0.24
drainage layer	0.3	0.21
mantle	0.5	0.13
top soil	0.15	0.11

Table 2: Calculation of radon emission on the top of capping system with GCL.

layer	depth of layer [m]	radon emission on the surface of layer [$Bq.m^{-2}.s^{-1}$]
deposited tailings – initial level		24
sub base layer	0.7	1.37
GCL – geosynthetic clay liner/ bentonite mattresses (NaBento)	0.009	1.18
drainage layer - geodrain	0.01	1.18
mantle	0.65	0.59
top soil	0.15	0.56

For the other tailing dams with radon emission on the surface 9 $Bq.m^{-2}.s^{-1}$ the final emission on the surface of the same cross section is lower than 0.04 $Bq.m^{-2}.s^{-1}$. For this case the following composition is sufficient to protect an environment against radon emission: sub base layer 0.4 m, GCL 0.005 m, geodrain 0.01 m, mantle 0.65 m and top soil 0.15 m. To summarize the differences between calculated 2 cross sections we can state:

- Total thickness of the capping system can be decreased because roughly 20 mm of geosynthetics can substitute 0.9 m of the classical sealing and draining layers (but not fully to guarantee the required values for gamma radiation - minimum 1.0 m - and for radon emission) and so to decrease the total settlement of the deposited tailings.
- Strict requirements for sealing and drainage layers are substituted by unsorted soil, which is cheaper and can be found in near surroundings.
- The geosynthetic clay liner is more flexible than the classical compacted clay where tensile cracks as the result of differential settlement can be a significant problem – Vanicek (2002).

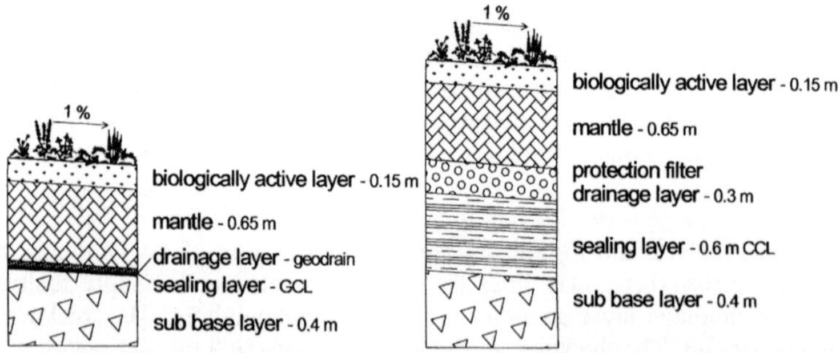

Figure 4: Composition of the capping system – a) with geosynthetic clay liner GCL and geodrain, b) with compacted clay liner and natural drain.

Field investigation and laboratory tests

To be able to predict the total settlement and consolidation in time t, the first phase of site investigation started. For the first phase of the investigation the light penetrometer with height of fall 0.5 m, drive head diameter 25.2 mm and falling weight 10 kg was used. The reason was a limited access on the tailing ponds. The tests were performed only very closely to the embankment, because the rest of the pond was flooded by water. Nevertheless this penetrometer device proved to be a very useful tool for this type of material, the resistance was not so high to protect penetration of the tip even for the coarser material which is usually around periphery of the pond. In the centre of the pond the deposited material is generally finer. Therefore the results are mostly on the optimistic side. Briefly summarize, this penetrometer, which is easily mobile, can recognise the total depth of deposited material and give us the first information about anisotropy, change of properties as a function of position and depth and also values of modulus of deformation as a function of observed profile. Grain size distribution tests were performed on samples collected for oedometric tests; results are showing that the material has a character of fine sand. As mentioned above, the grain size composition in the centre of the tailing pond is much finer in most cases, it contains higher percentage of particles lower than 0.06 mm. For the calculation of total settlement the deformation characteristics have to be determined and the following methods were used for a rough estimation:

- With the help of classical oedometric test on samples with dia 120 mm which were collected close to the place where penetrometer tests were performed. For a lower range of stresses the oedometric modulus was lower than 10 MPa, for a higher range of stresses between 10 and 20 MPa.
- From the penetrometer test the result of which is shown on Figure 6. It is evident that some layers are very soft, the modulus of deformation is close or even less than 1 Mpa, even at the place situated very close to the pond periphery.
- From the back analysis. For the first tailing pond K I, covered by protection layers with thickness roughly 2 m, the maximum settlement measured on the surface was up to 750 mm after 3.5 years Sedlacek,M (2003). From the consolidation curve it is obvious that the further settlement can be expected. For the maximum depth of the deposited material 23 m there, the expected oedometric modulus of deformation E_{oed} (for 1 D) is close to 1 MPa.

Note. Some other testing is recommended but now we are in the phase of determining the protective measures for laboratory handling with the material which is to a certain extent radioactive. Nevertheless the results from this first phase can be used for a rough estimation of the total settlement.

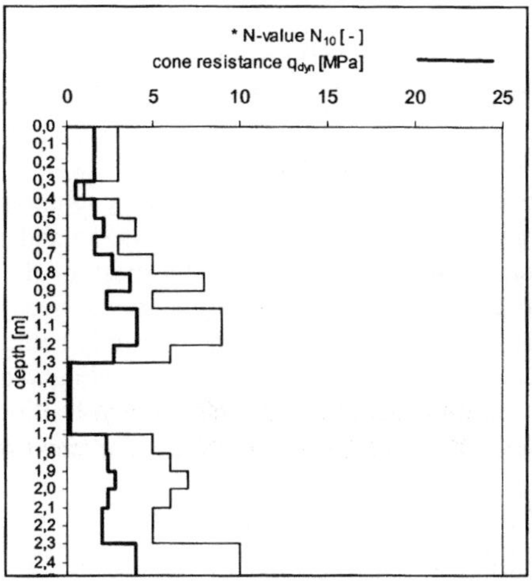

depth [m]	N_{10} [-]	q_{dyn} [MPa]
0,1	3	1,6
0,2	3	1,6
0,3	3	1,6
0,4	1	0,5
0,5	3	1,6
0,6	4	2,1
0,7	3	1,6
0,8	5	2,6
0,9	8	3,6
1,0	5	2,3
1,1	9	4,1
1,2	9	4,1
1,3	6	2,7
1,4	0	0,1
1,5	0	0,1
1,6	0	0,1
1,7	0	0,1
1,8	5	2,3
1,9	6	2,4
2,0	7	2,8
2,1	6	2,4
2,2	5	2,0
2,3	5	2,0
2,4	10	4,0

Figure 6: The result of penetration test.

Estimation of the total settlement

The first estimation was calculated for maximum depth of the deposited uranium tailings in the middle of the tailing pond 10 m, and for minimum one (1m) around periphery of the pond and for the width of 400 m. For the roof-shaped dewatering and demanded inclination of the surface only 1%, the thickness of the sub base layer will be changed from minimally assumed 0.4 m up to 2.4 m only for the dewatering, plus additional part due to the calculated settlement. The minimum total thickness of the capping system was previously calculated for 1.2 m assuming geosynthetics in cross section and radon emission on the top of deposited uranium tailings 9 $Bq.m^{-2}.s^{-1}$. The thickness of deposited material 1.0 m is assumed to be around the periphery of the tailing ponds. For expected modulus of deformation E_{oed} = 1 MPa, calculated settlement of deposited material is roughly 24 mm there. In the centre of the pond the total thickness (without additional settlement) is supposed to be 3.2 m and therefore the expected settlement of the deposit there is close to 640 mm. Therefore we must add another 640 mm to the sub base layer which will increase the total settlement to the value 728 mm. By this step by step method roughly 800 mm of additional sub base material have to be added in the centre of the tailing pond to eliminate the expected settlement and to guarantee the surface of capping system inclination on average, see Figure 7. The advantage is that the consolidation close to the periphery will be quicker due to more coarse (and permeable) character of deposited material, so the inclination of the surface during fist years will be a little bit steeper than 1%.

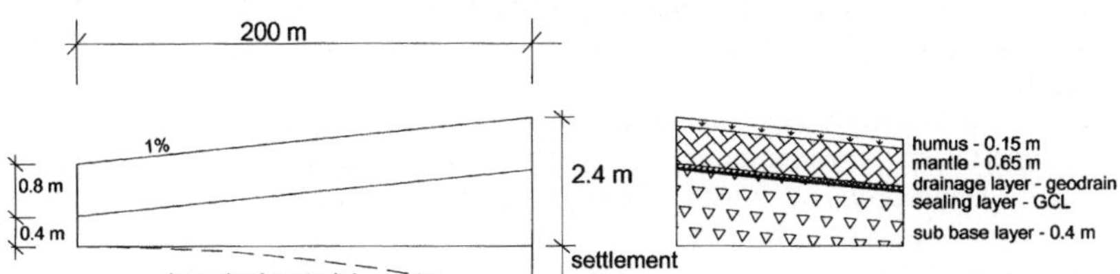

Figure 7: Expected capping profile taking into account the settlement.

Conclusion

Tailing ponds from uranium mill in Mydlovary represent very complicated problem, the solution of which will need a long time and a huge amount of material for capping system. The following results of the first phase of calculation and investigation, can be emphasized:

- The proposed capping profile utilising geosynthetics for sealing layer (geosynthetic clay liner) and for drainage layer (geodrain) was found as an optimum solution which is able to fulfill all required technical criteria, such as an aerial radon emission and gamma radiation, with minimum claims on the protection material. Total thickness of the capping system 1.2 m was found to be adequate for radon emission on the surface of tailing pond 9 $Bq.m^{-2}.s^{-1}$, resp. 1.5 m for 24 $Bq.m^{-2}.s^{-1}$. At the same time this solution can decrease the soil requirements with strictly determined characteristics of sealing and drainage layers. Geomembrane is also not as sensitive as the clay sealing layer is to the differential settlement elicited by anisotropy of the deposited material. For clay sealing layer this differential settlement can lead to tensile loading and finally to the tensile cracks.
- This high anisotropy was verified by penetrometer tests and it is the result of wet process of deposition. Light penetrometer also proved its usability for a quick and cheep estimation of the total depth of deposited material and for its deformation characteristics.
- A minimum thickness of sub base layer, which can be composed from unsorted partly permeable soils, is around pond periphery 0.4 m, but in the centre of the pond (the minimum width of which is 400 m) it should be much higher. Value of 3.2 m was calculated there as a summation of these 0.4 m plus 2.0 m to fulfill required inclination of the surface 1% and adding another 0.8 m as the result of calculated settlement.
- The volume of the material for sub base layer can be roughly estimated, but in all cases it will be very high. Therefore to decrease this volume we recommend return to the idea of transferring the deposited tailings from smaller and shallower pond to the bigger one. Nevertheless the environmental aspects of this transfer should be solved.
- The main phase of site and laboratory investigation has to start, but after the adoption of strict rules of radioactive material management.

Acknowledgements

This paper was prepared with the help of grant project No. MSM 6840770005 Sustainable construction.

References

Gens, A.: The role of Geotechnical Engineering for nuclear energy utilisation – Special Lecture. In: Proc. XIII ECSMGE, Vanicek et al.(eds.), Prague 2003, CGtS, Vol.3, p. 25- 67.

Kolarova, E. : Remediation of old mining activity. MSc. thesis, CTU, Prague. 154 p. (in Czech).

Leupold, D., Oswald,K.D., Heerten,G., Horstmann,J,: Geotechnical aspects concerning the long-term remediation of tailings management facilities as former uranium ore mining industry legacies. In: Proc. XIII ECSMGE, Vanicek et al.(eds.), Prague 2003, CGtS, Vol.3, p. 277-282.

Sedlacek, M.: Information report about the observation of the tailing dam KI. Report of VD TBD - Supervision Organization in Hydro Engineering. Prague, 2003, 25 p. (in Czech).

Vanicek, I.: Deposition of Waste Material on the Surface. Main Lecture. Session 4: Environmental Protection. In: Proc. XII Danube ECSMGE, Passau 2002, Verlag Gluckauf GmbH, Essen pp. 543-554.

Vanicek, I.: Remediation of landfills, old ecological burdens. Czech Technical University Press, Prague. 2002. 247 p. (in Czech).

The Water Balance of the BF4 Uranium Waste Pile, Poços de Caldas, Brazil, as a contribution for future remediation projects.

Adilson do Lago Leite, Federal University of Ouro Preto, Civil Engineering Department, Ouro Preto, Brazil
Junio René Toledo Fagundes, Federal University of Ouro Preto, Civil Engineering Department, Ouro Preto, Brazil
Carlos Eduardo Ferraz de Mello, Federal University of Ouro Preto, Civil Engineering Department, Ouro Preto, Brazil

Abstract
This paper focus on the water balance for the BF4 uranium waste rock pile, INB, Poços de Caldas, Brazil, as a first step to understand the groundwater flow regime at this site and the acid mine drainage (AMD) generation. This is essential for choosing and designing adequate remediation measures in the future. Firstly, a drainage basin that comprises the BF4 had to be delimitated, then all the available data from the site was retrieved and organized in specific files. These include historical meteorological data, flow measures and pumping rates at the BF4. Some statistics has been provided for estimating the infiltration, runnoff and evapotranspiration. To estimate evapotranspiration, the Penman Model is proposed. The results clearly reveals that the main contributor to the AMD generation at the BF4 is precipitation and that there is no significant contribution of the local groundwater to the discharge at the base of BF4. So, a cover system which reduces infiltration is suggested as a passive remediation measure.

Introduction

Acid Drainage (AD) or Acid Mine Drainage (AMD) is known worldwide as a side effect of mining operation. Basically, AMD is generated by the oxidation of sulfides, mainly pyrite, in the presence of water, oxigen and naturally occurring bacteria.

Large amount of AMD has been generated at the Mining-Industrial Complex of Poços de Caldas (CIPC), Brazil, since the beginning of its operation. The Osamu Utsumi Mine operated at the CIPC from 1981 to 1995 in a discontinuous way, where approximately 2,316,527 tons of uranium ore was produced (Cipriani, 2002). Nowadays, the CIPC belongs to the Industrias Nucleares do Brasil S.A. (INB), which is property of the Brazilian government.

Besides the ore mass, 108 million tons of waste rock came out as overburden, which, in turn, were deposited in waste rock piles. The main waste rock pile, here called BF4, covers an area of 560,000 m^2, with volume of 12.4 million m^3.

Since its disposal, BF4 has been generating AMD in its base, which nowadays is collected on a small dam, called BNF, to be later pumped in pipelines to the mine open pit reservoir. An average volume of 5 x 10^5 m^3 of AMD (pH of 3 to 3.5) is annually collected at the BNF dam, representing about 40% of the total acidic water generated at the CIPC (Cipirani, 2002).

All the CIPC acidic waters, including the discharge from BF4, are pumped and treated with hydrated lime and other chemicals on a water treatment plant. Later they are released on the environment. Some reports say that, only with chemicals, around US$ 150,000 is spent each year for acid neutralization (Cipriani, 2002). This is costly prohibitive, once AMD generation is expected to occur for centuries at this site. So, a more reasonable approach would be reducing the AMD generation, including the BF4 generation, using passive remediation technologies.

This paper aims to perform the water balance of the BF4 waste pile drainage basin, as a first step to understand the flow regime at this site. As a part of the chain of nuclear fuel production, this facility has been well monitored for many environmental parameters since the beginning of its operation. So, these are valuable data which shall not be wasted.

This water balance was intended to determine the BF4 water contributors in terms of volume. In this way, passive remediation measures could be suggested as an attempt to decrease the costs associated with the active treatment currently used at this site.

Background

Covering an area of about 15 km^2, the CIPC stands between 1300 and 1600 meters above the sea level, at the so-called Poços de Caldas Plateau. The climate there is typically tropical, with two very distinguished seasons: wet (March to October) and dry (April to September). The average temperature hangs around 19 °C, with minimum of 1°C and maximum of 36 °C. The 30-year average precipitation is 1700 mm. About 82% of this precipitation falls in the wet season.

Natural vegetation is basically composed of savannas and grass on the hills and plains, along with small stains of rain forest at the river valleys (Prado, 1994).

The CIPC stands at the great volcanic caldera of Poços de Caldas, Brazil. This caldera has approximately 30 km in diameter and is composed of alkaline rocks from the Later Cretaceous. Ring dikes can be found at its boundaries. Nepheline sienites, phonolites and foyaites are the main rock types of the plateau (Holmes et al, 1992). Faults, fractures and breccias can easily be found all over the caldera.

Figure 1 shows the CIPC plant. In the mine, the uranium exploration resulted in a huge open pit as a circular whole of about 176 m deep and 1200 m diameter, which is now filled with water. Nowadays, the open pit reservoir receives liquid contributions from direct precipitation, boundary runoff and pumping from the BF4 and from the CIPC water treatment plant.

Figure 2 presents a picture of the BF4 main slope (70° and approximately 90 m height). BF4 was dumped on the ancient Consulta stream valley, which, in turn, was deviated from its original course. The Consulta stream belongs to the Rio Verde river drainage basin. The bedrock geology and the waste deposit configuration makes it easier the groundwater to flow off the BF4 in a single point, located at its base. This water is highly acidic (pH 3 to 3.5) and is totally collected in a small dam called BNF. Since its construction, in 1989, the BNF acidic water has been pumped to the open pit reservoir.

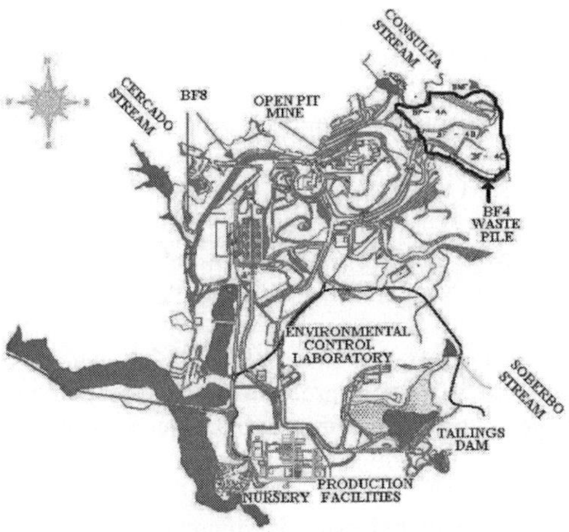

Figure 1 - CIPC schematic (Adapted from Cipriani, 2002).

Figure 2 - Partial view of the BF4 slope.

BF4 Water Balance Method

In general terms, a water balance involves the difference between the inflow and the outflow of an geometric element, such as shown by the following equation:

$$[\textit{outflow}] - [\textit{inflow}] = [\textit{variation inside the element}] \quad (1)$$

In this way, two approaches can be used in the water balance: (a) not considering the groundwater flow and (b) considering the groundwater flow.

In the case (a), the water balance can be expressed by the next equation.

$$[P + R_{in}] - [R_{out} + ET + I] = \Delta V_{ECS} \quad (2)$$

where P is the precipitation, R_{in} and R_{out} the surface runoff that respectively enters and leaves the element, ET the evapotranspiration, I the infiltration and ΔV_{ECS} the water variation (storage) at the close surface of the soil.
The case (b) involves groundwater flow, as can be seen in the equation (3).

$$[P + R_{in} + G_{in}] - [R_{out} + ET + G_{out}] = \Delta V_{ECG} \quad (3)$$

where G_{in} and G_{out} are respectively the groundwater inflow and outflow and ΔV_{ECG} the groundwater storage.

In a global scale, the water balance can be considered as a close system. However, in a regional scale this not the case (Tucci, 2002). To solve this problem, the concept of the drainage basin can be used. In this way, a geometric element is defined such as all the components of the water balance can be estimated.

Equation (3), case (b), was used for the BF4 water balance. Firstly, its drainage basin had to be delimitated and for that the surface water divides (drainage divides) were determined over a 1:7,500-scale topographic map. Since no significant surface drainage was observed on the BF4, the BNF dam was defined as the only outflow of its drainage basin.

The components of the water balance were calculated by using the average values from the available hydrological and meteorological data. These data was retrieved and organized in specific files, such as: precipitation, potential evaporation, temperatures, humidity, the water discharges of the Consulta stream and the pumping rates from the BNF to the open pit reservoir. Also some statistics has been operated on these data to provide monthly average input for estimating infiltration, runnoff and evapotranspiration.

Some hypotheses were assumed, based on the observations made at the site. They are as follows:

- as a long term analyses (annual), the term ΔV_{ECG} of equation (3) equals to zero;
- the R_{in} component is neglected, since the volume of surface water that comes from other drainage basins is very small;
- the R_{out} term was also neglected, since no erosion features nor surface drainage plants were observed over the BF4 area;
- potential evapotranspiration (ET_p) was considered to be the same as evapotranspiration (ET).

In this way, equation (3) turns into the following:

$$[P - ET]A + G_{in} = G_{out} \quad (4)$$

where A is the BF4 area, standing for turning the units of equation (4) to volume. In this equation, the component G_{out} is considered to be the volume of acidic waters collected on the BNF dam, at the base of BF4. G_{out} data corresponds to the monthly average discharge from 1998 to 2004.

According to equation (4), the G_{in} component is the difference between [P - ET]A and G_{out}. Once the runoff components R_{in} and R_{out} are neglected, all the precipitation must infiltrate into the soil towards groundwater. In this way, equation (4) is rearranged as:

$$[I]A - G_{out} = \pm G_{in} \tag{5}$$

Following the hypotheses formulated in equations (4) and (5), if G_{in} is positive, water is escaping from BF4 to adjacent drainage basins through other points besides the BNF dam discharge. If it is negative, water from other sources besides precipitation is infiltrating in BF4, probably from other groundwater sources.

Potential evapotranspiration (ET_p), the most complex term of the water balance, was estimated by using the Penman Model (Tucci, 2002). This model is based on the energy balance of the system along with the aerodynamic conditions of the site. Temperature, humidity, wind velocities, insolation and geographic position are some of the parameters considered in this model, as can be seen in equation (6) (Tucci, 2004).

$$ET_p = \frac{\dfrac{\Delta}{\gamma}\dfrac{R_{liq}}{\rho_w L} + E_i}{\left(\dfrac{\Delta}{\gamma} + 1\right)} \tag{6}$$

where:
ET_p = potential evapotranspiration (cm/day^{-1});
ρ_w = water unit weight (g/cm^{-3});
L = latent heat of vaporization (cal/g^{-1});
R_{liq} = efective radiation absorbed by the soil surface (cal/cm^2.day);
E_i = evaporation under isothermic conditions (cm/day);
Δ = rate of change of the saturated vapor pressure of water in air with respect to temperature (mm-Hg/^{o}C)
γ = Bowen Constant (0,49 mm-Hg/^{o}C).

The term R_{liq} was estimated through the following expression (Tucci, 2002):

$$R_{liq} = R_t\left(\alpha + \beta\frac{n}{N}\right)(1-a) - \sigma T^4\left(0.56 - 0.09\sqrt{e}\right)\left(b + c\frac{n}{N}\right) \tag{7}$$

where:
R_t = short waves radiation at the top of the earth atmosphere;
α and β = correction factors which account for the atmosphere vapor content and the height and thickness of the clouds;
n = average effective daily insolation;
N = maximum period of effective daily insolation;
a = albedo or reflective factor;
σ = Stefan-Boltzman constant (1.19 x 10^{-7} cal/(cm^{2}dayK^4);
T = absolute temperature;
e = vapor pressure;
b and c = correction factor considering the clouds effect on the long wave radiation;

For more details about the Penman equation, the reader should refer to basic hydrology texts.

In the absence of the factors α and β for Poços de Caldas, these parameters were obtained from the nearest cities of Lavras (0.23 and 0.49 respectively) and São Paulo (0.24 and 0.58 respectively).

The albedo (a) was obtained from Demattê et al (2004). They have analyzed the albedo of three different classes of soil from the state of São Paulo, Brazil, including a red-yellow podzol. Field observations at the CIPC site revealed that the BF4 surface soil is similar to aforementioned soil, so a values of 0,30 and 0,35 from Demattê et al (2004) were applied.

RESULTS AND DISCUSSIONS

Precipitation (P)

The Environmental Control Laboratory/INB have registered the monthly precipitation at the CIPC since 1976. The closest gauge station stands about 2.5 km from the BF4. The precipitation data from this station were used in the performed water balance.

The double mass curve method was used for checking the consistency among this gauge station and other three stations, standing no more than 7 km from the BF4. The coefficient of determination (R^2) determined in this analysis was 0.9992 , validating the precipitation data used here.

Table 1 shows the monthly average precipitation and the annual accumulated from 1977 to 2004 at the CIPC.

Table 1 - Monthly average and annual precipitation at the CIPC.

MONTH	AVERAGE (mm)
January	319.44
February	215.75
March	204.82
April	94.12
May	70.50
June	26.74
July	24.49
August	36.32
September	79.11
October	131.01
November	177.93
December	257.24
Total (Annual)	**1637.48**

It is clear that two distinguished seasons exist in the site: (1) dry season, from April to September, where the average precipitation is less than 100 mm/month and (2) wet season, from October to March. January offers the greatest precipitation, with more than 300 mm/month. At the same time, July is the driest month, with less than 25 mm/month.

Potential Evapotranspiration (ET_p)

The results of the Penman model application for ET_p estimation are presented in Table 2. In this table the different values of α, β and a are tested.

BNF Discharge (G_{out})

The monthly average discharges at the BNF dam, considered here as G_{out} , are given in Table 3. They represent weekly barrel measures at the base of BF4 from January/1998 until August/2004.

Table 2 - Potential Evapotranspiration (ET_p) at the CIPC.

Month	$\alpha = 0.23$ e $\beta = 0.49$		$\alpha = 0.24$ e $\beta = 0.58$	
	$a = 0.30$	$a = 0.35$	$a = 0.30$	$a = 0.35$
Jan	90.61	83.74	100.03	92.49
Feb	86.94	80.59	96.28	89.26
Mar	84.09	77.83	93.37	86.45
Apr	71.64	65.68	81.50	74.84
May	55.08	50.17	63.22	57.73
Jun	54.76	50.39	62.08	57.18
Jul	54.37	49.64	62.26	56.97
Aug	80.25	74.34	90.37	83.74
Sep	95.26	88.43	106.76	99.11
Oct	105.53	98.46	116.39	108.55
Nov	95.76	88.52	106.72	98.70
Dec	102.14	95.01	112.31	104.44

Table 3 - Discharge at the BNF Dam (01/1998 to 08/2004)

MONTH	VOLUME (m^3)
January	49986.17
February	48748.80
March	53621.14
April	49453.71
May	50007.43
June	45617.14
July	47541.60
August	43779.09
September	40488.00
October	40932.40
November	42312.00
December	46016.40
Total (annual)	**558503.89**

Water Balance

As mentioned before, equations (4) and (5), here considered to be the same, have been used to estimate the water contributions at the BF4 drainage basin. For that, monthly average values of P, Et_p and G_{out} were applied under different α, β and a Penman parameters. The analysis was performed on an annual basis and the results are presented in Table 4.

As can be seen in Table 4, in three of the four types of analysis, G_{in} values were negative. This means that water sources other than precipitation are effective in BF4. On the other hand, the difference between G_{out} and I goes from 1.2 to 18.36 % , which does not seem to be high considering all the possible errors involved in the estimation of the water balance. Thus, it is concluded that precipitation is by far the main contribution to the AMD collected at the BNF dam (G_{out}).

Table 4 - Water balance results for different α, β and a Penman parameters.

$I = [P - Et_p]A$ (m^3)	G_{out} (m^3)	G_{in} (m^3)	$\left\lvert \frac{G_{out} - I}{G_{out}} \right\rvert \times 100$ (%)
551,828.83[(1)]	558,503.89	-6,675.06	1.20
613,293.58[(2)]	Idem	54,789.69	9.81
455,946,17[(3)]	Idem	-102,558.00	18.36
524,256.10[(4)]	Idem	-34,247.80	6.13

[(1)] $\alpha = 0.23$; $\beta = 0{,}49$ and $a = 0.30$
[(2)] $\alpha = 0.23$; $\beta = 0{,}49$ and $a = 0.35$
[(3)] $\alpha = 0.24$; $\beta = 0{,}58$ and $a = 0.30$
[(3)] $\alpha = 0.24$; $\beta = 0{,}58$ and $a = 0.35$

The correlation between the precipitation (P) and the BNF discharge (G_{out}) was checked as shown in Figure 3. In both graphics, the numbers mean maximums and the letters mean minimums. The apostrophe (') on the numbers and letters refers to the BNF discharge curve. For example, the precipitation number 2 correlates to the BNF discharge number 2' and so on; at the same time, the letter A at the precipitation curve can be correlated to the letter A' at the BNF discharge curve.

It is quite clear that both curves in Figure 3 are correlated, particularly the minimums (letters). This correlation corroborates the results from the water balance, in which precipitation is the main water contributor at the BF4 drainage basin. The lag time estimated between the precipitation and its response on G_{out} is about 2 and 3 months.

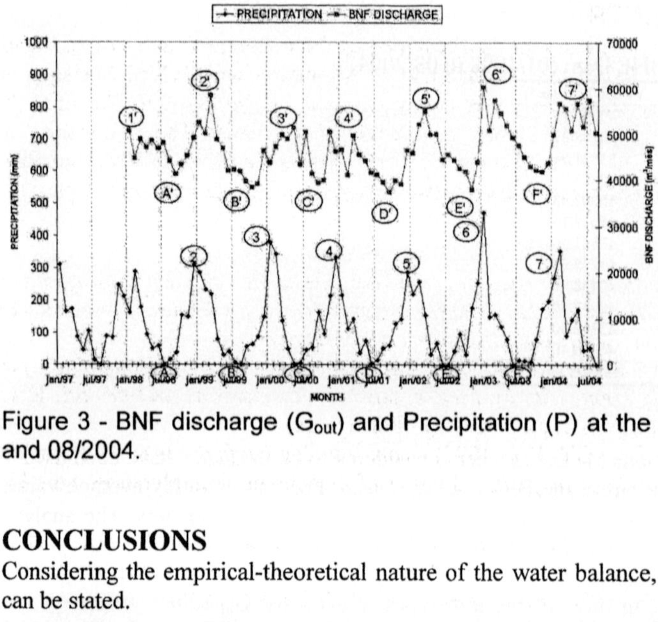

Figure 3 - BNF discharge (G_{out}) and Precipitation (P) at the CIPC, between 01/1997 and 08/2004.

CONCLUSIONS

Considering the empirical-theoretical nature of the water balance, the following conclusions can be stated.

The infiltration (I) estimated from the water balance is somewhat similar to the measured BNF discharge (G_{out}), with differences between 1.20 to 18.36 % . Considering all the parameters computed in the water balance, some of them estimated from the literature and some produced by statistics, and also taking into account the model used to estimate the potential evapotranspiration (ET_p), these differences can be due to intrinsic errors of the method.

Being I and G_{out} similar parameters, the following immediate conclusions are stated:

- Most of the precipitation infiltrates into BF4, later moving to the water table and flowing to the BNF dam as groundwater (G_{out});
- There is no significant groundwater outflow from BF4 (G_{out}) but the one aforementioned;
- There is no significant water contribution from sources other than precipitation;

Thus, it can be concluded that in long term, the interruption of infiltration on the BF4 surface would lead to an expressive decrease in the BNF discharge and consequently a reduction in the overall AMD generation at the CIPC (recall that 40% of the AMD treated at the CIPC comes from the BNF dam).

A cover system is suggested to be constructed on the top of the BF4 so as to reduce water infiltration and AMD generation. Further studies are needed to evaluate the technical details and the cost-benefit analysis of the proposed cover system.

The results show the significant contribution of the water balance as a method for understanding the flow regime at the BF4 drainage basin. This is important to take appropriate AMD remediation measures at this site. Water balance is considered an easy-to-use and low-cost tool once historic hydrological and meteorological data are available.

ACKNOWLEDGEMENTS

The authors appreciate the support and the data provided by the Industrias Nucleares do Brasil S/A (INB). The support and help from Dr. Mauricio de Almeida Ribeiro, CIPC Manager, is sincerely acknowledged. Insolation data was gently donated by the Instituto Nacional de Meteorologia (INMET), for which the authors extend their thankfulness. The scholarship for the second author was provided by the Coordenadoria de Aperfeiçoamento de Nível Superior (CAPES). Finally, the authors appreciate the reviewers suggestions for improving this paper.

REFERENCES

CIPRIANI, M. (2002). *Mitigation of the environmental and social impacts from uranium mine closure procedures.* Ph.D. Thesis. Instituto de Geociências, Universidade Estadual de Campinas. Campinas, SP, 332 p. (In Portuguese)

DEMATTÊ, J. A. M.; TOLEDO, A. M. A.; SIMOES, M. S. (2004) *Methodology for recognizing three different soils by remote sensing: laboratorial and orbital.* Rev. Bras. Ciênc. Solo, vol.28, n°5, p.877-889. (In Portuguese)

HOLMES, D.C.; PITTY, A.E.; NOY, D.J. (1992) *Geomorphological features of the Poços de Caldas Caldera analogue study sites.* Journal of Geochemical Exploration, vol. 45, p.215-247.

PRADO, V. C. S. (1994). *The impact of the uranium concentrated over the river water quality - a case study at the Poços de Caldas Mining-Industrial Complex.* Master Dissertation. UFRJ, COPPE. Rio de Janeiro, 195 p. (In Portuguese)

TUCCI, C. E. M. (editor) (2002). *Hydrology: Science and Application.* 3ª ed. Editora UFRGS/ABRH, Porto Alegre, RS, 943 p. (In Portuguese)

Geotechnical Aspects of Peat Dams on Bog Land

G. P. McInerney, Department of Civil, Structural and Environmental Engineering, Trinity College Dublin, Ireland.
B. C. O'Kelly, Department of Civil, Structural and Environmental Engineering, Trinity College Dublin, Ireland.
P. M. Johnston, Department of Civil, Structural and Environmental Engineering, Trinity College Dublin, Ireland.

Abstract
Boglands are wetlands formed by the gradual accumulation of decaying plant material. Many bogs have been damaged to varying degrees by anthropogenic activities such as drainage schemes, afforestation and harvesting of peat. The conservation of a bog requires the careful management of the groundwater to support plant growth and to prevent drying out of the peat material. Drainage and cutaway areas cause subsidence that tends to propagate across the bog. One conservation approach involves the construction of peat dams along the bog perimeter at an affected site. This paper reviews a case study at Raheenmore bog (Ireland) where a series of peat dams were constructed. A local slope failure of one of the peat dams occurred after construction. The dams were studied with the aim of identifying the cause of failure. On-site information was reviewed and a series of laboratory tests were conducted on undisturbed and remoulded peat specimens. Numerical back-analysis of the failed peat dam was carried out.

Introduction
Raised bogs, characterised by a domed relief, originated after the last glaciation over 10,000 years ago when the lakes formed by the glacial melt waters were colonised and gradually filled in with fen vegetation and decaying plant material. Raised bogs have been threatened by drainage schemes used to improve the land for agricultural use, afforestation or the harvesting of peat. Plant root activity ceases when the groundwater level falls about 10 cm below the ground surface. The bog surface subsides in response until the acrotelm (upper growing peat layer) reaches the new equilibrium groundwater level thereby allowing plant growth to recommence. However, there are also limits to the steepness of the slopes that will support sphagnum moss growth.

In the past 20 years, the conservation of semi-intact raised bogs has been highlighted by more stringent environmental policy and legislation, particularly concerning Special Areas of Conservation. For example, only about 19% of the 1.3 million hectares of peatland that originally covered Ireland remains relatively intact. Many engineering solutions have been considered to reverse the damage to protected raised bog ecosystems. One method that was used at Raheenmore bog (Ireland) was the construction of peat dams along the bog perimeter to pond surface water that would re-saturate and regenerate

5th ICEG Environmental Geotechnics, Thomas Telford, London 2006.

the vegetation in locally damaged areas. The purpose of the peat dams was also to arrest the propagation of the subsidence initiated by the perimeter drainage.

Case study

Raheenmore bog (Figure 1a) located in County Offaly, Ireland, is a designated Nature Reserve, 162 hectares in area with peat depths of up to 15 m, which are underlain by lacustrine clay. The bog has been effected by man-made drainage along sections of its perimeter.

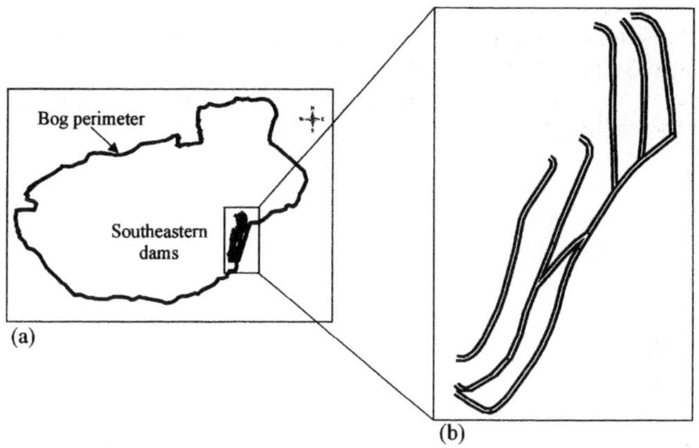

Figure 1 – (a) Raheenmore bog. (b) South-eastern peat dams.

Following a series of geological and hydrogeological studies completed by the Irish-Dutch Raised Bog Study (Streefkerk and Zandstra, 1994), a site was chosen at the south-eastern part of Raheenmore bog for the construction of a series of semi-concentric peat dams (Figure 1b). The dams, 0.5 to 5.0 m in height, were constructed during the summers of 1994 and 1995 using peat material that was at different stages of biodegradation, namely; very slightly decomposed (H2 to H3 on the von Post classification scale), slightly decomposed (H3 to H5), and highly decomposed (H7 to H8). The von Post classification system (von Post and Granlund, 1926) characterises the peat material in terms of its degree of biodegradation on a scale of H1 to H10, with H10 being the most strongly degraded, amorphous material. Figure 2 shows a typical cross-section of a dam constructed of slightly decomposed peat. Figure 3 shows the cross-section of a dam constructed of strongly decomposed peat.

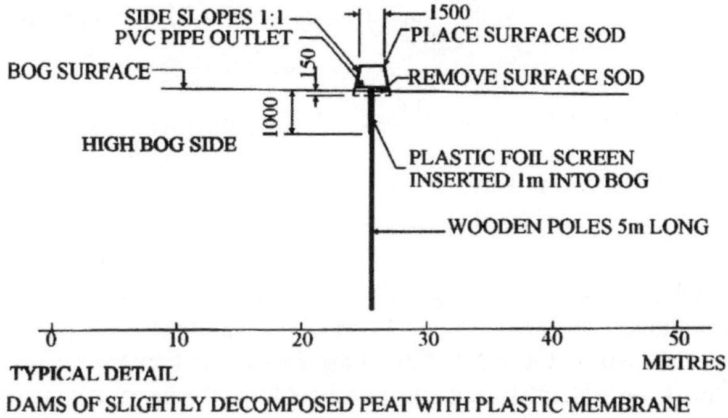

Figure 2 – Cross section of dam constructed of slightly decomposed peat (Bennett, 1998).

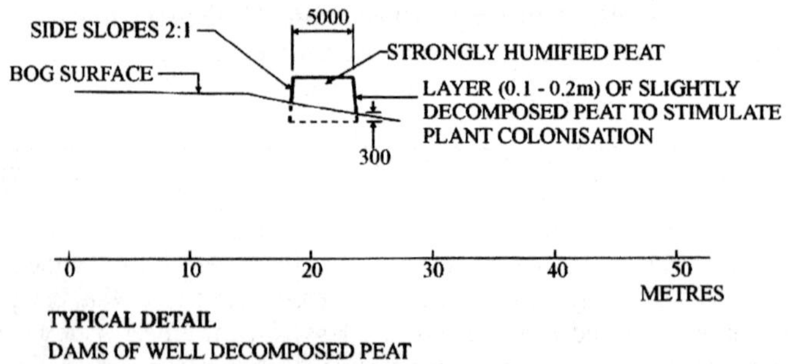

Figure 3 – Cross section of dam constructed of strongly decomposed peat (Bennett, 1998).

These dams were designed based on the experiences from the construction of similar dams at the Bargerveen Nature Reserve (The Netherlands). Wooden poles, 5 m in length, were driven vertically into the ground under the dams that were constructed of the slightly decomposed peat material to increase bearing stability.

Performance of dams at Raheenmore

A local slope failure of one of the dams occurred in December 1998, about four years after the completion of construction. A large flow of water and peat debris was discharged and covered an area of about 0.36 hectares (Figure 4).

Tension cracks were evident around the area of the breach. Large-scale vertical and horizontal displacements occurred with the crest of the dam having been displaced down-slope by 1.5 m. On-site monitoring ceased prior to the slope failure so it was difficult to identify with certainty the cause of failure.

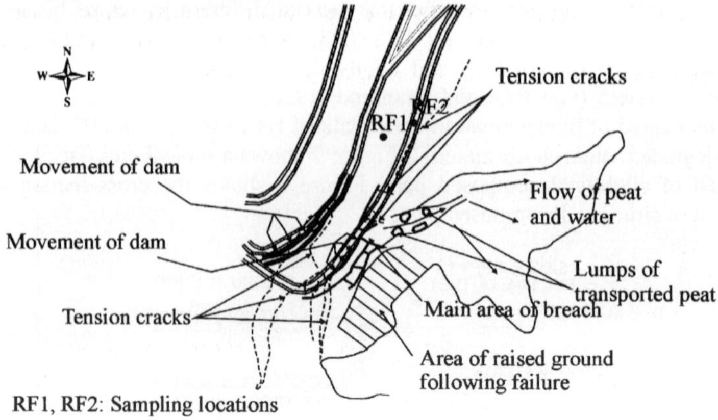

Figure 4 – Extent of dam failure.

Material properties and test programme

Peat samples were taken at locations RF1 (*in-situ* peat) and RF2 (remoulded used to construct the dam) shown in Figure 4. The samples were taken from about 1.0 m below ground level by pushing a thin-walled sampling tube, 100 mm in diameter and 1.0 m in length, into the ground. Some index properties of the peat material are listed in Table 1.

Table 1 –Index properties of the peat material.

Location	von Post classification	Water content (%)	Specific gravity	Mass loss on ignition (%)
RF1	H4	1170	1.42	98
RF2	H4	590	1.46	92

The water content values were determined by oven drying specimens of the peat material at 105°C. The mass loss on ignition values were determined by igniting dry peat material at 440°C. Location RF2 had a lower ignition loss since the peat material used in the construction of the dam was sourced from nearer the bog perimeter and included some clay material. The specific gravity (G) values were estimated using the equation G=3.8/(0.013N+1.4) where N (%) = the mass loss on ignition (Hobbs, 1986).

Landva and La Rochelle (1983) reported that the ring shear test was the most reliable in determining the effective stress shear strength parameters of peat. A programme of consolidated-drained shearbox and ring shear tests was carried out. The shearbox specimens were 101.6 mm in diameter and 20.0 mm in height. The ring shear specimens had an outer diameter of 50.0 mm and an inner diameter of 35.0 mm. The specimens were consolidated under applied vertical stresses of 10, 20 and 40 kPa. The specimens were then sheared slowly at a rate of 0.00975 mm/minute in the shearbox apparatus and 0.096 deg/minute in the ring shear apparatus, which allowed full dissipation of the excess porewater pressures.

Laboratory test results

The stress-strain behaviour of the peat specimens measured using the shearbox apparatus is shown in Figure 5. The results from the ring shear tests are shown as shear stress versus angular rotation in Figure 6. A Mohr-Coulomb analysis (Figure 7) indicated that the ring shear and shearbox apparatus gave similar effective stress shear strength parameters of c' = 0 and ϕ' = 38°. Farrell and Hebib (1998) reported similar values from shearbox and ring shear tests on the Raheenmore peat material.

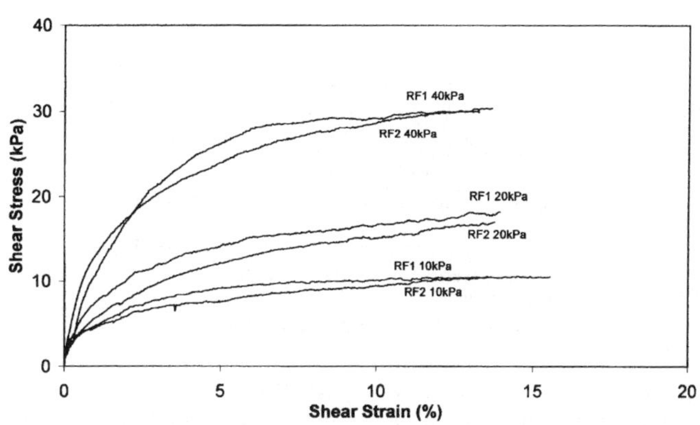

Figure 5 – Shear stress versus shear strain data from shearbox tests.

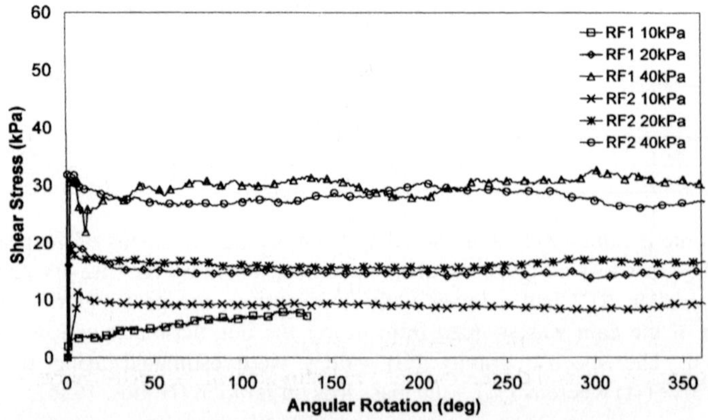

Figure 6 – Shear stress versus angular rotation from ring shear tests.

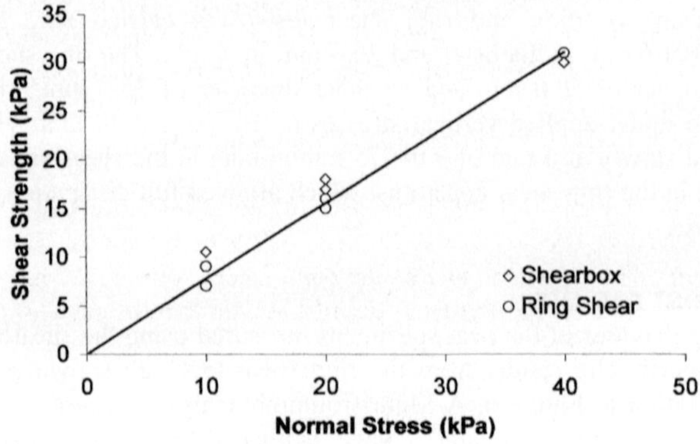

Figure 7 – Shear stresses versus effective normal stress plot.

Numerical modelling

The Raheenmore dam was modelled using the finite element package PLAXIS (version 8.2). Table 2 shows the soil input parameters used for the analysis. The analysis indicated that slope instability would not occur under these conditions (factor of safety of 1.12). However, the analysis predicted some vertical and lateral movements of the peat dam (Figure 8).

Table 2 – Soil properties for PLAXIS analysis.

Soil	Saturated unit weight (kN/m³) γ_{sat}	Horizontal permeability (m/day) k_x	Vertical permeability (m/day) k_y	Young's modulus (kN/m²) E_{ref}	Poisson's ratio ν	Cohesion (kN/m²) c_{ref}	Friction angle (deg) ϕ'
Dam peat (RF2)	11.4	10^{-4}	10^{-5}	420	0.2	0	38
Bog foundation (RF3)	10.8	10^{-4}	10^{-5}	420	0.2	0	38
Clay	21.0	10^{-8}	10^{-8}	2000	0.35	10	30

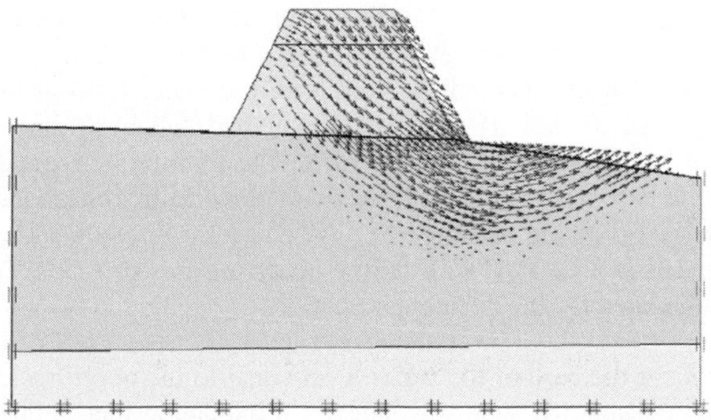

Figure 8 – Ground displacements predictions from PLAXIS analysis.

van Baars (2005) presented a case study involving the translational failure of a canal dyke constructed of peat material in The Netherlands. The failure occurred due to the decrease in the weight of the dyke as a result of drying out of the peat material near the crest of the dyke. At Raheenmore bog, the dam crest would also have dried out during extended periods of dry weather. The effect of the increase in the horizontal pressure that acted behind the dam during sudden rainfall events was considered in the horizontal and vertical stability analysis according to method of van Baars (2005).

Figure 9 shows a cross section of the dam close to the failure location. The maximum head of water retained by the dam was limited to 2.5 m by a pipe that allowed the overflow to discharge to a point in front of the dam. There were concerns that the water exiting this pipe may have eroded the front toe of the dam thereby further increasing the likelihood of slope instability.

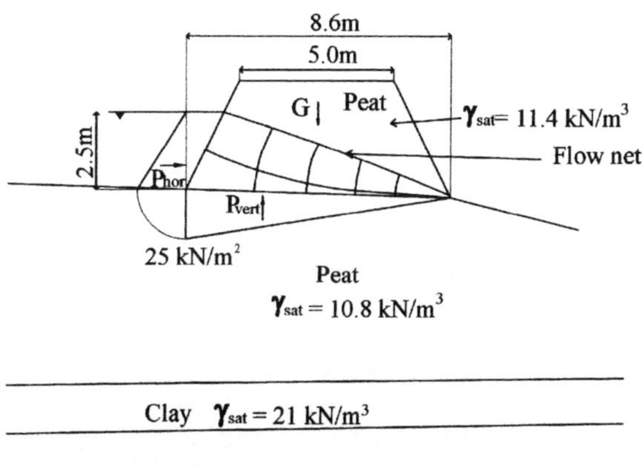

Figure 9 – Cross section of dam.

Horizontal and vertical stability

The total cross-sectional area of the dam (I_{tot}) was 24 m^2. If the peat material had been saturated over the full cross-sectional area, the weight of the dam ($G_{sat} = I_{tot} \times \gamma_{sat}$) would have equalled 274 kN/m run. The flow net in Figure 9 shows the general flow path through the dam and the zone above which drying out of the peat material could have occur during extended periods of dry weather. If this zone had dried out, the weight of the dam ($G_{unsat} = G_{sat} - l_{unsat} \times (\gamma_{sat} - \gamma_{dry})$) would have been reduced to 144 kN. The hydrostatic uplift beneath the dam ($P_{vert} = 0.5 \times p \times l$) was 108 kN. The hydrostatic force ($P_{hor} = 0.5 \times \gamma \times h^2$) acting behind the dam with the water level at the overflow level would have equalled 31 kN. If the water level had fallen by just 1.0 m, then the hydrostatic force would have reduced to 11 kN. The limiting horizontal shear resistance mobilised over the base of the peat dam was determined using the laboratory-measured c' and ϕ' values in the following equation: $F_{max} = l.c' + (G - P_{vert}) \times \tan \phi'$. The safety factor (SF) was calculated as F_{max}/P_{hor}, with failure occurring for SF < 1.0. Table 3 shows the safety factors calculated for the different scenarios.

The safety factor for the case of the water level equal to the overflow level and the crest of the dam in an unsaturated state was below 1.0 indicating that failure would occur. A large rainfall event did occur prior to failure of the Raheenmore dam in December 1998 that would have increased the hydrostatic force acting behind the dam. Tension cracks and the possibility of erosion and undermining of the front toe of the dam by the water discharging from the overflow pipe could have reduced the safety factor even further.

Table 3 – Safety factors for different scenarios.

Water level	Peat at crest	P_{hor}	P_{vert}	G	F_{max}	Safety factor
High	Saturated	31	108	274	129.7	4.2
High	Unsaturated	31	108	144	28.1	0.9
Low	Saturated	11	108	274	129.7	11.5
Low	Unsaturated	11	108	144	28.1	2.5

Summary and conclusions

Peat dams used in the conservation of raised bogs are generally constructed on the *in-situ* bog foundation. The dams at Raheenmore bog (Ireland) were designed from experience of dam construction at Bargerveen Nature Reserve (The Netherlands). Translational failure is of major concern and is not always considered in the design of peat dams. A local translational failure occurred at one of the Raheenmore dams about 4 years after the completion of construction. Shearbox and ring shear tests carried out on peat material indicated effective stress shear strength parameters c' = 0 and ϕ' = 38°. The ground model and soil parameters were inputted into a finite element package to determine the factor of safety on slope instability. A factor of safety on slope instability of 1.12 was obtained for a classical circular slip failure indicating that the dam should have been stable. However, further analysis using the same soil parameters in horizontal and vertical stability calculations after van Baars (2005) indicated a factor of safety of less than 1.0 for the case of a dried out zone near the dam crest and a high water level acting behind the dam. Tension cracks evident in the failure location and piping underneath the dam may have also been contributing factors. Translational failure is a principle mode of failure for peat dams built on bogland and should always be considered in design.

Acknowledgements

The authors would like to thank Michael Gill for his helpful comments during the preparation of this paper. Grainne McInerney would also like to acknowledge research scholarships from RPS Consulting Engineers and the Geotechnical Trust Fund of the Institution of Engineers of Ireland.

References

Bennett, J., 1998, Monitoring the hydrogeological and geotechnical performance of peat dams at Raheenmore bog, Co. Offaly, Ireland. MSc thesis, University of Dublin, Trinity College.

Farrell, E. R., and Hebib, S., 1983, The determination of geotechnical parameters of organic soils. Proceedings International Symposium on Problematic Soils, IS-TOHOKU 98, Sendai, Japan, 33–36.

Gill, M., 2005, The use of peat dams in wetland conservation. MSc thesis, University of Dublin, Trinity College.

Hobbs, N. B., 1986, Mire morphology and the properties of some British and foreign peats. Quarterly Journal of Engineering Geology. London. Vol. 19, 7–80.

Landva, A. O., and La Rochelle, P., 1983, Compressibility and shear characteristics of Radforth peats. Testing of peats and organic soils, ASTM STP820, P. M. Jarrett, Ed., American Society of Testing and Materials.

Streefkerk, J. G. and Zandstra R. J., 1994. Experimental management measures in the south-east corner of the Raheenmore raised bog reserve. Staatsbosbeheer, The Netherlands.

von Post, L., and Granlund, E., 1926, Sodra sveriges torvtillgangar. Sveriges geologiska undersokninh C. 335 yearbook Vol. 19, No. 2, Stockholm.

van Baars, S., 2005, The horizontal failure mechanism of the Wilnis peat dyke. Geotechnique, Vol. 55, No. 4, 319–323.

Discussion on Theoretical and Practical Aspects of Gas Fluxes in Mining Waste Deposits

Rose Mary Gondim Mendonça – Federal University of Tocantins, Palmas, Brazil
Maria Claudia Barbosa - Federal University of Rio de Janeiro, Rio de Janeiro, Brazil
Laura De Simone Borma - Federal University of Tocantins, Palmas, Brazil

Abstract

A capillary barrier cover will be used to minimize O_2 and water fluxes into pyritic waste piles of carbon mines in South Brazil. The objective is to control the generation of AMD inside the piles. As part of the project, a research on the gas flux by diffusion through the barrier materials is being realized in laboratory scale, and some results are presented herein. At first, a theoretical framework on gas flux through porous medium is presented, together with a brief discussion on the relative importance of advection and diffusion mechanisms in gas fluxes of usual practical situations. O_2 diffusion tests were performed in a double reservoir laboratory cell using two different testing procedures: constant concentration source and finite mass source. The O_2 consumption was also calculated and compared to the experimental results. The diffusion coefficient obtained was 21 or 28 m^2/y, depending on test procedure, and the measured reaction coefficients were inferior to the calculated values.

Introduction

The studies of Environmental Geotechnics in the beginning were mainly concerned with the liquid phase of the soil, through which the potential contaminants could be transported. To consider the soil saturated can be a conservative approach in the case of water-soluble species, since in those cases the mass flux is faster through the water than through the air phase.

But there are many practical situations that require a better understanding of the gas phase behavior in soils and solid wastes. So there are today an increasing number of studies being developed for:

- *Comprehension of natural phenomena* – examples: quantification of oxygen consumption by organic matter and soil minerals; degradation of uranium wastes with radon gas liberation; aerobic and anaerobic degradation processes of natural organic matter and organic compounds;
- *The gas as an instrument in soil remediation* – examples: the use of techniques to facilitate the contaminant oxidation and increase the produced gas removal; soil aeration to promote biological augmentation and increase the resulting aerobic degradation of contaminants; or the opposite, to isolate the system from the oxygen of the air, to promote anaerobic degradation; isolation of MSW and uranium mining waste from the atmosphere, to prevent gas emissions, respectively methane and radon gases; inhibition of oxygen flux into sulfide rich waste piles, to prevent acid mine drainage (AMD) generation;
- *The role of gas phase in the mechanical behavior of unsaturated soils and wastes.*

Mechanisms of Gas Transport in The Soil

The air is a dynamic constituent of the soil and the air movement may occur either in saturated condition (dissolved air flux in the water phase) or in unsaturated condition (air flux through the connected pores filled with air plus the dissolved air flux through the pores filled with water).

The flux of air in the soil occurs based on two major transport mechanisms: advection and diffusion. According to many authors, RENAULT & SIERRA (1994) and WEERTS *et al.* (2000), diffusion can be considered the dominant mechanism in soils, but the contribution of advection cannot be ignored (RENAULT & SIERRA, 1994), particularly for granular materials (HILLEL, 1998).

Gas diffusion in the soil or waste

There are two mechanisms of diffusion in unsaturated media. The first one involves the diffusion of air and/or water vapor through the soil pores filled with air. The second one involves the flux of the air as a dissolved constituent through the soil liquid phase (water) due to a chemical concentration gradient or to an osmotic suction gradient.

The amount of a particular gas in the soil is determined by the amount of soil pores filled with air and the proportion of the air volume that is occupied by the gas. The natural soil generally presents the same or slightly higher proportion of nytrogen than the atmospheric air, and less oxygen (YANFUL *et al.*, 1993). At high water contents, the percentage of oxygen is significantly reduced, prevailing the amount of dissolved oxygen in the soil water (HILLEL, 1998). In consequence, there is a seasonal variation of each gas concentration both in the soil and in the atmospheric air composition, due to variations in temperature and humidity as observed by DAVIS *et al.* (1996).

At the particular situation of an AMD generating mining waste pile, the cover design is mainly concerned with oxygen flux into the pile. Although advective fluxes related to wind currents, thermal effects and convection cannot be ignored, most authors agree that the major contribution in oxygen entrance in the waste in the case of high water content cover materials is the diffusion mechanism of transport. That is the reason of performing diffusion gas tests for the design of the waste pile cover reported herein.

Basic Equations

Gas diffusion coefficient in the soil or waste

TINDALL *et al* (1999) state that, historically, the Fick law has been used to model the gas diffusion flux in the soil. But many authors have shown that Fick law may underestimate the gas flux in soils because the model only considers the gas movement in response to partial pressure gradients. Four different mechanisms of gas diffusion in soils are recognized today: Fick law ordinary diffusion, Knudsen diffusion, non-equimolar diffusion and the solid surface diffusion.

Despite these different mechanisms, NICHOLSON *et al.*, (1989) and YANFUL (1993) state that one-dimensional diffusion of both soil phases, water and air, through non-reactive wastes and soils, obey the Fick first law, as presented in Table 1. In the same table are presented the equations for reactive materials, using a reaction coefficient (K and λ) that represent the decrease of concentration due to the consumption of the gas by the soil or the waste constituents. The advective flux is not included in these equations.

The gas properties can be separated from the porous medium properties, by considering the relationship between D/D_0 (D – diffusion coefficient of the gas in the porous medium; D_0 – diffusion coefficient of the gas in free solution, calculated and available in literature for many gases). This relationship is referred as tortuosity factor by many authors (WEERTS *et al.*, 2000). This approach is identical to the conventional approach adopted for transport of solutes through the liquid phase of the soils.

Table 1 – Mathematical formulations presented by YANFUL (1993) and by program POLLUTE v 6.0 for determination of the gas diffusion coefficient.

	YANFUL (1993)	POLLUTE
Fick 1st Law	$F(t) = -\theta_a D_e \frac{\partial C}{\partial t}$ $\theta_a = n\,(1\text{-}S)$	$F(t) = -nD_e \frac{\partial C}{\partial t}$
Fick 2nd Law plus kinetic reaction term	$\frac{\partial C}{\partial t} = D_e \frac{\partial^2 C}{\partial z^2} - KC$	$n\frac{\partial C}{\partial t} = nD_e \frac{\partial^2 C}{\partial z^2} - \lambda C$

Where:
F – mass flux ($ML^{-2}T^{-1}$);
θ_a – volumetric water content (L^3L^{-3});
n – total porosity (L^3L^{-3});
D_e – effective diffusion coefficient (L^2T^{-1});
C – concentration (ML^{-3});
t - time (T);
Z – depth (L);
K and λ – chemical equilibrium constant of oxygen consumption by sulfides (first order reaction) or radioactive or biologic decay (first order reaction).

To estimate the gas diffusion coefficient another expression can be used, equation 1 proposed by MBONIMPA *et al.* (2003).

$$D_e = \frac{1}{n^2}\left(D_a^0 \theta_a^{p_a} - H D_w^0 \theta_w^{p_w}\right) \quad (1)$$

Where: D_a^0 and D_w^0 are the oxygen diffusion coefficients in air and in water, respectively. The exponents p_a and p_w are the tortuosity factors for the air and water phases, respectively. In the interpretation of the tests presented herein, the approximation $p_a = p_w = 3.3$ was used, as proposed by MBONIMPA *et al.* (2003) for soils with total porosity of 0.4. And H is the Henry constant expressed by the ratio $H = C_w/C_a$, equal to 0.03 for oxygen at 20^0C.

Reaction coefficients for gas in soil or waste

The consumption of oxygen by the oxidation process is an attenuation factor for oxygen migration through waste piles, but the oxidation reaction produces unwanted chemical substances. In soils the bacteria and the organic matter are responsible for the attenuation of oxygen concentration in the soil pores. In the case of covered ADM generating waste piles, the oxygen that is available for consumption by the waste is the gas coming by diffusion from the soil cover, so it is important to determine the reaction coefficient for the cover materials.

According to NICHOLSON *et al.* (1989), at any depth within the waste pile the variation rate of oxygen concentration can be determined by the Second Fick Law combined to the kinetic reaction term (K) presented in Table 1. On the other hand, CABRAL (2000) states that oxygen consumption by bacteria existing in paper and cellulose residue can also be described by a first order reaction as in pyritic wastes. Consequently, the same equations can be used to represent oxygen consumption rates both in the soil cover and in the waste pile, at least as a first approximation.

COLLIN & RASMUSON (1988) proposed equation 2 to estimate the reaction coefficient, based on the kinetics of the reaction on the particle surface varying linearly with the percentage of sulfidic minerals present in waste composition. The equation considers the porosity and the surface area of an equivalent grain size D_H.

$$\lambda = \lambda' \frac{6}{D_H}(1-n)C_p \qquad (2)$$

Where: λ' is the reaction of pyrite with oxygen ($5x10^{-10}$ m^3O_2/m^2 pyrite surface); C_p is the pyrite content of the waste in dry weight (kg/kg). D_H can be estimated using grain size distribution curve parameters as presented by MBONIMPA *et al.* (2003) in equation 3:

$$D_H = \left[1 + 1.17\log\left(\frac{D_{60}}{D_{10}}\right)\right] D_{10} \qquad (3)$$

Where: D_{10} and D_{60} are the diameters corresponding to 10% and 60% of passing material, respectively.

In the case of interpretation of diffusion cell tests, equation 4 proposed by ELBERLING *et al.* (1994) can also be used to estimate oxygen consumption. This expression assumes that the oxygen consumed will be reflected in the decrease of concentration in the reservoir source.

$$\ln\left(\frac{C}{C_0}\right) = -t(\lambda D_e)^{0.5}\frac{A}{V} \qquad (4)$$

Where: A is the cell area, V is the volume of air above the soil or waste test sample, S is the inclination of the $\ln(C/C_0)$ versus time (t) graph. And D_e can be estimated by equation 1.

$$\lambda = \left(\frac{SV}{A}\right)^2 \frac{1}{D_e} \qquad (5)$$

Since the reaction is influenced by many different factors such as mineralogy, sulfide content, temperature, sulfide state of oxidation, bacterial activity, among others, it is expected that the estimated and the observed reaction coefficients are not coincident.

Experimental Results

A capillary barrier is being designed as the cover system of some existing waste piles in South Brazil. The cover objective is the control of the acid mine drainage (AMD) that is being leached from these piles for a long time, causing a significant environmental impact

(MENDONÇA *et al.*, 2003). One of the configurations designed that are being investigated is the use of local thermoelectric bottom ashes as the coarse layer in the capillary barrier (UBALDO *et al*, 2006, this congress). Table 2 summarizes the main geotechnical characteristics determined for the ashes, following the Brazilian standards.

Table 2 – Geotechnical characteristics of the thermoelectric bottom ashes

Parameter	G	K	w_L	w_P	IP	Gravel	Sand	Silt	Clay	D_{60}	D_{10}
Unit		(cm/s)	(%)			(%)				($x10^{-5}$ m)	
Test Standard	DNER ME 093/94	NBR 13292	NBR 6459/84	NBR 7180/84	NBR 7180/84	NBR 7181/84					
Value	2.043*	$2.45x10^{-4}$*	NP	NP	NP	2.0	58.0	36.0	4.0	24.0	5.0

* - UBALDO (2005)

The bottom ashes present approximately 2% of organic matter (chemical method as in EMBRAPA, 1996) and 0.05% of pyritic sulfur, determined with LECO equipment and also observed in the X-rays difractogram. Mineralogical analysis indicated the presence of quartz, muscovite, kaolinite, gypsite and pyrite.

Laboratory diffusion tests were realized in a double reservoir cell to determine the oxygen diffusion and reaction rate coefficients through a compacted sample of the bottom ashes. Tests results were interpreted by program POLLUTE v.6.0 and equations presented in the previous section.

Testing equipment

The apparatus developed for the research comprises a Plexiglas cell with the following segments: a collection reservoir at the bottom, a source reservoir at the top, the sample cell, and base and top blind plates with O-rings. At each collection and source reservoirs there are two symmetrical valves for the purging procedure with nitrogen gas, and one for the oxygen electronic sensor. The soil or waste sample tested is 3 cm high. The test is monitored by an automatic data acquisition system where both oxygen concentration and temperature are registered and plotted continuously with time. The sensor was previously calibrated and the calibration curve was inserted in the data acquisition program.

Testing procedure

Two procedures were used in the oxygen diffusion tests: constant concentration source and finite mass or decreasing concentration source. The experimental program described herein followed a three step routine. At first the cell was closed and the superior and bottom reservoirs were purged with humidified nitrogen gas for approximately 5 minutes at 3 mL/s rate. When oxygen readings stabilized in both reservoirs, the source reservoir was purged with humidified air until stabilization, the reservoir was closed and the finite mass test started. The test ended when equilibrium was attained in O_2 concentration between the two reservoirs. At this point the upper reservoir was opened to the atmosphere, and the bottom collecting reservoir was again purged with nitrogen gas. The constant concentration source test started when the oxygen sensors readings stabilized in 20.9% of O_2 in the upper reservoir and approximately 1% O_2 in the collecting bottom reservoir.

Results presentation and discussion

To determine the possible values of the effective diffusion coefficient and of the reaction rate constant, the experimental results were theoretically fitted by program POLLUTE v.6.0.

Figures 3 and 4 present the theoretical fitting curves and experimental data for the finite mass source test and the constant concentration source test, respectively. The same parameters were also estimated by equations 1, 2 and 5 for comparison. Table 3 summarizes the parameters altogether.

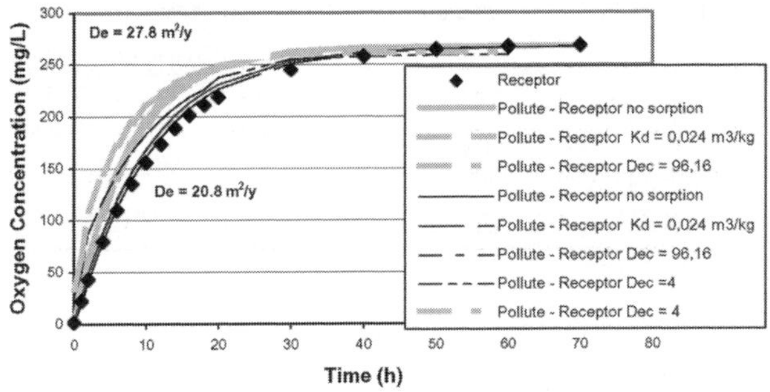

Figure 3 – Experimental results and theoretical fitting by POLLUTE v.6 – Constant concentration (20.9%) source O_2 diffusion test; compacted thermoelectric bottom ashes sample.

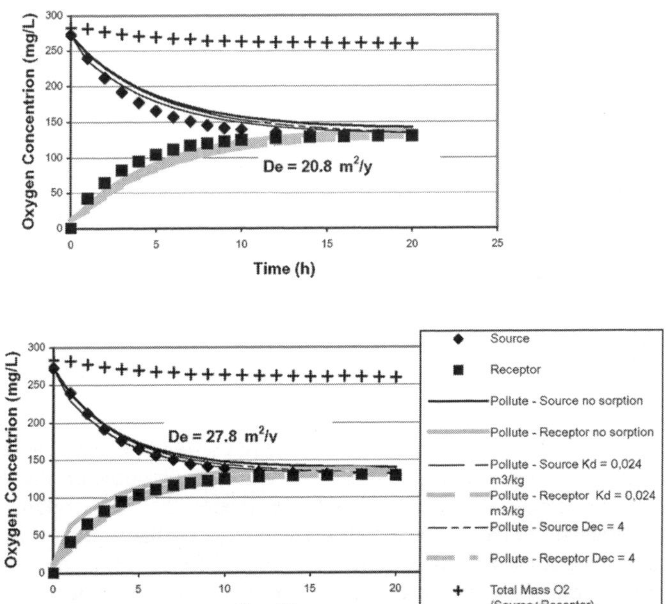

Figure 4 – Experimental results and theoretical fitting by POLLUTE v.6 – Finite mass source O_2 diffusion test; compacted thermoelectric bottom ashes sample.

The finite mass source test presented a loss of total O_2 mass with time, indicating oxygen consumption by the thermoelectric bottom ashes (Figure 4). This consumption can be explained by the presence of pyrite and organic matter in ashes composition, and also by biological action, since neither the sample nor the cell were sterilized prior to the test. Table 3 summarizes the cell and test sample characteristics.

Table 3 – Descriptive parameters of cell and sample tested

Characteristic			Value
Sample	γ (KN/m^3)		10,59
	w (%)		39.57
	n		0.48
	θ_w		0.42
	θa		0.06
	S (%)		86.7
Cell	Source	Height (m)	0.05
		Volume (m^3)	3.93×10^{-4}
	Collecting	Height (m)	0.05
		Volume (m^3)	3.93×10^{-4}

Estimated Values			
De	0.23 m^2/y		7.27×10^{-9} m^2/s
λ	Equation 2		Equation 5
	7.23×10^{-3} h^{-1}		3.12×10^{-5} h^{-1}
POLLUTE v.6 - Simulated			
De	Finite Mass	20.8 m^2/y	6.6×10^{-7} m^2/s
	Constant. Concentration.	27.8 m^2/y	8.8×10^{-7} m^2/s
λ		4 h^{-1}	
Kd		0.0024 m^3/kg	

The estimated value of the coefficient of diffusion from equation 1 (7.27×10^{-9} m^2/s or 2.62×10^{-5} m^2/h) is much smaller than that obtained from test interpretation by program POLLUTE v.6. The difference between the two values may be related to the estimation of the tortuosity factor, or to the fact that the sample tested present a porosity higher than 0.4, wich was the validity limit of the formulation adopted. The coefficient of reaction estimated from equations 2 and 5 is higher than the values obtained by adjusting the experimental results. The estimated values caused a sharp decrease in concentration of the oxygen that was not observed in the reservoirs at the end of the test. The experimental values of D_e and λ are close to the values determined by CABRAL *et al.* (2000) for the residue of paper (Table 4).

Tabela 4 – Typical results of effective diffusion coefficient (De) and coefficient of reaction (λ) reported in literature.

Cabral *et al.* (2000) – Pulp and Paper Residue				Mboninpa *et al.* (2003) - Tailings			
Sample	S (%)	De (m^2/s)	λ (h^{-1})	Sample	S (%)	De (m^2/s)	λ (h^{-1})
Nonsterilized	76	8.3×10^{-7}	2.31	Nonsterilized	84	2.1×10^{-9}	0.002
Sterilized	77	2.8×10^{-8}	2.17	Nonsterilized	90	3.4×10^{-10}	0.002

Conclusions

It can be concluded that the test procedure with finite mass source is faster than the one with constant concentration in the source and allows the determination of the coefficient of reaction easily. The kinetic coefficient obtained indicates that the bottom ash has low reactivity when compared to the residues with pyrite, and is similar to the values obtained for the residue of paper cellulose by CABRAL *et al.* (2000). These results confirm the viability of the use of this material as a constituent of dry cover system of mine waste piles. The estimation of parameters from empirical equations is important because it facilitates the use of the POLLUTE v.6. However, the empirical approach many times does not fit to the experimental data, as occurred in the present research with the equation of COLLIN & RASMUSON (1988) for D_e, and MBONINPA *et al.* (2003) and ELBERLING *et al.* (1994) for λ. For that reason the research continues to search for a reliable laboratory method of determining D_e to apply in the study of gas diffusion flow in the field. New studies are also necessary to investigate the influence of the bacteria, pyrite content, organic content and humidity on the decline the concentration of the oxygen with time in finite mass source analysis. The presented data and analysis are the initial information of in progress research.

Acknowledgements

Center of Mineral Technology and FINEP for allowing to the collection of samples, X ray diffraction analysis and sulphur text analysis. We wish to thank those who provided assistance to this study at COPPE-UFRJ and Federal University of the Tocantins.

References

CABRAL, A., RACINE, I., BURNOTTE, F., LEFEBVE, G., 2000, "Diffusion of oxygen through a pulp and paper residue barrier", *Canadian Geotechnical Journal*, v. 37, pp. 201-217.

COLLIN, M. RASMUSON, A., 1988, "A comparison of gas diffusivity models for unsaturated porous media", *Soil Science Society of America Journal*, v. 52, pp.1559-1565.

DAVIS, G.D., DOHERTY, G; RITCHIE, A.I.M., 1996b, "A model of oxidation in pyritic mine wastes: part 2 – comparison of numerical and approximate solutions". *Appli. Math. Modelling*, vol. 10, December, pp. 323-329.

ELBERLING, B., NICHOLSON, R.V., REARDON, E.J., TIBBLE, P., 1994, "Evaluation of sulphide oxidation rates: a laboratory study comparing oxygen fluxes and rates of oxidation product release", *Canadian Geotechnical Journal*, v. 31, pp. 375-383.

HILLEL, D., 1998, *Environmental soil physics*, London : Academic Press.

MBONIMPA, M., AUBERTIN, M., AACHIB, M., BUSSIÈRE, B., 2003, "Diffusion and consumption of oxygen in unsaturated cover materials", *Canadian Geotechnical Journal*, v. 40, n. 5 (October), pp. 916-932.

MENDONÇA, R.M.G., BORMA, L.D.S., SOARES, P.S.M., 2003, "Caracterização do Potencial de Geração de Acidez de Resíduos de Mineração de Carvão". In: Proceedings V Congresso Brasileiro de Geotecnia Ambiental – REGEO 2003, CD-ROM, Porto Alegre, Brasil, May.

NICHOLSON, R. V., GILLHAM, R. W., CHERRY, J. A., REARDON, E. J., 1989, "Reduction of acid generation in mine tailings through the use of moisture-retaining cover layers as oxygen barriers", *Canadian Geotechnical Journal*, v. 26, p.1-8.

RENAULT, P., SITENGEL, P., 1994, "Modeling oxygen diffusion in aggregated soils: I. Anaerobiosis inside aggregates", *Soil Science Society of America Journal*, v. 58, pp.1017-1023.

TINDALL, J. A., KUNKEL, J. R., ANDERSON, D. E., 1999, *Unsaturated zone hydrology for scientists engineers*, New Jersey, Prentice Hall.

UBALDO, M. O.; BARBOSA, M. C.; BORMA, L.S., 2006, "Technical and Environmental Viability Study of the use of Coal Ash as Covering Material for Mining" (this congress)

UBALDO, M. O., 2005, "Uso de cinzas de carvão na composição de uma cobertura de rejeitos de mineração". MSc. Thesis, Program of Civil Eng., COPPE, Federal University of Rio de Janeiro-Brazil.

WEERTS, A. H., FREIJER, J.I., BOUSTEN, W., 2000, "Modelling the gas diffusion coefficient in analogy to electrical conductivity using a capillary model", *Soil Science Society of America Journal*, v. 65, n. 6 (Nov-Dec), pp. 1577-1584.

YANFUL, E. K., 1993, "Oxygen diffusion through soil covers on sulphidic mine tailing", *Journal of Geotechnical Engineering ASCE*, v. 119, pp. 1207-1228.

Study on Dewatering of High Water Content Dredged Material Using Geotextiles

A. E. Muthukumaran, Former Post-Graduate Student, Division of Soil Mechanics and Foundation Engineering, College of Engineering, Anna University, Chennai, India 600 025
K. Ilamparuthi, Professor and Head, Division of Soil Mechanics and Foundation Engineering, College of Engineering, Anna University, Chennai, India 600 025

Abstract
In the recent past geotextile dewatering is being considered as a suitable disposal technique for sludge with high water content. Literature currently available validates this technique as a suitable alternative for conventional disposal techniques. However, not many works have been carried out to study the factors influencing the performance of this technique. In this study influence of initial water content of sludge, and aperture opening size of geotextile on the sludge geotextile behaviour has been studied through twenty-four filtration tests on dredge refuse of Chennai harbour. Flow response curves were used to understand the sludge geotextile behaviour during dewatering. Blinding of geotextile pores and formation of filter cake influenced the flow behaviour during dewatering. Variation in water content did not have appreciable influence on retention of solids but on the expulsion of water. 'Critical water content' was identified for the geotextiles tested at which retention of solids and expulsion of water was balanced.

Introduction
In the recent past, disposal of high water content sludge is turning out to be a menace for the environmentalists. Huge land resources are required for disposal, which become unfit for future applications. 'Geotextile dewatering' is a recent technique which has gained wide spread acceptance as a suitable alternative technique for disposal of high water content sludge. Geotextile is sewn to make geotextile tubes, which are subsequently filled with sludge to be dewatered. Solids in the sludge are retained in the geotextile tubes and the liquid part of sludge is drained out. The expelled liquid is disposed either on land or in water bodies with or without secondary treatment depending upon its quality after dewatering. Substantial literature is available on this technique, which validates the efficiency of this technique to dewater high water content sludge. Several real time projects were carried out to successfully dewater sludge. Geotextile bags, tubes and containers filled with a variety of fill materials can be used as submerged stability berms, groins, sill structures for controlling thalweg erosion, scour protection around piers, contraction dikes, dredged material containment and disposal of clean and contaminated materials.

Mori et al., 2002, studied suitability of geotubes made out of a single layer of geotextile for the containment of dioxin-contaminated sediment. The concentration of dioxins in the drainage from the geotubes decreased in the dewatering process. Formation of filter cake in the inner surface of the geotextile tube was responsible for the decrease in the concentration. Geotubes were 99% efficient in detaining the contaminants. Prototype high strength geotubes of various sizes and materials were used to study their effectiveness in dewatering tannery sludge (Hasbach, 1999). Primary slurry and other biological sludge were also tested. After two weeks solids content in primary slurry and biological sludge rose up to 25% and 18-30% respectively in the geotubes. The total time taken for dewatering 1070 cubic metres of sludge was three weeks. The dredged material, at an initial wet bulk density of 1.3 g/ml in the geotubes, consolidated 70% from an initial height of about 1.2 m to about 0.4 m in about two months (Fowler et al., 1997). Geotextile tubes filled hydraulically with fine-grained sand were used extensively on the northern shores of the Netherlands for barrier dikes for subsequent hydraulic fill behind the dikes (Krystian, 1994). Geotextile containers, which are dumped either from dump trucks or split hull, bottom dump hopper barges, have been used successfully to construct underwater stability berms, closures for repair of breached dikes, groins etc., (Duarte et al., 1995). U. S. Army Corps of Engineers (USACE) demonstrated that geotextile tubes could be filled with fine-grained dredged material for potential use for dike construction and wetland creation at Gaillard Island (Gaffney et al., 2001).

Current literature on geotextile dewatering of sludge mostly validate the efficiency of this technique. The various factors that influence the sludge-geotextile behaviour during dewatering have not been studied widely. The current study was intended to study the influence of initial water content of sludge and aperture opening size of geotextile on the sludge-geotextile behaviour. For this purpose, twenty-four laboratory filtration tests were carried out in the laboratory on dredge refuse using four different geotextiles. Results of the experimental investigations are reported in this paper.

Materials

Dredge Refuse

Dredge refuse used for the study was collected from a hopper during the dredging operations carried out in the Chennai Harbour. Care was taken to ensure that the collected samples were representative of the actual gradation. Tests were planned to carry out at different initial water contents of sludge. The liquid dredge refuse was carefully drained to remove all water without loss of fine materials. After complete removal of water the solids were properly air dried before use. Properties of the dredge refuse are listed in Table 1.

Table 1 Dredge refuse properties

Description	Value
Specific gravity, G	2.7
Portion finer than 75μ, (%)	82
Sand, (%)	18
Silt, (%)	72
Clay, (%)	10
d_{50} (mm)	0.012

Geotextile

Four different geotextiles based on aperture opening size were selected for the study. Properties of the geotextiles provided by the manufacturers are listed in Table 2. Aperture opening sizes of geotextiles were selected such that they are representative of commonly used opening sizes for filtration applications.

Table 2 Geotextile properties

Description	Geotextile			
	G1	G2	G3	G4
Aperture Opening Size, (mm)	0.090	0.225	0.425	0.600
Tensile strength, (kN/m)	100 x 100	72 x 72	70 x 105	70 x 70
Water flow rate, (l/min/m^2)	13	240	810	1830

Test Program and Methodology

Dry dredge refuse was mixed with water to prepare model sludge. Series of tests were carried out using the four geotextiles selected for the study. In each series different initial water contents of slurry were tested. A standard mix proportion of dry dredge refuse was followed to maintain the gradation of solids in the slurry. Details of the different water contents tested are listed in Table 3.

Table 3 Test program

Test legend	Geotextile	Initial Water Contents (%)
FG1	G1	80, 110, 140, 170, 200, 230, 290
FG2	G2	80, 110, 140, 170, 200, 230, 260, 290
FG3	G3	80, 110, 140, 170, 200, 230
FG4	G4	80, 140, 200

The laboratory tests were carried in a specially fabricated filtration mould made of Perspex (Fig. 1). Filtration mould comprised of two major chambers, the filtration chamber where slurry was filled in for dewatering, and the collection chamber where filtrate was collected during the test. Collection chamber was conical in shape with an angle of 60° at the vertex to avoid deposition of solids in the collection chamber. The sludge prepared at a specific water content was filled in the filtration chamber keeping the valve in the collection chamber closed to hold the flow until filling is complete. Filtrate was collected from the outlet continuously and measured simultaneously to observe the flow rate variation during the test. The filtrate was later tested for total suspended solids (TSS).

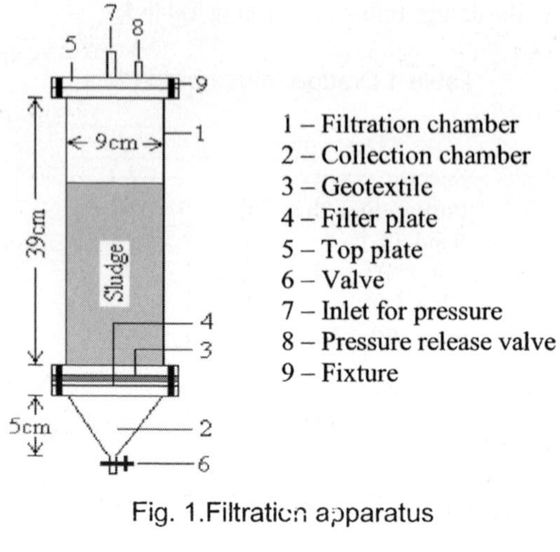

Fig. 1.Filtration apparatus

Results and Discussion

Flow Response

Sludge-geotextile behaviour during dewatering is complex and involves retention of solids and expulsion of sludge liquid that take place simultaneously. These two phenomenons are influenced by the flow behaviour during dewatering. In order to study the flow behaviour during dewatering and the factors that influence the flow behaviour, flow rate during the test was measured in predetermined intervals. From pilot tests it was found that flow rate varied rapidly during the first few minutes of the test. Hence, flow rate was measured in short intervals during the first few minutes of the test.

Influence of Initial Water Content

Variation of flow rate (flow rate was calculated based on the quantity of filtrate collected per unit time) with logarithm of time was plotted. The resulting curve was termed as 'flow response curve'. Flow response curves were used to understand the flow behaviour during dewatering and also to study the parameters that influence the sludge-geotextile behaviour. Figure 2 (a) shows the flow response curves for FG1 series for different initial water contents of sludge. Geotextile G1 was used in the model dewatering tests. The initial water contents of model sludge tested are listed in Table 3. Flow rate was observed to vary with time in all cases tested. Invariably in all tests flow rate did not decrease linearly with time. Bilinear response in flow rate with time was observed. It decreased from an initially high rate of flow and attaining a constant rate after sometime. This indicates that the flow behaviour comprised of two prominent stages. First stage of flow response curve is marked by the rapid decrease in flow rate and in the second stage stabilization of flow takes place. It was also observed that flow rate in the initial stage increased with increasing initial water content of sludge. However, for the initial water contents of 80% and 100% the bilinear flow response was insignificant.

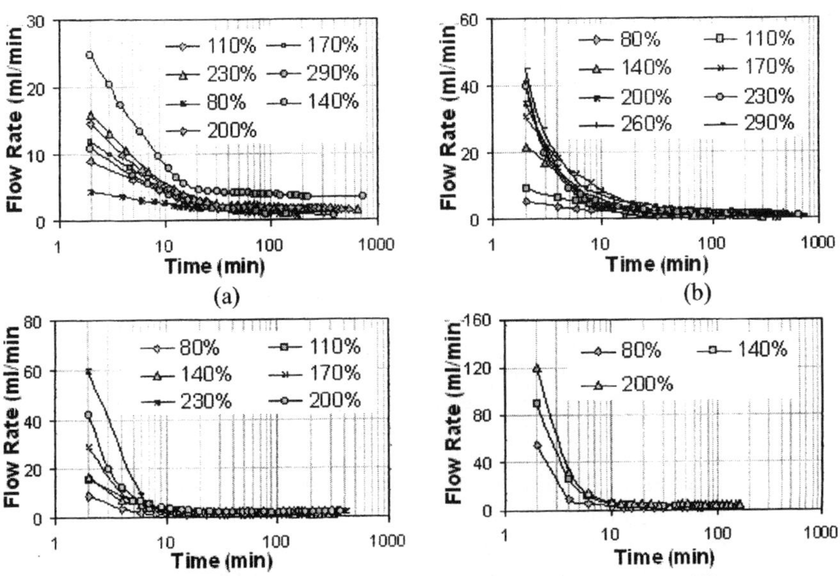

Fig. 2 Flow response curves (a) FG1 series (b) FG2 series (c) FG3 series (d) FG4 series

Flow rate practically remained the same through out the test for these two water contents. For water contents higher than 110% flow response took the bilinear form. Even though there was appreciable variation in the flow rate in the first stage after stabilization of the flow, the response curves of all the water contents congregated. Nevertheless, flow rate is influenced by initial water content of sludge and is increased with increase in initial water content of the sludge. The flow response of other series of tests (FG2, FG3, and FG4) on the geotextiles G2, G3, and G4 respectively at different initial water contents of sludge in each series are illustrated in Fig. 2 (b) to 2(d) respectively. Flow responses observed in all these tests follow the bilinear variation as observed in FG1 series. Apart from that, the flow rate increased with increase in initial water content of the sludge in the series of tests also.

Influence of Aperture Opening Size (AOS) of Geotextile

In order to study the influence of aperture opening size on the variation of instantaneous flow rate, the flow rate measured two minutes after the commencement of the test were compared for different initial water contents of sludge in Fig 3. There was no appreciable variation in the flow rate with increase in aperture opening size for the water contents of 80% and 140% in FG1, FG2, and FG3 series. However, considerable variation was observed among these series of tests with 200% water content. For the same set of water contents, flow rate in FG4 series increased drastically. Even at 80% water content flow rate was much higher than the flow rate in other series. Compared to FG1, FG2, and, FG3 series there was 3-6 fold increase in the flow rate in case of FG4 series for the range of water contents studied. The observations presented above indicated that AOS of geotextiles has influenced on the first phase of flow response curve particularly at higher initial water content of the sludge. However, AOS influence is very marginal in the second phase of flow among the geotextile tested. Further the geotextiles with AOS $\leq$ 0.425 mm showed almost identical behaviour on the sludge tested. Where as, the geotextile with AOS 0.6 mm performed differently from the rest and showed high rate of flow even for the lowest water content (80%) tested.

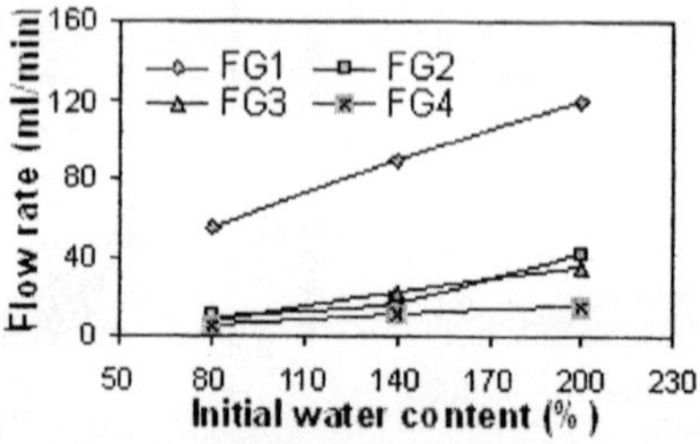

Fig. 3 Instantaneous flow rate

Predominant Mechanisms Governing Flow Rate

Flow behaviour during dewatering is predominantly governed by blinding of geotextiles and filter cake formation (filter cake is the thick layer formed due to deposition of solids on the surface of geotextile during dewatering). When dewatering begins simultaneous flow of sludge liquid and solids take place. Solids in the sludge start clogging the pores of geotextiles soon after the test begins, which hinders the flow of solids through the pores. Continuous and rapid deposition of solids takes place along with the flow, which is nothing but the progressive formation of filter cake. In this process the flow rate is decreased and flow of solids along with the liquid is also reduced drastically. At one stage there is no more migration of solids through geotextile, which indicates complete formation of filter cake. After filter cake formation flow will take place through two different mediums namely filter cake and geotextile. Compared to the thin geotextile layer, permeability of filter cake is very less. Since pores of the geotextile are clogged an average permeability is attained, which will be much less than the actual permeability of the geotextile. Once average permeability is attained, flow stabilizes and attains a constant rate. Formation of filter cake, therefore, plays a major role in the retention of solids.

Efficiency of Geotextile

Filtration Efficiency

Filtration efficiency is the measure of the capacity of a geotextile to retain maximum solids in the sludge without affecting the flow of sludge liquid. It is calculated using the formula proposed by Moo-Young et al., (2002), which is given below

$$\text{Filtration Efficiency, FE} = \frac{TS_{initial} - TSS_{final}}{TS_{initial}} \times 100 \qquad (1)$$

where, $TS_{initial}$ = Initial total solids in the sludge, mg/lit
TSS_{final} = Total suspended solids in the filtrate, mg/lit

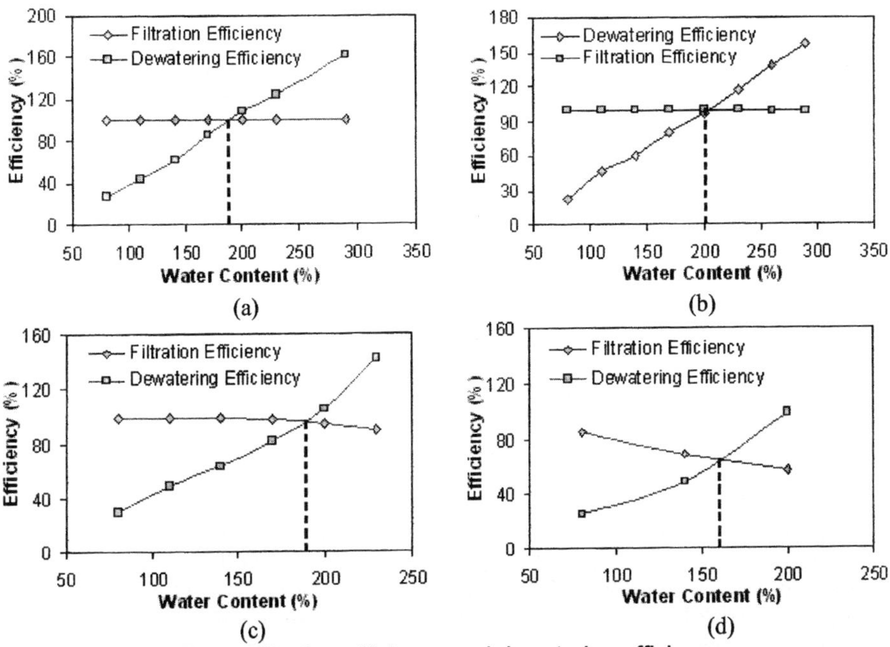

Fig. 4 Filtration efficiency and dewatering efficiency

Figures 4(a) to 4(d) show the variation of filtration efficiency with initial water content of sludge for the geotextiles tested. In FG1, FG2, and FG3 series filtration efficiency remained close to 100% for all initial water contents of sludge tested. Initial water content of sludge when increased did not have influence on filtration efficiency in these cases. On the other hand, in FG4 series, filtration efficiency drastically reduced with increase in initial water content of sludge. It dropped from approximately 85% at 80% initial water content to 57% at 200% initial water content. Geotextiles G1, G2, and G3 were found to perform better in terms of retention of solids, which is obvious through the high filtration efficiency values. Performance of geotextile G4 in terms of retention of solids was poor compared to the other geotextiles.

Dewatering Efficiency

Dewatering efficiency is the measure of the capacity of a geotextile to expel water without much loss of solids in the sludge. It is calculated using the formula given below, which is also proposed by Moo-Young et al., (2002).

$$\text{Dewatering Efficiency, DE} = \frac{PS_{final} - PS_{initial}}{PS_{initial}} \times 100 \qquad (2)$$

where, $PS_{initial}$ = Initial percent solids in the sludge, %
PS_{final} = Final percent solids in the sludge, %

Figures 4(a) to 4(d) also show the variation of dewatering efficiency with initial water content of sludge. In all cases, dewatering efficiency increased with increase in initial water content of sludge. The increase was almost linear with water content in all cases studied except FG4. In FG4 series dewatering efficiency increased exponentially. At low initial water contents of sludge, dewatering efficiency was approximately 30% in all cases tested. It attained a considerably high value for initial water content of sludge beyond 150%. At low initial water contents of sludge, even though geotextiles exhibited high retention of solids, performance in terms of expulsion of water was greatly low, irrespective of the geotextile used. Commendable performance in terms of balanced efficiency can be achieved when sludge with considerably high initial water content is dewatered.

Optimum Efficiency

It can be seen in Figures 4(a) to 4(d) that there is a water-content at which both filtration efficiency and dewatering efficiency have the same value. This water content is termed as 'Critical Water Content' and the corresponding efficiency is known as 'Optimum Efficiency'. It is the water content at which compatibility between retention of solids and expulsion of water is achieved. For water contents less than critical water content filtration efficiency is high and for water contents higher than critical water content dewatering efficiency dominates. Optimum efficiency was close to 100% for all cases except FG4 series, wherein the efficiency was only 60%. If sludge with water content less than critical water content is dewatered, it may result in good retention of solids but poor expulsion of sludge liquid. However, if sludge with higher water content than critical water content is dewatered the performance may not be affected as in the case of lesser water content. This is due to the fact that the decrease in filtration efficiency is not substantial even at high water contents for all geotextiles tested except G4. It obviates the fact that high water content sludge can be effectively dewatered through this technique.

Conclusions

Laboratory filtration tests were carried out on dredge refuse to study the influence of initial water content of sludge, and aperture opening size of geotextile on the sludge-geotextile behaviour. Based on the test results the following conclusions were drawn.

1. Flow behaviour during sludge dewatering takes a bilinear form. Flow takes place in stages – in the first stage it is high and in the second stage it stabilizes and attains a constant value. Increase in initial water content of sludge increased the flow rate. However, the influence of aperture opening size of geotextile was appreciable only in case of 0.600mm geotextile.
2. Clogging of geotextile pores and formation of filter cake are the predominant mechanisms that govern the flow behaviour during sludge dewatering. Filter cake induces an average permeability, which will be less than the permeability of the geotextile. It plays a major role in the retention of solids.
3. The influence of flow behaviour during sludge dewatering disappears when flow stabilizes. Flow rate was found to vary within a narrow range in all cases tested.
4. Filtration efficiency practically remained unaltered with increase in initial water content of sludge. Drastic decrease in filtration efficiency was observed only in 0.600mm geotextile. Conversely, dewatering efficiency increased with increase in water content. At low water contents very low dewatering efficiencies were observed.
5. There is a critical water content for each geotextiles tested and sludge used at which both filtration and dewatering efficiencies are the same and is close to 100% except for the geotextile opening size of 0.6mm.
6. Geotextiles with apertures sizes less than 0.425mm performed efficiently during dewatering, which can be extended to field situations after proper field tests.

Reference

Duarte, F., Joseph, J., Satterlee, G. 1995. Construction of geocontainer and geobag contraction dikes to eliminate dredging at Red Eye Crossing. In *Proceedings of Ports '95.* ASCE, New York, N.Y.

Fowler, J., Bagby, R. M., Trainer, E., 1997. Dewatering Sewage Sludge with Geotextile Tubes, *Geotechnical Fabrics Report*, pp 26-30.

Gaffney, D. A., Wells, L., Wickoren, D., Skelton, B.A., 2001. Environmental Benefits of Dredged Material Contained in Geotextile Tubes – Drakes Creek, TN. *Proceedings of Conference 32*, International Erosion Control Association, Las Vegas, NV.

Hasbach, A. (1999). Geotextile Tubes Handle Sludge Overload. *Pollution Engineering*, October 1999.

Krystian, W.P. 1994. Novel Systems in Coastal Engineering, Geotextile System and Other Methods, an Overview. Rijkswaterstaat, *Road and Hydraulic Engineering Division*, Delft, The Netherlands.

Moo-Young, H. K., Gaffney, D. A., and Mo, X., 2002. Testing Procedures to Assess the Viability of Dewatering with Geotextile Tubes. *Geotextiles and Geomembranes,* 20(5), 289-303.

Mori, H., Miki, H., Nobuyuki, T., 2002. The geo-tube method for dioxin-contaminated soil. *Geotextiles and Geomembranes,* 20(5), 289-303.

Effects of iron content due to the segregation pattern upon the strength parameters of iron ore wastes

Luís Fernando Martins Ribeiro, University of Brasília, Brasília, Brazil
Eleonardo Lucas Pereira, University of Brasília, Brasília, Brazil
Luiz Heleno Albuquerque Filho, University of Brasília, Brasília, Brazil
Ednelson da Silva Presotti, Companhia Vale do Rio Doce, Brazil
Alexandre Gonçalves Santos, Federal University of Ouro Preto

Abstract. The hydraulic fill technique is a common process used for waste disposal. According to this technique, the waste is placed randomly and a complex segregation pattern occurs due to the hydraulic parameters and the waste characteristics. The deposit behaviour is influenced by discharge parameters, presenting some heterogeneity that changes its geotechnical parameters. This research aims at studying the tailings behaviour regarding of the iron content and verifying the changes in the strength parameters. In order to verify the iron particle influence upon the tailings friction angle, some triaxial tests were carried out in samples with distinct iron content. The samples were collected from the hydraulic deposition simulation tests according with its distance to the discharge point. Through the obtained results, it was observed the relation among the hydraulic deposition variables and strength parameters considering the grain size distribution, densities and iron content. The most important aspect was the influence of the hydraulic segregation on the whole geotechnical behaviour of the iron tailing deposits, where the hydraulic selection is not only defined by size, but essentially influenced by iron particle weight.

Introduction

The waste disposal originated in the treatment process of iron ore and its use as construction material was started empirically. The construction methods applied to iron waste disposal presented low technical quality and the waste was placed in a randomly way. However, the waste disposal technologies have had some progress, there is still a great difficult in establishing the control of the variables and parameters that have some influence on the tailing deposits.

The conventional procedure adopted for waste deposition has been the hydraulic fill technique, normally linked to retaining structures such as tailing dams or tailing piles. This technique, basically, consists of hydraulic transport of waste mixed with water and its deposition is made in the form of a beach. The use of this technique applied to tailing dams is considered most attractive because of its low construction costs and the use of the waste as the construction material to build the body of the dam. The hydraulic fill has been especially used for the majority of mining companies as a main form of waste disposal. It allows the dam construction to be made in steps according to the mine production and the use of waste provides an increase in the storage capacity of the dam.

When applying the hydraulic fill technique the materials are transported and distributed to their final position by water (Hsu, 1988). Normally, this process use hydrocyclones, spigots, or discharge pipes. The use of pipes and/or spigots promotes a hydraulic segregation along the beach due to the discharge velocity and the solids concentration effects, while with the hydrocyclones the first segregation occurs before the placement.

The initial stage of the dam consists of building a starter dyke normally containing clay material and/or rock fills. After this stage, considering the upstream construction method, the iron waste is placed in the upstream direction forming the tailing beach that will become the foundation, and will eventually provide the construction material for the next rising (Figure 1).

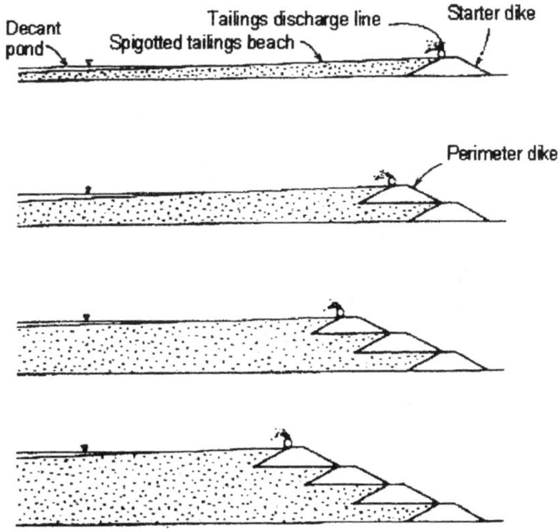

Figure 1: Sequential rising of hydraulic fill structure linked to upstream construction method (Vick, 1983).

Although the hydraulic fill structures linked to the upstream methods present some economic advantages, the waste behaviour tends to change significantly in function of the kind of ore, mineral exploration and mining process. In this context, the use of the waste as a construction material leads to an imperative analysis of the waste properties and its particularities viewing to reduce the uncertainties in the global behaviour of these structures. In the case of iron ore waste, the main aspect is the segregation caused by the weight of the iron particles which are present in the waste.

This research aims at studying the behaviour of waste based on the iron content and to verifying the changes in their geotechnical properties. Therefore, it presents analyses of iron particles percentage influence in the deposit density, as well as the effects in the strength parameters of iron waste. Some triaxial tests were carried out in order to evaluate the frictional angle for different samples with different iron content and porosities. The samples used were obtained from the hydraulic depositional simulation tests – HDST (Ribeiro, 2000). It is important to consider that the samples generated by HDST contain different percentages of iron content in a segregation process which is similar to the field one.

Experimental program

Tailing sample characteristics

The samples used in the experimental programme were obtained from the laboratory hydraulic simulation using HDST equipment (Ribeiro, 2000). The HDST equipment was developed to simulate, in a laboratory scale, the hydraulic deposition process considering the similar characteristics of the field hydraulic deposition. The HDST apparatus consists of a depositional channel, 6.0 m long, 0.40 m wide and 1,0 m high. The channel was built using steel profiles that support twelve panels of tempered glass (lateral walls). This kind of wall allows the observation of the evolution of the deposition process during the entire test. The slurry is formed in a reservoir equipped with a special mixing system of controlled velocity that keeps the slurry concentration constant during the period of test evolution. The discharge process is controlled by a maximum level of the main reservoir and by a valve that limits the flow rate. More details about the HDST equipment can be found in Ribeiro et al. (1999) and Ribeiro (2000). Figure 2 shows schematic details of hydraulic deposition simulation tests (HSDT).

Due to the HDST characteristics it was possible to simulate, qualitatively, the hydraulic segregation in function of the discharge parameters obtaining representative samples considering the similar effects in the field. In this way, the test methodologies consist of adopting different values of concentration and flow rates and their influence in the deposited iron waste. The samples were collected in regions along the channel each 40 cm from the discharge point (Figure 2). They were normally obtained 24 h after the end of the depositional phase.

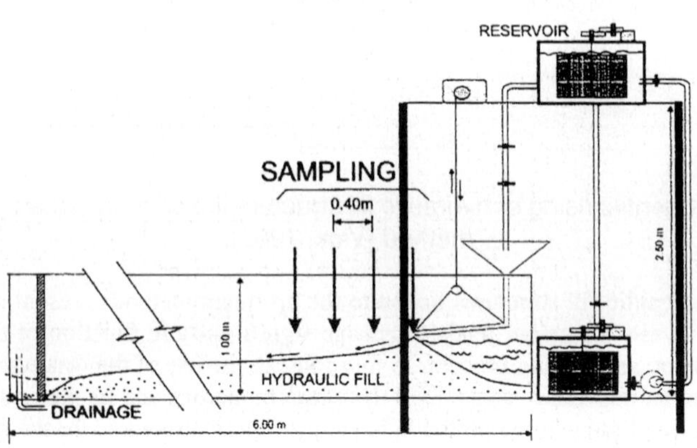

Figure 2: Schematic details of hydraulic deposition simulation tests (HSDT) and sample points. (Ribeiro, 2000)

Characterization of the waste

The hydraulic segregation in the HDST was analysed by grain size distribution, specific gravity of solids (Gs) and density measurements starting from the discharge point. The density determination was realised using a small cylinder (5.0 cm of diameter and 5.5 cm of height) and the grain size analyses and specific gravity of solids tests following standard recommendations. These data allows evaluating the iron content in function of the depositional process. It is important to consider the iron particle effects in the deposited material, this can be verified by the evaluation of Gs values related to the samples. In this iron waste the composition basically consists of iron and quartz particles. Therefore, the increase in the specific gravity of solids values represents the increase in the iron content. This fact reveals a particular segregation process linked to this material, in which the weight of the particle is more preponderant than the size. Table 1 shows the selection characteristics where the iron particles are concentrated near the discharge point. This represents the basis to evaluate the waste behaviour during the deposition and their changes due to the differences in the iron content in particular regions of the tailing beach.

In addition to the specific gravity of solids, other parameter could be used to identify the hydraulic sorting. In this research the grain size distribution was also used to evaluate size selection. The Figure 3 presents the grain size distribution related to the samples obtained in the HDST.

Table 1: Characteristics of Samples from HDST

Sample Dist. (m)	MA-8 (concentration = 10% e flow rate = 10 l/min)		
	Gs	(%) Fe	ρ_d (kN/m^3)
0.0	4.41	72.40	28.4
0.4	3.16	22.40	18.9
0.8	3.04	17.60	16.8
1.2	3.00	16.00	16.5

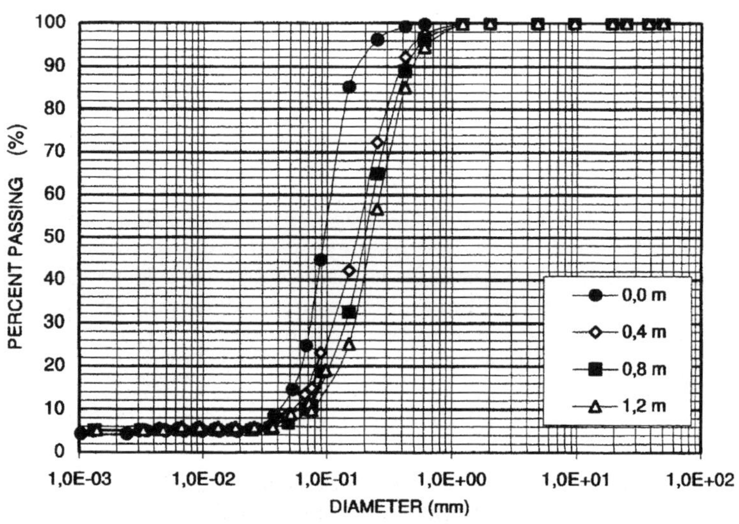

Figure 3: Grain size distribution of samples obtained by HDST

It can be observed that there is a decrease in the iron percentage as it gets distant from discharge point. This value is totally dependent of the specific gravity of solids (Gs) obtained for each analysed region along the deposit. This fact shows the applicability of the HDST in order to evaluate the changes in the deposit properties due to the hydraulic segregation and it also qualifies this technique to facilitate the sampling materials with different iron content in a similar way which occurs in the field. In Figure 3 it is possible to detect the mutual influence of size and weight particles, but in the specific case of this mine waste, the weight tends to dominate this process. It is also possible to observe, in the grain size distribution curves, the occurrence of a bigger concentration of small particles, although heavier ones near the discharge point may occur.

Additionally to the iron wastes characterization, the maximum and minimum void ratios were determined. These tests were carried out in order to analyse the density index of the deposit and were used in the remoulding samples technique. All these tests were performed considering the standard recommendations. The Table 2 shows the e_{max} and e_{min} values related to the different particle compositions in function of the distance to the discharge.

Table 2: Results of e_{min} and e_{max} values for the HDST samples.

Sample Dist. (m)	e_{min}	e_{max}
0.0	0.627	1.040
0.4	0.633	0.944
0.8	0.600	0.895
1.2	0.610	0.899

Strength characteristics obtained by CD triaxial tests

From the collected samples in four different regions along the channel, in relation to the discharge point, specimens were remoulded, using a simple compaction process in four different values of void ratio. All the samples were remoulded using the e_{max} and e_{min} values to define the required density index of each sample (Presotti, 2002). The remoulding method adopted was based on the suggestions and proposals presented by Head (1984) e Oliveira Filho (1987).

The reconstitution sample process was made using a special mould fixed in the lower part of the triaxial cell. The waste, in prefixed quantity, was transferred to the mould and the compaction occurred in four layers until the required density was achieved. Table 3 presents the void ratio adopted in the reconstituted triaxial samples. All the tests were performed using saturated samples with 50 mm of diameter and 110 mm of height sheared in drained conditions using a strain rate of 0.09 %/min

Table 3: Void ratios related to remoulding samples used in the triaxial tests.

Sample Dist. (m)	Void ratios of reconstituted samples			
0.0	0.78	0.82	0.88	0.95
0.4	0.74	0.82	0.83	0.90
0.8	0.67	0.77	0.82	0.87
1.2	0.67	0.77	0.82	0.87

The procedures used for the saturation and sample shear were based on the recommendations made by Head (1984). Specimens were saturated by application of slowly ascendant flow and back-pressure. Saturation was considered complete when the B value (Skempton, 1954) reaches approximately 0.98. During the tests, the axial and volumetric deformation and deviator stress was measured by an electronic measuring device connected to the automatic data acquisition system.

Results and discussions

The grain size analyses allow the obtaining a good correlation among the predefined diameters (D10, D50, D90) and iron content. Figure 4 shows the relation of these diameters and iron content. It is possible to observe the tendency of an increase in the iron content when there was a decrease in the particle diameter. Based on this analysis, one verifies that the iron particles are smaller than the quartz particles. As it was previously commented, the iron particles tend to be smaller, but heavier than the quartz particles and mainly concentrated near the discharge point. This consideration shows the importance of this kind of particles in the hydraulic segregation process which occurs during the tailing deposition.

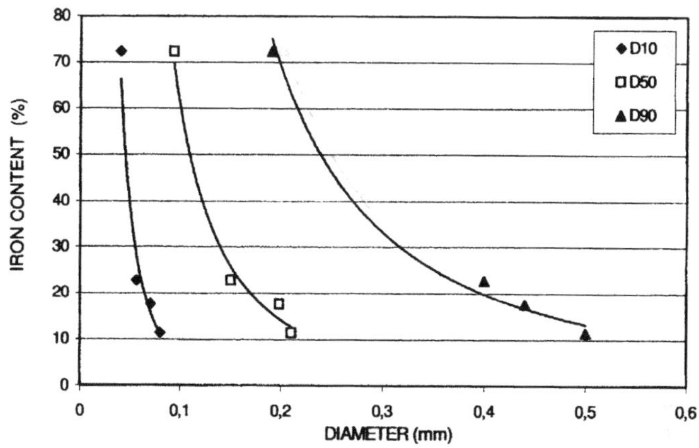

Figure 4: Relationship by iron particles and grain size.

The effects of iron content in the strength parameters of the iron waste were analysed considering the initial void ratio of the reconstituted specimens and the point of sampling using the HDST. The Figure 5 represents the relation between the iron percentage and the frictional angle. The influence of iron particles can be verified, where the increase in the iron particles concentration tends to generate an increase in the frictional angle. One can also observe that the effects of iron particles are more significant for dense conditions. Thus, for dense specimens the increase of 22 % in the iron content represents a gain of 4 degrees in the frictional angle. This situation is less significant in the loose state.

The Figure 6 shows the influence of porosity in the frictional angle, as expected, the frictional angle tends to reduce with the increase of the porosity. However, this figure shows the particularities of this material. The existence of different kinds of minerals alters the common behaviour related to the majority of conventional granular soils. In this context, the iron particles tend to generate a significant influence on the strength parameters. Thus, the obtained curves are not coincident because of the iron concentration effects in the material. Additionally, it can be inferred, from this analysis, that in spite of the small diameter presented by the iron particles, these generate a gain in the frictional angle in the deposited material.

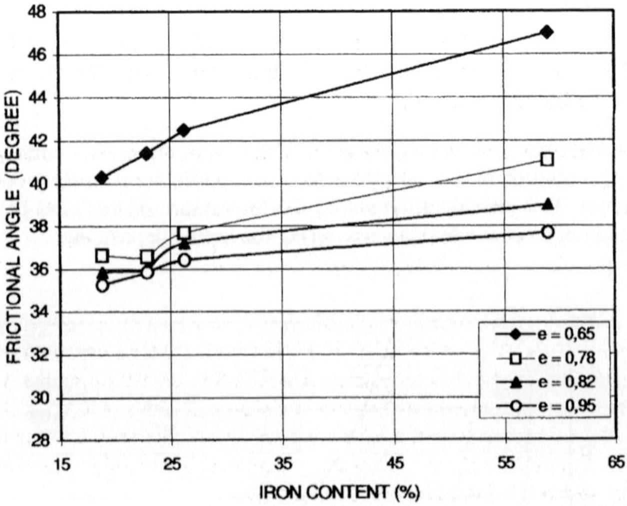

Figure 5: Relation between iron content and iron waste frictional angle

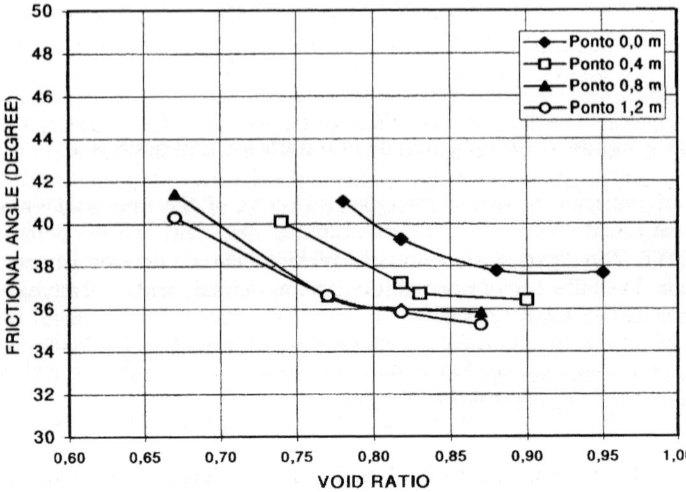

Figure 6: Relation between frictional angle and iron waste compactness

Conclusions

The general characterization of the samples shows the clear influence of the iron presence in the waste. In the tests which were used to determine the specific gravity of solids, the obtained values were approximately from 3.0 to 4.4, being strongly influenced by the iron content. From grain size distribution, it was possible to conclude that the iron particles are predominately smaller than quartz particles in this waste.

Considering the iron content effects in the compactness of the waste, it can be observed that the increase of iron content beside with the reduction of particles diameter tends to augment the maximum and minimum void ratio.

The results from triaxial tests demonstrated that the waste presents a similar behaviour of granular soils. Its strength is controlled by frictional effects among particles; however the presence of iron particles tends to introduce an important aspect in the waste behaviour considering mainly the strength. From these tests it was possible to quantify the iron particles influence in the friction angle, mainly for dense conditions.

This research also shows the applicability of HDST simulations in analyses related to the tailings dam behaviour. This simulation can generate samples characteristics which are similar the ones in the field. This represents a qualitative condition of the segregation that occurs in the field. The HDST aims at implementing an optimized project of waste deposition through the control of hydraulic variables during the waste disposal and has proved to be an important tool for the knowledge of iron waste properties. This study represents a part of this complex research line related to tailing dam behaviour.

References

Head, K.H. (1986). Manual of soil laboratory testing. London, UK: Pentech Press. 1240 p.

Hsu, S.J.C. (1988). Review of practice in hydraulic fill construction. Hydraulic Fill Structures, ASCE, Geotech Special Publication no 21, D.J.A. Van Zyl & S.G. Vick eds., pp.884-901.

Oliveira Filho, W.L. (1987). Considerations about triaxial tests in sands, MSc. Thesis., COPPE, Federal University of Rio de Janeiro, Rio de Janeiro, Brazil, 184 p. (In Portuguese).

Presotti, E.S., (2002). Influence of iron content into the strength parameters of a iron ore waste. MSc. Thesis. Civil Department, Federal University of Ouro Preto, Ouro Preto, Brazil, 153 p. (In Portuguese).

Ribeiro, L.F.M (2000). Physical simulation of the hydraulic fill formation process applied to tailings dams. Civil and Environmental Engineering Department, University of Brasilia, Brasília, Brasil, 232 p (In Portuguese).

Ribeiro, L.F.M. & Assis, A.P. (1999). Experimental simulation of the hydraulic deposition process in tailings dams. XI Pan-American Conference on Soil Mechanics and Geotechnical Engineering, ISSMGE, Foz do Iguaçu, Brazil, 3: 1113-1120.

Skempton, A.W. (1954) The pore pressure coefficients A and B. Géotechnique, v. 4, n. 4, p. 143-147.

Vick, S.G. (1983). Planning, Design and Analysis of Tailings Dams. Wiley International, New York, USA.

Stability and Settlement Characteristics of Hydraulically Placed Mine Tailings & Power Station Fly Ash

D.I. Stewart, School of Civil Engineering, University of Leeds, Leeds, UK
C.A. Charles-Cruz, School of Civil Engineering, University of Leeds, Leeds, UK
T.W. Cousens, School of Civil Engineering, University of Leeds, Leeds, UK

Abstract

Fine silt-sized waste particles are often disposed of by being mixed with water, pumped to a void and allowed to sediment. The resulting 'soil' has a very loose structure which makes it difficult to test. There is interest in these materials, e.g. for site redevelopment. This paper describes the preparation of very loose samples of pulverised fuel ash and mine tailings from fluorspar production, in an attempt to replicate the structure of hydraulically deposited fills. The results of oedometer and triaxial tests are also presented. Difficulties in sample preparation and measuring e_{max} are described. The use of Bolton's (1986) relative dilatancy index to describe the drained behaviour of such deposits is discussed. The materials display varying degrees of instability in undrained loading that may be dependent on stress level, particle shape and size distribution.

Introduction

Hydraulically placed fills are a common form of disposal for fine-grained inert solid wastes. Such wastes are mixed with water and pumped as slurry into impoundments. An inherent characteristic of the freshly sedimented material is its very open and loose structure. Interest arises in the behaviour of these deposits both when evaluating the stability of impoundment facilities and when considering the redevelopment of disposal sites. For this reason, there is great interest in the classification, shear strength and compressibility behaviour of such fills.

This paper presents the results of index, one dimensional consolidation and triaxial tests on two hydraulically placed fill materials: calcium fluoride (fluorspar) mine tailings from Glebe Mine, Chesterfield, Derbyshire and pulverised fuel ash (PFA) from Skelton Grange, Leeds, West Yorkshire. These materials were selected as representing two extremes of particle shape: angular and rounded. Due to the very loose nature of the deposits it was not possible to recover even notionally undisturbed samples for testing; therefore water pluviation of either initially dry material or slurry was used to prepare samples for consolidation and triaxial testing.

Background

The geotechnical behaviour of hydraulic fills is of interest in many areas, e.g. design of impoundment sites and site reuse. These materials are typically of silt size. Difficulty in investigating the geotechnical properties of hydraulic fills lies partly in their heterogeneity. Variations over time arise due to changes in the mineral composition of the parent rock, the minerals being extracted and the methods of separation (crushing, grinding, use of surfactants, etc.). In addition, the type of slurry deposition (single or multiple point discharge) influences spatial variations within the impoundment. This variability affects the analysis of single sites and also makes comparison between sites difficult.

The geotechnical properties of several different types of tailings can be found in the literature, including copper tailings (Qiu & Sego, 2001), gold tailings (Stone et al., 1994), and bauxite (Consoli & Sills, 2000). These papers focus on consolidation characteristics, hydraulic conductivity and liquefaction susceptibility, due to their importance in impoundment design.

The study of the engineering characteristics of PFA can be traced back to the work of Raymond (1958) and Knight (1960). Gray and Lin (1972), Leonards and Bailey (1982), McLaren and DiGioia (1987), Parylak (1992), and Okumura et al. (1996) also describe the use and engineering properties of PFA. There has also been work on the collapse behaviour of loose PFA deposits (Henau & Thijs, 1985; Trivedi & Sud, 1999).

It is very difficult to take undisturbed samples of hydraulically placed fill because any disturbance (including stress relief at the bottom of a cased borehole) can lead to strength loss in loose non-cohesive deposits, and instability of the borehole base. Where the fill is very loose it is impossible to recover even notionally undisturbed samples for laboratory testing. Correlation between field and laboratory tests has been attempted to overcome this limitation; e.g. seismic cone penetration testing (SCPT) by Woeller, et al. (1996), piezocone testing by Davies (1999), and cone penetration testing (CPT) by Cousens and Stewart (2003) and Stewart et al. (in press). Another approach to this difficulty has been to calibrate theoretical/ numerical models using either centrifugal model tests (Stone et al., 1994) or field studies (Consoli & Sills, 2000). Theoretical modelling studies using large strain consolidation theory have been conducted by Cargill (1984) and Seneviratne et al. (1996).

Hydraulically placed fills and embankments are susceptible to liquefaction under monotonic or cyclic loads (Fourie et al., 2000; Ishihara, 1993). As generally defined, liquefaction is a sudden loss of shear strength due to a rapid increase in pore water pressure. The study of liquefaction phenomena in sands has been extensive and there are a variety of empirical and theoretical approaches to assess liquefaction potential (Ng et al., 2004; Verdugo & Ishihara, 1996). However, there has been less research on the effect of fines in the liquefaction process and on the study of this phenomenon for silty or clayey soils. Finn et al. (1994) attempt to summarise the effects of fines content on liquefaction resistance for both non-plastic and plastic fine grained soils. According to their guidelines, for non-plastic fines, there is an increase in liquefaction resistance of a silty sand compared to clean sand with same normalised standard penetration resistance $(N_1)_{60}$, there is a decrease in liquefaction resistance with increasing fines content for materials with same gross void ratio, and materials with the same void ratio in the sand skeleton have the same liquefaction resistance. For the case of plastic fines, the

Table 1. Material Properties

	PL/ LL	G_s	D_{50} (μm)	D_{60}/D_{10}	Fines Content (%)	e_{min}	e_{max}	e_{max}^*	Morphology
Glebe Tailings	20.9/ 23.2	2.90	4	40	40	0.51	1.38	1.38	Angular flakes
Skelton Grange PFA	NP/ 41.4	2.30	15	7	5	0.55	0.98	1.62	Spherical or rounded

G_s – particle specific gravity, e_{min} – minimum voids ratio, e_{max} – maximum voids ratio, D_x – the particle size that x% of the sample is finer than, PL – plastic limit, LL – liquid limit. Note: e_{max}^* determined by water sedimentation, other tests conducted in accordance with BS1377(1990): Parts 2 &4.

modified Chinese criteria (after Finn et al., *op. cit.*) suggest that soils that satisfy four requirements (ie % finer than 0.005 mm $\leq$ 10%; liquid limit $\leq$ 36%; natural water content $\geq$ 0.92 LL; liquidity index, $I_L \leq 0.7$) are susceptible to liquefaction. Yamamuro and Lade (1998) suggest that unlike the behaviour of clean sands, where susceptibility to liquefaction increases with confining stress, it can initially decrease with confining stress for silty sands. This is attributed to the silt particles producing meta-stable structures at low confining stress. Guo and Prakash (1999) suggest that soil fabric and aging seem to slow down pore pressure generation and that a higher percent of plastic fines may increase the liquefaction resistance of silty soils.

Materials

Glebe Mine Tailings

Fluorspar is the only major source of fluorine and is also used in the production of a wide range of industrial products, such as pharmaceuticals, glass, plastics (Notholt, 1971). Glebe Mine is located in Stoney Middleton, Derbyshire. Operations on site consist of fluorspar mining, processing and tailings disposal. Processing involves crushing the parent rock, forming a slurry, and separating the different minerals gravimetrically using detergents. The resulting tailings are mainly composed of barites (40%), limestone (33%) and silica (25%). Table 1 gives a summary of the Geotechnical properties of Glebe Mine tailings. A scanning electron micrograph (SEM) is shown in Figure 1a, which shows occasional sub-angular coarse particles and the predominantly angular and plate-like finer particles (D_{50}=4μm).

In the mine's active disposal lagoon, one half is being used as a sluicing pond, while partially dewatered tailings are recovered from the other half. The excavated material is used as backfill in the opencast mine. This releases lagoon capacity and creates a sustainable waste cycle. The bulk tailings samples used in this study were taken from the latter area of the lagoon.

Skelton Grange PFA

PFA, or fly-ash, is a waste from pulverised coal combustion for electricity generation. It is the fine residue that is collected by mechanical or static precipitators. PFA is often disposed of locally by pumping a water-based slurry into sedimentation lagoons. In the lagoon, the PFA quickly settles out while the water drains away. PFA particles are typical spherical, ranging in diameter from 1μm to 150μm, and due to the rapid cooling of the flue gases, consist primarily of non-crystalline particles or glasses (about 80%) and only a small amount of crystalline material (Cabrera et al., 1986). Figure 1b shows an SEM image of the mainly spherical particles of Skelton Grange PFA (the amorphous particles are assumed to be unburned carbon).

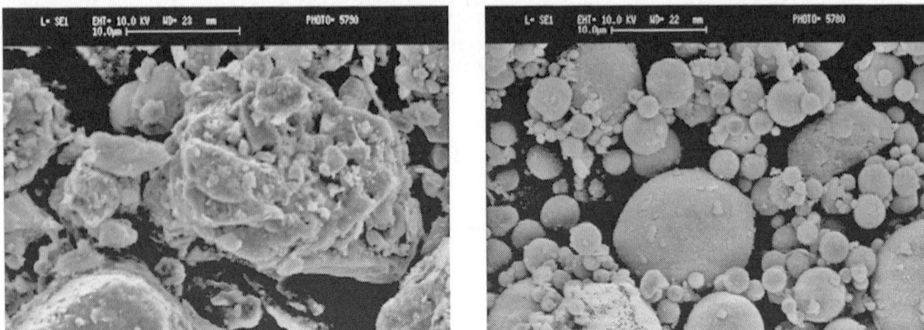

(a) (b)

Figure 1: SEM images of (a) Glebe Mine Tailings and (b) Skelton Grange PFA

For this study, PFA was obtained from the Skelton Grange lagoons in Leeds, West Yorkshire. Details of the site and previous field investigations are included in Cousens and Stewart (2003) and Stewart et al. (in press). Although in previous investigations this material was catalogued as a low plasticity silt (Stewart et al., in press), the material used in this investigation was classified as non-plastic silt. Skelton grange PFA is classified as type F (low carbonate) according to BS-BN450. A summary of PFA characteristics is included in Table 1.

Experimental Methods

Testing was conducted at 21±1°C. Classification, consolidation, undrained and drained triaxial strength tests were performed generally in accordance with BS1377(1990), sections 2, 4, 7 and 8, respectively, with minor variations so that high void ratio samples could be produced.

One-dimensional consolidation tests

One-dimensional consolidation tests were conducted in fixed ring oedometers with top and bottom drainage. To avoid the need to move the specimen, with the inherent risk of sample disturbance, sample preparation by water pluviation was undertaken with the oedometer cell within the loading apparatus. During sample preparation an extension tube made from acetate film and adhesive tape was fitted to the water-bath. Only a small head of water was used, and a perfectly adequate water seal was achieved by ensuring a close fit between the water bath and the extension tube and by coating the contact surfaces with silicon grease. A second, slightly shorter, extension tube was close-fitted inside the top of the oedometer collar. An equal water level was established in both the inner and outer tubes about 120 mm above the oedometer cell, and approximately 80g of material was introduced slowly into the inner tube.

When testing PFA, dry material was sprinkled slowly and evenly into the inner sleeve to avoid segregation. Trials indicated that this procedure was unsuitable for the finer Glebe tailings, which were added as a slurry. The sedimentation times were approximately 1 and 2.5 hours for PFA and Glebe tailings, respectively. A minor shortcoming of this technique is the difficulty of measuring the initial sample volume to directly determine the initial void ratio, which has to be back calculated from the final specimen height after extrusion from the cell.

The top-cap, with a porous stone and filter paper, was placed very carefully on the specimen and a seating load applied (giving a stress of 13 kPa). The sample was loaded in increments typically of 13, 25, 52, 104, 208, 415, 832 and 1788 kPa to a maximum vertical stress of about 3.5 MPa. Each loading step was monitored for 30 minutes, which was found experimentally to exceed the time for 90% consolidation for all loading increments (t_{90} was found by plotting the settlement against $\sqrt{\text{time}}$). After some loading increments the specimen was monitored for up to 3 days to observe any secondary (or creep) compression. Small stress increments were used initially on reloading after a creep period. Unloading was undertaken in two or three large increments because the rebound was very small and the response was essentially instantaneous. After unloading the specimen was extruded, its height and weight measured, and after oven drying its dry weight and moisture content were determined. Extrusion disturbance meant that the sample height could not always be determined accurately, so the final void ratio was also estimated from the final moisture content by assuming saturation.

Triaxial compression tests

Triaxial compression tests were conducted in standard Wykeham Farrance 38mm triaxial cells equipped with 3kN submersible load cells, and "Imperial College" type volume change units. The accuracy of individual measurements are estimated to be ±1 kPa for pressure, ±1 kPa deviator stress, ±0.005 mm for displacement, ±0.025ml for volume change. Hence the accuracy of the axial and volumetric strain measurements is better than ±0.03%.

Sample preparation was by sedimentation. A rubber membrane was attached to the pedestal and surrounded by a 38mm split-mould secured with a rubber band. Instead of adopting the common practice of rolling the membrane back onto the mould, a short 54mm ID metal tube was inserted inside the membrane to sit about 5mm above the top of the mould. Subsequently a 15 cm long plastic extension tube was inserted through the metal tube to lightly contact the split mould and make an adequate water seal against the short length of exposed membrane.

To avoid size segregation problems that can occur with incremental water pluviation (Vaid et al., 1999), samples were prepared as a slurry at a moisture content of ~52%. This was shaken and rapidly poured into the mould and allowed to settle. The volume of slurry required for the correct sample height was found by trial and error. The cap was very carefully placed and initially fixed by a thin rubber band so that a small suction could be applied (~10kPa). After ~20 minutes the specimens were sufficiently stable for rubber O-rings to be placed on the top-cap. Full width drainage was provided at the bottom of the specimen, and side drains were used with Glebe tailings. Trials showed that this was the most appropriate method for preparing loose triaxial specimens of these materials. Suetsugu et al. (2000) used a similar procedure to prepare loose PFA triaxial specimens. After testing, the final void ratio was calculated from specimen dry weight and initial geometry accounting for volume change during testing.

Specimen saturation was ensured by using a back pressure. A cell pressure lower than the intended minor principal stress was initially applied, and then the cell pressure and backpressure were slowly raised together to apply the desired back-pressure. These pressures were maintained until an acceptable B-value was measured (99% and 80% for undrained and drained tests, respectively). After saturation, the specimens were isotropically consolidated to the desired effective stress. After consolidation, the samples were axially compressed using strain rates of 1 and 0.16 %/min for the undrained and drained tests, respectively.

The void ratio of PFA specimens prepared for consolidation and shear testing significantly exceeded the maximum void ratio measured by the method specified in BS1377(1990): Part 4.4. Since the principal difference is that the standard test employs air sedimentation, it was

Table 2. Summary of 1-D Consolidation Results

Test ID	e_o	C_c	C_r	C_r/C_c	Surcharge Stress During Creep Stage (kPa)	C_α	C_α/C_c
GM4I-1	1.54	0.22	0.02	0.08	103	0.002	0.010
					3,406	0.005	0.021
GM4I-4	1.44	0.27	0.04	0.16	206	0.004	0.015
					3,844	0.005	0.018
GM4I-5	1.51	0.18	0.02	0.13	102	0.003	0.017
					3,323	0.003	0.014
PFA-3	1.62	0.25	0.04	0.16	-	-	-
PFA-4	1.47	0.24	0.04	0.15	204	0.001	0.006
					3,385	0.002	0.006
PFA-5	1.43	0.24	0.04	0.15	-	-	-
PFA-6	1.61	0.29	0.04	0.13	204	0.004	0.015
					3,388	0.004	0.012
PFA-7	1.52	0.30	0.04	0.14	204	0.003	0.009
					3,287	0.003	0.010
PFA-9	1.59	0.28	0.03	0.11	408	0.006	0.020
					3,256	0.003	0.012

repeated using water sedimentation (the measuring cylinder was completely filled with water). The result is reported as e_{max}^* in table 1.

Results

One-dimensional consolidation tests

Consolidation test results are reported in table 2. The compression index, C_c, of the two soils is similar; 0.18 to 0.27 for Glebe tailings and 0.24 to 0.30 for PFA. The rebound index, C_r, is about 0.03 for Glebe tailings and 0.04 for PFA. The compression and rebound indices are characteristic of low plasticity silty clay or silt (Lambe & Whitman, 1979). The shape of the consolidation curves is illustrated in Figure 3. These were typical tests in that the response of the PFA was almost linear on the semi-logarithmic consolidation plot, whereas the initial response of Glebe tailings was more variable. Secondary compression of both materials varied linearly with the log(time) for up to 3 days. The average C_α values are 0.004 and 0.003 for Glebe tailings and PFA, respectively, giving average C_α/C_c ratios of 0.016 and 0.011, which are only slightly less than would be anticipated for inorganic clays and silts (Mitchell, 1993).

Triaxial Testing

CD Triaxial tests

Table 3 reports the results of the drained triaxial tests. The relative density (I_D) values of the Glebe specimens (0.33, 0.41) indicate that they were all quite loose. Thus the maximum angle of shearing resistance should approximate to the critical state friction angle, suggesting ϕ_{crit} value of ~35° for angular Glebe tailings. The two PFA I_D values (0.30, 0.96) are representative of a loose and dense state. The former suggests the PFA has a critical state friction angle of ~33°. Figure 4 shows the triaxial data plotted using axes of stress ratio and rate of dilatancy (Rowe, 1962). Rowe proposed that failure is governed by the equation

$$\frac{\sigma_1'}{\sigma_3'} = \tan^2(45° - \phi_f/2)\left(1 - \frac{d\varepsilon_v}{d\varepsilon_1}\right) \qquad (1)$$

where $\phi_f \approx \phi_{crit}$ for loose materials where failure is associated with turbulent shear, but $\phi_f \approx \phi_\mu$ for denser materials where particle sliding is dominant at failure. Figure 4 indicates that failure of loose PFA specimen is associated with turbulent shear mobilising ϕ_{crit}=35°, whereas that of the dense PFA specimen is associated particle sliding mobilising ϕ_μ=28°. Failure of Glebe specimen GM-D2 (e=1.09, I_D=0.33) seems to be associated with turbulent shear mobilising ϕ_{crit}=34°, but the mode of failure of Glebe specimen GM-D11 (e=1.02, I_D=0.41), is less clear, possibly representing a transition between particle sliding and turbulent shear.

CU Triaxial tests

Table 4 reports the results of the undrained triaxial tests. These specimens were intended to be loose; the I_D values were 0.44, 0.53 and 0.58 for Glebe specimens, and 0.22, 0.33, 0.59, 0.63, 0.66 and 0.72 for PFA specimens. Thus some of the PFA specimens should be regarded as moderately dense (although there is some uncertainty about the e_{max} for the PFA).

The stress-paths in the undrained triaxial tests are shown in Figure 5a. The stress-paths followed by the three Glebe tailings specimens are typical of undrained response of loose sands and silts (Yamamuro & Lade, 1998). They start nearly vertical but increasingly curve back towards the q-axis, due to rapidly increasing pore water pressure. In two tests the stress path curves over with q decreasing slightly while p′ decreases rapidly (the stress ratio when dq/dp′=0 is given as η_{flat} in table 4 and is equivalent to $\phi \approx 20°$). At a moderately high stress ratio the stress-paths double back towards the direction of increasing q and p′ due to decreasing pore water pressure (the stress ratio here, given as η_{elbow} in table 4, was equivalent to

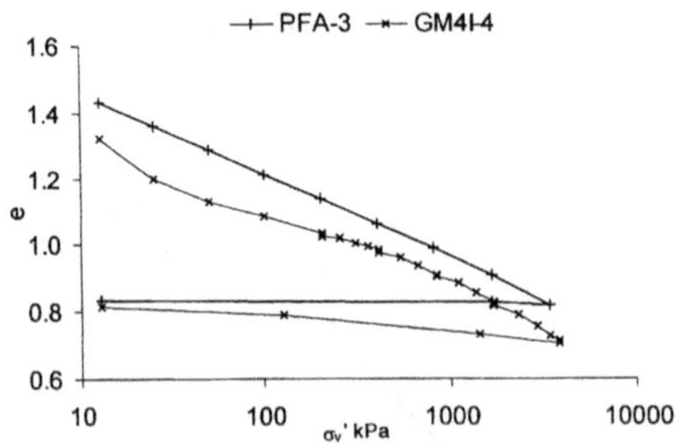

Figure 3. Compressibility curves for Glebe Tailings and PFA

$\phi \approx 28°$ in all three tests) to approach a critical state stress ratio asymptotically.

The stress-paths followed by the PFA specimens are similar to those followed by the Glebe tailings. However, the principle difference is that although both materials appear to have a critical state friction angle around 35°, η_{flat} is equivalent to $\phi \approx 28°$ and η_{elbow} to $\phi \approx 31°$, for the PFA. These are considerably higher than the stress ratios at equivalent points in the tests on Glebe tailings. With both materials the axial strain required to mobilise η_{flat} is relatively small (~1%), but more strain occurs as p′ decreases (typically $(\varepsilon_1)_{elbow} \approx 5\%$). Larger strains arise as the materials work-harden towards a critical state. The end points of the stress-paths shown in Figure 5 is where non-uniform sample deformation started to produced erratic data.

Discussion

The hydraulic method of disposal of fine materials can result in very loose deposits. The preparation of very loose test specimens, especially triaxial specimens, was very difficult, as immediately after sedimentation the samples are very sensitive to disturbance and readily collapse. Deposits in the field often settle through relatively large bodies of water and may have a looser structure than that produced by the techniques described here, although they may be subject to external vibration and slumping or flow slides which will affect the structure.

The two materials respond differently to deposition through water, rather than air. With air pluviation, e_{max} is achieved by using high deposition rates and low heights of fall (Kolbuszewski, 1948). High void ratios result from the low kinetic energy of the particles, and from individual particles not having the freedom to move to a closer packed structure due to the restricting effect of adjacent particles. With water pluviation, particles will tend to a limiting settling velocity, generally lower than in air, which is a function of shape, size and mass. The water sedimentation technique used to find e_{max}* has a large flow rate with a small height of fall, and so individual particle freedom to move to a closer packing will have been restricted. Using a water sedimentation method compared to air sedimentation e_{max} for PFA is

Table 3. Details of the consolidated drained triaxial tests.

Test ID	e_o	I_D	Cell pressure (kPa)	Back pressure (kPa)	$(\sigma_1\text{-}\sigma_3)_f$ (kPa)	$(\varepsilon_1)_f$ (%)	$(\varepsilon_v)_f$ (%)	$(-d\varepsilon_v/d\varepsilon_1)_f$	ϕ_{peak}
GM-D2	1.09	0.33	405	199	599	9.0	1.28	0.10	36.3°
GM-D11	1.02	0.41	250	150	260	16.0	0.00	0.07	34.1°
PFA-D1	1.29	0.30	311	161	360	8.9	2.82	-0.11	33.2°
PFA-D3	0.59	0.96	451	303	390	12.1	-1.84	0.17	34.9°

The relative density, I_D, of the PFA specimens has been calculated using the e_{max}^* (water sedimentation).

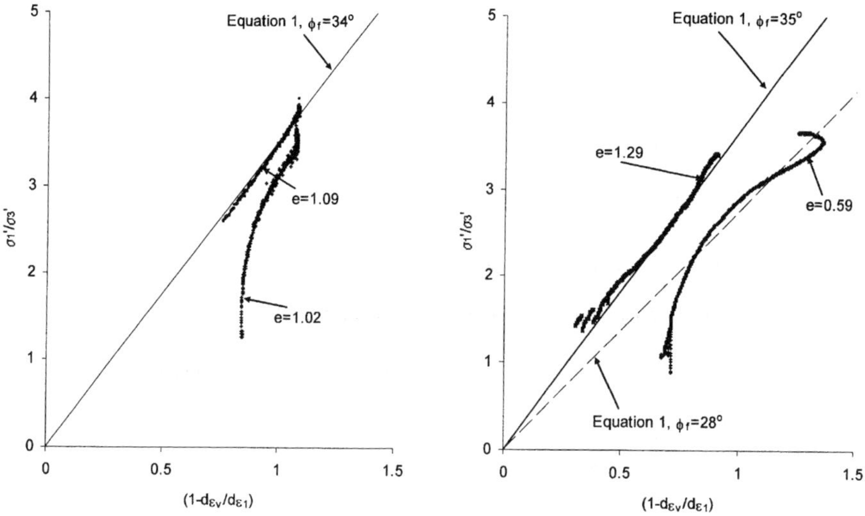

Figure 4. CD triaxial data for (a) Glebe Mine tailings and (b) Skelton Grange PFA

1.62 compared to 0.98, whereas for the Glebe Mine tailings the value is unchanged at 1.38. The lower specific gravity of the PFA may account for the large effect of water sedimentation on the PFA, but the reason for the unchanged e_{max} value for Glebe Mine tailings is unclear.

In the oedometer the PFA e_o values were slightly lower than e_{max}* ($e_{o,ave}$=1.52), as would be expected as a low flow rate was used. However, $e_{o,ave}$ for the Glebe Mine tailings was 1.50 using the water slurry method. Hence $I_{D\ average}$ for the PFA samples is 0.09 while for the Glebe Mine tailings it is -0.14. These data suggests that the structure of these materials is very sensitive to the method of preparation, and that measuring of e_{max} may be problematical.

The results of the drained triaxial tests suggests that the majority of the samples behaved in a loose fashion, tending to give a maximum mobilised angle of friction equal to the critical state angle, although large strains may be necessary to achieve this. The behaviour of a soil depends on its packing and stress state. Bolton (1986) proposed a relative dilatancy index, I_R;

$$I_R = I_D(Q - \ln p') - 1 \qquad (2)$$

where p′ is the stress at failure. The parameter Q (taken as 10 for quartz and feldspar sands) is

Table 4. Details of the consolidated undrained triaxial tests.

Test ID	e_o	I_D	Effective cell pressure (kPa)	η_{flat}	η_{elbow}	η_{max}	$(\varepsilon_1)_{flat}$ (%)	$(\varepsilon_1)_{elbow}$ (%)	$(\varepsilon_1)_{max}$ (%)
GM-U1	0.92	0.53	193	-	1.07	1.41	-	4.9	20.0
GM-U2	0.85	0.58	382	0.77	1.09	1.36	0.8	2.9	17.9
GM-US1	1.00	0.44	149	0.73	1.24	1.33	1.5	6.1	19.7
PFA-U2	0.94	0.63	395	1.02	1.34	1.46	1.5	5.1	14.3
PFA-U4	1.26	0.33	239	1.19	1.37	1.47	2.5	8.9	14.2
PFA-U6	1.38	0.22	49	1.08	1.17	1.58	0.8	5.6	8.3
PFA-U9	0.99	0.59	147	-	1.19	1.46	-	1.6	10.0
PFA-U11	0.91	0.66	149	1.10	1.34	1.72	1.1	5.4	10.0
PFA-U12	0.84	0.72	53	0.98	1.07	1.69	1.4	2.3	5.0

The relative density, I_D, of the PFA specimens has been calculated using the e_{max}* (water sedimentation).

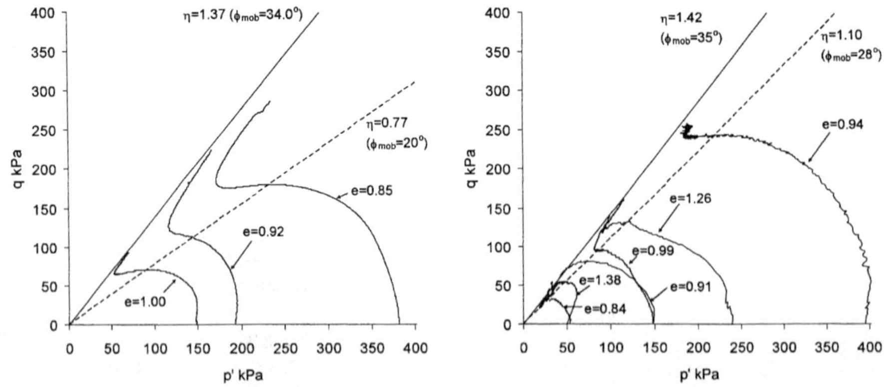

Figure 5. CU triaxial data for (a) Glebe Mine tailings and (b) Skelton Grange PFA

related to the mean effective stress required to suppress dilatancy, and is a function of the crushing strength of the particles (McDowell & Bolton, 1998). Bolton used I_R to account for the effect of stress on the measured angle of friction and the maximum rate of dilation, and for triaxial compression proposed that:

$$\phi_{max}-\phi_{crit} = 3I_R{}^{\circ} \quad \text{and} \quad (-d\varepsilon_v/d\varepsilon_1)_{max} = 0.3I_R \qquad (3)$$

Table 5 reports I_R values for the drained triaxial tests, together with predicted values of $\phi_{max}-\phi_{crit}$ and $(-d\varepsilon_v/d\varepsilon_1)_{max}$ assuming Q=10. Comparison between the values in tables 3 and 5 indicates generally good agreement between the data and equations (3). The exception is PFA-D3, where low measured ϕ_{peak} and $(-d\varepsilon_v/d\varepsilon_1)_f$ values must cast doubt on the void ratio.

The three undrained Glebe tailing specimens showed the expected trend of increasing instability during undrained loading with increasing void ratio. Most of the PFA specimens showed temporary instability during undrained loading. The greatest degree of instability (in terms of reduction in shear stress) was exhibited by specimens PFA-U6 and PFA-U11, where the stress ratio actually reduced slightly during instability. With the exception of PFA-U12 (which was relatively dense), these were the PFA tests where the effective stress was lowest at the onset of instability. It is interesting to note, but no inference is drawn from, the fact that the stress ratio at the onset of instability in PFA is approximately equal to the stress ratio for "particle sliding" in drained loading. In all the undrained tests, the maximum stress ratio approximates to critical state.

Conclusions

It is difficult to measure the maximum void ratio of silt sized materials, and particularly PFA. It is thus difficult to calculate absolute values of relative density. However trends in ϕ_{peak} and the maximum rate of dilation during drained loading generally correlate with dilatancy index. The limited undrained triaxial data show that these materials display varying degrees of instability dependent on stress level, with the pattern possibly influenced by particle shape and size distribution.

Table 5. Predictions based on I_R

Test ID	I_D	p'_f (kPa)	I_R	Predicted $\phi_{max}-\phi_{crit}$	Predicted $(-d\varepsilon_v/d\varepsilon_1)_{max}$	Test ID	I_D	p'_f (kPa)	I_R	Predicted $\phi_{max}-\phi_{crit}$	Predicted $(-d\varepsilon_v/d\varepsilon_1)_{max}$
GM-D2	0.33	402	0.32	1.0°	0.10	PFA-D1	0.30	271	0.42	1.0°	0.10
GM-D11	0.41	191	0.95	2.8°	0.28	PFA-D3	0.96	277	3.20	9.6°	0.96

References

Bolton, M.D., 1986. The strength and dilatancy of sands. Geotechnique, 35(1): 65-78.
BS1377, 1990. Methods of test for soils for civil engineering purposes. BSI.
Cabrera, J.G., Hopkins, C.J., Woolley, G.R. *et al.*, 1986. Evaluation of the properties of British pfa and their influence on the strength of concrete, ACI SP-91.
Cargill, K.W., 1984. Prediction of Consolidation on Very Soft Soil. ASCE J. Geot. Eng., 110(6): 775-96.
Consoli, N.C. & Sills, G.C., 2000. Soil formation from tailings: Comparison of predictions and field measurements. Geotechnique, 50(1): 25-33.
Cousens, T.W. & Stewart, D.I., 2003. Behaviour of a trial embankment on hydraulically placed pfa. Eng. Geology, 70(3-4): 293-303.
Davies, M.P., 1999. Piezocone Technology for the Geoenvironmental Characterization of Mine Tailings. PhD Thesis, Univ. British Columbia.
Finn, W.D.L., Ledbetter, R.H. & Wu, G., 1994. Liquefaction in Silty Soils: Design and Analysis, ASCE GSP 44, 51-76.
Fourie, A.B., G., P. & G.E., B., 2000. Static liquefaction as an explanation for two catastrophic tailings dam failures in SA, 7th Int. Conf. on Tailings and Mine Waste 2000, 149-58.
Gray, D.H. & Lin, Y.K., 1972. Eng. Properties of Compacted Fly Ash. ASCE J. Soil Mech. and Found. Div., 98(4): 361-80.
Guo, T. & Prakash, S., 1999. Liquefaction of Silts and Silt-Clay Mixtures. J. Geot. and Geoenv. Eng., 125(8): 706-10.
Henau, A.D. & Thijs, M., 1985. Controle de la compacite d'un remblai en cendres volantes, 11th ICSMFE. Balkema, San Francisco, 867-70.
Ishihara, K., 1993. Liquefaction and flow failure during earthquakes. Geotechnique, 43(3): 351-415.
Knight, P.G.K., 1960. Pulverised Fuel Ash as Construction Material. Proc. ICE, 16: 419-32.
Kolbuszewski, J.J., 1948. An experimental study of the Min. and Max. Porosities of Sand, Proc 2nd ICSMFE.
Lambe, T.W. & Whitman, R.V., 1979. Soil Mechanics. Wiley.
Leonards, G.A. & Bailey, B., 1982. Pulverised Coal Ash as Structural Fill. ASCE J. Geot. Eng., 108(4): 517-31.
McDowell, G.R. & Bolton, M.D., 1998. On the micromechanics of crushable aggregates. Geotechnique, 48(5): 667-79.
McLaren, R.J. & DiGioia, A.M., 1987. Typical Engineering Properties of Fly Ash. Geot. Practice for Waste Disposal '87, ASCE GSP13: 683-97.
Mitchell, J.K., 1993. Fundamentals of Soil Behaviour. Wiley.
Ng, C.W.W., Fung, W.T., Cheuk, C.Y. *et al.*, 2004. Influence of stress ratio and stress path on behavior of loose decomposed granite. J. Geot. and Geoenv. Eng., 130(1): 36-44.
Notholt, A.J.G., 1971. Fluorspar, Mineral Resources Consultative Committee, London.
Okumura, T., Takahashi, K., Tezuka, M. *et al.*, 1996. Geotechnical properties of fresh coal ashes from thermal power plants, 2nd Int. Conf. on Environ. Geotech., Lisbon, 869-74.
Parylak, K., 1992. Influence of Particle Structure on Properties of Fly Ash & Sand, ASCE GSP30, 1031-41.
Qiu, Y. & Sego, D.C., 2001. Laboratory Properties of Mine Tailings. Can. Geot. J., 38: 183-90.
Raymond, S., 1958. Utilization of Pulverised Fuel Ash, Civil Eng. and Public Works Review, 1513-6.
Rowe, P.W., 1962. The stress-dilatancy relation for staic equilibrium of an assembly of particles in contact. Proc. Roy. Soc. Lond. A, 269: 500-27.
Seneviratne, N.H., Fahey, M., Newson, T.A. *et al.*, 1996. Numerical modelling of consolidation and evaporation of slurried mine tailings. Int. J. Num.and Anal. Methods in Geomech., 20(9): 647-71.
Stewart, D.I., Cousens, T.W. & Charles-Cruz, C.A., in press. The interpretation of CPT data from hydraulically placed pfa. Eng. Geology.
Stone, K.J.L., Randolph, M.F., Toh, S. *et al.*, 1994. Evaluation of consolidation behavior of mine tailings. ASCE J. Geot. Eng., 120(3): 473-90.
Suetsugu, D., Miyata, Y. & Kogure, K., 2000. Mechanical property of submerged coal fly ash. In: Nakase and Tsuchida (Editors), Coastal Geotech. Eng. in Practice. Balkema, Rotterdam.
Trivedi, A. & Sud, V.K., 1999. Collapse Behavior of Coal Ash. J. Geot. and Geoenv. Eng., 130(4): 403-15.
Vaid, Y.P., Sivathalayan, S. & Stedman, D., 1999. Influence of Specimen-Reconstituting Method on the Undrained Response of Sand. ASTM Geot. Testing J., 22(3): 187-95.
Verdugo, R. & Ishihara, K., 1996. The steady state of sandy soils. Soils and Found., 36(2): 81-91.
Woeller, D.J., Robertson, P.K., Boyd, T.J. *et al.*, 1996. Seismic cone penetration testing for evaluating the compressibility and liquefaction potential of mine tailings, 3rd Int. Conf. on Tailings and Mine Waste, 101.
Yamamuro, J.A. & Lade, P.V., 1998. Steady-state concepts and static liquefaction of silty sands. J. Geot. and Geoenv. Eng., 124(9): 868-77.

Large-Scale Column Testing of Hypersaline Tailings

Daniel J. Stolberg, The University of Queensland, Brisbane, Australia
David J. Williams, The University of Queensland, Brisbane, Australia

Abstract

As part of a research project investigating cover systems on hypersaline tailings storage facilities in the arid climate of the Eastern Goldfields region of Western Australia, a large-scale column test was conducted under controlled laboratory conditions to aid understanding of the water balance of the tailings profile under tailings deposition, dewatering and desiccation; and subsequent re-wetting by fresh water simulating rainfall, followed by further desiccation. Under dewatering and desiccation of the tailings in the laboratory column following both deposition and re-wetting, about 60% of the total water loss from the tailings was attributed to seepage. Under the higher evaporation environment in the field, minimal seepage would occur from the tailings post deposition, with nearly all of the cumulative water loss from the tailings in the field being taken up by evaporation.

Test Design and Setup

The column is 2 m high and 300 mm in diameter. It was instrumented along its height by 10 sets of sensors to measure volumetric water content and matric suction. Potential evaporation was measured adjacent to the column and seepage from the column was measured directly. The water balance for the column allowed the actual evaporation to be calculated.

The original test strategy was to deposit the tailings slurry at 50% solids by mass into the column in 15 layers of about 100 mm thickness, at intervals of a few days between layers, and with the free water resulting from sedimentation and consolidation of the tailings removed prior to the placement of the next layer. This was intended to achieve a "layered" system similar to field conditions, whereby a thickness of tailings is placed and allowed to settle, consolidate and possibly desiccate prior to the next layer being deposited. However, this strategy proved impossible to achieve in the laboratory within a reasonable timeframe. As a result, it was decided to place the tailings slurry in a single lift to a height of about 1.5 m (Figure 1). The tailings slurry was wet up to the required 50% solids using the supplied saline tailings water (having an Electrical Conductivity of about 67,000 μS/cm). Beneath the base of the tailings, a graded sand filter was placed in order to prevent "blinding" by the tailings.

The instrumentation used for automated volumetric water content and matric suction determination comprised Enviroscan capacitive sensors for volumetric water content (Sentek, 1997, 2000) and CS229 heat dissipation sensors (Campbell Scientific, 1998) for matric suction. Only the volumetric water content data are presented herein. The volumetric water content sensor labels and their locations, which are referenced throughout this paper, are described in Table 1.

Table 1: Volumetric water content sensor labels and locations

Height from base of tailings (mm)	955	855	755	655	555	455	355	255	155	55
Enviroscan sensor ID	A1	A2	A3	A4	A5	A6	A7	A8	A9	A10

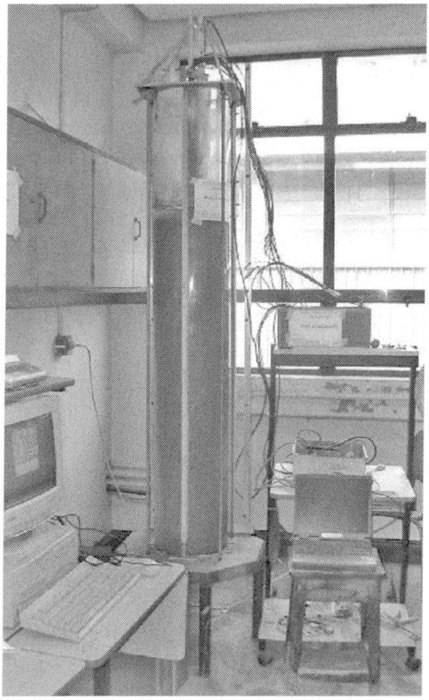

Figure 1: Column test setup immediately following tailings deposition

A 100 mm rainfall event was simulated to re-wet the desiccated tailings, involving the addition of 100 mm depth of deionised water, which was carefully placed on the desiccated tailings surface. Potential evaporation was measured adjacent to the column using a pan and automated balance.

Test Results and Interpretation

Sedimentation and consolidation of the tailings in the column test were monitored following the deposition of the tailings slurry from an initial 50% solids (specific gravity of 2.62, initial height of about 1,500 mm, initial gravimetric moisture content of 100%, wet density of 1.448 t/m^3 and dry density of 0.724 t/m^3). Monitoring the change in the height of the tailings column (shown on Figure 2, together with the locations of the volumetric water content sensors) enabled the average % solids and average dry density to be calculated as the test progressed. At any point in time, the Enviroscan sensors provided a profile of volumetric water content with depth. Accounting for the % solids and the dry density of the tailings column, the averaged volumetric water content was converted to the averaged gravimetric moisture content using the following equation.

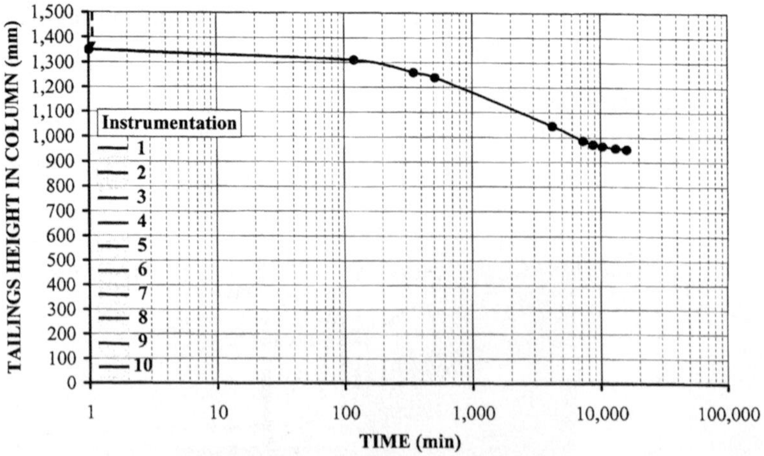

Figure 2: Sedimentation and consolidation of tailings in instrumented column

$$\theta_w = w\frac{\rho_d}{\rho_w} \quad [1]$$

where

θ_w = volumetric water content
w = gravimetric moisture content
ρ_d = dry density
ρ_w = density of water.

After 11.15 days, the height of the tailings column was 951 mm, its % solids had increased to 66.9%, its gravimetric moisture content had decreased to 49.4%, and its averaged wet and dry densities had increased to 1.706 t/m^3 and 1.142 t/m^3, respectively.

The volumetric water content readings from the Enviroscan system were used as a progressive means of monitoring the progress of the test. Following the initial deposition of the tailings, the drying period was determined from the time it took each sensor to level off to a "residual value". This did not apply to the base sensors, which were not expected to show a great deal of change during the course of each phase of the test, since the thickness and high water storage capacity of the tailings would prevent them from fully drying out, despite the free drainage provided at the base of the column.

Figure 3 shows the volumetric water content versus time plot for the first drying phase following the deposition of the tailings, and for the second drying phase following the simulated rainfall event. The sensor labels correspond with the depths given in Table 1, with the lowest number corresponding to the sensor nearest to the tailings surface. Figure 3 shows that the sensors towards the top surface generally displayed slightly higher saturated volumetric water contents, due to the lower self-weight experienced by the surficial tailings, but experienced the greatest desiccation.

Figure 3 also shows that the surficial tailings dried out to lower volumetric water contents following the simulated rainfall event compared with drying following tailings deposition, while at depth the water contents levelled off to relatively higher values following the simulated rainfall event. The different drying behaviour following the simulated rainfall event is most likely due to the development of surface cracks on initial desiccation, which acted like drains on re-wetting, limiting surficial re-wetting and directing water to greater depth. It is also possible that initial surficial drying following tailings deposition was reduced by the higher osmotic suctions that developed at surface under the influence of the saline tailings water.

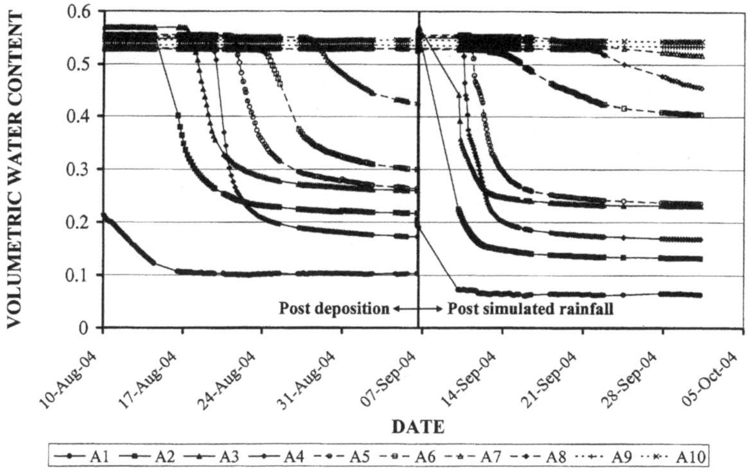

Figure 3: Volumetric water content versus time for drying phases of column test

Figure 4 shows plots of volumetric water content versus depth on initial tailings deposition, following desiccation of the tailings, on initial re-wetting and following the second drying phase, base on the Enviroscan sensor data. Superimposed on Figure 4 is the final water content profile measured by physical sampling and the oven drying method, which shows a similar trend to the corresponding Enviroscan sensor data. Figure 5 shows a series of plots of volumetric water content versus depth at weekly time intervals during the first drying phase following tailings deposition, which indicates that the tailings profile dried out most rapidly during the second week of the four-week drying period.

Figure 6 shows the progressive rate of evaporation measured during the course of the column test. A three-day "moving average" is also plotted, which reduces some of the scatter in the measured data. The linear trend line shown indicates a slight positive gradient, since the ambient temperature was rising over the course of the column test (as summer approached), with an average rate of potential evaporation of 3.5 mm/day.

The averaged gravimetric moisture contents for the tailings column, determined from Figure 2 and Equation [1], were used to calculate the water mass balance. Seepage was measured directly; hence actual evaporation may be back-calculated. Measured gravimetric moisture contents, determined at the end of the column test using the oven drying method (AS 1289.2.1.1, 1994), verified the Enviroscan sensor calibrations and confirmed the accuracy of the method used for deriving the actual evaporation.

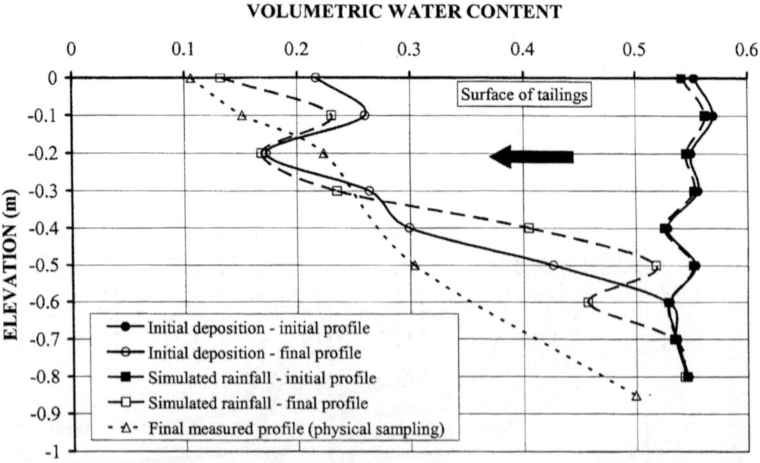

Figure 4: Volumetric water content versus depth plots for initial tailings deposition and drying, and for simulated rainfall followed by further drying

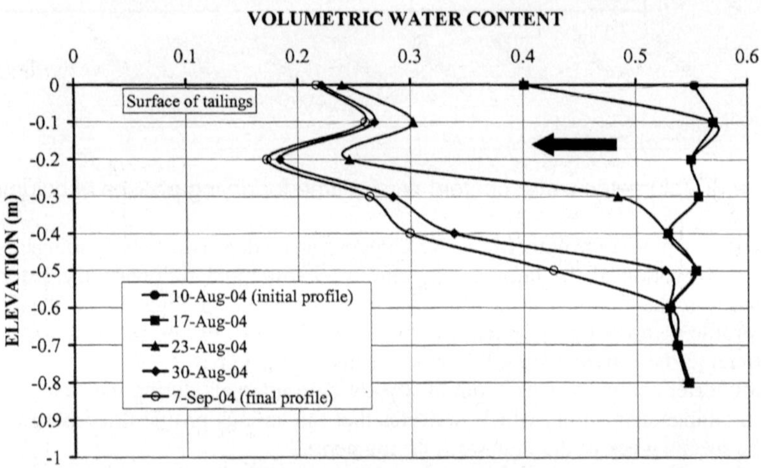

Figure 5: Progressive volumetric water content versus depth plots during drying following tailings deposition

Figures 7 and 8 show the results of the mass balance calculations following tailings deposition and simulated rainfall, respectively. Figures 7 and 8 include plots of column seepage loss, calculated (actual) evaporation loss, total water loss (the sum of seepage and evaporation losses), estimated potential evaporation loss from the column under laboratory conditions at The University of Queensland, and the approximated potential and actual evaporation losses from the column under field conditions.

The laboratory potential evaporation and field evaporation (potential and actual) were determined by the following means. Figure 9 is a hypothetical plot of the cumulative potential evaporation that would occur if the column contained a reservoir of freshwater, which was allowed to evaporate under the same laboratory conditions. These cumulative plots were then used in conjunction with the (actual) evaporation losses from the column, back-calculated from the mass balance calculations, to determine a running ratio of actual evaporation to potential evaporation (that is, AE/PE).

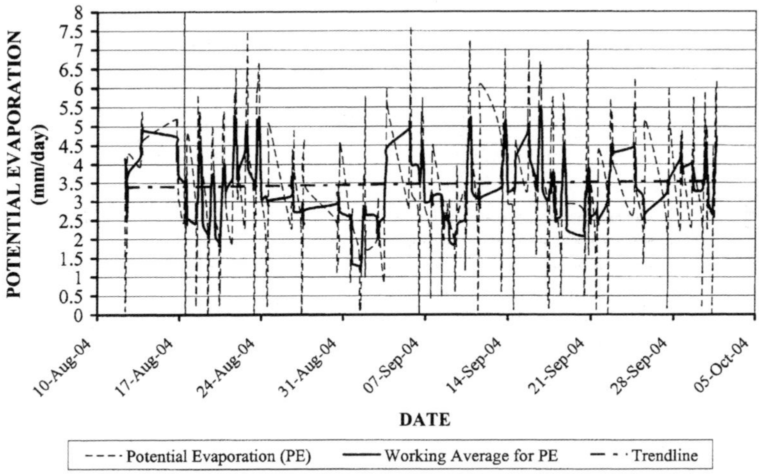

Figure 6: Measured rate of potential evaporation under laboratory conditions at The University of Queensland

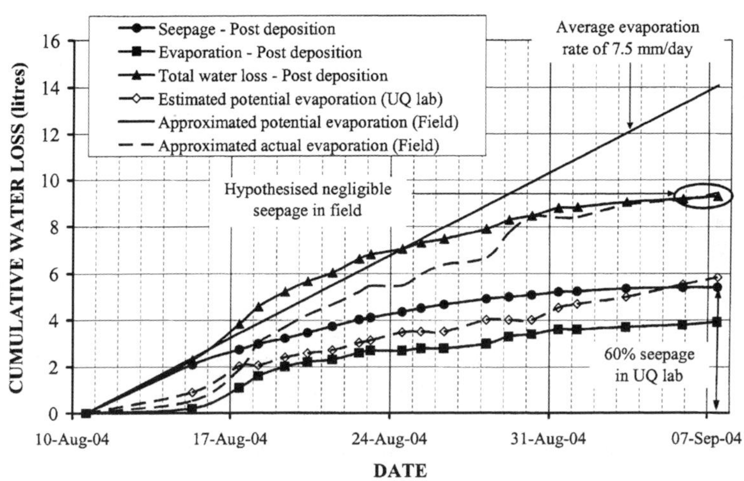

Figure 7: Column water balance versus time following tailings deposition

Since daily evaporation data were not available for the Yeelirrie/Mt Keith region, an average daily pan evaporation rate of 7.5 mm/day (Bureau of Meteorology, 2004c) was adopted for the generation of a cumulative evaporation plot for the field (the pan evaporation rate for Yeelirrie/Mt Keith varies from about 3 mm/day in June/July to 13 mm/day in January/February). The running AE/PE ratio determined for laboratory conditions was then used to estimate an actual cumulative evaporation plot for the field, which is plotted on Figures 7 and 8. By comparison, the average pan evaporation rate for Brisbane (that is, The

University of Queensland) during the test period was 4.9 mm/day (Bureau of Meteorology, 2004a, 2004b), while the annual average pan evaporation rate for Brisbane is 5.3 mm/day (Bureau of Meteorology, 2004d).

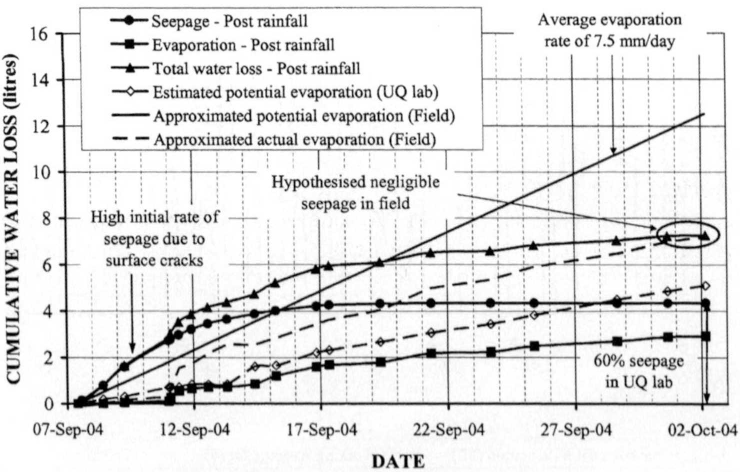

Figure 8: Column water balance versus time following simulated rainfall

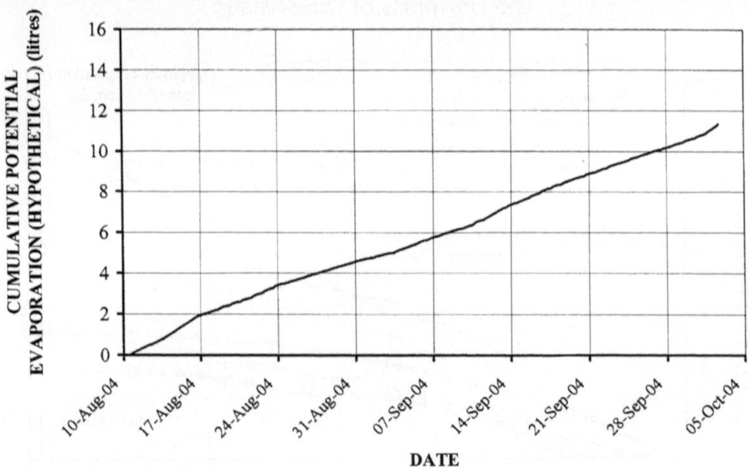

Figure 9: Hypothetical cumulative potential evaporation from column tests

The key outcome of the column testing, which is illustrated by Figures 7 and 8 is that although seepage very much dominated the moisture flux at the outset of each wetting cycle (especially following the simulated rainfall event, as a result of previous surface desiccation cracking), evaporative moisture flux takes over, accounting for about 40% of the column water loss by the end of each drying cycle. If the estimated actual evaporation plot for Mt Keith is adopted instead of The University of Queensland laboratory plot, then there would be negligible seepage loss, with essentially all of the cumulative water loss from the tailings column being taken up by evaporation.

Conclusion

The column testing provided a means of studying the behaviour of the Mt Keith tailings under controlled laboratory conditions and aided understanding of the water balance of the tailings profile under tailings deposition, dewatering and desiccation; and subsequent re-wetting by fresh water simulating rainfall, followed by desiccation. Under laboratory conditions, approximately 60% of the water loss from the tailings column was attributed to seepage. Under field conditions, minimal seepage would occur from the tailings post deposition, so that designing a cover system to pond rainfall runoff for removal by evaporation, rather than shedding runoff, has potential in this arid region.

Acknowledgements

The Australian Research Council and WMC Resources Limited are gratefully acknowledged for funding the research on which this paper is based. The management and staff of WMC Resources Limited are also acknowledged for assisting with the work described and for giving their permission for the work to be published.

References

AS 1289.2.1.1 (1994). *Australian Standard - Method 2.1.1: Soil moisture content tests - Determination of the moisture content of a soil - Oven drying method (standard method).* Standards Australia.

Bureau of Meteorology (2004a). *Brisbane, Queensland - August 2004 Daily Weather Observations.* Bureau of Meteorology, Australia.

Bureau of Meteorology (2004b). *Brisbane, Queensland - September 2004 Daily Weather Observations.* Bureau of Meteorology, Australia.

Bureau of Meteorology (2004c). *Climate averages for Station 012090: Yeelirrie, Western Australia.* Bureau of Meteorology, Australia.

Bureau of Meteorology (2004d). *Climate averages for Station 040223: Brisbane Aero, Queensland.* Bureau of Meteorology, Australia.

Campbell Scientific (1998). *Application Note: 229 Soil Water Potential Probe.* Campbell Scientific, Inc., Logan Utah.

Sentek (1997). *Enviroscan Hardware Manual.* Sentek Pty Ltd, Stepney, South Australia.

Sentek (2000). *Enviroscan Version 4.1 Manual.* Sentek Pty Ltd, Stepney, South Australia.

The capping of sludge and other weak deposits

Eur.-Ing. Paul Thurlwell, PSA Design, Preston, UK
Cliff Hall, Tensar International, Blackburn, UK
Dipl.-Ing. Omar Naciri, Tensar International, Bonn, Germany

Abstract

Arising from many different industrial and commercial processes, sludge and other weak deposits are often placed into on-site lagoons. There normally comes a time, commonly when a lagoon has reached its capacity, when these deposits have to be closed. The practice of closing, or capping over the weak deposits, has been made possible by the application of geosynthetic products generally and geogrids in particular.

At the commencement of the capping operation, often the biggest problem is the difficulty in gaining safe access onto the weak deposits for construction personnel, followed by the ability of the working platform to safely support the weight of the earthmoving plant. Finally integrity and robustness of the capping has to be provided for the long term utilisation of the surface.

This capping of tailing lagoons, sludge lagoons, dock silts and dredged deposits has been one of a developing methodology over a number of years, and the successful use of Tensar biaxial geogrids on a number of challenging projects has provided the opportunity to identify the main components of an appropriate technique.

The projects described in this paper illustrate important design and construction issues associated with the safe and ecologically acceptable capping of sludge and weak deposits. Emphasis is placed on the use of geogrids as part of a bespoke practical approach to each problem.

Introduction

A by-product of many industrial and commercial industries is a near-fluid effluent which is directed to settlement areas generally described as sludge lagoons. Eventually, as part of the owners' operating plans, which are normally heavily influenced by environmental considerations, the sludge lagoons reach their retention capacity and have to be permanently covered over. The covering, or capping, of lagoons is a civil engineering operation requiring a careful review of the available capping options.

The number of available techniques is relatively few and most capping projects will employ a soil covering in which a critical component is the use of mechanical stabilisation using geogrids. In the case of contaminated deposits the capping solution also has to function as a sealing layer to prevent losses, leading to pollution, and also minimise rain water ingress. Therefore for the long-term capping of sludge lagoons, the following list of engineering criteria needs to be fulfilled:

5th ICEG Environmental Geotechnics, Thomas Telford, London 2006.

- The losses and emissions from the sludge must be controlled
- Rainwater has to be collected and drained at the surface
- The penetration of the rainwater in the deposits has to be minimised
- A dewatering of the deposits has to be achieved
- Settlement of the capping layer must be controlled
- Erosion control of the surface must be provided
- A beneficial utilization of the surface, possibly recreational or commercial, must be achieved

General

At the commencement of the capping works, the primary problem is the weakness of the deposits. Any attempt to measure the shear strength on the near surface deposits often indicates strengths of not more than 2–3 kN/m^2, or no measurable strength at all. Under such conditions numerical analysis and dimensioning has a limited application. The safe and successful capping of lagoons is often related more to the designer's methodologies and the installer's techniques, allied to the use of observational methods and strict site control.

As part of both the methodology and attendant installation techniques, the use of monolithically produced and therefore stiff geogrids make it possible to provide:

- safe access for installation personnel (Figure1)
- construction access for fill delivery (Figure 2)
- bearing capacity for fill placing plant (Figure 3)
- bearing capacity for future uses (Figure 4)

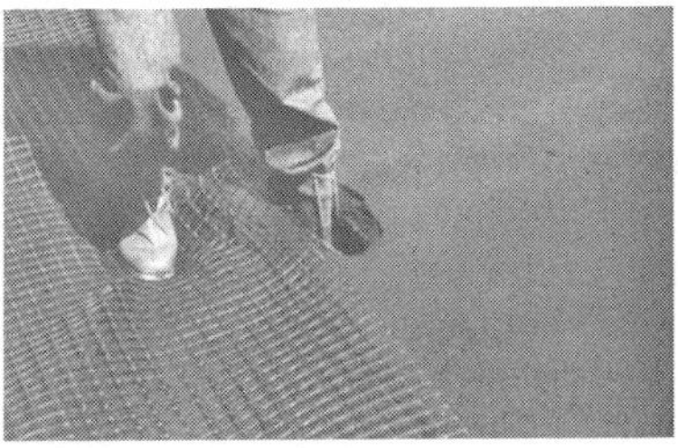

Figure 1: safe access for installation personnel – the snowshoe effect

Figure 2: initial construction access

Figure 3: bearing capacity for fill-placing plant

Figure 4: bearing capacity for future uses

The general methodology for the wider management of a capping project, using stiff and integral geogrids, has been previously reported, [1, 2]. Figure 5 concentrates on more detailed aspects, showing the range of sludge strength typical of capping projects, and the important design issues that need to be considered. Figure 5 also shows the sludge strength and considerations for the three projects described in this paper.

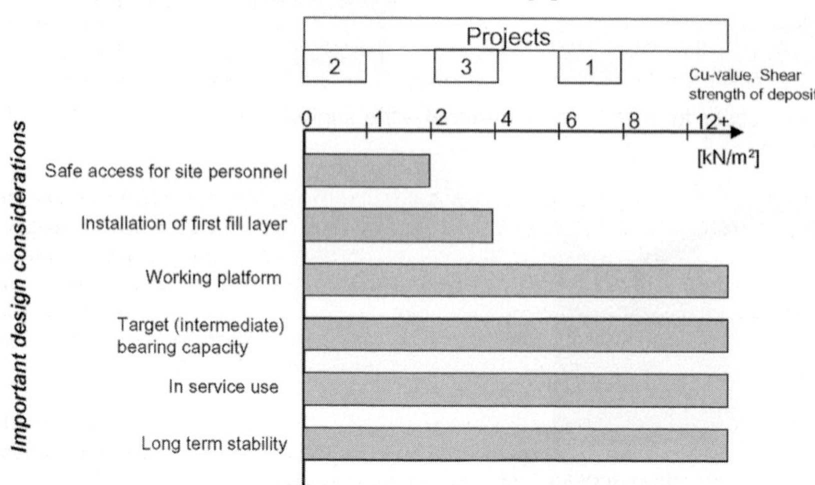

Figure 5: design considerations for the described projects

Once initial access and the placement of the first layer of fill is achieved the risk of failure reduces as the sensitivity to the sludge strength diminishes. To create an initial working platform, the contractor's interpretation of the designer's outline method statement, which is usually specified as part of the contract, needs to be transformed into a detailed method statement taking account of:

- the placement of the geogrid rolls, (sequence, orientation, pattern)
- the methods to be employed in placing and spreading the first and critical layer of fill
- the routes for delivering fill material
- the mode of operation for controlling sludge displacement, (bow waves)
- monitoring, (usually levelling for displacements and settlements)
- compaction requirements, if any
- the target bearing capacity, (often vertical drain installation rigs)
- the requirements of the final design
- observational trigger levels

Projects

Project 1: The capping of a sludge in North Germany near Hannover

Acidic gypsum sludge had been deposited over many years in an earth-lined basin surrounded by earth containment bunds within the grounds of a factory. The depth of the sludge was approximately 6m. The surface of the sludge was pulpy, Figure 6, and had no crust to facilitate safe access on foot but had various waste materials, (e.g. oil drums) which required removal to a regulated waste facility. The undrained shear strength, Cu, was measured at 6-8kN/m^2 below the weak, and fluid, surface and the water content was measured as 360-440%. Owing to the future required use of the ground and the requirements of the environmental authority, the surface of the sludge lagoon had to be capped.

First, various solutions to gain safe access to the surface, prior to the capping operation, were considered. Solidification of the surface by cement or chemical treatment was rejected due to the lack of assured installation method and uncertain outcome. A capping of woven geotextiles combined with an aggregate was abandoned, as it was not possible to develop an installation method. The client took the option to simply fill the basin with soil in order to minimise costs. However, this proved to be ineffectual as the filling material instantly sank to the bottom of the basin, displacing the sludge which rose to the surface. Therefore, a solution using multi-layers of stiff geogrid was chosen.

Using a very stiff and rigid geogrid, the sludge surface was made safe and accessible to construction personnel. The first layer was installed using wide rolls of a composite geogrid and geotextile. Fill materials were specified to provide both shear strength and drainage functions. The basal geogrid composite was combined with an additional two layers of geogrid to form a total construction thickness of 1.5m. The multi-layers of geogrid provided a flexurally stiff platform that not only maintained the integrity of the capping system but also accommodated and mitigated the effects of differential settlements.

Consolidation settlements were regularly measured on the capping surface. The measurement results showed a typical consolidation profile measured against time. After about six years settlements up to 120cm were recorded and consolidation had reached about 50%, [3].

Figure 6: Safe access for the clear-up operations at the start of the installation, [4]

Project 2: Bingley and Knostrop, UK

The influence of environmental considerations and health and safety concerns gave rise to the need to provide a capping to a disused sludge lagoon in Bingley. The final use of the area was to be dedicated to public amenity space and loading from light vehicles. However, the consistency of the sludge and a shallow depth of water meant that initial access and the installation of the capping were the most critical phases of the work.

The lagoon had been in use for approximately fifty years becoming disused in the late 1990's. The 7m deep deposits had not consolidated to any extent and as such no measurable shear strength had developed.

Three layers of geogrid were incorporated into a capping layer thickness of 1.0m. The first layer of geogrid was placed directly on the fluid deposits. The requirement of this layer was that its flexural stiffness was sufficient to provide safe access for the installation personnel. This is often called the snow-shoe effect as the physical nature of a geogrid resembles the load spreading ability that is a feature of snow shoes. Once installed, fill can then be placed in a controlled manner. Control is exercised on both the fill thickness and the type and operation of the mechanical plant. The overall aim is to observe the performance of the fill and sludge and to exercise control on the displacement of the sludge to ensure that the bow wave build-up does not flow back over the installation; as recovery from this situation is very difficult to achieve. The temptation to place too thick a layer must be avoided, as this could lead to a classical rotation failure at the edge of the platform. In the most difficult areas, usually towards the middle of a lagoon, manual work may be required such as installing barrow runs on timber planks and using wheel barrows, to import a thin layer of fill over a wide area. Normally however, fill can be placed using long-reach hydraulic excavators or cranes with strut jibs discharging the fill from skips.

Following the success of the Bingley capping, the client had a similar requirement for capping work at Knostrop. With the same construction company employed to undertake the capping works, the techniques devised for Bingley could be reviewed and refined. This was aided by the opportunity to carry out evaluation works in advance of the main contract using a range of alternative materials. Although the capping in these trials appeared to be functioning as intended, its integrity could not be assessed. Therefore, a method of examining integrity, (robustness and resilience), was devised by the installer. This involved a dynamic test comprising dropping a heavy weight on the trial capping sections and observing and measuring the response. A 1.0 cubic metre concrete block was dropped from a height of 1m and the integrity of the capping layer observed and the permanent deformation measured under the point of impact. For the chosen system, Figure 7, the deformation was in the order of 5mm.

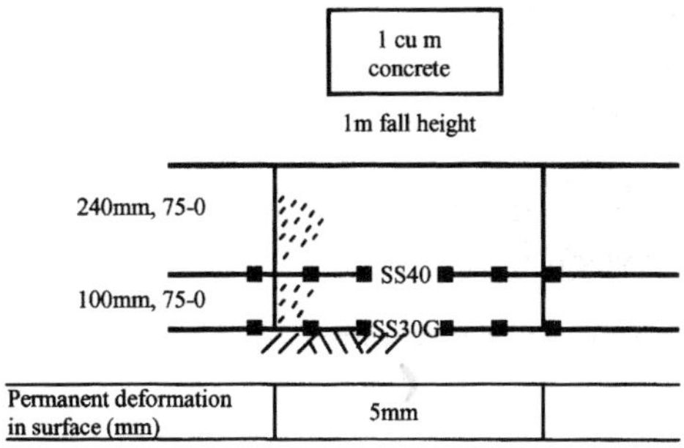

Figure 7: Investigation of the integrity of a capping system

Project 3: The Covering of Tailings ponds – Wismut – Germany

Over 2million m^2 of fine, weak and very wet uranium mill tailings had already been dewatered as part of a major remedial scheme of uranium mill tailings ponds located in Saxony and Thuringia in Germany. In a long-term project, Tensar geogrids were used to provide stable working platforms on the tailings surface. The geogrids enabled access for the installation of vertical geosynthetic (wick) drains, Figure 8, to aid consolidation, and also formed part of the final cover layer system.

Figure 8: Installation of geogrid and wick drains

The vast open areas of tailings, the biggest being the Culmitzsch TMF with a surface area of over 235ha and a total volume of 85million m^3, were visually and environmentally undesirable. The weak fine tailings which had settled below the water table had a near surface undrained shear strength of less than 5kN/m^2. This water had to be removed with minimum environmental impact prior to capping the lagoons safely.

Owing to the history of flushing of the lagoon the profile of the contents varied from the outer parts, which were predominantly sandy, to the central deeper areas, which were finer and weaker.

The initial step was to place a geotextile separator layer over the sludge, and cover it with a biaxial geogrid that enabled access for the wick drain rigs. The vertical wick drains were installed in order to accelerate the consolidation and increase the strength of the weak deposits. The geogrid was then covered with a 1m thickness of granular fill, which also created a stable reinforced working platform for the installation of the final cover layer.

In this way, the wick drains provided in-situ dewatering and this dramatically increased the rate of consolidation of the fine material. In six years, over 2million m^2 of surface treatment has been successfully installed over these tailings ponds.

Conclusion

This paper has reviewed the practical aspects of waste capping and reported on successful experiences of installing a mechanically stabilised soil layer using integral, stiff and rigid geogrids. The use of such geosynthetic products enables safe access onto very weak deposits as well as assisting in the creation of a long term robust and durable capping layer.

Reference

1. Daenecke, R. Johannsen, K. Kums, H. Oswald, K-D. & Merkel,G., 1999: The application of Geosynthetics in the Rehabilitation of Tailings Ponds at Wismut GmbH. Mine Water & Environment, IMWA Congress, Sevilla, Spain.
2. Leupold, D et al 2004: Practical experience with the covering of tailings and the use of geosynthetics in the framework of the rehabilitation of Wismut tailings ponds, EUROGEO 3 /2004, Munich
3. Biczok, E, & Scharf, K.: Abdeckung eines Schlammbeckens durch eine Geogitter -Vlies-Kombination, 8. Information- und Vortragstagung über „Kunststoffe in der Geotechnik" K-GEO 2003, München
4. Naciri, O., Pfannenschmidt, P., 1998: Überbauung von Schlammteichen, Straßen- und Tiefbau, Heft 6/1998

Capping/covering of Sludge lagoons and stabilisation of soft alluvial soils using Multi-layered geosynthetic/soil systems: Specific Case Histories

Laurence TOMLIN, Peter Brett Associates, Exeter, UK
Dimiter ALEXIEW, HUESKER Synthetic GmbH & Co.,Gescher, Germany
Iñaki AMIGOT, HUESKER geosintéticos S.A., Pamplona, Spain
Graham HORGAN, HUESKER Limited, Warrington, UK

Abstract
The use of multi-layered systems consisting either of (1) different mineral (soil) layers, (2) soil-geosynthetic "sandwiches" or (3) only of geosynthetics are widely used as basal and capping systems for waste disposals and remediation. Recently the combinations of soil & geosynthetic seem to be most popular and are becoming established as "standard solutions" in many countries worldwide. However the construction of these capping systems (incorporating a low permeability barrier) or covering systems (not incorporating a barrier) over very soft (undrained shear strength cu < 20 kN/m2) and extremely soft (cu < 10 kN/m2) soils poses additional design and constructions challenges. This paper identifies some of those critical issues associated with edge stability, bearing capacity and constructability of the covering systems and practical solutions adopted on specific sites.

The paper focuses on three cases histories differing in problem and adopted solution:
- **Stabilisation of soft alluvial clays beneath a new landfill cell at Docksway, Newport, South Wales, UK**
- **Capping system over a large basin with thixotropic industrial sludge, situated in Nordenham, Germany**
- **Capping of very soft industrial sludge in Navarra in Spain**

For each site the specific problems, the engineering solution and the construction details are presented

Design and construction considerations

Multi-layered systems consisting of different soils and/or geosynthetic layers are widely used as basal and capping systems for waste disposals. Soil-geosynthetic combinations are most popular and have become established as "standard solutions" in many countries, allowing the unique advantages of both materials to be realised. In some cases additional specific problems such as stability, bearing capacity and/or deformation problems have to be solved by such systems in addition to their primary barrier, drainage and protection functions.

Often very soft sludge's are too soft to even walk on and as such require unconventional construction measures to allow access. These measures include complete geosynthetic covers sewn together on site from individual panels and deployed from the periphery of the sludge lagoons or floating pontoons. Other measures include the use of non-conventional low bearing pressure construction plant, or utilization of ground improvement techniques to improve the engineering characteristics of these weak soils and hence allow the access of construction equipment.

Site 1: Stabilisation of Soft Alluvial Clays Beneath a New Landfill Cell at Docksway, Newport, South Wales

Background

The existing landfill facility at Docksway, Newport, South Wales comprises land raising on the flood plain of the River Severn estuary. This existing landfill is nearing capacity and a new landfill area is to be developed on ground to the south of the existing facility. The area of the proposed new landfill is sited within a former meander of the River Ebbw. This area was reclaimed by damming off each end of the meander when the river was straightened in 1990. The ground levels in the area of the new landfill range between about 1m OD in the centre of the former river channel to about 8m OD around the site perimeter.

Following the damming of the river, alluvial clay was removed from the river banks along the length of the meander. The firm crust was typically stripped to depths of up to 2m below the original ground level leaving soft alluvial clay exposed. This clay was used within the exiting landfill for day cover and capping works. Following the stripping of the firm crust, the meander was allowed to flood with surface water, forming three lakes. The new landfill cell is to be constructed in the area of one of the lakes. The clays remaining beneath the proposed area of the landfill cell after the firm crust had been removed are generally soft to very soft.

Due to the soft nature of the clays in the area to be developed, it was identified that stabilisation would need to be undertaken in order to permit an artificial clay liner to be constructed in the base of the cell.

Geology and Hydrology

The geology beneath the area of the proposed cell to be developed for the new landfill generally comprises alluvium overlying river gravels, which in turn rest on Mercia mudstone. The alluvium is generally a very soft to firm clay, ranging between approximately 3m to 9m thick with the base at an elevation varying between about -0.5m OD and -3.5m OD.

The river gravels below the alluvium comprise medium dense to very dense sandy gravels containing clay, cobbles and boulders. The thickness of these gravels range between about 4m and 9m. The Mercia mudstone beneath the River comprises weathered, very stiff, red brown clay and silt. Groundwater is present within the river gravels and is artesian with a maximum phreatic head of 4.15m OD. A typical section thorough the site indicating the ground profile and geological conditions is presented as Figure1.

Proposed Landfill Cell Construction

Based on a detailed appraisal of the ground conditions and geotechnical stability analysis undertaken by the consultant engineers, Peter Brett Associates (PBA) it was identified that considerable settlement could occur beneath the proposed landfill. In addition, the soft and very soft nature of the clays were such, that it would not be possible to form the proposed artificial clay liner unless these soft alluvial clays were stabilised effectively to allow plant movements across the site.

The proposed landfill cell construction includes a 1.2m thick artificial clay liner laid on the low permeable alluvial clay soils. The thickness and low permeability of the of the alluvial clay soils beneath the artificial clay liner would form an appropriate natural geological barrier. In this respect, it was important not to significantly compromise the effectiveness of this natural geological barrier with any stabilisation works that were to be undertaken.

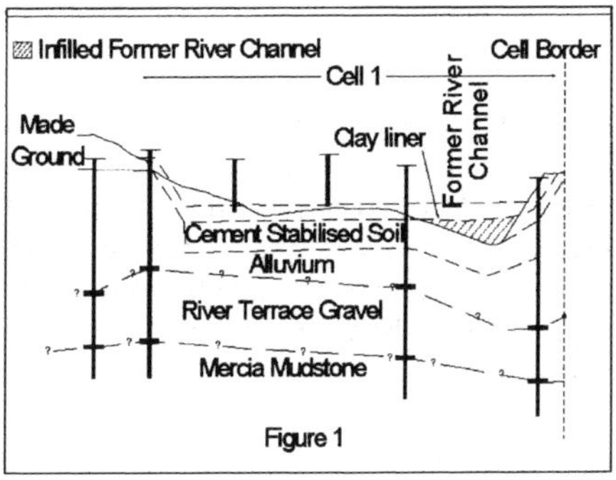

Figure 1. Ground Profile and Schematic Geological Section

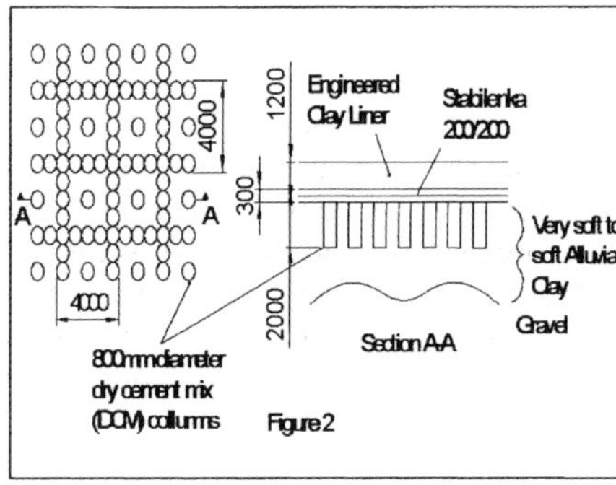

Figure 2. Schematic of the soil mix columns

In consideration of the need to form a surface on which the artificial clay liner could be formed and the importance of maintaining the low permeability of the alluvial clays, it was decided that near surface stabilization using a soil mixing technique would be most appropriate.

Soil Stabilization

The design of the stabilization was undertaken by specialist ground stabilization contractor in consultation with the main consultant. The design included for installation of soil/cement mixed columns 800mm in diameter formed in the soft to very soft clays to a depth of no more than 2m. This ensured that there would be at least 1m between the underside of any column and the top of the gravels. The stabilized columns were to be installed as a grid system as indicated on Figure 2.

The stabilized columns were designed to form an upper stabilized soil zone with an effective combined shear strength of 64KPa and a bearing capacity of 20 MPa. In order for the stabilized ground to support construction plant, a basal reinforced platform was incorporated over the dry soil mix columns incorporating a high strength biaxial polyester woven geotextile, (with 200 kN/m ultimate tensile strength in each direction), with a 300mm granular blanket above. The design also relied on the material above the geotextile being at least 0.5m thick to transfer loads effectively through the blanket and into the columns in order that the stabilized soil would act as a relatively uniform layer. In this respect, the lower part of the clay liner would also form part of the basal reinforced platform.

Site 2: Capping system for the sludge lagoon "Galing 1", Nordenham, Germany

The company Metaleurop in Nordenham, Germany is a producer of zinc. Although environmental-friendly technologies are used, some substances from the production process cannot be recycled and have to be stored as "waste": namely "jarosit sludge", which contains some percentage of heavy metals. At the beginning of the 1970's the industrial waste disposal site at Galing in Nordenham was built for this purpose consisting of two large basins (Galing I and II) with basal barrier and drainage system designed according to German standards at that time.

The main components of the basal barrier are a heavy 2.5 mm thick HDPE geomembrane and a coupled leachate control and drainage system. Galing I has an area of 105,000 m² (≅ 300 m x 350 m) and a volume of about 680,000 m³ with a depth of 6.0m to 6.5m. Pumped sludge began filling the site in 1974 and no additional surcharge was applied for consolidation at any time. The excess contaminated sludge water on the surface after sedimentation was pumped back to the plant. The Galing I lagoon was completely filled in 1989 and had to be covered after that by an appropriate capping system. In 1991, German standards concerning waste disposals became more rigorous. The Galing I-capping-system had to be completely water impermeable over hundreds of years and had to include a vegetated soil cover on top, for landscape purposes. The remediation project and capping system developed by German consultant "Wersche" Oldenburg were accepted by the regulatory officials in December 1993.

The most problematic issue was the behaviour of "jarosit" sludge. Due to the specific filling process, granulometry, plasticity etc. of "jarosit" and due to the period of about 4 years between end of filling and start of remediation, the sludge had no free water on the surface and a crust of about 0.2 m; the remaining sludge extending to a depth approximately 6 m below was not consolidated and still had a high water content, soft consistency and c_u-values generally less than 15 kN/m². Additionally, the sludge (even the crust) was thixotropic: a person could walk on the crust slowly, but jumping resulted in softening and the danger of sinking-through. The capping system was required to provide not only the classic water barrier and drainage functions but additional deal with specific problems: the local and overall stability of the capping system on the sludge as "subsoil", the compensation of consolidation settlements after installation of capping and the possible thixotropic softening by dynamic influence of equipment during construction.

Multi-layered capping system

The capping system design conformed to the new German standards at the beginning of the 1990's e.g. incorporating a combined soil-geomembrane barrier (Fig. 3). The consolidation settlement of sludge due to the added weight of capping system was expected to be up to 0.40 m over the 6 m thick sludge zones decreasing with decreasing sludge thickness e.g. near the lagoon borders. After this inhomogeneous settlement a sufficient inclination of ≥ 0.5% of the barrier and drainage layers had to be guaranteed. For this reason, a settlement compensation soil layer of varying thickness was foreseen (Fig. 3) in a transition zone around the periphery of the lagoon. The excess contaminated water had to be accumulated in the a non-cohesive granular "capillary layer" (Fig. 3), which should sink completely in the sludge during consolidation. The first geosynthetic layer between sludge surface and non-cohesive soil layer had to fulfill three functions: reinforcement to ensure local and overall stability during installation of capping layers on the sensitive thixotropic sludge, separation and filtration (free water permeation from sludge into 'capillary layer' during the consolidation 'settlement'). Design calculations resulted in required design tensile forces in both directions of ≥ 20 kN/m at ≤ 3% strain for at least one year, including at joints/overlaps and all material reduction factors. The effective opening size O_{90} had to be about limited to approximately 0.1 mm.

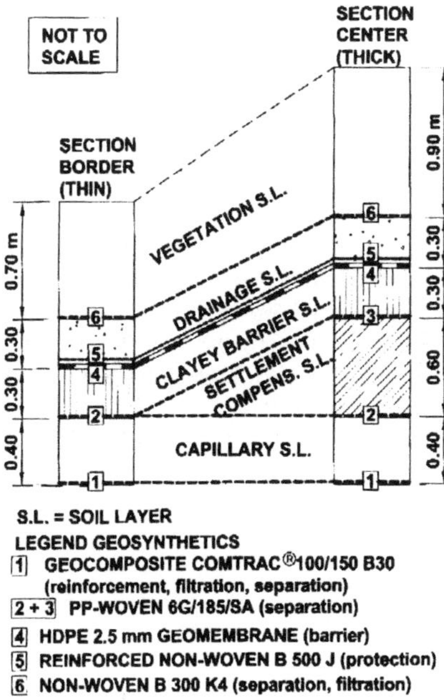

Figure 3. Multi-layered capping system for the sludge lagoon "Galing I"

Figure 4. Sewing the geocomposite geotextile rolls on site

A specific engineered geocomposite was produced by Huesker to meet all mechanical and hydraulic requirements. The 5 m wide geocomposite (Fig. 4) consists of a composite of a high-strength low-creep net providing the bearing component and a non-woven geotextile. The ultimate tensile strength in the longitudinal direction amounted to 100 kN/m and in the cross direction to 150 kN/m. The higher tensile strength of 150 kN/m in cross direction compensates the lower strength and higher strain in the sewn joints of parallel rolls (see Figure 4). The result is a nearly homogeneous continuous biaxial reinforcement over the entire area. For the in-situ joints special yarns were used. The five soil and six geosynthetic components and their main functions are shown in Figure 3. More detailed project information is given in elsewhere (e.g. Nüske 1995).

Construction

Construction started in 1994 and was completed in 1995 . Temporary embankments were built across the area for transportation, distribution and installation of soils and geosynthetics and for drainage pipes and shafts as well (not discussed herein). The embankments had an additional function of damping dynamic action of heavy trucks on the thixotropic sludge, which could otherwise fail due to softening. After completion of the capping system the overfill embankment soil in the upper part was removed and replaced by vegetation soil. Due to thixotropy the contractor used light 'snowcats' on rubber tracks for installation of the first two soil layers on the sludge, construction was completed without excessive deformations.

Site 3: Sealing/Capping system of sludge lagoon, Navarra, Spain

Introduction

Viscofan S.A. is a world leader in the supply of artificial coverings predominantly to the meat & food Industry. At its production plant in Caseda in Navarra, Spain the manufacturing process generates a liquid sludge which historically was pumped into an on-site storage lagoon. At the end of it's useful life in 2002, the surface of the lagoon was sealed in compliance with the provisions of the European directive 1999/31. The lagoon surface area was approximately 4,500 m^2 and was formed on side-long ground with a periphery retaining dyke to the eastern edge of the lagoon, (see Figure 5). The lagoon consisted of up to 4.5m of soft sludge The geotechnical properties of the sludge was determined and shear strength parameters applied to enable a limit equilibrium slope stability analysis to be undertaken for both static and seismic conditions.

Figure 5. surface of sludge lagoon and eastern dyke

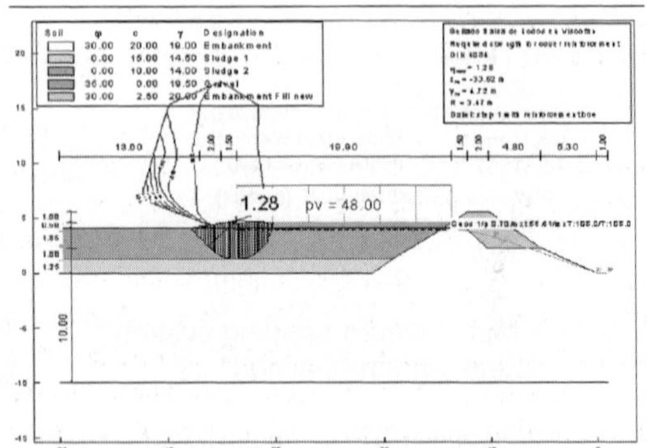

Figure 6. Output from stability analysis

The consistency of the liquid sludge was such that even pedestrian access across the lagoon was not possible. This meant that the construction traffic would not be able to install the lining system across the surface without some form of additional support. A geosynthetic reinforced layer was therefore adopted over the entire area of the lagoon to enable an effective liner system to be installed.

Lining System

The engineered layer at the base of the lining system had to be designed to cope with the weight of the components of the lining system as well as the construction traffic. The initial phasing and placement of layers of the fill were crucial to ensure adequate peripheral anchorage of the geosynthetic reinforcement. The reinforcement grade was determined using limit equilibrium analysis. This considered the in-situ soil properties and the placement of the 500mm gravel layer as well as the construction traffic loading acting as a uniformly distributed load (UDL) across the surface of the gravel. GGU slope stability software was used for the analysis, a typical graphical output form the software is shown. (Figure 6).

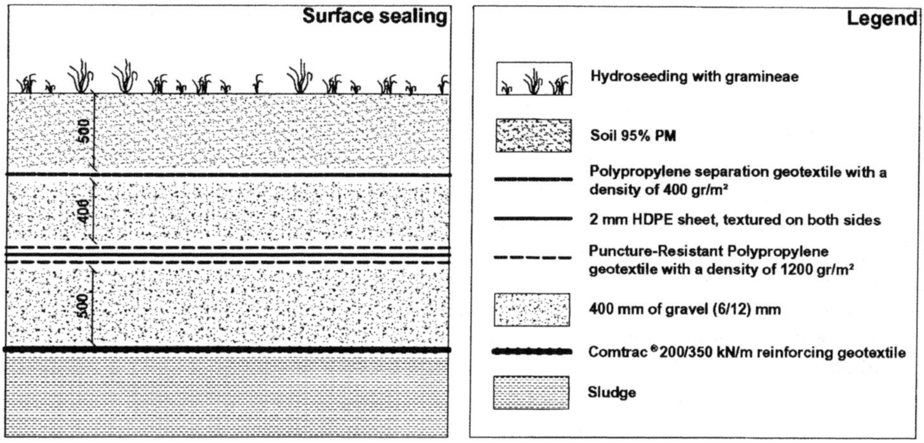

Figure 7: Multi-layered capping system for 'Viscofan' Lagoon

The adopted lining system is shown above, (Figure 7). The geosynthetic reinforcement was provided in three tailor made panels with lengths to meet the exact dimensions of the lagoon. The panels consisted of 5m wide sections of a factory sewn geocomposite. This geocomposite was produced to meet the mechanical and hydraulic requirements. The geocomposite has an ultimate tensile strength in longitudinal direction equivalent to 200 kN/m and in cross direction equivalent to 350 kN/m. This higher tensile strength of 350 kN/m in cross direction compensates for the lower strength and higher strain in the sewn joints of parallel rolls. This result is a nearly homogeneous continuous biaxial reinforcement (200/200) over the entire area. The first and largest panel was installed lengthways across the lagoon extending from the natural slope to the periphery dyke. The second and third panels were then orientated at an approximately 60° angle to the first panel extending a distance of 10 metres over the first panel. (See Figure 8).

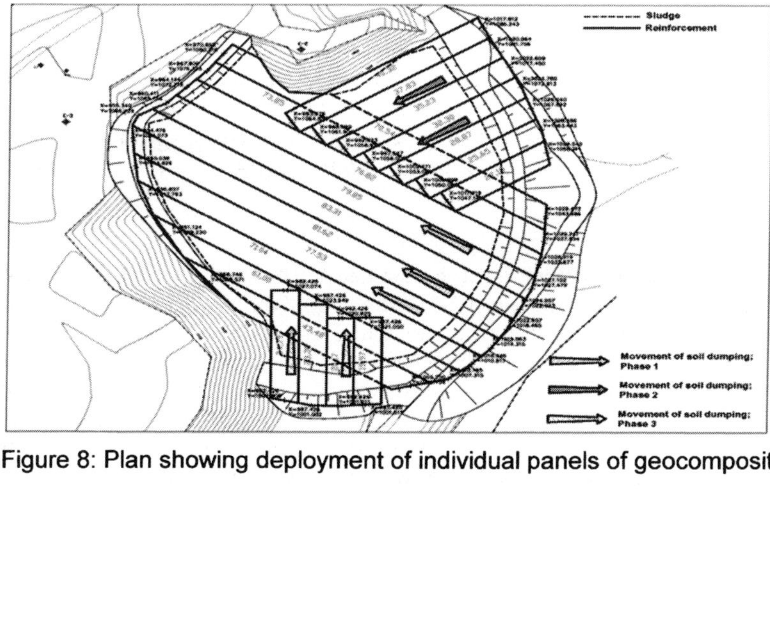

Figure 8: Plan showing deployment of individual panels of geocomposite

Anchorage was provided around the periphery of the lagoon, by extending the geocomposite across the crest of the dyke and continuing 3m down the down-stream face. Subsequently covering with 1.5 m depth of fill across the dyke crest and 1m depth of fill above the down-stream face, prior to placing the first gravel layer. In order to avoid/limit a progressive sludge wave the gravel was placed in a series of tongues approximately 2.5 m installed approximately 4m apart extending across the full width of the lagoon, once the tongues were installed the remaining fill was placed between the 'tongues', See Figure 9 Once the first gravel layers had been placed the fill depth across the periphery dyke was reduced by 1m level with the surface of the gravel and the subsequent geosynthetics installed.

Concluding remarks

Multi-layered capping systems consisting of different soils and specialised geosynthetics proved to be the optimal solution for sludge lagoons and site remediation. They can successfully combine barrier and bearing functions even over very soft subsoils.

Different measures including complete geosynthetic covers sewn together in large prefabricated panels, the use of non-conventional low bearing pressure construction plant and utilization of ground improvement techniques have all been successfully adopted to provide a suitable bearing layer on which to install a capping/covering layer over very soft soils. The common denominator in each of the reported projects is that the primary geosynthetic reinforcement was produced and supplied by Huesker Synthetic GmbH or its sister companies. The authors would like to acknowledge the efforts of everybody who took part in the projects described.

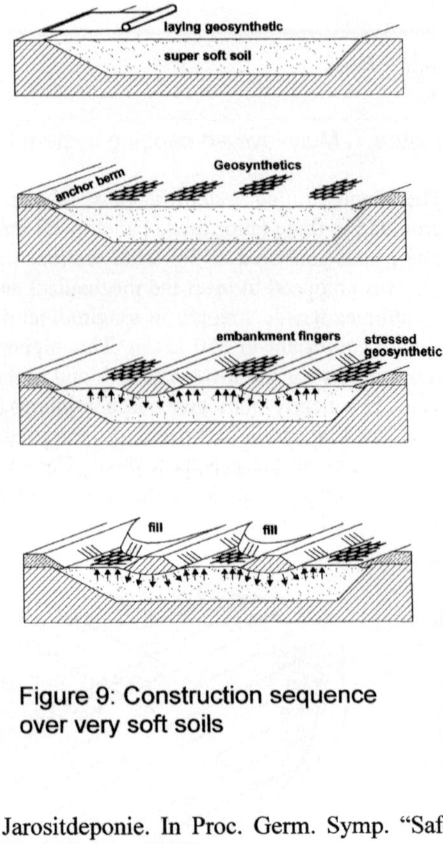

Figure 9: Construction sequence over very soft soils

References

Nüske, Cl. (1995). Oberflächendichtung einer Jarositdeponie. In Proc. Germ. Symp. "Safe waste Disposals", Würzburg, 16-17 February 1995. Würzburg, SKZ

Erosional Stability of Tailings Storage Facilities in an Arid Climate

David J. Williams, The University of Queensland, Brisbane, Australia
Daniel J. Stolberg, The University of Queensland, Brisbane, Australia

Abstract
As part of a research project investigating tailings storage facilities (TSFs) in the arid climate of the Eastern Goldfields region of Western Australia, erosion gullies in the outer slopes of TSFs and a natural slope analogue have been monitored for 3 years using high-resolution, digital, stereo-photography. The more pronounced erosion gullies were observed in the outer slopes of disused tailings storage facilities that had been capped to shed runoff, leading to overtopping of the wall at low points. More minor erosion was also observed on topsoil-covered outer slopes, resulting from rainfall incident only on the slope itself. Rock-covered outer slopes and the natural slope analogue appeared to have undergone negligible erosion gullying over the monitoring period.

Introduction
Three different mine operations were selected as test sites for the erosion studies, chosen because of their longer-term future, their geographical spread within the Eastern Goldfields, and to encompass a range of commodities and mine waste types. The test sites included Leinster Nickel Operations (LNO), which is located approximately 400 km north of Kalgoorlie; and St Ives Gold (SIG) and Kambalda Nickel operations (KNO), which are both located approximately 100 km south of Kalgoorlie.

Water-Induced Erosion in Relation to TSF Slopes
A natural watercourse is normally in a state of metastable equilibrium, with the size, shape gradient and surface cover of the channel being for the most part suitable to carry the flows imposed (Hudson, 1981; Williams, 2000). Gully erosion occurs when the flows are out of balance with the capacity of the topography and surface cover to handle them and continues until metastable equilibrium is re-established.

Water-induced erosion of an exposed surface will only occur if the velocity of flow is sufficient to pick up and transport particles, which will necessarily be lower for dispersive materials. For the relatively flat tailings and cover material slopes, the erosion loss will be limited by the ability of the flow to maintain particles in suspension, which is a strong function of slope steepness. This is referred to as transport limited erosion loss. On the other hand, for the steeper slopes of the outer wall of a TSF the erosion loss will be detachment limited; that is, limited by the ability of the flow to detach particles from the slope, as characterised by Figure 1, after Haan, *et al.* (1982), relatively independent of slope steepness.

Western Australian regulatory authorities, along with many of their counterparts in other Australian states, suggest a maximum slope for the outer slope of a TSF wall of 1 vertical to 3

5th ICEG Environmental Geotechnics, Thomas Telford, London 2006.

horizontal (about 18.5° or about half the angle of repose of waste rock) to allow the safe movement of dozers on the slope and to facilitate the topsoiling and revegetation of the face (Williams, 2000). Some TSF outer slopes are constructed using waste rock, with the outer face at the angle of repose of the material, requiring the slope to be flattened. In addition to this expense, most growth media are highly erodable unless a good vegetative cover can be established quickly and be sustainable. In the arid Eastern Goldfields, a vegetative cover is both difficult to establish and difficult to sustain, and any topsoil cover material is both scarce and likely to readily erode (Williams, 2000).

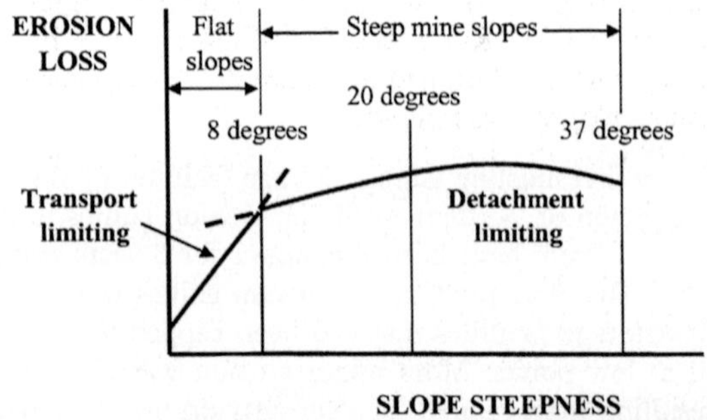

Figure 1: Erosion loss as a function of slope steepness

For a given TSF wall height with a given surface cover, flattening the outer face results in a longer slope and an enlargement of its catchment, leading to greater erosion loss overall. Furthermore, flattening the slope angle tends to crush and bury any coarse-grained particles on the face, increasing both runoff and erosion loss (Williams, 2000). The slope profile will also affect the erosion loss. Concave slope profiles, which are the most common shape found in nature, are inherently more stable than linear or convex slope profiles (Williams, 2000). At Normandy Mining's Wiluna Operation, which is located approximately 200 km north-west of Leinster, one of their TSFs has recently been rehabilitated using a concave slope design (Raynes, 2001). With a concave slope any sediment that erodes off the steeper upper part of the slope profile is potentially caught on the flatter lower part of the slope.

Geotechnical engineers have traditionally considered slope stability only in terms of mass translational or rotational movement of the slope. However, erosion is a very common problem for constructed slopes and must also be accounted for as part of a slope stability assessment (Blight and Costa, 1999; Southcott, 1997). A "non-erodable" cover on the outer slope of a TSF may be difficult to achieve, especially in the arid Eastern Goldfields. However, concentrated runoff down the face of a TSF wall, arising from the ponding of runoff against a low point of the wall or on inter-slope benches, leading to overtopping, should be avoided.

High-Resolution Digital Stereo-Photogrammetry

The traditional means of monitoring erosion off the outer slope of a TSF involves the isolation of an erosion plot and the installation of erosion collection troughs and instrumentation at the base of the plot. This approach faces a number of difficulties including the effective isolation of a section down the slope, the inevitable disturbance that results from the construction of the plot and collection troughs, the collection and removal of the high volumes of sediment involved, and the remoteness of many mine sites.

Photogrammetry is defined as "The art, science and technology of obtaining reliable quantitative information about physical objects and the environment throughout the process of recording, measuring and interpreting photographic images and patterns of radiant energy derived from sensor systems" (Slama, 1980). Stereo-photogrammetry is essentially a process of triangulation of angles measured from a pair of images.

High-resolution digital stereo-photogrammetry may be applied to monitoring erosion gullies on unvegetated TSF and natural slopes. The method described herein revolves around a system developed by CSIRO Division of Mineral Exploration in Brisbane, Australia, including the proprietary three-dimensional imaging computer program SIRO3D and the commercially available programs SURFER and Split-Desktop. The high-resolution digital stereo-photogrammetry approach is illustrated in Figure 2. The application of the method to unvegetated TSF and natural slopes is as follows.

- A baseline for positioning the digital camera is established, either in front of the toe of the slope for monitoring gullies, or at the crest of the slope for monitoring erosion sediment deltas (volume and extent) beyond the toe of the slope.
- A single target is located on the slope.
- The baseline and target are surveyed.
- The photographs are processed using the survey information.
- A three-dimensional image is produced using SIRO3D.
- The three-dimensional image is imported to a topographic program, such as SURFER, from which the volume of the erosion gully, or erosion sediment delta, can be determined.
- Subsequent photographs taken of the same slope, from the same baseline, enable the erosion over time to be monitored.
- Close-up photographs of the surface texture of the slope may be processed using Split-Desktop to give an estimate of the particle size distribution of the surface particles.

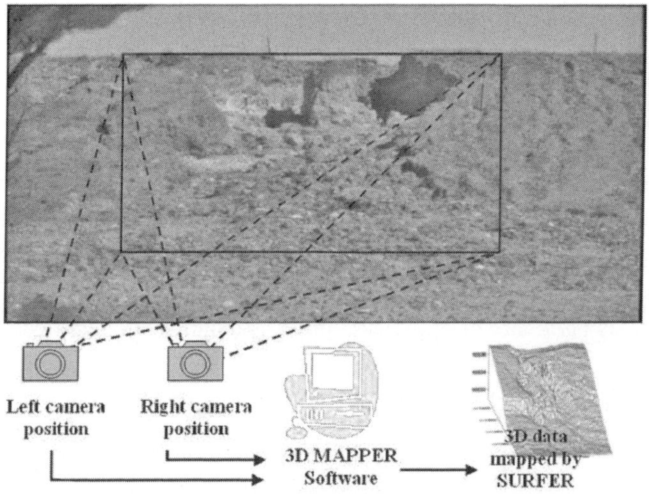

Figure 2: Application of high-resolution digital stereo-photogrammetry to monitoring an erosion gully on an unvegetated TSF slope

Processing of Digital Images

The stereo-photographs of the erosion gullies were processed using SIRO3D to produce three-dimensional images, from which x, y, z coordinates were derived. The coordinates were then imported into the commercially available contouring and three-dimensional surface mapping computer program SURFER (Golden Software, 1999) to enable estimates to be made of the total amount of erosion that the gullies had experienced.

The gully erosion volumes calculated from the processing of the first pass (July 2001) images are necessarily approximate only, since the as-constructed or original slope profile and age of the slopes were not known precisely. The original slopes were assumed to be planar. A comparison was also made between the estimated gully erosion loss from the outer slopes of various TSFs and that from a steep natural eroding slope, known as the "Breakaways" (located on the eastern perimeter of Leinster Nickel Operation), to place the two in perspective. The estimated total erosion losses for all gullies investigated in July 2001 are presented in Table 1.

Table 1: Estimated gully erosion to July 2001

Site/TSF/ Location	Sediment loss (m^3)	Slope length (m)	Slope height (m)	Average width (m)	Gully area (m^2)	Slope angle (°)	Normalised Sediment Loss (m^3/m^2)
LNO/1/North	21.6	12.9	3.9	2.2	28.4	17.6	**0.76**
LNO/2/East	139	25.0	6.1	6.0	150.0	14.1	**0.93**
LNO/2/South	391	35.3	5.0	5.8	204.7	8.1	**1.91**
Breakaways	1315	43.3	10.3	13.2	571.6	13.7	**2.30**
SIG/1/North	258	23.0	10.2	14.0	322.0	26.3	**0.80**
SIG/2/SE-major	57.5	25.8	7.0	1.4	36.1	15.8	**1.59**
SIG/2/SE-minor	28.9	24.3	6.3	0.5	12.1	15.0	**2.39**
KNO/2/East	35.5	11.7	6.0	2.5	29.3	31.0	**1.21**
KNO/2/NE	76.7	10.7	6.3	4.6	49.2	36.2	**1.56**
KNO/2/S *	-	-	-	-	-	-	-
KNO/2/SW	52.9	7.6	4.2	4.7	35.7	33.7	**1.48**

* These images could not be processed due to obscured tunnelling of the tailings at this location.

Figures 3 to 4 show the SURFER wireframe plots (produced from the SIRO3D data points) and the corresponding corrected (for distortion) left-hand images for two of the erosion gullies studied; on the mid north face of TSF 1 at Leinster, and on the north-east corner of TSF 2 at Kambalda.

Verification Using Total Station

A Leica Total Station, together with the computer program FACE SCANNING, was used to survey the gully on the mid-north face of TSF 1 at Leinster Nickel Operation. The survey results were imported into SURFER to enable the volume of the gully to be calculated. This gave an estimated volume of 30.3 m^3, compared with the photogrammetric estimate of 21.6 m^3. This result was considered acceptable given that a total station interpolates between larger intervals (that is, fewer points are surveyed per equivalent area), and therefore will in theory give a less accurate result to the method of high-resolution digital stereo photogrammetry.

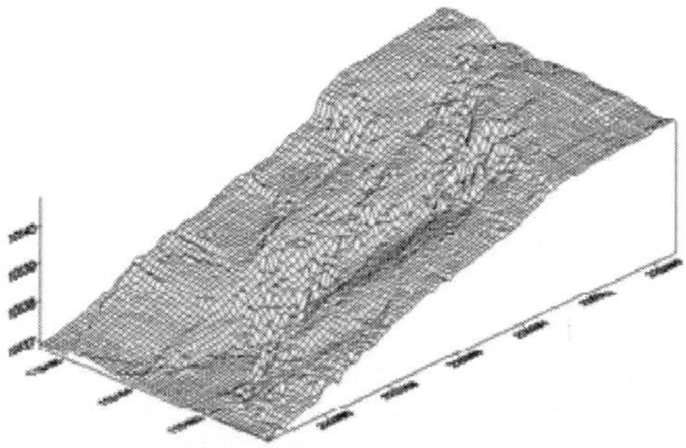

Figure 3: LNO TSF 1 (mid-north face) - SURFER wireframe map and corresponding left-hand corrected image (July 2001)

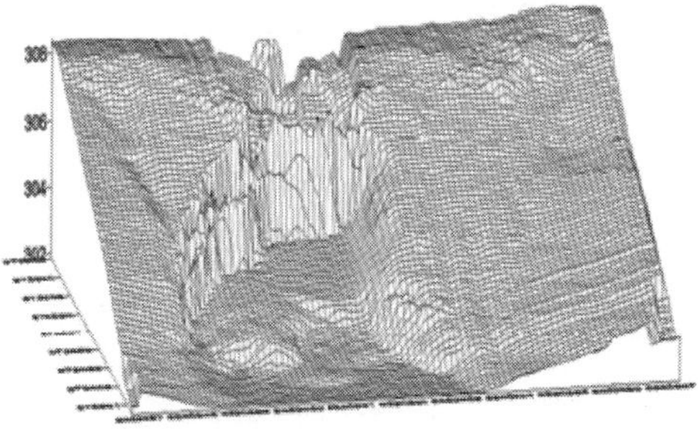

Figure 4: KNO TSF 2 (north-east corner) - SURFER wireframe map and corresponding left-hand corrected image (July 2001)

Discussion of Results and Implications for TSF Closure

In Table 1, an attempt was made to estimate the average depth to which each gully had eroded. For the TSF outer slopes, the gullying depths ranged from 0.76 to 2.30 m, with an average of 1.31 m. There appear to be no consistent differences in gully depth between the sites, despite the fact that the covers applied to the TSF outer slopes varied from predominantly steeply-sloped tailings with some waste rock at Kambalda, a very thin (50 mm) cover of gravel directly over the tailings on TSF 1 at Leinster, topsoiled waste rock at St Ives' TSF 2, weathered waste rock at Leinster's TSF 2, to durable waste rock at St Ives Gold's TSF 1, and the slopes are of varying age. However, once a gully has formed the eroded cover would not be expected to provide any further protection except against any potential widening of the gully.

For the natural Breakaways, the average gullying depth was calculated to be 2.30 m. However, this slope is believed to be thousands of years old, and the slope is curved, implying that assuming an initially planar slope was very subjective. The estimated average gully depths for the TSF and natural slopes appear to be of a similar order, notwithstanding the differences in the age of the slopes, and differences in the sizes of the catchments feeding the slopes, the geometries of the slopes, and the erodability of the face materials.

Table 2 shows the cumulative results obtained from the five monitoring times over the 3-year period from July 2001 to June 2004, while Table 3 shows the normalised rate of erosion loss in m^3/m^2 of area/year (or annualised gully deepening in m/year). The Breakaways showed negligible net erosion loss over this period. The variable cumulative erosion losses for the Breakaways are most likely due to inaccuracies in the interpretation of the photographs (attributed to the presence of vegetation and the subjective assumption of an initially planar slope), and suggest that the method is accurate to ±10% for the Breakaways.

The St Ives Gold Mine durable rock slope (the north face of TSF 1) would be expected to show negligible erosion loss over the monitoring period, and the measured erosion losses suggest an over-prediction of between 0% and 25%, attributable to the assumption of an initially planar slope and complicated by the presence of a partial bench mid-slope. At Leinster, the average rate of gully deepening (erosion loss) over the monitoring period was 0.068 to 0.115 m/year. At Kambalda, the average rate of gully deepening over the monitoring period was 0.075 m/year, on a steep, unprotected tailings face.

Table 2: Estimated gully erosion from July 2001 to June 2004

Site/TSF/ Location	**Cumulative erosion loss (m^3)**					**Erosion loss between investigations (m^3)**			
	July 2001	May 2002	Nov 2002	Oct 2003	Jun 2004	Jul 01 to May 02	May 02 to Nov 02	Nov 02 to Oct 03	Oct 03 to Jun 04
LNO/1/North	21.6	25.1	26.4	28.9	31.0	3.5	1.3	2.5	2.1
LNO/2/East	139.0	153.0	161.2	179.7	190.8	14.0	8.2	18.5	11.1
LNO/2/South	391.0	411.0	348.7	421.5	432.8	20.0	~0	10.5	11.3
Breakaways	1315	1205	1172	1299	1289	~0	~0	~0	~0
SIG/1/North	257.5	358.0	342.5	355.0	356.5	~0	~0	~0	~0
SIG/2/SE-major	57.5	N/A	N/A	N/A	N/A	N/A	N/A	N/A	N/A
SIG/2/SE-minor	28.9	N/A	N/A	N/A	N/A	N/A	N/A	N/A	N/A
KNO/2/NE	76.7	81.4	82.9	85.7	87.7	4.7	1.5	2.8	2.0

Table 3: Normalised rate of gully erosion from July 2001 to June 2004

Site/TSF/Location	**Normalised rate of erosion loss (m^3/m^2/year)**				
	Jul 01 to May 02	May 02 to Nov 02	Nov 02 to Oct 03	Oct 03 to Jun 04	**Average** (Jul 01 to Jun 04)
LNO/1/North	0.137	0.097	0.087	0.109	**0.110**
LNO/2/East	0.106	0.117	0.124	0.110	**0.115**
LNO/2/South	0.116	~0	0.054	0.082	**0.068**
Average for Leinster	**0.120**	**0.071**	**0.088**	**0/100**	**0.098**
Breakaways	~0	~0	~0	~0	**~0**
SIG/1/North	~0	~0	~0	~0	**~0**
SIG/2/SE-major	N/A	N/A	N/A	N/A	N/A
SIG/2/SE-minor	N/A	N/A	N/A	N/A	N/A
KNO/2/NE	0.096	0.058	0.051	0.060	**0.075**

The relationship between TSF outer slope gully deepening and rainfall was also investigated, however no relationship was found other than that it appears that a certain minimum amount and/or intensity of rainfall is required to initiate erosion. The average annual rainfall for the arid Eastern Goldfields region is about 257 mm, with significant variations due to periodic droughts and heavy rainfall associated with occasional cyclonic influences to the north. While the rainfall is well distributed throughout the year, it occurs mainly in intense events of short duration. The region experiences average maximum and minimum temperatures of 25°C and 11°C, respectively, and an average annual pan evaporation of 2,400 mm, with actual evapotranspiration taking up most of the annual rainfall.

The research outcomes suggest the need for TSF closure strategies in arid climates vastly different to "conventional" mine site practice and regulatory guidelines, which in the past have promoted the shedding of rainfall runoff from TSFs. Decants must effectively be closed down to prevent direct seepage to the foundation. Shedding rainfall runoff from the tops of TSFs inevitably leads to the concentrated overtopping of the outer slopes, causing erosion gullying. In an arid climate, rainfall incident on the top of a TSF may be removed through evaporation, and/or transpiration if revegetation is possible, provided that the rainfall runoff is spread over as large an area as possible.

Constructing concave-shaped TSF outer slopes that mimic natural slopes (rather than the conventional uniform, flatter slopes, often incorporating drainage benches that concentrate runoff) has the potential to substantially reduce erosion loss off the TSF outer slopes (Loch, 2002; Williams and Stolberg, 2004). Such slopes can be as steep as the angle of repose over the upper section if durable, inert rock is used, as supported by natural analogues. The "conventional" approach of a "topsoil" cover lacks the coarse elements necessary to resist erosion in the likely absence of sufficient vegetation cover due to the arid climate.

Conclusion

Erosion gullies in the outer slopes of TSFs and a natural analogue in the arid Eastern Goldfields region of Western Australia have been monitored over a period of approximately 3 years using high-resolution, digital, stereo-photogrammetry. The more pronounced erosion gullies were observed in the outer slopes of disused TSFs that had been capped to shed rainfall runoff, leading to overtopping of the walls at low points. More minor erosion was observed on topsoil-covered outer slopes, resulting from rainfall incident only on the slope itself. A rock-covered TSF outer slope and a natural slope analogue appeared to have undergone little erosion gullying over the 3-year monitoring period.

High-resolution digital stereo-photogrammetry proved to be an appropriate method for monitoring erosion loss off the outer slopes of TSFs and natural slopes in an arid climate, provided that the gullies were not shadowed by overhanging material or vegetation. The method could also be applied to monitoring erosion of waste rock dump slopes and the degradation over time of disused open pit slopes. It could also be applied to larger-scale aerial stereo-photographs taken from an elevation of about 1,000 m, which would capture an entire TSF, waste rock dump, open pit, or small mine site.

The research outcomes confirm that arid climate TSF closure strategies based on shedding rainfall runoff, and flattened and topsoiled outer slopes, lead to concentrated overtopping and the loss of fines offsite. Rather, rainfall runoff should be removed from the tops of TSFs by evapotranspiration, and concave-shaped outer slopes protected against erosion by coarse-grained, durable material should be employed.

The "conventional" approach of a "topsoil" cover on a flattened slope lacks the coarse elements to resist erosion in the (likely) absence of sufficient vegetation cover in an arid climate, and nature will "correct" this by eroding the excess fines offsite. Durable waste rock at its angle of repose lacks the fines to retain moisture and is unsuitable as a growth medium, however, it does not erode offsite. Gradually over time, nature will break down the rock producing fines that fill the voids, eventually retaining moisture and nutrients to support vegetation.

Acknowledgements

The Australian Research Council and WMC Resources Limited are gratefully acknowledged for funding the research on which this paper is based. The management and staff of WMC Resources Limited are also acknowledged for assisting with the work described and for giving their permission for the work to be published.

References

Blight, G.E. and Costa, F.A. (1999). In search of the 1000-year tailings dam slope. *Civil Engineering: October 1999*, 15-21.

Golden Software (1999). *Surfer (ver. 7) - User's Guide. Contouring and 3-D Surface Mapping for Scientists and Engineers*. Golden Software Inc. Colorado, USA.

Haan, C.T., Barfield, B.J. and Hayes, J.C. (1994). *Design Hydrology and Sedimentology for Small Catchments*. Academic Press, San Diego.

Hudson, N. (1981). *Soil Conservation*. Second Edition. Cornell University Press.

Loch, R.J. (2002). *Private communication.*

Raynes, J. (2001). *Directors Awards Wiluna Environmental Submission - Rehabilitation of Tailings Storage Facility "C" Using the Concave Slope Design*. Normandy Mining Limited.

Slama, C.C. (1980). *Manual of Photogrammetry*. Fourth Edition. American Society of Photogrammetry.

Southcott, P.H. (1997). Slope stability and soil erosion. *GeoEnvironment 97, Melbourne, Australia, 26-28 November 1997*, 481-486, A. Bouazza, J. Kodikara and R.J. Parker (eds). A.A. Balkema, Rotterdam.

Williams, D.J. (1996). Lateral thinking on mine site rehabilitation. *Proceedings of National Environmental Law Association Conference. Coolum, Australia, 8-12 May 1996, 12.2.1-12.2.9*, S. Blain and T. Slater (eds). National Environmental Law Association, Canberra.

Williams, D.J. (2000). *Predicting Long-Term Seepage and Runoff from Tailings Facilities to Facilitate Cost-Effective and Timely Closure and Lease Surrender - Proposed Study Scope.* Australian Research Council in collaboration with WMC Resources Limited, July 2000.

Williams, D.J. and Stolberg, D.J. (2004). Erosional stability of tailings storage facilities in an arid climate. *Tailings and Mine Waste '04, Vail, Colorado, 10-13 October 2004*, 55-61. A.A. Balkema, Leiden.

Alternative Design and Instrumentation of Covers Over Potentially Acid Forming Mine Wastes

David J. Williams, The University of Queensland, Brisbane, Australia
Timothy K Rohde, The University of Queensland, Brisbane, Australia
Daniel J. Stolberg, The University of Queensland, Brisbane, Australia
Graham Pope, The University of Queensland, Brisbane, Australia

Abstract
The design of earthen covers for potentially acid forming mine wastes is site specific, largely dictated by climatic conditions, the materials available for cover construction, and the expectations of stakeholders. The "store/release" cover system is the preferred approach in semi-arid conditions, such as those that exist over much of Australia. It typically comprises a compacted clayey sealing layer overlain by a vegetated, loose "rocky soil mulch". As an alternative, a thick mono-layer of overburden may be used. The performance of such a cover system in a coastal, sub-tropical, seasonal climate is being monitored using time domain reflectrometry to measure volumetric water content, thermocouple sensors to measure soil suction, and lysimeters to measure percolation rates at the cover/waste interface and the waste/foundation interface. The paper describes the site setting, laboratory and field testing of cover and waste materials, the instrumentation of the cover system, and preliminary monitoring data illustrating the successful performance of the trial cover to date.

Introduction
The "store/release" cover system for potentially acid forming mine wastes was first developed in Australia (Williams *et al.*, 1997) to manage acid rock drainage from mineralised waste rock dumps at Kidston Gold Mines' open pit operations in the semi-arid, seasonal, sub-tropical climate of North Queensland, Australia, where it has been monitored for the past 9 years. In Kidston's highly seasonal and variable semi-arid climate, a rainfall-shedding cover would not be a sustainable means of minimising rainfall percolation into the mineralised waste rock. During the long dry season, the oxide waste rock available for cover construction would desiccate leading to vegetation die-back. The subsequent summer storms would then erode the desiccated, poorly vegetated surface, with the likelihood of breaking through the cover. Kidston's store/release cover comprised a 0.5 m thick compacted clayey oxide waste rock sealing layer overlain by a minimum 1.5 m thick rocky soil mulch layer.

The store/release cover system has the following aims.

- To provide a cover with sufficient water storage capacity to "store" the bulk of the short, summer, wet season rainfall, limiting percolation through the cover to < 5% of annual rainfall.
- To "release" the stored water through evapotranspiration during the long dry season, while maintaining the compacted clayey layer at the base of the cover moist to preserve its integrity. The vegetative cover plays a key role in ensuring that this aim is met.

- To provide a cover system that cycles annually between wet and dry states, without progressively wetting up or drying out.

In coastal Central Queensland, Australia, the somewhat wetter, less evaporative climate, the soil-like cover construction materials available, and the closer proximity to centres of population influence the selection of strategies for covering potentially acid forming mine wastes. A 6 m thick mono-layer store/release cover comprising uncompacted overburden materials is being trialled at the Stuart Oil Shale Project in this region. Although no compaction was applied to the layer immediately overlying the mine wastes, the height of overburden above effects some compaction, and the greater thickness provides high water storage capacity to limit breakthrough.

Site, Climatic and Geological Settings

The Stuart Oil Shale Project is located about 15 km north of Gladstone on the Central Queensland coast. The site is located on a level to undulating coastal plain of low elevation (0 to 40 m Australian Height Datum [AHD]). The coastal plain rises gradually to the west towards undulating to rolling low hills of about 50 m relief, which abut the rolling to steep hills of Captain's Mountain and the Mount Larcom range complex (rising to 280 to 450 m AHD). The site is covered by pasture and native bush, and has been grazed by cattle.

The climatic setting of the site is sub-tropical, with annual rainfall typically varying between 600 mm and 1,400 mm, and an average annual rainfall of about 980 mm. The driest year on record experienced about 350 mm of rainfall, while the wettest year on record experienced about 2,200 mm of rainfall. Typically, rain falls in each month, with the highest falls in January and February. The average annual pan evaporation is about 1,700 mm, about 1.7 times the average annual rainfall. Average daily temperatures range from a maximum of 22 to 30°C to a minimum of 11 to 22°C. For most of the year winds are predominantly onshore (easterly), but there is typically a variation in wind speed and direction on a daily basis.

The Stuart Oil Shale deposit forms part of the preserved Tertiary sediment fill of the Narrows Graben, which is about 40 km long (aligned with the coast) and 5 km wide. The sequence is fault-bounded to the east and west. The strata comprising the infill sediments are up to 1,000 m thick, at least 400 m of which is oil shale. The strata comprise oil shale, claystone, carbonaceous oil shale and lignite, siltstone, sandstone and minor conglomerate. The overburden layers include Quaternary alluvium and colluvium, with some gravels; white clay (T1) and white/orange clay (T2); and partially weathered shale (T3). These materials cover oil shale, including waste shale and claystone of the Rundle Formation, and Curlew Formation (in places). The oil shale extends to depth of between 50 m and 190 m in the upper Kerosene Creek Member of the Rundle Formation, which is the focus of this study.

The mining of Stage 1 of the project commenced in July 1998. The Stage 1 Pit reached a depth of about 50 m by the end of 2004, producing about 11 Mm^3 of mine wastes, including processed shale, that are contained in an Ex-Pit Dump located to the west of the pit. It is proposed to rehabilitate the Ex-Pit Dump by placing a 2 to 6 m thick cover of inert overburden to encapsulate the processed shale, with a minimum 200 mm thickness of topsoil placed on the surface and revegetated with native trees.

Testing and Monitoring Program

The program undertaken to assist in the design and performance monitoring of the proposed cover for the Ex-Pit Dump for Stage 1 of the Stuart Oil Shale Project included the testing of selected samples recovered from the 14 boreholes drilled between August and December 2004 to further characterise the resource, the testing of surficial cover materials and processed shale, and the design and installation of instrumentation to monitor the performance of the placed cover. The testing was carried out in accordance with the test methods in AS 1289.

Testing of Resource Characterisation Borehole Samples

The laboratory testing of selected resource characterisation borehole samples included moisture content and matric suction determinations, Atterberg limits, sieving and full particle size distribution analyses (including specific gravity determinations), density determination, Emerson Class Number, Pinhole testing and saturated hydraulic conductivity testing.

Moisture State, Hydraulic Conductivity and Groundwater

The gravimetric moisture content data showed drier conditions towards the ground surface due to desiccation, approaching 15% at the ground surface, a peak in the moisture content of about 65% towards the base of the weathered shale at about 20 m depth (initially significant inflows into the Stage 1 Pit were observed at this depth on breakthrough), and a near-constant moisture content averaging 27% below that. The measured matric suctions appeared to oscillate with depth, suggesting the presence of perched water tables at about 20 m and 90 m depth, with the permanent groundwater table at about 130 m depth. This was borne out by the calculated degree of saturation, which indicated near-saturated conditions at 20 m and 90 m depth, and below 130 m depth. The matric suction profile with depth was generally consistent with a predominantly (upward) evaporative moisture flux.

The laboratory-determined saturated hydraulic conductivities were very low, ranging from 4×10^{-11} to 30×10^{-11} m/s (1.3 to 9.5 mm/year), the near-surface soils being slightly more permeable than most of the profile. The profile would pass minimal percolation if water were continuously ponded on the surface, and much less if rainfall runoff were drained, and/or the surface remained unsaturated.

The historical groundwater levels (based on standing water levels, SWLs, measured in exploration boreholes) coincided with the base of the weathered shale at 11 to 23 m below the ground surface. The resource characterisation boreholes showed SWLs from 16.5 to 38.7 m below the ground surface. The apparent drop in the water levels was attributed to drawdown on mining Stage 1, and was consistent with model predictions prior to the excavation of the Stage 1 Pit. The high measured SWLs may be due to the inflow of perched groundwater that is unable to drain due to the very low hydraulic conductivity of the ground.

Profile Characterisation

Over the upper 15 m of the profile, the field moisture content was found to coincide with the plastic limit of the materials encountered. Between 15 and 30 m depth, the field moisture content was above the plastic limit of the materials, and below 30 m depth the field moisture content was considerably drier than the plastic limit of the materials.

The near-surface soils were found to be fine-grained, overlying coarse-grained soils to about 20 m depth, with fine-grained soils or clay rocks below that. The measured specific gravities trended generally down with increasing depth, influenced by the lower specific gravity of the oil shale, with high and variable values over the upper 15 m depth, a distinct dip in value at the base of the weathered shale at about 20 m depth, and an average value of about 2.5 below about 60 m depth.

The measured wet and dry densities displayed dips at the base of the weathered shale at about 20 m depth, increases towards the surface due to desiccation effects, and reasonably constant values at depth, with wet and dry densities of about 1.88 t/m^3 and 1.45 t/m^3, respectively.

The sources of cover material are the near-surface Quaternary (Q) and the T2 overburden soils. The Q samples tested were found to be variably erodable, while the T2 samples tested were found to be generally erosion resistant.

Testing of Surficial Cover Materials and Processed Shale

The materials tested included surficial unvegetated new and one-year old, and vegetated, three to four-year old placed cover materials, and processed shale sampled immediately after placement and some time after its placement in the dump. The field testing included field density and field hydraulic conductivity testing. The laboratory testing included moisture content and matric suction determinations, Atterberg limits, full particle size distribution analyses (including specific gravity determinations), and Emerson Class Number, Standard compaction, saturated hydraulic conductivity and Soil Water Characteristic Curve (SWCC) testing. Unsaturated hydraulic conductivity functions were calculated from the saturated hydraulic conductivities and SWCCs.

Field Moisture State

The older, vegetated surficial cover materials were on average considerably wetter than the younger, unvegetated cover materials, and both were considerable drier (by up to two-fold) than the undisturbed surficial materials from which they were sourced. The field densities of the new and one-year old surficial cover materials were considerably higher than those of the older, vegetated cover materials, which would have been ripped, and both were well below the surficial undisturbed values, due to a combination of a lower gravimetric moisture content and a less dense state of packing. The calculated degree of saturation of the surficial cover materials ranged from 13 to 41%, with an average value of 23%, similar to that of the source materials at the ground surface. The degree of saturation of the surficial processed shale ranged from 33 to 50%, with an average value of 44%.

The matric suctions obtained for the surficial cover materials were of a similar order to those obtained for the surficial undisturbed Q and T2 overburden materials from which they were sourced, and are consistent with an evaporative moisture flux.

Material Characterisation

The Atterberg limits of the cover materials were similar to those of the source materials, while the processed shale had much higher Liquid and Plastic Limits (of 71% and 65%, respectively), but a low Plasticity Index (of 6%). The field moisture contents of the surficial cover materials (4.4 to 11%, averaging 7.8%) and processed shale (37.5%) were much lower than their respective Plastic Limits.

The three surficial cover materials had similar, reasonably well-graded, coarse-grained particle size distributions, but with decreasing clay content with increasing age, suggesting that any dispersive fine-grained particles may have been lost on exposure. The surficial cover materials were generally at the coarse-grained end of the range obtained for the Q and T2 overburden source materials, with all of the T2 source curves plotting much finer than the surficial cover materials sampled, suggesting that Q unit materials were the main source of the cover materials sampled and possibly that the finer-grained fractions have eroded from the covers (Figure 1).

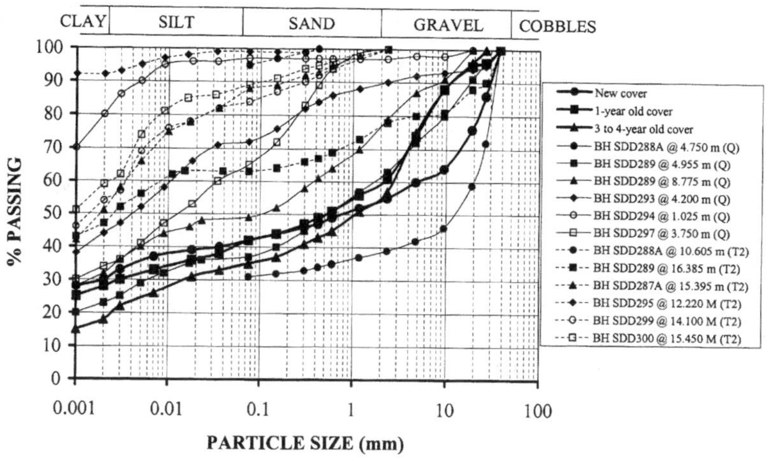

Figure 1: Comparison of particle size distribution curves for surficial cover materials and cover source materials

The surficial cover materials sampled as-placed were either completely erosion resistant or were potentially erodable if disturbed, based on their Emerson Class Numbers; and were less erodable than the previous testing of both Q and T2 overburden source materials obtained from boreholes suggested. The processed shale comprised gravelly, sandy, coarse-grained silt-sized particles, and was completely erosion resistant, based on its Emerson Class Number.

The Standard Maximum Dry Density (MDD) of the surficial cover materials was 1.86 t/m^3 at an Optimum (gravimetric) Moisture Content (OMC) of 14.5%. The field dry density of the trafficked and unvegetated surficial cover materials was in the range from 80 to 87% of the Standard MDD, with an average value of 83% (corresponding to an average dry density of about 1.54 t/m^3). The field dry density of the ripped and vegetated surficial cover materials was in the range from 59 to 63% of the Standard MDD, with an average value of 61% (an average dry density of 1.13 t/m^3). All of the field moisture contents of the surficial cover materials were well dry of the OMC.

The Standard MDD of the processed shale was 0.96 t/m^3 at an OMC of 41%. The field dry density of the trafficked processed shale was in the range from 88 to 95% of the Standard MDD, with an average value of 90% (an average dry density of 0.86 t/m^3), while that of the untrafficked processed shale was only 66% of the Standard MDD. All of the field moisture contents of the processed shale were dry of the OMC.

Hydraulic Conductivities and Soil Water Characteristics

The laboratory and field measures of saturated hydraulic conductivity obtained for undisturbed and disturbed cover materials are summarised on Figure 2. The average measurements of saturated hydraulic conductivity for the cover materials are seen to rise from 8.3×10^{-11} m/s in their undisturbed state (as tested in the laboratory); to 6.5×10^{-8} m/s compacted to 95% Standard MDD (as tested in the laboratory); to 3.7×10^{-7} m/s (as tested in the field on a trafficked, unvegetated); to 1.94×10^{-5} m/s (as tested in the field on a ripped, vegetated surface). The laboratory-determined saturated hydraulic conductivity values obtained for compacted cover materials, which averaged 6.5×10^{-8} m/s, are considered to be representative of the placed cover as a whole, since the minimum 6 m thick cover will be compacted by self-weight at depth.

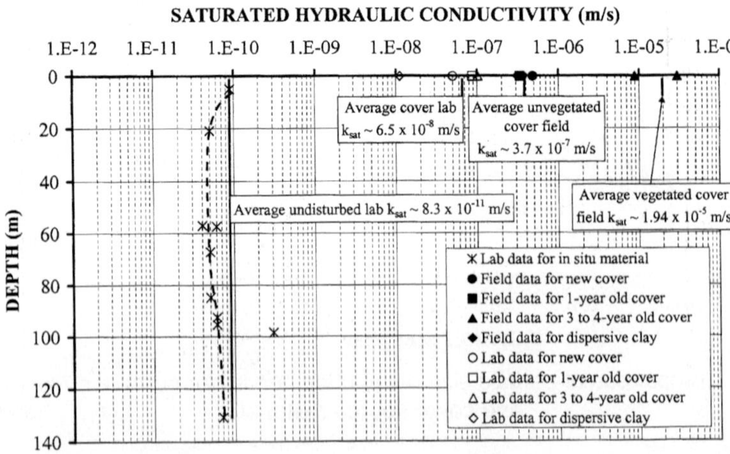

Figure 2: Comparison of laboratory and field measures of saturated hydraulic conductivity for undisturbed and disturbed cover materials

Laboratory Soil Water Characteristic Curves (SWCCs) were obtained for the as-placed cover materials using Tempe cells. Saturation of the samples largely destroyed any soil structure. The drying SWCCs are shown on Figure 3, fitted using the method of Fredlund *et al.* (1997) and the program *SoilVision*, with the vertical axis converted to gravimetric moisture content. The low Air Entry Values (AEVs) indicate that the cover materials would have a capillary rise in the range from 0.3 to 1 m. The flat post-AEV slopes of the SWCCs indicate that they have a large storage capacity for water.

Also shown on Figure 3 are the field SWCC data for undisturbed Q and T2 overburden materials, for the surficial cover materials, and from the instrumented cover. The data for the undisturbed overburden and surficial cover materials plot to the right (higher suction for a given moisture content) of the laboratory SWCC data. This can be explained by the well-developed structure of the undisturbed overburden materials. The dry surficial cover materials, while having a poorly-developed structure, exist at low moisture content and hence high matric suction, due to desiccation effects. The limited data obtained to date from the instrumented cover suggest that the recently-constructed cover is yet to develop structure, and is currently existing at relatively high moisture content and moderate matric suction. Over time, the cover material would be expected to develop structure due to seasonal wetting and drying cycles, with an accompanying increase in matric suction.

The cover materials are likely over time to develop matric suctions in the range from about 1,000 to 5,000 kPa over the full gravimetric moisture content range up to about 45%. This represents up to between 84% and 96% saturation for dry densities of between 1.56 t/m^3 and 1.77 t/m^3. This matric suction range suggests an unsaturated hydraulic conductivity in the range from 10^{-14} to 10^{-12} m/s (based on unsaturated hydraulic conductivity functions calculated using the method of Fredlund *et al.*, 1994 from the measured saturated hydraulic conductivities and SWCCs), indicating that the cover would be practically impermeable under likely unsaturated field conditions. The 6 m thick cover would typically have no more than about half its pore space filled with water from rainfall infiltration.

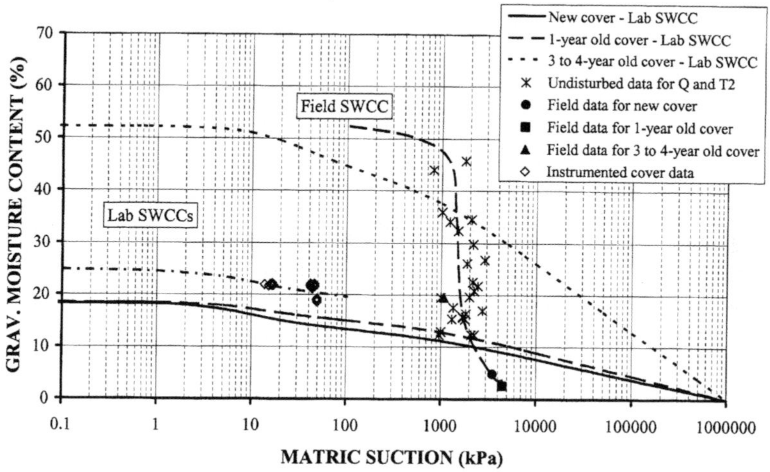

Figure 3: Comparison of laboratory and field SWCC data for undisturbed and disturbed cover materials

Since the characterisation of the cover placed on the dump has been limited to the field and laboratory testing of surficial cover materials only, it will be supplemented by the laboratory testing of samples from geotechnical boreholes through the cover and into the underlying processed shale.

Cover Monitoring

The monitoring equipment installed in the dump and cover included the following.

1. A base (processed shale) lysimeter, gravity draining to an automated tipping bucket, to monitor the seepage rate of any leachate from the placed processed shale, and allow the collection of leachate for possible subsequent chemical analysis.
2. Temperature sensors and gas monitoring tubes in the placed processed shale 10 m and 20 m in from the outside face of the dump.
3. A full weather station to provide the climatic data required for numerical modelling and prediction of dump cover performance.
4. A cover lysimeter, gravity draining to an automated tipping bucket, to monitor percolation rates through the 6 m overburden cover, and allow the collection of percolation for possible subsequent chemical analysis.
5. Moisture and suction sensors within the cover to demonstrate the capability of the cover to store rainfall infiltration and limit oxygen ingress into the underlying processed shale.
6. Erosion plots on a vegetated cover and on an undisturbed vegetated surface (not discussed in detail herein).

Preliminary Results

The processed shale is placed in the dump at a temperature of about 75°C. The temperature monitoring showed a 10 to 25°C drop in the temperature of the processed shale over a 6-month period, the temperature dropping at a faster rate closer to the face of the dump and at a diminishing rate with time. Over this period, the average ambient temperature also dropped from about 28 to 20°C.

The flow into both lysimeters has settled down to < 5% of cumulative incident rainfall. The cover sensors were only commissioned in August 2005, giving a profile of volumetric water content with depth ranging from 0.3 towards the surface to 0.4 at depth corresponds to a gravimetric moisture content range from about 16.5 to 22%. Adopting a field dry density of 1.54 t/m^3, this range of moisture content implies a range of saturation from 60 to 80%, considerably wetter than the average degree of saturation of 23% calculated for the surficial cover materials.

Conclusion

The key findings arising from the work to date are summarised in the following.

- The permanent groundwater table appears to be located at about 130 m depth below the ground surface, with perched water tables at about 20 m and 90 m depth.
- The hydraulic conductivity of the undisturbed ground is very low, which limited groundwater inflows into the pit during mining and would limit potential flows out of the pit post-mining.
- While their low density would result in a significant saturated hydraulic conductivity, the cover materials have a high storage capacity for rainfall infiltration prior to saturated breakthrough, and the trial cover is performing well to date.

Acknowledgements

The authors gratefully acknowledge the management and staff of Queensland Energy Resources Limited for providing access to the data on which this paper is based and for allowing publication of these data.

References

AS 1289 *Testing Soils for Engineering Purposes*. Standards Australia.

Fredlund, D.G., Xing, A. and Huang, S. (1994). Predicting the permeability function for unsaturated soils using the soil water characteristic curve. *Canadian Geotechnical Journal*, **31**, 533-546.

Fredlund, M.D., Fredlund, D.G. and Wilson, G.W. (1997). Prediction of the soil water characteristic curve from grain size distribution and volume mass properties. *Proceedings of 3rd Brazilian Symposium on Unsaturated Soils, Rio de Janeiro, Brazil, 22-25 April 1997*, 12 pp.

Williams, D.J., Wilson, G.W. and Currey, N.A. (1997). A cover system for a potentially acid forming waste rock dump in a dry climate. *Proceedings 4th International Conference on Tailings and Mine Waste '97, Fort Collins, Colorado, 13-17 January 1997*, 231-235. A.A. Balkema, Rotterdam.

Underground mining influence on an environment

Stanis E.V., Russian Peoples' Friendship University, Russia, Moscow

Abstract
This paper presents an attempt to analyze the modern influence of underground mining on geoecological environment. The coal mine, from the point of view of environmental impact, is a complex composed of overground and underground parts, each of them having influence on a biosphere.
The basic directions of negative influence are the following: withdrawal of land from usage and disturbance of lands; exhausting of water resources and disturbance of surface waters regimes; pollution of surface waters; pollution of a surface by wastes of a mining and coal preparation; relief deformation and creation of technogenic landscapes. Detailed monitoring of geoenvironment is necessary to avoid negative influence of underground coal mining industry.

In regions with developed mining industry, there is an essential reduction of resources of the geological environment of a surface part of the lithosphere. The underground method of coal excavation is rather capacious as regards geological space: mining of 1 million t of fuel results in alienation of about 8 hectares of land [1]. On the territory of Russia and Ukraine essential part of coal deposits is situated in agricultural areas with unique fertile soils – "chernozems". Alienation of such grounds without preliminary "evacuation" causes irreparable ecological and economic damage.

In regions of an underground coal mining, the considerable disturbance of territorial resources occurs at the result of subsidence in places of underground workings. The disturbances of a surface as a result of mining operations can be smooth or interrupted. They can originate during mining operations or many years after it.

Interrupted deformations are formed at a non-uniform subsidence of the ground in case of disturbance of surface layer structure. This process can result in hollows, deflections and breaks in breed. The similar phenomena can be connected with destruction of small depth underground coal cameras. Smooth deformations such as subsidence occur without disturbance of surface layer structure. The depth of such subsidence in the Moscow coal basin can reach 2 - 3 m on the area of 150 km^2, in Donbass (Donets Basin) - 5 - 7 m on the area of more than 20 km^2. The time of subsidence formation is different - from 3 to 20 years and sometimes such subsidence may have catastrophic character [2].

Mines in Lenin and Belovsk regions in the western part of Kuzbass are situated not far from the river Inja. In the spring period water rises on 4-5 meters and creates danger of flooding of mines and adjoining territories in places of a surface subsidence as a result of a coal mining. More than 60 km of dams were constructed during last 20 years in Kuzbass for protection of such territories from flooding.

Change of hydro-geological conditions causes the considerable injury of territories resource potential owing to water lowering and mine drainage. Formation of large depression cones can cause active karst and suffusion processes. In Donbass, near the city of Artemovsk, such subsidence has caused serious deformation of buildings; many houses became unsuitable for living. Depression cones can reach enormous size. For example, in Karatau foothills (Kazakhstan), near the city of Kentau, in 1978, in Devonian limestone, there was a karst falling accompanied by a mushroom-like column of a dust. The area of the cone's mouth is more then 1200 m^2 and the depth reaches 50 - 55 m [2].

The formation of colliery dumps of different shapes, sizes and composition will convert a surface and create elements of a technogenic positive relief. Alongside with it the rock constituents naked during the formation of colliery and unloaded in dumps are the reason of other phenomena at a surface part of lithosphere. The dirt of mine dumps is exposed to weathering (oxidation, dissolution, hydrolysis). In some cases such intensive oxidation results in flaring up of dumps.

Two types of dirt are formed as a result of coal mining: common and flotation. The surface transforming of a material of dumps occurs variously, depending on their composition, volume and hydrothermal conditions. Products of aeration and dust and clay material are brought out from the dumps as a result of erosive processes, plane and linear washout.

As the depth of coal mining becomes deeper, the salinization of dirt increases. In Tula region (Central Russia) in some areas the module of washout from dumps changes from 138400 up to 795900 m^3 / km^2 per one year. Carrying out of ions H^+, Fe^{2+}, SO_4^{2-} and others exerts negative influence on soils and productivity of agriculture on neighboring lands [3].

Atmospheric precipitation, after interaction with dumps, is enriched with soluble compounds. Waters brought out from the dumps are characterized by acid reaction (pH 1 - 3), high concentration of a sulfate - ion (up to 30 g/kg), iron (up to 8 g/kg), and high mineralization (up to 50 g/kg). The infiltration of sinks from dumps and mine waters to aeration zone influences chemical composition of underground, and, first of all, ground waters, physical-mechanical and filtration characteristics of grounds, results in a salinization of soils. The waters acquire aggressive nature as regards concrete, the geological processes become more active: suffusion, karst, landslides etc.

All kinds of dump deformations (landslips, taluses, etc.) are ecologically dangerous. The catastrophic landslip in mine dumps in Wales (Great Britain, October, 1966) is widely known. In Silesia coal basin dumps consist mainly of rubble, rotten stone, gravel. In such dumps two types of erosive forms are formed: gullies, cutting dump slopes, and the ruts cutting the borders of slopes. Growth of gullies results in the retreat of slopes at a speed of 1,73 cm/year, and ruts lead to smoothing of slope borders at a speed of 0,9 cm/year.

The accelerated accumulation of substance in channels of superficial water-currents, in ponds and natural reservoirs is connected both with the processes of washout, and with high turbidity of technological waters.

Ravine erosion is connected with an escape of mine waters. It is especially intensive at a high hydraulic gradient in areas with the poor or disturbed vegetation cover. In these regions, formation of ravines is widely distributed and usually has high speed about 10 - 20 m per one year and more.

Deformation of the earth surface results in river channel disturbance along with the formation of the drowned lands. The magnification of hydraulic gradient in regions of surface squatting provokes the erosion by surface outwash and disturbance of natural hydrological equal balance. Decrease of hydraulic gradient of river channel results in the formation of the drowned territories, high water in surroundings and marsh formation.

Water scoop and depressing of water level connected with the last results in a decrease of an underground sink in the rivers and pools. In case of water scoop from the zone of underground waters depression, all superficial sources of water (rivers, lakes, etc.) are exhausted, the level of rivers, lakes and water basins falls, bogs dry up, springs and small rivers disappear.

The water escape of mine and service waters in surface pools, on the contrary, frequently results in considerable increase of water content in strands and rivers.

The escape of mine waters results in strong pollution of superficial (and partly underground) waters. Investigation in Donbass have shown, that the escape of 55 million m^3 of waters from mines in the rivers brings more than 200 000 t of dissolved salts: sulfates about 22 000 t, strong acids (in recalculation on H_2SO_4) - 5 000 t annually. In case of mixture of mine and river waters, pH decreases from 7,9 to 3,6; the mineralization increases in 2 times, and the contents of sulfates - in 3 times, the contents of the weighted particles sharply increases, while transparency decreases [3].

Dumps are also the source of surface and underground waters pollution. Study of chemical composition of soil, ground and superficial waters on the territory of the coal enterprises of Moscow coal basin has shown that the increase in the mineralization of natural waters occurs due to water erosion on dump periphery, and owing to infiltration of acid sulphatic waters through the bottom horizons.

Researches of influence on an environment of waste banks of collieries in Donbass have established that rain waters bring out selenium, cobalt, copper and other heavy metals. Carrying out of selenium from slag is so intensive that 1 g of slag is enough for concentration in 1 l of water which has exceeded allowable concentration [4].

More than 11800 t per day of different salts are brought into rivers by mine waters in Silesia (Poland). Chlorides and sulfates give more than 5900 t per day from this quantity. These polluted waters are brought into the tributaries of Vistula and Oder. Many rivers became actually collectors of mine waters [5].

The agents influencing the soil cover are the mine waters, dumps, other phenomena connected with them.

It is also necessary to note the following kinds of influence upon soils: alienation of the fertile soils in the places of coal mining and destruction of a soil layer, salinization, pollution, development of erosive processes as a result of many factors connected with underground coal mining. Destruction of soils results in losses of unique ecosystems including plant and animal communities.

In conclusion it is necessary to emphasize that underground coal mining influences almost all the components of biosphere. Coordinated work of specialists in different areas of science, industry and nature protection and detailed monitoring of geoenvironment is necessary to avoid negative influence of underground coal mining industry.

Literature.

1. Galperin A.M., Ferster V.V., Shef H.Ju. (1997) Technogenic massives and environmental protection. M., 534 p. (in Russian).

2. Ecological functions of litosphere. (2000) M., Moscow State University, 430 p. (in Russian).

3. Stanis E.V. (2004) Hydroshere changing under effect of underground output coal//Bulletin of Russian Peoples' Friendship University, series Ecology and Life Safety, 2004, № 1(10), pp.99 – 104 (in Russian).

4. Gribanova L.P. (1990) Estimating of geological environment tolerance to the waste polygons influence// Prospecting and protection of entrails, № 3, pp. 29-33 (in Russian).

5. Rozhkovski A. (1989) Influence of mining on water environment in Upper Silesia coal basin (Poland)//Influence of mining upon environment. Part 1., M., pp. 98-111(in Russian).

SUSTAINABILITY, REGULATION AND RISK MANAGEMENT

Regulation and Behavior of Silty Clay Liners in Argentina

Gonzalo M. Aiassa Universidad Tecnológica Nacional, Córdoba, Argentina
Marcelo E. Zeballos Universidad Nacional de Córdoba, Córdoba, Argentina
Roberto E. Terzariol Universidad Nacional de Córdoba, Córdoba, Argentina

Abstract
Loess soils cover a large area in the center of Argentina. The Argentinian loess has an open structure made of fine sand and volcanic silt particles connected by clay binders. In its natural state loess can be considered as an unstable soil which develops deformations or collapses when moistened. In the center of Argentina loess soil is used in most geotechnical constructions, including liners for solid waste landfills. In small towns located in loessic areas the treatment of solid waste is made through sanitary landfills. Operation of landfills and seepage control toward watertable depend on the degree of compaction, liner thickness and watertable depth. This paper reports world landfill design regulation and summarizes the current status of hazardous and non-hazardous waste regulations in the center of Argentina. Compacted silty clay soils have been evaluated in laboratory tests between flow simulations in one-dimensional numerical computer program. Approaches of design based on modelling process are presented and proposed as future regulations for local landfill design.

Introduction
Sanitary landfills have been a common method for final waste disposal. The landfills areas are divided in small cells where waste is covered with local soils daily. In the past, sanitary landfills did not work in efficient ways to prevent leachate from waste. The detection of environmental problems have induced multiple research for the development of more appropriate design criteria.

The new technologies that let a close sanitary landfill and produce less environmental problems have more restricted regulation and legislation results. In the majority of the cases, sanitary landfill design is managed for regulation more than technical and scientific conditions. The liner systems of sanitary landfill are different for each country showing variations in the design of criteria used. To compare the design criteria to the local regulations and provide a view of the minimum requirements for liner design, it is appropriate to establish an overview of the main systems applied in different parts of the world.

Geotechnical component design requires an analysis of the hydraulic system behavior. Particularly, it is interesting to know the hydrologic cycle in the evaluation of contaminant migration. The rainfall over landfill can runoff, evaporate, or infiltrate through waste. The infiltrated water can take contaminants or increase its original contaminant levels in the way toward watertable. In front of fails or deficient landfill drainage systems, the leachate can move through liners toward watertable. The liner materials used in the landfill system are compacted clay, mixed clay and geosynthetic materials. Evans (1991) recommended multilayer liner design combining clay and synthetic materials. International research establishes some design parameters (Mitchell and Jaber, 1990; Benson et al., 1994). The adopted solution is conditioned by material available and construction size. Generally, for landfills of small towns, liners construction with local soils is procured and the use of synthetic materials is reduced. In the central part of Argentina, the principal soil belongs to the loess formation. The hydraulic conductivity of soils is structure dependent and presents anisotropy. Loess soils are macroporous in the natural state and undergo strong volumetric changes due to wetting collapse. In these conditions, foundation soil is strongly sensitive to water content change that places the landfill structure at risk.

The engineering properties of the world loess soils, particularly for Argentina, are reported in Rocca (1985). Uses of these soils as liners are a common solution in small and medium towns in the providence of Córdoba. Under these conditions, the soil is transformed in a compacted structure lowly sensitive to water content change. The sanitary landfill construction is not regulated by a clearly defined normative. The extended uses of these soils in geotechnical constructions establish the need for definitions of hydraulic parameters. The characteristics of affected and used soils in landfills construction established an interest in detailed seepage analysis. This work present recommendations for local regulations based on experimental and numerical simulations results about foundation soil and liner interaction.

International regulations

For a long time landfill construction was only a clay compacted liner. An alternative initiated in 1980 was the use of a simple geomembrane layer as liner. In this case the geomembrane was located above the natural foundation soil. Above the geomembrane a gravel layer and drilled pipe were located for remove leachate. A fine soil layer was situated between gravel and waste. Regulations of sanitary landfill liner system can differ in different countries and regions, still starting from the same general premises. The minimum requirements for hazardous waste landfills established the use of a double liner, leachate collection and a leakage detection system. For non-hazardous waste a simple liner and leachate collection system must be used.

The design of a landfill liner system is generally based on either a prescriptive or performance standard. Most of the present regulations around the world belong to the prescriptive design standard type. Only some countries, as Canada and USA, have introduced performance standards as an alternative to minimum requirements for a liner system. The liner design varies both, within and between countries, depending on waste management strategies and practice. Actually most of the modern landfills are constructed with a composite liner system, where a geomembrane is placed over a compacted clay layer (CCL). Leakage rates through well constructed composite liners are far lower than through geomembrane or compacted clay liners alone (Manassero et al., 2000).

Chai and Miura (2002) have reported a comparative analysis between minimum municipal solid waste landfill design requirements from Europe, America (Figure 1) and Japan (Figure 2). The liner systems were compared in terms of their ability to prevent advective and diffusive transport through liners. They concluded that, (a) Japan established systems have a variable behavior, (b) the retardation effect of the soil liner has very important effects on contaminant migration, (c) a geomembrane layer, even with a given number of defects, has a significant effect on reducing the peak concentration of contaminants in the ground beneath a landfill. Manassero et al. (2000) have reported regulations for landfill design in Oceania's countries (Figure 2).

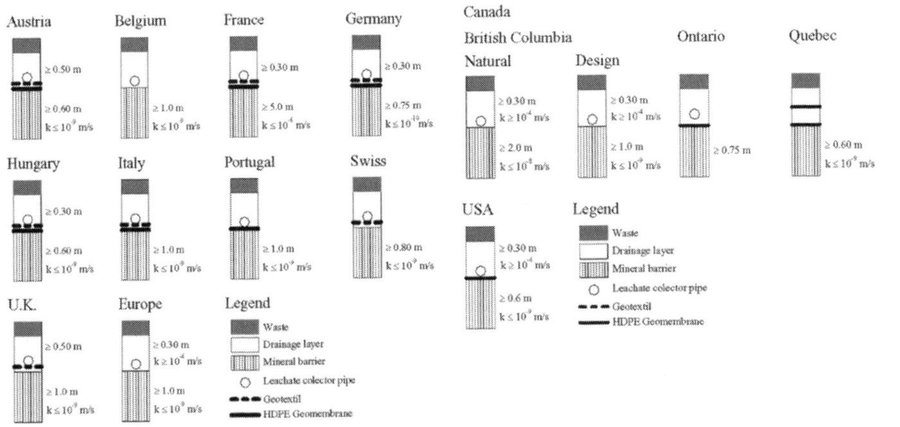

(a) (b)

Figure 1: Liner system for municipal solid waste in (a) Europe, (b) America (Chai and Miura, 2002; Manassero et al., 2000)

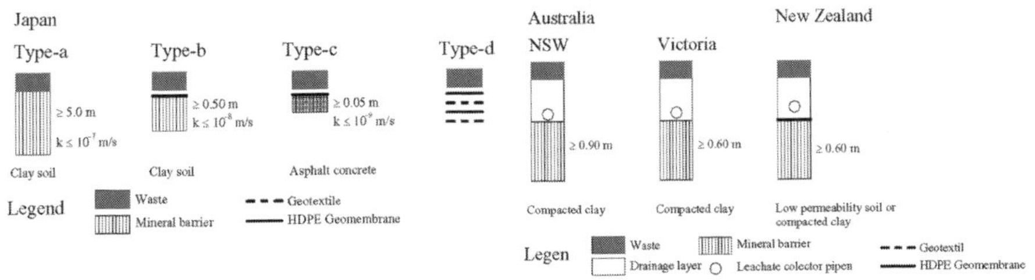

(a) (b)

Figure 2: Liner system for municipal solid waste in (a) Japan, (b) Oceania (Chai and Miura, 2002; Manassero et al., 2000)

Landfills design regulations in Argentina are not defined and justified clearly. The Health and Social Security Minister (1996) proposed instructions for landfill design and constructions. Here a permeability of 10^{-9} m/s is established. If this permeability can not be achieved with compacted soils, a geosynthetic of 250 microns must be used. A minimum requirement for the final cover is a compacted barrier of 0.60 m thickness. Local review shows that liners are single layers of compacted soil determining an environmental contaminant source.

Loess soil description

Loess deposits cover about a tenth part of the continental earth surface. They are usually located in arid and semiarid regions. The Argentinian deposit is one of the most important in the southern hemisphere with a thickness average of 30 meters. In Generally these soils have been transported by winds to the deposition place (Teruggi, 1957). Loess soils located in the central area of Argentina belong to the Pampean formation and cover about 600,000 Km^2. A large part of the providence of Córdoba is covered with loess deposits (Moll and Rocca, 1991). The main constituents of sand and silt fraction are plagioclases, quartz, volcanic glass and feldspars. In clay fractions, the main minerals are montmorillonite and illite. Calcium carbonate is found in two forms; as concretions or precipitated at particle contacts. Loess is alkaline in nature. (Teruggi, 1957; Moll and Rocca, 1991). The principal geotechnical characteristic of natural loess is the sudden volume change due to wetting, loading, or both combined (Moll and Rocca, 1991). Wet and load increase in natural loess can be generated by waste on the liner. The mechanical behavior of Argentinian loess can be explained from its open structure, characterized by silt and fine sand particles connected by clay bridges.

The soil used in the testing program is representative of soil in the central area of Argentina. It is a low plasticity silty clay soil designated as CL-ML in Unified Soil Classification System (USCS). Their properties are summarized in Table 1. Previous studies have been employed for some variable definitions. Saturated hydraulic conductivity for natural and compacted loess soils are contained in Francisca et al. (1998). For natural soil the saturated hydraulic conductivity (k_s) is 1.2×10^{-7} m/s. Saturated conductivity of compacted soils is around 1.5×10^{-9} m/s. Standard Proctor (AASHTO-T99) and optimum water content were used.

Table 1: Geotechnical soil properties

Variable	Unit	Value
Liquid limit	[%]	26.9
Plasticity index	[%]	3.9
Fines (particle < 0.074 mm)	[%]	95
Class (ASTM D-2487-69)		CL-ML
Dry unit weight, γ_d	[kN/m^3]	12.9
Max dry unit weight, $\gamma_{d\,max}$		17.4
Optimal moisture, ω_{op}		17.5
Specific gravity, G_s		2.65

Detailed analyses require knowledge of unsaturated soil behavior. Previous studies (Redolfi, 1990; Blundo et al., 1996; Rinaldi, 2002) have been employed for the definition of the relation suction-water content. Natural and compacted loess soil have been characterized by experimental laboratory tests with the use of a suction control cell and was compared with the data obtained from the Unsaturated soil hydraulic database UNSODA (Leij et al, 1996). Laboratory infiltration tests were done using undisturbed and remolded samples. During the tests, one-dimensional flow was evaluated under constant hydraulic gradient. Cumulative infiltration was measured until a steady state flow condition was achieved. Infiltration tests were conducted in a cylindrical rigid wall cell with an inside diameter of 102 mm and a height of 105 mm (Figure 3). A Similar cell has been employed for clay soils in Wang and Benson (1995). The top plate includes a port for venting gases and a port for inflow. A Mariotte bottle was installed in the inflow line to maintain a constant head. The bottom plate has a central port for outflow. Porous stones at the top and bottom complete the system. Undisturbed samples were cut in the cell. Soil-cell space was sealed to prevent preferential flow. Remolded samples were compacted in the cell. The constant top head applied was 0.15 kPa. The van Genutchen (1980)-Mualem (1976) unsaturated hydraulic conductivity function and soil water characteristic curve equation were used (Table 2).

$$\theta = \theta_r + (\theta_s - \theta_r)\left[1 + (p\psi)^n\right]^{-\left(1-\frac{1}{n}\right)} \tag{1}$$

$$k_\psi = k_s \frac{\left\{1 - (p\psi)^{nm}\left[1 + (p\psi)^n\right]^{-m}\right\}^2}{\left[1 + (p\psi)^n\right]^{0.5m}}; \; m = 1 - \frac{1}{n} \tag{2}$$

where, θ: volumetric water content; ψ: matric suction; k_ψ: unsaturated hydraulic conductivity; m n p: parameters

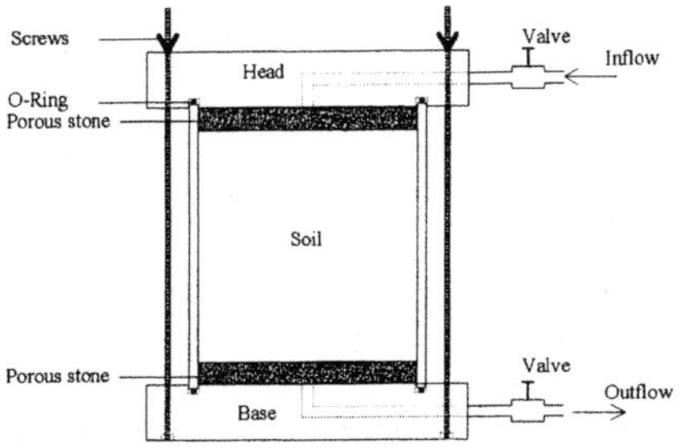

Figure 3: Cell for infiltration tests

Table 2: Soil water and permeability function parameters

Variable	Unit	Natural soil	Compacted soil for 100% T-99
Initial suction	cm	1000	1100
Saturated water content, θ_s	cm^3/cm^3	0.50	0.34
Residual water content, θ_r	cm^3/cm^3	0.10	0.12
Parameter, p	1/cm	0.0018	0.0009
Parameter, n		2.41	1.40
Saturated permeability, k_s	m/s	1.2×10^{-7}	1.5×10^{-9}
	cm/h	0.043	5.4×10^{-4}

Problem model

The case studied is a typical one-dimensional transient flow problem in unsaturated porous media. The problem is governed by Richard's partial differential equation (Richards, 1931),

$$\frac{\partial\theta}{\partial t}=\frac{\partial}{\partial z}\left[k_\psi\frac{\partial(\psi+z)}{\partial z}\right] \tag{3}$$

where, z: coordinate; t: time

In this study the Unsaturated Water and Heat Flow software, UNSAT-H, was employed. It was developed at Pacific Northwest Laboratory in Richland, Washington (Fayer, 2000). UNSAT-H is a one-dimensional model and does not compute lateral drainage. The software solves the equation (3) by a finite difference technique. Input data are soil properties and boundary conditions. Khire et al. (1999) have performed water-balance models on capillary barriers used as final covers of municipal solid waste landfill in semiarid regions. In loess soils, Zeballos and Terzariol (2002) analyze seepage under different compaction conditions and Terzariol et al. (2003) compare modelling with field tests. Results were very sensitive to the saturated hydraulic conductivity (k_s) between other variables. Complementarily, a sensitivity numerical analysis of principal variables has been made. The Range of variables belong to appropriate limits for this soil. Sensitivity analysis have shown that saturated hydraulic conductivity (k_s) affect the infiltration rate considerably. Initial matric suction and soil-water characteristic curve parameters, p and n, have a moderate influence. The constant top head almost does not affect the infiltration curve. This suggests that, for these tests, the matric suction dominates the infiltration process, being significantly more important than the hydrostatic pressure head or gravity head (Wang and Benson, 1995).

Liner behavior

The liner behavior has been analyzed by means of a numerical model. Typical compaction conditions and thickness of 1.0 meter in liners have been considered. The seepage conditions imposed were; a constant head top of 30 cm (3 kPa), a surface saturation and a non-evaporation conditions. A head top of 3 kPa is the maximum acceptable for actual international regulations (Qian et al., 2004). The watertable has been considered in two positions. In condition 1 it is located at 3.0 meters under the bottom liner. Here, a linear variation for matric suction was assumed. Boundary conditions imposed for matric suction were 0 kPa on water table and 110 kPa on natural soil liner interface. In condition 2, the watertable was considered 20 meters under the bottom liner. The variation of matric suction and boundary conditions were considered in a similar way than in condition 1. Evolutions of volumetric water content profile for both conditions are shown in Figure 4. Results are influenced by watertable positions. The infiltration simulation for condition 1 shows that a steady state flow was achieved in 175 days, with a cumulative infiltration of 4.8 centimeter (cm). For condition 2 it was achieved in 590 days with a cumulative infiltration of 14.5 cm. The medium rate of infiltration in condition 1 was 0.027 cm/day and in condition 2 it was 0.025 cm/day. A similar result obtained is attributed to a steady state flow (Table 3).

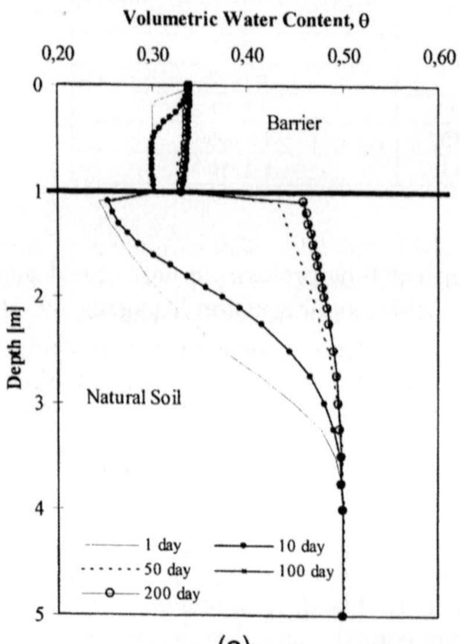

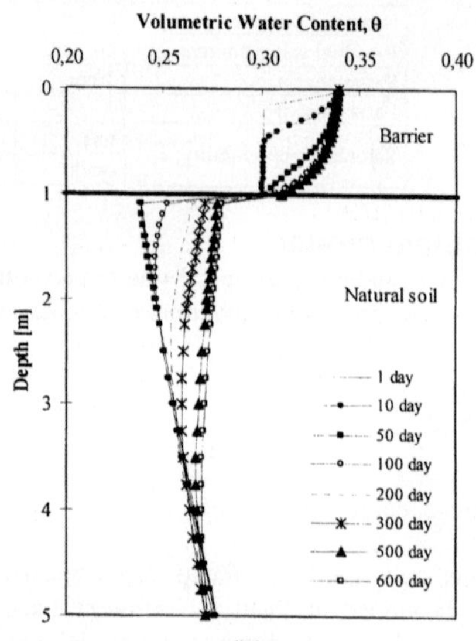

Figure 4: Water content profile for watertable conditions (a) 1, (b) 2

Table 3: Modelation results

Cond.	Average Rate (cm/day)	Cumulative Infiltration (cm)	Stability time (day)
1	0.027	4.8	175
2	0.025	14.5	590

For condition 1 natural soil saturation is achieved in around half a year. This condition combines environmental and collapse problems. Despite this, in condition 2 saturation levels are low, still for steady state flow. Macroporous and collapsible loess characteristic constituted a special topic in conformed models. If foundation soils undergo water content increase, an excessive settlement must happen and the structure can be damage. Hence, case 1 shows that a special caution must be taken into account, even in small constructions.

Results were valued, and compared to these constructive recommendations. Special interesting topics are; (a) a stable soil selection, well graded and medium plasticity, (b) a high compacted unit weight and a specific water content wet of optimum, (c) the use of a high compaction effort in the order of increase the shear strain, (d) The prevented analysis about potential environmental changes in time and its possible effects on the liner, (e) The evaluated compatibility between waste, leachate and clay liner. Considerations (d) and (e) are the most difficult problems. According to these results it is possible to establish recommendations for liners on loess soils as the following,

- An adequate configuration and size.
- A foundation preparation. A high level compaction, even at a lower unit weight than the liner, in order to reduce soil potential collapse.
- A Liner thickness of no less than 1.0 m, and hydraulic conductivity (k_s) no more than 1×10^{-9} m/s.
- A degree of compaction with a dry unit weight larger than 95% of Standar Proctor.
- The verification of the watertable depth and the increase of quality controls levels when it is near land surface.
- The consideration of collapse risk by water content changes. A tenso-deformational analysis must be employed.
- The compaction water content must be established on the wet side of optimum.

Conclusions

A numerical model for loess liner seepage analysis has been calibrated, taking the unsaturated soil mechanics behavior into account. Infiltration tests have been used in the calibration processes by means of comparison between time cumulative infiltration and wetting front advance both in natural and compacted conditions. Simulation results agree reasonably well with laboratory results. This validates the variable identification methods. Infiltration and advance of wetting front in natural and liner loess were evaluated. Modelling results have shown that the watertable has a strong influence on the behavior of these structures. The Watertable close to the liner reduces the time needed to obtain steady state flow conditions and an instantaneous infiltration rate. In this case, special attention of collapse behavior must be considered for high saturation. Water content increase to obtain steady state flow depending on the watertable position. When the watertable is depth, a steady state flow produces a low increase in the water content of natural soil. In this situation the risk of structural collapse of natural soil is reduced. Finally, it is important to consider the accurate control of compaction conditions due to the influences on soil structure and permeability. A deficient compaction produces a higher level of infiltration and a reduction in the time needed to achieve a steady state.

Acknowledgments

The writers like to thank the Department of Civil Engineering, National Technological University of Cordoba and National University of Cordoba. Gonzalo Aiassa thanks the postgraduate grant awarded by the National Technological University.

References

Benson, C. H.; Zhai, H. and Wang, X. 1994. Estimating hydraulic conductivity of compacted clay liners. *J. Geotech. Engrg.*, ASCE, 120(2), 366-387.

Blundo, M.; Redolfi, E. y Zeballos, M. 1996. Estimación de asentamientos inducidos por variaciones en el nivel freático. GT'96 Encuentro de Geotécnicos Argentinos, Córdoba, Argentina.

Chai, J. and Miura, N. 2002. Comparing the performance of landfill liner systems. *Journal of Material Cycles and Waste Management.* 4 (2), 135-142.

Evans, J. C. 1991. Geotechnics of hazardous waste control systems. *Foundation Engineering Handbook.* Second Edition. Edited by Hsai-Yang Fang. Van Nostrand Reinhold. New York. 750-777.

Fayer, M. J. 2000. *Unsaturated soil water and heat flow model.* UNSAT-H, ver. 3.0. U.S. Department of Energy. DE-AC06-76RLO 1830 Pacific Northwest National Laboratory, Richland, Wash.

Francisca, F. M., Cuestas, G. A. y Rinaldi, V. A. 1998. Estudio de permeabilidad en limos loéssicos. GT'98, Encuentro de Geotécnicos Argentinos. Córdoba, Argentina.

Khire, M. B., Benson, C. H. and Bosscher, P. J. 1999. Field data from a capillary barrier and model predictions with UNSAT-H. *J. Geotech. Geoenviron. Eng.*, ASCE, 125(6), 518-527.

Leij, F. J., Alves, W. J., van Genuchten, M. Th. and Williams, J. R. 1996. *Unsaturated soil hydraulic database UNSODA 1.0.* User's Manual. Report EPA/600/R-96/95. U.S. Environmental Protection Agency, Ada, OK 103 p.

Manassero, M.; Benson, C. H. and Bouazza, A. 2000. Solid waste containment systems. *Proceedings of GeoEng2000.* Melbourne, Australia. Technomic Publishing Company, Lancaster, PA, USA, 520-642.

Ministerio de Salud y Seguridad Social 1996. Destino de residuos sólidos urbanos. Manual de líneas operativas para los municipios. Gobierno de Córdoba, Argentina.

Mitchell, J. K. and Jaber, M. 1990. Factors controlling the long-term properties of clay liners. Waste Containment Systems: Construction, Regulations and Perfomance. Geotechnical Special Publication, ASCE (26), 84-105.

Moll, L. L. and Rocca, R. J. 1991. Properties of loess in the center of Argentina. Proc. of the IX Panamerican Conference on Soil Mechanics and Foundation Engineering, Chile, 1, 1-13.

Mualem, Y. 1976. A new model for predicting the hydraulic conductivity of unsaturated porous media. *Water Resources Research*, 12(3), 513-522.

Qian, X., Gray, D. H. and Koerner, R. M. 2004. Estimation of maximum liquid head over landfill barriers. *J. Geotech. Geoenviron. Eng.*, ASCE, 130(5), 488-497.

Redolfi, E. 1990. Comportamiento de pilotes en suelos colapsables. Tesis Doctoral, Universidad Politécnica de Madrid.

Richards, L. A. 1931. Capillary conduction of liquids through porous medium. *J. Physics*, 1, 318-333.

Rinaldi, V. A. 2002. Estudio de succión en suelos loéssicos compactados. Informe de Laboratorio del Área Geotecnia. Universidad Nacional de Córdoba. Argentina

Rocca, R. J. 1985. Review of engineering properties of loess. CE 299 Report. University of California, Berkeley.

Teruggi, M. E. 1957. The Nature and Origin of Argentinian Loess. *J. Sed. Petrol.* 27(3), 322-332

Terzariol, R. E., Redolfi, E., Rocca, R. J. and Zeballos, M. E. 2003. Unsaturated soil model to loessic soils. Proc. of the XII Panam. Conf. Soil Mech. and Geotech. Eng., USA, 1317-1332.

van Genuchten, M. T. 1980. A closed-form equation for predicting the hydraulic conductivity of unsaturated soils. *Soil Science Society of American Journal*. 44, 892-898.

Wang, X. and Benson, C. H. 1995. Infiltration and saturated hydraulic conductivity of compacted clay. *J. Geotech. Engrg.*, ASCE, 121(10), 713-722.

Zeballos, M. E. y Terzariol, R. E. 2002. Modelaciones de flujo unidimensional en suelos loessicos. Proc. of the "XVI Cong. Arg. Mec. de Suelos e Ing. Geot." Patagonia, Argentina.

Recent Developments in Ground Contamination — Basic Thoughts on Regulations

Koichi AKAI, Senior Advisor, Geo-Research Institute, Osaka, Japan
Makoto TANAKA, Senior Researcher, Tokyo Office, Geo-Research Institute, Tokyo, Japan

Abstract
This paper deals with basic thoughts of regulation policy and novel technologies about ground contamination. The authors believe that many techniques are developed depending on the regulation policy. Since each country has its own natural and historical background, it is not always a smart way that a government imports the idea of other governments when it issues a new regulation on environmental pollution. The Japanese government is currently discussing a new regulation on ground contamination by oil since no regulation on that but benzene has been established in Japan. The authors point out the necessary matters to issue such a new regulation, considering their former activity on contamination by oil and a case history. The authors also propose an appropriate application of the ideas of "precautionary principle" and "risk-based approach" to avoid a meaningless regulation on the reflection of Minamata disease.

Introduction
There is no doubt that novel technologies about ground contamination are developed depending on the regulation policy of each government. Recently, many techniques for characterization, analyses and remediation of ground contaminated with heavy metals, volatile organic compounds (VOCs) and oil. The developments of most of these techniques are motivated as they fulfil the performance required in the laws and the regulations by the governments.

In these three decades, each government established many kinds of laws and regulations, according to the developments in researches on ground contamination. These regulations depend on basic thought of each government, where great differences can be seen.

In 2000, the CleanSoil Consortium was organized in Japan. This consortium was a field of technical exchange between Japanese and Dutch researchers, organizations and companies concerning with soil pollution, mainly by oil. The Japanese side of the consortium were consisted of research, technical and steering adjustment divisions. About ten researchers and forty companies joined in the research and the technical divisions respectively. The Dutch side were consisted of seven companies having novel technologies on site characterization, analyses, remediation and risk management. This consortium had acted for two years and played a great role in technical exchange between Japan and the Netherlands. Not only technical relationships between Japanese and Dutch companies have founded but also the difference between Japanese and Dutch thought of regulations of ground contamination was brought out.

In addition, discussions about ground contamination by oil have at last started in Japan. Environmental regulations were discussed usually after a great problem on people's health once appeared, such as Minamata disease (pollution by methyl mercury included in waste water) that killed hundreds of inhabitants. Since there is no great disease or problem has occurred, no regulation on the contamination by oil but benzene has been established in Japan. The authors believe that there must be countermeasures against ground pollution by oil and oil-based substances such as toluene, xylene and polycyclic aromatic hydrocarbons (PAHs) depending on risk-based evaluations.

The authors, who respectively acted as the president and the secretary general of the CleanSoil consortium, report on recent developments in techniques on ground contamination

5th ICEG Environmental Geotechnics, Thomas Telford, London 2006.

by many researchers and companies, from the viewpoint of basic thoughts and policies of regulations on ground contamination by each government.

Histories and Purposes of Japanese Regulations on Ground Contamination

The Japanese government designates air, water and soil pollutions, odour, noise, vibration and ground settlement as the seven typical pollution and establishes environmental quality standards (EQS) for air, water and soil pollutions and noise in order to protect the human health and the living environment. The substantial policies and regulations by the government have started in 1967 establishing *Pollution Measures Fundamental Law*, which was replaced by *Environment Fundamental Law* in 1993 after *Rio Declaration on Environment and Development*, United Nations Conference on Environment and Development in 1992. The Japan Environmental Agency was founded in 1971 in order to play roles on those issues, which was reorganized as the Ministry of Environment.

The government has started acting on the environmental pollution from about 1964 after hundreds of people were killed by and much more suffered from Minamata disease found in 1953 (officially declared by the government in 1956).

In the four greatest pollutions for which EQS has been established, the measures against the soil pollution is the newest and started in 1991. The brief history of the regulations on the soil and groundwater pollution is as follows.

1991 Established, *EQS for Soil Pollutants* (10 items, heavy metals etc.)
1993 Revised, *EQS for Water Pollutants* (items added, intensified for lead and arsenic)
1994 Revised, *EQS for Soil Pollutants* (intensified for lead and arsenic, items added resulting 25)
1994 Established, *Guidelines on Investigation Methods and Measures against Soil Pollution by Heavy Metals and Volatile Organic Compounds (Chlorinated Organics)*
1997 Established, *EQS for Groundwater Pollutants* (23 items)
1997 Management of final waste disposals Enforced by Revised *Waste Management and Public Cleansing Law*
1999 Established, *Guidelines on Investigation Methods and Measures against Soil and Groundwater Pollution*
1999 Revised, *EQS for Groundwater Pollutants* (items added resulting 26)
1999 Established, *Law Concerning Special Measures against Dioxins*
2002 Established, *Soil Pollution Control Measures Law*
2004 Discussions about guidelines on investigation methods and measures against soil pollution by oil
2004 Studies on infiltration behaviour of DNAPL and measures

Each level of EQS has been decided after some analysis evaluating the risk of occurrence of cancers or fatal disease by the toxicity of each pollutant, only considering the worst-case scenario of intake to a human body. This cannot be said the true risk-based analysis. The rules should apply flexibly as their purposes are to protect the human health and the natural and living environment. This is very difficult under the existing circumstances since the levels described in EQS and emission and effluent standards stay so strictly once the rules are established, while the measurements of the nature have statistical variations. Many people may have a false impression to judge dualistically whether an environment is safe or not only seeing some measurement results. There is no doubt that it should not be judged under safe-unsafe dualism and risks should be analyzed. Of course it is very difficult to build a risk-based regulation system considering each risk and scenario. This does not mean that we do not have to make efforts to build such a system.

Recent Developments in Techniques on Ground Contamination

Recently the common contaminants found in ground are heavy metals, VOCs and oil (Domenico and Schwartz, 1990).

Investigation Methods

One of the features of investigation methods developed recently is using cone penetration test (CPT) machine. The CPT probe is connected to the tip cone and the rods and usually measure three components, the tip resistance, the pore water pressure and the sleeve friction. This type of probe can be replaced by or connected to other probe that can detect the contaminants. For example, the energy-dispersive X-ray fluorescence (EDXRF), the gas chromatograph (GC) and the ultraviolet fluorescence (UVF) techniques are used for heavy metals, VOCs and oil respectively (e. g. U. S. Army Engineering R&D Center Environmental Laboratory, 2000).

Countermeasures

The countermeasures against ground contamination can be divided into some groups: containment (walls and seals), withdrawal (pumping, venting and excavation), ex situ treatment (soil washing), in situ chemical treatment, bioremediation (biostimulation, bioaugmentation and phytoremediation) and others.
Novel techniques tend to be developed mainly to clear the EQS. We have to make a judgement that the technique is really "good" or "smart" in each case before applying it to the practical problem. In many cases the risk by the contamination will remain after the countermeasure is performed. Then we have to evaluate the risk concerning with each scenario and think what level of risk is acceptable. Risk-based approach is necessary when we perform any countermeasure against the ground contamination. Risk is also an index for the judgement if a certain technique is appropriate.

CleanSoil Project - Field of Discussions about Contamination by Oil

As described above, no regulation on the contamination by oil but benzene has been established in Japan. There are many discussions about measures against ground pollution by oil. The authors had acted to make a field of discussions and founded a consortium where Japanese and Dutch researchers and engineers joined, called CleanSoil Consortium.
The CleanSoil Consortium was organized in 2000, when the time was ripe for the discussions about ground contamination. This consortium was a field of technical exchange between Japanese and Dutch researchers, organizations and companies concerning with study, investigation and countermeasures on soil pollution, mainly by oil. The whole system of the consortium is shown in Figure 1. The Japanese side of the consortium were consisted of research, technical and steering adjustment divisions. About ten researchers and forty companies joined in the research and the technical divisions respectively. The Dutch side were consisted of seven companies having novel technologies on site characterization, analyses, remediation and risk management. This consortium had acted for two years from April 2000 to April 2002 and played a great role in technical exchange between Japan and the Netherlands.
Not only technical relationships between Japanese and Dutch companies have founded but also the difference between Japanese and Dutch thought of regulations of ground contamination was brought out. The Dutch government has turned its policy about the ground contamination from so to speak "complete remediation" to "risk assessment and natural attenuation" from 1995, as the recognition that former is almost impossible by the technical and financial reasons was widely spread. Many Japanese participants were interested in this basic thought about the ground contamination.
The brief history of the activities of the consortium is as follows.

April 2000	MOU exchanged between Japanese and Dutch sides Opening Conference
October 2000	1st CleanSoil Seminar in the Netherlands
April 2001	General Meeting
July 2001	Presentation by a Dutch engineer working in Japan
December 2001	2nd CleanSoil Seminar in Japan
April 2002	Closing Conference

The consortium held two seminars, one in October 2000 in the Netherlands and the other in December 2001 in Japan. The main topics in the seminars are listed in Table 1.

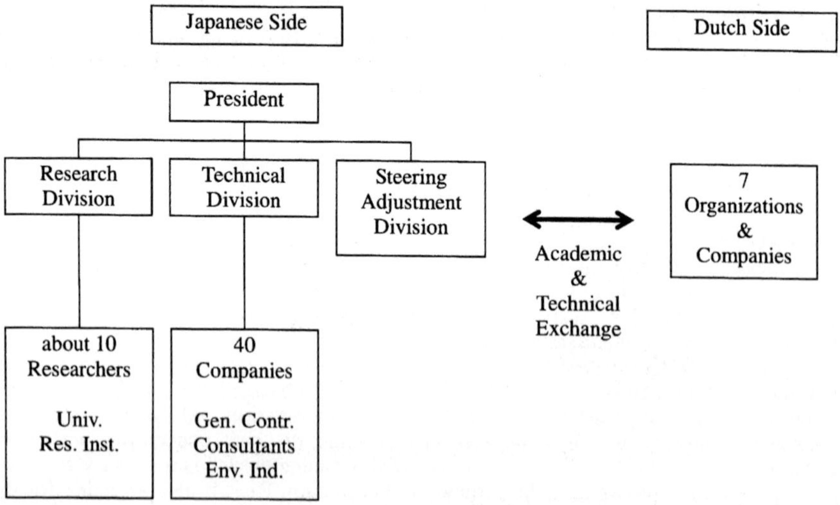

Figure 1: System of CleanSoil Consortium.

Table 1: Main topics in the seminar in CleanSoil Project.

Date	Place	Contents
Oct. 2000	The Netherlands	Workshop on soil contamination in coal gas factory Visit to analysis laboratory, chemical factory and soil washing plant Laboratory and investigation equipments Discussions about EQS for soil in the two countries
Dec. 2001	Japan	FFD (fuel fluorescence detector) Risk assessment in the Netherlands Novel technologies and policy changes in soil remediation Soil and groundwater remediation in Japan Small incineration plant contaminated with dioxins

One of the things Japanese side had the greatest interest in this project is the thought of risk-based approach that imports the risk analysis and the (monitored) natural attenuation. This is supported by the European common understanding that it is very difficult to remove the pollution completely because of the technical and the financial reasons.

Risk-based Approach to Ground Contamination

Reflection of Minamata Disease

In 1953, Minamata disease was found in the city of Minamata (see Figure 2). This disease was officially designated as the pollution-related illness by the government in 1968 and evaluated to be one of the four greatest pollution-related illnesses. The government has started acting on the environmental pollution after hundreds of people were killed by this disease.

Minamata disease is a pollution-related illness caused by methyl mercury included in waste water. This disease was first thought to be a contagious disease, which was wrong. Next, it was found that this was a pollution-related illness caused by waste water from a factory facing to the sea producing fertilizer, acetic acid, vinyl chloride, PVC and plasticizers. It was estimated that some pollutant included in the waste water was condensed in the food chain at the sea area next to the factory and taken by the inhabitants. Then the pollutant that causes this disease was studied at a national university, not at a governmental organization. The concluded pollutant was manganese in the first, selenium in the next, thallium and finally identified as alkyl mercury. In fact alkyl mercury was used in the factory as a catalyst when it produce acetaldehyde, a raw material to make acetic acid and plasticizers, from acetylene. There framed other hypotheses that explosives and amines were the pollutant. The further study identified the cause of the disease is that the suffering habitants ate fish taken in the sea next to the factory inside which methyl mercury condensed.

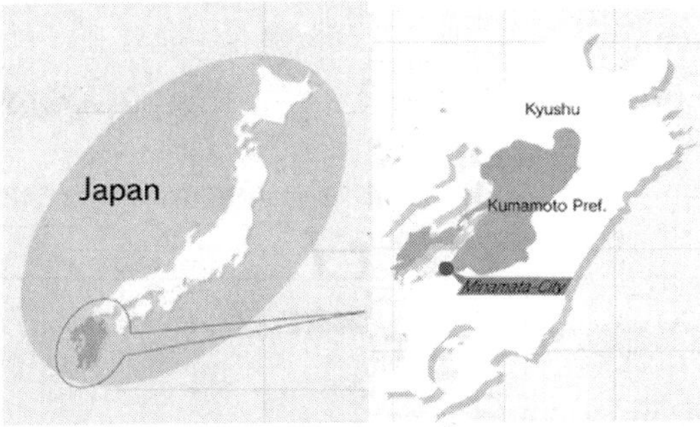

Figure 2: Location of Minamata City (after Minamata City Government).

The number of the casualties recognized by the government is 2,265 (1,517 of them were killed) as of 31 March 2004, and the number of the objective of the governmental relief fixed in 1997 is 11,540 (Minamata Disease Municipal Museum, 2000). As these numbers do not include that of the killed before the recognition started, there is no reliable data about the number of the casualties.

The other greatest pollution-related illnesses are the Niigata-Minamata disease (found 1964, designated 1968, caused by methyl mercury), *itai-itai* disease (1910s, 1968, cadmium) and Yokkaichi asthma (1962, 1972, oxidized sulphur). About all four illnesses administrative litigations were instituted and the court recognized the responsibility of the government and the enterprises that caused the illnesses.

Very similar problems have/had occurred at factories beside Songhuajiang River in Heilongjiang Province, China, Amazon River, Brazil, Canada, Indonesia, and so on. The knowledge from the case of Minamata should be utilized.

Precautionary Principle

There can be seen many insistence about the "precautionary principle" in Japan. In fact, not only Minamata disease but other pollution-related illnesses could have been prevented if the government had taken appropriate measures. The authors believe that this principle is useless for avoiding any disease caused by environmental pollution such as Minamata disease. If the government had regulated the manganese in waste water when the main cause was thought to be manganese, no effective countermeasure would have been established. If selenium? If thallium? It is clear that manganese, selenium and thallium are toxic and harmful for human health. At the same time, it is also clear that the regulations on these heavy metals result nothing about protecting the human health. We should never stop thinking what is "precautionary principle". If we cry A is dangerous, B is harmful, C is hazardous etc. without a reliable ground, no good result will occur. In the worst case, the phenomenon of "The Boy Who Cried Wolf" would occur. On the contrary, if we do nothing about the regulation, many people will suffer from this kind of disease. From this short thinking, it is understood that we must make an effort to *think* about the problem.

Case History — Osaka Amenity Park

Brief History

Osaka Amenity Park (OAP) is a complex consisted of offices, hotels, shops and residences in the middle of Osaka city, beside Okawa (Old Yodo River) constructed on a 500,000m^2 land where a metal refinery used to be. It was redeveloped from 1997 by Mitsubishi Estate Co., Ltd. (ME) and Mitsubishi Material Corporation (MM). The location of the redeveloped land is shown in Figure 3. The arrangement of the buildings on the redeveloped land is shown in Figure 4. Soil pollution by heavy metals were found, however, the developers hid the fact.

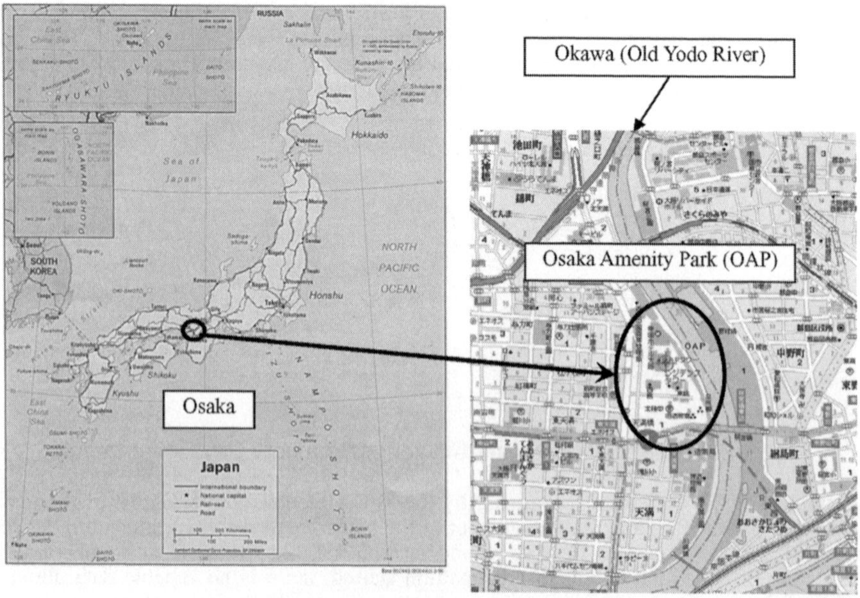

Figure 3: Location of Osaka Amenity Park.

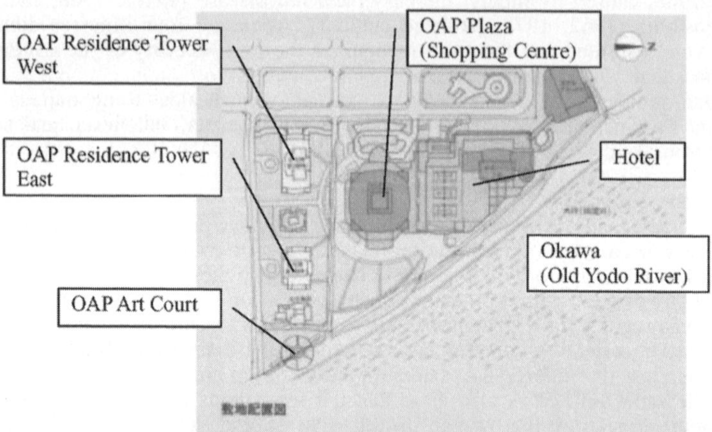

Figure 4: Arrangement of Osaka Amenity Park.

The brief history of the matter is as follows (after Mitsubishi Estate Co., Ltd., Mitsubishi Estate Residence Selling Co., Ltd., Mitsubishi Material Corporation, 2004-2005, Asahi, Mainichi and Yomiuri daily newspapers).

January 1989	Started boring investigation for the redevelopment
June 1995	Removed about 200,000m^3 contaminated soil
January 1997	Found heavy metals in groundwater
February 1997	Started selling East Building (February 1998 Completed)
September 1999	Started selling West Building (December 2000 Completed)
January 2001	Found Arsenic in 65 times concentration of EQS, etc.

November 2001	"The Guideline on Soil Contamination" was carried by the board of Real Estate Association
September 2002	ME and MM first noticed the fact of the contamination to the residents.
October 2004	Osaka Pref. Police investigated each Osaka Office of ME, MM and ME's daughter company Mitsubishi Estate Resident Selling Co., Ltd. (MERS).
November 2004	Osaka Pref. Police investigated each head office of the three.
December 2004	Osaka Pref. Police sent the papers pertaining to MMRS to the Public Prosecutor's Office
February 2005	ME proposed paying JPY 1,500M to the residents as the compensation and preparing JPY 4,500M for the countermeasure
March 2005	The Resident's Union basically accepted the proposal
May 2005	The Resident's Union agreed the amount of the compensation of 25% of the purchasing price.
June 2005	The Prosecutor's Office decided not to prosecute ME, MM and their board members.

Reactions by Mass Media

The three main newspapers in Japan, Asahi, Mainichi and Yomiuri issued many articles dealing with this problem. It is well known among those who are concerned with engineering that mass media do not always report the truth, however, no false article has been found up to now in these newspapers in this matter.

It is clear that the problem become larger and more difficult as the developers hid the fact. The developers sold the residences without notifying to the purchasers that the soil underneath the complex is contaminated with heavy metals. When OAP Residence sold, the law system on the ground contamination were not built in Japan, however, this is a violation of the *Building Lots and Buildings Transaction Business Law*. But the developers had not recognized that their behaviour is a violation, before they change their stance in 27 May 2005 at a press conference. As results, the presidents and some executives of the developers resigned. The Asahi estimates that the reason the Prosecutor's Office decided not to prosecute the developers is that they would be much damaged in their finance since the building lots and buildings transaction licence will be cancelled for 5 years and the payment of the compensation to the residents that is estimated to be no less than JPY 7,500M (25% of the total purchasing price) would be hindered in case they are prosecuted.

Investigation Committee

The developers and the residents have agreed to organize an investigation committee called "Technical Evaluation Conference on Countermeasures against Contamination of Soil and Groundwater in OAP", consisted of four researchers, four representatives of the residents and three representatives of the developers, with third-party secretaries by environmental investigators. Although official investigations and juristic activities finished when the Prosecutor's Office made the decision, some hot discussions will continue for some time. The first meeting of this conference was held in 30 July 2005. The authors are well interested in its activities since there are very sincere discussions between each party.

As shown in Figure 2, the location of OAP is next to a flood plain along Okawa (Old Yodo River). Since OAP Residence is a condominium and each resident is an owner of a unit of the land, responsibility for the migration of the contamination might occur on the residents under the current law system of Japan, if some soil, groundwater or water contamination is found and the Conference conclude that contamination results from the migration even if that is thought to be a very rare case. In this situation, any resident is forced to be serious about this problem.

Conclusions

The conclusions of this paper are as follows:

- Recently the common contaminants found in ground are heavy metals, VOCs and oil.
- Cone penetration test (CPT) machine is used in recent investigation methods. The energy-dispersive X-ray fluorescence, the gas chromatograph and the ultraviolet fluorescence techniques are used for heavy metals, VOCs and oil respectively.

- We have to make a judgement that the countermeasure against the ground contamination is really "good" or "smart" in each case before applying it to the practical problem. Risk-based approach is necessary when we perform any countermeasure against the ground contamination.
- The authors founded a consortium where Japanese and Dutch researchers and engineers joined, called CleanSoil Consortium. This consortium had acted for two years and played a great role in technical exchange between Japan and the Netherlands.

The authors insist that information is one of the most important thing when we deal with any environmental problem.

- In the case of Minamata disease, hundreds of people were killed before the correct information about the cause of the disease was made clear. In fact the government made a great mistake and failed to protect people. But on the contrary, much more people must have suffered if the government had established the regulations depending on a wrong information.
- In the case history described in this paper, the developers had a great damage on their corporate images and the presidents and some executives of them had to resign to avoid the prosecution. That can be said that those who manage these companies made a wrong choice about dealing with information of the soil and groundwater pollution.

We must think about the "precautionary principle" not for the regulation for the regulation but that for the protection of human health (Nakanishi, 1998).

References

Asahi (daily newspaper), Evening 19 March 2005, Evening 28 March 2005, Morning 9 May 2005, Morning 28 May 2005 and Morning 11 June 2005 (in Japanese).

Domenico, P. A. and Schwartz, F. W. (1990). "*Physical and Chemical Hydrogeology*", John Wiley & Sons.

Mainichi (daily newspaper), Morning 28 October 2004, Morning 29 October 2004, Morning 21 March 2005, Morning 19 April 2005 and Morning 21 April 2005 (in Japanese).

Minamata Disease Municipal Museum (2000), "Minamata Disease - Its History and Lessons -.

Mitsubishi Estate Co., Ltd., News Releases on 20 September 2002, 8 October 2002, 31 October 2003 and 30 October 2004 (in Japanese).

Mitsubishi Estate Residence Selling Co., Ltd., News Release on 22 November 2004 (in Japanese).

Mitsubishi Material Corporation, News Releases on 22 November 2004, 29 March 2005, 8 May 2005, 27 May 2005 and 10 June 2005 (in Japanese).

Nakanishi, J. (1998). "Much Ado about Nothing on Environmental Hormones", *Shincho 45*, December 1998 (in Japanese).

Tanaka, M. (2000). "Reports on the Japan - the Netherlands Exchange Meeting on CleanSoil Project and the Special Lecture", *Tsuchi-to-Kiso*, Proc. Japanese Geotechnical Society, Vol. 48, No. 10 (in Japanese).

U. S. Army Engineering R&D Center Environmental Laboratory (2000), The Tri-Service Site Characterization and Analysis Penetrometer System-SCAPS: Innovative Environmental Technology from Concept to Commercialization. Report Number: SFIM-AEC-ET-TR-99073.

Yomiuri (daily newspaper), Morning 9 October 2002 and Morning 30 October 2004 (in Japanese).

The evaluation of Soil and Groundwater Remediation at an Industrial Site Using Risk Analysis

Rosane Cunha Maia Nobre, Professor at Universidade Federal de Alagoas and Doctorate Student at UFRJ/COPPE, RJ, Brazil, rosanemaia@uol.com.br
Manoel de Melo Maia Nobre, Professor at Universidade Federal de Alagoas, UFAL/CCEN/GEO, Maceió, Brazil, manoelmaia@uol.com.br
Linice Cavalcante Bonaparte, Chemical Engineer, BRASKEM S.A., Maceió- AL, Brazil, linice.bonaparte@braskem.com.br

Abstract:

Risk analysis was applied at an industrial site to evaluate the performance of a second-phase groundwater remediation program for the migration control of a 1,2 dichloroethane groundwater plume. Monitored natural attenuation was observed to be an effective method for the remediation of the dissolved phase downgradient a physical barrier as source control was achieved. In spite of the complete absence of human settlements at the site, human health risk was assessed to a point of compliance 265 m from the source area. This protective assumption was motivated by many risk assessment procedures such as the ASTM RBCA (E2081-00). The downgradient concentrations were calculated by the analytical model Biochlor®, assuming a first order degradation rate for 1,2 dichloroethane. The calculated human risk was considered tolerable. The predicted concentration at the point of compliance was also compared with some ecological standards as a preliminary ecological assessment and the comparisons also suggested a tolerable ecological risk.

Introduction

Risk to human health is a concept not easily quantified and standardized. Similarly, ecological risks, designed to evaluate potential ecological effects, is not totally understood. Usually, the responsibility to define risk from contaminants in the environment and set acceptable standards is delegated to environmental agencies, which rely on updated information on toxicity and other sciences to meet their needs. This has been done in developed countries with some level of success, with the establishment of different protocols to deal with the subject. In Brazil, in particular, environmental agencies are demanding such analysis as a tool to enhance the decision making process. Although most methodologies assume lack of data for initial stages and increased understanding in higher levels in order to achieve remediation goals, the nature of transport phenomena in soils holds most of the uncertainties involved.

Risk involves both toxicity and exposure, and risk assessment is the process by which the total risk to human health and the environment is calculated (Brady et al., 1998). The widely recognized approach involves a four-stage process, as illustrated in Figure 1: i) data collection and hazard identification: the characterization of chemicals of interest from potential contaminant sources; ii) exposure assessment: pathways or environmental routes by which chemicals from the site can reach receptors; characterization of potentially exposed populations (receptors); and characterization of exposure routes: ingestion, inhalation and dermal contact (Figure 2); iii) toxicity assessment: the determination of numerical indices of toxicity to calculate risk; and iv) risk characterization: the estimation of the magnitude of risk as well as the uncertainty of the estimate.

Another application of risk assessment in site remediation is the establishment of cleanup standards. It is a way to define, objectively, "how clean is clean" and "what is safe". These standards may guide site remediation actions as well as promote site closure. It is an inverse procedure, that is, the process starts with a definition of acceptable risk and works back to the level of contamination that will produce the acceptable risk level (LaGrega et al., 1994). It is expected that cleanup decisions and site characterization reflect an equilibrium between cost and health-risk reduction.

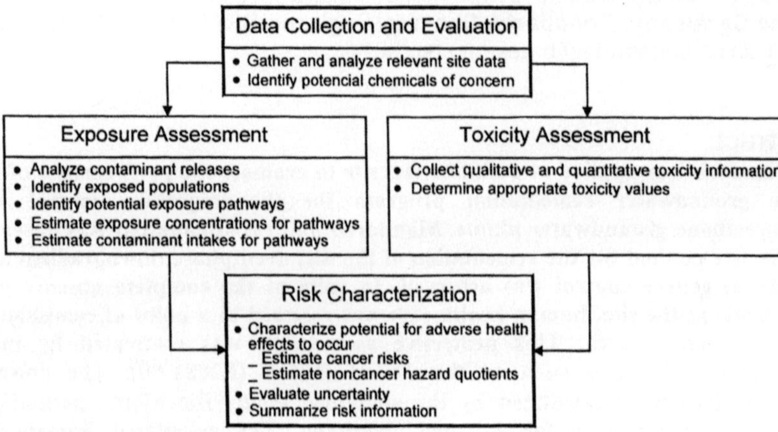

Figure 1: Four-stage process of risk assessment (modified from USEPA, 1989).

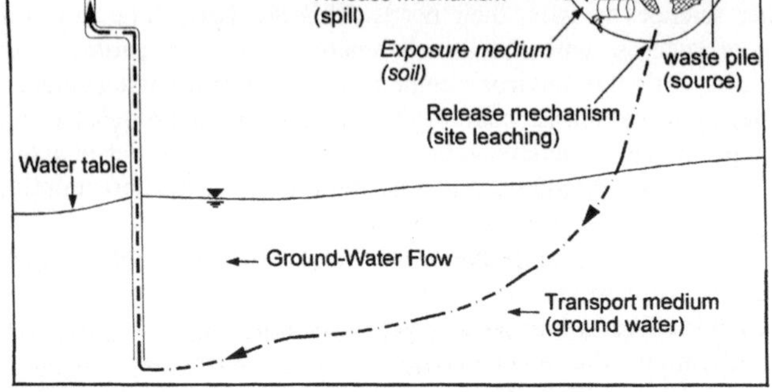

Figure 2: Examples of exposure pathways (modified from USEPA, 1989).

Ecological concerns and human health risks are usually assessed by different methodologies. In North America, a number of attempts have been made to define protocols for the clean-up decision making process. The most well known of these is the RBCA – *Risk Based Corrective Action* – approach. It was based on the theoretical principles of RAGS – Risk Assessment Guidance for Superfund, developed by the american EPA (USEPA, 1989), to be used in remedial investigation/feasibility study (RI/FS) processes at Superfund sites. The RBCA protocol establishes a tiered approach of gradual response in site characterization and remediation efforts to soil and groundwater remediation. Site characterization efforts are directed at collecting only the information that is needed for each stage or tier. In this manner, the lower-risk sites demand less expense from site characterization to achieve a desirable clean-up level. In this protocol, human health risks for carcinogens are calculated as the probability of developing cancer in a lifetime and for non-carcinogens as hazard indices. Details on the RBCA protocols are found at ASTM E2081-00 (ASTM, 2000) for chemical releases, in general, and ASTM E1739-95 (ASTM,1995) for petroleum release sites.

This work presents a remediation program in which risk analysis is being incorporated in an advanced stage to evaluate its performance. When a decision making process is incorporated into the remediation process, the result is a risk based corrective action. Hence, the remedial actions at the site have been implemented in order to satisfy the RBCA protocol criteria.

Case Study

Physical Setting: Chlor-Alkali Plant

A chlor-alkali plant located in a coastal lagoon region in northeastern Brazil was impacted by an accidental release of more than 135 metric tons of 1,2 DCA (Nobre and Nobre 1997, 1999, 2004). The product free phase (i.e. DNAPL secondary source) was vertically limited to a sandstone containing a high organic fraction that occurs at approximately 11m of depth. The contamination is currently being contained by a combined system of 12 extraction wells and a 300 m physical barrier, that consists of a cement-bentonite cut-off wall.

Although the system has already extracted over 110 metric tons of 1,2 DCA and controlled plume migration, the rate of mass removal has been reduced over the years which demanded the evaluation of alternative technologies to help accelerate the remediation process. Therefore, additional studies were carried out to investigate the potential for natural attenuation of the dissolved plume downgradient from the source, where sulfate-reducing conditions are predominant, as well as the aerobic degradation potential of 1,2 DCA and vinyl chloride, through the use of an air sparging system near the source. Natural attenuation processes were evaluated as a second-phase remedial program for the 1,2 DCA contaminated aquifer, including a risk analysis based on the RBCA protocol (ASTM, 2000).

The site is located at a local recharge zone area, between the Atlantic Ocean and a lagoon channel that constitutes the discharge zone of the contaminated aquifer. Recharge to the upper aquifer is directly through rainfall. The area is underlain by about 30 m of surficial deposits above a silty clay layer, followed by a massive limestone and a confined sandstone aquifer. The surficial unit consists of a well-graded sand of alluvial and marine origin from the Quaternary age. At about 11 m of depth, approximately, occurs 4 m of well-cemented sandstone, with high organic fraction, that has constituted the vertical limit for the migration of the free-phase product. The hydraulic conductivity (k) of the surficial unit is estimated to range from 10^{-5} to 10^{-4} m/s, with an anisotropy ratio (k_H/k_V) less than 10, except the organic sand layer with K values ranging from 10^{-9} to 10^{-7} m/s. The hydraulic gradient is about 0.5%

at the rainy season. The dissolved concentrations of 1,2 DCA are significant on the upper portion of the aquifer, in the order of mg/L closer to the free phase source.

The 1,2 DCA storage tanks, which are less than 300 m from the nearest open discharge zone, are surrounded by the physical barrier. Because of the confinement created by the cut-off wall and the confining layer underneath, extraction wells located in the vicinity of this tank have been removing a significant amount of mass, with 1,2 DCA concentration in the order of hundreds to thousands of ppm. It is expected that the air sparging/biosparging system will accelerate mass removal at this location.

Demonstrating Natural Attenuation

A specific groundwater sampling program is being carried out to assess the potential for relying on natural degradation for the portion of the aquifer downgradient the physical barrier. The parameters associated with intrinsic biodegradation were monitored in accordance with USEPA protocols (Wiedemeier et al. 1998), including dissolved oxygen, redox potential, biochemical and chemical oxygen demand, pH, water conductivity, sulfate, nitrate, iron, ethane, ethane, methane and volatile organic compounds. The degradation parameters have been estimated from a 10 years site-specific database.

Recent analytical results reveal that the process of abiotic hydrolysis, under alkaline pH, is possibly occurring in which part of the dissolved 1,2 DCA is being naturally transformed into vinyl chloride (VC). Furthermore, under the anaerobic-reducing conditions at the site, 1,2 DCA appears also to be degrading via dihalo-elimination to ethene. It is expected that 1,2 DCA and VC rapidly mineralize to CO_2, within the air sparging zone of influence, where groundwater redox conditions change to aerobic and oxidizing. Also, in the presence of oxygen and methane, methanotrophic bacteria appear to be using the methane produced within the anaerobic portion of the plume and to be co-metabolically degrading the contaminants in the aerobic zone to produce CO_2. High concentrations of ethene, VC and methane are always detected in association with high 1,2 DCA concentrations, with few exceptions (Nobre and Nobre, 2004). The dihalo-elimination of 1,2 dichloroethane to ethene seems to be related to methanogenic microorganisms. Ethene is the most prevalent by-product at this site. The organic matter present at the site is probably behaving as a substrate source - electron donor - which is stimulating the reductive dechlorination process of 1,2 DCA to occur.

Risk Analysis: Tier 1

A risk analysis was conducted at the chlor-alkali plant site based on the protocol RBCA (ASTM E2081-00 RBCA), including Tiers 1 and 2 and recent concentrations at the site. The estimation of exposure concentration is frequently based on the concentration of the source media. Once the concentration of the source media is determined, fate and transport models may be applied to estimate the exact amount of chemicals that could be taken by the receptor. The use of 95% upper confidence limits of the mean concentration (95% UCL or C_{95}) is accepted by most environmental agencies (including USEPA) and will be applied here.

The first procedure in conducting tier 1 of RBCA consisted in three steps: i) quantification of possible exposure pathways at the site; ii) definition of exposure routes and iii) comparison of analytical results (C_{95} medium concentration) to the *Preliminary Remediation Goals* - PRGs that must be achieved at the point of compliance. The exposure routes were considered to be inhalation and dermal contact since the waters from the lagoon channel, at the point of compliance, were not potable, neither was the groundwater at the site. However, a risk calculation involving oral ingestion (for the ingestion pathway) as well as the two other routes (inhalation and dermal pathways) was performed for comparison purposes and completeness.

The transport medium for the different exposure pathways included groundwater only. The PRGs were calculated from the OHIO-EPA Excel spreadsheets (OHIO-EPA, 2002), which were adjusted to comply with the Brazilian legislation for a Hazard Quotient of 1 and a Carcinogenic Risk of $1x10^{-5}$. The total calculated risk (carcinogenic and non-carcinogenic) is the summation of the risk values obtained for each exposure route and compound. In this case study, the only contaminant considered was 1,2 DCA. The non-carcinogenic risk was calculated from

$$Non-carcinogenic\ Risk = \left[\frac{C}{PRG(nc)_{DCA}}\right] \quad (1)$$

where C is the 95% upper confidence limits of the mean concentration of 1,2 DCA in groundwater (abbreviated as C_{95} or 95% UCL hereafter) and $PRG(nc)_{DCA}$ is the non-carcinogenic standard for that compound. The carcinogenic risk was calculated as follows

$$Carcinogenic\ Risk = \left[\frac{C}{PRG(c)_{DCA}}\right] x\ 10^{-5} \quad (2)$$

where $PRG(c)_{DCA}$ is the carcinogenic standard for 1,2 DCA in groundwater. The calculated PRGs are summarized in Table 1.

Table 1: Non-Carcinogenic and Carcinogenic PRGs for 1,2 Dichloroethane

Preliminary Remediation Goal - Non-Carcinogenic $PRG_{(NC)DCA}$ at point of exposure - STANDARS		Preliminary Remediation Goal - Carcinogenic $PRG_{(C)DCA}$ at point of exposure - STANDARS	
(WITH oral ingestion)		**(WITH oral ingestion)**	
Receptor	$PRG_{(NC)}$ (ppb)	**Receptor**	$PRG_{(C)}$ (ppb)
Adult receptor	**592.0**	Adult receptor	**8.9**
Child receptor	**205.0**	Child receptor	**14.8**
(WITHOUT oral ingestion)		**(WITHOUT oral ingestion)**	
Receptor	$PRG_{(NC)}$ (ppb)	**Receptor**	$PRG_{(C)}$ (ppb)
Adult receptor	**1290.0**	Adult receptor	**186.0**
Child receptor	**472.0**	Child receptor	**362.0**
Assumptions:Hazard Quotient =1		Assumptions:Cancer Risk =10E-5	

The mean concentration (95% UCL) of 1,2 dichloroethane was calculated for recent analytical results of 3 wells located immediately downgradient the physical barrier (only the detectable values). At least 15 values of concentrations were considered, resulting in C_{95} equal to 6000 ppb. The results of Table 1 indicate that the PRGs (either for oral ingestion or not) are smaller than the calculated 95% UCL at the site. Therefore, the conductance of tier 2 was needed, with groundwater flow and transport analysis in order to move the point of compliance far from the source area.

Risk Analysis: Tier 2

Tier 2 procedure requires the establishment of a point of exposure (receptor), which is, approximately, 265m far from the source at the site. It is represented by a stream that discharges its water into a larger channel, as illustrated in Figure 3. Concentrations upgradient the physical barrier are significant, greater than 100ppm. On the other hand, concentrations downgradient the barrier are smaller, generally below 1 ppm. However, these values are still greater then the minimum PRGs calculated for 1,2 dichloroethane (Table 1) during Tier 1. A maximum concentration of 18 ppm was detected at monitoring well PP-30. The EPA BIOCHLOR analytical model (USEPA, 2000) was chosen to evaluate the transport of 1,2 DCA contaminated dissolved plumes that are located downgradient the physical barrier. This model is capable of evaluating transport phenomena of chlorinated compounds in saturated media, with a chemical reaction term, represented by a first order decay model to simulate a biotic or abiotic degradation process. Transport simulations were performed for a period of 30 years, considering natural degradation processes for the 1,2 DCA a half life equal to 1.5 years, calculated with a first-order reaction kinetics.

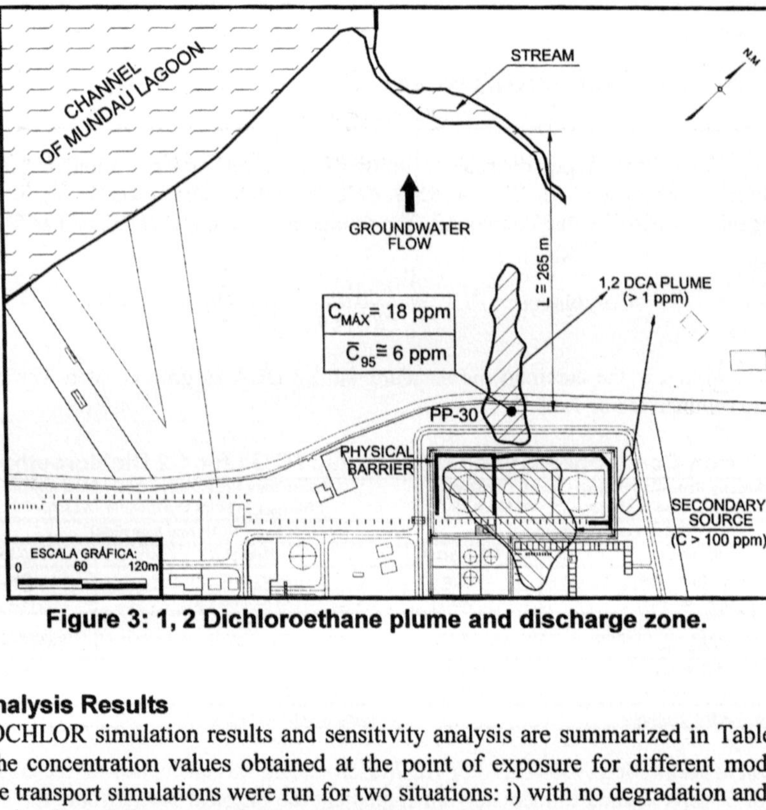

Figure 3: 1, 2 Dichloroethane plume and discharge zone.

Risk Analysis Results

The BIOCHLOR simulation results and sensitivity analysis are summarized in Table 2, that shows the concentration values obtained at the point of exposure for different model input data. The transport simulations were run for two situations: i) with no degradation and ii) with a first order degradation, assuming a concentration at the source equal to 6 ppm. For each value in Table 2, the non-carcinogenic risk (or hazard quotient - HQ) and carcinogenic risk (CR) were calculated using equations 1 and 2. Considering, for example, a concentration of C = 0.176 ppm (176 ppb) at the point of exposure, a HQ = 0.14 and CR = 0.95×10^{-5} were obtained (adult receptor; without groundwater ingestion). Considering water ingestion, however, these values increase considerably, resulting in a HQ = 0.30 and CR = 1.98×10^{-4}. In this example, the non-carcinogenic risk is not a problem (< 1 in both situations). However, CR is greater than 10^{-5} (with water ingestion), which is not allowed by local legislation.

Transport simulation results were very sensitive to the reaction term. If no degradation is considered, concentration levels at the point of exposure are much higher, comparatively. This is because groundwater flow is very slow and some sort of degradation is always expected with time. The results are also sensitive to the hydraulic conductivity term. As mentioned previously, the RBCA protocol is based on a reversed procedure, the process starts with a definition of acceptable risk and works back to the level of contamination that will produce the acceptable risk level. Therefore, site specific target levels – SSTL were calculated at the site in order to specify the maximum acceptable concentration of 1,2 DCA at the source, considering the attenuation process of Tier 2. Table 3 summarizes the results, including SSTLs, considering the most probable parameter values for the transport simulation problem ($k=1x10^{-4}$ m/s, $\alpha_L = 5$ m, $t_{1/2} = 1.5$ years). The results indicate that, considering oral ingestion at the point of exposure, a concentration of 0.3 ppm (C_{95}) at the source is acceptable. On the other hand, without oral ingestion, the acceptable level may be up to 6.3 ppm.

Table 2: Transport simulation results and sensitivity analysis

Concentration Values (ppm) at 265 m from the source (C_{95} = 6 ppm) BIOCHLOR Results				
Hydraulic Conductivity (k)	**Transport Model**	**Dispersivity (α_L=2 m)**	**Dispersivity (α_L=5 m)**	**Dispersivity (α_L=10 m)**
$5x10^{-5}$ m/s	(t ½) = 1.5 year*	0.006	0.008	0.013
	(t ½) = 3 years*	0.165	0.179	0.187
	no degradation	5.819	5.515	4.771
$1x10^{-4}$ m/s	(t ½) = 1.5 year*	0.163	**0.176**	0.185
	(t ½) = 3 years*	0.942	0.926	0.854
	no degradation	5.728	5.432	4.749
$5x10^{-4}$ m/s	(t ½) = 1.5 year*	2.731	2.605	2.300
	(t ½) = 3 years*	3.926	3.729	3.268
	no degradation	5.656	5.363	4.689

* 1st order decay (natural attenuation).

Based on the results, it was determined that the remediation program must continue in operation until concentration (C_{95}) drops to below 1 ppm at the source area (downgradient the physical barrier). Then, the pump-and-treat system may be discontinued and rely solely on monitored natural attenuation processes downgradient the physical barrier to achieve the risk criteria previously defined.

Table 3: Site Specific Target Levels for 1,2 DCA at the site

Site Specific Target Levels - SSTLs for 1,2 DCA using BIOCHLOR (at source)		
(WITH oral ingestion)		
Parameter		**Value**
Hydraulic Conductivity	k	$1x10^{-4}$ m/s
Longitudinal Dispersivity	(α_L)	5 m
1,2 DCA Half-life	($t_{1/2}$)	1.5 year
Retardation Coefficient	R	1
SSTL (adult receptor)	-	**0.3 ppm**
(WITHOUT oral ingestion)		
Parameter		**Value**
Hydraulic Conductivity	k	$1x10^{-4}$ m/s
Longitudinal Dispersivity	(α_L)	5 m
1,2 DCA Half-life	($t_{1/2}$)	1.5 year
Retardation Coefficient	R	1
SSTL (adult receptor)	-	**6.3 ppm**
Assumptions:Hazard Quotient =1 and Cancer Risk =10E-5		

Conclusions

In Brazil, risk assessment is starting to be used as a tool to enhance the decision making process for remediation programs in the environmental area. It provides decision-makers with a technical evaluation of the risks posed at a site, a characterization of the potential routes of exposure and an assessment of remedial alternatives.

The paper presents the results of a second-phase remediation program for a 1,2 DCA groundwater plume downgradient a containment area. Plume control was obtained as a result of combined efforts, which included the operation of an integrated system of a hydraulic barrier and a physical barrier. In addition, monitoring of natural degradation mechanisms was part of the overall remedial program. The calculated risks based on RBCA protocol provided guidance in the selection of alternatives for the management of a dissolved low concentration plume at the site, located downgradient the physical barrier. The results indicated that the risk based remediation program, based on the actual scenario, could not depend solely on monitored natural attenuation as a remediation remedy. Hydraulic pumping was still needed at the site in order to lower the concentrations to the site specific target levels (SSTLs) calculated during the risk analysis procedure.

References

ASTM, Method E2081-00 (2000). *Standard Guide for Risk-Based Corrective Action*. Annual Book of American Society of Testing and Materials Standards, Philadelphia, PA, USA.

ASTM, Method E1739-95 (1995). *Standard Guide for Risk-Based Corrective Action Applied at Petroleum Release Sites*. Annual Book of American Society of Testing and Materials Standards, Philadelphia, PA, USA.

Brady, P.V., Brady, M.V. and Borns, D. (1998). *Natural Attenuation*. Boca Raton, Fla., Lewis Publishers, 245p.

La Grega, M.D.; Buckingham, P.L.; Evens, J.C. (1994). *Hazardous Waste Management*. McGraw Hill Ed.

Nobre, M.M.M. and Nobre, R.C.M (1997). Uma Solução Geotécnica para Remediação de Águas Subterrâneas por Compostos Organoclorados. *Revista Solos e Rochas* da ABMS/ABGE, Volume 20, No. 02, 1997. p.121-134.

Nobre, M.M.M. and. Nobre, R.C.M. (1999) Field evidence of 1,2 dichloroethane degradation in a contaminated aquifer. 4° Congresso Brasileiro de Geotecnia Ambiental (REGEO'99). São José dos Campos, SP, 28-30 de nov99, p.400-409.

Nobre, R.C.M. and M.M.M. Nobre (2004). Natural Attenuation of Chlorinated Organics in a Shallow Sand Aquifer. *Journal of Hazardous Materials*, Vol 110/1-3, pp.129-137, 2004.

OHIO-EPA (2002). *Support Document for the Development of Generic Numerical Standards and Risk Assessment Procedures*. The Voluntary Action Program.

USEPA (1989). Risk Assessment Guidance for Superfund, Volume I: Human Health Evaluation Manual, Part A. EPA/600/1-89/002. Office of Emergency and Remedial Response. Washington DC.

Wiedemeier, T.H., Swanson, M.A., Moutoux, D.E., Gordon, E.K., Wilson, J.T., Wilson, B.H., Kampbell, D.H., Haas, P.E., Miller, R.N., Hansen, J.E. & Chapelle, F.H. (1998). *Technical Protocol for Evaluating Natural Attenuation of Chlorinated Solvents in Groundwater*. EPA/600/R-98/128. U.S.Environmental Prot. Agency, Washington, D.C.

Ground Surface Subsidence Risk Assessment by Underground Mining in Estonia

Jyri-Rivaldo Pastarus Tallinn University of Technology, Tallinn, Estonia

Abstract

The paper deals with risk assessment/management problems in Estonian oil shale mines, where room-and-pillar mining with blasting is used. The primary interest of this study concerns evaluating the usability of the applied risk assessment method at room-and-pillar mining. The study addresses risk associated with the collapse of mining blocks and resulting surface subsidence. Some of the various factors, relevant to Estonian oil shale mines, have been determined. For risk estimation the judgmental approach and event tree are used. Investigations in Estonian oil shale mines show that the likelihood of the surface subsidence is not acceptable. Mitigation methods will reduce the likelihood. The results of the investigation are of particular interest for practical purposes.

Introduction

The most important mineral resource in Estonia is a special kind of oil shale. It is located in a densely populated farming district of high soil fertility. Underground oil shale production is obtained by room-and-pillar method with blasting. This method is cheap, highly productive and easily mechanizable.

Underground mining in Estonian oil shale mines causes a large number of technical, economical, ecological and juridical problems. Conventional theoretical basis does not enable to solve these problems. The data which have become available in the last 40 - 50 years concern experience obtained at about 400 mining blocks on the area of 100 km^2. They give a good basis for elaboration of the concept and methods of risk assessment/management, and for using the results to solve the mining and environmental problems.

This study addresses risks associated with the collapse of mining blocks, which includes surface subsidence. The primary interest of this study concerns evaluating the usability of the methodology and estimating the probability of failure without detailed assessment of its consequences. The study is based on literature and Estonian experiences. As an application example the risk analysis of Estonian oil shale mines has been conducted.

Risk assessment/management involves making judgments about taking a risk, and all parties must recognize the possibility of adverse consequences which might materialize. Prevention of the hazardous situations is more moral, ethical and economical than facing the adverse consequences. In the stage of design, risk assessment gives the possibility to determine negative consequences and to work out the methods of their mitigation. Risk assessment method gives information about the influence of mining on the environment. Having received the information, the management of a mine can come to adequate political and strategic decisions. The concept and methods of risk assessment/management may be used for different purposes and at different levels: at the stage of mining block design; as the basis for decision-making when choosing between different remedial actions for mined-out area within temporal and financial restraints; to relate risk levels of ground surface subsidence to acceptable risk levels established by society for other activities. Risk assessment/management is a most powerful tool to solve the complicated mining problems.

Risk analysis is the use of available information to estimate the risk of ground surface subsidence from hazards. Some of the various factors relevant to Estonian oil shale mines are determined in the present study. Probabilistic risk analysis is the more rational basis for evaluation. For risk estimation the event tree is used. Having obtained the risk information and knowing the risk valuation criteria, we come to a decision. Investigations showed that in Estonian oil shale mines the likelihood and the consequences of the surface subsidence are not acceptable. The risk exceeds the limit 15 times. The risk mitigation process reduces the likelihood of mining block collapse and surface subsidence.

Analysis showed that the used risk assessment/management methods are applicable for Estonian oil shale mines. The elaborated risk assessment methods will be usable in different geological conditions, where the room-and-pillar mining system is used. The results of risk assessment are of particular interest for practical purposes.

Geology and mining

The Estonia oil shale deposit is located in the north-eastern part of the country. The commercially important oil shale layer stretches for 200 km from west to east, and for 30 km from north to south. The oil shale bed has a form of a flat bed slightly inclined (2 - 3 m per km) southward. The depth at which oil shale bed occurs varies significantly: the bed lies straight under the Quaternary sediments at the northern border of the field and it reaches 100 - 150 m at its southern rim. The oil shale reserves in Estonia are estimated to be approximately four thousand million tons.

The commercial oil shale bed and its immediate roof consist of oil shale and limestone seams. The main roof consists of carbonate rocks of varying thickness. Characteristics of the individual oil shale and limestone seams are rather different. Oil shale compressive strength is 20 - 40 MPa and that of limestone is 40 - 80 MPa. The strength of the rocks increases southward, and their volume density is 1.5 - 1.8 and 2.2 - 2.6 Mg/m^3, respectively. Calorific value of dry oil shale varies within 7.5 - 18.8 MJ/kg depending on the seam and deposit area.

In Estonian oil shale mines the room-and-pillar mining system with blasting is used (see Figure 1). It gives an extraction factor of 70 - 80 %. The field of the oil shale mine is divided into panels, subdivided into mining blocks approximately 300 - 350 m wide and 600...800 m long each. A mining block consists usually of two semi blocks. The oil shale bed is embedded at the depth of 40 - 70 m. The height of a block corresponds to the thickness of the commercial oil shale bed, approximately 2.8 m. The width of the room is determined by the stability of the immediate roof. Although the latter is very stable when it is 6 -10 m wide, bolting must still support the immediate roof. The pillars are arranged in a singular grid. Actual mining practice has shown that pillars with a square cross-section suit best. Intra-chamber pillars are left to secure the solidity of the whole rock mass laying above the chambers and to eliminate surface subsidence. The cross-sectional area of the pillars is 25 - 40 m^2, depending on the depth of the oil shale bed. Loading and transportation of blasted mined rock is carried out by powerful load-haul-dump (LHD) machines with diesel drive. A work cycle lasts over a week.

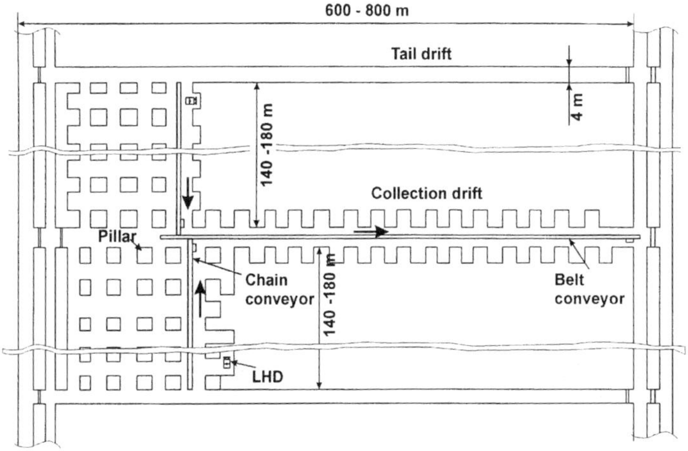

Figure 1. Schematic layout of room-and-pillar mining in Estonian oil-shale mines

Theoretical background

In Estonia, like in other countries, risk assessment/management methods are used in different branches of industry and for many different technical systems. In spite of the varied terminology, there is general agreement on the basic requirements (Calow 1998, Grahm & Bartsh 1995, Rowe 1980). The terminology and risk assessment/management methodology used in the frame of this project are presented below.

Risk management is the systematic application of management policies, procedures and practices to the task of identifying, analyzing, assessing, treating and monitoring risk (Calow 1998, Rowe 1980). Having obtained the risk information, a decision-maker must come to a decision. The primary steps of risk assessment/management are presented in Figure 2.

Risk assessment is the process of deciding whether existing risks are tolerable and risk control measures are adequate (Calow 1998, Grahm & Bartsh 1995, Rowe 1980). It involves making judgments about taking of the risk (whether the object or process is assessed as safe enough), and all parties must recognize that the adverse consequences might materialize and owners will be required to deal effectively with the consequences of the failure event. Risk assessment incorporates the risk analysis and risk evaluation phases.

Risk analysis is the process of determining how safe the object or process is. A risk analysis contains the following steps: scope definition, hazard identification, and risk estimation. The description of the system, scope and expectations of risk analysis should be defined at the outset. An iterative approach should be adopted with qualitative methods being employed at the early stages of the process. If more information becomes available, use of quantitative analyses is required.

Risk identification is the process of determining what can go wrong, why and how. Failure can be described on many different levels. Conceptualization of the different possible failure modes for a technical system is an important part of risk identification. One should first take into account as many types of failure as possible. The initial list can then be reduced by eliminating those types of failures which are considered implausible.

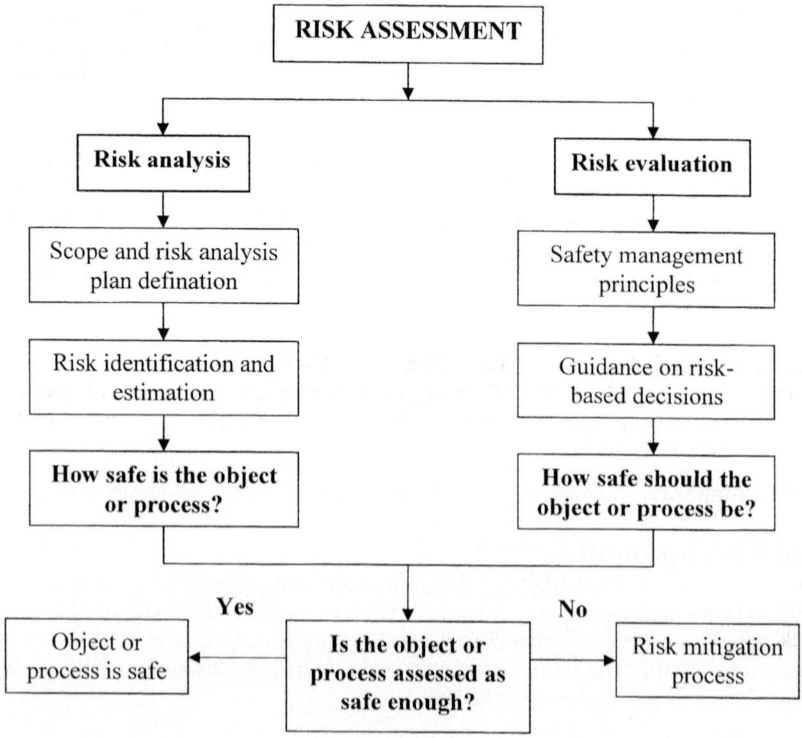

Figure 2. Risk assessment/management process

Risk estimation entails the assignment of probabilities to the events and responses identified under risk identification. The assessment of appropriate probability estimates is one of the most difficult tasks of the entire process. Tools that are often used to help in risk estimation are fault/event trees. Probability estimation can be performed according to three general approaches depending on the type and quality of the available data:

1. Analytical approach uses logical models for calculating probabilities.
2. Empirical approach uses existing databases to establish probability.
3. Judgmental approach uses experience of practicing engineers in guiding the estimation of probabilities.

Attaining an exact value of probability at examining technical systems and processes is not a realistic expectation.

Risk evaluation is the process of examining and judging the significance of risk. It must answer the question, how safe the process or object should be. It is based on the available information, including consideration of the importance of the estimated risks and the associated social, environmental and economic consequences. The principal role of risk evaluation in risk assessment is the generation of decision guidance against which the results of risk analysis can be assessed.

Risk acceptance is an informed decision to accept the likelihood and the consequences of a particular risk. In some countries, there is a certain risk level that is defined as the limit of unacceptable risk. For failure events with no potential fatalities or irreparable damage to the environment, the target failure probability may be decided exclusively basing on economic considerations and corresponding risk analysis. A target level of 10^{-3} - 10^{-2} for lifetime risk of the object or process may be a reasonable criterion (Hoeg 1996).

Risk mitigation is a selective application of appropriate techniques and management principles to reduce either likelihood of an occurrence or its consequences, or both (Calow 1998, Lind 2004). If the calculated risk of the existing system is judged to be too high, alternatives are proposed to reduce the risk of failure. After repeated study the decision-makers can be provided with suitable alternatives and their estimated costs for consideration in improving overall technical system safety.

Investigation results in Estonian oil shale mines

Technological and geological aspects of underground mining can influence collapse of a mining block and surface subsidence. Some of various factors relevant to Estonian oil shale mines are presented in Figure 3.

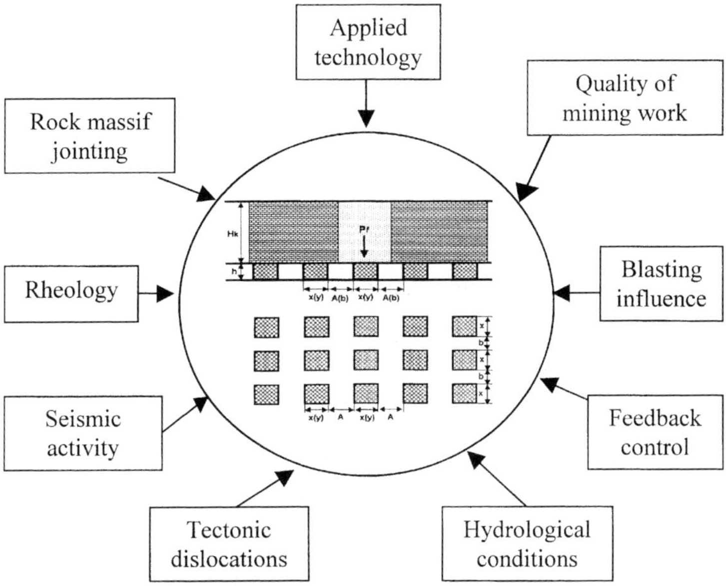

Figure 3. Factors contributing to mining block collapse and surface subsidence

Main technological aspects influencing the stability of a mining block concern quality of mining and blasting works. Investigations have shown that dimensions of rooms and pillars in Estonian oil shale mines may deviate from their project values. Maximum deviation is about ±1 m. It depends on quality of the applied machinery and skill of the miners. On the other hand, properties of the rock mass can change as a result of blasting works. In the roof and pillars there appear additional cracks, which decrease the strength of the rocks. Influence of geological parameters on mining block stability is significant. Rheological behavior of rocks was studied, and the results were considered at compiling the mining law (Mining-law, 1998).

Influence of karst and fissures (joints) on the stability of a mining block is evident. These factors have been determined for the Estonia oil shale deposit and are presented on the map of a mining block. Especial attention must be paid to hydrological conditions. Many old mines are completely flooded. It is not known, what will happen in the future, after 20 - 30 years. Seismic activity in Estonia is at such a low level, practically negligible, that it has been considered in this study only to a limited extent.

The above-mentioned technological and geological factors can lead to damages of the ground surface. Up to present, 73 collapses have been recorded. These complicated cases demand further investigations.

In the first stage of the project two factors were taken into consideration: quality of mining works and that of geological data. For probability determination the judgmental and subjective degree-of-belief approaches are used. The event tree presents risk estimation (Figure 4).

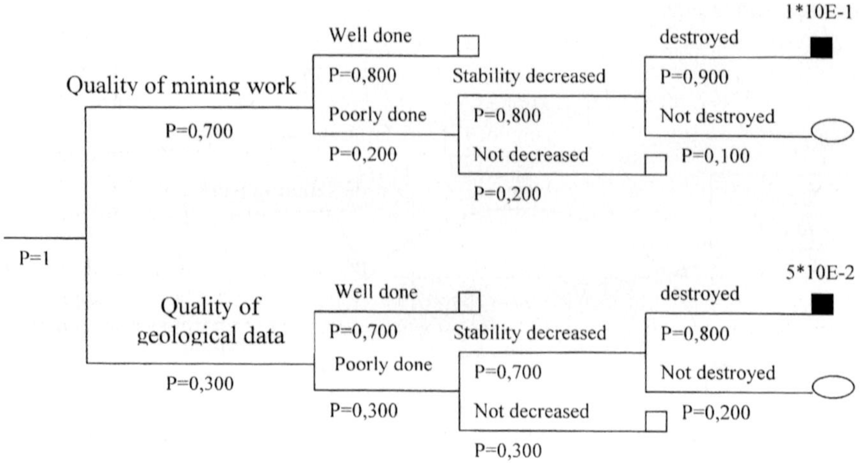

Figure 4. Event tree

The investigation showed that in the case of mining works the estimated probability of surface subsidence is 10^{-1}, in the case of quality of geological data - $5*10^{-2}$. Surface subsidence probability in the event tree is settled by "OR gate" (Williams et al. 2004). It requires that only one of these factors (the quality of mining works or the quality of geological data) need to exist for surface subsidence to occur. Calculations showed that the total probability of surface subsidence is $1.45*10^{-1}$. It is evident that the estimation probability of the surface subsidence exceeds the limit (10^{-3} - 10^{-2}) 15 times. Consequently, the risk for mining blocks is not acceptable. Risk mitigation methods allow to reduce the likelihood of the mining block collapse and surface subsidence.

Design of mining block parameters is based on the instruction used in Estonian oil shale mines (Mining-law, 1998). Analysis of the applicable design method of mining blocks showed that the factor of safety is very large (1.2 – 1.8). Practically, it is impossible to determine the adequate factor of safety for a mining block. Consequently, it contains unknown factors. Determination of the unknown factors and reducing the factor of safety is the task for the future.

On the other hand, the method of adaptive design can reduce the probability of collapse of a working mining block (Pastarus 2005). It is based on the monitoring system which allows to determine the potential collapse center in a mining block.

The above-mentioned risk mitigation methods reduce the likelihood of mining block collapse and surface subsidence. At application of these methods the risk will be acceptable. If not, alternatives will be proposed to reduce the probability of collapse and surface subsidence.

Acknowledgment

Estonian Science Foundation (Grant No. 5164, 2002-2005) supported the research.

Conclusions

As a result of this study, the following conclusions and recommendations can be made.

1. Underground production in Estonian oil shale mines is carried out by room-and-pillar method with blasting which causes a large number of technical, economical, ecological and juridical problems.
2. The present study addresses risk associated with collapse of mining blocks, including surface subsidence. The primary interest of this study concerns evaluating the usability of the methodology and evaluating the probability of the mining block collapse and surface subsidence without detailed assessment of the consequences.
3. Some of the various factors relevant Estonian oil shale mines have been determined. The event tree determines the probability of the surface subsidence. The investigation showed that the likelihood and the consequences of risk are not acceptable. The risk exceeds the limit 15 times. Risk mitigation process will reduce the likelihood.
4. Risk assessment/management method is a powerful tool to solve complicated mining problems. Analysis showed that the used risk assessment/management methods are applicable Estonian oil shale mines. The results of the investigation are of particular interest for practical purposes.

Bibliography

Calow, P. Handbook of environmental risk assessment and management. Oxford, Blackwell Science. 1998.

Graham, L. P., Bartsch, M. Risk analysis safety assessment for use at Swedish dams. Symposium: Research and development in the field of dams. Crans-Montana, Switzerland, 7-9 September 1995, p.12-19.

Hoeg, K. Performance evaluation, safety assessment and risk analysis for dams. Hydropower & Dams. Issue 6, 1996.

Lind, H. G. Mitigating the risk associated with underground coal pillar extraction in South Africa. Mining Technology. Transaction of the Institutions of Mining and Metallurgy. Vol 113, No 4, 2004. p. 215-224.

Mining-law and legal regulation acts. Ministry of Environment, Ministry of Economy. Part II. Tallinn, 1998 (in Estonian).

Pastarus, J. R., Sabanov, S. A method for securing working mining block stability in Estonian oil shale mines. Proc. of the Estonian Academy of Sciences, Engineering, 11/1. Tallinn, 2005, p.59-68.

Rowe, EW. D. Risk assessment approaches and methods. Society, Technology and Risk Assessment. (Conrad, J., ed.). Academic Press, London, 1980, p.3-29.

Williams, D. J., Loch, R. J., Vacher, C. Risk assessment applied to tunnel erosion of mine spoils. Proc. of the 11th Tailings and Mine Waste Conference. 10-13 October 2004, Vail, Colorado, USA, 2004, p. 63-70.

Sustainability in the context of vibro stone column ground improvement techniques

C.J. Serridge, Pennine Vibropiling Ltd., Bacup, Lancashire, UK

Abstract
Natural aggregate resources are not unlimited and their extraction causes increasingly unacceptable geo-environmental impacts. As part of achieving environmental sustainability in ground improvement, there is an increasing desire to utilise recycled aggregates in vibro stone column techniques. Spent railway track ballast and crushed concrete probably have the greatest potential for this application in the UK. However, where such materials are considered for use it is important that they are "fit-for-purpose". Some applications of recycled aggregate in the context of vibro stone columns are presented, including a case history for both spent railway track ballast and crushed concrete. Also presented is an innovative technique for avoiding pollutant linkages via stone columns using "vibro concrete plug" technology.

Introduction
At its heart sustainable development is the simple idea of ensuring a better quality of life for everyone, now and for generations to come. The "sun diagram" in Figure 1 (Leiper, 2003) illustrates very effectively how the specific impacts of the construction business (including specialist geotechnical contracting), are related to the UK government's four objectives for sustainable development:

- Social progress which recognises the needs of everyone
- Effective protection of the environment
- Prudent use of natural resources
- Economic growth and prosperity

This paper attempts to outline and summarise how the use of recycled aggregates in vibro stone columns and also the use of "vibro concrete plug technology", can contribute to sustainable development in environmental geotechnics. More detailed discussion is given elsewhere (Serridge, 2005).

Vibro stone columns
Vibro stone columns have been used as a ground improvement/treatment technique in the UK since the 1960's and are currently the most common form of ground improvement employed. Useful commentary and guidance on the vibro stone column technique and its applications are given in Moseley and Priebe (1993) and Building Research Establishment (BRE) Report BR391, (2000) among others. While the use of recycled aggregates in stone columns is not an entirely new concept, historically the technique has consumed predominantly freshly quarried primary aggregates.

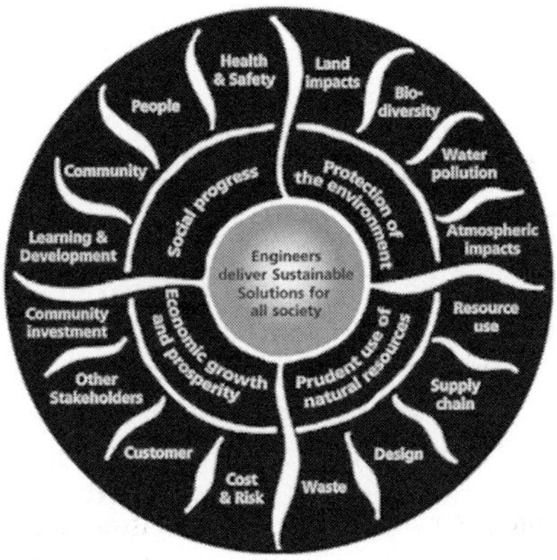

Figure 1: Sustainability "sun diagram" (Source: Leiper, 2003).

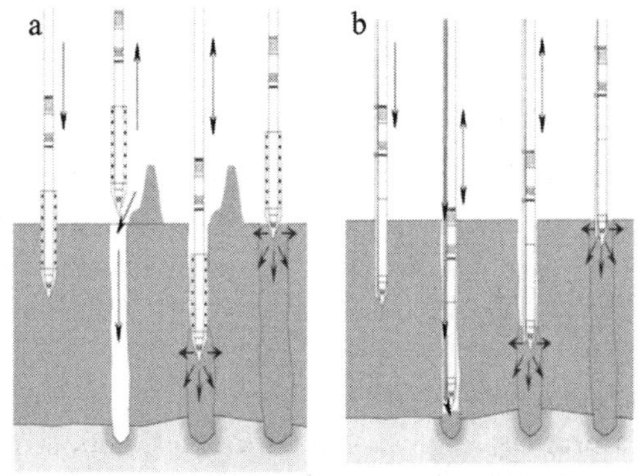

Figure 2: a) Vibro top feed technique and b) vibro bottom feed technique.

However, increasing awareness for sustainable development and construction is leading to a greater desire for use of recycled aggregates/materials in vibro stone column ground improvement (with utilisation in this sector currently estimated at around 25-30% and increasing). Recycling is of significant benefit through both reducing demand on natural (primary) aggregate resources (including associated environmental impact) and disposal of material to landfill. This contributes to the sustainability objectives of protection of the environment and prudent use of natural resources described above.

Within the UK stone column installation is generally carried out using the dry technique. The specific circumstances of bore stability (particularly in respect of fine-grained soils), and groundwater regime, determine whether a top feed (Figure 2a) or bottom feed (Figure 2b) technique is selected. Charges of stone are introduced (from the surface, in the case of the top feed technique, and by way of a tremie pipe in the case of the bottom feed technique) and compacted in stages by the vibroflot until a dense stone column is constructed to the surface. Within the UK the dry bottom feed technique (Figure 2b), has largely superseded the wet top feed technique on geo-environmental grounds.

Aggregate requirements for vibro stone columns and influence of stone column material on design

The material used to form stone columns should be "fit-for-purpose" and typically this requires it to be free draining, hard, inert and capable of forming stone columns with relatively high angles of friction; it must also comply with acceptable criteria in terms of material type, grading, particle shape, flakiness and chemical stability. Of particular importance is the Ten Percent Fines Value (TFV) (the test samples should be soaked where vibro treatment is anticipated to extend below groundwater level), BS 812, Part 111, (1992) and the Los Angeles Abrasion Value (LAA) which replaces the TFV in the new European Standards, BS EN 1097-2 (1998). Specifications for stone column material are included in the Institution of Civil Engineers (ICE) Specification for ground treatment (1987) and BRE Report BR391 (2000). Based upon vibroflot equipment currently operating in the UK the following aggregate grading specifications are normally adopted for the two main vibro stone column techniques:

(a) *Dry top feed technique:* a nominal single-sized material within the range 40-75mm or a graded material within these limits is appropriate with a maximum fines (clay/silt) content of 5% usually permitted.
(b) *Dry bottom feed technique:* a nominal single-sized material within the range 20-50mm is normally considered appropriate and the fines content should be less than 5%.

Where the wet top feed technique is utilised, a nominal single-sized material within the range 20-50 mm and with less than 5% fines is normally appropriate. Material selected to form stone columns should be able to withstand the impact forces of the vibrating poker and retain long-term integrity under the applied foundation loads. The fact that a compacted stone column is cohesionless implies that the load-carrying capacity of the column is a function of the angle of internal friction of the column material. The maximum adopted friction angle in the UK is 45 degrees, with 40 degrees being a more typical value used in design. Laboratory determined values which are higher than this would normally be downgraded to 45 degrees. Some further downgrading of friction angle may also be necessary when addressing stone columns in weak cohesive soils. The influence of friction angle on column capacity, Hughes and Withers (1974), and settlement, Priebe (1995) is clearly demonstrated in Table 1 and Figure 3.

Table 1: The ultimate carrying capacity of vibro stone columns (Hughes and Withers, 1974), for different angles of internal friction (φ´),normalised against 45°.

φ´	$\frac{(1 + \sin \varphi')}{(1 - \sin \varphi')}$	% (relative to 45°)
45°	5.828	100
40°	4.599	79
35°	3.690	63

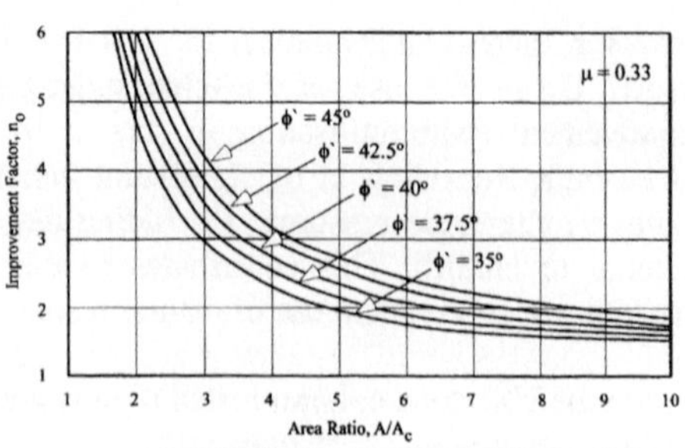

A - effective treatment area of stone column
μ - Poisson's ratio
A_c - cross sectional area of stone column
φ´- angle of friction of stone column

Figure 3: Settlement reduction factors (after Priebe,1995).

Opportunities exist for optimising the potential for use of recycled aggregates in vibro stone columns. The Specialist ground improvement/ground treatment contractor can contribute to this in the following ways:

(a) Requesting information on availability and cost of recycled and secondary aggregates from suppliers, when enquiries are made about primary aggregates for use in vibro stone columns. An increasing number of suppliers are now offering alternative sources to primary aggregates, dependent upon location and availability. Where appropriate, services such as *AggRegain*, among others, should be considered.
(b) Developing partnering approaches with suppliers offering recycled aggregates, to ensure consistency and quality of materials.
(c) Promoting use of recycled aggregates to clients, main contractors, architects, engineers and other relevant parties, by providing design options for recycled aggregates, particularly where there are clear potential environmental and economic gains (healthy 'green' profits) to be made, supported by appropriate quality control protocols and case histories. Encourage clients/stakeholders to permit options for recycling in conditions of contract and specifications.

(d) Adopting/establishing environmental management systems, to reinforce and further enhance the sustainability ethos/concept.
(e) Investigating the various possibilities for re-use of construction and demolition material (C&D) waste generated on redevelopment sites. This requires specialist ground improvement contractor involvement at an early stage to establish whether (with appropriate *in-situ* crushing, processing and quality control), these materials could be utilised in vibro stone columns.

To date, recycled railway track ballast and crushed concrete appear to have the greatest potential for use in vibro stone columns in the UK.

Recycled railway track ballast

Total arisings of spent railway ballast in England and Wales were around 1.30 million tonnes in 2001, of which some 95% was used as recycled aggregate, Office of the Deputy Prime Minister (2001). In Scotland the total arisings was around 0.1 million tonnes, 81% of which was recycled, Winter and Henderson (2001). While the majority of spent railway ballast is supplied to the construction industry from regional recycling and supply depots, as a fill or sub-base material, it is also proving an appropriate resource for a higher grade application, namely vibro stone columns. The properties of ballast render the material particularly suitable for use in stone column construction. Historically in the UK specifications have required ballast aggregate to have an upper size limit of 63 mm with not more than 0.8% finer than 1.18 mm, and with most of the material lying between 50 and 28 mm. The material was also required to be clean, hard, durable stone, angular in shape and free from dust. The flakiness index and elongation index was not to exceed 40 and 50% respectively and the Aggregate Crushing value (ACV) - BS 812, Part 111, (1992) was not to exceed 22% when the 14 to 10 mm fraction was tested. Within the railway network ballast is removed and replaced primarily because of inter-particle attrition and aggregate breakdown.

Use of spent railway ballast – case history

At a site in Streetley, West Midlands (UK) a 5.7ha brownfield site representing a large infilled former sand extraction pit has recently been re-developed for domestic housing. Site investigation showed the presence of up to a maximum of around 20 m of made ground comprising predominantly loose and loose-medium dense sand, ash, concrete and brick with some metal and minor constituents of wood, rubber and plastic. (However, the made ground was shown to be competent below around 10-12 m depth). The made ground deposits were underlain by 1-2 m thickness of clayey sand and sandy clay over sandstone bedrock. As a result, conventional shallow foundations were considered unsuitable without additional re-engineering or implementation of ground improvement techniques to provide a stable development plateau to mitigate the risk of experiencing unacceptable total and differential settlements. This was achieved by first undertaking the installation of vibro stone columns on a 2 m grid pattern and to a depth of approximately 10 m (locally 12 m along lines of proposed sewers and drainage). This involved the installation of approximately 14,500 columns to provide a minimum allowable bearing pressure of 75 kN/m^2 (following completion of stone column installation and construction of 1m thick geogrid reinforced granular layer) to accommodate raft foundations for the houses. Apart from screening and re-use of arisings resulting from regrading/re-contouring operations (to maximise on-site retention of materials and minimise off-site disposal), it was also decided to use recycled spent railway ballast as the principal aggregate source for construction of the stone columns.

The Bescot ballast recycling depot, near Wednesbury in the West Midlands, was selected as the source for the stone column aggregate, with in excess of 75,000 t required during the duration of the ground improvement contract.

Quality control procedures were implemented at source (and on site) to ensure that the aggregate had been processed, cleaned and graded to meet the stone column aggregate requirements outlined earlier in this paper. The Recycled spent railway ballast used was predominantly a 40-20 mm granite aggregate (occasionally dolerite sources were used), with a TFV - BS 812 Part 111,(1992) of 180 kN (soaked) and 220 kN (dry). Laboratory shear box testing confirmed the friction angle to be in excess of 43 degrees. It should be noted that the columns were also required to provide a secondary, passive gas pathway. Therefore it was essential that the aggregate had to be clean and free draining. Zone load testing of the treated ground (stone column-soil composite) recorded settlements of less than 10mm at two times the design bearing pressure. The 40-20 mm recycled ballast was transported to site over a distance of 9.5 km. Cheapest available suitable primary aggregate (around 20% higher than the cost of the recycled ballast) would have had to be transported from Rowley Regis over a distance of some 18 km. Recycled spent railway ballast therefore clearly proved a very sustainable solution.

Recycled crushed concrete aggregate

Apart from recycled spent railway ballast, crushed concrete has perhaps the greatest potential of all recycled aggregates for use in vibro stone columns, particularly in view of its availability, notably in urban or industrialised areas, where concrete structures are being demolished and floor slabs and concrete pavements broken up (Figure 4).

Using laboratory shear box testing, Sarsby (2000) reports angles of internal friction in the range 35-55 degrees for crushed concrete aggregate. Mc Kelvey *et al.* (2002) recorded a friction angle of 39 degrees for crushed concrete with particle sizes in the range 20-40 mm (values of this magnitude are acceptable and unlikely to result in significantly lower bearing capacity or higher settlement of any vibro stone column ground improvement).Of significance, however, is that this friction angle reduced to around 30 degrees when the crushed concrete was smeared with up to 20% kaolin slurry. Also, some particle crushing was observed at higher applied stresses, indicating that crushed concrete is slightly weaker than primary aggregates normally used in stone column construction. Hence in applications such as stone columns, it is important that designers and specifiers do consider these issues/aspects and ensure that the crushed concrete aggregate selected is "fit-for-purpose" and not contaminated with fines (clay/silt).

Use of recycled crushed concrete aggregate – case history

Between 4000 and 4500t of graded 75-40 mm recycled crushed concrete aggregate have been utilised for a vibro stone column ground improvement project in Coatbridge, Scotland. The site had a legacy of industrial activity, latterly having been occupied by an ironworks and was re-developed to accommodate a new large retail warehouse and garden centre (with associated access roads and car parking), following ground improvement using vibro stone columns. Ground conditions comprised a heterogeneous sequence of made ground (fill) deposits extending to depths of around 6.0 m including loose ash, slag and brick together with some clayey zones, underlain by natural more competent soils. Site investigation had also revealed the presence of remnant foundations (including chimney bases and floor slabs) and some significant zones of dense to very dense fused slag, which it was necessary to remediate before ground improvement and foundation construction. In order to implement the principle of "sustainable construction" it was decided by the main contractor to establish crushing and screening facilities on site to permit recycling of arisings from demolished structures, remnant foundations and concrete slabs. The objective of this was to permit production of crushed concrete and brick which was suitable for use as a granular infill associated with the remediation described above, in advance of vibro stone column installation and higher quality crushed concrete material (i.e. with less fines) for use in vibro stone column construction. Specifications and grading envelopes were provided by the specialist ground improvement contractor to confirm specification requirements for both granular backfill/infill to excavations associated with the remediation and stone column aggregate. Quality control procedures were also implemented on site, incorporating regular inspection, testing and monitoring to ensure consistency and quality of supply and therefore "fitness-for-purpose" (in line with stone column aggregate requirements discussed earlier in the paper).

The ground improvement was designed to improve bearing capacity and control settlements (total settlements were to be restricted to 25 mm and differential settlement, expressed as an angular distortion, to better than 1 in 500) in the made ground deposits (including loose granular backfill in remediated areas), to permit construction of shallow foundations and ground bearing floor slabs for the large retail warehouse and garden centre (and associated car parking and access). Stone columns were installed in closely spaced groups beneath pad and strip foundations (associated with the main portal frame structure, to provide a bearing pressure of 100 kN/m^2) and on a square grid arrangement beneath ground bearing floor slab areas and designated external hardstanding areas to provide bearing pressures in the range 30-50 kN/m^2. Stone column lengths typically ranged from 2.5 to 6.0 m to accommodate the variation in soil conditions beneath treatment areas. The crushed concrete aggregate proved very satisfactory for stone column construction. An average stone column deformation modulus of 45 MN/m^2 (Table 2) was determined from 600 mm diameter plate load tests on stone columns selected randomly for testing by the consulting engineer for the project.

Table 2: Calculated modulus (E) values from 600 mm diameter plate load tests on columns constructed with recycled crushed concrete.

Test Ref	Stone column length m	Test level m	Load kN/m^2	Poisson's Ratio μ	Settle-ment Mm	Modulus (E) MN/m^2
1	5.0	-0.55	318.3	0.3	2.94	46
2	3.8	-0.55	318.3	0.3	2.91	47
3	4.5	-0.55	176.8	0.3	1.58	48
4	3.6	-0.55	176.8	0.3	1.37	55
5	5.0	-0.55	318.3	0.3	3.30	41
6	3.6	-0.55	176.8	0.3	1.71	44
7	5.7	-0.55	176.8	0.3	1.56	49
8	6.0	-0.55	318.3	0.3	3.17	43
9	3.0	-0.55	176.8	0.3	2.15	35

Figure 4: Production of recycled crushed concrete.

Other potential sources of recycled aggregate

Other potential sources of reclaimed aggregate for use in vibro stone columns include inert general demolition materials and building debris (Recycled Aggregate-RA) and waste rock and steel slag (Secondary Aggregate-SA). However, the availability of secondary aggregates does depend on location. They are not as geographically widespread as recycled aggregates and there are various factors which may cause some limitations on their use. These aspects are discussed further in Serridge (2005).

Vibro concrete plug technology

Penetrative ground improvement techniques such as vibro stone columns can generate potential pathways for contaminant migration. This raises environmental concerns, particularly if sensitive groundwater or underlying aquifers are present due to potential pollutant linkages (i.e. source-pathway-receptor linkages). Some guidance on pollution prevention in this context is given in Environment Agency Report NC/99/73 (2001). However, an innovative approach to address this issue has been the development of "vibro concrete plug" technology, incorporating the introduction of lean mix concrete into the basal section (toe) of the stone column, thereby isolating any pathways for downward migration of contamination via the stone columns. The technique has been implemented successfully for a vibro stone column ground improvement/treatment project in Kidderminster, UK. Below potentially contaminated made ground (fill), essentially cohesive alluvial strata was present, but containing significant granular soil horizons with potential for hydraulic continuity with sensitive groundwater in the underlying Triassic sandstone aquifer. A schematic section through the ground treatment carried out and a typical section through a stone column-concrete plug composite is shown in Figure 5(a) and 5(b) respectively.

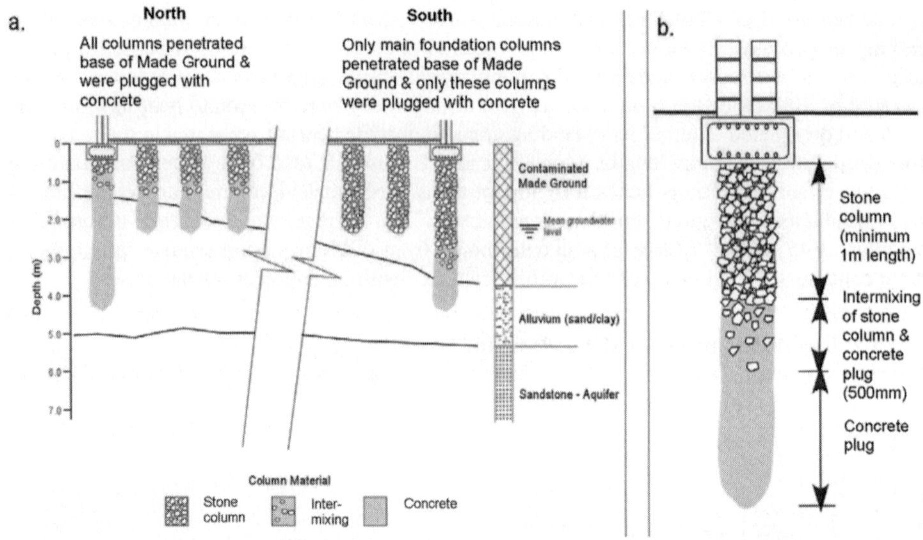

Figure : 5 a) Stone column/concrete plug arrangement on a site in Kidderminster (UK), b) Typical section through vibro stone column with concrete plug.

Adopting 'vibro concrete plug' technology in this way avoids pollutant linkages being formed from contamination/pollutants in the soils through which the vibro stone column ground improvement/treatment is installed.

Conclusions

Simple sustainability advances can be made through replacement of natural aggregates with recycled ones and greater use of *in-situ* and *ex-situ* recycling. To ensure effective utilisation it is important, particularly where vibro stone column techniques are being considered that there is specialist contractor involvement at as early a stage as possible.

Specifications and quality control systems/protocols are available which will assist in ensuring that recycled aggregates selected for application in stone columns are "fit-for-purpose". The fines content, in particular, has a significant effect on the angle of internal friction of the stone column material and therefore load carrying capacity and settlement potential of the treated ground.

Based upon the author's experience, the best opportunities to date for use of recycled aggregates in vibro stone columns in the UK have been with recycled spent railway ballast and crushed concrete. Brief case histories have been provided for both these materials. Further dissemination of this information within the industry, will ensure that more advances can be made with use of recycled aggregates in vibro stone column ground improvement techniques.

Vibro concrete plugs are an effective method of avoiding contaminant migration via stone columns on brownfield sites with underlying sensitive ground waters.

References

AggRegain web site (www.aggregain.org.uk)

British Standards Institution (1990). Methods for sampling and testing of mineral aggregates and fillers. *BS 812: 100 Series. Testing aggregates – Part 111: Method for determination of ten per cent fines (TFV). Part 110: Method for determination of aggregate crushing value (ACV). BSI, London.*

British Standards Institution (1998). BS EN 1097. Test methods - Physical and Mechanical Part 2 - *Methods for determination of resistance to fragmentation, Los Angeles Abrasion (LAA) test, BSI, London, BS EN 1097-2.*

Building Research Establishment (2000). Specifying vibro stone columns. *BRE Report BR391, CRC Ltd., Garston, UK.*

Environment Agency (2001). Piling and penetrative ground improvement methods on land affected by contamination: *Guidance on Pollution Prevention. NC/99/73.*

Hughes, J.M. and Withers, N.J. (1974). Reinforcing of soft cohesive soils with stone columns. *Ground Engineering, May, pp. 42-49.*

Institution of Civil Engineers (1987). *Specification for ground treatment, ICE, Thomas Telford, London.*

Leiper, Q. (2003). Editorial. *Journal of Engineering Sustainability, 156, June, issue ES2, pp. 67-68. Proceedings of the Institution of Civil Engineers, Thomas Telford, London.*

Mc Kelvey, D., Sivakumar, V., Bell, A. and Mc Laverty, G. (2002). Shear strength of recycled construction materials intended for use in vibro ground improvement, *Ground Improvement, Vol. 6, No. 2, pp. 59-68, Thomas Telford, London.*

Moseley, M.P. and Priebe, H.J.(1993). Vibro techniques, *Ground improvement, (Ed. Moseley, M.P.), pp. 1-19, Blackie Academic and Professional, London.*

Office of the Deputy Prime Minister (2001). *Survey of arisings and use of secondary materials as aggregate in England and Wales* www.planning.odpm.gov.uk/consdemo

Priebe, H.J. (1995). The design of vibro replacement. *Ground Engineering, December, pp. 31-37.*

Sarsby, R.W. (2000). *Environmental geotechnics. Thomas Telford, London.*

Serridge, C.J. (2005). Achieving sustainability in vibro stone column techniques. *Journal of Engineering Sustainability 158, December, issue ES4, pp. 211-222. Proceedings of the Institution of Civil Engineers, Thomas Telford, London.*

Winter, M.G. and Henderson, C. (2001). *Recycled aggregates in Scotland.* Edinburgh, Scottish Executive Central Research Unit.

Sustainable development: Durability of natural geotextiles as erosion control product

Singh, R. M., Geoenvironmental Research Centre, Cardiff School of Engineering, Cardiff University, Cardiff, UK
Venkatappa Rao, G., Indian Institute of Technology Delhi, Department of Civil Engineering, New Delhi 110016 India.

Abstract

Natural geotextiles made up of coconut husk is an emerging product for erosion control of soil slopes. Coconut is abundantly available in tropical countries and generates large quantities of organic waste, so use of it as an engineering material to preserve the environment is a good example of sustainable development. In view of the inadequate information on the biodegradability behaviour of natural geotextiles, a series of tensile tests were conducted on four types of woven coir geotextiles to understand the durability behaviour under the influence of different environmental conditions, viz. submerged in water, embedded in saturated kaolinite clay and moist manure. Basically accelerated degradation studies were done by keeping the specimens in adverse conditions. The results of the experiments established that most of the geotextiles remained unaffected in water and clay even after six months. However, the tensile strength of most of the geotextiles was reduced considerably in aerobically decomposed manure.

Introduction

Modern living compels the mankind to avoid indiscriminate use of plastics and such ecologically damaging materials. Industries that manufacture these products generate vast quantities of chemical wastes that pollute the air, land and water. The non-degradable plastics prevent recharge of ground water and chokes drainage. The consequent environmental degradation is threatening the very existence of all forms of life on this planet. Nevertheless, there is an increasing awareness around the world over the urgency to maintain the much needed biological balance. It is imperative that any developmental effort should go hand in hand with ecological equilibrium so as to sustain the equilibrium and a return to natural living has been accepted as a safe way to solve many environmental problems. Thus in the geosynthetic field coir geotextiles become an ecologically acceptable material.

Majority of geosynthetics, covering a wide range of woven and non-woven geotextiles, geogrids, geomembranes and geocomposties used in civil engineering applications are polymeric. These products generally have a long life and do not undergo biological degradation, but are liable to create environmental problems in the long run. Throughout the world growing awareness of the sustainable development to preserve the environment has led to the rehabilitation of areas damaged either by natural or industrial causes. Hence, the use of biodegradable natural materials is gaining popularity.

Coir geotextile, a generic member of the geosynthetic family is made from coconut fibre extracted from the husk of the coconut fruit. Like their polymeric counterpart, coir geotextiles are developed for specific application in civil engineering like erosion control, ground improvement etc. Coir fibre can be converted into fabric both by woven and non-woven (air laid, needle punched, adhesive bonded) processes. From these, a variety of composites like rolled erosion control products, drains and vegetation fascines can be manufactured to suit particular applications. It can be easily blended with man made fibres too. Its low cost makes it attractive for geotechnical applications. The major drawback is its biodegradability. However, coir has begun to be accepted as a geotextile material suitable for specific application in civil engineering. Field trials conducted in Australia, Germany and India have proved coir geotextiles as an excellent material for riverbank protection and embankment stabilisation (Cammack, 1988; Schurholz, 1991; Venkatarangaraju et al., 1988; Natarajan et al., 1988; Sudhakran, 1993). Coir as a biodegradable and environmentally friendly material is virtually irreplaceable by any of the modern synthetic substitutes. In view of this, a research was undertaken to study the feasibility of using coir geotextiles in geotechnical applications. This calls for a thorough understanding of the engineering and biodegradability behaviour of coir in isolation and in conjunction with soils.

The objective of the present paper is to study the durability of natural coir geotextile subjected to various adverse conditions. In view of the inadequate information on the engineering properties and biodegradability behaviour of natural geotextiles, a series of tensile tests were conducted on four types of woven coir geotextiles kept under different conditions. Three types of test conditions were adopted for the study, namely; (i) submerged in water, (ii) embedded in saturated kaolinite clay, and (iii) embedded in moist manure. The influence of these test conditions on the tensile strength of the above mentioned geotextiles were studied for a period of 6 months in machine (m/c) and cross-machine (x-m/c) directions.

Material

Four types of woven coir geotextiles namely type A, B, C and D has been used for durability studies. These geotextiles have different aperture size, thickness and density as shown in Fig.1. The coir has 54% cellulose and 46% lignin. The rate of decomposition of coir fibre is much less than that of jute due to the high content of lignin. The various physical properties measured as per ASTM specifications are given in Table 1. The thickness of the specimens was measured after loading it to a normal pressure of 2 kPa for 1 minute as per ASTM D5199-01. Density (mass per unit area) was determined on the specimens of size 27 cm x 20 cm according to ASTM D5261-92.

Table 1 Physical property of natural woven geotextiles

Property	Woven geotextile			
	A	B	C	D
Width of roll (m)	1.2	1.2	1.2	1.2
Length of roll (m)	25	26	25	24
Weight of roll (kg)	10	11	18	42
Aperture size (mm x mm)	45 x 30	25 x 25	6.25 x 6.25	7 x 4
Thickness (mm)	7.27	6.69	8.04	9.54
Mass per unit area (g/m^2)	335	361	608	1335

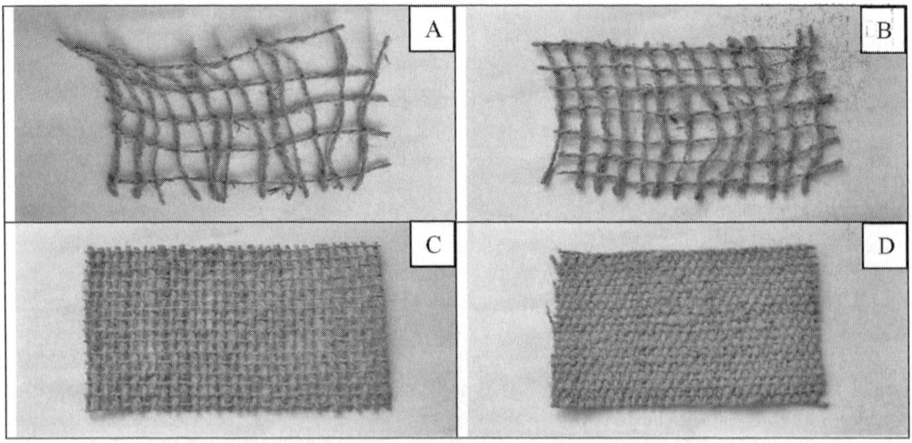

Fig. 1. View of coir geotextiles Type A, B, C and D

Methodology

The specimens of size 27cm x 20cm were cut out from the rolls of geotextiles in machine (m/c) direction and cross-machine (x-m/c) direction based on ASTM D4595-86. The tensile strength of the specimens has been measured in m/c direction and x-m/c direction for all four types A, B, C and D by wide width tensile test. The wide width tensile test has been carried out at 10 mm per minute strain rate as per the ASTM specifications on a computer controlled Hounsfield Universal Testing Machine as shown in Fig. 2. Aluminium gripping plates with serrated face were used to hold the specimens during the tensile strength tests as shown in Fig. 3. The specimens were gripped at both ends such that the un-gripped length of the specimens was 10 cm as per the ASTM specifications. The specimens should be properly gripped at both ends in order to prevent slipping during the tensile test. The specimens were placed in accelerated degradation conditions, namely, (i) submerged in water, (ii) embedded in saturated kaolinite clay, and (iii) embedded in moist manure for six months to investigate the influence of these factors on the durability of the coir fibres.

Five specimens of each type have been considered for the initial tensile tests and three specimens of each type have been considered for the accelerated degradation conditions as show in the attached testing programme (Table 2).

Table 2 Testing programme: Number of specimens used

Type of test	A		B		C		D	
	m/c	x-m/c	m/c	x-m/c	m/c	x-m/c	m/c	x-m/c
Initial tensile test	5	5	5	5	5	5	5	5
Submerged in water	3	3	3	3	3	3	3	3
Embedded in Saturated kaolinite clay	3	3	3	3	3	3	3	3
Embedded in moist manure	3	3	3	3	3	3	3	3

Fig. 2. Hounsfield testing machine

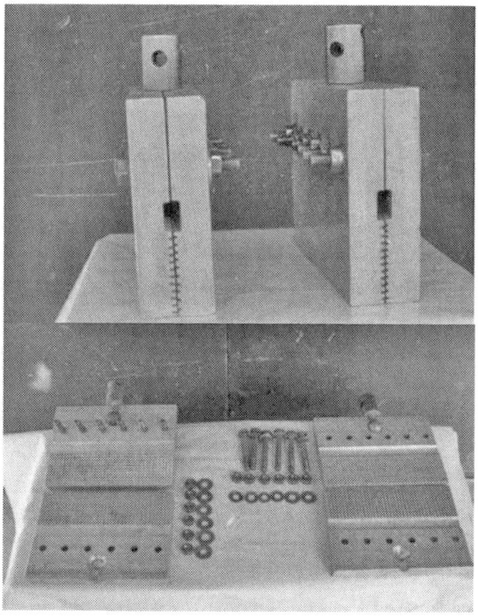

Fig. 3. Gripping plates

Submerged in water

Three specimens in machine direction (m/c) and three specimens in cross-machine (x-m/c) direction for all four varieties of coir geotextiles were submerged in tap water and kept at room temperature for 6 months. After six months, the specimens were taken out of water and dried up in an artificial sunbathe. The dried up specimens were tensile tested to examine the effect of water on the strength of the coir geotextiles.

Embedded in saturated kaolinite clay

Three specimens in machine (m/c) direction and three specimens in cross-machine (x-m/c) direction (of all the four types of coir geotextiles) were embedded in saturated kaolinite clay in laboratory trays as shown in Fig. 4. These trays were kept in a humidity chamber at a temperature of 30 ± 1°C and a relative humidity of 90 ± 1% for a duration of 6 months. At the end of 6 months, the specimens were retrieved, washed with water to remove the adsorbed clay particles and dried in an artificial sunbathe. The dry samples were then tensile tested to determine the effect of saturated clay on the tensile strength of the natural geotextiles.

Embedded in moist manure

The above test procedure was repeated for the moist manure case as well, in order to determine the effect of moist manure on the tensile strength of the natural geotextiles.

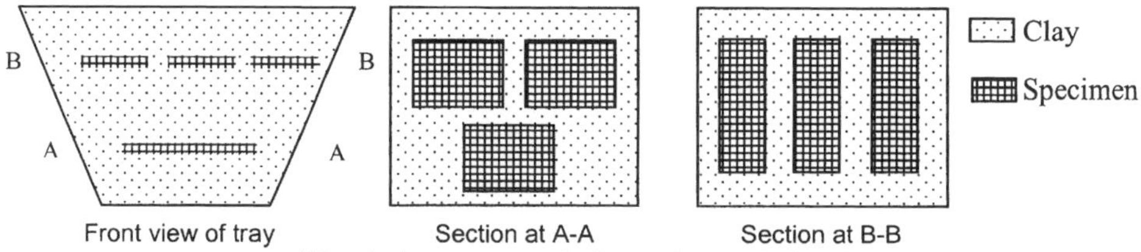

Fig. 4. Arrangement of specimens in tray

Results

The geotextiles exhibit high initial tensile strength and elongation in m/c direction compared to x-m/c, as shown in Fig. 4. The type D geotextile shows the highest tensile strength in both directions due to the fact that it has the highest density compared to other types as mentioned in Table 1.

Submerged in water

Fig. 5 and 6 show the effect of water on strength in both m/c and x-m/c directions respectively. The results suggest that water does not affect the tensile strength considerably for any of the geotextiles even after 6 month immersion, but elongation at failure is shown to increase which can be attributed to water absorption by the coir yarn resulting in increased elasticity. This statement is further reinforced by the observations made by Schurholz (1991) who found that when natural coirs were exposed to water continuously for 167 days (in order to simulate the traction effect while flooding), it had almost no damage.

Embedded in saturated kaolinite clay

The influence of saturated clay on tensile strength of the geotextiles in m/c and x-m/c direction is shown in Fig. 7 and 8. Similar to water, clay also does not show any considerable reduction in the strength of any of the geotextiles but elongation at failure is shown to increase which can be attributed to water absorption by the coir yarn from the saturated clay resulting in increased elasticity. This behaviour has been confirmed by Schurholz (1991) who reported that the natural coir he used retained approximately 80% of its strength after one year in an incubator test with soil (at 28-30°C and 90% humidity). Similarly, Balan (1995) reported that the natural coir he used also retained its initial strength for up to 3 months.

Embedded in moist manure

The influence of moist manure on tensile strength of the geotextiles in m/c and x-m/c direction is shown in Fig. 9 and 10. The results show a considerable reduction in the strength of the geotextiles in both directions (m/c and x-m/c). This can be attributed to the bacterial growth (present in the manure) which biodegrade the geotextiles. The elongation at failure also dropped marginally because of biodegradation of coir yarns. Similar results have been obtained by Balan (1995) and Rao et al. (2000).

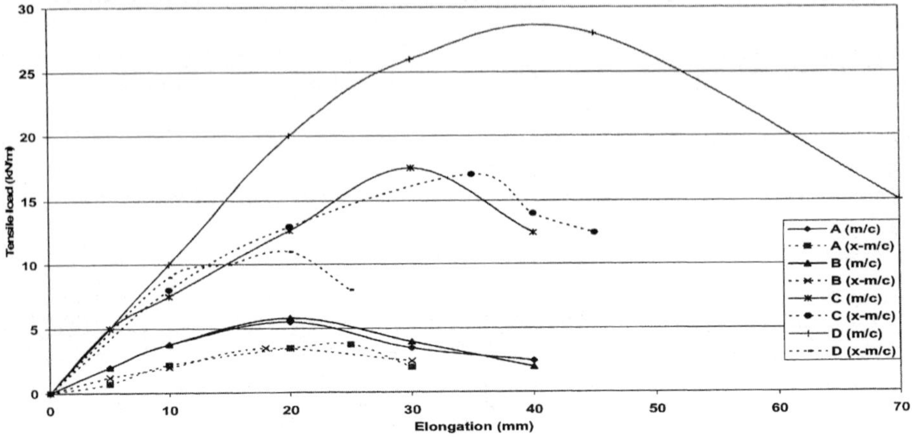

Fig. 4. Initial tensile strength in m/c and x-m/c direction

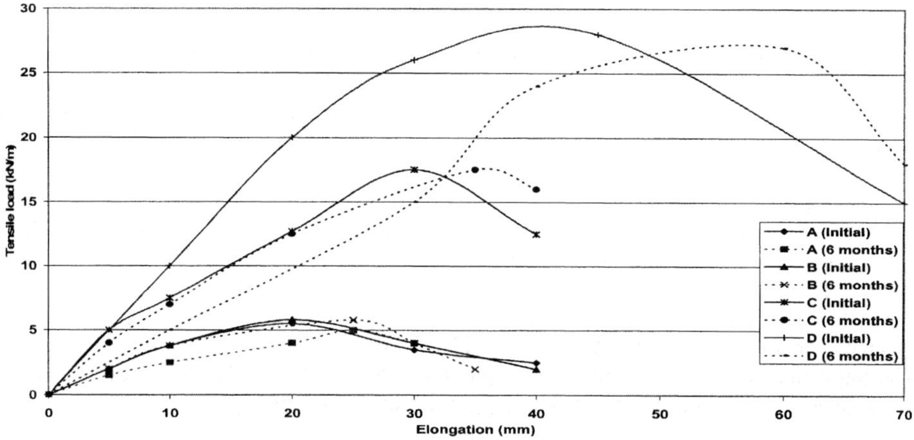

Fig. 5. Effect of water on tensile strength in m/c direction

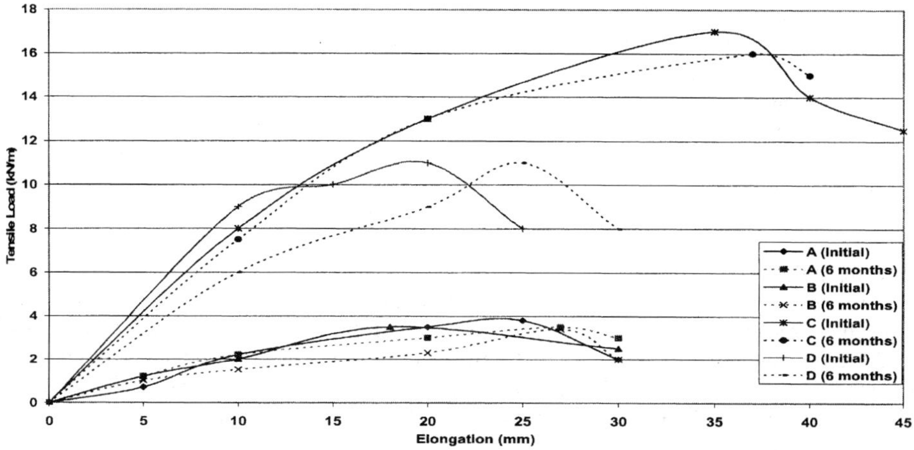

Fig. 6. Effect of water on tensile strength in x-m/c direction

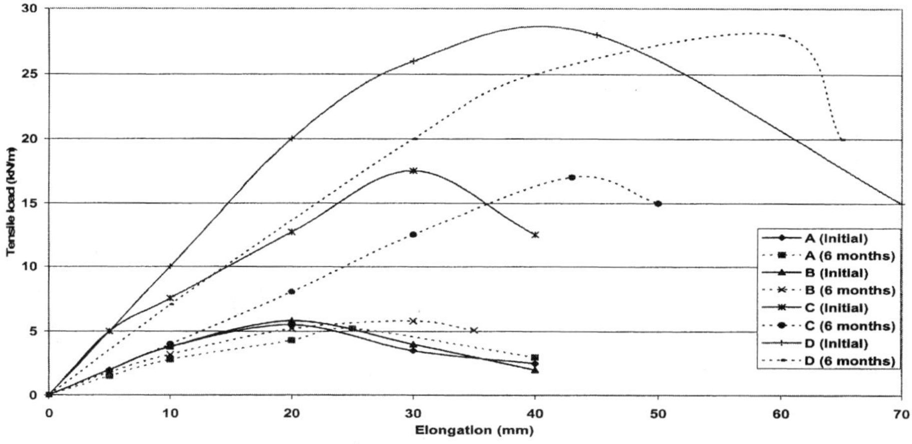

Fig. 7. Effect of clay on tensile strength in m/c direction

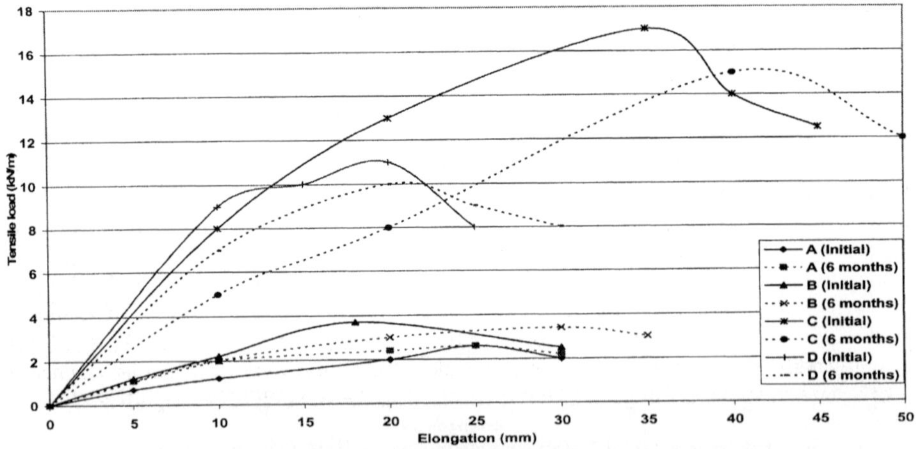

Fig. 8. Effect of clay on tensile strength in x-m/c direction

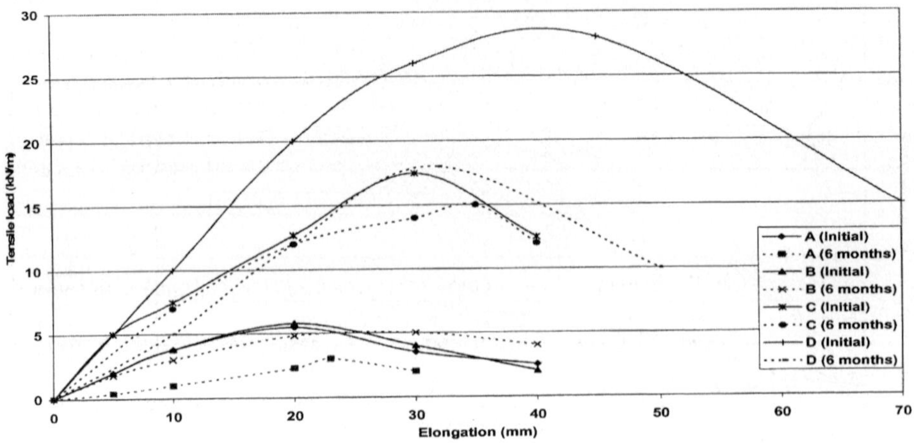

Fig. 9. Effect of manure on tensile strength in m/c direction

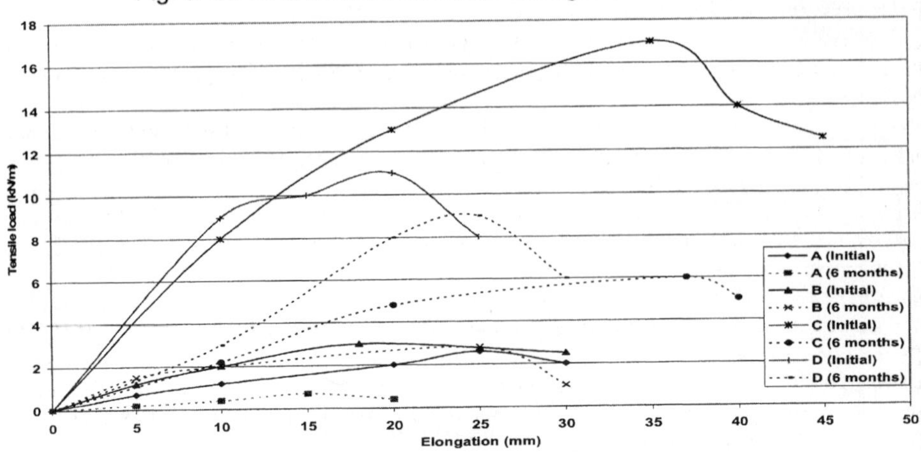

Fig. 10. Effect of manure on tensile strength in x-m/c direction

Discussion

From the tests, it is evident that both water and clay do not affect the strength of the geotextiles but adverse biodegradable condition of manure can weaken the natural coir yarns. Although natural coirs biodegrade very quickly in hostile conditions like moist manure, it can still be used as an erosion control product on natural soil slopes where such conditions (bacterial growth) do not exist. Natural coir geotextile can be used successfully in many applications where more expensive polymeric fibres are now used. Specifically, the type D geotextile has very high tensile strength and could be used as a reinforcement material in pavements and retaining walls. The observations made in this study have also been confirmed by many previous studies. For instance, the performance of coir geotextiles (of density 500 g/m^2) in reducing the water velocity and soil loss is reported to be 77.0% and 98.4% respectively relative to bare soil conditions (Cammack, 1988). Sarsby et al. (1992) report that a retaining wall reinforced with simple ropes made from natural fibres in a similar manner to that of a conventional steel strips can be used to enhance the strength of soil significantly.

Although natural fibre reinforcements used are fairly extensible and most of their extension will take place during construction period and thus will not affect their visual performance. Natural woven coir geotextile has many potential applications such as separation and reinforcement in road, small height embankments in soft soils, stabilisation of slopes, river bank protection and land reclamation. It is quite a cost effective solution as compared to polymeric geosynthetics to many soil engineering problems, so it can be used in various developing countries without posing any danger to environment. It is claimed that this work is an attempt to supplement the existing knowledge of the mechanical behaviour of the natural geotextiles.

References

ASTM D5199-01 (2001) Standard Test Method for Measuring the Nominal Thickness of Geosynthetics ASTM International

ASTM D5261-92 (1996) Standard Test Method for Measuring Mass per Unit Area of Geotextiles ASTM International

ASTM D4595-86 (2001) Standard Test Method for Tensile Properties of Geotextiles by the Wide-Width Strip Method ASTM International

Balan, K., 1995. Studies on engineering behaviour and uses of geotextiles with natural fibres. PhD thesis submitted to Indian Institute of Technology Delhi.

Cammack, A., 1988. A role for coir fibre geofabrics in soil stabilisation and erosion control. Proc. Workshop on coir geogrids and geofabrics in civil engineering practice, Coimbatore, India, 28-31.

Natarajan, T.K., Rao, P.J., and Murty, A.V.S.R., 1988. Coir netting for erosion treatment and surficial landslide correction. Proc. Workshop on coir geogrids and geofabrics in civil engineering practice, Coimbatore, India, 21-27.

Sarsby, R.W., Ali, M., Alwis, RDe., Khaffaf, J.H., and McDougall, J.M., 1992. Low cots oil reinforcement for developing countries. Int. Conf. on Non Woven, The textile Institute, North India Section, 297-310.

Schurholz, H., 1991. Use of woven coir geotextile in Europe. Coir Vol. XXXV, No. 2, 18-25.

Sudhakaran, P.M., 1993. Protection to the side slopes of kabini canal. Central Institute of Coir Technology, Bangalore.

Venkatappa Rao, G., Dutta, R.K., and Sivaramakrishnan, R., 2000. Evaluation of Indian made erosion control products. Coir geotextiles emerging trends, publisher Kerala State Coir Corporation Ltd. Kerala, India.

Predicting ecological environment for a residential community using intelligent hybrid system

Hong-Ci Wu, School of Civil & Architectural Eng., Guizhou Univ., Guiyang, China
B.-Y. Lu, School of Civil & Architectural Eng., Guizhou Univ., Guiyang, China
Ji-Bing Wei, School of Civil & Architectural Eng., Guizhou Univ., Guiyang, China
Bin-Jie Wu, School of Civil & Architectural Eng., Guizhou Univ., Guiyang, China

Abstract
In this paper, intelligent hybrid system (IHS), in which fuzzy reasoning, Genetic Algorithm (GA) and neural network based on fuzzy rule (NNFR) are combined, is employed to incorporate diverse residential environment aspects and predict ecological environment levels. Based on connotation of composite ecosystem, the affected ecological environment aspects (and here the input patterns for the IHS) in such predictions are extended to consider eight random factors (and here linguistic variables): afforested environment, atmospheric environment, water environment, sound environment, traffic environment, energy supply, waste disposal, and civilization environment. The predicted result (and here the output pattern for the IHS) is represented by one linguistic variable, i.e. ecological environment level for residential community. Consequently, the IHS enables us to predict successfully the ecological environment levels of a residential community in Guiyang city of China, and the effectiveness of the IHS is verified and compared with in situ measurement.

Introduction
Prediction of ecological environment has received increased attention in China in the last decade, the focus of concern is on the lack of a general code of practice on ecological environmental assessments, which has led to non-uniform approval practices. Despite such, current prediction for ecological environment in China is still often made autonomously and without fully investigating the influence of diverse environment aspects on the ecological prediction. Because many qualitative, fuzzy and uncertain aspects (and here linguistic variables) are involved in the procedure of predicting the most suitable ecological environment levels for a residential community, in this paper we present an initial attempt to consider more aspects of affected prediction of ecological environment, such as afforested environment, atmospheric environment, water environment, sound environment, traffic environment, energy supply, waste disposal, and civilization environment etc, and predict the suitable ecological environment levels by means of the IHS, consisting of fuzzy reasoning [Cho 1995; Czogala 2000; Grabish 1995; Klir 2000; Wu 1997a, 2001, 2002, 2003, 2004], the GA [Bäck 1996; Davis 1990; Grefenstette 1986, 1994; Wu 1998, 2001, 2002, 2003, 2004], and the NNFR [Cho, S. B. 1995; Cho, K.B. 1996; Czogala 2000; Haykin 1999; Pham 2000; Sundararajan 2000; Wu 1997b, 1999, 2001, 2002, 2003, 2004]. The domain of application of the IHS here is the preliminary ecological environment prediction of a residential community in Guiyang city of China, and the results of the study are pretty well.

Fuzzy reasoning for problem of environment prediction

The problem for the prediction of ecological environment levels of a residential community can be represented as a procedure of fuzzy reasoning [Cho 1995; Czogala 2000; Grabish 1995; Klir 2000; Wu 1997a, 2001, 2002, 2003, 2004]. Suppose that n number of m-input and one-output fuzzy rules or fuzzy conditional statements are given. The i-th rule in the if-then form, which describes input-output relation, may be written in the following form:

$$R_i\text{: if}\{(x_{1i}=A_{1i}) \wedge (x_{2i}=A_{2i}) \wedge \ldots \wedge (x_{mi}=A_{mi})\}\text{, Then }\{y=B_i\}\} \quad (1)$$

Where $x_{1i}, x_{2i}, \ldots, x_{mi}$ and y are respectively linguistic variables of the input and the output patterns, A_{ji} (j=1, 2, …, m; i=1, 2, …, n) and B_i are fuzzy subsets of x_{ji} and y, respectively.

When an observation $(x_{1i}, x_{2i}, \ldots, x_{mi})$ is given , a fuzzy inference output y can be obtained by using product-sum-gravity fuzzy reasoning method as follows:

$$h_i=A_{1i}(x_1)\,A_{2i}(x_2)\ldots A_{mi}(x_m)=\prod_{j=1}^{m} A_{ji}(x_j) \quad (2)$$

$$y=\left(\sum_{i=1}^{n} h_i s_i y_i\right)/\left(\sum_{i=1}^{n} h_i s_i\right) \quad (3)$$

Where h_i (i=1, 2, …, n) is an agreement of the antecedent of the i-th fuzzy rule, s_i is the area of Bi, and y_i is the center of gravity of Bi.

The input patterns here are denoted by means of eight linguistic variables: x_{1i}=afforested environment, x_{2i}=atmospheric environment, x_{3i}=water environment, x_{4i}=sound environment, x_{5i}=traffic environment, x_{6i}=energy supply, x_{7i}=waste disposal, and x_{8i}=civilization environment [Grabish 1995; Pham 2000; Tian 2004; Wu 1997a, 2001, 2002, 2003, 2004]. The output pattern is represented by one linguistic variable, y=ecological environment level [Tian 2004; Wu 1997a, 2001, 2002, 2003, 2004], these linguistic variables are normalized, thus taking values in the interval [0, 1]. Consequently, the proposed IHS enables us to model the relations (1).

A procedure of predicting ecological environment using The NNFR

The learning capability of the NNFR can be used for automatic fuzzy if-then rules generation. The NNFR can utilize linguistic information from the human expert and measurement data, in other words, linguistic prior knowledge from expert and in-situ can be incorporated into the NNFR to identify and predict the defined problem [Cho 1995; Czogala 2000; Grabish 1995; Klir 2000; Wu 1997a, 2001, 2002, 2003, 2004]. The NNFR, shown in Figure 1, possesses three layers: one input, two hidden and one output layer. In the input layer, eight linguistic variables, i.e. $(x_{1i}, x_{2i}, \ldots, x_{8i})$ are to be presented to the network. The output layer, with one neuron, is used to predict the alternatives i.e. the ecological environment levels. When the training input-output variables $(x_1, x_2, \ldots, x_m, y^*)$ are given, the following inference error function E is adopted to evaluate an error between y^* and y:

$$E=\sum_{i=1}^{n} (y^* - y)^2/n \quad (4)$$

Where y^* is the desired output, and y is the corresponding fuzzy inference output, and Guassian-type membership functions A_{ji} and B_i (j=1, 2, …, m; i=1, 2, …, n) of (1) are defined as:

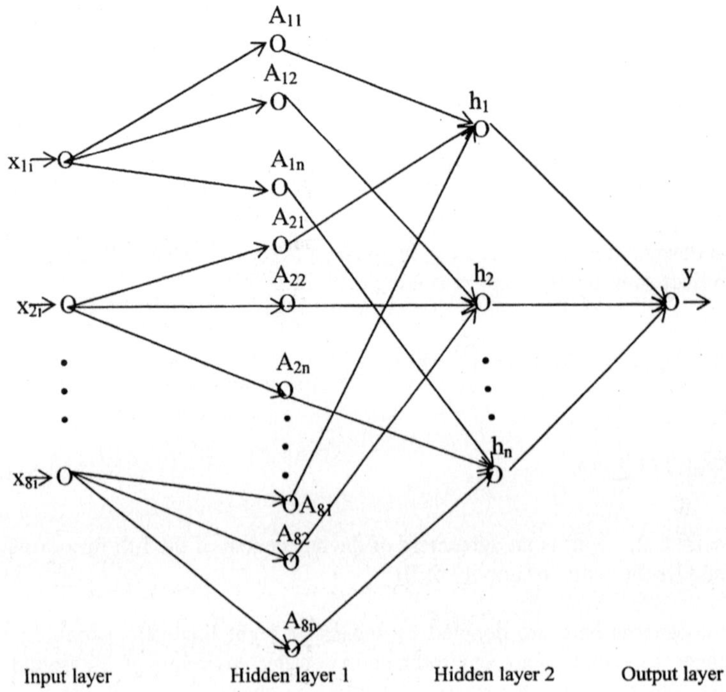

Figure 1. The NNFR

$$A_{ji}(x_j)=\exp(-(x_j - a_{ji})^2 / 2\sigma_{ji}^2) \quad (5)$$

$$B_i(y)=\exp(-(y - y_i)^2 / 2\sigma_i^2) \quad (6)$$

Where a_{ji} and σ_{ji} (j= 1, 2, …, m; i= 1, 2, …, n) are the center and width of A_{ji}, and y_i and σ_i (i=1, 2, …, n) are the center and width of Bi, respectively. In order to minimize the inference error function E, a self-tuning learning algorithm for tuning the parameters a_{ji}, σ_{ji} and y_i is adopted in the following forms [Cho, K.B. 1996; Cho, S.B. 1995; Haykin 1999; Pham 2000; Sundararajan 2000; Wu 1997A, 1997b, 1999, 2001, 2002, 2003, 2004]:

$$a_{ji}(t+1)=a_{ji}(t) - \alpha\,\partial E / \partial a_{ji}(t) = a_{ji}(t) + \alpha (y^* - y)(y_i - y)\, h_i\, (x_j - a_{ji}) / (\sigma_{ji}^2 \sum_{i=1}^{n} h_i) \quad (7)$$

$$\sigma_{ji}(t+1)=\sigma_{ji}(t) - \beta\,\partial E / \partial \sigma_{ji}(t) = \sigma_{ji}(t) + \beta(y^* - y)(y_i - y)\, h_i\, (x_j - a_{ji})^2 / (\sigma_{ji}^3 \sum_{i=1}^{n} h_i) \quad (8)$$

$$y_i(t+1)=y_i(t) - \gamma\,\partial E / \partial y_i(t) = y_i(t) + \gamma (y^* - y)\, h_i / (\sum_{i=1}^{n} h_i) \quad (9)$$

Where α, β and γ denote the learning rates, and t means the learning iteration. Two hidden layers are adopted to realize basis functions h_i and A_{ji} defined by equations (2) and (5), i.e. the membership functions and fuzzy rules are tuned by using a self-tuning learning algorithm. The learning procedure of the self-tuning algorithms is accomplished by minimizing the inference error function E of the network in (4). The self-tuning learning algorithms allow us to reach both an optimal approximation and save computational time.

Genetic training algorithm in the prediction

The NNFR is 'trained' by presentation of 32 training-case sets associated with the above input and output patterns. The GA here is applied to modify / learn the definition of the membership function shapes and / or the composition of the fuzzy rules [Bäck 1996; Davis 1990; Grefenstette 1986, 1994; Tian 2004; Wu 1998, 2001, 2002, 2003, 2004]. In order to do so, the linguistic variables here are further partitioned by fuzzy characteristics: x_{1i}={excellent, good, average, passable, poor}; x_{2i}={excellent, good, average, passable, poor}; x_{3i}={excellent, good, average, poor, very poor}; x_{4i}={very silent, silent, fair, noisy, very noisy}; x_{5i}={very convenient, convenient, common, inconvenient, very inconvenient}; x_{6i}={very sufficient, sufficient, fair, insufficient, very insufficient}; x_{7i}={very advanced, advanced, common, backward, very backward}; x_{8i}={very harmonious, harmonious, regular, inharmonious, very inharmonious}; y_i={excellent, good, average, passable, un-passable} [Grabish 1995; Tian 2004; Wu 1997a, 1997b, 1998, 1999, 2001, 2002, 2003, 2004], each one is composed of five linguistic labels (and here L_{MR}, M=1, 2, ..., 8, R=1, 2, ..., 5, for example, L_{34} denotes that the water environment in a residential community is poor) [Bäck 1996; Davis 1990; Grefenstette 1986, 1994; Tian 2004; Wu 1997a, 1997b, 1998, 1999, 2001, 2002, 2003, 2004], thereupon these fuzzy partitions can be obtained from a normalization process in which the universe of discourse of each variable is equally divided into 5 parts, and a triangular fuzzy set is associated to each one of them. An example of a fuzzy partition with five linguistic labels is shown diagrammatically in Figure 2.

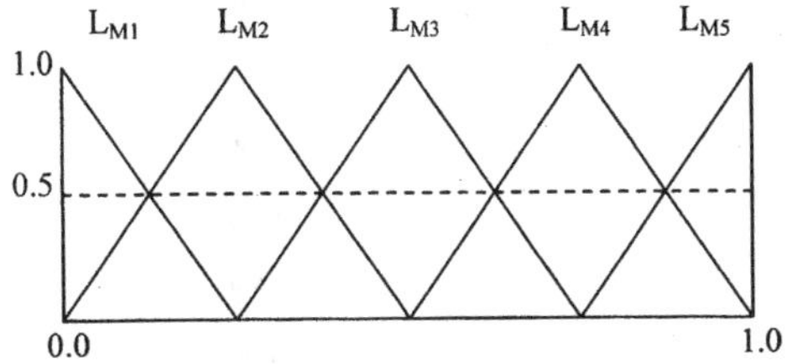

Figure 2. Example of fuzzy partition with five linguistic labels

The fuzzy partitions for each input variable in the process of genetic training are allowed to be different from other ones, the membership function can be used more than one time in the fuzzy rule base, and the representation of the fuzzy rule base is never destroyed even after the learning process, so that the case of weak-firing can be well avoided by the learning algorithms in general.

To overcome the permutation problem (the offspring will duplicate some elements of the permutation and will omit others), we here use a structure of parallelism in genetic learning procedure by dividing the global population into several sub-populations. Each sub-population is evolved on each neuron in the hidden layer and the input layer. Population size = 30; number of generation, g = 100; probability of crossover = 0.6; and probability of mutation = 0.01. The GA basically consists of three different genetic operators: selection, crossover and mutation.

The selection is a process in which a new 'generation' of population is formed by randomly selecting strings from an existing population, a two-stage selection scheme here is adopted. In the first stage (e.g. for the first 10 generations), all populations are conserved. That is, for each population, the best individual is first selected to be reproduced; the rest of the population is applied an extinctive selection with linear ranking. In the next stage, the worst predefined populations are rejected and new population conservation is maintained. The first stage increases the possibility of searching widely across the whole search space. The second stage reduces the number of search paths and increases the search efficiency, and this procedure can be repeated more than once if necessary.

The crossover operator is responsible for producing an offspring from the selected parents. Suppose that two parents from a population having 11 bits are chosen for crossover. The crossover site is decided randomly. If the crossover site is between bits 4 and 5, the first offspring will get the first four bits of chromosome from parent 1 and the last seven bits from parent 2; while the second offspring will get the first four bits of chromosome from parent 2 and the last seven bits of chromosome from parent 1. This procedure is shown diagrammatically in Figure 3.

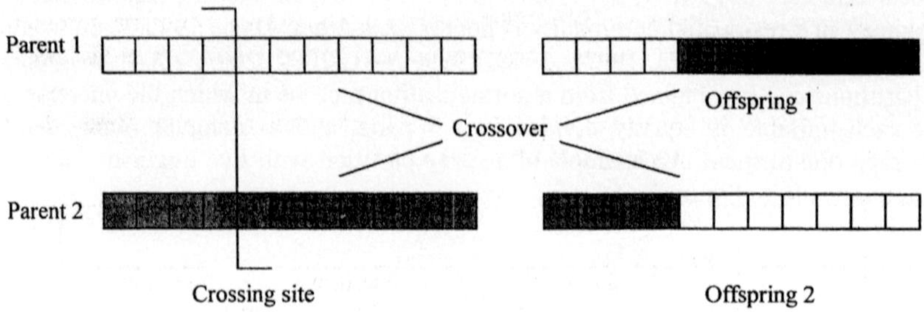

Figure 3. Graphic Scheme of One-point Crossover

In order to maintain the variability of the population, a mutation should be performed on certain individuals. According to the mutation probability p_m=0.01 together with selection and crossover operators, mutation acts as an insurance against total loss of any 'gene' coded by the numbers {0, 1} in a particular position throughout the population, using its ability to introduce a 'gene' which may not, initially, have existed due to limited population size or was lost through application of the operators. The mutation procedure is shown diagrammatically in Figure 4.

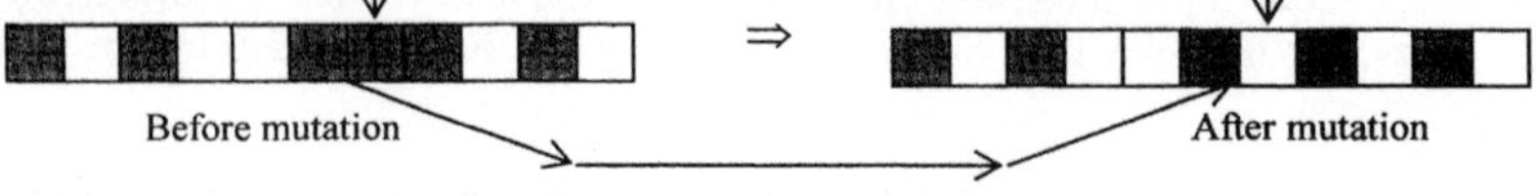

Figure 4. Mutation procedure

The process of the GA can be implemented as follows:

begin (1)

```
    g=0;
    initialize p(g);
    evaluate p(g);
    while (not termination-condition) do
    begin (2)
        g=g+1;
        select p(g) from p(g-1);
        recombine p(g);
        evaluate p(g);
    end (2)
end (1)
```

Where p(g) denotes the population at generation g.

In the training process, the learning rates are taken as $\alpha=0.01$, $\beta=0.05$ and $\gamma=0.65$. The algorithm stops the corresponding learning process when the inference error function E for identifying training data is less than the threshold δ (and here $\delta=0.0006$), and the final learning iteration t=350.

A comparison of the predicted levels to the site ones of ecological environment

The trained NNFR is 'tested' by presentation of 6 testing-case sets from the site measurement in 6 residential communities. Table 1 shows a comparison between the predicted levels and

Table 1. A comparison between the predicted levels and the site ones of ecological environment

Residential communities	L_{MR}	The predicted	The site
1	L_{11} L_{21} L_{31} L_{41} L_{51} L_{61} L_{71} L_{82}	Excellent	Excellent
2	L_{14} L_{24} L_{34} L_{44} L_{54} L_{65} L_{75} L_{84}	Passable	Passable
3	L_{12} L_{22} L_{32} L_{42} L_{52} L_{62} L_{73} L_{83}	Good	Good
4	L_{11} L_{21} L_{31} L_{41} L_{51} L_{61} L_{72} L_{81}	Excellent	Excellent
5	L_{13} L_{23} L_{33} L_{43} L_{53} L_{64} L_{73} L_{84}	Average	Average
6	L_{11} L_{21} L_{31} L_{41} L_{51} L_{61} L_{72} L_{81}	Excellent	Excellent

the site ones of ecological environments.

It seems that the developed IHS for the prediction problem of ecological environment in residential communities is altogether efficient, based on Table 1.

Conclusions

In this study, an initial progress is successfully made in considering more aspects of affected prediction of ecological environment and predicting the suitable ecological environment levels by means of the IHS. In this IHS, fuzzy logic, the GA and the NNFR are combined.

The developed IHS increases the efficiency for solving the qualitative, fuzzy and uncertain problems, and provides the possibility for incorporating more aspects of affected prediction of ecological environment and predicting the suitable ecological environment levels.

Applications to the prediction of most suitable ecological environment levels yield satisfactory results in close agreement with the site measurement.

References

Bäck, T. (1996). Evolutionary algorithms in theory and practice. Oxford University Press.

Cho, K. B. et al (1996). Radial basis function based adaptive fuzzy systems and their applications to system identification and prediction. Fuzzy Set and Systems, 83, 325-339.

Cho, S. B. et al (1995). Combining multiple neural networks by fuzzy integral for robust classification. IEEE Transactions on Systems, Man and Cybernetics, 25, 380-384.

Czogala, Ernest (2000). Fuzzy and Neuro-Fuzzy Intelligent Systems. Physica-Verlag Heidelberg, Germany.

Davis, L. (1990). A handbook of genetic algorithms, New York.

Grabish, H. T. and Nguyen M. (1995). Fundamentals of uncertainty calculation with application to fuzzy inference. Kluwer, Dordrecht, The Netherlands.

Grefenstette, J. J. (1986). Optimization of control parameters for genetic algorithms. IEEE Trans. on Systems, Man and Cybernetic, vol. SMC-16(1), 122-128.

Grefenstette, J. J. (Ed.) (1994). Genetic algorithms for machine learning. Kluwer Academic, The Netherlands.

Haykin, S. (1999). Neural networks: A comprehensive foundation. Second edition. Prentice-Hall International Inc., New Jersey.

Klir, George J. and Yuan, Bo (2000). Fuzzy sets and fuzzy logic. Prentice Hall PTR, New Jersey.

Pham, D. T. and Karaboga, D. (2000). Intelligent optimisation techniques. Springer-Verlag London Limited, Great Britain.

Sundararajan, N. et al. (2000). Radial basis function neural networks with sequential learning. World Scientific Publishing Co. Pte. Ltd., Singapore.

Tian, j. (2004). The research on ecological residential community and target system. Industrial Construction, Vol. 34, No. 2, 10-12.

Wu, Hong-Ci (2004). Updated evolutionary computation for parameter identification of RC structures. Proceedings of the 2004 Structures Congress, Nashville, Tennessee, USA, May 22-26, ASCE, 1583-1592.

Wu, Hong-Ci (2003). Modal genetic identification for RC structural systems. Engineering Mechanics, 20 (Supp.), 675-679

Wu, Hong-Ci (2001). Incorporated building design intelligent hybrid systems. J. of Guizhou University of Technology, 30 (1), 76-82

Wu, Hong-Ci (2002). Integrating GA with fuzzy rule system for site layout in building construction. J. of Guizhou University of Technology, 31 (1), 88-93.

Wu, Hong-Ci (1997a). Classifying composite strata using fuzzy mathematics-hierarchy analysis methods. J. of Guizhou Univ. of Tech., 26 (1): 18-24.

Wu, Hong-Ci (1999). Predicting stability for roof in a mine using neural networks. Ground Pressure and Roof Control, 1, 16—22.

Wu, Hong-Ci (1998). Scheduling excavated plan for slope using genetic algorithms. Chinese J. of Rock Mechanics and Engineering, 17 (5), 565-570.

Wu, Hong-Ci (1997b). Modelling constitutive relationship for soil using finite element-neural network hybrid algorithms. Proc. 9th Int. Conf. on Computer Methods and Advances in Geomechanics, 662--668

A Sustainable Geoenvironment - the Impact of Information and Communication Technologies and E-business

L. Yi Geoenvironmental Research Centre, Cardiff School of Engineering, Cardiff University, UK
H. R. Thomas Geoenvironmental Research Centre, Cardiff School of Engineering, Cardiff University, UK

Abstract
Sustainable geoenvironment development and e-business, as two fast growing themes, are affecting the world's social development and economic growth. This paper essentially considers whether these two factors interact with each other. The geoenvironmental impact of e-business and Information & Communication Technologies are discussed, considering various aspects such as land use, e-waste and resources. It is shown that the impacts appear to be mixed, with further research required in this area. A neural network approach is proposed as a possible mechanism for further understanding of the cross coupling effects.

Introduction
Since the first international Earth Summit in 1992 when more than 100 countries met in Rio de Janeiro, Brazil, sustainable development has been at the top of the agenda of most countries. For example in China, the economy has been growing at a high speed for years, but at a heavy cost to its environment. An area of 1.5 million hectares in northeast China is reported to have turned into desert. The soil erosion area has expanded to 17.6 million hectares in this region (People's Daily1). Over the period 1990 to 2001 oil consumption in China doubled, natural gas consumption increased by 95% and copper consumption by 189% (People's Daily2). Having realized the serious consequences of this development, China is injecting $US85 billion into its environmental protection campaign. This amounts to 1.3% of the country's GDP.

Within this general context, the maintenance of a sustainable geoenvironment has been considered in detail by Yong (2003), who states "the geoenvironment, which includes the terrestrial and in-land aquatic ecosystems, is a key component of the global environment. With proper attention to the protocols and requirements for management, preservation and enrichment of the geoenvironment's natural resources, many of the goals of a sustainable geoenvironment can be realized. Particular attention must be paid to the total geoenvironment because it is the resource base which serves as: (a) the engine that provides for human sustenance – food, shelter, clothing and resource exploitation, and (b) the habitat for various land and aquatic biota. Maintaining sustainability of the geoenvironment is a requisite for Sustainable Development. "

Running in parallel with these developments, e-business and digital technology have also been developing rapidly. The sector is of increasing importance to the world economy and has the potential to profoundly change society. Electronic commerce lies at the heart of the UK government's vision for building a modern, knowledge-driven, economy (PIU, 1999). Online sales in the UK almost doubled in 2003 to £39.5 billion, up from £19 billion in 2002, according to a report by the Office for National Statistics. Internet sales in the US reached $21.4 billion in the 4th quarter of 2004, an increase of 22.3% from the same period during 2003 (IMRG, 2004). Also the British government aimed to provide all government services online by 2005 – transacting and procuring as far as possible, and promised over 99% of the population will have broadband services available (Cabinet Office, 2005).

Some of the environmental impact of this activity can be highlighted as follows:

- According to DEFRA, powering 86 PCs for a year uses 23,000 kWh of electricity and can generate 10 tonnes of CO_2 equivalent emissions.
- BT's tele-working programme made a 50% reduction in floor space since mid 1990s;
- Sun Microsystems' iWork reduced office space needs by 25% in the past four years (Goodman et al, 2004)
- Making the average PC requires 10 times the weight of the product in chemicals and fossil fuels while making car or refrigerator only uses one or two times (Hirsch, 2004).

In view of all the above facts, it appears timely to attempt to address the impact of e-business and "Information and Communication Technologies" (ICT) on a sustainable geoenvironment.

Environmental Impact Indicators

In Gothenburg 2001, the European Council agreed on a sustainable development strategy that should be checked annually on the basis of a number of headline indicators, as follows:

- greenhouse gas emissions;
- the energy intensity of the economy;
- ratio of volume of transport to GDP;
- modal split of transport;
- urban air quality;
- municipal waste collected, landfilled and incinerated.

Taking these six indicators as frame of reference, Erdmann et al defined three levels of environmental effects, with reference to ICT (Erdmann, 2003). These are as follows:

- Primary effects: The impacts and opportunities created by the physical existence of ICT and the processes involved (energy consumption of PCs, hardware waste etc).
- Secondary effects: The impacts and opportunities created by the ongoing use and application of ICT (e.g. online shopping, satellite navigation systems in cars).

- Tertiary effects: The impacts and opportunities created by the aggregated effects of large numbers of people using ICT over the medium to long term (e.g. increased consumption due to lower prices).

Geoenvironmental Impact Assessment

Focussing specifically on the geoenvironment, major impacts include:

- decreased use of office, warehouse and retail space resulting from e-business and tele-working, hence reduced use of land
- wastes disposed to land, generated from e-business associated activities
- energy usage of ICT equipment, leading to Green House Gas (GHG) emission and climate change. The impact of GHG and climate change on the land environment can be felt for example in acid rain interaction with soil, and erosion of costal areas due to increasing water level (Yong and Mulligan, 2004).

Land use and contamination

The Internet Economy could in principle, render unnecessary as much as 3 billion square feet of buildings - some 5% of U.S. commercial floor space (Romm, 1999). Through the integrated supply chain management, commercial companies especially manufacturers could easily reduce the warehouse space, inventories and corresponding transport energy consumption.

ICT supported tele-working also helps to reduce office space and related energy consumption. The construction, heating and lighting of buildings in the UK accounts for half of all energy consumed, half of CO_2 emissions and 40% of raw materials. But office space is only actually used for 22% of the average week. Many organisations have seen huge savings through teleworking, e.g. BT's flexi-working programme made a 50% reduction in floor space since mid 1990s; Sun Microsystems' iWork reduced office space needs by 25% in the past four years (Goodman et al, 2004); At&T estimates a US $25 million per year saving in real estate through its virtual office programme (Allenby, 2001).

In contrast to traditional business using ICT as a complement, some pure online businesses offer even more potential to eliminate retail space completely. C2C (consumer to consumer) e-commerce, for instance EBay, serving 100 million users with more than 1.4 million items up for auction each day, is maintaining a virtual website rather than a huge retail store.

However, according to SVTC (Silicon Valley Toxics Coalition), the original Silicon Valley in Santa Clara County, California, the centre of computer industry, has 29 Superfund sites, which is more than any other county in the US. The process of chip manufacturing requires massive amounts of water – millions of gallons per day in the newest chip plants. For example, in the City of Santa Clara, the high tech electronics industry alone used almost 24% of the City's water in 1994/1995. Of the top waste water discharging companies in Santa Clara County in 1994, 65% were electronics companies. In addition, vast amounts of groundwater reserves are contaminated in the chip-making process. In Arizona, about 25-30% of the groundwater beneath and around Phoenix has been contaminated resulting in 15 mile long plume – 70% of this by the high tech electronics industry. The cost of cleaning the aquifer beneath Phoenix has been estimated as high as $800 million.

e-Waste

According to Hirstch (2004), making the average PC requires 10 times the weight of the product in chemicals and fossil fuels. Additionally many of these chemicals are toxic.

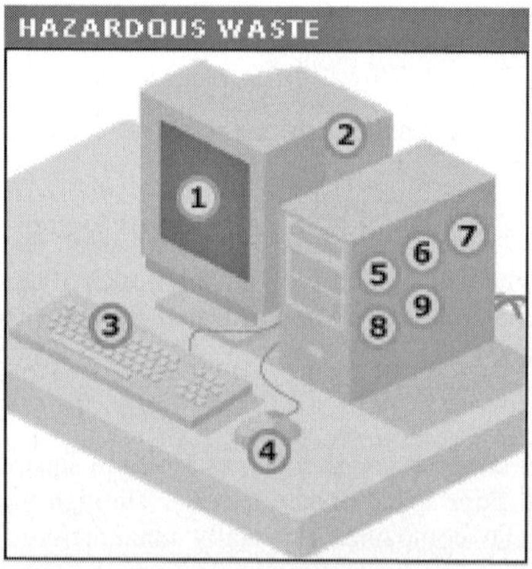

- **1: Lead** in cathode ray tube and solder
- **2: Arsenic** in older cathode ray tubes
- **5: Antimony trioxide** as flame retardant
- **4: Polybrominated flame retardants** in plastic casings, cables and circuit boards
- **3: Selenium** in circuit boards as power supply rectifier
- **6: Cadmium** in circuit boards and semiconductors
- **7: Chromium** in steel as corrosion protection
- **8: Cobalt** in steel for structure and magnetivity
- **9: Mercury** in switches and housing

Figure1: Hazardous Waste in a PC (Source: Hirstch, 2004, "Computers must be greener")

With a fast growing and constantly updating ICT sector, consumers are disposing of old electronic and electrical equipment as the latest products are purchased. In the UK for example there are more mobile phones than the population. Millions of PCs are thrown away each year, and until recently most of these e-wastes were simply disposed of or shredded rather than recycled.

Wastes generated from the various activities associated with the production of goods and services will ultimately find their way into one or all of three disposal media: (a) receiving waters, (b) atmosphere, (c) land. Of these, land disposal appears to be the most popular method for waste containment and management, resulting thereby in ground contamination by pollutants from waste leachates (Yong and Mulligan, 2004).

Waste Electronic and Electrical Equipment (WEEE) now constitutes one of the fastest growing waste streams in the EU, with current estimates indicating a rate of growth three times that of municipal waste. In 1998, 6 million tonnes of WEEE were generated (4% of the municipal waste stream). This is likely to have doubled by 2010 (European Commission, 2000).

Recognising the detrimental effects, more companies have started designing more easily recyclable products and introducing recycling services. HP offers end-of-life return programs for HP as well as other manufacturer's hardware in many countries, and developed state-of-the-art processes to ensure economical and environmental sound management of e-waste. Each month, HP's recycling centres around the world process roughly 1.8 million kilograms of computer related hardware.

Another waste resource is e-commerce induced extra packaging. On a single Saturday in July 2000, 100 airplanes and 9,000 trucks delivered more than 250,000 copies of "Harry

Potter and the Goblet of Fire" to Amazon.com customers all over the United States. These on-line shoppers got the year's hottest kid's book faster than local bookstores. The venture set a record for the online provision of goods in volume. It also probably set a record for the quantities of empty shipping boxes and packaging that would wind up in landfills, let alone the energy used and transport pollution caused in the deal (Matthews et al, 2001).

Products bought online are mostly packed and delivered individually, unlike a traditional retail approach where large amount of products are packed in bulk to be sent to stores and displayed on shelves. Thus e-commerce may bring more packaging waste to our life. On the other hand since supply chain management may well eliminate the middleman and reduce inventories, cardboard and crates may decline. A comprehensive cradle-to-grave approach may help find the net result.

Resources and energy use

As stated previously, making the average PC requires 10 times the weight of the product in chemicals and fossil fuels (additionally many of the chemicals are toxic, while the use of fossil fuels helps contribute to global warming). It needs at least 22kg of chemicals, 1.5 tonnes of water and 240kg of fossil fuels to provide the energy to manufacture 24kg of PC with monitor. Compare this with cars or refrigerators, which use only between one and two times their weight in fossil fuels, it is clear that making more than 130 million computers worldwide has a significant impact.

During the usage stage, Internet equipment also consumes significant amounts of energy from, for example, the constant running of Internet routers, switches, computers etc. According to DEFRA, powering 86 PCs for a year uses 23,000 kWh of electricity and can generate 10 tonnes of CO_2 equivalent emissions. As a more serious consequence resulting from all these energy consumptions may eventually lead to climate change and global warming.

However taking ICT resulted applications into consideration, resources and energy are widely saved in many areas. It is estimated that by 2010, e-materialization of paper, construction, and other activities could reduce US industrial energy and GHG emissions by more than 1.5% (Romm, 1999). Furthermore, the Internet has made virtualisation of products and digitisation of information possible. Taking the music industry as an example, physical products like CDs could easily be replaced with music files downloadable from the Internet. Production as well as related warehouse, transport etc could be eliminated. The digitisation of standard business transactions is also changing office paper use. For instance, e-procurement has saved BT 5.8 tons of paper from 100+ page catalogues which were previously circulated and 1 ton of paper from 24,500 order pads which were previously printed.

However, some rebound effects such as re-materialization may offset these contributions. Given more choices, some people may burn the music files to CDs, may print out emails, e-commerce may cause excessive packaging from products bought online.

The net effect of e-business and ICT on raw materials and energy consumption still remains to be seen. Taking into account that increased demand on natural resources and depletion of natural capital could lead to land and surface degradation (Yong, 2003), and energy consumption could produce GHG and cause climate change which may bring acid rain to soil and water, the changing consumption of resources resulting from e-business and ICT will definitely affect the geoenvironment.

Artificial Neural Network

A general picture of how e-business and ICT affects the geoenvironment has been presented above. However understanding of this rather complex picture is limited. Issues that need to be explained further include:

- What are the complete e-business and ICT factors that have an impact on the geoenvironment?
- What indicators should be chosen to represent the different aspects of the geoenvironment that are affected?
- How much impact does each e-business and ICT factor have on the different aspects of the geoenvironment?

An Artificial Neural Network approach is therefore proposed herein as a possible solution. This could be used, for example, by an individual company to analyse how its e-business operations influences the geoenvironment.

First of all a set of indicators must be chosen to represent e-business/ICT factors and geoenvironmental performance. For example how many computers does the company have and how much energy is consumed per year. A certain amount of data is to be collected in this industry sector, which would be fed into an artificial neural network with e-business side indicators as inputs and geoenvironmental side indicators as outputs. Appropriate amounts of neurons would be added to the hidden layers in the neural network with each neuron given an initial weight. The neural network would be presented with these data and trained in batch or incrementally. The initial weight of each neuron would be adjusted gradually to lower the network performance error. Once the neural network achieved a certain performance goal, which means the neural network has learned how much an e-business factor contributes to each geoenvironmental factor, it can be given an unseen dataset for prediction. The company can then foresee how changes in the e-business side will affect the geoenvironmental performance and can also simulate how to change the inputs to achieve certain outputs target.

Conclusions

E-business and ICT are obviously influencing the geoenvironment. The impacts appear to be mixed.

The main negative impacts include: more raw material and energy consumed during the production of the hardware facilities; significant amount of energy consumed during the usage of Internet equipment; toxic WEEE; extra packaging waste etc.

The main positive changes include: land and corresponding energy saved by virtualization of business operations and tele-working; paper saved by e-materialization; raw materials reduced by digitization of products etc.

In depth comparison of certain geoenvironmental indicators may help to reveal how e-business and ICT have been changing the geoenvironment. Though data is inadequate and there is no standard process available, further study seems necessary. An artificial neural network approach is proposed as a possible solution.

Acknowledgement

The authors would like to thank Prof R N Yong for his valuable contribution at the initial stage of this project. His guidance and support as the former supervisor of the first author is gratefully acknowledged.

References

Allenby B., 2001, "Testimony to Congress - Telework: The AT&T Experience" available at http://www.att.com/telework/article_library/congress_testimony.html accessed at 2005-04-05

Broadband Stakeholder Group, 2001, "Report and Strategic Recommendations", available at http://www.broadbanduk.org/reports/report04.htm accessed at 2005-04-05

Cabinet Office, 2005, "Connecting the UK: the Digital Strategy", available at http://www.strategy.gov.uk/downloads/work_areas/digital_strategy/digital_strategy.pdf accessed on 2005-04-01

Cohen, A., 2000, "Catching the green express from Euston", Financial Times, 2nd January 2000

Digital Economy 2000, US Department of Commerce (Economics & Statistics Administration) available at: www.esa.doc.gov/de2000.pdf accessed on 2005-04-05

Erdmann, L., Behrendt, S., 2003, "The Future Impact of ICT on Environmental sustainability, second interim report" Berlin: IZT

European Commission, 2000, "Directive on Waste Electrical and Electronic Equipment and Directive on the Restriction of the Use of Certain Hazardous Substances in Electrical and Electronic Equipment", Explanatory Memorandum, Brussels, 13 June 2000

Fulford, B., 2003, "Korea's Weird Wired World" in Forbes.com, accessed on 2005-04-05

Goodman, J., Alakeson, V., Jorgensen, B., 2004, "Encouraging Green Network", *forum for the future*

Hirsch, T., 2004, "Computers 'must be greener'", *BBC news*, available at http://news.bbc.co.uk/2/hi/technology/3541623.stm , accessed 2005-04-04

IMRG (Interactive media in retail group), 2004, "UK ecommerce sales doubled in 2003", available at http://www.imrg.org/IMRG/IMRGReports.nsf/(httpReportsByHotNews)/B0C6F22C5094 22B880256F5500392252 accessed on 2005-04-01

Matthews, H.S. et al, 2001, "How much did Harry Potter cost? E-commerce is growing and environmentalists should be happy, Or should they" available at http://www.oecdobserver.org/news/fullstory.php/aid/415/How_much_did_Harry_Potter_c ost_.html accessed on 2005-04-05

Net Impact Study, 2002, "The Net Impact Study overview" http://www.netimpactstudy.com/nis_2002.html accessed on 2005-04-01

Norton, J., 2001, "Winning in the race for e-business", *The Royal Academy of Engineering*, ISBN 1 871634 94 6, January 2001

Romm, J., 1999, "The Internet Economy and Global Warming - a scenario of the Impact of E-commerce on Energy and the Environment"

SVTC (Silicon Valley Toxics Coalition), 2005, www.svtc.org

People's Daily Online 1, "NE China faces serious environmental problem", available at http://www2.chinadaily.com.cn/en/doc/2003-11/20/content_283328.htm accessed on 2005-7-6

People's Daily Online 2, "Foreign vision vs. Chinese insight: money on environment", available at http://english.people.com.cn/200407/30/eng20040730_151352.html accessed on 2005-7-6

PIU - Performance and Innovation Unit, 1999, 'E-commerce @ its Best', UK Government

Yong, R.N., 2003, "Sustainable Development and Environmental Management: A Geoenvironmental Perspective", Proceedings of "The First Kyoto University Conference on International Cooperation to Integrate Diverse Environmental Studies in View of a Future Global Civilization – An experimental dialogue between the modern sciences and the humanities", September 29-30, 2003, p13-29

Yong, R.N., and Mulligan, C.N., 2004, "Natural attenuation of contaminants in soils", CRC Press, Boca Raton, 310p

THE LESSONS OF THE ECOLOGICAL DISASTER AT THE CHERNOBYL NPS

V.N. ZHIVODEROV, "OOO Transgidrostroi", Moscow, Russia

Abstract

When eliminating the consequences at the Chernobyl NPS (occurred in 1986 in the Ukraine in the territory of the former USSR) it was revealed that when designing the nuclear power station (NPS) there were not envisaged any geotechnical protection arrangements in its territory. Therefore the working design and arranging geotechnical measures at the Chernobyl NPS was carried out under conditions of strong radiational action. This circumstance had hard consequences and led to a death of many specialists.

As distinguished from systematic improvement of protective devices of the main technical installations, the geotechnical protections of NPS as an obligatory measure have not undergone any changes and are not envisaged by the Russian normative documents up to now. We can suppose that a similar situation takes place also when constructing nuclear power stations in other countries. There are more than 0.5 thousand reactors at the NPS all over the world. They are located as a rule near large waterways and reservoirs. Therefore the most endangered are riverside territories and water areas having a way out to inside or outside seas. In such cases a danger can arise for inhabitants of other countries. The geography of a radioactive contamination is vast and hardly predictable, but the most active its part is near the NPS territory, which requires arranging reliable geotechnical protections in good time.

Now many countries use electric power from nuclear power stations, therefore the questions of the safety and of creating geotechnical protections at NPS may be of a common interest for many countries. The hazard of occurring accidents grows additionally in recent years due to a threat of committing acts of terrorism, therefore the role of protective measures at NPS becomes still more urgent. The lessons of the large ecological disaster at the Chernobyl NPS must be taken into account in the modern life.

Introduction

To the end April 2006 20 years have passed since the moment of the largest technological disaster of the twentieth century - the accident at the Chernobyl nuclear power station.

At the night of 27 April 1986 on the territory of the former USSR at the distance of 130 km from the capital of the Ukraine Kiev a sudden explosion occurred in the reactor part of the Chernobyl nuclear power station (ChNPS).

The situation turned out

As a result of occurred non-controlled nuclear reaction the upper part of the reactor dome of the 4th power unit of the ChNPS, and with the steam and smoke thousands and tens of thousands Roentgen per hour rushed to the atmosphere. At the next day an increased radioactivity was revealed at an immense distance from the accident place, including the shores of Sweden, Finland and, Norway, which are at the distance of more than 1200 km from the seat of radiation.

As a result of extraordinary measurements urgently undertaken by the Leadership and the Government of the USSR the intensity of the output of radioactive substances was reduced during the first 7 – 8 days by a factor of hundreds and in some cases of 1000.

Geotechnical works during the period of eliminating the accident

Beginning from the first days after the explosion and the fire at the reactor specialists of a geotechnical branch took active part in the works for eliminating the consequences of the disaster. After the evacuation of civil population (which was carried out during twenty four hours) under heavy and dangerous conditions of restoration works with a high intensity of radioactive radiation thousands of specialists including geotechnicians were sent on a mission and a multitude of units of special equipment and transport.

The measurements for geotechnical protection of the environment were developed, designed, and realized at the workplaces in Chernobyl. The main of them are the following.

1. Arrangement of water-receiving wells in low places of the terrain for gathering radioactive-contaminated surface water on the NPS territory.
2. Arrangement of a deep antifiltration "wall in ground" (H is up to 32 m) on the path of movement of potentially contaminated underground water to the cooling pond and to Pripyat' river.
3. Arrangement of dikes and embankments for guarding water areas from a possible ingress into them of contaminated surface run-offs.
4. Arrangement in the base of an accidental reactor of an additional under-foundation reinforced concrete plate with coil pipes for supplying into them of a cooling reagent for the purpose of reducing the temperature of a melted inner mass existing in the reactor.
5. -Drilling and arrangement of holes for obtaining clean (potable) water and also for using it when radioactive decontamination.
6. Arrangement of geotechnical protections with using a system of drainage boreholes.

The terms of carrying out the works and their results.

The basic volumes of restoration works for eliminating the consequences of the accident, due to centralization and the possible optimum organization of works, were carried out with a high quality and in the shortest terms in 3 – 3.5 months, and already in September 1986 around the faulted reactor a monolith reinforced concrete sarcophagus was erected, enveloping hermetically the reactor broken down from all sides, including its underground part.

However, in doing so, irretrievable losses were suffered including considerable human victims and the broken down health of the accident liquidators, which manifests itself long ago after the disaster.

The problems posed by Chernobyl and possible ways of their solution

According to the information the author has, in all countries, which have or construct NPS including Russia, there is no Construction Norms and Rules (CNaR) or Technical Regulations, which would require an obligatory application of geotechnical protections when erecting nuclear power stations. In Russian normative documents as well as in the European norms the necessity of obligatory construction of geotechnical protections at the objects of nuclear power is not reflected up to now. The situation turned out puts questions requiring rapid decision: "How to ensure that the necessity of constructing geotechnical protections under conditions of the strongest radioactive fields would fall off?", "How to preserve human lives, human health , equipment and large material means?" Simpler speaking, is not it better to struggle with the causes of an effect, than with its consequences?

However, up to now, after 20 years, even in such a country as Russia, which itself has suffered (and is suffering) large-scale consequences of the Chernobyl tragedy, there is no clear comprehension and unambiguous scientific substantiation for the development of typical technical decisions on the arrangement of geotechnical protections at the NPS.

It is not clear yet for the specialists, for example, what kind of a contour of the "wall in ground" or a grouted curtain it is better to make around the NPS: closed, half-closed, with a drainage system or without it? Which combinations of an antifiltration curtain and drainage are preferable? What degree of a waterproof (density) must be ensured by an injection curtain and in what dependence on the radiation field? What to do with the radioactive contaminated ground mass and similar waters?

These and many-many other questions, in spite of the urgency of the theme touched on, have not yet concrete answers due to poor studying of these problems. As concerns the author's opinion on the question considered, the author is sure that to avoid (or to reduce) the victims, when constructing geotechnical protections in an accident period, it is necessary that the geotechnical protections at the NPS (and other objects ecologically dangerous) would be envisaged at the stage of designing, and the realization of the protective measurements would be carried out during the construction and not in an after-accident period.

Since for each NPS the conditions for the construction are individual, it would be objectively correct that the ways and forms of geotechnical protections would be chosen and assigned by a general designer in accordance with the general Provisions of the Unified International Technical Regulations. However, for the development of such Regulations there is no necessary basic data generalizing and considering the most probable combinations of radiation situations taking into account hydro-geological and hydrological characteristics of the region and other local conditions (the intensity of radioactive radiation, terrain of the locality, distances from a potential seat of radiation to a large populated locality, to the nearest water areas etc.).

Modern problems of geotechnical construction

When working over this material a new serious circumstance emerged, which as if confirmed the author's proposition on the urgent necessity of elaborating the Unified International Technical Regulations, and considerably increased its importance.

Non-controlled natural events occurring recently in the world and causing large-scale spontaneous calamities on various continents (hurricanes, tsunami, floods, earthquakes), obviously, can also provoke and make breaking down actions on the NPS being constructed and having been constructed.

These circumstances side by side with the experience of the liquidation works at the Chernobyl NPS and existing terrorism threats confirm once more the necessity of undertaking urgent measurements on the development of certain technical decisions obligatory when constructing NPS and fixed by an international practice of construction.

There is an impression that the main problem to-day is the lack of a centralized management of the safety of nuclear power stations, which are constructed in different countries of the world, and the main cause of such a situation is the absence of Unified requirements for the construction. The tragedy of the Chernobyl NPS has shown that the radioactive contamination can propagate with the velocity of a wind and can be carried over distances of several thousands kilometers during 2-3 days. Hence in a case of a large accident, when a considerable amount of radioactive substances enters the atmosphere, the consequences of this event become practically unpredictable for any country in the world.

The solution of this problem can be found in the development and subsequent affirmation of the Unified International Technical Regulations for the NPS construction obligatory for using in all countries of the world.

Not considering in this case organizational problems, we would like to dwell on the Regulations.

The Unified Technical Regulations on the construction of NPS

In the basis of such Regulations the obligatory for builders Propositions and Requirements may be, including, for instance, the following:

- to construct NPS of only underground type (or preferably of underground type),
- the NPS olcation must not be near large water areas and rivers (the minimum distance from them must be pointed in the Regulations),
- the distance of the NPS from large populated locality must ensure their maximum safety (the boundaries of the distance must be stated beforehand),
- an obligatory design and arrangement of geotechnical protections is envisaged when constructing the NPS (especially of an underground type), these protections must be developed in good time and then united into typical ones taking into account the hydrogeological and engineering-geological conditions,
- other important Propositions.

The main landmarks in the path to solve the problem

The Technical Regulations of such type would be able to be developed after carrying out scientific-research works in one of the most developed countries and after their discussion, addition and approval they could be sent for an agreement and confirmation to the International Atomic Energy Agency (IAEA).

After this the Regulations would acquire a status of an obligatory International document. A special inspecting organ, for example, under a leadership of the IAEA, could be charged to control their fulfillment in each country, which constructs NPS. (Or it is possible to carry out some other forms of a control accepted in the International practice). Undoubtedly the proposed sequence of actions can be considered as the main landmarks of one of versions on the path of solving the problem.

Conclusion

The development of the Unified International Technical Regulations for the construction of NPS will require certain, possibly considerable, material expense and also time expense. But taking into account the already existing Construction Norms and Rules, National Technical Regulations and also the accumulated experience (including at the Chernobyl NPS), facilitates somewhat this task. But it is obvious that it is highly difficult to find another alternative way of solving the problem. And one point remains clear: “The sooner this work is begun, the earlier its solution can be obtained”.

If the author’s propositions are approved, he would not like to hold himself aloof of this work. The author would be grateful for an invitation to take part in the work, for example, as a consultant having an experience at Chernobyl (by correspondence).

Inferences

1. Experience of the works for arrangement of geotechnical protections under conditions of high radiation fields at the Chernobyl NPS convinces us of the fact that such works should be obligatory included into the project and carry out them in

good time at the stage of the NPS construction, and not under conditions of a radiation contamination after an accident.

2. The disaster at the Chernobyl NPS have shown that the radiation contamination can propagate with the velocity of a wind to distances of thousands of kilometers in 36-48 hours and present a threat practically to any country in the world.
3. Taking into account the presence of terrorist threats, the technical causes and emerging natural cataclysms (hurricanes, tsunami, earthquakes, floods) it is considered as necessary to begin the development and issue of the Unified International Technical Regulations obligatory for the fulfillment in all countries constructing NPS.
4. It is evident that it is necessary to carry out the works for the development and issue of the Unified International Regulations under the supervision of the IAEA of other influential International Organization with a subsequent discussion and approval of specialists-geotechnicians of the leading countries.

References

1. Dmitriev N.V., Zhivoderov V.N. and Grachev Y.A. 1994. Elimination of consequences of the accident at Chernobyl atomic Station. Proc. XIII ICSMFE, New Delhi, vol. 4, 1657-1660
2. Dmitriev N.V and Zhivoderov V.N. 1994. Features in organization and fulfillment of the works for protection of underground water when eliminating the consequences of accident at Chernobyl AES. Bases, Foundations and Soil Mechanics, 1, 24-28 (in Russian language).
3. Dmitriev N.V. and Zhivoderov V.N. 1995. Post-disaster geotechnical works at Chernobyl. Proceedings of the Eleventh European Conference on Soil Mechanics and Foundation Engineering, The Interlay between Geotechnical Engineering and Engineering Geology: 8.195 – 8.200. Copenhagen, 28 May – 01 June 1995.
4. Zhivoderov V.N. 1997. Possible role of ground improvement measures in mitigating the Chernobyl environmental disaster. Ground Improvement, vol. 1, num. 3: 177-178.

FATE AND TRANSPORT OF POLLUTANTS

Leaching of Chromium, Copper and Arsenic from Utility Poles Treated with Chromated Copper Arsenate

Joseph A. Arisi, Faculty of Engineering and Applied Science, Memorial University of Newfoundland, St. John's, Canada
Cynthia A. Coles, Faculty of Engineering and Applied Science, Memorial University of Newfoundland, St. John's, Canada
Marion Organ, AMEC Earth & Environmental, St John's, Canada

Abstract

The objectives of this study were to examine leaching of chromated copper arsenate (CCA) from treated timber and to determine the cation exchange capacity (CEC), pH and total organic carbon (TOC) of the contaminated soil samples. A total of 114 surface soil samples were collected around 28 Douglas fir utility poles at distances of 0, 0.30, and 0.60 m and 3 background samples were collected at a distance of 7.0 m from the poles. Also a freshly treated, 1.50 m long utility pole section was suspended outside in a large cylinder to collect runoff. The Cu:As:Cr (molar) ratios of the average metal contents in the soil samples collected at a distance of 0 m, and in the runoff from the pole segment, were 100:46:56 and 100:44:57, respectively. At 0.60 m distance the metal concentration had reached background levels. The results suggested that As was more mobile than the Cr and that the metals were being continuously leached from the poles over time. A correlation between the soil CEC and TOC was observed.

Introduction

Timber utility poles in North America are treated with preservatives to reduce bacterial, fungi and insect attack and increase their life expectancy. The wood preservatives are either organic (and oil based) or inorganic (and waterborne). Pentachlorophenol, creosote, chromated copper arsenate (CCA) and ammoniacal copper arsenate (ACA) are the most widely used preservatives, while CCA is the most widely used in aquatic environments (Hingston et al., 2001). CCA is an inorganic wood preservative developed by an Indian scientist, Dr. Sonti Kamesam in 1933 (Lahiry, 1997) and since then has been used extensively although recently its use has been phased out in the United States (Rahman et al., 2004).

Three types of CCA (-A, -B, and -C) are commercially available and they have varying proportions of copper oxide (CuO), chromium oxide or anhydrous chromic acid (CrO_3), and arsenic pentoxide (As_2O_5) present. The CuO ≈ 19% by volume, the CrO_3 varies between 35 and 65% by volume and the As_2O_5 makes up between 46 and 16% by volume. Type CCA-C (with 18.5% CuO, 47.5% CrO_3 and 34% As_2O_5) is most commonly used (Cooper, 1994). The chromic acid (H_2CrO_4) dissociates in water to form the soluble chromate (CrO_4^{2-}) and dichromate ($Cr_2O_7^{2-}$) ions.

The copper (Cu) and arsenic (As) in CCA inhibit fungi, bacteria and insects whereas the chromium (Cr) promotes fixation of CCA with the timber (Lebow, 1996) and promotes precipitation. Successful fixation will minimize leaching of metals from the wood. Bull (2001) proposed a simple reaction for CCA fixation in wood and according to his model hexavalent chromium or Cr (VI) is reduced to Cr (III) as shown in the following equation.

$$Cr_2O_7^{2-} + 6e^- + 14H^+ \rightarrow 2Cr^{3+} + 7H_2O$$

The major fixation products in treated utility poles are comprised of the Cr (III) and they form as Cr (VI) groups are reduced (Hingston et al., 2001). The proposed products include chromium (III) arsenate ($CrAsO_4$), that forms complexes with the lignin, CrO_4^{2-} that also complexes with lignin, Cu (II) precipitates and complexes with lignin and cellulose, and chromium (III) hydroxide (Bull, 2001). X-ray absorption fine structure analysis has revealed that primary alcohol groups in the lignin and carbohydrate fractions in the wood are the electron donors in the reduction reaction (Bull, 2001).

The reduction process occurs in stages. The Cr(VI) and other electron acceptors first become sorbed to the carbohydrates and then they are reduced and form complexes of CrO_4 with lignin and cellulose, and CrO_4^{2-} with lignin. The changes in metal species during fixation and the long-term complex reactions influence the distribution of metals in the wood, the wood toxicity, and the species that may be leached (Hingston et al., 2001). It is believed that metals may be leached as Cu or Cr arsenates, as inorganic complexes, or as organo-metallic complexes bound to water-soluble wood extractives (Lebow, 1996) though little research has been undertaken in this area (Albuquerque and Cragg, 1995).

Chemical leaching of the treated timber also progresses in stages by "loss of surface deposits and unfixed components, penetration of water into the wood and hydrolysis or dissolution of preservatives from the surface of the wood" (Cooper, 1994). Leaching rates are affected by local climate, leaching media, wood properties and wood treatment techniques (Hingston et al., 2001).

CCA exhibits a broad-spectrum of toxicity and the toxic components bond strongly with the wood (Bull, 2001) though recent toxicity tests suggest that in aquatic environments the leaching of preservative components from wood is harmful to the environment (Hingston et al., 2001). Leaching and toxicity of CCA treated wood may be a problem during storage, use and particularly disposal (Riberiro et al., 2000). Approximately 5 million tons of treated timber are landfilled annually in the US (Falk, 1997). Disposal in landfills is expensive and may be prohibited in the future (Riberiro et al., 2000).

Nine different oxidation states exist for Cr (Massara et al., 2004) but only the Cr (VI) and Cr (III) are common in soils (Pagilla, 1999; Massara et al., 2004). Cr (VI) is more toxic and more mobile than Cr (III) which forms stable complexes and is retained in soils. Cr (VI) is reduced to Cr (III) by organic matter (humic and fulvic acids) and inorganic agents (Fe) (Sanders and Reidel, 1987; Pagilla, 1999). Ideally, CCA fixation in wood converts the Cr (VI) to Cr (III) but Nygren and Nilsson (1993) found that 20% of Cr was present in the hexavalent form or incomplete fixation was occurring. Cr (III) may also be mobilized by oxidation to the hexavalent form or its complexation with naturally occurring ligands (Massara et al., 2004). Dichromate, a major constituent of CCA, is generally weakly sorbed to soil and could pose the greatest potential immediate threat to groundwater supplies (Carey et al., 1996).

Cu (II) is an essential micronutrient but is toxic above trace levels. It tends to become bonded to organic material (especially humic acids) in soils (Hung et al., 1993). Cu sorption to soil increases with pH (Carey et al., 1996) and greater Cu leaching from treated wood occurs at lower pH (Hingston et al., 2001). Lebow et al. (1999) found Cu leaching from CCA treated timber was much greater in seawater than in freshwater and their study raised concern about the use of CCA treated wood in marine environments. One study of CCA treated Pine found that 25% of the Cu was leached in the first 6 months, and a maximum of 52% of the Cu was leached within 85 months (Hingston et al., 2001).

There are 4 oxidation states for As and arsenate (V) and arsenite (III) are the most common (Wang and Mulligan, 2004). As (III) is the more toxic and As (V) is the more abundant of the two states (Hingston et al., 2001). Woolson and Gjovik (1981) found As (III) comprised 3% of the As in treated wood although Nyren and Nilsson (1993) found no trivalent As present in commercial wood supplies. In a study on marine piles, leaching of As into the interstitial and overlying waters was found in all cases (Baldwin et al., 1996).

Very little research has dealt with in-situ leaching rates of CCA treated timber and research under actual and different environmental conditions is needed (Lebow, 1996; Hingston et al., 2001). Simple, standard and repeatable laboratory tests on small blocks of wood are not applicable to field conditions and so leaching rates from utility poles are not well defined (Hingston et al., 2001).

Mortimer (1991) investigated CCA/Polyethylene glycol (PEG) treated utility poles, 3 to 36 months in age, and found very low concentrations of Cu, Cr and As near the poles. Cooper et al. (1997) studied 50 CCA-C/PEG poles and found high Cu, Cr and As concentrations near the poles. Zagury et al. (2003) studied soil surrounding 6 CCA/PEG treated utility poles and measured 1460 ± 677 mg/kg of Cu, 410 ± 150 mg/kg of As and 287 ± 32 mg/kg of Cr.

The purpose of this study was to determine the Cu, Cr and As concentrations in soils adjacent to CCA treated utility poles, metal concentration variations with distance from the poles, approximately field scale leaching of metals, and soil properties including cation exchange capacity (CEC), pH and total organic carbon (TOC). Leaching tests were conducted on timber under conditions that closely resembled natural field conditions. Metal concentrations in runoff from an isolated log were determined to obtain an estimate of runoff in the field.

Materials and Methods

The transmission lines used in this study were selected according to criteria proposed by the Electric Power Research Institute (EPRI, 1997) for penta and creosote treated poles. The CCA treated poles in this study were all Douglas fir and of known age, came from readily accessible areas having variable soil properties, had not been exposed to pesticides and other contamination, were far from industrial areas, were at least 6 ft away from a roadway and were situated in a clearing of at least 6 ft in diameter.

Samples were collected around 28 poles along three different transmission lines (227, 259, 225) situated in Gros Morne National Park, at Rocky Harbour and at Deer Lake, respectively. The 28 poles were located at 24 sites and a total of 114 samples were collected.

For 21 sites with 23 poles, three samples were taken at distances of 0, 30 and 60 cm from the poles and 2 of these sites had double poles. At 2 other sites that had 1 pole each, samples were taken at distances of 0, 30 and 60 cm in north, south, east and west directions from the pole for a total of 12 samples per pole. At 1 site with 3 poles, 3 samples each were taken for 2 of the poles and 12 samples were taken for the other pole. At least 3 samples were collected at 3 distances from each pole to determine the variation in soil metal concentrations with distance from the pole. At 3 pole locations, soil samples were also collected at 7 m from the poles to obtain background soil conditions.

Surface soil samples were collected with a spoon, placed in clean Ziploc bags, sealed and refrigerated at 4°C. The spoon was cleaned between sampling locations. Each Ziploc bag was labelled with the date of collection, a code for each pole, the pole age, the distance, and where applicable, the direction from the pole.

Soil samples were analyzed for TOC by the modified Walkely-Black wet oxidation method and for CEC according to the $Ca(OAc)_2$ – $CaCl_2$ method (Sheldrick, ed. 1984). Soil pH was determined using a ratio of 1:2 soil:0.01M $CaCl_2$ (Sheldrick, ed. 1984). Soil samples were digested using US EPA method 3050 B (American Chemical Society, 1986). Samples were stored in the refrigerator at 4°C until analysis with a Perkin-Elmer SCIEX Elan 6100 inductively coupled plasma mass spectrometer (ICP MS). All pH measurements and 68% of TOC measurements for each sample were conducted in duplicate. CEC determinations were conducted in duplicate for only 10% of the samples since at least 3 samples were available for each pole. Only 10% of acid digestion experiments (for soil metal determination) were conducted in duplicate since it is known that metal concentration tends to decrease as distance from the pole increases (Cooper et al., 1997; Hingston et al., 2001; Zagury et al., 2003).

A freshly treated section of a utility pole, 1.50 m long and 0.30 m in diameter, was suspended vertically outside in a large plastic cylinder measuring 2.40 m high by 1.20 m in diameter. The bottom of the cylinder was connected with an outlet for runoff collection and the pole and cylinder were exposed to rain and snow. Runoff from the pole segment was collected regularly over a period of 2.5 years. The water samples were then acidified with HNO_3 to $<$ pH 2 and stored at room temperature until they were analyzed for Cr, Cu and As. One backgound water sample was also collected for analysis.

Results and Discussion

At all 28 poles metal concentrations were highest adjacent to the poles and decreased with distance from the poles. This was similarly observed in previous studies (Cooper et al., 1997; EPRI, 1997; Hingston et al., 2001; Zagury et al., 2003). This phenomenon at pole 1-32 is illustrated in Figure 1 and the trends seen in this figure are typical results for this study and are the same trends that were seen when measured values from all samples in this study were averaged.

For 27 of the 28 poles Cu concentrations were higher than the Cr and As concentrations and this was also observed by Cooper et al. (1997) and Zagury et al. (2003). For approximately 80% of the poles, the As concentration exceeded the Cr concentration immediately next to the pole but at 30 cm from the poles, the Cr concentration was greater than the As concentration (when concentrations are measured in mg/kg). These results suggest As is more mobile than Cr and could therefore be more likely to contaminate groundwater. Stilwell and Gorny (1997) and Zagury et al. (2003) also found As to be more mobile than Cr.

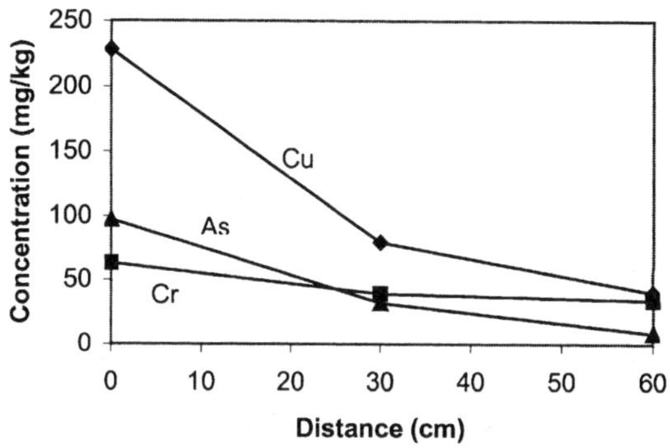

Figure 1. Reduction in metal concentration with distance from pole 1-32

An overall trend in soil immediately adjacent to the utility poles was observed whereby Cu had the highest average concentration of 574 ± 58 mg/kg, followed by As with an average concentration of 309 ± 33 mg/kg and finally by Cr with an average concentration of 261 ± 27 mg/kg. It can be seen from Table 1 below, these concentrations in soils are considerably higher than Canadian maximum acceptable levels. There can be significant variation in the relative amounts of As and Cr applied to the timber depending on the type of CCA used, and so these values for As and Cr might not necessarily be exactly comparable for studies using different types of CCA.

Table 1. Maximum acceptable levels of Cu, As and Cr (Environment Canada and Canadian Council of Ministers of Environment, (CCME), 2003)

Metals	Water (µg/L)					Soil (mg/kg)	
	Community	Aquatic life		Agriculture		Agriculture	Commercial
	MAC[a]/ IMAC[b]/AO[c]	Fresh-water	Marine	Irrigation	Livestock	Residential Parkland	Industrial
Cu	≤ 1000 (AO)	2-4	-	200-1000	500-1000	63	91
As	25 (IMAC)	5	12.5	100	25	12	12
Cr	50 (MAC)	-	-	-	-	64	87

[a] MAC = Maximum acceptable concentration.
[b] IMAC = Interim maximum acceptable concentration.
[c] AO = Aesthetic objective

The metal concentrations in soils at 7.0 m from the utility poles were comparable to metal concentrations in the soils at 0.60 m from the poles, as shown in Table 2. Therefore, metal background levels in surface soils were reached at about 0.60 m from the poles.

Average values of pH, CEC and TOC for soil at each pole were calculated and Figure 2 shows a direct correlation between the CEC and TOC of each soil. This might be explained by the fact that the TOC is a function of the total organic matter in the soil and the soil organic matter contains a large proportion of carboxyl groups that dissociate and contribute to negative charge and result in a high CEC. Zagury et al. (2003) compared different soils and found that organic soils had higher CEC values than clay soils. They concluded that an organic matter contribution to CEC would be greater than a clay contribution.

Table 2. Metal concentrations in surface soils at 3 pole locations

		Metal concentrations at four distances from poles (mg/kg)			
Pole No.	Metal	0 m	0.30 m	0.60 m	7.00 m
P 1-31(S)	Cu	1257	117	24.1	18.6
	As	487	42.4	2.43	24.2
	Cr	232	47.4	20.5	24.3
P 1-4 (S)	Cu	575	14.8	18.6	24.6
	As	375	4.38	3.68	6.77
	Cr	117	20.0	27.7	25.2
P 19L(S)	Cu	512	5.52	7.52	11.6
	As	79.8	1.00	2.25	4.11
	Cr	140	1.01	2.18	3.38

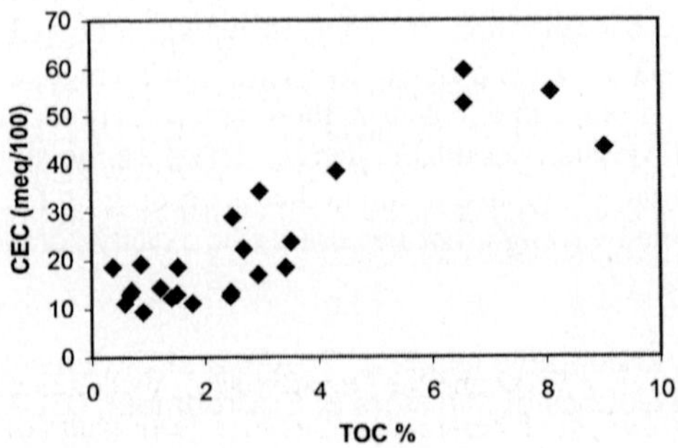

Figure 2. Average of all TOC site values against average of all CEC site values

The average pH values for all of the sites ranged between pH 4.1 and pH 7.2 and were therefore in the acid to neutral pH range for all soils. There was no direct relationship between the TOC and the pH or between CEC and pH, as can be seen from Figures 3 and 4 respectively. However, wherever there was a high TOC, and thus a high CEC, there was also a low pH. This may be explained by the fact organic matter can have a significant affect on the lowering of pH.

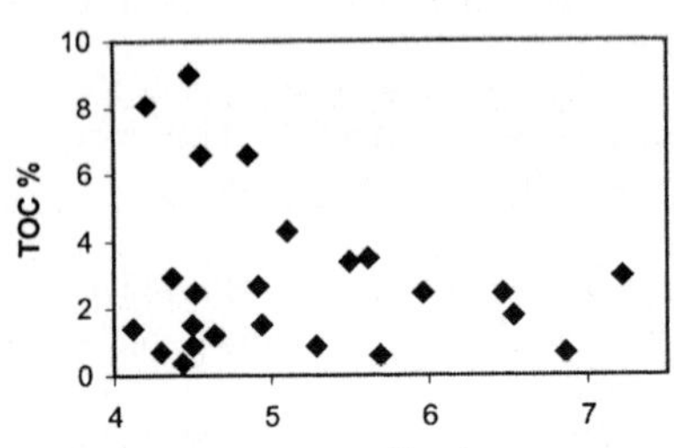

Figure 3. Average of pH values against average of TOC values

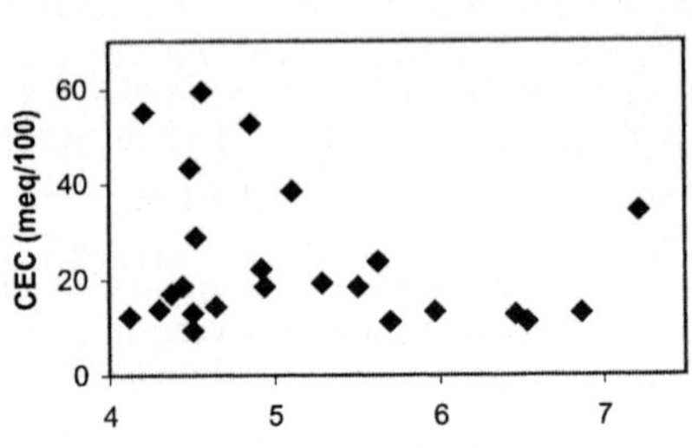

Figure 4. Average of pH values against average of CEC values

Between 5 September 2002 and 12 February 2004 water samples were collected from the utility log that had been suspended outside. Cu was found to be the most abundant metal to be leached. The average amount of Cu found in the water samples was 641 μg/L and the average amounts of As and Cr leached were 331 and 300 μg/L respectively. Based on the recommended maximum levels in Table 1, leaching of these metals could be unsafe, particularly in freshwater or marine environments, where toxic levels of Cu and As (and possibly Cr) could be released in the immediate vicinity of the treated timber. The trends and proportions of metal leaching into the water and into the soil immediately adjacent to the poles, were similar. The background concentrations of these metals in water were only 72 μg/L of Cu, 0.17 μg/L of As and 0.80 μg/L of Cr.

Conclusions

The average metal contents in soil collected immediately adjacent to the utility poles gave concentrations of 574 ± 58 mg/kg Cu, 309 ± 33 mg/kg As, and 261 ± 27 mg/kg Cr or a Cu:As:Cr molar ratio of 100:46:57. A similar trend of leaching of the metals from the suspended log was observed as the average concentrations in the water samples were 641 μg/L Cu, 331 μg/L As and 300 μg/L Cr or a Cu:As:Cr molar ratio of 100:44:57. The metal concentrations in soil samples immediately adjacent to the utility poles were accurately reflecting the same relative proportions of metal leaching from the suspended log. This could mean there is a continuous leaching of the metals over time since the results also suggested that As was more mobile than Cr. It appeared that soil background conditions in surface soils were reached at a distance of only 0.60 m from the utility poles.

Acknowledgements

The authors thank Dr. Geoff Veinott and the Department of Fisheries and Oceans for assisting with the use of their ICP-MS, the laboratory personnel in the Faculty of Engineering and Applied Science at Memorial University of Newfoundland (MUN) and Ms. Linda Windsor in the Chemistry Department at MUN, and Dr. A. S. J. Swamidas in the Faculty of Engineering and Applied Science at MUN. The authors also thank the MUN Faculty of Engineering and Applied Science and the VP Research for providing partial financial support.

References

Albuquerque, R.M. and Cragg, S.M., 1995, "Evaluation impact of CCA- treated wood on the marine environment, International Research Group on Wood Preservation, Stockholm.

American Chemical Society, (1986), "Reagent Chemicals", 7th Ed., American Chemical Society Specifications, Washington, DC

Baldwin, W.J., Pasek, E.A. and Osborne, P.D., (1996), "Sediment toxicity study of CCA-C treated marine pile", Forest Products Journal, 4:42-50

Bull, D.C., (2001), "The Chemistry of Chromated Copper Arsenate II; Preservative Wood Interactions, Wood Science and Technology", 34(6):450-466, Feb.

Carey, P.L., Mclaren, R. and Adams, J. A., (1996), "Sorption of Cupric, Dichromate, and Arsenate Ions in some New Zealand Soils", Water, Air and Soil Pollution, 87:189-203.

Cooper, P.A., (1994), "Leaching of CCA: Is it a problem? Environmental Considerations in the Manufacture, Use and Disposal of Pressure Treated Wood", Forest Products Society, Madison, WI

Cooper, P.A., Ung, Y. T. and Aucoin, J.P., (1997), "Environmental impact of CCA poles in service", in Proceedings of the 28th Annual meeting of the Int. Res. Group on Wood Preservation, Whistler, B.C, Canada, Section 5: Environmental Aspects, pp. 1-20.

EPRI, (1997), "Preservatives in Soils Adjacent to In Service Utility Poles in the United States", December.

Environment Canada and CCME, 2003, "Summary of Existing Canadian Environmental Quality Guidelines", http://www.ccme.ca/assets/pdf/e1_062.pdf

Falk, B., (1997), "Opportunities for the Wood Waste Resource"; Forest Products Journal", 47(6):17-22.

Hingston, J.A., Collins, C.D., Murphy, R. J. and Lester, J. N., (2001), "Leaching of chromated copper arsenate wood preservatives: a review", Environmental Pollution, 111:53-66.

Hung, T., Meng, P.J. and Wu, S., (1993), "Species of Copper and Zinc in Sediment Collected from the Antarctic Ocean and the Taiwan Erhjin Chi Coastal Area", Environmental Pollution, 80:223 –230.

Lahiry, A.K., (1997), "An introduction to environmental aspects of groundwater arsenic and CCA treated wood poles in Bangladesh", Section 5 (Environmental Aspects), Doc. IRG/WP 97 – 50081, Int. Res. Group on Wood Preservation, Stockholm.

Lebow, S., (1996), "Leaching of wood preservative components and their mobility in the environment. Summary of pertinent literature", General Technical Report, Forest Products Laboratory, USDA Forest Service, Madison, WI., FPL-GTR-93:36.

Lebow, S. T., Foster, D. O. and Lebow, P. K., (1999), "Release of Cooper, Chromium and Arsenic from Treated Southern Pine Exposed in Seawater and Freshwater", Forest Journal, 49(7-8):80-89.

Massara, H., Mulligan, C. N. and Hadjinicolaou, J., (2004), "Innovative Treatment for Chromium Contaminated Sites", 57th Canadian Geotechnical Conference, Montreal, QC

Mortimer, W.P., (1991), "The environmental persistence and migration of wood pole preservatives", Canadian Electrical Association, Montreal, QC, Canada, CEA 118 D 697

Nygren, O. and Nilsson, C.A, (1993), "Determination of chromium, copper and arsenic in wood and dust from CCA impregnated timber", Analusis, 21:83-89

Pagilla, K. R. and Canter, L.W., (1999), "Laboratory Studies on Remediation of Chromium Contaminated Soils", Journal of Environmental Engineering, 125:(3)243-248, March.

Rahman, F. A., Allan, D. L., Rosen, C. J. and Sadowsky, M. J., (2004), "Arsenic Availability from Chromated Copper Arsenate (CCA)-Treated Wood", Journal of Environmental Quality, 33:173-180.

Ribeiro, A. B., Mateus, E. P., Ottosen, L.M. and Bech-Nielsen, G., (2000), "Electrodialytic Removal of Cu, Cr and As from Chromated Copper Arsenate treated Timber Waste"; Environmental Science and Technology, 34(5):784-788.

Sheldrick, B. H., ed., (1984), "Analytical methods manual 1984", Land Resource Research Institute, Ottawa, Ontario

Sanders, J.G., and Riedel, G.F., (1987), "Control of trace element toxicity by phytoplankton", in "Phytochemical Effects on Environmental Compounds", Saunders, J.A., Kosak-Channing, L. and Conn, E. E., (Eds), Plenum Press, New York, pp. 131-149.

Stilwell, D.E., and Gorny, K.D, (1997), "Contamination of soil with copper, chromium and arsenic under decks built from pressure treated wood", Environmental Contamination and Toxicology, 58:22-29.

Wang, S.; and Mullingan, C.N.; 2004; "Arsenic in Canada"; 57th Canadian Geotechnical Conference; Montreal, QC, Canada

Woolson, E. E. and Gjovik, L. R., (1981), "The Valence State of Arsenic on Treated Wood", Proceedings of the American Wood Preservers Association, 77:15-22.

Zagury, G.J., Samson, R. and Deschênes, L., (2003), "Occurrence of Metals in Soil and Ground Water Near Chromated Copper Arsenate-Treated Utility Poles", Journal of Environmental Quality, 32:507-514.

Influence of secondary changes of pH value from motorway A 4 Katowice-Krakow side-space on cadmium, lead and zinc concentration

Arkadiusz Bauerek, Central Mining Institute, Katowice, POLAND
Michal Gwozdziewicz, Central Mining Institute, Katowice, POLAND

Abstract

Research carried out in 2004 and 2005 showed the occurrence of zone changes of cadmium, lead and zinc contents in soils situated along the one of main motorway A 4 between Chrzanow and Krakow in Poland. It was found out that high contents of examined heavy metals in soils in the western part of the examined section is associated with the occurrence in the region of Chrzanow positive anomaly of geochemical contents of cadmium, lead and zinc. Secondary soil alkalization in motorway surroundings was also observed which may cause gradual concentration of cadmium, lead and zinc in increased pH value of the soil. Increase of soil pH value is associated with the use of chemical substances for winter maintains of the road. Observed heavy metals contents in soils situated near the motorway are mainly associated with the emission of industrial pollutions.

Introduction

Investigation of cadmium, lead and zinc concentration in soil in the surroundings of A 4 motorway in the section between Chrzanow and Krakow was carried out in the Environment Monitoring Department of Central Mining Institute in Katowice. The tests of soil adjacent to the A 4 motorway were performed in 2004 and 2005

The operation of road may influence indirectly the concentration of heavy metals in such soils coming from diverse sources of pollution, also by change of chemistry of soils lying in the closest neighbourhood. An instance of such a phenomenon may be accumulation of cadmium, lead and zinc in soils located near the investigated section of A 4 motorway, running in its western part through the areas featuring well-developed mining industry and zinc and lead metallurgy.

Location and geological and soil conditions of the investigation area.

Motorway A 4 is one of the most important transport routes in the southern Poland. The section of A 4 motorway where samples have been taken is 35 km long. It is located in the western part of Malopolska voivodship between Chrzanow and Krakow (Fig. 1)

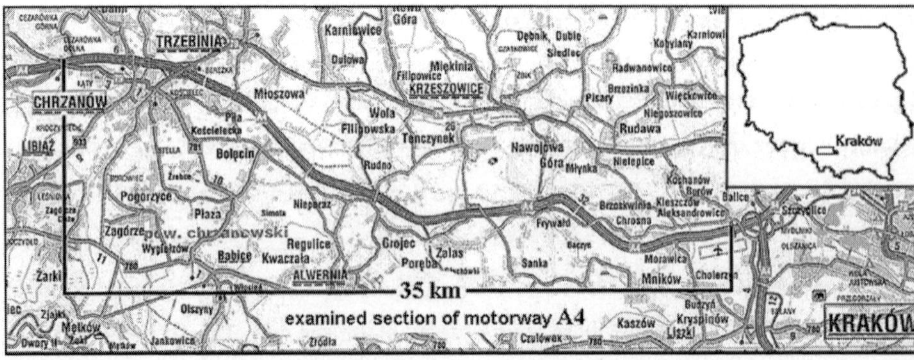

Fig. 1 Situation of examined section of motorway A 4 between Chrzanow and Krakow

Literature analysis has shown that the subsoil of the motorway features Carboniferous, Perm, Triassic, Jurassic, Tertiary and Quaternary formations [4].

The soil cover appearing along the motorway section under investigation was formed mainly on the Quaternary, upper Jura and middle Trias deposits. Soils formed from Quaternary deposits are represented mainly by podzols and brown soils. Podzol soil appears mainly in the western part of the area under investigation where sands of glacier and water-glacier accumulation are deposited of the Middle Poland Glaciation (Saalian) as well as sands of eolian origin. Such deposits fill the large depressions of Krzeszowice Trench. In its eastern part (from Brzoskwinia to Balice) the dominating part is constituted by brown soils formed upon the loess cover that has been formed during the North Poland Glaciation (Vistulian).

In the middle and extreme western part of the investigated area, where the motorway crosses the hills made of carbonate rocks of upper Jura and middle Trias, the predominating type is limestone soil. Upper Jura deposits formed as clear slab limestone and marls create outcrops between Grojec and Rudno, in the area of Zalas and near Krzemionki, Chrosna and Morawica. West from Chrzanow the land surface shows middle-Trias limestone and marl, and the ore-bearing dolomites constitute the surface part of the hills.

Methods and scope of research

Taking into account the linear character of the object under investigation, 35 soil samples have been taken in the A zone, at the distance of 3 to 10 m from the motorway curb. The sampling points were located uniformly along the entire section investigated, every 10 km. Subsequently, 10 soil samples have been taken from B-zone – at the distance of 10 – 15 m from the motorway, and 10 soil samples have been taken from C – zone, at the distance of 30 – 40 m from road border. In order to determine the geo-chemical background of indexes investigated, 12 equalizing soil samples have been take at the points located uniformly on both areas of the motorway, at the distance of 1000 – 2500 m (D zone). In order to perform a proper determination of heavy metals contents and pH, the soil samples have been taken at the level of humus accumulation, at the depth of 5 to 20 cm.

The reaction of soil (pH in KCl) was measured using the electrometric method, whereas Cd, Pb and Zn content were determined using emission spectrometry with inductively excited plasma (AES – ICP) after mineralization of samples in aqua regna and hydrofluoric acid.

Discussion of results

The impact of road operation on the increase of pH of soils located in the direct vicinity of the road is related mainly with use of chemicals used for the winter road maintenance [2]. Chlorides such as NaCl, $CaCl_2$, and $MgCl_2$ used for removal of snow and ice from the asphalt surfaces, impair the ionic balance of the environment, causing the alkalization of soil at the distance up to several meters from road border.

Transfer of soil alkalizing chemicals takes place through a direct spreading of mineral grains in solid form, splashing of saline water by cars and carrying the mist by wind

The process of gradual alkalization of soil was observed along the investigated section of the motorway A 4. The average values of pH_{KCl} in soils collected in A zone (at the distance of 3 to 10 m from motorway) taken from idle lands and forests was 7,25 at the values contained in the range 5,19 to 8,24 (Tab 1). At the larger distance from the highway in B zone (10–15 m) the average soil pH_{KCl} is lower and amounts to 6,68, and the values noted fall into the range from 4,44 to 8,10. pH_{KCl} of soil from C zone (30 – 40 m from the motorway) takes values from the range 5,08 to 7,99, and its average value is 6,46. Whereas average reaction of soil from reference samples representing the geo-chemical background was 5,47 at the noted range values from 4,00 to 7,74.

Table 1

Values of heavy metals contents and pH_{KCl} in the soil samples taken from the section of A 4 motorway between Chrzanow and Krakow

Sample location	Index		Western area Chrzanow - Bolecin		Central and eastern area Bolecin - Krakow	
			range	average	range	average
Zone A 3 – 10 m	pH_{KCl}		5,31 – 8,21	7,65	5,19 – 8,02	6,86
	Cd	mg/kg	< 1 - 15	5,3	< 1 – 2	1,2
	Pb	mg/kg	50 - 1035	366,0	17 – 86	40,6
	Zn	mg/kg	113 - 1667	553,9	15 - 224	80,1
Zone B 10 - 15 m	pH_{KCl}		4,97 – 8,10	7,08	4,44 – 7,42	6,29
	Cd	mg/kg	< 1 - 12	4,0	< 1 - 2	1,2
	Pb	mg/kg	31 - 573	209,6	25 - 58	35,8
	Zn	mg/kg	64 - 1311	442,8	47 - 91	63,4
Zone C 30 - 40 m	pH_{KCl}		5,08 – 7,99	6,52	5,41 – 7,03	6,40
	Cd	mg/kg	2 - 7	5,0	< 1 - 2	1,2
	Pb	mg/kg	39 - 281	158,6	25 - 64	35,2
	Zn	mg/kg	61 - 791	368,2	36 - 166	83,8
Zone D 1000 – 2500 m	pH_{KCl}		4,21 – 7,34	5,60	4,0 – 7,74	5,35
	Cd	mg/kg	2 - 4	2,8	< 1 - 3	1,4
	Pb	mg/kg	37 - 147	103,2	31 - 90	54,7
	Zn	mg/kg	30 - 281	170,0	14 - 215	82,7

In accordance with the division proposed by J. Siuta [7] a majority of soils taken along the motorway (from zone A) characterizes the following reaction: alkaline, medium alkaline or low alkaline. Whereas 18 from 32 soil samples taken from zones B, C and D – not covered by the impact of chemicals used for de-icing of roads characterizes the neutral, acidic, medium acidic and low acidic reaction. The significant reduction of pH of soil from the direct area of motorway takes place on the first 10 – 15 m from road border (Fig. 2). Among 10 samples of soil representing B zone (10 – 15 m) 9 featured the drop of pH_{KCl}, when compared with pH_{KCl} of A zone soil, amounting in average to 1,16 (the observed range is from 0,03 to 3,40).

The results presented indicate that on both sides of the highway there exist a zone of soil, several to ten-odd meters wide, featuring increased pH values in comparison to the soils not covered by secondary pH changes.

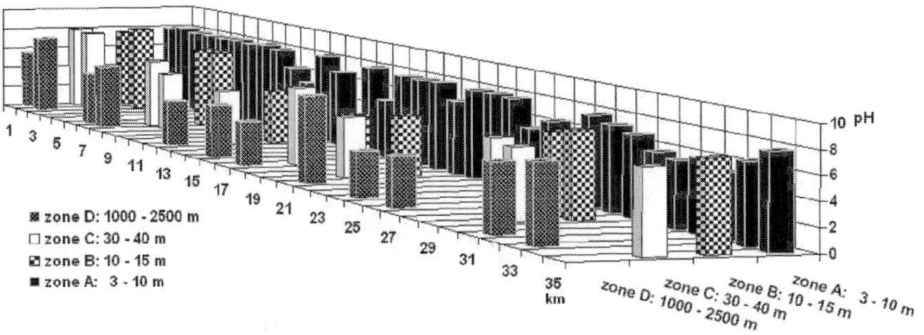

Fig. 2. pH_{KCl} of soils existing in the direct neighbourhood of A 4 highway and representing the geo-chemical background

The western part of investigated A 4 motorway area, ca 11 km long, (from Chrzanow to Bolecin) lies within the area of positive geotechnical anomaly of cadmium, zinc and lead contents in the soil [5]. This anomaly showing as occurrence of increased contents of the aforesaid heavy metals is related with mining and metallurgy of zinc and lead ores in the area of Chrzanow. Abnormal amounts of cadmium, zinc and lead in soil may be also of local origin related with appearance of ore-bearing dolomites in the soil profile as base rock.

Analysis of test results for soil samples taken in the A, B and C zones along the motorway and samples representing the geochemical background (D zone) has shown existence of a two-way changes of cadmium, zinc and lead contents in the soil of the area tested. The averaged test results are presented in Table 1.

Also very apparent is equatorial character of heavy metals contents changes (parallel to the course of motorway). The observed concentrations of cadmium, zinc and lead in the soil taken along the motorway and representing the background along the western part of the section in question (between Chrzanow and Bolecin) are several times higher than the average values of these metals for Upper Silesia soils amounting to 1,4 mg/kg for cadmium, 53 mg/kg for lead and 121 mg/kg for zinc [6].

Quantities of cadmium, lead and zinc both in the soil taken in A, B and C zones (at the distance from 3 to 40 from the highway) and in the soils taken in the areas located 1000 to 2500 m from it (C zone) drop drastically at the distance of 6 km east from Chrzanow near Bolecin town (Fig 3, 4 and 5).

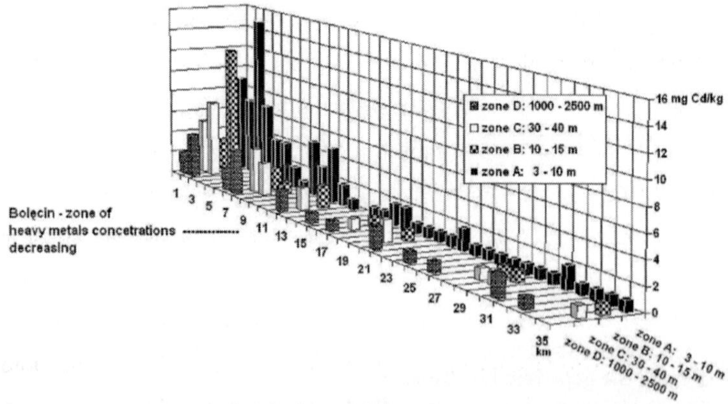

Fig. 3. Contents of cadmium in the soils taken from the direct surroundings of the motorway and those representing the geochemical background

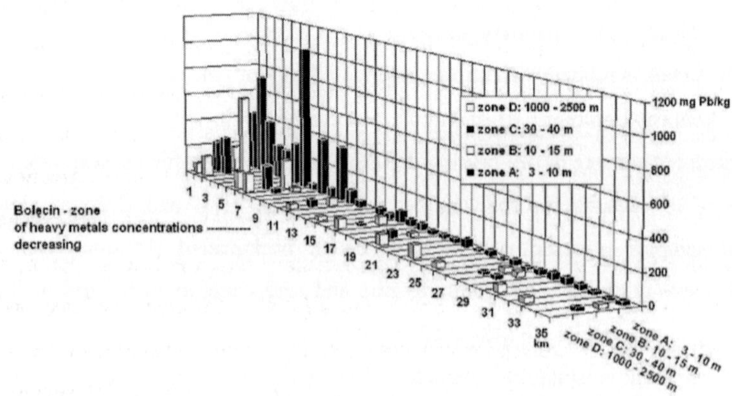

Fig. 4. Contents of lead in the soils taken from the direct surroundings of the motorway and those representing the geochemical background

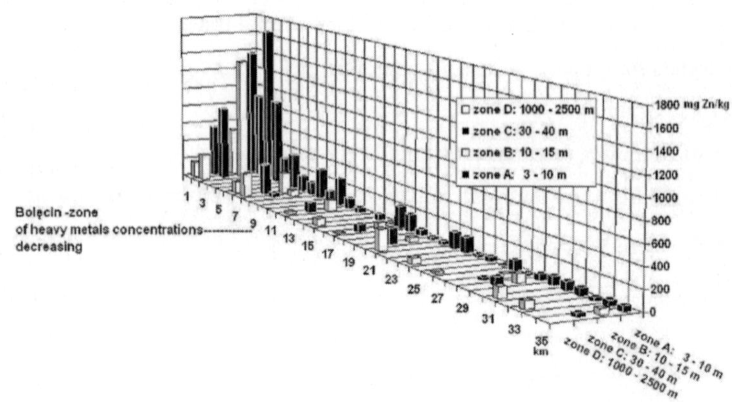

Fig. 5. Contents of zinc in the soils taken from the direct surroundings of the motorway and those representing the geochemical background

Location of Cd, Pb and Zn contents reduction zone in the soils is located at the distance of ca. 6 km east from the potential source of heavy metals emission in the Chrzanow and Trzebinia areas, as well as the outcrop of ore-bearing dolomites. Along the middle and eastern section (between Bolecin and Krakow) the average contents of lead and cadmium in the soils taken along the highway and those representing the background are close to the values stated for Upper Silesia soils, whereas the calculated average zinc contents in the soils of this area is lower than those in soils of Upper Silesia.

In the western part of motorway section under investigation it was also stated appearance of zone-oriented variability of heavy metal contents in the soil of motorway surroundings having meridian direction (perpendicular to the route of motorway). At the changed values of pH_{KCl} (in average 7,65) in soils lying directly along the motorway (in A zone) high concentration of cadmium (in average 5,3 mg/kg), lead (in average 366,0 mg/kg) and zinc (in average 553,9 mg/kg) are observed - Table 1. Concentration of the metals in question in A zone are considerably higher than the noted contents on soil taken between 10 and 15 m from the highway. A significant reduction of concentration of cadmium, lead and zinc in the soils located several to ten-odd meters from the highway (between Chrzanow and Bolecin) is correlated with reduction of soil pH_{KCl} by 0,99 in average. In zone C, contained between 30 and 40 m from the highway, the noted contents of zinc and lead as well as pH_{KCl} values are lower than those observed in A and B zones (Tab.1). Statement of lower indexes (except for cadmium whose concentration has increased) in C zone is related with the gradual disappearance of highway impact and transfer to the values characteristic for geochemical background. The observed concentrations of Cd, Pb and Zn in soil along the route are probably related with their secondary alkalizing. The metals under investigation are trapped in soil where pH exceeds values from 5,5 to 6,5 [3]. The zone of their concentration may be several to ten-odd meters wide from the highway. In the soils with lower pH, nearest the natural one, cadmium, lead and zinc do not accumulate as they appear in movable forms [1].

Along the middle and eastern section of the motorway in its direct neighbourhood no sub-concentration of heavy metals was observed, even if the alkalization of soil was stated. This is probably due to the fact that within the area between Bolecin and Krakow the pressure of industrial pollutants is much lower and the impact of motorway remaining in operation since mid-eighties is not remarkable where the Cd, Pb and Zn are concerned. In spite of stated significantly lower contents of heavy metals, the process of gradual reduction of pH values for soils in each zone is also observed along the middle and eastern motorway section, as well as related reduction of metals in soil.

Summary

The investigations carried out in 2004 and 2005 have shown that there exist zones of various concentration of cadmium, lead and zinc in soils existing along A 4 motorway between Chrzanow and Krakow. It was stated that the presence of high values of investigated heavy metals along the western part of the section in question is related with positive geotechnical anomaly occurring in the area of Chrzanow concerning the contents of cadmium, zinc, and lead. Also a secondary alkalization of soil was observed in the direct neighbourhood of the motorway. This may result in a gradual concentration of cadmium, zinc and lead. The increase of soil pH is related with use of chemicals during winter maintenance of road. The observed contents of heavy metals in soils appearing along the motorway are related mainly wit emission of industrial pollutants

LITERATURE

1. Alloway J., Heavy Metals in Soils, Blackie Academic and Professional, Glasgow 1995
2. Dobierska K., Waszkiewicz S., Wplyw srodkow chemicznych stosowanych w zimowym utrzymaniu drog na srodowisko gruntowe i szate roslinną, (Impact of chemicals used in the winter maintenance of roads on soil environment and vegetation) – The International Technical and Scientific Conference Miedzynarodowa Konferencja Naukowo – Techniczna, Krzyzowa 17 – 19 November 2004.
3. Kabata-Pendias A., Pendias H., Biogeochemia pierwiastkow sladowych, (Bio-geo-chemistry of trace elements) Wydawnictwo Naukowe PWN, Warszawa 1999.
4. Kaziuk H., Mapa geologiczna Polski w skali 1: 200 000 arkusz Krakow, (Geological map of Poland scaled 1: 200 000, sheet Krakow, Wydawnictwa Geologiczne, Warsaw 1980.
5. Lis J., Pasieczna A., Atlas geochemiczny Gornego Sląska 1:200 000, (Geo-chemical maps of Upper Silesia) Państw. Inst Geol., Warsaw 1995.
6. Lis J., Pasieczna A., Atlas geochemiczny Polski 1:2 500 000, (Geo-chemical maps of Poland) Państwowy. Inst Geol., Warsaw 1995.
7. Siuta J., Gleba diagnozowanie stanu i zagrozenia, (Soil – diagnostics of condition and hazard) Instytut Ochrony Srodowiska, Warsaw 1995.

Numerical model for leachate and gas flow in a sanitary landfill

Bernardo Caicedo Associated Professor Los Andes University, Bogotá, Colombia
Natalia Quintero Civil Engineer, Los Andes University, Bogotá, Colombia

Abstract: The gases and leachate flow in a sanitary landfill are a significant issue. On one hand, from the geotechnical point of view, the gas and liquid pressures that are developed during the decomposition process of organic matter could affect the stability of the landfill. On the other hand, the environmental aspects, the physicochemical characteristics of the leachate are important variables that depend on the efficiency of its treatment. The typical models of leachate generation usually estimate the amount of leachate that is produced by using water balance for the landfill, taking into account the concepts of infiltration and field capacity. This calculation methodology assumes that no gas pressures are developed inside the landfill, which disagrees with the direct observations in sanitary landfills. This methodology does not consider the leachates that can be produced. This paper covers a model which intends to correct these deficiencies by coupling the biological decomposition processes with the gas and liquid flow processes in partially saturated media.

Introduction

The control and knowledge of the gases and leachates produced by a sanitary landfill are of vital importance for the environmental health. Furthermore it has great influence on the stability of the waste and efficiency of the landfill. The objective of this document is to explain the development and to obtain a numerical model capable of calculating the production and flow of gas and leachates in a waste column for a specific landfill.

In order to solve this problem soil mechanics concepts related to partially saturated soils generating liquid and gas in the mass of the material are used. The gas and liquid generation depends on the kinetics of the biological processes, which are developed in the solid waste. Hydrology concepts are used to model the boundary conditions. The equations are solved using the explicit finite difference method. After presenting the numerical model, a column of sanitary landfill is simulated.

Generalities

Generation of gases and leachates in a sanitary landfill

A sanitary landfill is a media that receives waste and water, and expels leachates and gases. The velocity, composition and pressures generated by these components depend on the waste constitution, water infiltration and microbial growth, among others. The generation of leachates and gases in sanitary landfills are processes intimately related to one another. Whenever solid wastes are disposed on a sanitary landfill, a sequential decomposition process of the biodegradable organic fraction begins. These processes affect the generation of leachates and the formation of gases. It is important to understand the sequence followed in the production of gases and leachates, which is explained bellow (Tchobanoglous, 1993):

Initially, the micro organisms presented in solid wastes consume rapidly the oxygen of the air trapped in the voids, turning the environment anaerobic. Under these conditions, the decomposition of existing organic matter is accomplished by fermentation caused by micro organisms. These micro organisms degrading the organic matter dissolve the particles, forming large quantities of organic acids (e.g. acetic acid) and giving off substances (e.g. ammonia). This stage is named acidogenesis or stage of fermentation. Most of the organic acids produced in the first stages of waste decomposition escape in the leached liquids and are responsible to a large extent for the bad odour and the high pollutant power. This organic matter, which escapes with the leachates, is no longer available for the generation of gas in the sanitary landfill.

The generated acids, besides leaving mainly with the leachate, are also used by other micro organisms presented in the wastes to produce gas. These other micro organisms known as methanogenics are mainly producers of methane gas. Unfortunately, these micro organisms are rare in the original solid waste. Thus, the gas production is slow in the beginning while natural populations of these micro organisms are increasing. At first, there is a disruption between acid producer micro organisms and acid-consuming micro organisms. Eventually, the acid-consuming micro organisms grow and as their population increases, the gas production rates also increase considerably. Simultaneously, the acid concentrations in the leachate are diminished. The growth velocity of the methanogenic micro organisms is rather slow and different in each case. Typically, the gas production only reaches large volumes several years after the wastes have been disposed. In order to consider the formation of leachates, it is important to take into account that there are two origins, one biological associated with waste decomposition within the landfill, and other hydrological associated with the amount of water infiltrated into the waste layers.

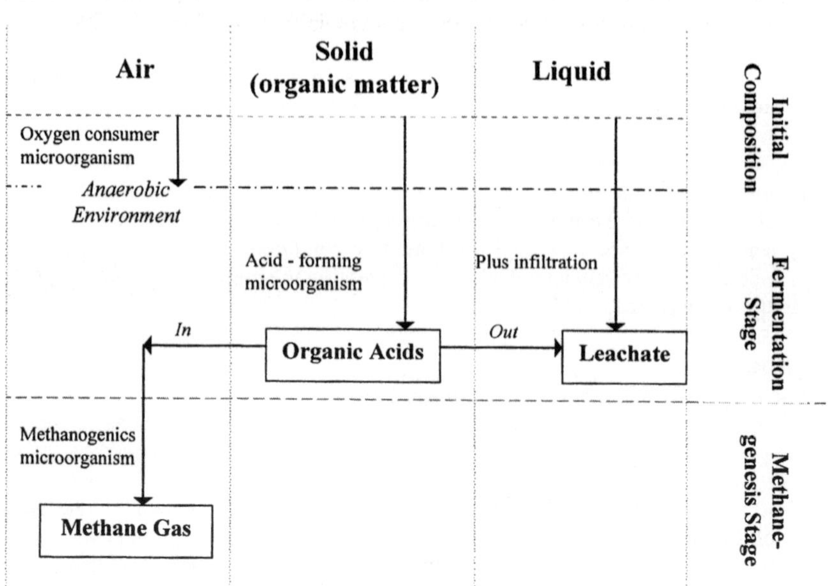

Figure 1. Gas and leachates generation process sketch

The model proposed in this work couple two processes. The first process is the decomposition of the organic matter in its two stages: initial fermentation and methane-genesis. The second process contemplates liquid and gas migration by using classical theories for partially saturated soils. In fact, the process of leachate and biogas migration in a sanitary landfill is a problem similar to the moisture flow in partially saturated soils. Several variables should be considered in the model: relative permeability of water and biogas according with its saturation degree, material water retention curve, and deformation laws of the material according to the applied stresses.

Description of the numerical model

The numerical model of sanitary landfills behaviour consists of three main modules functioning in a coupled manner: a biochemical module of gas and leachate generation, a gas and water flow module in a porous deformable partially saturated media and boundary condition module.

Microscopic Description of municipal solid waste

The municipal solid waste consists mainly in inert waste, biodegradable waste, leachate and gas. This composition occurs in three phases: solid, liquid and gas. Each one can be defined by its volumetric content, θ_1, θ_2 and θ_3. The summation of the three phases must equal unity; $\Sigma\theta_i=1$ and is composed of different items:

The solid phase is composed of three items: inert solids, θ_{11}; biodegradable solids, θ_{12}; and micro organisms. This last one is divided into two groups: acid-forming microorganisms,θ_{13}; and methanogenic micro organisms, θ_{14}. The summation of these components must equal the volume of the solid phase: $\theta_1=\Sigma\theta_{1j}$.

The liquid phase is composed of three items: water, θ_{21}; grease acids θ_{22}; and ammonia, θ_{23}. The summation of these three components equals the volume of the liquid phase: $\theta_2=\Sigma\theta_{2j}$.
The gas phase is composed of three items: air, θ_{31}; methane CH_4, θ_{32}; and carbon dioxide CO_2, θ_{33}. The summation of these three components equals the gas volume: $\theta_3=\Sigma\theta_{3j}$.

Similarly, it is possible to define the mass of each phase as w_i and the mass of each component as w_{ij}, where the subscripts ij correspond to the above definition.
The concentration of component j in phase i is defined as:

$$C_{ij} = \frac{w_{ij}}{w_i}$$

Considering an element of solid waste volume, in which there is no occurrence of gas and liquid flow, the time evolution of the mass of each of the 10 components of the solid waste can be expressed by the following phenomenological equations:

The inert solids, the biodegradable solids, the acid-forming micro organisms and the methanogenic micro organisms' variations, respectively:

$$\dot{w}_{11} = 0 \qquad \dot{w}_{12} = -k_1 w_{12} w_{13} \qquad \dot{w}_{13} = y_1 k_1 w_{12} w_{13} \qquad \dot{w}_{14} = y_2 k_2 w_{22} w_{14}$$

The water, grease acids and ammonia variations:

$\dot{w}_{21} = -y_3\dot{w}_{13} - y_4\dot{w}_{14} - \beta_w\dot{w}_{12}$ $\quad \dot{w}_{22} = \beta_1 k_1 w_{12} w_{13} - k_2 w_{22} w_{14}$ $\quad \dot{w}_{23} = \beta_2 k_1 w_{12} w_{13}$

The gas phase, the air, methane and the carbon dioxide variation:

$\dot{w}_{31} = 0 \quad \dot{w}_{32} = \beta_3 k_2 w_{22} w_{14} \quad \dot{w}_{33} = \beta_4 k_2 w_{22} w_{14}$

Where k_i are the constants that define the substrate consumption rates (solid biodegradable and grease acids), y_i are the constants that define the micro organisms' production (acids-forming and acid-consuming) and β_i are the stoichiometric acid-genesis and methane-genesis reactions constants, respectively:

$$C_aH_bO_cN_d + \beta_w H_2O \xrightarrow{Micro-organisms+Heat} \beta_1 CH_3COOH + \beta_2 NH_3$$

$$CH_3COOH \xrightarrow{Micro-organisms+Heat} \beta_3 CH_4 + \beta_4 CO_2$$

Flow in an unsaturated media (Not isothermal flow)

Water flow in a sanitary landfill is caused by climatic changes and microbial reactions. The three main state equations are presented; they describe the water, air and heat behaviour with regard to the soil's physical variations (saturation, porosity, etc) and the other states variation.

The air flow differential equation in *Y* direction, includes the influence of the water pressure and the temperature variation (Dakshanamurthy and Fredlund, 1981), the pressure contribute caused by the micro organisms' gases production should be added:

$$\frac{\partial u_a}{\partial t} = -C_a \frac{\partial u_w}{\partial t} + c_a^v \frac{\partial^2 u_a}{\partial y^2} + C_{at} \frac{\partial T}{\partial t} + gas$$

Where: u_a= air pressure, u_w= water pressure, T = temperature, C_a = interaction coefficient related with the partial differential equation of water phase, c_a^v = consolidation coefficient with regard to the air phase and C_{at} = interaction coefficient with regard to the temperature phase.

The differential equation of the water phase includes the water and the water vapour pressure due to the processes of advection and convection. In the *Y* direction (Wilson, 1990), plus the leachates pressure component:

$$\frac{\partial u_w}{\partial t} = -C_w \frac{\partial u_a}{\partial t} + c_v^w \frac{\partial^2 u_w}{\partial y^2} + c_v^{wv} \frac{\partial^2 \bar{u}_v}{\partial y^2} + leachates$$

Where: c_v^w = consolidation coefficient with regard to the water phase and c_v^{wv} = consolidation coefficient with regard to the water vapor phase.

$$c_v^w = \frac{k_w}{\rho_w g m_2^w} \qquad c_v^{wv} = \frac{\bar{u}_a + \bar{u}_v}{\bar{u}_a} \frac{D_V}{\rho_w m_2^w}$$

where: C_w = interaction coefficient related with the partial differential equation of air phase, $\bar{u}_v$= partial water vapor pressure, D_v = diffusion coefficient of water vapor, $\bar{u}_a$ = absolute atmospheric pressure , ρ_w= water density , g= gravitational constant, k_w = water permeability coefficient, and m_2^w = coefficient of volume change of water with regard to a suction change ($d\ (u_a - u_w)$) .

Finally, to describe the heat transfer of the soil the Fourier's diffusion equation will be used in the following way, Wilson (1990):

$$\zeta\frac{\partial T}{\partial t}=\frac{\partial}{\partial y}\left(\lambda\frac{\partial T}{\partial y}\right)-L_v\left(\frac{\bar{u}_a+\bar{u}_v}{\bar{u}_a}\right)\frac{\partial}{\partial y}\left(D_v\frac{\partial\bar{u}_v}{\partial y}\right)+heat$$

Where: ζ = volumetric heat capacity of the soil as a function of the water content, λ = thermal conductivity of the soil as a function of the water content and L_v = latent heat of vaporization of water.

Boundary conditions

For each one of the states, a boundary condition should be found, specifically, a condition that relates the hydrologic state of the atmosphere and the soil. This condition induces the system transformation.

Air pressure

Local atmospheric pressure can be taken as the air pressure boundary condition.

Water pressure

Water pressure is determined by the water quantity in the soil, it is directly related to the precipitation occurrence, which is part of the meteorological information. If there is a precipitation event, the capacity of the soil to infiltrate a quantity of water should be analyzed; and the volume of infiltration should be calculated. The volume of infiltration must be calculated using an imposed water pressure condition. In order to calculate the percentage of precipitation that the soil can absorb, we need to compare the absorbing capacity of the soil to the amount of precipitation.

Water vapour boundary condition development is acquired by applying the following statement: the evaporation quantity that occur (mm/day) depends on two parameters, the climatic conditions and availability of water in the soil. Each one of these parameters is limited by the other; the evaporation quantity under climatic conditions is calculated in an ideal state, which is the total saturation state. Total evaporation is reduced if the soil's surface saturation diminishes, because there is not evaporation potential.

Heat Flow

The soil's temperature at a surface level can be used as a boundary condition when the heat flow equation is solved (Wilson, 1990).

Numerical Results

A sanitary landfill's 10 meters height column is simulated, with a 7 m depth initial water table and dry climatic conditions. The results for different days since the initial day are shown. Days 15, 100 and 183 are denoted as initial, medial and final simulation days.

The suction values (figure 2) vary from a surface maximum value to a null value around the water table level. Through the simulation it can be observed how the suction maximum values begin to be concentrated on the unsaturated zone middle part.

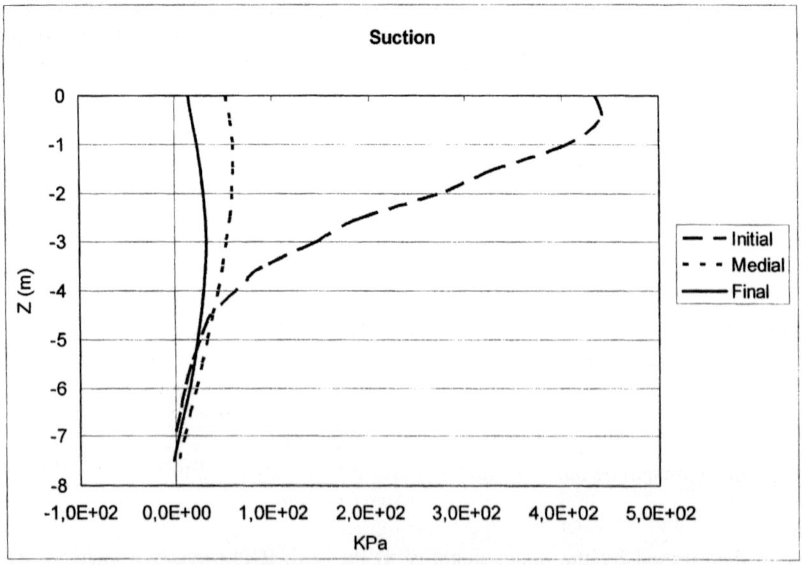

Figure 2. Suction vs. Depth

The saturation values as well as the suction values (figure 3) tend to take a parabolic form in the backfill middle unsaturated area. The saturation values become higher as the simulation proceeds and maximum values begin to be situated in the landfill's nucleus.

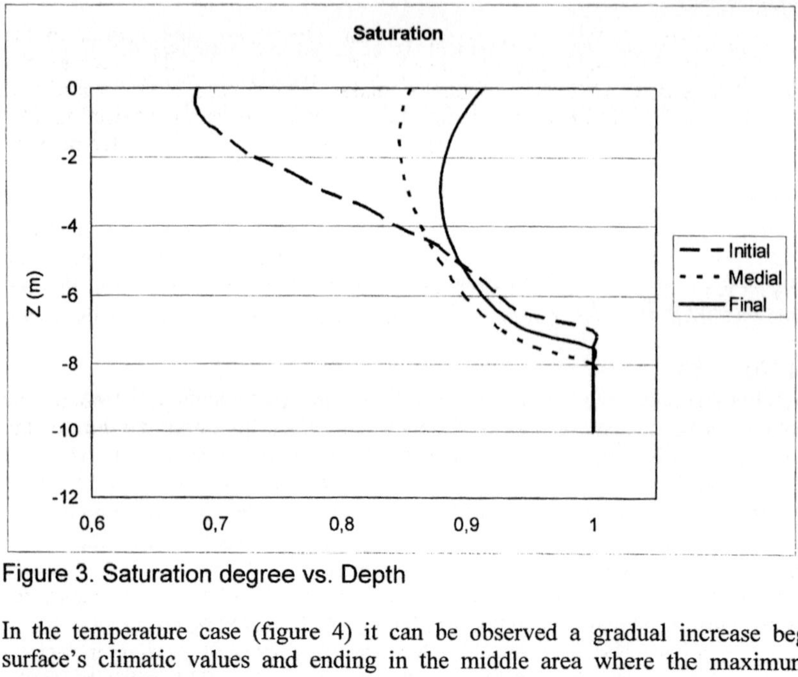

Figure 3. Saturation degree vs. Depth

In the temperature case (figure 4) it can be observed a gradual increase beginning with surface's climatic values and ending in the middle area where the maximum values are presented and those values are consequence of the exothermic microbial reactions. Final simulation shows a tendency to acquire a soft parabolic form getting a final value around the 332 Kelvin degrees. The biodegradable solid concentration values diminish gradually, while the acid forming and methanogenics micro organism's concentration values enlarge slowly.

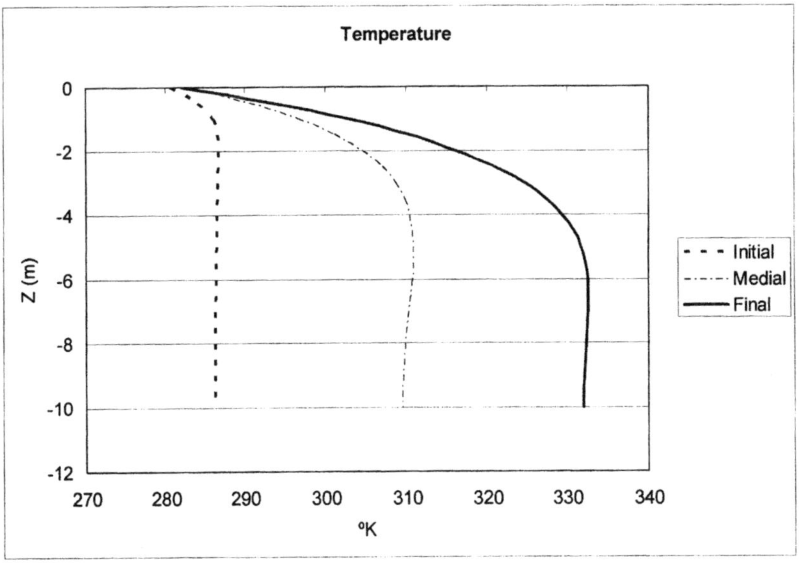

Figure 4. Temperature vs. Depth

Conclusions

The coupled model given in this paper presents great advantages with respect to the traditional models based on the field capacity concept. First of all, it is possible to foresee the pressures generated by gases and liquids, which are fundamental to compute the stability of the landfill. It is also possible to foresee the effect of the drainage pipes in the behaviour of a landfill. Biological kinetics are very important, the air-water pressure and temperature gradients caused by flow and hydrologic mechanisms.

The microbial growth and production is based on a series of main concepts that can be briefly simplified in the following way: change of an aerobic to anaerobic media, fermentation process accompanied by grease acids production due to the acid forming micro organisms and methane production due to the methanogenics micro organisms whose substrate are the grease acids.

In coupling the biological model, it is possible to attaint results over the amount and physicochemical characteristics of the leachates that will be obtained during the decomposition process. Finally, it is clear that the concepts of the partially saturated soil mechanics allow the unification of the decomposition phenomena of solid waste and the flow of liquids and gases in a sanitary landfill.

References

Bear J.; Bachmat Y.; 1984, Transport phenomena in porous media – basic equations. Fundamentals of Transport Phenomena in Porous Media. NATO ASI Series

Caicedo, B., Giraldo, E. and Behrentz, E. (2002) Modeling leachate and gas flow in sanitary landfills, Unsaturated Soils. Proc. 3rd Int. Conf. on Unsaturated Soils (UNSAT 2002), Recife,

Brazil (ed. Jucá, J.F.T., de Campos, T.M.P. and Marinho, F.A.M.), Lisse: Swets & Zeitlinger, Vol. 1, pp. 107-112.

Fredlund D.G.; 1984. Consolidation of unsaturated porous media. Fundamentals of Transport Phenomena in Porous Media. NATO ASI Series.

Fredlund D.G.; Rahardjo H; 1993, Soil Mechanics for Unsaturated Soils. John Wiley and Sons.

Giraldo E.; Behrentz E.; 2000, Modelo acoplado para la estimación de la generación de gases y produccion de lixiviados en rellenos sanitarios. VIII Colombian Geotechnical Congress.

Tchobanoglous, 1993. Integrated solid waste management, Ed McGraw Hill

Tindall J.A.; Kundel J.R.; Anderson D.E; 1999, Unsaturated zone hydrology for scientists and engineers. Prentice Hall.

Upscaling of micro-scale reactive transport equations using homogenization: iron adsorption during arsenic remediation

M.I.M. Darwish, Postdoctoral fellow, Delft University of Technology, Delft, The Netherlands.
J. Bruining, Associate Professor, Delft University of Technology, Delft, The Netherlands.

Abstract
Injection of aerated water is a remediation technique to prevent arsenic-containing water from entering in the drinking water tube wells. The idea is to develop a filtration zone around the well consisting of soil grains coated with metal (iron)-hydroxides, which adsorb arsenic (As). Under reduced conditions, abundant Fe(II) is dissolved in ground water. After injecting aerated water, Fe(II) is oxidized to iron-hydroxides precipitates that adsorb to the grains forming coating. In the subsequent production stage Fe(II) and As adsorb to the coated grains and deactivate the filter. Since the Fe(II) concentration in the produced water is much higher than As concentration we focus on Fe(II) adsorption neglecting the adsorption competition. On the pore scale we have diffusion, convection and reaction. The reaction occurs near the coated grains, but this detail needs to be upscaled to the scale of measurements. A powerful mathematical upscaling method is called homogenization. It starts with the micro-scale equations and uses well defined assumptions to obtain the upscaled equations. We follow the mathematics literature, but use a physical approach to derive the upscaled reactive transport model during resumption of the production stage. A second purpose of the paper is to give a physically comprehensive explanation of the homogenization technique. The upscaled equation shows that the convection and reaction are the dominant mechanisms, where as diffusional and dispersion effects play a minor role.

Introduction

In recent years, the presence of dissolved arsenic in contaminated groundwater from both natural and anthropogenic sources has emerged as a major concern on a global scale. Arsenic (As) is a ubiquitous element present in various compounds through out the earth's crust. Natural geochemical contamination through soil leaching is the primary contribution of dissolved arsenic in groundwater around the world.

Adsorption/co-precipitation with iron oxyhydroxides is the most commonly adopted method for arsenic removal from contaminated water because the method is both effective and cheap compared with other methods.

Intermittent introduction of aerated water in extraction wells can in principle reduce the As load in the pumped water due to oxidation of Fe(II) and Mn(II) followed by formation of metal(hydr)oxides which will contain As through adsorption/ coprecipitation phenomena. These reactions occur at the transition zone between oxic and anoxic groundwater. The main idea of the in-situ treatment method of groundwater is to transfer the physical, chemical and biological processes of conventional above ground treatment plants into the aquifer. The aquifer carries the function of natural subterranean reactor where oxidation and filtration processes for the removal of arsenic and other constituents take place. We distinguish three phases. In the first phase, the injection of oxygen-enriched water displaces the original groundwater containing iron, arsenic and other water constituents. Thus, an oxidation zone is built up in the aquifer where Fe(II) will be converted to Fe(III) compounds, which largely precipitate on the soil. During the following production phase (phase 2), flow is reversed and groundwater flows through the oxidation zone towards the well. The As(III) (and also As(V)) are adsorbed onto the Fe(III) compounds deposited on the grain. However, also the dissolved Fe(II) compounds are adsorbed to the soil grains and cover the grains that were previously coated with Fe(III). The Fe(II) compounds shield the Fe(III) compounds off and further adsorption of As compounds is inhibited (i.e. the filter is exhausted). Oxygenated water is re-injected (phase 3), thus the adsorbed Fe(II) is oxidized to ferric hydroxides by the oxygen transported with the injected water. This restores the filter capacity of the iron coated soil grains. In the ideal situation phase 2 and phase 3 can be repeated indefinitely.

In this paper we deal specifically with the second stage of the remediation process where the Fe(II) and As(III) (As V) are adsorbed to the soil grains, which are partially coated by previously deposited oxidation products. However, since the As concentration is negligible compared with the Fe(II) in ground water we focus on the Fe(II) adsorption. We choose the second stage because it is considered to be the simplest case from the modeling point of view. The purpose of this paper is to show how the homogenization upscaling procedure can be applied. Homogenization is developed in the seventies by mathematicians see Hornung (1992, 1997) for references. One considers large scale media with a periodic (or well defined random) microstructure and averages out the physical and chemical properties on a small scale to obtain average properties. The first step of applying homogenization to porous media was set by Tartar (1980). Bouddour et al. (1996) describe homogenization for particle erosion and deposition and we essentially follow their approach in this paper. In spite of the powerful possibilities of homogenization it is not widely applied by engineers (Smit et al. (2003)). The reason is that the mathematical framework is not always easy to grasp and that the physical aspects are not always made clear. By application of the method to the second stage of the As remediation process, we show that the method has a well defined physical basis and can be applied straightforwardly to numerous problems of engineering interest.
Modeling complex reactive transport processes in porous media consists of three parts. First a model at the pore scale needs to be developed. Secondly this model must be upscaled to the experimental (core) scale where measurements are possible. Thirdly the experiments are used to validate the upscaled model.

Foremost we describe the physical model on the pore scale. Also we give the procedure that is used for homogenization with the underlying assumptions. Then the pore scale model equations are presented. Moreover we give the non-dimensionalized equations including a separation of spatial differentiation on the global and local scale. Finally all dependent variables are separated into parts of decreasing significance and thus equations of various degrees of significance are obtained. By averaging these equations over the unit cell the upscaled equations are derived.

Physical model at pore scale

For reasons of illustration we will present a relatively simple 2-D example. The reader will find it easy to extend the procedure to geometrically more complex situations. This complexity is in practice only limited by computer power.

Consider a system of macroscopic (global or big scale) dimensions with a characteristic length L (see Figure 1). The macroscopic domain consists of a collection of periodic microscopic domains Ω (unit cells), on the microscopic (local or small) scale with characteristic length l. To ensure the separation of scales we need that $\varepsilon = l/L << 1$.

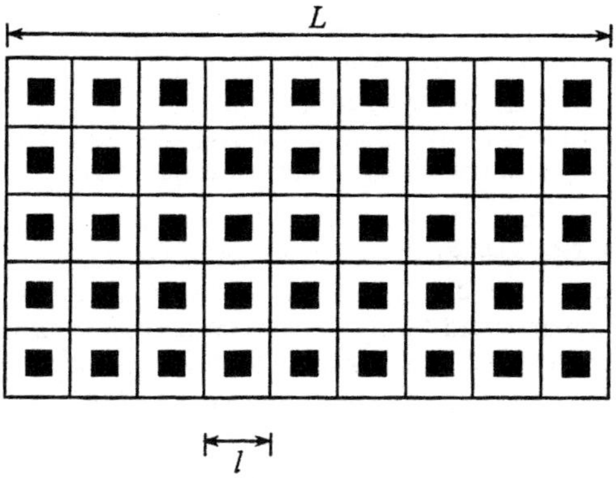

Figure 1: *a periodic lattice. In each cell we have a void space and an impermeable grain shown as a square. The length of the entire lattice is* L *and the length of a single unit cell is* l, *where* $\varepsilon = l/L << 1$.

In our example the periodic unit cell (PUC) consists of a (soil) square grain domain Ω_m with dimensions $a \times a$ in the center with a single phase fluid domain Ω_l flowing around it (see Figure 2). The boundary between Ω_l and Ω_m is denoted by Γ. However, the method can easily be extended using more complex unit cells. The PUC plays an analogous role as the representative elementary volume (REV) (Bear (1972)), but is usually smaller.

Fluid flow in Ω_l is described by Stokes law using periodic boundary conditions and the velocity is zero both perpendicular and along the grain surface Γ (impervious grains with no slip boundary conditions). It is also possible to use other conditions at the boundary of the unit cell (e.g. no transport of tracer across the side boundary) but the periodic boundary condition appears to be the most flexible. The fluid is considered incompressible.

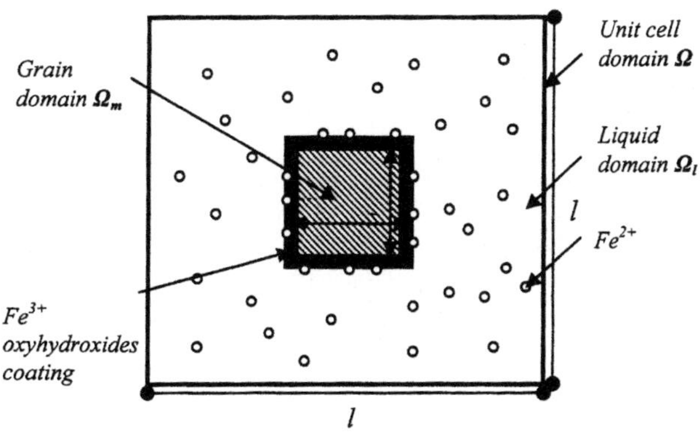

Figure 2: *Periodic unit cell and relevant domains. The picture shows the oxyhydroxide coating and the* Fe^{2+} *ions (not drawn to scale for reasons of illustration).*

Transport in the fluid phase is governed by convection, diffusion and reaction at the grain surface according to the surface reaction equation $k\,(c - c^{eq})$, where k is the first order kinetic rate coefficient, c is the Fe^{2+} concentration and c^{eq} is the Fe^{2+} equilibrium concentration near the precipitated Fe^{2+}. Hence the adsorption rate is determined partially by diffusion transport and partially by the kinetics of the surface reaction. The diffusion coefficient is independent of composition. The fluid of interest is water containing dissolved Fe(II) (or As) compounds flowing around the grains.
Fluid properties like density and viscosity are considered to be independent of the solute concentrations. Consequently also gravity effects can be disregarded.

Pore-scale model equations

Stokes equation

The Stokes equation, the continuity equation and the boundary condition at the grain read respectively

$$\mu\,\Delta v = \nabla\Phi \qquad [1] \qquad \text{and} \qquad \nabla \cdot v = 0 \text{ in } \Omega_l \qquad [2]$$

$$v = 0 \text{ on } \Gamma$$

where $\Phi[Pa] = P + \rho g z$ is the potential, where P is the pressure, ρ [kg/m³] is the density, g [m/s²] is the acceleration due to gravity and z [m] the coordinate in the vertical direction. Furthermore v [m/s] is the velocity vector. Within this formulation (μ, ρ are constant) the flow equation is completely decoupled from the solute transport equations and can thus be solved independently (Guceri et al. (1985)).

Solute transport

The Fe^{2+}, with a molar concentration c, transport is described by convection using the velocity field computed with the Stokes equation and diffusion. At the grain boundary Γ the Fe^{2+} concentration is c_s and the adsorption rate $r = dc_s/dt$ is given by $r = k\,(c - c^{eq})$.
The convection diffusion equation and the continuity equation inside the PUC i.e. in Ω_l can be written as:

$$\frac{\partial c}{\partial t} + div(vc) = div(D\ grad\, c) \text{ and } div\ v = 0 \qquad [3]$$

where c is the molar Fe^{2+} concentration, v the velocity field obtained from the solution of the Stokes equation and D is the molecular diffusion coefficient.
The boundary condition at the grain surface is written as:

$$(-D\ grad\ c)_n = k\,(c - c^{eq}) \qquad [4]$$

and $v = 0$ at Γ where n is the unit vector perpendicular to Γ boundary.

Procedure for homogenization

The procedure used for homogenization is as follows: The transport equations are non-dimensionalized (Shook et al. (1992)), where we split all spatial differentiation into a contribution on the local scale and on the global scale. We assume that these two contributions are of the same order of magnitude with respect to the ε scale. This means that in the non-dimensionalizing procedure L is the characteristic length for global differentiaiton and l is the characteristic length for local differentiation. Then we expand the dependent variables into contributions of decreasing significance with respect to ε. In other words we write a power series for each of the dependent variables in terms of ε. Substituting the series into the dimensionless equations makes that each of the equations consist of terms with different order with respect to ε. Because the upscaling procedure should be valid for a range of ε values of the same order of magnitude, it is stated that each of the terms with a specific order in ε constitutes an equation that is separately satisfied. First, the equation of the lowest order in ε is solved leading to solutions of the most significant contributions of the dependent variables. Subsequently the equation with one order higher in ε is solved using the previously obtained relations. More accurate (but also more complex) models can be obtained by adding the averaged equations with the next in the order of ε. We only derive the leading order model. This averaged equation over the PUC is the upscaled equation. Combining the results also the expressions for the upscaled parameters in the upscaled equation are obtained.

Splitting differentiation into global and local scale and non-dimensionalizing

Stokes equation

Equations [1] and [2] can be straightforwardly solved numerically in 2D by introducing the stream function and the vorticity (Guceri et al. (1985)). As a result we find in each periodic unit cell the same velocity distribution, because in each PUC the flow is subjected to the same potential gradient. As a result the velocity does not change on the global scale. Only local velocity variations i.e. within the PUC are relevant. It is convenient to separate the divergence ∇ into a global ∇_b and a local ∇_s contribution. We notice that $\nabla_b \cdot v = 0$ because of the periodicity assumption and therefore we obtain from $\nabla \cdot v = \nabla_b \cdot v + \nabla_s \cdot v = 0, \nabla_b \cdot v = 0$, and $\nabla_s \cdot v = 0$

The Stokes flow problem within the periodic unit cell can be solved such that we obtain the velocity field $v = v(r_s)$.

Solute transport

The differentiation is split into contributions at the local (small) and the global (big) scale. As stated above, we assume that the contribution of the global and local scale differentiation have the same order of magnitude with respect to ε. Subsequently we non-dimensionlize the equations (Shook et al. (1992)), using l and L as the characteristic length of the local (s) and global (b) scale respectively. Moreover we use the interstitial velocity v_R i.e. the Darcy velocity divided by the porosity as the characteristic velocity scale. This leads to

$$\begin{aligned} &\frac{L}{v_R t_R}\frac{\partial c}{\partial t} + div_b\,(vc) + \frac{L}{l}\,div_s\,(vc) = \frac{l}{L}\,div_b\,(\frac{D}{v_R l}\,grad_b\,c) + \\ &div_b(\frac{D}{v_R l}\,grad_s\,c) + div_s(\frac{D}{v_R l}\,grad_b\,c) + \frac{l}{L}\,div_s\,(\frac{D}{v_R l}\,grad_s\,c) \end{aligned}\quad , \text{in } \Omega \qquad [5]$$

where t_R is a reference time and $t_R = L/v_R$. We assume that $\partial c/\partial t$ is of the same order of magnitude as $div_b\,(vc)$, because we anticipate that the main concentration change $\partial c/\partial t$ is caused by global convection effects.

The reference velocity v_R is given a value of 10^{-5}m/s as a typical velocity occurring in the field and the diffusion constant $D \approx 10^{-9}$m²/s. For a size of the periodic unit cell of l=100μ=10^{-4}m we obtain for the Peclet number $Pe = v_R\, l/D$, i.e. of the order of one. Note that by defining the Peclet number we use l as a reference length. We substitute the Peclet number and use the abbreviation that $\varepsilon = l/L$. Hence we obtain the equation

$$\frac{\partial c}{\partial t}+div_b\,(vc)+\frac{1}{\varepsilon}\,div_s\,(vc)=\frac{\varepsilon}{Pe}\,div_b\,(grad_b\,c)+\frac{1}{Pe}div_b(grad_s\,c)+ \\ \frac{1}{Pe}div_s(grad_b\,c)+\frac{1}{\varepsilon}\frac{1}{Pe}\,div_s\,(grad_s\,c) \qquad [6]$$

The boundary conditions at the grain surface reads

$$-\left(\frac{D}{v_R L}\,grad_b\,c\right)_n-\left(\frac{D}{v_R L}\,grad_s\,c\right)_n=\frac{k}{v_R}(c-c^{eq}) \qquad [7]$$

and $v=0$ at Γ

The boundary condition can be written in dimensionless form as

$$-\left(\frac{\varepsilon}{Pe}\,grad_b\,c\right)_n-\left(\frac{1}{Pe}\,grad_s\,c\right)_n=N_{Da}\;(c-c^{eq})\text{ at }\Gamma, \qquad [8]$$

where N_{Da} is the Damkohler number. In this paper the case where $N_{Da}\;(c-c^{eq})$ is of the order of magnitude of ε, with respect to the left term, is considered i.e. where the transport dominates over the adsorption.

Expansion of dependent variables and equating the terms with the same order of magnitude

The concentration c can be developed into a series of terms with decreasing significance. This significance is expressed in terms of ε i.e. with respect to the upscaling parameter. Using concepts as in perturbation theory we equate the terms of the same order of magnitude. The order of magnitude is expressed in terms of the scaling parameter ε. The approach is valid for any value of ε in the range of our interest e.g. 10^{-5}-10^{-3}. We use the series expansion $c=c_o+\varepsilon\,c_1+\varepsilon\,c_2+\cdots$ where c_o, c_1, c_2 are of the same order of magnitude. The whole procedure until now had as purpose to know which terms are of the same order of magnitude with respect to the scaling i.e. of the order of $1/\varepsilon$, ε^o, ε etc. we need only two orders of ε to obtain the leading order upscaled equation. Hence we obtain

Order $\frac{1}{\varepsilon}$
$$div_s\,(vc_o)=\frac{1}{Pe}div_s\,grad_s\,c_o, \qquad [9]$$

Order ε^o
$$\frac{\partial c_o}{\partial t}+div_b\,(vc_o)+div_s\,(vc_1)+div_s\,(vc_o)=\frac{1}{Pe}div_b\,(grad_s\,c_o) \\ +\frac{1}{Pe}div_s\,(grad_b\,c_o)+\frac{1}{Pe}div_s\,(grad_s\,c_1), \qquad [10]$$

for the convection diffusion equation. For the continuity equation we get as shown before

$$div_s\,v=0 \qquad [11]$$

$$div_b\,v+div_s\,v=0, \qquad [12]$$

For the boundary condition we obtain

Order ε^o
$$\left(\frac{1}{Pe}grad_s\,c_o\right)_n=0,\quad v_0=0\text{ at }\Gamma, \qquad [13]$$

Order ε
$$-\frac{1}{Pe}\left(grad_b\,c_o+grad_s\,c_1\right)_n=\frac{N_{Da}}{\varepsilon}(c_o-c^{eq}),\quad v_1=0\text{ at }\Gamma. \qquad [14]$$

Solution of the model equations

In order to solve the equations we need to specify the periodic boundary conditions on the local scale. (i.e. $c_i(x_s, y_s) = c_i(x_s + 1, y_s)$ and $c_i(x_s, y_s) = c_i(x_s, y_s + 1)$ with $i = 0,1,...$) to allow the solution of the equation.

The order $\frac{1}{\varepsilon}$ equation $div_s\,(v c_o) = \frac{1}{Pe} div_s\; grad_s\; c_o$ has as boundary condition that c_o is Ω periodic on the PUC boundaries. Mathematics (Lions (1981)) dictates that equation [9], using the Ω periodicity of c_o and Eq. [13] has a unique solution i.e. c_o is independent on r_s and hence $c_o = c_o(r_b)$ only depends on the global scale coordinate r_b. In saying this we mean that on the scale of the unit cell c_o remains constant, but it does not remain constant on the global scale.

Using this and the arguments given below we can simplify order ε^o equation [10] to

$$\frac{\partial c_o}{\partial t} + div_s\,(v c_1) + v \cdot grad_b\; c_o = \frac{1}{Pe} div_s\,(grad_b\; c_o) + \frac{1}{Pe} div_s\,(grad_s\; c_1) \qquad [15]$$

where we use that

$div_b\,(v c_o) + div_s\,(v c_o) = c_o\,(div_b\,(v) + div_s\,(v)) + v\; grad_b\; c_o = v\; grad_b\; c_o$, and

$div_b\; v + div_s\; v = 0$ and from the fact that $\frac{1}{Pe} div_s(\,grad_s\; c_o) = 0$, because c_o does not depend on the local scale coordinate r_s (periodic unit cell) i.e. $grad_s\; c_o = 0$.

If we integrate Eq. [10] over the unit cell and divide by the volume of the unit cell Ω i.e. we multiply at the left with the operator $\frac{1}{|\Omega|}\int_{\Omega_l} dr_s$ we obtain

$$\left(\varphi\frac{\partial c_o}{\partial t} + \bar{u} \cdot grad_b\; c_o\right) = \frac{1}{Pe}\frac{1}{|\Omega|}\int_{\Gamma}((grad_b\; c_o) + (grad_s\; c_1)) \cdot dA \qquad [16]$$

where we apply Gauss's integral theorem and the periodicity of the dependent variables. Indeed for any Ω periodic vector Q we have that

$$\frac{1}{|\Omega|}\int_{\Omega_l} div_s\, Q\; dr_s = \frac{1}{|\Omega|}\int_{\Gamma} Q\; dA \qquad [17]$$

Note that we denote the average velocity by $\bar{u}$ because it is a Darcy velocity, where $\bar{u} = \frac{1}{|\Omega|}\int_{\Omega_l} v\; dr_s$ denotes the average in the periodic unit cell.

Hence it follows from equations [14] and [16] that

$$\varphi\frac{\partial c_o}{\partial t} + \bar{u} \cdot grad_b c_o = -\frac{N_{Da}}{\varepsilon}\,(c_o - c^{eq})\,A_v \qquad [18]$$

where A_v is the dimensionless specific surface area i.e. the area of the grain divided by the volume of the PUC. In full dimensional form, the first order macroscopic description obtained in a notation where now we use only full dimensional quantities reads:

$\varphi\frac{\partial c_o}{\partial t} + \bar{u} \cdot grad_b c_o = -k\,(c_o - c^{eq})\,A_v$ and the approximation of the macroscopic behavior within an error $O(\varepsilon)$ is $\varphi\frac{\partial <c>}{\partial t} + \bar{u} \cdot grad_b <c> = k\,(<c> - c^{eq})\,A_v$, where $<c>$ is the averaged concentration. This equation is valid for adsorption of any species on soil grains where the transport dominates over the adsorption. The practical implication is that in the upscaled equation all the quantities can be measured in contrary to the micro-scale equation.

However, in this research work we were inspired by the iron adsorption on coated grains during arsenic remediation.
In fact, Bouddour et al. (1996) got the same equation as well with their more mathematically oriented approach.

Conclusions

- We applied homogenization to derive the upscaled reactive transport equations for the Fe(II) and As precipitation on metal hydroxides-coated soil grains.
- Homogenization is a powerful technique to derive upscaled equations for reactive transport models in porous media starting from the micro-scale equations.
- We used a physically oriented approach for homogenization, but arrive at the same result as obtained from the purely mathematical approach.
- The upscaled equation shows that diffusion and dispersion play a minor role for the problem considered.
- The physically oriented approach makes it easy to use homogenization for problems of engineering interest.

Acknowledgments

This paper is based on the article of Bouddour et al. (1996). The ideas regarding the dimensional analysis originated during a discussion of the authors with Jacob Bear, Marjan Smit, Ruud Schotting and Andrea Cortis.

References

Bear, J., 1972, Dynamics of fluids in porous media, Dover, ISBN 048-65675-6.

Bouddour, A., J.-L. Auriault and M.Mhamdi-Alloui, 1996, Erosion and deposition of solid particles in porous media: homogenization analysis of formation damage, TIPM, 25:121-146.

Guceri, S., and Farouk, B., 1985, Numerical solutions in laminar and turbulent natural convection", in Kakac, S., W. Aung and R. Viskanta (eds). Natural Convection: Fundamentals and Applications, Washington, D.C.: Hemisphere.

Hornung, U, 1992, Applications of the homogenization method to flow and transport in porous media, Xiao Shutie (Editor), "Summer school on flow and transport in porous media", Beijing China, 8-26 August, World Scientific, ISBN 9971-50-934-2.

Hornung, U. (ed.): 1997, Homogenization and porous media, Vol. 6 of Interdisciplinary applied mathematics, Springer-Verlag, Berlin.

Lions, J. L.: 1981, Some methods in the mathematical analysis of systems and their control, Science Press, Gordon and Breach, New York

Shook, M., Li Dachang and Lake, L., 1992, Scaling Immiscible flow through permeable media by inspectional analysis, IN SITU 16 (4) 311-349, Appendix A

Smit, M., Bruining, J., and Schotting, R.J., 2003, Homogenization of the Stokes equation", XIV International conference on Computational Methods in Water Resources.

Tartar, L., 1980, Incompressible fluid flow in a porous medium-convergence of the homogenization process, E. Sanchez-Palencia (Ed.), Non-homogeneous media and vibration theory, Lecture notes in physics 127 Springer, Berlin, 368-377.

Micro/Macro Modeling of Electrolyte Transport through Semipermeable Bentonite Layers

Andrea Dominijanni, DITAG, Politecnico di Torino, Torino, Italy
Mario Manassero, DITAG, Politecnico di Torino, Torino, Italy
Daniele Vanni, Trevi SpA, Cesena, Italy

Abstract
Clay soils, characterized by high specific surface and high cation exchange capacity, as bentonite, are able to behave as semipermeable porous media that restrict the passage of charged solutes. This paper describes a microscopic model for the structure of bentonite. The ability of bentonite to generate ion partitioning is analysed using the diffuse double layer theory and the Donnan's equilibrium theory. The analogies and differences between these two approaches are analysed. Successively, the macroscopic transport flux equations are derived from spatial averaging of the microscopic Generalized Navier-Stokes equation and the Generalized Nernst-Planck equations. Some analytical solutions for specific problems are provided and the potentially significant beneficial aspect of the semipermeable properties for contaminant containment is discussed.

Bentonite Microstructure.
The bentonite is a clay soils containing at least 70% of the three layered (2:1) clay mineral montmorillonite. Montmorillonite particles are characterised by a negative net charge, which is a result of substitution of lower-valence cations for higher-valence cations within the structure The surface charge of montmorillonite, σ, is approximately equal to 0.114 C/m^2, the total specific surface, S, available for water adsorption, is generally assumed equal to 760 m^2/g and the solid density, ρ_s is typically equal to 2.65 g/cm^3.

The transport of a solution through a clay soil at the microscopic scale is well represented by the motion of a film between infinitely extended platy clay particles. The half distance, b, between the clay particles may be estimated from the total porosity, n, or the void index, e. The work by Norrish (1954) has shown that bentonite may have a dispersed structure, in which clay particles are present as well separated units, or an aggregated structure, that consists of packets of particles, or tactoids, within which several clay platelets are in parallel array, with interparticle distance of 9 Å. The tactoid formation has the net result of reducing the surface area of the montmorillonite, which than behaves like a much larger particle with the diffuse double layer only fully manifesting itself on the outside surfaces of the tactoids (see Fig. 1). The formation of tactoids is due to internal flocculation of clay platelets and depends on the concentration and the valence of the ions in the soil solution. The number, N, of clay platelets forming a tactoid increases increasing the ion concentration and the valence of cations in the soil solution. Unfortunately, the number of platelets in a tactoid cannot be predicted and has to be estimated from macroscopic measurements of the transport parameters (e.g. the permeability of clay at zero electric potential difference).

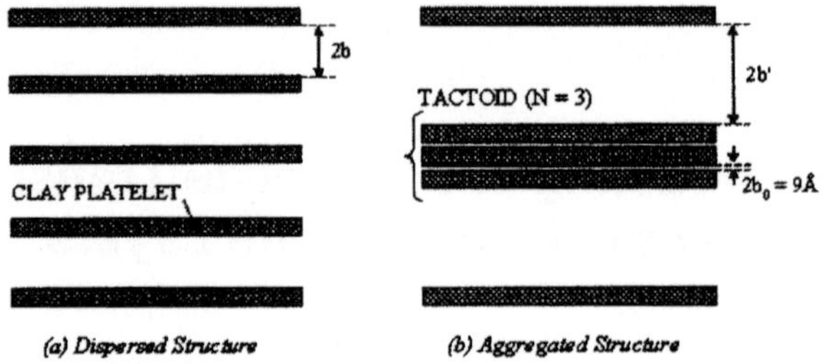

Figure 1. Microscopic structure of montmorillonite.

The half spacing, b, in dispersed clays may be estimated assuming a uniform distribution of clay platelets in parallel orientation, from the relationship

$$b = \frac{e}{\rho_s S} \tag{1}$$

If the clay has an aggregated structure, only the external surface of tactoids should be considered and the void space within the platelets in tactoids should be subtracted to the total void space (see Fig. 1). Assuming tactoids with N platelets, the internal specific surface, S', and the external specific surface, S'', are given by:

$$S' = S / N \tag{2}$$

$$S'' = S - S' = \frac{(N-1)}{N} S . \tag{3}$$

The average half spacing between the platelets in the tactoids as determined by X-ray measurements is $b_0 = 4.5$ Å (Shainberg et al., 1971). Thus the void index associated with the internal surfaces of the tactoid, e_0, is

$$e_0 = b_0 \rho_s S'' . \tag{4}$$

The corrected half spacing b' between the tactoids may be estimated from an equation analogous to equation (1):

$$b' = \frac{e'}{\rho_s S'} . \tag{5}$$

where $e' = e - e_0$. When the number, N, of clay platelets in tactoids increases, the external specific surface decreases and the half spacing, b, between tactoids increases.

The permeability at zero electric potential difference, k_ϕ, of the clay may be related to the half spacing, b', as follows:

$$k_\phi = \frac{n' \tau_m}{\mu} \frac{b'^2}{3} \gamma_w \tag{6}$$

where $n' = e'/(1+e')$ = effective porosity, γ_w = unit weight of the solvent, τ_m = matrix tortuosity factor, that accounts for the actual film pathways through the clay due to the geometry of the interconnected pores, and μ = solvent viscosity ($\mu = 10^{-3}$ Pa·s for water). k_ϕ represents the permeability of the clay measured at zero electric potential difference, that is

$$k_\phi = \gamma_w L \left(\frac{q}{\Delta P} \right)_{\Delta\varphi=0, \Delta C_i=0} \qquad i = 1,2, \ldots, N_{ions} \tag{7}$$

where q = volumetric flux of the solution or Darcy's velocity, ΔP = hydraulic pressure difference applied to the sample, ΔC_i = concentration difference of the i-th ion, L = length of the specimen, N_{ions} = number of ions in solution, $\Delta\varphi$ = electric potential difference.

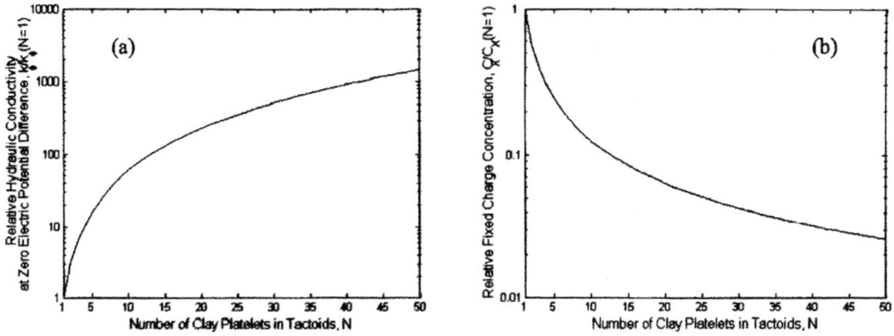

Figure 2. Dependency of the hydraulic conductivity at zero electric potential difference, k_ϕ, (a) and the fixed charge concentration, C_X, (b) on the number, N, of clay platelets in tactoids for a bentonite having a total porosity n = 0.8.

The hydraulic conductivity, k, as it is usually defined, is measured not at zero electric potential difference, but at zero electric current. The difference between k_ϕ and k is very important for charged porous media, as bentonites, because k_ϕ depends on the ion concentration and valence only through the half spacing between the clay platelets, whereas k depends explicitly on the ion concentration and valence through the interaction with the solvent molecules beyond that with the clay platelets. The relation between k and the ion concentration and valence will be discussed in a following paragraph.

From Fig.2a, the effect of an internal flocculation on the permeability at zero electric potential difference may be evaluated. If the number of clay platelets forming the tactoids increases, for example due to an increase of the concentration of divalent cations in solution, the characteristic half distance b' increases, inducing a large increase of k_ϕ. As a result, the dispersed structure of bentonite is associated to low permeabilities and satisfactory performances as contaminant barrier, whereas the aggregated structure is associated to high permeabilities and poor containment ability.

Diffuse Double Layer and Donnan's Equilibrium Theories.

The ability of clays to generate ion partitioning is generally analysed on the basis of the DDL theory. Sometimes, as an alternative, the Donnan's equilibrium of electrolyte solutions is used. The goal of this paragraph is to clarify the relation existing between these two approaches.

When a porous medium separates two solutions of different concentrations of an electrolyte, the internal surface of the pores acquires a charge and electric double layers develops within the pore. To describe the flow, a rectangular coordinate system (x,y) is used, where x is the position along the pore longitudinal axis, while y is the distance from the central axis in the transversal direction (see Fig. 3). Following Gross and Osterle (1968), the electric potential, $\overline{\varphi}(x,y)$ may be assumed composed by two parts: the first, $\psi(x,y)$ represents the electric potential due to the double layer, while the second, $\Phi(x)$, denotes the potential of a virtual bulk solution that is in thermodynamic equilibrium with the soil solution at the generic position x.

The total electric potential in the pore satisfies the Poisson equation:

Figure 3. Reference scheme for a typical pore of the bentonite.

$$\nabla^2 \overline{\varphi} = -\frac{F}{\varepsilon_r \varepsilon_0} \sum_{i=1}^{N_{ions}} z_i \overline{c}_i \tag{8}$$

where F is the Faraday's constant (96,485 C/mol), ε_r is the relative permittivity (for water, $\varepsilon_r = 78$), ε_0 is the permittivity of vacuum ($8.854 \cdot 10^{-12}$ C/(V·m)), z_i is the electo-chemical valence of the i-th ion and $\overline{c}_i$ is the molar concentration of the i-th ion.

Since the length of the pores is much longer than the transversal spacing (L >> b'), the axial variations of the potential may be neglected with respect to the transversal ones:

$$\frac{d^2\psi}{dy^2} = -\frac{F}{\varepsilon_r \varepsilon_0} \sum_{i=1}^{N_{ions}} z_i \overline{c}_i \,. \tag{9}$$

The Poisson equation is generally associated to the following boundary conditions:

$$\frac{d\psi}{dy}(y=0) = 0 \text{ at the centre of the pore } \quad (10a); \qquad \frac{d\psi}{dy}(y=b') = -\frac{\sigma_d}{\varepsilon_r \varepsilon_0} \text{ at the clay surface } \quad (10b)$$

where σ_d = electric surface charge associated to the diffuse double layer. σ_d is given by the total surface charge σ minus the surface charge in the Stern layer σ_s.

Using the boundary conditions (Eqs. 10), the area average of Eq. 9 may be expressed as follows (Cwirko and Carbonell, 1988; Moyne and Murad, 2002):

$$\sum_{i=1}^{N_{ions}} z_i \overline{C}_i - C_X = 0 \tag{11}$$

where $\overline{C}_i = \langle \overline{c}_i \rangle$ = macroscopic ion concentration of the i-th ion, $C_X = \sigma_d / (F \cdot b')$ = fixed charge concentration, i.e. the macroscopic concentration of the charges of the solid skeleton referred to the pore volume, the brackets mean averaging over the capillary cross section, i.e. $\langle \bullet \rangle = \frac{1}{b'} \int_0^{b'} \bullet \, dy$. Equation 11 corresponds to the statement of macroscopic electroneutrality in the pore. In Fig. 2b, the variation of the fixed charge concentration, C_X, as a function of the number, N, of clay platelets forming the tactoids is illustrated: when N increases, the half spacing, b', increases and, therefore, C_X decreases.

In the DDL theory, the Poisson equation is coupled with the equations that impose the equality between the electro-chemical potentials of the components of the soil solution and the electro-chemical potentials of the components of a virtual bulk solution:

$$\overline{\mu}_i^{ec} = \mu_i^{ec} \quad i = 1, 2, \ldots N_{ions} \tag{12a}$$

$$\overline{\mu}_w = \mu_w \tag{12b}$$

where $\overline{\mu}_i^{ec} = \mu_i^0 + RT\ln(\overline{c}_i) + z_i F \overline{\varphi}$ = electro-chemical potential of the i-th ion, $\mu_i^{ec} = \mu_i^0 + RT\ln(C_i) + z_i F \Phi$ = electro-chemical potential of the i-th ion in the virtual bulk solution, $\overline{\mu}_w = \mu_w^0 + (\overline{p} - \overline{\pi})/C_w$ = chemical potential of water, $\mu_w = \mu_w^0 + (P - \Pi)/C_w$ =

chemical potential of water in the virtual bulk solution, μ_i^0, μ_w^0 = constants of integration, $\bar{c}_i, C_i$ = concentrations of the i-th ion in the real and in the virtual solution, respectively, $\bar{\varphi}_i, \Phi_i$ = electric potentials in the real and in the virtual solution, respectively, $\bar{p}, P$ = pressures of the fluid in the real and in the virtual solution, respectively, $\bar{\pi} = RT\sum_{i=1}^{Nions} \bar{c}_i$ = osmotic pressure in the pore solution, $\Pi = RT\sum_{i=1}^{Nions} C_i$ = osmotic pressure in the virtual solution, R = universal gas constant (8.314 J/(mol·K)) and T = absolute temperature (K).

Using the definitions of electro-chemical potential, Eqs. 12 may be written as follows:

$$\bar{c}_i = C_i \exp\left(-z_i \frac{F}{RT}\psi\right) \quad (13a); \qquad \bar{p} = P - \Pi + \bar{\pi} \quad (13b)$$

where $\psi = \bar{\varphi} - \Phi$. The ion concentration distribution given by Eq. 13a is known as the Boltzmann distribution. Coupling Eq. 9 with Eq. 13a, the Poisson-Boltzmann equation is obtained.

At the boundaries of the porous medium (x = 0 and x = L), the virtual solution coincides with the external bulk solutions. Therefore, knowing the external ion concentrations, the electric potential distribution may be evaluated from the Poisson-Boltzmann equation.

All the microscopic variables, $\bar{c}_i, \bar{p}, \bar{\varphi}$ are constituted by an average or macroscopic part and a microscopic fluctuation. The Donnan's relations are obtained neglecting the microscopic fluctuations, i.e. assuming as a first approximation $\bar{c}_i \cong \langle \bar{c}_i \rangle = \overline{C}_i$, $\bar{p} \cong \langle \bar{p} \rangle = \overline{P}$ and $\bar{\varphi} \cong \langle \bar{\varphi} \rangle = \overline{\Phi}$. Based on such an assumption, Eqs. 13 give the following relations:

$$\overline{C}_i = C_i \exp\left(-z_i \frac{F}{RT}\Psi\right) \quad (14a); \qquad \overline{P} = P - \Pi + \overline{\Pi} \quad (14b)$$

where $\Psi = \overline{\Phi} - \Phi$ and $\overline{\Pi} = RT\sum_{i=1}^{Nions} \overline{C}_i$. The macroscopic electric potential Ψ is known as the *Donnan's potential* or *membrane potential.*

Eqs. 14 coupled with the electroneutrality condition (Eq. 11) constitute a set of equations that allows the macroscopic variables within the porous medium to be determined when the values of the virtual variables are known.

The presented analysis clarifies the relation existing between the DDL theory and the Donnan's equilibrium theory. The Donnan's equations may be considered to be a sort of first-order approximation of the exact solution of the DDL theory. If the local fluctuation of the variables is small the error due to the use of the Donnan's equilibrium equation instead of the Poisson-Boltzmann equation is also small. Considering the boundary condition Eq. 10b, it is possible to verify that the microscopic fluctuation are small if $Q = \frac{F}{RT}\frac{\sigma_d \cdot b}{\varepsilon_r \varepsilon_0} < 1$. However, an important difference between the two approaches is that the Donnan's equilibrium doesn't allow the influence of the solvent dielectric constant on the electric potential distribution to be evaluated. On the other hand, the use of the Donnan's equilibrium equations allows the integration of the highly non-linear Poisson-Boltzmann equation to be avoided.

Transport Equations.

Following Gross and Osterle (1968), Cwirko and Carbonell (1988), and Moyne and Murad (2002), the transport equations at the microscopic scale are the Generalized Nernst-Planck Equations for the ion transport and the Generalized Navier-Stokes Equation for the liquid transport. In the axial direction, these equations may be written as follows:

$$j_i = v_x \bar{c}_i - D_{0,i}\frac{\partial \bar{c}_i}{\partial x} - z_i \frac{F}{RT} D_{i,0} \bar{c}_i \frac{\partial \bar{\varphi}}{\partial x} \tag{15a}$$

$$v_x = -\frac{1}{\mu}\int_y^{b'} dy' \int_0^{y'} \left(\frac{\partial \bar{p}}{\partial x} + F \sum_{i=1}^{Nions} z_i \bar{c}_i \frac{\partial \bar{\varphi}}{\partial x} \right) dy'' \tag{15b}$$

where j_i = pore flux of the i-th ion, v_x = liquid velocity within the pore, $D_{0,i}$ = free-solution diffusion coefficient of the i-th ion. Assuming as a first approximation $\bar{c}_i \cong \langle \bar{c}_i \rangle = \bar{C}_i$, $\bar{p} \cong \langle \bar{p} \rangle = \bar{P}$ and $\bar{\varphi} \cong \langle \bar{\varphi} \rangle = \bar{\Phi}$, the following macroscopic equations may be obtained from Eqs. 15:

$$J_i = n\tau_m \langle j_i \rangle = q\bar{C}_i - nD_i^* \frac{\partial \bar{C}_i}{\partial x} - nz_i \frac{F}{RT} D_i^* \bar{C}_i \frac{\partial \bar{\Phi}}{\partial x} \tag{16a}$$

$$q = n\tau_m \langle v_x \rangle = -\frac{k_\phi}{\gamma_w}\left(\frac{\partial \bar{P}}{\partial x} + F \sum_{i=1}^{N} z_i \bar{C}_i \frac{\partial \bar{\Phi}}{\partial x} \right) \tag{16b}$$

where $D_i^* = \tau_m D_{0,i}$ = effective diffusion coefficient of the i-th ion. When a contaminant transport analysis is carried out, Eqs. 16 should be coupled with the condition of the absence of electric current:

$$I_e = F\left(\sum_{i=1}^{Nions} z_i J_i \right) = 0 \tag{17}$$

where I_e is the electric current density.

In the case of a single binary electrolyte in solution, Dominijanni and Manassero (2005a,b) derived the following equations:

$$q = -\frac{k_\lambda}{\gamma_w}\left[\frac{dP}{dx} - \omega_\lambda RT(\nu_1 + \nu_2)\frac{dC_s}{dx} \right] \tag{18a}$$

$$J_s = (1-\omega_\lambda) q C_s - nD^*_{\omega,\lambda}\frac{dC_s}{dx} \tag{18b}$$

where $J_s = J_1/\nu_1 = J_2/\nu_2$ = salt molar flux, J_1 = cation molar flux, J_2 = anion molar flux, ν_1 = cation stoichiometric coefficient; ν_2 = anion stoichiometric coefficient, $C_s = C_1/\nu_1 = C_2/\nu_2$ = virtual salt concentration, C_1 = cation concentration, C_2 = anion concentration, k_λ = local hydraulic conductivity, ω_λ = local chemico-osmotic efficiency coefficient and $D^*_{\omega,\lambda}$ = local osmotic effective diffusion coefficient. The local phenomenological parameters are given by the following expressions:

$$k_\lambda = \frac{k_\phi}{1 + \frac{k_\phi RT(\Gamma_1 - \Gamma_2)^2 \nu_1 \nu_2 C_s}{n\gamma_w(\nu_1 \Gamma_2 D_2^* + \nu_2 \Gamma_1 D_1^*)}} \ \text{(19a)}; \quad \omega_\lambda = 1 - \frac{\nu_1 D_{0,2} + \nu_2 D_{0,1}}{\nu_1 \Gamma_2 D_{0,2} + \nu_2 \Gamma_1 D_{0,1}} \Gamma_1 \Gamma_2 \ \text{(19b)}; \quad D^*_{\omega,\lambda} = (1-\omega_\lambda) D_s^* \ \text{(19c)}$$

where $\Gamma_1 = \bar{C}_1 / C_1$ = cation partition coefficient, $\Gamma_2 = \bar{C}_2 / C_2$ = anion partition coefficient, $D_s^* = \tau_m (D_{0,1} D_{0,2})(\nu_1 + \nu_2)/(D_{0,1} + D_{0,2})$ = salt effective diffusion coefficient. The ion partition coefficients may be related to each other by the Donnan's equilibrium relations and the electroneutrality condition:

$$\Gamma_1 = \Gamma_2^{-\nu_2/\nu_1} \quad (20a) \qquad z_1\nu_1\Gamma_1 C_s + z_2\nu_2\Gamma_2 C_s - C_X = 0 \quad (20b)$$

The ion partition coefficients are functions of the ion valence and the ratio C_X/C_s. The fixed charge concentration, C_X, is a function of the clay structure and the surface charge of clay platelets.

The local hydraulic conductivity, evaluated at zero electric current, depends explicitly on the salt concentration and the ion valence. This phenomenon is known as the electroviscous effect and corresponds to a reduction of hydraulic conductivity at low salt concentration (see Fig.4). However, generally the electroviscous effect is small compared to changes of hydraulic conductivity due to structural modifications, as tactoid formation, and soil deformations (see Fig. 2a).

When the soil solution contains a 2:1 electrolyte, the restrictive capacity of the membrane, expressed by $(1 - \omega_\lambda)$ is reduced. This phenomenon is clearly illustrated in Fig. 5, where ω_λ is plotted as a function of the relative salt concentration, $\chi = C_s/C_X$, for the case of KCl and $CaCl_2$ solution. In correspondence of the same ratio C_s/C_X, the osmotic efficiency is considerably lower in the presence of divalent Ca^{2+}.

For the special case of low Peclet number and equal mobility of the ions, i.e. $D_{0,1} = D_{0,2}$, an analytical solution for the exit salt flux, $J_s(x = L, t)$, may be obtained for (1:1) electrolytes, assuming at the exit boundary (") a perfect flushing condition ($C_s'' = 0$):

$$J_s(x=L,t) = \frac{nD_s^*}{2L}\left(-C_X + \sqrt{C_X^2 + 4C_0^2}\right) + \frac{nD_s^*}{L}\left(-C_X + \sqrt{C_X^2 + 4C_0^2}\right)\cdot\sum_{m=1}^{\infty}\cos(m\pi)\exp\left(-\frac{D_s^* m^2\pi^2 t}{L^2}\right) \quad (21)$$

where $C_0 = C_s'$ = source concentration.

In Fig. 6, the steady state salt flux is plotted as a function of the fixed charge concentration and the source concentration $C_0 = C_s'$. When the source salt concentration tends to zero, the local osmotic efficiency coefficient, ω_λ, tends to 1 and, therefore, the bentonite layer acts as a "perfect" barrier able to completely hinder the passage of the salt (Manassero and Dominijanni, 2003). The magnitude of the salt restriction is controlled not only by the source concentration, but also by the fixed charge concentration, C_X. If $C_X = 0$ M, the bentonite layer does not have any semipermeable property. Increasing the value of C_X, the ability to hinder the passage of the salt of the layer improves for a given value of the source concentration.

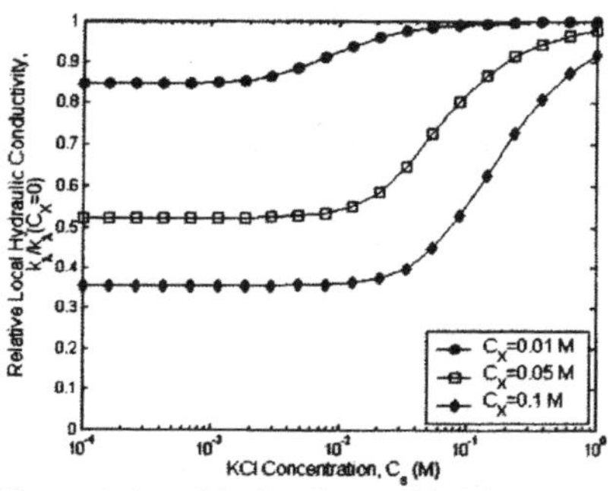

Figure 4. Local hydraulic conductivity, k_λ, as a function of the KCl concentration, C_s, and the fixed charge concentration, C_X.

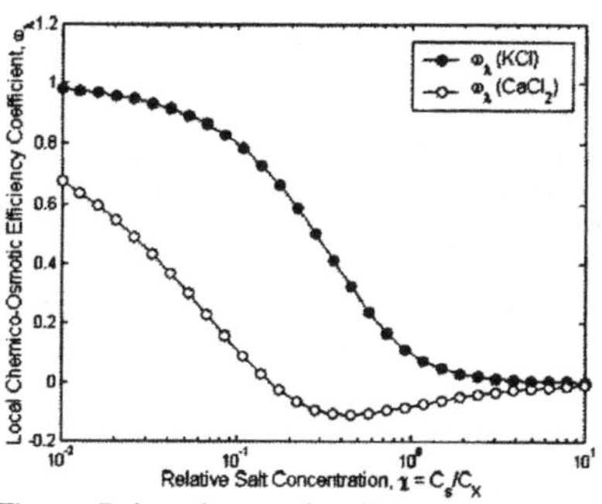

Figure 5. Local osmotic efficiency coefficient as a function of the ion concentration, C_s, the fixed charge concentration, C_X and the cation valence.

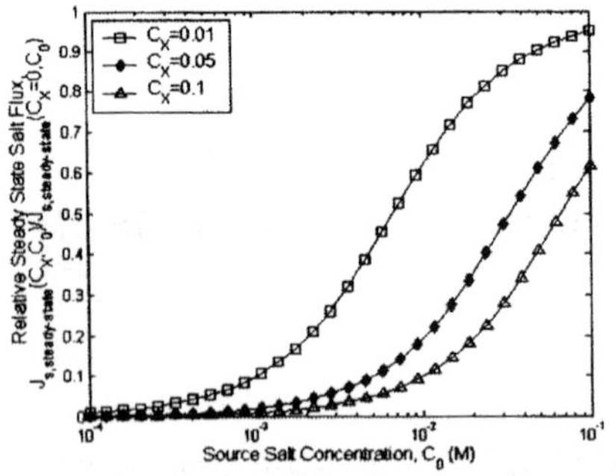

Figure 6. Relative steady state salt flux, $J_{s,steady\text{-}state}(C_X,C_0)/J_{s,steady\text{-}state}(C_X = 0,C_0)$, as a function of the source salt concentration, C_0, and the fixed charge concentration, C_X.

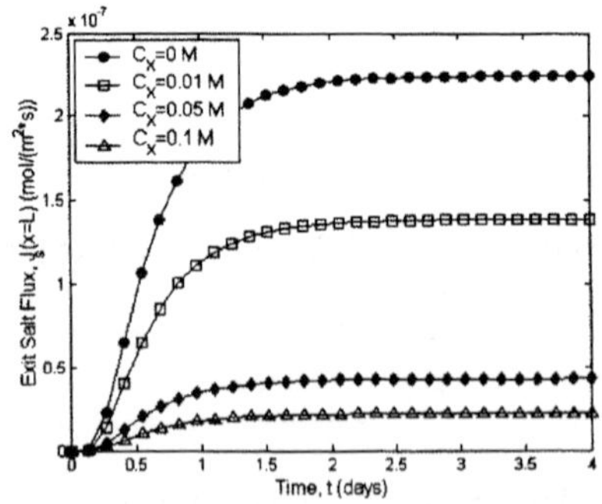

Figure 7. Exit salt flux, J_s (x=L,t) as a function of the time, t, and the fixed charge concentration, C_X. The source concentration, C_0, is equal to 0.01 M.

From the regression analysis of the experimental data of Malusis and Shackelford (2002a,b) on a GCL having porosity n = 0.79, the following parameters have been obtained: N = 4, τ_m = 0.1355, C_X = 0.0123 M. In the experiments of Malusis and Shackelford the soil solution was a KCl solution. In the analyses of Fig. 6 and 7, the following parameters have been adopted: n = 0.8, N = 4, τ_m = 0.14, $D_{0,K} = D_{0,Cl} = 2\cdot 10^{-9}$ m²/s.

Conclusions

Looking at Figs. 6 and 7, the reduction of the salt flux at steady state conditions due to semipermeable properties of bentonite may be evaluated. It should be stressed that also a small reduction in terms of steady state flux means a consistent reduction of the cumulative mass passing through the layer versus time. Therefore, the reduction of the salt flux at steady state has a significant role in the evaluation of the performance of a bentonite layer used as contaminant barrier.

References

Cwirko, E.H. & Carbonell, R.G. (1988). Transport of electrolytes in charged pores: analysis using the method of spatial averaging. Journal of Colloid and Interface Science, 129(2): 513-531.

Dominijanni, A. & Manassero, M. (2005a). Modelling Contaminant Transport Through Clay Membrane barriers. Proceedings of the 16th ICSMGE, Osaka, Japan, September 12-15, 2005.

Dominijanni, A. & Manassero,M. (2005b). Modelling osmosis and solute transport through clay membrane barriers. Waste Containment and Remediation (ASCE Geotechnical Special Publication No. 47), Alshawabkleh, A., et al. (Eds), ASCE, Reston, VA.

Gross, R. J. & Osterle, J. F. (1968). Membrane transport characteristics of ultrafine capillaries. The Journal of Chemical Physics, 49(1): 228-234.

Malusis, M.A. & Shackelford, C.D. (2002a). Chemico-osmotic efficiency of a geosynthetic clay liner. Journal of Geotechnical and Geoenvironmental Engineering 128(2): 97-106.

Malusis, M.A. & Shackelford, C.D. (2002b). Coupling effects during steady-state solute diffusion through a semi-permeable clay membrane. Environmental Science and Technology 36: 1312-1319.

Manassero, M. & Dominijanni, A. (2003). Modelling the osmosis effect on solute migration through porous media. Geotechnique, 53(5): 481-492.

Moyne, C. & Murad, M. A. (2002). Electro-chemo-mechanical couplings in swelling clays derived from a micro/macro-homogenization procedure. International Journal of Solids and Structures, 39: 6159-6190.

Norrish, K. (1954). The swelling of montmorillonite. Discussions of the Faraday Society, 18: 120-134.

Shainberg, I. Bresler, E. & Klausner, Y. (1971). Studies on Na/Ca montmorillonite systems. 1. The swelling pressure. Soil Science, 111(4): 214-219.

Acknowledgements

This project has been sponsored by the Italian Research Ministry (MIUR) and by the private companies Trevi SpA (Cesena) and Teseco SpA (Pisa). Their financial and technical support is gratefully acknowledged.

A mass-conservation boundary condition for the modeling of contaminant transport through leaking geo-membranes

Abbas El-Zein, Senior Lecturer, Centre for Geotechnical Research, Department of Civil Engineering, University of Sydney, NSW 2006, Australia, Email: aelzein@civil.usyd.edu.au

Abstract
Modelling contaminant migration through the intact or leaking geo-membrane of composite landfill liners is computationally expensive, owing to the contrast in dimensions and hydraulic conductivities of adjacent media. The paper proposes a new approach for modelling contaminant transport through leaking geo-membranes. The geo-membrane is treated as a boundary condition to the underlying clay layer, while conserving contaminant mass in the waste. The proposed approach works with any standard analytical or numerical solution method, such as the finite element, finite difference or boundary element methods, and has three advantages. First, it is computationally efficient because the geo-membrane is implicitly modelled as a boundary condition and the need to model two layers with very different characteristics is removed. Second, the method takes into account leakage through any number and locations of holes. Third, mass rather than concentration in conserved in the waste.

Introduction
Research over the last decade has shown that geo-membranes used in composite landfills undergo wear and tear, leading to leakage of leachate [1-3]. The formation of wrinkles increases the chances of dents and holes and widens the leakage area. Modelling contaminant migration through geo-membranes and the underlying clay layer is inherently expensive, given the contrast in dimensions and hydraulic conductivities of the two adjacent media. This difficulty is compounded, in the case of damaged geo-membranes, by fast transport through the leaks. Models of leaking geo-membranes have circumvent it this difficulty by making simplifying assumptions about the problem's dimensionality, combined effect of holes and boundary conditions in the landfill [4-8]. This paper proposes a new method which treats the geo-membrane as a boundary condition to the underlying clay layer, while conserving contaminant mass in the waste. The method has three advantages. First, it is computationally efficient because the geo-membrane is implicitly modelled as a boundary condition and the need to model two layers with very different characteristics is removed. Second, the method takes into account leakage through any number and locations of holes. Third, mass rather than concentration in conserved in the waste.

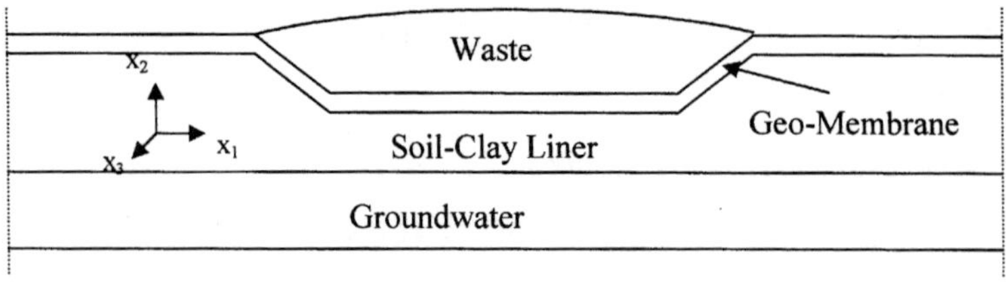

a. Cross-section of landfill

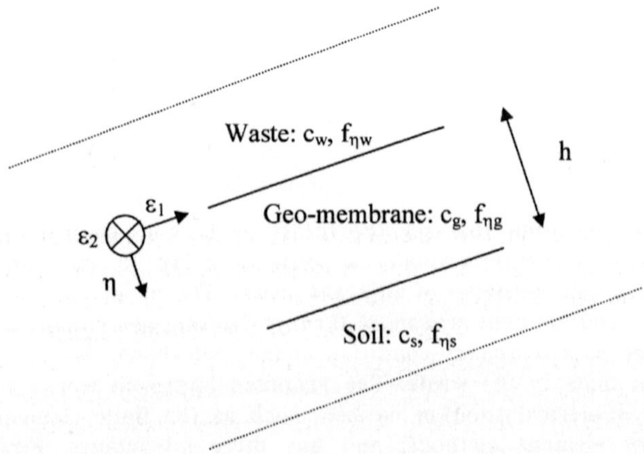

b. Local co-ordinate system (axis ε_2 is outward normal to the surface of the page)

Figure 1. Landfill with geo-membrane of thickness h

Contaminant Transport through Leaking Composite Liners

Figure 1a shows the layout of a landfill composite liner, with reference to a Cartesian co-ordinate system x_i (i=1,3). Figure 1b presents a local coordinate system (ε_1, ε_2, η), with $\eta=0$ at the top of the geo-membrane and $\eta=h$ at its base. Γ indicates the bottom surface of landfill. The geo-membrane, separating the waste from the mineral liner, has a total surface area of A_w. Three assumptions are made concerning mass transfer in the geo-membrane. First, given its small thickness compared to its other dimensions, transport in the geo-membrane is taken to be one-dimensional and governed by a Fickian diffusion process. Second, a jump in concentration occurs at the interface between the geo-membrane and any liquid contact medium, reflecting the molecular intrusion into the plastic material of contaminant dissolved in the water. Both of the above assumptions are supported by empirical evidence and are widely used in the literature on transport in geo-membranes [4-8]. Finally, there is instant mixing of contaminants in the waste and concentration in the waste is described by a single variable c_w, which may change in time but not in space. This is warranted by the fast hydraulic regime governing landfill leachate, in comparison with slow transport in the underlying liner.

The governing equations of contaminant migration in the geo-membrane and the soil, as well as continuity conditions of concentration and flux at the two interfaces (waste-geo-membrane and geo-membrane-soil) are written next. Mass-transfer and compatibility equations are shown successively as one moves down the system from the waste, through the geo-membrane to the clay layer.

$$c_w = c_{w0} - \frac{1}{R_w A_w H_f} \int_0^t \left(\int_\Gamma f_{\eta w} d\Gamma \right) d\tau - \frac{1}{R_w} \int_0^t \lambda_w c_w d\tau \qquad \text{(at } \eta=0) \qquad (1)$$

$$\frac{\partial}{\partial \eta}\left(D_g \frac{\partial c_g}{\partial \eta} \right) - \lambda_g c_g = \frac{\partial c_g}{\partial t} \qquad \text{(at } 0 \le \eta \le h) \qquad (2)$$

$$f_{\eta g} = -D_g \frac{\partial c_g}{\partial \eta} \qquad \text{(at } 0 \le \eta \le h) \qquad (3)$$

$$f_{\eta w} = f_{\eta g} \qquad \text{(at } \eta=0) \qquad (4)$$

$$c_g = k_{gw} c_w \qquad \text{(at } \eta=0) \qquad (5)$$

$$f_{\eta g} = f_{\eta s} \qquad \text{(at } \eta=h) \qquad (6)$$

$$c_g = k_{gs} c_s \qquad \text{(at } \eta=h) \qquad (7)$$

$$\nabla(\mathbf{D_s} \nabla c_s) - \mathbf{v_s} \nabla c_s - \lambda_s c_s = R_s \frac{\partial c_s}{\partial t} \qquad \text{(at } \eta>h) \qquad (8)$$

where t is time; w, g and s are indices referring to the waste, geo-membrane and soil, respectively; c is concentration; f_η is the flux in direction η; λ is a first-order decay coefficient; are the coefficients of diffusion and first-order decay in the geo-membrane, respectively; along direction η; k_{gw} and k_{gs} are constants that are determined experimentally, for various combinations of contaminant and contact media [9-11]; D_g is the diffusion coefficient in the geo-membrane; $\mathbf{D_s}$ (D_{11}, D_{12}, D_{22}) and $\mathbf{v_s}$ (v_1, v_2) are the hydrodynamic dispersion and advection tensors, respectively; R is a sorption retardation factor. Note that equation (1) conserves mass in the waste.

Equation (1) still applies in the case of a leaking geo-membrane. However, the integral of the flux over the surface area of the geo-membrane must now take into account leakage, as well as transport through the intact parts of the geo-membrane:

$$\int_\Gamma f_{\eta w} d\Gamma = \int_{\Gamma_i} f_{\eta wi} d\Gamma + \int_{\Gamma_l} f_{\eta wl} d\Gamma \qquad (9)$$

Where indices i and l indicate intact and leaking parts of the geo-membrane, respectively. Note that $f_{\eta wi}$ must be evaluated at η=0, while $f_{\eta wl}$ applies at the top of the soil. Hence, for leaking geo-membranes, equation (8) becomes:

$$c_w = c_{w0} - \frac{1}{R_w A_w H_f} \int_0^t \left(\int_{\Gamma_i} f_{\eta wi} d\Gamma + \int_{\Gamma_l} f_{\eta wl} d\Gamma \right) d\tau - \frac{1}{R_w} \int_0^t \lambda_w c_w d\tau \quad \text{(at } \eta=0) \quad (10)$$

The same equation applies over Γ_i and Γ_l. However, over Γ_l, c_w can be replaced with c_{sl}:

$$c_{sl} = c_{w0} - \frac{1}{R_w A_w H_f} \int_0^t \left(\int_{\Gamma_i} f_{\eta wi} d\Gamma + \int_{\Gamma_l} f_{\eta wl} d\Gamma \right) d\tau - \frac{1}{R_w} \int_0^t \lambda_w c_w d\tau \quad \text{(at } \eta=0) \quad (11)$$

For the case of steady-state fluid flow, the equations can be linearized with respect to time by applying a Laplace transform. An equivalent boundary condition, in terms of the concentration and flux at the top of the soil, can be then be derived. Hence, under the intact parts of geo-membrane:

$$\frac{E_3 E_0 k_{gs}}{E_1 e^{(\alpha_1+\alpha_2)h} k_{gw}} \bar{c}_{si} + \frac{E_2 E_0}{E_1 e^{(\alpha_1+\alpha_2)h} D_g k_{gw}} \bar{f}_{\eta si} + \frac{E_1}{E_3 R_w A_w H_f} \int_{\Gamma_i} \bar{f}_{\eta si} d\Gamma + \frac{1}{R_w A_w H_f} \int_{\Gamma_l} \bar{f}_{\eta sl} d\Gamma = c_{w0}$$

(at η=h) (12)

Under the leaking part:

$$E_0 \bar{c}_{sl} + \frac{E_1}{E_3 R_w A_w H_f} \int_{\Gamma_i} \bar{f}_{\eta si} d\Gamma + \frac{1}{R_w A_w H_f} \int_{\Gamma_l} \bar{f}_{\eta sl} d\Gamma = c_{w0} \quad \text{(at } \eta = h) \qquad (13)$$

This boundary condition can be used as a proxy for the geo-membrane, which conserves mass of contaminants in the waste.

Results

The method was validated by comparing its predictions to those of conventional approaches which model the geo-membrane explicitly. To illustrate the scope of the method, a landfill shown in figure 2a is simulated using the implicit approach in conjunction with a boundary element program CONTAN. Symmetry with respect to the x_2x_3 plane is used to reduce the size of the numerical model. The landfill incorporates a composite liner with a geo-membrane on top of a 1m-deep clay layer. The geo-membrane is leaking in five locations, as shown in the figure. Values of k_{gw} and k_{gs} for dichloromethane in HDPE geo-membranes, found in the literature, are used in the analysis [1]. Figure 2b shows a typical boundary element mesh generated for the simulations. landfill-related parameters: c_{w0}=100 mg.l^{-1}; H_f=0.01m; R_w=5.67; λ_w=0.01 year^{-1}; geo-membrane-related parameters: h=1.5mm; D_g=1e-7 m^2.year^{-1} ; k_{gw}=3.; k_{gs}=3.; λ_g=0; clay-related parameters: D_{x1}= 3.15e-4 m^2.year^{-1}; D_{x2}= 4.25e-4 m^2.year^{-1}; v_{x2}=-3.15e-4 m.year^{-1}; R_s= 9.01; λ_s=1e-4 year^{-1}. The base of the clay layer is taken as a zero-concentration surface, simulating fast groundwater transport; all other surfaces, apart from geo-membrane and leaks, are zero-flux surfaces

Figure 3a shows the cumulative flux through the leaks ($\int_0^t \left(\int_{\Gamma_l} f_{\eta wl} d\Gamma \right) d\tau$) and the intact parts of the geo-membrane ($\int_0^t \left(\int_{\Gamma_i} f_{\eta wi} d\Gamma \right) d\tau$) in the model, versus time. (Given the use of symmetry, the total mass transfer over the whole landfill is twice that shown in figures 3). Finally, figure 3b shows the cumulative flux through the leaks relative to the cumulative flux through the intact parts. Results are shown up to 500 and 1000 years in order to develop an understanding of the relationship between key parameters in the system even if, in reality, the geo-membrane's performance is expected to deteriorate over the long term. Figure 3 shows that, at a relatively early stage of transport (t=35 years), in the geo-membrane with 2m-wide leaks, the mass of contaminants leaking into the clay is similar in magnitude to the amount transferred through the intact parts. In geo-membranes with 10-cm leaks, on the other hand, the amount of leakage is one order of magnitude smaller than the mass migrating through the intact parts. In addition, as the contaminants diffuse through the geo-membrane and concentration builds up underneath it, and as c_w decreases in the waste, the direction of the flux through the leaks is reversed. This is clearly shown in figure 3a where the cumulative flux versus time reaches a maximum at the point where the flux reverses. Figure 3b shows

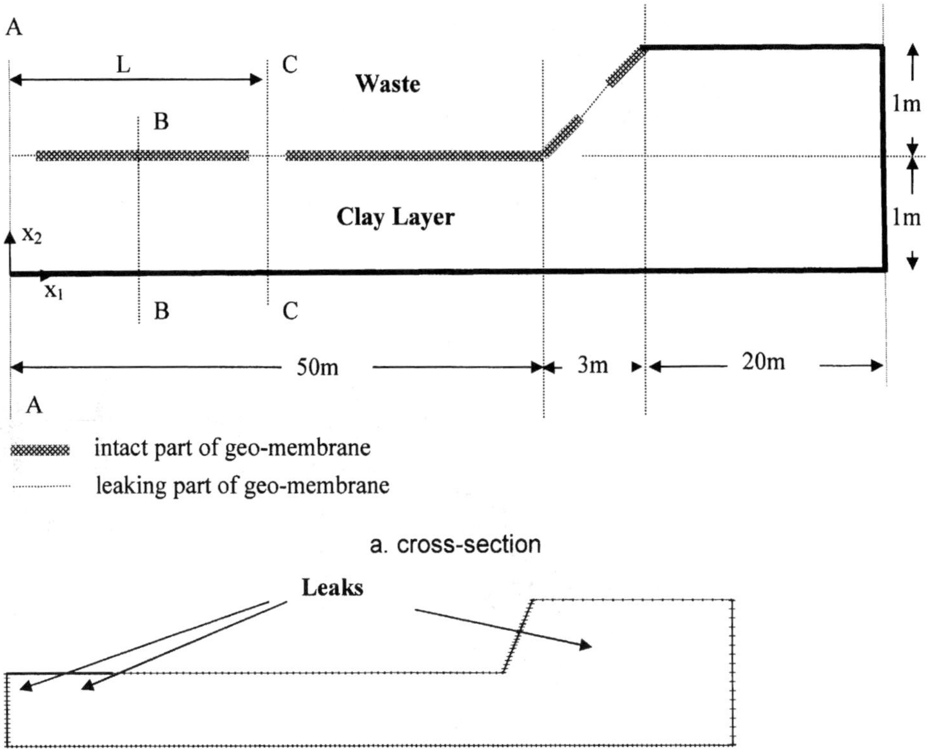

a. cross-section

b. typical boundary element mesh used in the analyses

Figure 2. Landfill with leaking geo-membrane; symmetry with respect to the x_2x_3 plane is used; five leakage areas of equal width; AA and CC run across the centres of the two horizontal leaks; BB runs across the centre of the intact area between the two leaks.

that the contribution of the leaks to contamination declines exponentially with time and that most of the damage caused by the leaks occurs at a relatively early stage of contamination. This is due to the retardation provided by the geo-membrane which allows contaminants to migrate over longer periods of time. Hence, early leakage rates may overestimate the total, long-term contamination caused by the defects in the geo-membrane.

Conclusion

The proposed method is an economical technique for modeling contaminant migration through intact or leaking geo-membranes. It introduces no additional computational cost associated with simulating the effect of the geo-membrane and, as shown in the paper, can give valuable insight into the short- and long-term effects of defects in landfill composite liners.

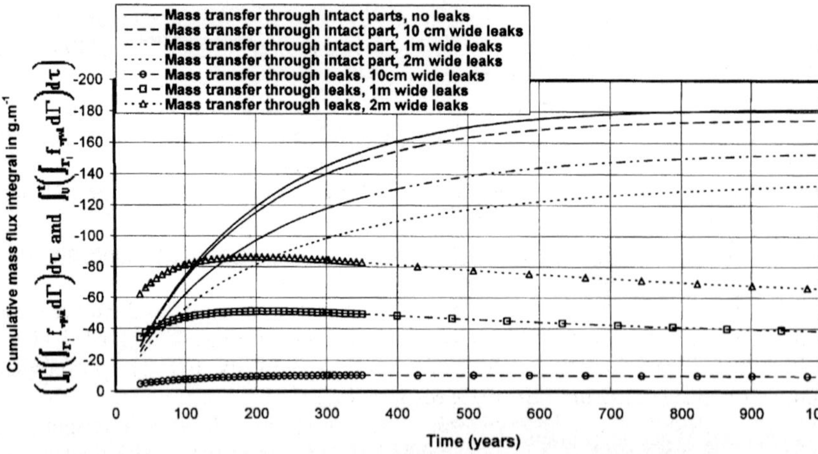

a. cumulative total mass transfer (positive flux indicates contaminant leaving the clay layer)

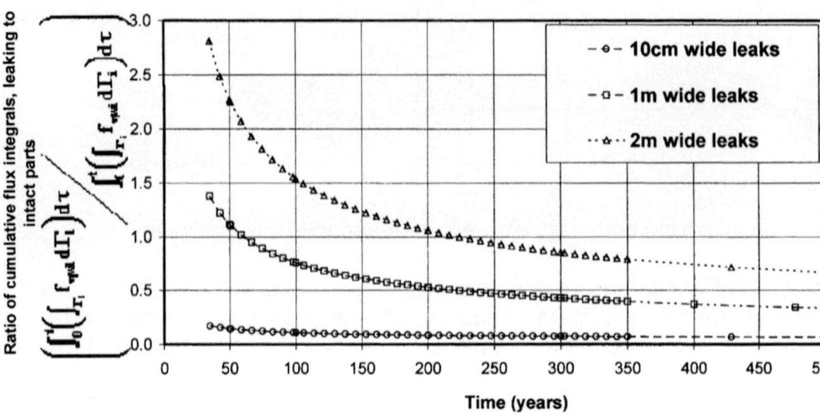

b. cumulative total mass transfer from leaking parts relative to intact parts

Figure 3. Cumulative total mass transfer versus time, across half the interface between the clay layer and the geo-membrane (L=10m)

References

1. Rowe RK, Sangam HP, Lake C. 2003. Evaluation of an HPDE geomembrane after 14 years as a leachate lagoon liner. *Canadian Goetechnical Journal* **40**:536-550.
2. Buss SE, Butler AP, Johnston PM, Sollars CJ and Perry R. Mechanisms of leakage through synthetic landfill liner materials; *Journal of the Chartered Institution of Water and Environment Management* 1995, **9**(4):353-359.
3. Bonaparte R. and Gross B. Field behaviour of double-liner systems. *Proc Symp Waste Contain Sys: Construc, Regulat Perform*, Nov 1990, San

4. Mueller W, Jakob I, Tatzky-Gerth R and August H. Solubilities, Diffusion and Partition Coefficients of Organic Pollutants in HDPE Geomembranes: Experimental Results ad Calculations. Sixth International Conference on Geosynthetics 1998, 239-248.
5. Brown KW and Thomas JC. A comparison of the convective and diffusive flux of contaminants through landfill liner systems. *Waste Management and Research* 1998; **16**(3):296-301.
6. Katsumi T, Benson CH, Foose GJ, Kamon M. 2001. Performance-based design of landfill liners. *Eng Geo* **60**:139-148.
7. Kalbe U, Muller WW, Berger W and Eckardt J. Transport of organic contaminants within composite liner systems. *Applied Clay Science* 2002; **21**:67-76
8. Foose GJ, Benson CH and Edil TB. Comparison of solute transport in three composite liners. *Journal of Geotechnical and Geoenvironmental Engineering* 2002; **128**:391-403.
9. Park JK and Nibras M. Mass flux of organic chemicals through polyethylene geomembranes. *Water Environ. Res.* 1993; **65**(3):227-237.
10. Muller W, Jakob I, Tatzky-Gerth R, August H. Stofftransport in Deponieabdichtungssystemen, Teil 1: Diffusionsund Verteilungskoeffizienten von Schadstoffen bei der Permeation in PEHD-Dichtungsbahnen. *Bautechnik* 1997; **74** (3):176–190.
11. Muller W, Buttgenbach B, Tatzky-Gerth R, August H. 1997. Stofftransport in Deponieabdichtungssystemen, Teil 2: Permeation in der Kombinationsdichtung. *Bautechnik* 1997; **74**(5):331–344.

A procedure for modeling coupled soil consolidation and solute migration processes in compressible soils

Roberto Lopes Ferraz, State University of Maringá, Paraná, Brazil
Maurício Ehrlich, Federal University of Rio de Janeiro, Rio de Janeiro, Brazil
Maria Claudia Barbosa, Federal University of Rio de Janeiro, Rio de Janeiro, Brazil

Abstract

In this paper is presented a procedure for modeling coupled soil consolidation and solute transport in compressible soils. The mathematical models used consider finite strains in soil consolidation, non-linear soil properties and variation in hydraulic conductivity and hydrodynamic dispersion coefficients in space and in time. This approach presents many advantages relative to current conventional ones, which generally consider a scenario of infinitesimal strains and constant parameters of soil consolidation and solute transport in the coupled process. The Finite Volume Method (FVM) is used to discretize the governing equations. Validation exercises have shown that the model correctly predicts the expected behavior. Parametric studies have shown that soil deformation may have a great influence on solute transport.

Introduction

Studies have shown that the deformation process of a porous medium through which the solutes are moving can reduce or increase the breakthrough times, depending on the problem boundary conditions. Some examples include waste disposal sites and their clay liners built on compressible substrates, where the consolidation process of the foundation soil can result in breakthrough curves very distinct from those predicted by models for a rigid porous medium.

Due to the growing need to deal with these situations, in recent years an increase in the number of research projects involved with the development of mathematical and numerical models addressing solute migration through deformable porous medium has been observed, i.e., Loroy et al., 1996, Smith, 2000, Smiles, 2000, Ferreira, 2001, Peters & Smith, 2002 and Alshawabkeh et al., 2005.

In terms of mathematical and numerical modeling, the common practice has been to solve the coupling of solute transport and the consolidation of the porous medium in two steps. In the first step the consolidation model is solved to generate the advective component of the solute transport process, which is solved in the second step.

Generally the used mathematical models consider a scenario of infinitesimal strains and constant parameters of soil consolidation and solute transport in the coupled process. However, Peters and Smith, 2002, showed that a non-linear large deformation model can greatly reduce solute breakthrough time compared to a standard transport analysis typically used by environmental engineers. Besides, they showed that it is important to include spatial and temporal void ratio variations in the large deformations analysis.

An example of field condition involving solute transport through a deformable porous medium is the foundation of Gramacho Municipal Solid Waste Landfill, Rio de Janeiro, Brazil, which is described in Ehrlich et al., 1994, Castro et al., 1994, Barbosa, 1994, Barbosa et al., 1996. This landfill has been in operation since 1978, and was built on a layer of organic-saline soft clay with thickness varying between 5 and 8 meters (Barbosa, 1994). The high compressibility and salinity of the landfill foundation soil produce a complex process of ion migration. Due to the concentration differences the ions migrate in both ways, through the landfill into the foundation and through the foundation into the landfill, as showed by Barbosa, 1994. In addition, the consolidation process of the soft clay induces an advective flux towards the landfill. This results in the scenario is sketched in Figure 1.

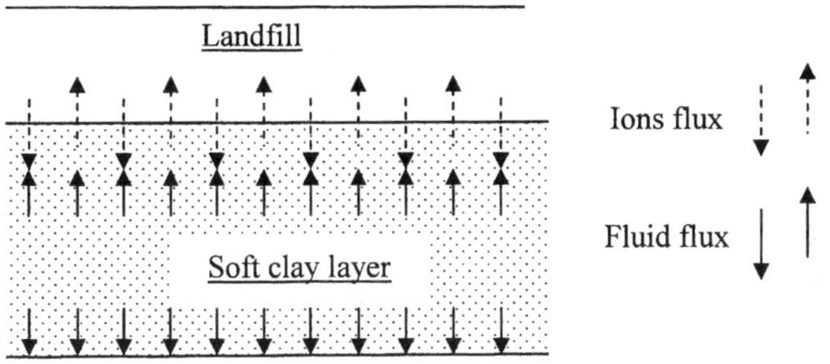

Figure 1: The existing flux in the foundation soil of Gramacho Municipal Landfill

The mathematical models used

The basic hypotheses for the formulation of the mathematical models used can be summarized as follows: (a) the soil is homogeneous and fully saturated, (b) the effective stress principle is valid, $\sigma' = \sigma - u_f$, where σ' and σ denotes effective and total stress, respectively, and u_f is the pore-water pressure, (c) solid particles and pore fluid are incompressible, (d) deformation and flows are one-dimensional (vertical) under isothermal conditions, (e) Darcy's and Fick's laws are valid for fluid and chemical species mass transport, respectively, (f) there is a unique relationship, independent of time, between void ratio, e, and effective stress, (g) the mass transfer between fluid and solid phases do not change the density of each phase, (h) there is no decay of solute mass, (i) osmotic effects due to the exposure of the clay to a saline solution are not considered. This will be considered in further developments.

Under the simplified hypotheses of the previous paragraph, the governing equation for solute migration processes in compressible soils can be derived by considering solute mass balance in the porous medium. This results in the following equation:

$$\frac{\partial}{\partial t}(e.C+\rho_d.S)=\frac{\partial}{\partial z}\left[\frac{e.D(e)}{(1+e)^2}\cdot\frac{\partial C}{\partial z}\right]-\frac{\partial}{\partial z}\left[\frac{e}{1+e}.(v_f-v_s).C\right] \qquad (1)$$

in which e = void ratio, C = solute concentration in the fluid phase, ρ_d = dry bulk density of the soil, S = mass of solute sorbed per unit mass of dry soil, D = coefficient of hydrodynamic dispersion, v_f and v_s = velocities of the fluid and solid phases, respectively, z = reduced coordinate (oriented towards gravity). The co-ordinate z is defined as the volume of soil particles (per unit area) lying between the datum plane and the Lagrangian (initial) co-ordinate point (McNabb, 1960).

The use of the co-ordinate z has an advantage over the Eulerian co-ordinate because it allows to treat a problem with moving boundaries as one with fixed boundaries. This eliminates the need of updating the space mesh at each time step. However, from a physical point of view, it is more convenient to express the dependent variables in terms of Eulerian co-ordinates x and time t. Therefore, once the solution has been obtained in reduced z co-ordinate, it is converted to the Eulerian co-ordinates by means of the co-ordinate transformation, expressed by following equation:

$$x(t) = \int_0^z [1 + e(z', t)] dz' \qquad (2)$$

The equation which describes the deformation of porous medium can be derived from the continuity equations for the solid and fluid phases, Darcy's law, and the equilibrium equation in terms of effective stress. For a homogeneous soil the resulting equation takes the form given by Gibson et al., 1967, (3), where ρ_s and ρ_f = density of solid and fluid phases, respectively, k = hydraulic conductivity in the vertical direction and other symbols are as previously defined.

$$\frac{\partial e}{\partial t} = \left(\frac{\rho_s}{\rho_f} - 1\right) \cdot \frac{d}{de}\left[\frac{k(e)}{1+e}\right] \cdot \frac{\partial e}{\partial z} - \frac{\partial}{\partial z}\left[\frac{k(e)}{\rho_f \cdot (1+e)} \cdot \frac{d\sigma'}{de} \cdot \frac{\partial e}{\partial z}\right] \qquad (3)$$

This equation governs non-linear finite strain consolidation of saturated clays and is based on the constitutive assumptions $k = k(e)$ and $\sigma' = \sigma'(e)$, where the latter function being restricted to a monotonic loading path.

The numerical solution of the mathematical models used

The governing differential equations of the mathematical models were adapted and discretized using the Finite Volume Method (FVM), whose fundamental aspects are described in Patankar, 1980, and Versteeg and Malalasekera, 1995.

The FVM is a class of discretization schemes that have proved highly successful in approximating the solution of a wide range of systems of conservation laws. In FVM the calculation domain is divided into a number of non-overlapping control volumes so that there is one control volume surrounding each grid point. The differential equation is integrated over each control volume. The most attractive feature of FVM is that the resulting solution will imply that the integral conservation of mass is satisfied over any group of control volumes and, of course, over the whole calculation domain. This characteristic exists for any number of grid points and not just in a limiting sense, when the number of grid points becomes large. Thus, even the coarse-grid solution exhibits exact integral balances (Patankar, 1980).

In the numerical model development the following possibilities were considered: non-linear soil properties, variation in hydraulic conductivity and in coefficient of hydrodynamic dispersion in space and time.

The resulting equations of the discretization process by FVM were implemented in a computer program. In this computer program a series of fast and robust routines were implemented for the automatic processing of the calculations, as well as an user friendly environment for data entry.

Model validation

In this section the uncoupled parts of the model are validated with data existing in literature and an available commercial software.

The finite strain consolidation process

Since analytical solutions do not exist for the general non-linear equation of finite strain consolidation, the validation of the consolidation part of the model was carried out referencing some results of the numerical simulations presented by Townsend and McVay, 1990.

The scenarios sketched in Figure 2, as described in Townsend and McVay, 1990, were used for validation.

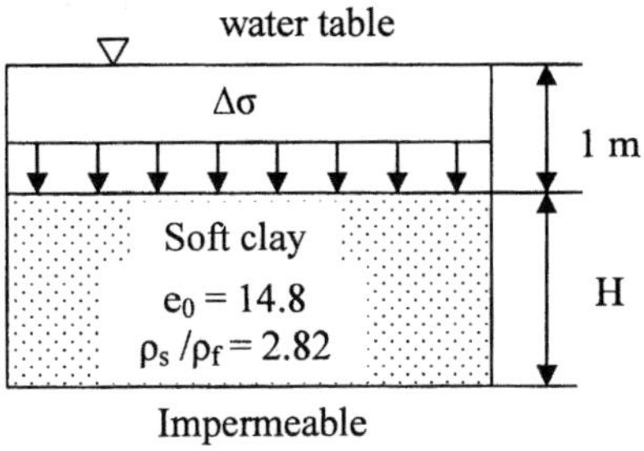

Scenario (a): Quiescent consolidation, uniform initial void ratio ($\Delta\sigma = 0$, H = 9.6 m)

Scenario (b): Quiescent consolidation, surcharge loading, uniform initial void ratio ($\Delta\sigma = 9.576$ kPa, H = 7.2 m)

Figure 2: Scenarios for validation of the consolidation part of the model

The effective stress-void ratio-permeability relationships that were used for the predictions are presented by the following equations:

$$e = 7.72.\sigma'^{\,(-0.22)} \quad (\sigma' \rightarrow \text{kPa}) \qquad (4)$$

$$k = (0.2532\text{x}10^{-6}).e^{4.65} \quad (k \rightarrow \text{m/day}) \qquad (5)$$

The results of the calculations for the two scenarios described and the summary of the results (average values) predicted by several numerical models for the proposed scenarios (Townsend and McVay, 1990) are presented in Table 1. The comparison of these values shows that the numerical model developed displays predictions within the average of other models for the analysis of finite strain consolidation.

Table 1: Comparison of results of the proposed procedure with the average values presented by Townsend and McVay, 1990

	Numerical model developed				Townsend and McVay, 1990			
Scenario	Final	After 1 year			Final	After 1 year		
	Height	Height	e(bottom)	u(bottom)	Height	Height	e(bottom)	u(bottom)
(a)	4.11 m	6.82 m	6.43	8.55 kPa	4.16 m	7.19 m	6.58	8.71 kPa
(b)	2.42 m	4.47 m	6.43	15.41 kPa	2.50 m	4.75 m	6.44	15.50 kPa

The advection-dispersion process

The solute transport part of the model was tested by comparing the numerical results obtained by the model and by the software POLLUTE© V6. For this, several hypothetical cases were used. One of the cases analyzed referred to a saturated soil layer, with height H = 5 m, porosity n = 0.60, dry density 1.2 g/cm^3, hydrodynamic dispersion coefficient D = 0.04 m^2/year and S = 0 (no sorption). This layer, initially free of contamination, was supposed to be in contact in the top with an infinite source of pollutants (C0 = 1000 mg/L). For this case, the predictions for the contamination profile in the layer were done for a period of 20 years. In order to evaluate model sensitivity to different values and directions of the advective velocity, va, the following cases were adopted for comparison: in case 1 (va = 0), case 2 (va = - 0.1 m/year) and case 3 (va = + 0.1 m/year).

The results of the simulations are presented in Figure 3, showing excellent agreement between the models.

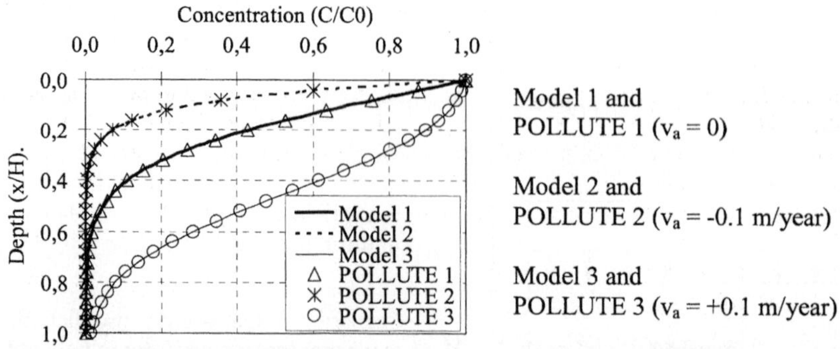

Figure 3: Concentration profiles for t = 20 years (for rigid porous medium).

Simulation of a coupled case

Since analytical solutions, or computer programs for analyses of the coupling of solute transport and consolidation were not available, a hypothetical example was used to evaluate the coherence of the results supplied by the model. In this case, the geometric configuration of scenario (b) (Figure 2) and the relationships (e x σ', k x e) expressed by equations 4 and 5 were used.

For the consolidation process the vertical stress increment Δσ = 9.576 kPa remained constant for a period of 5 years. After this time the vertical stress increment Δσ was suddenly increased to 50 kPa and maintained at this level thereafter.

The analysis of the evolution of the coupled process was done by obtaining the concentration profiles of the hypothetical pollutant for the times t = 2, 5, 5.2 and 10 years. The period of 5.2 years was used to verify if the changes in the advective flow, originated by the load increment in t = 5 years, were detected by the model.

For the modeling of the solute transport process, the soil is assumed initially to be free from contamination. The adsorption process was not considered (S = 0). In order to evaluate the interference of the hydrodynamic dispersion coefficient in the transport process, two cases were studied. In the first case (Case 1) a linear relationship was used between D and the void ratio given by equation (5). This relationship is based on results of diffusion test experiments in reconstituted soil samples performed by Ferreira, 2001.

$$D = 9.322*10^{-4}*e + 0.022 \quad (m^2/year) \qquad (5)$$

In the other case (Case 2) it was considered that the whole process occurs with an average value of D = 0.03 m^2/year. In both cases a contaminant source with infinite mass (constant concentration C0 = 5000 mg/L) was placed at the top of the layer.

The results of the simulations are presented in the Figure 4. As the thickness of the layers are different at the end of each analyzed period, the depth values were normalized to the final heights of the layer in each period (Ht).

As expected, for both studied cases, numerical results in Figure 4 reveals that the advective flux generated by the consolidation process delays the progress of the contamination front inside the soil layer. This happens because the advective and diffusive fluxes have opposite directions. In addition, the advection portion has a greater intensity at the beginning of the excess pore pressure dissipation process. That is perfectly described when the profiles corresponding to t = 5 and t = 5.2 years are compared. As observed in both graphics of Figure 4, the profile of t = 5.2 years is "delayed" (i. e., the contaminant flux goes in opposite direction) in relation to the observed flux before the vertical stress increment from 9.5 kPa to 50 kPa, that occurred at t = 5 years. Such behavior is due to the high value of the advective portion at the beginning of the new load increment (50 kPa).

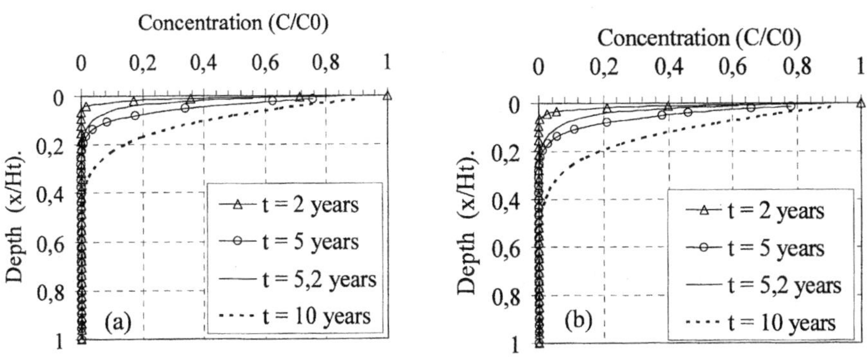

Figure 4: Concentration profiles for cases 1 and 2 (for coupled cases condition) (a) Case 1: D varies recording equation 7 and (b) Case 2: D = 0.03 m^2/year

In relation to the interference of the hydrodynamic dispersion coefficient, it was verified that assuming a constant average value for D during the process (case 2) causes a faster progress of the contamination front compared to the condition where it was considered a linear variation of D with the void ratio (case 1).

Conclusions

This paper presents a numerical study of coupled soil consolidation and solute migration in compressible soils. The numerical model described uses Finite Volume Method for discretization of the governing differential equations.

The resulting equations were implemented in a computer program, where it was possible to consider a finite strain consolidation process, non-linear properties and variation of the hydraulic conductivity and hydrodynamic dispersion coefficient in space and time. The process of validating each isolated part of the model showed that its predictions compare well with those already existing in literature, as well as with numerical results from available commercial software.

The simulation of a coupled case condition, where the soil consolidation and solute migration process occurs simultaneously, allowed the evaluation of this problem. The obtained results showed that the advective flux generated by the consolidation process may delay the progress of the contamination front inside the soil layer.

References

Alshwabkeh, A. N., Rahbar, N. and Sheahan, T., 2005. A model for contaminant mass flux in capped sediment under consolidation. Journal of Contaminant Hydrology. Vol. 17, Issue 3, pp. 147–165.

Barbosa, M. C., 1994. Geoenvironmental investigation of the organic soft clay deposit under Gramacho Municipal Landfill, DSc Thesis, COPPE/UFRJ, Rio de Janeiro (in Portuguese).

Barbosa, M. C., Almeira, M. S. S. and Ehrlich, M., 1996. Ion diffusion through an organic saline clay soil, Proc. 2nd International Congress on Environmental Geotechnics, Osaka, Japan, pp. 7-12.

Castro, F. C., Ehrlich, M., Barbosa, M. C. and Almeira, M. S. S., 1994. Site investigation for the rehabilitation of a municipal landfill, Proc. 1st International Congress on Environmental Geotechnics, Edmonton, Alberta, Canada, pp. 637-642.

Ehrlich, M., Almeira, M. S. S. and Barbosa, M. C., 1994. Pollution Control of Gramacho Municipal Landfill, Proc. 1st International Congress on Environmental Geotechnics, Edmonton, Alberta, Canada, pp. 657-664.

Ferreira, J. S., 2001. Coupled consolidation and contaminant transport soft clayed soils, MSc Thesis, COPPE/UFRJ, Rio de Janeiro (in Portuguese).

Gibson, R. E., England, G. L. and Hussey, M. J. L., 1967. The theory of one- dimensional consolidation of saturated clays. Géotechnique. Vol. 17, N° 3, pp. 261–273.

Loroy, J. J., Soga, K. and Brito, A. M., 1996. Finite element analysis of consolidation and contaminant transport in porous media. Second International Congress on Enviromental Geotechnics. Osaka, Japan, p. 263–268.

McNabb, A., 1960. A mathematical treatment of one-dimensional soil consolidation. Quarterly Journal of Applied Mathematics, 17, pp. 337-347.

Patankar, S. V., 1980. numerical heat transfer and fluid flow. New York. Hemisphere Publishing Corporation. 197p.

Peters, G. P. and Smith, D. W., 2002. Solute transport through a deforming porous medium. International Journal for numerical and Analytical Methods in Geomechanics. Vol. 26, p. 683 – 717.

Smiles, D. E., 2000. Material coordinates and solute movement in consolidating clay. Chemical Engineering Science. Vol. 55, N° 5, pp. 773–781.

Smith, D. W., 2000. One-dimensional contaminant transport through a deforming porous medium: theory and a solution for a quasi-steady-state problem. International Journal for numerical and Analytical Methods in Geomechanics, Vol. 24, pp. 693–722.

Townsend, F. C. and McVay, M. C., 1990. SOA: Large strain consolidation predictions. Journal of Geotechnical Engineering. Vol. 116, N° 2. pp. 222–243.

Versteeg, H. K. and Malalasekera, W., 1995. An introduction to Computational Fluid Dynamics: The Finite Volume Method. London. Longman Group Ltd. 257p.

Effect of Porosity Variability on Ganglia Entrapment during Immiscible Displacement

Franco M. Francisca, CONICET-Universidad Nacional de Córdoba, Córdoba, Argentina

Abstract

Ganglia are occluded blobs of non-aqueous phase liquids (NAPL) that arise during the displacement of two immiscible fluids in porous media. Their presence increases the residual saturation of organic fluids during the removal of contaminants. Immiscible flow has several uncertainties that govern the macroscopic behavior of fluids inside porous media. These uncertainties are essentially related to the porous media properties including anisotropy and spatial variability of pore size. The displacement of two immiscible fluids within soil pores is analyzed by means of a micromechanical pore network model. The model assumes that Young-Laplace's equation governs the interface instability (capillarity prevails over gravity and viscosity). The effect of porosity anisotropy, presence of preferential flow paths, and coefficient of variation of pore size is analyzed. Results show that pattern of ganglia and residual saturation of NAPL are directly related to porosity anisotropy and pore size spatial variability.

Introduction

Non-Aqueous Phase Liquids (NAPLs) are commonly discovered contaminating soil and groundwater (e.g. Fetter, 1993). Contamination is usually associated with the leakage of underground tanks and pipes, with industry activities and spills produced because of accidents during the transportation of hydrocarbons. Oil recovery and remediation of sites contaminated with NAPL are typical problems involving immiscible displacement of fluids.

When water and NAPL exist within soil pores an interface separate both immiscible phases. The displacement of immiscible fluids and interfaces inside porous media is known as immiscible fluid flow. The most relevant forces controlling fluid displacement are capillarity, gravity and viscosity. Capillary forces are responsible for the instability of interfaces between the displacing and displaced phase (Mercer and Cohen 1990, Dullien 1992, Adamson and Gast 1997, Hardisty et al. 1998), while gravity and viscosity have significant effect during steady state flow conditions (Corey 1986).

Pore network models are frequently used to simulate the displacement of immiscible fluids in porous media (Fatt, 1956, Lenormand et al. 1988, Dullien 1992, Vidales et al. 1998, Blunt

5th ICEG Environmental Geotechnics, Thomas Telford, London 2006.

2001). The pore space is characterized by a discrete network of capillaries which can capture properties of the porous media. The model simulates the displacement of interfaces by considering the pore-level displacement physics (Lenormand et al. 1988), wettability, spreading coefficient (Zhou and Blunt 1997, Mani and Mohanty 1997) and interconnectivity between pores (Dullien 1992).

In this work, a pore network model is developed to analyze the pattern formed by a fluid pushing a second immiscible fluid and considering that capillary effects are responsible for the interface movement and ganglia generation. Hence, capillary pressure given by the Young-Laplace's equation determines the position of interfaces which in turn define the advancing front.

The purpose of this work is to show that ganglia generation, NAPL residual saturation and the pressure-saturation response depend on porosity anisotropy, on pore size spatial variability and on pore size distribution.

Immiscible displacement in soils

Pore geometry and the pressure of both fluid phases determine the curvature of meniscus/interfaces. When capillarity prevails over inertial effects (e.g. fluid velocity tends to zero values) and gravity effects (e.g. both fluids have similar density), the physical law that controls the displacement of interfaces is the Young-Laplace´s equation (Mercer and Cohen 1990):

$$\Delta P = Pnw - Pw = \frac{2\,T_s \cos\theta}{r} \qquad (1)$$

where ΔP = capillary pressure, Pnw = pressure of the non-wetting phase, Pw = pressure of the wetting phase, r = pore radius, T_s = interfacial tension and θ = contact angle between the interface and the solid surface. Contact angle is a measure of wettability (Morrow 1990) and capture the physicochemical interaction between the two fluids and the solid phase. The higher the affinity between the solid phase and water the lower the contact angle and the easier to displace the non-wetting phase (Francisca et al. 2003).

The movement of interfaces defines the advancing front. After the displacement of every interface NAPL saturation changes and some of oil can form a ganglion (trapped oil completely surrounded by the displacing fluid).

Macro- and micro-scale

Porous media are inherently non-homogenous phase mixtures. At macro-scale and under certain circumstances, soils and rocks can be considered as homogeneous for several practical purposes even though they are heterogeneous at micro-scale (Sahimi 1993). Behavior of macroscopically homogeneous systems is well predicted by average properties while spatial variability must be considered in micro-scale studies.

Immiscible flow can be analyzed at macro-scale by means of equivalent continuous models based on the generalized Darcy's law and mass conservation equation (Corey 1986, Binning and Celia 1999). In this case, the fluid velocity of each phase depends on the relative permeability which in turns depends on saturation and capillary pressure.

On the other hand, evolution of the advancing front and ganglia generation can only be described by means of micro-scale analysis. The distribution of ganglia within the soil matrix can only be described by considering the soil heterogeneity given by the changes in capillary characteristics (pore-size distribution or wettability, Bradford et al. 1998).

Pore network model

A two-dimensional pore network is used to simulate displacement of immiscible fluids. This discrete model consists of a network of pores bodies interconnected to four neighbor pores by means of capillaries. The sizes of capillary tubes are generated by using a lognormal probability distribution function. The resulting mean (μ) and standard deviation (σ) of pore sizes of the pore network define the coefficient of variation (COV) of pores as:

$$COV = \frac{\sigma}{\mu} \tag{2}$$

The mean pore size was set as 15 μm. The size of each pore body is assumed equal to the larger pore throat connected with it. Hence, there is spontaneous flooding of pore bodies when the interface of any neighbor capillary tube moves. All capillaries have circular cross-section, the same length and no roughness.

The model is used to simulate the displacement of a non-wetting fluid by a wetting fluid (imbibition process). The initial state considers that the pore network is fully saturated with NAPL. The organic fluid is displaced as the pressure of the displacing phase increases. The instability and movement of each particular interface define the advancing front. Interfaces follow a piston displacement mechanism and they move when the pressure difference ΔP surpass the capillary forces (equation 1, Young-Laplace). Some NAPL can remain trapped within pores during the advance of the displacing front. This NAPL forms ganglia and determines the residual saturation of the displaced phase. The simulation ends when the percolating path is generated (there is not considered the subsequent hydrodynamic fluid displacement on the motion of ganglia).

Results and analysis

Figure 1 shows the influence of the COV of pore sizes on the normalized capillary pressure versus saturation curves. Pore sizes are randomly distributed in the pore network. Results show that the pressure for the initial displacement decreases as the COV rises. Porous media with almost uniform pore sizes (standard deviation and COV of pore sizes tend to zero values) show a less gradual variation of the pressure-saturation response. As a particular case, a pore network with null COV should have a unique entry pressure given by equation (1). The lower entry pressures are related with the large pore fraction of networks with higher standard deviation of pore sizes.

Figure 2 shows the influence of the COV on the residual saturation of NAPL which in turn is related to the number and size of ganglia. Simulated cases also consider a random distribution of pore sizes within the network. It can be observed that NAPL can be removed almost completely in case of a network with uniform pore sized (COV$\approx$0) and that residual saturation increases significantly for very low values of COV. This result suggests that the intrinsic anisotropy and heterogeneity of porous media are responsible for the ganglia generation, even though the system may be considered as homogenous at macro-scale.

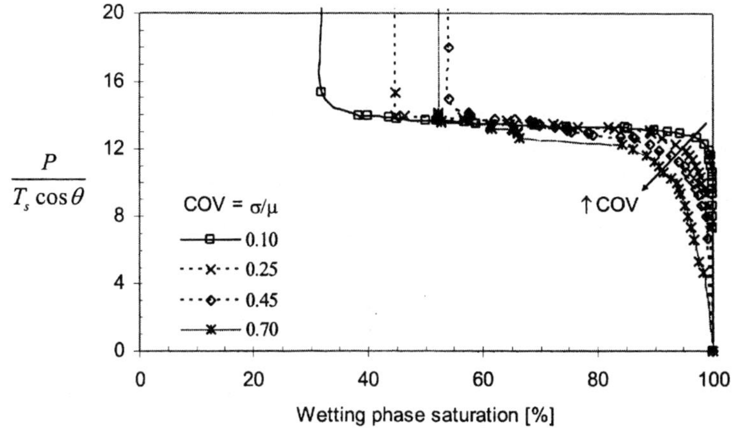

Figure 1: Influence of COV of pore sizes on the pressure-saturation curves.

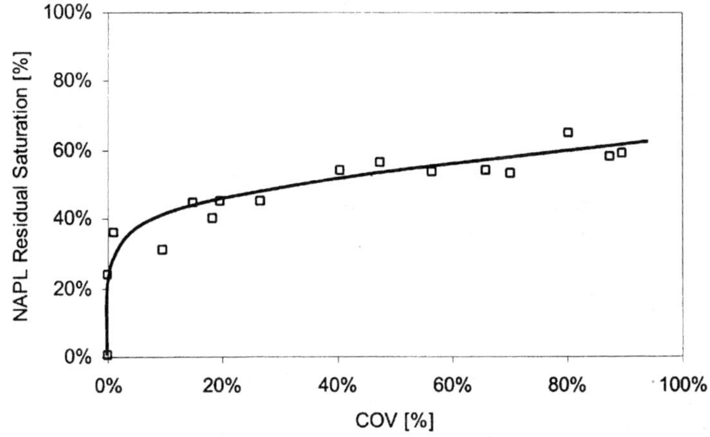

Figure 2: Influence of COV of pore sizes on the residual saturation of NAPL.

Figure 3 presents the influence of the spatial variability of porosity on ganglia generation. The mean and standard deviation of pore sizes is the same for all cases. Lateral borders are considered impervious, the displacing fluid is introduced from the top and the displaced fluid is eliminated from the bottom of the pore network. Case "a" shows a typical pattern of ganglia generated for a randomly distributed pore sizes. In cases "b", "c" and "d" the pore sizes increases toward the bottom, top and right respectively. Case "d" simulates a porous media with macropores or a preferential flow path. Finally, case "e" has a periodic distribution of pore sizes (e.g. simulating a fractured porous media, conglomerates or agglomerates of particles). Notice that even the macro-scale properties are the same in all cases (e.g. mean and

COV of pore sizes and porosity), the number and size of ganglia are very different. Obtained results suggest that pattern of ganglia and NAPL residual saturation are mainly influenced by the spatial variability of pore sizes. Hence, the spatial correlation of pore sizes should be considered in order to obtain a unique pattern of ganglia.

Figure 4 shows the pressure versus saturation curves for the same cases presented in Figure 3. Several results are shown for the randomly distributed pore sizes. Notice that there is a clear relationship between the evolution of the advancing front and the spatial distribution of pore sizes. In addition, the residual saturation (macro-scale property) is related to the spatial distribution and correlation of pore sizes (micro-scale material parameters). The heterogeneities at micro-scale level are responsible for the nonuniqueness of the capillary pressure–saturation relationships (Ataie-Ashtiani 2002).

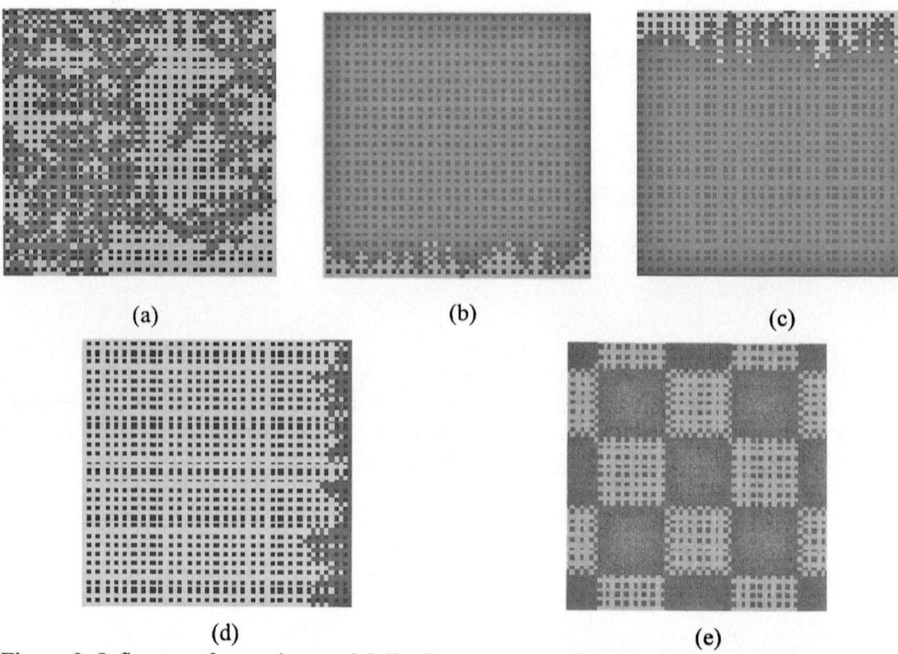

Figure 3: Influence of pore size spatial distribution on ganglia generation and NAPL residual saturation: (a) random distribution, Sr=57.1%, (b) pore size increases to the bottom, SR=1.6%, (c) pore size increases to the top, Sr= 2.8%, (d) pore size increases to the right, Sr=78.7% and (e) periodic spatial distribution of pore sizes, Sr=64.3%. Light color = NAPL, dark color = water, regular black squares = solid phase.

Conclusions

A pore network model has been developed to simulate immiscible displacement in porous media at micro-scale. The main conclusions of this research can be summarized as follows:

- Ganglia generation and NAPL residual saturation depend on micro-scale porous media properties. The coefficient of variation of pore sizes determines the initial displacement pressure and residual saturation. Lower entry pressures are related with the fraction of large pores of porous media with higher standard deviation and COV of pore sizes.

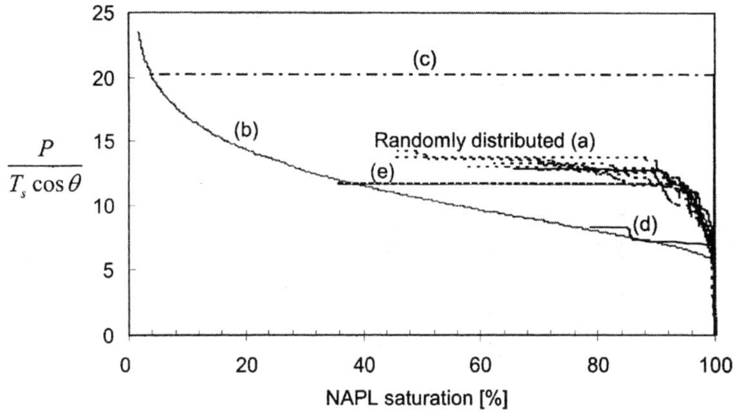

Figure 4: Influence of pore size spatial distribution on the normalized capillary pressure.

- Residual saturation increases with the variation (COV) of pore sizes. The intrinsic anisotropy and heterogeneity of porous media at micro-scale are responsible for the ganglia generation, even though the system may be considered as homogenous at macro-scale. The spatial variability of pore sizes determines the pattern of ganglia and residual saturation for the case of correlated porous structures.

- The number and size of ganglia are directly related with the spatial distribution of pore sizes. The spatial distribution of pore sizes and presence of preferential flow path (e.g. fractured porous media) determine the number and size of ganglia which in turn modify the residual saturation of NAPL.

Acknowledgements

This research was partially supported by CONICET and SECyT. The author thanks the motivation given by Professors J.C. Santamarina and V. Rinaldi.

References

Adamson A. W. and Gast A. P. (1997). “Physical Chemistry of Surfaces” Sixth Edition, *John Wiley and Sons*, 784p.

Ataie-Ashtiani B., Hassanizadeh S.M., Celia M.A. (2002). “Effects of heterogeneities on capillary pressure–saturation–relative permeability relationships”, *Journal of Contaminant Hydrology* 56, pp 175– 192

Binning P. and Celia M.A. (1999). "Practical Implementation of the Fractional Flow Approach to Multiphase Flow Simulation", *Advances in Water Resources* 22(5), pp. 461-478.

Corey A. T. (1986). "*Mechanics of Immiscible Fluids in Porous Media*", Water Resources Publications, 255p.

Blunt M.J. (2001). "Flow in porous media: pore-network models and multiphase flow", *Current Opinion in Colloid & Interface Science* 6, pp. 197-207.

Bradford S.A., Abriola L.M., Rathfelder K.M. (1998). "Flow and entrapment of dense nonaqueous phase liquids in physically and chemically heterogeneous aquifer formations", *Advances in Water Resources* 22(2), pp. 117-132

Dullien F.A.L. (1992). "*Porous media fluid transport and pore structure*", 2nd Ed., Academic Press, 574p.

Fatt I. (1956). "The network model of porous media I. Capillary pressure characteristics", *Trans. AIME* 207, pp. 144-181.

Fetter C.W. (1993). "Contaminant Hydrogeology", *Prentice-Hall*, 500p.

Francisca F.M., Rinaldi V.A. and Santamarina J.C. (2003). "Instability of Hydrocarbon Films over Mineral Surfaces-Microscale Experimental Studies-", *ASCE Journal of Environmental Engineering* 129(12), pp. 1120-1128.

Hardisty P.E., Wheater H.S., Johnston P.M., and Bracken R.A. (1998). "Behaviour of Light Immiscible Liquid Contaminants in Fractured Aquifers", *Geotechnique* 48(6), pp. 747-760.

Lenormand R., Touboul E. and Zarcone C. (1988). "Numerical models and experiments on immiscible displacement in porous media", *Journal of Fluid Mechanics* 189, pp. 165-187.

Mani V. and Mohanty K.K. (1997). "Effect of the spreading coefficient on Three-Phase Flow in Porous Media", *Journal of Colloids and Interface Science* 187, pp. 45-56.

Mercer J. W. and Cohen R. M. (1990). "A review of Immiscible Fluids in The Subsurface: Properties, Models, Characterization and Remediation", *Journal of Contaminant Hydrology* 6, pp. 107–163.

Morrow N. R. (1990). "Wettability and Its Effect on Oil Recovery", *Journal of Petroleum Technology* 42, pp. 1476-1484.

Sahimi M. (1993). "Flow phenomena in rocks: from continuum models to fractals, percolation, cellular automata, and simulated annealing", *Review of Modern Physics* 65(4), pp. 1393-1537.

Vidales A.M., Riccardo J.L. and Zgrablich G. (1998). "Pore-level modeling of wetting on correlated porous media", *J. Physics D: Applied Physics* 31, pp 2861-2868.

Zhou D. and Blunt M. (1997). "Effect of spreading coefficient on the distribution of light non-aqueous phase liquid in the subsurface", *Journal of Contaminant Hydrology* 25, pp. 1-19.

Hydro-Chemical Coupling in Unsaturated Geomaterials

Hasan Ghasemzadeh, CERMES, Ecole Nationale des Ponts et Chaussées (ENPC), Paris, France and KN Toosi university of Technology, Tehran, Iran
Behrouz Gatmiri, CERMES, Ecole Nationale des Ponts et Chaussées, Paris, France and University of Tehran, Iran

Abstract

The transport process of solute via advection, dispersion, diffusion and some basic chemical reactions in the liquid phase of water has been studied in the frame of a coupled numerical model in a multiphase medium. Water flow occurs in both vapour and liquid phases. The governing equations in the basic formulation are written in terms of soil displacements, water and air pressure, temperature and concentration of contaminant. These equations are solved by finite element method. In this paper some examples of transport process in isothermal and undeformable cases are presented to validate the developed model. The results are compared to the results of the other numerical model and experimental observation.

Introduction

The problems in the environmental geomechanics often involve the study of heat, mass and contaminant transport in a number of engineering situations (Charlier and Gens [1]). One of the most important issues in the contaminant transport is heavy metal transport in soil. Heavy metals are elements having atomic weights between 63.546 and 200.590 (Kennish [2]), and a specific gravity greater than 4.0 (Connell et Miller [3]). The contaminant transport normally takes place in the unsaturated soil and in the presence of heat due to climate change. This paper gives some application examples of the hydro-chemical coupling in order to estimate the variation of concentration of chemical elements like heavy metal in a multiphase medium.

A suction based-formulation of fully coupled behaviour of a deformable unsaturated porous medium under heating has been proposed and validated (Gatmiri et al [4], [5]). In the mentioned formulation the nonlinear behaviour and phase changes (vapor and liquid) are considered. The theory of Philip and de Vries [6], initially developed for undeformable media is adopted for transfer of humidity; furthermore this approach has modified in order to be consistent with deformability of skeleton. In the Philip and de Vries approach, water flow occurs in both vapor and liquid phases; and heat convection and conduction is taken into account. In the THM modeling of multiphase medium the coupling effects of skeleton, suction and temperature are integrated via the concept of state surfaces of void ratio and degree of saturation (Gatmiri [4]). Based on proposed formulation, a fully coupled numerical model for the behavior of soil deformation, water flow, air flow, heat flow in unsaturated soil is developed and integrated in a finite element code "θ-STOCK" (Gatmiri [4]).

5th ICEG Environmental Geotechnics, Thomas Telford, London 2006.

An extension of the pre-mentioned model is developed to include the chemical effects. In this extension the advection, dispersion, diffusion of contaminant and some basic chemical interactions between the liquid phases and soil are included. The governing equations in terms of soil displacements, water and air pressure, temperature and concentration of contaminant are given.

Field equations

Equations of suction-based approach of multiphase media as developed in (Gatmiri [3] and [4]) are briefly presented here. Moisture mass conservation and moisture movement are:

$$\frac{\partial \rho_m}{\partial t} = -\mathrm{div}\left(\rho_w (U + V)\right) \tag{1}$$

$$\rho_m = \theta\rho_w + (n - \theta)\rho_v = nS_r\rho_w + n(1 - S_r)\rho_v \tag{2}$$

$$U + V = -D_T \nabla T - D_\theta \nabla\theta - D_w \nabla Z \tag{3}$$

Where U is water velocity, V is vapor velocity, ρ_w is water density, ρ_v is vapor density, S_r is degree of saturation, D_T is thermal moisture diffusivity, D_θ is isothermal moisture diffusivity and D_w is gravitational diffusivity.

Equation of mass conservation and movement of gas in a control volume of unsaturated porous media is given as:

$$\frac{\partial}{\partial t}\left[n\rho_g(1 - S_r + HS_r)\right] = -\mathrm{div}(\rho_g V_g) - \mathrm{div}(\rho_g HU) \tag{4}$$

$$V_g = \frac{-K_g}{\gamma_g}\frac{\partial P_g}{\partial T}\nabla T - K_g\left(\nabla\left(\frac{P_g}{\gamma_g}\right) + \nabla Z\right). \tag{5}$$

Where H is Henry constant, Vg is vector of gas velocity and ρ_g is gas density.

Since the application examples presented in this paper treat only the hydrochemical process in isothermal and undeformable media, the description of equations of skeleton behaviour and heat transfer are omitted.

Chemical species transport and thier interactions with the other phases can be expressed in mass conservation equation as:

$$\frac{\partial(\theta C)}{\partial t} = \frac{\partial}{\partial x_i}\left(\theta D\frac{\partial C}{\partial x_j}\right) - \frac{\partial}{\partial x_i}(\theta v_i C) + q_r C_r + \sum R_n \tag{6}$$

$$\sum R_n = -\rho_b \frac{\partial C_s}{\partial t} - K_1 \theta C - K_2 \rho_b C_s \quad (7)$$

Where C is the concentration of contaminant, C_s: the concentration of contaminant in solid phase, v_i: water velocity in direction i, q_r and C_r : discharge and concentration of source, D : dispersion coefficient, ρ_b bulk density of soil, and K_1 & K_2 are decay coefficient of solved and adsorbed concentration. Different geochemical models used to express the relation between C_s and C are: Linear sorption, Freundlich sorption, Langmuir sorption and Langmuir two-surface sorption. In the present paper dispersion coefficient (*D*) is a linear function of molecular diffusion (D_m) and water velocity (v)[7]:

$$D = \tau(\theta) D_m + a(\theta) v \quad (8)$$

Where τ is the tortuosity factor and α is the water content dependent coefficient of dispersivity.

Numerical solution and integration in θ-STOCK

The integral form of the proposed model is obtained by using the weighted residual method and the Galerkin choice of weighted functions. The final matrix form is obtained by discretization in space and in time by applying the single step integration. This final matrix form of THCM model has been integrated in **θ-STOCK,** finite element code already developed at CERMES-ENPC [4].

Validation and verification

The robustness of the numerical schemes and THM models in **θ-STOCK** are illustrated by an extensive use of this code over a period of several years (1997-2004) by the different research activities in CERMES. The objective of this section is to show the validity of the developed formulation (THCM) and its numerical integration by comparison with the published experimental and numerical results reported in the literature. The experiments of Nutzmann et al [8] and Thomas and Cleall [11] have been chosen as the references of comparison.

Chloride solution transport

The first example shows the vigour of the developed model by evaluation of solute dispersion in unsaturated medium without any chemical reaction. For this purpose, the experiments of Nützmann et al [8] on unsaturated glass beads are used.

In these tests, a column of 100 cm height and 20.6 cm internal diameter filled of glass beads has been used. The glass beads had particle sizes ranging from 0.36 to 0.71 mm in diameter with the properties presented in Table 1. Each displacement experiment has been begun with a constant water infiltration. The tracer was injected when the steady state water flow through the column was reached; thus there is no hysteresis problem. Experiments have been carried out for different infiltration rates, where the calculated molecular Peclet numbers were between $40<Pe_{mol}<600$, indicating that mass flux was dominated by dispersion with negligible diffusion [9, 10]. All experiments have been repeated three times.

Table 1: Physical characteristics of the medium

Properties	Unit	Values
Porosity	n (cm^3/cm^3)	0.347
Particle density	ρ (kg/dm^3)	2.57
Saturated hydraulic conductivity	Ks (cm/s)	0.0794
Water content	θ (kg/dm^3)	0.14
Infiltration rate	q_{inf}(cm/s)	0.009867
Initial sorbent concentration	Cs (g/l)	0.06

Three platinum probes were placed at each depth with an angle of 120° between them to measure concentration. The breakthrough curves measured at 15 cm (z=85 cm) and 50 cm (z= 50 cm) distances from the top of the column are presented in figure 2. The difference among the results of different probes in the same level is due to high fluctuation of water velocity in the unsaturated medium [8]. The mean values of measured concentration in each level are presented in figure 3.

In the numerical analysis, a mesh of 2*50 rectangular elements was used which is presented in the figure 1. The tracer (Nacl) with a concentration of 1.02 g/l was injected to the top of column after a steady state infiltration of water. The duration of the experiment was approximately two hours and half. Dispersion coefficient was considered as [8]:

$$D = 0.003956\theta^{-2.89689} \tag{9}$$

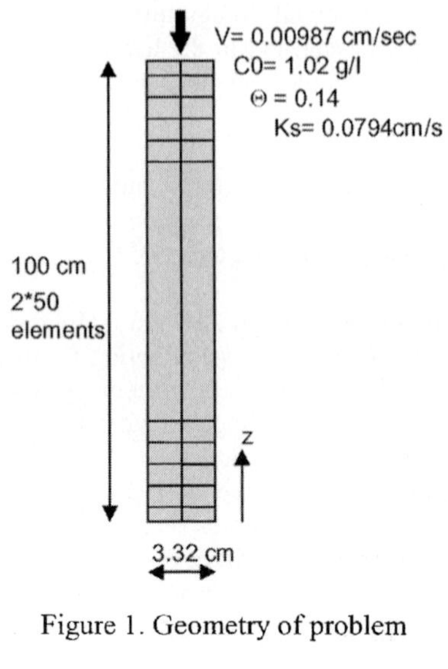

Figure 1. Geometry of problem

The calculated breakthrough curves at z=85 cm and z= 50 cm distances are given also in the figure 2. At z=0.85 m the concentration starts to increase from initial value (C_s= 0.06 g/l) after about 120 seconds and it reaches to entrance concentration (C=1.02 g/l) after about 500 seconds. At z=0.50 m the concentration starts to increase from initial value after about 450 seconds and it reaches to entrance concentration after about 1250 seconds. We observe that these calculated results in different depth are between the results obtained from different probes at the same depth. Figure 3 shows comparison of calculated results and average of measured tracer breakthrough curves. A good agreement between average of experimental results and numerical simulation can be observed.

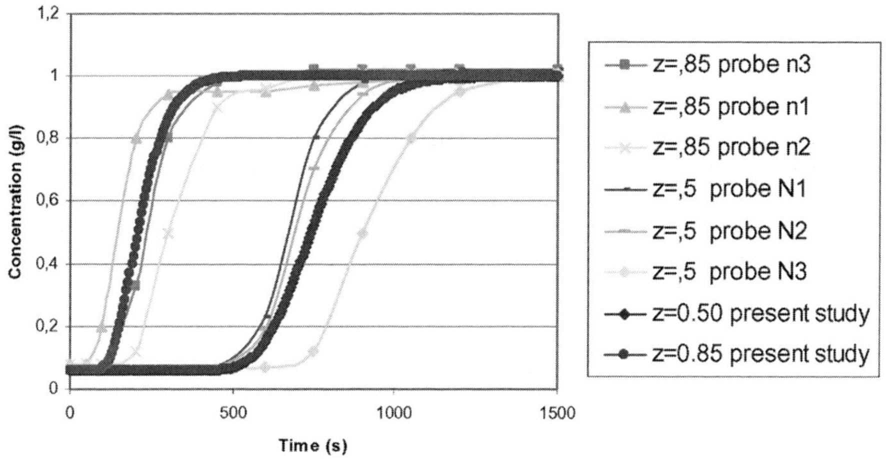

Figure 2: Measured tracer breakthrough curves in unsaturated glass beads for different probes from Nützmann et al [8] and result s of this calculation.

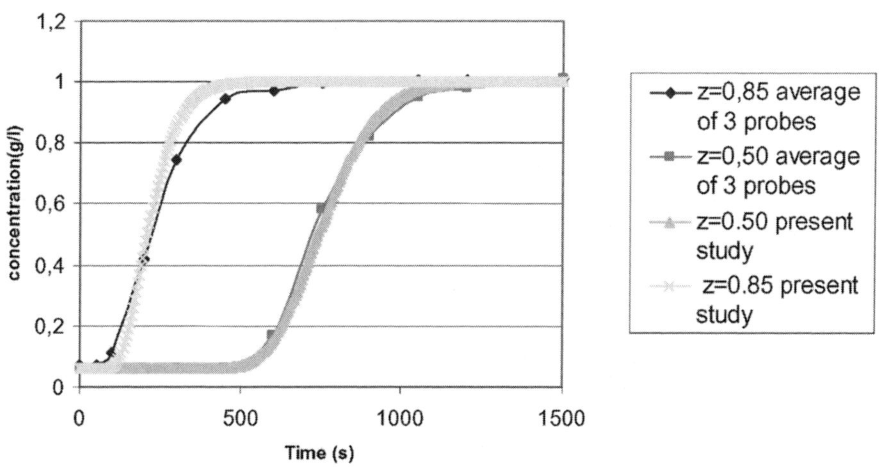

Figure 3. Comparison of present calculation results and average of measured tracer breakthrough curves in unsaturated glass beads from Nützmann et al [8].

Lead transport

The second example is horizontal leaching test in the unsaturated soil. A series of tests have been performed on the soil sample by Thomas and Cleall [11]. The properties of soil sample are presented in Table 2. In this example, a soil column of 300 mm was leached with 500 ppm lead solution at constant atmospheric pressure under zero head until the cell reached full saturation.

Table 2: Physical characteristics of the soil

Properties	Unit	Values
Porosity	n (cm^3/cm^3)	0.33
Dry density	ρ (kg/dm^3)	1.69
pH	Log(H^+)	7.6
Clay of soil	(cm^3/cm^3)	30%
Sand of soil	(cm^3/cm^3)	70%
Initial sorbent concentration	Csoil (mol/m^3)	0.66
Initial degree of saturation	Sr (cm^3/cm^3)	0.2
Initial water pressure	Pw (kPa)	-2560

In the numerical simulation, a mesh of 60 thick elements of 5 mm was used (Figure 4) with a time step size of 1200 seconds. The initial condition is presented in Table 2. The Dirichlet boundary condition in the left limit is: water pressure =0.0 and concentration =2.41 mol/ m^3. The Langmuir model is adapted to simulate sorption of lead on the surface of soil particles.

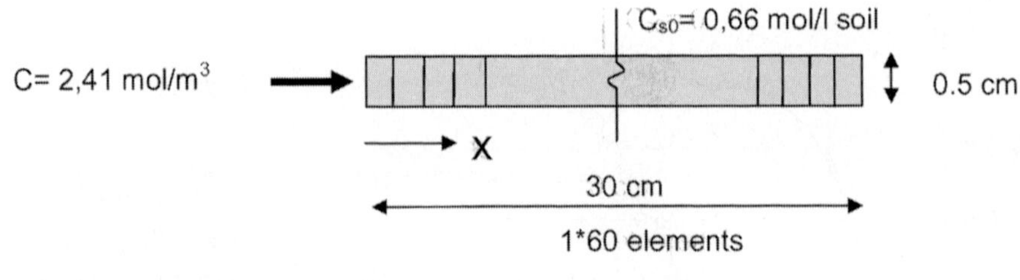

Figure 4: Geometry of problem

The concentration profile of solution is presented in the figure 5. The concentration is equal to 2.41 in the left limit and it decreases sharply to zero in a distance about four cm from source. In the figure 6, the results of numerical simulation are presented. The experimental results and numerical results with code COMPASS are also presented in this figure (Thoms et Cleall [11]). The results for sorbed concentration were calculated at 20mm intervals throughout the soil so each point in the figure represents the average over 4 elements. It can be seen a relatively good agreement between experimental results and numerical simulation although the calculated absorbed concentration is not enough satisfactory in relation to experimental result.

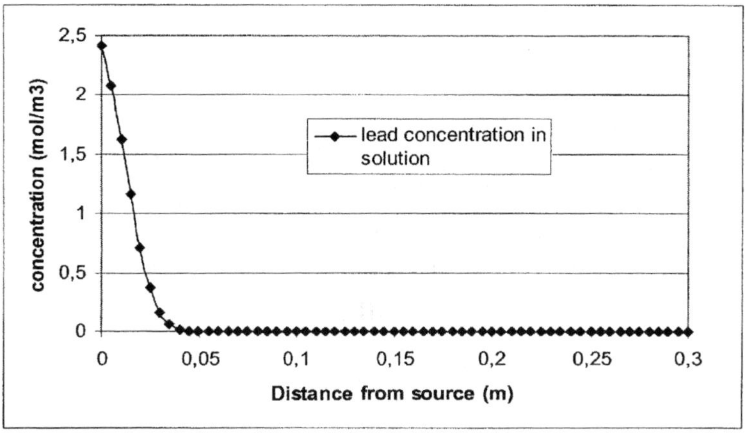

Figure 5: calculated profiles of concentration in solution.

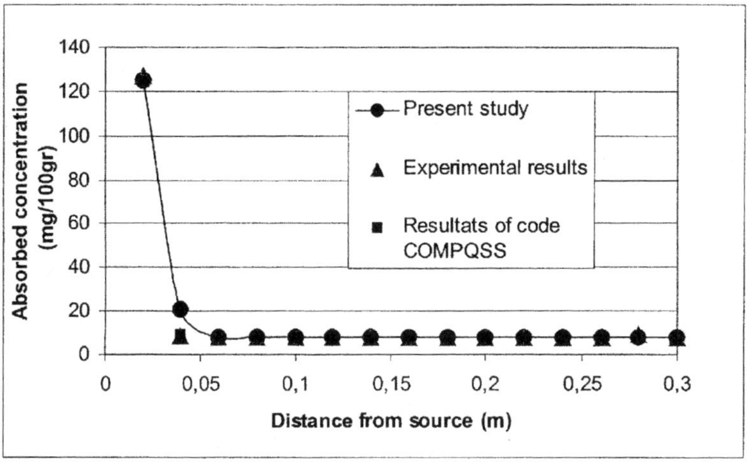

Figure 6: Numerical profiles of sorbed concentration compare to experimental ones (Thomas and Cleall [11]).

Conclusion

Solute transfer through unsaturated medium is studied and modeled in this paper. Two applications related to tracer and heavy metal transport in undeformable and unsaturated soil were presented in which chemical transport and geochemical interactions were simulated. The results of present study showed a relatively good agreement with the experimental ones. More development can be done to model solute transfer in unsaturated and deformable media in order to consider mechanical effects of solute concentration.

Reference

1 Charlier R., Gens A., 2001, Environmental Geomechanics. *Revue française de génie civil, volume 5 n°6/2001.* 743-775.

2 Kennish, M.J. 1992. *Ecology of Estuaries: anthropogenic effects.* Marine Science Series, CRC Press: Boca Raton,USA.

3 Connell, D.W., G.J. Miller. 1984. *Chemistry and Ecotoxicology of Pollution.* John Wiley & Sons, NY.

4 Gatmiri B., "Analysis of fully coupled behaviour of unsaturated porous media under stress, suction and temperature gradient", *Rapport final, 1997, CERMES-EDF.*

5 Gatmiri B., Jenab-Vossoughi B., Delage P. 1999 'Validation of θ -STOCK, A finite element software for the analysis of thermo-hydro-mechanical behaviour of engineering clay barriers', NAFEMS World congress '99 on effective engineering analysis, Rhode Island, USA.

6 Philip J.R & D.A. de Vries, 1957 'Moisture Movement in porous materials under temperature gradents', *Trans. Am. Geophys.* Un. 38, 222-232

7 Maciejewski, S., 1993. Numerical and experinmental study of solute transport in unsaturated soils. *J Contam Hydrol*, 14. 193–206.

8 Nützmann, G., Maciejewski, S., Joswing, K., 2002. Estimation of water saturation dependence of dispersion in unsaturated porous media: experiments and modelling analysis. *Advances in Water Resources*, 25. 565–576.

9 Pfannkuch, HO. 1963, Contribution a l'étude des déplacements de fluides miscibles dans un milieu poreux. Revue de l'institut Français du Pétrole. 18, 215-70.

10 Van Genuchten MTh, Wierenga PJ. Mass transfer studies in sorbing porous media. I: Analytical solutions. Soil Sci Soc Am J 1976; 40. 473-80.

11 Thomas H.R., Cleall P.J., 2001, Pollutant transport and chemical effects. *Revue française de génie civil, volume 5 n°6/2001.* 743-775.

Fate and Transport of Polybrominated Biphenyl Ethers from Biosolids

Tamer Alexan Gorgy, Department of Civil Engineering, University of British Columbia , Vancouver, Canada V6T 1Z4
Loretta Li , Department of Civil Engineering, University of British Columbia, Vancouver,Canada
John Grace, Department of Chemical and Biological Engineering, Vancouver, Canada

Abstract

Polybrominated diphenyl ethers (PBDEs) are popular flame-retardants which are highly hydrophobic and ubiquitous. This is a review of studies that have taken place in Europe and North America since 2001 on PBDE levels in biosolids and agricultural soils that received biosolids which were 0.7×10^{-3}–9.3 μg/g and 0.2×10^{-3}- 1.5 μg/g, dry weight basis, respectively. Also a preliminary study was carried out to examine the degree of contamination by various PBDEs in agricultural soils due to the application of biosolids, and to examine their mobility by conducting desorption tests on these soils. The levels of PBDEs in the soils that have received biosolids increased by 2-4 orders of magnitude from their background/control levels. Concentrations of PBDEs in the soil which underwent desorption were reduced to up to 51% of the pre-desorption concentrations. These results provide evidence that PBDEs can be mobile due to their desorption from the soil, which has received biosolids.

Introduction

Polybrominated diphenyl ethers (PBDEs) are used as flame-retardants in plastics, textiles, electronic circuitry and insulation material. There are 209 congeners of PBDEs, classified by the number of bromine atoms. These congeners have become emerging contaminants that are ubiquitous aided by their high octanol-water partitioning coefficients (K_{OW}) in the range of 10^4-10^{10}. As a result, PBDEs have been found in fish, birds, marine mammals, humans, plasma, sediments soils and vegetation [1]. Sewage sludge and biosolids have also been reported to be contaminated by PBDEs [2]. Although modelling studies have indicated that small proportions of PBDEs exist in air and sediments, they can undergo long-range transport via these compartments sorbed to particulate matter. However, studies regarding transport of these contaminants through soil, especially those that receive biosolids, are few in number.

Products containing PBDE compounds are ultimately disposed in landfills or incinerated. These compounds may escape due to volatilization or by leaching. The relatively low volatilities and aqueous solubilities of PBDEs suggest that the bulk of the environmental burden of these chemicals will eventually reside in either sediments or soils [2]. The environmental fate of PBDEs has been assessed based on the available properties for the groups of congeners. Gouin and Harner [3] modelled long-range atmospheric transport of PBDEs, evaluating their surface-air exchange behaviour, and concluded that these compounds will largely partition to organic carbon in soil and sediment. Other models have also been developed to predict environmental partitioning of individual PBDEs to aerosols and soils using aerosol-air and soil-air partitioning functions. As for the effect of leachate quality on the fate and transport of PBDEs, it has neither been studied nor quantified.

It is a common practice in North America to dispose of sludge produced from sewage treatment plants on agricultural lands. Information about the mobility of these contaminants from such soils is absent, and hence the mechanisms of their transport through the soil are also unknown. Examining the desorption potential from biosolids and agricultural soils that have received these biosolids can determine the extent of their mobility via the surrounding aqueous phases. This information should also make it possible to evaluate their risk of contaminating groundwater, and hence the risks of exposure by different receptors.

This paper presents a comprehensive examination of what is known about the distribution of PBDEs in biosolids and agricultural soils in North America and Europe in order to assist in understanding the extent of the problem. We include results from our preliminary study which examined the degree of contamination by PBDEs in agricultural soil due to the application of biosolids and their desorption potential from this soil. In addition, we discuss several issues that need to be investigated with regards to their distribution in biosolids and agricultural soils to achieve a better understanding of the fate of PBDEs in agricultural soils.

PBDEs in Biosolids

North America

Almost half of the sludge produced from sewage treatment plants in North America is disposed on agricultural lands. PBDEs are likely to be present in these high organic-content media. Samples from different regions of North America have verified the presence of this class of brominated flame-retardants, at levels almost two orders of magnitude higher than European levels.

Eleven biosolids samples from the Lake Superior and Lake Michigan regions in the U.S were collected prior to application on land [4]. This study found that PBDEs containing 4 to 6 bromine atoms are present in significant amounts in the biosolids (Figure 1). Concentrations ranged between 0.140 and 0.76 µg/g dry weight (d.w.). The total PBDEs concentration in these biosolids was 1.10.–2.29 µg/g d.w.. Hale et al. [4] suggest that these high concentrations are consistent, regardless of the region of origin or the pre-application treatment.

Biosolids samples from three different sewage treatment plants in Greater Cincinnati, Ohio were analysed for PBDEs [5] . The average concentrations of BDE-47, -99, and –100 BDE ranged between 7.49×10^{-3} and 16×10^{-3} µg/g d.w. (Figure 1). These primary congeners represented 40, 42 and 18%, respectively, of the total average concentration. In 2002, it was estimated that the State of Ohio used 247,364 tonnes sewage sludge-dry matter (DM) in land application, representing 62% of the annual sludge produced [6].

Highly elevated PBDE concentrations in biosolids have also been found in the Pacific Northwest region. Samples from two Sewage treatment plants (STPs) located in the Tacoma region, Washington, and three STPs along the Yakima River, again in Washington State. Concentrations from the Tacoma region biosolids had a total average concentration of 9.30 µg/g d.w. (Figure 1), having BDE-47,-99, -209 comprising 20%, 24%, and 44%, respectively, of the total concentration [7]. Average total concentration from activated and dewatered sludge from the Yakima plants were 2.43 and 3.49 µg/g d.w., respectively [7]. BDE-47, and-99, in activated sludge, contributed 32 and 35% of the total PBDE concentrations, respectively, and for the dewatered sludge their contributions were 19 and 22%, respectively.

A single study at the Paolo Alto California STP indicated that the levels of BDE-47, -99, and –209 were greater than the averages reported in the Eastern U.S. The mean concentrations of individual PBDE congeners in the sewage sludge presented in Figure 1 ranged between 0.2×10^{-3} and 1.18 µg/g d.w., with a total PBDE concentration of 3.38 µg/g d.w. [8]. The six congeners with the highest concentration were BDE-47, -99, -100, -153, -154, and –209, with percent contributions of 22, 28, 5, 3, 2, and 35%, respectively.

Europe

Three studies have reported concentrations of PBDEs in sludge or biosolids generated from Swedish, Dutch and Danish STPs. The Swedish results were reported on a wet weight basis rather than dry weight basis. 114 samples from 22 municipal STPs were collected and analyzed for single PBDE congeners. The concentrations ranged from non-detectable to 0.39 µg/g (Figure 1), with BDE-47, -99, and –209 contributing 22, 32 and 35% of the total PBDEs concentration [9].

Dutch sludge samples from the Ameland and St. Annaparochie STPs , were analyzed for PBDEs as part of a national study on estrogenic contaminants in aquatic environments. The concentrations of the BDE congeners ranged from 1.4×10^{-3} to 0.19 µg/g d.w. and 0.7×10^{-3} to 0.19 µg/g d.w. from the sludge generated from the each plant, respectively (Figure 1) [10]. BDE-47, -99, and –209 represented 4, 6 and 87% respectively, in Ameland STP sludge and 5, 6, and 86%, respectively, in St Annaparochi STPs of the total PBDE concentrations.

Sewage sludge was collected at Bjergmarken, a typical Danish provincial STP plant receiving both urban and agricultural wastewater [11]. Concentrations of the different congeners ranged from 2.0×10^{-3} to 0.25 µg/g d.w., with a total concentration of 0.48 µg/g d.w. (Figure 1). BDE-47, -99, -209 represented 22, 16 and 53% by mass of the total PBDEs concentrations respectively.

PBDEs in soil receiving biosolids

Soils in Sweden that have received sewage sludge have been analyzed for PBDEs as reported by Matscheko et al. [12]. Sewage sludge from two STPs in southern Sweden (Lund and Malmö) were applied at three different dosages: 2.3, 20 and 60 tonnes/ha dry mass basis on different fields. The average concentrations of PBDEs in the fields that received the same dosage were calculated and reported in Figure 2, and are 8.5×10^{-6}-3.3×10^{-5}, 1.5×10^{-6}-1.3×10^{-4} and 2.2×10^{-6} – 2.2×10^{-4} µg/g d.w., respectively . The total PBDEs concentrations were in the range 7.5×10^{-5} to 5.9×10^{-4} µg/g d.w. [12] . BDE-47 and BDE-99 were dominant, representing 27–43% and 29–49% of the total PBDE concentrations, respectively.

Dewatered sludge from one of the treatment plants located along the Yakima River in Washington State, USA, were used as a soil fertilizer in a hop field [7]. Samples from different locations in the field were analyzed for PBDEs. Three samples were taken from a hop field known to have received these dewatered biosolids. The control sample/reference, was taken from an area that had not received any biosolids. The other two samples were taken from disked land between hops and from an area where no disking had taken place [7].

The concentrations of PBDEs in the disked and non-disked soil samples were 3.1×10^{-3}–39×10^{-3} $\mu g\cdot g^{-1}$ d.w. and 0.12–1.50 μg/g d.w., respectively, whereas the concentrations in the control samples ranged from non-detectable to 0.15×10^{-3} μg/g d.w. (Figure 2). Total PBDE concentrations for the disked and non-disked samples were 0.14 and 7.6 μg/g d.w. (Figure 2). BDE 47, and BDE-99 were dominant representing 24 and 28% of the total concentration in the disked soil, and 17 and 20% in the non-disked soil, respectively.

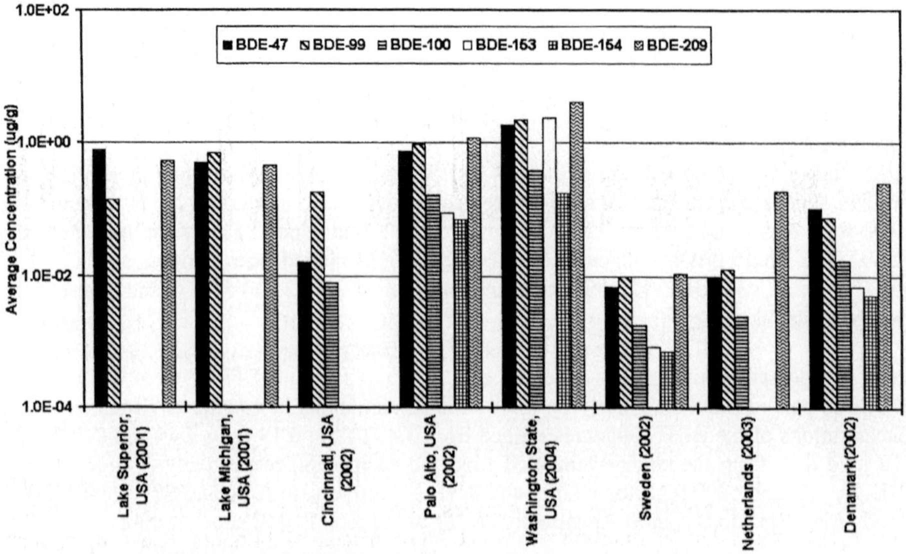

Figure 1: Average concentrations of BDE-47, -99, -153, -154 and –209 in North American and European biosolids. All concentrations are reported as dry weight basis except for the Swedish results which are reported as wet weight basis.

Preliminary study

A preliminary study was carried out by us to examine the degree of contamination by PBDEs in agricultural soils due to the application of biosolids and their desorption potential from these soils.

Material and Methods

Samples were collected from an experimental agricultural farm managed by Agriculture Canada located in the South-Central Interior of the Province of British Columbia, Canada. The general soil classification of this farm is Brown Chernozem (66% sand, 17% silt, 17% clay, 1.9% Corg). Three samples each were collected from: (1) background/control area which received no biosolids, (2) an area which had received 20 tonnes/hectare (T/ha) biosolids loading, and (3) an area which had received 60 T/ha biosolids loading. In both cases, the biosolids were applied in the late summer of 2002. The samples were collected from depths between 0.05 and 0.20 m below the surface. The extent of PBDE contamination due to the application of biosolids was determined for the three samples by measuring BDE-47, -85, -99, -100, -153, -154 and -183 according to the method described by the Canadian Technical Report of Fisheries and Aquatic Sciences [13]. This method was modified by extracting the contaminants using accelerated solvent extraction, instead of Soxhlet extraction.

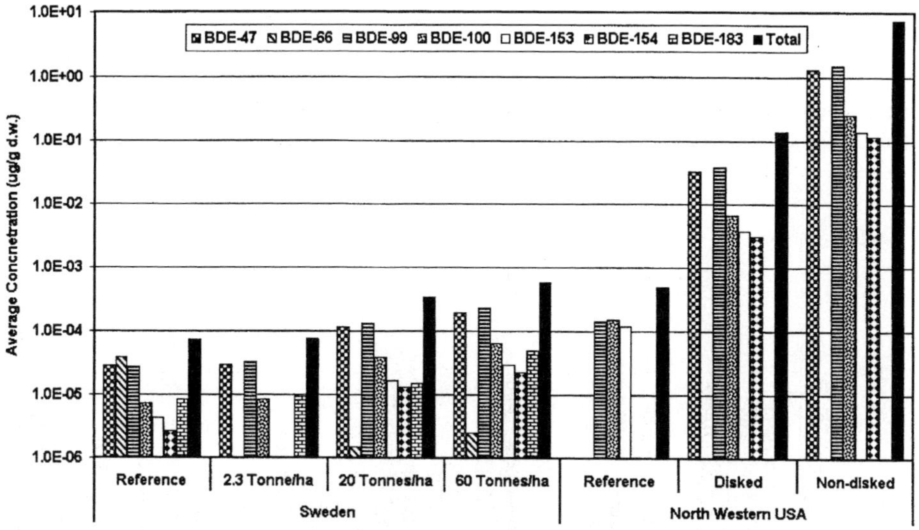

Figure 2: Average concentrations of BDE-47, -66, -99, -100, -153, -154 and total-BDE in Swedish and North West American soils that have received biosolids.

The samples from the 20 and 60 T/ha biosolids treatments were subjected to two sets of desorption tests: (a) desorption using distilled water, and (b) desorption using a leachate. The leachate used was produced by passing distilled water through a column packed with soil from the control soil sample. The desorption tests were similar to the batch equilibrium test method [14], except that the duration was set at 7 days instead of 24 hours. The samples were then analyzed using the same method as above.

Results

The degree of contamination by PBDES due to biosolids application is shown in Figure 3. The concentrations of PBDEs in the soils which received 20 T/ha and 60 T/ha were 2-4 and 3-4 orders of magnitude higher, respectively, than the concentrations in the control sample. Concentrations of PBDEs in the control, 20 T/ha and 60 T/ha samples were in the range of no-detection (ND)-0.48×10^{-6} μg/g d.w., 0.3×10^{-3}–8.6 ×10^{-3} μg/g d.w., and 2.0×10^{-2}–0.40 μg/g d.w.. These results show that PBDEs can be introduced from the biosolids to the underlying soil.

The desorption tests indicate that PBDEs can be desorbed from the soil (Figure 3). The levels of PBDEs in the 20 Mg/ha soil samples after desorption were in the range of 2.6×0^{-3}–6.7×10^{-3} μg/g d.w. and 2.9×10^{-3}–8.1×10^{-3} μg/g d.w, for leachate and distilled water desorption, respectively. As for the samples that had received 60 Mg/ha biosolids, desorption of all PBDEs was observed (Figure 3). The concentrations were in the range of 9.0×0^{-3}–0.240 μg/g d.w. and 9.0×10^{-3}–0.25 μg/g d.w. for leachate and distilled water desorption, respectively.

Discussion

The main objective of each of the studies reviewed above as well as our own preliminary measurements, was to report the levels of PBDEs in various biosolids and soils. The studies that reported levels of PBDEs in biosolids from North America and Europe have indicated a significant rise in the concentration of these contaminants over the last five years. The North American levels are typically almost 3 orders of magnitude higher than their European counterparts. None of the earlier studies published accounted for the temporal trends of PBDE levels at the given locations or their levels in the soils which have received these biosolids.

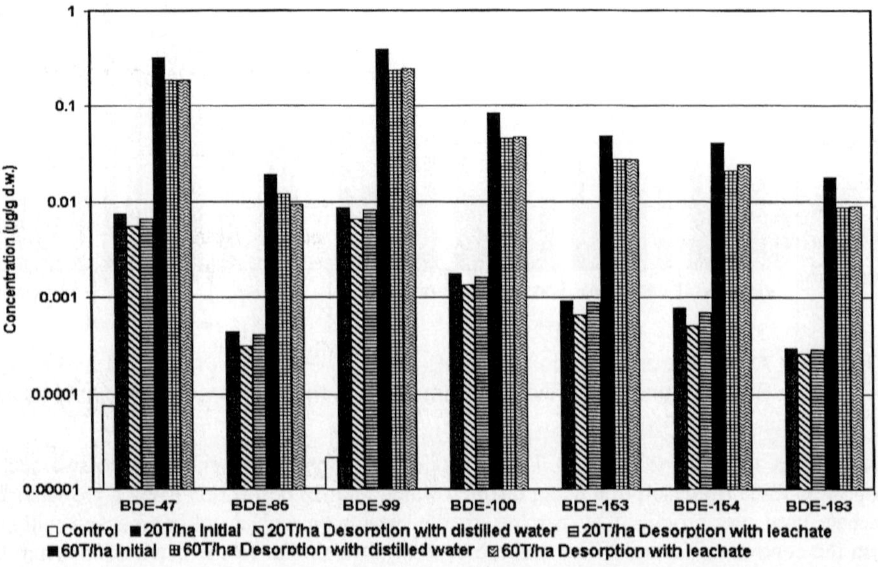

Figure 3: PBDEs levels in soil samples from control, 20 T/ha, and 60 T/ha areas in the agricultural field, as well as PBDE levels after desorption tests.

The studies from Sweden and Washington State [7, 12], show clearly that BDE-47, and BDE-99 are most abundant in soils that have received biosolids. The average contributions of BDE-47 and –99 in the total PBDEs are higher in the agricultural soils than in the biosolids, whereas BDE-209, only analyzed in biosolids, had the highest average contribution, accounting for 50% of the PBDEs. This comparison could indicate that BDE-47 and –99 are most mobile and are transferred from the biosolids to the agricultural soil, that more highly brominated congeners decompose producing these two congeners as products, or that both occur. The absence of a relationship between the levels in biosolids and the agricultural soils to which they have been applied makes it impossible to specify the mechanism by which these congeners end up having such high concentrations in the studies. It is therefore clear that further research is needed to determine the mobility and transfer of the different congeners between solid phase environments, especially between biosolids and agricultural soils.

The preliminary study results presented above confirms the Swedish and American studies [7, 12] that for the soils which had received biosolids, the PBDEs levels increased. In our preliminary study these levels increased by 2-4 orders of magnitude from their background/control. The preliminary study has also shown that these soils can also desorb these contaminants, and their levels can be reduced by 51% from their initial levels. Desorption using distilled water mainly resulted in greater reductions in PBDEs levels compared to the leachate. This observation needs further examination to determine the conditions which control the desorption kinetics between the soil and aqueous phases. This study provides insight with respect to the application of biosolids on the possible contamination of groundwater. PBDEs can be transferred from the biosolids applied to the agricultural land to the soil beneath. The water leaching through the soil, might cause desorption of these PBDEs from the soil, possibly resulting in transport to groundwater.

Further studies are needed to confirm these results, and to investigate factors controlling the desorption kinetics of PBDEs from solid to aqueous phases. Suggested factors include suspended matter in the aqueous phases and their pH values, as well as the chemical and physical properties of the soils. It is vital to understand the competition of the different PBDEs to sorb onto solid phases, and to be able to determine the concentration limits beyond which these contaminants become mobile. The findings of the recommended studies will provide a basis for assessing environmental risks from practices such as biosolids application to agricultural lands, and can further include other practices or sources such as municipal and industrial landfills contained by liners of different soil types.

Conclusion

Research in the last five years has focused on the levels of PBDEs in different environments, and has drawn attention to their rising levels in multiple organisms, including humans. Reported PBDE levels are often higher by several orders of magnitude in North America than in Europe. In addition, PBDEs concentrations are 3 orders of magnitude higher in agricultural soils that have received biosolids than in reference or control soils.

There is a strong indication from previous studies, that the application of biosolids to agricultural soils increases the PBDE levels in the latter. The contributions of the lower-brominated congeners in agricultural soils are greater than in the biosolids, and these lower-brominated congeners are present in greater proportion than more highly brominated ones. This could indicate that lower PBDEs are mobile and can be transferred from the biosolids to soils and/or that the higher brominated congeners undergo debromination to lower-brominated congeners.

The preliminary study has also shown that PBDEs have a potential to desorb from the agricultural soils. This desorption allows them to be mobile in the aqueous subsurface environments, and can therefore lead to the contamination of groundwater resources. Further research is necessary to investigate the various chemical and physical conditions of both the soils and aqueous environments, which allow these contaminants to be mobile. It is also important to study decomposition of more highly brominated congeners into lower-brominated compounds, as this could be playing a role in explaining the results of the experimental studies reported in Sweden, the U.S. ad now Canada.

References

1.Hites, R.A., *Polybrominated diphenyl ethers in the environment and in people: A meta-analysis of concentrations.* Environmental Science & Technology, 2004. **38**(4): p. 945-956.

2.Hale, R.C., et al., *Polybrominated diphenyl ether flame retardants in the North American environment.* Environment International, 2003. **29**(6): p. 771-779.

3.Gouin, T. and T. Harner, *Modelling the environmental fate of the polybrominated diphenyl ethers.* Environment International, 2003. **29**(6): p. 717-724.

4.Hale, R.C., et al., *Polybrominated diphenyl ether flame retardants in Virginia freshwater fishes (USA).* Environmental Science & Technology, 2001. **35**(23): p. 4585-4591.

5.Vonderheide, A.P., M. Montes-Bayon, and J.A. Caruso, *Development and application of a method for the analysis of brominated flame retardants by fast gas chromatography with inductively coupled plasma mass spectrometric detection.* Journal of Analytical Atomic Spectrometry, 2002. **17**(11): p. 1480-1485.

6.Ohio-EPA, *Sewage Sludge Use or Disposal in Ohio -Calendar Year 2002.* 2002, State of Ohio Environmental Protection Agency: Ohio, USA.

7.Reick, R.H., *Polybrominated Diphenyl Ethers Analysis in Fish Tissue and other Matrices by GC-ECD.* LCGC North America, 2004. **22**(9): p. 914-925.

8.North, K.D., *Tracking polybrominated diphenyl ether releases in a wastewater treatment plant effluent, Palo Alto, California.* Environmental Science & Technology, 2004. **38**(17): p. 4484-4488.

9.Oberg, K., K. Warman, and T. Oberg, *Distribution and levels of brominated flame retardants in sewage sludge.* Chemosphere, 2002. **48**(8): p. 805-809.

10.de Boer, J., et al., *Polybrominated diphenyl ethers in influents, suspended particulate matter, sediments, sewage treatment plant and effluents and biota from the Netherlands.* Environmental Pollution, 2003. **122**(1): p. 63-74.

11.Christensen, J.H., et al., *Polybrominated diphenyl ethers (PBDEs) in marine fish and blue mussels from southern Greenland.* Chemosphere, 2002. **47**(6): p. 631-638.

12.Matscheko, N., et al., *Application of sewage sludge to arable land-soil concentrations of polybrominated diphenyl ethers and polychlorinated dibenzo-p-dioxins, dibenzofurans, and biphenyls, and their accumulation in earthworms.* Environmental Toxicology and Chemistry, 2002. **21**(12): p. 2515-2525.

13.Ikonomou, M.G., et al., *A comprehensive multi-residue ultra-trace analytical method, based on HRGC/HRMS, for the determination of PCDDs, PCDFs, PCBs, PBDEs, PCDEs, and organochlorines pesticides in six different environmental matrices. Report 2389*, in *Canadian Technical Report of Fisheries and Aquatic Sciences*. 2001.

14.ASTM, *Standard Test Method for Determining a Sorption Constant (Koc) for an Organic Chemical in Soil and Sediments*, in *American Standards for Test Methods*. 1985.

Assessment of the pollutant transports parameters in barrier materials: Performance of GCL compared to clay liner.

Dr. ALIMI-ICHOLA Ibrahim. Laboratoire URGC-Géotechnique, INSA-LYON, VILLEURBANNE, FRANCE

Abstract

The manufactured sealing materials become current in environmental protection plant. The performance of these materials concerns specially their permeability. The tests described in this study show that the clay liner can have a better performance because of its retention capacity and its chemical stability. The long time flow through the GCL increases its permeability. It is shown that the barrier system made up of clay delays the flow of pollutants better and its effective dispersion coefficient has little variation during the leachate percolation. It is necessary to reduce the stay time on the GCL by efficient drainage system above it. The pollutant flux delay can be insured by increasing the barrier system thickness.

Introduction

The realization of the tight barriers saw the development of the manufactured products whose performance must be higher than natural materials. This performance is especially judged by only one parameter which is the coefficient of permeability of the product. But to prevent subsoil and ground water pollution, that not only the leachate flow which must be reduced, the reduction of the solute transported by the leachate must be taken into account. The solute flux reduction depends on the sorption-desorption phenomenon but it is shown that the barrier thickness delays its transport (Alimi-Ichola and Gaidi, 2005). So the material performance can be assessed by monitoring the solute concentration in the outflow leachate.

After the Geosynthetic Clay Liner (GCL) and the clay are installed between two layers of clayey sand (SC) as the barrier systems of waste containment, the obtained columns are percolated by tip water until saturation. The saturation process is controlled by TDR measurements. So, after the soil saturation, the columns are percolated by lead slag leachate.

The hydraulic performance of the two materials is examined by comparing the infiltration curves and by measuring the saturation delay during water percolation. To study the material permeability during leachate percolation, the outflow rate is determined. The influence of the material on the leachate is highlighted by measuring the electrical conductivity of the material and of the percolated leachate. The increase of the solute concentration is computed with the calibration relationship between scaled solute concentration and scaled leachate electrical conductivity (Gaidi and Alimi-Ichola, 2003).

Materials and experimental method

The infiltration tests are carried out on columns built with PEHD rings of 50mm of height and 100.6 mm of diameter. Each test column is composed by three rings and the clay and the GCL are placed in the mid ring as it is shown on the Figure 1. The moisture movement and the soil electrical conductivity are monitored with TDR probes in the two upper rings as it is shown on the schematic drawing (Fig.1). Three rod TDR probes are used.

The geotechnical characteristics of the clayey sand and the Gault clay are embedded on the Table 1. The GCL used is the "Saline 100" and the chemical properties of the used lead slag leachate are presented on Table 2.

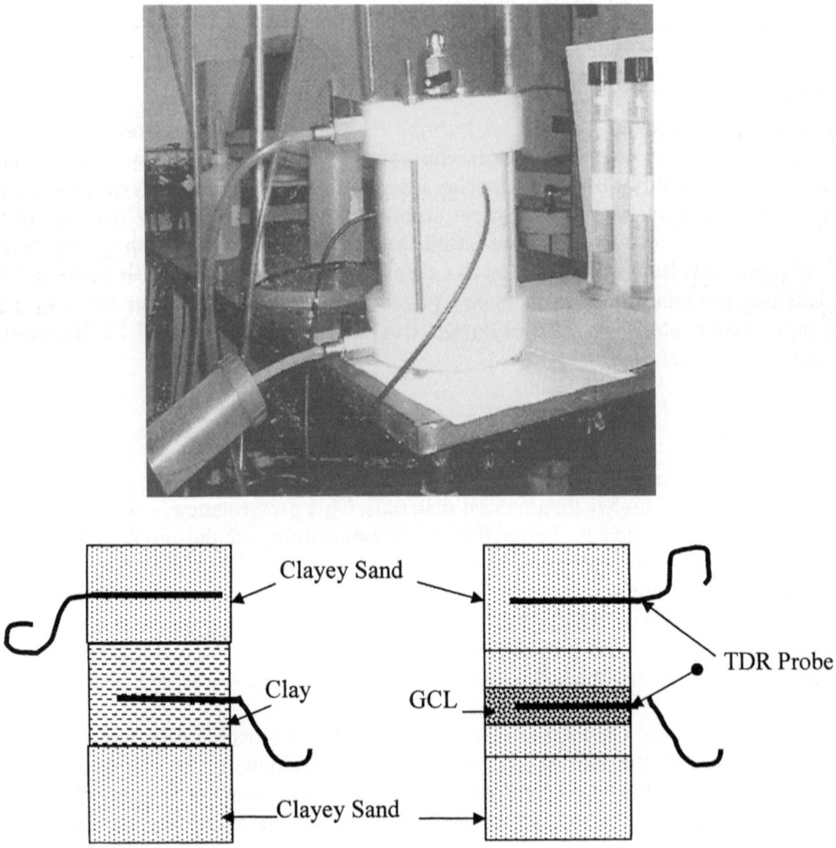

Figure 1 Infiltration column picture and schematic of the barrier systems (Clay and GCL)

Before leachate percolation tests are carried out, the two columns is saturated with tip water by infiltration under a pressure head of 1m of water given by a Mariotte burette. The saturation procedure is stopped when the outflow volume reaches three times the clay pore volume after the steady flow occurs. So, the tip water in the Mariotte burette is replaced by the leachate. The infiltration head pressure remains 1m of water. Infiltrated volume and TDR curves are collected to compute the infiltration curves, the soil water content and the soil electrical conductivity.

Table 1 Soil geotechnical characteristics

Soil	Fines (%<80μm) (%)	Clay (%<2μm) (%)	LL (%)	IP (%)	Optimal dry density	Optimal water content (%)
Clayey sand	71	22	30	19	1.77	12.5
Gault Clay	100	69	40	21	1.85	17.5

Table 2 Lead slag leachate chemical properties

pH	Electrical Conductivity	Solute Concentration
6.3	5330 mS/m	75446 mg/l

Experimental results

To evaluate pollutant reduction during leachate percolation, it is needed to compute the infiltration rate, the solute retention level and the outflow rate. It is noticed first that the saturation process is longer in the barrier system SC-Clay-SC than in the system SC-GCL-SC. The saturation duration is 69 days higher for SC-Clay-SC system than for SC-GCL-SC system. After the steady outflow rate is reached, it is found that the Darcy permeability is hardly the same for the two systems. So the leachate influence on the barrier systems can be observed on the infiltration curves.

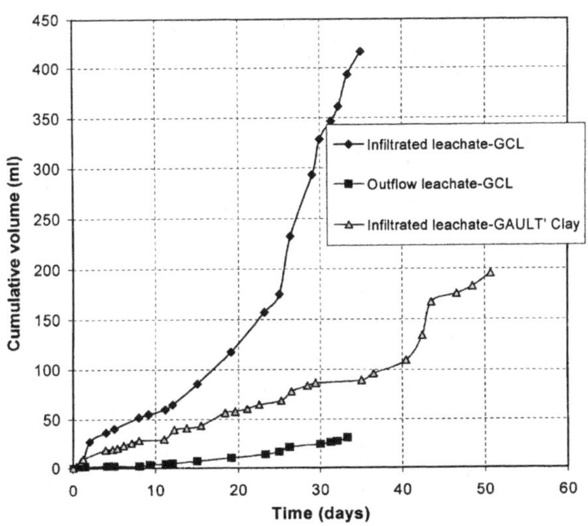

Figure 2 Leachate infiltration curves in the systems SC-GCL-SC and SC-Clay-SC

Infiltration curve analysis

It is shown in the Figure 2 that the infiltration rate is high in prior time in the system SC-GCL-SC and changes three times for the two systems during the leachate infiltration. The second change in infiltration rate occurs at the beginning of the outflow of the leachate. The rate reaches six times the value of the rate of the prior steady infiltration flow. The second rate change appears after 25 days in the system SC-GCL-SC and after 42 days in the system SC-Clay-SC. The steady outflow rate remains higher in the system SC-GCL-SC than in the system SC-Clay-SC (Table 3). The barrier system SC-Clay-SC reduces more the solute transport by reducing leachate flow. This increase of the infiltration flow proves that the chemical resistance of the system SC-Clay-SC is higher than one of the system SC-GCL-SC.

Table 3 Infiltration Rate change in the two barrier systems

Barrier System	1st Rate (m/s)	2nd Rate (m/s)	3rd Rate (m/s)
SC-GCL-SC	$5.09\ 10^{-9}$	$1.25\ 10^{-8}$	$3.12\ 10^{-8}$
SC-Clay-SC	$3.64\ 10^{-9}$	$2.63\ 10^{-8}$	$0.71\ 10^{-8}$

Water content evolution analysis and permeability assessment

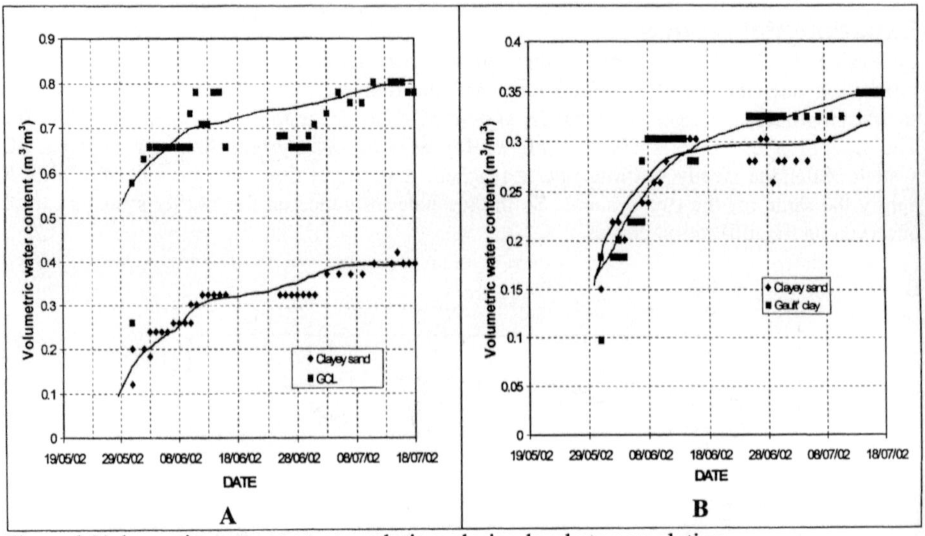

Figure 3 Volumetric water content evolutions during leachate percolation

For the accuracy of the TDR measurements, a new polynomial relationship is determined using five different soils and two bentonites (Gaidi and Alimi-Ichola, 2000). It is shown on Figure 3A, when the infiltration of the leachate began on June 28th , that the volumetric water content reached the equilibrium value in the clayey sand and in the GCL. The same state is observed in the SC-Clay-SC when the leachate infiltration started on July 16 (Fig.3B). The equilibrium water content is hardly the same in the clayey sand and in the Gault clay. Since the equilibrium water content is reached in the two systems, the infiltration rate can be used to compute the Darcy permeability coefficient.

Table 4 Permeability value in the two barrier systems

Barrier System	1st Coefficient (m/s)	2nd Coefficient (m/s)	3rd Coefficient (m/s)
SC-GCL-SC	$6.64\ 10^{-10}$	$1.63\ 10^{-9}$	$4.07\ 10^{-9}$
SC-Clay-SC	$4.75\ 10^{-10}$	$3.43\ 10^{-9}$	$0.92\ 10^{-9}$

It is noticed in table 4 that the coefficients of permeability of the two systems remain close during the prior infiltration time. The second change in the infiltration rate occurs at the beginning of the outflow of the leachate. The permeability of the system SC-GCL-SC increases during the infiltration test. One of the system SC-Clay-SC was not monotone. It increases and decreases during the leachate percolation. It seems that the first value of the coefficient of the permeability of the two systems concerns the pore water movement. The increase of the permeability coefficient can be attributed to the leachate chemical action on the soil system. It is noticed that the leachate action is more marked on the system SC-GCL-SC than on the system SC-Clay-SC since the SC-Clay-SC system permeability decreases after the 3rd change (Table 4).

Solute retention capacity of the GCL and of the clay in the two systems

It is shown by Meddahi et al. (1993) that the relative retained solute mass (ratio of retained solute and the total mass of solute in the leachate) can be estimate by measuring the soil impedance. The following relation is used:

$$M_{L,R}(t) = \left[\frac{\left(R_{L,t}\right)^{-1} - \left(R_{L,t_i}\right)^{-1}}{\left(R_{L,t_o}\right)^{-1} - \left(R_{L,t_i}\right)^{-1}} \right] \qquad (1)$$

where :

$R_{L,ti}$ the initial soil impedance over the depth L, $R_{L,to}$. the minimum value of the soil impedance and $R_{L,t}$ the soil impedance at the instant t. The values of $R_{L,ti}$ and $R_{L,to}$ are given in the table 5.

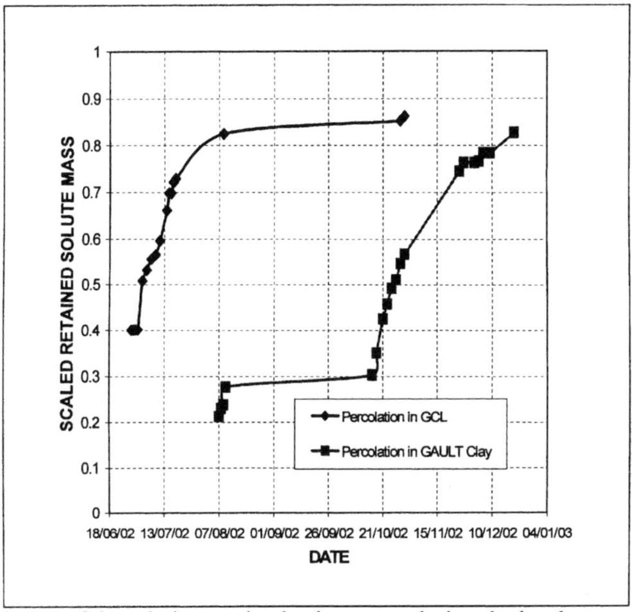

Figure 4 Evolution of the relative retained solute mass during the leachate percolation

In the Figure 4 it is shown that 86% of the capacity of the GCL to retain solute is reached after 120 days. The impedance change at the depth of 75mm starts after three days of the infiltration. No delay was noticed on solute migration in the SC-GCL-SC system. The evolution of the relative mass of solute on the depth of 75mm is fast during the first four days after the beginning the leachate infiltration in the system SC-Clay-SC. Then no variation of retained solute is observed during the following 67days until the outflow of the leachate begins. After this duration the retention process restarts with high rate. 65 days are needed to reach 83% of the retention capacity. This percentage (83%) of the retention capacity is reached in the system SC-GCL-SC in 39days. The shape of the two curves proves that the retention process continues in the system SC-Clay-SC while the retention threshold is reached in the system SC-GCL-SC.

Table 5 Initial and minimal values of the impedance

Soil	$R_{L,ti}$ (Ω)	$R_{L,to}$ (Ω)
GCL Bentonite	50	12.6
Gault Clay	136	19.5

The clay system delays more the leachate flux and the solute migration. This result is in accordance with the decrease of the infiltration rate in the system SC-Clay-SC during the leachate percolation. The solute migration delay will be confirmed by the evolution of the solute concentration in the percolated leachate.

Evolution of the solute concentration in percolated leachate : apparent dispersion coefficient

Gaidi and Alimi-Ichola (2003) have computed a relationship between leachate solute concentration and leachate electrical conductivity. The equation (2) gives the scaled value of the solute concentration (FS/FS_0) when the scaled value of the electrical conductivity of the leachate (σ/σ_0) is measured. FS_0 is the initial value of the solute concentration in the leachate and σ_0 its electrical conductivity. Measuring the electrical conductivity of the leachate during the percolation of the barrier system, the change in leachate concentration is computed using the relation (2).

$$\frac{FS}{FS_0}=5.3747\left(\frac{\sigma}{\sigma_0}\right)^4-8.8419\left(\frac{\sigma}{\sigma_0}\right)^3+4.1908\left(\frac{\sigma}{\sigma_0}\right)^2+0.4095\left(\frac{\sigma}{\sigma_0}\right)-0.0159 \quad (2)$$

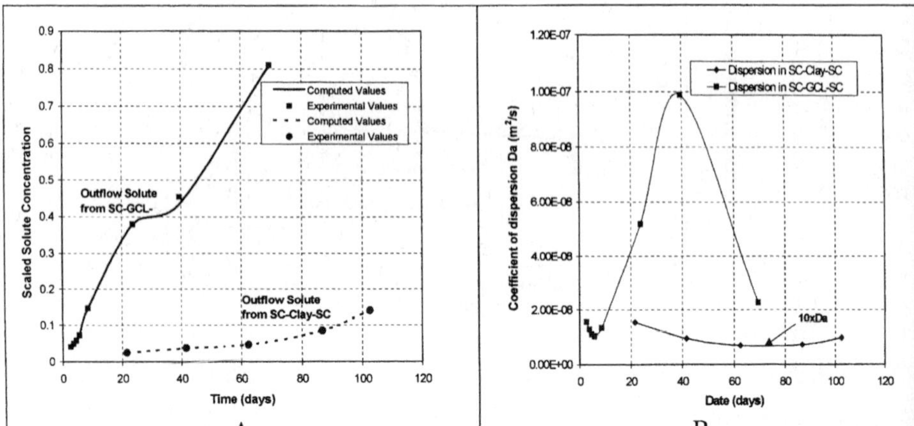

Figure 5 Solute concentrations in outflow leachate and estimated effective dispersion coefficient

Figure 5A presents the variation of the leachate solute concentration during the percolation of the two barrier systems. For the system SC-GCL-SC, 81% of the solute concentration appears in the percolated leachate after 66days. After 80days, the solute concentration in the leachate is only 14% the solute concentration is measured after the percolation of the system SC-Clay-SC. This result confirms the fact that the retention capacity of the system SC-Clay-SC is higher than one of the system SC-GCL-SC. Only 50mm of the clay of the Gault can retain more solute than the GCL bentonite.

Mermoud (1978) have shown that the solute concentration profile can be expressed by the relation (3) when the concentration is maintained constant at one end of a semi infinite column.

$$\frac{FS}{FS_0} = \frac{1}{2}\left[erfc\left(\frac{z-ut}{\sqrt{4Dt}} \right) \right] = \frac{1}{\sqrt{\pi}} \int_{\frac{z-ut}{\sqrt{4Dt}}}^{\infty} e^{-S^2} ds \qquad (3)$$

where FS is the solute concentration at the instant t, u is the Darcy pore rate and D the effective dispersion coefficient in non reactive porous media.

Gaidi and Alimi-Ichola (2002) have shown that the solute migration can be represented by the Cl^- migration with no delay (delay coefficient equal1). So using the relation (3), le variation of the effective dispersion coefficient is computed. The results are shown in the Figure 5B. It is noticed that the effective dispersion coefficient is 100 times higher in the system SC-GCL-SC than in the system SC-Clay-SC. The effective dispersion coefficient increases and decreases during the leachate percolation through the system SC-GCL-SC. In the system SC-Clay-SC, the effective dispersion coefficient has little variation during the percolation. The SC-Clay-SC barrier system is chemically stable.

In Figure5A are presented the curves of the solute concentration computed with the model of the effective dispersion coefficient represented on the figure 5B. It is noticed that the experimental points fitted well the computed curves.

Conclusion

Comparing clayey sand and clay layers behaviour (Alimi-Ichola and Gaidi, 2004), it is found that the increase of the layer thickness does not compensate the increase of the permeability coefficient of the sealing material. This study proves that, with the same thickness of barrier systems which have the coefficient of permeability on the same range, chemical action can reduce the barrier efficiency.

The retention capacity is more insured by the system SC-Clay-SC because of the sealing material thickness. It seems that the apparent height of the immobile water phase in the clay delays the solute migration. The tests performed on the two materials (GCL and clay) prove that the barrier thickness is an important parameter on which it must be taken into account for barrier design.

The chemical action on the barrier depends on the leachate stay time. It is necessary to reduce this time by the installation of efficiency drainage system above the sealing material. A natural material (loam or silt) which permeability is not low enough can be used on the GCL to delay the convective movement through the GCL.

Regarding to these results, long time percolation tests must be performed to highlight the chemical action of the leachate on the barrier system to evaluate sealing material performance. Indeed, the pollutant flux can be reduced during the first time of the life of the system and the increase of the pollutant flux can become non compatible with the environment if the thickness is not sufficient and if the drainage system is not efficiency.

References

ALIMI-ICHOLA I. and **GAIDI L.** Influence of the unsaturated zone of soil layer on solute migration, CR of 4th BGA Environmental Engineering Conference, Stratford Upon Avon UK,Thomas Telford, June 2004 p.3-10

GAIDI L. et ALIMI ICHOLA I., Etude des caractéristiques hydrodynamiques des sols par la méthode TDR (Bull. Eng. Geol.Env. (2000) vol.59 pp247-255) A48439

GAIDI L. I. ALIMI-ICHOLA, Study of the Hydraulic Behaviour of Clay Geosynthetic Barrier Subjected to the Leachate Infiltration, Proc.IS Nuremberg 2002 April 2002 Nurnberg Germany. pp. 233-246

GAIDI L. et ALIMI ICHOLA I., Utilisation des sondes TDR pour estimer la masse de soluté retenue par une couche d'argile lors d'une infiltration (Bull.Eng.Geol. Env.(2003) 62:47-56)

MEDDAHI M.E., D. MALLANTS, J. FEYEN, M. BADJI and H. VEREECKEN. 1993. Paramètres essentiels d'écoulement des solutés dans un milieu poreux saturé : Méthode de mesure et modèle d'interprétation, *Science du sol,* 31, (4), 233-250.

MERMOUD, A., (1978), "Propagation de substances miscibles en milieu poreux. Etude théorique et expérimentale", Bulletin Technique de la Suisse Romande, 23, p. 341-345.

Numerical modelling of contaminant transport in unsaturated soils

A. A. Javadi, Department of Engineering, University of Exeter, Exeter, U.K.
M. M. AL-Najjar, Department of Engineering, University of Exeter, Exeter, U.K.
A. S. I. Elkassas, Department of Engineering, University of Exeter, Exeter, U.K.

Abstract

This paper presents the development and validation of a numerical model for simulation of the flow of air and water and contaminant transport through unsaturated soils. The governing differential equations include two mass balance equations for the water phase and air phase together with a balance equation for contaminant transport through the two phases. In the model the nonlinear system of the governing differential equations is solved using a finite element method in the space domain and a finite difference scheme in the time domain. The governing equations of the miscible contaminant transport including advection, dispersion, and diffusion effects are presented. The mathematical framework and the numerical implementation of the model are described in detail. The model is validated by application to standard experiments on contaminant transport in unsaturated soils. Finally, the merits and limitations of the model are highlighted.

Keywords: contaminant transport, unsaturated soil, finite element method

Introduction

In recent years, interest in understanding the mechanisms and prediction of contaminant transport through soils has dramatically increased because of growing evidence and public concern that the quality of the subsurface environment is being adversely affected by industrial, municipal and agricultural waste. In assessing the environmental impacts of waste discharges, engineers seek to predicate the impact of emission on contaminant concentration in nearby air and water (Nazaroff and Alvarez-Cohen, 2001). The high costs, large time scales and lack of control over the boundary conditions have prevented the development of field scale experiments (Hellawell and Sawidou, 1994). In this paper, a numerical model is presented for simulation of contaminant transport in unsaturated soils. The main governing phenomena of the miscible contaminant transport including advection, mechanical dispersion, and molecular diffusion are considered. The contaminant transport equation together with the balance equations for flow of air and water are solved numerically using the finite element method, subject to prescribed initial and boundary conditions. The developed model is validated by comparing the model prediction results with those available in the literature.

Governing equations of fluid flow and contaminant transport in soil

The introduction outlined a broad range of issues that are of interest in the unsaturated zone. Unsaturated zone is a multiphase system, because at least two fluid phases are present: air and water. The governing equations that describe fluid flow and contaminant transport in the unsaturated zone will be presented in this section.

Modelling of water flow

The governing differential equation for water flow is based on the conservation of mass of the ground water, leading to:

$$C_{ww}\frac{\partial u_w}{\partial t}+C_{wa}\frac{\partial u_a}{\partial t}=\nabla[K_{ww}\nabla u_w]+\nabla[K_{wa}\nabla u_a]+J_w \tag{1}$$

where,

$$C_{ww}=C_{fw}+C_{vw}\ ;\ C_{wa}=C_{fa}+C_{va}\quad C_{vw}=nS_aK_{fw}\ ;\ c_{va}=nS_aK_{fa}\ ;\ C_{fw}=-n(\rho_w-\rho_v)\frac{\partial S_w}{\partial s}\ ;$$

$$C_{fa}=n(\rho_w-\rho_v)\frac{\partial S_w}{\partial s}\quad ;\quad K_{ww}=\frac{\rho_w K_w}{\gamma_w}+K_{vw}\rho_w\quad ;\quad K_{wa}=\rho_v K_a+\rho_w K_{va}\quad ;$$

$$K_{fa}=\rho_0\frac{\partial h}{\partial\psi}\frac{\partial\psi}{\partial s}\nabla u_a\quad ;\quad K_{fw}=-\rho_0\frac{\partial h}{\partial\psi}\frac{\partial\psi}{\partial s}\nabla u_w\quad ;\quad K_{vw}=-\frac{D_{atms}V_v n}{\rho_w}K_{fw}\quad ;$$

$$K_{va}=-\frac{D_{atms}V_v n}{\rho_w}K_{fa}\ ;\ J_w=\rho_w\nabla(K_w\nabla z)$$

in which n is the porosity of the soil, K_w is the conductivity of water $[L][T]^{-1}$, K_a is the conductivity of air $[L][T]^{-1}$, S_w is the degree of saturation of water, S_a is the degree of saturation of air, ρ_w is the density of water $[M][L]^{-3}$, ρ_v is the density of water vapor $[M][L]^{-3}$, ρ_0 is the density of saturated soil vapor $[M][L]^{-3}$, s is the soil suction $[M][L]^{-1}[T]^{-2}$, V_v is the mass flow factor, u_w is the pore-water pressure $[M][L]^{-1}[T]^{-2}$, u_a is the pore-air pressure $[M][L]^{-1}[T]^{-2}$, D_{atms} is the molecular diffusivity of vapor through air, γ_w is the unit weight of water, ψ is the capillary potential, h is the relative humidity and ∇z is the unit normal oriented downwards in the direction of the force of gravity.

Modelling of air flow

The governing differential equation for air flow is based on the conservation of mass of the ground air, leading to:

$$C_{aw}\frac{\partial u_w}{\partial t}+C_{aa}\frac{\partial u_a}{\partial t}=\nabla[K_{aw}\nabla u_w]+\nabla(K_{aa}\nabla u_a)+J_a \tag{2}$$

where,

$$C_{aw}=C_{aw1}+C_{aw2}\ ;\ C_{aa}=C_{aa1}+C_{aa2}\ ;\ C_{aw1}=-n\rho_{da}(H_a-1)\frac{\partial S_w}{\partial s}\ ;\ C_{aa1}=n\rho_{da}(H_a-1)\frac{\partial S_w}{\partial s}\ ;$$

$$C_{aw2}=n(S_a+H_aS_w)C_{daw}\quad ;\quad C_{aa2}=n(S_a+H_aS_w)c_{daa}\quad ;\quad C_{daw}=-\frac{R_v}{R_{da}}K_{fw}\quad ;$$

$$C_{daa}=\frac{1}{R_{da}T}-\frac{R_v}{R_{da}}K_{fa}\ ;\ K_{aw}=\frac{H_a\rho_{da}}{\gamma_w}K_w\ ;\ K_{aa}=K_a\nabla u_w\ ;\ J_a=H_a\rho_{da}\nabla(K_w\nabla z)$$

in which H_a is the Henry's volumetric coefficient of solubility, ρ_{da} is the density of dry air $[M][L]^{-3}$, R_{da} is the specific gas constant for dry air and R_v is the specific gas constant for water vapor.

Modelling of Contaminant Transport

In porous media, contaminant transport occurs mainly by various processes including advection, diffusion and mechanical dispersion.

Advection

Advection is the transport of material caused by the net flow of the fluid in which that material is suspended. Whenever a fluid is in motion, all contaminants in the flowing fluid, including both molecules and particles, are advected along with the fluid (Nazaroff and Alvarez-Cohen, 2001). The rate of contaminant transport that occurs by advection is given by the product of contaminant concentration c and the component of apparent groundwater velocity like v_x, v_y, etc. For two dimensional cases with two components of the contaminant transport in the x and y directions, the rate of contaminant transport due to advection will be (Sethi, 1995):

$$F_{x,advection} = v_{w_x} c \tag{3a}$$

$$F_{y,advection} = v_{w_y} c \tag{3b}$$

Diffusion

The process by which contaminants are transported by the random thermal motion of contaminant molecules is called diffusion (Yong at al., 1992). The rate of contaminant transport that occurs by diffusion is given by Fick's law. In terms of two components of the contaminant transported in the x and y directions, the rate of contaminant transported by diffusion will be (Sethi, 1995):

$$F_{x,diffusion} = -D_m \frac{\partial c}{\partial x} \tag{4a}$$

$$F_{y,diffusion} = -D_m \frac{\partial c}{\partial y} \tag{4b}$$

where, D_m is the molecular diffusion coefficient in the porous medium $[L]^2[T]^{-1}$.

Mechanical dispersion

Mechanical dispersion is a mixing or spreading process caused by small scale fluctuation in groundwater velocity along the tortuous flow paths within individual pores. In terms of two components of contaminant transport in the x and y directions, the rate of contaminant transported by mechanical dispersion is given by:

$$F_{x,dispersion} = -D_{w_{xx}} \frac{\partial}{\partial x}(\theta_w c) - D_{w_{xy}} \frac{\partial}{\partial y}(\theta_w c) \tag{5a}$$

$$F_{y,dispersion} = -D_{w_{yx}} \frac{\partial}{\partial x}(\theta_w c) - D_{w_{yy}} \frac{\partial}{\partial y}(\theta_w c) \tag{5b}$$

where $D_{w_{xx}}$, $D_{w_{xy}}$, $D_{w_{yx}}$, $D_{w_{yy}}$ are coefficients of dispersivity tensor $[L][T]^{-1}$ for the water phase and θ_w is the volumetric water content. The amount of contaminant in each phase (α) is denoted by the mass fraction, which is defined as:

$$w_\alpha = \text{Mass of contaminant in phase } (\alpha) \,/\, \text{Mass of phase } (\alpha) \tag{6}$$

The standard mass balance equation of contaminant transport can then be written as:

$$\frac{\partial(\theta_w \rho_w \omega_w)}{\partial t} + F_{,advection} - F_{,dispersion-diffusion} + \lambda_w \theta_w \rho_w \omega_w = F^w \tag{7}$$

where λ_w is the reaction rate for water $[T]^{-1}$ and F^w is the source/sink term for water $[M][L]^{-3}[T]^{-1}$. Rather than using the contaminant mass fraction it is often more convenient to use the volumetric concentration (c_α) which is defined as:

c_α = Mass of contaminant in phase (α) / Volume of phase (α) $= \rho_\alpha \omega_\alpha$ (8)

The governing equation can then be rewritten in terms of c by substituting the above relation into the governing equation:

$$\frac{\partial(\theta c)}{\partial t} + \nabla(vc) - \nabla(D\nabla c) + \lambda c = F^w \tag{9}$$

Numerical solution

Numerical solution of governing differential equation for water and air flow

The governing differential equations for water flow (1) and air flow (2), as defined in the previous section, have two variables u_w and u_a; these variables are primary unknowns. The primary unknowns can be approximated using the shape function approach for an element as:

$$u_w = \hat{u}_w = \sum_1^n N_s u_{ws} \; ; \; u_a = \hat{u}_a = \sum_1^n N_s u_{as}$$

where, N_s is the shape function, u_{ws} is the nodal pore-water pressure, u_{as} is the nodal pore-air pressure and n is the number of nodes in element. Replacing the primary unknowns with shape functions approximations of above equations, equation (1) and (2) can be written as:

$$\nabla[K_{ww}\nabla\hat{u}_w] + \nabla[K_{wa}\nabla\hat{u}_a] + J_w - C_{ww}\frac{\partial\hat{u}_w}{\partial t} - C_{wa}\frac{\partial\hat{u}_a}{\partial t} = R_{\Omega w} \tag{10}$$

$$\nabla(K_{aw}\nabla\hat{u}_w) + \nabla(K_{aa}\nabla\hat{u}_a) + J_a - C_{aw}\frac{\partial\hat{u}_w}{\partial t} - C_{aa}\frac{\partial\hat{u}_a}{\partial t} = R_{\Omega a} \tag{11}$$

where, $R_{\Omega w}$ and $R_{\Omega a}$ are the residual errors introduced by the approximation functions for water and air respectively. A Galerkin finite element scheme is applied to the spatial terms. Employing the Galerkin weighted residual approach to minimise the residual error represented by equation (10) or (11) and integrating the equation over the spatial domain (Ω^e), the spatial discretisation of governing differential equation for water flow can be written as:

$$C_{ww}\frac{\partial u_{ws}}{\partial t} + C_{wa}\frac{\partial u_{as}}{\partial t} + K_{ww}u_{ws} + K_{wa}u_{wa} = f_w \tag{12}$$

where,

$$C_{ww} = \sum_{e=1}^n \int_{\Omega_e} [N^T C_{ww} N] d\Omega_e \; ; \; C_{wa} = \sum_{e=1}^n \int_{\Omega_e} [N^T C_{wa} N] d\Omega_e \; ; \; K_{ww} = \sum_{e=1}^n \int_{\Omega_e} [\nabla N^T (K_{ww} N)] d\Omega_e \; ;$$

$$K_{wa} = \sum_{e=1}^n \int_{\Omega_e} [\nabla N^T (K_{wa} N)] d\Omega_e \; ;$$

$$f_w = \sum_{e=1}^n \int_{\Omega_e} [\nabla N^T (K_w \rho_w \nabla z)] d\Omega_e - \sum_{e=1}^n \int_{\Gamma_e} N_r \{\rho_w \hat{v}_{wn} + \rho_w \hat{v}_{vd} + \rho_w \hat{v}_{va}\} . d\Gamma^e$$

in which $\hat{v}_{wn}$ is the approximate water velocity normal to the boundary surface, $\hat{v}_{vd}$ is the approximated diffusive vapour velocity normal to the boundary surface, $\hat{v}_{va}$ is the approximate pressure vapour velocity normal to the boundary surface and Γ^e is the element boundary surface. Similarly, the spatial discretisation of governing differential equation for air flow leads to:

$$C_{aw}\frac{\partial u_{ws}}{\partial t} + C_{aa}\frac{\partial u_{as}}{\partial t} + K_{aw}u_{ws} + K_{aa}u_{as} = f_a \tag{13}$$

where,

$$C_{aw} = \sum_{e=1}^{n} \int_{\Omega_e} \left[N^T C_{aw} N \right] d\Omega_e \quad ; \quad C_{aa} = \sum_{e=1}^{n} \int_{\Omega_e} \left[N^T C_{aa} N \right] d\Omega_e \quad ; \quad K_{aw} = \sum_{e=1}^{n} \int_{\Omega_e} \left[\nabla N^T (K_{aw} \nabla N) \right] d\Omega_e \quad ;$$

$$K_{aa} = \sum_{e=1}^{n} \int_{\Omega_e} \left[\nabla N^T (K_{aa} \nabla N) \right] d\Omega_e \quad ;$$

$$f_a = \sum_{e=1}^{n} \int_{\Omega_e} \left[\nabla N^T (K_w \rho_{da} H_a \nabla z) \right] d\Omega_e - \sum_{e=1}^{n} \int_{\Gamma_e} N^T \rho_{da} \left\{ \hat{v}_{fn} + \hat{v}_{an} \right\} . d\Gamma^e$$

in which $\hat{v}_{fn}$ is the approximate velocity of free dry air and $\hat{v}_{an}$ is the approximate velocity of dissolved dry air. Spatially discretised equations for coupled flow of water and air can be combined in a matrix form as:

$$\begin{bmatrix} K_{ww} & K_{wa} \\ K_{aw} & K_{aa} \end{bmatrix} . \begin{bmatrix} u_{ws} \\ u_{as} \end{bmatrix} + \begin{bmatrix} C_{ww} & C_{wa} \\ C_{aw} & C_{aa} \end{bmatrix} . \begin{bmatrix} \dot{u}_{ws} \\ \dot{u}_{as} \end{bmatrix} - \begin{bmatrix} f_w \\ f_a \end{bmatrix} = 0 \tag{14}$$

where,

$$\dot{u}_{ws} = \frac{\partial u_{ws}}{\partial t} \ ; \ \dot{u}_{as} = \frac{\partial u_{as}}{\partial t}$$

A time discretisation of equation (14) is achieved here by application of a fully implicit mid interval backward difference algorithm. Applying a finite difference scheme to equation (14) will result in:

$$A^{\left(\frac{n+1}{2}\right)} (\phi^{n+1}) + B^{\left(\frac{n+1}{2}\right)} \left[\frac{\phi^{n+1} - \phi^n}{\Delta t} \right] + C^{\left(\frac{n+1}{2}\right)} = 0 \tag{15}$$

where, $A = \begin{bmatrix} K_{ww} & K_{wa} \\ K_{aw} & K_{aa} \end{bmatrix}$; $B = \begin{bmatrix} C_{ww} & C_{wa} \\ C_{aw} & C_{aa} \end{bmatrix}$; $C = \begin{bmatrix} f_W \\ f_a \end{bmatrix}$; $\phi = \begin{bmatrix} u_{ws} \\ u_{as} \end{bmatrix}$

Equation (15) can be further simplified to:

$$\phi^{n+1} = \left[A^{\left(\frac{n+1}{2}\right)} + \frac{B^{\left(\frac{n+1}{2}\right)}}{\Delta t} \right] \left[\frac{B^{\left(\frac{n+1}{2}\right)} (\phi^n)}{\Delta t} - C^{\left(\frac{n+1}{2}\right)} \right]$$

Numerical Solution of the Contaminant Transport Governing Equation

Ignoring the source and sink terms (F^w) equation (9) reduces to:

$$\frac{\partial(\theta c)}{\partial t} + \nabla(vc) - \nabla(D\nabla c) + \lambda c = 0 \tag{16}$$

The primary unknown can be approximated using the shape function approach as:

$$\theta c = \theta \hat{c} = \sum_{1}^{n} N_s \theta c_s \ ; \ c = \hat{c} = \sum_{1}^{n} N_s c_s$$

Replacing the primary unknowns with shape function approximation of, equation (16) can be written as:

$$\frac{\partial(\theta \hat{c})}{\partial t} + \nabla(v\hat{c}) - \nabla(D\nabla \hat{c}) + \lambda \hat{c} = R_{\Omega c} \tag{17}$$

where $R_{\Omega c}$ are the residual errors introduced by the approximation function. Employing the Galerkin weighted residual approach to minimise the residual error represented by equation (17) the discretised global finite element equation for single component of contaminant takes the form:

$$M\frac{dc}{dt}+Hc+F=0 \tag{18}$$

Applying a finite difference scheme to equation (18) will result in:

$$M\frac{(\theta c)^{n+1}-(\theta c)^{n}}{\Delta t}+H\left[(1-\gamma)c^{n}+\gamma c^{n+1}\right]+F^{n+1}=0 \tag{19}$$

where, Δt is the time step, γ is a parameter between 0 and 1 and n and $n+1$ stand for time levels (t_n and $t_{n+1}=t_n+\Delta t$). Solution of equations (15) and (19) will give the distribution of the contaminant concentrations at various points within the soil and at different times, taking into account the effects of air and water flow.

Numerical Example

The developed finite element model is applied to two contaminant transport problems from the literature (Li et al. 1999 and Karkuri et al. 1997). In the first example the transport of a contaminant through a bar, 30 m long and 1 m high, is considered. The bar is subjected to an initial Gaussian distribution of contaminant concentration of amplitude c=1 centered at x = 5.0 m (see Figure 1). A steady and uniform intrinsic velocity field with $v_w=1$ is assumed over the bar.

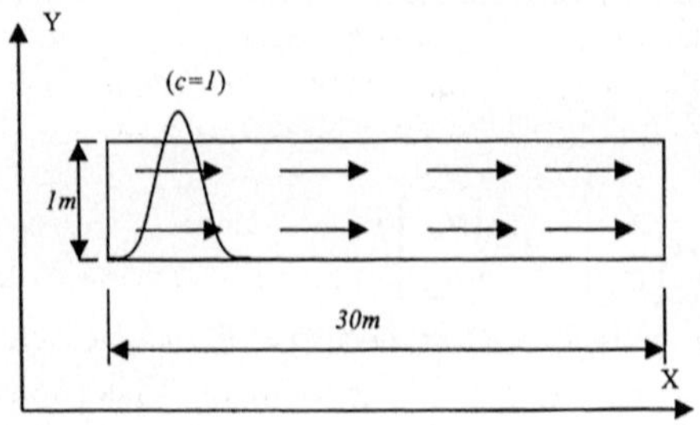

Figure 1. Problem definition.

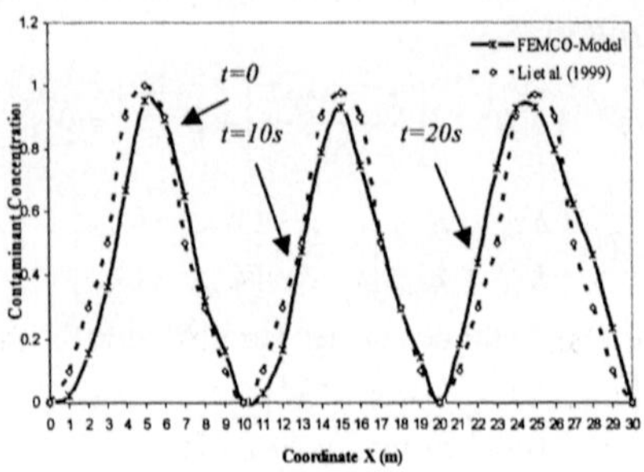

Figure 2. Concentration distribution at *t=0,* 10 and 20 *s* for advection.

In the first case, only the transport of the contaminant by advection is considered. Figure 2 shows the distribution of contaminant concentration along the bar at times *t=0, 10* and *20* seconds. It is shown that the results of the developed model are in close agreement with those reported by Li et al. 1999 for this example. To illustrate the capabilities of the model to predict the effects of other mechanisms of contaminant transport, another case involving the combined effect of advection and dispersion is considered and the result are shown in Figure 3. In this case, the same initial Gaussian distribution of concentration as in the previous case is considered together with additional dispersion and diffusion parameters including the longitudinal dispersivity for water phase $\alpha_{L_w}=0.5$, the coefficient of water molecular diffusion $D_{m_w}=0$ and the reaction rate for water $\lambda_w=0$. Figure 3 shows the distribution of contaminant concentration a long the bar due to the combined effect of advection and dispersion at times (t = 0, 5, 10 *and* 20 *seconds*). It is seen that the distribution of the contaminant concentration continuously decreased from the initial amplitude of ($c=1$) due to the effect of the addition of the dispersion mechanism. Figure 4 shows the distribution of contaminant concentration along the bar due to the combined effect of advection and dispersion at time *t=10* seconds. Again, it is shown that the results of the model are in very good agreement with those reported by Li et al. 1999 for this case.

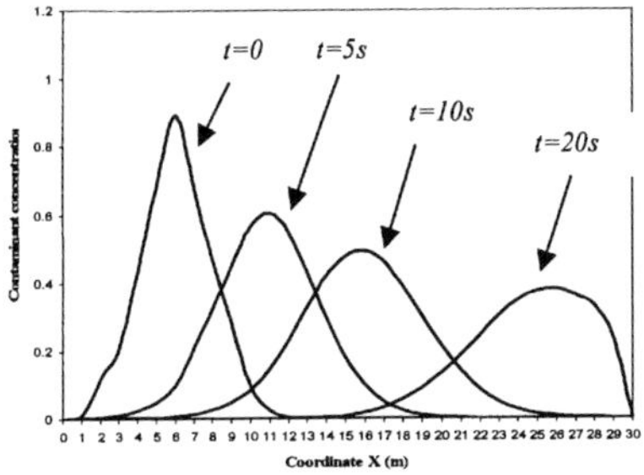

Figure 3. Contaminant concentration distribution at (t = 0, 5, 10 *and* 20s) for advection and dispersion.

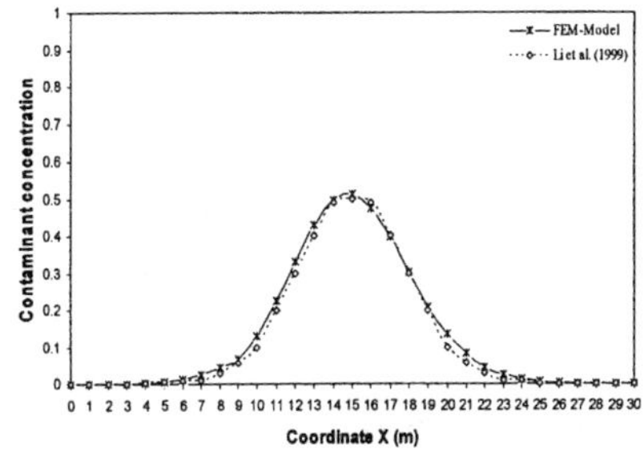

Figure 4. Concentration distribution at *t=10 s* for advection, dispersion and diffusion.

The second example involves application of the developed finite element model to another case involving pollutant transport through a saturated sandy soil column with a height of (2m) and a width of (0.02m) (Karkuri et al. 1997). The results are compared with those reported by Karkuri et al. 1997 as well as with those obtained using an analytical solution (Appelo and Postma, 1993). The loading system in this analysis involves the concentration of pollutant, *c*, picked up by the flow from the base of the column and distributed with time in the vertical direction. The pollutant influx has the form of a pulse entering at the bottom of the soil column and moving upward from the base under the effect of vertical groundwater advective velocity $v_w = 10^{-5}\, m/\sec$. The soil has a porosity of $n = 0.35$. The numerical analysis of the advection and dispersion of the pollutant in the vertical column with $D = 10^{-7}\, m^2/\sec$ and a time increment of $\Delta t = 2\times10^4$ sec is illustrated in Figure 5 in the form of concentration profiles at different times ($t = 2\times10^4$, 4×10^4, 6×10^4 to 1.8×10^5 sec). This figure indicates that the pollutant is flowing upwards in the column with the groundwater. The maximum concentration is decreasing gradually from its initial value of the infiltrating mixture ($c_o = 1 \quad mole/m^3$) due to the effect of advection and dispersion while the width of the pulse increases.

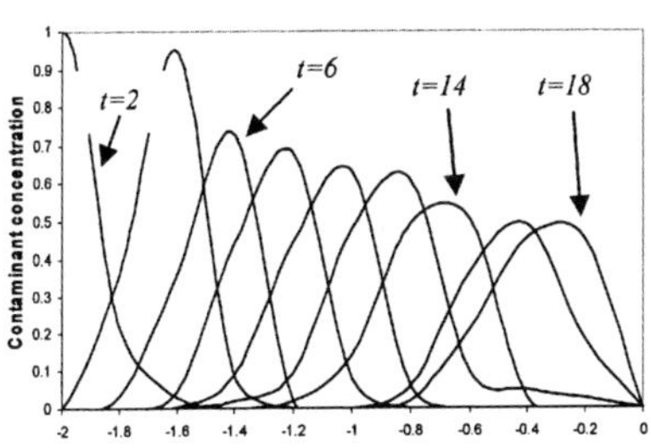

Figure 5. numerical advection-dispersion results

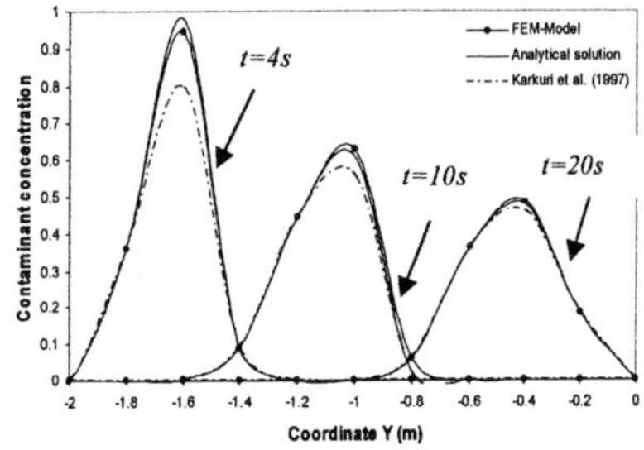

Figure 6. Comparison of numerical and analytical solutions

The results of the finite element analysis are compared with those of an analytical solution proposed by Appelo and Postma (1993) as:

$$c = \frac{M}{\sqrt{\pi D t}} \exp^{-\left(\frac{(y-y_o)^2}{4Dt}\right)} \tag{19}$$

where c is the tracer concentration in ($moles/m^3$), M, in moles per (m^2) of cross sectional area normal to the flow, is half of the total mass of the pollutant entering at ($y = 0$) during ($t \geq 0$), D is the coefficient of hydrodynamic dispersion in ($m^2/\sec$) and y_o is the location of the maximum concentration which is ($y_o = 0$) at ($t = 2\times10^4$) (Karkuri et al. 1997). The numerical and analytical solutions are shown for three times steps of ($t = 4\times10^4$, 1×10^5 and 1.6×10^5 sec) in Figure 6. Comparison of the results shows a very good agreement between the results of the numerical model and the analytical solution.

Conclusions

This paper presented a numerical model for predicting contaminant transport through soils. The model is capable of simulating several phenomena governing miscible contaminant transport in the soils including advection, dispersion and diffusion. A two dimensional finite element model was developed to solve the governing equation of contaminant transport together with the equations for air and water flow. The numerical results illustrated the performance of the presented model in simulating the effects of different phenomena governing the contaminant transport on the concentration distribution. The numerical model was validated by application to a number of case studies from literature. The results show that the developed numerical model is capable of predicting, with a very good accuracy, the effects of various mechanisms of contaminant transport through soils. This work is a part of a research project on developing a finite element model to simulate various aspects of contaminant transport in unsaturated soils.

References

1- Appelo,C. A. J. and Postma, D. 1993. Geochemistry, Groundeater and Pollution. A.A.Balkema, Rotterdam, Netherlands.

2- Hellawell, E. E. and Sawidou, C. 1994. A study of contaminant transport involving density driven flow and hydrodynamic clean up, Proc., of the Centrifuge '94 Conf., University of Cambridge, UK.

3- Karkuri, H.M. and Molenkamp, F. 1997. Analysis of advection dispersion of pollutant transport through a layered porous media. Proceeding of the International Conference on Geoenvironmental Engineering, Cardiff University, UK 193-198.

4- Li, X., Cescotto, S. and Thomas, H. R. 1999. Finite element method for contaminant transport in unsaturated soils. Journal of Hydrologic Engineering, 4(3), 265-274.

5- Nazaroff, W. W. and Alvarez-Cohen, L. 2001. Environmental Engineering Science. John Wiley & Sons, Inc. USA.

6- Radu, J. P., Biver, P., Charlier, R., and Cescotto, S. 1994. 2D and 3D finite element modeling of miscible pollutant transport in groundwater, below the unsaturated zone. Proc., Int. Conf. of Hydrodyn., ICHD'94, China Ocean Press, Beijing, China.

7- Sethi, H. S. 1995. A 3-dimensional finite element method for groundwater flow and contaminant transport. PhD Dissertation, Department of Mechanical Engineering, University of Nevada, Las Vegas, USA.

8- Yong, R.N., Mohamed, A.M.O and Warkentin, B.P. 1992. Principles of Contaminant Transport in Soils. Elsevier Science, Amsterdam, The Netherlands.

Variations In Lead (Pb) Content In Soils Downslope And Upslope Of Victoria Falls Municipal Waste Dump In Zimbabwe

Mhosisi Masocha, University of Zimbabwe, Harare, Zimbabwe
Daniel Silas Tevera, University of Zimbabwe, Harare, Zimbabwe

Abstract
This study examines lead (Pb) content in soils downslope and upslope of the Victoria Falls town municipal waste dump. Fourteen soil samples were collected in October 2003 from 20*20m plots located along two linear transects (one downslope and the other upslope of the waste dump) and analysed for Pb using the atomic absorption method. It was hypothesised that there is a significant relationship between distance and direction from the municipal waste dump and soil Pb content. The study found that soils that are closer to and downslope of the waste dump had significantly higher ($p<0.05$) Pb content than those further from and upslope, respectively. It was concluded that the disposal of urban solid wastes in Victoria Falls town increases the concentration of Pb in the soil.

Introduction
Widespread crude tipping of wastes at open unprotected urban waste dumps is one of the principal causes of environmental pollution in Zimbabwe (Tevera, 1991). The World Resources Institute (1996), Abreau (1997) and Iqbal *et al.*, (2000) have noted with concern that crude dumping of mixed urban solid wastes contributes to soil chemical pollution because many components of the urban waste stream undergo chemical transformation that results in the release of various contaminants. For instance, some industrial and commercial wastes contained in the urban solid waste stream such as coloured plastics, glass, residues from metal shredding, lead-acid batteries, mercury and cadmium batteries as well as paint and pesticide residues contain heavy metals notably cadmium (Cd), lead (Pb) and zinc (Zn). During processes such as rusting, overland flow and leaching these heavy metals escape into urban soils where they suppress soil microbial activity, inhibit vegetation growth and disrupt ecosystem processes like food webs. Also, heavy metals can contaminate ground and surface waters thereby damaging aquatic life (UNEP, 1994). Heavy metals are metallic chemical elements that have a relatively high density of 5g/cm^3 and more. They tend to bioaccumulate, are generally toxic to plant and animal life even at low concentrations (AMAP, 1991; SEPA, 2003).

Although it is widely accepted by many researchers that heavy metals are natural components of soil released during natural geo-chemical processes like rock weathering, most of them (e.g. Cd and Pb) are only present in eco-toxicologically insignificant quantities. However, when land is used as a repository for urban wastes, their concentration levels in soils usually rise. It is

5th ICEG Environmental Geotechnics, Thomas Telford, London 2006.

hypothesised in this study that most significant increases occur in areas closer to and downslope of urban waste dumps than those further and upslope, respectively. Iqbal *et al.*, (2000) argued that although urban solid wastes may not have high concentrations of heavy metals per tonne however, because of their sheer huge volumes, they can increase heavy metal budgets for urban soils considerably. Although research done in other regions such as Western Europe (European Commission, 2002) shows that poor urban solid waste disposal contributes to soil contamination, to date in Zimbabwe, very little is known about the impact of urban crude waste dumps on heavy metal loads in soils.

Against the above backdrop, this study measures and compares Pb content in soils upslope and downslope of the Victoria Falls municipal waste dump. The study also examines the relationship between Pb content in the soil and distance from the waste dump. The hypotheses formulated to guide the study are that (a) there is a significant difference in the mean Pb content in soils upslope or downslope of the waste dump; and (b) there is a significant relationship between distance and direction from the waste dump and soil Pb content. As of October 2003, there were more than 14 open waste dumps that have become quasi-permanent features of Victoria Falls' urban landscape (Masocha, 2004). This study focuses on the municipal waste dump because it receives the bulk of the wastes produced in the town.

Materials and Methods

Study Area

The study was undertaken in Zimbabwe's northwestern town of Victoria Falls. Victoria Falls town is located 17^0 55′ south of the equator and 25^0 50′ east of the Greenwich meridian. The town is bordered to the West by Zambezi National Park while to the East lies the Victoria Falls National Park. The town receives an average of 600mm of rainfall per year and its long term mean monthly maximum temperature is 29.8^0C. Layers of chalcedony, sandstone and Kalahari sands overlie the basalt rocks of the Karoo group that dominate the geological base of the town. The Kalahari sands are deep regosols and loosely textured reddish brown sands with very high infiltration rates. In contrast, the basalt-derived clays are shallow but compact lithosols (about 25cm thick) hence there are less permeable. The climate and soils of the town sustain a savannah type of vegetation that is dominated by *Colophospermum mopane* woodland.

Population Growth Trends

Since 1969 when the first comprehensive national census was conducted, the population of Victoria Falls town has been doubling every decade making it the fastest growing urban centre in Zimbabwe. The total population of the town was 3 450 in 1969. It rose sharply to 8 126 and 16 826 by 1982 and 1992, respectively. Results of the 2002 national census show the population of the town was 31 375 (CSO, 2002). Apart from the resident population, Victoria Falls town also caters for about 30 000 visitors who comprise domestic, regional and international holiday makers as well as cross-border traders mostly from Zambia every month (Plan Afric, 2001). The natural beauty of the waterfalls and the rich biodiversity of the Zambezi River and surrounding national parks attract thousands of visitors to the town. The fast pace of urbanisation being experienced in the town exposes the administrative inadequacies of Victoria Falls Municipality (VFM) regarding timely removal and safe disposal of wastes generated.

Waste Management Issues

Victoria Falls town consists of a fairly dense road network, less than 10 000 formal housing units, light industries and several hotels, which generate copious amounts of wastes. Waste management is currently an enormous challenge facing the local authority (Sanyanga and Masundire, 1999; Plan Afric, 2001; Velma and Masocha, 2003). Under the Urban Councils Act (*Chapter 29:15*), Victoria Falls Municipality, through its Department of Housing and Community Services (DHCS), is solely responsible for providing basic sanitation including waste collection and disposal. Like most local urban authorities in the country, the Municipality charges a flat refuse charge that covers all aspects of waste management ranging from provision of waste bins and skips, collection, transportation and disposal.

Waste management in the town is very simplistic and focuses mostly on waste collection and crude dumping. Out of an estimated 8 000 tonnes produced every month, about 60% are temporarily stored in waste bins, in skips and other waste receptacles from where they are collected using curbside collection method and transported to the municipal dump where they are crudely tipped and left uncovered (Masocha, 2004). The major waste generation areas in the town are residential areas, factories, warehouses, retail shops, offices and institutions like hospitals and hotels, vegetable and fruit markets as well as construction sites. The composition of the MSW stream by weight was determined through physical analysis of samples collected. The wastes have very high proportions of paper and cardboard (about 34%) and organic materials (about 26%). Plastics constitute the third largest component of the waste stream (about 15%). Metals account for a fairly small proportion (about 5%). The rest of the wastes are glass and ceramics, clothes and textiles, leather and rubber as well as inert materials (Masocha, 2004).

Normally, the frequency of waste collection ranges from once per week in the low-income areas (e.g. Chinotimba township), to twice per week in the industrial areas and high-income areas (e.g. Baobab suburb), to daily (except on Sunday) in the Central Business District. However, like in all other urban centres in the country, which are faced with a crippling fuel crisis and shortage of adequate refuse vehicles waste collection has become very irregular and erratic in recent years and this largely accounts for widespread indiscriminate open waste dumping being experienced in the town. Since 1967, when Victoria Falls was accorded town status, crude dumping remains the most common and cheapest disposal method in spite of the odours, leachate and pest problems associated with this elementary method no wonder it has been outlawed in many countries.

Transect Surveys And Systematic Plot Sampling

In order to measure Pb content, soil samples were carefully collected in October 2003 from 14 study plots (each measuring 20 by 20 metres) which were positioned at a regular interval of 40m along two linear transects. Distance between plots was measured using a 100m tape measure. Transect 1 was downslope of the municipal dump while transect 2, which was upslope of the waste dump, was used as the control. The rationale for this is that along transect 2 all plots were upgradient of the municipal waste dump hence they were considered less contaminated by heavy metals, which are mostly distributed in the soil by runoff and that are gravity controlled hydrological processes. In choosing the above sampling strategy and plot size, the intention was to capture small-scale variations in Pb content with changing distance from the waste dump.

Soil samples were augured at a depth of 25cm from close to the corner and centre of each study plot, mixed thoroughly to form a representative sample and preserved in sealed polythene bags. The vegetative layer comprising decaying leaves and branches was carefully removed before collection. The study therefore examined total Pb count in surface and near surface soils. In that regard, it does not show variability of Pb content with changing soil depth.

Measuring Soil Pb Content

Atomic absorption flame spectroscopy using air-acetylene was used to detect Pb content in the soil. Laboratory analysis was done in the Institute of Mining Research at the University of Zimbabwe. Since heavy metal concentration levels can be affected by particle size, samples were first dried and then sieved (using a 0.21mm mesh) to remove larger particles. Perchloric acid ($HClO_4$) was added to each sample to destroy the organic matrix (acid attack). For safety reasons, nitric acid (HNO_3) was added to counter the effects of $HClO_4$, which reacts explosively when heated. The ratio of $HClO_4$ to HNO_3 used was 1:1 because the samples had low organic matter content. Finally, Pb content was measured using an atomic absorption (AA) spectrometer (VARIAN SPEC 10).

Comparing Mean Pb Content

In order to compare the differences in the mean soil Pb content downslope and upslope of the municipal waste dump, box plots were first plotted against each other to visualise and discern any differences. Data were then tested for normality using the Kolmogorov-Sminorov test in order to decide whether to use parametric or non-parametric methods. Thereafter, a paired samples *t*-test was used to test if the mean soil Pb content downslope and upslope of the municipal dump were significantly different.

Testing And Modelling Relationships

The relationships between distance and direction from the municipal waste dump and soil Pb content were determined through Pearson's correlation analysis and further analysed using simple regression. For curve estimation, the logarithm method was used because it best fitted the data. This was evidenced by the highest coefficient of determination (R^2) it yielded for the parameters regressed when compared to linear, quadratic and exponential methods.

Results And Discussion

Variations In Soil Pb Content

The results of soil Pb content measured for each study plot are shown in Table 1. It was observed that generally along both transects, soils close to the municipal dump had higher Pb content than those further. The results show that downslope of the waste dump soil Pb content decreases steadily as distance from the dump increases while upslope soil Pb content fluctuated but decreased marginally with increasing distance. The mean soil Pb content was 44 mg/kg and 22 mg/kg downslope and upslope of the municipal waste dump, respectively. Results of the paired samples *t*-test show that soils downslope of the waste dump had significantly ($p<0.05$) more Pb content than those upslope (Figure 2).

Table 1: Soil Pb content downslope and upslope of the municipal dump

Study plot	distance from dump (m)	Soil Pb content (in mg/kg)	
		downslope (transect 1)	upslope (transect 2)
1	20	78	28
2	60	52	18
3	100	49	19
4	140	24	21
5	180	39	26
6	220	32	22
7	260	34	20

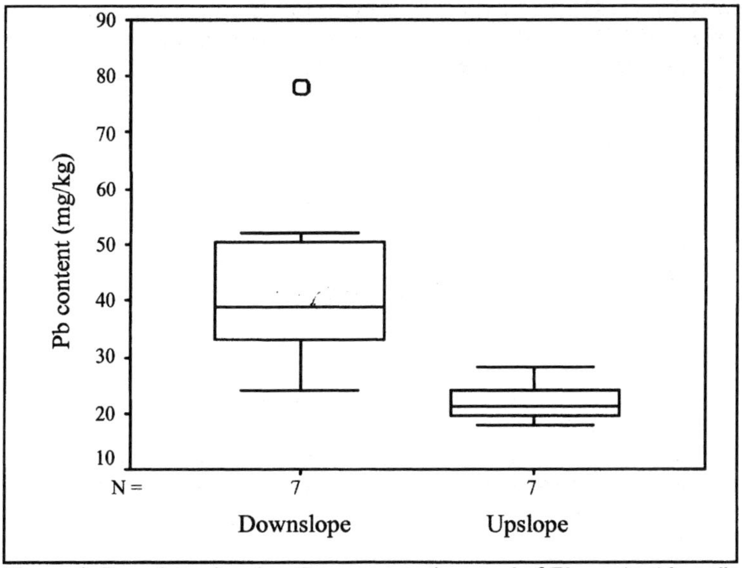

Figure 1: Box plots showing the mean and spread of Pb content in soils downslope and upslope of the municipal waste dump

The concentrations levels of Pb content were compared with Swedish Environmental Protection Agency (SEPA) guideline values for polluted soils and most sensitive land use types (SEPA, 2003), which meet international standards because presently Zimbabwe does not have any national guidelines/standards in place. The results of Pb measurements (Table 1) were below the Swedish guideline value of 80mg/kg. The average amounts of 44mg/kg and 22mg/kg for transects 1 and 2, respectively generally show that Pb levels at the municipal waste dump are below the eco-toxicological limits. However, the highest amount of 78mg/kg recorded for plot 1 (downslope of the waste dump) is quite close to the recommended limit and indicates that the level of Pb in soils downslope of the dump is reaching critical levels.

Relationships Between Distance And Variations In Soil Pb Content

Results of Pearson' correlation analysis (Table 2) show that contrary to what was hypothesised, downslope there was a strong and significant negative correlation between soil Pb content and distance from the waste dump. However, upslope soil Pb content was not significantly correlated ($p<0.05$) with distance from the dump.

Table 2: Pearson's correlation coefficients between distance from the waste dump and soil Pb content

Parameters	Soil Pb content (mg/kg)	
distance from the dump (m)	downslope	upslope
correlation coefficient (*r*)	-0.786*	-0.188
p value	0.036	0.687

* correlation is significant at the 0.05 level (2-tailed)

In order to quantify how soil Pb content is related to distance from the municipal dump simple regression was used. The regression equation computed (Figure 2), which was statistically significant ($p<0.05$), show that downslope of the municipal waste dump most of the variation in soil Pb content (R^2=0.83) can be explained by changes in distance. Upslope of the waste dump there was no significant relationship between distance from the waste dump and soil Pb content (Table 2) hence no regression model was developed. From the preceding analysis, it is clear the municipal waste dump is a source of Pb that contaminates soils especially those downslope. However, its effect on other heavy metals such as zinc (Zn) and cadmium (Cd) still needs to be investigated further.

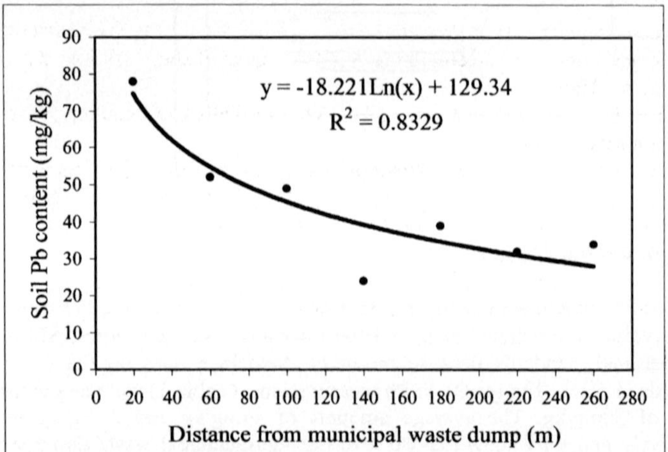

Figure 2: Model showing relationship between distance from the municipal dump and Pb content downslope of the dump (transect 1).

Conclusion

From the above discussion, it was concluded that the disposal of solid wastes through crude tipping at the municipal waste dump causes soil contamination by lead (Pb) with soils closer to and downslope of Victoria Falls municipal waste dump being worse affected than those further and upslope. The bulk of solid wastes generated in urban centres throughout Zimbabwe are crudely dumped at open waste dumps all of which have no environmental controls (e.g. geosynthetic membranes). In addition, no attempts are made to treat wastes prior to crude dumping to make them environmentally benign. Therefore, it is very likely that urban soils throughout the country are under stress from waste dump-related chemical pollutants although more conclusive evidence needs to be gathered.

References

Abreau, V. (1997) *Urban Water Quality and Waste Management*, University of Michigan, Earthscape, Michigan.

AMAP (1991) *Arctic Pollution Issues: A State of the Arctic Environment Report,* The Arctic Monitoring and Assessment Programme (AMAP) Working Group, http://www.amap.no/assess/soaer-cn.htm.

CSO (2002) *Census 2002 Zimbabwe Preliminary Results Summary*, Central Statistical Office, Harare.

European Commission (2002) *Heavy Metals in Waste Final Report February 2002*, European Commission DG ENV. E3, Project ENV.E.3/ETU/2000/0058, Environmental Expert, S.L, http://europa.eu.int/comm/environment.

Iqbal, M, Z., M. Rukhsana and S. Muhammad (2000) "The Effect of Lead and Cadmium on Trees", *Journal of the Society of Municipal Arborists,* Vol. 36, Number 1, January/February 2000.

Masocha, M. (2004) Solid waste disposal in Victoria Falls: spatial dynamics, environmental impacts, health threats and socio-economic benefits, Unpublished MPhil thesis, University of Zimbabwe, Harare.

Plan Afric (2001) *Victoria Falls Combination Master Plan*, Victoria Falls Combination Master Plan Preparation Authority, Bulawayo.

Sanyanga, R. A. and Masundire, H. M. (1999) "Waste Management in the Major Population Centres of the Zambezi Valley-Botswana, Zambia and Zimbabwe", In Yap, N. T. (ed.) *Cleaner Production and Consumption: Challenges and opportunities in East and Southern Africa,* Weaver Press, Harare.

SEPA (2003) *Pollutants: Heavy Metals*, Swedish Environmental Protection Agency (SEPA), Stockholm, www.internat.naturvardsverket.se/index.php3.

Tevera, D.S. (1991) "Solid Waste Disposal in Harare and its Effects on The Environment: Some Preliminary Observations", *The Zimbabwe Science News,* Vol. 25, No.1/3, pp. 9-13.

UNEP (1994) *Environmental Data Report 1993*, United Nations Environment Programme, Nairobi.

World Resources Institute (1996) *A Guide to the Global Environment: The Urban Environment*, Oxford University Press, New York, http://www.wri.org/wri/wr-96-97/ee_txt4.html.

The challenge to adapt soil properties

Gerard van Meurs, GeoDelft, Delft, the Netherlands
Willem van der Zon, GeoDelft, Delft, the Netherlands
John Lambert, GeoDelft, Delft, the Netherlands
Derk van Ree, GeoDelft, Delft, the Netherlands
Victoria Whiffin, GeoDelft, Delft, the Netherlands
Waldo Molendijk, GeoDelft, Delft, the Netherlands

Abstract

Whenever people want to build in, onto or with soft soils like sand, clay or peat, soil properties strongly influence the design and construction. Safety standards and building codes also require that the 'weakest' value for properties is used. This means that soil properties have a significant cost impact and can lead to unwanted limitations in the utilisation of space in densely populated areas. The challenge from a geo-engineering point of view is to develop technology that mitigates this impact on society.

SmartSoils®, is a strategic research program initiated by GeoDelft, aimed at the development of technologies to adapt soil properties and obtain more favourable values to be used in the process of design and construct. Various innovative technologies can be identified. Among these are BioGrout (stiffness and strength improvement) and BioSealing (reduction of permeability).

In this contribution, the relevant phenomena are explained, selected results of current research are presented and an outline of future work is described.

Introduction

More than half of the world population inhabit deltaic areas where rivers flow into lakes or the sea. Examples can be found in Europe (e.g. the Netherlands formed by the rivers Meeuse and Rhine, the Po-area in Italy and the Donau-area in Central Europe), the United States of America and Asia. These densely populated areas mostly exhibit shallow groundwater tables and a subsurface consisting of unconsolidated, compressive materials like peat, clays and sand, deposited in a perimarine or fluvial environments. Both the economic activity and the necessity of flood protection lead to specific building requirements.

With respect to construction in deltaic areas, the natural soil conditions and inherent uncertainty of its properties have a strong impact on suitable technologies and costs. Looking at the life-time costs of a construction, some 40 to 60 % of these costs are for maintenance and repair. The costs of failures in soil, road and hydraulic engineering, and in commercial and industrial buildings amount to about 25 % of the total costs. Half of these costs are related to the subsoil. Common practice is to accept the soil as it is, and incorporate the uncertainty by means of a safety factor or more recently, a probabilistic approach. As a consequence, the overall design is adapted to the weakest property of the soil volume present. In cases where the soil is unsuitable for the desired use, soil improvement technologies have been developed, e.g. grouting techniques and in some cases the soil has even been replaced by synthetic lightweight materials to obtain the chosen properties. In addition to these issues, adverse soil conditions, lack of space and cultural heritage protection, lead to increasing pressure on existing construction technologies and thus higher costs in densely populated regions

An important breakthrough could be achieved whenever the soil can be treated as an industrial material with properties that can be produced according to the specifications of the construction engineer. This implies a complete change in perception, a true paradigm shift. Our strategic aim is to adapt the soil properties using in situ processes to obtain high quality and sustainable structural material with specific and required performance: so-called "soil on demand".

Geological and soil forming processes are physiochemically and (micro-) biologically very complex and occur over a long time span. Some aspects can however be controlled to generate a desired effect in the treated area. An example is the BioGrout-process in which in situ calcium carbonate formation is regulated by means of a biological process with the result of transforming non-cohesive sands into a cemented sandstone-like material.

The significant benefit of this breakthrough from a construction point of view is that natural soils will become a more industrialised material. This means that soils can be generated with processes that are fully quality controlled, irrespective of its complexities and natural variations and corresponding uncertainties. Current practice, which commonly adopts a large factor of safety to account for unforeseen circumstances, will change drastically and a broader range of construction options will evolve with a significant reduction of life cycle cost and offering building opportunities that did not exist before.

SmartSoils®, an initiative of GeoDelft, focuses on the controlled adaptation of soil mechanical properties [Barends, 2005]. Research is currently underway to develop a better understanding of the relevant phenomena. Among these are the complex interactions between the soil (groundwater and solid particles), the growth of the micro-organisms and the interaction with the geo-chemistry [Whiffin et al., 2005].

Areas of application

Existing soil improvement technologies such as chemical injection and grouting with cement-based grouts have a limited field of application and are relatively expensive. A wider field of application of SmartSoils®-technologies is sought after, based on development needs that can be identified in geotechnical engineering. All of the applications presented here are taken from the field of geo-engineering.

BioGrout is aimed at the improvement of soil strength and stiffness, whilst largely maintaining soil permeability. The latter is also important for the implementation of the process in situ over distances of 10 meters and more. Various areas of application can be identified in which differential settlement, erosion damage or lack of space lead to significant expenditures in maintenance and reconstruction. Amongst these is the reinforcement of coastal defences such as dikes and dunes that may need to be upgraded as a result of climate change due to global warming. The current Dutch policy of dynamic coastal defence is to allow dune-shores to recede during storms. At some specific high-risk locations, this policy may not be adequate any more, requiring improvement to maintain safety but without destruction of natural areas. Treating dikes with sand cores with the BioGrout process could also lead to avoiding the costly replacement of existing covers. In areas with a high groundwater tables dynamic loading can result in construction damage and collision risks as has been observed in practice. This in turn results in maintenance costs and restrictions in the rail- and transport infrastructure. Significant social and economic benefit can be derived from in situ improvement of the strength and stiffness of a track without closing it.

Another SmartSoils® technology is BioSealing, which aims to reduce the local permeability of the soil. The first application is related to common civil engineering in deltaic areas. Existing groundwater management policies lead to strict requirements for water retaining barriers whenever soil excavation takes place in a situation with a relative high groundwater table. Leaks can have a significant impact on the surroundings and are difficult and costly to repair due to difficulties in localisation and accessibility. BioSealing has shown to be applicable in this area. The technology involves a number of physical, geochemical and microbial processes which lead to a reduction of the permeability of the soil in the area a leak. The second possible application is related to polder areas, which are common in the western regions of the Netherlands. The surface of a polder is below sea level. As a consequence, groundwater flows upward from deeper sandy formations and seeps into the ditches of the polder. Frequently, the upcoming groundwater is brackish and thus deteriorates the quality of the shallow groundwater. This flow phenomenon is not always uniform and there are often several spots where the flow is more intense than elsewhere. The introduction of a higher hydraulic resistance at such a spot, such as that which can be achieved with BioSealing, will strongly improve the quality of the shallow groundwater.

Principle of formation

SmartSoils® processes are based on adapting the soil properties in ways that simulate nature. However, in nature under normal conditions these phenomena take place over a geological time frame. The challenge is to speed up the rate of formation so that it becomes possible to meet the time frame in which a construction must be realised. The durability of the process should also meet construction needs.

Technologies such as BioGrout and BioSealing require nutrients. Bacteria and reactants (chemical species) need to be brought together. Whenever one of these components is not present in the porous medium like sand, there is a need for infiltration or injection. The design of injection points is based upon the characteristics of the flow of groundwater, the transport of dissolved species, the growth of bacteria and the rate of the reactions. Beside the intended species, also residual products are formed. Care needs to be taken for the environmental effects of these residuals, thus in addition there may be a need for extraction or conversion.

BioGrout

Sandstone is formed on a geological time frame due to several mechanisms including chemical and biological processes. In BioGrout specific bacteria mediate a rection which reuslts in the precipitation of calcium carbonate in the form of crystalline calcite (Figure 2) on the surface of the sandy particles [Molendijk et al., 2004] (Figure 1, 1a). When this reaction occurs under the right conditions, crystals are formed between two adjacent particles producing a connection between them. This connection increases the strength and stiffness of the porous material. The longer the nutrients, bacteria and reactants are present, the thicker the layer of mineralization (1b). An important characteristic of this process of cementation is that the permeability of the porous material only reduces slightly. Preservation of permeability is required for many applications on a field scale.

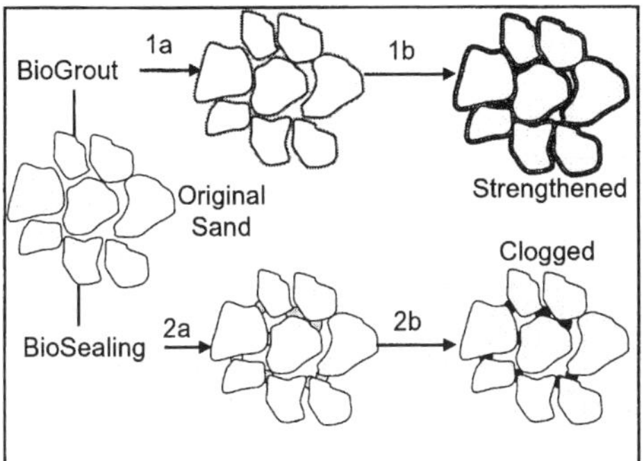

Figure 1: Schematisation of the principle of formation of BioGrout and BioSealing.

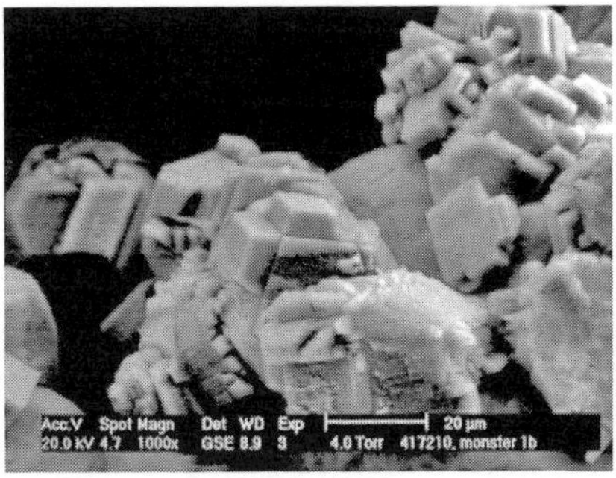

Figure 2: Scanning Electron Micrograph (SEM) to show the mineralization of calcite onto sandy grains.

As already mentioned before, the biomineralization process progresses slowly in natural circumstances. Fortunately, the rate of mineralization can be enhanced by stimulating the conditions. The ongoing research is focused at up scaling and process-control for specific applications. Both laboratory experiments and implementation in the field are executed.

Table 1 gives an indication of the values for the various soil mechanical properties which may be the result when the soil is treated with the technology BioGrout and BioSealing.

Table 1: Indication of various geotechnical parameters: strength (σ), e-modulus (E_{50}) and permeability (k).

	σ (kPa)	E_{50} (MPa)	k (m/s)
Original sand	460	80	$2.2 * 10^{-5}$
Strengthened	2,860	270	$1.0 * 10^{-5}$
Clogged	460	80	$2.0 * 10^{-7}$

BioSealing

For BioSealing the growth of bacteria is stimulated by the injection of nutrients. Bacteria flow with the water and are attached at surfaces of the sandy particles especially near the pore throat (Figure 1, 2a). Sometimes, the chemistry of the water changes and precipitates occur (2b). As a consequence, the permeability of the porous medium reduces.

Clogging processes have been studied over many years and are described in literature. In the past the drawback of clogging during groundwater extraction was illustrated together with solutions describing how to handle the clogging [McLaughlan, 2002]. [Vandevivere and Baveye, 1992] showed with laboratory tests that under aerobic conditions bacteria were capable of reducing the hydraulic conductivity of a sandy column. [Hill and Sleep, 2001] and [Thullner et al., 2002] showed that the biomass has a minor effect on the porosity and a major effect on the permeability. [Thullner et al., 2002] carried out experiments in a two-dimensional set-up filled with glass beads. They showed that bioclogging takes place due to the injection of nutrients. Gradually, the initial uniform flow pattern changes and is affected by the spatial distribution of the biomass. Ultimately, the water flows around the clogged area.

In most practical circumstances there is only a localised area that is responsible for most to the inflow of groundwater. A solution may be to inject nutrients in the groundwater at an upstream position to the leak. Due to the flow regime of the groundwater the nutrients are displaced in the direction of the appointed area, thus the leak is self-locating. In our research program we investigated the use of anaerobic BioSealing technology to reduce the permeability in areas through which the major amount of groundwater flows.

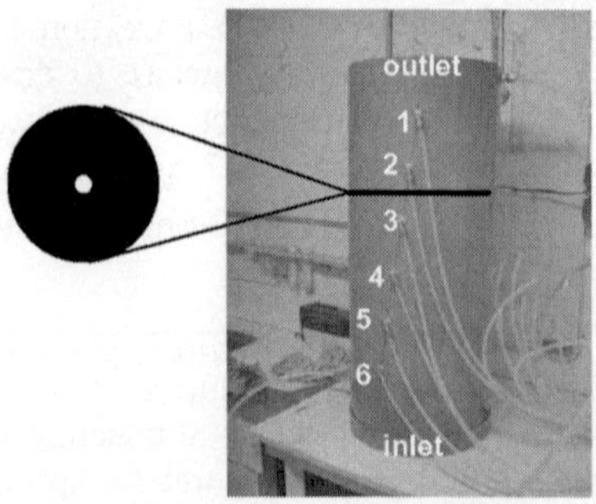

Figure 3: Experimental set-up of column experiment for BioSealing.

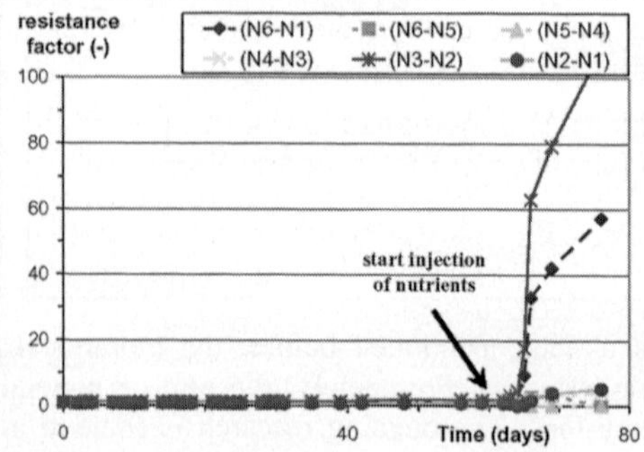

Figure 4: Increase in hydraulic resistance as a result of injection of nutrients.

Column tests were carried out in the laboratory to investigate the impact of clogging on the reduction of the permeability of sandy material at a desired location. Figure 3 gives a schematic view of the experimental set-up. A column, made of PVC, was filled with sand. The dimensions of the column were: height 0.73 m and diameter 0.30 m. At various vertical positions along the column, six piezometric tubes were installed numbered from top (N1) to bottom (N6). Each piezometric tube measured the water pressure at different positions along the column. A change in the difference in water pressure between two (adjacent) piezometric tubes indicated a change in hydraulic resistance and thus extent of clogging.

The water flow in the columns was in upward direction; from the bottom to the top. In each column, a separation plate was placed at about two third from the inlet (Figure 3). The location of the plate was between piezometric tube Number 2 and Number 3. A hole, with a diameter of 0.03 m was made in the centre of the separation plate and served as the location of the leak. At the inlet, inflow of water takes place. After a certain time period, nutrients were injected with the inflow of water. Several column tests were in operation each with a different flow regime and / or a different composition of nutrients. The sand and the water were taken from a Pleistocene stratum in order to represent a typical composition. The Pleistocene sand that was used contained about 1 to 3 % of iron oxide. The nutrition consisted of sugar syrup or a mixture of residual products from the potato starch industry.

Several parameters were measured. Among them were: acidity (pH), glucose concentration, electrical conductivity (Ec) and redox potential. The piezometric head is measured at the locations of the piezometric tubes. These results were used to determine the difference in piezometric head between two (adjacent) locations. Division of this value by the initial difference normalises to a dimensionless value: the so-called resistance factor. This factor is a measure for the change in hydraulic resistance. Beside across the plate, at the start of the experiment almost equal hydraulic resistance between adjacent piezometric tubes. The initial hydraulic resistance across the plate is about four times as much as the hydraulic resistance of only the sandy material.

During a time period of about sixty days only water was flowing through the column. No change in hydraulic resistance occurred. Then, nutrients were injected at the inlet. The resistance factor of hardly changed when only water was flowing through the column. However, the hydraulic resistance increases at some time after the injection of nutrients at the inlet started (Figure 4).

The main increase in the resistance factor is between N2 and N3. At this location the hydraulic resistance increased with more than a factor 100 within approximately 10 days. There appeared to be very little change in hydraulic resistance between each other pair of adjacent piezometric tubes. The overall hydraulic resistance of the column (N6-N1) increased by a factor of more than 50. The overall resistance as well as the resistance around the hole in the separation plate were still increasing at the end of the experiment.

All the column tests showed that clogging took place at the hole of the separation plate independently of the flow or the amount of nutrients. Further analyses of the results showed that biological growth had already started around the injection point.

Given the positive results in the laboratory trials, a field test was performed to demonstrate BioSealing under large-scale conditions. For this experiment, three sea transport shipping-containers were buried vertically. Each container was made to simulate a building pit with leaking sheet piles therefore each container had a different configuration of holes at its bottom end (the leakage). The surface areas of the various leakage configurations were comparable.

Figure 5: The used 20 ft sea transport shipping container is buried in a vertical position at the Maasvlakte near Rotterdam. The right drawing gives the locations of equipment put into the soil and the elevation due to sealevel (NAP).

Three injection tubes were placed approximately 2.5 m away from each (hole of the) container (Figure 5). Nutrients were injected at three different depths. Monitoring wells were installed between the point of injection and the hole to measure several parameters. The containers were partially filled with sand and the water was pumped out of the container with a drain. The water level in the drain was kept constant and the rate of water, extracted from each container to maintain the water level, was measured. This value was an indicator of the amount of clogging (lower water flow means more clogging). Further details about this experiment can be found in [Veenbergen et al., 2005].

The pilot test started on March 1st 2004. Initially, the addition of the nutrients took place once a week, later on, followed by five times a week. It took about six weeks before the hydraulic resistance around the inflow of Container 2 was enhanced by a factor five, which was the objective during the first session of the experiment (Figure 6). Afterwards, the resistance factor increased further to a value of about thirty. Even more then three months after the last injection of nutrients, the hydraulic resistance remained stable.

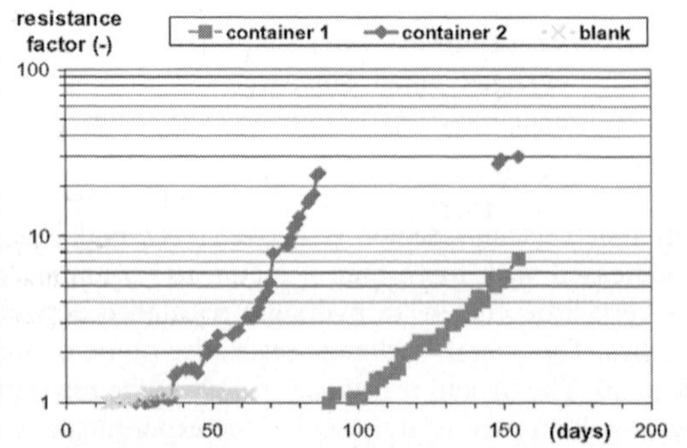

Figure 6: Increase in hydraulic resistance due to nutrition.

At the same time, also Container 3 was drained off. However, no nutrients were injected. Therefore, this container can be seen as 'the blank' (Figure 6). It was found that the flow rate, and therefore the resistance, did not vary. Therefore, it can be concluded that enhancement in hydraulic resistance around Container 2 is due to the mechanisms activated by the injection of nutrients. The daily injection with nutrients in the vicinity of Container 1 started at day 90 and ended at day 155. The resistance factor gradually increased to a value of about seven at the end.

In July 2004 the containers were removed from the soil. Several interesting observations were made. Among others, the soil between the injection points and the holes of the containers was coloured black. This means that precipitation took place not only near the holes, but also in the area between the container and the injection point. However, black colour of the soil is most intense near the leak.

Concluding remarks and future work

Several tests were carried out as part of our research program. The results indicate that it is possible to adapt soil mechanical properties in situ. Consequently, a start is made to consider the soil as a starting material that can be adapted to suit a desired use. Then the soil may be treated as an industrial material. In future, possible processes that can change the properties of other materials like clay and peat will be considered. The goal is to change soils to meet the requirements of construction: soil on demand.

Both BioGrout and BioSealing have been tested in the laboratory. To date, BioSealing has been applied at a pilot scale and full field scale. BioSealing was capable of reducing the local hydraulic conductivity where excessive inflow of water occurred BioGrout is currently progressing toward scaling up the process to a field situation.

References

Barends, F. B.J (2005)
Associating with advancing insight,
Terzaghi Oration, 16th Int. Conf. on Soil Mech. and Geot. Eng., Osaka, Japan.

Hill, D.D. and Sleep, B.E. (2001)
Effects of bio film growth on flow and transport through a glass parallel plate fracture,
Journal of Contaminant Hydrology, vol. 56, pp. 227-246.

McLaughlan, R.G. (2002)
Managing water well deterioration,
Int. Cont. Hydrogeology, Vol. 22: 128 S.; Lisse (Balkema).

Molendijk, W.O., Admiraal, B.J. and VanPaassen, L.A. (2004)
SmartSoils®: soil properties on demand (in Dutch),
Geotechniek, Vol. 8, Nr. 5, pp. 30-33.

Huller, M., Mauclaire, L., Schroth, M.H., Kinzelbach, W. and Zeyer, J. (2002)
Interaction between water flow and spatial distribution of microbial growth in a two-dimensional flow field in saturated porous media, Journal of Contaminant Hydrogeology, vol. 58, pp. 169-189.

Vandevivere, P. and Baveye, P. (1992)
Saturated Hydraulic Conductivity Reduction Caused by Aerobic Bacteria in Sand Columns, Soil Science Society of America Journal, vol. 56, no. 1, pp. 1-13.

Veenbergen, V., Lambert, J.W.M., VanderHoek, E.E., VanTol, A.F., Weersma, S.I. (2005)
Underground Space Use, Analysis of the Past and Lessons for the Future,
Proceedings ITA-AITES World Tunnel Congress, 7-12 May 2005 Istanbul

Whiffin, V.S., Lambert, J.W.M., and Van Ree, C.C.D.F. (2005)
BioGrout and BioSealing - Pore-space engineering with bacteria.
Geostrata - Bio-Geo Insitu processes. American society of civil engineers

Acknowledgements

The Biogrout and BioSealing research projects are carried out in cooperation with Volker Wessels Stevin Geotechniek, Visser and Smit Bouw B.V. and the Technical University Delft. The research is in part supported by the Ministry of Economic Affairs.

Correlation Between 1, 2 and 3-D LNAPL Migration Patterns at Equilibrium in Unsaturated and Saturated Sand

Mostafa H. A. Mohamed, School of Engineering, Design and Technology, University of Bradford, Bradford, UK.
Radhey S. Sharma, Department of Civil and Environmental Engineering, Louisiana State University Baton Rouge, USA

Abstract
This paper presents results of an experimental investigation for Light Non-Aqueous Phase Liquids (LNAPLs) migration patterns in unsaturated and saturated sand. Field and 3-D experimental investigations for LNAPL migration in variably saturated soils are generally avoided because of limited resources, excessive time requirements, and potential adverse environmental effects at the site. With the objective of exploring patterns of LNAPL migration, an extensive experimental investigation was undertaken. The migration patterns of LNAPL obtained using 1-D (column), 2-D and 3-D tanks filled with unsaturated and saturated sand were critically examined to illustrate the salient features of major differences between various LNAPL patterns. Digital imaging was employed to record the advancing front of LNAPL during infiltration as well as the final distribution of LNAPL at equilibrium. The results suggest that 1-D column test results should be treated with caution because of the major implication of the boundary conditions. In addition, results of 2-D LNAPL migration pattern were found to be in good correlation with those obtained from 3-D tank. However, a scale factor should be applied on 2-D LNAPL distribution at equilibrium to account for the restriction of LNAPL migration in the third direction.

Introduction
Groundwater aquifers are a major resource for domestic water supply in Europe as well as in many areas around the world. Light Non-Aqueous Phase Liquids (LNAPLs) such as gasoline and kerosene are a major source of pollution to groundwater aquifers, which would result in serious environment risks if no action is implemented. With the increasing number of contamination incidents, a considerable attention has been given to understand the processes that control migration of LNAPL in unsaturated and saturated soils with view of designing successful and economic remediation schemes. Several numerical codes have been developed for the simulation of LNAPLs in unsaturated and saturated zones (see for example, Abriola and Pinder, 1985 and Steffy et. al., 1998). Proper validation of such a numerical code requires

high quality field and/or laboratory data simulating various scenarios of LNAPL migration patterns.

Large-scale field and 3-D experiments are not favourable because of the possible adverse environmental effects at site, limited resources and excessive time requirement. Alternatively, small-scale laboratory investigations using 1 and 2-D set-ups were conducted. Several studies have been undertaken to investigate LNAPL migration in homogenous 1-D samples (see, for example, Eckberg and Sunada, 1984, Abdul, 1988 and Sharma and Mohamed, 2001). The outcomes of 1-D investigations were quite useful in improving our knowledge about the migration aspects of LNAPL in the vadose zone. Deeper understanding is obtained using 2-D laboratory set-ups, which partly eliminates the influence of boundary condition. A number of studies have been undertaken to investigate the migration of LNAPL in 2-D homogenous samples (see for example, Schieg and McBride, 1987, Van Geel and Sykes, 1994 and Sharma and Mohamed, 2003). Distribution of LNAPL in the unsaturated and saturated zones was captured along with measuring the degree of saturation and suction head. However, limited results were presented by Schieg and McBride (1987). The results presented by Van Geel and Sykes (1994) indicated that the relationship between suction head and degree of saturation is dynamic before the arrival of LNAPL as the measured water pressure head increased prior to the arrival of LNAPL front. This outcome needs further investigation, as it was not picked up by other investigators. Sharma and Mohamed (2003) investigated LNAPL migration patterns in homogenous sand samples under different spill conditions. It will be useful if similar studies are carried out in heterogeneous samples. A small number of experiments were carried out using 3-D experiments for LNAPL migration in unsaturated and saturated sand (Mohamed, 2003).

The primary objective of this paper is to examine the major differences between various LNAPL patterns in 1, 2 and 3-D tanks. The results presented in this paper are part of an extensive experimental investigation carried out at the University of Bradford for LNAPL migration in 1-D (column), 2-D and 3-D tanks filled with unsaturated and saturated sand.

Experimental Techniques and Materials

In this investigation, three different set-ups including 1-D (column), 2-D and 3-D tanks were used to obtain the migration patterns of LNAPL in unsaturated and saturated sands. Silica sands were used to construct unsaturated and saturated samples and mineral oil was used as a LNAPL.

Experimental set-ups

Column, 1-D set-up comprising of a Perspex column with an internal diameter of 9 cm and with a height of 40 cm was used. The column was connected with a water tank and a burette. The burette was used to monitor the water level inside the column. Water level inside the column was controlled by adjusting the height of an overflow valve. A fine mesh was placed at the bottom of the column in order for water to be distributed uniformly throughout the whole cross-sectional area and to prevent leaching of fine particles. Further details are in Sharma and Mohamed (2001).

2-D and 3-D LNAPL flow patterns were generated using specially designed relatively large-scale fully instrumented tanks. The 2-D tank has internal dimensions of 120 cm x 120 cm x 10 cm and the internal dimensions of the 3-D tank are 60 cm x 75 cm x 60 cm. The front wall of 2-D tank was made of glass for visual inspection and recording of the LNAPL flow pattern using electronic imaging. It should be noted that in case of 3-D tank two sides were made of glass. In both set-ups, a drainage system consisting of perforated tube, connectors, water tank, well and peristaltic pump, was designed to maintain uniform flow into and from the tank and to keep the water level inside the tank at a given height according to the test stage. The rate of inflow and outflow of water from the tanks was selected in such a way to change the water level by a rate of 10 cm/hour. The pumping rate was kept constant in all tests. Further details about the description and function of various components of 2 and 3-D tanks are in Mohamed (2003). All tests were conducted in a controlled temperature room. The temperature was kept at 20 ± 1 °C.

Materials and sample preparation

Mineral oil, which is lighter than and insoluble with water and safe to work with, was used as a LNAPL. The measured specific gravity of mineral oil is 0.85. Of note, mineral oil is colourless and therefore, a biological Red O dye was used to dye the oil red in order to enhance the contrast between different fluids. This enhances the visual inspection and detection of the advancing LNAPL front. The amount of dye was kept constant at 0.015 % by weight of oil in all tests. 1-D experiments were conducted using poorly graded coarse sand (C_u = 1.35 and C_c = 0.92) with mean diameter range between 0.80 and 0.85 mm. However, poorly graded fine sand (C_u = 1.23 and C_c = 0.98) was used to construct 2 and 3-D sand samples. The fine sand has mean diameter of 0.36 mm.

Fully saturated samples were initially prepared by pouring dry sand into tanks, which were partially filled with water. During this stage, controlled flow of water into the tanks was maintained to produce ~ 5.0 cm head of water above the sand level. Since the dry sand was always poured into the water, it is unlikely that air would be trapped in the pore voids. Upon filling of the tank with saturated sand, the sand surface was carefully levelled. The sand samples were then left for 24 hours to ensure reaching full saturation. Stratification of sand layers was unlikely to occur provided that poorly graded sand was dropped from a constant height and a constant water head was maintained above the sand surface.

Experimental procedure

Following the preparation of saturated sand sample, the water level was gradually lowered to a predetermined height at a constant rate. The sand samples was then left for 48 hours to make sure that water distribution within the sand samples reached an equilibrium. Then a given volume of LNAPL was spilled over an area using LNAPL tanks. In case of 1-D column set-up the LNAPL tank covered the whole sand surface whereas for 2 and 3-D experiments the dimensions of LNAPL tank are 10 cm x 10 cm. In tests involving 2 and 3-D tanks, the bottom of LNAPL tank was placed at about 7 cm below sand surface to prevent pooling of LNAPL above the sand surface. To ensure uniform distribution of LNAPL over the spill area, a filter sheet was fastened to the bottom of the LNAPL tanks. LNAPL was allowed to infiltrate and distribute in unsaturated and saturated zones for 48 hours from the commencement of the spill. It should be noted that during this stage, water level was kept constant. Finally, two cycles of wetting and drying after an initial lowering of water was performed to investigate the influence of water table fluctuation on the LNAPL migration pattern.

Results and discussion

Results of LNAPL migration patterns obtained using 1, 2 and 3-D tank set-ups are presented in this section to highlight the major differences between various LNAPL patterns at equilibrium as well as after fluctuating the water level. Figure 1 shows the 1-D distribution of LNAPL in unsaturated and saturated coarse sand as well as the redistribution of LNAPL under a fluctuating water level. In this test 50 cm^3 of LNAPL was spilled above the sand surface. The LNAPL zones are delineated by the dashed lines. Once LNAPL was introduced on the sand surface, it started migrating downwards under the influence of gravity through visible and invisible pathways. Eventually, it distributed itself in 1.5 cm thick zone above the top of the capillary fringe. It was observed that an irregular bottom of LNAPL zone was formed. The thickness of LNAPL zone was measured after equilibrium condition was established between water and LNAPL distribution, which is marked by no further outflow flow of water. Lowering water table from 8.5 cm to 3.4 cm caused LNAPL to redistribute itself in a relatively larger area (~ 4 cm thick) than the initial zone (1.5 cm thick). Finally, with rising water table to 11 cm, LNAPL redistributed itself under and above the water table. LNAPL zone became 9.0 cm thick after 6 days. These results clearly show that fluctuations of water level would result in LNAPL being distributed over a large area.

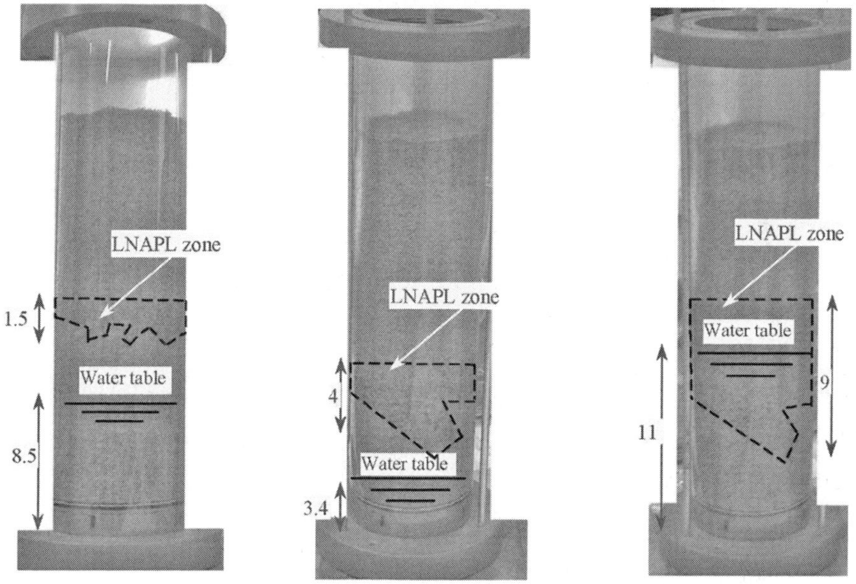

Figure 1: distribution and redistribution of LNAPL in 1-D column

Results of LNAPL patterns in 2-D tank are presented in Figures 2 and 3. Of note, the boundaries given in those figures correspond to the glass panel of the 2-D tank. In this test 4.5 l of LNAPL was spilled above an area of 10 cm x 10 cm while the water level within the sand sample was kept at 45 cm height. It can be seen from Figure 2 that LNAPL migrated continuously and uniformly during LNAPL spill period, which lasted for 135 minutes. With further downward migration, continuous increase in the lateral spread of LNAPL in the unsaturated zone was recorded. It is clear that LNAPL front took the shape of an arc of circle during the infiltration period. In addition, the advancing LNAPL front travelled with constant velocity measured along the vertical centreline of the tank in the unsaturated zone. Once LNAPL reached the saturated zone, slower vertical downward migration and increased lateral migration were observed (compare LNAPL distribution at 100 minutes and 120 minutes in Figure 2). With further accumulation of LNAPL on top of the capillary fringe, large LNAPL pool was obtained. Eventually, a bowl-shape pool with flat-top was observed. The dimension of the LNAPL pool (thickness x length) was measured and found to be 18.0 x 105.0 cm.

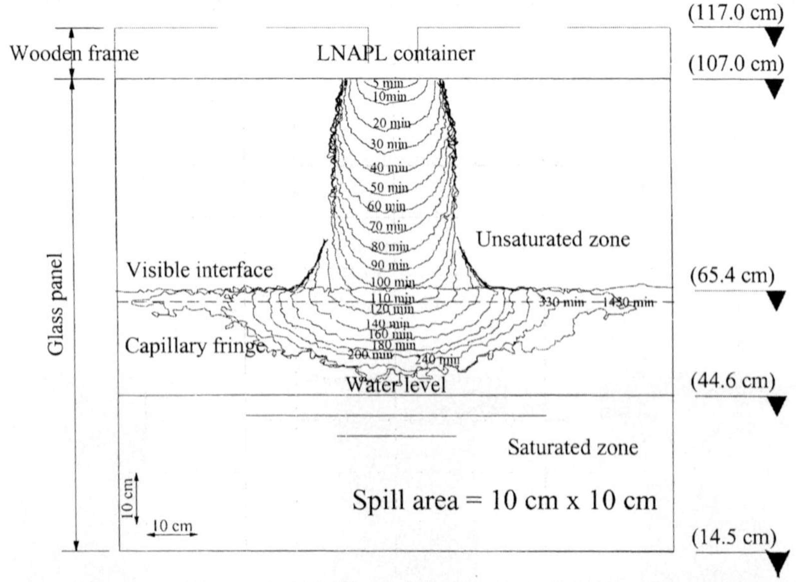

Figure 2: LNAPL distribution in 2-D tank

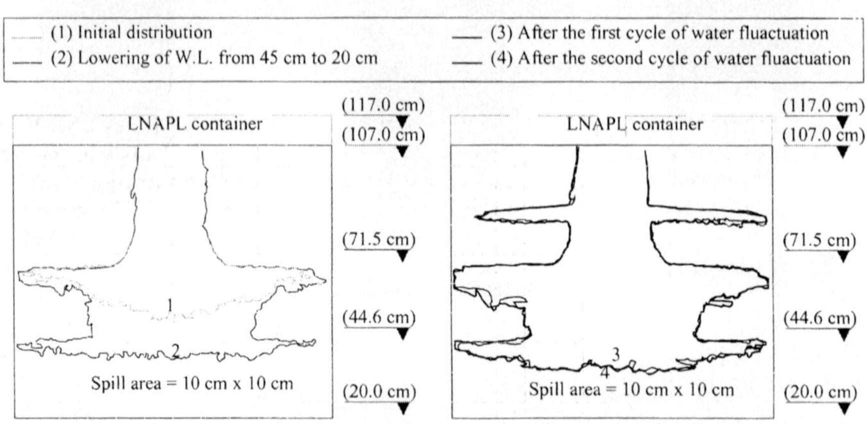

Figure 3: Redistribution of LNAPL in 2-D tank

Redistribution of LNAPL was investigated by fluctuating water table between 70 cm and 20 cm heights after an initial lowering of water level from 45 cm to 20 cm height. Figure 3 shows redistribution pattern of LNAPL in 2-D tank. It can be seen from Figure 3 that during the initial lowering of water level from 45 cm to 20 cm, LNAPL migrated in a trapezium shape till reaching the capillary fringe corresponding to a water level of 20 cm and ther distributing above it. During elevation process of water level from 20 cm to 70 cm with a rate of 10 cm/hour, it was observed that no further lateral migration of LNAPL occurred indicating that LNAPL migrated through the previously invaded pore voids. However, during the equalisation period which lasted for 48 hours, continuous accumulation of LNAPL above the top of the capillary fringe corresponding to a water level of 70 cm resulted in a third poo being formed as shown in Figure 3. The new pool took the shape of convex upper surface with flat bottom.

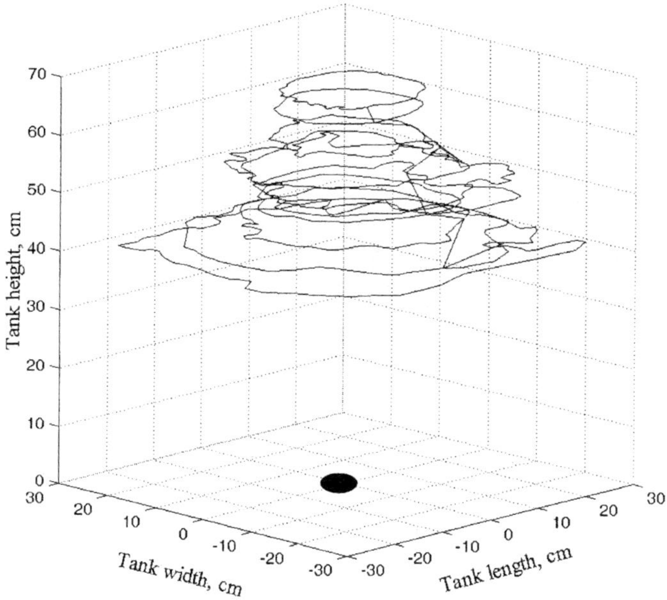

Figure 4: Distribution and redistribution of LNAPL in 3-D tank

Figure 4 presents the results of LNAPL distribution in 3-D tank when 4.5 l of LNAPL was spilled over an area of 10 cm x 10 cm. It should be noted that the 3-D LNAPL migration pattern was obtained at the end of the test by taking images of LNAPL distribution at arbitrary horizontal planes, when it was observed that large variation in terms of LNAPL distribution occurred. The axes in Figure 4 indicate the internal boundary of the 3-D tank. Of note the water level was kept at 20 cm height during the infiltration period and then water was fluctuated between 20 cm and 40 cm heights. Inspection of Figure 4 reveals that the first 20 cm of LNAPL migration (from 67 cm to 47 cm heights) corresponds to the initial distribution of LNAPL. It can be depicted that LNAPL migrated in an incomplete conical shape. The measured diameter of the LNAPL migration cone underneath the LNAPL tank was 19 cm. The dimension of the cross-sectional area increased with further downward migration. It was observed that there was a LNAPL pool at height of 54 ~ 56 cm. This would be due to the elevation of water level to 40 cm. The maximum diameter of the upper pool was found to be 45 cm. The thickness of the upper pool could not be measured accurately but it was estimated to be ~5 cm. Below 50 cm height, there was a reduction in the equivalent diameter of the cross section of the cone representing LNAPL migration suggesting that this part belongs to the initial distribution but it might have suffered slight enlargement during fluctuations of water level. Accurate measurements of the bottom LNAPL pool was practically difficult because of the alteration of capillary pressure head as the sand was removed below the top of the capillary fringe causing faster spreading of LNAPL above the saturated zone. It should be noted that the distribution of LNAPL at some levels inside the 3-D tank was not symmetrically around the vertical axis. This could be due to the presence of the transducers inside the sand sample, which resulted in preferential pathways of LNAPL increasing macroscopic heterogeneity of sand sample.

A link between the laboratory results obtained using 1-D, 2-D and 3-D models has not been clearly identified and/or established in the literature. Most researchers have carried out 1 or 2-D tests and generalised the results to field scale where 3-D flow is encountered. It should be noted that in 1-D column tests and in 2-D tanks, flow of water and LNAPL is constrained in two directions and in one direction respectively. Generally, in 1-D column set-up, it is assumed that the diameter of column is small enough so that flow of water and LNAPL occurs uniformly and parallel to the vertical axis whereas in 2-D flow, it is assumed that flow of fluids is parallel to the tank walls and is uniform throughout the third direction. In 3-D flow, LNAPL flows radially without boundary constraints. It is therefore worth examining the linkage between 1-D to 2-D then to 3-D LNAPL flow patterns.

The results of LNAPL distribution in 1-D model show that LNAPL distributed like a strip above the saturated zone (see, Figure 1), which is similar to observations from 2 and 3-D experiments. However, with limiting lateral spread in 1-D columns tests, there was no indication for the shape of the LNAPL pool such as those recorded in 2 and 3-D experiments. In 2-D test, a bowl-shape pool was clearly observed. Generally, the redistribution of LNAPL in 1-D model under a fluctuating water level causes LNAPL to redistribute itself below and above the water table and occupying a larger area than the initial one. In contrast the results from 2 and 3-D models showed that no significant change in the migration pattern between the water levels is recorded. In addition, second and third LNAPL pools were observed above the capillary fringes corresponding to the highest and lowest water level. These results are very descriptive and show the consistency between 2-D and 3-D model results. However, limiting the thickness of 2-D tank to 10 cm resulted in huge LNAPL pool being obtained, which is significantly greater than this obtained in 3-D tank for the same spill volume. Comparing the cross sectional area of LNAPL migration in the unsaturated and saturated zones shows that LNAPL migrated within almost the same area. This suggests that a good estimate of 3-D LNAPL contaminated area could be deduced if accurate 2-D data are available. It seem reasonable to conclude that 1-D results should be considered as an indicative only whereas 2-D patterns of LNAPL migration can be extended to describe different spill cases in 3-D models. In addition, extrapolation of 1-D results to simulate field scenarios should be undertaken with caution due to the great influence of boundary condition on the observed migration pattern. However, 2-D results can be used to simulate field situation but with the application of a scale factor as the measured lateral distribution of LNAPL in 2-D tanks was larger than that in 3-D tank for the same spill conditions.

Summary and Conclusions

A detailed laboratory investigation using 1, 2 and 3-D models was carried out to investigate various patterns of LNAPL migration as well as the influence of groundwater fluctuations. Main conclusions are as follows:

1- The column results indicated that LNAPL would spread as a thin layer above the capillary fringe zone. Fluctuations of groundwater caused LNAPL to spread out in large area.
2- The results obtained from 2-D tank showed that LNAPL migrated in a regular shape and the advancing LNAPL front formed an arc of circle. It was observed that LNAPL travelled with almost constant velocity in the unsaturated zone. LNAPL took longer to migrate through the saturated zone than in the unsaturated zone. Eventually, a bowl-shape LNAPL pool was observed.

3- Results suggest that lowering of water level causes LNAPL to migrate downwards following the water level but with a decreasing width taking the shape of trapezium. As a result of continuous migration of LNAPL, the amount of free LNAPL declines resulting in a smaller LNAPL pool floating over the new capillary fringe. Additionally, raising the water level led to a third pool to be formed above the capillary fringe with convex upper boundary.

4- Results obtained from 1-D (column) tests can be only used for general purpose since the boundary condition was found to have a significant influence controlling the distribution of LNAPL. However, results of 2-D and 3-D tank matched well indicating that 2-D model results would be good enough to model the flow of 3-phase system in unsaturated/saturated sand. The consistency between the 2-D results and those obtained in 3-D gives credibility to 2-D results.

5- It is suggested that a scale factor should be applied on 2-D LNAPL distribution at equilibrium to account for the restriction of LNAPL migration in the third direction. By the use of scale factor, a good estimate of the 3-D pattern can be deduced for the same spill conditions.

References

Abdul, S. (1988). Migration of petroleum products through sandy hydogeologic systems. Ground Water Monitoring Review, 8(4), 73-81.

Abriola, L.M. and Pinder, G.F. (1985). A multiphase approach to the modelling of porous media contamination by organic compounds 2. numerical simulation. Water Resources Research, 21(1), 19-26.

Eckberg, D.K. and Sunada, D.K. (1984). Nonsteady three-phase immiscible fluid distribution in porous media. Water Resources Research, 20(12), 1891-97.

Mohamed, M.H.A. (2003). Migration of light non-aqueous phase liquids in unsaturated and saturated sand. DPhil thesis, University of Bradford, U.K.

Schiegg, H.O. and McBride, J.F. (1987). Laboratory set-up to study two-dimensional multiphase flow in porous media. In Proceedings, Petroleum Hydrocarbons and Organic Chemicals in Ground Water Conference, National Water Well Association, Houston, 371-395.

Sharma, R.S. and Mohamed, M.H.A. (2001). Migration of LNAPL in an unsaturated/saturated Sand. Proceedings of the third Conference organised by the British Geotechnical Association and Cardiff School of Engineering.

Sharma, R.S. and Mohamed, M.H.A. (2003). Patterns and mechanisms of light non-aqueous phase liquid in unsaturated sand. Geotechnique, 53(2), 225-239.

Steffy, D.A., Johnston, C.D. and Barry, D.A., "Numerical simulations and long-column tests of LNAPL displacement and trapping by a fluctuating water table" Journal of Soil Contamination, 7(3), 325-356, 1998.

Van Geel, P.J. and Sykes, J.F., "Laboratory and model simulations of a LNAPL spill in a variably-saturated sand, 1. laboratory experiment and image analysis techniques", Journal of Contaminant Hydrology, 17:1-25, 1994.

Evaluation of Heavy Metal Contamination for Urban Soils

Catherine N. Mulligan, Concordia University, Montreal, Canada
Yi-Ting Huang, Concordia University, Montreal, Canada

Abstract

Since an increasing number of urban dwellers have come to recognize the close relationship between urban soils and water and air quality, as well as the growing popularity of urban agriculture, concern regarding soil contamination is on the increase. The objective of this research was to investigate the mobility of the heavy metals (lead and zinc) in contaminated urban soils. Soil samples from downtown Montreal, Canada were collected. The concentrations of lead and zinc of some soil samples were highly elevated, with the highest concentration being 2000 mg/kg (the standard limits for both metals are 300 mg/kg). Soil column leaching tests were carried out and they indicated that zinc has a higher mobility in the soils than lead. With this knowledge of heavy metal mobility in soils, the soil decontamination schemes can be designed or the potential impact of these soils can be evaluated.

Introduction

With increasing human population, cities are expanding into rural areas. Although urban lands contain significant areas which are impermeable, exposed land areas located within cities in park areas, recreational areas, building sites, golf courses, community gardens, and lawns, are highly important to city life. Due to a variety of human activities, the quality of urban soils is decreasing. Sources of contamination of urban soils by heavy metals include atmospheric depositions, runoff, waste disposal, excessive fertilizer, and herbicide use and chemical spills. Other signs of degradation are soil acidification, loss of organic matter and soil erosion. Thus, the importance of maintaining the quality of urban soils should not be minimized.

Heavy metals, unlike other organic contaminants, tend to remain in the soil system for a long period of time. For example, after the heavy metals are allowed to pass through the municipal waste treatment facility, they can return to the environment where they are persistent, cannot be biodegraded and can thus follow a number of different pathways. The metals can adsorb onto the soil, run off into rivers or lakes or leach in the groundwater, an important source of drinking water. Exposure to the heavy metals through ingestion or ingestion of drinking water and foods can lead to accumulation in animals, plants and humans. Children are also susceptible to exposure to surface soils and dusts containing the toxic heavy metals.

The mobility of the heavy metals depends on numerous factors. Understanding the factors influencing the mobility is required to design remediation technologies and evaluate the risk of exposure to humans and the ecosystem. Although an analysis of the heavy metal content in soils provides an indication of potential risk to the environment, determining the affinity of the heavy metals to the various fractions in the soil can provide a better indication of heavy metal mobility. The objectives of this paper were to determine the distribution of heavy metals among the various soil fractions and factors influencing the mobility of heavy metals in the soils. Urban soils from downtown Montreal, Canada containing lead and zinc contamination were studied.

Materials and Methods

Soil Sampling

Thirty-four soil samples were collected in the greater Montreal area. Ten of the soil samples were collected in a communal backyard garden and nearby area, as heavy traffic was a concern of the residents. All of the samples were taken within a 20 cm depth (topsoil). A systematic sampling scheme was used where every sampling location (usually a plot of 170 cm × 125 cm) was divided into 4 to 6 sub-areas. A standard steel hand auger was used for the collection of soil samples here. After samples were collected from each sub-area, they were mixed thoroughly to obtain a uniform soil sample representing each specific sample location. These moist, well-mixed soil samples were stored in several plastic centrifuge tubes before being transferred to the laboratory for various analyses. Soil samples were dried in an oven at 105°C for 24 hours before analysis with a series of sieves for soil particle size distribution.

Soil Digestion

Soil samples were digested for analysis of metal content with 4 M hydrochloric acid (HCl) (Roundhill 2001). Twenty ml of 4 M HCl were added to 2 g of each oven-dried soil sample. The acid-soil slurry was placed into centrifuge tubes on a shaker and shaken for 24 h. After this period, all samples were filtered with a No. 41 filter paper and then the clear supernatants were analyzed by a Perkins Elmer Model AAnalyst 100 Atomic Adsorption Spectrometer (AAS) for its heavy metal concentrations.

Cation Exchange Capacity (CEC)

The CEC measurement procedure used was according to the method of Chapman (1965). The cation exchange capacity ranged from 51.0 to 70.5 meq/100g for all soil samples tested with the exception of sample no. 9 which was 20.0 meq/100g.

Particle Size Determination

Soil samples were dried in an oven at 105°C for 24 hours and then a series of sieves (from 0.075 mm to 1.18 mm aperture openings) were utilized for analyzing soil particle size distribution. Most of the soil samples have the majority of the particle sizes ranging from 0.6 mm (sieve No. 30) to 0.15 mm (sieve No. 100); the fractions of soil particles bigger than 0.6 mm (those would not pass sieve No. 30) were discarded. The fraction of soil particles smaller than 0.15 mm (those pass through sieve No. 100) was kept because smaller size particles, especially clay-size particles, influence greatly the soil surface properties. From then on, all the experiments were performed on soil fractions that pass through sieve No. 30. The soils were thus in the range of coarse to fine sand.

Selective Sequential Extraction

The sequential extraction by Yong et al. (1993) was used for selective sequential extraction. 1.5 g of soil was used for each step. After each digestion, the supernatant was filtered and the clear supernatant was collected from each fraction. These supernatants were analyzed by atomic absorption spectrometry.

Specific Surface Area (SSA)

The SSA measuring method was based on wetting soil samples with a polar reagent (ethylene glycol monoethyl ether (EGME)) and subsequent equilibration (Heilman 1965). Excess liquid was removed until only a monomolecular layer of reagent was adsorbed by the soil surface. Measurements were taken every 1 hour 45min, until a constant weight was reached. The SSA was calculated by converting the mass of retained liquid to surface area using the formulae of Cerato (2002).

Permeability and Leaching Tests

To minimally disturb the soil structure, soil columns were utilized with intact soil samples to simulate field conditions. Soil permeability and leaching tests were carried out in the soil column. Water was leached through soil columns for permeability tests and the effluent flow was measured. From Darcy's law, the hydraulic conductivity for the soil samples was calculated by:

$$K = -\ (\Delta h/L)/q$$

where q is the specific flow rate (flow rate/cross sectional area), $\Delta h/L$ is the hydraulic gradient and K is the hydraulic conductivity.

Using soil columns, three types of leachates were evaluated: a 300 mg/L solution of lead chloride or zinc nitrate or a solution of mixed lead and zinc, each with a concentration of 300 mg/L. A continuous solution was passed through the soil column, effluents were collected every 2 to 10 pore volumes (PV) of solution passed through the column and the heavy metals were analyzed with an atomic adsorption spectrophotometer. The relative concentration of the effluent to influent was plotted (C/Co) versus the number of pore volumes.

Results and Discussion

Heavy Metal Analysis

Seven soil samples from the communal garden in area I, were digested with acid to test six concerned heavy metals concentrations: lead (Pb), zinc (Zn), chromium (Cr), cadmium (Cd), copper (Cu) and nickel (Ni). Four out of the six heavy metals tested were within their background concentration levels, thus they do not pose a threat to human health by the standard according to the Environmental Protection Agency (EPA) and Quebec soil quality criteria. However, lead and zinc concentrations (Fig. 1), especially for sample No. 6, were significantly higher than the threshold concentrations (Sittig 1994). The rest of the experiments thus focused on these two heavy metals in terms of their fate and transport in soil.

It can be seen from Figure 1 that the concentration of lead and zinc in urban soils varies greatly. The lead concentration ranges from almost zero (5 mg/kg) to almost 2000 mg/kg; the zinc concentration ranges from 100 mg/kg to over 2,000 mg/kg. According to the Ministry of Environment, Quebec soil quality criteria, the Pb and Zn contents of the collected soil samples are within B to C range (requiring monitoring, but no remediation and the soil is acceptable for residential/commercial use). The results by Murray (1999) showed the same level of heavy metal contamination for urban soils from the Montreal area.

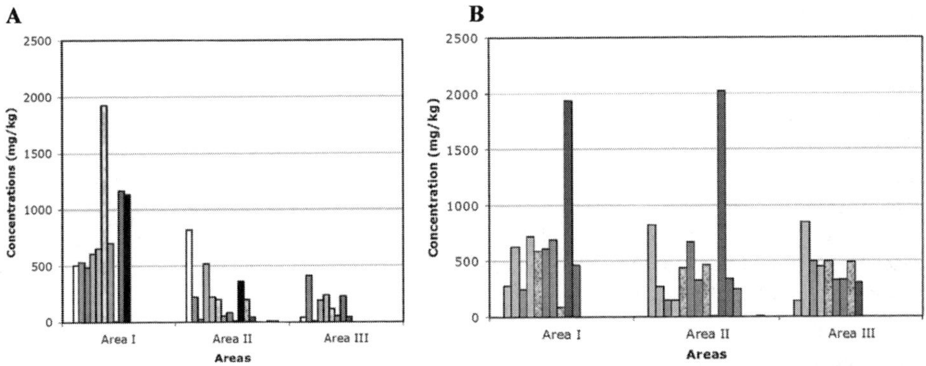

Figure 1: Lead (A) and zinc (B) concentrations in soil samples taken from various locations of three areas (Area I: 10 samples from a communal garden and locations near it, Area II: 15 samples near Concordia University. Area III: 9 samples from near Metro Vendome.

Depending on the history of land uses and the degree of leaching by rainfall and urban runoff in the areas, huge variations of heavy metal concentrations are expected. The concentrations of both lead and zinc in area I were substantially higher than the rest of the samples (zinc concentrations in area II were also high). Urban communal gardens are relatively small compared to agricultural land, but they are likely to be subjected to more concentrated pollution, since all of the gardens are easily contaminated from atmospheric and particulate material deposition. The area where this communal garden is situated used to be a dumping ground for industrial activities and is only a few blocks away from two major traffic streets. The metals from the industrial waste, traffic emissions and other contamination sources such as paint chips, bonfire ash, and the lead from children toys all contribute to the higher concentrations of lead and zinc.

Sequential Extraction Tests

Sequential extraction tests were performed on several of the soil samples. The general trends are presented in Table 1. It can be seen that both lead and zinc have different affinities towards different soil fractions. The affinity order agrees completely with the study of

Table 1. Heavy metals affinity for various soil fractions

Heavy Metal	Affinity towards soil fraction (average % of total metal content for all soil samples tested)
Lead (Pb)	Soil organic matter (39.0%) > Oxides (36.1%) > Carbonate (14.8%) > Residual (9.8%) > Exchangeable (0.4%)
Zinc (Zn)	Oxides (50.9 %) >> Residual (19.7%) > Soil organic matter (16.1%) > Carbonate (12.8%) > Exchangeable (0.5%)

Mulligan and Dahrazma (2003) for sediments. Both Pb and Zn have a higher affinity towards soil fractions of soil organic matter and oxides; while small amounts of both metals are attached to the exchangeable fraction. The humus in soil organic matter has a large specific surface area and contains various functional groups; therefore, a series of chemical and physical interactions between metals cations and soil organic surfaces, including complexation, adsorption and chelation, take place there. Therefore, heavy metals would be bound to soil organic matter more easily than other soil fractions. Metals bound to the exchangeable fraction of soil are mostly physically adsorbed (by electrostatic force) to the soil surfaces, thus the bonding is weaker compared to other binding mechanisms. The moderate to high degree of leaching by rainfall and the competition from other cations present in the leachate solution possibly explains why only a limited amount of Pb and Zn were retained by this soil fraction. The results agree with those of Lu et al. (2003), who showed that significant association of Pb and Zn with soil oxides and organic matter is partially due to the sources of these heavy metals. Oxide is the most abundant soil fraction that contains zinc.

Permeability Tests

Permeability tests were performed on ten soil samples. Table 2 indicates the permeability coefficients calculated from Darcy's Law which ranged from 10^{-3} to 10^{-1} cm/sec. This is within the range of silty (fine) sand.

Table 2. Soil permeability tests

Sample no.	No. 10	No. 11	No. 17	No. 23	No. 24	No. 25	No. 26	No. 28	No. 30	No. 31
Permeability coefficient (cm/sec)×10^{-2}	2.3	2.4	0.9	3.9	2.1	1.6	0.6	1.9	0.6	3.5

By considering soil permeating fluids as well as soil structural properties, the permeability coefficient (K) can be expressed according to the Poiseuille and Kozeny-Carman (K-C) model:

$$K = (C_S n^3 \gamma) / (\eta T^2 S_w^{\ 2})$$

From this model, Figure 2 can be obtained. The soil permeability tests resulting from Table 2 can be plotted in this graph to determine the corresponding wetted surface area for each soil column test. In comparison to a study by Yong and Mulligan (2003), the soil samples have a much higher soil permeability coefficient than those in that study. In this study it was in the range of 10^{-4} to 10 compared to the range of 10^{-9} to 10^{-6}. A much smaller soil wetted surface area is thus expected.

Leaching Tests

Five soil samples with low initial heavy metal contents were chosen for the leaching tests. A sixth sample with a much higher metal content (sample no. 10) was also chosen for comparison. They were divided into two tests: (I) soils were leached with single heavy metal solutions, e.g. lead and zinc, of concentration of 300 mg/(L water) (II) soils were leached with a mixed lead and zinc solution, each heavy metal with a concentration of 300 mg/L. Table 3 summarizes the results of the breakthrough pore volumes for each sample, under each condition (single or mixed metal leachate).

It can be seen that for all of the soil samples, it takes fewer pore volumes of zinc solution to saturate the soils in comparison to the lead solution. This is consistent with the study of Phadungchewit (1990), which concluded that in soil solution, zinc has higher mobility than lead. From the results of the selective sequential extraction (Table 1), it also shows that lead has the highest affinity towards the soil organic matter fraction, while zinc towards the soil oxides fraction. Therefore, zinc would be more mobile in soils than lead. This is also consistent with the result obtained in Table 3.

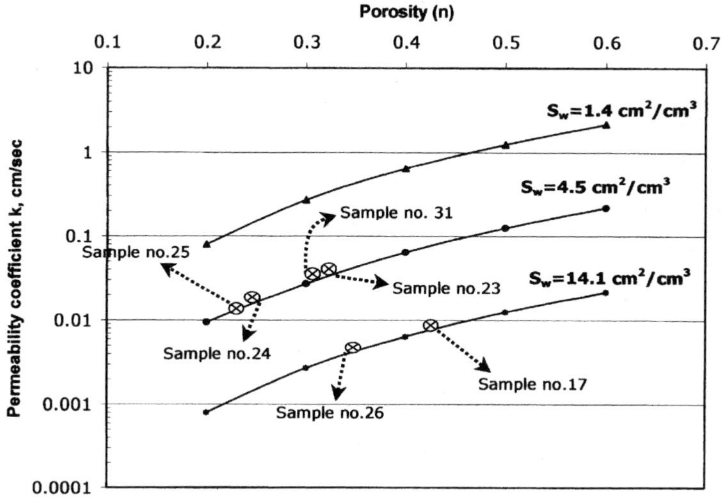

Figure 2. Relationship of the soil wetted surface area (S_w) to the permeability coefficient (K, cm/sec) and porosity of the soil. Experimental points are marked by ⊗

Table 3. Summary of breakthrough curve results

Sample No.	No. 8	No.10	No. 19	No. 22	No. 25	No. 26
Initial heavy metal content before leaching (mg/kg)	Pb (5) Zn (89)	Pb (1132.5) Zn (462.5)	Pb (10) Zn (10)	Pb (45) Zn (250)	Pb (10) Zn (10)	Pb (42) Zn (138)
Pb leachate breakthrough (PV)	162	100	168	156	220	121
Zn leachate breakthrough (PV).	115	17	156	72	115	53
Pb and Zn leachate breakthrough (PV)	Pb (156) Zn (100)	Pb (107) Zn (15)	Pb (160) Zn (68)	Pb (126) Zn (60)	Pb (199) Zn (129)	Pb (116) Zn (26)

For a mixed zinc and lead solution, the pore volumes for breakthrough to occur were reduced slightly for both Pb and Zn, due to the competition for adsorption sites on the soil surfaces. Table 3 also shows that although sample no. 10 contained high concentrations of Pb (1132.5 mg/kg) and Zn (462.5 mg/kg) before the leaching test, accumulation of additional amounts of heavy metals was possible but this soil sample became saturated more quickly than the other soil samples. For the soil samples with lower concentrations of Pb and Zn, the pore volumes for breakthrough ranged from 121 PV to 220 PV for Pb, and 53 PV to 156 PV for Zn.

Three soil samples originally contaminated with approximately the same level of Pb and Zn were chosen for comparison of their heavy metal retention capacities and to evaluate the dominant factors for the breakthrough pore volumes. Sample no. 8 (5 mg Pb/kg, 89 mg Zn /kg), sample no. 25 (10 mg Pb /kg, 10 mg Zn /kg) and sample no. 26 (42 Pb mg/kg; 138 mg Zn /kg) were chosen. For the lead leaching tests, sample no. 25 has the highest Pb retention capacity, with breakthrough pore volumes of 220 PV, compared to 162 PV for sample no. 8 and 121 PV for sample no. 26. Sample no. 25 as well as sample no. 8 has the same Zn retention capacity of the breakthrough pore volume of 115 PV while sample no. 26 has a lower retention capacity of 53 PV. Among these three soil samples, soil sample no. 25 showed the highest metal accumulation capacity for both lead and zinc, while soil sample no. 26 had the lowest capacity.

The chemical and surface properties of soils which including the cation exchange capacity (CEC) and the specific surface area (SSA) contribute greatly to soil metal retention capacity (breakthrough pore volume). The CEC and SSA can be related through the equation below (Yong and Mulligan 2003):

$$CEC = 10^5\ \sigma_s\ (SSA)$$

where CEC is the cationic exchange capacity (meq/100 g soil), σ_s is the soil surface charge density (keq/m^2) and SSA is the specific surface area (m^2/kg soil). Taking the values of both CEC and SSA for the three soil samples, Fig. 3 shows that soil sample no.25 has the highest value of CEC/SSA (7.06), while soil sample no. 26 has the lowest value (1.28). Figure 3 also shows the influence of this property of the soil surface properties with metal retention capacity. The soil sample with the highest retention capacity has the highest CEC/SSA value.

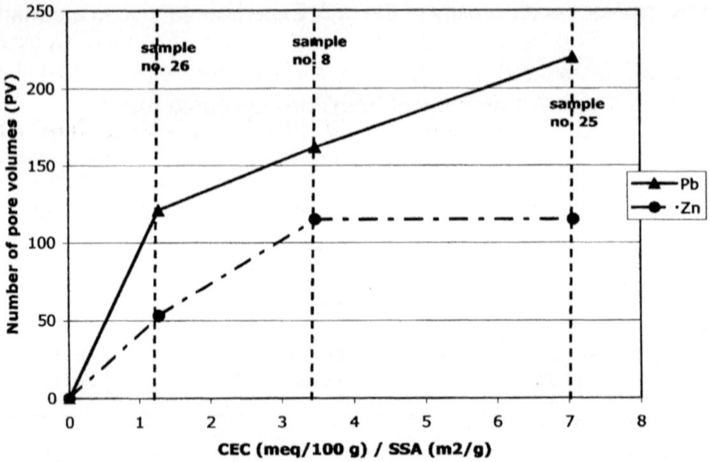

Figure 3: Number of breakthrough pore volumes vs CEC / SSA

Conclusions

Urban soil properties in relation to heavy metal contamination were investigated in this study.

Elevated lead and zinc concentrations were found in some urban soil samples up to 2000 mg/kg for lead and zinc which is much higher than the standard limit for both metals of 300 mg/kg. The majority of soil samples had a texture of fine sand (loamy and silty). From the selective sequential extraction tests, it can be determined that Pb has the highest affinity toward the organic soil fraction while Zn has the highest affinity toward soil oxides. From the leaching tests, it was determined that zinc has a higher mobility in soil than lead. Thus, lead is retained preferentially in the soil in comparison to zinc. When both metals are leached through soil columns, it took fewer pore volumes for both lead and zinc solutions to saturate the soil columns, due to the competition of retention sites. The soil sample with the highest retention capacity has the highest CEC/SSA value which further demonstrates that the understanding and knowledge of the soil surface properties in retaining heavy metal contaminants is highly important in determining mobility and risk of exposure.

Acknowledgements

The authors would like to acknowledge the financial support of NSERC and the Faculty of Engineering and Computer Science of Concordia University.

References

Chapman, H. D. (1965) Determination of CEC at pH 7 with ammonium acetate. Methods of soil analysis. *American Society of Agronomy*. 9: 159-165.

Cerato, A. B. and Lutenegger, A. J. (2002) Determination of Surface Area of Fine-Grained Soils by the Ethylene Glycol Monoethy Ether (EGME) Method. *Geotechnical Testing Journal,* 25: 314-320.

Heilman, M.D., Carter, D.L., and Gonzalez, C.L. (1965) The ethylene glycol monoethyl ether (EGME) technique for determining soil-surface area. *Soil Science* 100: 409-413.

Lu, Y., Gong, Z., Zhang, G., and Burghardt, W. (2003) Concentrations and chemical speciation of Cu, Zn, Pb and Cr of urban soils in Nanjing, China. *Geoderma,* 115: 101 – 111.

Mulligan, C. N. and Dahrazma, B. (2003) Use of selective sequential extraction for the remediation of contaminated sediments. *2nd International Symposium on Contaminated Sediments*, Quebec, Canada, May 26.

Murray, P. Ge, Y. and Hendershot, W. H. (1999) Evaluating three trace metals contaminated sites: a field and laboratory investigation *Environmental Pollution*, 107: 127 – 135.

Phadungchewit, Y. (1990) *The role of pH and soil buffer capacity in heavy metal retention in clay soils.* Ph.D thesis, Department of Civil Engineering and Applied Mechanics, McGill University.

Roundhill, D. M. (2001) *Extraction of Metals from Soils and Water.*" Kluwer Academic/Plenum Publishers, New York.

Sittig, M. (1994) *World-wide Limits for Toxic and Hazardous Chemicals in Air, Water and Soil.* Noyes Park Ridge, N.J.

Yong, R.N., Galvez-Cloutier, R., and Phadingchewit, Y. (1993) Selective extraction analysis of heavy-metal retention in soil, *Canadian Geotechnical Journal,* 30(5): 834-847.

Yong, R.N and Mulligan, C.N. (2003) The impact of clay microstructural features on the natural attenuation of contaminants. *Applied Clay Science* 23: 179-186.

Diffusion and Retardation of Benzene and Toluene Vapours inside a Low Moisture Layer of the Unsaturated Zone

Panagiotis C. Nerantzis, University of Strathclyde, Glasgow, United Kingdom (panagiotisnerantzis@yahoo.co.uk)
Mark R. Dyer, University of Strathclyde, Glasgow, United Kingdom

Abstract

The pore water in the unsaturated zone of the ground effects both abiotic and biotic processes concerning the fate and gaseous transport of volatile organic compounds (VOC). The diffusion and the abiotic retardation due to presence of pore water of Benzene and Toluene vapours were investigated with five soil column experiments using silica uniform medium coarse Leighton Buzzard sand with relative low field capacity. The results show that Benzene and Toluene vapours migrated at same rate in case of multicomponent and single component systems, even if the vapour concentration at the source is lower in the first case. However, both Benzene and Toluene vapours migrate faster inside the low moisture content (1.4%) than the air dried sand even if the effective diffusion coefficient is smaller and the retardation is higher due to the presence of water in the first case. The results cast doubt on the validity of current theory for migration of pollution vapours in granular soils at low moisture content.

Introduction

In the vadose zone, the effective gas phase diffusion coefficient of a chemical compound is estimated typically using the chemical-specific diffusion coefficient in air and the soil tortuosity factor ($D_e = \tau D_a$). The most widely used approximation for the tortuosity factor is that usually attributed to Millington and Quirk (1961) ($\tau = \theta_a^{7/3}/\theta^2$). The presence of the water inside the soil pores reduces the tortuosity factor. Hers et al. (2000) measured in situ tortuosity factors based on push pull tests and found that the tortuosity factor decreases (range 0.22-0.039) as the moisture content of the ground increases (range 5-12%) for a relative uniform soil. As a consequence, a smaller effective diffusion coefficient results slow vapour migration inside the soil. Jury et al. (1990) discussed the effect of the moisture content on the effective diffusion coefficient and suggested that increasing the water content of a layer significantly decreases the volatilization flux through this layer.

Furthermore, the existence of soil water inside the ground has as result the retardation of VOC during the gaseous transport. Sorption of VOC vapours at the soil water and at air-water interface are the main mechanisms. Corley et al. (1996) derived experimentally retardation factors for vapour transport indicated that retardation increases with moisture content for four VOC including Benzene.

In addition, Nerantzis and Dyer (2005) showed using soil column experiments that soil water induces biodegradation activating indigenous soil microbes which start to utilize BTEX vapours to gain energy for growth.

The aim of this study is to investigate the diffusion and associated retardation of Benzene and Toluene vapours within a ground with low moisture content (1.4%) compare to an air-dried. The migration of vapours in a multicomponent system, where a mixture of BTEX compounds is used, and single component systems, where pure compounds are used as source of vapours, were also investigated. The results of five laboratory soil column experiments are presented here. Breakthrough curves for multicomponent and single component systems are compared as well as curves from single component systems with air dried and low moisture content sand.

Materials and Methods

Chemicals

Pure products (purity > 99%) of Benzene, Toluene, Ethylbenzene and m-Xylene are obtained from BDH Laboratory Supplies and Acros Organics. Table 1 shows some properties of BTEX compounds.

Table 1. Properties of BTEX compounds.

Compound	Formula	Molecular Weight (g/mol)	Density (g/ml)	Vapour Pressure (mmHg) (20-25 °C)	Diffusion Coefficients in Air, D_a (cm^2/s)
Benzene	C_6H_6	78.1	0.8787	95.2	0.093
Toluene	C_7H_8	92.4	0.8669	30	0.085
Ethylbenzene	C_8H_{10}	106.2	0.867	10	0.076
m-Xylene	C_8H_{10}	106.16	0.8642	8	0.070

Characterization of the Sand

The sand used in this study is silica uniform medium-coarse Leighton Buzzard sand (DA 16/30) (WBB Minerals Ltd, UK). The Particle Size Distribution using the Dry Sieving Method (BS1377 1990: Part 2:9.3) is: < 1.18 mm: 100 %, < 0.6 mm: 9 % and < 0.425 mm: 0 %. The Field Capacity using the Pressure Plate Method is 1.6%. The Particle Density using the Small Pyknometer Method (BS1377 1990: Part 2:8.3) is 2.65 g/cm^3. The Fraction Organic Carbon is 0.001452 mg organic carbon per mg of sand. Initially, the sand was air-dried. The microbial population was measured using the Plate Count Method and no viable microbes were found.

Soil Column Experiments

Five one-dimensional vertical soil column experiments were carried out at room temperature (23±3 °C) using stainless steel (SS316) gas-tight columns. The sand is accommodated inside the main part of the column tube with 50 cm height and 10 cm internal diameter. A media support screen (0.063 mm mesh) at the bottom of the main part of column retains the sand. A stainless steel fuel reservoir is attached at the bottom end of the main column. The reservoir is partially filled with pollutants. A headspace was left in every case to facilitate the formation of pollutant vapours. The bottom end of the sand column is in direct contact with the fuel headspace. Figure 1 shows the columns set up. Every column is equipped with 2 sampling ports. The first is positioned at the reservoir and the second one at the top end of the column. These ports are made from stainless steel (SS 316) fittings (Swagelok). GC septa (SGE) fitted into a hole of 10 mm diameter of this fitting allow the sampling with a syringe. Gas samples were taken with gas-tight syringes equipped with Teflon® tip plungers and stainless steel needles (length 70 mm, outside diameter 0.50 mm) (SGE).

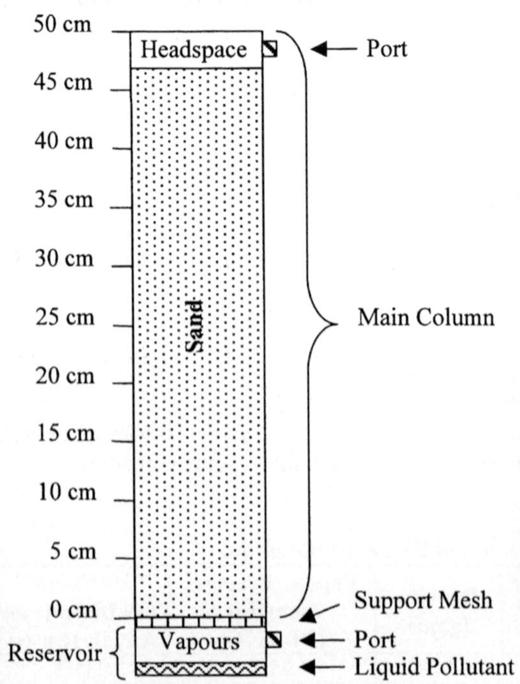

Figure 1. Columns Set Up

The 1st column was packed homogeneously with air-dried sand and the reservoir was filled with a 60 ml liquid mixture of BTEX (15 ml of each compound). The 2nd column was packed with air-dried sand and the reservoir was filled with 15 ml pure liquid Benzene. The 3rd column was packed with moisturized (1.4 % moisture content) sand and the reservoir was filled with 30 ml pure liquid Benzene. The 4th column was packed with air-dried sand and the reservoir was filled with 30 ml pure liquid Toluene. The 5th column was packed with moisturized (1.4 % moisture content) sand and the reservoir was filled with 30 ml pure liquid Toluene. Table 2 and 3 show the soil properties that achieved. Special attention was paid to achieve similar densities among the columns and also the distribution of moisture content to be homogeneous. The soil water is considered immobile among the soil particles. No fresh fuel was added in any of the experiments until the end. At the end of all the experiments fuel still existed in the reservoirs. Table 4 shows the mole fractions and the initial vapour concentrations at multicomponent and single component systems.

Table 2. Soil properties after packing.

	Bulk Density (g/cm^3)	Dry Density (g/cm^3)	Moisture Content (%)	Porosity, θ	Void Ratio
1st Column with Air-dried Sand and pure BTEX	1.71	1.71	0.0	0.36	0.55
2nd Column with Air-dried Sand and pure Benzene	1.72	1.72	0.0	0.35	0.54
3rd Column with Moisturized Sand and pure Benzene	1.71	1.68	1.4	0.36	0.57
4th Column with Air-dried Sand and pure Toluene	1.70	1.70	0.0	0.36	0.56
5th Column with Moisturized Sand and pure Toluene	1.71	1.69	1.4	0.36	0.57

The concentrations of BTEX vapours were measured at the two sampling ports. At the end of each experiment the concentrations of CO_2, O_2 and N_2 were measured to confirm that they were at atmospheric levels and no biodegradation occurred inside the columns.

Table 3. Soil properties after packing.

	Volumetric Water Content, θ_w	Volumetric Air Content, θ_a	Theoretical Tortuosity Factor, τ	Effective Diffusion Coefficients, D_e (cm^2/s)
1st Column with Air-dried Sand and pure BTEX	0.00	0.36	0.71	Benzene: 0.066 Toluene: 0.060
2nd Column with Air-dried Sand and pure Benzene	0.00	0.35	0.71	Benzene: 0.066
3rd Column with Moisturized Sand and pure Benzene	0.02	0.34	0.61	Benzene: 0.057
4th Column with Air-dried Sand and pure Toluene	0.00	0.36	0.71	Toluene: 0.060
5th Column with Moisturized Sand and pure Toluene	0.02	0.34	0.61	Toluene: 0.052

Table 4. Mole fractions and initial vapour concentrations of fuel mixtures.

	Multi Component System		Single Component System
Compound	**Mole Fraction**	**Initial Vapour Concentration (23 °C) (g/m^3)**	**Initial Vapour Concentration (23 °C) (g/m^3)**
Benzene	0.305	122.8	403.1
Toluene	0.254	38.2	150.3
Ethylbenzene	0.221	12.7	-
m-Xylene	0.220	10.2	-

Gas Analysis

Gas concentrations of BTEX compounds are analysed by injecting 50 µl of gas into a Unicam ProGC Gas Chromatograph (GC) (Unicam, UK) equipped with a CP - Poraplot Q (Varian) capillary column (column length 27.5 m, internal diameter 0.25 mm, film thickness 8 µm) and a Flame Ionisation Detector (FID) (GC - FID). The following chromatography method is used: Detector Temperature = 245 °C, Injector Temperature = 245 °C, Column Temperature = 215 °C (Isothermal), Split = 10 ml/min, Carrier Gas Pressure = 28 psi (Helium) and Make Up Gas Pressure= 5 psi (Helium). The detection limits are: Benzene = 0.1 g/m^3, Toluene = 0.6 g/m^3, Ethylbenzene = 0.7 g/m^3 and m-Xylene = 0.7 g/m^3 in air. Calibration standards of BTEX are used to quantify the unknown samples. The standards are prepared by vaporising known quantities of BTEX in certain volume of air.

The concentration of CO_2 in soil air is analysed by injecting 250 µl of gas into a Unicam ProGC Gas Chromatograph (GC) (Unicam, UK) equipped with a CP-CarboPLOT P7 (Varian) capillary column (column length 27.5 m, internal diameter 0.53 mm, film thickness 25 μm) and a Thermal Conductivity Detector (TCD) (GC - TCD). The following chromatography method is used: Detector Temperature = 110 °C, Injector Temperature = 80 °C, Column Temperature = 85 °C (Isothermal), Split = 5 ml/min, Carrier Gas Pressure=10 psi (Helium) (Carrier Gas Flow=8.5 ml/min) and Make Up Gas Pressure=5.5 psi (Helium) (Make Up Flow= 8.5 ml/min). The detection limit is 0.04% of CO_2 in air. Certificated calibration standards of CO_2 in air are used to quantify the unknown samples (Greyhound Chromatography & Allied Chemicals, UK).

The concentrations of O_2 and N_2 in soil air are analysed by injecting 10 μl of gas into a Unicam ProGC Gas Chromatograph (GC) (Unicam, UK) equipped with a CP-CarboPLOT P7 (Varian) capillary column (column length 27.5 m, internal diameter 0.53 mm, film thickness 25 μm) and a Thermal Conductivity Detector (TCD) (GC - TCD). The following chromatography method is used: Detector Temperature = 110 °C, Injector Temperature = 80 °C, Column Temperature = 30 °C (Isothermal), Split = 15 ml/min, Carrier Gas Pressure=2.5 psi (Helium) (Carrier Gas Flow=3.0 ml/min) and Make Up Gas Pressure=2 psi (Helium) (Make Up Flow=3.0 ml/min). The detection limit was 5 % of O_2 in air. Certificated calibration standards of O_2 in air are used to quantify the unknown samples (Greyhound Chromatography & Allied Chemicals, UK).

Results

The ratio (C/C_o) of concentrations between the source (C_o) and the top headspace (C) against time is plotted for each column experiment (Figures 2, 3, 4, 5, 6 and 7). In every experiment this ratio initially increases at a great rate and then remains relative constant increasing slowly and reaching asymptotic the ratio 1.0. The concentrations of Carbon Dioxide and Oxygen were measured after the end of each experiment and in all cases were at the atmospheric levels, which mean that there was no degradation of pollutant vapours inside the gas-tight columns. The shape of these breakthrough curves is affected only by the diffusion and the abiotic retardation mechanisms. Sorption at soil particles, at soil water and at air-water interface tend to retardant the transport of pollutant vapours inside the columns. Volatilisation of compounds from liquid source and the formation of vapours also influence the shape of breakthrough curves.

The gaseous concentration of Benzene at the headspace above the liquid source in the multicomponent system with air-dried sand (122.8 g/m^3) is lower than in the single component system (403.1 g/m^3). This occurrence does not seem to affect the shape of the breakthrough curve. In both cases 48 hours are needed to reach the ratio of 0.8 (Figures 2 and 3).

Toluene vapours migrate with similar rate inside the air dried sand in both single and multicomponent systems, even if the vapour concentration at the source is higher in the first system (150.3 g/m^3 compare to 38.2 g/m^3). The ratio of 0.7 was reached after 144 hours in both systems (Figure 5 and 6).

Comparing the breakthrough curves for Benzene vapours between air-dried and moisturised sand (1.4%) in single component systems, the 0.8 ratio was reached in about 48 hours compare to 16 hours respectively (Figures 3 and 4). Hence, Benzene vapours travel 3 times faster inside the moisturised sand.

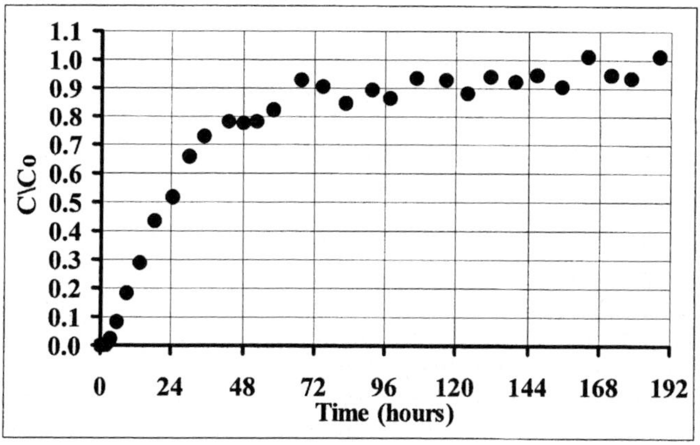

Figure 2. Breakthrough curve of Benzene in multicomponent system with air-dried sand.

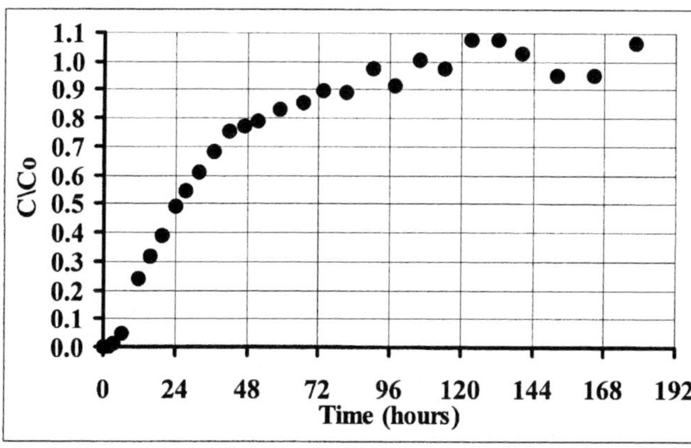

Figure 3. Breakthrough curve of Benzene in single component system with air-dried sand.

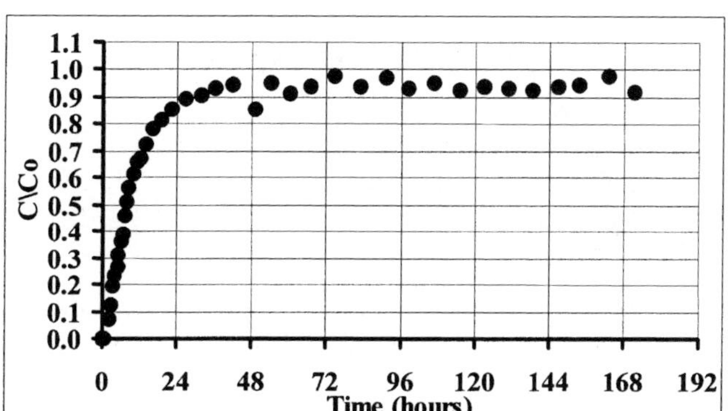

Figure 4. Breakthrough curve of Benzene in single component system with moisturized sand.

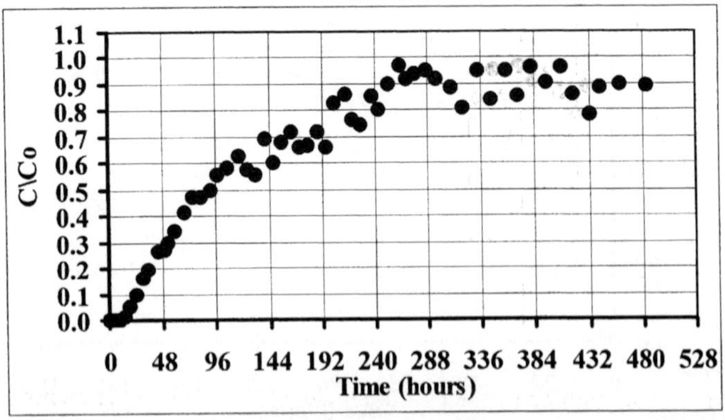

Figure 5. Breakthrough curve of Toluene in multicomponent system with air-dried sand.

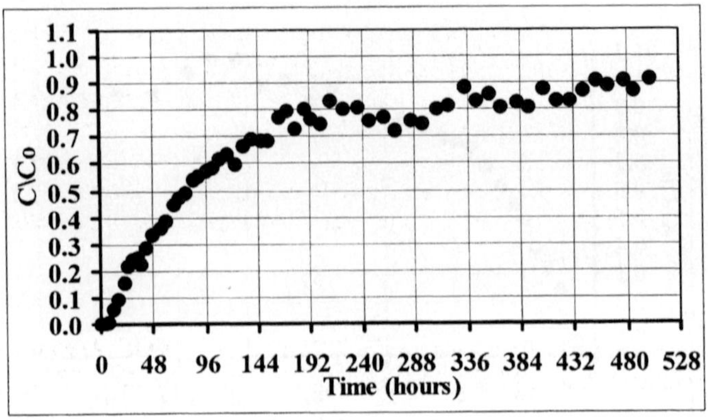

Figure 6. Breakthrough curve of Toluene in single component system with air-dried sand.

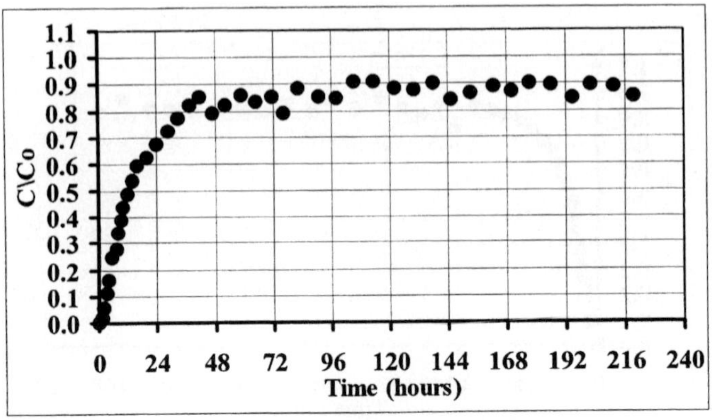

Figure 7. Breakthrough curve of Toluene in single component system with moisturized sand.

In both cases the initial vapour concentration is the same (403.1 g/m^3) but there is difference in theoretical tortuosity factor 0.61 in moisturised compare to 0.71 in air-dried sand and as well in effective diffusion coefficient 0.057 cm^2/sec compare to 0.066 cm^2/sec respectively.

Toluene vapours show similar migration attributes. They migrate faster inside the moisturised sand (1.4%) compare to air-dried; 24 hours are needed to reach the 0.7 compare to 144 hours (Figures 6 and 7). Hence, Toluene vapours travel 6 times faster inside the moisturized than in air-dried sand. In both cases the initial vapour concentration is the same (150.3 g/m^3) but there is difference in tortuosity factor 0.61 in moisturised compare to 0.71 in air-dried sand and as well in effective diffusion coefficient 0.052 cm^2/sec compare to 0.060 cm^2/sec respectively.

Conclusions

The results from the multicomponent and single component experiments show that the migration of Benzene and Toluene vapours inside the air-dried sand are independent of the vapour concentration at the source. As a consequence Benzene and Toluene vapours migrate at the same rate whether the source is a petroleum fuel mixture or pure product.

Benzene and Toluene vapours diffused faster inside the moisturised sand even though the theoretical effective diffusion coefficient is smaller than the air-dried sand. Even though, the retardation of vapours in moisturised sand caused by sorption on soil water and on air-water interface is expected to be higher than air-dried silica sand. Where as, according to the existing theory the Benzene and Toluene vapours should migrate faster inside the air-dried than in moisturised sand. As a consequence these experimental results cast doubt on the validity of existing theory for migration of pollution vapours in unsaturated granular soils wetted to low moisture content.

References

Corley T.L., J. Farrell, B. Hong and M.H. Conklin 1996. "VOC Accumulation and Pore Filling in Unsaturated Porous Media." Environ. Sci. Technol., 30, 2884-2891.

Hers I., R. Zapf-Gilje, L. Li and J. Atwater 2000. "Measurement of In Situ Gas-Phase Diffusion Coefficients." Environmental Technology, Vol. 21, 631-640.

Jury W.A., D. Russo, G. Streile and H. El Abd 1990. "Evaluation of Volatilization by Organic Chemicals Residing Below the Soil Surface." Water Resources Research, Vol. 26, No. 1, Pages 13–20.

Millington R.J. and Quirk J.P. 1961. "Permeability of porous solids." Trans. Faraday Soc, Vol. 57, Pages 1200–1207.

Nerantzis P. C. and Dyer M. R. 2005. "The effect of a soil layer of high moisture to the transport of BTEX vapours in the unsaturated zone." Proceedings paper of In Situ and On-Site Bioremediation, The Eighth International Symposium, Baltimore, Maryland, U.S.A – 2005.

Sodium Diffusion in Compacted Lateritic Soil

Osinubi, K. J., M. ASCE Assoc. Prof. Dept of Civil Engrg Ahmadu Bello Univ., Zaria 810001, Nigeria

Nwaiwu, C. M. O., AM.SCE Snr. Lect.Dept. of Civil & Water Res. Engrg., Univ. Of Maduguri, Maiduguri, Nigeria.

Abstract

The effective diffusion coefficients, D*, of a monovalent cation (Na^+) diffusing in three compacted lateritic soil samples were measured in the laboratory. The sodium ion is contained in municipal solid waste leachate. The combined effects of soil plasticity and grading represented by the soil plasticity product on the measured D* values were examined. The calculated D* values varied from 3.27×10^{-6} to 3.35×10^{-6} cm^2/s and are sensitive to variations in the plasticity product. In general, the retardation factors increased with higher values of plasticity product while the partitioning coefficients, effective diffusion coefficients, tortuosity factors as well as apparent diffusion coefficients decreased with higher values of plasticity product. The measured D* values are in good agreement with corresponding values from previous studies. The D* values determined from reservoir concentrations are higher than the D* values determined from soil concentration profiles.

Introduction

Hydraulic conductivity is regarded as the primary parameter of concern in the design of waste containment barriers. Traditionally, the design of these engineered barriers that retard the release of contaminants from land disposal facilities has been solely based on the hydraulic conductivity of the material (Daniel and Shackelford, 1988).

The design of earthen barriers has often been based on the assumption that the hydraulic conductivity controls the rate of leachate migration. However, a number of field studies have indicated that diffusion is an important, if not dominant, transport process in fine-grained soils (Quigley and Rowe, 1986). For compacted clayey soils, it is also necessary to consider that diffusion, and not advection governs the transport of contaminants through them. Recognition of the potential contribution of diffusion to contaminant transport through low-permeability barrier materials has stimulated substantial research aimed at measuring the effective diffusion coefficients D* of potential contaminants with potential soil barrier materials. Laboratory diffusion testing will become a necessary component of the overall design and evaluation processes associated with waste disposal practice.

Researchers that studied diffusion of contaminants under laboratory conditions mainly used compacted clay soils of temperate zones (Shackelford and Daniel, 1991) or sand - bentonite mixtures (Gillham *et al.*, 1984) or soil - bentonite mixtures (Mott and Weber, 1991). However, not much work has been reported on fine - grained lateritic soils. The objectives of this study are to evaluate the diffusion of sodium (Na^+) in compacted lateritic soils and to show the dependence of this diffusional transport on soil index properties.

Materials and Methods

Soils

Three samples of fine-grained lateritic soils obtained from Shika, near Zaria (latitude 11° 15' N and longitude 7 45' E) in Northern Nigeria and designated as SZL1, SZL2 and SZl3 were used in this study. The soils are reddish-brown in colour and belong to the group of ferruginous tropical soils derived from acid igneous and metamorphic rocks (Osinubi, 1998a). The mineralogical composition of these soils was investigated qualitatively with the use of the differential thermal analysis (DTA) method. The DTA thermograms of soils SZL1 and SZL2 suggest that they contain kaolinite as the predominant clay mineral while soil SZL3 contains kaolinite with some quartz. These results are in agreement with those reported by Osinubi (1998b) in an earlier study of material from the same borrow pit. Physical properties as well as compaction characteristics of the soils obtained at the energy of the British Standard heavy (BSH) effort in accordance with procedures outlined in BS 1377 (1990) are summarised in Table 1.

Table 1: Physical properties and compaction characteristics of the lateritic-soils

Property	Designation of soil samples		
	SZL-1	SZL-2	SZL-3
Liquid limit, %	38	40	46
Plasticity index, %	14	18	17
Specific gravity	2.67	2.68	2.69
Linear shrinkage, %	6	9	8
Gravel, (< 40 76mm), %)	0.44	0.38	0.53
Sand, (0.075 - 4.76mm),%	32	29	32
Fines, (< 0.075mm)	68	71	68
Clay (< 2μm) %	36	34	31
USCS classification	CL	CL	CL
Plasticity product	952	1278	1156
[a]Maximum dry unit weight, kN/m^3	18	18	18
Optimum water content, %	14	14	15

a - BSH compaction

Batch Equilibrium Absorption Tests

Batch equilibrium adsorption tests (BEATs) were performed with the soil samples and municipal solid waste (MSW) leachate to quantify the potential adsorption sodium ion. The procedures employed for BEATs generally followed those used by Shackelford and Daniel (1991) and Cotten *et al* (1998).

A 1:4 soil - solution ratio (by weight of dry soil) which is the highest recommended ratio ("Batch - Type" 1987) was used in the BEATs. This 1:4 ratio was maintained by adding 50 g of soil and 200 g of solution into each of eight different plastic containers previously rinsed with distilled water. The concentration of the leachate in the containers were made to vary from one another. The mixtures were agitated from time to time for about 48 hours before obtaining the supernatant by filtering into plastic bottles using filter papers. Sodium ion concentrations were measured by using UNICAM 969 Atomic Absorption Spectrophotometer (AAS).

The results of the chemical analyses were plotted in the form of adsorption isotherms, or sorbed concentration, C_s, versus dissolved equilibrium concentration, C_e, of solute. The sorbed concentrations were determined by mass balance using the equation:

$$C_s = \frac{(C_o - C_e)V}{M_s} \tag{1}$$

where C_o = initial concentration of solute; M_s = mass of soil solids; and V = volume of solution. The slope of the adsorption isotherm (plot of sorbed concentration versus equilibrium), called the partition or partitioning coefficient, K_p, is given by the equation:

$$K_p = \frac{dc_s}{dc_e} \tag{2}$$

Diffusion Test

The diffusion tests followed the procedure outlined by Shackelford and Daniel (1991) for single-reservoir, decreasing source tests. The diffusion test set-ups used in this study consisted of transparent plastic pipes containing the compacted soils with leachate ponding over each of the three compacted soils. The cells were sealed from the top and leachate was introduced through a short aluminum pipe that was fitted to the top cover. Small diameter hoses were connected to the top of the aluminum pipes to aid the introduction of leachate of the desired quantity into the diffusion cells.

The soil samples were mixed with tap water by an amount that corresponds to 2% on the wet side of optimum water content, and allowed to hydrate for 4 days prior to compaction to the same volume and density as with the British Standard compaction mould. In order to achieve saturation and hence minimize mass transport of contaminants due to suction, soil saturation was accomplished by allowing distilled waste to pond over the compacted soils for at least 42 days.

The diffusion stage of the tests was initiated by gradually introducing the leachate into the cell and ensuring that there were no air bubbles on top of the solution. Diffusion continued for 126 days. The cation concentrations of the leachate before diffusion testing are given in Table 2. At the end of the diffusion stage, each cell was disassembled and the soils extruded before sectioning in order to obtain specimens and chemical extracts. Sodium ion concentrations were measured by using the UNICAM 969 Atomic Absorption Spectrophotometer.

Table 2: Chemical characteristics of leachate used for diffusion testing

Chemical Parameters	Concentrations
Electrical conductivity, µmhos/cm	1000
Chemical oxygen demand, mg/l	1050
Total dissolved solids, mg/l	4100
Hardness	10.1
Chloride, mg/l	9.99
Sulphate, mg/L	460.0
Phosphate, mg/L	0.03
Calcium, ca	28.31
Sodium, Na^+	37.67
Potassium, K+	71.42
Magnesium, Mg+	-
Iron, Fe	0.78
Pb^{2+}, Zn^{2t}, Cd^{t}	Trace

The diffusion test data were evaluated by using the analytical solution to Fick's second law for one-dimensional diffusion through saturated soil for the case of a decreasing source concentration and finite cell length (Crank, 1975):

$$\frac{C(x \geq 0,t)}{C_o} = \frac{\alpha}{1+\alpha} + \sum_{m=1}^{\infty} \frac{2\alpha}{1+\alpha+\alpha^2 q_m^2} \exp\left(\frac{-D^* q^2_{m} t}{R_d L^2}\right) \frac{\cos[q_m(1-(\frac{x}{L})]}{\cos(q_m)} \tag{3}$$

Where all q_m = non zero positive roots of the following function

$$\tan(q_m) = -\alpha q_m \tag{4}$$

and α = a constant defined as

$$\alpha = \frac{H_L}{nR_d L} \tag{5}$$

and the retardation factor is given as:

$$R_d = 1 + \frac{\rho_d}{n_e} K_p \tag{6}$$

where n and n_e are the porosity and effective porosity of the in-situ soil, respectively; :ρ_d is the dry or bulk density; K_p is the partition coefficient; H_L and L are height of the liquid in the source reservoir and length of the soil, respectively.

Results and Discussion

Adsorption Behaviour

Measured adsorption isotherms using the three lateritic soil samples for Na^+ are shown in Fig. 1. The sorption of Na^+ into lateritic soils tends to occur more at higher concentrations. The empirically derived Freundlich isotherm was used to describe the adsorption characteristics of the lateritic soils and is defined as:

$$\frac{x}{m} = K_f C_e^{1/n} \tag{7}$$

where x/m is the amount of adsorbate adsorbed per unit weight of adsorbent, t; C_e is the equilibrium concentration of adsorbate in solution after adsorption while K_f and n are empirical constants obtained by the using eq. (7) that was rewritten as eq. (8):

$$\log(x/m) = \log K_f + \left(\tfrac{1}{n}\right)\log C_c \tag{8}$$

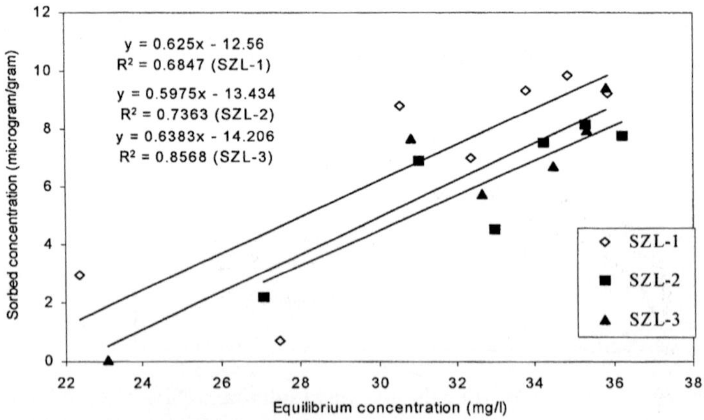

Fig. 1 Adsorption isotherms for the lateritic soils

A linear adsorption equation is given as:

$$x/m = K_d C_c \,(or\ K_p C_c) \tag{9}$$

where K_d is distribution coefficient and K_p is partitioning coefficient. The partitioning coefficient K_p was obtained from eqs.(7) and (9) as:

$$K_p = K_f C_c^{\frac{1}{n}-1} \tag{10}$$

Thus, utilizing the equilibrium concentration for the soil samples and the initial concentration of Na^+, the following K_p values were obtained: 0.3 cm^3/g for soil SZL1; 0.26 cm^3/g for soil SZL2 and 0.26 cm^3/g for soil SZL3. These K_p values that were obtained by using eq. (10) are small, thus implying that the corresponding values of retardation factor (R_d) will be small and therefore contaminant transport will be more rapid. The sorption behaviour of lateritic soils to Na^+ is of significance because high sodium concentration levels can affect (or alter) the physico-chemical properties of clays (Edil *et al.*, 1992).

Diffusion

The water contents determined for each slice of the three soil samples represent the weighted averages of the water contents after diffusion testing. The variations in water content with depth are shown for all three lateritic soils in Fig. 2. The concentration profiles for Na^+ are shown in Fig. 3 for the three lateritic soils.

The retardation factor (R_d) summarised in Table 3 for each soil sample were evaluated with the aid of eq. (6). The values of K_p and R_d tend to suggest that they decrease and increase, respectively, with higher values of plasticity product defined as:

$$Plasticity\ product = plastic\ index\ x \quad percentage\ pas \sin g\ 75\mu m \tag{11}$$

The plasticity product represents the effective contribution of the plasticity of the fines to the performance of the whole material. As R_d values are greater than 1.0, it is clear that Na^+ exhibited retardation. Eq. (3) was used to calculate values of the effective diffusion coefficient given in Table 3. The analytical solution employed was based on two assumptions: namely: (1) that the effective diffusion coefficient, D*, is a constant and (2) that the sorption (ion exchange) process is linear, which may not always be valid with respect to diffusion testing for waste disposal.

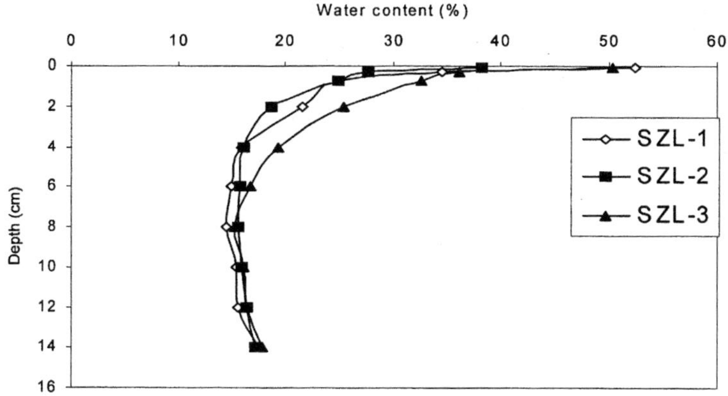

Fig. 2 Observed w ater content versus depth after diffusion testing for the lateritic soils

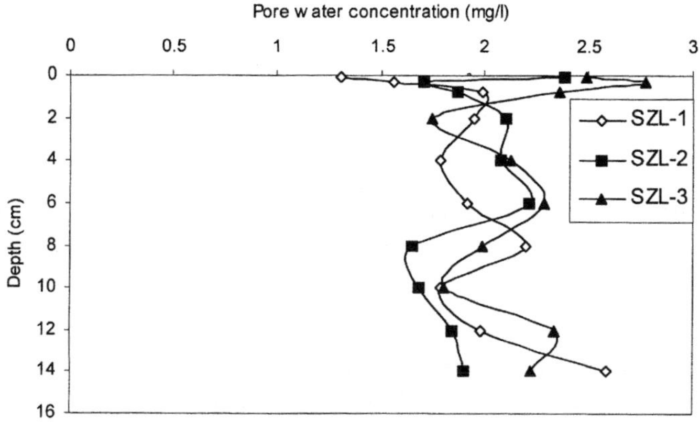

Fig. 3 Observed sodium ion concentrations after diffusion testing for the lateritic soils

Table 3: Transport parameters for sodium (Na^+) diffusion into lateritic soils.

Contaminant transport parameters	Designation of soil samples		
	SZL-1	SZL-2	SZL-3
Plasticity product, PP	952	1278	1156
Partitioning coefficient, K_p	0.3	0.26	0.26
Retardation factor, R_d	2.44	2.54	2.49
α	0.40	0.43	0.44
Reservoir diffusion coefficient, cm^2/s	13.23 x 10^{-6}	13.83 x 10^{-6}	13.86 x 10^{-6}
Soil profile diffusion coefficient, cm^2/s	3.34 x 10^{-6}	3.27 x 10^{-6}	3.29 x 10^{-6}
Tortuosity factor,	0.25	0.25	0.25
Apparent diffusion coefficient, cm^2/s	1.37 x 10^{-6}	1.29 x 10^{-6}	1.32 x 10^{-6}

The values for the transport parameters (effective diffusion coefficient for reservoir concentration data (D_{Res}), effective diffusion coefficient for the soil profile data (D*), tortuosity factor ($\tau_{a,soill}$) and apparent diffusion coefficient (D^*_A or D_s), respectively, for sodium (Na^+) diffusion are summarised in Table 2: . The values of D^*_{Res} are visibly higher than those of D^*_{soill}. The D^*_{Res} values for soil SZL2 and SZL3 are higher than the corresponding free solution diffusion coefficient, D_o (= 13.3 x 10^{-6} cm^2/s) for Na^+ (Daniel and Shackelford, 1988; Shackelford, 1989). The D^*_{soill} values that were used to evaluate the tortuosity factors (τ_a) and D^*_A values that ranged from 2.11 x 10^{-6} to 8.75 x 10^{-6} cm^2/s reported for Na^+ by Shackelford and Redmond (1995). The retardation factors as well as the D^*_A (or D^*_{soil}/R_d) values obtained for Na^+ in this study also ranged from 1.98 ± 0.11 to 9.53 ± 6.69 for R_d and 0.92 x 10^{-6} cm^2/s for D^*_A as reported by Shackelford and Redmond (1995) for Na^+ with respect to contaminant transport in processed Kaolin.

The tortuosity factors (τ_a) varied from 0.24 to 0.25 and fall within the range of 0.1 to 0.58 for natural soils (Daniel and Shackelford 1988). According to Shackelford and Rowe (1998), since τ_a values are less than 1.0 for diffusion in soil, an upper limit on D* for most inorganic chemical species is 2 x 10^{-5} cm^2/s based on the equation:

$$D^* = D_o \tau_o \qquad (12)$$

The values of the reservoir diffusion coefficient as well as those of the soil profile diffusion coefficient are less than 2 x $10^{-5}$$cm^2/s$.

The values of soil profile diffusion coefficient, tortuosity factor and apparent diffusion coefficient are affected by the plasticity product. The values of these transport parameters decreased with higher plasticity product, thus suggesting that the compositional factors of lateritic soils have some effects on the transport of inorganic chemical species in them. The incidence of reservoir diffusion coefficients being higher than the free-solution diffusion coefficients is in agreement with the findings of Shackelford and Daniel (1991)

Conclusion

Laboratory adsorption and diffusion tests on three lateritic soils were conducted to evaluate the nature of sodium diffusion in compacted lateritic soils. Soil composition was characterized by plasticity product. The partitioning coefficients as well as the retardation factors are low and indicate more rapid contaminant transport for sodium (Na^+). The effective diffusion coefficients, tortuosity factors and apparent diffusion coefficients determined from diffusion test results are within the normal range of values reported in literature. The study showed that the sorption behaviour of and contaminant transport in lateritic soils is a function of soil compositional factors as characterized by the plasticity product. For example, the retardation factors increased with higher values of plasticity product, while the partitioning coefficients, effective diffusion coefficients, the tortuosity factors as well as apparent diffusion coefficients decreased with higher values of plasticity product.

REFERENCES

Batch-type (1987). 'Batch-type adsorption procedures for estimating soil attenuation of chemicals, EPA/530 - SW - 006: Draft Technical Resource Document for Public comment (NTLS - 146 - 155).' US Environmental Protection Agency, Cincinnati, Ohio.

BS 1377 (1990). 'Methods of tests for soils for civil engineering purposes.' British Standards Institution, London.

Cotten, T. E., Davis, M. M., and Shackelford, C. D. (1998). 'Effects of test duration and specimen length on diffusion testing of unconfined specimens.' Geotechnical Testing Journal, 21(2), 79 - 94.

Crank, J. (1975). The Mathematics of Diffusion, Second Edition, Oxford University Press, Oxford, England, 414p.

Daniel, D. E. and Shackelford, C. D. (1988). ' Disposal barriers that release contaminants only by molecular diffusion.' Nuclear and Chemical Waste Management, 8: 299 - 305.

Edil, T. B., Sandstrom, K, and Berthouex, P. M. (1992). 'Interaction of inorganic leachate with compacted pozzolanic fly ash.' Journal of Geotechnical Engineering, ASCE, 118(9), 1410-1430.

Gillham, R.W., Robin, M. L. J., Dytynyshyn, D. J. and Johnston, H. M. (1984). 'Diffusion of non-reactive and reactive solutes through fine-grained barrier materials.' Canadian Geotechnical Journal, 21(3), 541-550.

Mott, H. V. and Weber, W. J. Jr. (1991). 'Factors influencing organic contaminants through soil-bentonite cut-off barriers.' Environmental Science and Technology, 25(10), 1708-1715.

Osinubi, K. J. (1998a). 'Influence of compactive efforts and compaction delays on lime-treated soil.' Journal of Transportation Engineering, ASCE, 124(2), 149 - 155.

Osinubi, K. J. (1998b). 'Permeability of lime-treated lateritic soil.' Journal of Transportation Engineering, ASCE, 124(5), 465-469.

Quigley, R. M., and Rowe, R.K. (1986).'Leachate migration through clay below a domestic waste landfill, Sarnia , Ontario, Canada: Chemical interpretation and modeling philosophies.' Hazardous and Industrial Solid Waste Testing and Disposal STP 933, D. Lorenzen, et al. (eds). American Society for Testing and Materials (ASTM), Philadelphia, Pa, 93 -103.

Shackelford, C. D. (1989). 'Diffusion of contaminant through waste containment barriers.' Transportation Research Record, 1219, 169-182.

Shackelford, C. D. and Daniel, D. E. (1991). 'Diffusion in Saturated soil. II: Results for compacted clay.' Journal of Geotechnical Engineering, ASCE, 117(3), 485 - 506.

Shackelford, C. D. and Redmond, P. L (1995). 'Solute breakthrough curves for processed kaolin at low flow rates.' Journal of Geotechnical Engineering, ASCE, 121 (1), 17-32.

Shackelford, C. D. and Rowe, R. K. (1998). 'Contaminant transport modeling.' Proceedings of the Third International Congress on Environmental Geotechnics, Lisbon, Portugal, 7-11 September, 939 - 956.

Retention curves and hydraulic properties of a soil contaminated by NAPL

Rabozzi Cristina Politecnico di Torino - DITAG, Torino,Italy
Ribotta Laura Politecnico di Torino - DITAG, Torino, Italy
Giacomo Gremigni Teseco SpA, Pisa, Italy

Abstract

This paper presents some experimental results on retention and transport properties of a LNAPL in silty sand samples. Water and oil retention curves for two-phase and three-phase systems were determined in order to predict the relative permeability of different fluids in partially saturated conditions. Some considerations on the structural changes on the soil sample when LNAPL was used either as a permeation fluid or during soil sample preparation are given.
The basic objective of this research was to clarify some aspects of LNAPL transport and oil-soil interaction at different wetting conditions. The experimental results obtained from this study can be considered of practical interest to predict NAPL movements within a contaminated soil in the presence of different phases of immiscible fluids.

Introduction

Recent developments in the area of unsaturated soils offer an appropriate framework and laboratory equipment for a detailed study of seepage of hydrocarbon through the vadose zone (Cui et al. 2003, Delage et. al. 2000).

Successful remediation strategies associated to hydrocarbon contamination of soil requires modelling of the multi-fluid flow and transport. Flow and transport in porous media are controlled by interfacial processes between fluid-fluid and fluid-solid properties. One of the most important parameters is the relative permeability related to the different saturation degrees and to the different pore fluids.

For a porous medium with an incompressible structure, the permeability function can be theoretically expressed in three different forms:

- As empirical functions of degree of saturation, volumetric water content, or soil suction.
- As macroscopic models based on a mechanistic view that the fluid-filled pores could be represented as bundles of capillary tubes of various sizes.
- As statistical models in which the relative permeability is calculated from the soil-water retention curve.

From an experimental view point, the relative permeability of a soil to each fluid phase can be directly measured in a sample by either steady-state or unsteady-state methods. In the steady-state method, a fixed fluid ratio is forced through the sample until the saturation and pressure equilibria are established. In unsteady-state method only one of the phases is injected into the sample. Unsteady-state relative permeability measurements can be made more rapidly than steady-state measurements, but the mathematical analysis is more difficult.

Some experimental results obtained in a two-phase system (oil-water) are shown in this paper in order to evaluate the fraction oil recovery. This test has been interpreted by simplified analytical solutions referring to unsteady-state conditions.

The relative permeability curves, for the same two-phase systems (i.e. oil-water) and for air-water system, were preliminarily estimated on the basis of the experimental retention curves. Moreover, a prediction of the hydraulic behaviour of a three-phase system (air-NAPL-water) was provided. Finally, a cross comparison of the results was carried out and the related considerations were proposed.

Relative permeabilites

Two-phase systems

Laboratory tests were performed on a synthetic mixture of Ticino sand and Pontida silt (50%-50% by weight). Samples were prepared by means of static compaction. First, equal dry weights of sand and silt were mixed at laboratory conditions. Up to a final content of 10%, deionized water was added in stages while mixing. Specimens were then loaded in oedometric conditions until reaching the desired dry density, $\rho_d = 1.70$ Mg/m^3. In the air-oil experiments, the dry sample was directly compacted in the oedometer then fully saturated with oil. Further experimental details are reported in Manassero et al. (2005) and Rabozzi and Ribotta (2005).
For the air-water system, the hydraulic behaviour in unsaturated conditions, was predicted from the experimental retention curve (Figure 1). Two different techniques were used: vapour equilibrium for the high suction range (463-330000 kPa) and axis translation for the low suction range (5-95 kPa). The experimental data were interpreted using the van Genuchten function modified by Parker (Parker et al., 1987).

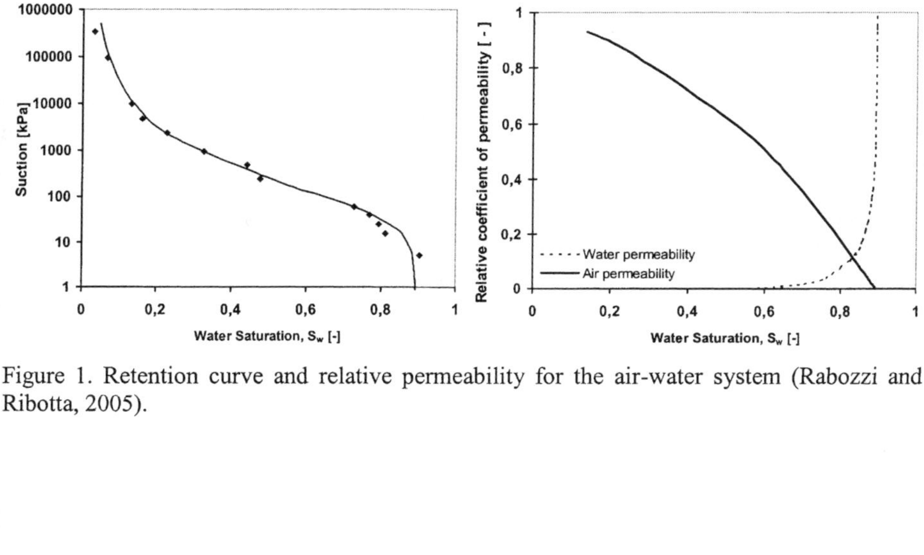

Figure 1. Retention curve and relative permeability for the air-water system (Rabozzi and Ribotta, 2005).

The two-phase oil and air system was investigated by performing two laboratory experiments. The NAPL used in this experimental study is a non toxic liquid called Soltrol 170. This fluid is a mixture of C14-C16 isoalkanes without polycyclic aromatic hydrocarbons (Delage et al. 2000, Cui et al. 2003, Manassero et al., 2005).
A suction controlled oedometer was used and as a result it was possible to control the air and oil pressure, P_a and P_o, as well as the suction, $\Psi=P_a-P_o$. The first test was conducted in four steps corresponding to four suction values ranging from 5 to 40 kPa (Rabozzi and Ribotta, 2005). A second test was performed in order to confirm the results. In this case, a larger number of suction values was applied. The experimental procedure of the second experiment is similar to the first one, but it was possible to describe the retention curve more accurately. The suction range imposed during these tests is larger than most values reported in literature. The soil samples were preliminary saturated with oil then a drying cycle was applied in order to assess the retention properties of the contaminated soil. Tests with suction over 40 kPa pointed out an air leakage flowing with the NAPL from the bottom of the sample. The ceramic porous plate used in the bottom of the oedometer cell has an air entry value (AEV) equal to 100 kPa. This AEV was confirmed through preliminary tests in an air-water system. In the air-oil system, where the ceramic plate was saturated with Soltrol, the measured air entry pressure, ranged between 40 to 50 kPa. Figure 2 shows the retention curve relative to the oil-air system.

The equations, describing the oil functions of Figure 2, are:

$$\theta_o = \frac{\theta_{so} - \theta_{ro}}{\left[1+(a\cdot\Psi)^b\right]^c} + \theta_{ro} \tag{1}$$

$$k_{r,o} = \sqrt{\frac{\theta_o - \theta_{ro}}{\theta_{so} - \theta_{ro}}}\left\{1-\left[1-\left(\frac{\theta_o - \theta_{ro}}{\theta_{so} - \theta_{ro}}\right)^{1/c}\right]^c\right\}^2 \tag{2}$$

where θ_o is the volumetric oil content, θ_{so} is the volumetric oil content in saturated conditions, θ_{ro} is the residual volumetric oil content, Ψ is the total suction expressed in kPa and $k_{r,o}$ is the relative permeability coefficient for the oil. Only the results of the second test were used to obtain the van Genuchten coefficients (a,b,c).

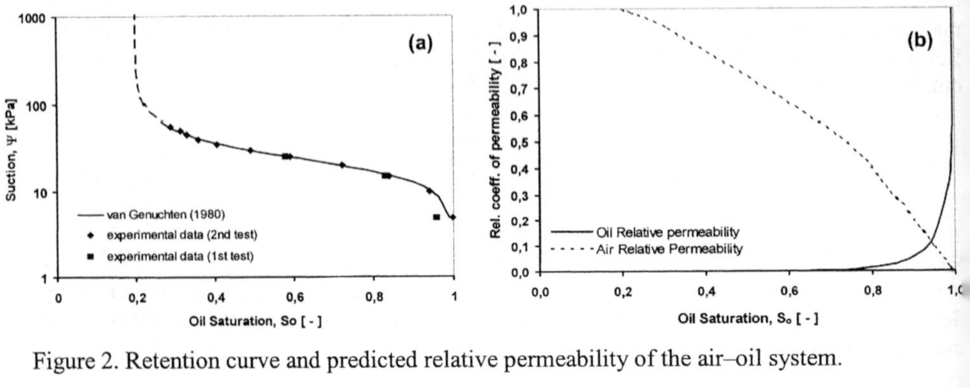

Figure 2. Retention curve and predicted relative permeability of the air–oil system.

As expected, it is possible to observe that, the oil permeability coefficient increases with the volumetric oil content and decreases with suction, while the air relative permeability k_{ra} undergoes the opposite process.

The curve in Figure 2a also shows that most oil desaturation occurs at a low suction of 40 kPa. In fact, the oil saturation, S_o, is equal to 36% at 40 kPa of suction. The oil-air retention curve in Figure 2 shows that in order to reach low oil content very high values of suction are necessary.

The direct evaluation of water and oil permeability in fully saturated conditions was carried out with a flexible wall permeameter. In order to allow comparisons, the data were expressed in terms of intrinsic permeability. As seen in Figure 3, the concept of intrinsic permeability does not stand, since two different results were obtained with each two fluid. This could be due to the different interfacial tensions when soil sample is permeated with oil and/or to a possible incomplete oil saturation of the specimen.
This experimental data means that the porous medium through which the fluids were permeated is not inert to the fluids, at least for water and oil.

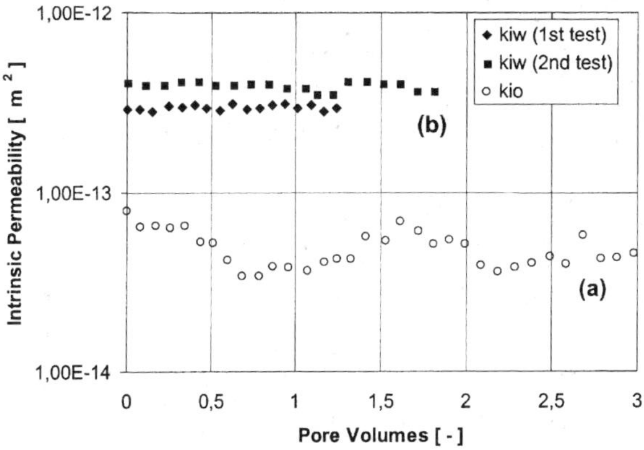

Figure 3. Intrinsic permeability obtained on: a) fully saturated sample permeated with water and b) oil.

The relative permeability of the oil-water system was experimentally obtained performing a permeability test in unsteady-state conditions. The test results were interpreted on the basis of the theory developed by Buckley and Leverett (1942) and extended by Welge (1952). The mathematical basis for the interpretation of the test data may be summarized as follows. Leverett combined Darcy's law with a definition of capillary pressure in differential form to obtain:

$$f_{w2} = \frac{1 + \dfrac{k_o}{q_t \mu_o}\left(\dfrac{\partial \Psi}{\partial x} - g\ \Delta\rho \sin\theta\right)}{1 + \dfrac{k_o}{k_w}\dfrac{\mu_w}{\mu_o}} \tag{3}$$

where f_{w2} is the fraction water in outlet stream, q_t is the superficial velocity of total fluid leaving the sample, θ is the angle between direction x and the horizontal, and $\Delta\rho$ is the density difference between displacing and displaced fluids. By definition of f_{o2} (fraction oil in outlet stream):

$$f_{o2} = \frac{q_o}{q_o + q_w} \tag{4}$$

and combining this equation with Darcy's law, it can be shown that:

$$f_{o2} = \frac{1}{1 + \dfrac{\mu_o / k_{ro}}{\mu_w / k_{rw}}} \tag{5}$$

Since μ_o and μ_w are known, the relative permeability ratio k_{ro}/k_{rw} can be determined from Equation (5).

Note that for any two immiscible fluids, e.g. water and oil, the fraction of the water flowing at any point in the porous media is given by:

$$f_w = 1 - f_o \tag{6}$$

The above expression indicates that during the displacement of oil by water flushing, an increase in f_o will cause a proportional decrease of f_w and water mobility.

The permeability tests in unsteady-state conditions consisted of the following steps: 1) saturation of the sample with water, 2) drainage (NAPL displaces water) until the fraction water in the outlet stream was equal to zero, 3) second imbibition (water displaces NAPL) until the fraction oil in the outlet stream was equal to zero.
The trend of f_o, during the primary drainage and secondary imbibition, are shown in Figure 4a.

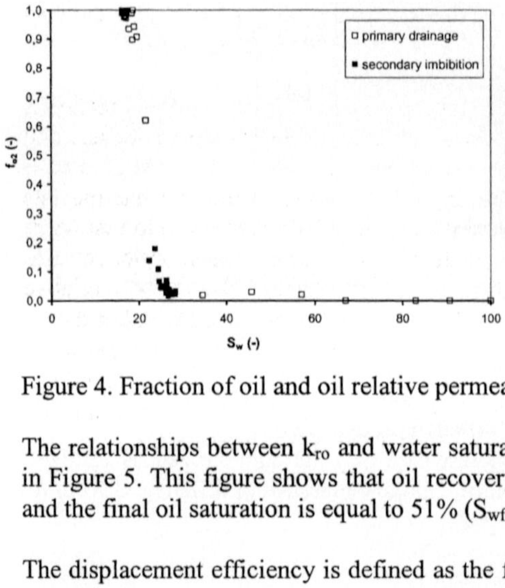

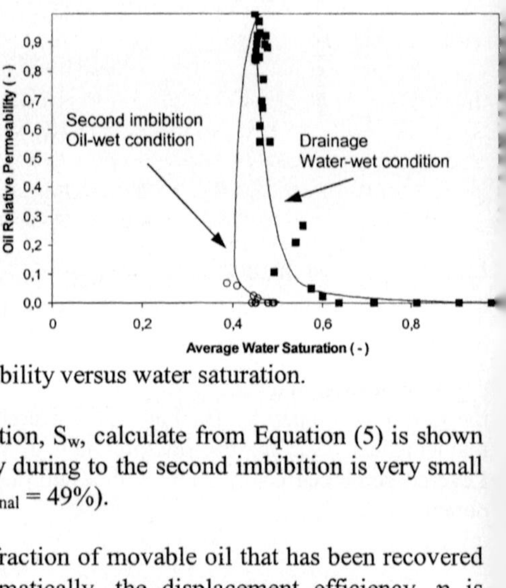

Figure 4. Fraction of oil and oil relative permeability versus water saturation.

The relationships between k_{ro} and water saturation, S_w, calculate from Equation (5) is shown in Figure 5. This figure shows that oil recovery during to the second imbibition is very small and the final oil saturation is equal to 51% ($S_{wfinal} = 49\%$).

The displacement efficiency is defined as the fraction of movable oil that has been recovered from the sample at any given time. Mathematically, the displacement efficiency, η is expressed as:

$$\eta = \frac{Initial\ volume\ of\ oil - Remaining\ oil\ volume}{Initial\ volume\ of\ oil} = \frac{S_w - S_{wi}}{1 - S_{wi}} \tag{7}$$

where S_w is the water saturation at a given time and S_{wi} is the initial water saturation. Figure 4b shows the trend of the displacement efficiency during the secondary imbibition.

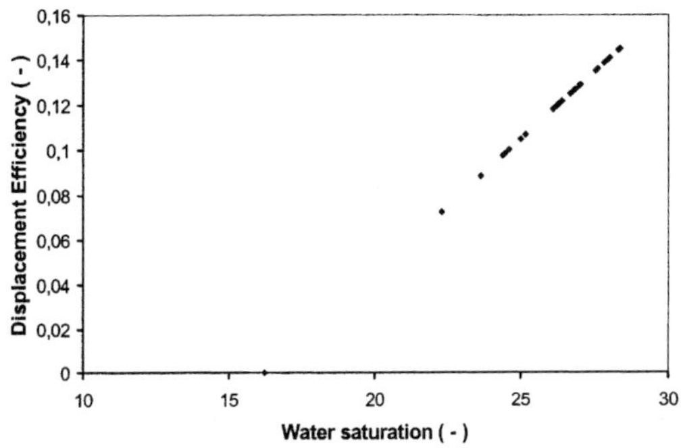

Figure 5. Displacement efficiency versus water saturation.

Several authors have studied the hysteretic behaviour of two-phase systems pointed out by the experimental data shown in Figure 4. Al-Fataisi and Patzek (2004) proposed a pore network model where each pore is idealized as a triangular duct. Initially, the duct is completely filled with water, and the solid walls are naturally water-wet. Primary drainage starts by increasing capillary pressure, which allows the NAPL to migrate into the duct when the oil entry pressure is exceeded. When NAPL invades, it occupies the center of the duct, and water resides in the corners of the ducts. After the NAPL recedes some distance toward the duct corners, the water film rupture occurs and parts of the duct wall become NAPL-wet.

In Figure 6 are reported oil and water relative permeabilities in different wetting conditions simulated by the Al-Fataisi and Patzek model. In the NAPL-wet condition, Al-Fataisi and Patzek observe that the water relative permeability is the highest and the NAPL relative permeability is the lowest. In this wetting condition, NAPL moves in the thin intermediate layers, whereas water moves along the corners and the centers of the ducts. The lowest water relative permeability is observed for the intermediate-wet condition, because water remains pinned near the duct corners without invading the duct centres. The NAPL relative permeability in the latter case is at first similar to that in the water-wet condition, but then it becomes lower because of significant NAPL trapping. For the water-wet condition, the water relative permeability is between those in the intermediate-wet and NAPL-wet conditions, while the NAPL relative permeability is the highest because the NAPL flows predominantly in the duct centres. The Al-Fataisi and Patzek model confirms, at least qualitatively, the experimental results obtained with Soltrol in sandy-silt specimens.

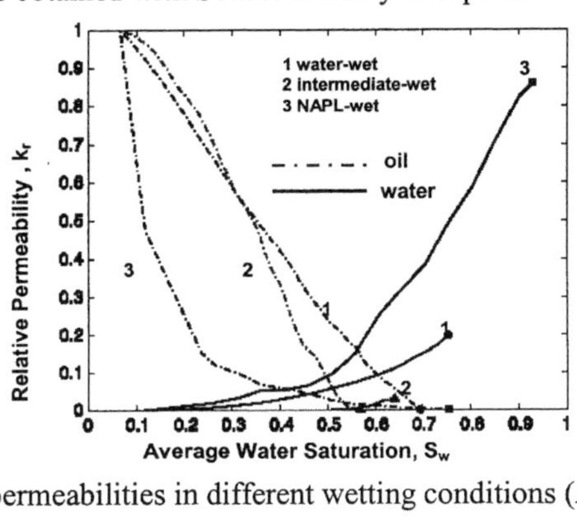

Figure 6. Relative permeabilities in different wetting conditions (Al-Fataisi and Patzek, 2004).

Three-phase system

A retention curve for a three phase system was obtained for a soil sample with an initial water content of 10% and 5% of Soltrol (LNAPL). Significant changes of soil structure were observed when Soltrol was used since the sample preparation phase (Manassero et al., 2005).

The data obtained with suction controlled oedometer and vapour equilibrium were jointly interpreted in order to define the suction-water saturation relationship.
The relative permeability of the water, air and Soltrol were preliminary estimated for the three-phase system using the test results available at this initial stage of the research programme. The original van Genuchten empirical equation, based on water retention curve has been modified assuming that the oil occupies the intermediate pore size range (Cui et al. 2003, Charbeneau, 2000). Therefore, the relative permeability values to water ($k_{r,w}$), air ($k_{r,a}$) and Soltrol, at 5% content by weight ($k_{r,o}$), are related to the volumetric content of the liquid phases through the following expressions (Manassero et al., 2005):

$$k_{r,w} = \sqrt{\frac{\theta_w - \theta_r}{\theta_s - \theta_r}} \left\{ 1 - \left[1 - \left(\frac{\theta_w - \theta_r}{\theta_s - \theta_r} \right)^{1/c} \right]^c \right\}^2 \tag{9a}$$

$$k_{r,a} = \sqrt{1 - \frac{\theta_w + \theta_o - \theta_r}{\theta_s - \theta_r}} \left[1 - \left(\frac{\theta_w + \theta_o - \theta_r}{\theta_s - \theta_r} \right)^{1/c} \right]^{2c} \tag{9b}$$

$$k_{r,o} = \sqrt{\frac{\theta_o}{\theta_s - \theta_r}} \left[\left[1 - \left(\frac{\theta_w - \theta_r}{\theta_s - \theta_r} \right)^{1/c} \right]^c - \left[1 - \left(\frac{\theta_w + \theta_o - \theta_r}{\theta_s - \theta_r} \right)^{1/c} \right]^c \right]^2 \tag{9c}$$

The results of the preliminary estimation of the relative permeability for the considered soil mixture are shown in Figure 7. Looking at this figure, it is interesting to point out the significant Soltrol relative permeability increase, at high water saturation values, for a given fixed content of 5% by weight. On the other hand, the air relative permeability remains rather high over most of the water saturation range. The opposite trend can be observed for the water relative permeability. Therefore, as expected for the considered soil, remediation techniques based on a simple water flushing seem to be hopeless, the only alternatives being the use of washing liquid different from water and/or the Soltrol mobilization by soil venting and/or air sparging techniques inducing phase changes of the considered hydrocarbon.

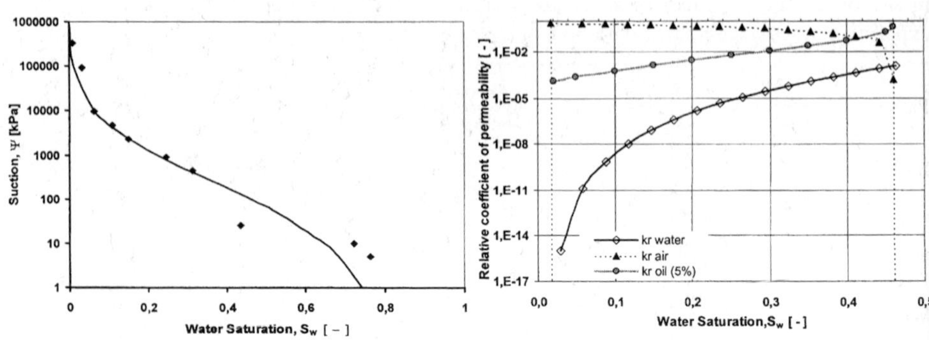

Figure 7. Retention curve and relative permeability according to equations (9) for the three-phase system (air-NAPL-water).

Conclusions

The retention properties and the hydraulic behaviour of two-phase and three-phase systems, consisting of air, water and oil within a silty sand soil, have been experimentally studied in this paper. In particular, laboratory data relative to oil-water and air-oil-water systems show that remediation techniques based on a simple water flushing seem to be hopeless. This low efficiency is due to the low value of water permeability and to high percentages of oil trapped in the pore volume. In particular, if the soil is oil-wet, the displacement efficiency is low and high values of residual oil content in soil are present as theoretically demonstrated by Al-Fataisi and Patzek (2004) and experimentally confirmed by the results shown in this paper.
Significant changes in the soil structure were observed in particular when the oil (Soltrol) was used since the sample preparation phase and also when Soltrol was infiltrated after the soil sample preparation some soil structure modification was apparent.
In order to obtain a more comprehensive and consistent framework the presented results, which were obtained during the initial phase of a more comprehensive research programme, must be extended by further experimental tests concerning the three phase system (air-oil-water) under both steady-state and unsteady-state conditions.

Acknowledgements

This project has been sponsored by the Italian Research Ministry (MIUR) and by the private companies Trevi SpA (Cesena) and Teseco SpA (Pisa). Their financial and technical support is gratefully acknowledged.

References

Al-Fataisi, A., Patzek, T.W. (2004). "Secondary imbibition in NAPL-invaded mixed-wet sediments". Journal of Contaminant Hydrology 74, pp. 61-81.

Buckley, S.E., Leverett, M.C. (1942). "Mechanism of fluid displacement in sands". Trans AIME, 146.

Charbeneau, R.J. (2000). Groundwater Hydraulics and Pollutant Transport. Upper Saddle River, New Jersey. Prentice Hall.

Cui, Y.J., Delage, P., Alzoghbi, P. (2003). "Retention and transport of hydrocarbon in a silt". Géotechnique 53 (1), pp.83-91.

Delage, P., Cui, Y.J , Alzoghbi, P. (2000). "Air, water and oil transfers in an unsaturated compacted silt". Proc. Int. Symp. On Physical Modelling and Testing in Environmental Geotechnic, La Baule, France, 15-17 May. Ed. by J. Garnier, L. Thorel, E. Haza, pp. 61-68.

Manassero, M., Musso, G., Rabozzi, C., Ribotta, L., "Retention curves for a polluted soil". Proc. of International Symposium, Advanced Experimental Unsaturated Soil Mechanics, Trento, 27-29 June 2005, Ed. Balkema, pp. 459-465.

Parker, J.C., Lenhard, R.J., Kuppusamy, T. (1987). "A parametric model for constitutive proprieties governing multiphase flow in porous media". Water Resources Research. 23, pp. 618-624.

Rabozzi, C., Ribotta, L. (2005). "Experimental study on a partially saturated polluted soil". Proc. of International Conference on Problematic Soils, 25-27 May 2005, Eastern Mediterranean University, Famagusta, N. Cyprus, pp.343-349.

van Genuchten, M.T. (1980). "A closed-form equation for predicting the hydraulic conductivity of unsaturated soils". Soil Sci. Soc. Am. J. 44, pp. 379-392.

Weldge, H.J. (1956). "A simplified method for computing recovery by gas or water drive". Trans AIME, 195.

Soil Contamination Profiles at Marambaia M.S.W. Landfill, Brazil

Elisabeth Ritter, Assistant Professor, Rio de Janeiro State University (UERJ), Brazil
Manoel Fernando Pereira da Mota, Engineer, Municipal Solid Waste Management Company (COMLURB), Rio de Janeiro, Brazil
Juacyara Carbonelli Campos, Assistant Professor, Rio de Janeiro Federal University, Brazil
Marcus Antonio Ventura, Undergraduate Student, Rio de Janeiro State University, Brazil

Abstract

The contamination profile was measured in the soil at three points on the Marambaia MSW. The various ions studied were chloride, sodium, calcium, magnesium, potassium, ammonium and nickel. On the South size of the area, there are signs of a slight shifting process. On the East side, contamination has occurred due to high concentrations of chloride, sodium, calcium, potassium and magnesium at depth. On the North side, next to the Iguaçu River, ammonium concentrations of about 140 mg/L at a depth of 6.5m also indicate a possible process of contamination in this area.

Introduction

Most Brazilian cities do not yet have facilities for waste disposal in landfills. However, some cities have started to make more concentrated efforts to set up sanitary landfills. But problems caused by this deficient disposal still persist even after the closure of the landfills; leachate continues to be produced and consequently its migration within the soil has not been interrupted. It is important to understand and evaluate the degree of soil contamination at the dumpsite in order to analyse the migration process and advance of contaminants in the leachate. Nova Iguaçu county in Rio de Janeiro State, Brazil, had used the Marambaia landfill for final disposal of collected urban solid waste between the late 1980s and February 2003. In

2001 a recovery process of the area began in which geotechnical soundings and chemical analyses of the leachate were carried out as a basis for the local remediation project.

The purpose was to do in-depth assessment of the degree of soil contamination using three sounding points on the boundary of the old landfill (Mota, 2005), in samples taken from the area during the remediation work.

The landfill occupied an area of approximately 200,000m² and each day for more than twenty years it received about a thousand tons of urban solid waste, including household, public, commercial, hospital and industrial refuse from the towns of Nova Iguaçu and Mesquita. The local relief consists of low hills covered with the trees and shrubs remaining from the native vegetation. An old brick factory still exists next to the landfill, a sign that the area was explored by pottery activities, hence the dips in the local contour. To the North, the region is bound by the Iguaçu River about 600m away, and between the waste disposal area and the river there are some depressions in which small leachate deposits were found. Next to the entrance to the landfill, there is a vertical slope on the edge of the landfill area and at its base a pool for storing percolated liquids was implemented in July 2003. Figure 1 shows an overview of the area in 2003.

Soundings carried out in 2001 show a predominantly clayey, slightly sandy substratum of medium to hard consistency in the fertile southern valley on the other side from the Iguaçu River. Thus, although the soil in the northern direction towards the river is clayey, it also has some sandy characteristics. The water level on the South side is at a depth of 8.5m and on the North side at a depth of 2.60m, in a region showing features of a fertile plain close to the road alongside the Iguaçu River.

The remediation process of the Marambaia area began in January 2001. The first step regulated platforms and created berms and slopes. On the boundary of the landfill, retention levees were built using clay material and a line of peripheral drains built at its base for percolates that were later taken to the accumulation pools. In addition to the percolate capture system, it was planned to install a drainage system for gases and a rainwater drainage system. The entire waste dump was coated with a 60cm clay layer.

The in-depth assessment of the degree of contamination required a collection of undisturbed soil samples and later extraction of pore water in a laboratory. It was necessary to identify the contamination source and description of the leachate. Thus, field, soil and leachate samplings, and laboratory procedures with extraction of the pore water from the soil were carried out to determine physical and chemical parameters. The leachate was collected from different points in the landfill and analysed in the Sanitary Engineering Laboratory at UERJ and in the Environmental Technology Centre (CTA) at FIRJAN. The selected points were chosen based on soundings conducted in April 2001. The chosen points covered three sides of the landfill - North, South and East – in order to be able to check the possible shift of the contamination point. The West side was not analysed because it is situated close to a main access road where waste disposal was negligible. Sampling points SP-B3, SP-L2 and SM-277 are located around the landfill in adjacent areas at the foot of the retention levee (Figure 1). The holes were drilled using boring or drilling equipment where the waste barrier or land did not permit water flush drilling. Special care was taken to minimize the flushing next to the sampling point. Five undisturbed samples were collected from each borehole using 8"-diameter Shelby and Denisson samplers at different depths. Two-inch piezometers were installed in the boreholes.

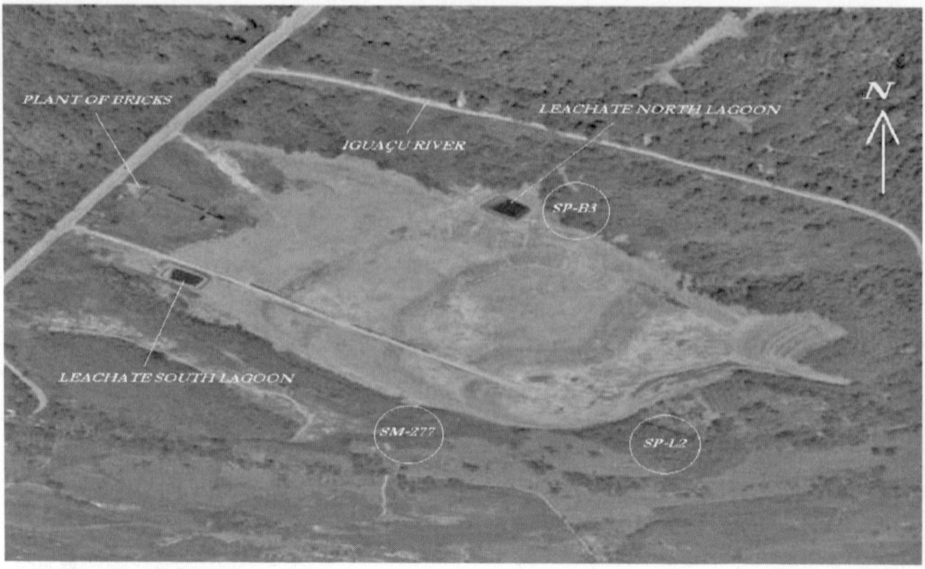

Figure 1 Overview of Marambaia MSW

Leachate samples were collected during three different periods: November 2003, July 2004, and October 2004. The physicochemical analyses were performed using the Standard Methods (APHA1998). The analysed parameters were conductivity, pH, COD, BOD, chloride, suspended solids, ammonium, sodium, calcium, magnesium, potassium, lead and cadmium. Pore water from the soil was extracted using a pneumatic press, and a visual-tactile method was used to classify a small portion of the soil. The chemical analyses were chloride, sodium, calcium, magnesium, potassium, ammonium, lead, chromium, iron, nickel and cadmium. The analytical methodology adopted for analysing the metals was by using a flame atomic absorption spectrometer. The ammonium was analysed using the potentiometric method (specific ion electrode).

Test Results and Discussion

An analysis performed on the Marambaia landfill on the South side was presented in Table 1. This table includes earlier analytical data from the Engineering School of São Carlos, University of São Paulo (EESC-USP) in 2001, in order to make comparative analyses of the values at the different times. It was found from the results of this analysis that the more recent leachate samples collected after closure of the Marambaia landfill contained significantly lower values of BOD and COD than the samples observed in May 2001 when the landfill was in operation. On the South side, the COD showed a decrease of up to 71.7% compared to the value first found. The BOD and COD ratio is normally used as a parameter for biodegradable leachate; it can range between zero and one (zero considered not fully biodegradable and one fully biodegradable). At the C1 point in the first analysis performed, the BOD and COD ratio presented values equal to 0.09, thus indicating a low level of biodegradability. In the analysis performed in October 2004, this figure was about 0.89, therefore showing a good level of biodegradability. The analyses on the North side show values of the BOD-COD ratio that also indicated good biodegradable conditions. Ammonium concentrations were compatible with other values found in different derivations of leachate from other Brazilian landfills; as shown in the following comparative data: 500 mg/L in the Brasilia landfill (Junqueira, 1999) and 1500 mg/L in the Gramacho Metropolitan landfill (Ritter et al, 2004). These values are higher than that permitted for effluent discharge standards, 20 mg/L (CONAMA 357).

In SP -L2 and SP-B3 boreholes, it was not possible to remove the pore water from the deeper samples; so there was not enough to perform the physicochemical analysis. The metals analysed (lead, chromium, iron, nickel and cadmium), with the exception of nickel, gave a very low concentration, lower than the standard fixed by the 2004 Government Decree no. 518, on the potability of water (Health Ministry). Figure 2 presents the results for potassium, ammonium, calcium, magnesium, sodium and nickel for the three boreholes. Chloride measures were possible only in a few samples.

Table 1 Leachate South Pool

Parameter	Samples			
	C – 1 EESC-USP 7 May 2001	C – 1 LES-UERJ 3 November 2003	South Pool LES-UERJ 7 July 2004	South Pool LES-UERJ 30 November 2004
pH	8.0	7.8	8.0	8.7
COD (mg O_2/L)	5680	1206	2825	1605
BOD (mg O_2/L)	494	220	116	1429
BOD/COD ratio	0.09	0.18	0.04	0.89
Conductivity (mS/cm)	*	5.0	17.5	7.2
Chloride (mg Cl^-/L)	*	953	2312	1413
Total Suspended Solids (mg/L)	313	220	264	56
Ammonia-nitrogen (mg N-NH_4^+/L)	939	214	1465	173
Magnesium (mg Mg^{+2}/L)	*	*	*	35.0
Potassium (mg K^+/L)	*	*	*	684
Sodium (mg Na^+/L)	*	*	*	930
Lead (mg Pb/L)	0.07	0.3	*	*
Cadmium (mg Cd/L)	< 0.0006	< 0.01	*	*
Calcium (mg Ca^{+2}/L)	*	*	*	33.0

* indeterminate

In the sounding of the SM-277 borehole on the South side of the landfill opposite the Iguaçu River, the soil was silty clay and little sand, with consistently hard to even harder as the depth increased; the water level was at 8.50m. It was possible to measure the chloride concentration between 2.5 and 4m in depth, and indicated values of 27 and 45 mg/L, respectively. Ammonium and potassium concentrations measured 2.5m in depth, which indicated a possible diffusion process. Calcium and magnesium concentrations were very low. Sodium concentrations revealed a more irregular behaviour but indicated that migration could possibly occur to a depth of seven meters. However, the sodium concentration at this depth was relatively low when compared to those of leachate (930 mg/L) present in the nearest pool.

After analysing the chloride values in the two samples (about 2.5-4m in depth), extremely low values were found compared to the 1413 mg/L concentration of the leachate pool. Among the heavy metals, nickel had a higher value than the standard requirement (2004 Government Decree no. 518, Health Ministry) at a depth of seven meters. It does, therefore, slightly indicate a migration process, although the water concentration measured (Mattos, 2005) at a depth of 6.5m had indicated no contamination.

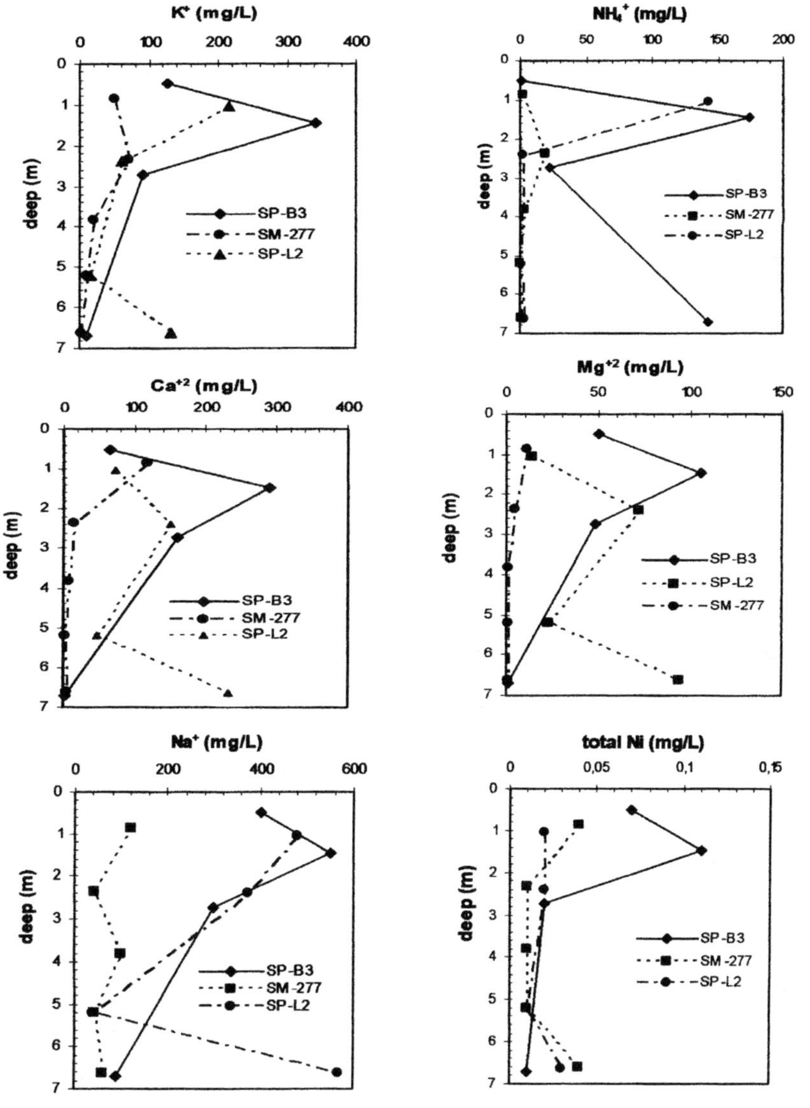

Figure 2 Chemical species concentrations of pore water

In the area of borehole SP-L2 on the East side of the landfill, close to a small farm where citric fruit and coconuts are cultivated, the sounding reached 19.40m, which found the water level to be around .60m from the surface. The soil shows characteristics of fine to medium clayey sand and young residual silty-clayey soil in the continuity of the borehole. Chloride concentration measures could only be found in the samples at 5.0 and 6.5m, giving values of about 72 and 405 mg/L, respectively. There are significant sodium and potassium concentrations in the first two samplings at 2.5m in depth, and at 7m sodium was found in concentrations of more than 500mg/L while potassium in concentrations of over 200 mg/L. The chloride concentration was about 405 mg/L at that same depth, and calcium and magnesium follow the same trend. Therefore, a contamination process does exist, through the long migration time of the leachate close to the foot of the landfill slope, which is about 40m high in waste. On the other hand, ammonium values are high only on the surface. Of all the heavy metals, nickel has higher values at a depth of 7m than permitted by environmental standards. An important aspect on this site near a small coconut plantation is that the groundwater is at the surface and chemical measures taken in water from the piezometer installed in this borehole (Mattos, 2005) showed contamination.

Borehole SP-B3 is situated on the northern face of the landfill area close to the Iguaçu River. Given the fact that the SP-B3 point was moved 25m from the original point, the analysis adopted the visual-tactile method. Organic clay was identified as far as 2.5m, and clayey sand with coarse material to 7.0m with silty clay as far as 8.3m. The water level is 1.5m from the surface. Chloride concentration could be measured in the samples of about 0.5m (400 mg/L), 1.5m (under 10 mg/L), and 6.5m (18 mg/L) in depth.

The sodium, potassium, calcium, and magnesium concentrations are significant as far as 2.5m and drop sharply in the layer of sand and silt-sandy clay containing coarse material. The exception is sodium that, although it showed a drop as depth increased, it still shows values of about 88 mg/L at 6.5m in depth. The chloride concentrations in the three measured samples accompany the noted tendency. Considering the proximity of the river, it is possible that the underground flow is occurring through this layer towards the river. However, the concentration of ammonium offers differentiated values, since it is extremely low at the surface (the test was repeated, but some error may have occurred during sampling), reaching 174mg/L at 2.5m in depth, and again with 142 mg/L at 6.0m in depth, which clearly indicated leachate contamination. Apparently, the more permeable layer facilitated the ammonium migration to reach the residual soil. The water level is at a depth of 1.5m, but measured concentrations proved it had no significant value in the piezometer installed in that area.

Conclusions

Soil samples were taken in depth from the boundary of the old MSW landfill, whose pore water was analysed. In the South side of the area, there is a sign of a migration process that may be occurring due to the sodium concentrations, although low when compared to levels reached by the leachate at 7.0m in depth, including ammonium and potassium at 2.5m in depth. On the East side, a contamination process is underway due to high concentrations of chloride, sodium, calcium, potassium and magnesium at a depth of seven meters, while ammonium is only high in the surface layer; the groundwater level is at the surface and is already contaminated. On the North side, next to the Iguaçu River, there may be a flow through a more permeable layer towards the river. However, ammonium concentrations of around 140mg/L at 6.5m in depth also indicate a possible process of contamination in that area.

Acknowledgements

The authors are grateful to S.A. Paulista Company, for the soil sampling and for the financial support to some analyses, and also to Maria da Gloria Marcondes from Environmental Geotechnical Laboratory from COPPE/UFRJ.

References

APHA, AWWA, WPCF. Standard Methods for the Examination of Water and Wastewater, 20th edition, New York, 1998.

Junqueira, F.F. & Palmeira, E.M. (1999), "*Monitoramento do comportamento de lixo em células geotecnicamente preparadas*" [Monitoring waste behaviour in geotechnically prepared cells], REGEO'99. Brazilian Congress on Environmental Geotechnics. pp. 428-433.

Mattos, T.P. (2005), "*Avaliação da contaminação hídrica no entorno do antigo lixão da Marambaia – RJ*". [Assessment of water contamination around the former Marambaia dumpsite, Rio de Janeiro] Master's dissertation. Sanitary and Environmental Engineering Department, UERJ.

Mota, M.F.P. (2005), "*Avaliação da contaminação do solo e o transporte de contaminantes no antigo lixão da Marambaia – Nova Iguaçu/RJ*" [Assessment of soil contamination and transport of contaminants in the former Marambaia dumpsite – Nova Iguaçu, Rio de Janeiro] Master's dissertation. Sanitary and Environmental Engineering Department, UERJ.

Ritter, E., Campos, J.C & Gatto, R.L. (2004), "The contamination level through an organic soil in Gramacho MSW landfill", ISC-2 Proceedings on Geotechnical and Geophysical Site Characterization, Viana da Fonseca & Mayne (eds.), pp. 1339-1343.

Physical and Numerical Modelling of Groundwater and Surface-Water Interactions with a Conservative Tracer.

Tim Sparks, PhD student, Cardiff University, Cardiff, UK (sparkstd1@cf.ac.uk)
Diana Kountcheva, MSc student, Cardiff University, Cardiff, UK
Dr. Bettina Bockelmann, Cardiff University, Cardiff, UK
Professor Roger Falconer, Cardiff University, Cardiff, UK

Abstract
This paper outlines recent and on-going physical and numerical modelling work carried out at Cardiff University, the novel approaches used, some of the initial results obtained, the problems encountered and the envisaged solutions.

A laboratory flume has been adapted to simulate the interaction between surface water and groundwater. At the same time an integrated surface-water/groundwater model has been developed from an existing hydrodynamic two-dimensional surface water model (DIVAST). The flume uses 60 ppi open-cell foam as 'river-bank' material to allow flow and solute transport through the aquifer. The computer model extends the surface-water model horizontally into the groundwater by discretising the equations of flow for porous media in the same way as the surface water equations. The interactions between the two domains can then be modelled together, in the same numerical model.

The laboratory setup is described, together with problems encountered. Initial results from the laboratory are compared with the computer model in general terms, although there are significant hurdles to overcome in the physical modelling before it can be used to refine the numerical model.

Introduction
Managing groundwater associated with contaminated land is vital in order to control and reduce the spread of contamination. Contaminated groundwater will cause problems when it reaches the surface, either through direct abstraction, or through seepage into surface water bodies (diffuse pollution). A new type of numerical groundwater model has been developed from an existing 2D hydrodynamic surface-water model in order to simulate the latter situation where groundwater and surface water are closely linked, and a conservative tracer is used to predict the transport of contaminants.

This computer model was compared to results from laboratory experiments in a tidal flume, simulating a tidal river reach. River 'banks' were included as permeable 60ppi foam, and a conservative tracer was used to measure flow in the river banks and the seepage to the main channel. Using foam allows a large area of groundwater to be simulated without a reinforced flume (necessary when large masses of sand are used instead). Foam is also relatively easy to work with, and retains its shape in a strong flow (unlike sand).

Numerical Modelling

The extended DIVAST model has been described in recent conference papers (*Sparks et al. 2004; Sparks et al. 2005*) so this paper will only consider a brief overview of the principle, and the application to the laboratory work will be the main focus of this paper.

DIVAST (Depth Integrated Velocities And Solute Transport) is a well regarded 2-D hydrodynamic and water quality model. The 2-D nature makes it particularly suitable for modelling estuaries, wide rivers and coastal waters. The first incarnation of the model was developed by Prof. Falconer (*Falconer 1992*) and has since been refined by various researchers at Bradford and in Cardiff University. The original model simulates two-dimensional distributions of surface water currents, elevations and various water quality parameters as functions of time. The new extended model takes this existing model and extends the domain into the adjacent groundwater horizontally whilst retaining the computational procedure as before. Thus the hydrodynamic equations for groundwater and surface water are solved together in the same mathematical 'sweep', rather than solving each separately and linking the results with source/sink terms, as is commonly done (*Sparks et al. 2005*).

This allows the interactions between groundwater and surface water to be included in the model on a cell by cell basis and transport processes from the sub-surface to the river/estuary system and vice/versa can be modelled accurately. The fate of the contaminants can be predicted, and potential problems highlighted. The model is capable of modelling the distribution of conservative tracers at present. This can be extended to solutes once the processes at the biochemical sw/gw interface have been included in the model. Research is going on in this area and will be included in the future.

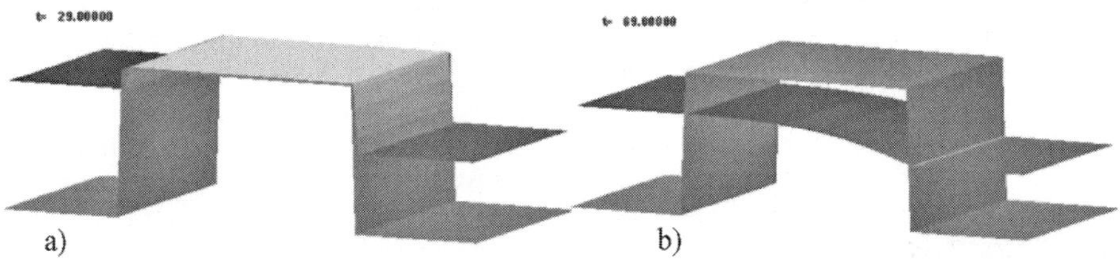

Figure 1 - a) Original DIVAST model with negligible depth assumed over the ground surface. Tidal water body on right has no influence on water body on left; and b) Extended model where flow continues through the aquifer. Tidal water body on right influences water body on left.

Hydrodynamic modelling is based on the physical equations of fluid flow. Conservation of mass and conservation of momentum are the two principles on which the original DIVAST model is based. The equations for conservation of mass and conservation of momentum are discretised in a finite difference format. The model domain is divided into square cells, and an equation created for each cell. The ADI (Alternating Direction Implicit) method (*Peaceman and Rachford-Jr. 1955*) is used in DIVAST to solve these equations simultaneously.

To extend the original model to include groundwater flow processes, we can discretise the equations of flow in porous media (conservation of mass, and Darcy's Law) in a similar fashion. We find that the equations that must be solved are identical to the surface-water equations in terms of the variables involved. The difference lies in the explicit coefficients for each variable. Therefore we can define the 'groundwater' coefficients and the 'surface-water' coefficients, and use them according to the status of each cell, solving the surface-water cells together with the groundwater ones, just as if they were all 'surface-water'. Figure 1 shows a comparison between the results from the original model and the extended model.

Modelling the laboratory conditions

A specific model designed to simulate the conditions in the laboratory flume was set up (Figure 2). The model includes an 'outfall' at the injection point (see Figure 3 and Figure 4), that introduces a conservative tracer into the aquifer at a specified rate and time. The model can record data from specific points as well as for the domain as a whole. Each laboratory experiment described below was mirrored in a numerical simulation.

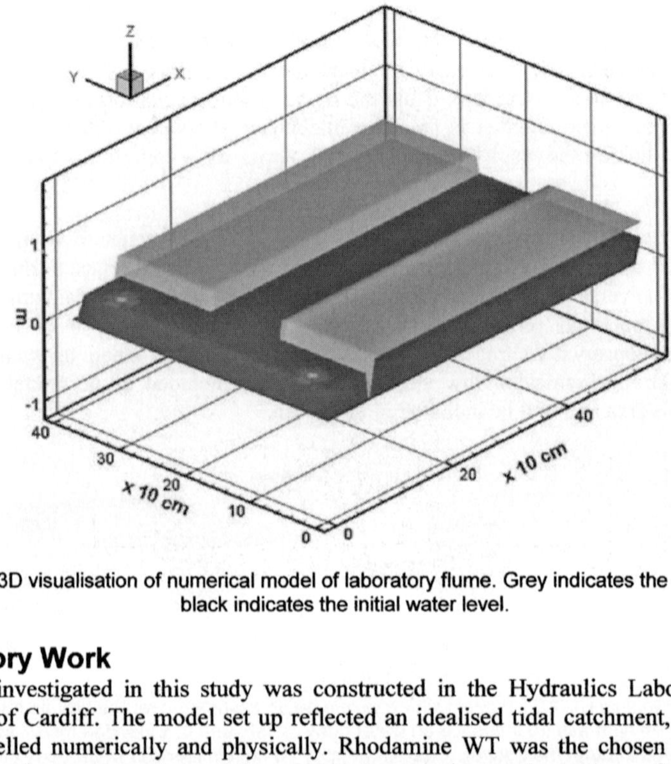

Figure 2 – 3D visualisation of numerical model of laboratory flume. Grey indicates the foam 'banks', black indicates the initial water level.

Laboratory Work

The basin investigated in this study was constructed in the Hydraulics Laboratory of the University of Cardiff. The model set up reflected an idealised tidal catchment, that could be easily modelled numerically and physically. Rhodamine WT was the chosen tracer for the simulations. The concentrations of dye used in the experiment, the settings of the measuring equipment, and the amounts of dye solute used in the model were predetermined with the help of the computer model. The results obtained were then corrected for delays and background concentration, and subsequently compared to the computer model output.

Flume construction

The channel was constructed out of 60ppi foam blocks with dimensions as shown in Figure 3. The foam is intended as porous media through which water and solute can flow as if through a river bank. The foam blocks were initially glued to the base of the flume, but during experimentation, weighted boards were added on top of the foam to stop it from floating (see Figure 5).

A rectangular tank with a suspended base was used for the tidal basin. Water was supplied from pipes connected to the main re-circulation tank. Water enters the basin through a large perforated pipe, and accumulates underneath the suspended base of the model. Holes in the suspended floor of the basin allowed the water level to rise to a predefined point within the flume. The tidal water fluctuation was simulated by increasing and decreasing the water elevation at the south end of the domain using a moveable weir. A perforated wall between the tide generation pool and channel area reduces the turbulence of the flow entering the flume.

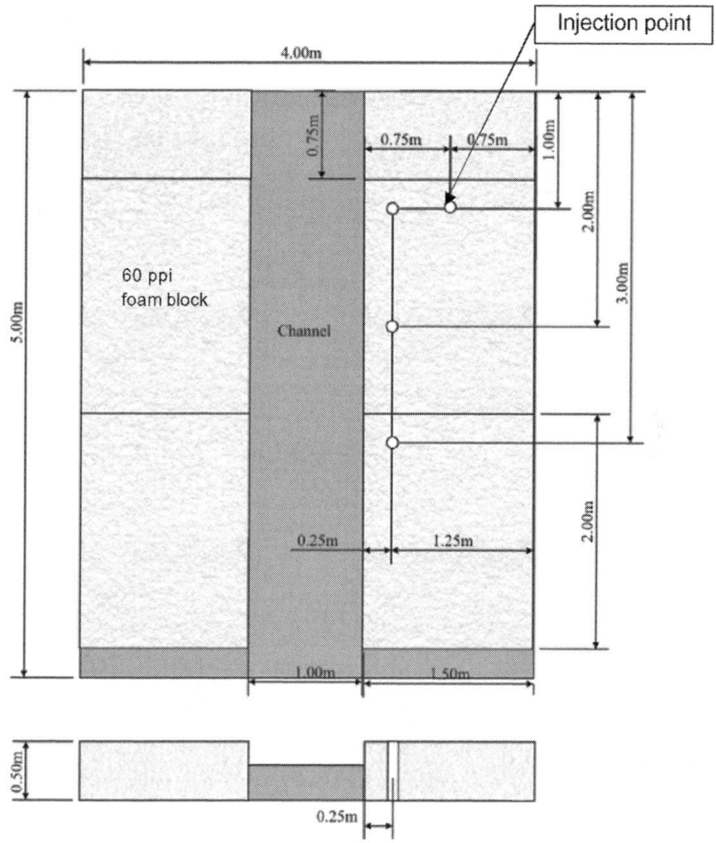

Figure 3 - Flume Design

Monitoring/injection points were cut out of the foam with a copper tube of 100 mm diameter. The holes went through the entire depth of the foam, and the extracted core was retained for reinsertion if necessary. The position of the injection point was selected so that it is some distance from the top and side boundary to reduce interference with contaminant accumulating by the walls. This allows for the migration of tracer along the channel and the bank to be investigated, while minimising unwanted effects of conducting the experiments in the lab with walls, rather than in a real river. The flume and monitoring/injection points are shown in Figure 4.

Two monitoring points were selected both from practical and analytical point of view to enable detection of the tracer plume. Point 'A' was selected along the perpendicular line from the outfall location to the channel. Point 'B' was selected to help investigate the spread of contaminant due to diffusion along the foam (see Figure 3 and Figure 4).

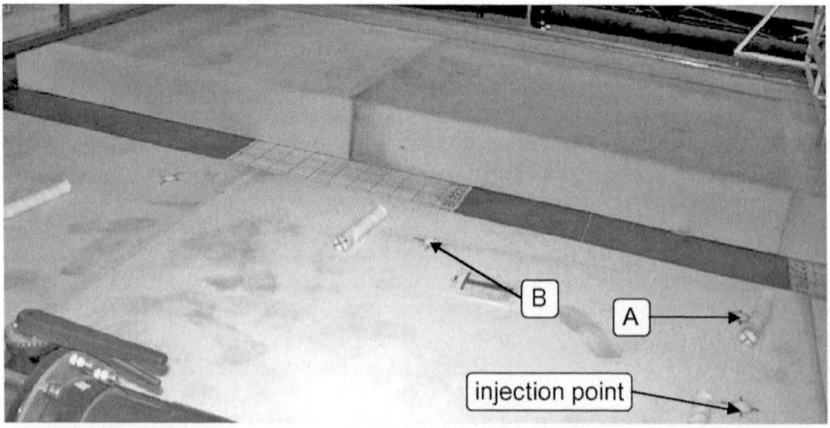

Figure 4 - Flume with no water, showing location of injection and monitoring points (A + B).

Figure 5 - weighted boards to prevent foam rising with tide.

Instruments

Two 10-AU fluorometers were used to measure the concentration of the dye in the two monitoring points simultaneously. The fluorometers were calibrated to ensure that they were measuring the correct values corresponding to the concentration of tracer in the water. A peristaltic pump was used to obtain the water sample from the monitoring locations, set to a low pump rate to keep the groundwater elevation stable. The travel time for flow through the pipe was measured, and the results adjusted accordingly.

The sampling point was positioned at an elevation just below the lowest water surface elevation (100mm from the floor of the flume) in order to keep it permanently submerged.

Experiments

The flume had a sinusoidal water elevation boundary varying from 290mm to 110mm and back to 290mm over a 30 minute period to simulate a tidal boundary. Rhodamine WT tracer was injected at the injection point over a period of two minutes at a concentration of 1 g/l.

The dye was injected at t=1800s, at the peak of a tidal cycle. The volumes injected were 50ml, 75ml and 100ml. Initial experiments of 100ml injections showed the concentrations were too high to measure, so subsequent experiments were performed with 75ml and 50ml injections. Selected results are show below in Figure 6.

Discussion

Figure 6 shows the data for monitoring point A. It appears to show a reasonably good correlation for the timing of the peaks between the numerical and physical models, although there are a few points that indicate that improvements need to be made in the physical model.

The predicted peaks are much higher than the measured peaks. The maximum concentration is theoretically 7.5 x 10^{-4} g/l, (or 750 ppb). The values measured are approximately an order of magnitude below this. Straight-away this indicates that much of the dye has been 'lost' from the foam. Some of the dye is probably lost due to adsorption to the foam, but it seems likely that for the effect to be this large, a lot of the dye has probably entered the surface water via a short-circuit underneath the foam. The initial experiments conducted support this theory; these were done while the foam was still firmly attached to the base of the flume and when the concentrations were measurable they were much closer to the predicted order of magnitude. (i.e. time 9000-9900s, predicted 250-300ppb, measured range 175-450ppb)

The peak shown in Figure 6 at about 1000 seconds is before the dye injection, and so must be due to background dye from previous tests. The similar shape of the peaks from 6300 to the end may indicate a similar origin, rather than a gradual dispersion from the injection point. The peaks occur during the rising of the tide, when the rising water re-enters from the channel into the foam and towards the monitoring points. This suggests that after t=7200s most of the dye detected originates from the channel, with small amounts from the foam. It is not certain to what extent the peaks are caused by water flowing through the foam, along the suspended floor of the basin, or rising from the larger basin underneath the suspended floor. (The sharp peaks occurs at a particular level of the rising tide, which could be associated with the critical amount of water necessary to cause enough buoyancy within the foam to lift it enough to allow water flow underneath.). In contrast, the numerical model predicts a decrease in concentration as the tide rises, as the cleaner water from the channel enters the foam.

The flow of water within the foam is slow and facilitates tortuosity of the flow paths of particles, which aids diffusion and mechanical dispersion of the dye, causing more scattering of the particles. Therefore the shape of the peak is a significant clue to its origin. A round peak indicates that the contaminant is more scattered within the water and it is likely that this reflects dye flowing through the foam. A sharp spiked peak is more characteristic of a source where the dye remains as a well defined "slug", indicating a short-circuit from the injection point or other dye-rich area. Hence it is likely that the rounded peaks which occur just after low water are from a diffuse source (i.e. have travelled through the foam). The timing of these rounded peaks matches up well with the predicted peaks in the numerical model.

In between these peaks the concentration drops back down almost to the background level very quickly. Again, this indicates that the monitoring holes are connected to the main channel by short-circuits. If the dye was leaving through the foam, the concentration would reduce much more gradually.

The numerical model predicted that the concentration at point 'B' would slowly increase with time as the dye diffused longitudinally through the foam, but the measured values showed no sign of this, only the same increasing background peaks as seen in Figure 6. The dye isn't reaching this second monitoring point as expected; another sign of short-circuiting of the tracer.

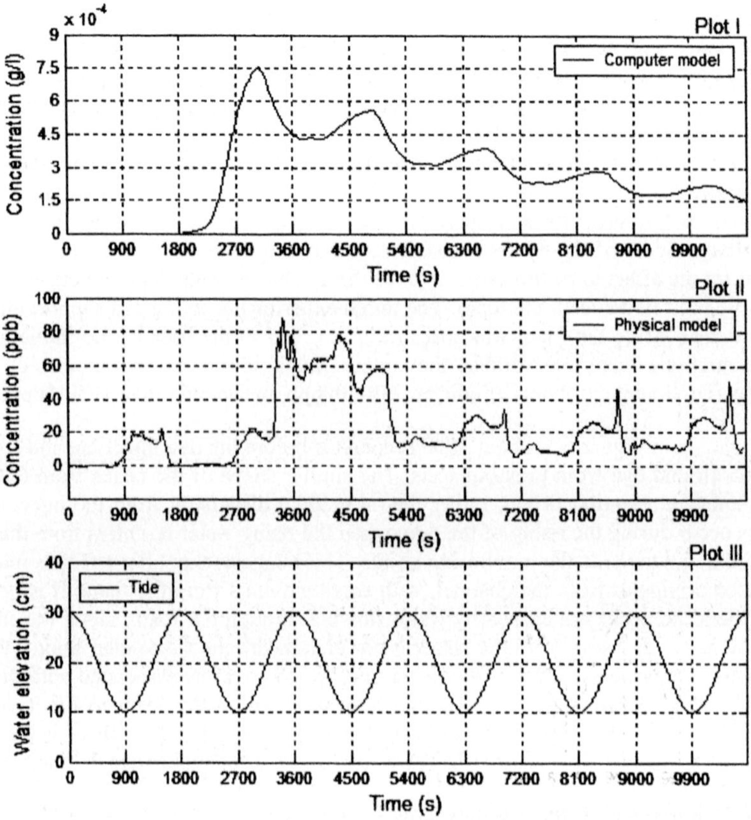

Figure 6 – Plot I – Graph of the numerical prediction of change of concentration of Rhodamine WT with time at point A for 75ml of 1g/l dye injected over two minutes. Plot II – the concentration of Rhodamine WT recorded at point A during physical experiment 4. Plot III – the respective tidal phase.

Problems, Solutions and Conclusions

The foam is buoyant, so in order to stop it rising it was glued to the base of the flume. However, after the first experiment, sections of the foam had lifted away from the flume base. Weighted boards were placed across the foam, but were insufficient to prevent short-circuiting of the water underneath the foam.

Stronger adhesive will be used to re-attach the foam to the base, or an impermeable board attached to the underside of the foam. The monitoring holes will also have the base of the 'core' reinserted, to prevent direct flow out of the base of the foam layer. Joints between blocks may have to be sealed with duct tape or similar. These methods or similar should prevent water flowing along unwanted pathways.

While it has caused some difficulties, the block structure of the foam will make it easy to rearrange the flume to simulate as many different scenarios as necessary. It is hoped to include a flow in the channel at the upstream boundary to more accurately represent a tidal river.

The initial results discussed briefly above, are more indicative of problems with the physical model that need to be addressed, rather than of how well the numerical model is performing. However, some of the observations do match up well with the numerical predictions in terms of timing, which is encouraging. Experimental work will continue when the modifications mentioned above are put in place. Once the short-circuiting problem is solved, it will be possible to compare the physical model with the numerical results, and refine the computer model accordingly.

References

(*Falconer 1992*)
Falconer, R. A. (1992). **Flow and Water Quality Modelling in Coastal and Inland Waters** *Journal of Hydraulic Research* Vol. 30 (4): 437-452pp.

(*Peaceman and Rachford-Jr. 1955*)
Peaceman, D. W. and Rachford-Jr., H. H. (1955). **The numerical solution of parabolic and elliptic differential equations** *Journal of the Society for Industrial and Applied Mathematics* Vol. 3: pp. 28–41pp.

(*Sparks et al. 2004*)
Sparks, T., Namin, M., Bockelmann, B. and Falconer, R. A. (2004). **Integrated River Basin Modelling - Finite Difference Modelling Of Surface Water And Groundwater Using Divast.** *Proceedings of "12th Association of Computational Mechanics in Engineering Annual Conference, 2004"*, Cardiff School of Engineering ACME 2004.

(*Sparks et al. 2005*)
Sparks, T., Bockelmann, B., Namin, M. and Falconer, R. A. (2005). **Integrated Modelling of 2-D Surface Water and Groundwater Flow with Contaminant Transport**. *Proceedings of "XXXI International Association of Hydraulic Engineering and Research Congress (IAHR Congress) "Water Engineering for the Future - Choices and Challenges""*, Seoul, Korea, (conference still in session at time of writing).

Nitrogen distribution in vadose zone under some Bulgarian soils

Dimitranka Stoicheva, Institute of Soil Science "Nikola Poushkarov", Sofia, Bulgaria
Milena Kercheva, Institute of Soil Science "Nikola Poushkarov", Sofia, Bulgaria
Dimitar Stoichev, Institute of Soil Science "Nikola Poushkarov", Sofia, Bulgaria

Abstract
The objective of this study is to estimate and compare the nitrogen distribution in geological materials underlying soils under natural and anthropogenic loading in different agro-climatic regions in Bulgaria. The obtained experimental results show that the applied fertilizer rates in excess of the nitrogen uptake by agricultural crops has led to considerable enrichment with NO_3-N of the entire profile (0-5 m) under the Fluvisol and the Chromic Luvisol. At the same conditions on the Vertisol nitrate nitrogen content is found till the 2 m depth and there is no direct relation between the fertilizer rates and contaminated shallow groundwater. The deep drillings completed in Haplic Chernozems show that independent of land use NO_3-N is detected in all samples from the soil surface to the groundwater table and varies from 0.2 to 33.0 mg.kg^{-1}. The area of the big dairy farm could be classified as a point source of contamination as about 400 kg.ha^{-1} NO_3-N is accumulated in the 2-meter layer.

Introduction
There are different approaches for studying the risk for enrichment of groundwater with nutrients and particularly with nitrate. In Bulgaria the identification of hot spots is realized through monitoring of the quality of groundwater organized by the Ministry of Environment and Waters (Drinking Water Quality in the Country, 1979) and by other research institutes in the frame of agro-ecological projects (Stoichev et al., 2001). More than 70% of the drinking water in the centralized water supply system in Bulgaria is from groundwater sources (Technical report N 27, 1988). The nitrate pollution of drinking water arose as a significant ecological problem in Bulgaria at the beginning of the 1980s when the contaminated sources of drinking water have increased to 20% (Gitsova et al., 1991) despite the accepted in 1983 higher value (50 ppm) of MPCL.

The sources for surplus of soil nitrogen such are fertilization, farm waste disposal and improper agrotechnical management are studied in control field experiments (Dinchev 1983; Stoichev et al., 1996; Stoichev, 1997; Stoicheva et al., 2004) and in real life farming practice (Stoichev et al., 2001). The fate of nutrients in the vadose zone is investigated rarely (Mateva et al., 1982; Stoichev et al., 1980, 1983). The efforts to link the different locations of nitrogen and trace its movement towards the groundwater are impeded by the complicate character of the processes and lack of enough information.

The objective of this study is to estimate and compare the nitrogen distribution in geological materials underlying soils under natural and anthropogenic loading in different agro-climatic regions in Bulgaria.

Material and methods
The main aquifers in the country consist of alluvial and deluvial quaternerry deposits in the plains of the Danube and Maritsa Rivers and their tributaries. It has been established that the

polluted groundwater is situated mainly in the zone of Vertisols, Luvisols, and Fluvisols in Southern Bulgaria and of Fluvisols and adjacent territories in Northern Bulgaria.

Sites layout

Location

Four areas situated in different agroecological regions of the country are studied for nitrogen distribution in soil and vadose zone at different land use (Table 1). Deep drillings were completed at three of the experimental stations of the Institute of Soil Science 'Nikola Poushkarov' near the villages of Gorni Lozen, Tsalapitsa and Sredetz, Southern Bulgaria under different fertilization treatments of irrigated maize grown as monoculrure. The drillings at the fourth site were made along the profile of a pilot watershed on the territory of the village of Pavomaitsi, Northern Bulgaria. The terrain in the studied area is flat with slight slope (0.7-2.0%) and groundwater gradient (0.28-0.82%) towards the Yantra River bed.

Climatic characteristics of sites

Climatic conditions are important factors, which influence the soil nitrogen transformations and nitrate leaching with water percolation. The regions differ in respect of the annual temperature amplitude and the distribution of precipitation (Table 1). Two of the sites (v. Parvomaitsi and v. Gorni Lozen) belong to the European moderate-continental zone with summer maximum of precipitation, cold winter and hot summer. The others (v. Tsalapitsa and v. Sredetz) bear the traits of the transitory subzone between European and Mediterranean continental zones with more evenly distributed precipitation, hot and dry summer and mild winter. The conditions for the high soil water percolation are created in wet years and at improper scheduled irrigation in summer.

The long-term records with meteorological data (Meteorological references, 1932-2002) allow to calculated appropriate indices to indicate the wet years. The Palmer Drought Severity Index – PDSI (Palmer, 1965) and the Leaching Index - LI (Williams and Kissel, 1991) are used in this study. The LI is ranged in Pierce et al. (1991) as low - 0-127 mm; moderate 127-254 mm; high 254-508 mm; excessive >508 mm.

Table 1. Geographical and main climatic parameters of the regions.

Site (nearest MTO station)	Lg.	Lat.	Alt. m	T°C Jan.	T°C July	Precipitation, mm		
						Apr-Sept	Oct-Mar	Annual
v. Parvomaitzi			94-110					
(G. Oryahovitsa)	25°36'	43°10'	87	-1.4	22.5	320	238	561
v. Gorni Lozen	23°30'	42°37'	565			355	219	574
(Sofia)	23°20'	42°39'	588	-1,3	20.3	367	278	647
v. Tsalapitsa	24°35'	42°13'	180					
(Plovdiv)	24°45'	42°09'	160	0,1	23.4	247	267	512
v. Sredetz	25°40'	42°16'	170					
(Stara Zagora)	25°39'	42°25'	229	1,0	23.8	355	273	628

Soil and geological description

The main characteristics of the studied soils are presented on Table 2. The most vulnerable to nitrate leaching are Fluvisols. They are coarse textured soils, which determines a high amount of drainage aeration pores and conditions for fast water movement downward the profile. The risk for nitrate pollution in the regions of these soils increases also due to their locality to the river bed and shallow groundwater table. These soils have one humic-accumulative horizon overlaying parent rock or deposition layers. The studied Fluvisols are sandy loam. Water holding capacity of 0-100 cm soil layer of Fluvisols is about 150 mm. The hydraulic conductivity is high - 26 cm/h for 30-180 cm soil layer in the village of Tsalapitsa (Mateva et al., 1982). The Fluvisols belong to the hydrological group A. Fine textured soils (Haplic Chernozem, Vertisol) and illuvial B horizon of Chromic Luvisol impede the fast water moving. The Haplic Chernozem and Chromic Luvisol could be classified as moderately well drained (hydrologic group B), while the Vertisol is poorly drained (hydrologic group C). Nevertheless fast water moving could occur through the cracks due to the high swelling of this soil.

Table 2. Main properties of studied soils (Stoichev 1997, Stoichev et al, 2001).

Soil horizons	Depth cm	Humus %	pH in H_2O	CEC mequ/100 g	Distribution (%) of soil particles size (mm)			d_b, g/cm^3	FC %v/v	BD %v/v
					1-0.05	0.05-0.001	<0.001			
Chromic Luvisol, v. Gorni Lozen										
Ap	0-25	1.95	6.6	27.9	21	37	42	1.28	37	18
A_2	25-46	1.60	6.8	30.5	21	36	43	1.37	37	21
B_1	46-76	1.49	7.2	31.3	15	34	51	1.41	39	23
B_2	76-105	1.32	7.2	30.9	15	32	53	1.49	40	26
B_3C_1	105-125	1.50	7.5	33.3	18	39	43			
Fluvisol, v. Tsalapitsa										
Ap	0-35	0.70	6.0	7.9				1.66	25	18
A_2	35-60	0.55	6.4	18.2	35	32	33	1.59	26	20
A_3C_1	60-87	0.42	6.5	22.8	27	37	36	1.45	32	16
C_2	87-118	0.38	6.5	23.1	26	47	27	1.61	25	17
Vertisol, v. Sredetz										
Ap_I	0-25	3.34	7.9	64.6	15	35	50	1.18	47	24
A_{II}	25-46	2.58	7.5	63.6	15	35	50	1.20	46	28
A	46-62	2.29	7.3	61.1	13	35	52	1.27	43	29
B_1	62-90	1.90	7.3	42 .0	10	38	52	1.36	40	32
B_2	90-113	0.93		43.7	10	36	54	1.35		
Fluvisol, v. Parvomaitzi										
A_1	0-24	1.50	7.5	16.1	64	26	10	1.38	-	-
A_2	24-50	0.88	7.5	16.2	60	28	11	1.46	-	-
A_3	50-80	0.29	7.9	15.2	55	41	4	1.46	-	-
C	80-170	0.32	8.0	17.1	64	33	3	1.46	-	-
Haplic Chernozem, v. Parvomaitzi										
A_I	0-23	3.31	6.7	35	12	54	34	1.29	37	19
A_{II}	23-36	2.07	6.7	35	10	53	36	1.42	39	22
AB	36-66	1.21	7	35	11	50	39	1.39	36	24
B	66-122	0.67	7.1	37	10	51	39	1.56	40	26
BC	122-140	0.49	7.5	38	11	57	33	-	-	-

The main characteristics of the geological profiles are presented on Table 3. Groundwater tables are situated at 3-4 m under Fluvisol in the v. Tsalapitsa (Stoichev, 1997), at about 7 m under Chromic Luvisol in the v. Gorni Lozen, 5 m – under Vertisol in v. Sredetz (Mateva et al., 1982), 4-5 m under Fluvisol and 11-17 m under Haplic Chernozen in v. Parvomaitsi (Stoichev et a., 2001).

Table 3. Vadose zone characteristics and depth of water table.

Site	Parvomaitsi		Gorni Lozen	Tsalapitsa	Sredetz
Soil	Fluvisol	Haplic Chernozem	Chromic Luvisol	Fluvisol	Vertisol
Litological description	Sand, fine granular (1.5-2 m); gravel, coarse sand (4-6 m); marl	Clay, light brown compact (15-17 m), limestone gravel(1-2 m),boulders, coarse granular gravel; marl	Clay, light brown; clay, brown, small gravel; gravel and sand	Clays, light brown; sand gray, coarse; gravel	Clays, silty brown; clay, brown compacted with limestone; clay
K_f, m/d	-	-	6.25	7.25	5.00
Water table depth, m	2-5	10-17	6-7	3-5	3.5-4.6

Experimental layout

The field trials with irrigated maize (Zea mays, L., FAO group 600) grown as monoculture in the experimental stations Gorni Lozen, Tsalapitsa and Sredetz involved four treatments with different nitrogen and phosphorous fertilizer rates compared with non-treated plots (V_1). In the optimum treatment (V_3) the rate was calculated for full compensation of the nitrogen uptake by the crop production. The rates for the other treatments were 125%(V_2) and 50%(V_5) from the amounts of optimum treatment. Sprinklers irrigation was scheduled to ensure a non-deficit water regime for the crop growth. Nitrogen content in precipitation, irrigation, leachate water, groundwater and residual soil nitrogen content in all treatments were also monitored. Deep drilling were completed in each of the described above treatments in 3 replicates.

The scheme of deep drilling in v. Parvomaitsi includes: virgin land (No 1), crop lands (No 2,3,4), big dairy farm area (No 5), natural pasture (No 6), and first flooded terrace of the Yantra River, near to a household garden (No 7). The drills No 1 to 6 are situated on the high terrace in the area of Haplic Chernozem, only No 7 is drilled in the area of Fluvisol.

The fate of nitrogen leached below the one-meter soil depth was studied using the samples of geological materials of vadose zone. The drilling of geological profiles were completed with automatic rotating machine F 220 mm. Disturbed samples are taken at the time of the drilling from every 30 cm layer to 1.5 m depth and from every 50 cm down to the groundwater level.

The nitrogen content is determined in 1% KCl (1:2.5) and measured by ADM-300 analyser (Methods of soil analysis, 1982).

Statistical analyses

The nitrogen data of geological samples along the longest geological profile under Haplic Chernozem were statistically analyzed for determining periodicity of the peaks. The statistical spectral analysis is used for determining the lengths of prevailing fluctuations (Shumway, 1988). The procedure calculates the peridiogram which is an estimation of the spectral function. The statistical software package ASTSA (Applied Statistical Time Series Analysis) is used.

Results and discussion

The analyses of soil samples and geological materials indicate that under Fluvisol in Southern Bulgaria, deep nitrogen leaching exists even in the case of no fertilizer applications (Fig. 1). The experiments carried out on a Vertisol (Southern Bulgaria) show that among the tested treatments (N_0, N_{110}, N_{220}, N_{276}), only the maximum rate results in a statistically proved nitrogen accumulation in 0-0.6 m layer. Detectable nitrogen amounts are found till the 2.0 m depth and there is no direct relation between the fertilizer rate and contaminated shallow (3.5-4.6 m) groundwater. Almost the same results are found in the field trial carried out on Luvisol, Southern Bulgaria.

The deep drilling completed in Haplic Chernozems and Fluvisols in Northern Bulgaria in the village of Parvomaitsi showed that independent of land use NO_3-N was detected in all samples from the soil surface to the groundwater table and varied from 0.2 to 33.0 mg.kg^{-1} (Figure 2). The area of the big dairy farm could be classified as a point source of contamination as about 400 kg.ha^{-1} NO_3-N was accumulated in the 2-meter layer. The whole geological profile from the soil surface to the groundwater level contained about 1200 kg.ha^{-1} residual nitrogen. Fluvisols on the Yantra River terrace are more vulnerable, because of their narrow soil and geological profile, low water storage capacity, high water permeability, shallow groundwater, and intensive anthropogenic loading.

Peaks values are measured in all long drillings of Haplic Chernozem and in some cases they could be found at very high depths even at the end of the aeration zone (Figure 2). They have different periodicity, which could be explained with the different properties of the geological materials of the layers from one hand, and with the water percolation regime in the geological profile from the other. The calculated peridiograms discern the most significant fluctuations in the profiles No1 (virgin land) – 4 m length and in No5 (dairy farm) – 5 m length (Table 4). These periods differs from the thickness of geological profiles (Table 3) which suggest to explain them with climate variability. Periodicity of wet years is irregular as it can be seen from the long-term records of PDSI (Figure 3) and leaching index (LI) (Figure 4). It should be mentioned that extremely wet years rarely (1%) occur in the region of Parvomaitsi (Table 5) and the peaks could originate also from other reasons – e.g. fluctuation of water table, internal nitrogen transformation. The first consideration could be rejected especially in the case of drill No 1, because it is situated at the top of the watershed and it was found no water above the marls.

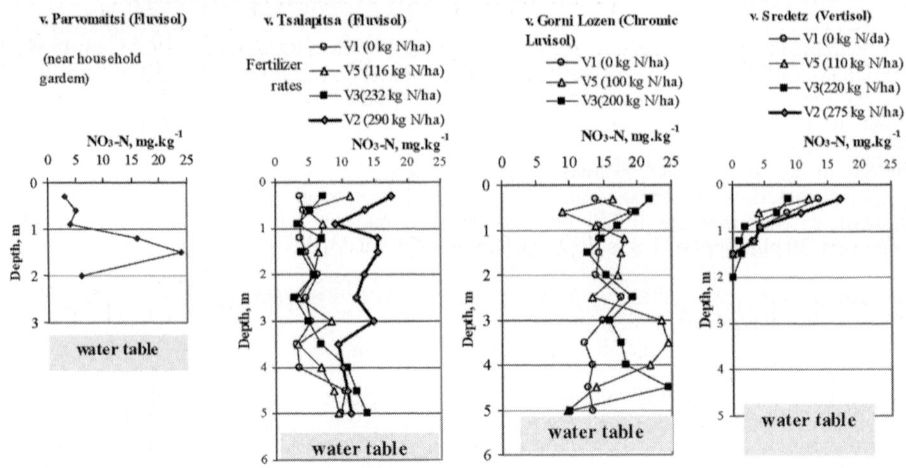

Figure 1. Distribution of nitrate-nitrogen (NO_3-N) in Fluvisols, Chromic Luvisols and Vertisols, and underlying geological profiles

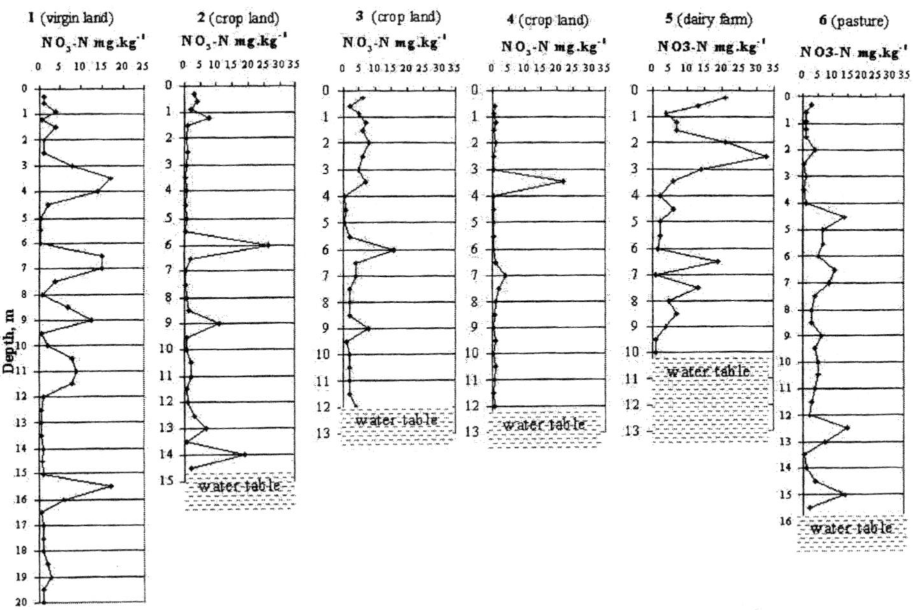

Figure 2. Distribution of nitrate-nitrogen (NO_3-N) in Haplic Chernozen at different land use and underlying geological profiles in the village of Parvomaitsi.

Table 4. Statistical parameters of nitrate-nitrogen distribution in the vadoze zone under Haplic Chernozem.

Parameters	Drillings, No					
	1	2	3	4	5	6
Thickness of vadose zone, m	20.0	14.5	12.0	12.0	10.5	15.5
Number of samples	40	29	24	23	20	30
Profile mean NO_3-N, mg/kg	4.3	3.3	4.0	1.6	8.5	4.9
Profile stdev NO_3-N, mg/kg	5.2	5.8	3.4	4.3	8.2	4.2
Main period, m	4.0	3.2	3.2	3.2	5.0	2.6
Second period, m	2.3				2.0	

Table 5. Mean values and distribution of LI for long-term period

Site (MTO data record length)	Soil hydrological group	LI, mm mean	Frequency (%) of years with LI 0-127 mm low	127-254 mm moderate	254-508 mm high
v. Parvomaitzi	A	103	68	31	1
(1931/32-2002)	B	39	99	1	0
v. Gorni Lozen	A	106	69	27	4
(1931/32-2002)	B	41	96	4	0
v. Tsalapitsa	A	79	80	18	1
(1931/32-2002)	B	27	99	1	0
v. Sredetz	A	111	63	30	7
(1954/55-1985)	B	43	90	10	0

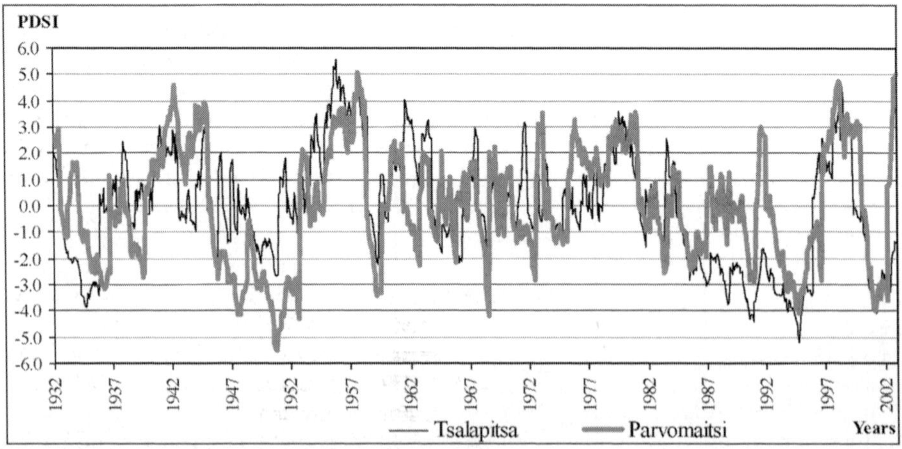

Figure 3. Palmer drought severity index (PDSI), v. Tslapitsa, Southern Bulgaria and v. Parvomaitsi , Northern Bulgaria

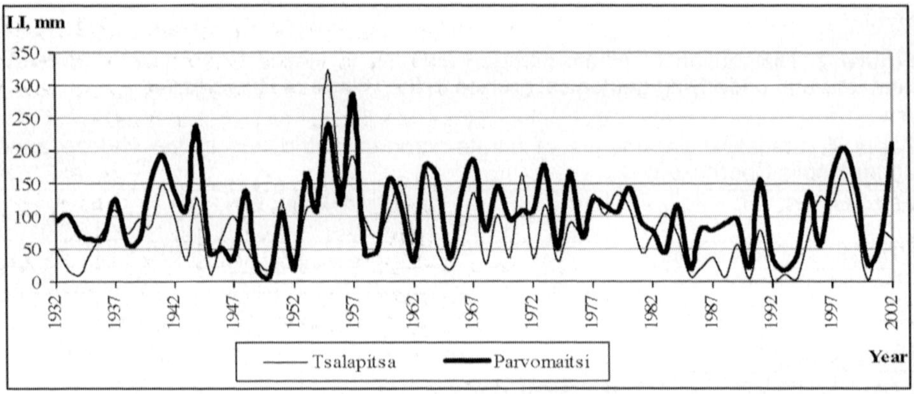

Figure 4. Leaching index (LI) for soils of A hydrological group in the regions of Tslapitsa (Southern Bulgaria) and Parvomaitsi (Northern Bulgaria)

Conclusion

The risk assessment for nitrate contamination of groundwater by diffusion and point agricultural sources is an important aspect of the environmental management. The obtained experimental results show that the applied fertilizer rates in excess of the nitrogen uptake by agricultural crops, have led to considerable enrichment with NO_3-N of the entire profile (0-500 cm) of the Fluvisol and Chromic Luvisol. At the same conditions on the Vertisol nitrate nitrogen content is found till the 200 cm depth and there is no direct relation between the fertilizer rates and contaminated shallow groundwater.

The deep drilling completed in Haplic Chernozems shows that independent of land use NO_3-N is detected in all samples from the soil surface to the groundwater table and varies from 0.2 to 33.0 $mg.kg^{-1}$. The area of the big dairy farm could be classified as a point source of contamination as about 400 $kg.ha^{-1}$ NO_3-N is accumulated in the 2-meter layer.

The efforts in this research should be directed to link the impact of different land use management practices on regional scale with climatic conditions provoking deep leaching.

References

Dinchev D., 1983. Soil nitrogen budget of Bulgarian soils. Zemizdat 136 pp (Bg),

Drinking Water Quality in the Country, 1979. Medicine and Sport, Sofia. (Bg)

Gitsova, S., Matev, B., Stoichev, D. 1991. On some aspects of water quality management and water pollution problems in Bulgaria. Report in Water Quality Workshop, Poznan, Poland,1-6.

Mateva, Ch., Stoichev, D., and Ahchiiski, P. 1982. Studing of nitrogen movement under intensive agriculture at different soils, Symposium Der Technische Fertschritt in der Wasserversorgung und in der Wasseraufbereittun, Band , 1, Varna, pp. 28-40.

Meteorological Annual References. 1932-2002. HMS, Sofia.

Methods of soil analysis, 1982. In: A.L.Page, R.H. Miller and D.R. Keeney (Editors) Part 2 – Chemical and microbiological properties, Second edition, Madison, Wisconsin (USA).

Palmer, W.C., 1965. Meteorological drought. U.S. Weather Bureau, Res. Pap. No. 45, 58 p.

Pierce, F.J., Shaffer, M., Halvorson, A.D., 1991. Screening procedure for estimating potentially leachable nitrate-nitrogen below the root-zone. In: R.F. Folett et al. (Eds.) Managing Nitrogen for Groundwater Quality and Farm Profitability, Soil Sci. Soc. Am., Madison, WI, chap. 12, 259-285.

Shumway, R.H., 1988. Applied statistical time series analysis. Englewood Cliffs Prentice Hall.

Stoichev, D., Atanassov, I., Glogov, L., 1980. Vertical movement of nitrogen in Alluvial-Meadow Soil. Soil Science and Agrochemistry, vol. XV, N 4: 10-17.

Stoichev, D., Stoicheva, D., Mateva, Ch., Ahchiiski, P., 1983. Distribution of nitrates in the profile of a Leached Smolnitza and their relationship to underground water. Soil, Sci. Agrochem., 18: 30-40.

Stoichev, D., Stoicheva, D., Angelov, G., and Dimitrova, K., 1996. Influence of long-term fertilization on nitrogen leaching. In: Van Cleemput et al. (Editors) Progress in Nitrogen Studies, Kluwer Academic Publisher, The Netherlands, 689-693.

Stoichev, D., 1997. Ecological aspects of anthropogenic loading of soils. Dr.Sc.Thesis, Sofia, 312 pp. (Bg).

Stoichev D., Kercheva M., Stoicheva D., 2001. NLEAP water quality applications in Bulgaria. In: Shaffer, M., Ma L., Hansen S. (eds.) Modeling Carbon and Nitrogen Dynamics for Soil Management. Lewis Publishers, Boca Raton, FL,.333-354.

Stoicheva D., Kercheva M., Koleva V., 2004. Assessment of nitrate leaching under different nitrogen supply of irrigated Maize. In Bieganowski, A., G. Jozefaciuk, R. Walczak (Eds.) "Modern Physical and Physicochemical Methods and their Applications in Agroecological Research". Institute of AGROPHYSICS, PAS, Lublin, 208-218.

Technical Report N 27, 1988. Nitrate and Drinking Water, Copyright (European Chemical Industry, Ecology and Toxicology Centre), ISSN 0773-8072-27. (Bg).

Williams, J.R., Kissel, D.E. 1991. Water Percolation: An indicator of nitrogen-leacjing potential. In R.F. Folett, D.R. Keeney and R.M. Cruse (Editors), Managing Nitrogen for Groundwater Quality and Farm Profitability, Ch. 4. Soil Sci. Soc. Am., Madison, WI, 59-83.

Time Integration for Numerical Solution of Advection-Diffusion Contaminant Transport

Guoping Tang, Northeastern University, Boston, USA
Akram N. Alshawabkeh, Northeastern University, Boston, USA

Abstract

A semi-analytical time integration (SATI) is proposed for numerical solution of the advection-diffusion contaminant transport equation. Compared with traditional finite difference method, SATI is stable, accurate and is not affected by the time step sizes. The computation costs for a single time step using SATI can be more expensive than using the finite difference method. However, the total computation cost may not be as expensive as that in finite difference methods since an accurate solution can be achieved by SATI using larger time step sizes and without any restrictions. Two examples from the literature are provided to demonstrate the effectiveness of SATI for 1D and 2D applications.

Introduction

Contaminant transport is often simulated by the advection-diffusion equation (ADE) that is given by (Fetter 1999),

$$R\frac{\partial C}{\partial t} = \nabla \cdot (D\nabla C + vC) + s \tag{1}$$

where R is retardation factor; C is solute concentration; D is diffusion coefficient; s is reaction or source; v is advection velocity; t is time. Analytical solutions are limited to simple geometry and in general require that the aquifer be homogeneous (Fetter 1999). Numerical solutions are more powerful than analytical solutions in the sense that the aquifers of any geometry can be analyzed and aquifer heterogeneities can be accommodated (Fetter 1999). However, numerical solutions often face difficulties of oscillation or instability, accuracy and prohibitive computation costs.

Numerical solutions of ADE usually involve space discretization to approximate the ADE by a system of ordinary differential equations (ODE) and time discretization to further approximate the ODE by a system of algebraic equations (AE). For the approximation to be accurate and stable, the grids in both space and time domain need to be fine enough. Too fine grids can make the computation costs prohibitive and truncation errors can accumulate to cause error and stability problem. This paper is devoted to the time integration of the ODE produced by the spatial approximation of the ADE with boundary conditions as

$$M\dot{x} + Kx = q \tag{2}$$

This is a semi-discrete advection-diffusion equation with M as capacity matrix, K as conductivity matrix, q as source vector due to boundary conditions and sources, x as the unknown vector of solute concentration at nodes. Details about space discretization are presented in Reddy (1993).

Traditionally, time integration of Eq. (2) is achieved by finite difference method that is described by the equation

$$(1-\theta)\dot{x}_k + \theta\dot{x}_{k+1} = \frac{x_{k+1} - x_k}{\Delta t} \tag{3}$$

Using this equation to approximate the derivative, Eq. (2) can be reduced to a system of algebraic equations as follows,

$$(M + \theta\Delta t K)x_{k+1} = [M - (1-\theta)\Delta t K]x_k + \Delta t[\theta q_{k+1} + (1-\theta)q_k] \tag{4}$$

If $(M + \theta\Delta t K)$ is nonsingular and stress q is constant during time interval k, Eq. (4) can be simplified to

$$x_{k+1} = A_d x_k + B_d q_k \tag{5}$$

where

$$A_d = (M + \theta\Delta t K)^{-1}[M - (1-\theta)\Delta t K]x_k \tag{6}$$

$$B_d = (M + \theta\Delta t K)^{-1}\Delta t \tag{7}$$

When $\theta < 1/2$, the θ method is stable only when the time step size Δt is very small. When $1/2 \le \theta \le 1$, θ method is unconditionally stable, but from θ = 1 to 1/2, numerical oscillation usually becomes significant as accuracy increases from the first order to the second order. Backward difference method (θ= 1) is very stable, but the accuracy is of the first order, so for accuracy, the time step sizes need to be small enough.

A stabilized second-order integration method–Runge-Kutta-Chebyshev (RKC) method (Verwer et al. 2004)–has been proposed for the time integration of Eq. (2). It is suitable for integrating a wide class of advection-diffusion-reaction problems. The advection-diffusion computations are explicit and the reaction computations are decoupled over the space grid. But finding critical time step sizes for use in actual solvers is a stability problem on its own (Verwer et al. 2004).

In this paper a semi-analytical time integration is proposed to solve the ODE produced by the spatial approximation of ADE. The stability, accuracy, time stepping and computation costs are analyzed. Two examples are given to illustrate the advantages and drawbacks of SATI

A Semi-Analytical Time Integration

Eq. (2) can be reduced to

$$\dot{x} = A_c x + B_c q \tag{8}$$

with $A_c = -M^{-1}K$, $B_c = M^{-1}$ because M is positive definite. The analytical solution for Eq. (8) is (Greenberg 1998)

$$x(t) = e^{A_c t} x(0) + \int_0^t e^{A_c(t-\tau)} B_c q(\tau) d\tau \tag{9}$$

Therefore, for a time step, if stress q can be viewed as a constant, then the solution can be discretized as

$$x_{k+1} = e^{A_c \Delta t} x_k + (I - e^{A_c \Delta t}) K^{-1} q_k \tag{10}$$

The physical meaning of the solution is clear: the first term on the right side describes the decay of the states, $K^{-1} q_k$ can be seen as the steady state solution of Eq. (2), so the second term is the remaining part of the response to the stress q. Usually, K is nonsingular, otherwise, the definition of matrix exponential as

$$e^A = \sum_{i=0}^{\infty} \frac{A^i}{i!} \tag{11}$$

can be used to calculate the second term without inversing K.

From physical meaning, and analysis as well, the solution of Eq. (10) as a time integration is unconditionally stable. When the stress is piece-wise constant, no error is produced in the time integration. Actually, if the stress is a linear, polynomial, exponential, trigonometric, or product of these functions of time t, the integration in Eq. (9) can be obtained analytically (Zhong 2004). Eq. (10) is the simplest form and the accuracy is dependent on the approximation of the stress as piece-wise constant. In groundwater flow and contaminant transport modeling, the time domain is usually divided into a number of stress periods in which the stress is constant. Therefore using SATI with time step sizes less than or equal to the stress period produces no errors in time integration. Traditionally when finite difference method is used for the temporal approximation, each stress period is further divided into a large number of time steps to achieve stable and accurate solution, thus prohibitive computation costs are usually involved.

SATI can be written in the same form as Eq. (5) but with

$$A_d = e^{A_c \Delta t} \tag{12}$$

$$B_d = (I - A_d) K^{-1} \tag{13}$$

In terms of computation costs, the differences between θ method and SATI are in the computation of A_d and B_d. For that θ method requires around $3n^3$ operations if only $O(n^3)$ computation (matrix inversion, matrix multiplication) is counted. SATI needs $4n^3$ operations plus a computation costs for a matrix exponential. Now the key issue for SATI is the computation of matrix exponential, which is challenging (Moler and Van Loan 2003) particularly when the scale becomes large. In general, the theoretical computation costs for matrix exponential is around 10~20 n^3. The matrices in the ODE that are obtained from spatial discretization are usually symmetric, positive definite and sparse. Employing these properties can significantly reduce the complexity, and thus the computation time for matrix exponential. The matrix exponential needs to be calculated only once if constant time size is used for linear case. Therefore the total computation costs with an accurate simulation can be saved using large constant time step sizes. Matrix exponential has a wide variety of applications in control theory as state space form (Brogan 1991) and structural dynamics (Fung 1997) and heat transfer (Chen et al. 2001), thus has the potential to be used for the groundwater flow and contaminant transport.

Examples

Two examples are selected from Tsai and Chen (1995) to demonstrate the effectiveness of the SATI method for time integration.

1D contaminant transport

Consider the following 1D governing equation

$$\frac{\partial C}{\partial t} = D\frac{\partial^2 C}{\partial x^2} - u\frac{\partial C}{\partial x}$$

with initial condition $C(x,0) = 0$ and boundary conditions $C(0,t) = C_0$, $C(\infty,t) = 0$. Assume $u = 1$ m/day and $D = 0.2$ m^2/day. Linear finite element is used with $\Delta x = 1$ m. The ranges of eigenvalues and spectral radius are presented in Fig. 1, which shows that the θ method will not be stable when $\Delta t > 1$ day for forward difference method, and when $\Delta t > 2$ day for θ = 1/3, because the minimum eigenvalues of A_d are less than -1, in another words, spectral radius of A_d exceed one. While SATI and $1/2 \le \theta \le 1$ are stable, oscillation can be expected as θ decreases from 1 to 1/2 for θ method due to the negative eigenvalues of A_d when Δt exceeds two days.

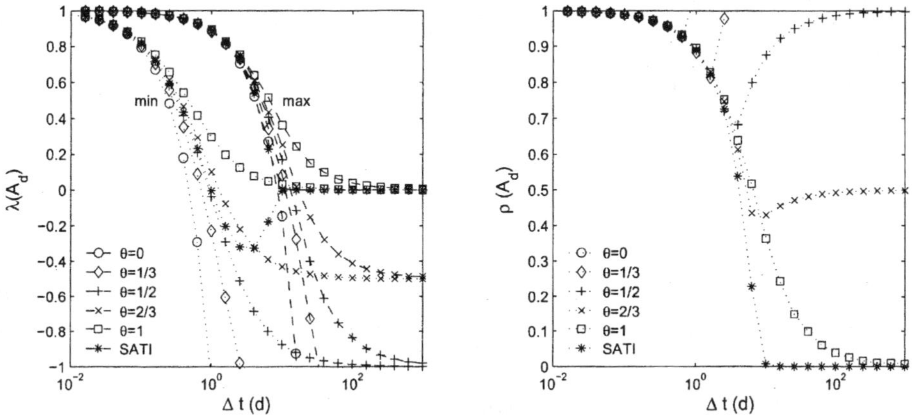

Fig. 1: Ranges of eigenvalues and spectral radius of A_d for 1-d example

The results by SATI and θ methods are compared in Fig. 2. When Δt is small (1 day), the solutions match the analytical solution. As Δt increases, the results obtained by Crank-Nicolson (θ=1/2) show oscillation, the results obtained by backward difference method (θ=1) are stable but with increasing errors, and the results obtained by Galerkin method (θ=2/3) are show a behavior that is somewhere between the first two methods. On the other hand, the results obtained by SATI are consistent regardless of Δt. Because the stress (here it is the prescribed boundary conditions) is constant, SATI produces no error in temporal approximation and there is no limitation on the time step size at all.

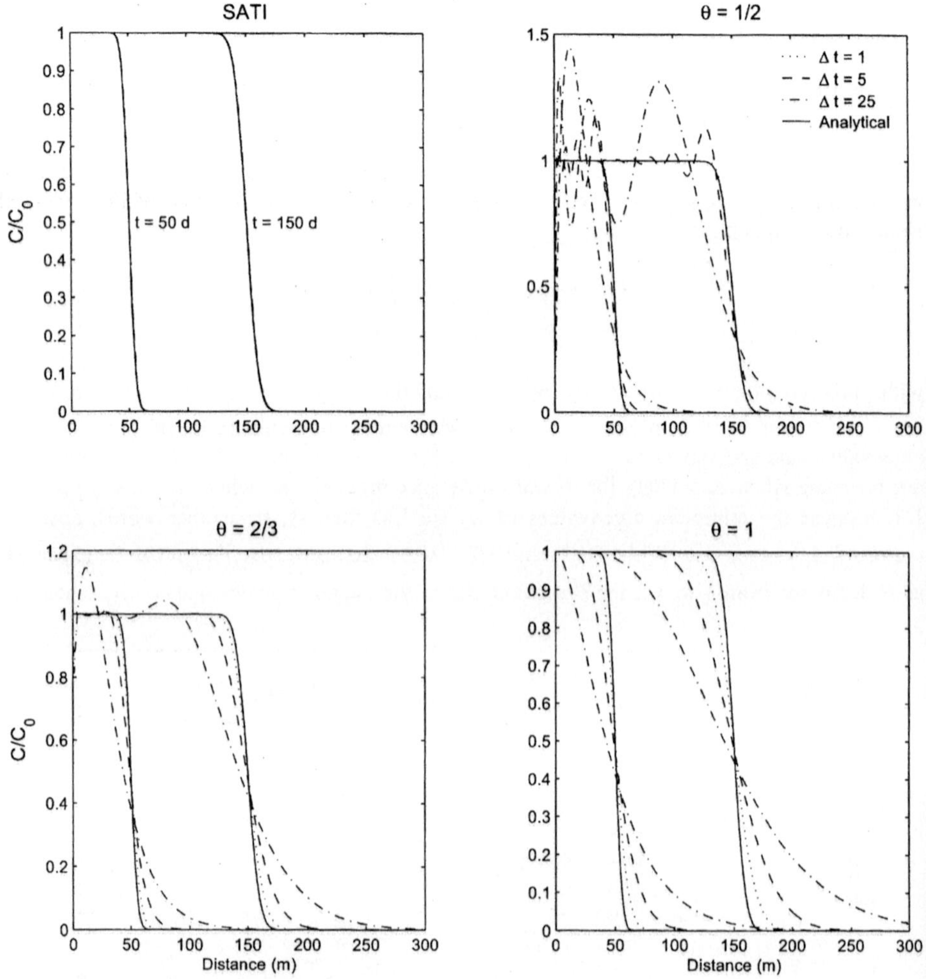

Fig 2: Concentration profiles at t = 50 day and 150 day for 1 d example

Two dimensional contaminant transport

The governing equation for two dimensional contaminant transport is

$$R\frac{\partial C}{\partial t}=\frac{\partial}{\partial x}\left(D_{xx}\frac{\partial C}{\partial x}\right)+\frac{\partial}{\partial y}\left(D_{yy}\frac{\partial C}{\partial y}\right)-\frac{\partial v_x C}{\partial x}-\frac{\partial v_y C}{\partial y}-kRC$$

A rectangular domain of 1000 m width and 800 m height is assumed with an initial condition of $C(x,y,0)=0$. The boundary conditions are no flux at the bottom, top and right boundaries, and prescribed solute concentration at the left boundary defined by a Gaussian distribution as

$$C(0,y,t)=e^{-\frac{(y-300)^2}{2\sigma^2}}$$

where $\sigma = 65$ m, R = 1, $D_{xx} = 32$ m^2/day, $D_{yy} = 4$ m^2/day, $v_x = 4$ m/day, $v_y = 1$ m/day, and k =0.01 1/day is the first order decay coefficient of chemical reaction. Linear rectangular element is used with $\Delta x = 40$ m and $\Delta y = 20$ m.

The concentration contours at t = 150 day using different Δt s and time integrations are presented in Fig. 3. The SATI does not introduce any error as Δt increases while θ method shows increasing errors.

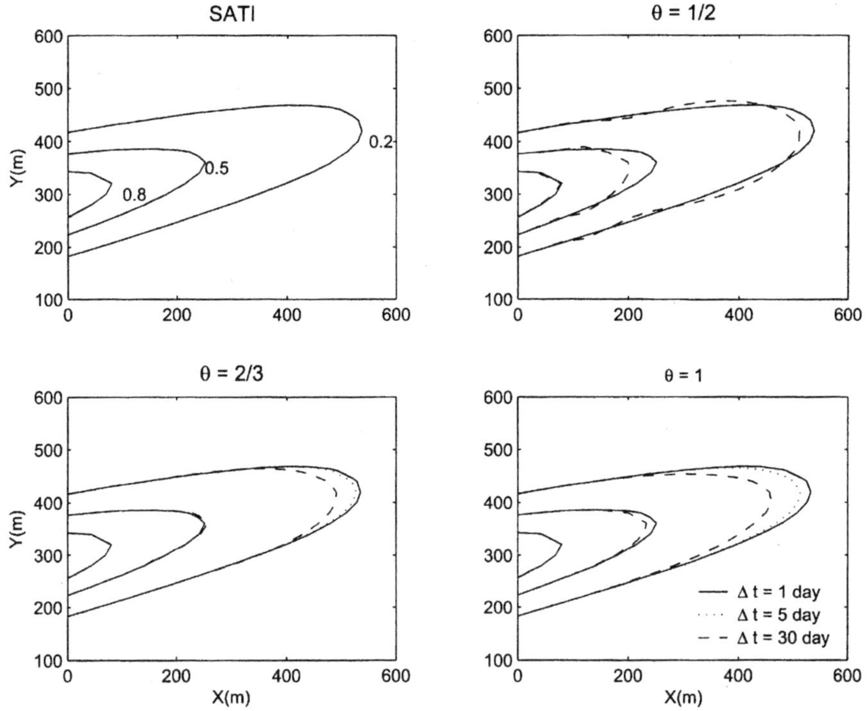

Fig. 3: Contours at t = 150 day using different Δt s and time integrations

Conclusions

A semi-analytical time integration (SATI) is proposed for the time integration of ordinary differential equations (ODE) produced by spatial approximation of advection-diffusion equation. Instead of approximating the time derivative as the traditional finite difference method, SATI approximates the analytical solution of the ODE. SATI is found to be unconditional stable, accurate but with expensive computation costs. The accuracy is dependent on the approximation of the stress. Two examples are given to show that it is possible to save computation costs while obtaining an accurate solution with using large time step sizes.

References

[1] Brogan, W. L. (1991), Modern control theory, 3rd Ed., Prentice Hall, Upper Saddle River, NJ.

[2] Chen B., Gu Y., Guan Z. and Zhang H. (2001), Nonlinear transient heat conduction analysis with precise time integration method, Numerical Heat Transfer, Part B;40:325–341.

[3] Fetter, C. W. (1999). Contaminant Hydrogeology, 2nd Ed. Prentice Hall, Upper Saddle River, NJ.

[4] Fung, T. C. (1997), A Precise Time-Step Integration Method by Step-Response and Impulsive-Response Matrices For Dynamic Problems, *Inter. J. Numer. Meth. Engng*, 40, 4501-4527.

[5] Greenberg, M. D. (1998), Advanced Engineering Mathematics, 2nd Ed. Prentice Hall, Upper Saddle River, NJ.

[6] Moler, C. and Van Loan, C. (2003), Nineteen Dubious Ways to Compute the Exponential of a Matrix, Twenty-Five Years Later. *SIAM REVIEW*, 45(1):3-49.

[7] Reddy, J. N. (1993) An Introduction to the Finite Element Method, 2nd Ed., McGraw-Hill.

[8] Tsai, W.-F. and Chen, C.-J. (1995) Unsteady finite-analytic method for solute transport in ground-water flow, *J. Eng. Mech.* 121(2): 230–243

[9] Verwer, J. G., Sommeijer, B. P. and Hundsdorfer, W (2004), RKC time stepping for advection-diffusion–reaction problems, *J. Comput. Physics*, 201(1):61–79.

[10] Zhong, W.X. (2004). On Precise Integration Method, *J. Comput. Appl. Math.* 163:59-78.

Parametric study of hydraulic barrier for contaminant transport under transient conditions

Martin Vaníček, Czech Technical University, Prague, Czech Republic

Abstract

Hydraulic barriers are very often used as short-term remediation measure to protect the endangered environment from different accidents in more permeable strata where the pollution will migrate very quickly. The parametric study to be presented was prepared as part of the development of finite element code to model contaminant transport under transient ground water flow during PhD research at the Czech Technical University in Prague.

At first a brief description of the program development will be presented and than the actual modelling of the parametric study will be discussed. The study concentrates on a protection of controlled waters from a failure of landfill liner in highly permeable strata. Different scenarios in terms of extraction rates in extraction wells lasting two 18-month periods were studied to determine the most effective strategy to protect the environment. In this situation the hydraulic barrier proposed is designed just as short-term solution until more permanent will be found and be able to be implemented.

Introduction

The computer program developed during the research is for contaminant transport modelling in a horizontal aquifer under transient ground water flow with free surface using finite element technique. During the development of the program FENST2DTAD two of university's developed independent codes have been utilised.

The parametric study models different scenarios for temporary remediation measures of the landfill liner failure in order to protect controlled waters in the vicinity. Mainly the influence of the transient conditions from the beginning of the extraction well operation has been studied together with the effect of different extraction rates.

FENST2DTAD program development

The development of the program consisted from three stages, introduction with the independent models for contaminant transport (Kazda & Valenta, 1994) and transient ground-water flow, coding of the final program and verification and evaluation of the program.

The program development consisted mainly of coding and merging the existing codes together. Most difficult and time consuming task was to differentiate between those two codes in terms of variables and proper sequencing throughout the time of transport. While the program has been coded, also verification of the program has being carried out so the program will calculate proper results. Verification of the code has been done against one-dimensional analytical solution developed by Ogata (1970) with only steady groundwater flow, as there is not available analytical solution with the transient groundwater flow.

5th ICEG Environmental Geotechnics, Thomas Telford, London 2006.

The program is using simple input from a text file, where all the information for the given problem is entered. This includes the discretisation of the modelled area into elements with coordinates of its nodes, soil properties for both hydraulic and transport part, initial conditions for both parts and boundary conditions, which can vary in time (defined time steps) and the total time (number of time steps) of the solving problem.

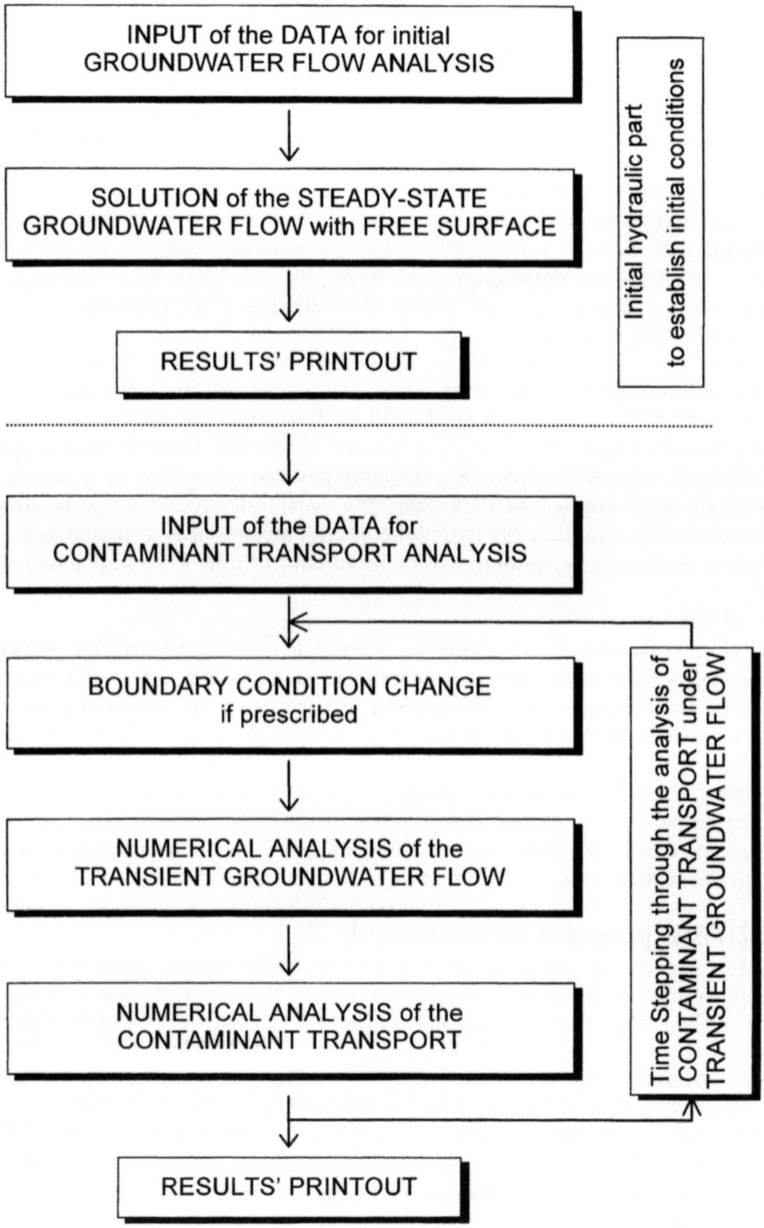

Figure 1: Schematic of the solution scheme used in the FENST2DTAD program

At the beginning of the analysis, steady state groundwater flow based on initial and boundary conditions is solved. Afterwards contaminant transport calculation is analysed in given amount of time steps. In each time step, first the hydraulic part is solved for the change of groundwater levels and / or flow rates in boundary nodes and therefore groundwater velocities. Groundwater velocities are then used in the determination of contaminant spreading in that time step accounting for advection, diffusion, dispersion and sorption transport mechanisms and also based on the change of the contaminant concentration at boundary nodes. The schematic diagram of the algorithm is shown in Figure 1.

To present the results from the numerical analysis, program uses output to the text data file and to the defined DXF files, which show the groundwater table in the modelled area in terms of contours of prescribed spacing and contours of the contaminant concentration in the modelled area.

Parametric study

The parametric study, which will be presented hereafter, was used also as evaluation exercise of the newly developed program.

This parametric study is based on a simplified realistic scenario and will present the influence of transient groundwater flow on the design of hydraulic barrier (see Figure 2). Hypothetic scenario of leakage from the crack in the liner of the landfill within a very permeable aquifer (coefficient of permeability 10 m/day) close to the controlled waters (river and lake) has been studied. Hydraulic barrier to protect the controlled waters seemed to be a solution of the given problem at least in temporary situation. The question than was, when to start and how much of the contaminated water to extract from the wells. An assumption has been made that the time, when any leakage from the analysed crack will be detected on site, will be one year (12 months).

After this time, remediation using extraction wells will start. Remediation measures are modelled as different water extraction rates from the wells in two stages of 18 months each. Rate 1 will be modelled between 12th and 30th month after start of simulation and rate 2 will be modelled between 30th and 48th month after the start of the simulation. The simulation itself will be finished after 48 months. At this time it is assumed that the crack would be sealed or other more sustainable remediation option developed and agreed.

Unfortunately in the limited space available for this paper, it is not possible to present and discuss all the analysed scenarios of the parametric study. Only those with the most significant influence on the results are presented here. The full results were presented in the PhD. Thesis of the author (Vaníček, 2003).

Comparisons of the results have been made with the case with no hydraulic barrier (see Figure 3) to show the effectiveness of each proposal. Graphical results presented here are only results at the end of each selected analysis. In the Figures, where the analysed results are presented, there are shown contours of groundwater levels every 0.5 m and contours of relative contaminant concentrations of 0.5, 5, 10, 25, 50, 75 and 100%. Case presented on Figure 4 is one of the preferred options (rate 1 of 1.16 l/s and rate 2 of 5 l/s). On the other hand the case presented on Figure 5 is one that represents one of the worst cases (rate 1 of 5 l/s and rate 2 of 1.16 l/s).

The case on Figure 4 allows for slightly higher spreading of lower concentrations mainly due to dispersion, but reduces the area affected by higher contaminant concentrations. The low rate of extraction attracts the water flow direction (advection process) towards the wells, but does not accelerate the spreading of contamination by dispersion. The higher rate in the later stage of remediation allows capture of contamination of higher concentrations that is present in that time around the extraction wells and does not allow contamination of higher concentrations to reach the controlled waters as the groundwater flow is from controlled waters towards the extraction wells (see Figure 4).

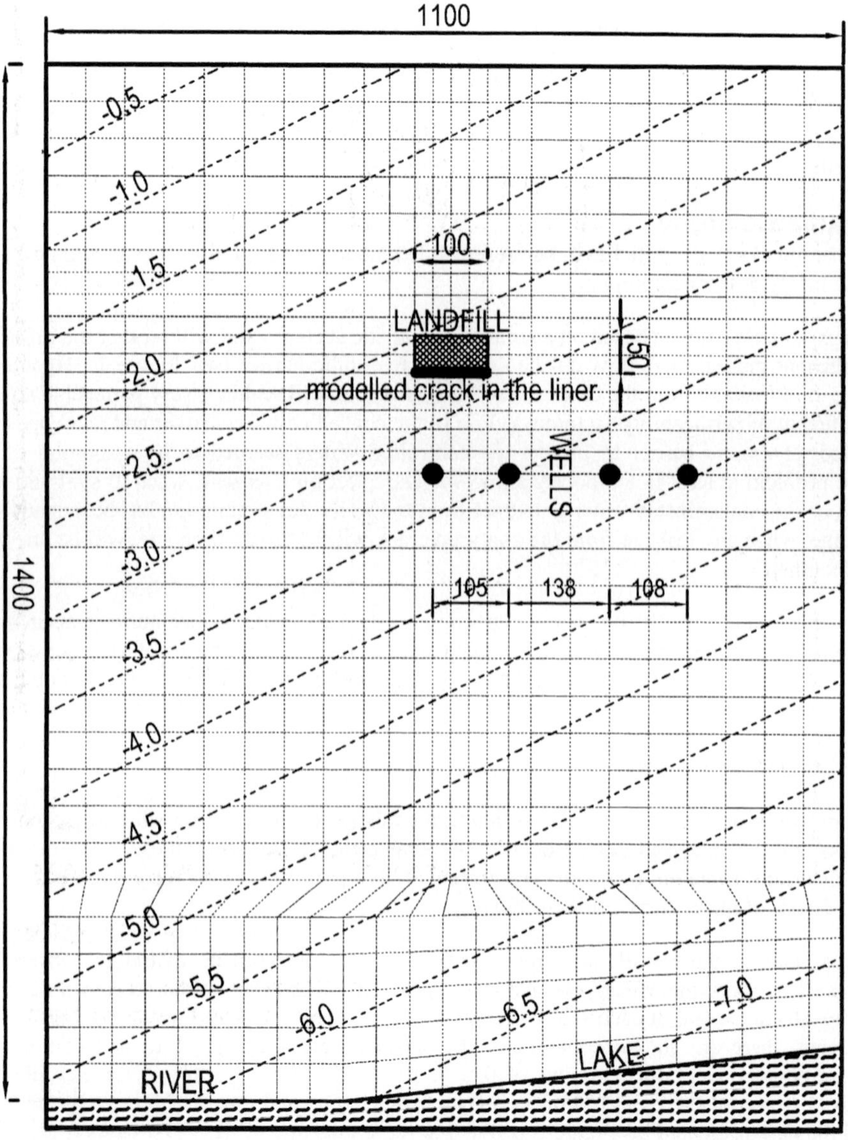

Figure 2: Modelled area with its dimensions, location of landfill, extraction wells, controlled waters (lake and river) and levels of impermeable base

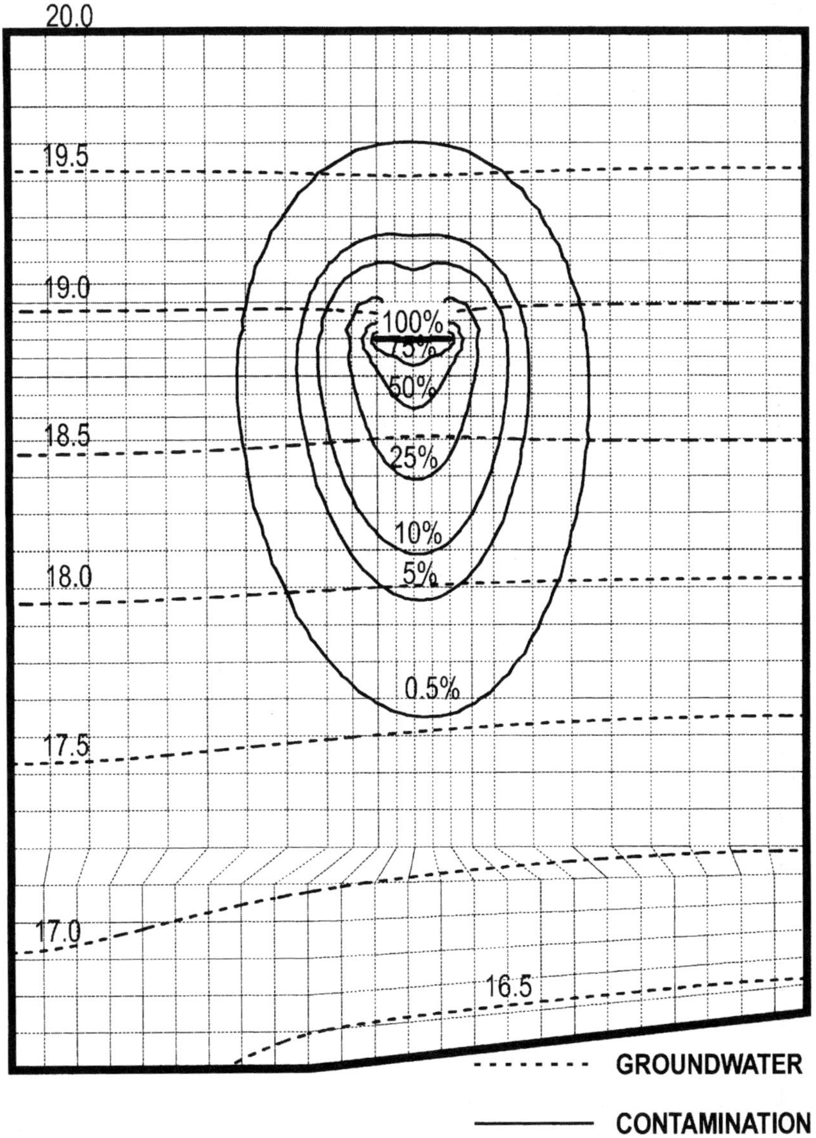

Figure 3: Contaminant spreading without any remediation measures after 48 months, contaminant concentration and ground water level contours

The case on Figure 5 presents the maximum contaminated area, where the spreading of contaminants goes furthest towards the controlled waters. This is caused by very high extraction rate at the beginning of the remediation process, when the contaminants are accelerated from the crack towards the wells, but also dispersed due to mixing (dispersion process) from high velocity of groundwater flow. The later stage of extraction with low rate

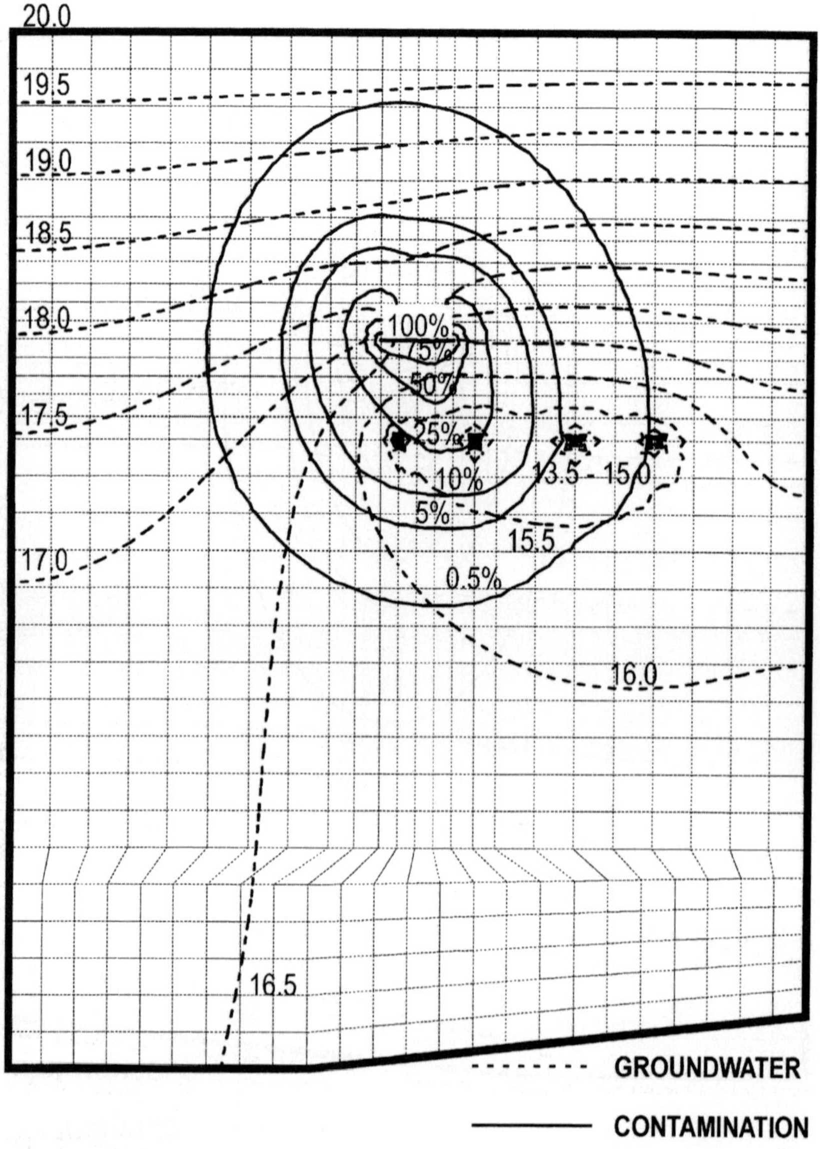

Figure 4: Contaminant spreading after 48 months with 1.16 l/s extraction between 12 and 30 months and 5.0 l/s extraction between 30 and 48 months (one of the preferred remediation options)

is not high enough to constrain the contamination in the area of extraction wells, but allows the plume to spread towards the controlled water as it can be seen from groundwater level contours on Figure 5.

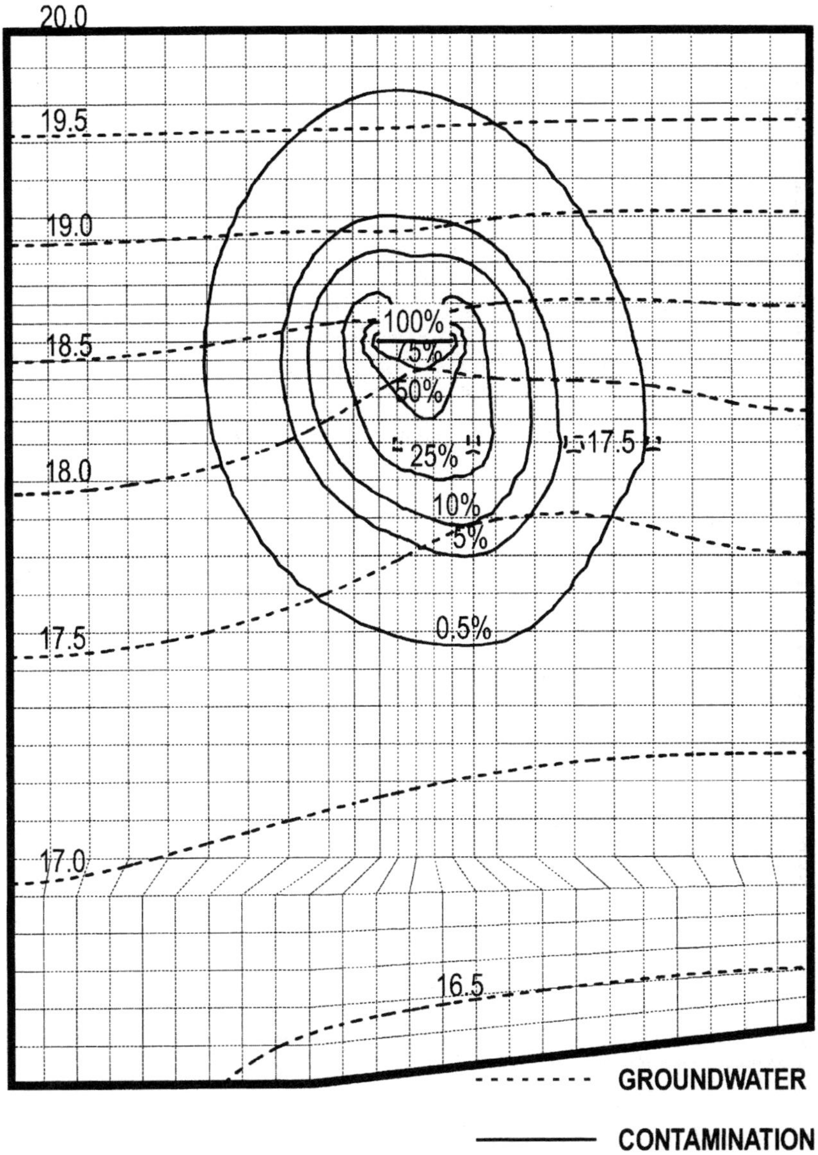

Figure 5: Contaminant spreading after 48 months with 5.0 l/s extraction between 12 and 30 months and 1.16 l/s extraction between 30 and 48 months – maximum advanced spreading

Conclusions

All modelled scenarios clearly showed that extraction wells can be used to protect controlled waters, at least temporarily, before more robust remediation options can be considered and the most appropriate will be put in place. But the rate of extraction and the timing of these extractions have to be also considered carefully.

Four different remediation strategies have been compared and it is obvious that it is important to know the area affected by contamination and the concentrations in relation to the wells location before any extraction from the wells began. The results from analyses (unfortunately not presented here for space reasons) indicate that the time needed to reach almost steady ground water flow conditions after the extraction rate change is around 9 months for the given site conditions. This finding shows how important role transient groundwater flow plays in the design of remediation strategies using numerical modelling of hydraulic barriers mainly at the beginning of the remediation process.

The presented results indicate that the newly developed program is capable of modelling the contaminant transport spreading under the transient groundwater flow. The developed program is a valuable addition to the research programs available at the Czech Technical University.

References

Kazda I.: Finite element techniques in groundwater flow studies. Development in Geotechnical Engineering, 61. Elsevier, Amsterdam; 1990.

Kazda I., Valenta P.: Program FE2DTAD 94. Manual, Czech Technical University; 1994.

Kazda I.: Underground hydraulics in ecological and engineering applications. Academia, Prague, 1997. (In Czech).

Ogata A.: Theory of Dispersion in a Granular Medium. U.S. Geological survey paper 411-1; 1970.

Vaníček M.: Contaminant transport in the host rock. Diploma Thesis, Czech Technical University; 1996.

Vaníček M.: Contaminant transport in the host rock – Numerical Modelling and Laboratory experiments. PhD Thesis, Czech Technical University; 2003.

Vaníček M.: Parametric study of numerical modelling of contaminant spreading under transient ground water flow. In: Proc. of XIIIth ECSMGE, Prague: CGS, 2003, Vol. 1, pp. 503-508.

WASTE REUSE AND WASTE MANAGEMENT

Management of waste sludge in trench cutting and re-mixing deep wall method

Minoru Aoi Ryobi Shinei Co., Ltd., Osaka, Japan
Junzo Fukunaga Obayashi Corporation, Osaka, Japan
Hiroki Yoshizaki Obayashi Corporation, Osaka, Japan
Fumio Kinoshita TRD Association, Tokyo, Japan
Masashi Kamon Kyoto University, Kyoto, Japan
Kazuyoshi Tateyama Ritsumeikan University, Shiga, Japan

Abstract
The trench cutting and re-mixing deep wall method (TRD method) is often used for retaining walls, for containment of contaminated soil, etc., because it can create not only vertical diaphragm walls, but also inclined ones of a highly homogeneous quality and without any defects. One of the technical problems encountered with the TRD method involves a substantial amount of waste sludge generated during the construction of such walls. The objective of this report is to determine an effective method to reduce the amount of waste sludge generated to the minimum. Although the simultaneous mixing method has much potential for reducing the volume of waste sludge, when compared to the subsequent mixing method, there is risk of cutter restriction by the solidified soil-cement mixture. This is because no additional bentonite slurry is injected into the trench to protect it against a collapse. Methods for evaluating the precise applicability of the simultaneous mixing method are discussed in this paper so as to maximize the potential use of simultaneous mixing method positively and in return minimize the impact to the environment by reducing the waste sludge.

Introducing the TRD Method
The trench cutting and re-mixing deep wall method (TRD method) is one of the methods for constructing diaphragm walls in the ground. The TRD method involves the excavation of deep trenches into which a solidifying agent such as cement slurry is simultaneously injected, agitated, and then mixed with in-situ soil. The method also involves the construction of continuous walls in the ground by a chain-saw type of cutter which travels horizontally in the

ground after being penetrated to a designated depth, as shown in Figure 1. When diaphragm walls are used as part of underground structures, and thus, a designated strength is required of the walls, core beams such as steel H beams are inserted into the trench before the soil-cement mixture begins to solidify, as shown in Figure 1. The TRD method enables high quality homogenous diaphragm walls of zero defects to be constructed which are of paramount importance when used for retaining walls, for containment of contaminated soil, etc.

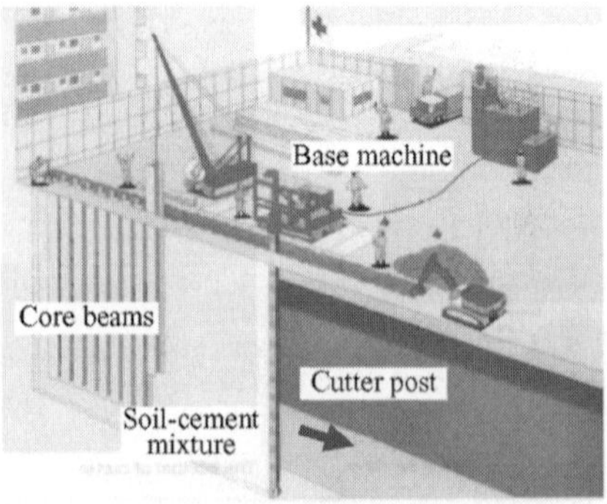

Figure 1 Construction of a diaphragm wall by the TRD method

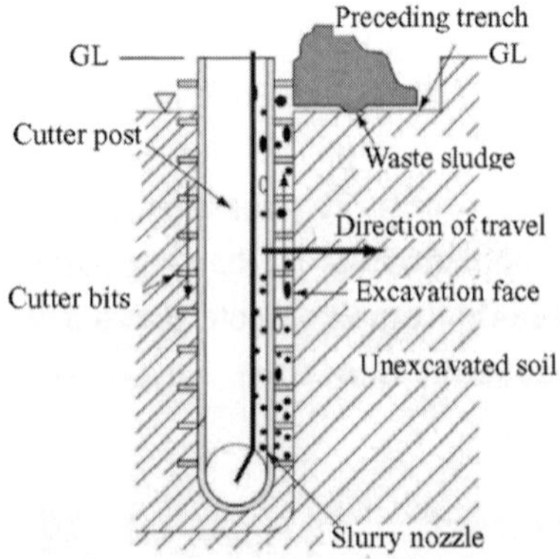

Figure 2 Movement of the cutter in the TRD method

In conventional construction methods, a rod equipped with a screw or a blade is penetrated perpendicularly into the ground. While the rod is rotated, a solidifying agent is injected into the ground through nozzles on the blade and mixed with in-situ soil. Each penetration of the rod forms a cylindrical solidification column, and thus, a diaphragm wall is constructed by placing series of columns, overlapping each other in the ground. However, defective mutual overlapping of the cylindrical columns can occur when deep construction is encountered. This is because of the irregular bend of the shaft during penetration, which brings about water leakage during or after construction. The TRD method can eliminate this defect by creating a continuous and uniform diaphragm wall. The cutter excavates the whole layer of soil vertically from its bottom to the surface of the trench, mixing and agitating the entire layer of soil uniformly to ensure the forming of a homogenous mix as shown in Figure 2.

Technical Problems with the TRD Method

The construction process using TRD method is classified into two different methods; namely the simultaneous mixing method and the subsequent mixing method. In the simultaneous mixing method, the excavation and mixing of soil are carried out simultaneously in a single operation as shown in Figure 3. This method is employed when the ground is soft enough and/or the construction depth is comparatively shallow, which enables high-speed excavations to be carried out. When the ground is hard and/or the construction depth is deep, the speed of the excavation is reduced permitting the solidification of the soil-cement mixture to take effect. In the case of an excavation in a hard ground or a deep construction, the subsequent mixing method is applied. In this method, the injection of the agent and its mixing with in-situ soil are carried out after the trench has been excavated, as shown in Figure 4. Bentonite slurry temporarily filled the trench upon excavation to stabilize the soil and protect the trench from collapsing. The cutter post then returns to its initial position before traveling again in the excavated trench, mixing the soil and the agent to form the diaphragm wall.

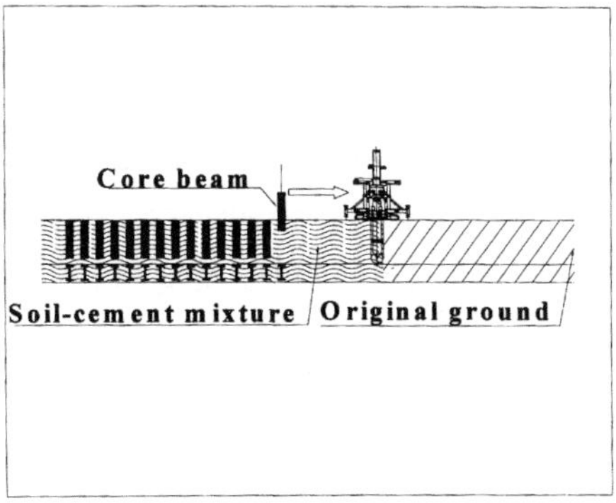

Figure 3 Simultaneous mixing method

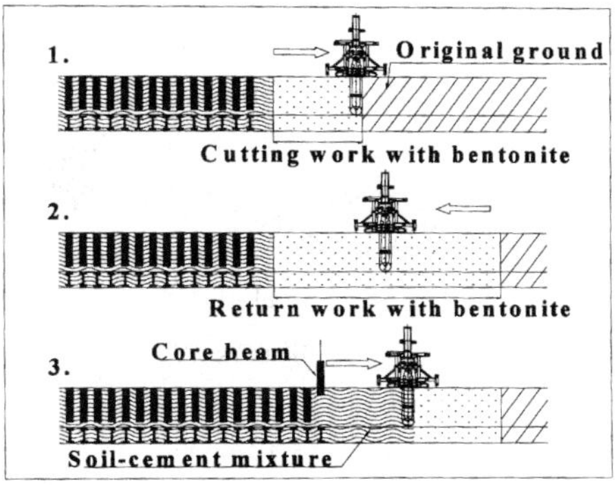

Figure 4 Subsequent mixing method

In the TRD method, a solidifying agent such as cement slurry is injected into the ground in the construction process. The volumetric increase in the trench will result in some volume of muck to be discharged out of the ground, as shown in Figure 2. When the subsequent mixing method is employed, the bentonite slurry is used to fill the trench to prevent collapsing prior to the injection of the cement slurry, and hence, the volume of waste sludge is more than that in the simultaneous mixing method. Figure 5 shows past records of a TRD construction, where the mucking ratio is plotted against the depth of the walls. The mucking ratio is defined by the ratio of the volume of the waste sludge to the total volume of the excavated trench.

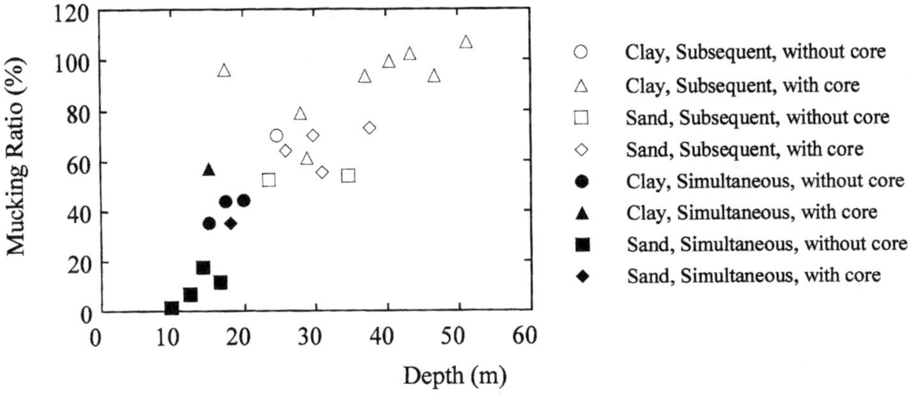

Figure 5 Construction records of the mucking ratio in the TRD method

It is clear from the figure that the mucking ratio in the subsequent mixing method ranged from 70 to 100% for clay grounds and from 60 to 70% for sandy grounds. In contrast, the ratio obtained in the simultaneous mixing method is from 30 to 60% for clay grounds and from 0 to 40% for sandy grounds. The volume of waste sludge is obviously less with the simultaneous mixing method than with the subsequent mixing method.

The volume of discharged waste sludge should be reduced not only to save costs related to the treatment of waste sludge, but also to minimize the environmental impact resulted from the construction. Thus, the applicability of the simultaneous mixing method is expected to be widely adopted in the TRD construction. When the simultaneous mixing method is employed in deep construction or in hard strata, the method's usefulness for reducing waste sludge cannot be undermined. To ensure the successful applicability of this method, the risk and difficulties brought about by the restriction of the cutter post from the progressing solidification of the soil-cement mixture must be precisely evaluated. In this paper, we discuss the mechanism for discharging waste sludge in the TRD method and judge the applicability of the TRD method to deep constructions or excavations in hard grounds.

Judging the Applicability of the TRD Method

The solidification of the soil-cement mixture generally begins in about two hours, and thus, the soil-cement mixing and the penetration of the core beams into the trench must be finished within two hours when the simultaneous mixing method is employed. This means that the traveling velocity of the cutter post should be fast enough to ensure the completion of these jobs within the specified two-hour period.

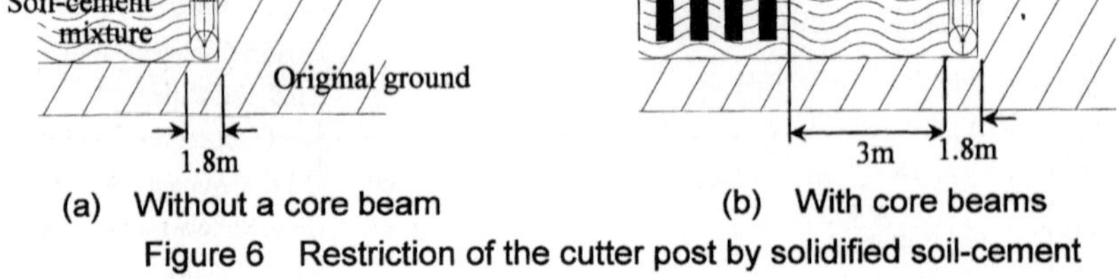

(a) Without a core beam (b) With core beams

Figure 6 Restriction of the cutter post by solidified soil-cement

Supposing the value of 1.8 m for the length of a popular cutter post, including the height of the bit, the horizontal traveling velocity of the cutter is required to be larger than 1.8 m per two hours in order to avoid the restriction of the cutter post from the solidifying soil-cement mixture, as shown in Figure 6(a), when core beams are not inserted. This means that the cutter post should travel horizontally in the ground with a velocity of more than 15 mm/min.

When core beams are inserted into a trench to reinforce the walls, another machine needs to be deployed for insertion of core beams into the trench. Space must be made available behind the TRD machine for deployment of this machine for core beams insertion before the solidification of the soil-cement mixture begins. When additional space of 3 m is required for the other machine to insert the core beams, the cutter post must travel in the ground at a velocity greater than 4.8 m per two hours, as shown in Figure 6(b). This translates to a traveling velocity of more than 40 mm/min for the cutter post.

The velocity of the cutter post can be determined by the relation between the cutting ability of the machine and the cutting resistance acting on the cutter from the ground. The former includes both factors of the mechanical power for cutting and that for traveling, while the latter includes the strength of the ground, the depth and the width of the wall, etc. Kinoshita and Mizutani developed a practical method for predicting the traveling velocity of a cutter post by combining the results of a theoretical study on the excavation mechanism in the TRD method and the records obtained at actual construction sites. Figure 7 shows the diagram that is used for predicting the velocity. The TRD value is an index expressing the resistance acting on the cutter from the ground. It can be determined from such construction conditions as the depth and the width of the wall, the machine parameters, and the ground conditions[1).

The applicability of the simultaneous mixing method can be judged by using Figure 7. When the traveling velocity predicted from Figure 7 is larger than 40 mm/min and core beams are used together or when the velocity is 15 mm/min and no core beams are used, the simultaneous mixing method can be applied for the construction.

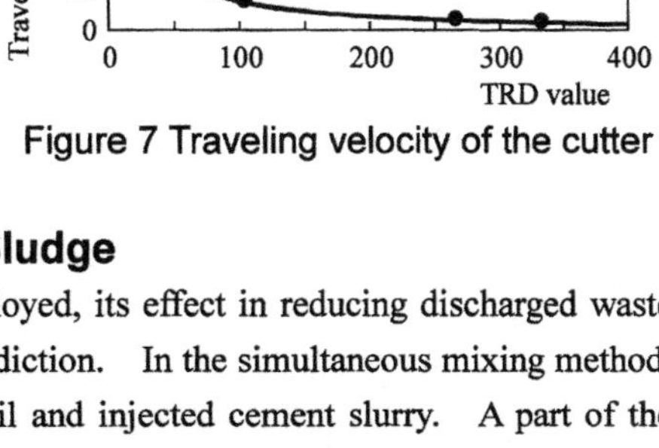

Figure 7 Traveling velocity of the cutter

Evaluating the Discharged Waste Sludge

When the simultaneous mixing method is employed, its effect in reducing discharged waste sludge must be confirmed through a precise prediction. In the simultaneous mixing method, a trench is filled by a mixture of excavated soil and injected cement slurry. A part of the cement slurry will permeate from the trench to the surrounding ground, and conversely, the water in the ground is drained to the trench. This means that the ground water is replaced by the permeating cement slurry. The volume of the permeating slurry depends on the type of soil composing in the surrounding ground. A comparatively larger volume of slurry will permeate into the surrounding ground in a sandy ground and a smaller volume will permeate in a clayey ground. The waste sludge discharged from the trench during construction will be reduced by the volume of permeating slurry, because it diffuses in the surrounding ground.

In the subsequent mixing method, bentonite slurry is filled once after the trench cutting to prevent the collapse of the excavated trench. The fine-grained fraction included in the bentonite slurry permeates into the ground and fills up all the voids in the soil to block any additional permeation of the slurry to the surrounding ground. Since the cement slurry hardly permeates from the trench to the surrounding ground, the volume of waste sludge will not be reduced by the permeation of the slurry.

In principle, the volume of waste sludge is equal to the volume of the cement slurry injected into the trench. For all practical purposes, however, it is smaller than the injected slurry. This is because of its permeation and diffusion in the surrounding ground, as mentioned above. We introduce the rate of replacement κ, defined by the ratio of the volume of ground water replaced by the slurry permeating into the void of the ground. The value of the rate depends on the type of soil, and its popular value for each type of soil is experimentally described in Table 1. In the simultaneous mixing method, a part of the water included in the void of the ground will be replaced by cement slurry with the rate in Table 1 according to the type of soil composing the surrounding ground. In the subsequent mixing method, the cement slurry is not able to permeate into the surrounding ground due to the sealed surface of bentonite slurry. The effect of reducing the waste sludge can be predicted by estimating the volume of permeating slurry with the rate of replacement κ.

Table 1 Rate of replacement depending on the type of soil

Clay	Sandy silt	Silty sand	Sand	Sandy gravel	Gravel
0.0	0.2	0.5	0.8	0.9	1.0

Judging the Applicability of the Simultaneous Mixing Method

The simultaneous mixing method was employed at a construction site after a precise investigation based on the above discussion. The construction site is located at the sandbank of a river in Osaka City where the ground is mainly composed of silt and sand, as shown in Figure 8(b). The construction was a renewal of an average building, and diaphragm walls were built in the ground for the repair and the expansion of the basement of the building, as shown in Figure 8(a). The size of the ground plan for the basement is 75.75 m x 47.31 m. The width and the depth of the diaphragm walls are 550 mm and 21.8 m, respectively. The south and the west sections of the basement were designed to be impermeable walls without core beams, while the north and the east sections were retaining walls with core beams. The rate of the cement slurry injection was designed corresponding to the ground properties and the designated strength of the walls.

Prior evaluation of the post-traveling velocity of the cutter

The applicability of the simultaneous mixing method is discussed by the method mentioned in the previous chapter. Although it will be expressed in the following chapter with the record measured at the construction site, the traveling velocity was estimated to be 51.6 mm/min, which is sufficiently larger than the threshold velocity of 40 mm/min.

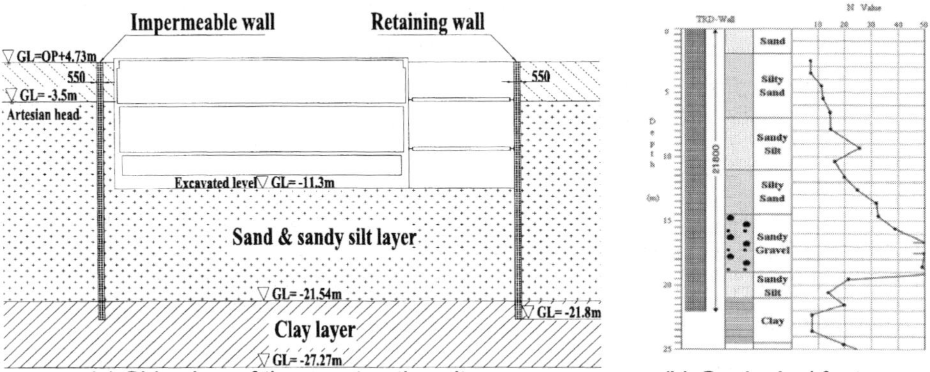

(a) Side view of the construction site (b) Geological features

Figure 8 Construction of the diaphragm walls by the TRD method

Prior evaluation of the volume of waste sludge

The volume of waste sludge was predicted using the geological features in Figure 8(b) and the rate of replacement κ in Table 1. The volume of the waste sludge can be calculated by the injected cement slurry subtracted by the volume of the slurry permeating into the surrounding ground. The volume of the slurry permeating into the surrounding ground is equal to the volume of void in each stratum replaced by the slurry, which can be calculated by multiplying the rate of replacement in Table 1 with the void in each stratum. Table 2 shows the calculation of the volume of the slurry permeation. The void ratio of each stratum is calculated from its bulk density, and the volume of the permeating slurry is calculated by $\boldsymbol{h_r = h\kappa e/(1+e)}$, being converted to the equivalent thickness in stratum, $\boldsymbol{h_r}$. The volume of the waste sludge, calculated through the above-mentioned process, will be expressed in the following chapter with the measurements recorded at the construction site.

Table 2 Volume of the slurry-ground water replacement calculated for each stratum

Stratum	Thickness of each stratum $\boldsymbol{h}$ (m)	N value	Bulk density γ (t/m^3)	Void ratio $\boldsymbol{e}$	Rate of permeation $\boldsymbol{\kappa}$	Replaced volume $\boldsymbol{h_r}$ (m)
Sand	1.5	15	1.65	1.46	0.8	0.71
Silty sand (1)	5	10	1.60	1.67	0.5	1.56
Sandy silt (1)	4	20	1.70	1.29	0.2	0.45
Silty sand (2)	3.5	25	1.75	1.13	0.5	0.93
Sandy gravel	4.5	50	2.00	0.60	0.9	1.52
Sandy silt (2)	2	15	1.65	1.46	0.2	0.24
Clay	1	10	1.60	1.67	0.0	0.00
Total	21.8					5.41

Results of the Actual Construction

Figure 9 shows the prediction and the record for the volume of discharged waste sludge at the south and the west sections of the construction site. The prediction for the case in which the subsequent mixing method is supposed to be employed is also included in the figure. It is clear from the figure that the predicted volume of the discharged waste sludge is almost equal to the measured volume, and that they equal about half the volume of the subsequent mixing method. Figure 10 shows the prediction and the record for the traveling velocity. The construction was carried out at about a 10 - 20% higher velocity than the predicted one. The difference is thought to be induced by the dispersion of the soil properties at the site. These results indicate that the predicted method for the volume of discharged sludge and the traveling velocity can be estimated through a prior examination, and that the simultaneous mixing method has enough potential for reducing the environmental impact that the waste sludge induces in TRD constructions.

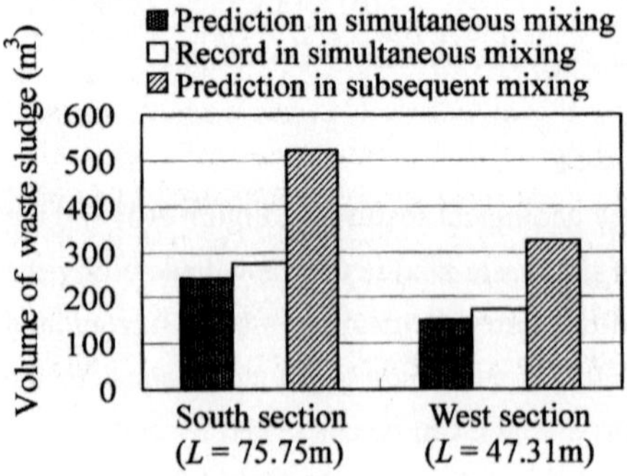

Figure 9 Prediction and record for the volume of waste sludge

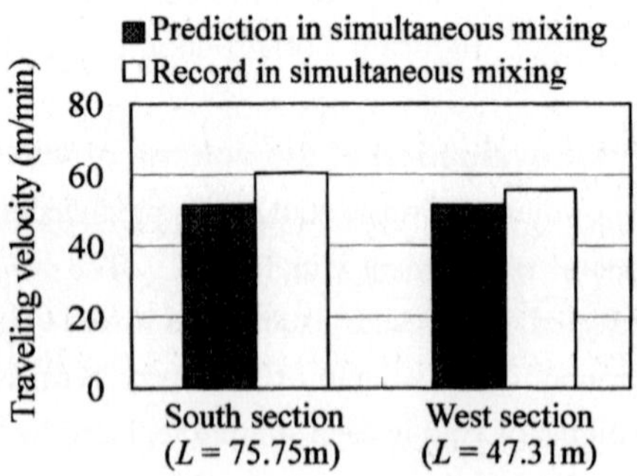

Figure 10 Prediction and record for the traveling speed

Conclusion

A practical method for evaluating the applicability of the simultaneous mixing method was developed in a TRD construction. In the method, the traveling velocity of a cutter post and the volume reduction of waste sludge discharged during construction were predicted in order to judge the applicability and the effect of employing the simultaneous mixing method in place of the subsequent mixing method. They are compared with the measured ones in the actual construction of a building, and the results showed that the predicted method for the volume of the discharged sludge and the traveling velocity can be estimated through a prior examination, and that the simultaneous mixing method has sufficient potential for reducing the environmental impact that the waste sludge induces in TRD constructions.

References

1) Kinoshita and Mizutani: Technical Report for TRD method (in Japanese), TRD-method Association: 2002.

Field and Laboratory Evaluation of Artificial Soil as an Evapotranspirative Barrier

Ramesh Bobba, Nicholson Construction Company, Sparta, NJ, USA
William J. Likos, University of Missouri – Columbia, Columbia, MO, USA
John J. Bowders, University of Missouri – Columbia, Columbia, MO, USA
R. David Hammer, USDA-NRCS, National Soil Survey Center, Lincoln, NE, USA

Abstract
An "artificial soil" created from industrial and municipal by-products has been evaluated as an evapotranspirative (ET) barrier via a full-scale field monitoring study and laboratory testing program. Two full-scale test sections were constructed and instrumented with moisture sensors, temperature sensors, lysimeters, and a surface weather station to quantify the rate and depth of moisture infiltration under natural precipitation events. Laboratory tests were conducted to quantify relevant engineering properties including moisture-density relationships, specific gravity, saturated hydraulic conductivity, and field capacity. The test sections were monitored for one year to observe infiltration behavior over a full range of environmental extremes. Results indicate that the artificial soil acts as an effective cover system by restricting wetting front propagation and causing water to either evaporate, transpirate, or flow laterally.

Introduction
Artificial soils engineered by blending various industrial and municipal by-products (e.g., bio-solids, fly ash, blast furnace slag) have received increasing attention as a sustainable land reclamation and beneficial reuse strategy (e.g., Reynolds et al., 1999; Cox and Whelan, 2000; Chu et al., 2005). An innovative application using an artificial soil as an evapotranspirative (ET) cover system was implemented for full-scale field evaluation in 2002 by a cement-manufacturing plant in Missouri, USA. Municipal yard waste (e.g., grass clippings, wood chips) and bio-solids are blended and placed as an organic-rich layer approximately 1.5-m-thick (5 ft) over a mineral-rich layer approximately 3-m-thick (10 ft) and consisting of blended coal ash and cement kiln dust (CKD). The layers are referred to as an "O" and a "C" horizon, respectively, in accordance with the US Natural Resources Conservation Service standards for describing soil. As illustrated in Figure 1, the O- and C-horizons form a stratified sequence that simulates a natural soil column (i.e., mineral-rich at depth and organics-rich near the surface). This "artificial soil" column is placed over an underlying CKD layer and then seeded with native grass and tree species to begin the biological processes associated with naturally-occurring soils. The system is intended to be an evapotranspirative barrier that limits percolation into the underlying CKD by storing and subsequently removing water under natural evaporation and transpiration processes. An additional role of the vegetation is to stabilize the soil surface both by above-ground biomass and by development of root systems, and to begin a biological cycle that will return organic matter to the system. By July 2005, more than 100,000 metric tonnes of artificial soil were placed and more than 16,000 m^2 (~4 acres) of land have been reclaimed (Table 1).

This paper presents results from a 12-month study that has been conducted by a research team at the University of Missouri Columbia (UMC) to assess the general effectiveness and behavior of artificial soil as an evapotranspirative barrier to fluid flow. Two full-scale artificial soil test sections were constructed and instrumented with moisture sensors, temperature sensors, lysimeters, and a surface weather station (rain gauge, ambient temperature, relative humidity). The sensors were installed at various depths throughout the soil profiles to quantify the rate and depth of moisture infiltration under natural precipitation and evaporation events. The sections were monitored for one year to observe infiltration behavior over a cycle of environmental extremes. A concurrent series of laboratory tests was conducted on artificial soil samples to quantify relevant engineering properties, including compaction behavior (moisture-density relationships), specific gravity, saturated hydraulic conductivity, and field capacity.

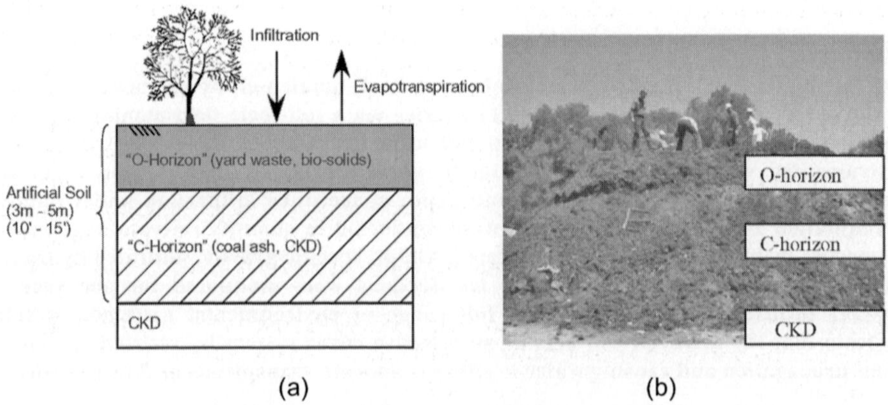

Figure 1. (a) Schematic diagram of artificial soil profile; (b) photo of Test Section 1.

Table 1. Summary of materials placed as artificial soil through July 2005.

Material	Source	Approx. Tonnes Placed
Cement Kiln Dust	Continental Cement Company	35,000
Yard Waste	Cities of Quincy, IL, Hannibal, MO Paris, MO, Springfield, IL	30,000
Bio-solids	Metro Sewer District: Hannibal, MO St. Louis, MO	22,000
Power Plant (Coal) Ash	University of Missouri – Columbia	15,000

Field Monitoring Program

Test Section No. 1 was constructed in late July and instrumented in early August of 2004. As illustrated by the photo in Figure 1b, the test section consisted of about 1.5 to 2.5 m of CKD placed on the surface of a quarry floor and overlain by about 4.5 m of artificial soil (3 m of C-Horizon and 1.5 m of O-Horizon). Dielectric moisture content sensors were placed in boreholes at various depths from the surface (Table 2). Two additional moisture sensors were installed in the uncovered CKD adjacent to the test section to serve as control sensors, i.e., to record infiltration into the bare (uncovered) CKD. Temperature probes were placed at various depths in the test section (Table 2) and at the ground surface. The latter was housed in a radiation shield located adjacent to the test section. A relative humidity (RH) probe was also housed in the radiation shield to record ambient RH. Finally, an electronic tipping-bucket rain gauge was installed adjacent to the test section for quantifying precipitation and for comparison to the moisture penetration recorded by the subsurface sensors. Boreholes containing the subsurface sensors were backfilled with artificial soil and compacted using a tamper to ensure good contact with the buried sensors.

Test Section No. 2 was constructed and simultaneously instrumented on October 5, 2004. Instrumentation was similar to that used in Test Section No. 1 and included twelve moisture sensors and three temperature sensors (Table 2). Unlike Test Section No. 1, however, each sensor was installed as the section was being constructed rather than placed in a post-construction borehole. This method of sensor installation was chosen to minimize the likelihood of creating a preferential flow zone by using a soil probe or auger to create a hole after material emplacement. In addition, four under-drain lysimeters were installed for capturing leachate at the following depths from the test section surface: 46 cm, 122 cm, 213 cm, and 366 cm (18, 48, 84, and 144 in). The lysimeters, which were constructed of a geomembrane, were drained through 4-cm-diameter PVC pipe connected to a center drain hole in the geomembrane (Figure 2). The drain pipes extended laterally through the test section and were individually capped to collect flow over a prescribed amount of time. The volume of water from each lysimeter was recorded during site visits on an approximately monthly basis. The temperature sensors, moisture sensors and rain gauge were connected to an automatic data logger programmed to record readings every 10 to 20 min.

Table 2. Summary of instrumentation locations in Test Sections No. 1 and No. 2; Sensor depths are given in centimeters (and inches) from the surface

Test Section 1		Test Section 2		
Moisture	Temp.	Moisture	Temp.	Lysimeter
23 (9)	56 (22)	15 (6)**	61 (24)	46 (18)
66 (26)	160 (63)	61 (24)**	122 (48)	122 (48)
102 (40)	274 (108)	122 (48)**	305 (120)	213 (84)
163 (64)		244 (96)**		366 (144)
206 (81)		305 (120)**		
274 (108)		376 (148)**		
409 (161)				
38 (15)*				
61 (24)*				

* Sensor placed in CKD adjacent to the test section (depth is noted from surface of bare CKD).
** Two sensors placed at same depth.

Laboratory Testing Program

Samples representative of the three primary materials at the field site (O-horizon, C-horizon, and CKD) were collected and returned to the laboratory for physical testing. Properties determined included compaction behavior (moisture-density relationships), long term flow behavior, saturated hydraulic conductivity, field moisture capacity, specific gravity, and water retention characteristics. The in-situ (placement) density of each layer comprising the test sections (O, C, and CKD) was also determined in the field using a sand cone apparatus. Results are shown in Table 3 along with the relevant test methods used.

(a) (b)

Figure 2. (a) 1m x 1m (3.3 ft x 3.3 ft) geomembrane lysimeter; (b) PVC pipes extending laterally from the lysimeters through Test Section 2 for leachate collection.

Table 3. Properties of artificial soil measured from laboratory investigation.

Property	Units	O-Horizon	C-Horizon	CKD	Test Method
Field Capacity	m^3/m^3	0.15	0.17	ND	column test
Specific Gravity	---	1.71	2.12	2.46	ASTM D854
In-situ moisture*	m^3/m^3	0.21	0.20	0.26	ASTM D4643
In-situ bulk dens	kN/m^3 (pcf)	4.9 (31)	7.1 (45)	7.1 (45)	ASTM D1556
In-situ dry dens	kN/m^3 (pcf)	2.5 (16)	5.0 (32)	4.9 (31)	ASTM D1556
Opt. Proct. dens	kN/m^3 (pcf)	5.3 (34)	10.8 (69)	ND	ASTM D698
Opt. moist.	g/g (%)	124	45	ND	ASTM D698
Sat. hyd. cond.	cm/s	7.0-8.3 X 10^{-3}	8.5-9.3 X 10^{-3}	ND	ASTM D 2434
at density	kN/m^3 (pcf)	2.6 (16.5)	6.1 (39)	ND	NA
Porosity	---	0.85 – 0.90	0.76 – 0.80	ND	NA

* Measured at time of sensor installation.

Monitoring Results and Discussion

Cumulative precipitation (Fig. 3a) recorded by the field rain gauge from August 6, 2004 through June 18, 2005 was 117 cm (46 in), which is typical for northeast Missouri (90-130 cm/yr). The maximum "single" event (Fig. 3b) was 16 cm (6.3 in), which occurred over a period of about 7 hours on August 27, 2004. The majority of single precipitation events were less than 5 cm (2 in). Ambient temperature ranged from a low of about -15°C (5°F) to a high of about 35°C (95 °F). RH ranged from about 20% to 100% over the measurement cycle.

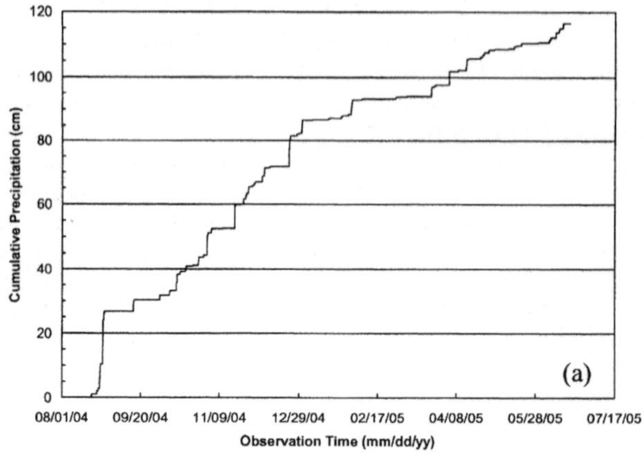

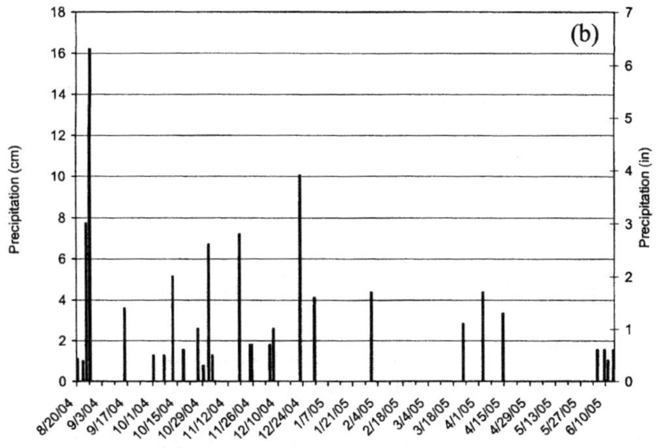

Figure 3. Cumulative (a) and incremental (b) precipitation records.

Subsurface temperature profiles for Test Sections No. 1 and 2 were similar (Figure 4). The maximum temperature in the O – horizon was 71°C (160°F) and in the C – horizon was 64 °C (148°F). Soil temperatures both on the surface and at depth greatly exceed values for natural soils in this part of Missouri. The depth of constant annual soil temperature in soils in the mesic soil temperature regime (Soil Survey Staff, 1975) is from approximately one to 1.3 meters depending upon vegetative cover and slope attributes (Hammer, 1986). Elevated temperatures are attributed to biological reactions (organic degradation) in the O-horizon and chemical reactions (hydration) in the C-horizon.

Similar subsurface moisture profiles were observed for both test sections. Figures 5a and 5b show the range of moisture contents observed over the complete monitoring cycle. Maximum volumetric moisture content observed in the O-horizon was about 40% and 85% in Test Sections 1 and 2, respectively (equal to gravimetric moisture content of 104% and 225%). Maximum volumetric moisture content in the C-horizon was about 60% and 40% (72% and 125% gravimetric) in Test Sections 1 and 2, respectively.

The largest range of moisture variation occurs in both test sections at or above the interface of the O- and C-Horizons. An upper limit corresponding to full saturation is reached in the O-Horizon (Test Section No. 2), but not in the C-horizon. Moisture variations near the bottom of the sections are relatively small and generally attenuate with increasing depth, indicating that the wetting front propagation to the bottom of the sections is limited. Sensors installed in the C-horizon show far smaller moisture content values (and corresponding degrees of saturation) than those in the O-horizon and at the O-C interface, indicating that water flow is restricted and is likely temporarily ponding at the interface. No water was observed in Test Section No. 2 in any of the lysimeters except during a period when water was artificially pumped onto the test section surface (1.7 Ml (450,000 gallons) over a period of 30 hours). These observations suggest that the artificial soil is acting as an effective hydraulic barrier by restricting wetting front propagation beyond the O-horizon. The water ponding at the O-C interface is likely evaporating under high ambient temperatures or flowing laterally, which was supported by visual observations made at the site.

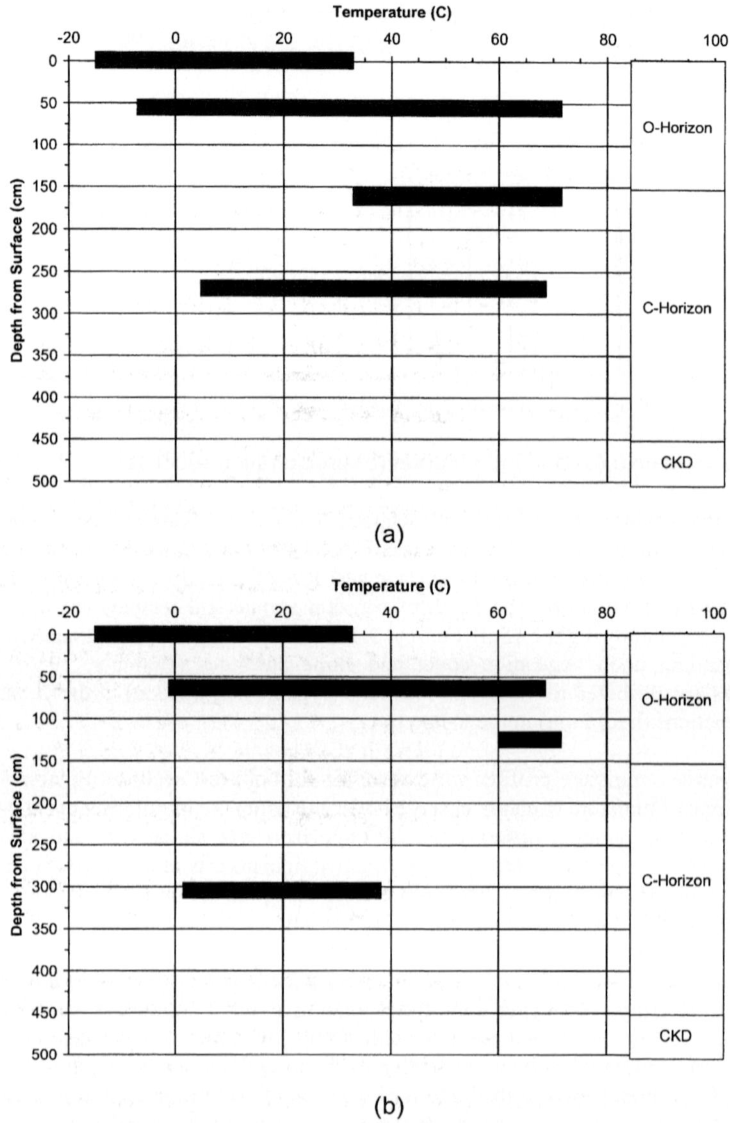

Figure 4. Subsurface temperature ranges: (a) Test Section 1 and (b) Test Section 2

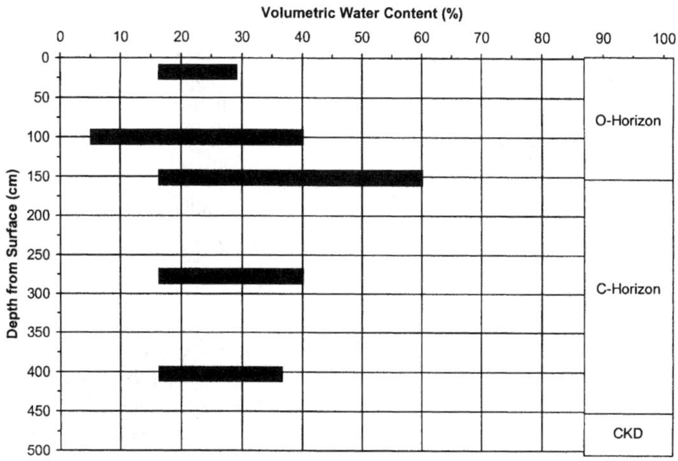

(a)

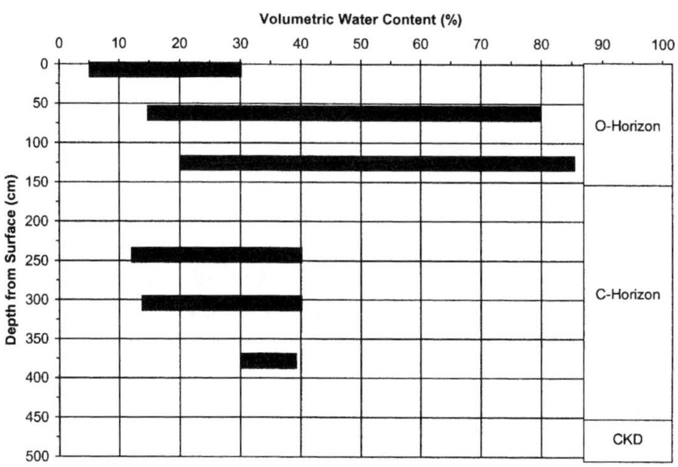

(b)

Figure 5. Subsurface moisture ranges: (a) Test Section 1 and (b) Test Section 2.

Summary and Conclusions

The "artificial soil" concept for creating a sustainable evapotranspirative cover system is an innovative technology aimed at controlling infiltration into underlying materials while establishing a permanent, natural landscape. Municipal yard waste and bio-solids are placed as an organic-rich "O-horizon" overlying a mineral-rich "C-horizon" consisting of power plant (coal) ash and cement kiln dust. The surface of "O-horizon" is seeded with native grasses and trees and the resulting system provides an infiltrative barrier, limiting vertical migration of water into the underlying materials. A 12-month field performance monitoring program was conducted and complemented with laboratory physical testing. The findings provide data for assessing the general effectiveness of the artificial soil barrier for capping CKD deposits and reclaiming surface mined areas.

Field monitoring results suggest that the artificial soil is acting as an effective cover system by restricting wetting front propagation beyond the interface of the O- and C-horizons, causing water to either evaporate, transpirate, or flow laterally along that interface. Lysimeter data collected over a 12-month period indicate that no water infiltrated completely through the barrier. The cover has been effective at limiting vertical infiltration even in Missouri's humid environment. Elevated temperatures measured in the system (as high as 71°C (160°F)) are attributed to biological reactions (organic degradation) in the O-horizon and chemical reactions (hydration) in the C-horizon. Continued monitoring will be necessary to determine the fates of materials weathering in and leaching from the organic materials into the C materials.

Acknowledgements

This program has been supported through a grant by Continental Cement Company, Hannibal Missouri. Their support is gratefully acknowledged. The authors are deeply indebted to Mr. Greg Haug, president of Resource Institute, Jefferson City Missouri, for his inception of the artificial soil program and for his numerous contributions to the authors' study. The efforts of Drexel Rothweiler, Mark Twain Solid Waste District, and Lennie Rosenkrans, Continental Cement Company, are gratefully acknowledged. Finally, we greatly appreciate the assistance of the students who helped with the project, including: Alexandra Wayllace, Benjamin Niesen, Daniel Huaco, Deepak Manjunath, Deepak Neupane, Devi Tulasi, Elizabeth Freeman, Jeff Zimmershied, Rafael Baltodano, and Ryan Mueller.

References

Chu, J., Goi, M.H., and Lim, T.T. 2005. Consolidation of cement-treated sewage sludge using vertical drains. *Canadian Geotechnical Journal*, 42, 528-540.

Cox, J.A., and Whelan, R.J. 2000. Soil development of an artificial soil mix: nutrient dynamics, plant growth, and initial physical changes. *Australian Journal of Soil Research*, 38, 465-477.

Hammer, R.D. 1986. Soil morphology, soil water and forest tree growth on three Mid-Cumberland Plateau landtypes Ph.D. Diss., University of Tennessee, Knoxville.

Soil Survey Staff. 1975. Soil Taxonomy: A basic system of soil classification for making and interpreting soil surveys. Agriculture Handbook 436. Soil Conservation Service, U.S. Department of Agriculture.

Reynolds, K., Kruger, R., and Rethman, N. 1999. The manufacture and evaluation of an artificial soil (SLASH) prepared from fly ash and sewage sludge. *1999 International Fly Ash Utilization Symposium*, Center for Applied Energy Research, University of Kentucky, Paper No. 1.

Using Disused Rails for Slope Stabilisation

Daniel. Castro, Universidad de Cantabria, Santander, Spain
Luis López, Universidad Católica de Valparaíso, Valparaíso, Chile.

Abstract

The use of recycled materials for slope stabilization is increasing. This is because such materials have advantages of speed, economy, simplicity and behaviour. The main applications to date have been in temporary works.

In this paper a methodology is described for the use of disused rails in slope stabilization. These rails take the form of metallic piles, such as are used in "soil nailing" techniques, the principle of which is that the piles intercept the slip surface of the slope, thus increasing the global resistance the failure plane.

Practical experiences and theoretical approaches for laterally loaded piles have been considered and an assessment has been made of the suitability of existing design methodologies. A new design procedure and appropriate charts have been developed for the use of rails in these circumstances. These are required because of the lack of bending resistance at the joints of these rail systems.

Introduction.

This paper is concerned with increasing global soil resistance using inserted structures. Slope stability is a recurrent problem in civil engineering. To solve the problem, a global analysis is required that enables clear identification of the source of the problem, the consideration of appropriate alternatives including construction methods, and the development of maintenance systems that will prevent future instability.

External inclusion methods have been employed to increase slope resistance for some time. These so called "soil nailing" methods employ continuous structures to resist shear and bending strengths. Railroad rails can work well in resisting shear and bending strengths when a single rail is driven, but are limited to supporting shear resistance only when there is more than one driven longitudinally. This is because the joints between the rails do not create a continuous structure are unable to transmit bending resistance from one rail to another.

This leads to new hypotheses which are not usually used in this inclusion design, where the earth pressure resistance diagrams decrease in the rail failure zones.

Figure 1 shows the earth pressure diagrams over a stabilisation pile according to NAVFAC (NAVFAC, 1986). The left hand side image represents the earth pressure diagram appropriate for rails, taken as a single continuous pile. The right hand image represents the theoretical situation for the several longitudinal rails. It can be seen that the continuous pile can retain more earth pressure than the separate rails. Because of this, one pile cannot retain a large soil pressure and thus limits the use of the system.

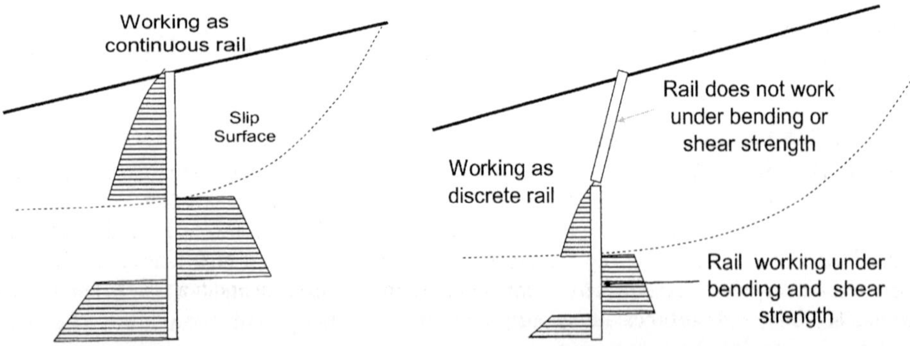

Figure 1. Earth pressure resistance diagrams for rails working continuously and discretely.

Thus soil nailing with rails involves driving the rails into the unstable slope to intercept the slip surface of the soil, thus increasing its shear resistance in order to stop or decrease the slope movement. Once installed, the system will not work until the soil start to move. The nail is then activated and the interaction of the elements commences (passive system).

The system has been used successfully in the North of Spain for stabilising areas with relatively small movements. Since the rails are recycled and are driven deep underground and therefore cannot be seen, it has low environmental impact. One potential problem is that the rails can corrode quickly, therefore it is important not to locate them close to groundwater. Generally they have been used adjacent to highways where the soil is shallow and there are no significant water runoffs. Where the technique has been applied in most cases the slip surface has already appeared. Analysis has then been necessary to determine the magnitude of movement and the type of slip surface and the soil characteristics, so that the initial hypothesis can be verified and the behaviour of the slope assessed.

The paper reports on investigations that have been carried out to determine how the rails work in practice and how possible problems can be resolved. The outcome of this work has been the development of appropriate design charts, as displayed below. When developing and applying these charts it is important to understand all the aspects involved in creating an effective design, which are the usual aspects of slope stability analysis. We need to select an analysis method and know the factor of safety within which the slope works. This will indicate how unstable the slope is and allow us determine the necessary force to increase its shear resistance. These aspects, together with others, such as geometrical design (position, length and separation), earth pressure, stabilization force and arch effects are discussed in the following paragraphs.

Models of Global Stability

There are various models that may be applied to analyse global stability. The Limit Equilibrium method is preferable, since it allows simple varying its formulation and its limitations are known.

Two calculation methods were used to develop the design charts: planar failure, when the recurrent variables are known; and the Modified Bishop Model, for detailed analysis.

Factor of Safety and Slip Surface

The failure surface can be obtained by the Bishop Method. The method generally starts by generating low values of factors of safety, given the initial soil conditions where the failure has occurred. After the rails have been driven, the values are increased to obtain a stable slope.

Hada et al., (1988), Sommer (1977) and the 'Guía de Cimentaciones en Carreteras', by the Ministry of Public Works of Spain (Ministerio de Fomento, 2002), provide safety factors between 1.05 and 2. For the cases studied, factors of safety were used for planar failure and finite slope values between 1.5 and 1.7, being considered appropriate for this kind of stabilisation, where the lack of precise soil data requires an increase in the factor of safety. In addition, stabilisation work with rails often needs to be executed quickly.

Locations of the rails

Using the method of slices the driven rail positions are given by the slope analysis. Each case will have a different position depending on its minimum failure circle. In the case of planar failure, the calculation is made by unit area; a rail size is calculated that will develop its shear strength toward the soil over a certain area. This leads to the installation of a certain number of rails per slope square metre, thus distributing the resistant strength homogeneously. In practice, however, this distribution may not always be achieved. For example, operational problems with machinery may prevent the ascension of the whole slope.

For small or localised slips all the inclusion resistance is considered to be mobilised in small areas that cut the slip surface. Recommendations are available from experience for the driven position in the planar failure case. For example, the slope may be divided longitudinally into three equal width tracks, and the rails are driven in the second track, beginning at the foot. The more unstable areas should obviously be intercepted. With reference to mass distribution, the lower slope areas have a larger weight, so here the stabilisation has to be decisive. The upper areas have smaller load, but their stabilisation does not affect the lower zones, therefore, slips could appear below them. In the *Manual para el Proyecto y Ejecución de Suelo Reforzado* (Ministerio de Fomento, 1998), recommendations are made that include nailing in the lower part of the central third.

Design

Obtaining the Stabilizing Force

Once the slip surface has been determined the required force to stabilise it must be calculated. This force is provided by the inclusions, in this case by the rails.

In the Bishop Method case, the general equation of the factor of safety with which the two analyses are carried out must be modified: the first analysis must be made without external forces, where an initial shear force is obtained; the second, entering the factor of safety required for the same conditions, when a final stabilising force is obtained. By subtracting one from the other the value of the force that should be mobilised by the rails to the soil is found. With this analysis, the stabilising force magnitude will be obtained, as well as the rails position on the slope.
The calculations begin with one rail for one metre of width and one line. If more than one line is needed, the necessary force should be divided by the number of rails per metre, thus obtaining the spacing implicitly. Strictly, the separation should generate a difference in the pressure that acts on

the rail; however, the models give limits of maximum and minimal spacing from 2 to 6 times the diameter of the pile. This leads to varying the spacing for values from 30 to 90 cm for a section of the rail. In Northern Spain (Cantabria), the normal spacing is between 0.5 to 1 metre over clays with a density of 1.9 t/m^3 and cohesion of 10 kPa, which seems to work well in the long and short term. Having this in mind, it can be stated that after the necessary stabilising force is obtained, that value should be divided by ultimate shear resistance of the rail and the result will be the number of rails per metre required.

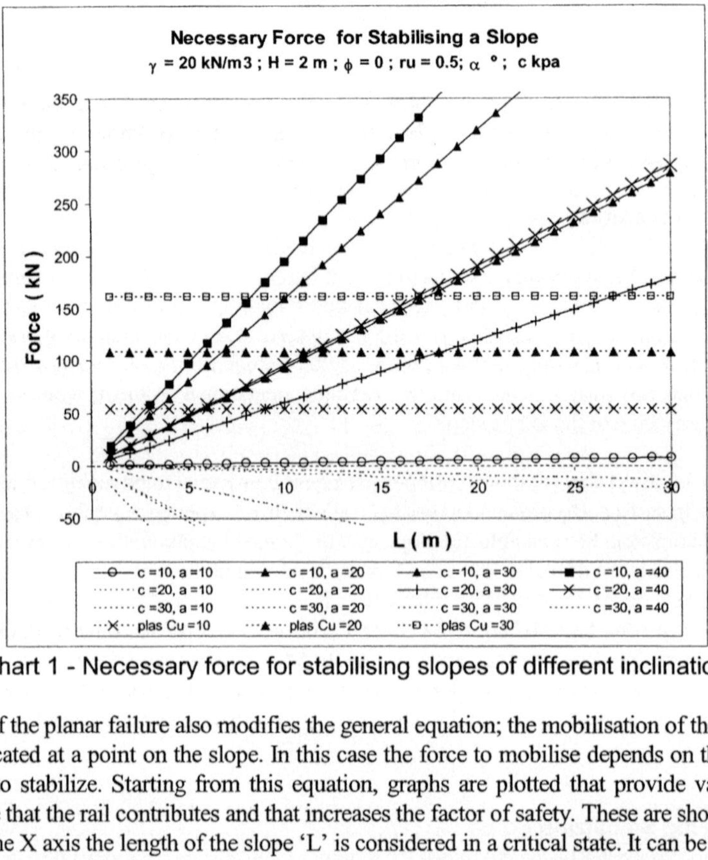

Chart 1 - Necessary force for stabilising slopes of different inclination.

The case of the planar failure also modifies the general equation; the mobilisation of the stabilising force is located at a point on the slope. In this case the force to mobilise depends on the length of the slope to stabilize. Starting from this equation, graphs are plotted that provide values of the shear force that the rail contributes and that increases the factor of safety. These are shown in Chart 1. Along the X axis the length of the slope 'L' is considered in a critical state. It can be appreciated that when extending the length, the force required to contain the slope will increase. Along the ordinates axis is the necessary force to contain the slope of assumed length 'L'. Each line on the chart is a function of cohesion 'c' and slope inclination angle 'α' (which appears as 'a' on the chart). Both vary, generating a series of lines. There are two types of lines, those that take negative and those that take positive values. Those of negative values have not been represented completely. The negative values indicate that for those slope conditions, an external force is not required to obtain a factor of safety 1.5, because the slope is stable. Contrarily, the positive values indicate that a force is required with the ordinate magnitudes shown (for that reason all the negative lines have been drawn with the same segmented line).

For example, a soil with cohesion of 10 kPa and a slope inclination angle of 40º corresponds to the "black square" series in the graph (the upper series). If we have a slope of 5 m. length, we need to retain 98 kN in shear. This value is relatively small for the rail, as it can resist 400 kN of admissible shear and ultimate value approximately of 600 kN, but has the soil enough resistance for retain this force? Soil can plastify before all resistance is mobilized from the rail to the soil. The lines which show the maximum resistances of the soil are horizontal, and they depend on cohesion. They are derived from Broms theory (Broms, 1964). For the example, a maximum of 52 kN per metre is resisted by a soil of c=10 kPa (lower horizontal line series). We then use two rails for metre, as each one can mobilize 98 kN/2 = 49 kN and it value is under the plastification line of c = 10 kPa.

This explains in simple terms how to calculate a structure that works like a pin over the slope, nailing the soil in the slip surface. However, is important to know where the various simplifications are made because many parameters are assumed. The geometrical configuration of the problem makes it impossible to apply any other system and a discrete system like this is easy to apply when pin joints have been used.

Earth Pressure Coefficients

The earth pressure coefficient and the separation of the rails are two of the most questioned parameters in rails systems design because the analysis is based on different types of data and different authors have produced different results, the most common being test on walls, piles and slopes, in plastic and cohesionless soils. A conclusion from all these studies, however, is that the earth pressure coefficient is always larger than the active coefficient. The maximum values are similar to a passive magnitude and, in some cases, are higher. In the NAVFAC Method (NAVFAC 1986), the values of the coefficients for lateral pressure are also larger than those given by Terzaghi. In cases of passive pressure, this method is based on the Brinch Hansen theory (Brinch Hansen 1961).

Arch Effects

The concept of the "arch effect" refers to the reaction of soil enclosed between two closely aligned rails. This produces a clear increase in soil resistance. There are analyses available for the increase and decrease of parameters such as angle of friction, cohesion, moisture and density of the soil, which show the variation of the soil state between the rails, in comparison to that in nearby free zones.

There are also analytical and empirical studies that determine the influence area of the rails. For elastic soils, studies generally use Rankine´s passive wedges which act between the rails and the soil. All the studies have determined a distance of between 1 to 7 diameters for the influence zone, with 12 diameters as a maximum. Thus, in the case of the rails which have a width of 15 cm, this means a separation of between 15 to 105 cm.

Earth Pressure Magnitude.

Earth pressure magnitude is used to analyse the equilibrium forces over the rails. This analysis takes into account the embedded forces, which are the passive zone forces that stabilize the rails, by reacting to an external force.

Where there is a single alignment of pile lines, it is acceptable to distribute the total earth pressure, calculated in equal parts, among the piles. Where there are several alignments or inclusion lines, the situation is more complex. If the load distribution problem becomes serious, this may require a specific analysis of interaction, using numerical models. The piles most exposed to the horizontal forces are those next to the loads and, as a general rule, it may be assumed that the total loads previously calculated are distributed among the successive alignments of rail screens with values equivalent to half of the previous ones (Guía de Cimentaciones en Carreteras', by the Ministry of Public Works of Spain (Ministerio de Fomento, 2002).

To determine the earth pressure magnitude, there are theories that quantify the rails pressures in relation to spacing. For example, Broms arrived at an equation for cohesion soils with $\phi = 0$ (Broms, 1964). He derives his equation assuming the pressure on these elements is as occurs in a foundation; that is through Terzaghi`s general formula. Analysing his research, it can be concluded that in varying the spacing from 0 to 1 m, the magnitude does not increase considerably, unless cohesion varies. It is concluded, therefore, that the biggest load variation on the rail is defined more by cohesion than by spacing. In Ito and Matsui's theory (Ito and Matsui, 1975), cohesion cannot vary the force magnitude in significant form in the analysed ranks, the pressure values result being similar. If the earth pressures caused by the different theories commonly employed are drawn in the same place, it can be verified that Broms` theory encircles them all, and that the pressure on the rail reaches larger values than Ito`s & Matsui`s theory and Rankine`s active and passive. Only the NAVFAC Method gives larger values than Broms, and therefore this was used to calculate the earth pressure and develop the design chart for embedded length.

Embedded length

There are various ways of approaching the calculation of embedded length in an element subject to loads for unstable mass methods. In this case the NAVFAC Method (NAVFAC, 1986) is chosen. The earth pressures assumed by this method are fairly conservative, which are supplemented by the factor of safety used (fos = 1.5). The embeddings values are between 1.4 to 1.6 times the depth of the unstable stratum for homogeneous soils; this provides reliable results for the case studied. Another advantage that the method presents is that the soil characteristics may be modified more easily, enabling them to vary from homogenous to heterogeneous soil. This is important when establishing embeddings in soils of variable quality, the better the soil quality, the smaller the necessary embedding.

In order to calculate the depth of embedding we need the characteristics of the soils and the rails. With these necessary parameters we can calculate the earth pressure and develop a force equilibrium analysis. The outcome of this analysis is the depth required to stabilize the rail. An example of this is shown in Chart 2 below. This was developed for a soil which has a variable friction angle between 0º to 10º a density of 2 t/m^3. These parameters are constants for the graph and represent a normal poor soil founded in Northern Spain (Cantabria). The "X" axis represents cohesion of the soil. On the "Y" axis we obtain the embedded length required. The lines represented are a function of the height of the rail and the angle of friction of the soil.

For example, if we have slip surface at 4 m. of depth, a soil with density of 20 kN/m^3 and an angle of friction between 0 to 10º, we can use the Chart 2. And if we have a clay on short term analysis ($\phi = 0$ obtained), and a cohesion of 30 kPa we use the "black square" series on the graph (H=4m; f=0) and against 30 kPa on the X axes we read an embedded length of 6,4 ms on the "Y" axis. To support the acting movement totally we add 4,0 m plus 6,4 m to obtain 10,4 m in order to stabilise the slope. In a normal work we should install a 10.4 m of piles but in our case the rails have a maximum length of 6 m. Then we need to install one stable rail plus a length to drive the pile, we can take a pile of 6 m length and put it in the failure zone 3 m under the slip surface and 3 m over the slip surface and we need 1 m of rail to drive to complete the depth of stabilization only with 7 m. If we follow a normal design, only driving one rail, then it will fail, because the embedded is poor or the last rail is not stable. Similar graphs need to be made for other parameter values for the problem, using the methodology described above, taking NAVFAC Manual (NAVFAC, 1986). If the soil parameters are improved, the required depth of embedding will be reduced.

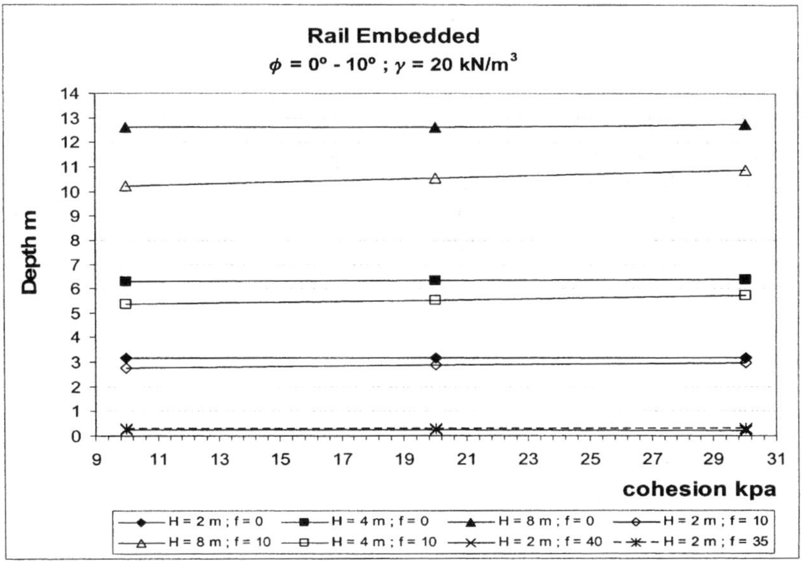

Chart 2. Embedded Curves for Different Kinds of Soils and Depth of Driven Rails.

Structural rail behaviour

Two situations should be verified for the earth pressure calculations to be effective. We need to know if the rail to be driven will act as a rigid or flexible structure.

If the rail acts flexibly (that is to say, it admits deformation), then the potential earth pressures acting must be studied to see if they will plasticize the rail, making it fail structurally as by the Broms method. If the rail acts like a rigid or short pile (that is to say, it does not admit deformation), it should be determined whether or not it can plasticize the soil around it. This means that the ultimate load capacity of the soil may be exceeded, resulting in big horizontal deformations.

The relative rigidity between the pile and soil surrounding it allows the examination of the failure devices described initially exposed in simple form in Chart 1 where its examples give the horizontal lines for the maximum limits of soil resistance. These lines were made according the US Federal Highway Administration, Department of Transportation (FHWA HI, 1998). Considering their shape and geometrical condition, the rails will act like short piles. There is a fundamental difference between rails and other methods of stabilisation using passive piles in that the kneecap formed in the joint between two rails breaks the system apart. This makes the piles that cut the slip surface behave indirectly like short piles in their characteristics of slenderness and module of elasticity. Hence the calculation is relatively easy and it may be possible to simplify the graphs still further. Generally, once it is known whether a pile works as a rigid or a flexible pile, the earth pressure and moments created on it can be calculated, in order to verify possible soil or pile failures. In the case of rails, both problems can be resolved by reference Design Charts 1 and 2.

Summary and Conclusions

Unlike other stabilisation systems with insertions, using driven rails means that there are joints that do not transmit bending strength. This makes traditional calculation methods ineffective because, on the one hand, the system fractures and the rails generate less resistance and, on the other hand, embedded lengths are obtained that do not apply because some rails do not receive bending stress as they are above or below the slip plane. In these cases only the rails positioned in the failure surface are significant.

When analysing plastification, it can be concluded that the rails will always will act as 'short piles' for the dimensions and characteristics presented. This will always cause plastification in the soil before rail plastification, making the design of hypotheses easier.

Two design charts design are easily obtained for constants parameters. If the user wishes to change this initial condition, the methodology should be followed for adjusting the graphs to the specific situations encountered.

Finally, the design sequence can be summarised as follows: 1) obtaining the failure surface, 2) calculation of embedded length, inclusions spacing and plastification resistance analysis, 3) revision of piles joint positions, and 4) making the design work in practice.

Acknowledgements

We gratefully acknowledge the cooperation of SIEC S.A. of Cantabria for providing case information and valuable experience in the use of rails for soil stabilization. Also we acknowledge the assistance of SODERCAN in funding this project and of KPC Consultancy Ltd for providing additional services.

References

Broms, Bengt. 1964. Lateral resistance of piles in cohesive soils. Journal Of The Soil Mechanics And Foundation Division. ASCE. USA. 90, 27-63

Brinch Hansen, J. 1961. The ultimate resistance of rigid piles against transversal forces. Geoteknisk Institut. Bull. Copenhagen. Nº 12.

Estaire, José. 2001. Dimensionamiento de pasadores estructurales para la estabilización de deslizamientos. V Simposio Nacional Sobre Taludes y Laderas Inestables. Spain. 1071-1103.

Ito, T and Matsui, T. 1975. Methods to estimate lateral force acting on stabilizing piles. Soil and Foundations, 15, no.4, pp. 43-49.

Ministerio de Fomento. 2002. Guía de Cimentaciones en Carreteras. Madrid, Spain.

Ministerio de Fomento.1998. Manual para el proyecto y ejecución de suelo reforzado. Madrid, Spain.

Naval Facilities Engineering Command (Navfac). 1986. Soil mechanics, design manual 7.01. USA.

Rodríguez Ortiz, José Maria. 1997. La utilización de pilotes y pasadores en la corrección de deslizamientos. Teoría y aplicaciones. IV Simposio Nacional Sobre Taludes y Laderas Inestables. Spain. pp. 1013-1049.

US Federal Highway Administration, Department Of Transportation. 1998. FHWA HI 97-013, Design and construction of driven pile foundations, Workshop Manual. USA. Volume I.

Geoenvironmental and mechanical properties of a fly ash

J.-C. Chai, Department of Civil Engineering, Saga University, Saga, Japan
S. Hayashi, Institute of Lowland Technology, Saga University, Saga, Japan
K. Onitsuka, Department of Civil Engineering, Saga University, Saga, Japan

Abstract
The geoenvironmental and mechanical properties of a fly ash (coal ash) in Japan were investigated experimentally. Geoenvironmental consideration is focused on the concentration of C_r^{6+} in solution/leakage. Batch contact test and column percolation test results indicate that mixing the fly ash with Ariake clay with a clay/fly ash ratio (C/F ratio) of larger than 20:80 by dry weight, or mixing the fly ash with 5% cement and cured for 1 week, the concentration of C_r^{6+} in the solution/leakage was less than 0.05 mg/*l* (Japanese environmental standard). The strength and modulus of compacted fly ash and the fly ash-cement-Ariake clay mixture were investigated by unconfined compression tests. For the fly ash alone or adding 5% cement to the fly ash and mixed with Ariake clay with a C/F ratio of 20:80, their unconfined compressive strength (q_u) was more than 100 kPa and the secant modulus corresponding to a tress level of 50% of the strength (E_{50}) was more than 10.0 MPa. For the samples of the fly ash plus 5% cement, q_u and E_{50} values were more than 10 times of that the fly ash alone.

Introduction
In Japan, the amount of fly ashes generated from thermal power stations was about 54 million Ton/year (RCUWGCM, 1999). How to treat and/or effectively use the fly ashes is an important environmental problem. One of the ways is to use them as construction material, such as for embankment construction. However, fly ashes normally contain ions of heavy metals such as Cr^{6+}, Cd^{2+}, Pb^{2+}, etc, which may have a negative effect/impact on environment (Kamon and Katsumi, 2001). It has to be clarified that under what kind of conditions the use of a fly ash as construction material will not cause environmental problem. As a construction material, the requirements on its mechanical properties, such as strength and modulus, have to be satisfied also. Several researchers (e.g. Yudhbir and Honjo, 1991; Shibuya et al., 2004) had investigated the strength of some fly ashes and fly ashes mixed with cement/lime. However, there is very few test results reported on fly ashes mixed with cement and clayey soil.

In this study, the environmental impact of a fly ash has been investigated by batch contact tests and column percolation tests focusing on Cr^{6+} concentration in the solutions/leakages. The effect of adding cement and Ariake clay (a soft clay deposit in Japan) into the fly ash on Cr^{6+} concentration was investigated experimentally. The conditions under which the Cr^{6+} concentration can satisfy the Japanese environmental standard have been specified. The strength and modulus of compacted samples of the fly ash, the fly ash plus 2 and 5% of the

5th ICEG Environmental Geotechnics, Thomas Telford, London 2006.

cement, the fly ash plus 2 and 5% of the cement and mixed with the Ariake clay with a clay/fly ash ratio (C/F ratio) of 20:80 by dry weight had been studied by unconfined compression tests. This paper reports the test conditions and the test results. Based on the test results, the effective ways of using the fly ash are discussed.

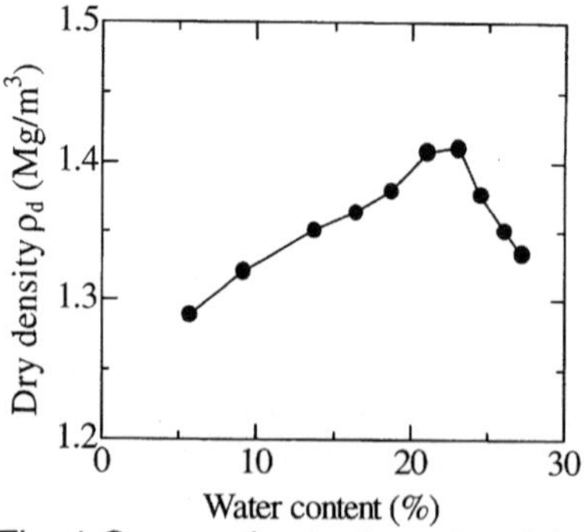

Fig. 1 Compaction test results of the fly ash

Materials used in this study

The fly ash tested has a brand name of Blair Athol (from Blair Athol coal). Some indexes of the fly ash and the corresponding standards are listed in Table 1. The chemical components of the fly ash are not tested. Table 2 lists the reference values of the same brand (JFAA, 2000). Compaction test (JIS A 1210 (JSA, 1999a)) resulted in a maximum dry density of 1.41 Mg/m^3 and optimum water content of about 23% (Fig. 1). The clayey soil adopted in this study was Ariake clay with a liquid limit (w_l) of 116.6% and plastic limit (w_p) of 57.5% and clay content of 31.0%. The dominant clay mineral in Ariake clay is smectic one (Ohtsubo et al., 1995). The pH value of the solution from the clay/water mixture with a ratio of 1:10 was 7.64. The cement used is a kind of Portland cement, which is a typical one for cement deep mixing ground improvement constructions in Japan.

Table 1 Index properties of the fly ash and corresponding standards

Item	Standards (JIS A 6201 (JSA, 1999b))	Test value
SiO_2 (%)	>45.0	59.9
Moisture (%)	<1.0	0.05
Ignition loss (%)	<5.0	0.9
Density (Mg/m^3)	>1.95	2.30
Remain on 45 μm sieve (%)	<40	14.0
Specific surface (m^2/g)	>0.25	0.346

Table 2 Reference chemical components (after JFAA, 2000)

Chemical component	SiO_2	Al_2O_3	Fe_2O_3	TiO_2	CaO	MgO	Na_2O	K_2O	P_2O_5	SO_3
Content (%)	62.6	26.6	5.8	1.4	0.7	0.6	0.5	0.5	0.4	0.3

Batch contact tests and column percolation tests

Batch contact test and column percolation test were employed to investigate the environmental impact of using the fly ash as a construction material focused on the concentration of C_r^{6+} in the solution/leakage.

Test methods and conditions

Batch contact test

The tests were conducted according to the regulation of the Japanese Environment Agency (1991). The solid/liquid ratio adopted was 1:10. The mixtures were put into 1*l* glass bottles and mixed for 24 hours by rotating the bottles with a speed of 30 rpm. Then the solid was separated from the liquid by a centrifuge machine. The concentration of Cr^{6+} and pH value of the solution were measured. For some tests, the electric conductivity of the solution was measured following the procedure of JGS 0212 (JGS, 2000).

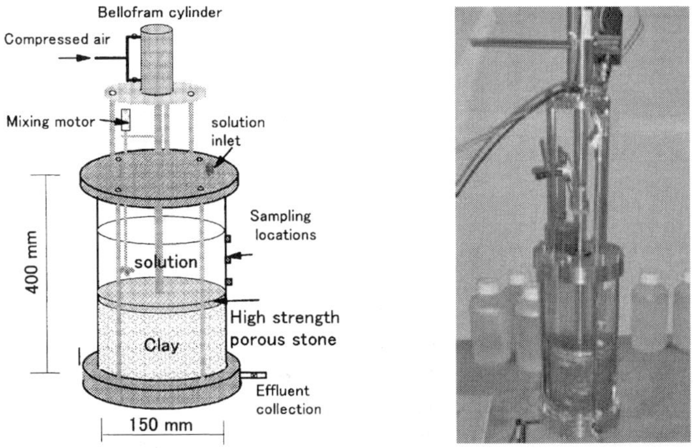

(a) Sketch of the device (b) Actual device

Fig. 2 Column percolation test device

Column percolation test

A column used was 150 mm in diameter and 300 mm in height. The samples were typically 20 mm thick. The samples were formed by compacting the materials inside the column. For the fly ash and the fly ash adding 2 and 5% the cement cases, the samples were compacted at a water content of about 23% and the dry density of the samples was about 1.34 Mg/m^3. In cases of C/F ratio of 20:80, the fly ash and the clay mixture was directly compacted without adding additional water. The resulting water content was about 26% and dry density was about 1.16 Mg/m^3. In case of C/F ratio of more than 20:80, the initial water content was too high for an effective compaction, and the samples were made by consolidating the material under a pressure of 50 kPa. The device used is shown in Fig. 2. During the tests, the depth of distilled water on the top of the samples was kept as 150 mm and the leakage was collected and the concentration of Cr^{6+} was measured periodically.

For both the batch contact tests and the column percolation tests, the conditions tested are listed in Table 3.

Test results

Batch contact test results

Cr^{6+} concentration in the solution is summarized in Fig. 3. Although there are scatters on the test results, generally the following two (2) points can be made.

(1) Cr^{6+} concentration reduced with the increase of the Ariake clay content. It shows that if the Ariake clay content was more than 20%, the Cr^{6+} concentration in the solution was less than the Japanese environmental standard (0.05 mg/*l*). As shown in Fig. 4, with the increase of the Ariake clay content in the fly ash/Ariake clay mixture, the pH value of the solution was reduced. Figure 5 shows the electric conductivities (*Ec*) of the solutions from some of the tests. As a general tendency, *Ec* increased with the increase of the clay content in the mixture. Reducing pH value means the increase of H^+ concentration, and increasing

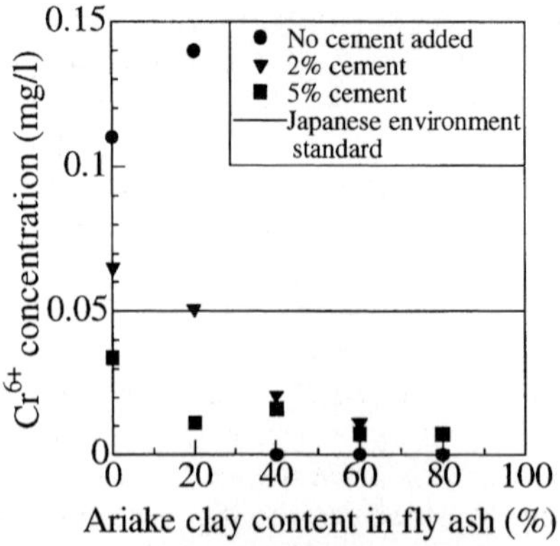

Fig. 3 Cr^{6+} concentration in solutions (batch tests)

Table 3 List of all tests conducted

Clay/fly ash ratio (C/F)	Cement mixed (%)	
	Column test	Batch contact test
0:100	0, 2, 5	0, 2, 5
20:80	2, 5	0, 2, 5
40:60	2, 5	0, 2, 5
60:40	-	0, 2, 5
80:20	-	0, 2, 5

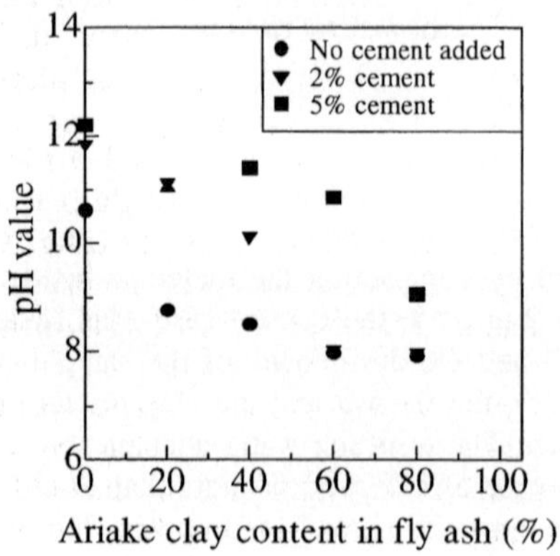

Fig. 4 pH value of solution (batch tests)

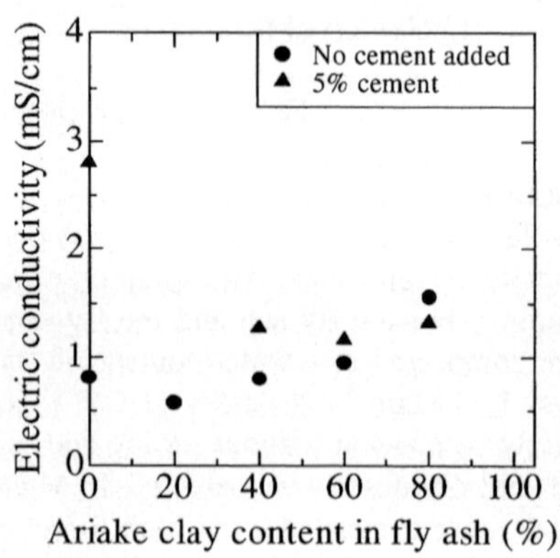

Fig. 5 Electric conductivity of the solutions (batch tests)

E_c value indicates the increase of the concentration of ions (including H^+) in the solution. Therefore, if only consider the concentration of H^+, the test results on pH and E_c are consistent.

Chai et al. (2004) reported that pH value had a significant effect on the adsorption capacity of Ariake clay to chromium. The Ariake clay tested had a plasticity limit of 47.6%, and liquid limit of 109.9%. The pH value of the solution from the clay/water mixture with a ratio of 1:10 was about 8.0. During the test, the pH values were adjusted by adding hydrochloric acid. The mechanism proposed is that the reduction of pH value promoted the following reduction reaction, which reduced Cr^{6+} (existed in the form of $Cr_2O_7^{2-}$) to Cr^{3+} and adsorbed by the clay mineral. It is considered that this mechanism is at least partially applicable here.

$$Cr_2O_7^{2-} + 14H^+ + 6e^- \leftrightarrow 2Cr^{3+} + 7H_2O \quad (1)$$

(2) For the Ariake clay content less than or equal to 20% cases, adding 2 and 5% of the cement reduced Cr^{6+} concentration considerably. As shown in Fig. 4 that adding the cement increased pH value of the solutions. Therefore, the mechanism expressed by Eq. (1) cannot

be applied to explain the reduction of Cr^{6+} concentration for adding the cement. The exact mechanism for this reduction has not been explored and it is considered mainly due to the cementation effect immobilized some of Cr^{6+} ions. However, for the cases of the Ariake clay content larger than 20%, reversely the Cr^{6+} concentration was increased after adding the cement. Although the reasons are not clear yet, it is considered partially due to the scatter of the data and partially may due to that the cement itself contains Cr^{6+} (Kamon and Katsumi, 2001).

Column percolation test results

Numbers of pore volume (NPV) versus Cr^{6+} concentration relationships are given in Figs 6 (a) and (b). NPV means the volume of the leakage divided by the pore volume of a sample. For most cases, after the sample was prepared, the percolation test was commenced and no curing time was given. Only one series, i.e. adding 5% of the cement, the percolation tests were started after 1 week curing time. For cases of C/F ratio of 40:60, no Cr^{6+} was detected in the leakage and the results were excluded. Figure 6 (a) shows that in cases of the fly ash alone and the fly ash plus 2 and 5% of the cement, the Cr^{6+} concentrations reduced with the increase of NPV. However, for no curing time cases, up to about 3 NPV, the Cr^{6+} concentrations were higher than the value from the batch contact test of the fly ash alone and much higher than the Japanese environmental standard. Comparing the batch contact test with the column percolation test, the amount of water in contact with solid particles is different. At a given time, there is more water in the batch contact test, which tends to reduce the concentration of a target chemical material. However, in the batch contact test, a chemical material can be more easily dissolved into the water than that in the column percolation test, which tends to increase the concentration of a target material in the solution. For the fly ash or the fly ash plus 2 and 5% of the cement cases, it seems that the amount of water in content with the fly ash particles played a more important role. The sample had a void ratio of about 0.7, and the ratio of the mass of the sample to 4 NPV of leakage by weight is approximately 1:1, which is much lower than the solid/liquid ratio of 1:10 adopted for the batch contact test. This partially explains the higher Cr^{6+} concentration from the column tests. Adding 2 and 5% of the cement reduced the Cr^{6+} concentrations, and at about 1 NPV, the value was about 70% of the corresponding value of the fly ash alone case. However, the difference between the 2 and 5% of the cement cases is not obvious. These results indicate that using the fly ash alone or the fly ash plus 2 to 5% of the cement as embankment material, the leakage from it may cause an environmental problem and it is unacceptable.

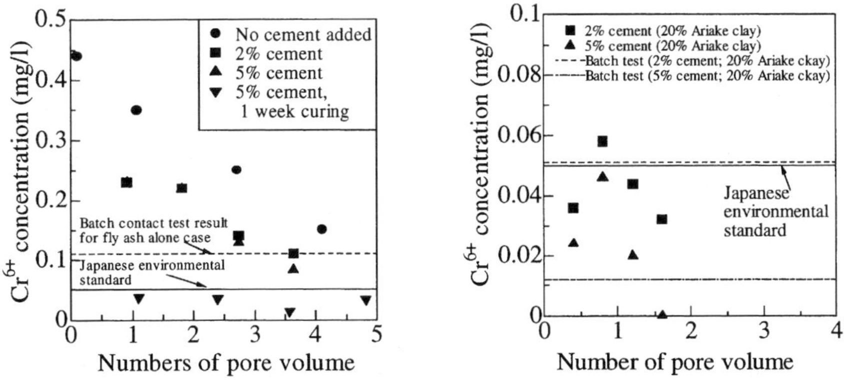

(a) Fly ash alone and fly ash plus cement (b) Fly ash, cement and Ariake clay mixture

Fig. 6 Cr^{6+} concentration in the leakage (column tests)

As shown in Fig. 6(a), in case of the fly ash plus 5% of the cement case, the curing time before starting the percolation test had a significant effect on the Cr^{6+} concentration in the leakage. With 1 week curing time, Cr^{6+} concentration was less than the Japanese environmental standard. This result has an important practical impact. It indicates that after constructing an embankment or other earth structure with the fly ash plus 5% the cement, covering the structures by a water proof sheet or by other means to prevent the infiltration of water for 1 week, the Cr^{6+} concentration in the leakage can satisfy the Japanese environmental standard.

For adding 2 and 5% of the cement into the fly ash and then mixed with the Ariake clay with a C/F ratio of 20:80 cases, the column percolation test results are given in Fig. 6 (b). For these cases, the Cr^{6+} concentrations from the column tests are comparable with those of from the batch contact tests at about 1 NPV and then reduced with the increase of NPV. Both the batch contact test and the column percolation test results indicate that adding the Ariake clay had a significant effect on reducing the Cr^{6+} concentration in the solution/leakage.

Based on the test results presented above, for the conditions evaluated, when adding 2 to 5% of the cement into the fly ash and then mixed with the Ariake clay with a C/F ratio of larger than 20:80, the concentration of Cr^{6+} in the solution/leakage can satisfy the environmental requirement. In Japan, there is a lot of waste clayey material generated from construction sites, as well as dredged sludges. There is a possibility of reusing the fly ash with that kind of waste clayey soil together. Another possible way to use the fly ash is to compact it with 5% cement and preventing water penetration for more than 1 week.

Unconfined compression test

Sample preparation and the conditions tested

The strength of the compacted fly ash and the fly ash-cement-Ariake clay mixture was investigated by unconfined compression tests. The samples were formed by compacting the material in disposable plastic molds which had a diameter of 50 mm and height of 100 mm. The mass of the rammer used is 1339 g and falling height was 112 mm. The compaction was made in three layers and each layer had 12 blows. The compaction water contents and resulting dry densities of the samples were about the same as of the samples for column percolation test as explained previously. The tests were conducted according to JIS A 1216 (JSA, 1998) and the compressive strain rate adopted was 1%/min. The conditions tested were (a) the fly ash alone; (b) the fly ash plus 2 and 5% of the cement; and (c) the fly ash plus 2 and 5% of the cement and mixed with the Ariake clay with a C/F ratio of 20:80. The curing times were 0, 2, 7, 14 and 28 days.

Test results

The fly ash alone

Figure 7 shows unconfined compressive stress (q) versus axial strain (ε_a) curves for different curing time. For each condition, three (3) samples were tested and it has been confirmed that the scatter was small. For clarity only one representative curve is plotted for each condition. It indicates that both unconfined compressive strength (q_u) and the secant modulus corresponding to a stress level of 50% of the strength (E_{50}) increased with curing time. q_u increased from about 60 kPa at 0 curing time to about 130 kPa at 28 days curing time, and the corresponding E_{50} values were about 5.5 MPa and16.0 MPa respectively.

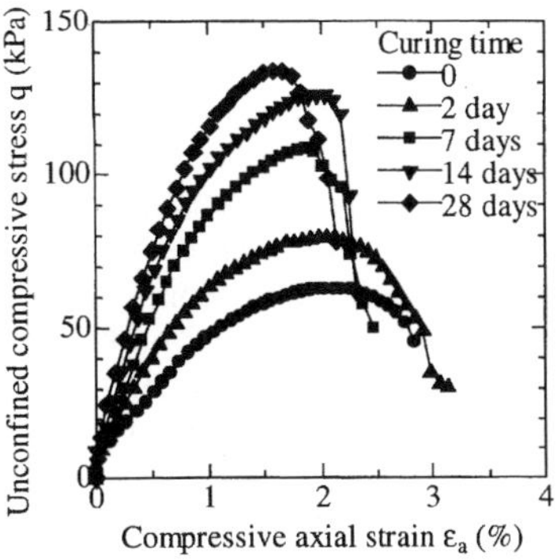

Fig. 7 q-ε_a curves of the fly ash alone

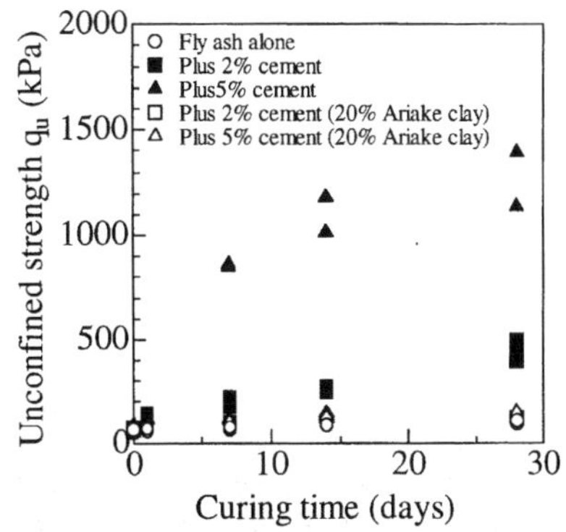

Fig. 8 Comparison of q_u values at different conditions

The fly ash mixed with the cement and the Ariake clay

Comparison of q_u values at different curing time and admixtures is given in Fig. 8. Except the general tendency of q_u increased with the curing time as already demonstrated in Fig. 7, Fig. 8 clearly indicates that adding the cement increased q_u values significantly. The q_u values of adding 2 and 5% of the cement at 28 days were about 500 kPa and 1300 kPa respectively, and about 3.8 and 10.0 times of that of the fly ash alone. Since adding the Ariake clay had a significant effect on reducing Cr^{6+} concentration in the leakage, the unconfined compression tests for the samples of adding 2 and 5% of the cement into the fly ash and then mixed with the Ariake clay with a C/F ratio of 20:80 were also conducted. Their q_u values at 28 days were about 125 kPa and 150 kPa for 2 and 5% of the cement cases respectively, which are comparable with that of the sample of the fly ash alone. E_{50} of the fly ash plus 5% the cement was about 200 MPa which is about 12.5 times of that of the fly ash alone. In case of mixed with Ariake clay with a C/F ratio of 20:80, the E_{50} values of 2% and 5% of the cement cases were about 8.2 MPa and 10.2 MPa respectively, which are lower than that of the fly ash alone (about 16.0 MPa).

Conclusion

Some of geoenvironmental and mechanical aspects of a fly ash or the fly ash-cement-Ariake clay mixture were investigated experimentally. Based on the test results, the following conclusions can be drawn.

On the geoenvironmental aspects

(1) In case of the fly ash alone, Cr^{6+} concentration in the solution/leakage was higher than Japanese environmental standard (0.05 mg/*l*). Adding 2 and 5% of cement without curing reduced the Cr^{6+} concentration, but still not satisfied the environmental requirement. For the fly ash plus 5% the cement case, curing the samples for 1 week before the column percolation test, the Cr^{6+} concentration in the leakage was less than Japanese environmental standard.

(2) Mixing the fly ash plus 2 and 5 % of the cement with the Ariake clay reduced Cr^{6+} concentration significantly. When clay/fly ash ratio (C/F ratio) was larger than 20:80, Cr^{6+} concentration satisfied the Japanese environmental standard.

(3) Comparing the results of the batch contact tests with that of the column percolation tests indicates that there is no general tendency on which test will yield a higher Cr^{6+} concentration.

It implies that only use either the batch contact test or the column percolation test may result in an unsafe result. For a field application, checking the concentration of a target chemical material by both the batch contact and the column percolation tests is recommended.

On the unconfined compressive strength (q_u) and the secant modulus (E_{50})

(1) For the fly ash samples compacted at optimum water content to 95 degree of compaction, the q_u value at 28 days was about 130 kPa and the secant modulus corresponding to a stress level of 50% of the strength (E_{50}) was about 16.0 MPa. Adding 2 and 5% of the cement increased q_u and E_{50} values significantly. For 5% of the cement cases, q_u was about 10.0 times and E_{50} was about 12.5 times of the corresponding values of the fly ash alone.

(2) For the cases of the fly ash plus 2 and 5% of the cement and mixed with the Ariake clay with a C/F ratio of 20:80, the q_u values were 125 and 150 kPa respectively, which is comparable with that of the fly ash alone, and E_{50} values were 8.2 and 10.2 MPa and less than that of the fly ash alone.

Acknowledgement

This study has been partially funded by Saga Branch, Kyusyu Regional Development Bureau, Ministry of Land, Infrastructure and Transport, Japan. The laboratory tests were conducted by Mr. H. Eguchi, a graduate of Saga University, Japan.

Reference

Chai, J.-C., Miura, N. and Onitsuka K. (2004). Adsorption characteristics of Ariake clays to chromium and the effect of pH-value. *Proc. 9th Australia New Zealand Conference on Geomechanics*. Auckland, pp.467-472.

Japanese Fly Ash Association (2000). Fly ash handbook.

Japanese Geotechnical Society (JGS) (2000). Test method for electric conductivity of suspended soils, JGS 0212-2000, Tokyo, Japan.

Japanese Standard Association (JSA) (1998). Method for unconfined compression test of soils, JIS A 1216, Tokyo, Japan.

Japan Standard Association (JSA) (1999a). The method for soil compaction using a rammer, JIS A 1210. Tokyo, Japan.

Japan Standard Association (JSA) (1999b). Fly ash for use in concrete, JIS A 6201. Tokyo, Japan.

Kamon, M. and Katsumi, T. (2001). Geotechnical utilization of solid waste materials and its environmental impacts. *Journal of Japanese Waste Material Society*, Vol. 12, No. 3, pp. 140-149.

Ohtsubo, K., Egashira, K. and Kashima K. (1995) Depositional and post-depositional geochemistry, and its correlation with the geotechnical properties of marine clays in Ariake Bay, Japan. *Geotechnique*, Vol. 45, No. 3, pp. 509-523.

RCUWGCM (1999). Activity report, Research Committee for Using Wastes as Geotechnical Construction Material (RCUWGCM), Japan.

Shibuya, S., Mitachi, T. and Kawaguchi, T. (2004). Engineering properties of some hardened fly ashes. *Geotechnical Engineering*, Southeast Asian Geotechnical Society, Vol. 35, No. 1, pp. 47-50.

The Japanese Environment Agency (1991). Environment standards regarding to soil contamination. Public Notice No. 46.

Yudhbir and Honjo, Y. (1991). Applications of geotechnical engineering to environmental control. *Proc. 9th Asian Regional Conference on Soil Mechanics and Geotechnical Engineering*, Bangkok, Thailand, pp. 431-469.

Stabilization of Sewage Sludge by a Combined Chemical and Mechanical Method

J. Chu and **T. T. Lim**
School of Civil and Environmental Engineering, Nanyang Technological University, Singapore 639798

M. H. Goi
CPG Consultants, Pte Ltd, Singapore 307685

Abstract
A large quantity of dewatered sewage sludge and other waste materials are produced each year in Singapore. The disposal of these waste materials has become a problem. On the other hand, a large amount of fill materials are required for land reclamation. A study to use the sewage sludge and other waste materials for land reclamation was carried out. Portland cement was used as a binder to stabilize the sludge. Another waste material, copper slag, was used as filler. The study shows that the stabilization effect of cement can be greatly enhanced if the cement mixed sludge can be consolidated by mechanical means. Therefore, it is cost-effective to adopt a combined chemical and mechanical method to treat the sludge. Leaching tests were also conducted to show that heavy metals were largely retained in the residual fraction, rendering the stabilized sludge to be more resistant in leaching.

Introduction

The disposal of sewage sludge and other waste materials in Singapore has become increasingly a problem due to a shortage of dumping ground. On the other hand, there is a great demand for fill materials for land reclamation. Therefore, it would be ideal if sewage sludge and other waste materials can be used as fill materials for land reclamation. Toward this purpose, a study on the feasibility of using binder treated sewage sludge and other waste materials for land reclamation was carried out. One of the methods studied is to mix the sludge and other waste materials with a small portion of binders and then consolidate the mixture on site after it is disposed. Depending on the usage of the reclaimed land, the sludge mixture layer can be capped with a layer of granular fill. In this method, the sludge needs not to be converted into a hard granular material before placement. It is only required to be treated into a material comparable to the consolidated seabed marine clay on which the sludge mixture is placed. In this way, the costs involved in treating and disposing sludge can be much lower compared with other disposing methods. The demand for granular fill can also be reduced.

5th ICEG Environmental Geotechnics, Thomas Telford, London 2006.

In this study, Portland cement and/or hydrated lime[1] were used as binders. Another waste, copper slag, was used as a filler. In this paper, the general geotechnical and geochemical properties of the sewage sludge and the copper slag are characterised. A preliminary study is presented to show that sewage sludge cannot be treated by binders or by consolidation alone. Both chemical and mechanical means need to be used together in order to improve the geotechnical properties of sludge in a cost effective way. As the consolidation behaviour of the binder treated sludge and copper slag mixtures can be quite different from that of soils, some laboratory model tests were conducted using a fully instrumented consolidation tank to study the consolidation process of the binder treated sludge mixture around vertical drains. Testing arrangement for the model tests is described. The geotechnical properties of the consolidated sludge mixtures are evaluated. Leaching tests were also conducted. The assessment on the leaching of various toxic metals from the cement-treated sludge mixtures is presented.

Materials tested

The anaerobically-digested sewage sludge used in this study was taken from a wastewater treatment plant in Singapore. The sludge had been filter-pressed for dewatering at the plant. Even so, the so-called dewatered sludge was still very soft, very high in water content and had little shear strength. It contained a considerable proportion of organic matters and some heavy metals. The physical and geochemical properties of the sludge are summarized in Tables 1 and 2 respectively. The heavy metal contents in the sludge were determined by microwave-assisted acid digestion of their oven-dried sub-samples followed by analysis of the digestates with the Inductively Coupled Plasma Optical Emission Spectrometer (ICP-OES). The results are presented in Table 3. Calcium and Silicon existed in high concentrations because quartz and calcite or calcium carbonate was the main mineral found in sludge, as revealed by the X-ray diffraction (XRD) tests conducted on the sludge sample (Goi 2004). Concentrations of Zn and Cu were relatively high. Zn could be introduced by factories producing galvanized zinc, whereas Cu might come from copper production plants and corroded copper pipe system. The coefficients of variations (CV) for Zn and Cu concentrations in the sludge were 56% and 36% respectively. These high CV values suggest that there were considerable spatial variations in the heavy metal concentrations in the sludge. Owing to the potential environmental concerns, the sludge needs to be treated to meet both the geotechnical and environmental requirements.

Table 1 Physical properties of sewage sludge

Property	Value
Moisture Content (%)	400-514
Total Solids (TS) Content (%)	16.3-20
Bulk Density (Mg/m^3)	1.02
Specific Gravity	1.634
Liquid Limit, W_L (%)	397
Plastic Limit, W_P (%)	63
Plasticity Index, PI (%)	334
Liquidity Index, LI	1.35

[1] Quicklime was not used as it was not practical to use it on a large scale due to the weather condition and the health regulations in Singapore.

Table 2 Geochemical properties of sewage sludge

Property	Value
pH at 25°C ± 2 (1:6 solid: water)	8.42
Electric Conductivity (EC) at 25°C ± 2 (1:6 solid: water) (mS/cm)	39.4
Redox Potential (Eh) at 25°C ± 2 (1:6 solid: water) (mV)	323
Dissolved Organic Carbon (DOC)* (mg/L)	1581
Phosphates* (mg/L)	66

* Based on the extraction of 1:5 solid: water ratio

Table 3 Total heavy metal concentrations in sewage sludge

Element	Concentration (mg/kg)
Sodium – Na	2,070 [705][a]
Calcium – Ca	32,600 [5,000]
Magnesium – Mg	4,820 [540]
Potassium – K	2,960 [290]
Aluminum – Al	12,700 [2490]
Iron – Fe	22,300 [2,670]
Silicon – Si	66,200 [4,440]
Barium – Ba	427 [76]
Arsenic – As	< 36
Selenium – Se	< 24
Cadmium – Cd	< 1
Chromium – Cr	694 (102)[b]
Copper – Cu	1,491 (533)
Lead – Pb	146 (24)
Nickel – Ni	314 (111)
Zinc – Zn	1,844 (1,032)

[a] Values in [ ] are standard deviations based on 6 determinations
[b] Values in ( ) are standard deviations based on 12 determinations

Table 4 Physical and geotechnical properties of copper slag

Property	Value
Bulk Density (Mg/m^3)	2.0
Specific Gravity	3.57
Void Ratio	0.264 - 0.651
D_{50} (mm)	0.5
Particle Distribution	Well graded
Particle Shape	Subrounded to
Permeability (m/s)	$10^{-4} \sim 10^{-5}$
Friction Angle (°)	29 ~ 47

Copper slag is a non-toxic waste product after it has been used as a blast-cleaning agent to remove rust and marine deposits on ships. It is generated in substantial quantities every year in Singapore and its disposal also becomes increasingly a problem. It is well-graded with a mean grain size of 0.5 mm. The geotechnical and geochemical properties of copper slag are given in Table 4. As copper slag contains heavy metals such as Pb, Zn, Cu, Cr and Ni, it cannot be used directly as a fill material for land reclamation. Ordinary Portland cement or/and hydrated lime were used as binders.

Preliminary study

Tests were conducted by mixing sludge with cement or lime or/and copper slag in various proportions. Some examples are given in Fig. 1. The unconfined compression (UC) test results versus curing time are plotted in Fig. 2. It can be seen that even with a cement ratio of 25%, the UC strengths are still low. Therefore, it is not effective to use chemical treatment alone to improvement the geotechnical properties of sewage sludge.

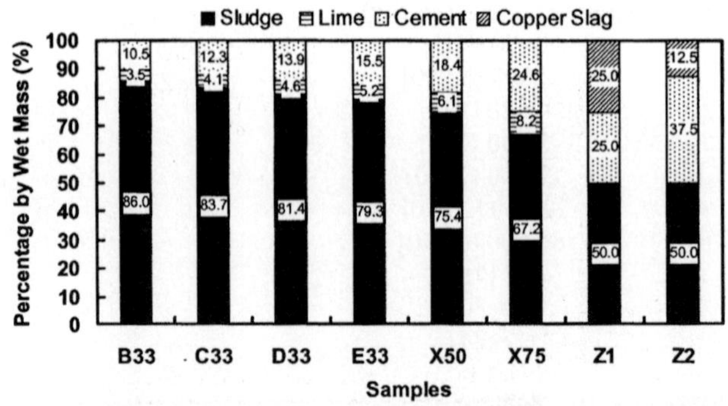

Figure 1 Proportions of constituents in each mix

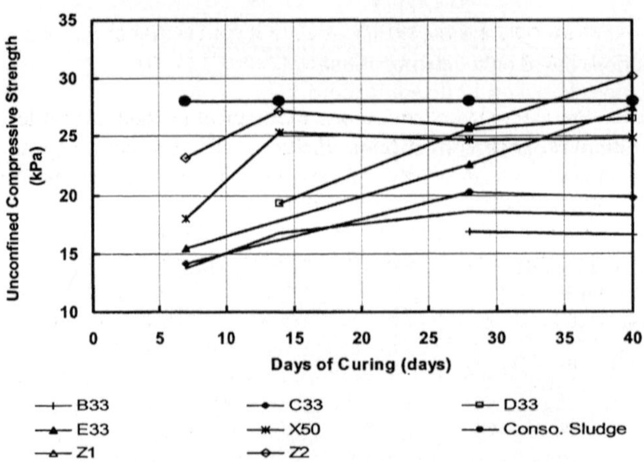

Figure 2 Results of unconfined compression test results

It is not effective either to treat sludge by consolidation or other mechanical means alone. According to Vesilind (2001), there are four distinct types of water that are associated with sludge: free (or bulk) water, interstitial water, vicinal water and water of hydration, and only the free water can be removed with relative ease by mechanical means. Interstitial water can also be removed, but the extra mechanical effort required to do so will not lead to a substantial reduction in the final water content that is efficient and cost-effective. Vicinal and water of hydration cannot be dewatered by any mechanical effort. The UC strength of a sludge sample consolidated under 80 kPa is also shown in Fig. 2. It can be seen that the strength of consolidated sludge was still very low. Therefore, a combined chemical and mechanical method was adopted in this study.

Model tests

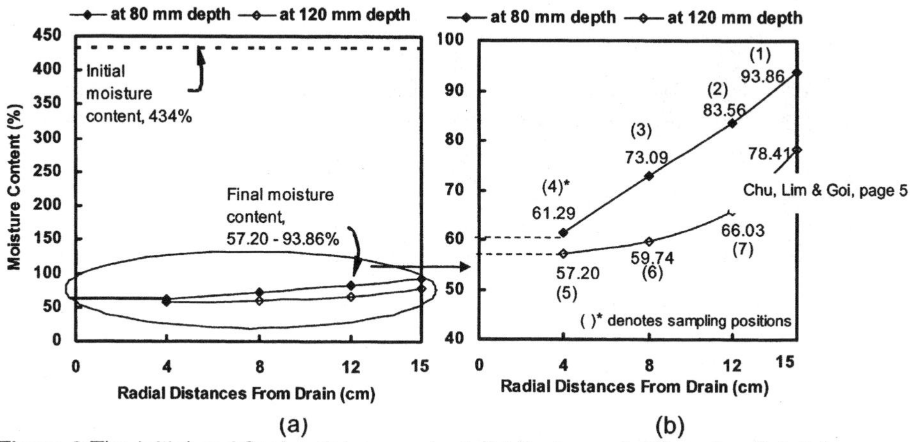

Figure 3 The initial and final moisture content distributions at different radial distances from the vertical drain: (a) distribution curves; (b) a local enlargement of (a).

To study the effectiveness of using the combined chemical and mechanical method, model tests using a 300 mm consolidation tank were conducted (Chu *et al.* 2005). The sewage sludge was first mixed with binders and the mixture was then placed in the consolidate tank for consolidation. The moisture content distribution along the radial distance obtained in one of the model tests is presented in Fig. 3. This test was conduced on a mixture of 77% of sludge, 15% of copper slag and 8% of cement. The moisture content of the mixed slurry was 434%. A prefabricated vertical drain was used for this test. The drain was installed before the slurry was placed. The consolidation pressure adopted was 80 kPa. The average degree of consolidation achieved at the end of the test was 89.1% (Goi 2004). The moisture content had reduced from the initial 434% to the final 57 to 94%. The moisture content of the soil closer to the vertical drain was lower than that further away from the drain. Similar observations were also made by Tanaka (1997) during the consolidation of slurry involving radial drainage. However, the variation in moisture content in sludge mixture is larger than that in natural clay. Therefore, for consolidation of sludge, closer drain spacing or longer duration of consolidation should be used. The UC strength obtained from the consolidated sludge varied from 45 to 160 kPa (Chu *et al.* 2005), which corresponds to an undrained shear strength to the effective vertical stress ratio, c_u/σ_{v0}', of 0.36 to 1.14. This range of value is higher than the typical c_u/σ_{v0}' ratio for normally consolidated remoulded soils (Terzaghi *et al.* 1996). The higher c_u/σ_{v0}' ratio reflects the effect of cementation due to the use of cement. As higher c_u/σ_{v0}' ratios were achieved at low water contents, it may imply that the lower the moisture content, the higher the cementation effect. In other words, consolidation enhances the effect of binders. This explains why the combined chemical and mechanical methods are more effective. The other model tests conducted by Goi (2004) also show that when a higher cement ratio of 10 to 12.5% is used, the water content of the consolidated sludge mixture can be much reduced and the shear strength can be much higher.

The surface morphologies and microstructures of the untreated sludge and the cement treated sludge mixture were examined using a Leica Stereoscan 420 scanning electron microscope (SEM). The samples were freeze-dried prior to the SEM test to remove moisture, thus minimizing alteration of their microstructure during drying. Their SEM images are shown in Fig. 4. The untreated sludge (Fig. 4a) had a sponge-like microstructure featuring large void spaces. It, therefore, would experience large volume change under compression. The pore spaces were initially occupied by water, which made up more than 80% of the sludge mass. After the sludge was mixed with copper slag and cement and consolidated, the resulting mixture possessed a compact, interlocking matrix that showed significantly reduced void space, as shown in Fig. 4b. These microstructures were formed possibly through filling of the voids by the copper slag particles and also cementation by cement hydration products such as calcium-silicate-hydrates which bound the sludge and copper slag particles together. The difference in the microstructures of the raw and stabilized sludge explains why there is a considerable increase in the undrained shear strength and a substantial reduction in compressibility of the treated sludge (Chu *et al.* 2002, 2005).

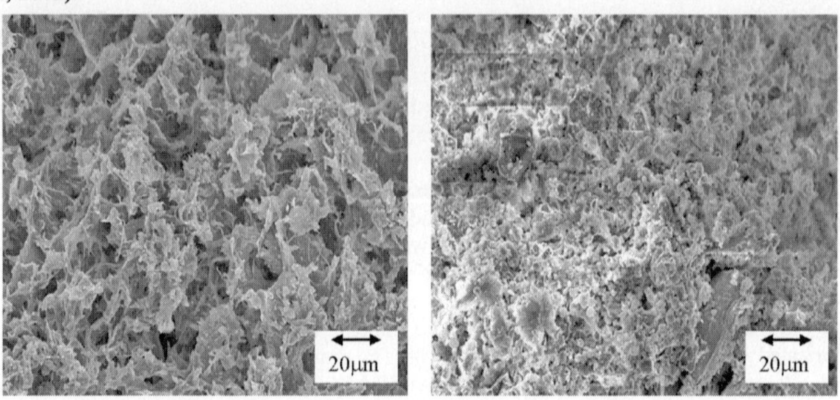

(a) (b)

Figure 4 Microstructures of (a) raw sludge and (b) Consolidated sludge, cement and copper slag mixture

Leaching properties of stabilized sludge

The leaching characteristics of both the raw and treated sludge were studied. The results for one treated sludge sample, Z2 (see Fig. 1), are shown in Fig. 5. The tests were conducted using the modified column leaching tests with distilled water as the leachant. Over the 50 days of leaching duration, a total of 60 PV of leachate was collected. Here PV refers to as pore volume. Fig. 5

indicates that the concentrations of heavy metals leached out from the treated sludge. These values were very much lower than those obtained from tests on untreated sludge (Chu *et al.* 2002; Lim *et al.* 2004; 2005). The maximum concentration of Cu in the leachate of Mix Z2 was 3.2 mg/L at 4 PV, and the concentration decreased gradually to consistently less than 1 mg/L thereafter. For Zn and Ni, their concentrations were the highest at 1 PV, and decreased significantly thereafter. Their maximum concentrations were 9.3 mg/L and 7.0 mg/L, respectively. The maximum concentration of Pb was 0.3 mg/L at 1 PV, and the Pb level began to decrease thereafter. For Cr and Cd, their concentrations leached were negligible, in the order of μg/L (ppb). Generally the concentrations of heavy metals leached in Mix Z2 were very low, well below their toxicity limits. This study shows that Mix Z2, which is sludge-cement-copper slag matrix, was significantly more resistant to leaching than sludge. This implies that addition of 12.5% cement into the sludge-copper slag mixture could significantly reduce the potential environmental impact due to leaching of heavy metal compounds in the sludge and copper slag.

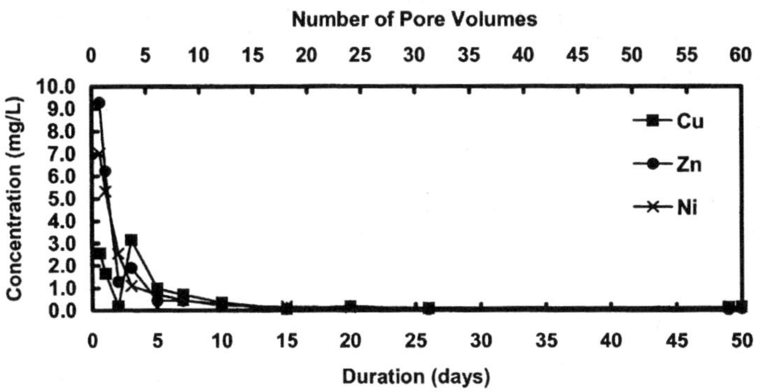

(a)

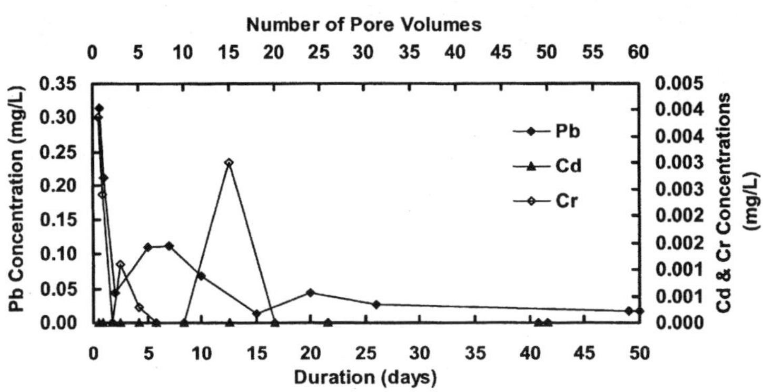

(b)

Figure 5: Leaching of heavy metals conducted on sample in Mix Z2

Conclusions

The following conclusions can be drawn from this study:

1) The geotechnical properties of sewage sludge cannot be improved effectively by using either chemical or mechanical method alone. A combined chemical and mechanical method is required in order to stabilise the sludge in a cost-effective way.
2) By mixing sludge with cement and copper slag and consolidating the mixture under a pressure of 80 kPa, the physical and geotechnical properties of the sludge can be significantly improved. The moisture content of the sludge can be reduced from more than 500% to a level of 57 to 94%. The cementation effect can be clearly identified.
3) The microstructures of the treated sludge mixtures are more densely packed and have much lesser pore spaces as compared to that of the raw sludge. This explains why the treated sludge has a considerable increase in undrained shear strength and a substantial reduction in compressibility.
4) The results obtained from the modified column leaching tests with distilled water as the leachant indicate that the concentration of Cu leached from sludge was significantly reduced when cement and copper slag were added. In general, the concentrations of heavy metals in the leachate decreased with leaching time.

References

Chu, J., Goi, M.H. and Lim, T.T. (2005). Consolidation of cement treated sewage sludge using vertical drains. *Canadian Geotechnical Journal,* 42(2), 528-540.

Chu. J, Lim T.T. and Goi. M. H. (2002). Potential Use of Stabilized Sewage Sludge for Land Reclamation, *Journal of The Institution of Engineers Singapore,* 42(6), 36-42.

Goi, M.H. (2004). Use of stabilised sewage sludge for land reclamation, *MEng Thesis*, Nanyang Technological University, Singapore.

Lim, T.T., Chu, J. and Goi, M.H. (2004). Assessment of heavy metals leachability in clay-amended sewage sludge stabilized with cement for use as fill material. *Journal of Residuals Science & Technology*, 1(3), 157-164.

Lim, T.T., Chu, J. and Goi, M.H. (2005). Effect of cement on redistribution of trace metals and dissolution of organics in sewage and its inorganic water-amended products. *Waste Management* (in press).

Tanaka, Y. (1997). Non-uniformly consolidated ground around a vertical drain. *Proc. 30th Anniversary Symp of Southeast Asian Geotechnical Society*, **1**, 232-243.

Terzaghi, K., Peck, R.B. and Mesri, G. (1996). *Soil Mechanics in Engineering Practice.* John Wiley & Sons, New York.

Vesilind. P.A. (2001). Sludge Dewatering: Why the Water Wins Every Time. *Conference Proceedings on Sludge Management Entering the 3rd Millennium,* 252-257.

Geotechnical Properties of Marble Dust and its Reuse in Slurry-Pond Embankment

Manoj Datta Indian Institute of Technology Delhi, India
J.T. Shahu Indian Institute of Technology Delhi, India
Saradindu Mondal Indian Institute of Technology Delhi, India

Abstract

In India, about 4000 marble mines and 1100 marble cutting units are operating in the small to medium scale industry in the northern state of Rajasthan. About 5 to 6 million tonnes of marble slurry is generated every year as by-product of cutting of marble. The slurry is disposed off in an uncontrolled manner in open spaces and low-lying areas. Upon drying, the marble dust causes serious air pollution problem. As a part of environmental control measures, a marble slurry-pond is being designed with 5 m high embankments all around the pond to contain the slurry and control the air pollution. The embankment height is to be raised to 25 m in stages. This paper describes the results of a laboratory investigation into the geotechnical properties of marble dust. It also reports the design undertaken to maximise the use of existing marble dust in the slurry-pond embankment and thereby reduce the cost of procuring local soil for embankment construction.

Introduction

The state of Rajasthan occupies a special position in India as the largest producer of marble. It contributes almost 60% to the production of marble in India. Consequently marble processing industries exist in large number for processing the mined marble blocks. Around 1100 gang saws are operating in Rajasthan for cutting mined marble blocks into the slabs. In the processing of marble by gang saws, around 25 to 30 % waste is generated in the form of marble slurry and it is estimated that around 5 to 6 million tonnes of marble slurry is generated annually in Rajasthan. The marble slurry composes of a lean mixture of water and marble dust produced during cutting of slabs.

To control the adverse impact of uncontrolled dumping on the geoenvironment, a slurry-pond is proposed at a site (500 m × 500 m in plan) in the region having the maximum number of marble processing units. All slurry will be transported and deposited in this pond. The marble dust will be kept immersed under water to prevent its dust particles from being air borne. An embankment along the periphery of slurry pond would be constructed using marble dust accumulated in existing dumps. The embankment is to be raised to a height of 25 m in increments of 5 m each.

Geoenvironmental Impact

Uncontrolled disposal of marble slurry imposes serious threats to the physical, chemical and biological components of environment. The problems encountered are given below.

5th ICEG Environmental Geotechnics, Thomas Telford, London 2006.

- When dumped on land, the marble dust adversely affects the productivity of land due to decreased porosity, water absorption, water percolation, etc.
- Slurry dumped areas cannot support any vegetation and remain degraded.
- When dried, the fine particles of marble dust become air borne and cause severe air pollution. Apart from occupational health problems, these particles also affect machinery and instruments installed in industrial areas.
- During rainy seasons, marble dust particles are carried away to rivers, drains and water bodies affecting quality of water, reducing storage capacities and damaging aquatic life.
- The heaps of marble dust that remain scattered all around the industrial estates are an eye sore and spoil aesthetics of entire region. Consequently tourism and industrial potential of the state is adversely affected.

Scope and Methodology

Disposal of marble slurry in a slurry-pond created by constructing an embankment along the perimeter of a 500 m x 500 m plot of flat ground is proposed as a remedial measure for such problems. The present study aims at designing the embankment using marble dust as the main construction material. As geotechnical properties of marble dust are needed for design, hence this study also includes the determination of geotechnical properties of marble dust.

This paper covers the following aspects:

- Determination of Geotechnical Properties: Determination of grain size distribution, specific gravity, Atterberg limits, compaction characteristics, shear strength characteristics, permeability and compressibility.
- Design of Slurry-Pond Embankments: Different types of embankments and different methods of construction are compared to arrive at the best possible solution.

Literature Review

There is no significant literature available on geotechnical properties of marble dust and its use in slurry-pond embankments. Literature regarding the formation and disposal of marble slurry in India (Natani, 2003; Pareek, 2002; Science Tech Entrepreneur, 2001; and Singh and Vijayalakshmi, 1999) is briefly discussed here. While marble blocks are cut by gang saws, water is used as a coolant. The blade thickness of the saws is about 5 mm and normally the blocks are cut to produce 20 mm thick sheets. Therefore, out of every 25 mm thickness of marble block, 5 mm are converted into powder while cutting. This powder flows along with the water as marble slurry. Nearly 20% of the total weight of the marble processed results into marble slurry. The marble slurry has high water content with solids concentration in the range of 20 percent. The chemical analysis (Table 1) of the marble dust shows that it is a

Table 1: Chemical analysis of marble dust

Constituents no.	Test carried out	Test value (in %)
1	Calcium Carbonate ($CaCO_3$)	92.53
2	Magnesia (MgO)	4.47
3	Silica (SiO_2)	1.73
4	Alumina (Al_2O_3)	1.04
5	Iron Oxide (Fe_2O_3)	0.21
6	Soda	< 0.01
7	Potash	< 0.01

mixture of calcium and magnesium carbonates. Minor impurities in the marble slurry include quartz, chert, flint, hematite, limonite, graphite, mica etc.

Experimental Investigations

Disturbed representative samples were obtained from an existing dump of marble dust to evaluate grain size distribution, Atterberg limits, specific gravity, plasticity, compaction characteristics, permeability, shear strength parameters and consolidation characteristics of the material.

Grain size distribution was carried out by hydrometer analysis. Specific gravity of marble dust was determined by density bottle method subjecting the sample for boiling and application of vacuum for 10 to 15 minutes for removing the entrapped air into it. Minimum dry density of the material was obtained using static settlement through water column test for solids concentration of 10 and 20 percent. Maximum dry density was achieved by vibrating table test subjecting the sample to vibration for 5 minutes. To evaluate the dry density and moisture content relationship, the sample was tested in the standard Proctor compaction test. Falling head permeability test on as-compacted-then-saturated samples was carried out for the determination of coefficient of permeability of marble dust. As-compacted-then-saturated samples were prepared by compaction at OMC and then saturated by immersion in water. Total shear strength parameters were obtained by unconsolidated undrained triaxial tests on statically as-compacted samples under cell pressure of 50 to 150 kPa. Direct shear tests were carried out on loose and as-compacted-then-saturated samples for determining the effective shear strength parameters under the normal stress of 50 to 200 kPa with a very slow strain rate of testing. To determine the consolidation and compressibility characteristics of marble dust, consolidation test on as-compacted-then-saturated samples were carried out. All the above tests were done according to relevant Indian Standard codes.

GEOTECHNICAL PROPERTIES

The laboratory test results on disturbed sample of marble dust show that the material is a non-plastic fine grained material with particle size in the range of medium to coarse silt sized particles and specific gravity of 2.83. The grain size distribution curves of marble dust are shown in Fig.1. The minimum dry density obtained was 9.8 kN/m^3 and maximum dry density was 18.5 kN/m^3. Table 2 shows the variation of minimum dry density at different solids concentration of slurry.

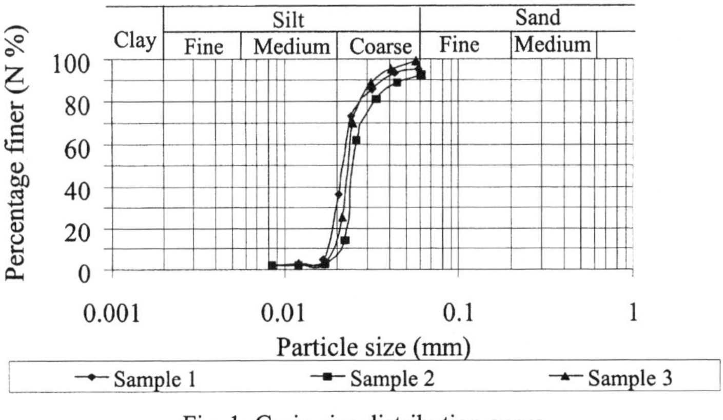

Fig. 1: Grain size distribution curve

Table 2: Variation of minimum dry density

Capacity of cylinder	10% slurry	20% slurry
1000 ml	9.81 kN/m^3	10.71 kN/m^3
100 ml	9.85 kN/m^3	10.23 kN/m^3

In the standard Proctor test, the optimum moisture content was observed to be 15.5% and the maximum dry density to be 18.3 kN/m^3. Fig. 2 shows the standard Proctor compaction curve with zero air void line for marble dust. Falling head permeability test on as-compacted-then-saturated samples revealed that the coefficient of permeability of the material is in the range of 1.2 to 1.9 × 10^{-8} m/sec. Unconsolidated undrained triaxial tests on as-compacted samples under cell pressure of 50 to 150 kPa yielded a cohesion intercept of 60 kPa and an angle of shearing resistance of 30°. Direct shear tests were performed under slow strain rates on as-compacted-then-saturated and loosely deposited saturated samples under normal stress of 50 to 200 kPa for the effective shear strength parameters. Fig. 3 shows the Mohr's envelopes for both the loose and dense states. Effective shear strength parameters are listed in Table 3. Triaxial test results are different from direct shear test results because the samples in the former were partially saturated and in the latter, these were saturated. The shear box results were used in stability analysis because these yielded the lower strength parameters as compared to triaxial tests and these were valid for the steady state seepage condition.

Slurry-Pond Embankment

A slurry-pond is proposed at a site, 500 m x 500 m in plan (Fig. 4). The pond will have an embankment (dyke) all along its perimeter, which would be raised incrementally, by 5 m in each stage, to reach a final height of 25 m. Marble slurry (a mixture of marble dust and water in 1:8 ratio) would be deposited in the pond. Excess water, after settlement of the marble dust, is to be pumped back for reuse. To construct the embankment, it was decided to maximize the use of marble dust available free of cost from an old waste dump. This is so because marble dust is not rated as hazardous and can thus be used as a construction material and also because acidic conditions are not expected to develop in the slurry pond which could otherwise cause dissolution reaction.

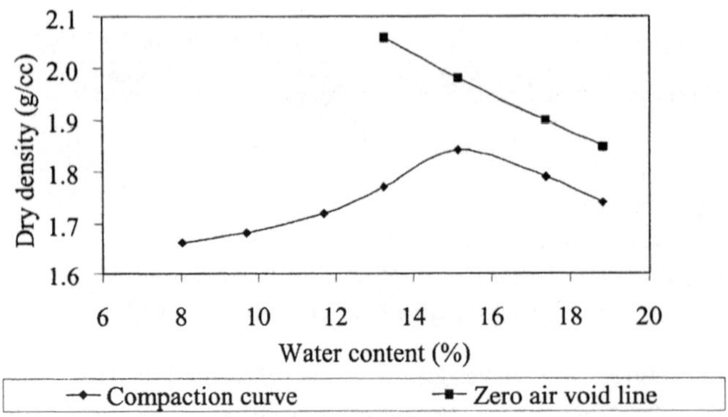

Fig. 2: Standard Proctor compaction curve

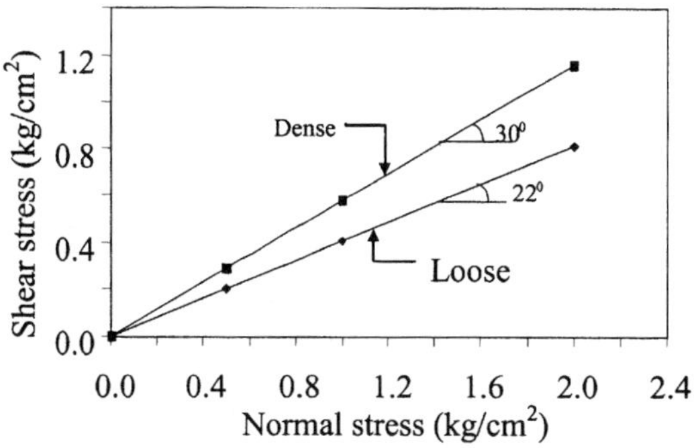

Fig. 3: Mohr's envelopes for both dense and loose state of marble dust

Table 3: Effective shear strength parameters

Parameters	State of samples	
	Loose	Dense
Dry density, γ_d (kN/m^3)	11.0	18.3
Cohesion (c′) (kPa)	0	0
Effective angle of shearing resistance, ϕ' (degrees)	22	30

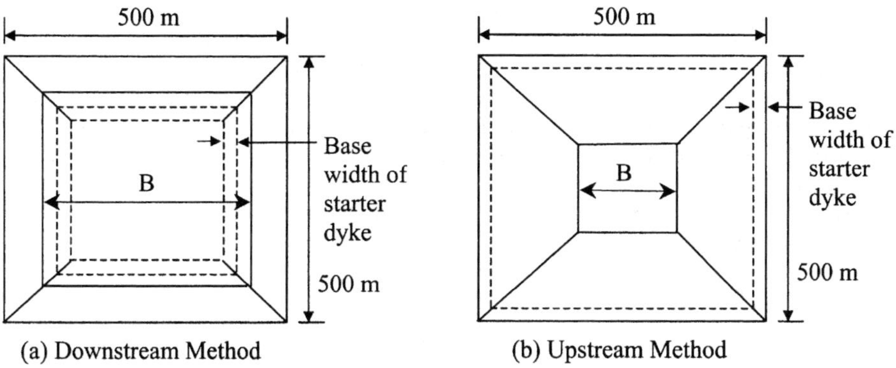

B – Top width of the slurry pond after final construction

Fig. 4: Plan of slurry ponds

Three alternatives were considered:

- embankment made of mixture of local soil and marble dust,
- embankment made of alternating layers of local soil and marble dust,
- embankment made of marble dust with a final cover of local soil.

The first alternative was expensive because of high cost of mixing. The third alternative was most desirable from the point of view of maximizing the use of marble dust and minimizing the use of soil from local borrow area. However, it was felt that dust emission could occur from dry surfaces of layers of marble dust during compaction in such a case. Hence it was decided that sprinkling of water coupled with provision of soil cover (0.15 m to 0.30 m thick) after every 2 to 3 m thickness of marble dust would be undertaken to offset dust emissions.

Upstream versus Downstream Method

To raise the embankment height in increments of 5 m, two methods were considered: the upstream method and the downstream method of construction (Figs. 5 and 6). Stability analysis was performed using Geoslope software and the steady-state condition was taken as the critical condition for safety of the embankment (ICOLD, 1996; U.S. Environmental Protection Agency, 1994).

The top width of the embankment was taken as 3 m and the berm width (at every 5 m height) was varied between 0 m to 6 m (Fig. 7). The shear strength parameters for the slurry deposited marble dust in the pond were taken as those corresponding to the loose state listed in Table 3 and for the compacted marble dust in the embankment as those in the dense state. The strength of thin layers of intermediate soil covers was ignored.

Table 4 shows the results of the stability analysis. It is observed that for the upstream method of construction, the downstream slope of the embankment would have to be in the range of 5H:1V to 6H:1V whereas for the downstream method of construction, the same would be 2.75H:1V. The stability analysis was carried out by three different methods available in the software, namely: (a) Ordinary method of slices, (b) Bishop method, and (c) Janbu method. For upstream method of construction, the slope could not be steeper because of two reasons, namely: (a) deep seated failure surface passing through loose slurry-deposited marble dust of low strength (Fig. 8), and (b) high pore water pressure due to phreatic line being close to the downstream surface of the slope as the location of the vertical drain was in the middle of the embankment (Fig. 5). In the case of downstream method of construction, the failure surface is located in compacted marble dust and significant pore water pressures are not encountered since the drain is well away from the downstream face of the embankment (Fig. 6).

Final Choice

The choice of the method of construction was governed by the volume of marble dust that could be placed in the slurry pond as well as within the body of the embankment. It was estimated that this would be in the range of 3.2×10^6 m^3 for the upstream method and 4.7×10^6 m^3 for the downstream method. The latter method of construction was taken as the final choice because it yielded 47 percent additional storage volume for marble dust despite a higher cost of construction. Adequately designed transition filters are placed between drains and the surrounding material to prevent blocking of the drains. The downstream slope will be protected by a cover of local soil of thickness 0.5 m and vegetative growth on the soil. The embankment is in planning stage. The construction is expected to be taken up in near future.

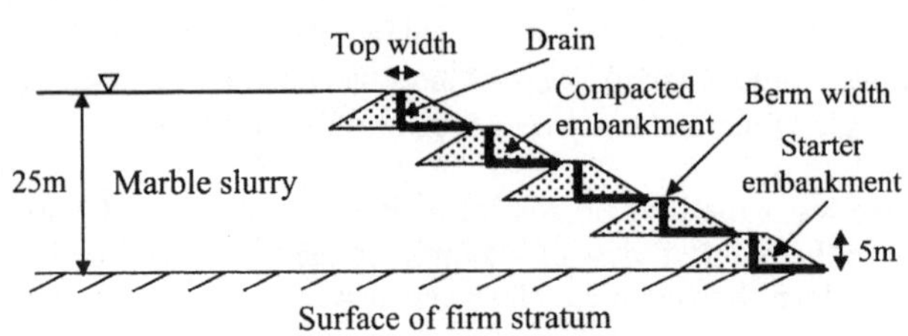

Fig. 5 Upstream method of construction

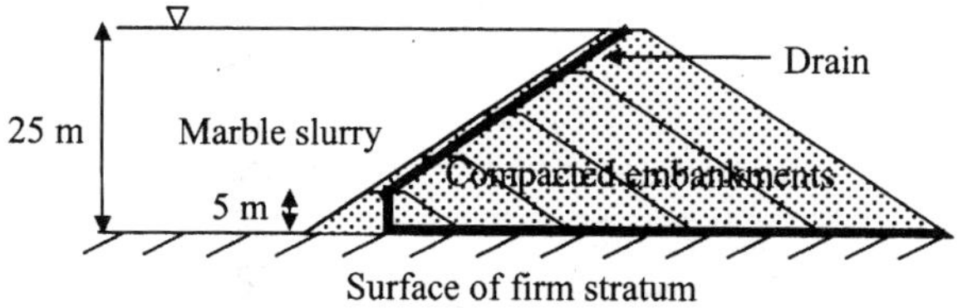

Fig. 6 Downstream method of construction

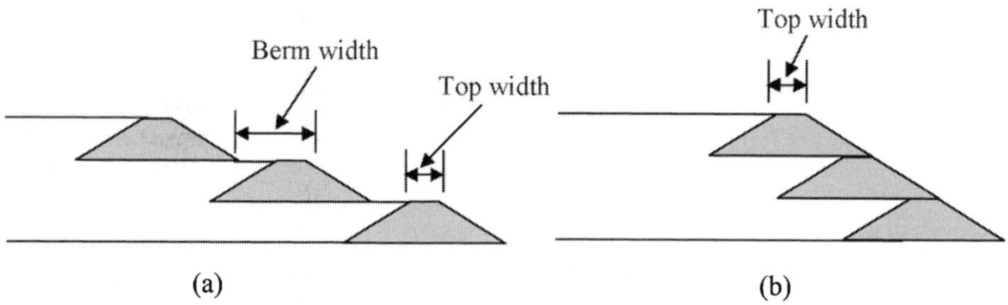

Fig. 7: Raised embankment (a) with berm and (b) without berm

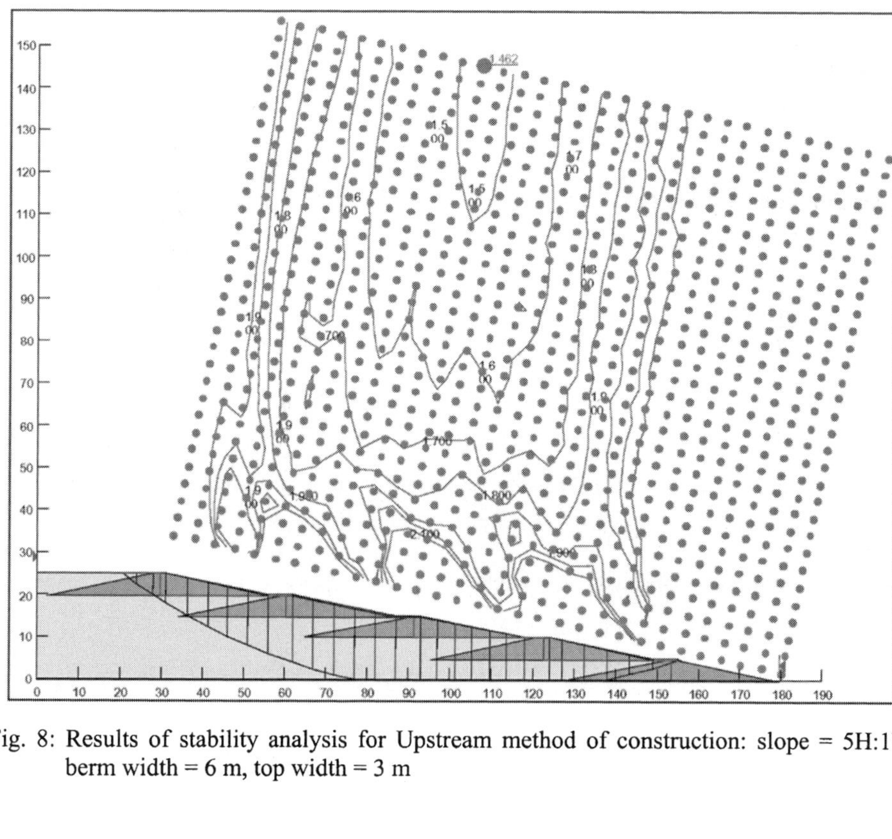

Fig. 8: Results of stability analysis for Upstream method of construction: slope = 5H:1V, berm width = 6 m, top width = 3 m

Table 4: Results of stability analysis

Method of construction	Parameters	Factor of Safety
Upstream	BW = 0, TW = 3m, D/s = 6H:1V	1.55
	BW = 3, TW = 3m, D/s = 5.5H:1V	1.51
	BW = 6, TW = 3m, D/s = 5H:1V	1.56
Downstream	BW = 0, TW = 3m, D/s = 2.75H:1V	1.59

BW - Berm width, TW – Top width, D/s – Downstream slope

Conclusions

The study of geotechnical properties of marble dust indicate that the material behaves similar to soils lying in the silt-size range. This waste material can be used for the construction of embankments of slurry ponds after proper compaction and provision of suitable soil cover.

References

ICOLD (1996), "A Guide to Tailings Dams and Impoundments: Design, Construction, Use and Rehabilitation", International Congress on Large Dams, Bulletin 106.

Natani, J.V (2003), "Geological and Environmental Status of Makrana Marble Mining Area of Nagpur District, Rajasthan and Strategy for Sustainable Development", Geological Survey of India, Jaipur.

Pareek S. (2002), "Gainful Utilization of Marble Waste: An effort towards protection of Ecology & Environment", Geological Survey of India, Jaipur.

Science Tech Entrepreneur (2001), "Marble Slurry – A new resource material for Entrepreneurs", DST-FICCI Project, http://www.techno-preneur.net/timeis/technology/sctechFeb-March/mb20.html.

Singh, S. and Vijayalakshmi, V. (1999), "Marble Slurry-A new building material", http://www.tifac.org.in/news/marble.html.

U.S. Environmental Protection Agency (1994), "Design and Evaluation of Tailings Dams", Technical Report, Washington.

Reuse of refractory bauxite ore waste as aggregate in asphalt concrete paving mixtures

G T P Fabbri University of Sao Paulo, Sao Carlos, Brazil
C A P Silva Jr. Unioeste, Cascavel, Brazil
A C Gigante University of Sao Paulo, Sao Carlos, Brazil
A P Furlan University of Sao Paulo, Sao Carlos, Brazil
H B Fontenele Assis Gurgacz College, Sao Carlos, Brazil

Abstract

This paper presents an evaluation of the potential use of a waste produced by exploration of refractory bauxite ore as aggregate in asphalt concrete mixtures. In a first phase, classification tests were accomplished in this waste by means of solubilization and leaching tests, in order to evaluate the hazard, in long term, of asphalt mixtures with this waste to the environment and to classify the materials/waste. Besides, the waste characterization as aggregate was based on gradation test results, Los Angeles abrasion and Marshall mixture design properties. In a second phase, mechanical property and durability tests were carried out in asphaltic mixtures with the waste. According to the results of environmental and mechanical property tests, it can be concluded that the use of refractory bauxite ore waste (ferruginous waste) is viable to compose the gradation in asphalt concrete.

INTRODUCTION

In some regions of Brazil, the production of asphalt mixtures for use in flexible pavements is difficult by the lacking of conventional aggregates (for example: basalt, granite or calcareous rocks). In order to supply this lack, it can be used alternative materials that are found in the neighborhoods of the road, or experimentally, industrial by-products as wastes or slag. The use of waste as a substitute for conventional aggregates in asphalt mixtures is matter of many researches. Hosking (1976) evaluated the strength and raveling susceptibility of some kinds of aggregates and oven slag of steel industry, and concluded that the use of the calcined bauxite improved the adherence between wheel and pavement.

In the north region of Brazil, at the city of Almerim, State of Para (Figure 1), the exploration of bauxite ore generates calcined bauxite, which is used in the manufacture of refractory materials for the steel industries. During the exploration and beneficiating of the refractory bauxite ore some residues are originated. Among them, a ferruginous one, called calcined ferruginous bauxite is generated and has no use or market value. Thus, this waste is disposed in the environment without potential pollution concerns (Mineracao Santa Lucrecia, 2000).

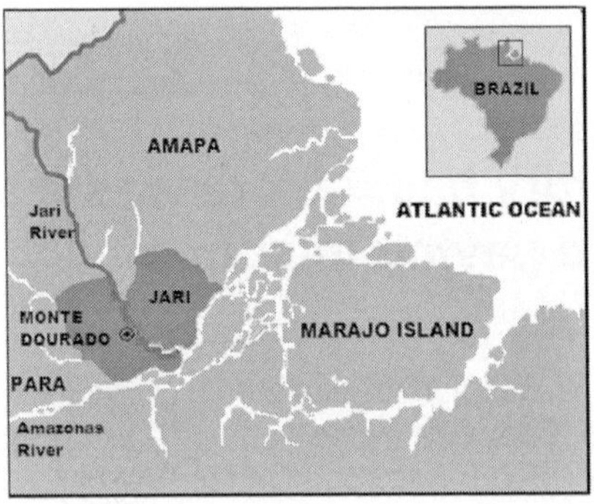

Figure 1: Bauxite ore exploration, north of Brazil

In the past, as an experimental test, one asphalt mixture with calcined ferruginous bauxite was used in the urban area of Almerin city. However the production of this mixture was not controlled in accordance to technical criteria of asphalt mixtures design.

Considering that the bauxite waste should have a friendly-environmental disposal and the lack of course aggregates in the region, this research evaluated the possibility of the use of this waste as aggregate in asphalt mixtures. The research considered two evaluation phases, the former focused a laboratorial study of mechanical and durability properties of an asphalt mixture, and the later an evaluation of the environmental risks

BAUXITE WASTE CHARACTERISTICS

The bauxite waste is composed basically of mulite, corindon, tielite and hematite and its specific gravity varies from 3.20 g/cm^3 to 3.30 g/cm^3 and the natural moisture content from 5% to 8%. Figure 2 shows the appearance of the bauxite waste.

Figure 2: Bauxite waste appearance

MATERIALS

Aggregate and Binder

The bauxite waste was evaluated by means of the specific gravity, Los Angeles abrasion and sieving analysis. The Los Angeles abrasion test result was 17.5% and the apparent specific gravity 3.202 g/cm^3. To compose a dense asphalt mixture, the bauxite waste was mixed with other materials (typical sand and a soil from Monte Dourado region) and the resultant gradation is presented in Table 1. The asphalt cement used in the research was a CAP 20, whose characteristics are summarized in Table 2.

Table 1: Gradation of asphalt mixture

Sieve (mm)	Gradation	Gradation C (Brazilian standard)
19.1		100
12.7		85 – 100
9.52	100	75 – 100
4.76	83	50 – 85
2.00	61	30 – 75
0.42	31	15 – 40
0.177	12	8 – 30
0.074	8	5 – 10

Table 2: Binder characteristics

Test		Results
	135°C	160
Saybolt-Furol	155°C	50
Viscosity (sSF)	175°C	26
Absolute Viscosity at 60°C (Poise)		2975
Penetration, 25°C, 100g, 5s (0.1 mm)		48
Softening Point (°C)		52
Flash Point (°C)		280
Specific Gravity		1.018

EXPERIMENTAL PROGRAM

Mechanical Characteristics of Asphalt Mixtures

A comprehensive laboratory program was performed on the asphalt mixture. First of all, Marshall mixture design was performed to establish the asphalt optimum content that was determined in compliance with the parameters of Marshall design (ABNT NBR 12891, 1993).

Indirect tensile tests were performed, in accordance with the DNER-ME 138/94, at 25°C, with a displacement speed equal to 0.8 ± 0.1 mm/s until specimen failure, which was characterized by splitting it into two parts in the vertical plane.

In agreement to standard test method DNER-ME 133/94, resilient modulus test was performed at 25°C, with a conditioning phase before the test, with 200 and 500 applications of dynamic load, respectively. Dynamic load is applied each second for 0.1s, resting for 0.9s, with stress level lower than 30% of the indirect tensile strength.

Moisture susceptibility tests were performed to evaluate if the mixtures with bauxite waste would show any problem related to adhesiveness and moisture damage, and if it would require the addition of an antistripping agent. In the ASTM D 4867 procedure, the asphalt mixture specimens are stabilized at compaction temperature for 2 hours and compacted up to 7% of air void or field void. A group is submitted to partial vacuum saturation and immersion in water at 60°C during 24 hours, then the temperature (25°C) is stabilized and ITS test is performed. The moisture susceptibility of asphalt mixture is represented by the ratio of tensile strengths, which is obtained from the relation between indirect tensile strength of the conditioning and the non-conditioning groups. The same procedure was assumed to resilient modulus test results.

Cantabro test was performed to evaluate the raveling potential of the asphalt mixture. The test method was applied in specimens in two conditions, dry and moisture. As NLT 325/86 states, the test consists in submit a specimen of asphalt mixture to 300 revolutions in Los Angeles abrasion equipment (without metal spheres) and after this is calculated the loss of mass. The test in moisture condition is similar to the dry one, however the specimen is kept on a thermal conditioning bath, which consists in the immersion of specimen in water at 60°C during 24 hours followed by immersion in water at 25°C during 24 hours. In the two test conditions, the asphalt mixture raveling potential is expressed as loss of mass due to abrasion. In most cases the moisture conditioning produces more damages to the asphalt mixtures, leading to higher loss of mass.

In order to evaluate if the presence of bauxite waste in asphalt mixture would contaminate the environment and cause health risks, the waste was classified by means of leaching, solubilization and metal determination tests.

Leaching test (ABNT NBR 10005, 2004) removes substances and compounds through washing or percolation of distilled water in a sample material. A water sample is collected for chemical analysis, which identifies the presence of substances with contaminant potential. In the leaching test, the sample is shaken in distilled water for 24 to 28h. The pH of water is monitored in specified periods to keep it equal to 5; sometimes, it is necessary to add acetic acid to control the pH solution. The resulting solution is filtered and submitted to chemical analysis.

In the solubilization test (ABNT NBR 10006, 2004), the sample is shaken in distilled water for 5 minutes, the pH is measured, and then the mixture rests for 7 days. After this period, the resulting solution is filtered and submitted to chemical analysis. Leaching test differs from solubilization test in testing time and pH control.

Metal determination test (ABNT NBR 10005, 2004; ABNT NBR 10006, 2004) allows the determination of metal content in materials. In this test, a sample is digested with nitro-perchloric acid; afterwards, the resulting solution is mixed to distilled water and then submitted to chemical analysis.

FINDINGS

Marshall design

The specimens were compacted in accordance with the Marshall method (ABNT NBR 12891, 1993) with binder content varying from 6.0 to 8.0%. Table 3 presents the average results of the Marshall design and the values of resilient modulus and tensile strength tests of the asphalt mixture with bauxite waste (BW).

If it had been considered all of the Marshall parameters, the asphalt content of mixture would be too high. Therefore, in this research, the optimum content was defined considering the bulk density and stability of the asphalt mixture. In order to clarify, if the Marshall air void interval (3 to 5 %) was considered, the mixture would exhibit an asphalt content higher than 8.0%, which appears to be too high and could be negative to the mixture mechanical behavior, as can be seem from the values of stability and the ITS presented in Table 3. Thus, it was used 7.0% as binder content in the evaluation of the mixture susceptibility to stripping and raveling.

Table 3: Results of Marshall design and mechanical properties of the asphalt mixture

Asphalt content	Bulk density	Air voids (%)	S (N)	F (0,1 mm)	RM (MPa)	ITS (MPa)
6.0	2.372	10.4	8692	20	5388	0.88
6.5	2.392	8.8	8823	20	5186	1.03
7.0	2.409	7.4	9328	23	4528	1.07
7.5	2.404	6.8	8736	25	3645	0.94
8.0	2.396	6.4	8343	28	3565	0.93

S-Stability; F-Fluency; RM-Resilient modulus; ITS-Indirect Tensile strength

Indirect Tensile strength

The indirect tensile strength was determined to all Marshall design asphalt contents and the maximum value matched to the optimum content assumed for mixture in this study, as it is showed in the Figure 3.

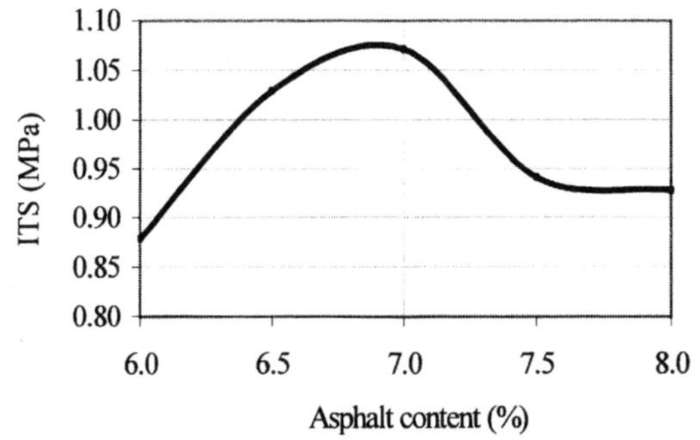

Figure 3: Asphalt mixture tensile strength *versus* asphalt content

Resilient Modulus

Resilient modulus variation *versus* of asphalt content of mixture are presented in Figure 4, which shows a RM's reduction with the increase of the asphalt content, indicating that the binder contributed with some flexibility gain of the asphalt mixture.

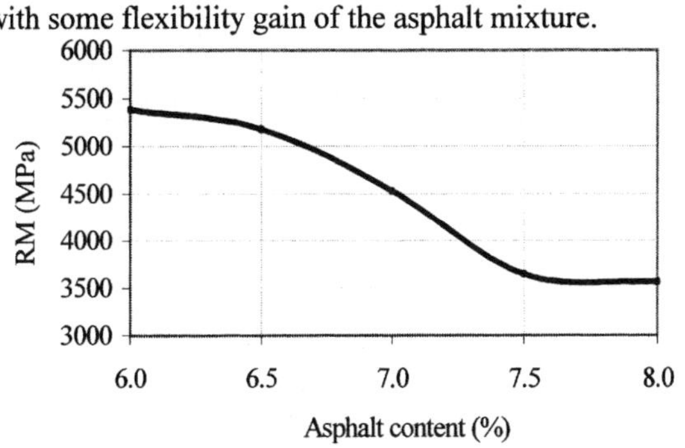

Figure 4: Asphalt mixture resilient modulus *versus* asphalt content

Moisture sensitivity (ASTM D 4867/84)

The acceptance criterion of ASTM D 4867/84 standard states that asphalt mixtures are susceptible to water if they present TSR (Tensile Strength Ratio) values lower than 70%. The same acceptance criterion was assumed to RMR (Resilient Modulus Ratio). Figure 5 shows the averages of TSR and RMR and the results indicated that the conditioning causes no significant damages, and that the asphalt mixture exhibited low level of sensitivity to moisture effects. This good behavior to water action is supposed to be due to the high air voids of the asphalt mixture. The average air voids was 8.9%. It is believed that asphalt mixtures with high air voids are less susceptible to water action due its high permeability, in the other words; there is not water retaining in the asphalt mixture air voids.

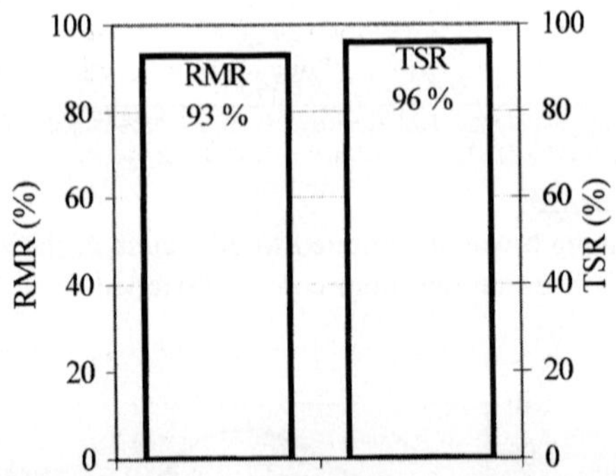

Figure 5: Tensile strength and resilient modulus average ratios

Table 4 shows a summary of the specimens' characteristics as the results of tensile strength and resilient modulus of control (G_1) and conditioning (G_2) groups of asphalt mixture. First of all, it can be noted that there are a significant difference between Marshall design (7.4%) and ASTM test specimens (8.9%) air voids. It was expected that the increasing in the air voids would produce some decrease in mechanical properties, and this was not verified.

Table 4: Characteristics and properties of the G_1 and G_2 groups

	Air Voids (%)	Sr (%)	RM (MPa)	RMR (%)	ITS (MPa)	TSR (%)
G_1	8.9	-	6975	93	1.30	96
G_2	8.9	68.3	6497		1.25	

Absorption

Different specimens production procedures (Marshall and ASTM) lead to different air void contents. Therefore, this difference was investigated by means of theoretical maximum and bulk specific gravities, apparent density and asphalt absorption according to Rice method (ASTM D 2041/95), as presented in table 5.

Analyzing the theoretical maximum specific gravity and absorption values, it is noted that the difficulty in obtaining the same air voids can be related to different asphalt absorption levels, because the ASTM procedure led to values ten times greater than the Marshall procedure.

Table 5: Characteristics of Marshall and ASTM asphalt mixtures in the optimum binder content

	Marshall	**ASTM**
Theorical maximum specific gravity	2.607	2.630
Aparent density	2.953	2.985
Bulk specific gravity	2.949	2.948
Absorption (%)	0.04	0.41

The ASTM method for the moisture damage evaluation states that the specimens' compaction should be preceded of temperature stabilization of loose asphalt mixture for 2 hours at compaction temperature. Asphalt in compaction temperature becomes less viscous and penetrates in aggregate porous; consequently, there is an increase in the air voids and a decrease in the asphalt coating thickness. This explains partially the resilient modulus and the tensile strength increasing. However, some effect of binder and/or mixture ageing does not have to be left behind, although not evaluated in this research.

Cantabro

The Cantabro test results are related to the cohesion of the asphalt mixture and its raveling strength. The characterization of the mixture with BW was performed in two groups of test (dry and wet) and the results are presented in Table 6. The results indicated that the loss of mass of G_1 and G_2 were of 12 % and 16 %, respectively. Therefore, in accordance with the acceptance criterion of the Spanish standard, which states that the maximum loss of mass must be lower than 25 %, the mixture with BW would be acceptable in both conditions.

Table 6: Cantabro loss of mass of the G_1 and G_2 groups

	G_1	G_2
Air voids (%)	9.4	9.4
Loss of mass (%)	12	16

Environmental Risks

The environmental risks evaluation was based on leaching, solubilization and metal determination tests. These tests were performed in a sample of bauxite waste. The results of chemical analysis classify the waste sample as inert or non-inert. Non-inert class is subdivided in other classes, depending on chemical element content, in accordance with Annex H - List n° 8 of ABNT NBR 10004 (2004). The results of all environmental tests (BIOAGRI Ambiental Ltda, 2005) indicated that the bauxite waste was classified as *Class II B Waste: Inert.*

CONCLUSIONS

From the results presented in this paper it can be concluded that the use of bauxite waste as asphalt mixture aggregate is feasible.

This can be stated because the asphalt mixture Marshall and mechanical properties are of the same range of the regular material ones. Considering durability properties (moisture sensitivity and Cantabro), the results also indicated a good behavior. However, the kind of aggregate led to high asphalt consumption, what could lead to rutting and exudation, not studied in present work.

So, the use of bauxite waste in asphalt mixtures can be a good alternative to north region of Brazil considering the lack of natural rocks resources and the low traffic level that the pavements would be exposed. Besides, this waste would have a friendly environmental disposal because it would be "encapsulated" within the asphalt mixture and not air exposed.

REFERENCES

ABNT NBR 10004. *Resíduos sólidos – Classificação.* Associação Brasileira de Normas Técnicas, Rio de Janeiro, 2004.

ABNT NBR 10005. *Procedimento para obtenção de extrato lixiviado de resíduos sólidos.* Associação Brasileira de Normas Técnicas, Rio de Janeiro, 2004.

ABNT NBR 10006. *Procedimento para obtenção de extrato solubilizado de resíduos sólidos.* Associação Brasileira de Normas Técnicas, Rio de Janeiro, 2004.

ABNT NBR 10007. *Amostragem de resíduos sólidos.* Associação Brasileira de Normas Técnicas, Rio de Janeiro, 2004.

ABNT NBR 12891. *Dosagem de Misturas Betuminosas pelo Método Marshall.* Associação Brasileira de Normas Técnicas, Rio de Janeiro, 1993.

ASTM D 2041. *Standard Test Method for Theoretical Maximum Specific Gravity and Density of Bituminous Paving Mixtures.* American Society for Testing and Materials, Philadelphia, 1995.

ASTM D 4867. *Standard Test Method for Effect of Moisture on Asphalt Concrete Paving Mixtures.* American Society for Testing and Materials, Philadelphia, 1984.

BIOAGRI AMBIENTAL Ltda, *Caracterização de resíduos segundo NBR 10004 – Resíduo de bauxita refratária. Boletim de análise nº 5004640.* Piracicaba, São Paulo, 2005.

NLT 325. *Proposta de Norma de Ensayo CANTABRO: Determinación de la perdida por desgaste de Mesclas el Empleo de La Maquina de Los Angeles.* Centro de Estudos de Carreteras. Madri, Espanha, 1986.

DNER ME 133. *Misturas Betuminosas – Determinação do Módulo de Resiliência.* Método de Ensaio, Departamento Nacional de Estradas de Rodagem, Rio de Janeiro, DNER, 1994.

DNER ME 138. *Misturas Betuminosas – Determinação da Resistência à Tração por Compressão Diametral,* Método de Ensaio. Departamento Nacional de Estradas de Rodagem, Rio de Janeiro, DNER, 1994.

HOSKING, J. R.. *Aggregates for skid em dash resistant roads.* Transport & Road Research Laboratory (Great Britain), TRRL Report, nº 693, 32p, 1976.

MINERAÇÃO SANTA LUCRÉCIA. *Histórico, informações de exploração, beneficiamento e produção de bauxita refratária.* Monte Dourado-Almerim-Pará, 2000.

GEOTECHNICAL PROPERTIES OF MINING WASTES AND MIXTURE OF ASH AND SLAG FOR CIVIL ENGINEERING

Piotr Filipowicz, Magdalena Borys, Department of Land Reclamation Engineering, Institute for Land Reclamation and Grassland Farming at Falenty, Mazowieckie province, 05 – 090 Raszyn, Poland

Abstract

Mining industry and Electric Power Stations produce a large number of waste materials. Utilization of the wastes in civil and harbour engineering allows to restrict building new waste dumps and expansion of the existing one. This way the existing waste dumps could be used for agricultural or forestry purposes.

The article presents the properties of mining wastes and the mix of ash and slag. Most important geotechnical parameters, which are necessary in engineering embankments designing, are discussed in detail. These parameters are: grain size distribution, water content and optimal moisture content, maximum dry density of solid particles and coefficient of permeability, cohesion and apparent angle of internal friction. These parameters will be characterised in relation to the different store time for mining wastes. Some of these parameters will be compared with parameters of ash and slag mixture.

Introduction

Generally, the main rocks of the mining wastes from Poland are claystone and siltstone. It could be found over 40% of those rocks in the mining wastes. The other rocks are sandstone and carbon rocks. They contribute in less than 33% to those rocks [Gazda et al., 1988; Bzowski et al., 2000; Skarżyńska, 1997]. In the mining wastes from Spain the main rocks are sandstone and claystone. The main rocks in the material from France are carbonaceous shale and sandstone. Thus we see that petrographical and mineral compositions of mining wastes are related to a type of rock, which accompany a coal deposit.

The main minerals found in mining wastes from Poland are silt minerals like illite, kaolinite. Next mineral is quartz and it occurs within the range from 15% to 37%. The rest of the minerals occur in less than 10% [Gazda et al., 1988; Bzowski et al., 2000; Skarżyńska, 1997].

The results of tests of basic chemicals and trace elements in tested mining wastes allow us to conclude that the material is not hazardous. It does not contain toxic elements. The results of monitoring of basic chemicals and trace elements prove the chemical composition of mining wastes does not change in time [Gazda et al., 1988; Bzowski et al., 2000].

Generally speaking properties of ash and slag mixture depends on type of coal (hard or brown), type of a furnace, technology of coal burning and a way of dropping an ash and a slag. The material we tested is a product of burning hard coal. Results of chemical tests of ash and slag mixture allow us to conclude that the main ingredients are aluminosilicate, iron and calcium. The material is not radioactive. Contain of radioactive elements and heavy metals in mixture of ash and slag do not exceed the chemical element limit in accordance with Polish standard.

This paper presents some geotechnical properties of mining wastes from a Coal Mine. It will discuss, in detail, some geotechnical properties for fresh wastes and those stored for 7 year. The article presents also the ash and slag mixture geotechnical properties from storage yards next to Electric Power Station. There were determined the main geotechnical properties, like physical, mechanical and permeability parameters.

Descriptions of geotechnical investigations

Grain size distribution of the mining wastes and the mixture of ash and slag were determined by sieving. The test was carried out for several samples of fresh and old mining wastes and fresh mixture of ash and slag. Obtained results of the test are presented in figure 1. The old mining wastes in comparison with the fresh material contain less gravel fraction. Fresh mining wastes contain over 70% of gravel fraction. The mining wastes stored for 7 year contain more sand and silt fractions than the fresh ones. The mixture of ash and slag contain less gravel fraction in comparison with the sand and silt fractions. The material contain up to 5% of gravel fraction and over 65% of sand fraction. Obtained results allow us to conclude that the grain size distribution of the mixture of ash and slag is similar to fine ground like fine sand and silt sand. The mixture of ash and slag and 7 year old mining wastes contribute up to 30% of silt with clay fractions. In comparison with the 7 year old mining wastes, the mixture contain much more sand fraction. Uniformity coefficient for the mixture of ash and slag was 10 and for the mining wastes was over 100.

Maximum dry density of solid particles and optimal moisture content were determined for the fresh and 7 year old mining wastes and the mixture of ash and slag. The tests were carried out with Proctor apparatus with II method in large cylinder for the mining wastes. The tests for the mixture was carried out with Proctor apparatus with I method in small cylinder [PN-88/B-04481]. The results of the tests are presented in figure 2.

Obtained results for the fresh mining wastes feature quite high maximum dry density of solid particles. The 7 year old material has lower maximum dry density of solid particles and higher optimal moisture content in comparison with the fresh material. Decreasing the maximum dry density of solid particles come from a disintegration of the cobble and gravel fractions of the mining wastes in time. Although these parameters undergo changes in time, the mining wastes have a good compactibility. The mixture of ash and slag has very low maximum dry density in comparison with 7 year old mining wastes. Obtained results indicated that new embankments built up from the mixtures of ash and slag would be lighter in comparison with the embankments built up from mineral soil. Optimal moisture content for the mixture of ash and slag is higher in comparison with the optimal moisture content for the mining wastes.

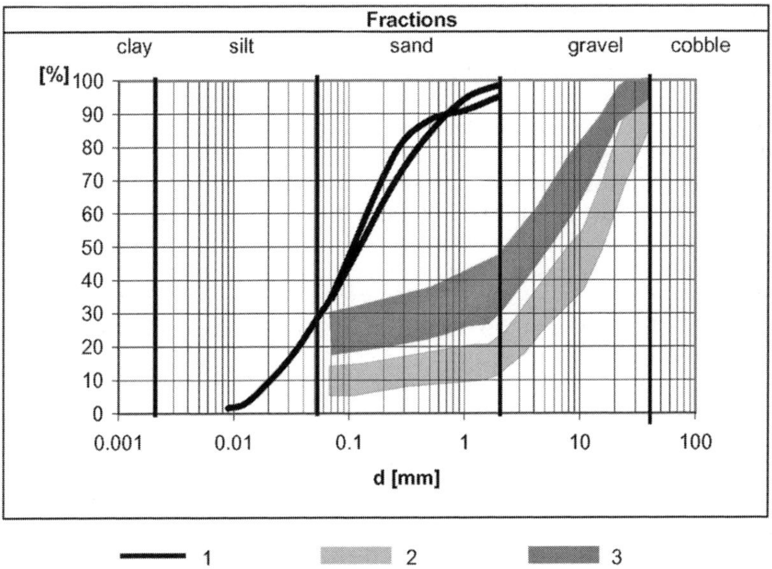

Figure 1. Range of grain size distribution of the tested material: 1 – the mixture of ash and slag, 2 – fresh mining wastes, 3 – 7 year old mining wastes.

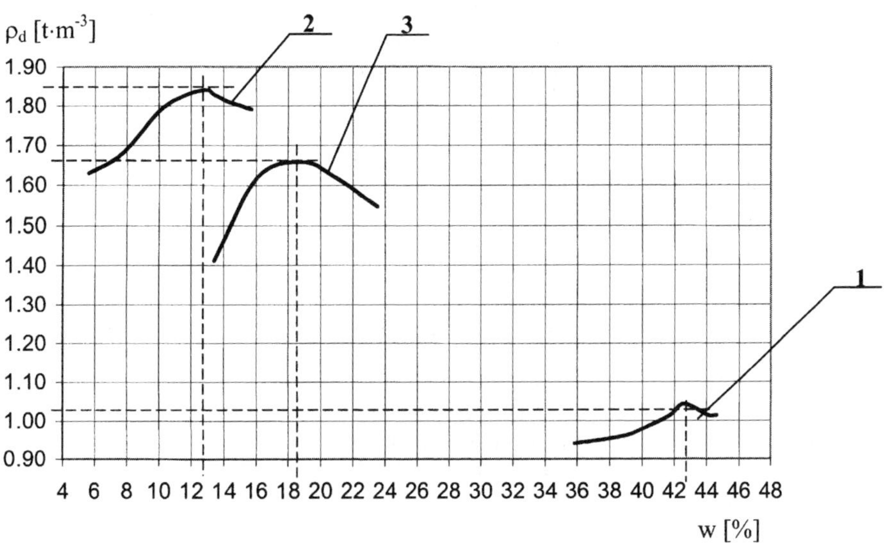

Figure 2. Curves of the compacted material: 1 – the mixture of ash and slag, 2 – fresh mining wastes, 3 – stored for 7 years mining wastes.

Coefficient of permeability was determined in a filtration column (foto 1). Each sample was compacted to a degree of compaction over 0.92. Obtained results of the tests for the mining wastes were compared with coefficient of permeability of the mixtures of ash and slag. The results are presented in table 1.

Obtained results show that the coefficient of permeability of the mining wastes changes in time. The old material has lower permeability in comparison with the fresh ones. For instance, permeability of the fresh material is similar to non-cohesive soil like sand. While, the mining wastes stored for 7 year has coefficient of permeability similar to cohesive soil like silt. Lower coefficient of permeability of the mining wastes in time come from disintegration of the cobble and gravel fractions. It was found, that the mining wastes are silting-up themselves in time [Pieczyrak J., 1982]. Range of changes of coefficient of permeability of the mixtures of ash and slag was similar to that obtained for the old mining wastes.

Table 1. Range of changes of the coefficient of permeability of the mining wastes and the mixture of ash and slag.

Type of the material	Coefficient of permeability [$m \cdot s^{-1}$]
Fresh mining wastes	10^{-4}
7 year old mining wastes	$10^{-6} - 10^{-7}$
The mixture of ash and slag	$10^{-6} - 10^{-7}$

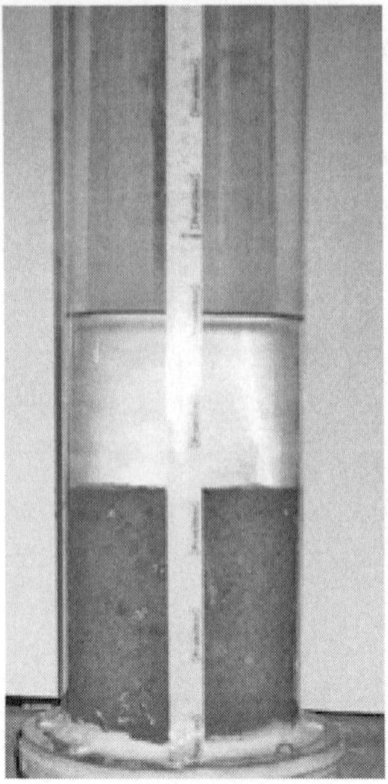

Foto. 1. Filtration column with a compacted sample of the tested material to a degree of compaction over 0,92

Shear strength of the mining wastes and the mixture of ash and slag were determined with direct shear apparatus. Obtained results of the tests are presented in table 2.

Table 2. Range of shear strength of the mining wastes and the mixture of ash and slag.

Type of the material	Cohesion [kPa]	Apparent angle of internal friction [°]
Fresh mining wastes	20 – 32	32 – 55
7 year old mining wastes	25 – 40	27 – 39
The mixture of ash and slag	18 – 44	28 – 32

To conclude, shear strength of the mining wastes undergoes changes in time. It was found that cohesion increased in time and was larger for the 7 years old mining wastes as compared with the fresh mining wastes. The apparent angle of internal friction decreased for the old mining wastes in comparison with the fresh mining wastes. Obtained high values of apparent angle of internal friction could come from high in gravel fraction in the samples, which were 12cm long and wide and 6cm high [Filipowicz P., Borys M. 2004a]. But the main aim of presented results was to show change of mechanical parameters of the mining waste in time. Also we want to compare obtained results for the mining wastes with the results for the mixture of ash and slag.

Obtained results for the mixture of ash and slag show that the material has quite high cohesion. Range of change of cohesion for the material was 18–44kPa. Obtained apparent angle of internal friction for the mixture of ash and slag was 28-32°. It was found also that obtained results for the material was similar to apparent angle of internal friction for fine soil like fine sand.

Settlement of the mixture of ash and slag was determined in consolidometer. Obtained results show that consolidation of the material was very fast. Settlement stabilization for the samples of the mixture of ash and slag last less than 24 hours after loading the samples. Range of change of compression modulus was 3-37MPa for range of load from 12,5kPa to 100kPa. Obtained results were similar for cohesive soil in plastic state.

For the mixture of ash and slag was determined collapsing coefficient. The collapsing coefficient was determined for the material after settlement stabilization for 50kPa and 100kPa stresses. Obtained results for the mixtures of ash and slag were within the range 0,0002-0,0003 for both stresses. Obtained collapsing coefficients were less from allowable coefficient (0,02). Generally speaking, the mixture of ash and slag could be qualified to unsubsidence soil, stable structure and non sensitive to water.

Settlement of the samples of the mixtures of ash and slag was carrier out also under the influence of water. Prepared samples were loaded with two different stresses and after settlement stabilization the samples were watered. Obtained results are presented in figure 3. Presented results show that settlement increase a little if the sample is watered. Increment of the settlement is not high for the mixture of ash and slag. For instance if the build in mixture of ash and slag would has a moisture content approach optimal moisture content, settlement would be a few millimetres only after watering the mixture. For the material of moisture content less than optimal moisture content, the settlement would be a few centimetres.

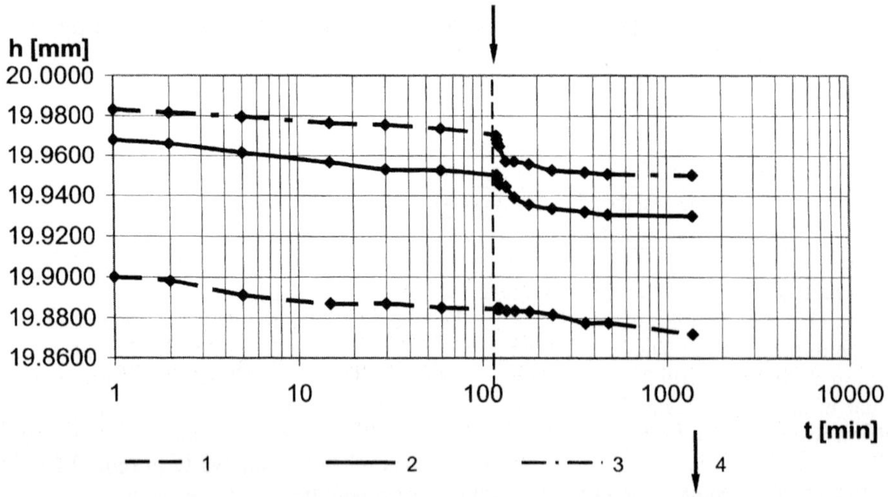

Figure 3. Obtained results for the mixture of ash and slag in consolidometer test with watering: 1 – for 12,5kPa stress, 2 – for 25kPa stress, 3 – for 50kPa stress, 4 – moment of watering.

Summary

Presented results of the basic chemical elements of the mixture of ash and slag and the mining wastes lead us to conclude that the material is not radioactive and contain neither toxic chemical elements nor toxic substances. Contain of radioactive elements and heavy metals in mixture of ash and slag do not exceed the chemical element limit in accordance with Polish standard.

The mining wastes and the mixture of ash and slag conform to the requirements of building materials that are used for hydrotechnical embankments. Obtained results indicate that their grain size distribution is similar to clay-gravel mixture for the mining wastes and to fine soil for the mixture of ash and slag. Uniformity coefficient for both tested materials was over 5. This is very positive feature and it is possible to utilize this material for hydrotechnical structures.

Both materials the mixtures of ash and slag and the mining wastes show good physical properties as a building material. The mining wastes show a good compactness and good mechanical properties because of a range of water contents which allow to easily compact the material to a degree of compaction Is $\geq$ 0.92. Although the apparent angle of internal friction for the mining wastes decreases in time, the material features good mechanical properties. The mixtureof ash and slag has a high cohesion and apparent angle of internal friction in comparison with the mining wastes and a fine soil. The material is not sensitive to water.

Obtained maximum dry density for the mixture of ash and slag were very low in comparison with the mining wastes. Findings lead to a conclusion that new embankments built up from the mixtures of ash and slag would be lighter in comparison with the embankments built up from mineral soil. It is very positive because the embankments built up from the mixture of ash and slag could be built on soft soil. In this case we have to remember to check the displacement of the embankment under the influence of water pressure.

Geotechnical parameters like grain size distribution, maximum dry density of solid particles, optimal moisture content, shear strength and coefficient of permeability could differ if the material is taken from other mining waste dumps and storage yards.

The mining wastes are not resistant to weather. Geotechnical properties are better in older mining wastes than in the fresh ones. The old mining wastes are low permeable. To conclude, this material could be used instead of building grounds. It is very important to observe regulations, if we construct or modernise hydrotechnical structures with the waste material like the mining wastes or the mixture of ash and slag [Filipowicz P., Borys M., 2004b].

Obtained results of the mixtures of ash and slag in consolidometer show that the settlement is not high and the material could be qualified to unsubsidence soil, stable structure and non sensitive to water.

References

1. Bzowski Z., Dawidowsk A., Zarębski K., Bernat A., Gwoździewicz M. 2000. Badania własności fizyko-chemicznych skał karbońskich lokowanych na składowisku w Bogdance i ocena ich przydatności do rekultywacji. POMIAR-GIG Lublin (typescript).
2. Filipowicz P., Borys M. 2004a. Geotechnical parameters of mine wastes from The Lubelskie Coal Basin as a material for hydrotechnical embankments, Journal of Water and Land Development. No 8, 163-170.
3. Filipowicz P., Borys M. 2004b. Possible utilization of mining wastes from the Lubelskie Coal Basin for hydrotechnical structures, Polish Journal of Environmental Studies, vol. 3, supplement III, 146-148.
4. Gazda L., Oleszczyński B., Pollo I., 1988. Charakterystyka mineralno-chemiczna oraz analiza możliwości wykorzystania przeróbczych odpadów przywęglowych w kopalni w Bogdance. Przegląd Górniczy (11-12), 16-18.
5. Pieczyrak J., 1982, Właściwości fizyczne oraz skład chemiczny i mineralny przepalonych odpadów kopalnianych, Ochrona Terenów Górniczych, 62, 30-38.
6. PN-88/B-04481, Building soils. Laboratory tests.
7. Skarżyńska K.M. 1997. Odpady powęglowe i ich zastosowanie w inżynierii lądowej i wodnej. AR Kraków Publishing House.

Mechanical and Leaching Properties of the Air-Cement Treated Soil Using Incineration Residue from MSW

Takuro Fujikawa, Technician, Department of Civil Engineering, Fukuoka University, Fukuoka, Japan
Kenichi Sato, Associate professor, Department of Civil Engineering, Fukuoka University, Fukuoka, Japan
Shoutarou Yamada, Research associate, Department of Civil Engineering, Fukuoka University, Fukuoka, Japan

ABSTRACT

In Japan, the remaining life of municipal waste landfill sites are decreasing and, to minimize the amount of waste going to landfill, much of it is first incinerated. Main component of land filled solid wastes are incineration residue such as bottom ash and fly ash. Recently, the new type incinerator factories for molten slag were built, which is melted a bottom ash and fly ash. On the other hand, the natural aggregates for construction are exhausted. Therefore, the solid wastes should be reused and recycled in order to prolong the life of landfill site and to prevent the environment.
This study carried out to investigate the mechanical characteristics of the air-cement treated soil using the incineration residue (bottom ash and molten slag) and surplus soil. Additionally, it is important to investigate the safety of these improved materials. Therefore, it carried out to investigate the leaching properties of this material for the influence evaluation of ground environment.

1. INTRODUCTION

Shortage of the natural aggregates used for embankment and backfilling materials is a big social problem. Therefore, the positive effective use of various waste materials is important due to the resources circulation.

Air-cement treated soil is a lightweight geomaterials. This is effective utilization methods of the surplus soils (Hayashi et al., 1998, Hayashi et al., 2000). Since the wet unit weight and strength set-up are possible for this material by each construction. This material can be reduced the load on the bearing stratum, and the pressure on the earth. The study of effective utilization of air-cement treated soil using industrial waste is performed centering on the coal fly ashes from thermal power plant (Horiuchi et al., 1991, Yasuhara et al., 2000). This study carried out the investigation of the mechanical characteristics of air-cement treated soil using incineration residue and surplus soil.

2. TESTING PROCEDURE

2.1 Incineration residue

The incineration residue samples used in this study were incineration bottom ash (IBA) and molten slag (Slag) for municipal solid waste. This study used two types IBA (IBA (A), IBA (B)). The particle sizes of these samples were 2mm or less. The photographs of the scanning type electron microscope of IBA and Slag are shown in Figure 1 (a) and (b). These photographs indicate that the particle surface of IBA is rough and porous. On the other hand, the particle surface of Slag looks like a glass. This study carried out to investigate the grain size effect for Slag sample. The grind powder Slag is shown in Figure 2. Grind Slag surface is very rough. Figure 3 shows the particle-size curves for IBA and Slag. The particle size of each sample is as same as sand. Table 1 shows that the physics properties and mechanical characteristic of IBA and Slag.

Figure 1(a): SEM of IBA

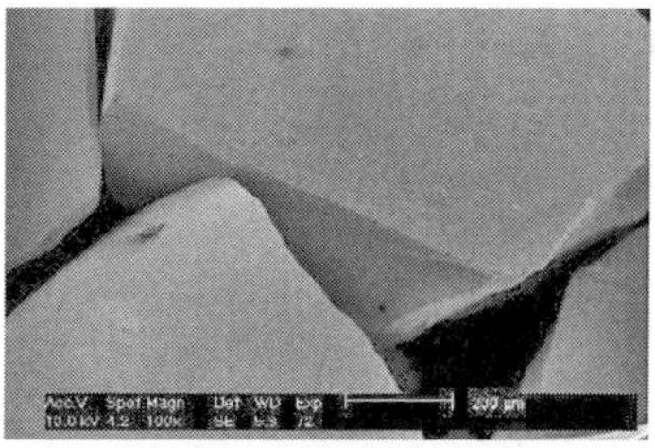

Figure 1(b): SEM of Slag

Figure 2: SEM of Grind Slag

2.2 Sample preparation

The air foam was prepared from kinds of synthetic surface active agents foaming material with a density of 0.05g/cm^3. The cement for the experiment is the blast-furnace Slag cement B (JISR5211). The air-cement treated mixing soil samples were preparing for the condition of flow value of F=180±20mm. The specimens were prepared by pouring the slurry into the chloride moulds (50mm in diameter and 100mm in height) and were then cured for several days in a thermostatic room (20±3℃). The specimens were remolds from the moulds and trimmed for each experiment. They were wrapped in polyethylene film to protect them from drying and were then cured in the same room. The unconfined compression tests were carried out to investigate the effect of the mixture and curing time of specimens.

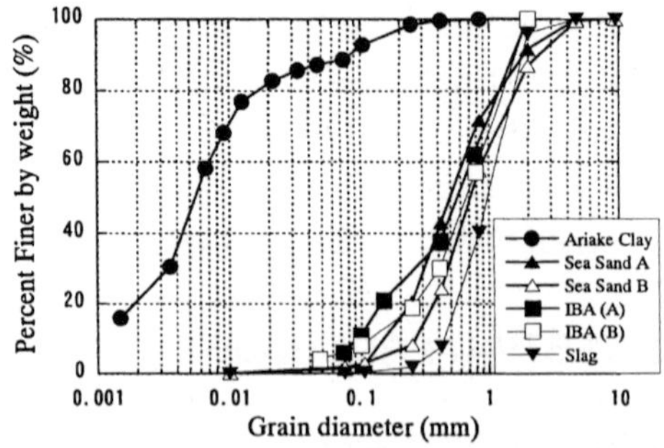

Figure 3: Particle-size curves for IBA and Slag

Table 1: Physical property of Soil, IBA and Slag

Sample name	Ariake Clay	Sand A	Sand B	IBA (A)	IBA (B)	Slag	Grind slag
ρ_s(g/cm^3)	2.55	2.65	2.66	2.47	2.14	2.73	2.73
w(%)	119.4	2.70	0.2	12.80	13.64	0	0
D_{10}(mm)	—	0.15	0.27	0.15	0.15	0.45	-
D50(mm)	0.0055	0.50	0.73	0.50	0.75	1	-
U_c	—	4.13	3.33	4.13	6.27	2.67	-
U_c'	—	0.97	0.91	0.97	1.13	0.91	-

Table 2: Condition 1 IBA (A)

Sample name	S/C	Rr (%)	Cement (kg/m^3)	Sample (kg/m^3)	IBA (A) (kg/m^3)	Water (kg/m^3)	Air (L/m^3)
Ariake Clay	1.0	0	168	369	0	453	224.9
		50	237	261	134	353	317.3
		75	276	151	233	320	370.6
		100	343	0	387	248	456.6
Sand A	1.0	0	365	375	0	235	497.7
		50	352	181	199	243	475.9
		75	346	89	294	249	464.1
		100	343	0	387	248	456.6

Each sample weight is wet condition.

2.3 Testing Condition

Table 2 and 3 show the mixing condition of the air-treated mixing soil samples for IBA (A) and IBA (B). Two type tests carried out to investigate the effect of main soil material (Sand and Clay) and cement volume of sample. Mixing condition 1 for IBA (A) was assumed to be the soil-cement (S/C) ratio 1.0 and there were four kinds of mixing rates of IBA (0, 50, 75, 100%). Curing days were 6 kinds of day (7, 28, 91, 365, 730,

Table 3: Condition 2 IBA (B)

Sample name	S/C	Rr (%)	Cement (kg/m^3)	Sample (kg/m^3)	IBA (B) (kg/m^3)	Water (kg/m^3)	Air (L/m^3)
Ariake Clay	1.0	0	205	375	0	408	274.4
		75	286	132	245	320	368.1
		100	323	0	367	289	410.2
	2.0	0	118	432	0	441	231.8
		75	183	168	314	322	338.9
		100	220	0	500	261	401.3
	3.0	0	82	450	0	455	217.6
		75	135	185	344	321	328.5
		100	160	0	545	276	382.2
Sand B	1.0	0	375	376	0	224	511
		75	338	85	290	266	437.0
		100	323	0	367	289	410.2
	2.0	0	267	535	0	170	540.7
		75	229	115	392	246	427.5
		100	220	0	500	261	401.3
	3.0	0	196	589	0	187	526.6
		75	168	126	430	255	413.9
		100	160	0	545	276	382.2

Each sample weight is wet condition.

1095 days). Mixing condition 2 for IBA (B) was assumed to be three kinds of soil-cement (S/C) ratio (1.0, 2.0, 3.0) and there were four kinds of mixing rates of bottom ash (0, 50, 75, 100%). Curing day was 28 days.

Table 4: Mixing condition for Slag

Soil Name	Type	Rr (%)	Cement (kg/m^3)	Soil (kg/m^3)	Slag (kg/m^3)	Water (kg/m^3)	Air (l/m^3)
Ariake Clay	Slag	0	168	369	0	453	225
		50	234	257	117	377	317
		75	290	160	219	313	396
		100	380	0	380	213	523
	Grind Slag	75	269	147	201	362	370
Sea Sand	Slag	0	365	375	0	235	498
		50	373	192	187	223	511
		75	376	97	282	219	516
		100	380	0	380	213	523

Each sample weight is wet condition.

These tests purpose that it thought about cost reduction measures according to a decrease in cement by changing S/C. Table 4 shows the mixing condition for Slag. Two type tests carried out to investigate the effect of main soil material (sand and clay) and the grain size of Slag. Mixing condition was assumed to be two kinds of materials (Ariake Clay and Sea Sand), four kinds of mixing rates (0, 50, 75, 100%), three kinds of curing day (7, 28, 91days) and two different kinds of grain diameter. This study performed an unconfined compression tests in order to find the strength and deformation characteristics of each testing conditions.

3 RESULTS AND DISCUSSIONS

3.1 Mechanical characteristics for IBA samples

3.1.1 Effect of IBA mixing ratio and curing term

Figure 4 shows the results of unconfined compression test of each mixing rate in air-cement treated soil (28days curing) using Ariake Clay for mixing condition 1. Unconfined compressive strength q_u increases with increasing of mixing rate (Rr) of IBA. These data indicate that the strength of improved soil is influenced by IBA mixed volume.

Figure 5 shows the long-term durability of air-cement treated soils using IBA for mixing condition 1. Unconfined strength of each improved soils are gradually steady after 91 days. These data are indicated that the durability of improved soil using IBA has sufficient strength for JH (Japan Highway Corporation) standard. (JH standard is general design strength of air-cement treated soil provided in Japan.)

Figure 6 shows the results of unconfined compression test of improved soils with Sand and Clay, which used IBA(B), for Rr=75% and curing time of 28days. The peak strength for Sand samples is larger than Clay samples. It is thought that Sand was used because W/C (water-cement ratio) has become small. This data appears the effect of different Soil materials.

3.1.2 Effect of cement contents and S/C (Soil - Cement ratio)

Figure 7 shows the relation between cement content (C) and q_u (28days curing). These data

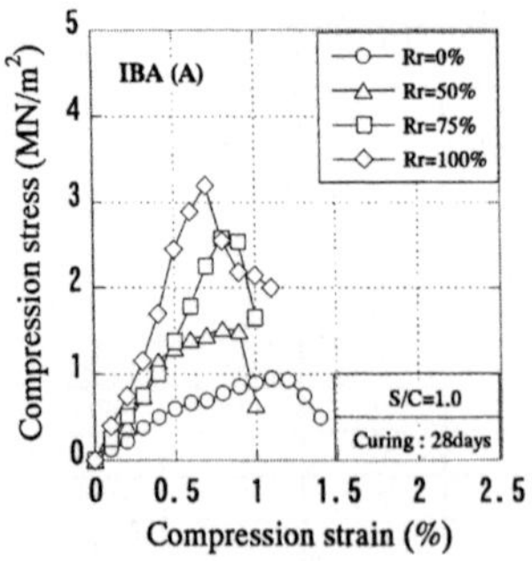

Figure 4: Results of unconfined compression

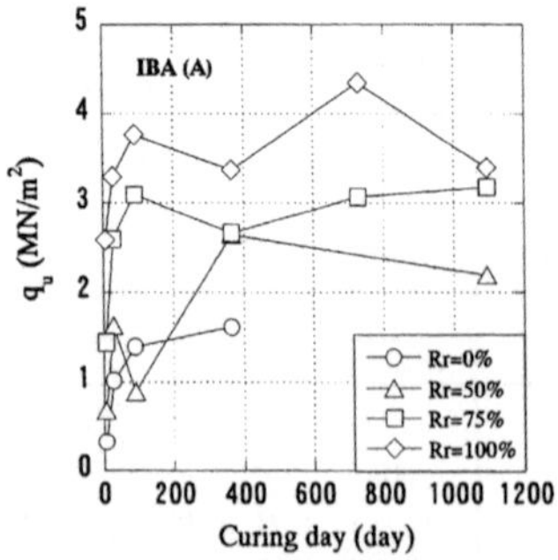

Figure 5: Long term curing strengths for IBA (A)

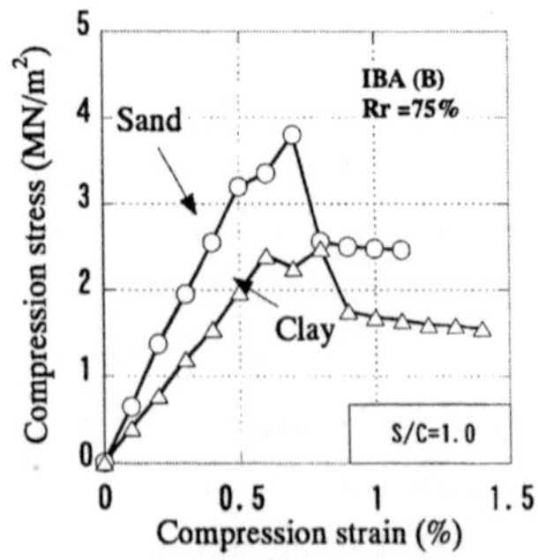

Figure 6: Influence of soil types

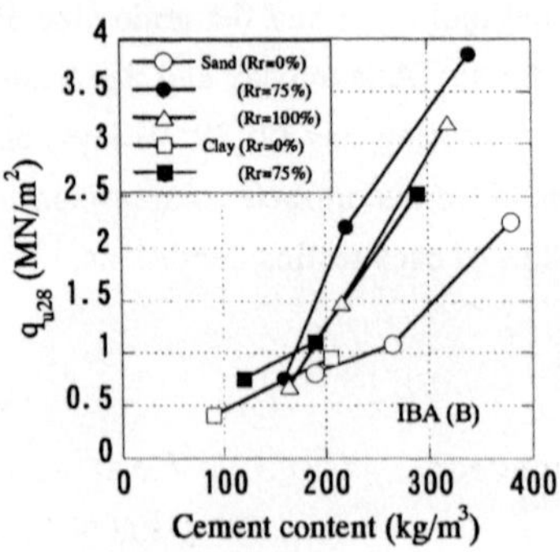

Figure 7: Relation between cement content and q_{u28}

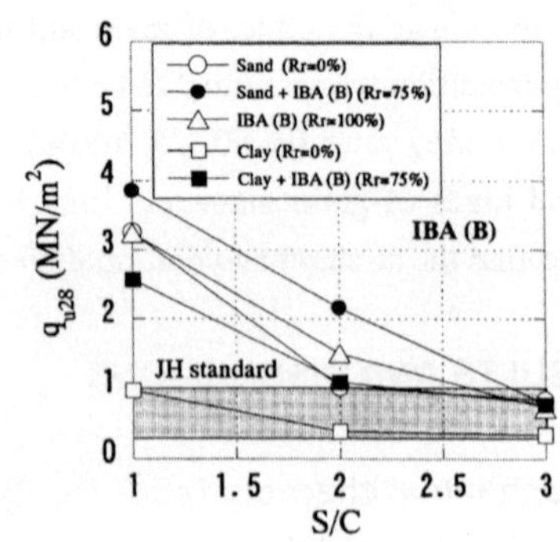

Figure 8: Relation between S/C and q_{u28}

indicate that the strength of improved soil increase with increasing Rr as same cement content. IBA can be increased the strength of improved soil. This data shows the possibility effective reuse of IBA. Figure 8 shows the relation between soil-cement ratio (S/C) and q_u. The improved soil strength q_u decreases with increasing S/C.

However, the strengths of improved soil for each testing condition are satisfied with JH standard. These results indicate that it reduces the amount of cement for improved soil. Thus, these results indicate that it can be reduce the manufacturing cost.

3.2 Mechanical characteristics for Slag samples

3.2.1 Effect of molten Slag mixing ratio and curing term

Figure 9 shows unconfined compression test results of each mixing rate in air-cement treated soil (with curing time of 28days) that used Ariake Clay. These results indicate that it can be

q_u increased with increasing the mixing rate of Slag. The failure strain of each tests strain decreased with increas-ing Slag mixing rate increased. It indicates that the rigid of improved soil become high.

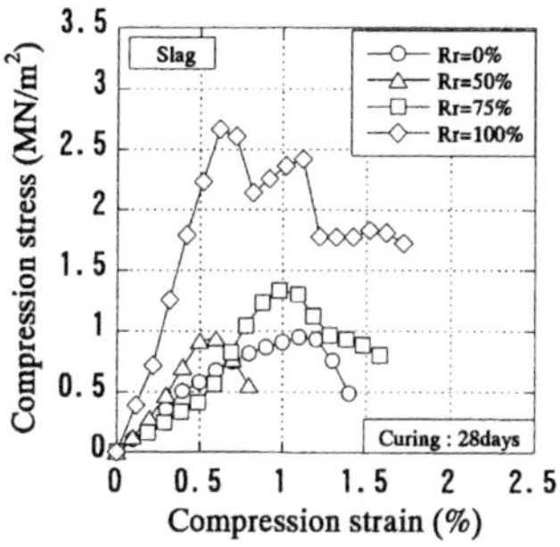

Figure 9: Results of unconfined compression

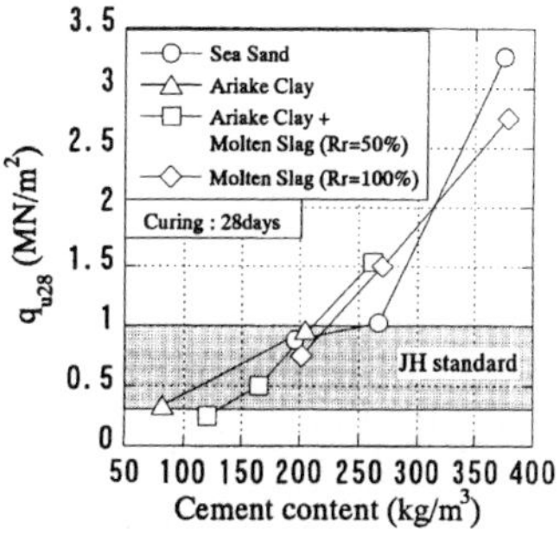

Figure 10: Relation between cement content and q_{u28}

3.2.2 Effect of cement content and S/C (soil-cement ratio)

Figure 10 shows the relation between cement content and q_u. The strength q_u increases with increasing cement content. Figure 11 shows the relation between soil-cement ratio (S/C) and q_u. It indicates that q_u decreases with increasing of S/C as well as IBA. This, a large amount of use for Slag can be hoped for in S/C=3.0 because JH standard is filled.

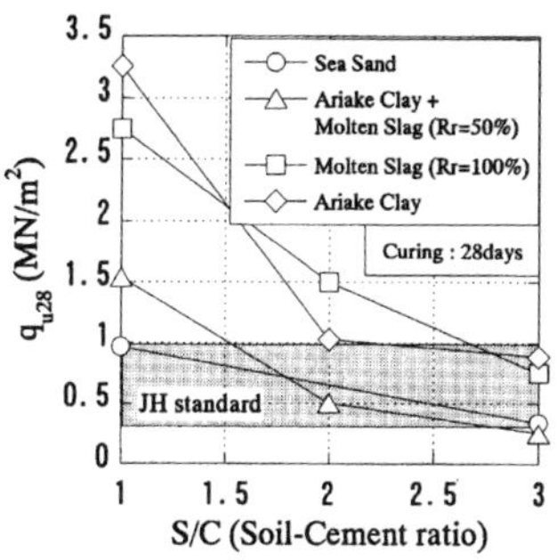

Figure11: Relation between S/C and q_{u28}

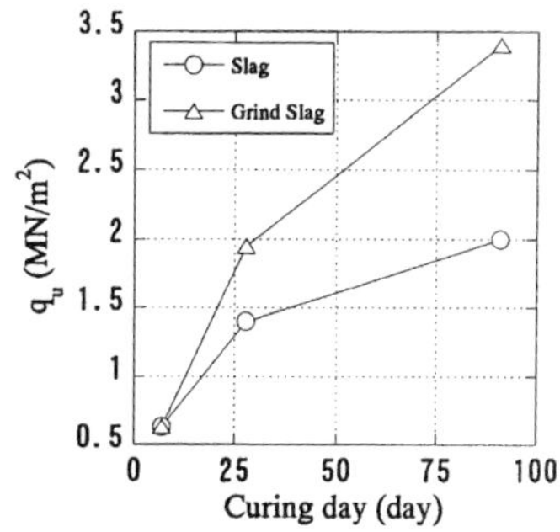

Figure 12: Relation between curing day and q_{u28}

3.2.3 Effect of grind size of Slag

Figure 12 shows the result of unconfined compression tests results for effect of particle size. With a curing time of 7 days, the differences in the strength due to particle size could not be seen. However with a curing time of 28 days or 91 days, the strength of Grind Slag is higher than Slag. The difference of strength was increasing with the increasing of each curing day. It is thought that this cause is dispersibility improves by grinding Slag and it can be made uniform specimen. Moreover, it is thought that the adhesion with cement has increased by crushing Slag. Figure 13 shows long-term durability of air-cement treated soils using Slag. The strength of each sample is gradually steady with increasing curing day.

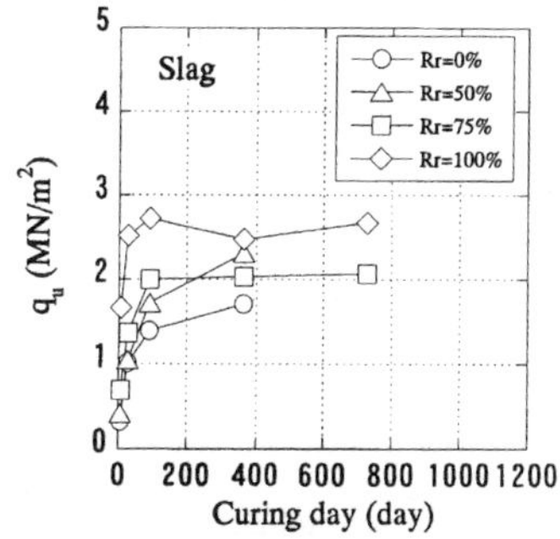

Figure13: Long term curing (Molten Slag)

4. LEACHING CHARACTERISTICS

In this study, the leaching (JLT 46 and Tank Leaching test (TL test) tests are carried out to investigate the geo-environment effect due to utilization of improved soil. The leaching sample is 28 days curing using surplus soil mixing bottom ash. A mixing ratio is 100%. In TL test, a sample was soaked in the water for 28 days. After soaking, a concentration was investigated by the leachate. In these tests, the measurement of concentration was 6 heavy metals (Zn, Cd, Fe, T-Cr, Cu, Pb). As a result, Cu and Pb leached from the bottom ash. However, the concentration of Cu was very low as compared with Japan geo-environment standard (12.5mg/l). Table 5 shows Pb leached from each sample. A concentration of Pb from bottom ash is over value as compared with Japan standard. On the other hand, the leaching properties of air-cement treated mixing soils is different depend on the testing method. In the case of JLT 46, the sample used the under 2mm. In the TL case, the sample of leaching test is just solid. The pH of JLT 46 is higher than TL. Therefore, when the mass of improved soil is in the ground, the possibility of leaching from improved soil using incineration residue is very low. TL test is very useful testing method for the real field condition.

Table 5: Concentration of samples (mg/l)

	Pb	pH
Blast-furnace slag cement B (JLT46)	N.D	11.92
Incineration bottom ash (JLT46)	0.02	9.89
Improved soils (JLT46)	0.05	11.92
Improved soil (TL)	N.D	8.30
Japan standard	0.01	-
Min.Limit of determination Value	0.01	–

5. CONCLUSION

(1) Air-cement treated soil that mixes IBA or molten Slag is satisfactory enough to be an engineering works material, on strength and the rigidity side.

(2) The strength of improved soil with poor mixing of cement is enough of the design standard. Therefore, it is able to reduce the cost of construction.

(3) The use of Grind Slag is effective for improved soil.

(4) When the mass of air-cement treated mixing soils is in the ground, the possibility of leaching from improved soil using incineration residue is very low. Tank leaching test is very useful testing method for the real field condition.

6. REFERENCES

Hayashi, Y., Suzuki, A. and Kitazono, Y. (1998)"Effect of soil properties on the improvement with form and cement mix", Proc. of the third International Conference Environmental Geotechics, pp.637-642.

Hayashi, Y. and Suzuki, A. (2000) "Influence of the void ratio in soil treated with air foam and cement", Costal Geotechnical Engineering in Practice, pp.453-458, 2000.

Horiuchi, S., Tamaoki, K. and Yasuhara, K. (1991) "Coal ash slurry for effective underwater disposal, Soils and Foundations, JGS, Vol. 35, No1, pp.1-19.

Yasuhara, K., Murakami, S., Kanazawa, T., Iikubo, T., Yoshino, H. and Horiuchi, S., "Microscopic aspect of a light-weight geomaterial using fly ashes", Costal Geotechnical Engineering in Practice, pp.771-776, 2000.

Seismic Response of Landfills

Behrouz Gatmiri, University of Tehran, Iran and Ecole National des Ponts et Chaussees Paris, France

Sara Navidi, University of Tehran, Iran

Abstract

Despite the lack of well documented case histories of seismic response of landfills, some researchers have derived solid waste properties such as unit weight and initial shear wave velocity gradients by back-calculating the seismic response of some existent landfills. In this paper the results of the seismic response analyses using a finite element computer program, ADINA, are used to examine the response of solid waste landfills to seismic loading. Thus, a synthetic landfill which its geometry and dynamic properties of its material are selected based on existent information in literature is analyzed and the effects of three factors on its seismic response are studied. These factors include: 1- The peak acceleration of the input motion. 2- The width of the landfill. 3- The shear strength parameters of solid waste. The results of the analyses demonstrate that increasing the peak acceleration of the input motion, increases the values of calculated acceleration spectra in different points of the landfill. Moreover seismic response of landfills can be affected by the width of the landfill. The values of acceleration spectra decrease when the width of the landfill increases. On the other hand changing the shear strength parameters of solid waste (c,φ) don't influence seismic response of landfills.

Introduction

Landfills are geometrically similar to earth embankments, but their materials are entirely different. Thus stability analysis and studying their seismic response are necessary and considered relatively new subject in geotechnical engineering.

One of the major limitations associated with analyzing landfills is uncertainty with respect to dynamic properties of solid waste. The operating industries, Inc. (OII) landfill is located 16 Km east of Los Angeles within the city of Monterey Park, California. Shortly after the 1987 Whittier-Narrows earthquake, two strong motion recording stations were placed at this landfill. One station was placed at the toe of the landfill and the other at the top of the landfill. Since their installation, the two instruments have recorded ground motions during 40 earthquakes. Strong motion records obtained during recent earthquakes provide for the first time a rational basis for back calculation of the dynamic properties of solid waste for seismic response analyses. By considering profiles of the unit weight and initial shear wave velocity of solid waste developed from field observations and measurements (Kavazanjian et al 1994) with observations of seismic response at OII, a calibrated set of material properties for seismic site response analyses of solid waste has been developed.

5th ICEG Environmental Geotechnics, Thomas Telford, London 2006.

Analyses employing these calibrated solid waste properties are used herein to examine seismic response of solid waste landfills and to study the effects of three factors on them. These factors include: 1-The peak acceleration of the input motion. 2- The width of the landfill.3-The shear strength parameters of solid waste.

The Geometry of the Synthetic Landfill and Its Material Properties

The first step in analyzing any structure using computer program ADINA is defining its geometry. Since the length of a landfill is extremely larger than its width, a plain strain 2D model is appropriate for its analysis. The dimensions of this landfill are selected based on those of existent landfills. The geometry of the landfill has shown in figure 1.

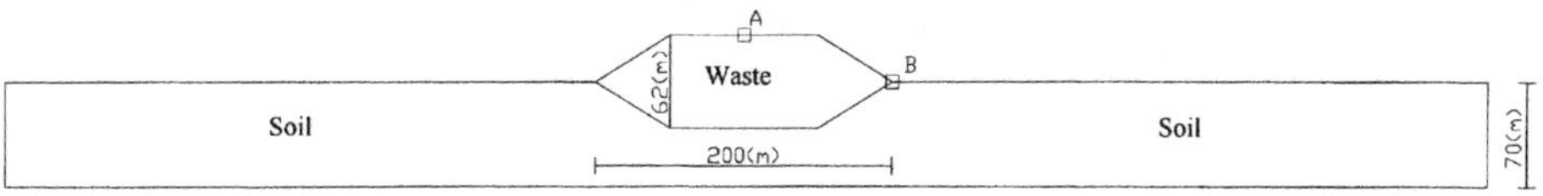

FIG.1. The Geometry of the Landfill

As mentioned some researchers have derived solid waste properties that are accessible in literature. Although obtained properties by various researchers are different, it is possible to consider a range of variation for these properties. In table 1 the solid waste properties used in this paper are summarized.

Table1-Used solid waste properties in this paper

Unit weight (KN/m^3)	8-12.75*
Elastic Modulus (MPa)	6-15*
Poisson's Ratio	0.33
Cohesion (KPa)	30
Angle of Internal Friction (°)	20

*Unit weight and Elastic modulus of solid waste are functions of the height of the landfill. The lower values indicate these parameters at the top of the landfill and the upper values indicate them at the base of the landfill.

2D Seismic Analysis

The landfill with mentioned properties is analyzed under 1990 ABBAR, Iran, earthquake which is scaled to peak acceleration of 0.35g.

Figure 2 shows the acceleration spectrum of the input motion and the calculated acceleration spectra of two points of the landfill. One of these points is located at the top of the landfill (point A) and the other at the toe of the landfill (point B).

As figure 2 shows the input motion is amplified by the landfill. Comparison of peak of these spectra demonstrates the amount of amplification. As observed in Figure 2 the peak of the input motion spectrum is 1.42g and the peak of points A and B acceleration spectra are 3.30 and 1.90. Thus the amplification value at point A is more than that of at point B.

Moreover comparison of the input motion spectrum and the spectrum of point B shows that the frequency content of the rock motion has changed.

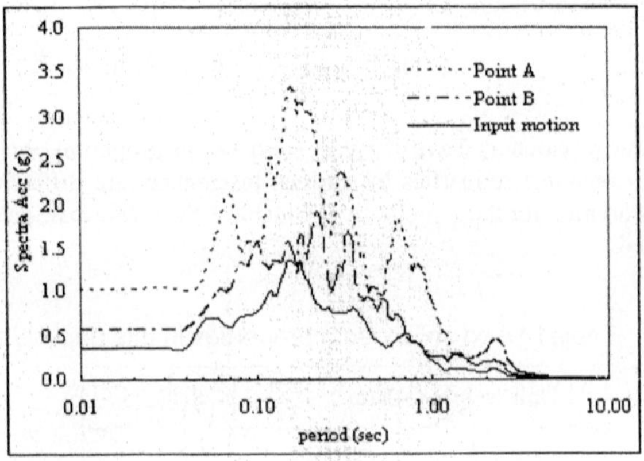

FIG.2.Calculated Response at points A and B of the landfill

Factors Affected the Seismic Response of Landfills

The Peak Acceleration of the Input Motion

One of the major characteristics of an input accelerogram is its peak acceleration. In this paper to investigate the effects of peak acceleration of the input motion on seismic response of landfills, the Abbar earthquake accelerogram is scaled to peak acceleration of 0.1g, 0.2g, 0.3g and 0.4g.

Figures 3 and 4 demonstrate calculated acceleration spectra at points A and B of the landfill in different analyses using various scaled accelerograms. Figures 5 and 6 also show the ratio of these spectra to the related acceleration spectrum of the input motion.

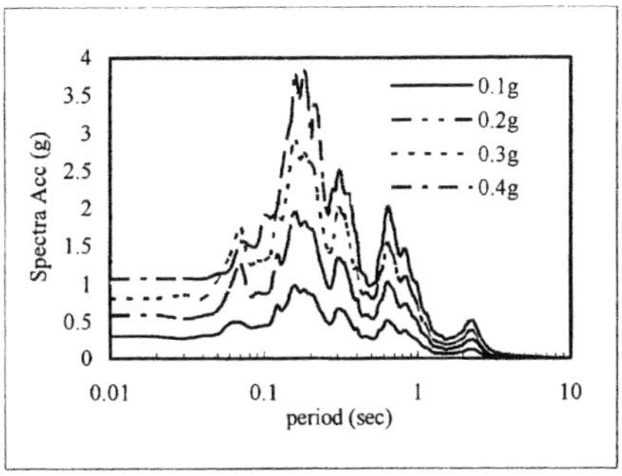

FIG.3. Calculated Response at point A of the landfill in different analyses with various peak acceleration of the input motion

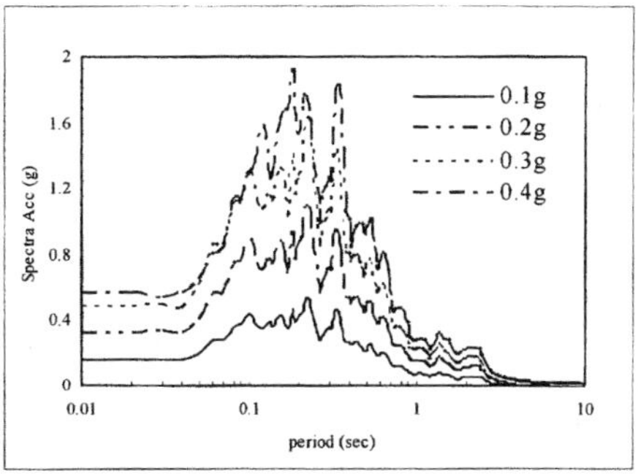

FIG.4. Calculated Response at point B of the landfill in different analyses with various peak acceleration of the input motion

As shown in Figs 3 and 4 for the more intense motion (larger peak ground acceleration) of the Abbar event, acceleration spectra show larger values in comparison with less intense motion. On the other hand as Figs 5 and 6 demonstrate increasing the peak ground acceleration has slight effect on response spectra ratio especially in frequency range between 1 to 10 Hz.

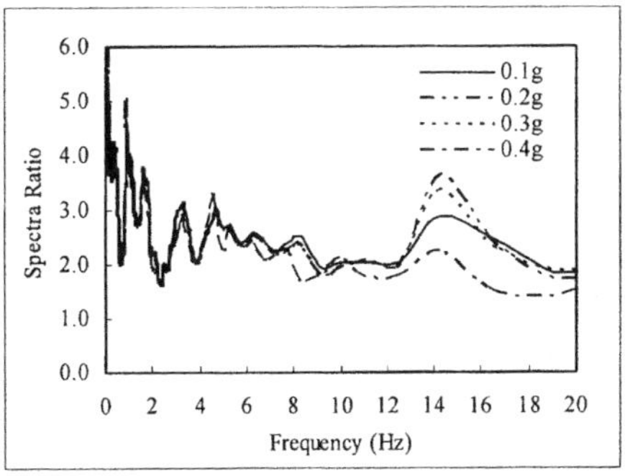

FIG.5. Response Ratio at point A of the landfill in different analyses with various peak acceleration of the input motion

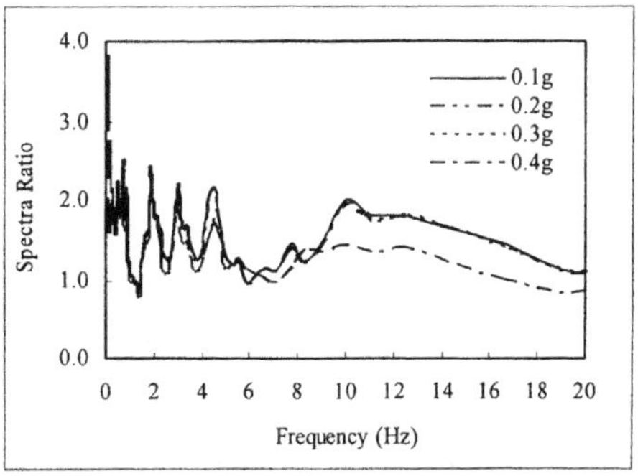

FIG.6. Response Ratio at point B of the landfill in different analyses with various peak acceleration of the input motion

The Width of the Landfill

To examine the effect of landfills' width on their seismic response, three landfills with 100, 200 and 250 m width are analyzed.

Figures 7 and 8 show that increasing the width of the landfill decreases values of acceleration spectra at points A and B. This decreasing trend is also observed in Figs 9 and 10 that demonstrate the ratio of these spectra to the acceleration spectrum of the input motion. In addition increasing the width of the landfill, changes the frequency content of the acceleration-time history of the input motion.

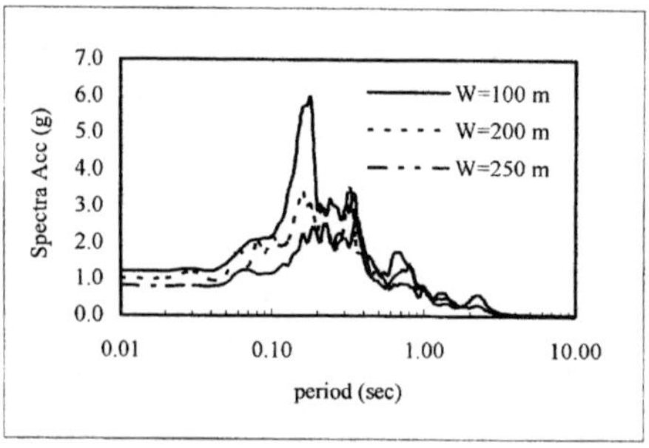

FIG.7. Calculated Response at point A of the landfills with different width

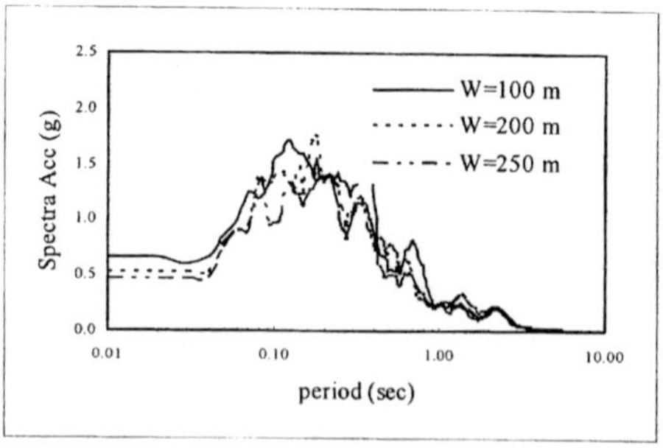

FIG.8. Calculated Response at point B of the landfills with different width

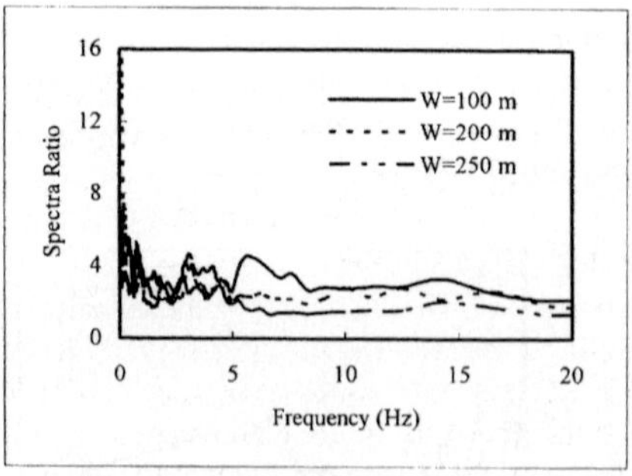

FIG.9. Response Ratio at point A of the landfills with different width

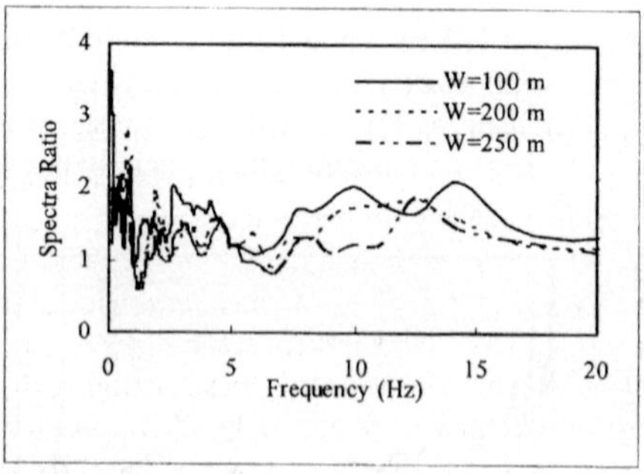

FIG.10. Response Ratio at point B of the landfills with different width

The Shear Strength Parameters of Solid Waste

The shear parameters of solid waste are usually determined from large scale triaxial or direct shear test, in-situ tests or back calculations from field trials or site performance records. There is an infinite number of shear strength combinations which is obtained from various authors.

In this paper to examine the effect of shear parameters of solid waste on seismic response of landfills, three combinations of shear parameters (c,φ) are used in analyzing seismic response of landfills. These combinations include: 1- C=30 kPa, φ=20°, 2- C=10 KPa, φ=24° and 3- C=0 KPa, φ=27°.

Figures 11 and 12 demonstrate acceleration spectra at points A and B of the landfills with three different shear parameters. As shown landfills with various combinations of shear parameters have similar response to seismic loading. Since shear parameters are related to failure conditions, same acceleration spectra in landfills with different combinations of shear parameters show that analyzed landfills aren't in failure condition.

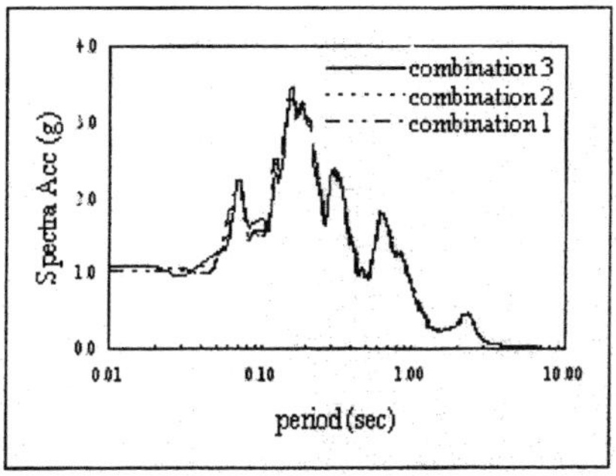

FIG.11. Calculated Response at point A of landfills with different shear parameters

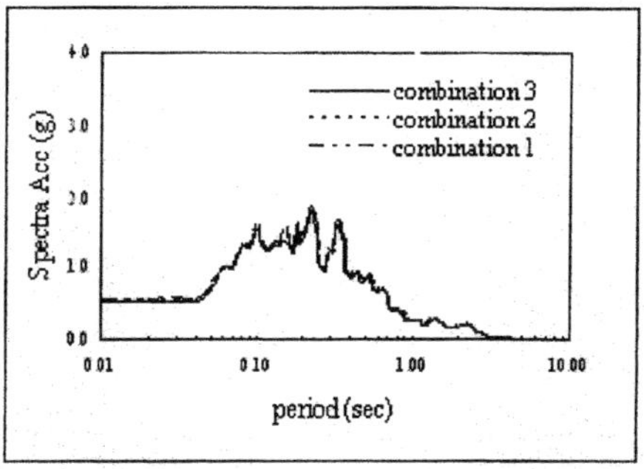

FIG.12. Calculated Response at point B of the landfills with different shear parameters

Conclusion

In this paper the effects of three factors on the seismic response of landfills are examined. These factors include: The peak acceleration of the input motion, the width of the landfill and the shear strength parameters of solid waste. As demonstrated, increasing the peak acceleration of the input motion increases the values of acceleration spectra but it has slight effect on spectra ratio especially in frequency range between 1 to 10 Hz. In addition increasing the width of the landfill decreases values of calculated acceleration spectra in different points of the landfill. This decreasing trend is observed in the ratio of these spectra to the acceleration spectrum of the input motion as well. On the other hand changing the shear parameters of solid waste (c,φ) don't affect landfills' seismic response of landfills. As a result since these parameters are related to failure conditions, the similar responses of landfills with different shear strength parameters show that analyzed landfills aren't in failure condition.

References

Japanese Society of Civil Engineers (2000), "Dynamic Analysis and Earthquake Resistant Design", edited by Tsueo Katayama et al.

Report of the ISSMFE Technical Committee TC5 on Environmental Geotechnics.

U.Aswathanarayana(1995), "Geoenvironment", A.A.Balkema Publishers.

Edward Kavazanjian, Neven Matasovic. (2000), "Seismic Analysis of Solid Waste Landfills", Geoenvironment.

Anthony.J.Augello, Jonathan.D. Bray, Gerald.A.Leonards, Pedro.C.Repetto, R.John.Byrne(2000) "Response of Landfills to Seismic Loading", Geoenvironment.

Jonathan.D. Bray, Anthony.J. Augello, Gerald.A. Leonards, Pedro.C. Repetto., R.John. Byrne "Seismic Stability Procedures for Solid Waste Landfills", Journal of Geotechnical Engineering.

V.Morochnik, J.P.Bardet, B.Hushmand.(1998), "Identification of Dynamic Properties of OII Landfill", Journal of Geotechnical and Geoenvironmental Engineering.No.3, Vol.124, pp186-196

Anthony.J. Augello, Jonathan.D. Bray, Norma.A.A brahamson, Raymond.B.Seed.(1998), "Dynamic Properties of Solid Waste Based on Back Analysis of OII Landfill", Journal of Geotechnical and Geoenvironmental Engineering. No. 3, Vol.124, pp211-222

I.M.Idriss, Martin.B.Hudson, Peter.K.Mundy, Roy Herzing. (1995), Seismic Response of the Operating Industries Landfill", Proceeding, ASCE Annual Convention, San Diego.

Anthony.J.Augello, NevenMatasovic, Jonathan.D.Bray, EdwardKavazanjian, Raymond.B.Seed. (1995), "Evaluation of Solid Waste Landfill Performance During the Northridge Earthquake", ASCE Geotechnical Special Publication No.54, pp.17-50.

Edward Kavazanjian, Neven Matasovic, Rudolph Bonaparte, Gray.R. Schmertman (1995), "Evaluation of MSW Properties for Seismic Analysis", In: Geoenvironment 2000, ASCE Geotechnical Special Publication No.46, Vol.2, pp.1126.

V.Cuellar, J.L.Monte, J.Valerio. (1998), "Static and Dynamic Elastic Model for Waste Landfills", Environmental Geotechnics, Seco e Pinto.

Neven Matasovic, Edward Kavazanjian. (1998), "Cyclic Characterization of OII Landfill Solid Waste", Journal of Geotechnical and Geoenvironmental Engineering ASCE, Vol.124, No.3, pp197-210.

Edward Kavazanjian, Neven Matasovic, (2001), "Seismic Design of Mixed and Hazardous Waste Landfills", State of the Art Paper No.11,Proc.4th International Conference on Recent Advances in Geotechnical Earthquake Engineering, University of Missouri,Rolla,27-31 March, San Diego, California.

Dean.K.Wall, Chris Zeiss.(1995), "Municipal Landfill Biodegration and Settlement", Journal of Engineering , Vol.121, No.3.

M.K.Yegian, A.M.Lahlaf. (1992), "Dynamic Interface Shear Strength Properties of Geomembranes and Geotextiles", Journal of Geotechnical Engineering , Vol.118, No.5.

Hisham.T.Eid,Timothy.D.Stark,W.Dougla Evans,Paul.E.Sherry .(2000),"Municipal Solid Waste Slope Failure.I: Waste and Foundation Soil Ptoperties", Journal of Geotechnical and Geoenvironmental Engineering.No.5,Vol.126,pp397-407

A.O. Lavanda, A.J. Valsangkar, S.G. Pelkey. (2000), "Lateral Earthquake Pressure at Rest and Compressibility of Municipal Solid Waste",Canadian Geotechnical Journal,Vol.37, pp1157-1165.

Venkaata Ramana Gunturi, Ahmed. W.M. Elgamal. (1998), "A Class of Inhomogeneous Shear models for Seismic Analysis of Landfills", Soil Dynamic and Earthquake Engineering, Vol.17, pp.197-209.

Use of steel slag and limestone quarry dust for remediation of acidity arising from runoff from colliery spoil – pilot study

Gurmel S Ghataora University of Birmingham, Edgbaston, Birmingham, B15 2TT
Nizar Ghazireh Tarmac Ltd, Ettingshal, Wolverhampton, WV4 6TJ
Brendan Burns University of Birmingham, Edgbaston, Birmingham, B15 2TT

Abstract
Runoff water from colliery spoils can be very acidic with a pH of perhaps as low as 2.5. Iron pyrites in the colliery spoil react with the oxygen and water to form iron and sulphates together with hydrogen, giving rise to high acidity. During research at the University of Birmingham, basic oxygen steel slag was found to release alkalinity gradually. BOS slag was used to treat colliery spoil and it was found that elevated levels of pH could be maintained in the field. However, pH of the treated spoil in the pilot study was too high to support vegetation. Nevertheless, the concept was proven and a further trial was initiated with a range of mixtures, which included BOS slag and limestone quarry fines together with compost.

Introduction
Weathered colliery spoils pose a problem as runoff water from them can be highly acidic with a pH of perhaps as low a 2.5 (Robinson *et al.*, 2005). Acidic arises because of the reaction between oxygen (from air and water) and iron pyrites, which is a naturally occurring compound found in carboniferous rock. This reaction is well understood and takes the form (CLAIRE, 2002):

$FeS_2 + 3.5\ O_2 + H_2O \Rightarrow Fe^{2+} + 2SO_4^{2-} + 2H^+$

Metal Sulphide oxidises to release ferrous iron, sulphate and acidity in the form of hydrogen

$Fe^{2+} + 0.25\ O_2 + H^+ \Rightarrow Fe^{3+} + 0.5\ H_2O$

Ferric iron may be oxidised to ferric iron

$Fe^{3+} + 3\ H_2O \Rightarrow Fe(OH)_3\ (s) + 3\ H^+$

Ferric iron may be converted to ferric hydroxide and further acidity

$FeS_2(s) + 14\ Fe^{3+} + 8\ H_2O \Rightarrow 15\ Fe^{2+} + 2SO_4^{2-} + 16\ H^+$

Additional iron may be produced via oxidation of pyrite by iron, thus perpetuating cycle.

A number of techniques, including the addition of lime, have been investigated in order to reduce the acidity in water leached from colliery spoil. However, most have limitations. For example, lime results in a "flash" release of alkalinity. Therefore, a slow release treatment is needed, so that re-dosing is not required.

Poh (2005), during his investigation into the use of Basic Oxygen Steel (BOS) slag, found that pH of the treated soil remained elevated for over 112 days. A comparison of change in pH levels with respect to time for BOS slag from three different sources and lime is shown in Figure 1. After about 7 days, the pH of soil treated with lime starts to drop as alkalinity is consumed in the pozzolanic reactions. In contrast to this the pH of soil treated with BOS slag remained elevated (with pH above 10) for at least about 4 months. This concept of slow release of alkalinity for treating acidity in colliery spoil was investigated by Jessic (2003). It was followed by a field trial at a colliery spoil site at Carr Wood near Aspatria, in Cumbria.

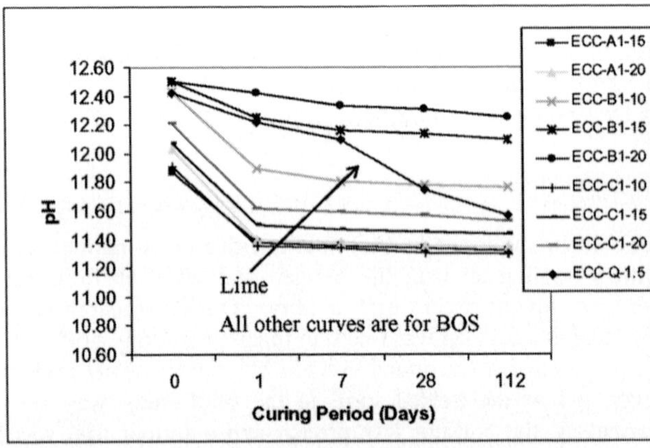

Figure 1. Relationship between pH levels and time for Lime and BOS slag from three different sources with Kaolin clay (Poh, 2005).

The main aim of this investigation was to treat acidity in the run-off water. Key findings of the field trial at Carr Wood and outlines of the construction of the second more extensive trial are summarised in this paper.

Pilot trial

The trial site contained colliery spoil from a coal pit, which was mined from about 1860 to 1930. It occupies an area of approximately 7 hectares. Deciduous and coniferous trees were planted to blend in with the surrounding woods at the cessation of the spoil dumping operation. The site was then used for commercial forestry. In the 1970s a section of the site was cleared of trees and was ploughed to make ready for re-planting. Soon after this, the landowner went into liquidation and new trees were never planted. The reworked area started to generate acidic runoff water with a pH of about 2.7 (Law, 2002). In the area downstream of the ploughed section, acidic water has destroyed vegetation. Annual monitoring of the surrounding watercourses has shown an increase in both the level and the period of pollution over consecutive years.

This colliery spoil contained approximately 25% sub 75 micron material. The coarser fraction was essentially well graded, ranging in size from sand to coarse gravel, see Figure 2. Steel slag from Teesport was used in this investigation. It was well graded with particle size distribution ranging from dust to 6mm (see Figure 2) and typical constituents of slag are shown in Table 1.

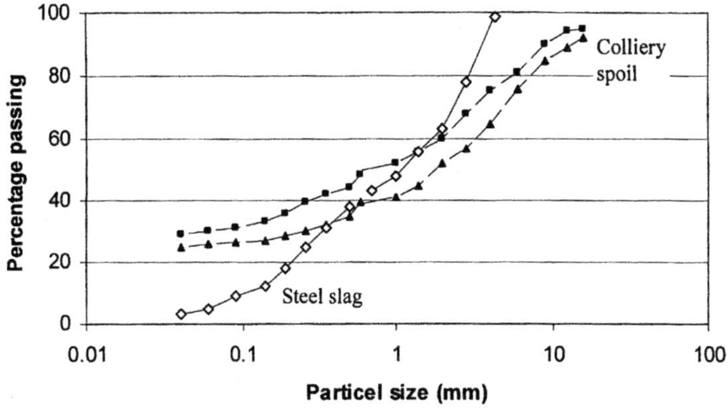

Figure 2. Particle size distribution curves for colliery spoil and BOS slag for trial site.

Table 1. Typical chemical and mineralogical composition of air-cooled blastfurnace slag and ground granulated blastfurnace slag (Poh, 2005).

	Air-cooled blastfurnace slag (National Slag Association, 2001a)	Air-cooled blastfurnace slag (Robinson, 2000a)	Ground granulated blastfurnace slag (Civil and Marine Slag Cement Limited, 2002)
Chemical constituent			
Calcium as CaO	32 – 45	41	40
Silicon as SiO_2	32 – 42	35	35
Aluminium as Al_2O_3	7 – 16	14	14
Magnesium as MgO	5 – 15	7	9
Iron as Fe_2O_3	1 – 1.5	-	0.4
Sulphide as S	1 – 2	0.8	0.8
Mineralogical phases	- Melilite - Wollastonite - Merwinite	- Melilite (solid solution series of gehlenite and akermanite - Calcium sulphide (oldhamite) - Merwinite - Dicalcium silicate	- Predominantly as glassy phase

Methodology for pilot tests

Leachate tests were conducted on colliery spoil and colliery spoil plus BOS slag in accordance with BS EN 1744-3, (2000). Chemical analyses of leachate arising from spoil and spoil plus 10% steel slag and the Environment Agency threshold values are shown in Table 2. With 10% BOS slag, pH of about 7.6 was achieved. Results also show that spoil treated with steel slag released lesser quantities of metals into the eluate compared to the untreated spoil. In all instances the metallic compounds in the eluate of the treated spoil were below the threshold limits set by the Environment Agency. Based on these results and a successful outcome of a small scale grass growing trial at the University, a larger trial was constructed at the Carr Wood colliery spoil site. The main aim of the trial was to increase and maintain pH in the range of approximately 5.5 to 9.0 for a prolonged period, to enable vegetation growth.

TABLE 2. Chemical Analysis of Leachate and EA contamination thresholds

Analyte	Untreated Spoil	Spoil + 10% Steel Slag Fines	EA Contamination Thresholds
COD as O_2	16.0 mg/l	15.0 mg/l	30.0 mg/l
pH	4.74	7.55	5.5 – 9.5
Copper as Cu	85.2 μg/l	9.03 μg/l	20.0 μg/l
Nickel as Ni	62.4 μg/l	<5.00 μg/l	50.0 μg/l
Zinc as Zn	166 μg/l	6.45 μg/l	500 μg/l
Aluminium as Al	792 μg/l	<10.0 μg/l	-
Iron as Fe	74.4 μg/l	<30.0 μg/l	100 μg/l
Manganese as Mn	1600 μg/l	<10.0 μg/l	-
Sulphate as SO_4	1540 μg/l	1570 μg/l	1500 μg/l
Conductivity	2020 μS/cm	2080 μS/cm	1000 μS/cm
Sulphide as S	<0.015 mg/l	<0.015 mg/l	150 mg/l
Alkalinity $CaCO_3$	<5.00 mg/l	63.1 mg/l	-
Chloride ion as Cl	3.92 mg/l	3.54 mg/l	200 mg/l

Although the laboratory work showed that 10% slag would perhaps be enough to maintain a pH of about 7.6. It was anticipated that the spoil may be variable and so it was decided to use 15% slag.

On an area cleared of vegetation, two 3.0m x 3.0m plots were constructed. Both were excavated to 0.75m in depth. For the control plot, spoil was mixed and replaced into the excavation. In the case of the treated plot, 15% slag was mixed with the spoil and replaced into the excavation. On completion of the plots, 500g of grass seed was sown in each area. The construction of the plots was completed in September 2003.

Chemical testing

After completion of the trial, samples were taken at regular intervals. Three specimens were obtained from each site from 0.1m and 0.3m depths. The three samples from each depth from the plot were mixed to form one sample for testing. Thus, for each round of tests, four samples were tested (two from 0.1m depth and two from 0.3m depth).

Specimens were tested for pH and a range of compounds. Relationship between pH and time (Figure 3) shows that alkalinity has remained at a high level (pH ranging from 8.0 to 11.0) for the area treated with slag, whereas pH has ranged from about 4.8 to 8.0 for the untreated area. It is anticipated that over time the pH level will drop in both areas. It was however clear that the pH of the treated area was too high for grass to propagate.

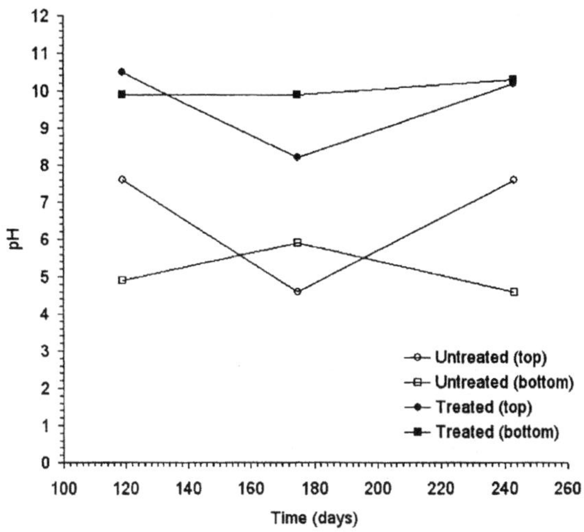

Figure 3. Relationship between pH and time for trial site.

Sulphate, and lead levels showed little change between treated and untreated areas. There was a distinct difference in both calcium and chromium contents between the treated and untreated areas. The treated area contains about four times the levels of calcium compared to untreated area. Chromium levels rose to about 5 times the initial value of about 50mg/kg. Six months later the level had dropped to about three times the initial level.

Conclusions from the pilot trial

Although the aim was to maintain pH level of from about 6.5 to 8.5, the pH levels remained elevated about 8 months after the start of the trial. The spoil was dosed with 15% BOS slag to overcome variability of the soil. Although the pH levels remained too high for the grass to grow, the concept of using steel slag to reduce acidity was proven.

Follow-up trial

Following the finding of the pilot scheme, a further trial was constructed at Carr Wood. This area was situated near to the pilot trial. The key aim of this research was to maintain elevated pH in the range of 5.5 to 9.0. Furthermore, the secondary aim was to grow grass (or grasses) that would be self-sustaining in the long term. Third aim was to explore the possibility of using hypo-accumulator vegetation to reduce heavy metals from near the surface.

A number of leachate tests were conducted on combinations of mixes: 5% slag, 5% limestone dust, and 5% slag plus 5% limestone dust. The addition of 5% slag to the mixes was considered to be the optimum dosage. This was lower than the findings at the pilot stage. This is most probably due the variability in the slag.

Analyses of eluates with 0%, 5% and 10% added slag are shown in Table 3. Results show that there was a general decrease in both the metallic and non-metallic compounds with increase in slag content. PH increased from 3.9 to 10.2 for the 0% to 5% respectively. For additional 5% slag (from 10% to 15%) there was no significant change as pH (increased only by 0.7).

Table 3. Chemical Analysis of Leachate of varying amounts of added slag.

Analyte (*ppm*)	0% Slag	5% Slag	10% Slag
Silicon as SiO_2	9.9	18.3	16.4
Calcium as CaO	700	539	457
Aluminium as Al_2O_3	0.5	<0.1	<0.1
Total Iron	0.4	<0.1	<0.1
Magnesium as MgO	3.1	2.1	2.5
Titanium as TiO_2	<0.1	<0.1	<0.1
Phosphorous as P_2O_5	4.4	1.2	2.1
Manganese as MnO	0.6	<0.1	<0.1
Sodium as Na_2O	6.6	4.3	4.2
Potassium as K_2O	12.5	10.0	6.6
Zinc as ZnO	<0.1	<0.1	<0.1
Chloride as Cl	4	<1	<1
Total Sulphur	442	299	237
COD as O_2	48	16	24
pH	3.93	10.17	10.85

Field trial comprised the mixtures shown in Figure 4.

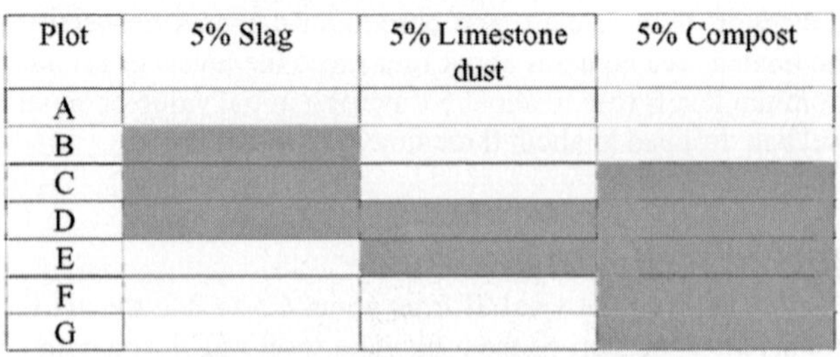

Figure 4 Mixtures used in the follow–up trial

Each plot measuring 2.5m x 2.5m was excavated to a depth of 0.75m. The excavated spoil was mixed with slag, limestone and compost as noted above and the mixture was re-placed into the excavation.

About 15mm to 20mm of mulch was added to surface of all the finished areas except the control plot. A 50:50 mix (in terms of seed numbers) of *Tall Fecue* and *Creeping Bent* seeds were planted on plots A, B, C, D, E and G. Hypo-accumulator grass (*Arabidopsis thaliana*) and shrubs (*Alyssum argenteum*) were planted in plots F. All areas except F were seeded in July at the time of construction of the trial. Plot F was seeded in September due to delayed delivery of seeds. The seven plots were repeated 3 times, as shown in Figure 5, with the view to getting repeatable results.

Spoil from all the seven plots was analysed in terms of pH and conductivity. Changes in pH and conductivity are shown in Figures 6 and 7 respectively. The level of acidity reduced significantly in plots B (slag only), C (slag plus compost), D (slag plus limestone dust and compost) and E (limestone dust and compost) compared to areas treated with slag alone compost alone. For all the areas pH reduced by a small amount between day zero and 77 days. The least reduction in pH was observed for area treated with slag, limestone dust and compost.

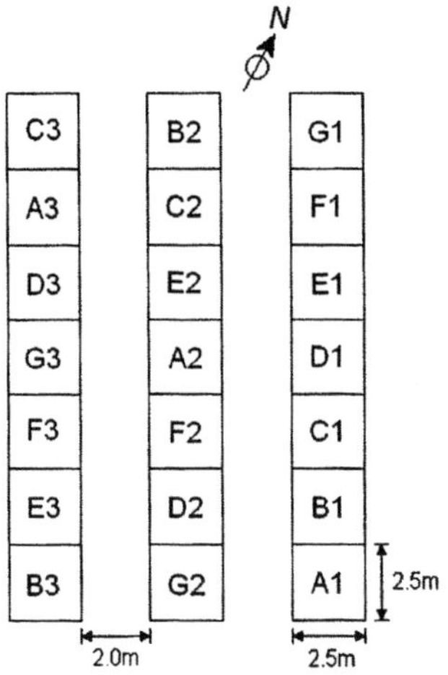

Figure 5. Layout of plots for field trial.

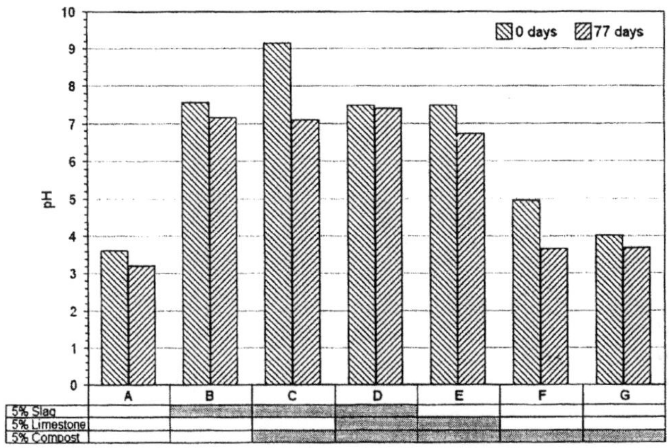

Figure 6. Laboratory results of pH tests on eluates.

Conductivity decreased in all the mixes compared to the control area (see Figure 7). By about 77 days increases in conductivity was recorded in plots D, F and G.

Early results of grasses grown in the laboratory showed that slag alone seem not to be adequate for growing grass. Plots C, D and E performed the best. This was also the case in the field. Control plots were found not to be able to support any vegetation. This was most probably because of both the low pH and high conductivity in the untreated areas.

Further investigations are underway and future trends will be ascertained from the planned monitoring programme, which also includes determinations of range on metallic and non-metallic compounds together with conductivity and pH.

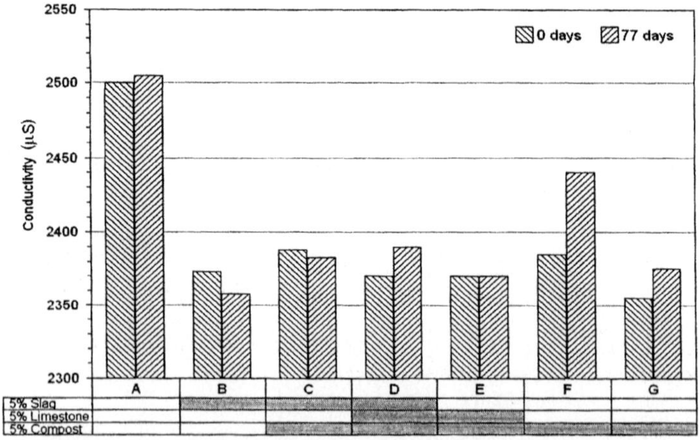

Figure 7. Laboratory results of conductivity tests on eluates.

Conclusions

Although the second trail has only been in place for about 11 weeks, following conclusions can be drawn.

- 5% concentration steel slag significantly reduced the acidity of the colliery spoil.
- Steel slag, steel slag plus compost and slag plus limestone dust and compost seemed to give the best the best results with pH of just over 7.
- Mixes incorporating steel slag gave the largest reduction in conductivity.
- Initial results of field trial and laboratory experiments show that grasses are growing best on plots C, D and E while the control plot is not sustaining growth.
- Further monitoring is necessary to assess the long terms trends.

References

BS EN 1744-3, 2000. Tests for chemical properties of aggregates, Part 3: Preparation of eluates by leaching. British Standards Institute, London.

CL:AIRE, 2002. Case Study Bulletin; CSB2, http://www.claire.co.uk/casestudy.php

Jessic J., 2003. The utilisation of reclaimed asphalt and steel slag fines. PhD Thesis, The University of Birmingham, Birmingham, UK.

Law, I., 2002. Carr Wood surface water survey, Environment Agency, unpublished, Cumbria.

Poh, D., 2005. Soil stabilisation using basic oxygen steel (BOS) slag fines. PhD Thesis, the University of Birmingham

Robinson, H., Ghazireh, N., and Ghataora, G.S., 2005. Use of Basic Oxygen Steel Slag and Limestone Quarry Dust for Colliery Spoil Remediation – a Pilot Case Study, Euro Slag Conference, Oulu, Finland.

Experimental Investigation of Compressibility of Cohesive Soil Mixed with Waste Tire Chips

Mahmoud Ghazavi, Assistant Professor, Civil Engineering Department, K.N.Toosi University of Technology, Tehran, Iran, Tel: +98 21 8779623; Fax: +98 21 8779476; email: ghazavi_ma@kntu.ac.ir
Amin Barmaki, Graduate Student, Geotechnical Engineering, Civil Engineering Department, K.N.Toosi University of Technology, Tehran, Iran

Abstract

A large number of scrap tires become out of use each year and enter environment. Dumping of waste tires in engineered designed landfills may not economically be justified. Therefore finding applications for such materials is environmentally sound. On the other hand, some soft clays are found in practice, which make problems. Although, geotechnical engineering has dealt with such problems by the use of appropriate foundations such as piles, caissons, etc., the soil improvement may also be considered in some cases. Such consideration becomes more valuable when environment issues are present. This paper presents the results of a series of consolidation tests performed on a cohesive soil mixed with various percentages of waste tire chips. Similar mixtures have also been prepared using a sand of the same size distribution as tire chips. This helps to distinguish the effectiveness of using tire chips for fine soil improvement. The volume of sand and tire chips in a certain mixture was kept the same for comparison. The results show that adding sand to the clay can improve the compressibility as expected. The addition of tire chips to the clay can also reduce the compressibility. Having the same volumes of sand and tire chips in each mixture, it will be shown that the compressibility of tire chip-clay mixtures is comparable to that of sand-clay mixture containing the same sand volume does. This is encouraging since the use of waste tires in engineering projects can reduce environmental impact.

Keywords: environment, cohesive soil, consolidation, waste tire chips.

Introduction

Soil improvement has been of great concern since many years ago. This has been undertaken to enhance mechanical properties of soft or problematic soils. A traditional solution for this improvement has been addition of some sound element to the soil. These materials themselves need to have sufficient strength and much durability. The aim of adding these elements to soil is to improve the soil for load carrying capacity and allowable settlement.

From environmental point of view, materials used for soil improvement must have no impact. Since waste tires don't go back to the environment cycle, engineers may make good ideas to re-use them. The use of waste tires in civil engineering applications is increasing such that it is anticipated that in the future, 65 percent of waste tires will be used in civil engineering projects. This means that each year there will be 29 to 50 percent increase in using them.

With the banning of deposing vehicle tires in sanitary landfills, many waste tires enter the environment, causing serious problems, especially with the development of societies. The number of waste tires in the United States averagely accounts for about 270 million. In Canada, this figure is about 28 million. Only 30% of the above waste tires are deposited in landfills (CCME 1994). Such wastes cannot generally be stocked in landfills since they require large spaces. In England, the number of waste tires is about 25 million per annum (the first author heard from local TV news). In England, 20% of total fuel supplied to produce electricity, is initiated from waste tires (the first writer heard from local TV news).

Waste tires are used in various applications in civil engineering, for instance, for reinforcing soft soil in road construction (Bosscher et al. 1997; Nightingale et al. 1997), for earthwork projects and stabilizing slopes (Poh and Broms 1995; O'Shaughnessy and Garga 2000a), as lightweight material for backfilling in retaining structures (Bosscher et al. 1997; Lee, et al. 1999; Sumanarathna et al. 1997; Tweedie et al. 1998; Garga and O'Shaughnessy 2000; O'Shaughnessy and Garga 2000a; Allman and Simundic 1998) etc.

The shear strength of tire shred-sand mixtures can also be enhanced significantly (Edil and Bosscher 1993; Foose et al. 1996; Ghazavi and Amel-Sakhi 2005a). The values of CBR of tire shred-sand mixtures are also of great importance (Ghazavi and Amel-Sakhi, 2005b).

The use of waste tires in forms of whole, shreds, and chips has attracted researchers and engineers in the last decade. Much research work has been undertaken on sand mixed with chips, grains, and shreds made of scrap tires. There is little research work on possible improvement of clay properties by using tire shreds, chips, or grains. This paper describes the results of a number of consolidation tests carried out on a cohesive soil mixed with various percentages of waste tire chips. Similar mixtures have also been prepared and tested using a sand of the same size distribution as tire chips. This helps to distinguish the effectiveness of using two additives of the same volume on soil compressibility.

Materials

In this research, a plastic bentonite soil was used. The rubber grains were obtained from scrap tires. The density of rubber grains was 1.32. The size distribution of the rubber grains In order to compare the compressibility of rubber-clay mixtures with sand-clay mixtures, the same size distribution has been used for the sand (Figure 1).

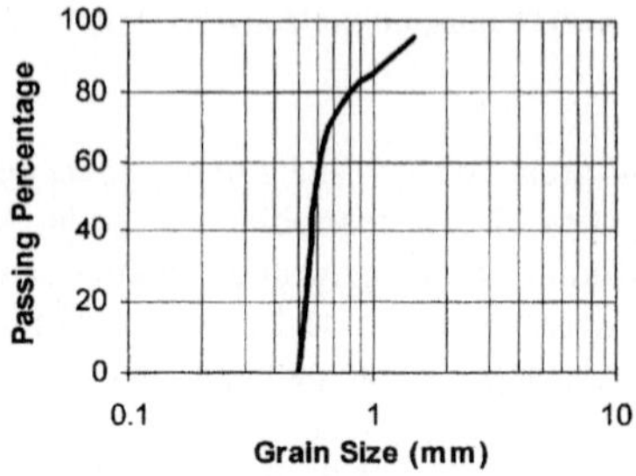

Figure 1: Particle size distribution of rubber and sand

Testing Procedure

A series of consolidation tests has been performed on rubber-clay and sand-clay mixtures according to ASTM D2435-93. To this aim, bentonite with moisture content more than liquid limit with various volumetric percentages of rubber particles and sand were mixed separately. The uniform sand-rubber and clay-rubber mixtures were completely saturated. The mixtures were located in the test mold, the sample was de-aired, and test apparatus was calibrated. Then the load was applied and reading was started till the final stage of settlement was reached.

Materials

Bentonite

In this research, a kind of bentonite soil of factory "Barite Flate Iran" with liquid limit of 280%, plastic limit of 180%, plasticity index of 100%, and grain density of 2.45 was used.

Rubber Aggregates

These aggregates were made from scrap tire in Yazd Tire Factory. Density of this aggregate is 1.32. and sieve analysis curve for it in shown in Figure 1. Various volumetric percentages of 10, 20, 30, and 40% were chosen for waste tire particles and mixed uniformly with clay.

Sand

To compare the impact of rubber particles on compressibility of bentonite-rubber mixtures, sand with similar grading and size distribution to rubber particles was used. For the sand also various volumetric percentages of 10, 20, 30, and 40% were chosen and mixed uniformly with clay. As mentioned, a comparative study between the use of sand and rubber grains for clay improvement was the main objective in this research. The size distribution obtained from the sieve analysis is shown in Figure 1.

Tests

Normal consolidation test was performed on all bentonite–rubber particle and bentonite-sand mixtures according to the standard ASTM D2435-93. For this purpose, first bentonite with moisture content of 56% with various volumetric percentages of rubber particles and sand with homogenous grading were mixed. Each mixture mixed gently such that became uniform. The mixture was poured in test mould and then de-aired. The loading was then started. Reading was done continuously until the end of consolidation settlement under each surcharge was reached. Three weights of 49.1, 98.1, and 196.2 kPa were used as surcharges.

Results

This section presents a summary of the results obtained from the performed tests. Figure 2 illustrates the final consolidation settlement of various clay-rubber mixtures under each surcharge. Figure 3 shows final settlement ratio after applying each surcharge.

Figure 4 compares the final consolidation settlements of clay-40% rubber and clay-40% sand mixtures. These mixtures contain 40% rubber or sand by volume as additives to the clay.

Figure 5 illustrates the variation of consolidation settlement versus time square for clay-rubber and clay-sand mixtures containing 40% rubber and 40% sand by volume.

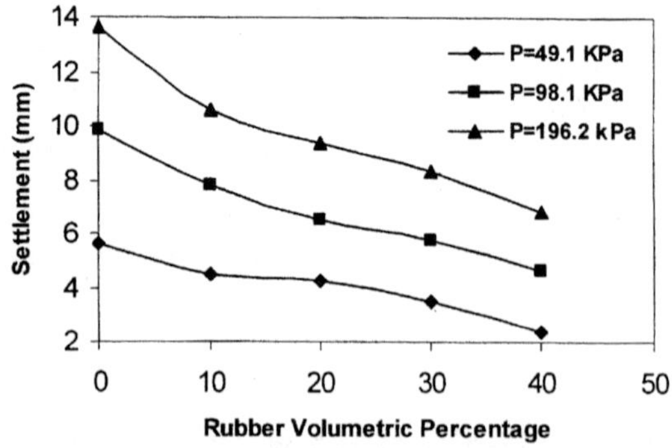

Figure 2. Final consolidation settlement of rubber-clay mixtures under various surcharges

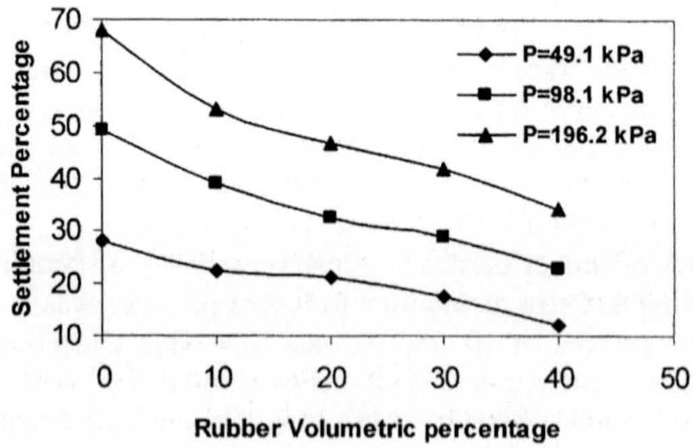

Figure 3: Final consolidation settlement ratio of rubber-clay mixtures under various surcharges

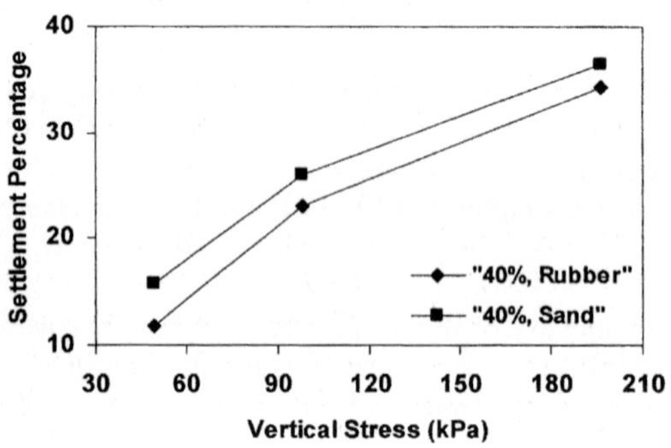

Figure 4: Comparison of final consolidation settlements of clay-40% rubber and clay-40% sand mixtures (40% by volume)

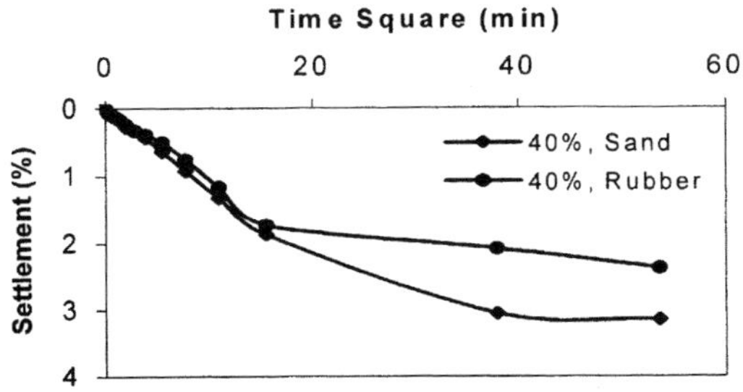

Figure 5: Plot of consolidation settlement with respect to time square for clay-40% rubber and clay-40% sand mixtures (40% by volume)

A number of points can be extracted from the above figures. These may be summarized as:
a) As expected, the compressibility of clay-rubber mixtures decreases with increasing rubber volume, as observed in clay-sand mixtures (Salehi, et al., 2002).

b) Rubber-clay mixtures in consolidation tests show the same trend as sand-clay mixtures do. There is no significant deviation between these trends when two types of mixtures contain the same rubber and sand by volume. Rubber-clay mixtures had slightly lower settlement than sand-clay mixtures of the same volumetric sand. From consolidation settlement point of view, clay-rubber mixtures can have an interesting function.

c) The clay-40% rubber mixture reaches 90% consolidation settlement in shorter time than the clay-40% sand mixture does. This is also an interesting feature.

Based on the above findings and adding other interesting features of clay-rubber mixtures such as low weight (Ghazavi, 2004) and acceptable unconfined compressive strength (Ghazavi and Barmaki, 2005), it may be said that clay-rubber mixtures may be of importance concern in comparison with clay-sand mixtures. In particular, in countries where sand and gravel are expensive, rubber grains may be used instead. More importantly, rubber-clay mixtures may be preferred due to environmental consideration, since a large number of scrap tires may be used in the form of chips and mixed with clay to enhance clay properties.

Conclusions

In this research, bentonite was mixed with various volumetric percentages of sand and also with the same volume of rubber grains. Normal consolidation tests were performed on all fully saturated mixtures. The values of consolidation settlements were determined under three surcharges. Values of settlement were determined for each mixture and compared. This comparison showed that by adding rubber particles or sand to the clay, the settlement generally deceased. Moreover, rubber-clay mixtures in consolidation tests showed slightly smaller settlement than sand-clay mixtures of the same volumetric sand. However, this was insignificant. Thus, clay-rubber mixtures can have an interesting function in compressibility. In addition, the clay-40% rubber mixture reached 90% consolidation settlement in shorter time than the clay-40% sand mixture does. Based on the limited findings in this research and other interesting features such as acceptable strength, light weight material, no apparent environmental impact, it may be said that rubber-clay mixtures may be comparative to sand-clay mixtures and thus may be recommended to use in practice.

References

Allman, M. A., and Simundic, G., 1998. "Testing of a retaining wall constructed of waste tyres." *Proceedings Third International Congress on Environmental Geotechnics*, Vol. 2, pp. 655-660, Lisoa, Portugal.

Bosscher, J., Edill, T. B., and Kuraoke, S. 1997. "Design of highway embankments using tire chips." *Journal of Geotechnical and Geoenviromental Engineering, ASCE*, Vol. 123, No. 4, pp. 297-304.

Council of Ministers for The Environment (CCME) 1994. "Harmonized economic instruments for used tires." *Final Rep. Prepared for the Canadian Council of Ministers for the Environment, Apogee Ref. 363 CCME, No. CCME-SPC-EITG-92E.*, Canada.

Edil, T. B., and Bosscher, P. J., 1994. "Engineering properties of tire chips and soil mixtures." *Geotechnical Testing Journal*, Vol. 17, No. 4, pp. 453-464.

Foose, J., Benson, H., and Bosscher, J. 1996. "Sand reinforced with shredded waste tires." *Journal of Geotechnical Engineering, ASCE*, Vol. 122, No. 9, pp. 760–767.

Garga, V. K., and O'Shaughnessy, V., 2000. "Tire–reinforced earthfill. Part 1: Construction of a test fill, performance, and retaining wall design." *Canadian Geotechnical Journal*, No. 37, pp. 75-96.

Ghazavi, M. 2004. "Shear strength characteristics of sand-mixed with granular rubber." *Geotechnical and Geological Engineering, An International Journal*, Vol. 22, No. 3, pp. 401-416.

Ghazavi, M., and Amel-Sakhi, M. 2005a. "Influence of optimized tire shreds on shear strength parameters of sand." *International Journal of Geomechanics, ASCE*, Vol. 5, No. 1, pp. 58-65.

Ghazavi, M., and Amel-Sakhi, M. 2005b. "Size optimization of waste tire shreds in sand-shred mixtures using CBR tests." *Geotechnical Testing Journal (ASTM)*, Vol. 28, No. 6, pp. 564-569.

Ghazavi, M. and Barmaki, A. 2005. "Strength enhancement of cohesive soil using waste tire chips." *8th International Conference on Cohesive Sediment Transport (INTERCOH2005)*, Institute of Lowland Technology, Saga University, Japan, pp. 26-29.

Humphrey, D.N., and Manion, W.P., 1992. "Properties of tire chips for lightweight fill." *Grouting, Soil Improvement, and Geosynthetics*, Vol. 2, ASCE, New York, N.Y., pp. 1344-1355.

Humphrey, D. N., Katz, L. E., and Blumenthal, M., 1997." Water quality effects of tire chip fill placed above the groundwater table." *Testing Soil Mixed with Waste or Recycled Materials*, ASTM STP 1275, M. A. Wasemiller , K .B . Hoddinott, Eds. pp. 299-313.

Lee, J. H., Salgado, R., Bernal, A., and Lovell, C. W., 1999. "Shredded tires and rubber–sand as lightweight backfill." *Journal of Geotechnical and Geoenviromental Engineering, ASCE*, Vol . 125, No 2, pp. 132-141.

Nightingale, D. E. B., and Green, W. P., 1997. "An unsolved riddle: Tire chips, two roadbeds, and spontaneous reactions." *Testing Soil Mixed with Waste or Recycled Materials*, ASTM STP 1275, M. A. Wasemiller , K. B. Hoddinott, Eds. , pp. 265-285.

O' Shaughnessy, V., and Garga, V. K., 2000a. "Tire–reinforced earthfill. Part 2: Pull–out behavior and reinforced slope design." *Canadian Geotechnical Journal*, No. 37, pp. 97-116.

O' Shaughnessy, V., and Garga, V. K., 2000b. "Tire–reinforced earthfill. Part 3: Environmental assessment." *Canadian Geotechnical Journal*, No. 37, pp. 117-131.

Poh, P. S. H., and Broms, B. B., 1995. "Slope stabilization using old rubber tires and geotextiles." *Journal of Performance of Constructed Facilities*, Vol. 9, No. 1, pp. 76-80, 1995.

Salehi, S., Ghazavi, M., and Aflaki, E. 2002. "Improvement of clay compressibility by use of sand and its effects on clay permeability." *Proceedings of the First National Conference on Soil Improvement*, Amir Kabir University of Technology, Tehran, Iran, pp. 213-223, (in Persian).

Sumanarathna, I. H. D., Mallawarctchie, D. P., and Kulathilaka, S. A. S., 1997. "Stabilization of slopes by anchored type retaining structures." *Proceedings 14th International Conference of Soil Mechanics and Foundation Engineering*, New Delhi, India, pp. 1261-1264.

Tweedie J.J, Humphrey, D.N., and Sanford, T.C., 1998. "Tire-shreds as lightweight retaining wall backfill." *Journal of Geotechnical and Geoenvironmental Engineering, ASCE*, 124(11), pp. 1061-1070.

Determination of Shear Strength Parameters of Tire Chips and Grains

Mahmoud Ghazavi, Assistant Professor, Civil Engineering Department, K.N.Toosi University of Technology, Tehran, Iran, Tel: +98 21 88779623; Fax: +98 21 88779476; email: ghazavi_ma@kntu.ac.ir
Mehdi Torbati, Undergraduate Student, Civil Engineering Department, K.N.Toosi University of Technology, Tehran, Iran
Masoud Kazem, Undergraduate Student, Civil Engineering Department, K.N.Toosi University of Technology, Tehran, Iran

Abstract

Each year large amount of waster tires enter the environment, causing serious problems. Although some of them may be dumped or used in various case forms, a very feasible application of them is in geotechnical projects. For example they may be used as lightweight backfill material behind retaining walls. For such applications, shear strength parameters of tire chips and grains are to be precisely determined. For this purpose, in this paper CD triaxial tests were performed on tire chips and grains. The performed experiments showed that friction angles obtained from tests were consistent with most of research work done on the subject. Also cohesion intercept was also obtained for tire grains. With the shear strength parameters and low unit weights which were obtained for waste tire grains and chips, it is certain that they are lightweight and thus the use of such material as backfill is encouraging. This is also sound since the use of waste tires in engineering projects can reduce environmental impact.
Keywords: environment, waste tire particles, triaxial test, friction angle, cohesion, geotechnical engineering

Introduction

In the United Kingdom, about 25 million car tires and 3 million truck tires become out of use each year and thus are discarded. According to Bridgwater and Mumford (1979), 8% is used for retreading, 6% for regeneration (e.g. granulation, carbon/oil extraction and steel recovery) and 9% for incineration purposes. Therefore, 67% of waste ties are stockpiled, landfilled or dumped illegally. In England, 20% of total fuel supplied to produce electricity, is initiated from waste tires. In the United States, it is estimated that 279 million scrap tires become out of use annually (House Bill S2462, 1990). Of these, about 85% are stockpiled, landfilled or illegally dumped. In Canada, the quantity of waste tires figures about 28 million each year. Generally speaking, only 30% of the waste tires may be deposited landfills, since such wastes capture large spaces and thus it is not reasonable to dump such a great number of scrap tires in engineered landfills.

To alleviate the environmental impact of scrap tires, they may be used in geotechnical projects. They have been used successfully on playgrounds and football fields and in paving materials. Waste tires in the forms of complete, shreds, chips, and/or grains are used in various applications in civil engineering, for instance, for reinforcing soft soil in road construction (Bosscher et al. 1997; Nightingale et al. 1997), for earthwork projects and stabilizing slopes (Poh and Broms 1995; O'Shaughnessy and Garga 2000a), as lightweight material for backfilling in retaining structures (Humphery and Manion, 1992; Bosscher et al. 1997; Lee, et al. 1999; Sumanarathna et al. 1997; Tweedie et al. 1998; Garga and O'Shaughnessy 2000; O'Shaughnessy and Garga 2000a; Allman and Simundic 1998), as soil-rubber mixtures in engineering applications (Black and Shakoor, 1997; Baykal et al., 1995; Baykal and Alpatli, 1995; Lyons et al., 1995; Al-Tabbaa, and Aravinthan, 1997; Al-Tabbaa, et al., 1997; Al-Tabbaa, and Aravinthan, 1998; Al-Tabbaa et al., 1997; Ghazavi and Amel Sakhi, 2005a; Ghazavi and Amel Sakhi, 2005b; Ghazavi and Barmaki, 2005) etc.

Wu et al. (1997) carried out triaxial tests on tire chips with sizes in the range of 2–38 mm. They obtained friction angles of 40–60° for chips. Dilation behavior was also observed in samples. Ghazavi (2004) performed direct shear tests on sand-rubber mixtures and obtained friction angles of 31-38° for sand-rubber particle mixtures. He reported an apparent cohesion for the mixtures. From environmental point of view and due to lightweight material for the sand-rubber particles, it may be useful to use such materials in corresponding geotechnical projects (Ghazavi, 2004).

This paper describes the results of a number of CD triaxial compression tests carried out on pure rubber particles.

Material

In this research, rubber grains were obtained from scrap tires. The specific density of rubber grains was 1.32. The size distribution of the rubber grains is shown in Figure 1.

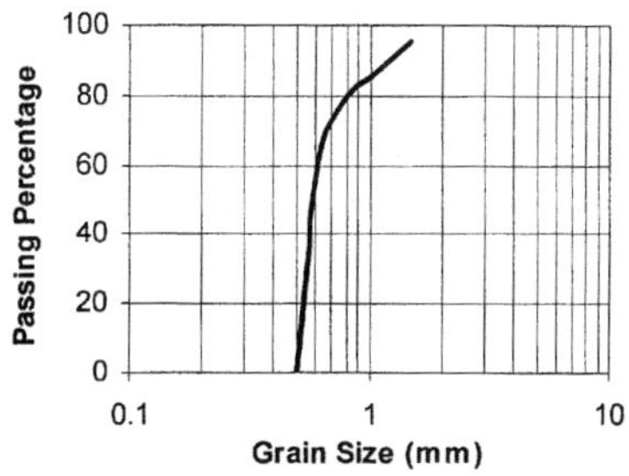

Figure 1: Particle size distribution of rubber

Testing Procedure

A series of triaxial compression tests has been performed on saturated rubber grains at two compaction levels according to the procedure described by ASTM. The material was poured in the test mold and compacted to the desired level. The compaction level was controlled by a certain weight and volume. The sample was then taken out and covered with the test membrane. It is noted that two cell pressures of 200 and 400 kPa have been used. The all-around pressure was first applied to consolidate the specimen. Axial load was then exerted to the sample and continued to the failure stage or to a certain axial strain. In this research work, no distinguished peak for deviator stress was reached, therefore the stress corresponding to 20% axial strain was considered as for the failure stage. The same trend was observed and therefore the same assumption was also made by Wu et al. (1997).

Results

Figures 2 and 3 illustrate the deviator stress versus the axial strain for two compaction efforts corresponding to void ratios of 0.614 and 0.624, respectively. As shown, for the cell pressure of 200 kPa, a relatively constant deviator stress was reached, however, for higher all-around pressure of 400 kPa, a distinguished deviator stress was not accessed. This is in good agreement with the finding of Wu et al. (1997).

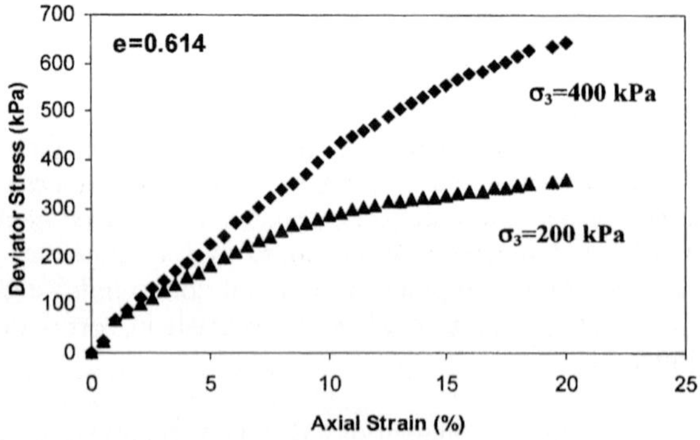

Figure 2: Variation of deviator stress versus axial strain for tire grains with void ratio of 0.614

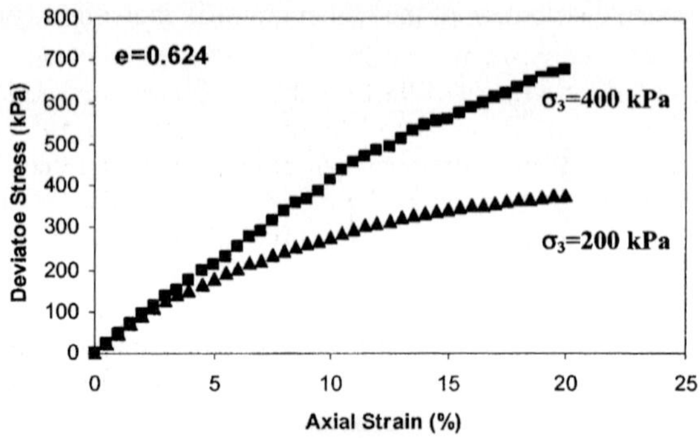

Figure 3: Variation of deviator stress versus axial strain for tire grains with void ratio of 0.622

Table 1 summarizes the friction angle and apparent cohesion obtained for rubber particles. As shown, the friction angle of tire grains is small. Also, an apparent cohesion intercept has been obtained.

As mentioned before, Wu et al. (1997) carried out triaxial tests on 5 types of tire chips with maximum sizes in the range of 2–38 mm and obtained friction angles between 40–60°. However, this range was not attained in this research. Here friction angles between 22-23° were obtained for tire grains having sizes between 0.5-2 mm. It seems that obtained friction angles in this research are reasonable due to supporting evidence. Performing small direct shear tests, Ghazavi (2004) obtained $\phi=30^{\circ}$ for rubber grains with sizes ranging 1-7 mm. Also according to a report from the University of Wisconsin-Madison (1995), a value of $\varphi=31^o$ was obtained for tire chips using direct ring shear test apparatus. Humphery et al. (1993) also tested pure tire chips using a direct shear apparatus and obtained $\varphi=19\text{-}26^o$. Significant friction angle may be achieved by the use of waste tire shreds mixed with sand. Thus tire shreds can enhance the friction angle of sands remarkably (Edil and Bosscher 1994; Foose et al. 1996; Ghazavi and Amel-Sakhi 2005a). For example, Ghazavi and Amel Sakhi (2005a) performed large direct shear tests on sand-tire shreds and obtained friction angle up to 67° for sand-shred mixtures containing optimized shred size.

As shown in Table 1, the values of cohesion intercept were to be in the range of about 19.7 to 20.9 kPa. Humphery et al. (1993) also found cohesion intercept between 4.3 to 11.5 kPa for pure tire chips. Wu et al. (1997) also found cohesion intercept on five types of pure tire chips. They used a consolidation stress of 55 kPa in their experiments. The authors believe that the cohesion obtained in the present tests may be attributed to large consolidation pressures used. This cohesion possibly is not dominant under low all-around pressures.

Table 1: Shear strength parameters of rubber particles obtained from triaxial tests

Parameter	Lower compaction (e=0.614)	Higher compaction (e=0.622)
Friction Angle, ϕ^o	22	23
Apparent Cohesion (kPa)	20	21

Conclusions

The present paper contributes to determine shear strength parameters of waste tire grains. Tire chips with sizes ranging between 0.5-2 mm have been tested using triaxial compression test apparatus. Two initial void ratios of 0.614 and 0.633 have been used to demonstrate the effectiveness of compaction efforts on shear strength of chips. The values of friction angles and cohesion intercept have been determined. The results show that the friction angles range $22\text{-}23^o$. The cohesion intercept are also in the range of 20-21 kPa. Based on the chip sizes used in this research and other research work reported in the literature, the obtained data in this paper are well within the ranges reported by others. Based on the limited findings in this research and other interesting features such as light weight material, no apparent environmental impact, etc., it may be said that tire chips may be used confidently instead of sandy and/or gravelly soils in appropriate applications. It is recommended to perform more research work to generalize the findings in this paper.

References

Allman, M. A., and Simundic, G., 1998. "Testing of a retaining wall constructed of waste tires." *Proceedings Third International Congress on Environmental Geotechnics*, Vol. 2, pp. 655-660, Lisoa, Portugal.

Al-Tabbaa A, Aravinthan T. 1997. "Engineered clay-shredded tire mix-tires as barrier materials." *Proceedings of the International Conference on Containment Technology*, Florida State Uni-versity, 621-627.

Al-Tabbaa, A., and Aravinthan, T. 1998. "Natural clay-shredded tire mixtures as landfill barrier materials." *Waste Management,* 18, pp. 9-16.

Al-Tabbaa A, Blackwell, O, Porter S.A. 1997. "An investigation into the use of soil-tire mixtures as construction material." *Journal of Environmental Technology*, 18(9), 913-920.

Baykal G, Kaval A, Alpatli H.M. 1995. "Rubber-kaolinite and rubber-bentonite liners." In: Sarsby W, editor. *Waste Disposal by Landfill-Green'93*. Balkema, 390-404.

Baykal G, Alpatli H.M. 1995. "Permeability of rubber-soil liners under confinement." Acar YB, Daniel DE, editors. *ASCE Geotechnical Special Publication No. 46. ASCE*, 718-731.

Black B.A, Shakoor A.A. 1997. "Geotechnical investigation of soil-tire mixtures for engineering applications." *Proceedings of 1st International Conference on Environmental Geotechnics*. Bitech Publications, 617-623.

Bosscher, J., Edill, T.B., and Kuraoke, S. 1997. "Design of highway embankments using tire chips." *Journal of Geotechnical and Geoenviromental Engineering, ASCE*, Vol. 123, No. 4, pp. 297-304.

Bridgwater A.V, Mumford C.J. 1979. *Waste recycling and pollution control handbook.* UK, George Godwin Ltd.

Edil, T.B., and Bosscher, P.J., 1994. "Engineering properties of tire chips and soil mixtures." *Geotechnical Testing Journal*, Vol. 17, No. 4, pp. 453-464.

Foose, J., Benson, H., and Bosscher, J. 1996. "Sand Reinforced with Shredded Waste Tires." *Journal of Geotechnical Engineering, ASCE*, Vol. 122, No. 9, pp. 760–767.

Garga, V.K., and O'Shaughnessy, V. 2000. "Tire–reinforced earthfill. Part 1: Construction of a test fill, performance, and retaining wall design." *Canadian Geotechnical Journal*, No. 37, pp. 75-96.

Ghazavi, M. 2004. "Shear strength characteristics of sand-mixed with granular rubber." *Geotechnical and Geological Engineering, An Internatinal Journal*, Vol. 22, No. 3, pp. 401-416.

Ghazavi, M., and Amel-Sakhi, M. 2005a. "Influence of optimized tire shreds on shear strength parameters of sand." *International Journal of Geomechanics, ASCE*, Vol. 5, No. 1, March, pp. 58-65.

Ghazavi, M., and Amel-Sakhi, M. 2005b. "Size optimization of waste tire shreds in sand-shred mixtures using CBR tests." *Geotechnical Testing Journal (ASTM),* Vol. 28, No. 6, pp. 564-569.

Ghazavi, M. and Barmaki, A. 2005. "Strength enhancement of cohesive soil using waste tire chips." *8th International Conference on Cohesive Sediment Transport (INTERCOH2005)*, Institute of Lowland Technology, Saga University, Japan, pp. 26-29.

House Bill S2462. 1990. "Tire recycling Incentive Act of 1990." 1st US Congress, Washington, DC.

Humphrey, D.N., and Manion, W.P., 1992. "Properties of tire chips for lightweight fill." *Grouting, Soil Improvement, and Geosynthetics*, Vol. 2, ASCE, New York, N.Y., pp. 1344-1355.

Humphery, D.N., Sanford, D.C., Cribbs, M.M., and Manion, W.P. 1993. "Shear strength and compressibility of tire chips for use as retaining wall backfill." *Transp. Res. Rec. 1422*, National Research Council, Transp., Res. Board, Washington, D.C., 29-35.

Lee, J. H., Salgado, R., Bernal, A., and Lovell, C. W. 1999. "Shredded tires and rubber–sand as lightweight backfill." *Journal of Geotechnical and Geoenviromental Engineering, ASCE*, Vol . 125, No 2, pp. 132-141.

Lyons J.C., Hitchens D.G., and Monticello DA. "Recovery of scrap rub-ber tires from landfill for construction uses." In: Sarsby W, editor.Waste disposal by landfill-green'93, Balkema, 1995, 327-334.

Nightingale, D.E.B., and Green, W.P., 1997. "An unsolved riddle: Tire chips, two roadbeds, and spontaneous reactions." *Testing Soil Mixed with Waste or Recycled Materials*, ASTM STP 1275, M. A. Wasemiller , K. B. Hoddinott, Eds. , pp. 265-285.

O' Shaughnessy, V., and Garga, V.K., 2000a. "Tire–reinforced earthfill. Part 2: Pull–out behavior and reinforced slope design." *Canadian Geotechnical Journal*, No. 37, pp. 97-116.

O' Shaughnessy, V., and Garga, V.K., 2000b. "Tire–reinforced earthfill. Part 3: Environmental assessment." *Canadian Geotechnical Journal*, No. 37, pp. 117-131.

Poh, P.S. H., and Broms, B.B., 1995. "Slope stabilization using old rubber tires and geotextiles." *Journal of Performance of Constructed Facilities*, Vol. 9, No. 1, pp. 76-80, 1995.

Sumanarathna, I.H.D., Mallawarctchie, D.P., and Kulathilaka, S.A. S., 1997. "Stabilization of slopes by anchored type retaining structures." *Proceedings 14th International Conference of Soil Mechanics and Foundation Engineering*, New Delhi, India, pp. 1261-1264.

Tweedie J.J, Humphrey, D.N., and Sanford, T.C., 1998. "Tire-shreds as lightweight retaining wall backfill." *Journal of Geotechnical and Geoenvironmental Engineering, ASCE*, 124(11), pp. 1061-1070.

University of Wisconsin-Madison. 1995. "Use of scrap tires in civil and environmental construction." *Environmental Geotechnics Rep. No. 95-2*, (Selected Publications from the University of Wisconsin-Madison), Madison, Wis.engineering

Wu, W.Y., Benda, C.C., and Cauley, R.F. 1997. "Triaxial determination of shear strength of tire chips." *Journal of Geotechnical and Geoenvironmental Engineering, ASCE*, 123 (5), 479–482.

Alternative Materials Used As Embankment Fill In Sustainable Engineering Landscape Applications

Dr. Richard Griffiths TRL, Cardiff, Wales
Martin Lamb TRL, Cardiff, Wales
Dr. John Lewis TRL, Cardiff, Wales

Abstract

A suitability study of the use of alternative materials in the construction of a vegetated bund has been carried out by the authors. Basic Oxygen Steel Slag (BOS) and Construction, Demolition and Excavation Waste (CD&EW) were selected as bund core materials based on their local availability and production volume as industrial by-products. A third core section of local sub-soil was also used as a control. The bund was covered with a 200mm thick topsoil layer, mulch, a standard grass mix and a selected range of indigenous shrubs and trees. The key areas studied were the ability of each core material to sustain vegetation, moisture retention and the potential for leachate generation, and potential effects on the local ground and surface waters. The results from tensiometer data and vegetation assessments indicate that BOS and CD&EW are suitable construction materials when used as unbound fill in the core of vegetated bunds. There are no clear discrepancies in the health or growth of vegetation on these alternative materials over conventional sub-soil. An assessment of the leaching characteristics of the materials utilised in the bund also indicate that they present no significant environmental risk in this application.

Introduction

This study provides a qualitative assessment of vegetation growth on alternative/secondary fill materials – BOS and CD&EW. An assessment of the suitability of alternative materials to act as a medium for vegetation growth was conducted on a purpose built bund at the Baglan Energy Park (BEP), near Port Talbot. The bund comprised three sections, each with a different core material; BOS, CD&EW and a control section of locally sourced sub-soil. These core materials received a 200mm covering of topsoil and a spent mushroom compost mix. To minimise the number of test variables, the bund was divided into uniform sections, each planted to an identical design specification consisting of mulched and un-mulched panels stocked with grass, shrubs and trees. Waste cellulose (waste paper) was used as the mulching compound.

Moisture retention characteristics of all materials were monitored by tensiometers installed within the bund at the topsoil/core-material interface, and within the core material. Monitoring of suction pressures (via the tensiometers), vegetation health, height and spread was carried out at regular intervals throughout the duration of the trial. Further to these tests, a leachate characterisation study of the BOS and CD&EW used in the bund was undertaken as part of an MSc project at Cardiff University, the results of which are summarised in this paper.

The Site

The bund (Figure 1) was designed as a modified flood defence structure, capable of directing overflow produced by a 125 year storm from an adjacent attenuation pond to the Baglan Brook via a lagoon. The structure crosses a former railway and access road, and CD&EW from these structures was incorporated within the core of the bund.

Figure 1 Vegetated bund at November 2002 (left) and June 2003 (right)

Design and Construction of the Bund

The structure is approximately 90m long by 10m wide with a 2m wide crest, and the design provides a slope angle of approximately 1 in 2 for both sides of the bund, thus providing a sufficient test area for controlled planting trials. To permit a comparison between the performance of the conventional and secondary bund core materials, the design of the bund was split into three sections, each 24 metres long with an approximately 4m long transition zone between adjacent core sections. The core of each section was constructed as follows:

a) Core Section S1 - composed of a silty clay sub-soil from Llanelli (control section).
b) Core Section S2 - composed of BOS sourced from the Corus steelworks at Margam.
c) Core Section S3 - composed of CD&EW locally sourced from the BEP.

The structure was constructed in 250 mm layers, each layer compacted using a vibrating roller. Once the core of the bund was complete, the slopes were trimmed to the specified design angle and height, and a 200mm layer of topsoil and spent mushroom compost mix was applied. Fencing, partially buried, was installed around the bund to prevent burrowing animals from gaining access.

During construction of the bund, the Environment Agency determined that both the BOS and CD&EW required no exemption from Waste Management Licensing due the bund being a research project. However, if recycled/secondary aggregates are produced in accordance with the Waste Resource Action Programme's (WRAP) quality protocol for the production of aggregates from inert waste (WRAP, 2005), the current Specification for Highway Works [MCHW Volume 1] (Highways Agency, 2005) and the Design Manual for Roads and Bridges [HD 35/04] (Highways Agency, 2004), permits their use as embankment fill without requiring exemption from Waste Management Licensing.

Planting Scheme for the Bund Vegetation

When planting a structure of this type, vegetation appropriate to the locality, soil and atmospheric conditions should be chosen. These considerations are also relevant when planting on alternative materials. However, the unorthodox nature of the alternative materials used in this study meant that the standard growing conditions and behaviour of the selected vegetation could not be assumed. Vegetation growing on a bund constructed with alternative materials may, for example, also encounter issues with drainage, gas (oxygen) permeability, pH, nutrient availability and, potentially, phytotoxins.

Within each of the three bund sections, the central 24 metres was used as a test area with approximately 4 metres at either end as a transition zone between adjacent sections. This removed the impact of varying soil conditions at the interface of the different core materials and any tolerances required during the construction phase. Each of the three 24 metre test sections was then split into twelve panels, six on each slope face, creating a total of 36 panels (Figure 2). With the bund orientated generally east to west, this allowed for relative growth assessments to be made on both south and north facing slopes.

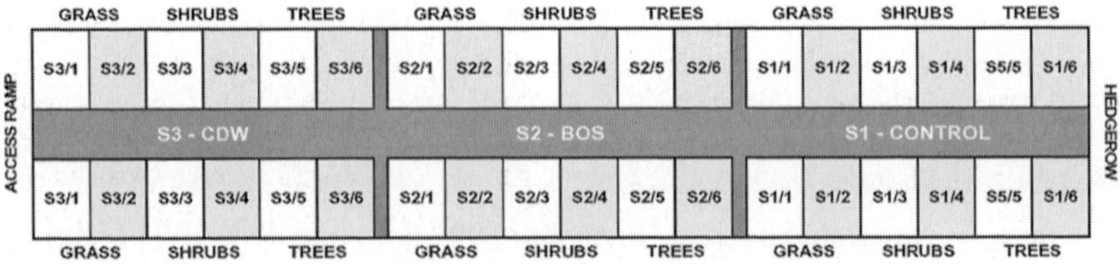

Figure 2 Bund sections and layout of panels

Each of the six panels was assigned specific types of vegetation; trees, shrubs or grasses. To limit the number of test variables, the layout of the panels on each slope was replicated across each section with, from west to east, two panels of grass, two panels of shrubs and two panels of trees (Figure 2). As part of the study, the performance of cellulose (waste paper) mulch, inoculated with rhizo bacteria and mycorrhizal fungi and produced by Excel Fibre Technology Ltd, was also assessed in terms of initial vegetation (grass) establishment. To enable an assessment to be made of the effect of the cellulose mulch on the establishment and growth of the vegetation, alternate panels of vegetation were mulched.

The size of the test areas within each section of the bund allowed for a number of selected tree and shrub species to be monitored in a controlled manner, along with grass. With the assistance of Neath Port Talbot County Borough Council's landscape specialists, appropriate native species of trees and shrubs specific to the maritime area of South Wales were identified and selected for planting on the bund. Figure 3 shows the arrangement of the trees and shrubs planted in December 2002.

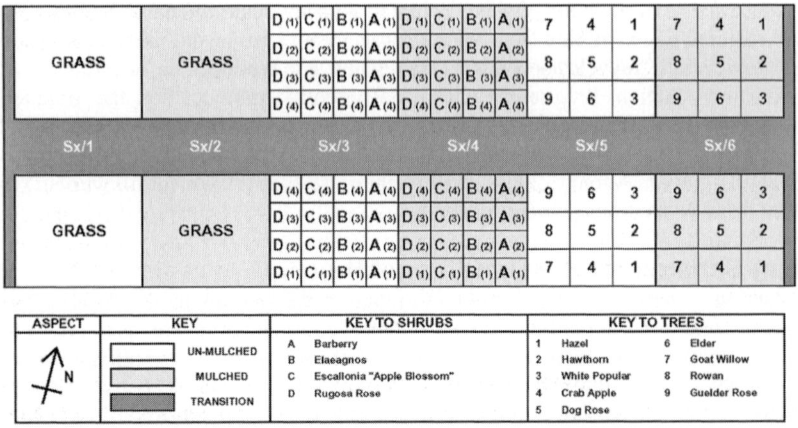

Figure 3 Vegetation selection and location

Prior to planting, the structure was hydroseeded with a DOT Pro 85 seed mix, as commonly used on highway embankments, in September 2002 across all 36 panels of the bund. The crest of the bund and the transition zones were seeded with a wildflower mix to provide seasonal colour and to enhance biodiversity. Mulch mats were placed around the base of each shrub and tree specimen to reduce localised transpiration of moisture from the grass.

Bund Monitoring

In combination with in situ density tests, disturbed samples of each core material were taken during construction, along with samples of the topsoil layer and the underlying made ground. Samples were subjected to laboratory tests including bulk density, particle density, moisture content, particle size distribution and chemical analysis. To provide an ongoing comparison of the moisture retention of the core materials and their effect on the upper soil layer, regular moisture content tests were also carried out on fresh core material samples across the bund. The effect of vegetation type on the moisture content within the differing bund materials, accompanied by the effectiveness of the recycled cellulose mulch to retain moisture, was measured using tensiometers. Two tensiometers were installed in each panel of the bund, one at the surface of the core material (T1), and another placed 150 mm in to the core material (T2). Monitoring was conducted on a weekly basis to accurately assess variations in moisture suction across the three core material sections.

Each tensiometer comprised an impermeable septum at the upper end and a permeable ceramic tip at the lower end, connected by a central trunk section. Once inserted into the bund, each tensiometer was filled with de-ionised water (leaving an air space) and the septum replaced to provide an impermeable seal. Water is drawn out of the permeable ceramic tip into the surrounding material at a rate dependant on the moisture content of the surrounding environment. Liquid exchange through the ceramic tip causes the pressure in the airspace to vary. In this way, by measuring pressure in the tensiometer airspace, it is possible to gain a measure of the hydrostatic pore pressure in the bund materials and, over time, how these may differ between materials and with changes in temperature and rainfall.

Vegetation was monitored for health, growth rate and spread, and formed two distinct stages. The first, lasting from seeding in September until mid December 2002 when the trees and shrubs were planted, focused solely on the establishment performance of the grass cover and grass shoot lengths. A photographic record of each panel was maintained which allowed comparisons to be drawn between either adjacent mulched and un-mulched panels, panels on the north and south sides of the bund, or across different sections of the bund. An assessment of vegetation growth on the transition or crest areas was not undertaken.

Following planting of the trees and shrubs a second stage of monitoring was commenced. A base line data set was taken in January 2003 which included a qualitative health assessment of each individual shrub or tree accompanied by measurements of height and spread. In each subsequent month (reduced to every three months as the project progressed) a health assessment was undertaken using a five-point scale (see Results section). In association with the photographic record of the bund vegetation, these techniques enabled accurate assessments to be made of the performance of the core materials and vegetation used, or of particular combinations of soil type and plant species.

Due to the associations between vegetation growth and weather, it was also necessary to monitor the weather conditions to which the bund was subjected, with daily datasets provided by the Meteorological Office weather station at the Mumbles, Swansea. In particular, the moisture suction measurements from the tensiometer monitoring needed to be considered in relation to rainfall and temperature readings.

Results

Geotechnical Data

The bund materials were mainly characterised by their Particle Size Distribution (Figure 4):

- With the exception of the topsoil, all materials used in construction of the bund were granular in nature and provided good strength and high permeability.
- The high fines content of the topsoil indicated that it would have a low permeability and would retain moisture and nutrients appropriately, thus providing a robust environment for root growth. This was confirmed, in part, by a measured moisture content of 27%.
- CD&EW and the control sub-soil were relatively variable materials.
- Due to industrial processing at source, BOS is a reasonably consistent granular material containing mainly gravel sized particles. Despite this, the moisture content of the material was higher than expected due primarily to the lime content of BOS and its ability to absorb water.
- As anticipated, samples of the underlying made ground indicated an inconsistent, poorly graded, gravelly sand with a low fines, coarse sand and fine gravel fraction. For clarity, grading curves for the made ground are not presented in Figure 4 due to the spread of results across the footprint of the bund.

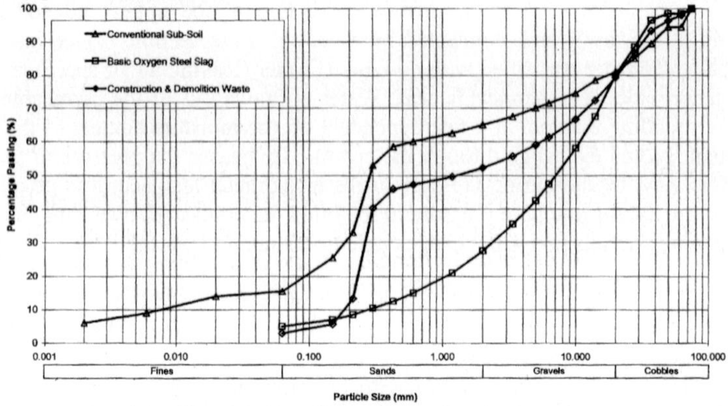

Figure 4 Particle Size Distribution curves for the bund core materials

Chemical Data

Chemical analyses of the bund core materials have been compared against Soil Guideline Values – risk-derived values based on intake routes including, for example, inhalation, ingestion and dermal contact. These results indicate that none of the materials contain any significant organic contamination, with all materials showing slightly elevated concentrations of metals, predominantly arsenic and chromium. Although all three core materials are alkaline, the CD&EW and BOS have a higher alkalinity with an average pH greater than 10. The BOS and CD&EW contain lower levels of nitrogen and organic materials when compared to the control sub-soil, although there is no significant difference between the available nutrients for the control soil and CD&EW. The BOS showed much higher concentrations of available magnesium and potassium than the other two materials, although the phosphorous concentrations were the same. The results suggested that with higher organic and nitrogen values the control sub-soil should provide the best medium for supporting vegetation, whilst the high alkalinity of the BOS and CD&EW may impair root growth.

Leaching

Leaching tests were performed on samples of BOS and CD&EW excavated from the bund. Two standard tests and an availability test were performed, and the BOS sample also underwent a series of column leaching trials simulating a range of environmental conditions. Results from the tests indicate that, based on the European standard CEN test (CEN, 2004), CD&EW and BOS (Type 2, as used in the bund) can be classified as Group 1 materials in compliance with CIRIA Report 167 leaching criteria. Group 1 materials require no restrictions on their use, based on their ability to potentially affect water quality.

In the BOS test, Mercury concentrations in leachate samples decreased rapidly with an increasing liquid/solid ratio and with time. The maximum Mercury concentration determined during the analysis of all test leachates was 0.004mg/l; requiring 4xdilution to satisfy the relevant environmental quality standard (EQS) and water quality standard (WQS) of 0.001mg/l. For the worse case scenario of a 125 year storm taking place on the uncovered core materials, without shielding by topsoil, suitable attenuation of contaminants by dilution will occur to reduce contaminants of concern (COC) to within EQS / WQS levels on reaching the adjacent water course (Baglan Brook). Additionally, in reality most of the water in such an event would not percolate through the bund but would run off directly into the Baglan Brook, via the adjacent lagoon. It should also be noted that under normal conditions only part of the rainfall is available to produce leachates as a fraction will be lost through evaporation and transpiration before it has a chance to interact with the bund materials.

Tensiometers

The results of the moisture suction monitoring of the 72 tensiometers across the bund, when compared with the controlled planting layout, enabled variations due to core materials, aspect, vegetation type and presence of the cellulose mulch to be observed and considered. Figure 5 shows the mean moisture suction results for the three core materials on each side of the bund. From Figure 5 it can be seen that from mid October 2002 until early March 2003 there was minimal moisture suction across any of the core material sections due to abundant precipitation. Prior to October and after March, clear variations in moisture suction were observed between the three core materials. Although, variations were apparent in moisture suction between the different core materials on the southern side of the bund, there remained no clear difference between sections on the northern side. For the shallower (T1) tensiometers on the southern side of the bund, from March 2003 to the end of the monitoring period it was apparent that the moisture suction was considerably higher within the CD&EW and BOS sections, suggesting that there is less available moisture than within the control soil. This supports the earlier PSD results that suggested that the BOS and CD&EW would be more permeable than the control soil.

The deeper (T2) tensiometers indicate that there is far less variation between moisture suction within the three core materials. Each material at some point has shown the highest moisture suction, but the levels arc variable. This is likely to be a result of air drying of the upper soil layer in combination with the higher permeability of the granular CD&EW and BOS materials. Although not shown in Figure 5, tensiometer data in April 2003 has shown variations in available moisture due to the type of vegetation present. Across the three core materials it has been observed that the tree panels provide the highest moisture suction, followed by the shrubs and then the grass. This is despite the grass having a substantially greater period of time to establish a comprehensive root system.

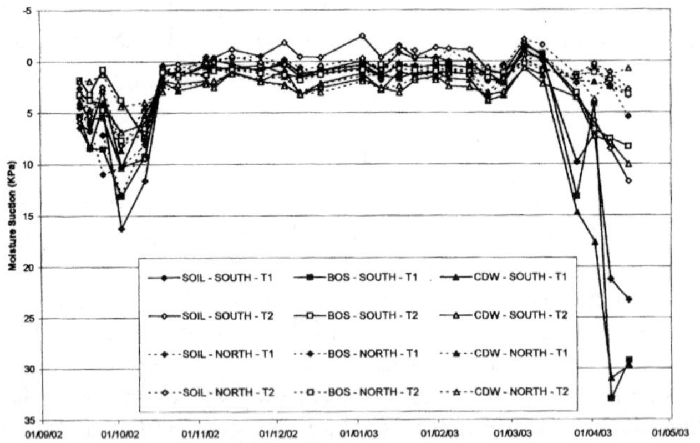

Figure 5 Tensiometer results

Assessment of Vegetation Growth

The results of the bund inspections clearly showed better grass establishment on panels treated with the waste paper cellulose mulch. Of the 18 pairs of panels, one month after grass seeding 15 of the mulched panels were outperforming their adjoining un-mulched panels. This trend remained until December 2002 by which point the un-mulched and mulched panels were all performing equally well. A general observation during the grass establishment period was that following heavy rain some surface material wash-out occurred on the un-mulched panels, whilst the mulched panels suffered significantly less. There is, therefore, an additional potential benefit in using mulch as it helps to prevent soil wash-out in the early period of vegetation establishment.

Following establishment of the grass and planting of the trees and shrubs, each individual tree, shrub and grass panel was given a rating from 1 (no signs of life) to 5 (thriving). This qualitative assessment, based on measured height and spread versus the original size of the shrubs/trees, assessed the subject's health and performance relative to the alternate growing mediums. A summary of the results is presented in Table 1 and shows almost no variation between growth performance on the three core materials. The majority of trees and shrubs were performing very well with average ratings between 'good' and 'excellent to thriving'.

Table 1 Mean vegetation assessment results

Vegetation Type	Control Soil	BOS	CD&EW
Trees	4.50	4.42	4.44
Shrubs	4.42	4.36	4.34

With the exception of Rugosa Rose, shrub health on all materials is classed as good or better. Unlike the case for grass, shrubs and trees were planted deeper into the bund, with their root systems becoming established at the top of the core materials. Barberry, Elaeagnos and Escallonia performed very well on all materials, with no clear pattern indicating that any one medium was better than another in supporting growth. Whilst performing acceptably, it is unclear why Rugosa was not as successful as the other shrubs; one possibility is that it suffered from competition from the other, larger shrubs. On the final inspection in October 2004, 13 of 108 trees showed no signs of life. It is considered that a loss of 13 trees over a 3 year period, however, is not exceptional for a vegetated structure subjected to a maritime environment. Additionally, a 'no sign of life' value should not be taken as being solely due to the growing conditions presented by the bund, as one tree is recorded to have been damaged during bund maintenance and another by animal foraging. Overall, the trees showed the ability to become established and thrive in all three growing mediums, with Hazel, Dog Rose, Elder and Goat Willow proving to be the most successful species and Hawthorn and Rowan being least successful.

Conclusions

To date the results from the tensiometer data and vegetation assessments from this study indicate that BOS and CD&EW are suitable core materials when used as unbound fill in bund construction for engineering or landscape applications. There are no clear discrepancies in the long term health or growth of vegetation on these alternative materials over conventional sub-soil. Across the structure, durability of the vegetation has been excellent, with only minimal vegetation failure. These results indicate success in both the selection of indigenous species and of the type of alternative materials used for sustaining vegetation establishment. Despite their granular nature, BOS and CD&EW were shown to retain sufficient moisture to support the vegetation. Tensiometer monitoring has revealed that moisture retention within the bund varies with weather patterns, as might be expected, and does not differ significantly between the alternative materials used and the control test section. Although the control sub-soil showed marginally better results, it is clear that all three core materials successfully support vegetation growth irrespective of their differences in chemical and physical properties. This study clearly indicates that through the vegetation establishment phase no growth impairment has been exhibited due to the core materials selected for the trials.

The trials also demonstrated that rapid establishment of grass on the bund was achieved when using cellulose mulch. Overall, mulched panels were shown to vegetate more quickly and generally outperformed the un-mulched panels in terms of initial growth and vegetation health. Evidence also exists that the mulched panels lose less material due to rain run off, thus preventing soil wash-out before grass growth. Results also show that the alternative materials used in construction of the bund do not produce leachates that could pose a significant risk to local controlled waters.

Acknowledgments

The bund study was a collaborative venture between TRL Limited, the Welsh Development Agency, Neath Port Talbot County Borough Council and BP Chemicals. It was undertaken as a work package within the Geoenvironmental Research Park project lead by Cardiff University's Geoenvironmental Research Centre.

References

Baldwin, G., Addis, R., Clark, J., and **Rosevear, A.,** (1997). Use of industrial by-products in road construction – Water quality effects. CIRIA Report 167. Construction Industry Research and Information Association, London. 151 pp.

CEN. (2004). Characterization of waste - Leaching behaviour tests - Influence of pH on leaching with continuous pH-control. Draft European Standard PrEN 14997. European committee for standardisation, Brussels.

Highways Agency. (2004). Design Manual for Roads and Bridges: Volume 7, Section 1, Part 2, HD 35/04. The Stationery Office, London.

Highways Agency. (2005) Manual of contract documents for highway works: Volume 1, Specification for highway works, The Stationery Office, London.

Waste and Resources Action Programme (WRAP). (2005). The quality protocol for the production of aggregates from inert waste, 2nd Edition. WRAP, Banbury.

Select Engineering Properties of Crushed Glass-Dredged Material (CG-DM) Blends

Dennis G. Grubb, Ph.D., P.E., Schnabel Engineering, West Chester, PA USA
Michael Carnivale, III, P.E., US Army Corps of Engineers, Philadelphia, PA, USA
Joseph Wartman, Ph.D., P.E., Drexel University, Philadelphia, PA, USA
Patricia M. Gallagher, Ph.D., P.E., Drexel University, Philadelphia, PA, USA

Abstract

To bolster recycling in the urban environment and to avoid disposal of potentially suitable fill materials, crushed curbside collected glass (CG) and dredged material (DM) classifying as organic silt (OH) were blended to evaluate their geotechnical properties. This paper reports on the basic soil parameters, modified Proctor compaction indices, strength, consolidation and hydraulic conductivity properties. The addition of as little as 20% CG led to important changes to DM, as summarized herein. The range of properties available from CG-DM blends allows for geotechnical designers to create "recycled material" fill for a wide range of uses.

Introduction

This paper reports on a laboratory evaluation of blending crushed glass (CG) and dredged material (DM) and the suitability of the blended products as landscaping, embankment and structural fill materials for transportation, airport, building and maritime construction, land reclamation, and brownfields and portfields redevelopment in urban areas. This study was motivated by the need for a pragmatic solution for two long-term compelling problems: the disposal of DM and beneficial use options for curbside-collected glass. The United States Army Corps of Engineers (USACE) has major challenges maintaining long-term disposal capacity for DM in its confined disposal facilities (CDFs) in or near many metropolitan areas, and other upland disposal options are limited, cost prohibitive or difficult to permit. In addition, state Departments of the Environment have overemphasized the curbside collection of glass without making comparable investments in beneficial use markets such as the use of CG as an aggregate substitute (Wartman et al. 2004a). This has resulted in a glut of glass (especially glass not suitable for bottling applications) at Material Recovery Facilities (MRFs) in non-bottle bill states. A joint solution for the beneficial use of DM and CG is to create blends using these materials to make synthetic fill materials for urban applications as the cost of acquiring fill materials is historically expensive. This paper is an abbreviated version of Grubb et al. (in press).

Laboratory Investigation

Materials and Basic Physical Properties

City of Philadelphia curbside-collected glass was the source of glass materials (CG) for this study. The glass was crushed and sieved through a 9.5 mm (3/8") sieve, a size that does not generally represent a physical handling hazard. Material dredged from the Schuylkill River federal navigation project, composed primarily of silts (MH, OH), and hydraulically placed in Basin A of the USACE's Fort Mifflin CDF (Philadelphia, PA) was the source of DM. All of the materials were collected and sealed in 55-gallon steel drums prior to shipment to the testing laboratory. The following CG-DM blends were evaluated

(CG/DM ratio by % dry wt.): 20/80, 40/60, 50/50, 60/40 and 80/20. Bulk samples of CG and DM were oven- or air-dried and subsequently blended to the specified proportions either by hand in a mixing bowl or in a cement mixer until they visually appeared to be of uniform consistency. A series of laboratory tests were performed to evaluate the basic physical properties of the CG, DM and CG-DM blends, including moisture content, specific gravity (G_s), loss on ignition (LOI), grain size distribution and Atterberg limits. The soils were then classified according to the Unified Soil Classification System (USCS).

Table 1 summarizes the basic physical properties of the CG, DM and the CG-DM blends and the applicable ASTM standards. The moisture contents presented reflect the "as-received" moisture of the CG and DM, and in the case of the CG-DM blends, immediately after mixing. The G_s of the CG was determined to be 2.48, which was consistent with the results of previous studies (Wartman et al. 2004a). The G_s of the DM was 2.40, a value that is consistent with the values determined in regional CDFs (SAIC 2001, Weston 2002).

The organic matter content, determined by the loss on ignition (LOI) [ASTM D 2974 (Method D)], of the 100% CG and DM were 3 and 11%, respectively, with the CG-DM blends containing slightly less organic matter than would be predicted by weight averaging. The organic content of the CG is attributed to debris such as paper, glue and plastic cap fragments whereas the organic matter in DM is comprised of decaying leaves, roots, branches and other detritus.

The grain size distributions were determined in accordance with ASTM D 421 and ASTM D 422 (mechanical sieving only) and are presented in Figure 1. Atterberg limit determinations were performed using material passing the 0.425 mm (U.S. Standard No. 40) sieve. The CG does not strongly influence the Atterberg limits of fine-grained soils through blending chiefly because the fraction of the non-plastic CG passing the 0.425 mm sieve is generally on the order of 5% or less (Wartman et al. 2004a).

Compaction Characteristics

Laboratory moisture-density relationships were developed for CG, DM, and CG-DM blends following the modified (ASTM D 1557) Proctor methods using 5 or 6 moisture-density points. Figure 2 shows the compaction curves for modified Proctor efforts. Zero air voids (ZAV) curves for specific gravities of 2.40 (DM), 2.48 (CG) and 2.65 (typical soil solids) are shown for comparative purposes. The moisture-density curves for the 100% DM exhibit the characteristic convex shape typical of OH soils. With increased CG content, the optimum water content (w_{opt}) decreased, the maximum dry density ($\gamma_{d,max}$) increased, and the shape of the compaction curve trended toward those associated with conventional coarse soils and aggregates. The impact of CG on the compaction characteristics of 100% DM is clearly evident—the addition of 20% CG results in a 20-point reduction in w_{opt} while increasing the dry density by approximately 3 kN/m^3 (18 lb/ft^3) for modified levels of compaction.

Strength Testing

Isotropically consolidated, undrained triaxial ($CI\bar{U}$) shear tests with pore pressure measurements were performed on CG, DM and CG-DM blend samples in general accordance with ASTM D 4757. The specimens were placed in 5 to 7 lifts using a modified undercompaction procedure (Ladd 1978) and compacted by placing a rubber stopper on top of the wet soil and tamping the lift with a Harvard miniature compactor to a minimum of 95% of their $\gamma_{d,max}$ values and 6% wet of optimum based on ASTM D 1557. After mounting the samples in the triaxial device, they were flushed with CO_2 for about 20 minutes and de-aired water for another 60 minutes. Finally, backpressure saturation was applied to achieve a B-value of at least 0.95. Confining pressures of 70, 210 and 345 kPa (10, 30 and 50 psi) were selected because they exceeded the compaction stresses induced during sample preparation, placing the samples in the normally consolidated range.

Table 1 - Physical Properties and Classification of CG, DM, and CG-DM Blends

Media Tested/ Method		Water Content	Loss on Ignition	Plasticity Indices			USCS	1-D Consolidation			
		D 2216	D 2974	D 4318			D 2487	D 2435			
		(%)	(%)	*LL*	*PL*	*PI*	--	C_c	C_r	c_v (cm^2/sec) @ 400 kPa	C_α
Crushed Glass (CG)		1.7	3.1	NP	NP	NP	SP	0.0420	0.005	0.1451	0.0016
Blends	80/20 CG-DM	11.9	2.9	71	48	23	SM	0.0820	0.011	0.0731	0.0009
Blends	60/40 CG-DM	18.5	4.0	74	49	25	SM	0.1380	0.006	0.0755	0.0038
Blends	50/50 CG-DM	24.5	5.7	73	46	27	SM	0.1050	0.014	0.0275	0.0065
Blends	40/60 CG-DM	27.0	5.6	75	44	31	OH	0.1650	0.022	0.0601	0.0179
Blends	20/80 CG-DM	32.9	8.7	80	51	29	OH	0.1450	0.019	0.0270	0.0191
Dredged Material (DM)		45.8	11.0	81	53	28	OH	0.2600	0.017	0.0611	0.024

ASTM Designation shown where relevant

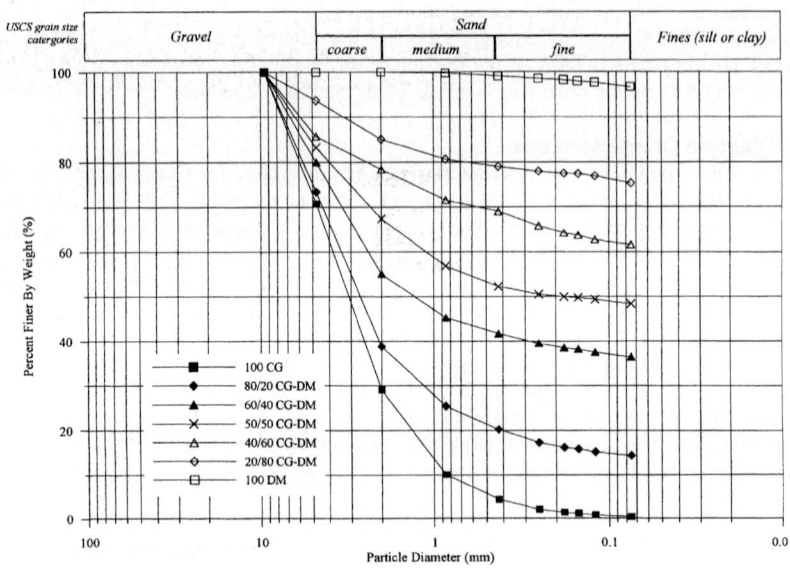

Figure 1 – Grain Size Distribution Curves for CG, DM and CG-DM Blends (by ASTM D 422)

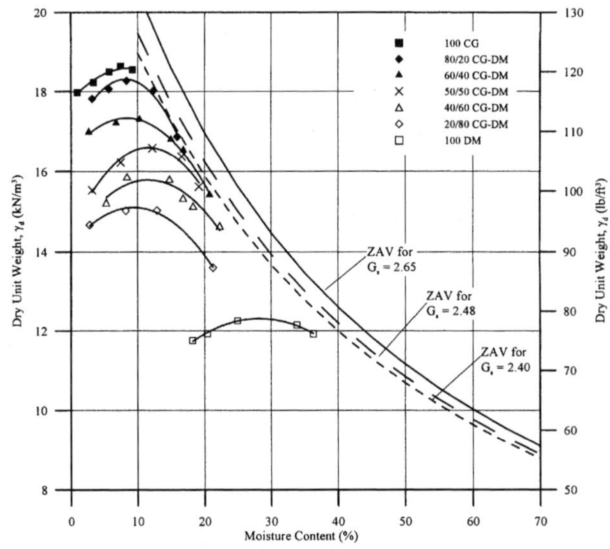

Figure 2 – Modified Proctor Compaction (ASTM D 1557) Curves for CG, DM and CG-DM Blends

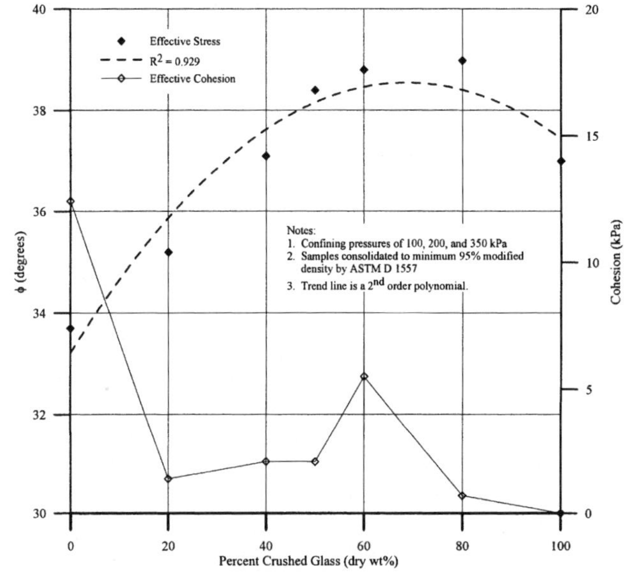

Figure 3 – Isotropically Consolidated Undrained (CIU) Triaxial Results for Compacted CG, DM and CG-DM Blends (by ASTM D 4757)

A strain rate of 2.5% was selected based on the t_{50} values which generally were less than 10 minutes. Time to failure during the tests ranged from 120 to 180 minutes. There was no definitive peak stress, so failure was defined as the shear stress corresponding to the largest ratio of peak stress to normal stress (max. stress obliquity). The effective stress strength parameters are summarized in Figure 3. The 100% DM had a $c_{CI\bar{U}}'$ of about 12 kPa. The addition of CG to the DM caused a gradual increase in $\phi_{CI\bar{U}}'$ from 34° (100% DM) to a high of about 39° for the 80/20 CG-DM blend. The $\phi_{CI\bar{U}}'$ of the 100% CG was determined to be 37° using confining pressures of 70, 210 and 345 kPa (10, 30 and 50 psi).

Consolidation Properties

One-dimensional consolidation properties of the CG and DM materials were determined in accordance with ASTM D 2435, as summarized in Table 1. The e-log P curves for the CG-DM blends are shown in Figure 4, using weight-averaged specific gravities for the blends. The response of the 100% DM to loading conditions is consistent with that of OH soils. Figure 4 illustrates a substantial reduction in settlement with only 20% CG with the addition of glass particles potentially corresponding to the emergence of a skeleton or network of relatively incompressible particles that significantly decreased the potential for settlement. After 40% CG, the e-log P curves appear to plot on top of each other, indicating a diminishing return on CG addition for settlement reduction. As expected, C_c of DM decreased significantly as the percentage of CG increased (Table 1). The largest drop, approximately 50%, occurred with the addition of 20% CG. C_c continued to decrease as the percentage of CG increased to a low of 0.08, but the most dramatic decrease occurred for the increment between the 100% DM and the 20/80 CG-DM blend. There is a general decrease in C_r with increasing percentage of CG. For most data points, $C_c/C_r \sim 7$, in general agreement with the critical state model that predicts $C_c/C_r \sim 5$ (Wroth and Wood (1978), as cited in Bardet (1997)).

Hydraulic Conductivity

The hydraulic conductivities of the 100% DM and CG-DM blends were determined in accordance with ASTM D 5084 while ASTM D2434 was used for the 100% CG. All tests were conducted at 20°C with a confining pressure of approximately 35 kPa (5 psi) with the results summarized in Figure 5. Figure 5 illustrates that the 20/80 and 40/60 CG-DM blends had lower hydraulic conductivity values than the 100% DM, which is likely due to the increased density that occurred prior to reaching a CG content (~40%) corresponding to a more open and interconnected pore space. Shakoor and Cook (1990) reported a similar trend in hydraulic conductivity for silty clay (CL) of glacial origin mixed with gravel particles (13- to 19-mm diameter) in increments of 10 to 80% gravel by weight. In their evaluation of the hydraulic conductivity behavior of mine spoil and kaolinite mixed separately with gravel, Shelley and Daniel (1993) found that the addition of gravel had little impact on the hydraulic conductivity of the blends at gravel contents less than 60%. However, when the gravel content increased above 60% there was only an order of magnitude increase for the mine spoil-gravel and kaolinite-gravel blends.

Discussion

The results of the laboratory testing program indicate that the addition of CG caused significant changes in the physical properties of DM, including reductions in moisture content, organic content and plasticity index, as well as coarsening of the grain size distribution. There are significant improvements in γ_{dmax} (increased) and w_{opt} (reduced) of all of the CG-DM blends. The addition of 20% CG produced a decrease in w_{opt} of 20 points (67%) and increased $\gamma_{d,max}$ by 23% at modified levels of compaction.

What is misleading about the CG-DM blends is that triaxial friction angle results suggest that CG addition to DM does not significantly improve the overall strength of the

100% DM. Figure 6 shows a maximum increase only on the order of 6° (max stress obliquity), whereas Wartman et al. (2004b) showed substantial increases (~12°) in the

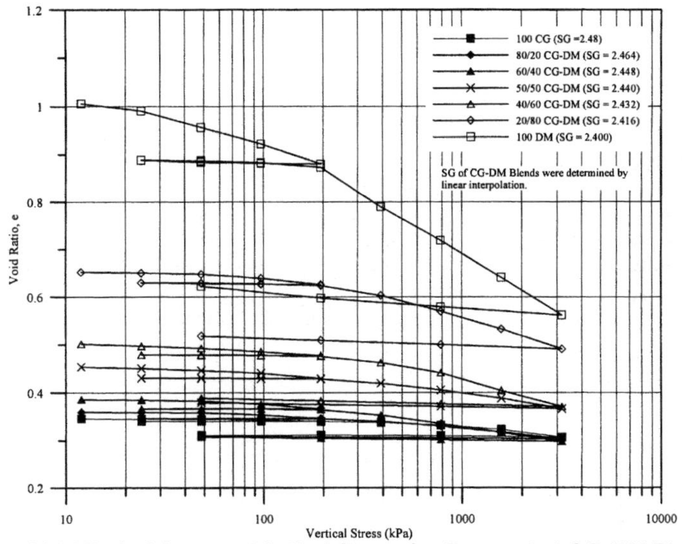

Figure 4 – Void Ratio (e) versus Vertical Stress for Compacted CG-DM Blends (by ASTM D 2435)

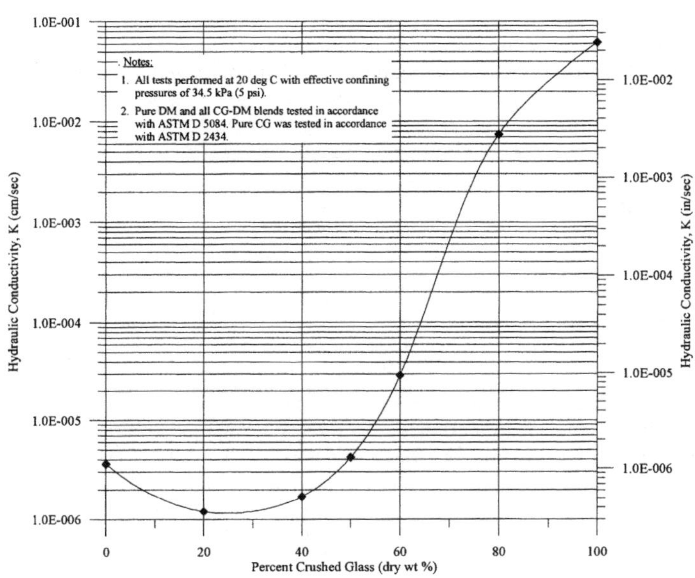

Figure 5 – Hydraulic Conductivity versus Percent Crushed Glass (dry wt %)

friction angles of CG-K blend when blending 50% CG with kaolinite failure defined as 10% axial strain). A more complete picture of the improvements offered by CG addition to DM becomes evident when the shear strength of the CG-DM is evaluated (not just ϕ) which incorporates the influence of cohesion, density and the friction angle. The shear strength of the soil is defined as:

$$\tau_i = c_i + (\gamma_i H_{fill}) \tan \phi_i$$

Where: τ_i is the soil shear strength; c_i is the cohesion; γ_i is the soil unit weight; H_{fill} is the thickness/height of a simulated fill; and ϕ_i is the internal friction angle and i is the blend. Figure 6 presents a contour plot of the normalized shear strength of CG-DM blends versus 100% DM (τ_{blend} / τ_{DM}) as a function of DM content and simulated fill depth (H_{fill} = 1 to 15 m) based on the CIŪ triaxial effective strength values and material properties corresponding to a minimum of 90% modified Proctor compaction for the 100% DM and 20% CG-DM blend, and a minimum of 95% modified Proctor compaction for blends containing greater than 20% CG. Figure 6 clearly shows that 100% DM is stronger than most CG-DM blends at shallow depth (c more important than H), and that blending can increase the strength of DM by up to a factor of 1.73 (for a 15 m fill project).

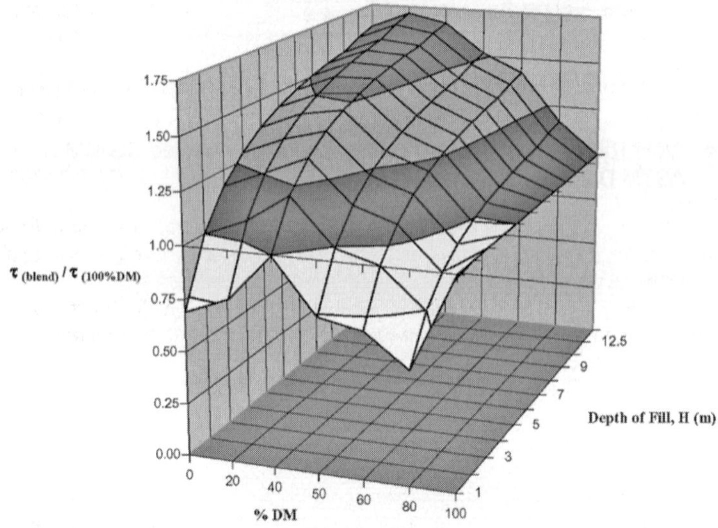

Figure 6 – Normalized Strength of CG-DM Blends as a Function of DM Content and Fill Depth, H

Conclusions

The significant changes in the properties of the CG-DM blends offers the geotechnical engineer the opportunity to optimize on several design considerations that can be evaluated/balanced for construction, the first of which is the ratio of available CG to DM. This may drive many choices, but it appears the major changes in DM behavior begin at 20% CG. For example, decreased compressibility occurs with the addition of just 20% CG, which reduces the compression index by 50%. Twenty to forty percent CG decreases the hydraulic conductivity value below 100% DM, whereas a 60 to 80% addition increased the hydraulic conductivity by 1 to 3 orders of magnitude, respectively. These characteristics allows the designer to hand-pick blends for construction, even to the point they may be varied on the same site depending on the functionality of the fill (inside/outside footprint of structures).

Acknowledgements

The USACE-Philadelphia District provided the funds and dredged material to support this research under contract DACW61-03-C-0021 to Apex Environmental, Inc. (Malvern, PA). Blue Mountain Recycling LLC (Philadelphia, PA) provided the crushed glass for this study. The views expressed here are solely those of the authors and endorsement is not implied by the project sponsors. With permission from ASCE.

References

ASTM. (1968). "Standard test method for permeability of granular soils (constant head)." *ASTM D2434-68*, West Conshohocken, PA.

ASTM. (1988). "Standard test method for specific gravity and absorption of coarse aggregate." *ASTM C127-88*, West Conshohocken, PA.

ASTM (1998a). "Standard test method for laboratory determination of water (moisture) content of soil and rock by mass," *ASTM D2216-98*, American Society for Testing Materials, West Conshohocken, PA.

ASTM (1998b). "Standard test method for particle size analysis of soils," *ASTM D422-63 (Re-approved 1998)*, American Society for Testing Materials, West Conshohocken, PA.

ASTM (2000). "Standard test methods for laboratory compaction characteristics of soil using modified effort," *ASTM D1557-00*, American Society for Testing Materials, West Conshohocken, PA.

ASTM (2003). "Standard Classification for Sizes of Aggregate for Road and Bridge Construction," *ASTM D448-03*, American Society for Testing Materials, West Conshohocken, PA.

Bardet, J.P. (1997). *Experimental Soil Mechanics,* Prentice-Hall, Inc., Upper Saddle River, NJ, 583 pp.

Grubb, D. G., Davis, A., Sands, S. C., Carnivale, M., and Wartman, J. (in press). Field evaluation of crushed glass-dredged material blends, *Journal of Geotechnical and Geoenvironmental Engineering*, ASCE.

Grubb, D.G., Gallagher, P.M., Wartman J., Carnivale III, M. and Liu, Y. (in press). Laboratory evaluation of crushed glass-dredged material blends, *Journal of Geotechnical and Geoenvironmental Engineering*, ASCE.

Ladd, R.S. (1978). Specimen preparation and liquefaction of sands, *Journal of the Geotechnical Engineering Division*, ASCE, 100(10), 1180-1184.

SAIC (2002). *Geotechnical Investigation Report, Fort Mifflin Disposal Area, Cells B and C, Philadelphia, PA,* Contract No. DACW61-00-D-007, Task No. 14, Prepared for: US Corps of Engineers-Philadelphia District, Philadelphia, PA, by: Science Applications International Corporation, 916 Springdale Drive, Exton, PA, 19341, January.

Shakoor, A. and Cook, B.D. (1990). The effect of stone content, size, and shape of the engineering properties of a compacted silty clay, *Bulletin of Association of Engrg. Geologists*, XXVII(2), 245-253, as cited in Shelley and Daniel (1993).

Shelley, T.L. and Daniel, D.E. (1993). Effect of gravel on hydraulic conductivity of compacted soil liners. *Journal of Geotechnical Engineering*, ASCE, 119(1), 54-68.

Wartman, J., Grubb, D.G., Nasim, A.S.M. (2004a). Select engineering characteristics of crushed glass, *Journal of Materials in Civil Engineering*, 16(6), 526-539.

Wartman, J., Grubb, D. G., and Strenk, P. (2004b) "Engineering properties of crushed glass-soil blends," *Geotechnical Engineering for Transportation Projects*, Yegian, M. K. and Kavazanjian, E., Eds., ASCE, 732-739.

Weston (2002). *Dredged Material Characterization Report, Fort Mifflin Confined Disposal Facility – Area A, Philadelphia, Pennsylvania*, Prepared for: Sunoco, Inc., 10 Penn Center, 1801 Market Street, Philadelphia, PA by: Roy F. Weston, Inc., 1400 Weston Way, West Chester, PA, February 1.

Application of molten slag from municipal waste as liquefied stabilized soil

Yasuhiro Hayashi, Kinki University Technical College, Kumano, Japan
Tomonori Taguchi, Token Corporation, Nagoya, Japan
Atsumi Suzuki, Kumanoto University, Kumamoto, Japan

Abstract
Most combustible municipal solid waste is incinerated in Japan, and some waste is treated at fusion resource facilities for safety reasons and volume reduction. Molten slag produced during this process is discarded into waste disposal sites. Some techniques to use the slag as construction material have been studied. Feasible uses are aggregates for concrete or subgrades. In order to develop a use for the slag, an attempt is made to use it as a component for liquefied stabilized soil. In this study, samples of molten slag taken from three different sources were used in two series of mix proportion tests. Dry density, cement content, and cement-clay slurry ratio affected the unconfined compression strength of the mix, but it was found the difference in strength was not primarily caused by the molten slag source.

Introduction
There is insufficient disposal space for the increasing amount of municipal solid waste (MSW) in Japan. In order to save space, combustible MSW has been incinerated to reduce its volume. The incinerated MSW has additional risks, containing hazardous materials such as harmful volatiles and heavy metals. Chemical non-hazardous treatment and molten techniques are possible solutions. Molten slag produced by the molten technique is a granular material that can be used as geomaterial.

The present study deals with the molten slag, which is produced after the cooling subsequent to the molten process. At least 127,000 ton of molten slag was produced in Japan in 2001, but 38% of the slag was discarded at disposal sites (e. g. Eco-slag Center). The molten slag is primarily used as an aggregate for concrete and subgrades. There is a limitation to the use molten slag because it is a hyaline material, and therefore the strength is lower than that of ordinary geomaterial.

In this study a technique to produce liquefied stabilized soil (LSS) is applied. LSS has a number of advantages; decreased labor requirements, higher construction speed, and lower in-place costs (Janardhanam, R. et al., 1992). Waste surplus soil can be used to produce LSS (e. g. Kuno (ed.), 1997). The required strength of the LSS is lower than that of concrete, and therefore the required strength for the aggregate may need to be lowered. It is considered that molten slag can be used as an alternative material to sand in LSS. Two series of mix tests are conducted, and the possible use of molten slag as component of LSS is studied.

Methods and Materials

Since cement develops soil strength, the amount of added cement was evaluated by the cement content per unit volume of treated soil for practical purposes. Kuno et al. (2003) reported that unconfined compression strength of LSS developed by the chemical bonding between cement and clay with water. Therefore, the strength is determined by the cement content against the clay slurry content. The consideration agrees with a two-phase mixture model. In the two-phase mixture model, the strength of the matrix and amount of the inclusion determines the strength of the mixture. But when the amount of the inclusion is sufficiently large to constitute a skeleton, the strength is determined by the strength of the matrix. In the case of LSS, cemented clay was considered as the matrix.

LSS in the present study was made from MSW molten slag, stabilizer and clay slurry. The molten slag is treated as an inclusion. A description of the materials follows.

Municipal solid waste molten slag

MSW or MSW-incinerated ash are molten in a furnace with a temperature over 1300 °C. In the molten process harmful volatiles in the ash are removed, but heavy metals remain. The slag remained after being cooled with water or air. The properties of molten slag depend on the source of the waste, sort of furnace, cooling method, and so on.

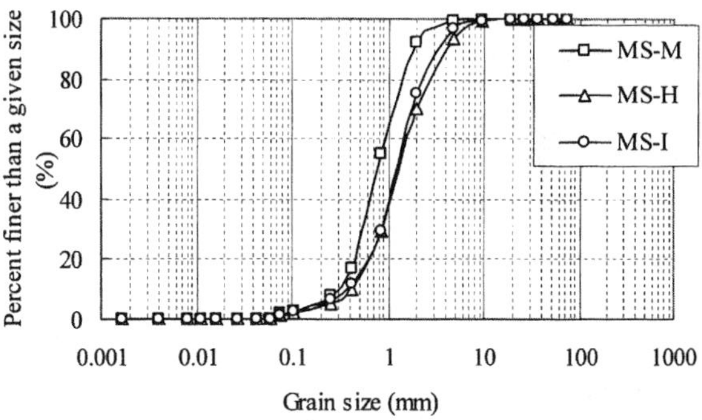

Fig. 1. Grain size distribution

Three kinds of water-cooled molten slag (MS-M, MS-H, MS-I) produced from different sources were prepared. The grain size distributions are shown in Fig.1 and their properties listed in Table 1. They fulfilled the dissolution and contents standards of heavy materials constituted by the Environmental Quality Standard for Soil Contamination, Japan (Ministry of the Environment, 2003).

Table 1. Properties of molten slag

	MS-M	MS-H	MS-I
Perticle density (g/cm^3)	2.83	2.77	2.78
Uniformity coefficient	3.5	3.6	3.1
Coeficient of curvature	1.3	1.2	1.0
Maximum void ratio	1.019	0.980	1.168
Minimum void ratio	0.626	0.613	0.718

Stabilizer and clay slurry

Stabilizer used was blast furnace slag cement which includes 40-60% slag. In order to avoid the segregating of the cement and molten slag, clay slurry, made of kaolin clay mined in Georgia, USA and tap water were used.

Experimental procedures

Properties required of LSS are fluidity in the fresh state and strength after curing. To check these properties, the flow value, bleeding ratio and compression strength were measured. The flow value was obtained with the cylinder method regulated by JHS A 313-1992. The LSS is workable if the flow value is between 110 and 300. The bleeding ratio was regulated by JSCE-F522-1994, and values below 1 % or 3 %, depending on the application, are required to avoid segregation of the materials. Unconfined compression tests confirmed by JIS-A1216 were applied to evaluate the strength after 7 days curing.

Two series of mix proportions were designed with five different slag contents. Important parameters were cement content, slag content, cement-clay ratio and flow value. For Series-I, the cement content was kept at 150 kg/m3, and the flow value was controlled at 170±5 mm per slag amount. Cement-clay slurry ratio was set at 0.12 in Series-II.

Right after the materials were mixed, the flow value was measured. Fresh LSS was subsequently poured into polyethylene sacks of 50 mm in diameter and 500 mm in length. The amount of dissociated water was measured after 20 hours. Other fresh LSS was poured into a plastic mould (50 mm in diameter and 100 mm in length), and subsequently cured for 7 days. After curing, the solidified LSS was taken from the mould and the unconfined compression test conducted. The sack and the mould with LSS were placed in a thermostatic room controlled at 20±3 °C during curing.

Results and Discussion

Mix proportion and bleeding ratio

Mix proportions of LSS with MS-M are listed in Table 2. In Series-I, wet and dry densities increased, and the water-clay ratio reduced as the molten slag content increased. In Series-II, the cement content decreased, densities increased, and flow value showed a peak value as the molten slag increased. The bleeding ratio was almost zero for all mix proportions. The slag might not constitute a skeleton because the maximum content of the slag was below the maximum void ratio.

Table 2. Mix proportions of LSS with MS-M

Series	Components				Wet density	Dry density	Flow value
	Clay slurry		Molten slag	Cement			
	clay	water					
	kg/m^3				t/m^3	t/m^3	mm
I&II	401	802	0	150	1.35	0.55	175
I	430	753	108	150	1.46	0.69	175
	403	705	268	150	1.53	0.82	175
	341	640	512	150	1.64	1.00	175
	220	551	882	150	1.82	1.25	175
II	388	775	97	143	1.40	0.63	195
	367	732	244	136	1.48	0.75	210
	331	659	495	122	1.61	0.95	180
	256	508	1017	94	1.89	1.37	125

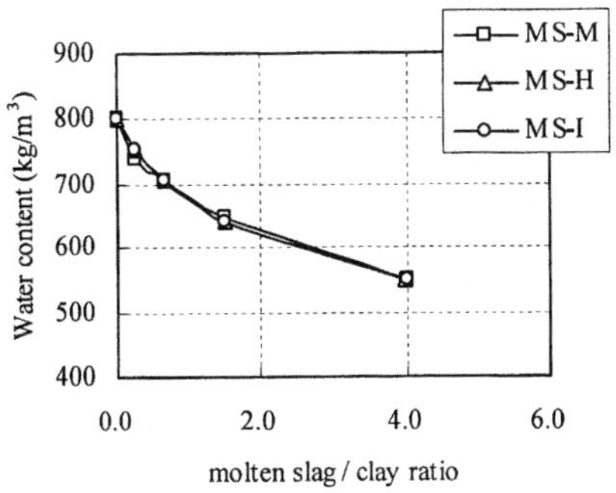

Fig. 2. Water content in series-I

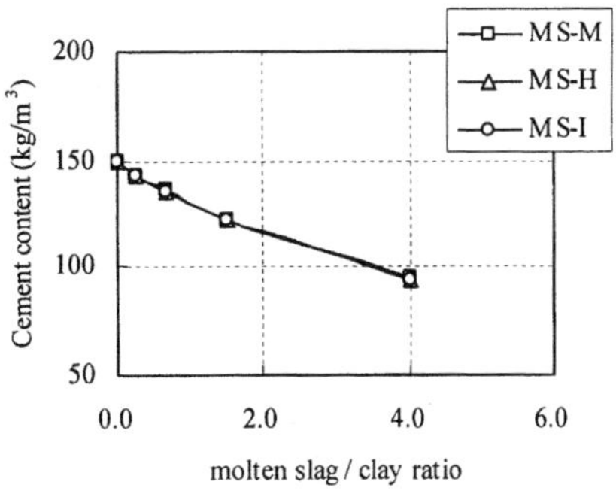

Fig. 3. Cement content in series-II

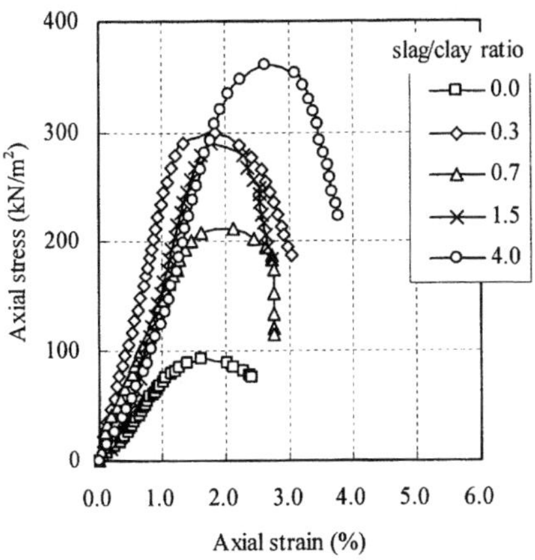

Fig. 4. Stress and strain curves for Series-I

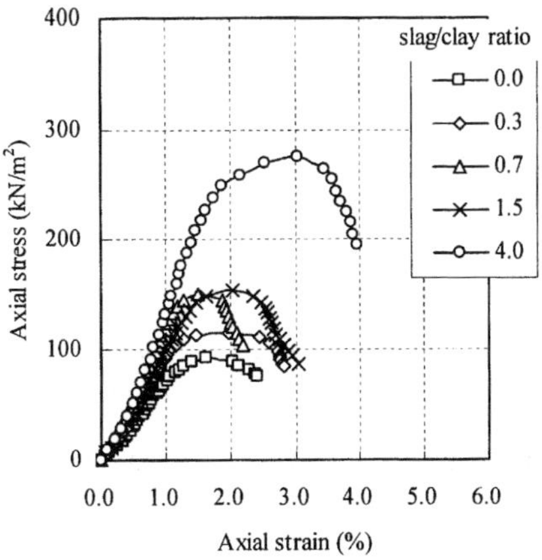

Fig. 5. Stress and strain curves for Series-II

Water content in Series-I and cement content in Series-II are shown in Figs. 2 and 3 against the molten slag-clay ratio. No remarkable differences among the sort of molten slag were observed. It is assumed that the differences in slag content properties are small enough to mask the effects of the varying amounts.

Unconfined compression strength

Stress and strain curves of the LSS containing MS-H are shown in Figs. 4 and 5. Both peak stress and failure strain, which is the strain at the stress, developed as the molten slag content increased.

The unconfined compression strengths are plotted against dry density in Fig. 6. Hollow and solid symbols indicate Series-I and Series-II. A significant difference was not apparent among the types of molten slag. The compression strength versus increasing dry densities in Series-II is smaller than in Series-I. This is caused because the cement content in Series-II decreased as the amount of molten slag increased.

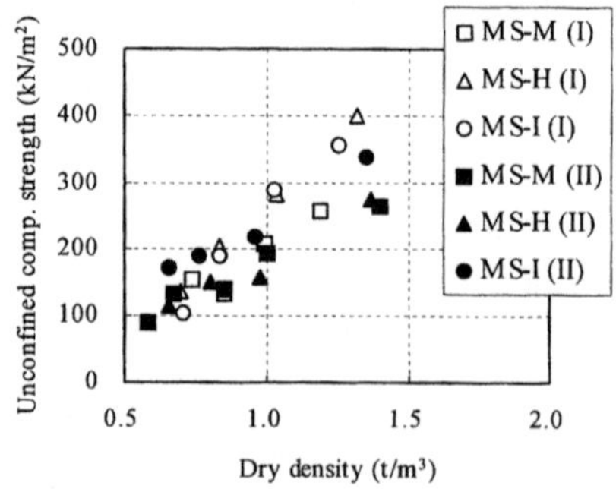

Fig. 6. Unconfined compression strength with dry density

The unconfined compression strength is evaluated by two important factors affecting the strength, the cement-clay slurry ratio and unit cement content in Figs. 7 and 8. It is found that the strength developed with increasing cement-clay slurry ratio and amount of slag, as shown in Fig. 7. The strength developed with increasing amounts of slag, even if the cement content was reduced (Fig. 8).

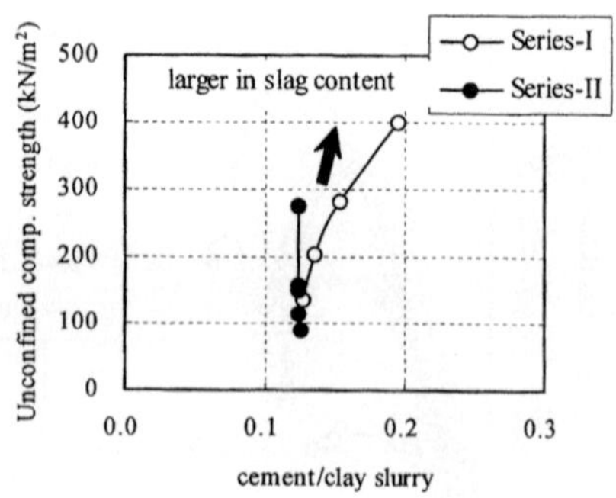

Fig. 7. Unconfined compression strength with cement/clay ratio

According to the study by Kuno et al. (2003) and the two-phase mixture model, the strength would not increase with increasing amount of slag in Series-II, since the cement-clay slurry ratio decreased and the amount of slag was insufficient to constitute a skeleton by itself. It is assumed that the interaction between the matrix and inclusion developed the strength due to chemical bonding by cement. Therefore, the strength of LSS is affected by both the matrix and inclusion content.

Conclusion

MSW molten slag was applied as a component of LSS. Two series of mix proportion tests, the cement content and flow value constant test, and cement and clay slurry ratio constant test, were conducted. For samples of LSS, this satisfied the qualification standard in flow value and bleeding ratio, the unconfined compression strength increased with increasing amount of molten slag and cement-clay slurry ratio. No clear differences were found among the types of slag.

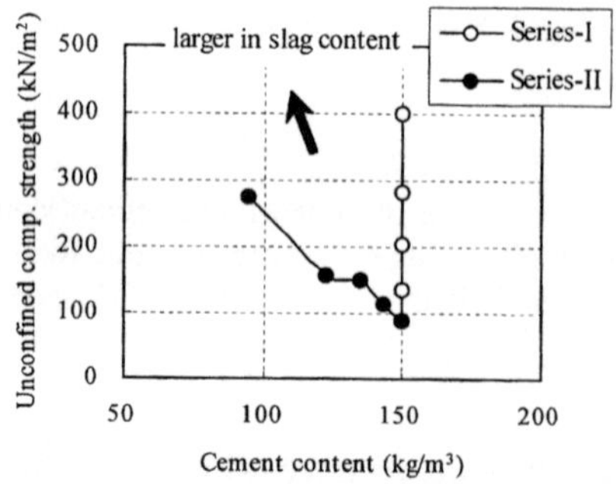

Fig. 8. Unconfined compression strength with cement content

As a result, the use of molten slag as a component for LSS is encouraging. The strength of the slag particles is relatively low in aggregates for concrete, and these effects require further study.

Acknowledgements
The study was conducted with the financial support of the Japanese Institution of Construction Engineering. Samples of molten slag were proffered by clean plaza in Minamata, Hitoyoshi, and Iizuka, Japan. The authors wish to thank them for their support.

References
Eco-slag Center: http://203.141.129.132/ecosla-center/ekosura.htm (in Japanese).

Kuno, G., Iwabuchi, J.and Ichihara, M. (2003): Experimental study on mechanical characteristics of liquefied stabilized soil, Journal of Geotechnical Engineering, JSCE, No.750/ III-65, pp.66-113 (in Japanese).

Kuno, G. (ed.) (1997): Liquefied stabilization of soils, Gihodo Shuppan (in Japanese).

Ministry of the Environment, Japam (2003): Environmental Quality Standard for Soil (in Japanese).

Scrap tire chips as a geomaterial to improve seismic performance of structures

H. Hazarika, Port and Airport Research Institute, Yokosuka, Japan
T. Sugano, Port and Airport Research Institute, Yokosuka, Japan
E. Kohama, Port and Airport Research Institute, Yokosuka, Japan
Y. Suzuki, Niigata Port and Airport Authority, Niigata, Japan

Abstract

This research attempts to exploit the potential of tire chips as a tool to reduce the earthquake induced load against offshore and coastal structures, and thus improving the seismic performance of such important facilities. An underwater shaking table test was conducted on a gravity type quay wall, behind which a compressible layer of tire chips was placed as cushion. The results show that the dynamic earth pressure against the wall could be significantly reduced using the tire chips cushion. In addition, as compared to conventional quay wall with no protective measures, the presence of such sandwiched cushion could reduce the earthquake-induced residual displacement of the quay wall.

Introduction

In many industrialized countries roughly one passenger car tire is generated per population in a year. With growing concern for environmental protection, new and promising technologies are emerging to turn industrial waste products, such as scrap tires, into a valuable resource. In response to the environmental problems caused by countless illegal scrap tires around the globe, most industrialized countries have instigated legal guidelines addressing this issue. Such regulations have led to an increasing use of tire derived recycled products.

Japan boasts of more than 90% recycling of the scrap tires, that she produces per year, for thermal, material and retreading purposes. However, the share of recycling as materials is still far from satisfactory. A major share of scrap tire recycling goes to the thermal recycling, a process that generates carbon dioxide (CO_2), and thus is not environmentally friendly. The amount of CO_2 emission by material recycling, on the other hand, is only 1/4 that of thermal recycling. In order to achieve a sustainable development, tire makers, scrap tire associations and construction industries in Japan are making strenuous efforts to increase the share of material recycling, and thereby cutting the share of thermal recycling and reducing the unaccounted scrap tires. This research attempts to contribute to this share-shifting process by utilizing tire chips as an earthquake resistant geomaterials.

Scrap tire derived products can be classified into two categories depending on the size of the grains: tire chips (size ranges from 12 mm to 76 mm) and tire shreds (size ranges from 76 mm to 305 mm). Fig. 1 shows the tire chips made out of shredding scrap tires. Increasing attention has been paid on the technology of using tire shreds and tire chips in geotechnical construction with the purpose of utilizing scrap tires in material recycling. With characteristics

5th ICEG Environmental Geotechnics, Thomas Telford, London 2006.

such as lightweight, compressible, permeable, durable and thermally insulating, this new class of geomaterial has a myriad of applications in civil engineering.

The objective of this research is to exploit the potential of tire chips as a tool to reduce the load against offshore and coastal structures during earthquake loading. As mentioned earlier, compressibility is a salient feature of tire chips. Many tend to label this characteristic of tire chips as a demerit. This particular characteristics of tire chips, however, can be utilized towards advantages rather than disadvantages, by using such material as a compressible layer behind the massive structures as shown in Fig. 2. Function of the sandwiched layer is to reduce the load on the structure by acting as a cushion, which provide flexibility, and thereby stability to a structure during earthquake by absorbing the energy.

Such technique has also proven to be effective in reducing the seismic load on rigid non-yielding structures (Hazarika et al, 2003) using other compressible and lightweight materials such as EPS geofoam (Horvath, 1997). Small size laboratory experiments under static loading conditions (Tsukamoto et al., 2002) have demonstrated that the use of compressible EPS geofoam behind rigid non-yielding retaining wall can significantly reduce the load against structure by bringing the soil-structure system to a quasi-active state. A field test (Hazarika and Sugano, 2004) and its numerical simulation (Hazarika, 2005) using tire chips as compressible and drainage enhancing materials have demonstrated that the static at rest pressure against a rigid non-yielding structure could be reduced substantially, thus proving the compressibility of such materials as a beneficial element in construction.

In this research, a series of shaking table tests using the underwater shaking table assembly of Port and Airport Research Institute (PARI) was performed. An actual earthquake loading was imparted to the soil-structure system and the response acceleration, the seismic load on the wall, the dynamic increment of the earth pressure acting along the wall and the water pressures in various locations of the backfill were investigated.

Figure 1. Scrap tire and derived product (tire chips)

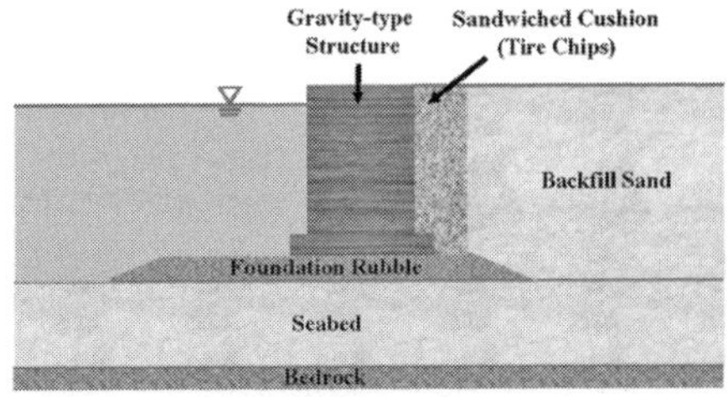

Figure 2. Sandwiched cushion made of tire chips

Test program and procedures

Shaking table

The three dimensional underwater shaking table assembly used in the testing program consists of a circular shake table of 5.65 m in diameter installed on a 15 square meter and 2.0 m deep water pool and a data acquisition system. The details of the shaking table can be found in Sugano et al (1996).

Model construction

A caisson type quay wall was used in the testing. Fig. 3 shows the cross section of the model caisson and the positions of various installed devices (Load cells, earth pressure cells, pore water pressure cells, accelerometers and displacement gauges). The model caisson consists of three parts; the central part (width 50 cm) and two dummy parts (width 35 cm each) as shown in Fig. 4. All the monitoring devices were installed at the central caisson to eliminate the effect of sidewall friction on the measurements. The soil box was made of a steel container 4.0 m long, 1.25 m wide and 1.5 m deep. As shown in Fig. 3, the model consists of bedrock layer, seabed layer, foundation rubble and the backfill.

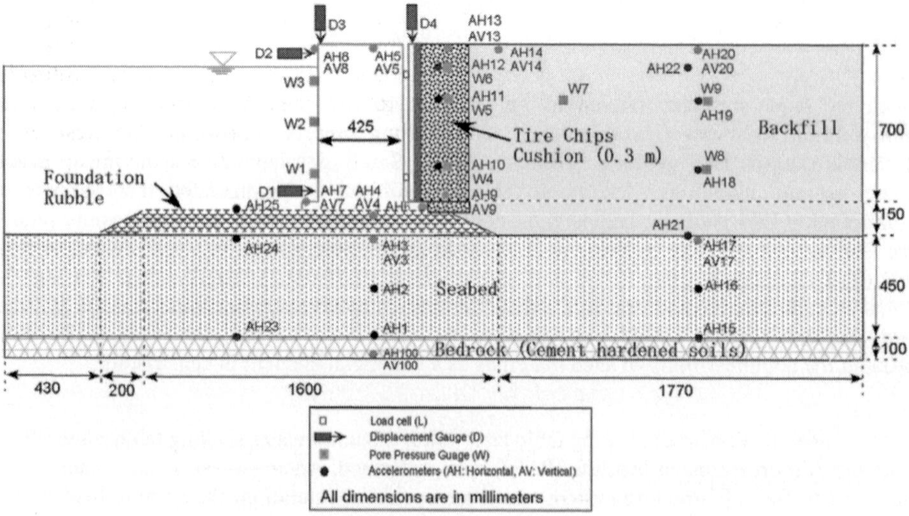

Figure 3. Cross section of the test model

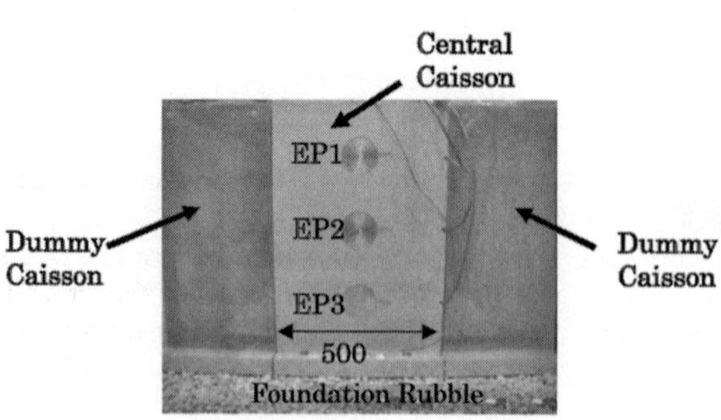

Figure 4. Front view of the model caisson

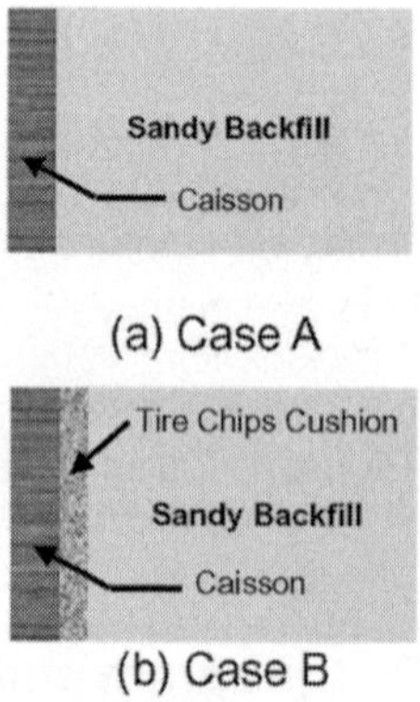

Figure 5. Test cases

The dense foundation soil (Sohma sand) representing the seabed layer was prepared in two layers using vibratory compaction technique. After preparing each layer, the whole assembly was shaken with 300 gals of vibration starting with a frequency of 5 Hz and increasing up to 50 Hz. Backfill soil (Sohma sand) was also prepared in stages using free falling technique and then compacting using vibratory soil compactor. After constructing the foundation and the backfill, and setting up of the instrumentations, the soil-structure assembly was filled with water up to a depth of 1.3 m. This submerged condition was maintained for two days so that the backfill attains a complete saturation stage. The initial ground surface heights were measured at 18 different points over the entire backfill. They were also subsequently measured after each oscillation to monitor the surface settlement during earthquake.

Test procedures

As shown in Fig. 5, two sets of tests were conducted. In one set (Case A) the caisson had the conventional sandy backfill with a backfill density of 1.452 t/m^3. In another set (Case B), behind the caisson, a 0.3 m thick layer of tire chips (average grain size 20 mm) was placed vertically down as cushion. The average density of the tire chips was 0.675 t/m^3. The cushion layer was prepared by filling tire chips inside a bag made from geotextile product. Geotextiles are required to wrap the tire chips so that they do not mix with the surrounding soils. Such confinement also makes the execution of backfilling easier. Furthermore, the presence of geotextiles also prevents flowing of sand particles into the chips structure, and thus prevents clogging and mixing, which may affect the compressibility and permeability of the chips.

Shaking table test in 1g gravitational field requires special consideration to define appropriate scaling law between prototype and model. The similitude of various parameters for the soil-structure-fluid system was calculated using the equations proposed by Iai (1989), and is shown in Table 1 for a model to prototype ratio (λ) of 1/10.

Table 1. Similitude for 1g field

Items	Prototype／Model	Scale factors
Length	λ	10
Time	$\lambda^{0.75}$	5.62
Density	1	1
Stress	λ	10
Water Pressure	λ	10
Displacement	$\lambda^{1.5}$	31.62
Acceleration	1	1

The input motion used in the test was the N-S component of the earthquake motion recorded at the Hachinohe port, during the 1968 Tokachi-Oki earthquake (M 7.9). The time axis of the accelerograms was reduced by a factor of 5.62 according to the similitude shown in Table 1. For the given seismic excitation, the response accelerations in various parts of the structure and the backfill were monitored, along with measurements of the seismic load on the wall, the residual displacement of the wall and the water pressures in various locations of the backfill.

Test results and analysis

Dynamic thrust against the caisson

Figs.6(a)~(c) show the measured horizontal seismic thrust acting on the top, middle and the bottom of the caisson for both the test cases (i.e. Case A and Case B). It can be observed that as compared to conventional caisson, the use of tire chips cushion yields a significant reduction of the seismic earth pressure acting on the quay wall. While the caisson without any protective cushion experiences high fluctuation of the earth pressure with a predominant peak, the earth pressure on the cushion-protected caisson stabilizes soon after reaching the peak. The maximum amplitude of the latter is also much lower than that of the former. This implies that the stability of the caisson improves with the use of the sandwiched cushion.

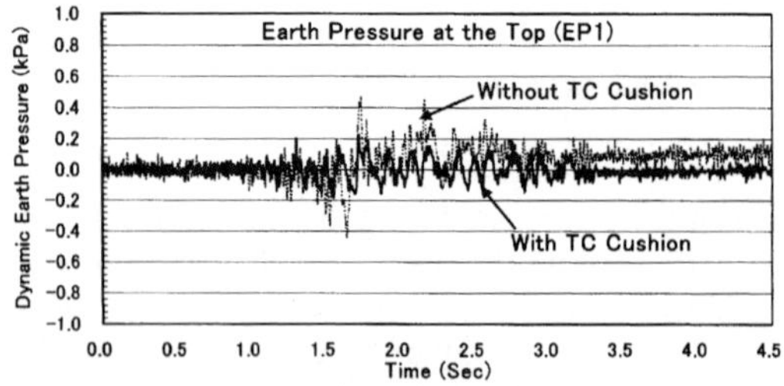

(a) Top of caisson

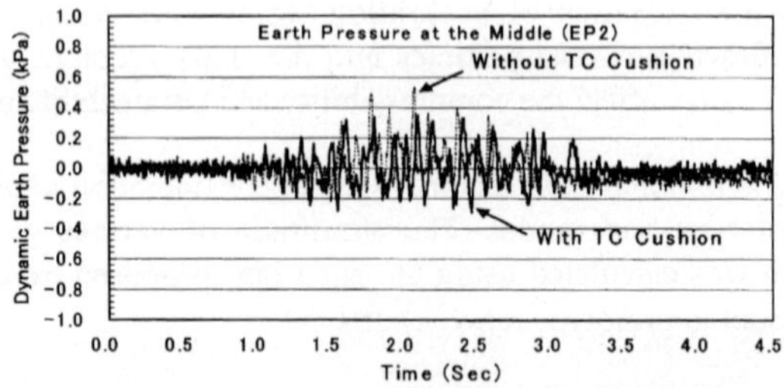

(b) Middle of caisson

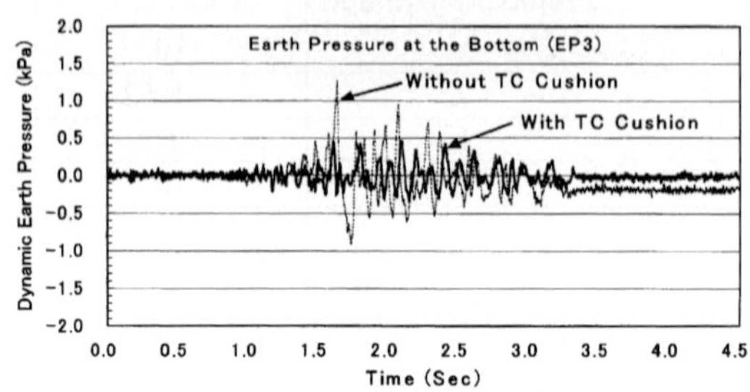

(c) Bottom of caisson

Figure 6. Increment of seismic thrust

Displacement of the caisson

Figs. 7(a)~(b) show the displacements experienced by the top (Fig. 7a) and the bottom (Fig. 7b) of the caisson during the shaking. The results show that the residual displacements were reduced to almost half for the caisson protected with the sandwiched cushion.

Shear deformation in the backfill

In order to observe the shear deformation behavior of the backfill during earthquake which may lead to the liquefaction of the backfill, the pore water pressure measured by the installed water pressure gauges are plotted in Figs. 8~10. Figs. 8(a) and (b) show the time history of the excess pore water pressure recorded by the gauges installed very close to the model caisson (refer to Fig. 3). It can be seen that while the test case A (without the protective cushion) exhibits an increase in pore water pressure (Fig. 8a), the test case B (with tire chips cushion), at the same location, however, exhibits no such increase. The highly granular tire chips layer prevents any development of such pressure, except for the little increase in the form of dynamic water pressure.

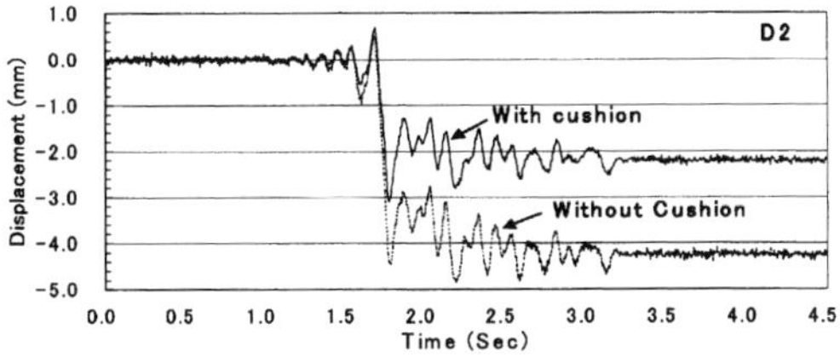

(a) At the top

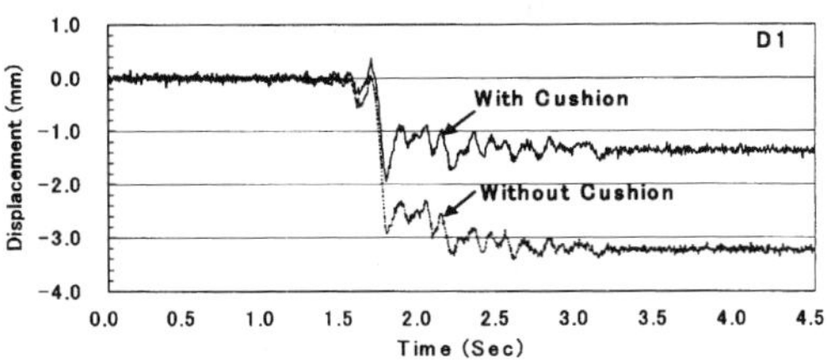

(b) At the bottom

Figure 7. Earthquake induced displacements of the caisson

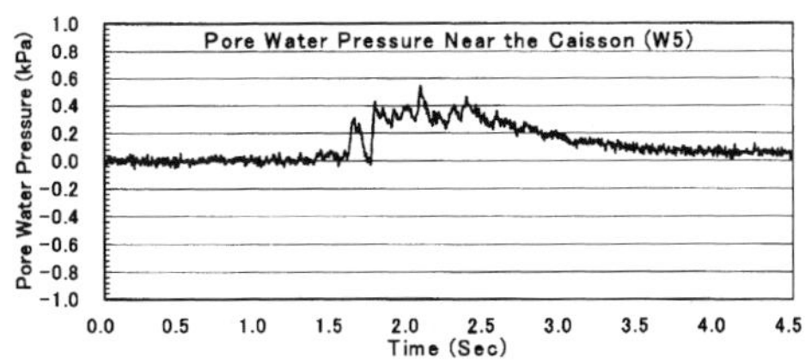

(a) Test case A

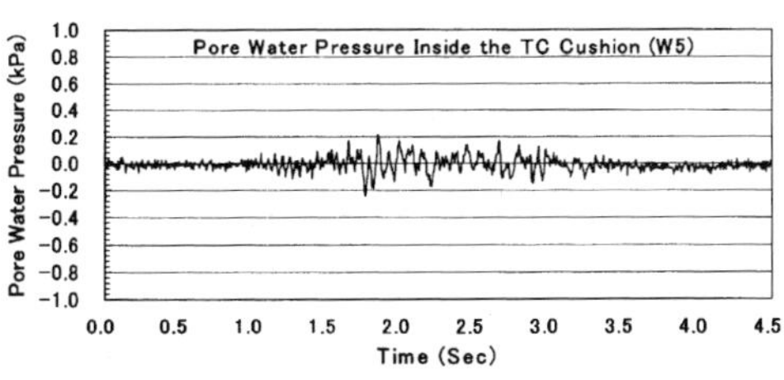

(b) Test case B

Figure 8. Pore water pressure (close to the caisson)

Figs. 9(a)~(b) show the time history of the excess pore water pressure developed at a location 0.65m from the caisson. While the unprotected caisson (Fig. 9a) backfill shows a significant development of the excess pore water pressure, and thus may experience liquefaction induced failure, the cushion protected caisson backfill (Fig. 9b) does not experience appreciable pore water pressure. Granular and highly permeable tire chips give the pore water pressure a chance to dissipate and consequently, there is a less chance of the increase of pore water pressure in the vicinity of the caisson. Compressibility of the tire chips also plays its role here.

The presence of highly compressible tire chips cushion can control the shear yielding of the sand particles, and thus increasing the cyclic mobility of the backfill soil. Consequently, there is less chance of increase of pore water pressure in the vicinity of the cushion.

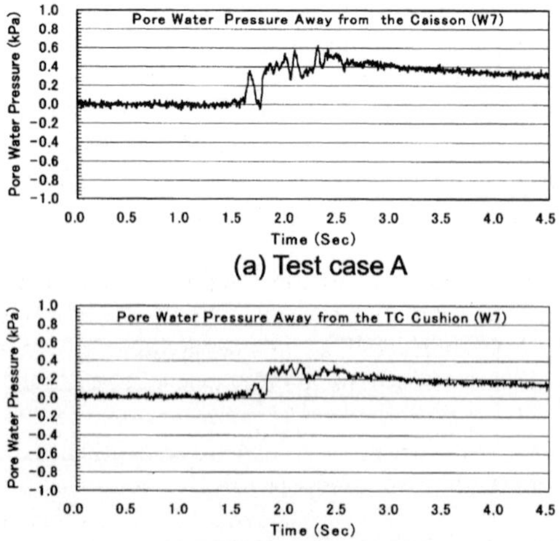

(a) Test case A

(b) Test case B

Figure 9. Pore water pressure (0.65 m away from the caisson)

Figs. 10(a)~(b) show the time history of the excess pore water pressure developed at a location 1.25 m away from the caisson. In this case, the pore water pressure develops appreciably in both the cases, which may lead to the onset of liquefaction. Thus, it can be concluded that the beyond the influence zone of the tire chips cushion, there is likelihood of liquefaction unless some protective measures are taken to prevent such occurrences.

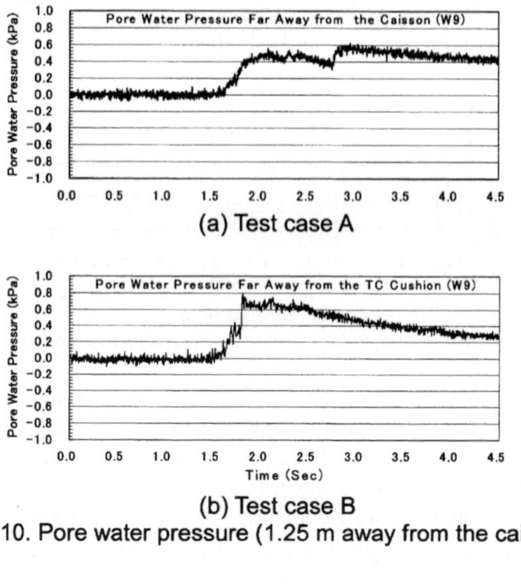

(a) Test case A

(b) Test case B

Figure 10. Pore water pressure (1.25 m away from the caisson)

Concluding remarks

Use of sandwiched cushion (made out of recycled tire chips) behind rigid and massive structures such as caisson quay wall, leads to reduction of the seismic load and the residual deformation during the earthquake. Reduction of the load against structure implies lowering of the design seismic load, which in turn yields a slim structure with reduced material cost. Tire chips, which are cheap and readily available thus not only leads to an economic design, but also contribute to the satisfactory performance of structures by rendering flexibility and stability to structures during earthquake. The benefit of the technique described here is not restricted to the newly constructed structures alone. The technique can also be applied for upgrading (reinforcing) the existing structures, which do not satisfy the current seismic design criteria and, thus, run the risk of damages during earthquakes.

This research is a first step towards establishing the effectiveness of the performance enhancement measure of geotechnical structures. Therefore, only a simple case of constant thickness cushion layer was considered in the testing process. Researches are ongoing to clarify the effects of the layer thickness as well as the size of the chips particles on the reduced load.

Recycling of scrap tire not only relieves the burden on our environment, but also can considerably reduce the execution and construction cost of a project. With growing emphasis on the industrial by-product in construction, and problem caused by the disposal of scrap tires, the performance enhancement measure described in this paper, thereby, will be a big boon for developing a sustainable environment.

Acknowledgments

This research was financially supported by the Port and Airport Research Institute (PARI) under the special grant for budding research. The authors gratefully acknowledge the support. The authors also would like to express their sincere gratitude to Mr. H. Eda of Alpha Engineering Co. Ltd., Tokyo, Japan for the technical help provided during the testing.

References

Hazarika, H., Okuzono, S. and Matsuo, Y. (2003): Seismic Stability Enhancement of Rigid Nonyielding Structures, *Intl. Society of Offshore and Polar Engineers (ISOPE) Transactions*, **2**, pp. 697-702.

Hazarika, H. and Sugano, T. (2004): Combined Use of EPS and Tire Chips As Compressible Cushion Behind Retaining Structure - A Field Test, *Intl. Conf. on Geotechnical and Geoenvironmental Engineering (ICGGE-2004),* Mumbai (India), pp. 229-234.

Hazarika, H. (2005): A Soil-structure Interaction Model with Multiple Participating Components, *11th International Conference on International Association for Computer Methods and Advances in Geomechanics (IACMAG)*, Torino, Italy, pp. 659-666.

Horvath, J.S. 1997. The Compressible Inclusion Function of EPS Geofoam, *Geotextiles and Geomembranes*, **15**, pp. 77-120.

Iai, S. (1989): Similitude for shaking table tests on soil-structure-fluid model in 1g gravitational field, *Soils and Foundations*, **29(1)**, pp. 105-118.

Sugano, T., Morita, T., Mito, M., Sasaki, T. and Inagaki, H. (1996) : Case Studies of Caisson Type Quay Wall Damage by 1995 Hyogoken-Nanbu Earthquake, *11th World Conference on Earthquake Engineering, WCEE*, Paper No.765.

Tsukamoto, Y., Ishihara, K., Kon, H., and Masuo, T. (2002): Use of Compressible Expanded Polystyrene Blocks and Geogrids for Retaining Wall Structures, *Soils and Foundations*, **42(4)**, pp. 29-41.

Issues of dissipated mercury wastes

Jan Jaremski, Rzeszow University of Technology, Rzeszow, Poland

Abstract
Landfill sites containing different wastes should be treated as dangerous bioreactors, in which very toxic and active compounds can form. Possible results of mercury wastes depositing together with the organic mass of municipal wastes have been presented in the paper. Mercury dissipation to the environment is harmful – this is a real danger. The author has considered possibilities of landfill protection and conception of separation of the waste deposit site from the geological environment for Rzeszow town in Poland. The conclusions can be used for construction of waste disposal sites in some other regions. A way of landfill protection should be selected after suitable tests in situ as well as analyses of the known technologies of utilization of volatile and soluble mercury compounds (by combustion of gas mercury, pure mercury intake and purification of sewage, containing soluble mercury).

Introduction
The contribution's aim is to pay attention to necessity of interdisciplinary research works and cooperation of many research workers in solving the problems connected with waste disposal sites where industrial waste products have been deposited.

In previous papers [4][9][10] the author discussed geological aspects of waste disposal sites structure, necessity of a new outlook for the old landfill sites where for example mercury had been deposited and now compounds of this heavy metal are present in the ground. Relatively small amounts of mercury can be connected with a mineralogical structure of the soils matrix and rock matrix. In many areas, mercury compounds are of anthropogenic origin, they come mainly from solid fuel combustion. Mercury and its compounds are also a result of industrial activity of man. In industry mercury is widely used. Additionally, mercury compounds can come from municipal waste sides. For example, in the vicinity of Rzeszow in Poland, large amounts of mercury from the Polam factory were deposited. Especially one landfill site in the town seems to be the most interesting for investigation purposes, where a relatively large amount of mercury was deposited. The existing closed and new waste disposal sites are bioreactors based on bioremediating and physicochemical processes. At these places liquid and gas impurities form occur and we sometimes are not able to predict their formation. Such processes, occurring in soils and deposits, are often well known but we do not know much about them when they exist in landfills conditions. In many countries specialists in geotechnology play a very important role in designing, protection and monitoring of the waste disposal sites.

Biomethylation of heavy metals belongs to very important processes, especialy biomethylation of mercury is very dangerous process. In the mentioned landfill, biomethylation of the deposited mercury seems to be a special threat to environment. We can expect, that microbes present in the deposited municipal wastes, containing much organic compounds, will participate in intense biomethylation processes. During bioremediation the microbes participating in the methylation process carry the alkyle group to the metal and we obtain dimethylmercury and methylmercury. These compounds are much more toxic than mercury in a metallic form. It is important that mechanisms of alkylation is not well known yet, but we know that methylation proceeds very well under oxygen-free conditions, occurring at the sites of wastes being a putrid organic matter. We can find the fact testifying toxicity of methylmercury. The chemistry professor Karen Wetterhahn from american university, died after 15-seconds contact with some millilitres of that compound by a latex glove [2].

In the fifties, in Japan emission of mercury compounds into the sea caused poisoning and death of people who had eaten fish containing methylmercury [12]. Children are especially sensitive to mercury. Vapours of mercury and methylmercury are the most harmful because they easily penetrate to the brain. Constant contact with high concentrations of mercury causes injuries of the nervous system, kidneys and the fetus. Mercury compounds join with enzymes, so harmful operation of mercury is long-lasting. Permissible dosage of mercury have been determined, but tolerance limits for low concentrations are not precisely known [1][11]. Since we know more and more about mercury and its compounds, mainly about their toxic properties, their consumption and dissipation are less and less every year.

Possibilities of heavy metals biomethylation in the closed landfills

When utilization of mercury wastes as well as wastes of other heavy metals was not performed, such wastes usually reached the municipal waste disposal sites. Thus, the suggested in [3][6] balance of toxic wastes deposited in the landfills (based on the data from factories) seems a very important problem.

Mercury biomethylation under natural conditions was discovered by Jansen and Jornelov in 1979. At present we know that also other heavy metals (for example As, Pb, Cd or Sn) are subjected to biomethylation, too. The results of process of biomethylation in the disposal sites of municipal wastes confirmed Lindberg's investigations [11]. Steve E. Lindberg and his coworkers found methylmercury in the water vapor that condensed out of the gas emanating from a Florida landfill. Concentrations were at least 100 times those typically seen in water. However, methylmercury comes in two forms - mono- and dimethylmercury - with the latter being the more toxic. To probe which form is made in landfills, Lindberg and his research group collected gases destined for flaring. They reported finding some 50 nanograms of dimethylmercury per cubic of landfill gas.

The further investigations are necessary to determine possible results of mercury wastes depositing together with the organic mass of municipal wastes. The proposed way of protection of old waste site by the baffles (cavity walls) separates the yard from the geological environment and obligatory gas intaking reduces harmful effects. This solution allows to obtain time necessary for complex and complementary research works leading to new bioremdiation technologies.

Characteristics of the Rzeszow waste disposal site

In 1970-1990 in Rzeszow was Polam factory producing discharge lamps. Metallic mercury was one of the raw materials used for production of these lamps.

The considered landfill site was operated since 1982 till 1990. About 2621550m^3 municipal and industrial wastes were deposited there. The site area was about 2-3 ha (the data are different in various sources), and the landfill height was 10 m. The wastes were thickened with use of a track bulldozer. After operation, wastes were covered with the soil from the scarps surrounding the waste site. The landfill has not got any additional sealings and its soil is permeable. While operating, grinded glass breakage was deposited (according to the data obtained from the Polam factory). Assuming the minimum estimated amounts of mercury in the wastes [5], we can say that about 2600 kg of mercury were deposited in the considered place.

Content of municipal wastes in Rzeszow differs from contents of such wastes in other Polish cities – in Rzeszow a participation of ashes is low (the city is heated from the central source or gas installations). Food wastes of organic or animal origin are dominating (54% - 75 % all the wastes). The mean moisture content is from 24% to 43 %. In Rzeszow, content of organic substances in the wastes is from 40.5% to 53.9%. It is important that the present law and the practice take into account only content of eluates. Such approach seems not to be right – also a balance of the deposited toxic substances and a prediction of changes in their chemical composition during bioremediation should be also included. In the considered landfill, mercury biomethylation is possible. It is a process of changes of inorganic forms of mercury into metaloorganic forms by microbes present in the deposited mass of wastes during bioremediation. We can expect that during methylation the microbes transfer the alkyle group to the metal, forming connections $(CH_3)_2Hg$, $(CH_3)Hg^+$. Mechanisms of alkylation are not well known yet. However, it is clear that methylation proceeds very well under oxygen-free conditions occurring at the landfillss containing the wastes being the putrescible organic matter. The methylmercury forms are much more toxic than the metallic mercury [5][8]. Methylmercury $(CH_3)Hg^+$ is readily soluble in water and has bioaccumulative properties, and dimethylmercury $(CH_3)_2Hg$ is sparingly soluble in water but quite volatile. Methylmercury and ethylmercury connections are resistant to biochemical degradation processes.

Geological and water conditions at the landfill site

Some landfills where mercury wastes were deposited are located at the Przedkarpacka Pradolina. This primeval valley contains pleistocene and holocene deposits. At the bottom, there are fluviglacial formations, next silty clays and silts being redeposited products of older loess weathering. The considered proglacial stream valley is located near the Podgorze Rzeszowskie, a region where considered landfill is located. The area of this landfill is an old excavation of the clay mine for and brickyard. At present excavation contains municipal wastes and glass breakage containing mercury (probably in forms of milled wastes). Under silty clays and silts there are interbeddings of clayey gravel with admixtures of boulders with thickness of 5 m. A range of interbeddings has not been known so far. The subsoil does not satisfy the requirements for geological barriers under the waste site bottoms. Below the quaternary formations there are illite clays which may be applied in the assumed idea of the landfill protection, described in the next paragraph.

The idea of the Rzeszow landfill site protection

There are some regulations and standards concerning disposal of wastes concerning mercury which say how to treat inefficient lamps and how to utilize the wastes. Even small amounts of mercury coming from the deposited fluorescent lamps, batteries etc. can lead to biomethylation [5][7] in municipal landfill sites. In the author's opinion, all the old waste disposal sites must be treated as sources of dangerous impurities which should be covered by suitable geotechnical barriers with installations for gas driving. The installations should make gas combustion possible. Gas products from the old landfill site should not be carried to atmosphere. The author proposes to burn such substances of low energy value at some yards with use of fuel gas. Mercury can be removed from gases with use of the Medisorbon process, applied for mercury removal from waste gases.

In order to protect hydrogeological environment we can consider an idea of encasing the Rzeszow waste disposal site by an screen against filtration with a diaphragm protecting against ionic diffusion. Such protection is good if there is an impermeable layer under the given area. The author also proposed how to make a bioremediator [7]. The landfill site area should be separated by the screen, i.e. a deep diaphragm reaching the cohesive soil layer and satisfying requirements as for the filtration coefficient. The screen should be done as a cavity wall in thixotropic suspension and with an additional membrane. Such screen should be 15-17 m in height. It would protect the geological environment against ionic diffusion, heavy metal washing out from industrial wastes and the heavy metals could not penetrate to underground waters. Such separated area including all the deposited wastes is going to be a bioremediator. Infiltrating water from precipitations will be intaken by wells (Fig.1).

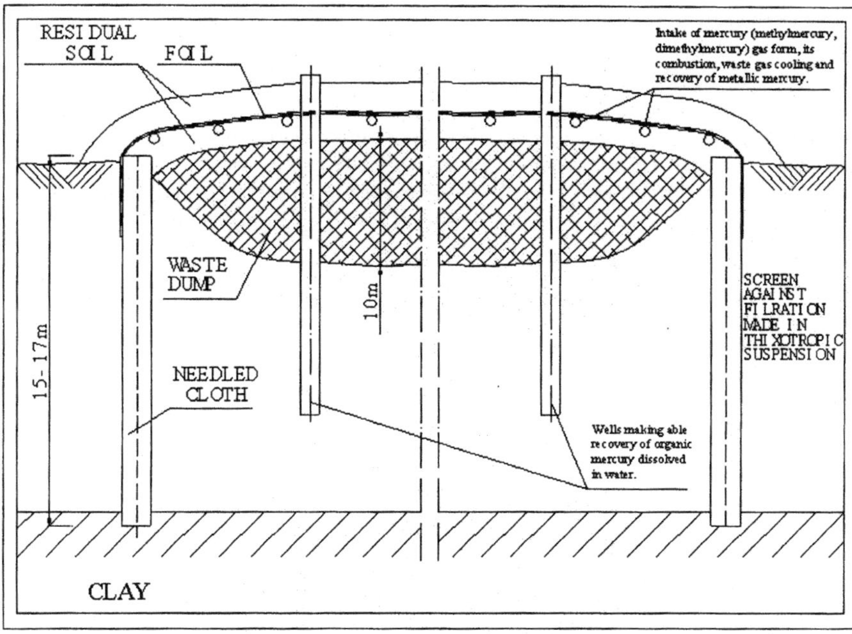

Fig.1. Scheme of the proposed protection of the Rzeszow waste disposal site.

Coclusions

Many methods of waste utilization were analysed within works on the considered problems. All the undertaken actions have to prevent a possible ecological catastrophe in the Rzeszow region in Poland. We must also remember that mercury wastes were deposited at many landfill sites, located in the region where the only water-bearing layer satisfying the requirements for drinking water occurs and the released compounds of deposited mercury may be transported in subsoil for even very long distances. Now the city authorities try to make a decision which protection method should be chosen.

All the problems connected with industrial and municipal wastes and the landfills sites should be considered and tested by research workers – specialists in various fields. Influence of the deposited toxic substances on geological environment should be determined. The problem of changes in sorption capacity of particular soil layers below the site bottom should be also taken into account. In the past, problems of protection of the mixed wastes (municipal and industrial) were often solved without geotechnics. Intensity of remediation was not taken into account and participation of deposited wastes coming from local industry in the deposited municipal wastes was not analysed.

Development of new techniques, chemistry and other fields of knowledge will help us in tests of toxic substances, we also will be able to calculate sorption of impurities by geological and geotechnical barriers more precisely.

References

[1] Eccles, C. U. and Aman, Z., eds., 1987. The Toxicity of Methylmercury, The John Hopkins University, the Press of Baltimore

[2] Gulanowska B., Cieślak-Golonka M., 1998. Karen E. Wetterhahn (1948-1997), Orbital 1

[3] Jaremski J., 1990. The analysis of influence waste from power station deposited in the open casts the composition of ground water. Proc. of 6th Int. Cong. IAEG, Amsterdam, A. A. Balkema

[4] Jaremski J., 1994. Some aspects of waste deposit in the marl open-air casts. Proc. of 7th Int. Cong. IAEG, Lisboa

[5] Jaremski J., 2000. About some results of bioremediation in waste yards. Scientific Books of Silesian Technical University, z. 45, Gliwice

[6] Jaremski J., 2002. Protection of the geological environment, especially ground water, against the influence of contaminants from waste materials containing heavy metals. Proc. of 9th Int. Cong. IAEG, Durban, South Africa

[7] Jaremski J.,1996. Identification of nitrogen compounds and organic coal and evaluation of excavation usability for waste deposits. Int. Conf. on Env. Geology, Spain

[8] Jaremski J., 2000. Some criteria for waste depositing in excavations. Proc. of 8th Int. Cong. IAEG, Vancouver, A. A. Balkema

[9] Jaremski, J., 2003. Some results of bioremediation in waste yards and their influence on the contaminated soils. Proc. of Int. Conf. on Problematic Soils, Nottingham, United Kingdom

[10] Jaremski, J., 2005. Some aspects of issues of dissipated mercury wastes. Int. Symp. on Latest Natural Disasters – New Challenges for Engineering Geology, Geotechnics and Civil Protection, Sofia, Bulgaria

[11] Raloff, J., 2001. Dimethylmercury in Landfills. Science News Week of July, 7, Vol. 160

[12] Smith, W.E. and Smith A.M., 1975. Minamata. New York: Holt, Rinehart and Winston,

Slag Bound Mixtures with Recycled Aggregates for Flexible Pavements

H A Khalid Dept. of Civil Engineering, Univ. of Liverpool, U.K.
Z L Soo
I S Nquot
I Artamendi

Abstract

This paper describes a laboratory study on the behaviour and mechanical properties of slag bound mixtures, including usage of recycled aggregates, with a view to assessing their adequacy to serve as base layers in flexible pavement construction. The study addresses the influence of a number of material variables, e.g. mixture composition, moisture content and moisture ingress, on the indirect tensile strength properties. The paper then considers the changes to the material's structural support properties, at the early stages, that occur at different curing durations and briefly assesses the impact of these changes on pavement design. The cyclic triaxial test was conducted to evaluate the resilient modulus, which was used to portray the material's structural support. An attempt was made to depict the non-linear behaviour of slag bound mixtures by expressing their stress dependency characteristics in terms of experimental models, mainly used to describe the behaviour of unbound granular materials.

Introduction

The increasing demand for high quality primary aggregates for use in road pavement construction has necessitated the search for alternative sources, such as slag, to mitigate the potential impact of resource depletion. Blastfurnace slag is an industrial by-product from the making of pig iron and is formed by the combination of the siliceous components of the iron ore with the limestone flux for smelting iron. In the UK, about 4 Mt are produced annually and are processed, mainly by water cooling, to produce granulated blastfurnace slag (GBS) for use as construction aggregates, or, by subsequent grinding, to produce ground granulated blastfurnace slag for use as cementitious binder. The latter is in huge demand as cement replacement in concrete production and, thus, accentuates its production from blastfurnace slag.

GBS is a latent hydraulic binder whose chemical composition is similar to cement. However, it is only soluble in alkaline solution, unlike cement which is soluble in water. GBS is not, by itself, cementitious, but its hydraulic binding properties can be developed by the addition of a catalyst, or activator, such as lime or other alkaline material. Basic oxygen steel slag (BOS) is one such material that could be used to produce the alkalinity condition that would assist in the hydrate formation reactions. BOS is an industrial by-product of the conversion process of iron to steel, with an estimated annual UK production of 1 Mt.

Aggregate such as crushed limestone and construction and demolition waste (CDW) are mixed with GBS, BOS and water to produce slag bound mixtures (SBM) for use as base layer in flexible pavements. Adverse environmental impacts associated with the use of SBM are considered rare (1) and can be attributed to bad practice or inappropriate use. The potential leachable lime or sulphur compounds in the GBS and BOS can be contained by proper material selection and pavement design. BFS typically contains <1% sulphur. Only a very small part of this is leachable when in contact with groundwater as most of it is in a bound condition within the pavement layer. In the UK, > 100Mt of unbound BFS have been used in road construction projects (2) with very few and insignificant cases of environmental pollution.

Due to their aforementioned latent hydraulic binding properties, SBM exhibit structural support behaviour akin to unbound aggregates at the initial service stages. Therefore, the principal structural distress mode would be excessive deformation that would lead to rutting at the pavement surface. As SBM's strength develops with curing time, their behaviour changes to that akin to cement bound materials, which are prone to flexural cracking as a result of the repeated wheel load applications at the pavement surface. This apparent change in behaviour with time is of major interest to civil engineers involved in the design, construction and maintenance of the transport infrastructure and, hence, forms the basis of the present research. This paper addresses some of the factors affecting SBM's mechanical properties at the early stages and considers the inherent complexity involved in their use in analytical pavement design caused by their stress dependency characteristics. The paper also provides results on recycled CDW aggregates replacing virgin limestone in SBM blends to show whether the waste material could produce equivalent properties to natural aggregates.

Materials and Test Procedures

Blends of GBS, BOS and aggregate were produced to satisfy the grading requirements of a 20mm nominal maximum aggregate size for a dense macadam binder course layer in accordance with BS4987 (3). The graded aggregates were either limestone or CDW, both of 20mm max size. The limestone was available in six sizes: 20, 14, 10, 6, 3mm - dust and filler, whereas the CDW came in two sizes, namely 20 and 5mm graded aggregates. Visual examination of the CDW revealed a composition comprising mainly two types of stone, one crushed and the other rounded, masonry, concrete, gypsum and others (~3%), e.g. glass, metal, timber, ceramic and plaster board. Optimum Moisture Content and Maximum Dry Density were determined for the chosen blends by vibratory compaction in 100mm cylindrical moulds. Compaction was conducted in accordance with BS EN 13286-51 (4), except for the vibration time being 30 seconds per specimen side. Optimum Moisture Content values for the limestone and CDW blends were 5.5 and 9.5% respectively. After compaction, specimens were left in the split moulds for up to 1 week wrapped in polythene bags for curing. After extraction from the moulds, specimens were either tested or stored for further curing wrapped in the polythene bags as before extrusion.

Two types of cylindrical specimen were produced for testing in this work: 150mm diameter by 70mm thick and 100mm diameter by 150mm thick specimens. The former were used for indirect tensile strength tests, while the latter were for conducting the cyclic triaxial test. The indirect tensile strength test was carried out monotonically until failure at different curing periods in accordance with BS EN 13286-42 (5) in a compression machine at a rate of 200 kPa/s.

Cyclic Triaxial Test Protocol

The cyclic triaxial test was carried out in a pneumatically controlled dynamic testing machine capable of applying load pulses of rectangular waveform for 1s duration followed by 1s rest period. Constant confining pressure was provided by means of a vacuum pump connected to the bottom platen, with the specimen being wrapped by a rubber membrane and the seal was achieved by the use of O-rings. Figure 1 shows the triaxial test assembly used in this work.

The resilient modulus, M_r, was determined as the ratio of cyclic deviator stress and the resilient strain, as in equation 1.

$$M_r = \sigma_d / \varepsilon_r \qquad (1)$$

where σ_d is the deviator stress and ε_r is the resilient vertical strain. M_r was determined by initially conditioning the sample under 100 pulses of 30 kPa, which was followed by 50 pulses of 30, 50 and 70 kPa of axial stress levels, without confinement. The resilient strain at the 50th pulse was measured and used to calculate M_r and the average value obtained from the three stress-level determinations. At least three samples were tested and the M_r value reported represents the average of these samples.

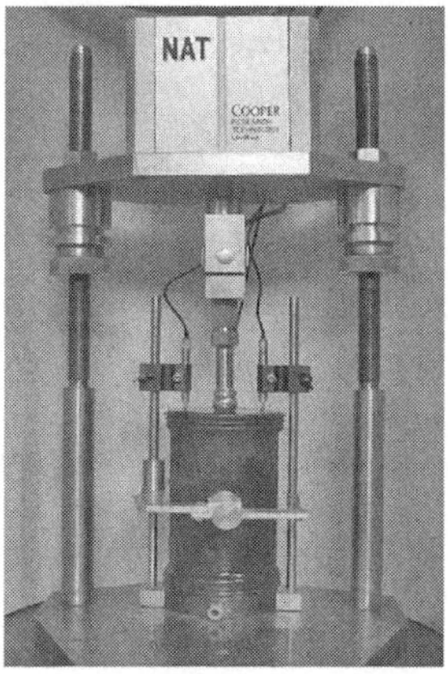

Figure 1. Cyclic triaxial test assembly

In determining the stress dependency of the material's resilient behaviour, the specimens were subjected to five confining stress levels, σ_3, and, at each confinement level, to 50 pulses of five σ_d cyclic axial stresses in accordance with the protocol given in Table 1. The stress levels used are very similar to Method B in BS EN 13286-7 (2004) for resilient testing in the cyclic load triaxial test and any deviations thereof were due to instrument limitations.

Table 1. Protocol for confined cyclic triaxial test

σ_3 (kPa)	σ_d / σ_3 Ratio
20	2; 2.5; 3.0; 3.5; 4.0
40	2; 2.5; 3.0; 3.5; 4.0
60	2; 2.5; 3.0; 3.5; 4.0
70	2; 2.5; 3.0; 3.5; 4.0
80	2; 2.5; 3.0; 3.5; 4.0

Effect of Slag-Aggregate Blend Composition

In the context of material variables, the aim is to minimise the use of finite aggregate resources and maximise the use of waste and by-product materials, provided that the service performance is at an acceptable level. This being so, there are market-driven constraints on the amount of GBS permitted to be used as hydraulic binder for road construction materials as there is an increasing pressure to produce ground granulated blastfurnace slag for use as cement replacement in concrete production underpinning wider construction demands by industry. In this work, a number of blends were made in which the proportion of slag to aggregate was increased from 10 to 50% by mass. In all these blends, the ratio of GBS to BOS was maintained at 4:1. The South African specifications (7) reported this to be the optimum ratio of binder to activator for stabilised road construction materials.

Figure 2 shows the indirect tensile strength results of the limestone and CDW blends with various slag contents at 7 days curing. Large proportions of slag in the blend would lead to too high a demand on the GBS and, as a result, would lead to high volume expansion if BOS were used as activator (2). The results indicate that any slag content higher than 30% would not be beneficial to the limestone blend and would not be viable for the CDW material. It was decided to adopt the 75/25 blend in this work to further limit the BOS content, thus avoiding any significant volume expansion potential, although the 70/30 blend is also recommendable, based on the data in Figure 2. For comparison purposes, Atkinson et al (8) obtained an ITS value of 0.3MPa at 28 days curing for a blend containing 70% blastfurnace slag aggregate, 20% GBS binder and 10% BOS activator.

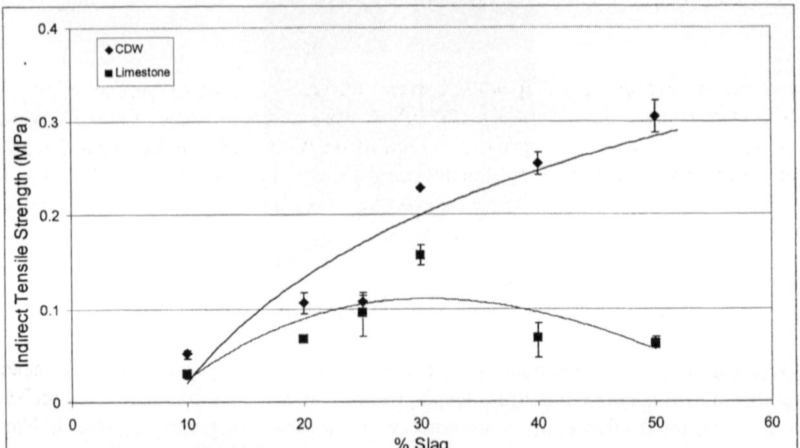

Figure 2. Effect of blend composition on the indirect tensile strength of SBM

Effect of Moisture Content

Figure 3 shows the influence of moisture content at compaction on the development of the indirect tensile strength for the limestone (dotted lines) and CDW blends (solid lines). Each blend was compacted at their respective optimum moisture contents and at optimum moisture content ± 1%. Within the adopted moisture content range, it can be seen that the limestone blend is less sensitive to compaction moisture content than the CDW blend.

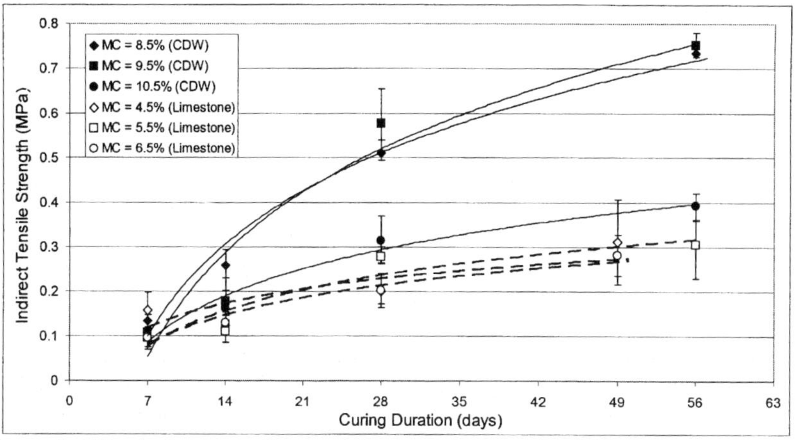

Figure 3. Effect of compaction moisture content on the SBM's tensile strength

Effect of Moisture on Resilient Behaviour

The resilient modulus of unbound and hydraulically bound materials, used as base layer, is the parameter sought in analytical pavement design to indicate the structural support contribution of that layer to the overall pavement structure. One of the aims of this study was to monitor the changes to the M_r value with curing time, thereby reflecting the improvement to the SBM's structural support with time, and to assess the influence of moisture ingress on this support.

After compaction, two groups of specimens were produced, one was wrapped and cured in a dry condition at room temperature and the other was soaked in water. Figure 4 shows the evolution of M_r up to an age of approx. two months for both blends under dry and wet curing conditions. In terms of sensitivity to moisture ingress, the retained M_r value, i.e. ratio of wet to dry moduli, is seen to increase from 78 to 82% for the CDW blend between 7 and 56 days curing. This shows that some improvement at the early stages is observed in the material's moisture sensitivity characteristics. It is difficult to comment on the limestone blend's moisture sensitivity as both dry and soaked results are inherently low at the early stages and, thus, very close to one another.

A power function was fitted to the data in Figure 4 to predict the M_r values at advanced stages of service. This prediction has produced 12-month M_r values of approx. 570 and 266 MPa for the CDW and limestone blends respectively. An analytical model of a typical flexible pavement structure was then developed using a common first-response analytical design program, with the SBM figuring as the base layer. The analysis has shown that, as a result of the resilient modulus increase, pavement design life exemplified by the number of standard axle passes to failure is increase by 4.4 and 2.3 times for the CDW and limestone blends respectively. The analysis carried out here, however, is purely for comparison purposes and involves an over-simplification of the nature and behaviour of SBM, which is discussed in more detail in the next section.

Stress Dependency

SBM behave almost like unbound to lightly bound materials, especially at the early stages of service. This is principally due to the slow rate reactions taking place between the slag and the lime from BOS, in the presence of water, that produce crystallisations of hydrated silicates and aluminates of calcium, just like the reactions occurring in the cement hydration process. The resilient behaviour of SBM is non-linear and is best described as stress-dependent, in the same vein as granular materials, which have often been modelled using the classical K-θ principle (9).

$$M_r = K_1\theta^{K_2} \tag{2}$$

where θ = sum of principal stresses and K_1, K_2 = constants that depend on the material and can be determined from linear regression.

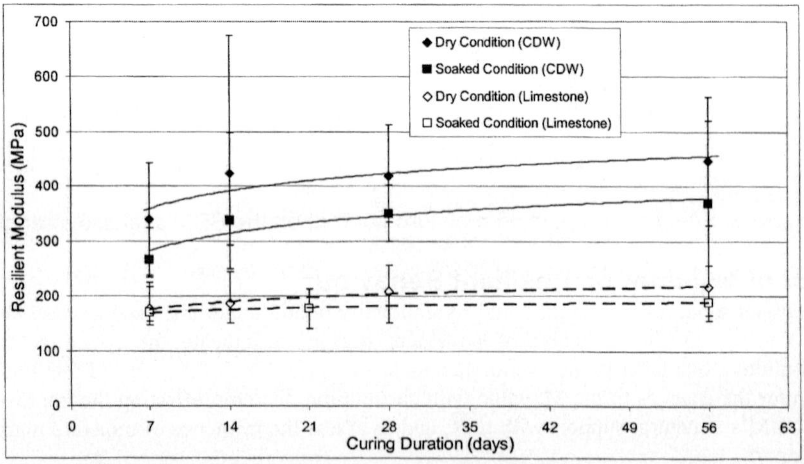

Figure 4. Effect of curing duration and moisture ingress on the resilient modulus

Although very simple and useful in analytical pavement design, the K-θ model has been reported to have a number of limitations (10, 11), the most notable being that it does not include other, possibly more influential, parameters affecting the resilient modulus. Consequently, a number of alternative models have been proposed, which consensually include the cyclic deviator stress, σ_d, and the confining stress, σ_3. A typical model of this genre is that developed by Uzan (12) and given here in equation 3.

$$M_r = K_1\left(\frac{\sigma_3}{\sigma_{3,0}}\right)^{K_2}\left(\frac{\sigma_d}{\sigma_{d,0}}\right)^{K_3} \tag{3}$$

where $\sigma_{3,0}$, $\sigma_{d,0}$ are reference confining and deviator stress values, and K_1, K_2, K_3 are material constants.

Figure 5 shows the K-θ model for the limestone and CDW blends after one week curing. At high confining stress levels, it can be seen that the materials undergo permanent deformation, which is portrayed by a decrease in M_r with increase in σ_d/σ_3 ratio, refer to Table 1. For comparison, another K-θ model was included in Figure 5 for an unbound 20mm recycled crushed concrete (RCC) base material, after Nataatmadja and Tan (10), from cyclic triaxial tests conducted by these researchers using a similar range of stress levels to those used in this work. It is quite clear how the figure ranks the three materials in terms of stress dependency and structural support value.

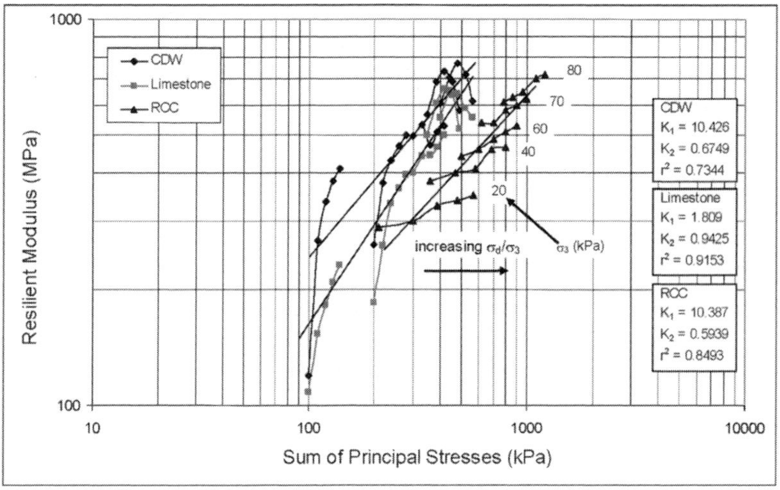

Figure 5. Comparison of K-θ models for different base materials

Figure 6 shows M_r as a function of the cyclic deviator stress and gives the Uzan model coefficients, K_1, K_2 and K_3 for the limestone and CDW blends, which were determined using multiple regression. K_1 in the Uzan model, i.e. equation 3, takes the units MPa, whereas K_1 in the K-θ model takes the units $(MPa)^x$, where $x = 1 - K_2$.

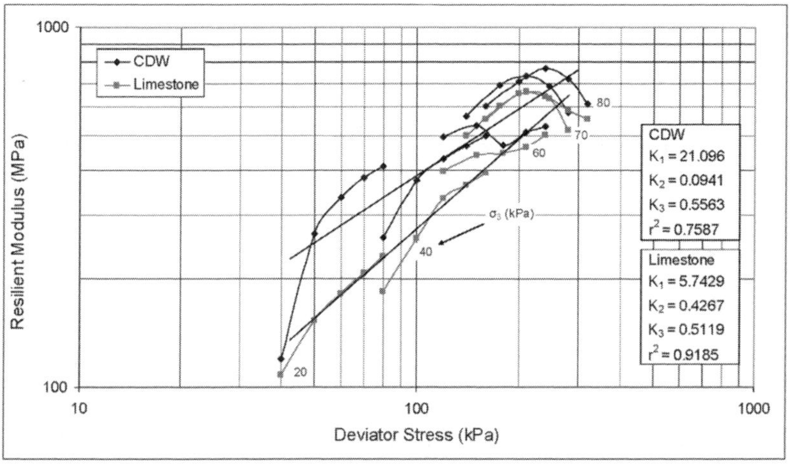

Figure 6. Effect of deviator stress on the resilient modulus with the Uzan model coefficients

Conclusions

To achieve a balance between maximising construction and demolition waste aggregate content and limiting potential volume expansion due to BOS, a 75/25 aggregate to slag blend was adopted, which gave 0.1 MPa indirect tensile strength after one week curing. SBM's resilient modulus increases with curing duration and their sensitivity to moisture ingress showed a slight improvement at the early stages. The predicted pavement performance showed that there is clear advantage in using SBM, which leads to increased design life by a factor of more than 4 and 2 for the CDW and limestone respectively. The CDW blend showed higher strength and resilient modulus properties than the limestone blend. This is mainly attributed to the inherent weakness of limestone and the presence of high toughness constituents in the construction and demolition waste aggregates. SBM exhibited stress dependent behaviour after one week curing, with the CDW showing less dependency than the limestone blend. No results were available to show whether any significant changes to the observed stress dependency occur at later service stages. The Uzan model provides a more stable equation with regard to the units of the K_1 coefficient.

Acknowledgement

The authors are indebted to Tarmac Ltd. for their technical and financial support of this work. Special thanks are to Drs Howard Robinson and Nizar Ghazireh for their continued encouragement, and to Mr Alan Banks for his help with experimental procedures.

References

(1) Dunster AM (2002). The use of blastfurnace and steel slag as aggregates. Proc. 4th European Symposium on Performance of Bituminous and Hydraulic Materials in Pavements, (eds. Zoorob, Collop and Brown), Balkema Publishers, 257-60.

(2) Dunster AM (2001). Blastfurnace slag and steel slag: their use as aggregates. Building Research Establishment Info. Paper No. IP18/01, BRE, Watford, UK.

(3) British Standards Institution (2001). Coated Macadam for roads and other paved areas. BS4987-1, BSI, London, UK.

(4) British Standards Institution (2004). Unbound and hydraulically bound mixtures – method for making test specimens: making specimens by vibrating hammer compaction. BS EN 13286-51: 2004, BSI, London, UK.

(5) British Standards Institution (2003). Unbound and hydraulically bound mixtures – test method for the determination of the indirect tensile strength of hydraulically bound mixtures. BS EN 13286-42:2003, BSI, London, UK.

(6) British Standards Institution (2004). Unbound and hydraulically bound mixtures – Cyclic load triaxial test for unbound mixtures. BS EN 13286-7:2004, BSI, London, UK.

(7) National Institute for Transport and Road Research (1986). Cementitious stabilisers for road construction, TRH13, NITRR, CSIR, South Africa.

(8) Atkinson VM, Chadock BC and Dawson AR (1999). Enabling the use of secondary aggregates and binders in pavement foundations, Transport Research Laboratory Report 408, TRL, Crowthorne, Berks., UK.

(9) Brown SF (1978). Material characteristics for analytical pavement design. In Developments in Highway Pavement Engineering - 1, (ed. PS Pell), Appl. Science Publishers, 41-92.

(10) Nataatmadja A and Tan YL (2000). The performance of recycled crushed concrete aggregates. Proc. Unbound Aggregates in Road Construction, (ed. AR Dawson), Balkema Publishers, 109-115.

(11) Van Niekerk AA and van Scheers J (2000). Effect of compaction on the mechanical behaviour of mix granulate base course materials and on pavement performance. Proc. Unbound Aggregates in Road Construction, (ed. AR Dawson), Balkema Publishers, 125-136.

(12) Uzan J (1985). Characterisation of granular materials. Transportation Research Record 1022, Transp. Res. Bd., Nat. Res. Council, Washington DC, 25-59.

Engineering property evaluation of cement treated soil with tire chips using X-ray CT Scanner

Y.Kikuchi, Port and Airport Research Institute, Yokosuka, Japan
T.Nagatome and **Y.Mitarai**, Toa Research & Development Center, Yokohama, Japan
J.Otani, Kumamoto University, Kumamoto, Japan

Abstract

The purpose of this study is to evaluate both the failure and permeability properties of cement treated soil with tire chips using a micro-focus X-ray CT scanner. The specimens used in this study were cement treated soil both with and without tire chips. First, the effect of fraction of tire chips on the physical properties of cement treated soil is discussed based on the results of CT scanning, considering the density distribution and content of tire chips in the soil. Next, a series of unconfined compression tests was conducted with these specimens, and the failure mechanism was investigated by CT scanning the specimens during unconfined compression. A series of permeability tests using a triaxial apparatus was also performed with these specimens during the process of unconfined compression, and their permeability change was examined. Based on the results, the engineering properties of cement treated soil with tire chips were evaluated.

Introduction

In Japan, almost 90% of used tires are recycled, but nearly half of this amount is used in thermal recycling, which produces heat for power generation and other applications but also generates carbon dioxide. On the other hand, material recycling, which reduces consumption of virgin resources and does not cause the CO_2 emissions associated with combustion, is a far more effective recycling method from the viewpoint of environmental sustainability. Development of a material recycling method for reusing waste tires in civil works would make it possible to recycle a large number of tires at even a single construction site. Thus, a reuse method of this type would have great potential for recycling.

Many studies have been conducted on barrier design for marine disposal fills in recent years. One important point in designing a barrier is to minimize barrier breakage in the event of serious deformation of the revetments. Cement treated soil is a promising sealing material for disposal fills due to its low permeability, but because it is also a brittle material, it cannot follow revetment deformation. The authors considered that addition of tire chips to cement treated soil offers an effective means of improving its ductility and maintaining its low permeability, enabling use in structures where deformation is anticipated (Mitarai et al. 2004).

The X-ray computed tomography (CT) scanner, which is used in medical diagnosis, has also been used for engineering purposes as a nondestructive method (Otani et al. 2002, Otani and Obara 2003, Kikuchi et al. 2005a, Kikuchi et al. 2005b), and a micro-focus X-ray CT scanner (SMX-225CT) was installed at the Port and Airport Research Institute, Japan in 2004.
The objective of this study is to evaluate both the failure and permeability properties of

cement treated soil with tire chips using a micro-focus X-ray CT scanner. The specimens used in this study were cement treated soil both with and without tire chips. First, the effect of fraction of tire chips on the physical properties of the cement treated soil was investigated based on the results of CT scanning with image processing analysis. A series of unconfined compression tests was then conducted with the same specimens. In this paper, density changes and the failure mechanism under unconfined compression are discussed as a part of an evaluation of the failure properties of the specimens using the CT data. A series of permeability tests using a triaxial apparatus was also performed with the specimens during the process of unconfined compression, and permeability change was examined. The effect of tire chips on the engineering properties of cement treated soil was then evaluated.

X-ray CT data

Although X-ray computed tomography (CT) is widely known as a medical technique, it is also an effective nondestructive inspection method which makes it possible to visualize the internal structure of specimens. The basis of formation of an X-ray CT image is the attenuation of X-rays transmitted through the specimen. CT image data are evaluated quantitatively using the so-called "gray level," which is a numerical number and proportional to the material density. Figure 1 shows the relationship between the average gray level and average density of specimens of cement treated soil. These results show a linear relationship, which means that the gray level can be used as a parameter for evaluating density distribution. In other words, the gray level enables a quantitative examination of material structural features based on this relationship with CT images. Using the results, a quantitative discussion of density changes in soil under compression is also considered possible. In CT images as shown in Figure 4, darker regions represent lower density zones, while lighter regions represent higher density zones.

Figure 2 illustrates an X-ray CT image of the object with CT scanning. The voxel (volumetric pixel) size can be changed as required by the purpose. For example, in this study, the scanning space was divided into 512 x 512 or 2048 x 2048 voxels. In scans of the entire cross section of specimens, 512 x 512 voxels were used, and the size of one voxel was 0.12 mm x 0.12 mm x 1 mm. Microscopic scans of the cross section were performed using 2048 x 2048 voxels, giving a voxel size of 0.01 mm x 0.01 mm x 0.01 mm. Three-dimensional images are reconstructed by superposing 2-dimensional images.

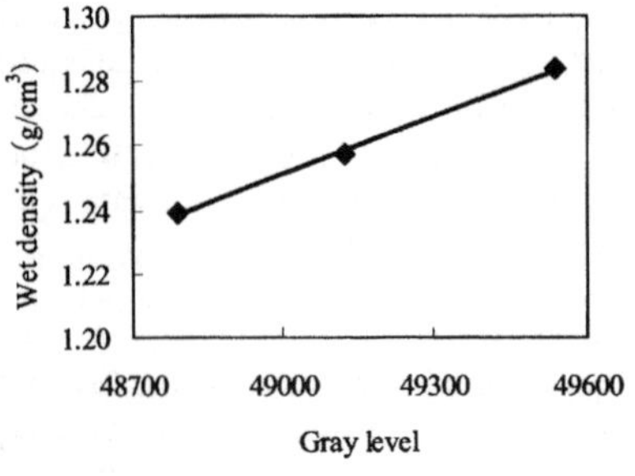

Figure 1. Relationship between the gray level and wet density for cement treated soil

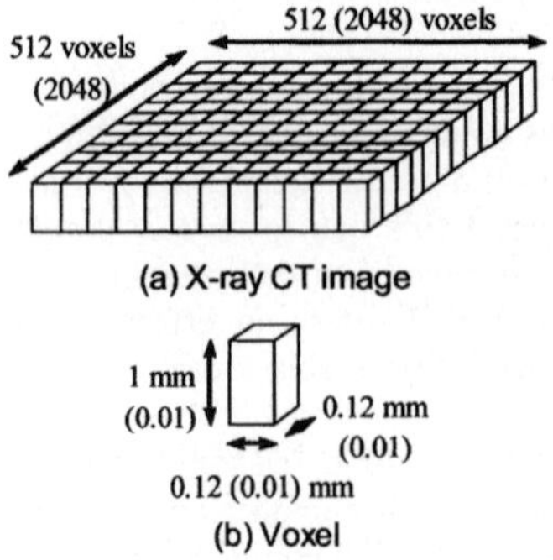

Figure 2. Concept of X-ray CT image and voxel

Physical properties

Specimens

Dredged slurry and cement and tire chips were used to make the cement treated soil. The dredged slurry was Tokyo Bay clay (ρ_s=2.716 g/cm^3, w_L=100%). The water content of it was controlled to 2.5 w_L with adding sea water. Cement was normal portland cement, the particle density of which was 3.16 g/cm^3. The tire chips were cut in pieces from used automobile tires to a separate size of approximately 2 mm as shown in Figure 3, the particle density of which was 1.15 g/cm^3.

The cement treated soil was made by mixing three materials in designated mixing condition as shown in Table 1. The target of unconfined compressive strength of the cement treated soil at 28 days of curing time was 300 kN/m^2. Molds of 5 cm in diameter and 10 cm in height were filled up with the mixture. The molds were placed in a high moisture closed box (more than 95% of relative humidity).

The specimens were trimmed to 50 mmϕ x 100 mm. The measured wet densities of cured specimen were 1.284 g/cm^3, 1.257 g/cm^3 and 1.239 g/cm^3 for Sample-A, Sample-B and Sample-C, respectively.

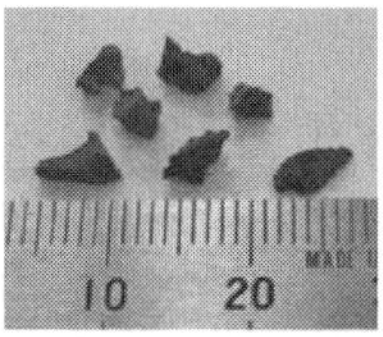

Figure 3. Tire chips

Table 1. Mixing condition

	Specimen-A (tire chips: 0%)		Specimen-B (tire chips: 16.7%)		Specimen-C (tire chips: 28.6%)	
	Mass (kg/m^3)	Volume (L/m^3)	Mass (kg/m^3)	Volume (L/m^3)	Mass (kg/m^3)	Volume (L/m^3)
Controlled clay	1225.7	980.5	1021.4	817.1	875.5	700.4
Cement	61.5	19.5	51.2	16.2	43.9	13.9
Tire chips	—	—	191.7	166.7	328.6	285.7
Total	1287.2	1000.0	1264.3	1000.0	1248.0	1000.0

Investigation of internal structure

CT images of each cross section are obtained based on the distribution of the gray level after scanning the specimen. Figure 4 shows an example of cross-sectional images of each specimen. These were chosen around the middle height in each specimen (50 mm). As described above, the CT images were drawn darker for lower densities and lighter for higher densities. The parts shown in black are air voids which formed in the material during specimen preparation, pure white indicates the presence of a cement lump or piece of shell, and gray parts approximately 1-3 mm in size are tire chips. Figure 5 shows high magnification images of the cross section of each specimen around the middle height. The multiple concentric circles in these images are attributed to image noise. In the images, isolation of the tire chips can be observed in both the macro-range and micro-range. From this, it is considered that the tire chips are suspended in the matrix of the treated soil.

Here, fluctuations in specimen density were evaluated quantitatively using the CT scanning results. As described above, the CT images are constructed using the spatial distribution of gray levels and the gray level can be converted to material density. The following equation was used for the characteristic curve, which expresses the relationship between the gray level and the density of the cement treated soil shown in Fig. 1. :

$$\rho_t=6.03\text{x}10^{-5}GL-1.702 \qquad (1)$$

(a) Specimen-A (b) Specimen-B (c) Specimen-C

Figure 4. Cross-sectional images

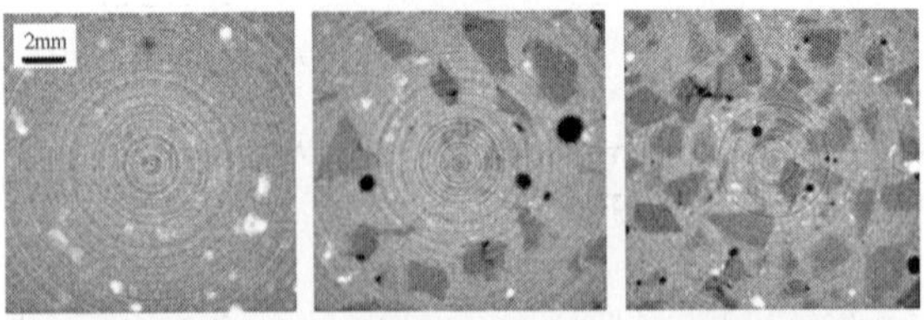

(a) Specimen-A (b) Specimen-B (c) Specimen-C

Figure 5. Cross-sectional high magnification images

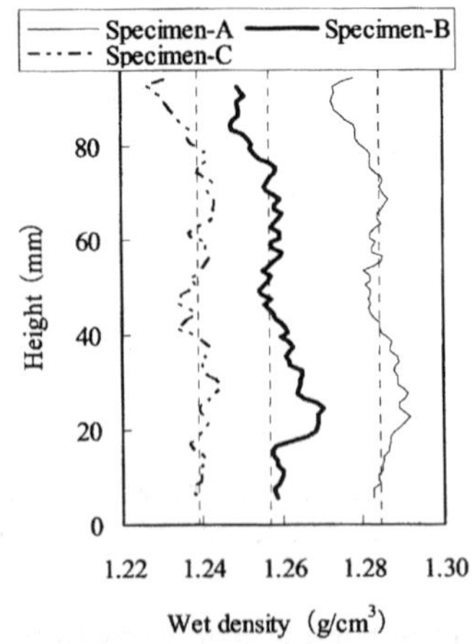

Figure 6. Density distribution

Where ρ_t is the average wet density and GL is the average gray level in each cross section whose height is 1.0 mm as shown in Fig. 2.

Figure 6 shows the estimated density distributions of each specimen using equation (1). Dotted lines in this figure are the average wet density of the specimens from volume and mass measurements. The density fluctuations of Specimen-A, Specimen-B and Specimen-C were 1.272-1.292 g/cm^3, 1.247-1.270g/cm^3 and 1.227-1.244 g/cm^3, respectively. From profile of the estimated densities, density variances are not changed even if mixing the tire chips and it means the tire chips are uniformly mixed in the specimens.

Failure properties

Specimens and test procedure

The specimens used in this study were the same specimens as those used in the investigation of physical properties (Sample-A, Sample-B and Sample-C). A series of unconfined compression tests was conducted with these specimens using a triaxial compression apparatus for the X-ray CT scanner. This apparatus was designed in such a way that the triaxial tests could be performed in the X-ray shield box. CT scanning was conducted before compression and at 0.5% increments of axial strains. When nondestructive inspection was conducted with the X-ray CT scanner during unconfined compression tests, compression was stopped momentarily while scanning was performed.

Results of unconfined compression test

Figure 7 shows the relationship between the stress and axial strain with Sample-A, Sample-B and Sample-C. The results show that the axial strain required to reach peak strength increases with increment of the fraction of the tire chips. The points marked by ◆ in the figure show the points where CT scanning were conducted. As mentioned above, compression was interrupted at each CT scanning point. Compressive stress shows a clear drop in each scanning point, it is due to the relaxation of the specimen at this time.

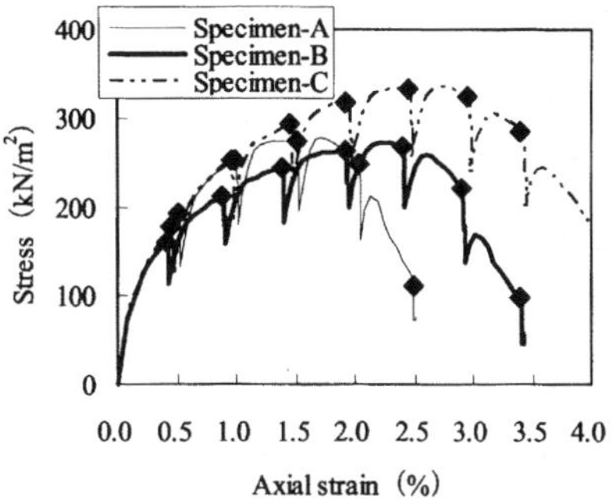

Figure 7. Relationship between the stress and axial strain

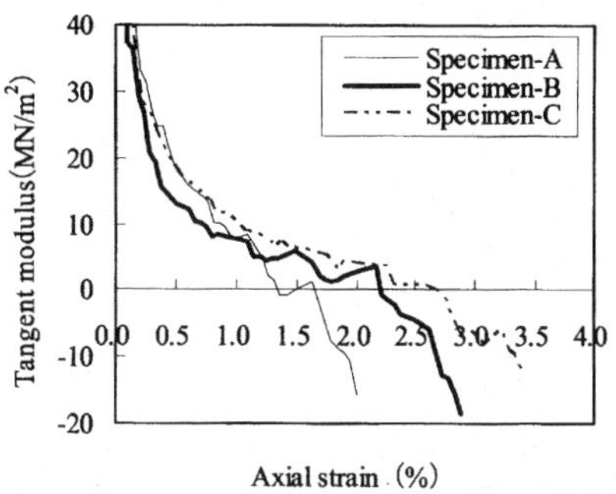

Figure 8. Relationship between the tangent modulus and axial strain

Figure 8 shows the relationship between the tangent modulus and axial strain of each specimen, excluding the scanning duration. At the beginning of shear deformation, each sample had approximately the same tangent modulus (>30 MN/m^2 in all cases). As shear deformation progressed, the tangent moduli of the cement treated soil with added tire chips (Sample-B and Sample-C) showed slighter decreases than that of the cement treated soil (Sample-A) after about 1.0% of axial strain. At that time, the tangent moduli of the specimens containing tire chips were about 10 MN/m^2, and approached that of rubber (1-5 MN/m^2). Based on these facts, it is considered that the tire chips in the cement treated soil did not contribute to shear resistance at the beginning of shear. However, as shear progressed, the tangent modulus of the cement treated soil decreased, and the ductility of the rubber tire chips contributed to the shear resistance of the material. From this, it can be understood that addition of tire chips to cement treated soil, which is essentially a brittle material, is effective for preventing a sudden drop in the tangent modulus.

Figure 9 shows the cross-sectional images of each specimen at initial state and axial strain of approximately 1.5%, 2.0%, 2.5%, 3.0% and 3.5%. These images were chosen at the middle height of each specimen (about 50 mm height). As a whole, there were no large changes before and at the peak condition of shear stress (the axial strain at failure of Specimen-A, Specimen-B and Specimen-C was 1.5%, 1.9% and 2.4%, respectively), but cracks occurred in each specimen after the peak. However, the appearance of the cracks in the respective cross sections was different in the simple cement treated soil (Specimen-A) and in the samples containing tire chips (Specimen-B and Specimen-C). The cracks in the cement treated soil without tire chips were almost straight and their thickness widened as axial strain increased. In contrast, the cracks in the cement treated soil with added tire chips were thin and developed successively in a reticulated pattern as axial strain increased. Figure 10 shows 3-dimensional images of the cracks after the peak condition from outside and inside in all specimens. The cracks in the simple cement treated soil (Specimen-A) were straight, while those in the specimens containing tire chips (Specimen-B and Specimen-C) were not straight. Thus, the mode of crack formation was changed by the presence of tire chips.

Failure propertise of cement treated soil with tire chips

The failure mechanism of cement treated soil with tire chips can be considered as follows: At the beginning of shear, the shear stress is borne by the part of the cement treated soil. As shear

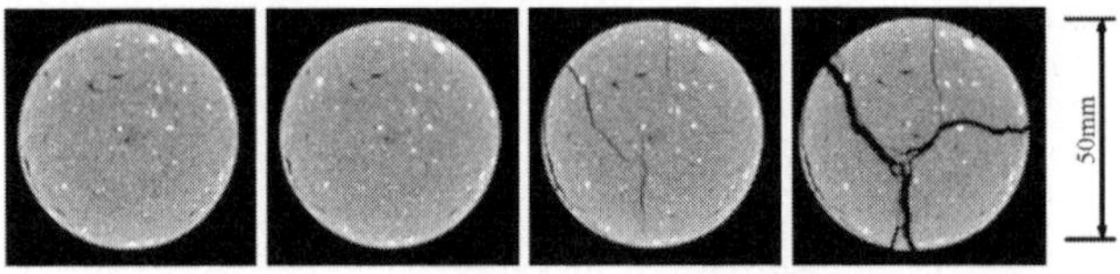

ε_a=0% ε_a=1.5% ε_a=2.0% ε_a=2.5%

a) Specimen-A

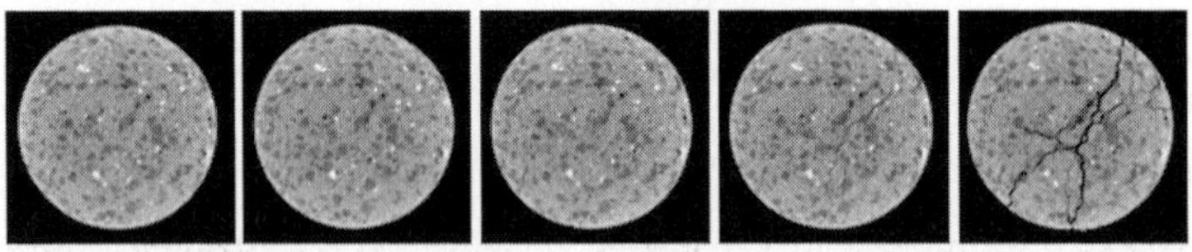

ε_a=0% ε_a=1.4% ε_a=1.9% ε_a=2.4% ε_a=2.9%

b) Specimen-B

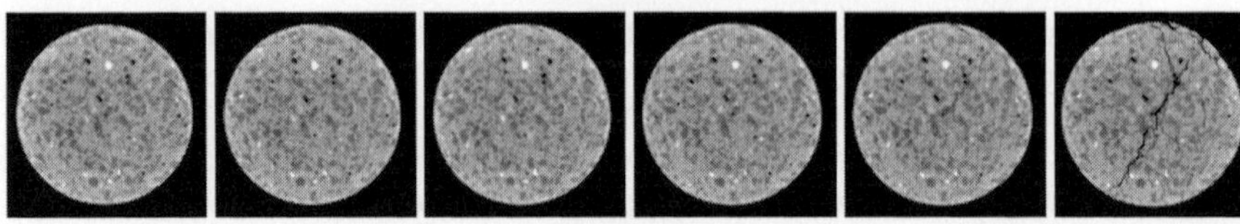

ε_a=0% ε_a=1.5% ε_a=1.9% ε_a=2.4% ε_a=3.0% ε_a=3.4%

c) Specimen-C

Figure 9. Cross-sectional images

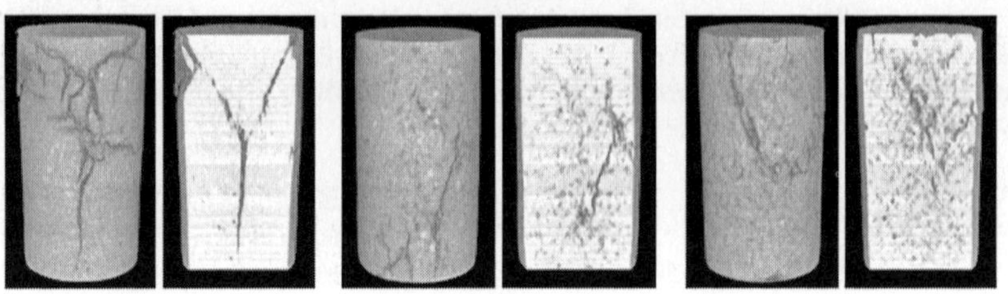

Outside Inside Outside Inside Outside Inside

a) Specimen-A (ε_a=2.5%) b) Specimen-B (ε_a=2.9%) c) Specimen-C (ε_a=3.4%)

Figure 10. 3-dimensional images

stress increases, the tangent modulus of the part of the cement treated soil decreases gradually. It causes the tangent modulus of tire chips is relatively getting higher than that of the part of the cement treated soil and some amount of shear resistance is transferred to the tire chips.

Permeability properties

Specimens and test procedure

The specimens used in this study were cement treated soil with tire chips, with two different percentages of tire chips relative to the volume of the specimen (0% and 16.7%). Table 2 shows the mixing conditions of these specimens. The materials used and procedure prepared were the same as described above (Specimen-A, Specimen-B and Specimen-C). And the water content ratio was 2.8 w_L. The cement content were slightly higher than Specimen-A,

Table 2. Mixing condition

	Specimen-D (tire chips: 0%)		Specimen-E (tire chips: 16.7%)	
	Mass (kg/m^3)	Volume (L/m^3)	Mass (kg/m^3)	Volume (L/m^3)
Controlled clay	1199.0	974.8	999.2	812.3
Cement	79.6	25.2	66.4	21.0
Tire chips	—	—	191.7	166.7
Total	1278.6	1000.0	1257.3	1000.0

Specimen-B and Specimen-C.
The specimens were trimmed to 50 mmϕ x 100 mm. The measured wet densities of cured specimen were 1.270 g/cm^3 and 1.251 g/cm^3 for Sample-D and Sample-E, respectively.

A series of permeability tests using the triaxial compression apparatus was performed some points during the process of unconfined compression. The compression tests were conducted under the unconfined condition. The permeability tests were conducted under confined condition, because of water pressure was applied to the specimen bottom to create a water pressure gradient. That is, first, the specimen was scanned with X-ray CT scanner. Then the permeability test was performed during the specimen was consolidated with confining pressure. After the permeability test the specimen was released from consolidation pressure and compressed in an unconfined condition until the planned axial strain was achieved. This procedure was repeated at the designed axial strain.

Results of permeability tests during process of unconfined compression

Figure 11 shows the relationship between the stress and axial strain with Specimen-D and Specimen-E. The points marked by ◆ in the figure are the CT scanning points. Qualitatively, however, the relationship between stress and axial strain showed a tendency similar to that in Fig. 7., in that the axial strain required to reach peak strength increased with the content of tire chips.

Figure 12 shows 3-dimensional images of Specimen-D and Specimen-E for the initial condition, the stress peak condition and axial strain of about 4% and 6% from inside of these specimens. Crack formation in the cement treated soil both with and without tire chips showed essentially the same tendency as in Fig. 10.

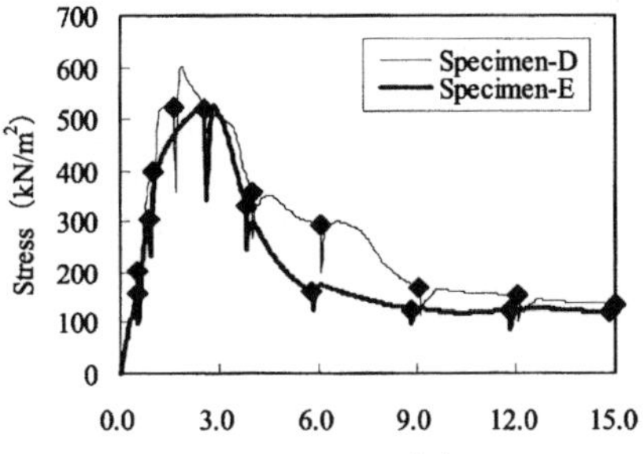

Figure 11. Relationship between the stress and axial strain

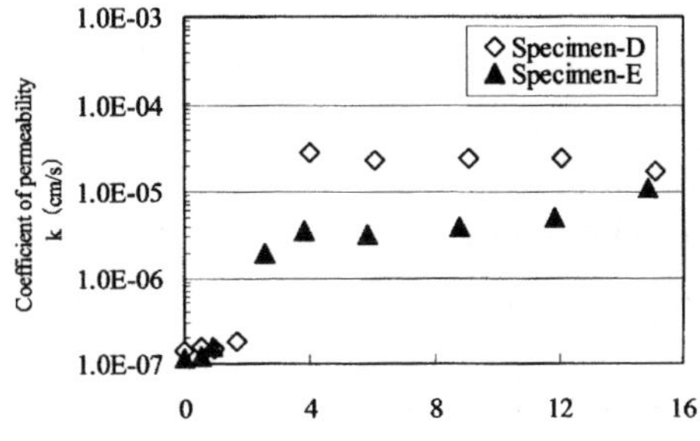

Figure 13. Relationship between the coefficient of permeability and axial strain

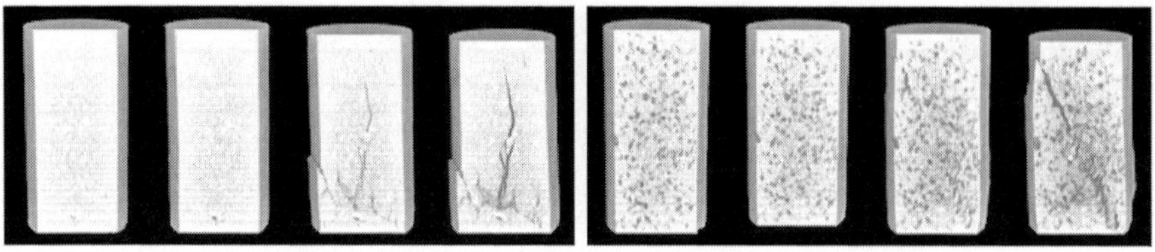

ε_a=0% ε_a=1.6% ε_a=4.0% ε_a=6.0% ε_a=0% ε_a=2.6% ε_a=3.8% ε_a=5.8%

a) Specimen-D b) Specimen-E

Figure 12. 3-dimensional images

Figure 13 shows the relationship between the coefficient of permeability and axial strain for Specimen-D and Specimen-E. With cement treated soil without tire chips (Specimen-D), the coefficient of permeability increased readily after crack initiation. In contrast, with the cement treated soil with tire chips (Specimen-E), the coefficient of permeability is about one order smaller than that of Specimen-D at this time. It is considered that large cracks formed in the cement treated soil without tire chips created large water-bleeding channels, and changed to the high coefficient of permeability. While, the cracks in cement treated soil with tire chips were thin and created little water-bleeding channels, and maintained the low coefficient of permeability.

Conclusion

In this study, the failure and permeability properties of cement treated soil with tire chips are discussed based on unconfined compression tests and permeability tests using a micro-focus X-ray CT scanner. The internal structure and failure mechanism during shear deformation were investigated visually and also evaluated quantitatively. The following conclusions were drawn from this study:

(1) It was observed that tire chips are isolated and suspended in the matrix of the cement treated soil.
(2) Tire chips are found to be well mixed in cement treated soil from the observation of the density distribution by using CT scanner.
(3) The failure mechanism of cement treated soil with tire chips under unconfined compression is different from that of cement treated soil without tire chips.
(4) The water interception capability of cement treated soil was significantly degraded by the appearance of cracks after peak strength, while crack growth after peak strength was inhibited in cement treated soil with tire chips and the coefficient of permeability is one order smaller than that of without tire chips. This is attributed to differences in crack formation and development under shear deformation.

References

Kikuchi, Y., Nagatome, T., Otani, J. and Mitarai, Y. (2005a): Engineering property of cement treated soil with tire-chips using X-ray CT scanner, *Proc. of the Sixth National Symposium on Environmental Geotechnology*, 343-350 (in Japanese).

Kikuchi, Y., Nagatome, T., Mitarai, Y. and Otani, J. (2005b): Failure properties of cement treated clay with added tire chips using X-ray CT scanner, *Proc. of the 2nd China-Japan Joint Symposium*, (in printing).

Otani, J., Mukunoki, T. and Kikuchi, Y. (2002): Visualization for engineering property of in-situ light weight soils with air foams, *Soils and Foundations*, **42** (3), 93-105.

Otani, J. and Obara, Y. (2003): *International Workshop on X-ray CT for Geomaterials –GeoX2003 –*

Mitarai, Y., Kawai, H., Yasuhara, K., Kikuchi, Y. and Karrmokar, A.K. (2004): The mechanical properties of composite geo-material mixing with scraped tire-chips and application example to harbor construction works, *Proc. of the 49th Symposium on Geotechnical Engineering*, 77-84 (in Japanese).

The Development of New Applications for Sandstone Quarry Sand in South Wales

Martin Lamb, TRL Ltd, Cardiff, Wales
Dr Murray Reid, TRL Ltd, Wokingham, England

Abstract

Approximately 1 million tonnes of sandstone quarry sand (SQS) are produced annually as a by-product of the South Wales quarrying industry. Around half of the material is used in various applications, but stockpiles of the remaining material are growing. The authors undertook a study to develop innovative applications for this potentially valuable material. SQS consists of approximately 85% sand-sized material and 15% silt and clay; it is well-graded, charcoal grey and very consistent in its properties. A literature review and market assessment were undertaken to identify applications. Laboratory tests were undertaken on promising applications and field trials on those where the laboratory tests were successful, including cycle path surfacing, block paving bedding sand and as a manufactured topsoil when mixed with compost. The material performed well in these applications and a design guide for the utilisation of SQS was produced to enable the results of the research to be implemented.

Introduction

The Minerals Technical Advice Note (Wales) – Aggregates (Welsh Assembly Government, 2004) identified that approximately 3.8 million tonnes of crushed rock 'fines' are produced annually in South Wales, a significant proportion of which is from sandstone quarries. The amounts of this material are increasing because modern methods of road surfacing use a higher proportion of the coarse aggregate from these quarries. There are markets for some of this material as a replacement for natural sand in construction, but this is insufficient to utilise all the material and large amounts are accumulating in sandstone quarries in South Wales.

Sandstone is quarried at five locations in South Wales, and is used predominantly as a high friction coarse aggregate for road surface courses. The Pennant Sandstone Series in South Wales comprise sandstones of the Upper Coal Measures of Upper Carboniferous age (Namurian or Westphalian), and the geographic area is essentially that of the South Wales coalfield, stretching from Swansea to the west, Cardiff to the east and Merthyr Tydfil to the north. The Pennant Sandstone is particularly prized as it is one of the few natural aggregates in the UK with a high polished-stone value (PSV); this is a measure of the resistance of a roadstone to the polishing action of a vehicle tyre.

A study of potential applications for the sandstone quarry sand (SQS) was undertaken by the authors between 2003 and 2004. A review of existing information and a market study were carried out to identify potential applications, and additional laboratory testing was carried out where necessary. A number of applications were selected for more detailed study by means of field trials. Based on the results of the study, a Design Guide for SQS was produced (Lamb, 2005).

5th ICEG Environmental Geotechnics, Thomas Telford, London 2006.

Laboratory Analysis

A selection of grading, physical, chemical and petrographic data was provided by CEMEX and Hanson Aggregates (Hanson) with further testing undertaken for specific applications or where there were gaps in the knowledge. Samples from three of the five active quarries in the Pennant Sandstone were tested. In total, thirty samples of SQS were subjected to particle size distribution (PSD) tests to BS 1377: Part 2: 1990. The PSD results presented in Figure 1 show the SQS to be extremely consistent and classified as a well graded silty sand. Around 15% of the material is less than 75µm, i.e. silt and clay with the remaining material of sand size. The fines increase moisture retention and enhance the susceptibility of the material to frost heave, limiting the potential uses of the material. Liquid and Plastic Limit tests carried out to BS 1377: Part 2: 1990 show that the SQS has a Plasticity Index of about 7%. This means it cannot be classed as non-plastic.

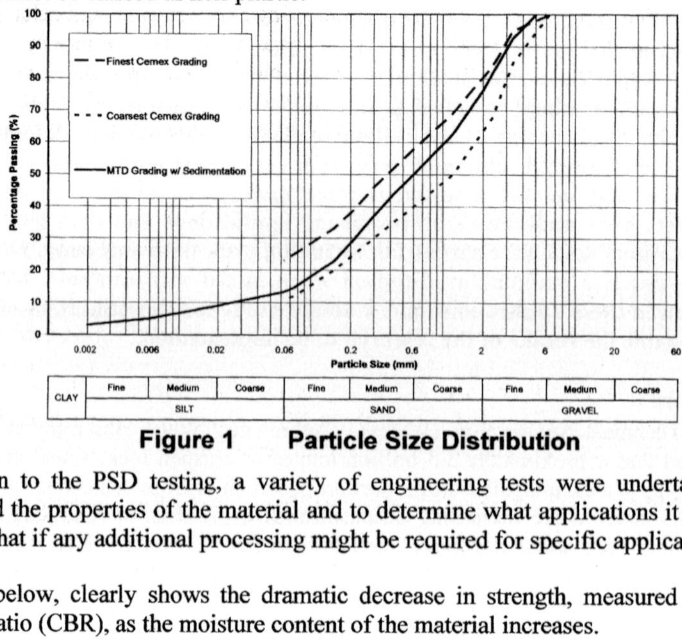

Figure 1 Particle Size Distribution

In addition to the PSD testing, a variety of engineering tests were undertaken to better understand the properties of the material and to determine what applications it may be suited for, and what if any additional processing might be required for specific applications.

Figure 2 below, clearly shows the dramatic decrease in strength, measured by California Bearing Ratio (CBR), as the moisture content of the material increases.

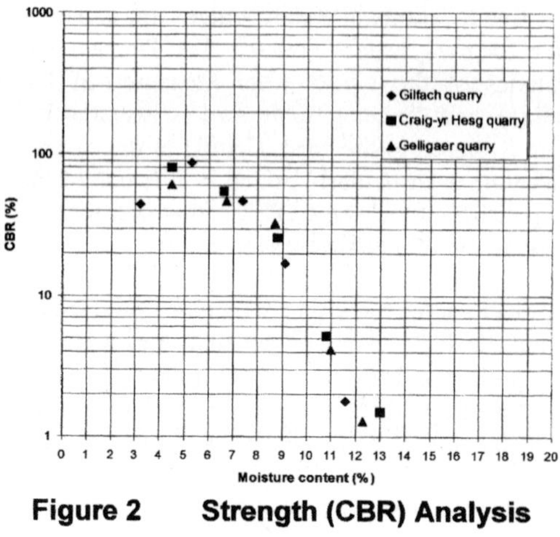

Figure 2 Strength (CBR) Analysis

The mean optimum moisture content was 8.6%; this equates to a laboratory CBR of around 30%. This strength is higher than required for capping applications (15% laboratory CBR) and similar to the lower limit recommended for sub-base (at least 30% laboratory CBR). As such, the strength data indicates that SQS can contribute good load bearing characteristics when at approximately optimum moisture content. However, the susceptibility of the SQS to increased moisture content is clearly demonstrated by the laboratory testing, with a sharp reduction in strength with minimal added moisture beyond optimum. The CBR decreases to 8% at 10% moisture content and less than 2% at a moisture content of 13%. This means that care must be taken in ensuring that SQS is close to, or below, the optimum moisture content when used in full-scale applications.

Physical Testing Conclusion

The testing undertaken demonstrated that the engineering characteristics and behaviour of the SQS from the three individual quarries were virtually identical. This feature of SQS can be viewed as a positive advantage in terms of developing general specifications for niche applications, as consideration of significant variances between local sources may be discounted. Whilst the results of most of the tests have been positive, potential problems with moisture susceptibility and frost heave were identified.

Selection of applications for further investigation

As part of the literature review and market research undertaken for this study, over forty potential applications were suggested for the material in either its 'natural' state or after further refinement and separation. Of the potential uses identified, twelve were selected for further investigation. These applications are listed in Table 1. These include applications for the unseparated material and for the separated sand (> 75μm) and filler (< 75μm) fractions. Discussions with quarry operators reveal that they have some confidence that markets could be secured for the sand (> 75 μm) fraction of the SQS. It has also been reported that they consider that separation of the sand would be relatively easy.

Table 1 Applications Selected for Further Investigation

Application	Reason for Investigation	Fraction
1. Surface course for paths	Identified by market review	Unseparated or sand
2. Structural concrete	Identified by literature review	Sand
3. High friction surfacing	Identified by market review	Sand
4. Tile manufacture	Possible trial available with local manufacturer	Filler
5. Cement extenders	Identified by market review	Filler
6. Manufactured topsoil	Local infrastructure projects that require landscaping.	Unseparated
7. Road sub-base (as cement bound material)	Local road construction & road widening schemes in S Wales	Unseparated
8. Decorative grit	Identified by market review. Awareness raising via report	Sand
9. Abrasives	Identified by market review	Unseparated or sand
10 Concrete block manufacture	Used historically. Awareness raising via report & case study	Unseparated
11. Block paving bedding sand	Used historically. Awareness raising via report & case study	Unseparated
12. Bentonite enhanced sand in landfill liners	Used historically in S Wales	Unseparated

Of the applications outlined above, a number were relatively quickly discounted. The quantity of fines in the material would make it unsuitable for applications involving mixing with asphalt, whereas abrasives applications such as shot-blasting would not be permissible due to the silica content, and inherent risk of silicosis. Initial trials were carried out for high friction surfacing applications, which proved to be extremely encouraging in terms of skid resistance, but subsequent tests indicated that the material would not be sufficiently durable for a highway application. In addition, both structural (C20-C50) concrete and concrete blocks were known to have been successfully manufactured with the incorporation of SQS at differing quantities, which was reported and highlighted through case studies.

Four of the more promising applications identified are discussed in more detail below.

Cycle Paths and Footways

The potential for use of SQS in a cycle path was considered, and following discussions with Neath Port Talbot County Borough Council a trial was arranged at Afan Argoed Country Park. Two trial areas were identified, the first a flat section, measuring 2 m wide by 60 m in length and the second a tight bend on a slope that was approximately 11 m in length and subject to harsh braking and water run-off. Construction was undertaken in February 2004.

Trial area 1 was divided into three sections:

- Unbound SQS,
- SQS stabilised with ordinary Portland cement (OPC) which was further subdivided into four, 5 m lengths containing 1%, 1.5%, 2% and 2.5% OPC and,
- Unbound SQS laid on 0 – 40 mm size scalpings.

All the sections were laid to a compacted depth of around 50 mm above original ground level and compacted in accordance with the Specification for Highways Works (Stationary Office, 2001). Control of the moisture content was found to be important, as the fresh SQS material was significantly dry of optimum moisture content.

Trial area 2 had 4.2% OPC mixed to take account of the extra stresses that would be placed on this corner. As with trial area 1, it was compacted in accordance with the SHW.

Monitoring was carried out regularly using visual inspection and taking photographic records. Nine months following construction it was observed that on trial area 1, all of the trial sections were performing well, with the unbound section performing the best of all. The 2.5% OPC bound section took on a concrete-like appearance quite quickly after construction, with the 2% OPC section appearing the same towards the end of the trial.

A levelling survey undertaken at the end of the trial revealed that there had been little or no erosion in the central area of the path, i.e. the area subjected to the most cycle and pedestrian traffic. Despite the addition of OPC to trial area 2, within two weeks it had taken on the appearance of an 'unbound' material, but performed very well. A clear divide between the trial material and the original surface remained until the end of the trial, suggesting that there had not been any significant transport of material as a result of cycles braking, or during periods of rainfall. The rangers at the country park were delighted with the performance of the material.

Manufactured Topsoil

Previous studies, (Keeling and Marchant, 2001) indicated that when mixed 70:30 by weight with a variety of composts, quarry fines form a suitable growing medium. It was decided to undertake a trial using SQS and filler at TRL's project office at the Geoenvironmental Research Park (GRP) near Port Talbot. Six beds were constructed at the site measuring 3 m by 3 m, and 450 mm tall. Green waste compost supplied by Swansea Waste was used as the organic component.

Four of the six test beds comprised a 2:1 mineral (either SQS or the filler fraction) to organic mix ratio by volume, with a fifth test bed a control bed divided longitudinally into a natural topsoil material and a local 'topsoil' manufactured from subsoil and construction, demolition and excavation waste; in this manner growth was compared between materials with similar specifications. A sixth bed was constructed and divided into SQS and filler only, i.e. no organic material to demonstrate the effect of the addition of the organic fraction as a growing medium. Half of each bed was compacted with a vibrating plate compacter and the other half was lightly compacted by foot by walking over a board. Each bed was sown with a standard DoT Pro85 grass and clover (5%) mix at a rate of 10 g/m^2, i.e. 90 g per test bed, as per the supplier's instructions.

Monitoring was undertaken on a weekly basis by the resident TRL staff at the GRP. This comprised an assessment of grass growth using a modified Braun-Blanquet scale and also through taking a photographic record.

British Standard BS 3882:1994 identifies three grades of soil; economy, general purpose grade and premium grade. Analysis of the manufactured soil mixes and of the topsoil for comparison with the general purpose grade target in BS 3882 was undertaken at the beginning of the trial. Six composite samples were taken from the test beds at the end of the trial in late October to assess the effect of vegetation establishment on the various soils. At the end of the trial, twenty two weeks following seeding, analysis indicated that all of the soils comfortably exceed the nutrient requirements and loss on ignition requirements, with all soils also comfortably meeting the pH requirements. The locally manufactured topsoil bed was the poorest performing with generally lower nutrients than the natural or manufactured soils, with exception of phosphorous which were similar for all mixes.

In terms of nitrogen, two of the six beds, the natural topsoil and one of the filler and compost beds, met the general grade standard, although none met the target initially. Overall, however, nitrogen levels did not appear to be directly related to vegetation performance. Bed 1, which recorded the lowest levels of nitrogen, supported good growth, whereas the locally manufactured soil which had higher total nitrogen, recorded relatively poor vegetation spread and limited growth. Results showed that with the exception of Bed 6, which contained SQS and filler alone, all beds showed the ability to grow vegetation at a reasonable rate and sustain vegetation successfully. Figure 3 presents the average grass growth on the test beds, based on observations using the Braun-Blanquet scale for grass vegetation cover.

The natural topsoil (Bed 5B) was the most successful of all the test beds in terms of initial vegetation cover, with the locally manufactured soil also performing well initially, which was not unexpected as they contained a natural seed bank resulting in increased vegetation cover and faster germination times. The disadvantage of this natural seed bank however, is that both soils contained numerous weeds. Initial performance, however, did not reflect overall growth throughout the trial. With the exception of the mineral only bed, the locally manufactured soil was the poorest performing bed in terms of overall growth and spread. This bed and the natural soil also exhibited a yellowing of the vegetation, something not noted on Beds 1 to 4.

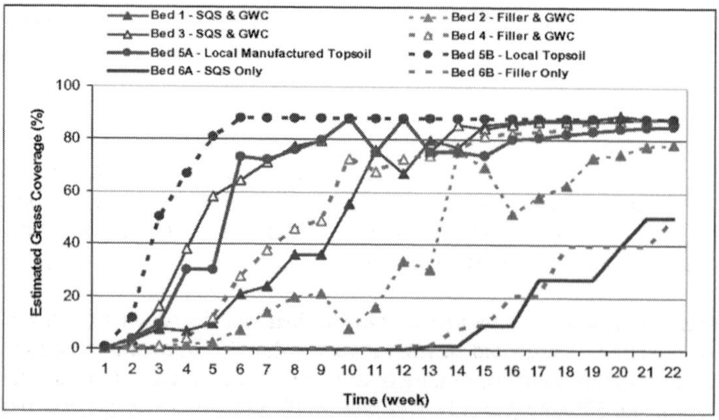

Figure 3 Average Grass Growth on Test Beds

The study has clearly demonstrated that either SQS or filler are ideal mineral components of a manufactured soil, and their use should be encouraged.

Cement Bound Sub-Base

Due to the potential for frost heave, plasticity and grading the material would not be suitable as unbound sub-base, but would fit within the specifications for a cement bound sub-base. A simple trial was undertaken to determine the cement content required at 100% compaction to meet the strength requirements for CBM1 in the Specification for Highway Works. Tests were carried out in accordance with BS1924: Part2: 1990. Mixes were carried out at 25, 50, 75, 100 and 125 kg/m^3 cement contents. Test cubes were compacted at a moisture content of 9% by mass of aggregate, which is the OMC of the material.

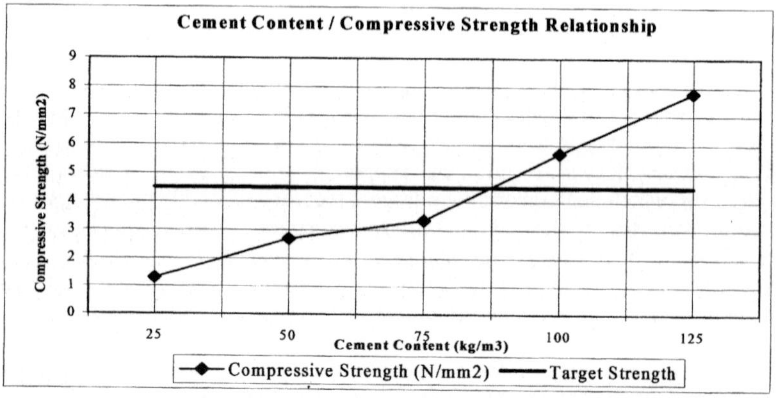

Figure 4 Cement content vs 7 day compressive strength

The results are shown on Figure 4. There is a steady increase in 7 day compressive strength with cement content, with the 4.5 N/mm^2 requirement for CBM1 falling between cement contents of 75 and 100 kg/m^3, at around 87.5 kg/m^3. In practice, the higher cement content (100 kg/m^3) would be used to ensure the strength was achieved. With a cube density of approximately 2.3 Mg/m^3, a cement content of 25 kg/m^3 represents approximately 1.1% by weight. Therefore, to achieve the CBM1 7 day compressive strength requirement of 4.5 N/mm^2, when using 100 kg/m^3, around 4.3% of cement would need to be added to the SQS.

Block Paving Bedding Sand

SQS does not fit the grading required in BS 7533: Part3: 1997 for block paving bedding sand due to the high fines content, although it is understood that some local contractors have used the material and report that it can perform better than the standard marine sand. It is likely that this is due to the well graded nature of the material giving excellent compaction characteristics at the optimum moisture content.

A local contractor agreed to perform a trial at their premises where three block paving parking bays would be constructed in accordance with TRL Application Guide AG26 (TRL Limited, 2003), on three different materials, a control marine sand, SQS and SQS with the filler fraction removed (>75μm SQS). The removal of the filler brings the material in line with the gradings detailed in the British Standard. The bays were constructed between December 2003 and January 2004 and were monitored for ten months.

The location of the trial was particularly useful as the bays would be in use between five and six days per week. A total of 128 survey points were measured before and after the trial comprising eight around the kerbs and 40 on each of the three test beds. Apart from a shallow dip in a corner of the control sand bay there was no visible difference between the three test bays. Assessment of the survey data revealed that all of the bays had settled slightly, but that the control bay had the single greatest and highest average settlement (-3.13 mm) followed by the SQS bay (-2.05 mm) and the >75μm SQS bay (-1.83 mm). This shows that unseparated SQS outperformed the control material and suggests that if used carefully, could form an ideal base for block paving.

Conclusions

A two year research project has been undertaken on SQS which has comprised a literature review, market assessment, laboratory trials and field trials of promising applications.

Three of the most promising applications identified were use as a cycle path, as a bedding sand for block paving and as the mineral component of a manufactured topsoil. Field trials were undertaken for all three of the above applications. The trials determined that whilst care was required in ensuring that the SQS was compacted at or close to the OMC of around 9%, the material was ideal for use as a cycle path, either unbound or combined with OPC, and actually outperformed the control sand for the block paving application. The field trials for the manufactured topsoil demonstrated that both SQS and the filler only fraction when mixed with green waste compost were successful in establishing vegetation and once established, outperformed the natural topsoil in terms of quantity of grass and of having fewer weeds.

The results show that SQS is an extremely consistent yet versatile material that has many benefits when used on its own or in combination with other materials for a wide variety of applications. It is hoped that the results of this research will lead to the wider use of this material. Although based on experience in South Wales, the results will be applicable to similar materials in other areas.

Acknowledgements

The project was funded by the RMC Community Fund via the Landfill Tax Credit Scheme, with contributing third party funding from the Welsh Assembly Government's Aggregates Levy Sustainability Fund, via the Environmental Body Viridis. The project was supported by a Steering Group comprising representatives from CEMEX, Hanson Aggregates, Neath Port Talbot County Borough Council, the Geoenvironmental Research Centre of Cardiff University, Minton Treharne and Davies, Jacobs Babtie, the Welsh Development Agency, Richards Moorehead and Laing, TRL and Viridis. The assistance of Gerald Davies Limited with the site trials is gratefully acknowledged.

References

Application Guide AG26 Version 2 (2003). "Footway and Cycle Route Design, Construction and Maintenance Guide". TRL Limited, Crowthorne.

British Standard BS 812-4 (1989). "Testing Aggregates. Part 4 Methods for determination of frost heave". British Standards Institution, London.

British Standard BS 1377-2 (1990). "Methods of test for soils for civil engineering purposes. Part 2 Classification Tests. ISBN: 0 580 17867 6

British Standard BS 1377-4 (1990). "Methods of test for soils for civil engineering purposes. Part 4 Compaction Related Tests". British Standards Institution, London.

British Standard BS 1377-9 (1990). "Methods of test for soils for civil engineering purposes. Part 9 In-situ tests". British Standards Institution, London.

British Standard BS 1924-2 (1990). "Methods of test for cement stabilised and lime stabilised materials. Part 2". British Standards Institution, London.

British Standard BS 3882 (1994). "Specification for Topsoil" British Standards Institution, London.

British Standard BS 7533-3 (1997). "Pavements Constructed with Clay, Natural Stone or Concrete Pavers: Part 3. Code of Practice for Laying Precast Concrete Paving Blocks and Clay Pavers for Flexible Pavements". British Standards Institution, London.

Keeling, A.A; Marchant, L (2001). "Grass and Tree Growth in Synthetic Soils - Early Results". *Presented as Appendix 4, Brite Euram Project Report BE97-5078 "Refill"*, Final Technical Report by Tarmac Fleming November 2001

Lamb, M (2005). "Design Guide for Applications of Sandstone Quarry Sand in South Wales". *Viridis Report VR8*. TRL Limited, Crowthorne. Available at www.viridis.co.uk.

Manual of Contract Documents for Highway Works (2001). "Volume 1: Specification for Highway Works (MCHW 1) (May 2001)". The Stationery Office, London.

Welsh Assembly Government (2004). "Minerals Planning Policy (Wales). Minerals Technical Advice Note (Wales) 1. Aggregates March 2004". ISBN 07504 3314 0.

Aging effect on MSW mechanical behaviour

S. L. Machado, Federal University of Bahia, Salvador, Brazil
M. F. Carvalho, Catholic University of Salvador, Salvador, Brazil
J. C. F. Nascimento, Federal University of Bahia, Salvador, Brazil
K. A. Dourado, Federal University of Bahia, Salvador, Brazil

Abstract

This paper presents results of drained conventional and stress path triaxial tests performed on MSW great dimension samples (200mm of diameter). The waste studied comes from Metropolitan Center Sanitary Landfill (Salvador-Brazil). In order to evaluate the influence of time since disposal on MSW mechanical behaviour, samples were collected in the disposal front (new residue) and on places where a time of 4 years was expected since disposal. MSW samples were molded with an unit weight of about 10 KN/m^3, what means a dry unit weight of about 5 KN/m^3, considering an average water content of 100% (dry basis). Typical friction angle and cohesion intercept *versus* axial strain curves are showed, allowing to evaluate the influence of the time since disposal on MSW mechanical behaviour. Besides that, are assessed the influence of the stress trajectory on the MSW behaviour and the effect of the time since disposal on the fibres mechanical properties.

Introduction

Despite of the growing interest in other destination alternatives for MSW, as selective collect, recycling, burning and reuse or even the prevention of waste generation, large amounts of waste must be still disposed in the environment and the main alternative for safe disposition of MSW continues to be the sanitary landfill, an engineered structure that comprises many fundamental devices such as the liner, the leachate and gas collection and removal systems, the waste cells and temporary and final cover. These engineering components interacts each other and the knowledge of the waste mechanical properties is necessary in order to answer many of the questions that arise from landfill management tasks. As part of the MSW components degrades with time, the knowledge of the MSW time differed behaviour is also required and its mechanical properties must be preferentially expressed as a function of time or as a function of the biodegadation process itself. In this way, the mechanical behaviour of MSW is a very complex subject. Different material types such as paper, metal, glass, wood and garden waste, food and plastic, with variable dimensions are usually present. Each one of these components has its particular mechanical behaviour with different strength and compressibility properties that can vary with time. The proportion of each component is highly variable and is related to cultural and economic aspects of the region where the refuse is generated. This probably imparts particular characteristics for MSW generated in different parts of the world.

These aspects and other difficulties related with obtainment of representative samples of in situ conditions, adequate testing procedures both in field and in laboratory, and the lack of reliable models to represent waste behaviour have led to study MSW considering some typical concepts of Soil Mechanics, sometimes adapted to take into account the features of such a peculiar material (Grisolia et al. 1995, Jessberger et al.1995). For instance, the shear strength of MSW is usually determined from large scale triaxial compression or direct shear tests, in situ tests or back analysis results that are interpreted according to the Mohr-Coulomb shear strength criteria. A typical feature of stress-strain behaviour of MSW is that failure cannot be observed even at high sample compression (Singh & Murphy 1990, Jesseberger & Kockel 1993, Jesseberger et al 1995, Grisolia et al 1995, Grisolia & Napoleoni 1996, Carvalho 1999). If Mohr-Coulomb theory is used, the shear strength parameters cannot be adequately defined, unless they are considered as strain dependent. The mobilized shear parameters tend to increase with the deformation in a process that seems to be largely commanded by the reinforcing effect of fibre materials (plastic, textiles, etc) as shown by Kockel & Jessberger (1995). The fibrous materials of the waste contribute to the increase the MSW cohesion, since they are able to withstand tension stress. Figure 1 shows results obtained by Kockel & Jessberger (1995) when testing a waste with different plastic contents.

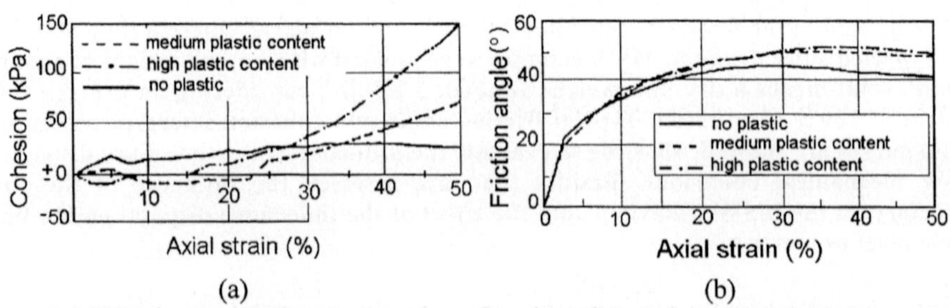

Figure 1. MSW shear strength parameters as a function of the axial strain. a) friction angle; b) cohesion intercept (Kockel & Jessberger; 1995).

As it can be seen in Figure 1, friction angle and the cohesion intercept are mobilized following different patterns. The friction angle tends to be fully mobilized at large strain values, while the fibre cohesion starts to be mobilized at strains of 10% and even more, and a limit value cannot usually be observed. Shear strength values of MSW reported in the literature vary widely, with friction angles between 10° to 53° and cohesion values varying from 0 to 67 kPa (Kavazanjian et al. 1995; Knochenmus et al. 1998). This variability can be credited to the heterogeneity of sample, the strain level used to define shear strength parameters, the choice of representative samples and the tests used. In addition, it is possible that other variables such as the age, composition, density and moisture content of the waste can interfere on the shear strength of MSW.

Based on limited data and calling attention to the wide scattering in the data, some authors such as Singh & Murphy (1990), Kavazanjian et al. (1995), Van Impe (1998) and Manassero et al. (1996) have proposed shear strength envelopes for practical purposes. Manassero et al (1996) proposed a shear strength envelope that depends on the normal stresses (σ). For very low normal stresses (σ<20 kPa) it is considered that MSW behaves as a cohesive soil-like material, thus showing a cohesion intercept (c) of about 20kPa. For low to moderate normal stresses, that 20< σ <60kPa, MSW is considered to be a purely frictional material with a shear strength angle (φ) of 38°. Finally, for larger normal stresses, it is suggested c ≥20kPa and φ ≈ 33°. These values are comparable to those suggested Kavazanjian et al. (1995).

A more comprehensive model that encompasses stress-strain behaviour was proposed by Machado et al. (2002). The proposal considers that two different effects command MSW mechanical behaviour, the reinforcement of MSW by synthetic fibres and the MSW paste, that is, the other non-fibrous materials component of the waste. The influence of the fibrous material in the MSW stress/strain curves is based on the idea that part of the deviator stress, q, is carried out by the fibrous components (plastic, textile and vegetable's fibres) present in the waste body. The axisymmetric mean stress, p, is the same for fibres and paste. Fibres elastic shear strain increments are computed as a function of the paste elastic shear strain increments by means of a mobilization function. In order better represent the reinforcement effect of the fibrous components, the Young modulus of the fibres is made dependent of the radial stress.

The fiber yield locus is given by the Von Mises criterion. Equation 1 presents the paste yield locus adopted by the model and equation 2 presents the flow rule adopted. In this equations, $\psi = n/(1+n)$, which corresponds to the p/p_o ratio in a critical state condition and $\beta = \ln(0.5)/\ln(\psi)$. M corresponds to the paste Critical State Line (CSL) slope on (q:p) plane, p_o corresponds to the paste isotropic pre-consolidation stress and n is a paste yield locus shape parameter, adopted as n = 2 by Machado et al. (2002). Although not presented here, Machado et al. (2002) model has been used in order to simulate results obtained from the stress path triaxial tests performed, showing good results.

$$f_p = q_p - M\cdot\left[p^n\cdot(p_o - p)\right]^{\left(\frac{1}{1+n}\right)}\cdot n^{\left(\frac{1}{1+n}\right)} = 0 \quad (1)$$

$$\frac{d\varepsilon^p_{vp}}{d\varepsilon^p_{sp}} = \left(\frac{1}{1+n}\right)\cdot M\cdot\left[p^n\cdot(p_o-p)\right]^{\left(\frac{1}{1+n}-1\right)}\cdot\left[p^{n-1}\cdot n\cdot(p_o-p)-p^n\right]\cdot\left(1+\sin\left[\frac{\pi}{2\cdot\psi^\beta}\left(\frac{p}{p_o}\right)^\beta\right]\cdot 2\cdot n\right)^{-1}$$
(2)

In this paper, some results of drained triaxial tests performed on MSW great dimension samples are presented and discussed as well as results of simple tension tests performed in the samples plastic material. New samples, coming from disposal front and samples 4 years old were tested in order to evaluate the effect of the elapsed time since disposal on MSW (and its fibres) mechanical properties.

Materials and methods

MSW characterization

The waste studied comes from Metropolitan Center Sanitary Landfill, located in city of Salvador, Bahia, Brazil. Around 2,500 tons of refuse are disposed daily in this sanitary landfill. New samples, coming from disposal front and samples 4 years old were collected in order to perform the proposed tests. Table 1 shows the physical composition of the collected waste samples (dry basis). Paste component includes food waste and other non easily separable materials. It is observed a reduction in almost all components with time, in contrast to paste fraction, which tends to increase its percentage. This is due the fact that significant amount of others components are incorporated in paste fraction after disposition, making difficult components segregation. Thus, part of sealing soil, paper, plastic, glass and textile, was incorporated in the paste fraction during landfill operation and after disposition and was computed as paste in the 4 years old sample. When the paste organic content is evaluated, however, there is a decrease (fast decrease) of its percentage with time. In this case, the Volatile Solid Content was used to measure the amount of organic matter in the MSW paste. An average volatile content of only 20% was obtained to 4 years samples, much less than 53.4% obtained to new ones. Paste organic matter decreases from about 16.4% to about 9.9% after 4 years since disposition.

Tests performed

Triaxial compression tests (CD) were carried out on MSW samples statically compacted in the base of triaxial cell. MSW samples were molded with an unit weight of about $10kN/m^3$, what means a dry unit weight of about $5kN/m^3$, considering an average water content of 100% (dry basis). Initially, the specimens were saturated by base to top percolation of water and after that the back pressure technique was used in order to saturate samples, accepting values of B greater than 0.90. The strain rate adopted was 0.8mm/min and effective radial stresses, σ_r, of 50, 100 and 200 kPa were used. Results of one test, performed increasing both radial stress, σ_r, and deviator stress, q (stress paths 1:1) and other performed decreasing radial stress while increasing deviator stress (stress paths 1:50), will be shown below. The strain rate adopted was the same of conventional tests.

Table 1- Physical composition of waste studied.

Components	Percentage of occurrence (%) (dry basis)	
	New waste samples	Samples 4 years old
Wood	5.8	5.6
Stone/ ceramic	14.5	13.4
Textile	3.5	2.5
Rubber	0.1	0.2
Plastic	19.8	13.8
Glass	4.8	4.1
Metal	5.0	5.0
Paper/ cardboard	15.6	5.2
Paste	30.8	50.2
Paste Volatile Solid Content	53.42	19.76
MSW paste organic matter	16.45	9.92

Simple tension tests were performed in plastic material token from MSW samples in order to evaluate the effect of the elapsed time since disposal on fibres mechanical properties. Plastic material 4 years old was obtained of the waste samples used in triaxial tests, using soft plastic derived from bags, knapsacks, etc. Samples were constructed washing and cutting plastic material in order to form bands with nominal size of 30 x 250 mm. Bands were putted together in order to form stacks with no less than 30 units. New plastics samples were obtained of new bags, knapsacks, etc. Samples equivalent transversal section was obtained as function of the stack mass and length (M and L respectively) and the average fibres density, ρ_f. An average value of $\rho_f = 1.0$ g/cm^3 was adopted.

$$A = \frac{M}{\rho_f \cdot L} \quad (3)$$

Obtained results and analysis

Figure 2 presents the results obtained from conventional triaxial tests performed. Typical stress-strain curves were obtained, with q increasing continuously, without reaching any peak or an ultimate value. Another interesting feature is the decrease in volumetric strains as the confining stresses increase. Similar behavior was obtained by Grisolia et al. (1993), Jessberger & Kockel (1993) and Carvalho (1999). Although presenting similar results, samples 4 years old tend to present higher values of deviator stress, particularly for higher levels of confining stress and axial strain. This is considered as an indirect influence of MSW mass loss, since it increases the proportion of fibres in the residue. Biodegradation process has also another effect that acts in the opposite way, reducing fibres young modulus and the values of tensile stress at failure, as will be see forward. It can be said, however, considering the samples tested, that the reduction in the mechanical fibres properties was less significative than increase in the plastic fraction. As failure usually cannot be clearly defined, shear strength parameters are obtained considering some values of strain.

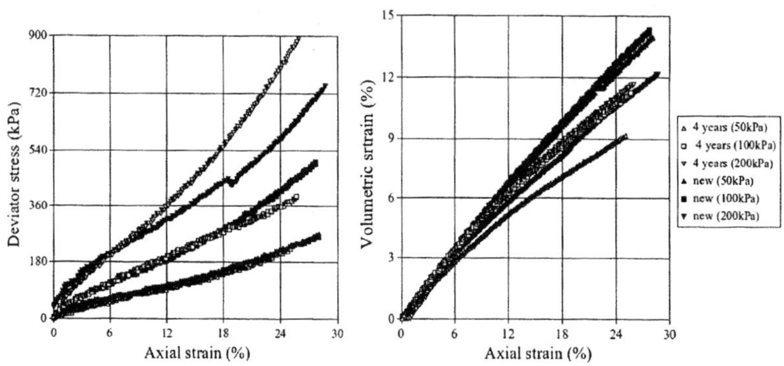

Figure 2. Conventional drained triaxial test results obtained on new samples and on samples 4 years old. a) Deviator stress-axial strain curve; b) Volumetric strain-axial strain curve.

Figure 3 shows the MSW cohesion intercept and friction angle variations with axial stress. As presented in figure 3, due the high values obtained to deviator stress in the test performed with 4 years samples, 200 KPa (figure 2), a significant difference can be observed in the shear strength mobilization curves. Other tests were performed using a confining stress of 200 KPa, showing similar results. New tests will be performed using confining stress values of 400 KPa in order to check if this tendency is maintained for higher values of confinement. Figure 4 presents some results of the stress path triaxial tests performed on MSW 4 years old samples. It is presented results of one conventional test (the same already presented before, stress path 1:3), performed with a constant confining stress of 100 kPa, and two other tests performed having initially 100 kPa of confining stress and stress paths of 1:1 and 1:50. In the 1:1 trajectory there is an increase of of the confining stress as the deviator stress increases, the opposite occurring with the 1:50 (almost vertical trajectory). As it can be seen, in the 1:2 trajectory there is a gain in the MSW rigidity, and more concave curves are obtained.

Figure 5 shows typical results obtained for plastic samples with 4 years old of disposition and new ones studied. It can be seen, there are relatively small changes in the initial young modulus and ultimate tensile stress, although samples 4 years old show a behaviour less plastic, where the stress decrease after to reach the pick value. This behaviour could lead to an instable behaviour of mass of waste, causing waste mass sliding in some circunstantes. Table 2 summarizes the results obtained. It was obtained a decrease of about 24% in the average young modulus and of about 9% in the ultimate tensile stress plastic samples after 4 years since disposition.

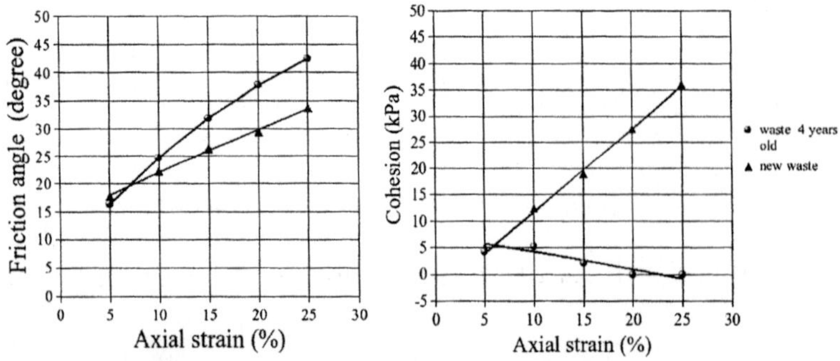

Figure 3. MSW shear strength parameters mobilization as a function of axial strain.

Table 2 – Mechanical properties of plastic new and 4 years of disposition.

Samples	Tests	Eu (MPa)	$q_{máx}$ (kPa)
Fibres of waste with 4 years of disposition	1	53.9	6359.09
	2	68.6	7617.46
	3	62.4	6439.42
	4	63.4	7610.49
Average value		62.10	7006.62
Standard deviation		60.98	702.13
New fibres	1	72.85	7321.85
	2	111.66	8218.90
	3	85.69	7829.92
	4	70.62	8507.33
	5	80.82	7497.10
	6	72.82	6889.22
Average value		82.41	7710.72
Standard deviation		15.43	596.45

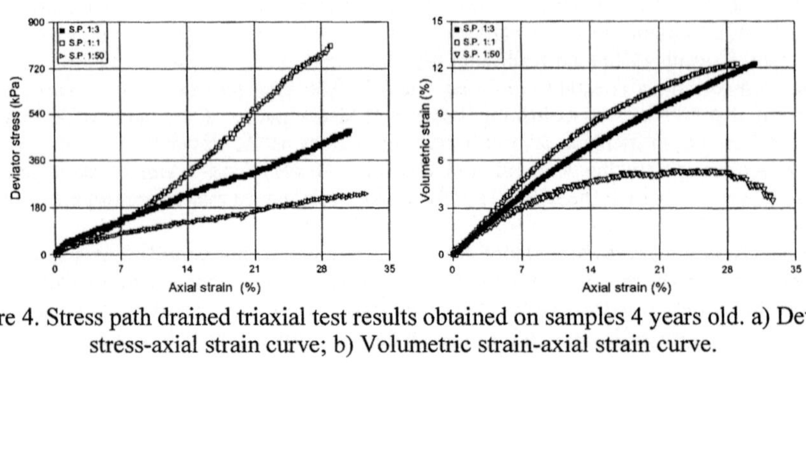

Figure 4. Stress path drained triaxial test results obtained on samples 4 years old. a) Deviator stress-axial strain curve; b) Volumetric strain-axial strain curve.

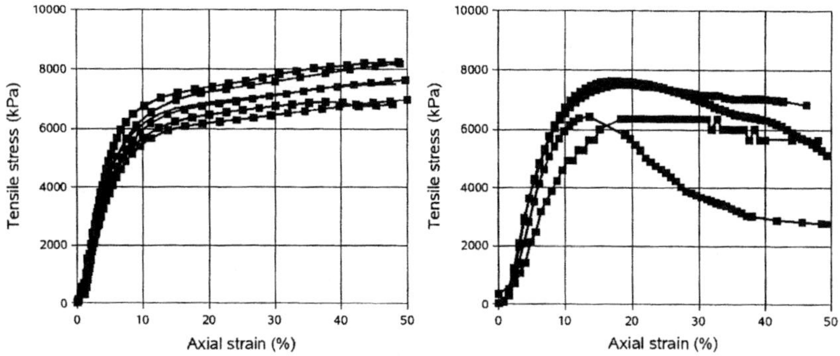

(a) (b)

Figure 5. Stress -strain curves obtained for plastic materials. (a) New samples; (b) 4 years old samples

Final remarks

Results of triaxial compression tests on large remolded specimens of waste with 4 years old and new samples were presented and discussed. Also were showed results of simple tension test in fibre of waste.

Drained triaxial compression tests showed stress-strain curves that were concave upwards and failure or an ultimate value of deviator stress could not be reached even for large values of strain. The shear strength parameters are strain dependent and friction angle tended to increase with the deformation. However, for the waste with 4 years of disposition, the cohesion intercept not follow this patterns.

Simple tension test in fibre shown a decrease of 24% in average elasticity moduli and about 9% in ultimate strength of plastic after 4 years old of disposition.

Acknowledgements

The authors are grateful to VEGA Bahia tratamento de Resíduos S.A. and FAPESB (Bahia state Research Funding Agency) for financing part of this research.

References

Carvalho, M. F. (1999). Comportamento Geotécnico de Resíduos Sólidos Urbanos. Tese de doutorado-Escola de Engenharia de São Carlos-Universidade de São Paulo.

Grisolia, M. e Napoleoni. (1996). Geotechnical Characterization of Municipal Solid Waste: Choice of Design Parameters. *Proc. Of The Second International Congress on Environmental Geotechnics,* Osaka, Japan, A.A.Balkema, v.2, p. 641-646.

Grisolia, M.; Napoleoni, Q. & Tancredi, G. (1995). The Use of Triaxial Tests for the Mechanical Characterization of MSW. *Proceedings Sardinia 95, Fifth International Landfill Symposium, S. Margherita di Pula,* Cagliari, Italy, october 1995, p.761-768.

Jessberger, H. L. & Kockel, R. (1993). Determination and Assessment of the Mechanical Properties of Waste Materials. *Proceedings Sardinia 93, Fourth International Landfill Symposium, S. Margherita di Pula,* Cagliari, Italy, October 1993, p.1383-1392.

Jessberger, H. L.; Syllwasschy, O. & Kockel, R. (1995). Investigation of Waste Body-Behaviour and Waste-Structure-Interaction. *Proceedings Sardinia 95, Fifth International Landfill Symposium, S. Margherita di Pula,* Cagliari, Italy, october 1995, p.731-743.

Kavazanjian, E.; Matasovic, N.; Bonaparte, R. & Schmertmam, G. R. (1995). Evaluation of MSW Properties for Seismic Analysis. *Geoenviromental 2000.* Geotechnical Special Publication n° 46, ASCE, vol.2, p.1126 - 1142. Ed. Yalcin B. Acar and David E. Daniel, New Orleans.

Knochenmus, G.; Wojnarowicz, M. & Van Impe W. F. (1998). Stability of Municipal Solid Wastes. *In: Proc. of the Third International Congress on Environmental Geotechnics,* Lisboa, Portugal, Sêco e Pinto (ed.), Balkema, Rotterdam, ISBN 90 5809 006x, p. 977-1000.

Kochel, R. Jessberger, H. L. (1995). Stability Evaluation of Municipal Solid Waste Slopes. Proc. 11th ECSMFE, Copenhagen, Denmark, Danish Geotechnical Society, Bulletin, 11, vol 2.

König, D. & Jessberger, H.L. (1997). Waste Mechanics. *In: ISSMFE Technical Committee TC5 on Environmental Geotechnics* pg. 35 – 76.

Machado, S. L.; Carvalho, M. F; Vilar, O. M. Constitutive Model for Municipal Solid Waste. *Journal Of Geotechnical And Geoenvironmental Engineering,* ASCE, New York, v. 128, n. 11, p. 940-951, 2002.

Manassero, M.; Van Impe, W.F. & Bouazza, A. (1996). Waste Disposal and Containment. *In: Proc. of the Second International Congress on Environmental Geotechnics*, Preprint of special lectures, Osaka, Japan, A.A.Balkema, v.3, p. 1425-1474.

Sánchez-Alciturri, J. M.; Palma, J.; Sagaseta, C. & Canizal, J. (1993). Mechanical Properties of Wastes in a Sanitary Landfill. *Proccedings of the Symposium Green'93 - Waste Disposal by Landfill.* Sarsby (ed.) 1995 Balkema, Rotterdam, ISBN 90 5410 356 6.

Sharma, H. D.; Dukes, M. T. & Olsen, D. M. (1990). Field Measurements of Dynamic Moduli and Poisson's Ratios of Refuse and Underlying Soils at a Landfill Site. *Geotechnics of Waste Fills - Theory and Practice*, ASTM STP 1070, Arvid Landva and G. David Knowles, Eds., American Society for Testing and Materials, Philadelphia, 1990, p.57-70.

Van Impe, W. F. 1998. Environmental Geotechnics: ITC5 activities – State of the art. *In: Proc. of the Third International Congress on Environmental Geotechnics,* Lisboa, Portugal, Sêco e Pinto (ed.), Balkema, Rotterdam, ISBN 90 5809 006x, p. 1163-1187.

Study of the MSW time differed compressibility in the Metropolitan Center Landfill

S. L. Machado, Federal University of Bahia, Salvador, Brazil,
M. F Carvalho, Catholic University of Salvador, Salvador, Brazil
K. A. Dourado, Federal University of Bahia, Salvador, Brazil
M. B. Rocha, Federal University of Bahia, Salvador, Brazil
J. C. F. Nascimento, Federal University of Bahia, Salvador, Brazil,

Abstract

Are presented results of confined compression and characterization tests performed on samples of Municipal Solid Waste, MSW, coming from Metropolitan Center Sanitary Landfill (Salvador/Brazil), collected in the disposal front (new residue) and in landfill points having 4 years since disposal. It is showed that time since disposal tends to reduct the relative amount of paper/cardboard in the MSW, whereas it tends to increase the MSW paste percentage. On the contrary, the amount of paste organic content represents about 9.9% of total MSW mass after 4 years since disposal, just 60% of the initial value (16.4%). It were obtained values of Cα around 0.0286 and average values of C'α, [C'α = Cα/(1+eo)] of about 0.0114. MSW primary compression index, Cc, showed to be depend of the MSW initial void ratio.

Introduction

Nowadays, the management and disposition of all kind of refuse, including municipal solid waste (MSW), is an important task faced by society. Many community have adopted management policies aiming to minimize waste production, to re-cycle or re-use some of its materials. However, these policies, together with other refuse reduction techniques are unable to avoid some kind of final deposition system such as sanitary landfill. A sanitary landfill is an engineering structure which is designed considering sanitary and environmental principles. In this sense, the mechanical properties of waste are of paramount importance in order to deal with questions such as stability against failure, deformation, stress imposed to foundation soil and the behaviour of accessory elements, such as liner and gas and leachate drains, among others. As could be expected, the mechanical behaviour of MSW is a very complex subject. Different material types such as paper, metal, glass, wood and garden waste, food and plastic, with variable dimensions are usually present. So, besides inert materials and soil of daily covering there are also degradable components, such as organic matter. Each one of these components has particular mechanical behaviour with different strength and compressibility properties that can vary with time. The proportion of each component is highly variable and is related to cultural and economic aspects of the region where the refuse is generated. This probably imparts particular characteristics for MSW generated in different parts of the world.

Regarding the compressibility of MSW, many mechanisms contribute to it. It is known that MSW exhibits a pronounced secondary compression. This is related to material creep and to biological and chemical decomposition which are time-dependent processes that extends for many years, depending on environmental and fill managing factors, such as climate and leachate recirculation. In addition to these important processes, it must be considered the almost instantaneous compression (credited to void ratio reduction and compression and distortion of some components) due to applied loads. Propositions about the mechanisms that command MSW compression and how they are distributed with time can be found for instance in Grisolia and Napoleoni (1996), Manassero et al. (1996), which is schematically shown in Figure 1.

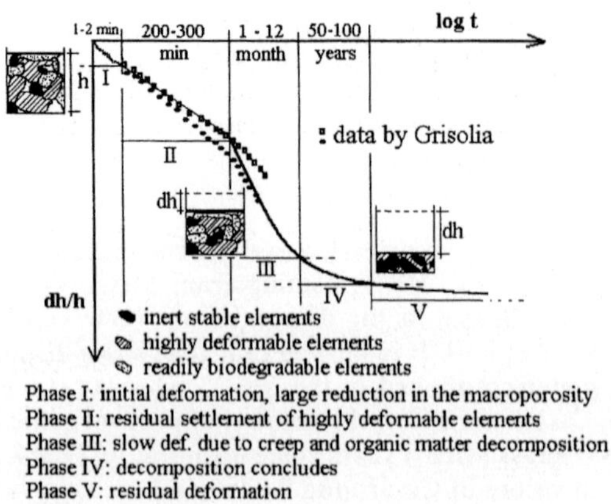

Figure 1. Qualitative model proposed by Grisolia & Napoleoni (1996) to reproduce the time differed compression behaviour of the MSW.

According to these authors, the compression of waste is governed by the following mechanisms: a) physical compression, governed by mechanical distortion, bending, crushing and reorientation of waste components; b) ravelling settlements due to migration of small particles into voids among large particles; c) viscous behaviour and consolidation phenomena involving both solid skeleton and single particles or components; d) decomposition settlement due to the biodegradation of the organic components and e) collapse of components due to physico-chemical changes such as corrosion, oxidation and degradation of inorganic components. The time line in Figure 1 must be seen as highly qualitative, since many of the aforementioned processes depend upon environmental and management characteristics. For instance, if the landfill is located in an arid region the degradation process can take much more time than if it is located in a temperate region and is subjected to an appreciable infiltration rate or if the leachate is recirculated through the landfill.

As far as modelling of MSW mechanical behaviour is concerned, compression is the topic that has deserved more attention from the researchers. For instance, Sowers (1973) extended the classical soil consolidation model, plus a secondary compression component, to study settlement of landfills. To deal with the accentuated secondary compression of waste some authors have used the Gibson and Lo (1961) rheological model or a power creep law (Edil et al. 1990). The biological degradation and its influence on waste settlement have been dealt with by some authors such as Edgers et al. (1992), Sanchez Alciturri et al. (1993) and Park and Lee (1997). Marques et al. (2003) presented a composite model that takes into account the initial mechanical compression in response to applied load and the secondary compression due to mechanical creep and biodegradation. A more comprehensive model that encompasses stress-strain behaviour and time dependent deformation was proposed by Machado et al. (2002).

Materials, methods and results

The waste studied comes from Metropolitan Center Sanitary Landfill, located in city of Salvador – Brazil, which is the greatest area of urban solid waste disposition in the city. Around 2,500 tons of refuse are disposed daily in this sanitary landfill. Waste samples 4 years old, collected in a square cave 1.5m width, and new waste samples, collected in the front of disposition, were studied. MSW samples were subjected to tests in order to measure their physical and chemical properties (natural moisture content, gravimetric composition, STV) and their mechanical properties, such as shear strength, in triaxial compression and compressibility, in confined compression tests. Table 1 shows the physical dry basis composition of the collected waste samples. In this table, paste term means mainly organic matter and all other components non easily separated by hand. It is observed an reduction in paper/cardboard percentage with the increase of the time since disposal, in the same time that paste fraction suffer an increase on its percentage. This occurs probably because a significant amount of the paper/cardboard fraction is incorporated in paste fraction during the degradation process. Besides that, part of sealing soil is incorporated in the paste fraction during the landfill operation, what contributes to increase the relative amount of MSW paste.

However, in despite of the fact that paste fraction percentage increases with time, the amount of remaining degradable organic matter decreases. This is shown in the Table 2. The volatile content was used to measure the organic matter of the MSW paste. The procedure adopted consisted in drying small shredding samples in stove at 70°C and after that calcine then in muffle at 600°C for two hours.

Table 1. Physical composition of waste studied.

Components	Percentage occurrence (%)	
	New waste	Waste 4 years old
Wood	5.8	5.6
Stone/ ceramic	14.5	13.4
Textile	3.5	2.5
Rubber	0.1	0.2
Plastic	19.8	13.8
Glass	4.8	4.1
Metal	5.0	5.0
Paper/ cardboard	15.6	5.2
paste Organic	30.8	50.2

Table 2. Paste volatile content values obtained.

Parameters	New waste	Waste 4 years old
Paste fraction (%)	30.80	50.2
Paste volatile solid content (%)	53.42	19.76
MSW paste organic matter (%) (dry basis)	16.45	9.92

Volatile content was obtained by the sample difference of mass after drying in stove (mineral plus organic material) and calcination in muffle (mineral material). It can be seen that, the average volatile content is about 20% for 4 years old samples and about 53.4% for new waste. Paste organic content represents about 9.9% of total MSW mass in 4 years old samples and about 16.4% in the new ones.

In order to characterize grain size waste materials it was determined a gradation curve, through sieving of the dry waste (dried under 70°C of temperature) using pre-selected sieves with openings of 50.8; 38.1; 31.7; 25.4; 19.1; 12.7; 9.4; 4.8; 2.38; 1.0; 0.50; 0.35; 0.25; 0.18; 0.125; 0.090 and 0.075mm. Components larger than 50.8mm had their dimensions determined directly. Figure 2 presents typical grain size distribution curves of the samples studied. Plastic components were not included. As can be seen, the largest particle diameter was about 400mm. 4 years old samples presented about 50% finer than 10mm, much more than 22% found in the new samples. In general, the grain size distribution showed a reduction of the MSW texture with time. The MSW solid particles unit weight was determined using the prescriptions of applicable standards, substituting the pyncnometer for a 2000 ml glass container calibrated for temperature of 15 – 40°C. The MSW solid particles unit weight was of about 22.2kN/m^3, considering the samples 4 years old.

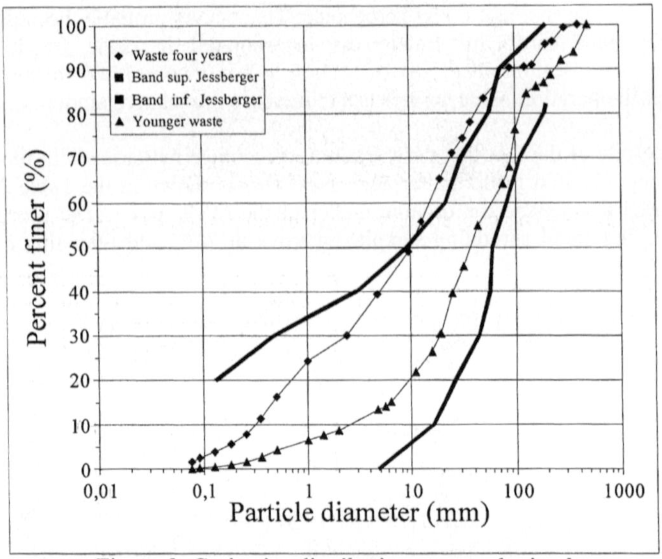

Figure 2. Grain size distribution curves obtained

Confined compression tests were carried out in a large oedometer (548,3 mm of diameter and 496,8 mm of height). Figure 3 shows the oedometer constructed, which has an pressure chamber for application of compressed air and porous stones in the sample top and base, for drainage of the liquid produced during the consolidation process. The waste was placed in the oedometer in layers (4 layers at whole) and was lightly pressed to reach the initial unit weight specified (10 and 12kN/m^3). Loading was applied in stages. At each new stage, the early vertical stress applied was duplicated. Each loading stage lasted about 30 days, although some stages were prolonged for more days in order to define adequately the creep characteristics of the waste. Table 3 shows the characteristics of the waste samples studied before and after of the confined compression tests. At this time, were executed two confined compression tests using samples 4 years old. Confined compression tests using new waste are in progress. Test executed with an initial nominal unit weight of 12kN/m^3 was finished in the stage of vertical stress of 160kPa, due to problems of air leakage in the chamber. The second test, executed with a nominal unit weight of 10kN/m^3, was initially compressed to 80kPa, unloaded to 20kPa and reloaded again up to 640 kPa. Figure 4 shows typical void ratio time differed curves obtained from the confined compression tests performed.

Figure 3- Oedometer used in the confined compression tests performed

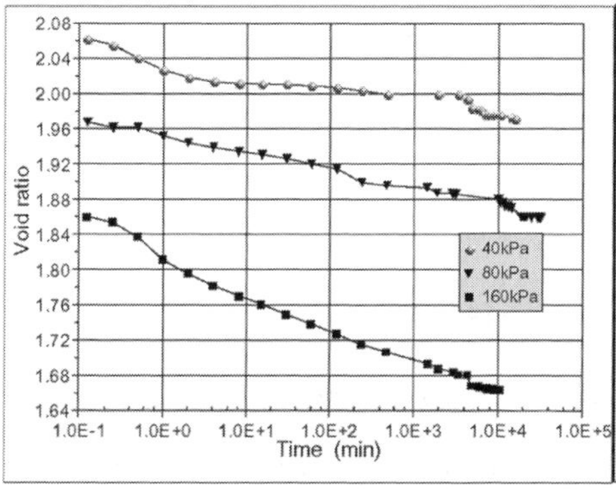

(a)

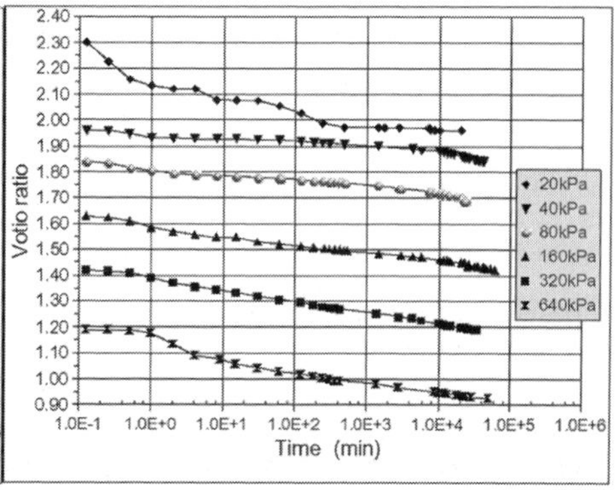

(b)

Figure 4. Typical time differed void ratio curves obtained. a) nominal unit weight of 12kN/m^3. b) nominal unit weight of 10kN/m^3.

Just after load application there was an almost instantaneous compression, followed by gradual time differed compression, characterizing a process of MSW secondary compression or creep. As it can be noted, the time differed void ratio curves presented behaviour approximately linear as a function of the logarithm of time. These data allowed to calculate secondary compression indexes C_α (C_α= -Δe/Δlog(t)) and secondary compression coefficients C'_α ($C'_\alpha = C_\alpha/(1+ e_o)$) for each stage of loading, which are showed in Table 4. Discarding the value related to the first stages of loading, more affected by accommodation of the waste inside the chamber, it can be observed that the C_α values obtained ranged from 0.0141 up to 0.0431, with an average value of 0.0286, and the values of C'_α varied between 0.0047 and 0.0201, with an average value of 0.0114. The values of C_α found are lower than those presented by Sowers (1973) and by Gabr & Valero (1995), but they are of the same order of magnitude of the waste tested by Landva & Clark (1990), and of the waste waste tested by Vilar & Carvalho (2004).

Table 3. Characteristics of the samples submitted to confined compression tests.

Parameters	waste with four years old			
	test 1		test 2	
	Initial	final	Initial	final
w (%)	64.73	-	53.72	23.82
γ (kN/m^3)	11.96	-	10.254	14.52
γ_d (kN/m^3)	7.258	8.340	6.67	11.52
γ_s (kN/m^3)	22.22			
e	2,0616	1.664	2.3312	0.9280
Sr (%)	69.76	-	51.2	57.03

Table 4. C_α and C'_α values obtained for waste studied.

Sample	γ kN/m^3	Average void ratio in stage	σ_v kPa	C_α	C'_α	R^2
Waste 4 years old	12.00		20	-	-	
		2.016	40	0.0141	0.0047	0.915
		1.916	80	0.0202	0.0069	0.988
		1.762	160	0.0386	0.0140	0.989
	10.00	2.146	20	0.0604	0.0192	0.950
		1.901	40	0.0155	0.0053	0.955
		1.764	80	0.0219	0.0079	0.964
		1.527	160	0.0339	0.0134	0.988
		1.306	320	0.0431	0.0187	0.998
		1.059	640	0.0413	0.0201	0.972

In despite of data scattering, it is possible to observe a tendency to increase C_α values with void ratio. In the range between 160 and 640kPa, the influence of overburden stress on C_α and C'_α was less pronounced, although a light increase C'_α was observed. Figure 5 shows compressibility curves for waste studied. When the sample compacted with a nominal unit weight of 12kN/m^3 is considered, it can be seen an pronounced plastic/viscous behaviour during the loading unloading cycle, which is probability related with the MSW secondary compression. Primary compression index, C_c (C_c= -Δe/Δlog(σ_v), varied between 0.51 and 0.65, as show in Table 5. Data from table 5 indicate that C_c is dependent on initial void ratio, since C_c decrease with then. If C_c is normalized through the specif volume (1+e_o), it can be seen that the compression coefficients C'_c [$C'_c=C_c/(1+e_o)$], obtained are very close, averaging 0.18. C_c values reported are among the lowest values reported by Sowers (1973) and are similar to data from Gabr and valer (1995) and Vilar & Carvalho (2004). Otherwise, C'_c was near to the lowest value reported by Landva & Clark (1990), when testing MSW from Canada in a large compression cell.

Table 5. Cc and C'c average values of MSW

Sample	g kN/m^3	Cc	C'c	R^2
Waste with	12.00	0.510	0.167	0.975
four years old	10.00	0.651	0.195	0.976

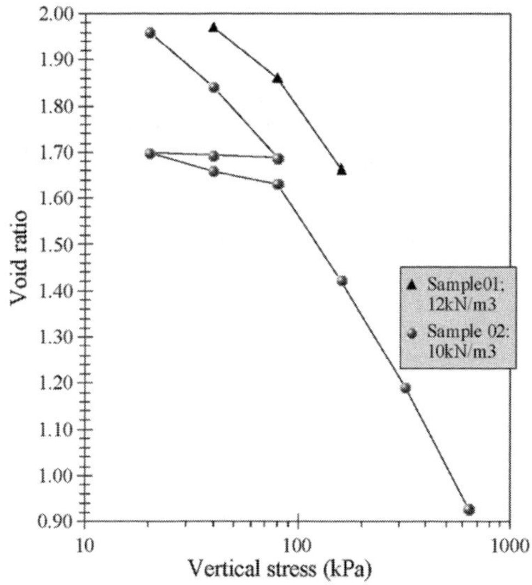

Figure 5 - Void ratio - vertical stress curve obtained from waste studied.

Final remarks

In this paper were presented some results of characterization tests performed on waste samples collected in the Metropolitan Center Sanitary Landfill- Brazil. It is observed a reduction in paper/cardboard percentage with time since disposal, in the same time that paste fraction suffered an increase on its percentage, due probably the migration of small particles into the the paste during decomposition process. Besides that, part of sealing soil is incorporated in the paste fraction during the landfill operation, what contributes to increase the relative amount of MSW paste mass. On the contrary, the amount of degradable organic matter remaining in the paste decreases with time. Paste total volatile solids after 4 years since disposal is about 20% (dry mass). This value is of about 53.4% in the new samples. Paste organic content represents about 9.9% of total MSW mass after 4 years since disposal, just 60% of the initial value (16.4%).

Confined compression tests performed showed the high compressibility of the waste, reflected in the high values of primary and secondary compression indexes. It was obtained values of C_α around 0.0286 and average values of C'_α, [$C'_\alpha = C_\alpha/(1+e_o)$] of about 0.0114. MSW primary compression index, C_c, showed to be depend of the MSW initial void ratio, since the lower is the void ratio, the lower are the obtained values of C_c. When this index is normalized by the MSW specif volume, producing the primary compression parameter C'_c [$C'_c = C_c/(1+e_o)$], more stable values are obtained.

Acknowledgements

The authors are grateful to VEGA Bahia tratamento de Resíduos S.A. and FAPESB (Bahia state Research Funding Agency) for financing part of this research.

References

Edgers, L.; Noble, J. J.; Williams, E. (1992). A Biologic Model for Long Term Settlement in Landfills. *In: Proc. of the Mediterranean Conference on Environmental Geotechnology,* Usmen & Acar (eds), Balkema, p.177 – 184.

Edil, T. B.; Ranguette, V. J. e Wuellner, W. W. (1990). Settlement of Municipal Refuse. *Geotechnics of Waste Fills-Theory and Practive, ASTM STP 1070*, Arvid Landva and G. David Knowles, Eds., American Society for Testing and Materials, Philadelphia, 1990, p.225-239.

Gabr, M. A. & Valero, S. N. (1995). Geotechnical Properties of Municipal Solid Waste. *Geotechnical Testing Journal*, GTJODJ, v.18, nº2, June 1995, p.241-251.

Gibson, R.E. & Lo, K.Y. (1961) A Theory of Soils Exhibiting Secondary Compression. Acta Polytechnica Scandinavica, C10, 296, p. 1-15 (apud Edil et al, 1990

Grisolia, M. e Napoleoni. (1996). Geotechnical Characterization of Municipal Solid Waste: Choice of Design Parameters. *Proc. Of The Second International Congress on Environmental Geotechnics,* Osaka, Japan, A.A.Balkema, v.2, p. 641-646.

Landva, A. O. & Clark, J. I. (1990). Geotechnics of Waste Fill. *Geotechnics of Waste Fills - Theory and Practice*, ASTM STP 1070, Arvid Landva, G. David Knowles, editors, American Society for Testing and Materials, Philadelphia, 1990, p.86-103.

Machado, S. L.; Carvalho, M. F; Vilar, O. M. Constitutive Model for Municipal Solid Waste. *Journal Of Geotechnical And Geoenvironmental Engineering,* ASCE, New York, v. 128, n. 11, p. 940-951, 2002.

Manassero, M.; Van Impe, W.F. & Bouazza, A. (1996). Waste Disposal and Containment. *In: Proc. of the Second International Congress on Environmental Geotechnics*, Preprint of special lectures, Osaka, Japan, A.A.Balkema, v.3, p. 1425-1474.

Marques, A. C. M. ; FILZ, G. M. ; VILAR, O. M. . Composite Compressibility Model for Municipal Solid Waste. Journal Of Geotechnical And Geoenvironmental Engineering, New York, v. 129, n. 4, 2003.

Park, H. L.; Lee, S. R. long Term settlement behaviour of landfills with refuse decomposition. Journal of solid waste technical management, vol 24, no. 4, 1997, pp 159-165.

Sánchez-Alciturri, J. M.; Palma, J.; Sagaseta, C. & Canizal, J. (1993). Mechanical Properties of Wastes in a Sanitary Landfill. *Proccedings of the Symposium Green'93 - Waste Disposal by Landfill.* Sarsby (ed.) 1995 Balkema, Rotterdam, ISBN 90 5410 356 6.

Sowers, G. F. (1973). Settlement of Waste Disposal Fills. *In: Eighth International Conf. on Soil Mechanics and Foundation Engineerings*, v.2, part 2, Moscow-URSS, 1973; p.207-210.

Vilar, O. M. ; Carvalho, M. F. Mechanical Properties of Municipal Solid Waste. *Geotechnical Testing Journal,* ASTM, v. 32, n. 6, 2004.

Sabkha soils mixed with oil lake residues – Reuse of a waste material.

Mahdi T.A., Al-Otaibi F. and Thomas H.R.
Geoenvironmental Research Centre,
Cardiff School of Engineering, Cardiff University, Cardiff, UK

Abstract

The work presented in this paper forms part of an experimental project to assess the viability of stabilising Sabkha soils with oil lake residues. The Sabkha soils investigated were from south Kuwait. Specimens were prepared by mixing the dry Sabkha soil samples with different percentages of oil residues by weight. The results presented include basic physical properties such as grain size distribution, consistency limits, compaction and permeability. Results indicated that the addition of oil residues improves the performance of the Sabkha soils. Maximum dry density increases with the addition of up to 8% oil residues and reduces with more addition. Soil gradations become coarser as the percentage of oil lake residues increases. Permeability and plasticity decreases and plasticity index was slightly affected.

Introduction

Sabkha is an Arabic term used to describe a large, flat, salt-encrusted and evaporative terrain. Sabkha is composed of clays, silts, sands and salty mixtures generally associated with high moisture content and a high groundwater table (1). Sabkha flats predominate in large areas along the coast of the Arabian Gulf. Sabkhas vary in both horizontal and vertical directions in textural characteristics, chemical and mineralogical compositions and the chemistry of the interstitial brine (2). Horizontal variations are expected to be related to the proximity to the shoreline, while vertical variations are related to the repeated drying of inundation waters rich in dissolved salts (3). The horizontal and vertical variations, in addition to the presence of highly concentrated brines, are the distinguishing features that can lead to highly variable and problematic geotechnical properties, as well as excessive settlement (3).

Sabkha soils are known for their low bearing strength at their natural conditions and therefore improvement of these unfavourable properties is deemed desirable (1). Due to the high cost and limited sources of soil construction materials, upgrading locally available Sabkha soils is of interest.

On the other hand, large amounts of oil residues accumulated in hundreds of oil lakes of different shapes and sizes exist in Kuwait. These oil lakes are the result of the destruction of more than 800 oil wells during the Gulf War (4). More oil lake residues are expected from accidental oil leakage. These oil lakes are major physical and environmental hazards. Oil lake residues could possibly be used, if by mixing with Sabkha soils, the soil's unfavourable properties could be stabilised and improved.

The main objective of this paper is to investigate the possibility that the physical properties of representative Sabkha soil samples collected from different locations in south Kuwait can be improved when mixed with different percentages of oil lake residues.

Materials and methods

Natural Sabkha soil

The experimental work was performed on Sabkha soil samples obtained from an area located approximately 65 km south of Kuwait city. Ten soil samples of about 50 kg each were collected from several excavated pits at depths ranging from 50 to 100 cm below ground level, covering an area of about 2.5 km by 12 km. The selection of the sampling area and the sampling locations was based on engineering judgment and previous knowledge of the studied area.

Soil samples were subjected to characterisation tests according to BS1377 to determine their engineering properties including grain size distribution, consistency limits, compaction, and specific gravity. Permeability was measured using the column leaching test described by Yong et al. (5).

Figure 1 shows the grain size distributions of different Sabkha soil samples. The figure shows a similarity in the distribution curves for the collected soil samples. It can be seen that sand, with its coarse, medium and fine size fractions is the most common size and comprises about 83% of the material. The Sabkha soil samples are classified as poorly graded sand (SPu) to poorly grade silty sand (SM) according to the BSI and USC systems. The tested soil samples have an average specific gravity of 2.7.

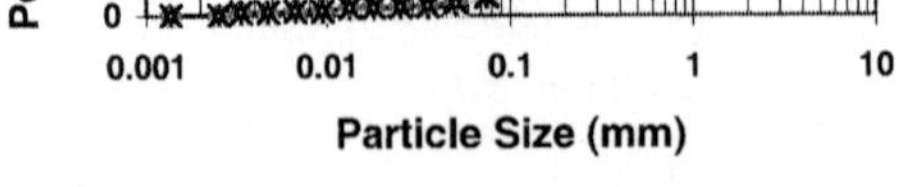

Figure 1 Grain size distributions of the different Sabkha soils

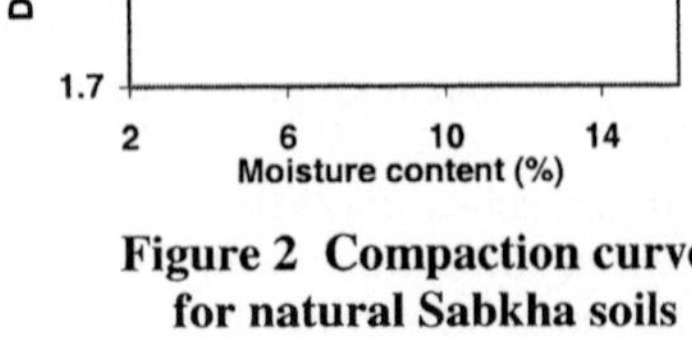

Figure 2 Compaction curves for natural Sabkha soils

Results of the modified proctor compaction tests for some of the soil samples are shown in figure 2. The maximum dry densities for the collected samples range from 1.895 g/cm^3 to 1.943 g/cm^3 and the optimum moisture contents range from 9.3% to 11.5%. Narrow ranges of variability of test results reflect the similarity of the collected soil samples.

Chemical analysis indicates that silica content is 54%, which is considered the main component. The carbonates content is 27.8%, mostly in the form of calcite. The sulphate and the chloride content were 11.2% and 1.0% respectively. Iron oxide and aluminium oxides were 1.3% and 4.1% respectively. The total organic content was very low. XRD results were in good agreement with the chemical analysis. The tested soil sample was composed of different constituents, identified as quartz, gypsum, calcite and halite.

Oil lakes residues

The oil residues used in this research was Atmospheric Bottom provided by the Kuwait Institute for Scientific Research - Petroleum Research, which is similar in properties to that found in the oil lakes as suggested by Khan et al (6).

The oil residue used is characterised by a boiling point of greater than 350 °C, high viscosity, a density of 0.997 g/ml and very low solubility of less than 1 mg/l. Since the density of the oil residue was lower than that of water and its solubility is very low, it can be considered as a light non-aqueous phase liquid (LNAPL).

Sample preparation

Representative soil samples from the selected location were mixed with 0 %, 2%, 3%, 4%, 5%, 6%, 7%, 8% and 10% of oil residue (by dry soil weight). The method of mixing followed the approach described by Basma et al. (7).

Results

Particle Size Distribution

Particle size distribution curves for Sabkha soils mixed with different percentages of oil residues are shown in figure 3. The figure shows that the percentages of particles passing different sieve sizes are reduced with oil addition. Moreover the grain size distribution of each tested soil sample is affected by oil residue addition at different rates. This is due to particle aggregation resulting from the addition of viscous oil residues. A close examination of figure 3 shows that fines (<63μ) reduces to less than 50% of its content in the natural Sabkha at an oil residues content of 4%, and almost disappears with further oil addition. Similar findings have been reported by Meegoda and Ratnaweera (8), in the context of clays and sands.

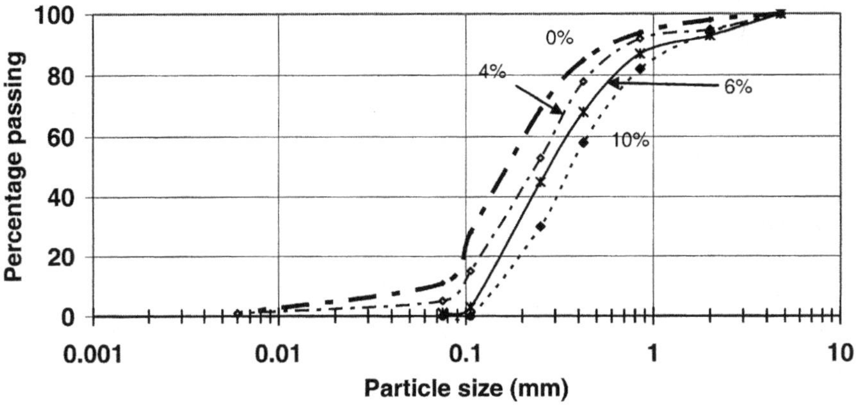

Figure 3 Grain size distributions of Sabkha soil mixed with different percentage of oil residues.

Visual inspection of Sabkha soils mixed with oil residues revealed that they were aggregated and completely covered with oil. The oil residues are expected to cover the surface of the aggregates and fill the accessible interaggregate spaces. The viscous non-aqueous phase liquid (NAPL) material acts as a gluing agent to hold the fine aggregate particles together (9).

Consistency Limits and Classification:

The variation of the liquid limit (LL), plastic limit (PL) and plasticity index (PI) of natural Sabkha soil samples and those mixed with oil residues are presented in Figure 4. Liquid limit shows an increasing trend with increase of oil content, since their values increased from 22 to 24.5 as oil content increased from 0% to 5%. Plastic limit shows a decreasing trend with increase in oil content.

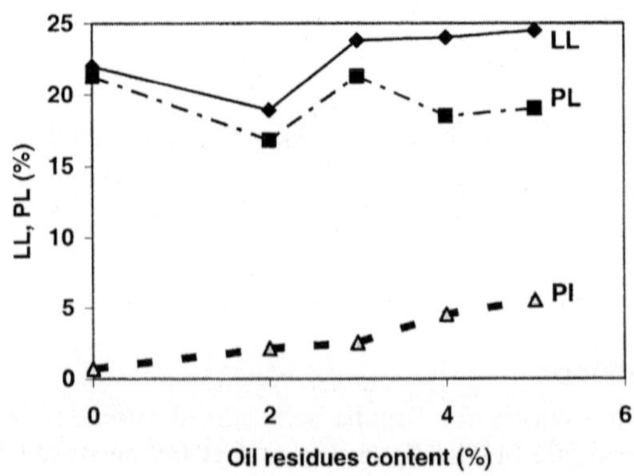

Figure 4 Variation of consistency limits with oil content

Plasticity index increases as oil residue content increases. Generally, natural Sabkha and those mixed with oil fall in the low plasticity chart according to BS5930: 1981, which suggests that the addition of oil residues has a marginal effect on the plasticity of the Sabkha soil. The reduction in the plasticity of the Sabkha soil may be attributed to the aggregation of fine grains due to oil residue. Similar findings were also reported by Meegoda and Ratnaweera (8), in the context of clays and silty sands.

Since the soil gradation and the consistency limits were affected by the oil residues addition, soil classification is expected to vary from that of the natural soil. Sabkha soils mixed with oil residues are classified as poorly graded sand (SP) according to the British Soil Classification System and the Unified Soil Classification System.

Specific gravity

There were some difficulties in carrying out the specific gravity experiment on the stabilised Sabkha soils. Due to the entrapment of air, particles of Sabkha soils mixed with oil residues were floating at the top of the pycnometer with air bubbles attached to the surfaces, which affect the test results.

Results show that specific gravity decreased from 2.68 for natural Sabkha to 2.58 for Sabkha soil mixed with 10% oil residues. The reduction in specific gravity can be attributed mainly to the low specific gravity of the oil residues, and to some extent, to the possible entrapment of air bubbles in the interior aggregate pores by oil residues. A similar trend was observed by

Srivastava and Pandey (10) during their investigation of oil contaminated alluvial soils and sands from India.

Compaction Test

Results of the modified proctor tests are shown in figure 5. The figure shows that the addition of oil at different percentages resulted in different compaction curves, each with a different optimum moisture content (OMC) and maximum dry density (MDD).

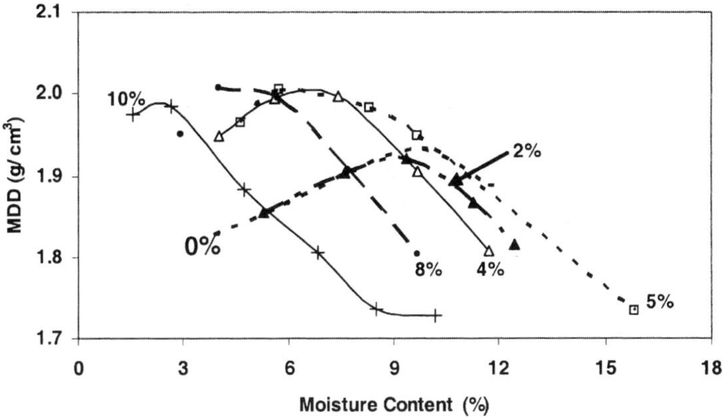

Figure 5 Compaction curves for Sabkha soils mixed with different oil contents

The variation in the maximum dry density with the oil contents is shown in figure 6. The figure shows a general trend of increasing maximum dry density with the increase in the amount of oil residues up to 8%. With oil addition beyond 8%, the maximum dry density starts to decrease. The reduction in the maximum dry density at higher oil content (>8%) could be attributed to the reduction in the specific gravity and the replacement of some soil particles with oil residues. Difficulty of compaction at high oil residue percentages (8% and 10%) was also observed, which may indicate that field compaction is expected to be extremely difficult at high oil residues content. The high percentage of oil residue will create a thick layer around the soil particles, which will prevent contact between the particles. Al-Sanad et al. (12) reported the same effects, in their study of oil contaminated sands.

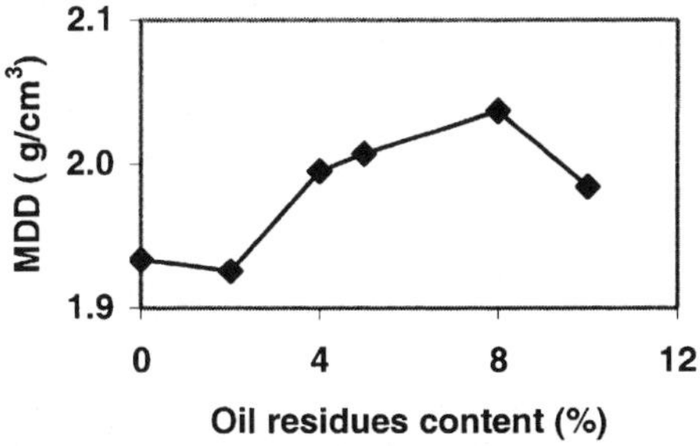

Figure 6 Maximum dry density variations with oil content

Variations in the optimum moisture and fluid contents (FC) of Sabkha soils mixed with different percentages of oil residues are presented in figure 7. The figure shows a general trend of reduction in optimum moisture content with the increase in the percentage of oil addition. However the rate of reduction increases with the increase in oil residues addition. The increase in oil residue reduces the amount of water required to achieve good compaction. The reduction in OMC for Sabkha soil mixed with oil residues is an advantage in areas of very arid condition due to the shortage of water (13).

Although the OMC decreases with oil residue addition, the FC, which is the sum of optimum moisture and oil residues contents, increases with the oil residues content.

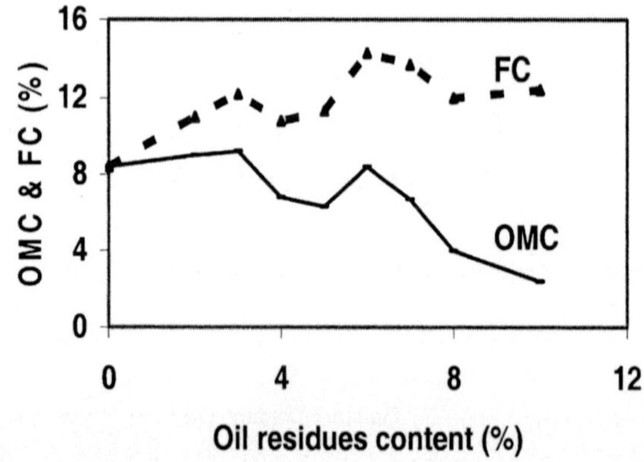

Figure 7 Variations of optimum moisture content and fluid content with oil content

The increase in maximum dry density and the reduction in the moisture content with the addition of oil residues can be attributed to the adsorbed oil residue on soil particles, which act as lubricants yielding a closer rearrangement of solid particles in the vicinity of maximum dry density and reduce the need for water.

Coefficient of Permeability

The coefficient of permeability (K) results for natural and oil mixed Sabkha soils are shown in figure 8. Results are for samples compacted to their maximum dry density at their optimum moisture content. The coefficient of permeability of the natural Sabkha soil is 7.26×10^{-8} m/sec indicating low to very low soil permeability. The measured permeability values fell within the range of values reported by different studies investigating oil contaminated sand in Kuwait (12).

The results indicate that the coefficient of permeability for the stabilised soils decreased with oil addition. The reduction percentages were 17%, 82%, 85% for the 2%, 5%, and 10% oil residue respectively. The percentage reduction was particularly noticeable for 5% oil residue,

The variation of void ratio with oil residue addition is plotted on the same figure. The figure indicates that hydraulic conductivity for the stabilised Sabkha soil samples follows a similar pattern to that of void ratio. The high reduction in the permeability with oil residue addition of 2% to 5% can be attributed to the higher reduction in the void ratio in this range. The appreciation higher oil residue percentage means more voids are occupied with oil residues and hence lower permeability (14). Similar findings were reported by Al-Sanad et al. (12), during their investigation of contaminated sand in Kuwait.

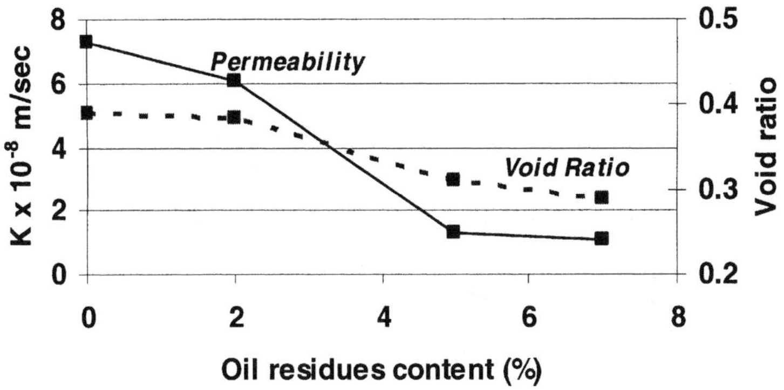

Figure 8 Variation of permeability and void ratio with the oil residue content

Conclusions

A laboratory testing programme, carried out to investigate the effect of oil residue addition on the physical properties of Sabkha soils, has been described. The investigated properties were grain size distribution, consistency limits, specific gravity, compaction parameters and permeability. The amount of oil residues used varied from 2% to 10%. The following conclusions can be drawn.

With oil addition, Sabkha soils show significant particle aggregation, altering the particle size distribution curve and the classification of the soil. The fines content is halved at 5% oil addition and almost disappears at higher percentages. Soil classification is changed from poorly graded sand to silty sand (SPu – SPM) for the natural Sabkha to poorly graded sand (SPu) for the Sabkha soil mixed with oil residues. Consistency limits are slightly affected by oil residues addition, while the specific gravity is reduced.

Compaction characteristics are modified with oil residue addition. The maximum dry density value increases and the optimum moisture content value decreases. Beyond 8% oil residue addition, the maximum dry density value decreases.

The coefficient of permeability of Sabkha soil mixed with oil residues decreases with the addition of oil. The reduction is very noticeable at 5% oil residue addition.

The results reported are obviously essentially preliminary characterisation tests. The outcome is sufficiently encouraging to merit further development of the experimental programme. Further investigations are ongoing. These include strength tests; i.e. unconfined compressive strength, direct shear and California Bearing Ratio tests; consolidation tests; i.e. compressibility, collapse potential and swelling tests; and leaching characteristics. Field studies are also ongoing. If proven viable this approach will reduce the environmental hazards of the oil lake residues, utilise a waste material and reduce the amount of material that needs to be replaced when constructing on Sabkha soil flats.

Acknowledgments

The authors wish to express their sincere gratitude to the Ministry of Defence in Kuwait for their support of the second author's Ph.D studies.

References

1. Al-Shamrani, M. A. and Dhowian, A. W., 1997, "Preloading for reduction compressibility characteristics of Sabkha profiles", Engineering Geology (48), pp.19-41.
2. Butler, G.P., 1969,"Modern evaporate deposition and geochemistry of coexisting brines, the Sabkha. trucial coast, Arabian Gulf.", Journal of Sedimentary Petrology, 39 (1), 70-89.
3. Akili, W. and Torrance, J.E., 1981, "The development and geotechnical problems of Sabkha with preliminary experiments on the static penetration resistance of cement sands", Quarterly Journal of Engineering Geology,(14), 59-73.
4. Al-Besharah, J., 1992, "The Kuwait oil fires and oil lakes facts and numbers", The environmental and health impact of Kuwait Oil Fires, Proceedings of an International Symposium, The University of Birmingham, pp.12-15.
5. Yong,R.N., Mohamed,A.M.O., and Warkentin,B.P.,1996, "Principles of contaminant transport in soils", Elsevier Science , Amsterdam,327 p.
6. Khan, Z.H., AbuSaeedo, F., Al-Besharah, J., and Salman, M., 1995, "Improvement of the quality of heavily weathered crude oils", Fuel, (74), pp.1375-1381.
7. Basma, A. A., Al-Homod , A.S., Khedaywi, T. S., and Al-Ajlouni , 1994, "Effect of bitumen content and curing condition on strength characteristics of asphalt stabilised soils", Geot. Eng. (25), pp.93-104.
8. Meegoda, N.J., and Ratnaweera, P., 1995, "Treatment of oil-contaminated soils for identification and classification", Geotechnical Testing J, 18(1),pp 41-49.
9. Lotfabad, S.K., and Gary M.R., 2000, "Characterisation of contaminated soil using Confocal Laser Scanning Microscopy and Cryogenic-Scanning Electron Microscopy", Environ. Sci. Tech., (34), pp.3408-3414.
10. Srivastava, R.K., & Pandey, V.D., 1998, "Geotechnical evaluation of oil contaminated soil", Green 2, Contaminated and derelict land. Thomas Telford, London.
11. Tarefder R.A., Ruckgaber, M.E., Zaman, M., and Patton, D., 2003, "Formulation of mix design for asphalt incorporation of hydrocarbon contaminated soil", J. Materials in Civil Eng. ASCE, pp 166-173.
12. Al-Sanad, H.A., Eid, W.K. and Ismael, N.F., 1995, "Geotechnical properties of oil-contaminated Kuwait sand", Journal of Geotechnical Engineering, 121(5), May 1995 ASCE, pp. 407-412.
13. Ingles, O.G. & Metcalf, J.B., 1972, "Soil stabilisation principles and practice butter worth's", Sydney, N.S.W.374 PP
14. Slattery, J.C.,1990, "Interfacial transport phenomena", Springer-Verlag , New York, 1159pp.

Fibre reinforced waste based artificial soil for grass covered areas subjected to occasional vehicular traffic

Mahdi T.A., Francis R.W., Thomas H.R. and Poon R.
Geoenvironmental Research Centre,
Cardiff School of Engineering, Cardiff University, Cardiff, UK

Abstract

This research is aimed at investigating the improvement in load bearing characteristics of a waste based artificial soil reinforced with polypropylene fibre. The soil, manufactured by mixing Sandstone Quarry Sand (SQS) and Green Waste Compost (GWC) at a ratio of 2:1 by volume, is to be used as a top layer in the construction of grassed covered areas subjected to vehicular traffic. A series of laboratory tests, including California Bearing Ratio (CBR), model footing test, and accelerated pavement testing were undertaken on reinforced and unreinforced artificial soil. Results suggest that the load bearing characteristics of the artificial soil can be greatly improved by reinforcing the soil with polypropylene fibre, under all studied conditions. Results of the model footing tests and accelerated pavement testing suggest that the polypropylene reinforced waste based artificial soil can be used as a top layer in grass covered areas subjected to occasional vehicular traffic.

Introduction

The problem of rising cost and decreasing availability of land in the UK has forced the use of large grass covered areas as access roads, overflow or occasional car parks, emergency access roads and heavy pedestrian usage areas. However, these grass areas often possess weak underlying soil and might be damaged by unexpected vehicular traffic.

One of the possible ways of improving the ability of grass covered areas to sustain occasional vehicular traffic while maintaining 100% grass cover is by fibre reinforcing the soil. Fibre soil reinforcement has been known for centuries. Different investigators have used different materials for soil reinforcement including the use of natural and synthetic fibre. A recent study (Jones 2003) highlighted the advantages of using polypropylene crimped fibre in the reinforcement of manufactured soil made from mixing a blend of subsoil with sand. The study concluded that the addition of fibre at 0.3% and 0.6% (by dry weight of soil) improved the load bearing characteristic of the soil making it fit to be used as a top layer for emergency exit vehicles.

The objective of this study is to investigate the effect of crimped polypropylene fibre reinforcement on the load bearing characteristics of waste based manufactured soil made from mixing sandstone quarry sand, a waste by product produced in huge quantities in South Wales from crushing the sandstones to produce aggregates for different applications (Arup 2004), and green waste compost at a ratio of 2:1 by volume.

Materials

Sandstone Quarry Sand (SQS): The SQS was supplied by a South Wales sandstone quarry. The chemical analysis provided by the quarry revealed that the SQS comprised of 78-80% Silica, 10% Alumina, 4% Ferric Oxide, 2.5% Potash, 1% Magnesia, 2-3% (Loss Of Ignition) and other traces. Results of particle size distribution undertaken by Transport Research Laboratory (TRL) Ltd (Lamb 2005) indicated that the SQS is well-graded silty sand with 23% fine gravel, 62.5% sand size particles and 14.5% fines (passing 0.063mm), of which 20% are clay and 80% are silt. Other properties including particle density, absorption, atterberg limit, compaction and CBR at different moisture content can be found in the TRL report (Lamb 2005).

Green Waste Compost (GWC): The GWC was obtained from a composting facility in Swansea. Tests on the compost revealed that the GWC complied with the requirement of BSI PAS 100 in terms of available nutrients, loss of ignition (41%), pH (7.4), potentially toxic elements and physical contaminants. The bulk density of the GWC is 0.53 Mg/m^3.

Fibre: fibres used in this research to reinforce the manufactured soil are named LokSand crimped mono-filament polypropylene fibres, which were developed and produced by Drake Extrusion Ltd, Leeds, UK. The fibres are 0.09mm diameter, crimped every 10mm and cut into a length of 35mm.

Manufactured soil: The manufactured soil was made from mixing SQS and GWC at 2:1 ratio by volume using a concrete drum mixer. The selection of the mix proportion followed previous work undertaken by Harper Adam college(Keeling and Marchant 2002) and TRL Ltd (Lamb 2005) to satisfy the nutrition, textural, pH, stone content and other requirements of BS 3882: 1994 (Specification for top soil). However the produced soil can satisfy all the requirement of the General purpose grade top soil of BS3882 apart from the Nitrogen content of 0.16% which was less than the specified amount of 0.2% minimum. Grass growing field trials using compacted 2:1 (SQS:GWC) mixes undertaken by TRL (Lamb 2005) concluded that a significant grass cover can be established on this soil. Bulk density and the specific gravity of the manufactured soil were 1.078 Mg/m^3 and 2.24 respectively. Manufactured soil particle size distribution obtained using the wet sieving method described in BS1377: Part 2:1990 showed that the manufactured soil consists of 27% gravel, 59% sand size particles and 14% Fines (Fines <0.63). The soil can be classified as well-graded silty/clayey SAND (SWM/SWC) according to the British soil classification system detailed in BS5930:1981. Different mixes are prepared by adding 0%, 0.3% and 0.6% of fibres (by weight of dry soil). The fibres are added to the manufactured soil during mixing in the concrete mixer and left to be homogenised for 15 minutes. Water was added as necessary during this process.

Sub-base: Sub-base is required to provide support to the top layer of grass covered soils. The selected sub base is a grade 1 slate sub-base produced by crushing slate waste. The slate sub-base used was supplied by Alfred McAlpine from their quarry in Bethedsa, North Wales.

Compaction Test

Compaction tests were undertaken to assess the effect of fibre addition on the moisture content dry density relationship and to obtain the value of the optimum moisture content to be used in the wheel loading tests and model footing test. The compaction test was carried out in accordance with BS1377:1900: Part 3.

The results for unreinforced soil samples and those reinforced with 0.3% and 0.6% fibres are presented in Table 1. The results show a clear trend of reduced maximum dry density with the amount of fibre addition. However the optimum moisture content has no trend and fall into a very narrow range of 14.01 – 14.05%. It was decided on the basis of these experiments to use the 14% moisture content as the base for the wheel loading test and the model footing test.

Table 1. Dry density and optimum moisture content for the tested materials

Soil	**Maximum Dry Density (Mg/m^3)**	**OMC (%)**
Unreinforced Manufactured Soil	1.76	14.05
Manufactured Soil with 0.3% Fibres	1.71	14.01
Manufactured Soil with 0.6% Fibres	1.66	14.04

CBR

The CBR tests were undertaken on remoulded samples. The design of the experiment was to assess the effect of fibre reinforcement of artificial soil at different moisture content. Results of the CBR tests presented in figure 1 show that the addition of the polypropylene fibre increases the CBR values of the unreinforced artificial soil. The amount of increase in CBR is proportional to the percentage of fibre addition, at 15 % moisture content CBR has almost doubled by the addition of 0.3% fibre and become 5 times the unreinforced value when 0.6% fibre is added to the soil. The figure also shows that CBR values are very sensitive to changes in moisture content. The CBR of the unreinforced and reinforced soil decreases sharply as the moisture content increases.

Model footing test

A model footing test was carried out to determine the ultimate bearing capacity of a compacted layered samples consisting of a subgrade layer (150mm) of certain CBR, in this case 5%, a sub-base layer (150mm) and a top layer (150mm) of manufactured soil. 3 samples, one unreinforced and two fibre reinforced manufactured soils were tested at optimum moisture content. Samples were compacted to the same compaction effort as that of the CBR test.

The model footing test uses a small circular footing (100mm diameter) placed on the surface of a soil sample that is loaded at constant penetration of 1mm/minute until the failure point is reached. The soil sample is placed in a cylindrical container of 400mm diameter and 500mm height. Full description of the test is given by Poon (Poon 2004). Typical results of the test are displayed as a stress-penetration diagram, the peak of which represents the ultimate bearing capacity.

According to the building regulations (DETR, 1990) the ultimate bearing capacity of layered design pavement should be greater than 815 kN/m^2 in order to meet the strength requirement for emergency access routes. The values of the ultimate bearing capacity for the three tests together with the corresponding penetration are presented in Table 2. The Table also includes the penetration at 815 kN/m^2. Results for the ultimate bearing capacity are presented graphically in figure 2.

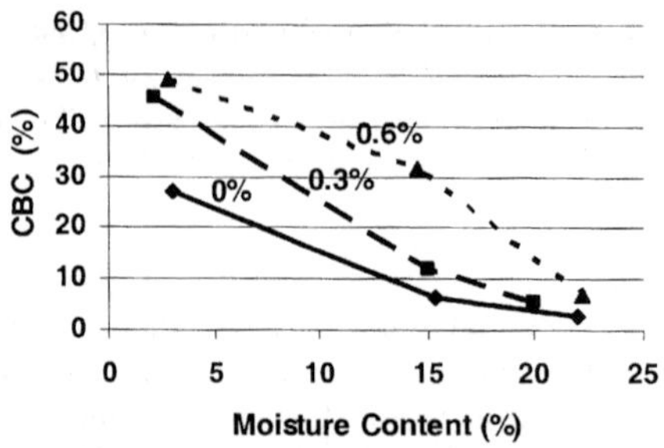

Figure 1. Variations of CBR with moisture content for different fibre content

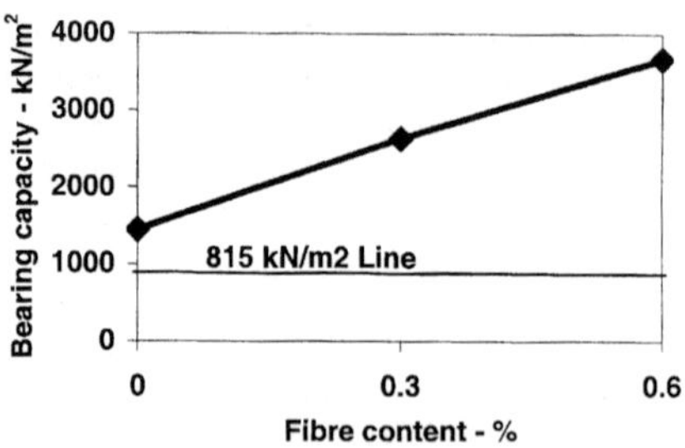

Figure 2. Ultimate bearing capacity for different fibre content

Table 2 and figure 2 show an increase in the ultimate bearing capacity with the fibre content. Figure 2 also shows that the layered-approach ultimate bearing capacity of the reinforced and unreinforced waste based artificial soils is much higher than the 815 kN/m2 required to satisfy the strength requirement of the building regulation. From this, it can be concluded that the unreinforced and fibre reinforced waste based artificial soil satisfy the strength requirement of the building regulation for emergency access routes and can be used as a top layer for these routes.

Table 2. Model footing test results

Fibre Content (%)	Ultimate Bearing Capacity (kN/m^2)	Penetration at Ultimate Bearing Capacity (mm)	Penetration at $815kN/m^2$ (mm)
0	1442	28.4	10.85
0.3	2633	37.5	15.85
0.6	3672	44.1	13.85

The results of the waste based artificial soil are much higher than those obtained by Jones (Jones 2003) for a blended recycled soil made from blending 30% of green waste soil with 70% sand). This could be attributed to the differences in the composition and moisture content between the two tests. It is worth mentioning that the unreinforced soil used in the previous study did not satisfy the strength requirement of the building regulation.

Results for the penetration at ultimate bearing capacity presented in Table 2 showed a similar trend of increase with the fibre content as that of the bearing capacity. The higher values of penetration at ultimate bearing capacity of the waste based soil can be considered as another advantage of this soil. The soil is able to sustain higher deformation before reaching the failure point. Values of the penetration at ultimate bearing capacity of this study are again much higher than those obtained by Jones (Jones 2003) for a blend of recycled and sand soil.

Wheel Loading Test

To ensure that the fibre-reinforced soils are suitable to be used as a top layer for grass covered areas subjected to vehicular traffic, an accelerated wheel loading pavement tests called the Wheel Loading Test (WLT) was undertaken in the laboratory. Full description of the WLT is detailed by Jones (Jones 2003). The test is to simulate the dynamic loading of the soil by passing traffic and was undertaken for three loading conditions, namely; cars (240 kPa), LGV (345 kPa), and HGV (620 kPa). The wheel loading machine works automatically and is connected to a computerised data logger, which records the rut depth of the sample at the mid point of each pass.

Samples are prepared by compacting a bottom layer of 150mm slate sub-base and a top layer of 150mm manufactured soil in a steel container 1400mm square by 300mm deep, using 8 passes of a 3kW, 60 kg Wacker plate (Model VP1B5A) for each layer. The thickness of the layers is similar to those used by Jones (Jones 2003) and considered to be a good representation of an actual layered design. Tests were undertaken on un-grassed samples due to time limitation.

In total 15 different tests where undertaken on soil samples of different conditions, namely; fibre content (0, 0.3%, 0.6%), moisture content (OMC, wet soil) and wheel loading (240kPa, 345kPa, 620kPa). To simulate the wet condition, the samples were prepared and compacted at optimum moisture content, and then watered with the equivalent of 25mm of water. Before starting the test, the sample was allowed to stand for 45 minutes to enable the water to penetrate to the bottom of the soil layer. This was intended to simulate a heavy rain condition, and to determine the effect of a high moisture content on the performance of the waste based manufactured soil.

Results

Typical test results, as obtained from the computerised data logger system, are presented in figure 3. The figure shows a plot of rut depth versus number of passes of the wheel.

Table 3 summarises the results of the tests undertaken on soil under different conditions of loading, moisture content and fibre content.

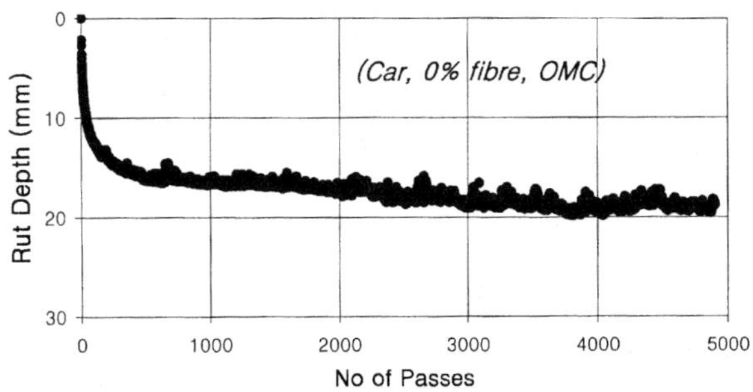

Figure 3. Typical test results of the simulated wheel loading test system.

Table 3. Wheel loading test results on soils at optimum moisture content and under wet condition.

Fibre Content (%)	Rut Depth of (mm)										
	Optimum Moisture Content					*Wet Soil Condition*					
	50 passes	100 passes	500 passes	1000 passes	4000 passes	5 passes	25 passes	50 passes	100 passes	150 passes	250 passes
	Car wheel Loading (240kPa)										
0	10.7	12.5	16.2	16.4	19.1	19.8	35	47.5	47.8	/	/
0.3	5.8	7.2	9.1	10.0	11.1	10.2	30.4	40	43.1	/	/
0.6	6.5	7.9	10.9	12.5	16.0	9.5	26.1	32.9	39.7	42	45
	LGV wheel loading (345kPa)										
0	13.5	15.4	21.7	23.3	26.2	13.5	15.4	21.7	13.5	15.4	21.7
0.3	12.5	14.0	18.8	21.2	25.1	12.5	14.0	18.8	12.5	14.0	18.8
0.6	9.0	10.5	16.4	18.6	24.7	9.0	10.5	16.4	9.0	10.5	16.4
	HGV wheel loading (620kPa)										
0	22.9	27.8	34.2	38.6	/	22.9	27.8	34.2	22.9	27.8	34.2
0.3	22.0	24.7	24.9	29.9	38.0	22.0	24.7	24.9	22.0	24.7	24.9
0.6	18.4	21.2	29.6	30.5	39.4	18.4	21.2	29.6	18.4	21.2	29.6

The results indicate a significant reduction in rut depth with the addition of fibres for all loading conditions. However, the amount and rate of reduction varies with the loading, number of passes, and moisture content of the soil.

OMC Results

Figure 4 presents the results of tests undertaken on reinforced and fibre reinforced soils at OMC. At 100 passes, car wheel loading, the measured rut depth was 12.5mm for unreinforced soil, whereas the reinforced soil with 0.3% of fibre needed to take up to 5000 passes to achieve this rut depth. The benefit from the addition of fibre is clear in the case of the 0.3% fibre content. However, the addition of extra fibre reinforcement (0.6%) failed to achieve the expected improvement compared to the 0.3% fibre addition, for the case of car loading and HGV loading at high number of passes. The reason for this is not clear as all other strength test results show a clear trend of improvement with the addition of fibre. Further investigation is required to check this point and to make sure that this behaviour is not related to experimental error.

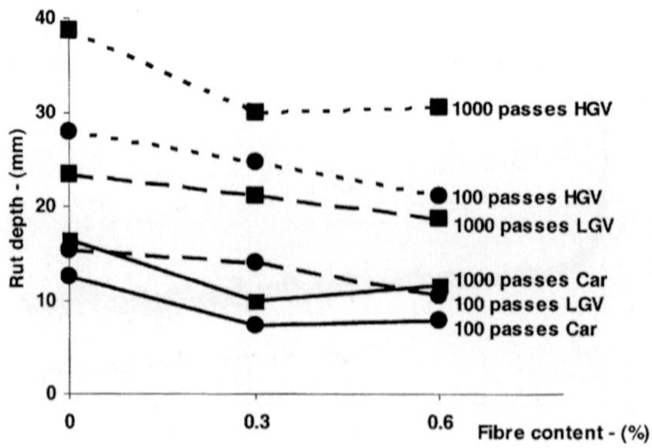

Figure 4. Variation of rut depth with fibre content for soils with OMC

The results for tests undertaken using the 345kPa (LGV) wheel indicate that the deformation of the soil is greater than that of 240kPa (Car) wheel for the same number of passes. At 50 passes the rut depth is 13.5 mm for the unreinforced soil sample, and it tended to stay almost constant after 2000 passes. The addition of 0.3% fibre to the soil samples shows a significant reduction in the rut depth of the soil. At 5000 passes, the rut depth was measured to be 24.7 mm given an improvement of 2.1 mm in rut depth. Figure 4 shows that the addition of 0.6% fibre to the manufactured soil has significantly reduced the rut depth compared to the unreinforced and reinforced with 0.3% of fibre soils. The improvement is more pronounce at lower traffic volume than higher traffic volume (above 4000 passes).

Test results undertaken using the 620kPa (HGV) wheel show that high rut formation occur much quicker than tests when cars and LGV are used. This is expected due to the large increase in the amount of load. For unreinforced soil sample, at 1000 passes rut depth is 38.6 mm. The same rut depth is achieved at 4000 passes for the 0.3% and 3800 passes for the 0.6% Fibre addition. This indicates a significant improvement of the load carrying capability of the reinforced soil compared to the unreinforced soil.

The overall conclusion of this section is that there is a clear trend of improvement with the addition of fibre in the rut depth for all types of loadings (cars, LGV, HGV) and under all traffic volumes. The relation between the fibre content and rut depth depends on the type of loading and traffic volume.

Wet Soil results

In a country like the UK, with high rainfall, the effect of wet conditions combined with traffic loading should be examined. Table 3 summarises the results of the tests undertaken on wet soils under different conditions of loading and fibre content. Some of the results are presented graphically in figure 5. Results show a large increase in the initial rut formation for all combination of loading and fibre content compare to samples with OMC.

For car loading, the rut depth for unreinforced soil is 35mm at 25 passes and almost levelled at 47.5mm after 50 passes. The test is stopped at 140 passes after reaching more than 50mm rut depth. When the soil is reinforced with 0.3% crimped polypropylene fibre, significantly less rut depth is achieved for the same number of passes. The rut depth is 30.4 mm at 25 passes. The test is stopped at 140 passes again, but the rut depth is 43.1 mm at 100 passes. This indicates a reduction of 4.7 mm in rut depth compared to the unreinforced sample. Soil with high fibre content of 0.6% performed better than those of 0.3%. The test is stopped at 250 passes, 110 passes more than the samples of unreinforced soil and soil reinforced with 0.3% fibre. Figure 5 indicates that the improvement in rut depth with the addition of fibre is almost linear for samples loaded with car loading (240kPa) at higher number of passes.

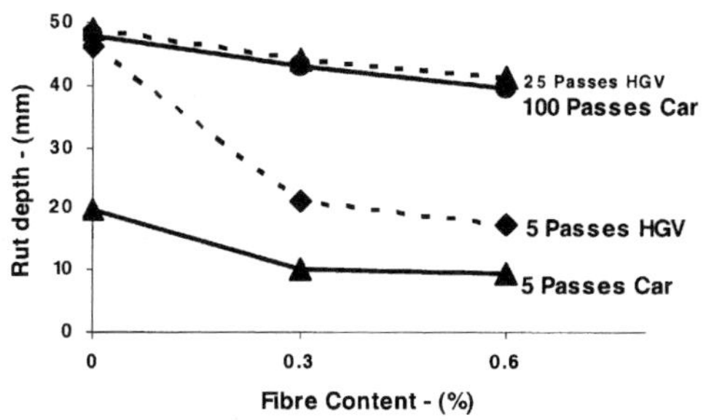

Figure 5. Variation of rut depth with fibre content for wet soils

Tests undertaken using the 620kPa (HGV) wheel on samples with 0% and 0.3% fibre were stopped at around 40 passes, while the test on the 0.6% fibre was stopped at 92 passes. A significant rut formation of 46.1mm was observed on the unreinforced sample at 5 passes, then it became almost stay constant on about 49 mm. With addition of 0.3% and 0.6% fibre, rut depth at 5 passes was reduced by 24.8mm and 28.9mm respectively compared to the unreinforced soil.

Figure 5 and Table 3 show that the addition of fibre achieves a good reduction of the rut depth when the soil is wet. However the amount of reduction in rut depth with fibre content is not linear and decreases as the amount of fibre increases and as the number of passes increases. The addition of 0.6% fibre has not shown a big improvement from that of the 0.3% fibre addition.

From the above discussion, it can be suggested that the increase of moisture content of the waste based soil has a significant effect on its ability to sustain vehicular traffic. However, the addition of fibre improved the soil considerably in terms of reduced rut depth and increased number of passes before failure.

Conclusions

The following conclusions can be drawn from this research;

1. Waste based manufactured soil used in this project is a mixture of 2:1 by volume of Sandstone Quarry Sand (SQS) and Green Waste Compost (GWC) and can be classified according to the British Soil Classification System as well-graded silty/clayey SAND.

2. The OMC and the maximum dry density for the waste based manufactured soil are 14% and 1.76 Mg/m^3 respectively. The additional of 0.3% and 0.6% of crimped polypropylene fibres caused a slight decrease in its maximum dry density.

3. CBR test showed a clear improvement in load bearing characteristics of artificial soil with the addition of fibre reinforcement. CBR of unreinforced and reinforced soil decreases sharply as the moisture content increases.

4. There is a clear trend of improvement in the rut depth with the addition of fibre to the soil, for all types of loadings (cars, LGV, HGV) and under all traffic conditions. The relation between the fibre content and rut depth depends on the type of loading (Cars, LGV, HGV) and the traffic volume.

5. Increase in moisture content of the waste based soil has a significant effect on its ability to sustain vehicular traffic. However, the addition of fibre improved the soil considerably in terms of reduced rut depth and increased number of passes before failure.

6. Waste based artificial soil, unreinforced or reinforced with fibre can satisfy the strength requirement of the building regulation regarding the use of the soil as a top layer for emergency access routes. The other advantage of the waste based artificial soil is its ability to sustain higher deformation before failure. The addition of fibre to the artificial soils has improved its CBR, resistance to rutting under different traffic loading and moisture content, ultimate bearing capacity and penetration at the ultimate bearing capacity.

References

Arup (2004), Improving the Information Base on Secondary Minerals/C&D Waste for Use as Aggregates, Arup, January 2004

BSI. (1994), BS3882:1994, Specification for topsoil. BSI.

BSI. (2002), BSI PAS100:2002, Specification for composted material. BSI

DETR (1991). The Building Regulations, Approved Document B: Fire Safety Department for the Environmental, Transport and Regions. London

Jones M (2003), The Properties of Crimped Fibre Reinforced soils and Their Use in the Provision of Grass Surfaced Roads. PhD thesis, Cardiff University.

Keeling, A.A. and Marchant, L. (2002), Producing synthetic soils from quarry wastes and green wastes. In: Proceedings of the 1st International Conference Addressing Government Sustainability Targets for Construction and Related Wastes, 104-109, Ed. H Al Nageim, Liverpool John Moores University.

Lamb M (2005), Application for sandstone quarry fines in South Wales. TRL - Viridis Report VR8.

Poon R (2004), Soil Reinforcement of Grass Covered Area Subjected to Occasional Vehicular Traffic, MSc dissertation, Cardiff University.

BSI (1990). BS 1377:1990: Methods of Test for Soils for Civil Engineering Purposes. BSI, London.

Surface Disposal of Paste Tailings

Vincent Martin, Ecole Polytechnique, Montreal, Canada
Michel Aubertin, Ecole Polytechnique, Montreal, Canada
Jacques McMullen, Barrick Gold Corporation, Toronto, Canada

Abstract

Underground disposal of tailings as a cemented paste backfill is becoming the state-of-the-art in many mining operations. Paste tailings also constitute a promising solution for surface storage. Due to the higher density and strength of paste tailings, their use reduces some of the risks linked with surface disposal of tailings. The densified state also increases the amount of water being re-circulated at the process plant. Moreover, because of its propensity to retain more water under unsaturated conditions, paste tailings containing sulphidic minerals are less likely to be exposed to an oxidizing environment, which reduces the potential for generation of acid mine drainage (AMD). Presently, the storage of paste tailings in surface impoundments is limited to a very small number of mining operations around the world. Barrick's Bulyanhulu Mine in Tanzania is considered a pioneer of this technology. This paper presents some of the developments of this technology, and discusses the advantages and limitations of surface storage of paste tailings using the Bulyanhulu Mine experience as a generic case.

Introduction

Mine tailings are typically produced as a slurry at the mill with a pulp density P varying between 30% - 45% w/w solids (e.g.Vick 1990, Aubertin *et al.* 2002). The slurry is usually deposited in a surface impoundment serving as a tailings storage facility (TSF). In these facilities, the tailings are generally deposited from discharge points located on the crest of a dyke. Following end pipe deposition, the tailings slurry flows toward a low point in the basin, with coarser particles settling closer to the discharge point forming a beach. This causes a segregation of tailings in the TSF, which affects their hydraulic and mechanical properties.

This classical deposition technique may cause some geotechnical and environmental problems. For example, as the coarser tailings become unsaturated along the beach, they may generate acid mine drainage (AMD) if sulphidic minerals are present. This issue, which affects many mining operations around the world, occurs when mine tailings containing reactive minerals, such as pyrite and pyrrhotite are oxidised when they come into contact with water and oxygen (e.g. Ritcey 1989; Aubertin *et al.* 2002). In order to better control this potential problem, the mining industry has been re-evaluating disposal methods. This reflection takes into account factors such as:

- The decrease in the cutoff grade of ore bodies, which increases the amount of tailings currently being produced by modern mines;
- The increased legal responsibility related to the possible failure of tailings dams and dykes, and the subsequent contamination of the surrounding environment;
- The significant increase in the construction and maintenance costs of TSFs;
- An increased environmental awareness of the industry;
- The significant costs of process water management;
- The difficulties brought by the rehabilitation of TSFs, due in part to the slow consolidation of the tailings (which are initially saturated and loose);

- More stringent environmental laws and regulations.

Professionals from the mining industry around the world have been studying various methods in order to engineer the properties of mine tailings to optimise disposal methods (e.g. Robinsky 1999; Aubertin *et al.* 2002; Bois *et al.* 2005) This paper presents some features of a novel deposition technique that has been inspired by the work of Robinsky (1999) on surface disposal of densified tailings and by the underground disposal of cemented paste backfill (e.g. Landriault *et al.* 2005). The core principle is to increase the slurry density before disposal in the TSF, so that the tailings can be less susceptible to geotechnical instabilities and differential settlements.

Densified Tailings

Various terms are used in the industry to identify densified tailings, based on the thickening technique and nature of the end product. A convenient parameter which distinguishes between various types of tailings is the pulp density P ($P(\%) = (M_S/M_T)100$). The following definitions, adapted from Jewell *et al.* (2002) and Crowder (2004), are adopted for this project:

i. *Slurried tailings*: The pulp of such commonly produced tailings has an elevated water content which translates into a critical flow velocity below which the solids contained in the slurry settle in a segregated fashion (with the coarser particles deposited first). P is usually between 30% and 45%, and the displacement measured during a slump test is at its highest. The water content ($w(\%) = (M_w/M_s)100 = (100 - P)/P$) is between 120% and 230%.

ii. *Thickened tailings*: These tailings have a relatively low water content due to the densification process at the mill, but they nevertheless show a critical flow velocity and do not segregate during sedimentation. Bleed water can be observed from thickened tailings following a slump test or after disposal in a TSF. In this case, P is usually between 45% and 70% (i.e. w between 40% and 120%). The slump can only be measured for tailings approaching a slurry density of 70%. Thickened tailings can be deposited from a central discharge point to form a conical stack (e.g. Kidd Creek - Robinsky 1999) or in layers in a dry stacking technique (Halle *et al.* 2000, Sofra and Boger 2002). Once the selected layer thickness is attained, the discharge point is moved to another location around the facility, while the fresh tailings are allowed to drain, settle, and dry. Typically, the whole cycle begins again once the whole facility has been covered with a fresh layer. This technique allows for synchronisation and control between the duration of deposition and the thickness of the deposited layers.

iii. *Paste tailings*: In a paste, the water content is decreased even further. Paste tailings do not have a critical flow velocity. In a pipeline, the tailings move according to a plug flow mechanism (e.g. Li *et al.* 2002). No bleed water is observed from paste tailings following deposition or after a slump test. This material can be deposited in TSF in a similar fashion as for thickened tailings, or in underground excavations (with a binder added). P is typically between 70% and 85% (w is roughly between 20% and 40%), depending on the properties of the solids (e.g. Benzaazoua *et al.* 2004; Potvin *et al.* 2005).

iv. *Filtered tailings*: When the water content is reduced even further ($P > 85\%$, $w < 20\%$), the material forms a filter cake. At this point, it is not possible to pump the tailings and it must therefore be transported by trucks or conveyors. The material produces a negligible displacement following a slump test.

Paste Tailings Application

Of the approaches presented above, paste tailings have interesting properties, as shown by the experience gained at Barrick Gold Corporation's Bulyanhulu Mine (Tanzania), where this technique has been used since 2001 (Grabinsky *et al.* 2002; Crowder 2004; Theron *et al.* 2005).

Paste tailings are usually deposited at the end of a pipe, forming a cone with a gentle outward slope. Once in place, the slurry can then flow freely along the interface to create a lobe with a texture approaching that of concrete cement (or toothpaste). The thickness of the tailings layer varies according to the deposition sequence defined by the operator, and depends on the rheological properties of the material. The first objective is to allow the freshly deposited tailings to consolidate and stabilise. However, the operator should ensure that the tailings will not desaturate too much if they are potentially acid generating. Figure 1 shows a section of a storage facility where potentially acid generating tailings have been stored, with various factors affecting the properties of the materials under potentially acid forming conditions.

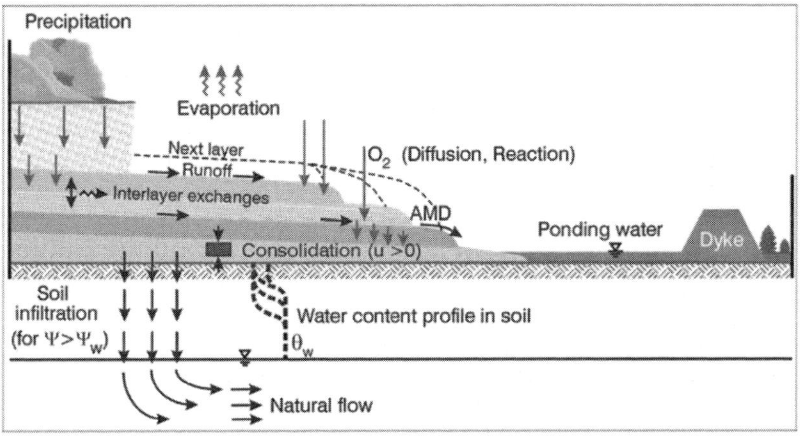

Figure 1. Schematic representation of reactive paste tailings deposited in successive layers (adapted from Aubertin 2004).

The schematic view shown in Figure 1 is based loosely on the actual practice at the Bulyanhulu mine paste tailings facility. In Figure 1, the deposition point would be located to the left of the figure, which represents a section of a deposition cone. The layers are usually thicker near the cone's central axis. Similarly, the slope angle becomes smaller near the outer limits of the facility. This is due to the induced shear stresses during free flow on the surface, which causes the viscosity of the pulp to decrease. This phenomenon is sometimes called shear thinning. The slope formed by the tailings cone favours flow of runoff water towards the outside of the facility. Accumulated water can then be captured and recirculated to the mill or pumped into a polishing pond.

Ideally, the tailings should have time to stabilise and consolidate before the next layer is added. However, when the tailings are potentially acid generating, it is preferable to keep the layers fairly saturated, to minimise the development of oxidation reactions. It is therefore necessary to adapt the deposition cycle in such a way that the tailings are covered with a fresh layer of paste before significant desaturation occurs. The length of this cycle depends on many factors that can influence infiltration and flow of water in the paste tailings including (Figure 1):

- Intensity, frequency and duration of precipitation events;
- Evaporation at the tailings surface;
- Settlement, consolidation, and hydraulic conductivity of the tailings;
- Unsaturated properties of the tailings and evolution of the degree of saturation, and of the pore pressure (positive or negative) profile in the various layers;
- Presence, size, and density of the desiccation cracks;
- Depth to the water table and hydrogeotechnical properties of the underlying soils;
- Tailings production rate, size of the tailings storage facility and time of exposure to meteorological elements;
- Reactivity of the tailings, which control the rate at which oxygen may be consumed and produce acid; this aspect is related directly to the mineralogical composition of the tailings (and their acid generating potential).

All of these factors make the analysis and design fairly complex. It is thus necessary to develop appropriate tools to manage reactive paste tailings deposited in TSF.

Particular Characteristics of Paste Tailings

As mentioned above, there are some potential advantages in using paste tailings for surface disposal. For instance, the lower water content means that more process water can be recirculated at the mill, and this can potentially reduce the treatment costs of the water from the tailings. Also, the storage volume may be somewhat decreased, because of the material's higher initial density. This can be illustrated with the mass-volume relationship used to estimate the storage volume for a pulp; it can be written as (e.g. Halle *et al.* 2000):

$$V_{st} = M_s \frac{[P(\rho_w - G_s) + G_s]}{PG_s\rho_w} \qquad [3]$$

where V_{st} = Storage volume (m^3).
M_s = Mass of solid particles (taken here as 1 tonne)
P = Pulp density (% solids w/w).
ρ_w = Density of water (1 tonne/m^3).
G_s = Grain density (2.93 for Bulyanhulu mine tailings collected in October 2004).

Using equation 3 for the Bulyanhulu case ($P \approx 75\%$) shows that the required storage volume is between 2.2 and 3.7 times lower than for typical slurried tailings (P between 30% and 45%). This reduction is based on the total volume of the tailings, and does not take into account consolidation and drainage that occurs after deposition. The actual storage volume is influenced by the final water content after deposition and upon closure. Because tailings consolidate fairly slowly (e.g. Vick 1990), it can take quite some time for slurried tailings to attain the same volume (density) as paste tailings. Also, the excess water generated by slurried tailings at the storage facility has to be managed. This may require the construction of larger collecting and polishing ponds. On the other hand, the large retention dykes typically constructed for slurried tailings, which create a bathtub shaped TSF, may allow for a greater volume of tailings to be stored for the same surface area (Blight 2003).

The increased density of paste tailings, compared to slurried tailings, influences the mechanical and hydraulic properties. For instance, in the course of this ongoing program, the saturated hydraulic conductivity k_{sat} and water retention curve (WRC) of the Bulyanhulu tailings have been measured in the lab, on reconstituted samples. Figure 2 shows the k_{sat} values obtained here, and those measured by Crowder (2004) in a complementary study being conducted at the University of Toronto (U of T). The data shown in Figure 2 are compared to k_{sat} values predicted by the Modified Kozeny-Carman (MKC) model developed at the École

Polytechnique, Montreal (EPM) (Mbonimpa *et al.* 2002); this model uses basic geotechnical properties (i.e. grain size and void ratio) to predict the saturated hydraulic conductivity of soils, tailings and other particulate media.

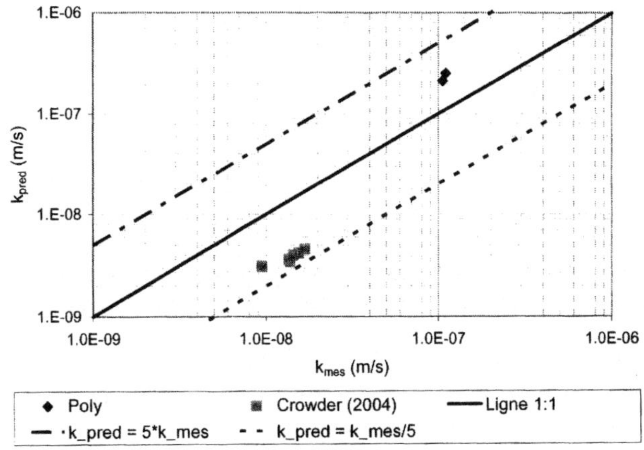

Figure 2. Hydraulic conductivity predicted and measured in the laboratory on reconstituted samples from the Bulyanhulu site.

The k_{sat} values measured at the École Polytechnique are higher than those obtained at the University of Toronto, in part because the grain size of the tailings was finer for the samples tested at U of T. Also, the latter were obtained from oedometer (consolidation) tests, which are known to give somewhat lower k_{sat} values. Nonetheless, in both cases, the measured values are in line with those measured on other similar types of tailings (e.g. Vick 1990).

Water retention curves were also measured on Bulyanhulu tailings samples. Representative results are shown in Figure 3; again, results obtained at EPM and at U of T are shown. Tempe cells were used for the tests performed by the Authors while a pressure plate apparatus was used at U of T (results presented by Simms and Grabinsky 2004). The curves shown in Figure 3 have been fitted to the test results, using the well known van Genuchten (1980) equation.

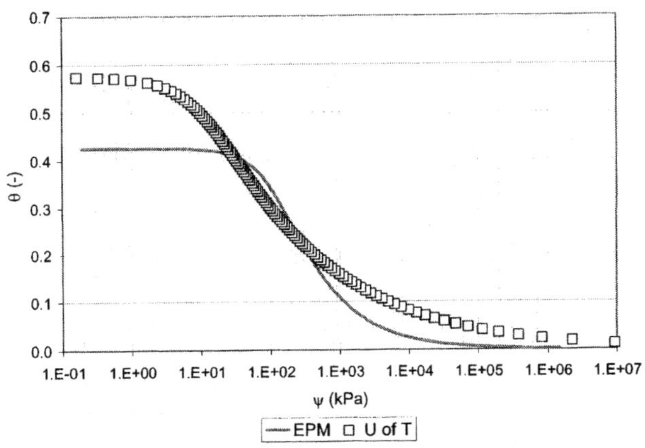

Figure 3. Water retention curves of the Bulyanhulu tailings at two initial void ratios.

The difference in grain size and porosity of the samples explains some of the differences observed with the two curves shown in Figure 3. The results indicate that a denser material induces a larger air entry value (AEV, or ψ_a), which is in agreement with other observations and with theoretical analyses of the water retention curve (e.g. Aubertin *et al.* 2003).

These and other unique properties of paste tailings can thus provide other advantages, which depending on the site, may include the following (e.g. Jewell *et al.* 2002):

- An increased strength upon deposition;
- A decreased demand for borrow materials for the construction of impoundment structures;
- A significant reduction (or elimination) of ponding;
- A more rapid creation of a workable surface, therefore decreasing the time at which reclamation of the facility can begin.

There are nonetheless some uncertainties and possible disadvantages linked with the use of paste tailings for surface disposal. For instance, because of their higher viscosity, paste tailings are not as easy to transport to the TSF as conventional slurries. Centrifugal pumps cannot be used; positive displacement pumps must be available for paste transportation in pipelines. Moreover, the equipment (thickeners and/or disc filters) used to decrease the water content can be costly to own and operate. These may significantly increase the operating costs of the facility. The planning of the tailings management system should involve a comparison between the higher operating costs of a paste tailings facility and the higher capital costs (linked to the construction of large dykes) of a traditional slurried tailings facility.

Other factors should also be considered when choosing a deposition method. For example, the lower water content of paste tailings can increase the potential for wind erosion, so special means must be adopted to minimise this problem. Additionally, the use of paste tailings for surface disposal is a fairly recent technology and it still requires an optimisation of the wetting/drying cycles in order to minimise the desaturation of reactive tailings, while maximising consolidation. When the tailings are prone to produce AMD, it is necessary to insure that they do not desaturate to a point where the oxidation reactions become important.

Final Remarks

As environmental rules and regulations become more stringent, the use of paste tailings technology for surface management is becoming a solution worth studying in greater detail. The environmental and economical advantages could counterbalance the operational and technical challenges that operators have to face today.

The experience acquired at the Bulyanhulu mine site shows that this disposal approach creates a stable deposit. However, greater characterisation efforts are necessary to better understand the various influence factors such as the soil-atmosphere exchanges and the hydrogeology of paste tailings. With the possible development of other paste tailings surface storage facilities (e.g. Verburg *et al.* 2003; Johnson *et al.* 2005), the mining industry is progressively acquiring a better understanding of this technology, and therefore gets access to another tailings disposal alternative. This study is ongoing, and will help define the key properties used to maximize the benefit of this technology.

Acknowledgements

The authors would like to thank Julie Shuttleworth, Josephine Sum, Andrew Cranley and Brett Thomson from the Bulyanhulu Mine for the information given on surface paste tailings. We would also like to extend our gratitude to Murray Grabinsky, Paul Simms and Jason J. Crowder at the University of Toronto. This work was completed as part of the program of the NSERC Polytechnique-UQAT Research Chair in Environment and Mine Wastes Management (http://www.polymtl.ca/enviro-geremi/).

References

AUBERTIN, M. (2004). "Paste tailings disposal and management at the Bulyanhulu mine : Proposed investigation". Document prepared for Barrick (unpublished), May 2004.

AUBERTIN, M., MBONIMPA, M., BUSSIÈRE, B., CHAPUIS, R.P. (2003). "A model to predict the water retention curve from basic geotechnical properties". Can. Geo. J. 40. 1104-1122.

AUBERTIN, M., BUSSIÈRE, B., BERNIER, L. (2002). "Environnement et gestion des rejets miniers". [CD-ROM] Montreal, Quebec : Les Presses Internationales de Polytechnique.

BENZAAZOUA, M., PEREZ, P., BELEM, T., FALL, M. (2004). "A laboratory study of the behaviour of surface paste disposal". In Proceedings Minefill 2004, Beijing, 1 : 180-192.

BLIGHT, G.E. "Quantified comparisons of disposal of thickened and unthickened tailings". In Tailings and Mine Waste '03. 63-71.

BOIS, D., BENZAAZOUA, M., BUSSIÈRE, B., KONGOLO, M., POIRIER, P. (2005). "A feasibility study on the use of desulphurized tailings to control acid mine drainage". CIM Bull. 98:1087. 74.

CROWDER, J.J. (2004). "Deposition, consolidation, and strength of a non-plastic tailings paste for surface disposal". 162p. Ph.D. Thesis in civil engineering, University of Toronto.

GRABINSKY, M.W., THERIAULT, J., WELCH, D. (2002). "An overview of paste and thickened tailings disposal on surface". In Symposium 2002 on environment and mines. November 3-5, 2002, Rouyn-Noranda, Canada. [CD-ROM].

HALLE, G., O'CALLAGHAN, W.B., HARTNEY, T.F. (2000). "Experience with thickened red mud tailings (paste) stacking in tropical and temperate climates and applicability of the technology to other tailings". In 53rd Can. Geo. Conf., Montreal, 1 : 683-690.

JEWELL, R.J., FOURIE, A.B., LORD, E.R. (2002). "Paste and thickened tailings – A guide". Nedlands, Australie : The Australian Center for Geomecanics. 173p.

JOHNSON, J.M., VIALPANDO, J., LEE, C. (2005). "Paste tailings management alternative – study results for Molycorp's lanthanide group operations in Mountain Pass, California". Mining Engineering. 57:2. 50-56.

LANDRIAULT, D., JOHNSON, J.M., PALKOVITS, F. (2005). "Thickened tailings and paste technology : The future of industrial waste disposal". SME Annual Meeting, 05-111.

LI, M., BERNIER, L., BOUCHER, J.-F. (2002). "Rheology of mineral pastes and its implications on underground pipeline delivery". In Symposium 2002 on environment and mines. November 3-5, 2002, Rouyn-Noranda, Canada. [CD-ROM].

MBONIMPA, M., AUBERTIN, M., CHAPUIS, R.P., BUSSIÈRE, B. (2002). "Practical pedotransfer functions for estimating the saturated hydraulic conductivity". Geotech. and Geol. Eng. 20. 235–259.

POTVIN, Y., THOMAS, E., FOURIE, A. (2005). "Handbook on mine fill". Nedlands, Western Australia : Australian Centre for Geomechanics. 179p.

RITCEY, G.M. (1989). "Tailings Management, Problems and Solutions in the Mining Industry". Amsterdam : Elsevier.

ROBINSKY, E.I. (1999). "Thickened tailings disposal in the mining industry". Toronto, Ontario : E.I. Robinsky Associates Ltd. 210p.

SIMMS, P., GRABINSKY, M.W. (2004). "A simple method for estimating rates of drying and desaturatoin of paste tailings during surface deposition". In Tails and Mine Waste '04, Fort Collins, Colorado.

SOFRA, F., BOGER, D.V. (2002). "Environmental rheology for waste minimisation in the minerals industry". Chem. Eng. J. 86. 319-330.

THERON, M., ADDIS, P.C., WATES, J.A., MARTIN, V. (2005). "Bulyanhulu mine (Tanzania) paste tailings facility : Relating the unsaturated properties of gold tailings to rate of rise". In Paste 2005, Santiago, Chile, 1 : 219-229.

VERBURG, R., JOHNSON, B., FORDHAM, M., LOGSDON, M. (2003). "A rapid and cost-effective method for bench screening of geochemical performance and disposal options for high-sulfide tailings". In 6th ICARD, Cairns, 1 : 739-749. [CD-ROM].

VICK, S.G. (1990). "Planning, Design, and Analysis of Tailings Dams". Vancouver, British-Columbia : Bitech. 369p.

Sulfate induced heaving in chromite ore processing residue

Deok Hyun Moon, Dimitris Dermatas, Maria Chrysochoou, Adriana M. Sanchez, Mahmoud Wazne, W. M. Keck Geoenvironmental Laboratory Center for Environmental Systems, Stevens Institute of Technology, Castle Point on Hudson, Hoboken, NJ 07030, USA
Dennis G. Grubb, Schnabel Engineering North, LLC, 510 East Gay Street, West Chester, PA 19380, USA

Abstract
Extensive heaving has been observed at sites in New Jersey and Maryland, USA, where chromite ore processing residue (COPR) was deposited. The heaving mechanisms were investigated using confined swell tests on COPR samples collected from deposition sites. The swell tests were conducted using a one-dimensional confined swell test apparatus employing wet and dry cycles with the introduction of different permeants such as tap water, sulfate, chromate and carbonate solutions. These swell tests were completed to evaluate the heaving related to hydrogarnets, ettringite, calcium aluminum chromium oxide hydrate (CAC) and calcite formation. The swell test results showed that significant swell (3%) occurred only upon sulfate addition at 2 wt%. The XRPD analyses of the swelled sample indicated a significant increase in ettringite content which demonstrated that the swell development was strongly associated with the increased ettringite content.

Introduction
Chromite ore processing residue (COPR) is a solid waste which is produced during the extraction process of chromium (Cr) from its ore (Hiller et al., 2003). In the process the chromite ore was mixed with soda ash and lime, and heated in a rotary kiln at about 1,200°C to produce sodium chromate. The primary phases that exist in COPR are brownmillerite (Ca_2FeAlO_5), periclase (MgO) and quicklime (CaO). Hydrogarnets (katoite $(CaO)_3Al_2O_3(H_2O)_6$ is the main phase), hydrotalcites ($Al_2Mg_4(OH)_{12}(CO_3)(H_2O)_3$ is a representative compound) and Calcium Aluminum Chromium Oxide Hydrates (CACs - $Ca_4Al_2O_6(CrO_4)\cdot nH_2O$) are the main pozzolanic products derived from brownmillerite reactions. Brucite ($Mg(OH)_2$) is the main periclase hydration product (besides hydrotalcites) that scavenges a portion of the magnesium. Hydrated lime ($Ca(OH)_2$) is the product of quicklime hydration. Ettringite ($Ca_6Al_2(SO_4)_3(OH)_{12}\cdot 26H_2O$) and calcite ($CaCO_3$) also form upon sulfate and carbonate influxes.

COPR materials were extensively used as structural fill material in several countries including England, Japan, West Germany and United States (Mckee, 1988). In the US, the major COPR deposition sites were Maryland, New Jersey, Ohio, and New York (Public Health Service, 1953). In New Jersey, about 1.5 million tons of COPR were deposited over a 34-acre area at Study Area 7 (SA7), which is located in Jersey City, Hudson Country. In Maryland, a significantly larger amount was deposited over an 85-acre area at Dundalk Marine Terminal (DMT), located in Baltimore. The use of these COPR deposition sites was very limited due to extensive heaving. Specifically, in the case of SA7, the site was not used at all.

The hydration of brownmillerite and carbonation reactions may induce bulking and heaving of the COPR material. Upon hydration, brownmillerite transforms into hydrogarnets and other byproducts that have lower specific gravity than the parent mineral, whereas the carbonation of COPR result in increased COPR mass and therefore bulking. However, the aforementioned reactions were not reported to cause swell in cementitious material. On the other hand, ettringite is widely known to cause heave in both the concrete and soil-stabilized matrices (Dermatas, 1995; Taylor, 2001; Puppala et al., 2005). In addition, the formation of ettringite was extensively identified in numerous COPR samples collected from SA7 and DMT deposition (Chrysochoou et al., 2005; Dermatas et al., 2005). Monosulfates, which have similar structure as CACs, were also reported to cause swell in concrete (Chatterji and Jeffery, 1963).

In this study, all of the aforementioned mechanisms were investigated for swell development in COPR using the confined swell apparatus; however, ettringite formation was considered as the most probable heaving mechanism.

Experimental methodology

COPR samples

In both the SA7 and DMT sites, two types of COPR materials were identified in terms of physical and mineralogical properties. One type of COPR is a granular sand-size, gray black color material (designated as GB), while the other type of COPR is a fine grained, orange to reddish brown color, with a high degree of lithification (designated as HB). The GB materials were found to be in an initial COPR reaction state, i.e., containing a high brownmillerite content. Brownmillerite is produced during the roasting process and it is thermodynamically unstable in the presence of water. Conversely, HB materials had very low brownmillerite content, indicating a highly advanced reaction stage. A total of four samples from SA7 and DMT sites were used for confined swell experiments. Three out of four samples, GB1, GB2 and HB, were collected from the SA7 site and one sample, GB3, was obtained from the DMT site. Pitcher tube samples, GB1 and HB, were extruded directly into the testing chamber using a hydraulically-operated piston. Sample GB2 was trimmed to fit the cylindrical shape of the testing chamber and the disturbed sample (GB3) was kneaded by hand to reach a density similar to the field. The initial descriptions of all COPR samples tested are presented in Table 1.

Table 1. Description of COPR field samples.

Sample ID	Site	Depth	Sample description
HB	SA7	6-8.7 ft	Pitcher barrel sample, high in hydrogarnets and hydrotalcites
GB1	SA7	3-5.8 ft	Pitcher barrel sample, rich in portlandite
GB2	SA7	1.5-2.5 ft	SA7 trench block sample, medium brownmillerite
GB3	DMT	3-4 ft	DMT disturbed sample, high brownmillerite

Physicochemical analyses

Key characteristics of the COPR samples are shown in Table 2. The pH measurements of the COPR materials were performed in accordance with ASTM D4972-89 and the water content was determined by ASTM D2216-92 (ASTM, 2000).

Table 2. COPR sample properties

Sample ID	pH	Sample weight (g)	Volume (cm^3)	Moist unit weight (g/cm^3)	Initial water content (%)
HB	10.98	410.8	202.9	2.02	36
GB1	11.75	335.6	184.8	1.82	28
GB2	11.26	305.5	269.5	1.57	4.8
GB3	13.0	294.1	176.2	1.67	19

Mineralogical analyses, testing conditions and mechanisms tested

X-ray Powder Diffraction (XRPD) was performed using a Rigaku DXR-3000 computer-automated diffractometer to assess the COPR mineralogy. Prior to XRPD analyses, samples were air dried for 24 hours and pulverized to pass a U.S. standard #400 sieve (37 μm). Step-scanned X-ray diffraction patterns were collected at 40kV and 40mA using a diffracted beam with Bragg-Brentano geometry. The XRPD data were collected in the range of 2θ values between 5° and 65° with a step size of 0.05° and a count time of 5 seconds per step. XRPD patterns were qualitatively analyzed by the Jade software ver. 7.1 (MDI, 2004) and reference to the patterns of the International Centre for Diffraction Data (ICDD) database. The Whole Pattern Fitting function of Jade which is based on the Rietveld method (Rietveld, 1969) was applied with Inorganic Crystal Structure Database in order to quantify the presence of the crystalline mineral phases.

The determination of the swell mechanism was completed according to the initial mineralogical composition of the samples. The swell test matrix, the testing conditions and mechanisms tests for all the samples are presented in Table 3.

Sodium sulfate (Na_2SO_4), sodium carbonate (Na_2CO_3) and potassium chromate (K_2CrO_4) were used in order to introduce sulfate, chromate and carbonate solutions into the COPR materials, respectively to assess which suite of reactions were most likely heave (Table 3). These chemicals were obtained from Fisher Scientific (Suwanee, GA).

Table 3: Mineralogical findings, testing conditions and mechanisms tested

Sample ID	Temperature	Chemical addition	Heave mechanism tested
HB	25°C	K_2CrO_4 (2 wt%)	Hydrogarnets to CACs
GB1	25°C	Na_2CO_3 (3 wt%)	Calcite formation
GB2	25°C, 50°C	Na_2SO_4 (2 wt%)	Ettringite formation and hydration/dehydration. Temperature effect
GB3	50°C	Water	Browmillerite to hydrogarnets

Due to its high content of hydrogarnets, sample HB was selected to evaluate CAC formation. A K_2CrO_4 solution was added to trigger the transformation of hydrogarnets to CACs. In addition, hydration and dehydration of CACs was also studied. Sample GB1 was found to contain a relatively high amount of portlandite ($Ca(OH)_2$). It was therefore selected to evaluate carbonation reactions since portlandite readily converts to calcite upon CO_2 exposure. A sodium sulfate solution was added to sample GB2 in order to investigate ettringite formation from brownmillerite and hydrogarnet dissolution. This sample, was also tested under two different temperatures to assess the role of temperature effects on ettringite formation. Finally, sample GB3 was at a less advanced reaction state, as indicated by the higher brownmillerite content, therefore hydration of brownmillerite to hydrogarnets was evaluated by the addition of tap water to this sample.

Test apparatus and methodology

The one dimensional (1D) confined swell test apparatus was used. Samples were placed into stainless steel cylindrical testing chambers fully open at the top and bottom. These testing chambers had an average effective diameter of 7.32 cm and a height of 7.62 cm. Sample enclosure was achieved by placing the bottom of the sample over a porous stone with a bottom support forming the base plate of the test system and then securing the bottom support to a top ring using nuts and bolts. Paper filters were placed in between the sample and each porous stone to prevent clogging. The steel frame transferred the applied dead load (5 lb) onto the tested specimen. Bottom and top inlets of the chamber were used for the injection of aqueous permeates and air.

Vertical height change of the samples was measured by linear vertical displacement transducer (LVDT). All the readings obtained during swell tests were automatically recorded by a Geolog Box and computer. The percent swelling was calculated by dividing the height increase of the COPR sample by its original height multiplied by 100.

Cycles and chemical addition

During wet cycles, the permeant was continuously introduced to the sample through the bottom inlet of the chamber until fluid was observed at the top of the sample to attain full saturation. The bottom inlet was then closed and the saturated sample was considered to be a closed system for the rest of the cycle. During the dry cycle, air was continuously introduced to the specimen from the top inlet of the device at the pressure of 14 kPa.

Variations were made to the swell test equipment when sample GB2 was tested. Sample GB2 was dried by raising the temperature inside the test console to 50°C instead of using an air injection to mitigate carbonation reactions that may obscure ettringite formation. Also, the sample was tested at two different wet cycle temperatures (25°C and 50°C). The initial temperature during the wet cycle was 25°C, but after the 4th cycle, the temperature was elevated to 50°C for two cycles.

Operating Conditions

Samples HB and GB1 were exposed to five and seven wet cycles respectively, GB2 underwent a total of six wet cycles, and GB3 underwent a total of two wet cycles. Carbonate was introduced in the form of soluble sodium carbonate to sample GB1 during the 5th and 7th wetting cycle at application rates 2 wt% and 1 wt%, respectively. Chromate was introduced as soluble K_2CrO_4 to sample HB at 1 wt% during the 5th wetting cycle. For sample GB2, 2 wt% sulfate was added in the form of Na_2SO_4 during the 1st wetting cycle. During wetting cycles when no chemical addition was performed, only water was added to keep the samples in saturation stage.

Results and discussion

Height change results

Samples HB, GB1 and GB3 consolidated as a result of exposure to their respective permeants. This suggests that under the testing conditions, brownmillerite reaction pathways to hydrogarnet, CAC, and calcite are not the principal mechanisms promoting swell. The swell behavior of sample GB2 is presented in Figure 2.

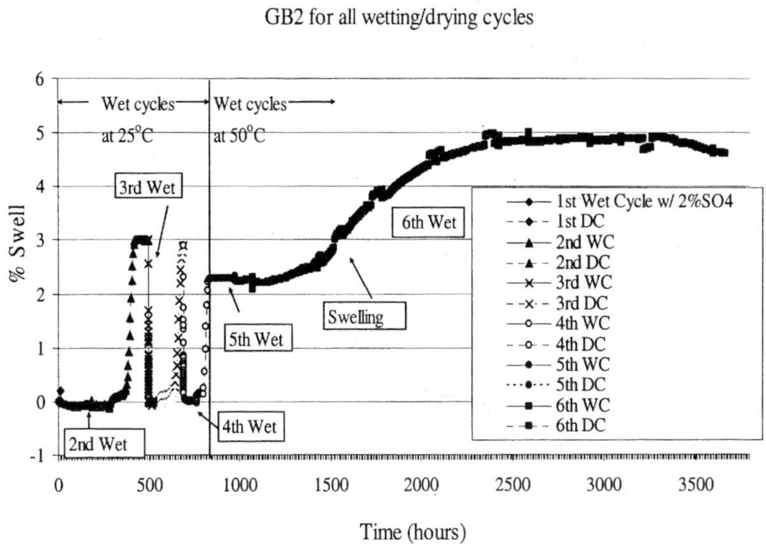

Figure 2. Height behavior for COPR sample GB2

Thermal expansion occurred in sample GB2 during the dry cycles at 50°C, and permanent swell was not observed during the wet cycles despite sulfate addition. This was confirmed by the immediate consolidation of this sample upon introduction of a sulfate solution. However, after the temperature of the wet cycle was increased to 50°C, the LVDT readings indicated approximately 3% swelling by the end of the 6th cycle. The readings for sample GB2 remained stable after swelling was observed (6th dry cycle, Figure 2). Inspection of this sample following swelling showed that the sample was very lithified and cracks were observed on the surface (Figure 3).

After testing, all samples were subdivided into two layers (top and bottom) and these layers were analyzed using XRPD to evaluate the phase transformations which occurred during the test. Specifically for sample GB2, XRPD analyses were conducted for three layers (top, middle and bottom) to identify potential mineralogical changes and their relationship to swell behavior. The sub-samples collected from each layer were mixed together to measure the final pH of each sample.

Figure 3. The surface of sample GB2 showing cracks after swelling

XRPD results of tested samples

In general, carbonation was the major reaction in all tested samples, except for sample GB2. It seems that the air influx during the dry cycle resulted in significant carbonation of the top layers, as the highly alkaline pH of the COPR samples enables rapid CO_2 sequestration. Carbonation reactions seemed to compete for the available calcium, resulting in the partial dissolution of hydrogarnets in some samples. Carbonate reactions, however, though they could not be controlled nevertheless appear unrelated to swell.

For sample GB1, calcite formation led to a reduction of portlandite (Table 4). Also, carbonation reactions seemed to "attack" the CACs, resulting in their partial dissolution. Portlandite dissolved in the top layer of the sample but it was still present in the lower layer, indicating that carbonation reactions were still ongoing at the end of the 5th cycle. Brownmillerite also dissolved partially in sample GB3 as a result of carbonation, while no appreciable hydrogarnets formed according to XRPD. No significant changes were observed in CACs content upon chromate influx. Carbonation also seemed to dominate as the prevalent geochemical mechanism in sample HB. These results indicated that, although calcite formation was a pronounced COPR reaction, it did not lead to volume increase. On the contrary, it led to consolidation depending on the initial geotechnical properties of the sample (lithification, void ratio).

XRPD data (Table 5) for sample GB2 after the 4th dry cycle showed the presence of sodium sulfate as thernardite (Na_2SO_4) with some monophases and ettringite at the bottom of the sample, indicating that most of the added sulfate had not reacted by the end of 4th cycle. Low pore concentrations of sulfate favored the formation of monophases whereas high sulfate concentrations favored the formation of ettringite (Thomas et al., 2003). Shortly after increasing the temperature of the wet cycle, swell of about 3% was observed. XRPD data confirmed the presence of ettringite throughout the entire sample and disappearance of thernardite and monophases. Ettringite formation seemed strongly favorable in sample GB2 once sulfate was readily available. Significant formation of ettringite was observed despite the competitive formation of calcite (with the partial dissolution of hydrogarnets acting as the likely calcium and aluminum source). Since calcite formation was ruled out as a heaving mechanism on the basis of the HB, GB1, and GB3 sample behavior, ettringite formation seems to be the only possible heaving mechanism for sample GB2.

Table 4. Comparison of COPR sample mineralogy (GB1, HB and GB3) before and after swell tests

	GB1 Initial pH: 11.75 Final pH: 11.85			HB Initial pH:10.98 Final pH: 11.76			GB3 Initial pH: 13 .0 Final pH: 12.02		
Mineral	Initial	Top final	Bottom final	Initial	Top final	Bottom final	Initial	Top final	Bottom final
Brownmillerite	11	19.7	32.7	1.8	6.6	3.8	58.4	37	41.2
Periclase	ND	ND	ND	3.3	4.2	2.8	3.6	2.3	2.2
Brucite	2.1	2.9	4	11.1	9.4	8.1	3.8	3	2.4
Portlandite	20	ND	1.9	ND	ND	ND	6.9	2.9	ND
Calcite	24	56.6	37.2	ND	27.6	20.2	8.4	24.1	28.7
Hydrotalcite	3.8	4.1	3.3	29.9	13.5	20.5	3.9	4.7	4.6
Hydrogarnets	18	7.9	13.5	41.2	26.5	30.3	12.1	17.2	13.8

Ettringite	2.8	2.9	1.5	ND	ND	ND	ND	1.9	ND
CACs (monophase)	14	2	1.3	12.7	12.2	14.3	2.8	2.2	3
Afwillite	4.1	4	4.5	ND	ND	ND	ND	4.6	4

Note: ND-not detected

Table 5. Comparison of COPR sample mineralogy (GB2) before and after swell tests

	GB2 (Initial pH: 11.26, Final pH: 12.01)					
Mineral	Initial	Top end of 4th dry cycle	Bottom end of 4th dry cycle	Top end of 6th dry cycle	Middle end of 6th dry cycle	Bottom end of 6th dry cycle
Brownmillerite	35	27.3	36.8	39	43.8	44.7
Periclase	0.6	1.6	1.5	1.4	1.5	2.8
Brucite	8.5	6.6	10.2	7.5	7.0	9.7
Portlandite	1.6	ND	ND	ND	ND	ND
Calcite	9.7	35.5	18.4	16.4	7.2	12.9
Hydrotalcite	5	2.2	3.5	3.0	4.2	4.3
Hydrogarnets	13	6.5	9.2	7.5	7.0	4.8
Ettringite	3.3	ND	6.9	25.2	29.3	20.7
CACs (monophase)	12	ND	5.0	ND	ND	ND
Afwillite	12	ND	ND	ND	ND	ND
Thernardite	-	20.2	8.4	ND	ND	ND

Note: ND-not detected

Conclusions

Confined swell tests were performed to investigate heaving phenomena in COPR. The COPR expansion mechanisms tested in this investigation were: a) brownmillerite hydration to hydrogarnets; b) carbonation reactions; and, c) ettringite and Calcium Aluminum Chromium Oxide Hydrates (CACs) formation. The observations can be summarized as follows:

- No apparent CAC formation was observed upon the addition of K_2CrO_4 to COPR sample HB, as carbonation dominated the geochemical reactions. The viability of the CAC-related expansion mechanism remained inconclusive.
- Significant calcite formation was observed in all samples tested and in particular sample GB1 due to the rapid CO_2 sequestration during the dry cycles. However, calcite formation in the existing voids produced no net volume increase and ultimately led to consolidation.
- No significant hydrogarnet formation was observed in the sample GB3 tested at 50°C. This elevated temperature was employed in order to accelerate the slow dissolution of brownmillerite. No correlation between hydrogarnets formation could, therefore, be clearly established between brownmillerite hydration and volume expansion.
- Upon the addition of a 2% sulfate solution, up to 3% swelling was observed in sample GB2 under the confinement imposed by a 5 lb dead load. In this sample the quantity of ettringite significantly increased (more than 17% from its initial content) upon sulfate influx.
- The present investigation verified that ettringite is a viable expansion and heaving mechanism in COPR and its formation is thermodynamically favorable, when enough sulfate is introduced into the system.

- The heave manifestation is associated with the overall geotechnical properties of the matrix. Carbonation was excluded as an expansion mechanism in COPR. On the other hand, the contribution of other hydration and pozzolanic products remained inconclusive since their formation was suppressed by carbonation reactions enabling ettringite reaction to prevail.

Acknowledgements

This study was supported by Honeywell Inc.

References

ASTM, 2000. Annual Book of ASTM Standards, American Society for Testing and Materials, Soil and Rock, Vol. 4.08.

Chatterji, S., and J.W. Jeffery. 1963. A New Hypothesis of Sulphate Expansion. Magazine of Concrete Research, 15(44), 83-86

Dermatas, D., "Ettringite-Induced Swelling in Soils: State-of-the-Art," Appl Mech Rev, Vol. 48, No10, 1995, pp. 659-673.

Dermatas, D., Chrysochoou, M., Moon, D.H.,.Christodoulatos, C., Lazarte, C.A., Pendleton, C., Bonaparte, R., Briggs, R., Myers, M., French, C., Morris, J., Kaouris, M., 2005. Mineralogical characterization of COPR at Dundalk Marine Terminal area 1800 to elucidate heave mechanism, Proceedings of the 8th International In Situ and On-Site Bioremediation Symposium, Baltimore, Maryland, USA, June 6-9, 2005.

Hiller, S., Roe, M.J., Geelhoed, J.S., Fraser, A.R., Farmer, J.G., and Parterson, E., 2003. Role of Quantitative Mineralogical Analysis in the Investigation of Sites Contaminated by Chromite Ore Processing Residue. the Science of the Total Environment, 308, 195-210.

Chrysochoou, M., Moon, D.H., Dermatas, D., Wazne, M., Christodoulatos C., Meng, X., Kaouris, M., Morris, J., French, C., Mineralogical Analysis of Chromite Ore Processing Residue by X-Ray Powder Diffraction, Proceedings of the 8th International In Situ and On-Site Bioremediation Symposium, Baltimore, Maryland, USA, June 6-9, 2005.

McKee, T., 1988. Chromate Chemical Production Industry: Waste Treatment Past and Present. Report to the New Jersey Department of Environmental Protection, Trenton, NJ, 1988.

Materials Data Inc, 2004, Jade Version, 7.1., California, USA.

Public Health Service, 1953. The Chromate-Producing Industry. In: Health of Workers in Chromate Producing Industry-A Study. Federal Security Agency, Public Health Service, Publication No. 192.

Puppala, A.J., Intharasombat, N., and Vempati, R.K., "Experimental Studies on Ettringite-Induced Heaving in Soils," Journal of Geotechnical and Geoenvironmental Engineering, 131, 3, 2005, pp. 325-337.

Rietveld, H.M., 1969. A Profile Refinement Method for Nuclear and Magnetic Structures. Journal of Applied Crystallography, 2, 65-71.

Taylor, H.F.W., Famy, C., and Scrivener, K.L., "Review Delayed Ettringite Formation," Cement and Concrete Research, 31, 2001, pp. 683-693.

Thomas, Jeffrey J., Rothstein, David, Jennings, Hamlin M., Christensen, Bruce J., 2003. Effect of Hydration Temperature on the Solubility of Ca-, S-, Al-, and Si- bearing solid phases in Portland Cement Pastes. Cement and Concrete Research, (33) pp 2037-2047.

Xeidakis, G.S., 1996. Stabilization of Swelling Clays by $Mg(OH)_2$: Changes in Clay Properties after Addition of Mg-Hydroxide. Engineering Geology, 44, 107-120.

Application of artificial granular material made from construction sludge to sand compaction piles

Yasuyuki Nabeshima, Osaka University, Osaka, Japan
Tamotsu Matsui, Fukui University of Technology, Fukui, Japan
Hideyuki Sugiura, Toa Corporation, Osaka, Japan
Kenji Ochi, Osaka Bentonite Coop, Osaka, Japan

Abstract
Applicability of the artificial granular material made from construction sludge was investigated as a filler for sand compaction piles through the full-scale SCP installation tests. It was confirmed that the artificial granular material can be used in the standard SCP installation procedures with a comparable execution period to the conventional sand. The ground improvement performances of the artificial granular material, such acceleration effect of consolidation and reduction of consolidation settlement, were firmly established in the clayey ground. For sandy ground, the artificial granular material piles provide definite ground improvement effects due to having higher stiffness than conventional sand piles.

Introduction
About eight million tons of construction sludge is generated every year in Japan according to the reports on the construction waste investigated by the Ministry of Land, Infrastructure and Transport (2001 & 2004), this means that the construction sludge occupied about 9% of the total amount of the construction waste.

Table 1. Actual conditions of construction waste recycle in 2000 and 2002.

	Recycling rates	
	Achievements in 2000	Achievements in 2002
Whole construction wastes	85%	92%
Asphalt concrete	98%	99%
Concrete	96%	98%
Construction sludge	41%	69%
Construction woods	83%	89%

Though the recycling rate of construction sludge increases from 41% to 61% as illustrated in Table 1, about 39% of the sludge is finally disposed without being recycled. This rate is still lower than those of other construction wastes and the increase in recycling rates of construction wastes, especially construction sludge, are necessary for prolonging the capacity of final disposal sites and the effective use of natural resources. Also, the domestic natural sand is exhausted and tightened regulations or imposed a ban on the extraction of natural sand in several regions. As an alternative material, the authors developed and proposed artificial granular material made from construction sludge. The manufacturing process of artificial granular material is follows: The construction sludge without soil contaminants is mixed with cement and polymer, and pressed into ingots, and then they are crushed to granular material after the curing. It does not return to sludge when it is immersed in water.

5th ICEG Environmental Geotechnics, Thomas Telford, London 2006.

This paper evaluates the characteristics of the artificial granular material and reports the results of full-scale tests conducted in Shimonoseki, Yamaguchi in July 2003 (Ochi, 2003) to confirm the applicability of the artificial granular material as a filler for sand compaction piles (SCP). This paper aims to contribute to the reduction of environmental disruption by extracting natural sand and the promotion for recycling construction sludge.

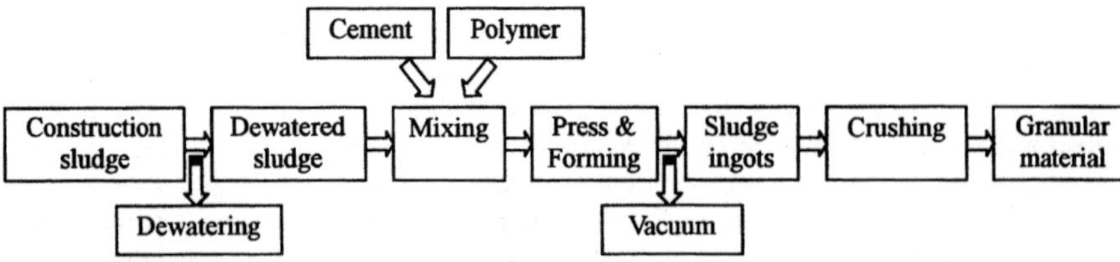

Figure 1. Production process of artificial granular material.

Manufacturing process of artificial granular material

Figure 1 illustrates the manufacturing process of the artificial granular material. This treatment is classified into "an advanced stabilization (Hamano and Ochi, 1998)," which is distinguished from conventional "stabilization." In the advanced stabilization, to obtain high strength solid material, the press or the autoclaving is conducted in addition to the stabilization of construction sludge by adding cement or lime. The grain size distribution of the artificial granular material can be arranged from sand to gravel for the required distributions. Construction sludge originally consists of clay and/or sand and contains no toxic substances. Using a safe hardening adjunct, the artificial granular material does not cause any secondary pollution. Furthermore, the advanced stabilization has advantages on the treatment cost compared to the burned and the melting treatments to produce granular materials.

Table 2 Soil properties of construction sludge treated by advanced stabilization.

Measured items	unit	data
Specific gravity	-	2.08
Dry density	g/cm^3	1.51
Water absorption ratio	%	27

Figure 2. Artificial granular materials.

As construction sludge usually has high water content, construction sludge is dewatered, then passed through a sifter and removed inclusions such as iron scraps by using a magnetic separator. The treated sludge is mixed with cement and polymer, and pressed and formed into ingots. To remove air in the sludge, the forming process is conducted under a vacuum condition. The sludge ingots are cured for one month and crushed into granular and sieved to each grain size as shown in Figure 2. The artificial granular material can be arranged for any grain size distributions from gravel, sand to fine powder, which allows a wide range of applications. The artificial granular material does not return to sludge when it is immersed in water because it is fully solidified. Table 2 shows typical soil properties of the artificial granular material (Hamano and Ochi, 1998). Advantages of the artificial granular material are follows; high unconfined compressive strength compared to the products of ordinary cement stabilization, good compaction characteristics, high water retention capacity, lighter weight than natural stones, selectable grain size distribution, relatively low treatment cost and stable material qualities.

Evaluation aspects of artificial granular material as a filler for SCP

The following six aspects of the artificial granular material were investigated to evaluate the applicability and usability as a filler for SCP.

Shear strength

The shear strength of SCP installed in clayey ground is required the higher shear angle (ϕ_d) than 30°, and the shear strength of SCP installed in sandy ground is required to exceed that of the soil surrounding SCP. To confirm them, specimens were made by arranging same grain size, density, and water content with the core samples obtained from the full-scale test piles, and direct shear box tests were carried out under a constant normal stress using a medium-sized shear box which has a hole of 20 cm in diameter and 10 cm in height.

Permeability

Permeability of SCP installed in clayey ground is required to exceed 1×10^{-4} cm/s. Standard permeability tests were carried out for core samples obtained from the full-scale test piles.

Execution period

The grain sizes of the artificial granular material distributes within the range of those of conventional sand for SCP, so that the conventional SCP machine can handle the artificial granular material. The execution period is required to keep a similar time compared to that using the conventional sand. Grain size distributions of core samples are measured, and then execution periods of SCP installation using conventional machine were compared for the artificial granular material and conventional sand to evaluate usability of the artificial granular material.

Ground improvement performances

The artificial granular material is required to show comparable ground improvement performances to the conventional sand, which are the increase in shear strength, acceleration effect of consolidation, reduction of settlement for clayey ground and the ground improvement effect for sandy ground. The acceleration effect of consolidation and reduction of settlement are compared on the basis of final settlement and consolidation degree in clayey grounds improved by SCP using the artificial granular material and conventional sand. To confirm the ground improvement performance in sandy ground, standard penetration tests were performed at pile centers and between piles, and the compaction effect of SCP using the artificial granular material and conventional sand was compared.

Durability

The artificial granular material is required to be stable for a long period and resistant to fracture. Resistance to fracturing under an alternating dry and wet condition due to changing the ground water level was estimated by monitoring grain size distribution in the alternating dry and wet tests. In addition, a long time monitoring of grain size distribution in the water was also performed to confirm the resistance for the fracture of grains.

Safety for geoenvironment

Leaching tests were carried out according to the environmental criteria concerning the soil pollution to confirm that the artificial granular material was safe as a filler for SCP from a viewpoint of geoenvironment.

All evaluations for six aspects listed above were obtained satisfactory results and confirmed that the artificial granular material met every requirements as a filler for SCP and had comparable performance to the conventional sand. Because of a page limitation, detailed descriptions about only the execution period and the ground improvement performances are here described.

Full-scale SCP installation test using the artificial granular material

To confirm the usability of the artificial granular material, it was necessary to perform a full-scale SCP installation test using the artificial granular material. Full-scale tests were performed at Chofu-Ougimachi in Shimonoseki city, Yamaguchi prefecture in July 2003 (Ochi, 2003). Figure 3 shows the locations of the test site. A conventional SCP machine was used to install every 25 piles (1.5 m x 1.5 m, L= 16 m) of the artificial granular material and the conventional sand. The execution periods for the artificial granular material and the sand were compared. Embankments with 3.0 m high were constructed to investigate acceleration effect of consolidation and reduction of settlement in clayey ground. The arrangement of SCP and embankments are illustrated in Figure 4. Figure 5 shows a typical ground condition at the test site.

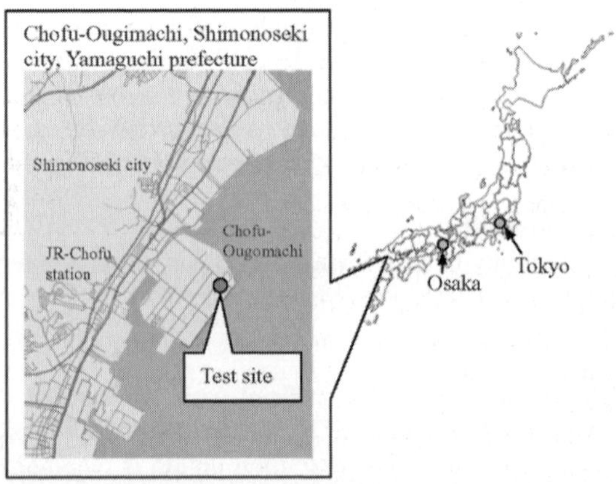

Figure 3. Full-scale SCP installation test site.

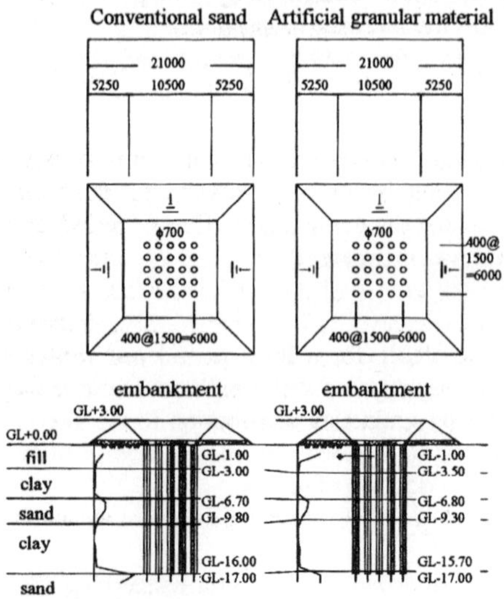

Figure 4. Arrangement of SCP and embankments.

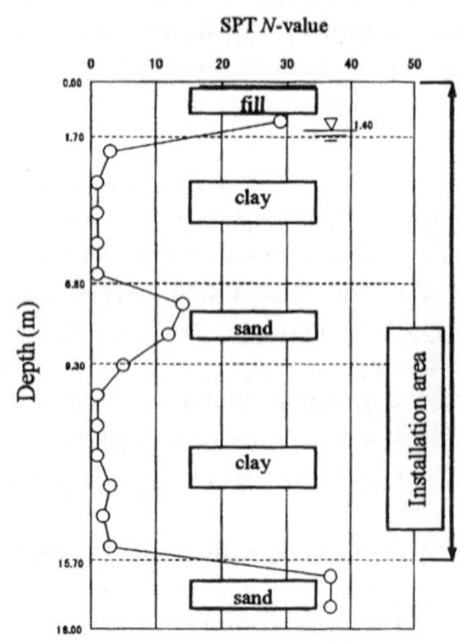

Figure 5. Ground condition.

Evaluation of the applicability of SCP using the artificial granular material through the full-scale test

Execution period

Usability of the artificial granular material in SCP installation was evaluated based on its grain size distribution. The grain size distributions were measured for the SCP core samples of the artificial granular material. Figure 6 shows the grain size distributions and they correspond to sandy gravel with fines in the standard classification in soil engineering. Representative cumulative grain size distributions of the artificial granular material piles are shown in Figure 6. Their grading curves are within the ranges of grain size used for conventional sand and gravel compaction piles (JGS, 1998). Execution periods for SCP installation of the artificial granular material and the conventional sand are summarized in Figure 7. The execution periods of primarily ten artificial granular material piles are longer than those of conventional sand piles as shown in Figure 7. This was because the new material required the observation of material losses during pile installation as well as the verification of resistance during the diameter expansion process of piles. From the eleventh pile on, the average execution period of 18.4 minutes for the artificial granular material was comparable to 17.3 minutes for the conventional sand. Figure 8 shows the installation scenes of the artificial granular material piles. The artificial granular material pile using standard SCP machine had a diameter of about 700 mm, which was the standard value for the conventional sand piles. It was thus confirmed that the artificial granular material can be used in SCP method in the same way as the conventional sand without significant increase in the execution period.

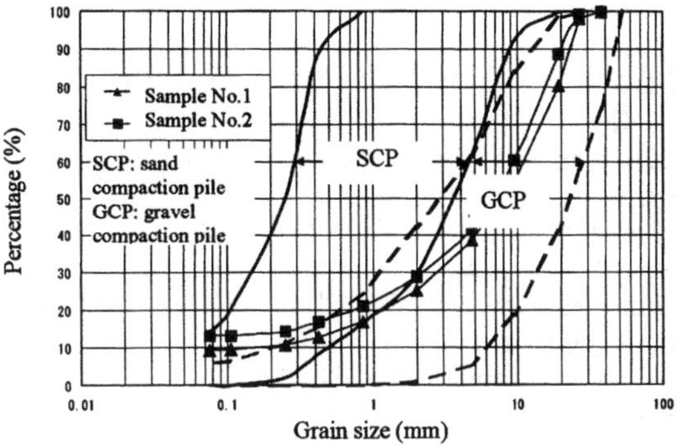

Figure 6. Grain size distribution.

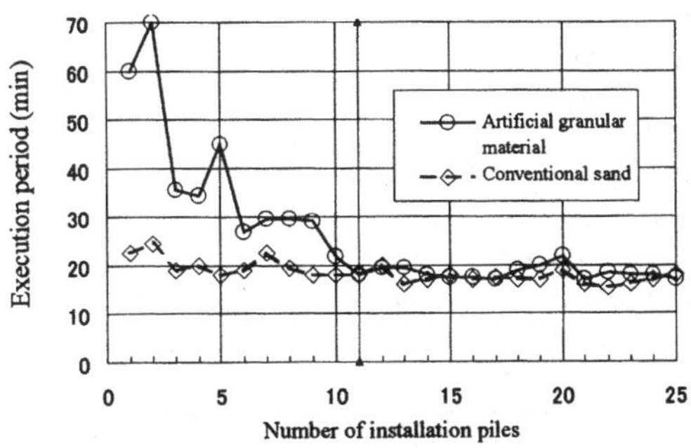

Figure 7. Execution periods of SCP installation.

Ground improvement performances

Ground improvement performances of the artificial granular material piles were evaluated for clayey and sandy grounds, respectively.

(a) Installation of SCP (b) Artificial granular material (c) Artificial granular material pile

Figure 8. Installation scenes of the artificial granular material piles.

Ground improvement effect for clayey ground

To discuss the ground improvement performances for clayey ground, the acceleration effect of consolidation and the reduction of settlement were investigated by measuring settlement after the embankments were constructed. Final settlement and degree of consolidation were estimated on the basis of the time-settlement curve. After installing the artificial granular material piles and conventional sand piles in the test site as shown in Figure 4, settlements were measured at the center of each pile after the completion of the embankments. Final settlement was estimated using the hyperbolic method, and calculated the consolidation degree, U (%) as a function of time. Figure 9 shows the measured settlements for the grounds improved by the artificial granular material piles and the conventional sand piles. Two grounds give similar consolidation curves, at 153 days after the completion of embankments, the consolidation degrees were 93.6% for the ground improved by the artificial granular material piles and 93.3% by the sand piles. Thus the artificial granular material was proved to have comparable acceleration effect of consolidation to the conventional sand. The curves in Figure 9 also show that the final settlement of 94 mm for the ground improved by the artificial granular material piles was much smaller than that of

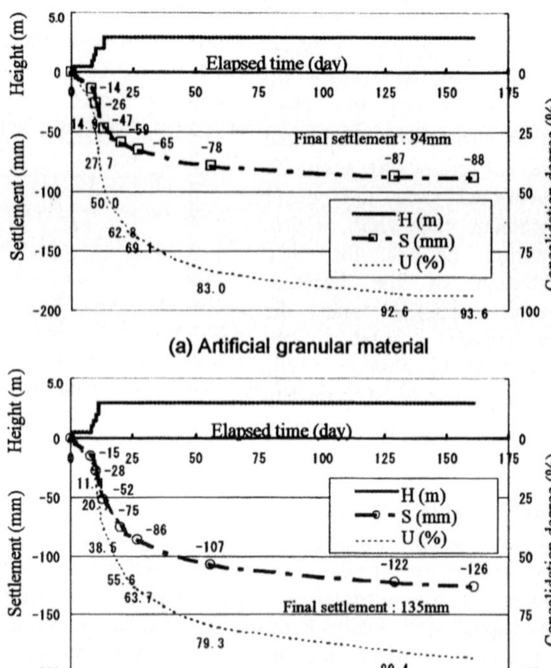

Figure 9. Consolidation settlement and consolidation degree with time.

135 mm for the conventional sand. The artificial granular material thus demonstrates a clearly greater reduction of consolidation settlement than the conventional sand. Similar results had been obtained in previous model experiments (Nabeshima et al., 2000). The ground improvement performances of the artificial granular material in the clayey ground were thus firmly established.

Since the clayey ground improved by SCP formed a composite ground consisting of clayey soil and sand piles, it is expected to increase the shear strength and the bearing capacity of the clayey ground around piles, in addition to ground improvement performances discussed above Standard penetration tests (SPT) were performed at the center of piles and between piles, SPT results are shown in Figures 10 and 11. Figure 9 shows that the artificial granular material provides a similar ground improvement performance and a comparable or greater reduction of settlement than that of conventional sand, the strengthening effect of an the artificial granular material pile on the surrounding ground was somewhat less than that of conventional sand. This is suggested by the *N*-value between the piles for the upper and lower clay layers shown in Figure 11, which were somewhat lower than those for sand piles. As for bearing capacity, however, Figure 10 shows much higher *N*-values for the artificial granular material samples at the center of piles than those for conventional sands, suggesting that the bearing capacity of the ground was sufficiently improved, even with only a small strengthening effect. This is due to well grading and angular particle shape of the artificial granular material. This conclusion was also supported by the greater settlement reduction effect of the artificial granular material piles shown by the present tests as well as the model experiments.

In summary, the full-scale SCP installation test showed that the artificial granular material provided lower strengthening and comparable consolidation acceleration effects on surrounding clayey ground in comparison with the conventional sand piles. The artificial granular material piles, which are more rigid than the conventional sand piles, have greater reduction of settlement and bigger bearing capacity. Therefore, the artificial granular material piles are equal or superior to the conventional sand piles as the ground improvement method if the strengthening by consolidation may be not required.

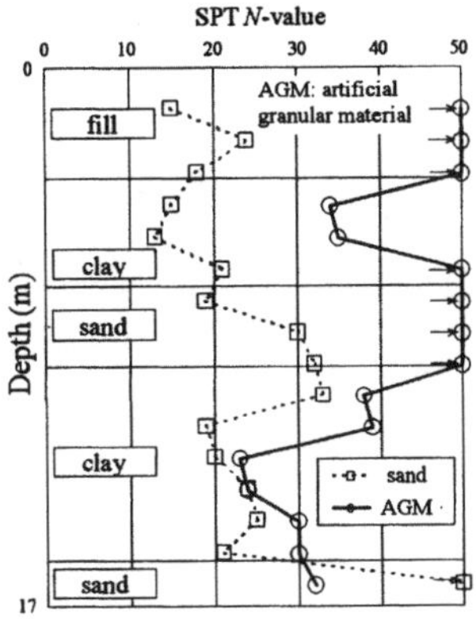

Figure 10. SPT *N*-values at the center of piles.

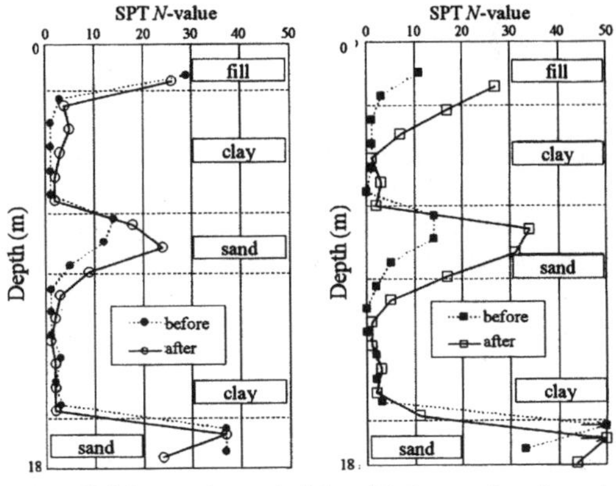

(a) Artificial granular material (b) Conventional sand

Figure 11. SPT *N*-values between piles.

Ground improvement effect for sandy ground

The sandy ground improved by SCP should have greater strength than the surrounding ground. This requirement was satisfied as shown in Figures 10 and 11, the *N*-values at center of the artificial granular material pile exceed 50 while those between the piles are about 20. The *N*-values for conventional sand piles are about 30 for both samples. In other words, the artificial granular material piles give higher *N*-values than the conventional sand at the center of piles and slightly lower between the piles.

Thus it was concluded that the artificial granular material piles have definite ground improvement effects of their high stiffness, in spite of slightly lower *N*-value between the piles.

Conclusions

Applicability of the artificial granular material made from construction sludge was evaluated as a filler for sand compaction piles through the full-scale SCP installation tests. The major results are summarized below:

1) The artificial granular material can be used in the standard SCP installation procedures with a comparable execution period to the conventional sand.
2) The artificial granular material piles have an acceleration effect of consolidation similar to standard SCP in clayey ground, however a slightly lower strengthening effect on the surrounding ground. The artificial granular material piles show greater reduction of settlement and bearing capacity increases comparable to or higher than conventional sand because they are more rigid.
3) For sandy ground, the artificial granular material piles provide definite ground improvement effects. They have higher stiffness than the conventional sand piles, however *N*-value between the piles is slightly lower.

References

Ministry of Land, Infrastructure and Transport, Policy Bureau (2001): "The report on actual conditions of construction wastes in 2000."

Ministry of Land, Infrastructure and Transport, Policy Bureau (2004): "The report on actual conditions of construction wastes in 2002."

Nabeshima, Y., Matsui, T. and Hamano, H. (2000): "Application of crushed cement-stabilized sludge as a filler of sand drain," Material Science, Vol.49, No.1, pp.18~21.

Hamano, H. and Ochi, K. (1998): "Advanced stabilization technique of construction sludge," New Technology of Construction Recycle, pp.45~48.

Ochi, K. (2003): "Application of recycled gravel made from construction sludge to sand compaction pile," Construction Recycle, Vol.25, pp.46~49.

Japan Geotechnical Society (1998): "Countermeasures for soft soil ground –from surveying, design to construction-," Soils and Foundations for Field Engineers, No.17.

Proposed geophysical method for investigating the seabed for a landfill project in Okayama, Japan

Makoto Nishigaki, Professor, Faculty of Environmental Science, Okayama University, Okayama, Japan (email:n_makoto@cc.okayama-u.ac.jp)

Mitsuru Komatsu, *Research Associate, Faculty of Environmental Science, Okayama University, Okayama, Japan*

Makoto Inoue, *Global Infor-Tech, Japan*

Kwabena Kankam-Yeboah, Senior Research Scientist, Water Research Institute, *Post Office Box M 32 Accra, Ghana.*

Stephan Mattersteig, Graduate Student, Faculty of Environmental Science, Okayama University, Okayama, Japan

John Apambilla Akudago, Graduate student, Faculty of Environmental Science, Okayama University, Japan

ABSTRACT

In Japan, wastes are dumped in mountainous areas due to the unavailability of land. The sea, thus, offers a good potential as landfill sites. The seabed along part of the coast of Okayama was first investigated by means of drilled boreholes. Information from the borehole logs showed variable clay depths of 1-3 m. In order to determine the continuity of the clay layer, an Induced Polarisation (IP) resistivity survey was carried out in the sea. The geophysical measurements showed that there is a continuous clay layer in the seabed; with a thickness of not less than 1 m. Further tests showed that the clay layer has a permeability of 1.0E-06 cm/s which meets the requirement of continuous clay liners of at least 5 m thick and permeability, *k*<1.0E-05 cm/s for landfill sites in Japan. Therefore, the clay layer in the sea bed along the coast of Okayama could serve as an effective liner and the site is suitable for a landfill, without any threat to the sea. The applicability of the method in underwater geotechnical investigations is discussed in this paper.

INTRODUCTION

Japan has a population of about 124 million people and an area of about 378,000 km^2 (PWMI, 1991). The population density is about 328 inhabitants per km^2. Mountains and forests occupy more than 70 % of Japan's land. Only 21 % of the land in Japan can be used as living space. Based on this 21 %, it has a population density of about 1,550 inhabitants per km^2, one of the highest in the world.

Due to the unavailability of land space for general land use and the high generation of domestic and industrial waste, the wastes are commonly disposed-off in mountainous parts of the country. These waste disposal sites are not adequate and there is a danger of pollution and contamination of the groundwater and the environment as a whole. Therefore, the use of the marine environment for waste disposal is being explored to complement the landfill sites in the mountains.

The Okayama Prefecture Environmental Conservation Corporation has constructed three landfill sites in the sea, bordering the southern section of the Mizushima Industrial Complex since 1979 (OPECC, 2003). The first and second landfills are already filled-up, reclaimed and the sites are being used as golf course. The designed life span of the third one is estimated to come to an end by April 2008. To ensure that disposal sites are available for the industries of Okayama Prefecture, a fourth landfill is to be constructed near the old ones.

A relatively impermeable and continuous clay liner, at least 5 m thick and a hydraulic conductivity, k <1.0E-05 cm/s is required for the new site. To ascertain this in the Okayama seabed, twenty boreholes were drilled on the planned site. The different types of soil and geological materials in each borehole were logged and analysed. The results indicated that a clay layer is located beneath a sand layer at a depth of about 15 m in the seabed. However, the thickness of the clay liner did not meet the minimum required depth of 5 m; it rather varied from 1 to 3 m with a hydraulic conductivity of 1.0E-6. At the critical location in the seabed where the thickness of the clay liner was about 1 m, the presence of big boulders did not make it possible to drill more boreholes to confirm the morphology of the seabed. It was therefore necessary to determine whether the clay layer was continuous over the whole site. A geophysical survey was therefore carried out on the seabed in Okayama in order to investigate the characteristics of the fourth Mizushima landfill site.

The objective of this paper is to compare the results obtained from the borehole investigations that had been previously carried out and the suitability of the Induced Polarisation (IP) method for the location of landfill sites.

STUDY AREA

The proposed landfill site is located at the Mizushima Disposal Facility, which is located on the southern edge of the Mizushima Industrial Complex near Kurashiki in Okayama as shown in Fig. 1. The planned landfill site has a dimension of 600 m long, 430 m wide and it is estimated that about 2,750,000 m^3 of industrial wastes could be stored in it. The density of industrial waste of the Okayama Prefecture is estimated to be 1.1 Mg/m^3. Thus, about 3,025,000 Mg of waste can be stored in the new landfill. The commissioning of the new landfill is set for 2008 and the design life is estimated to be 11 years.

GEOPHYSICAL SURVEY

Geophysical methods detect differences, or anomalies, of physical properties within the earth's crust (Todd, 1959) and electrical resistivity is one of such properties. In geophysical surveys, resistivities are determined from apparent resistivities, which are computed from the measurements of current and potential differences between pairs of electrodes placed on the surface. By this, a trace of the variation of resistivity with depth or lateral distance can be obtained and hence a resistivity log. In water resources and environmental studies, the logs are interpreted with respect to the geological formation and structure, porosity, water content and water quality

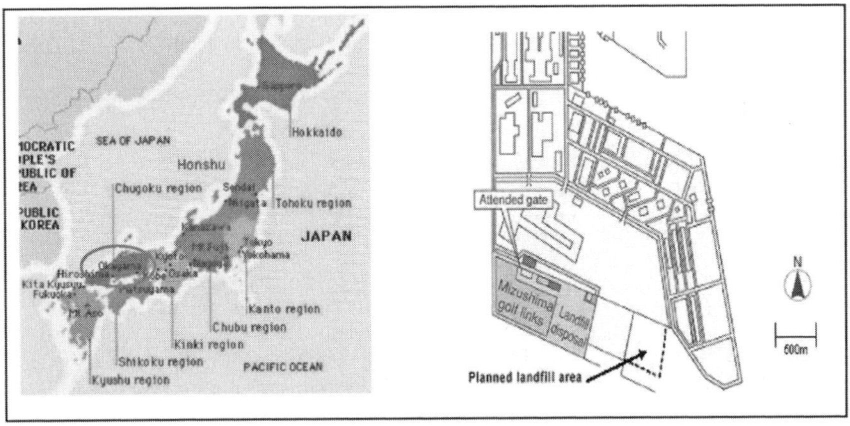

Figure 1: Location of the Mizushima Disposal Facility

The resistivity of a formation such as rock or clay limits the amount of current that passes through the formation when an electrical potential is applied. It is defined as the resistance in ohms between opposite faces of a unit cube of the material (Todd, 1959). If a material of resistance R has a cross-sectional area A and a length L, then its resistivity can be expressed by equation (1).

$$\rho = \frac{R * A}{L} \quad (\Omega\text{m}). \qquad (1)$$

Resistivity depends on the geological structure of the ground or underground material. Different kinds of rocks and soils exhibit different porosities, textures, water contents and thus resistivities. From gravel, sand, silt to clay, there is a decrease in the size of pores and an increase in the specific surface area (DUT, 2001). The resistivities decrease as the porosity and surface area of material increase. Solid rocks have the lowest porosity and the highest resistivity.

The IP method was first developed for mineral investigations but has been applied to environmental, water resources, geotechnical and geological projects in many parts of the world (Ward, 1990). The method can also be used to detect the migration of small concentrations of minerals. In recent years, the IP method has been applied to the marine environment; in the exploration of petroleum. Its applications in other fields are continuously being explored.

An electric current is passed into the ground from a source in all applications of the resistivity and IP methods. However, in IP survey, an alternating current is used and cut off from the source after some time, and the apparent resistivity resulting from the decay current is measured. The presence of clay helps in increasing the period of the current decay because the fine clay particles become charged which tend to conduct the current. This lowers the resistivity. The introduction of the current can be made through point electrodes or long line contacts. More often, point electrodes are used. With the aid of potential electrodes, the electrical current and the apparent resistivity of the ground are measured (Telford et al., 1990).

The natural occurrence of a homogeneous ground is rare. Most sub-surfaces exhibit two- or more layers. Different kinds of curves can be obtained if the ground exhibits several strata. Some typical curves are shown in Fig. 2.

Type H and K curves show the minimum and maximum resistivities respectively. The order of resistivity in type H and K is $\rho_1 > \rho_2 < \rho_3$ and $\rho_1 < \rho_2 > \rho_3$, respectively. Types A and Q occur in increasing and decreasing resistivities (Type A: $\rho_1 < \rho_2 < \rho_3$, Type Q: $\rho_1 > \rho_2 > \rho_3$), respectively. With the help of these standard curves, it is possible to identify the different kinds of layers in a sub-surface.

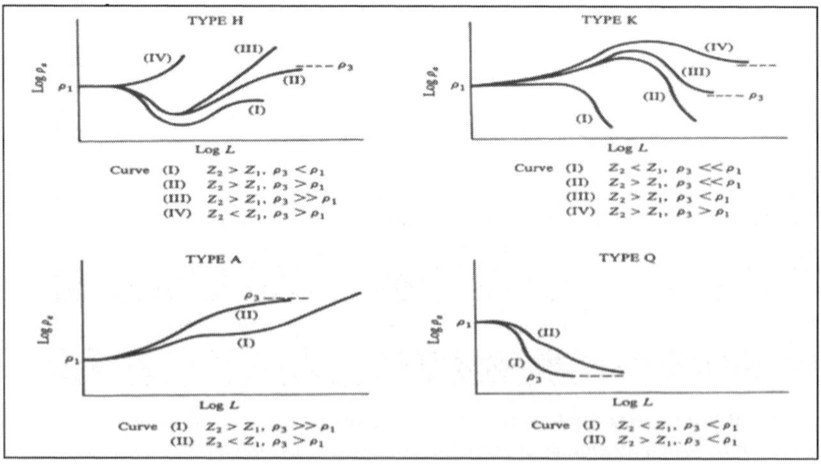

Figure 2: Types of curves in a three-layer sub-surface (Source: Telford et al., 1990)

It is good to have other resistivity survey data recorded with induced Polarization data. This enriches the interpretation of the resistivity data. The IP method is based on the changes in resistivity as a function of frequency. The resistivity and IP methods are used at low frequencies. Detailed review of the basic theory, the different kinds of arrays and the interpretation of resistivity survey has been done by Telford et al., (1990).

FIELD DATA COLLECTION

The geophysical survey was carried out using a survey vessel. On-board the survey vessel, were the IP transmitter and receiver, current source and electrodes. The electrodes were placed on a cable (streamer), which was dragged behind the survey vessel on the floor of the seabed as shown in Fig. 3. Three transponders were used. The transponder t_1 was located beside the survey vessel with the aid of a Differential Global Positioning System (DGPS). Transponders t_2 and t_3 were located on the streamer, to determine the exact position of the streamer and they were navigated from the survey vessel. The sea water level was also continuously measured.

Electrical current was passed into the sea through the current electrodes C1 and C2 and the potential difference between two potential electrodes recorded and stored automatically by a computer data logging system. The Pole-Dipole array was used with C2 fixed at 100 m from C1. The arrangement of the electrodes are shown in Table 1.

Measurements were taken along 3 lines. Lines 1 and 2 were parallel to each other. Line 1 was placed along *Bor.No. 1, B-No.1, B-No.2, Bor.No.6,* and *Bor.No.12.* Line 2 was placed along *Bor.No. 9, Bor.No. 8,* and *Bor.No. 7.* Thus, it was possible to make comparisons between the boreholes and their corresponding geophysical measurements. Line 3 was placed perpendicular to lines 1 and 2. The position of line 3 was chosen, because *Bor.No.5* and *Bor.No.7* showed a critical thickness of about 1 m. Line 3 was therefore chosen to provide more information on the extent, continuity and thickness of the clay layer. The orange (dash) line in Fig. 4 is the position of the streamer and the blue line is the position of the survey vessel during the measurements.

Table 1: Intervals between the electrodes

Electrodes	P_1	P_2	P_3	P_4	P_5	P_6	P_7	P_8	P_9	P_{10}	P_{11}	P_{12}	P_{13}	P_{14}	P_{15}	P_{16}	C_2
Distance from C_1 (m)	0.5	1	2	3	4	5	6	8	10	12	16	20	24	32	40	48	100

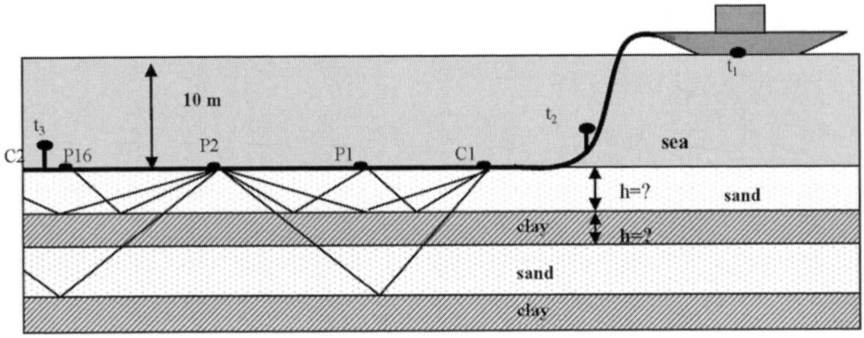

Figure 3: The cable streamer

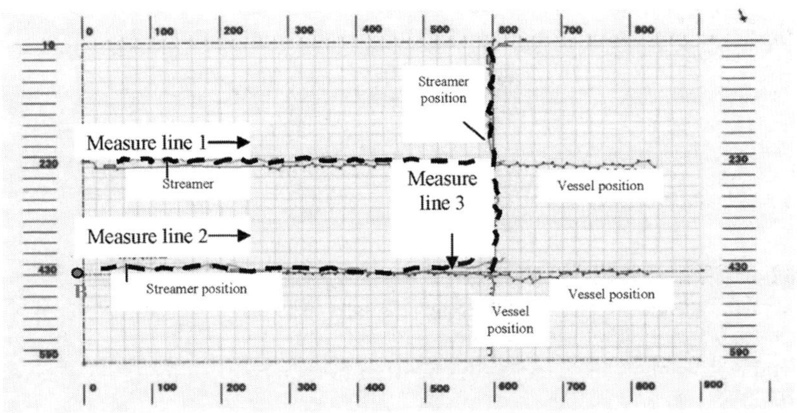

Figure 4: Simplified measuring grid – survey vessel and streamer location during the measurements (m).

DATA ANALYSIS

The data that were logged on the instruments of the survey vessel were analysed. The level of sea water was considered in the analysis of the geophysical data. A linear filter, a mathematical process that reduces undesirable effects ("noises") in the geophysical measurements, was incorporated into the analysis. Figure 5 shows the flowchart for analysing the geophysical resistivity data and Fig. 6 shows the resistivity change with depth. Figures 7, 8 and 9 show the filtered data.

During the measurements, it was possible to observe these curves on real-time basis. The shape in Fig.6 reflects Type H curve. At first, the resistivity decreased. Then, at a depth of 2 m the resistivity started to increase. The rise of the curve is sharp between 10 and 50m depth. A resistivity of > 2 Ω-m was measured at a depth of 50 m.

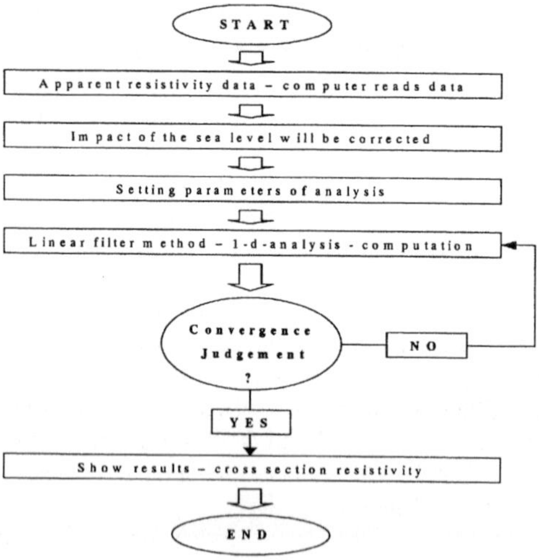

Figure 5: Flowchart of linear filtering method

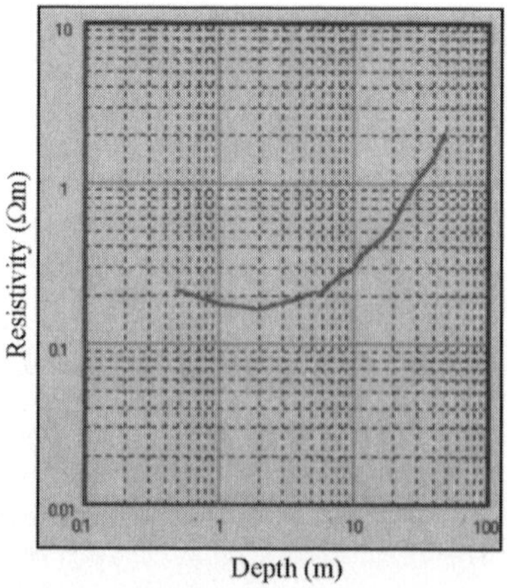

Figure 6: Variation of resistivity with depth

RESULTS AND DISCUSSION

The borehole logs indicated that a lower clay layer *Ac2* was found under a sand layer at a depth of about 15 m within the planned landfill area. In almost all the boreholes, the thickness of the clay layer *Ac2* was 1.5 - 3 m. At *Bor.No.5* and *Bor.No.7* however, the thickness of the clay layer *Ac2* was just about 1 m. For this reason, four more boreholes were drilled around *Bor.No.7*. All of them confirmed a thickness of 1 m. Due to the presence of extensive outcrops, no further boreholes could be drilled around *Bor. No. 5*. Using the filtered data from Figs. 7, 8 and 9, the resistivity of the lower clay layer *Ac2* ranged from 0.8 to 1.0 Ω-m. The results of lines 1 and 2 showed that the clay layer is continuous at a depth of about 15 m and has a thickness of 1 to 3.5 m.

Figures 7 and 8 again show the comparisons between the borehole and resistivity data of lines 1 and 2, respectively. The thickness of the clay layer *Ac2* was confirmed by the borehole and resistivity data. Along line 1, the four boreholes (*Bor.No.1, B-No.1, B-No.2* and *Bor.No.6*) showed that the clay layer *Ac2* occurs at a depth of about 15 m. It was found that the thickness of the clay layer *Ac2* ranged from 1.5 to 3.5 m. *Bor.No.1* showed that the layer was 1.5 m thick. However, the geophysical measurements indicated the layer to have a thickness of 4 m. *B-No.1, B-No.2* and *Bor.No.6* showed the layer's thickness to be 3.5 m, 2.5 m and 2.7 m, respectively. The geophysical survey results agreed with the thickness of the layer in *B-No.1, B-No.2* and *Bor-No.6* respectively.

Four boreholes (*Bor.No.12, Bor.No.9, Bor.No.8* and *Bor.No.7*) were drilled along line 2. The borehole logs showed that the clay layer *Ac2* was at a depth of about 15 m. The clay layer *Ac2* thickness ranged from 1 to 2.5 m. *Bor.No.7* showed that the clay layer *Ac2* had a thickness of only 1 m. All four boreholes that were drilled around *Bor.No.7* confirmed a thickness of 1 m. Borehole and resistivity data showed different results for lines 1 and 2. In spite of this, the measurements showed that a continuous clay layer *Ac2* is present at a depth of about 15 m. A comparison between the borehole and resistivity data of line 3 is shown in Fig. 9. There are three boreholes (*Bor.No.5, Bor.No.6* and *Bor.No.7*) along line 3. The borehole logs showed that the clay layer *Ac 2* was at a depth of about 15 m and has a thickness of 1 m at *Bor.No.5* and *Bor.No.7*. The drilling of additional boreholes around

Bor.No.5 could not be carried out because of the presence of big boulders in its neighbourhood. The thickness of the clay layer *Ac2* is about 1 m and 2.7 m at *Bor.No.5* and *Bor.No.6*, respectively. The geophysical measurements could not locate a clay layer along the first 250 m of line 3. From 250 to 400 m, however, clay layer *Ac2* was continuously located at a depth of 15 m. This could be explained by the fact that an electric power station discharges hot water into the sea. The discharge was close to the beginning of measured line 3. Since the conductivity of the sea water is dependent on the temperature of the sea, the resistivity data obtained might have been influenced by the hot water-sea water interactions. Figure 10 is a 3-dimensional section of lines 1, 2 and 3.

However, high conductivity of seawater served as a complete return path for the transmitted current detected at the potential electrodes P1 to P16. This property makes marine IP measurements very mobile. On the other hand, the high conductivity of seawater has an undesirable effect, because, in seawater, transmitted signals between two current electrodes form a short circuit. It was therefore necessary that the current electrode C_1 should have direct contact with the sea floor. If the current electrode C_1 were not to have such a contact, then the penetration depth into the seabed would be drastically reduced.

The electrodes P_8, P_9, P_{10}, P_{11} and P_{12} explored the seabed at a depth of approximately 15 m. The submarine geophysical measurements showed an accuracy of ± 1 m, due to the large intervals between the electrodes. In future investigations, the intervals between the electrodes could be reduced so that more measuring points can be taken. This would improve the accuracy of the results.

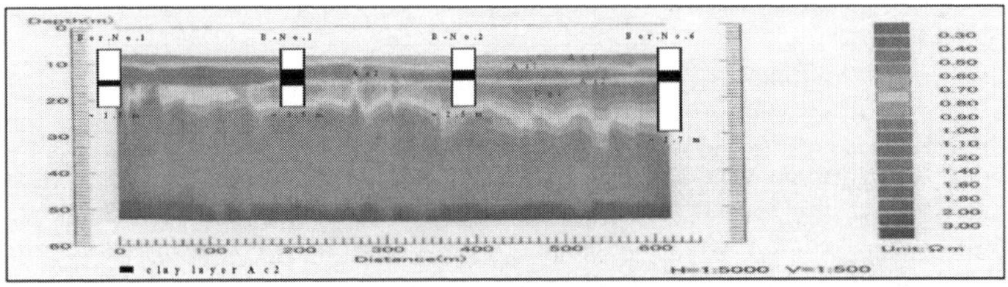

Figure 7: Comparison between the borehole and filtered resistivity data – Line1

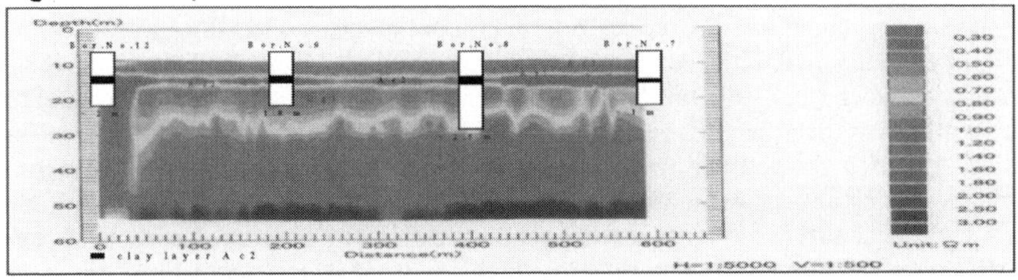

Figure 8: Comparison between the borehole and filtered resistivity data – Line 2

Test results showed that the hydraulic conductivity of the clay layer *Ac2* was 1.0E-06 cm/*s* which satisfes the requirement of k<1.0E-5 cm/s. The thickness of the clay was rather comparatively small (1-3 m). According to Bilitewski et al. (1997), if the natural soil liner in a landfill site is less than 3 m thick, it is necessary to install clay or synthetic layer of hydraulic conductivity of about 1.0E-7 cm/s. Comparing the standards in Japan with the result from the investigation, placing waste material on the site will cause reduction in thickness and the hydraulic conductivity.l. The equivalent velocity of leachate flow in 5 m thick clay with hydraulic conductivity of 1.0E-5 cm/s is equal to that with the clay of 0.5 m thick and hydraulic conductivity of 1.0E-6 cm/s. However, the time taken for the leachate to drain to the bottom part of the clay layer may be faster in the 0.5m thick clay but the quantity of leachate will be smaller due to the smaller hydraulic conductivity value.

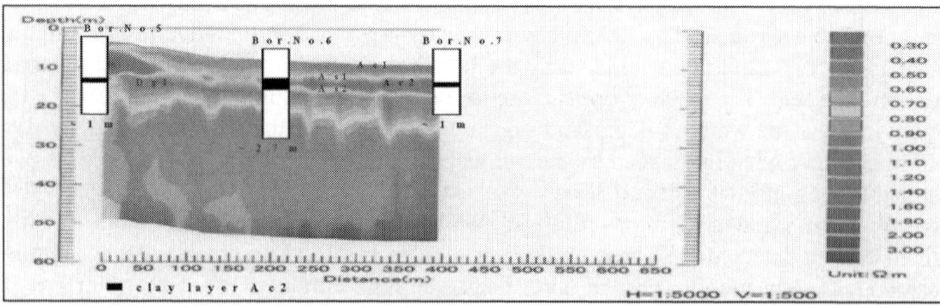

Figure 9: Comparison between the borehole and filtered resistivity data – Line 3

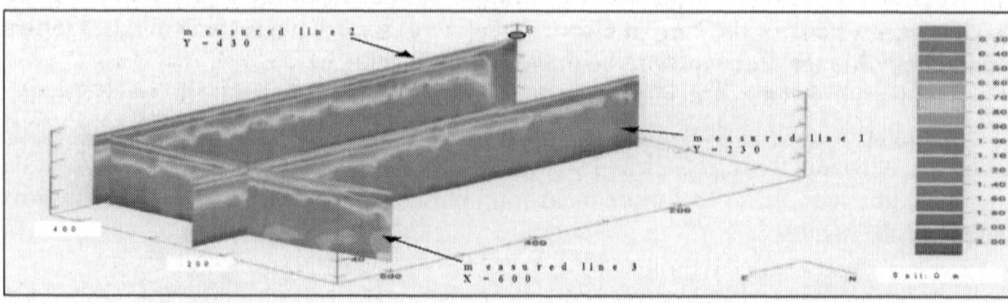

Figure 10: Three – dimensional illustration of measured lines 1, 2 and 3

CONCLUSION

Geophysical investigations were carried out along the coast of Okayama in order to study the characteristics of the seabed. The purpose of this research was to determine the morphology of the seabed; to locate a continuous clay layer that can serve as a good liner for the planned landfill site along the coast of Okayama.

The geophysical measurements agreed perfectly with the borehole logs and showed that there is a clay layer at about 15 m in the seabed; that is continuous and has thickness of 1-3 m. Tests showed that the hydraulic conductivity of the clay layer was 1.0E-06 cm/s.These properties of the seabed satisfy the minimum equivalent requirement of a 5-m clay layer and a hydraulic conductivity of 1.0E-05 cm/s, which is required for landfill liners in Japan. The results of this research showed that the Mizushima site could therefore be used as a marine landfill without posing any danger to the environment.

Unlike geophysical survey on land, marine geophysical survey uses advanced technology of GPS, streamers and equipment vessels. It is therefore rapid and is a reliable method for determining the morphology and other characteristics of the seabeds for geotechnical projects.

REFERENCES

Bilitewski B., Hardtle G., Marek K., Weissbach A., Boedddicker H. (1997). Waste Management, Springer Verlag, Berlin Heidelberg, Germany

Mizushima Disposal Facility, Okayama Prefecture, Company Brochure, Japan.

Okayama Prefectural Environmental Conservation Corporation (2003). Japan, (in Japanese).

Telford W.M., Geldart L.P., Sheriff R.E. (1990). *Applied Geophysics*, 2nd Edition, Cambridge University Press.

Todd, D.K. (1959). *Ground Water Hydrology*, John Wiley and Sons, Inc., USA.

Ward, S. H. (1990). *Geotechnical and Environmental Geophysics*, USA.

Mechanical Properties of Cohesive Soils Mixed with Carbide Possessing Water Absorption Property

Kiyoshi OMINE, Hidetoshi OCHIAI, Noriyuki YASUFUKU and Taizo KOBAYASHI
Department of Civil Engineering, Kyushu University, Fukuoka, Japan

Abstract
This study investigated the mechanical properties of cohesive soils mixed with carbide possessing water absorption property. The ability of charcoal to adsorb hexavalent chromium was also confirmed. The amount of Cr(IV) released from the cement-treated soil can be reduced by mixing charcoal. It was observed that the unconfined compressive strength of cement-treated soil increases by mixing charcoal or carbides prepared from waste materials. An improvement in the strength of cement-treated soil and the trafficability of volcanic cohesive soil by mixing water absorptive materials can be estimated from the reducing ratio of water content.

Introduction
Several types of cohesive soils are found in the Kyushu region of Japan. Although these soils are stabilized by adding hardening agents such as cement or lime for improving their mechanical properties, some soils are unsuitable for soil stabilization or require a large amount of hardening agents. This often causes problems such as an increase in the construction cost and leaching of hexavalent chromium from cement. A large amount of mown grass is generated from embankments and river areas, most of which is either burnt or disposed. A technique for carbonizing the compressed mown grass has been developed (Kuroki et al., 2005). Fowl droppings have also been carbonized by this technique. Since these carbides have a high capacity for water absorption, they can be used for water purification, soil improvement, as a deodorant, etc. The effect of immobilization due to charcoal mixing has also been investigated (Omine et al., 2005).

In this study, the improvement of the strength in cement-treated soils mixed with water absorptive materials is investigated by the unconfined compression test. The immobilization of hexavalent chromium in cement-treated soils mixed with charcoal is also investigated by the leaching test. Furthermore, the cone penetration test is performed on the volcanic cohesive soil mixed with various types of recycling materials, e.g., mown grass carbide, fowl droppings carbide, and used paper; thus, the improvement in trafficability is clarified. An evaluation method of the water absorption capability for the materials is discussed based on the test results.

Mixing Materials used

Manufacturing Methods
Mixing materials used in this study are charcoal, mown grass carbides (molded or uncompressed), fowl droppings carbides (pellet or powder), used paper (flocculated) and Toyoura sand. The shape and SEM micrographs of some water absorptive materials are shown in Figures 1 and 2. These materials are mixed with cohesive soil, and unconfined compression and cone penetration tests are performed on the mixtures. Table 1 shows the types of mixing materials used and the laboratory tests performed.

The used charcoal was prepared according to the following procedure. Disposable chopsticks were wrapped in an aluminum foil, and the charcoal was obtained by carbonizing for

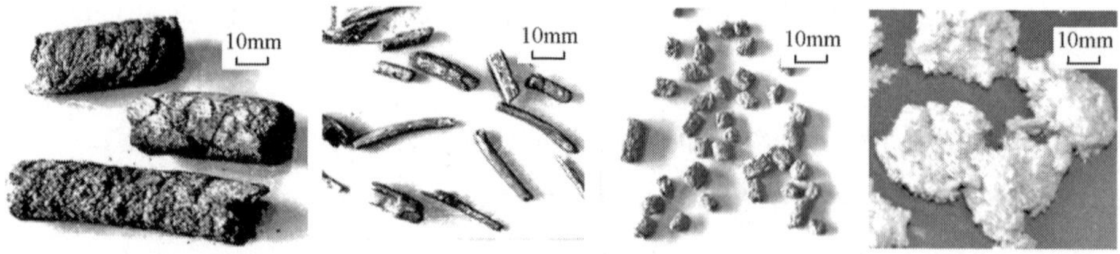

Mown grass carbide (molded) | Mown grass carbide (uncompressed) | Fowl droppings carbide (pellet) | Used paper (flocculated)

Figure 1: Shape of used water absorptive materials

Mown grass carbide (molded) | Mown grass carbide (uncompressed) | Fowl droppings carbide (pellet) | Used paper (flocculated)

Figure 2: SEM micrographs of the used water absorptive materials

Table 1: Types of mixing materials and laboratory tests performed

Types of mixing materials	Unconfined compression test	Cone penetration test
Charcoal	O	O
Mown grass carbide (molded)	O	O
Mown grass carbide (uncompressed)	O	O
Fowl droppings carbide (pellet)	O	-
Fowl droppings carbide (powder)	O	-
Used paper (flocculated)	-	O
Toyoura sand	O	O

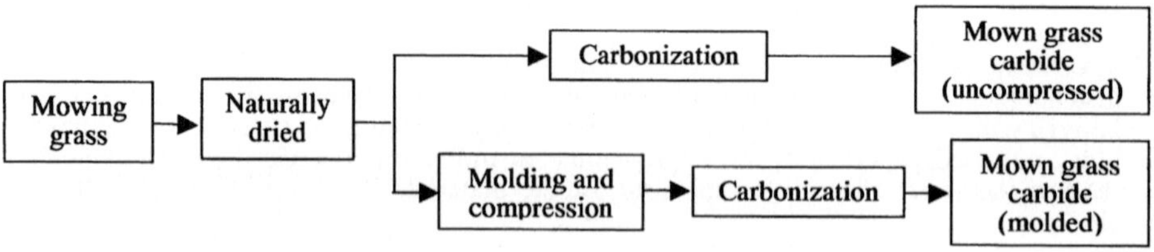

Figure 3: Flowchart of manufacture of mown grass carbides

approximately an hour in a table electric furnace at a temperature of 400°C. The obtained charcoal was crushed to a size of less than 2 mm.

Flowchart of the manufacture of the mown grass carbides is shown in Fig. 3. The mown grass carbide (molded) was obtained by the following procedure. Naturally dried mown grass was molded and compressed into a cylindrical form with a diameter of 2~3 cm and height of 1~10 cm. Carbide is then produced by carbonizing the compressed mown grass. Another type of mown grass carbide (uncompressed) is obtained by the procedure without mold and compression. These mown grass carbides crushed to less than 4.75 mm are used as mixing materials.

Fowl droppings carbide was obtained by carbonizing granulated fowl droppings. Although these are pellets, powdered fowl droppings carbide is also obtained in the manufacturing process.

Flocculated used paper was prepared according to the following procedure. Waste newspaper was mixed with water and unthreaded using a food processor. After it was dried in an oven, it was unthreaded again using the food processor, thereby obtaining flocculated paper. In addition, Toyoura sand was also used for comparison with other water absorptive materials.

Evaluation of Water Absorption Capability

In order to investigate the water absorption capability of the mixing materials, the water absorption ratio was measured by the following procedure: 1) a sample was submerged in water for a day, 2) it was then removed and the surface was wiped to remove water, and 3) its water content was measured after drying in an oven. The value thus obtained is defined as a maximum water absorption ratio. It is considered that the actual water absorption ratio in soil is lower than that in water. Table 2 shows the maximum water absorption ratio of the mixing materials. It is evident that the value for the maximum water absorption ratio of the mown grass carbide (uncompressed) is large, whereas the value for the mown grass carbide (molded) is small because of molding and compression.

It is considered that the water content of cohesive soil can be reduced by mixing it with a water absorptive material. The reducing ratio of the water content because of the mixing with the water absorptive material is defined as follows:

$$A = \left(\Delta m_w / m_{w0}\right) \times 100 \ (\%) \qquad (1)$$

where Δm_w is a change in the water content in a soil and represents the water absorption mass of the mixing material; further, m_{w0} is the initial mass of water in the soil. It is difficult to precisely measure the water absorption mass of the mixing material in a soil. It is assumed that the aspect of maximum water absorption capacity is introduced in a soil in the same manner as in water. Under such a condition, the reducing ratio of the water content for Eq.(1) is rewritten using a water absorption ratio W_{max}, content of mixing material M, and initial water content w_0 as follows:

$$A = \frac{W_{max} \times M}{w_0} = \frac{\left(\Delta m_w / m_c\right) \times \left(m_c / m_s\right)}{\left(m_{w0} / m_s\right)} \qquad (2)$$

where m_c is the mass of the mixing material and m_s is the mass of soil particles. Figure 4 shows the phase composition of a soil mixed with a water absorptive material.

Table 2: Water absorption ratio of mixing materials

Mixing materials	Maximum water absorption ratio (%)
Charcoal	188
Mown grass carbide (molded)	80
Mown grass carbide (uncompressed)	449
Fowl droppings carbide (Pellet)	82
Fowl droppings carbide (Powder)	101
Used paper (Flocculated)	677
Toyoura sand	21

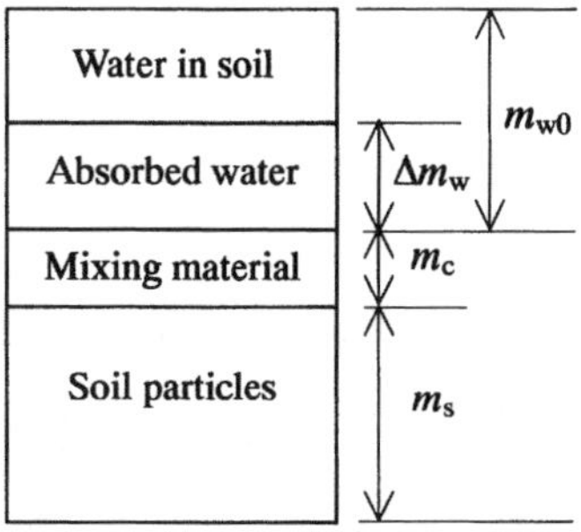

Figure 4: Phase composition of soil mixed with a water absorptive material

Improvement in Cement-treated Soil After Mixing Charcoal

Samples and Test Method

In order to investigate the fundamental improvements due to mixing carbide, unconfined compression and leaching tests were performed on cement-treated soils mixed with charcoal. Kaolin clay (I_P = 31.0, w_L = 50.6%) in 100% water content was used as a soil sample. A prescribed amount of charcoal was mixed with the clay in a slurry form for about 5 min followed by the addition of 200 kg/m^3 of Portland cement. The slurry was stirred for about 5 min. Specimens were prepared by pouring the slurry into a mold of diameter 50 mm and height 100 mm. They were cured at a temperature of 20°C in a humidity room for 7 or 28 days. The cement-treated soil mixed with Toyoura sand was also prepared under the same conditions. An unconfined compression test was then performed on the specimens.

For clarifying the adsorption effect of charcoal, the charcoal mixed cement-treated soil was crushed to a size of less than 2 mm. The leaching test was then performed according to Notification No. 46 of the Japan Environmental Agency. The test fluid was prepared with a solid-liquid ratio of 1:10 and agitated for 6 h to leach the contaminants in the specimen. The mass of the hexavalent chromium released from the cement was measured.

Test results

The relationship between the compressive stress and axial strain of the cement-treated soils for different charcoal contents is shown in Fig. 5. As indicated in this figure, two types of charcoal contents are defined, a dry weight ratio of charcoal and clay (%) and a dry weight of charcoal per unit volume of the specimen (kg/m^3). The maximum stress of the treated soil increases with an increase in the charcoal content, and the residual stress increases with the mixing of charcoal. Figure 6 shows the variation in the strength ratio R (=q_u/q_{u0}), where q_{u0} is the strength of the treated soil without charcoal mixing. The strength ratio increases with the charcoal content, and the increase in the curing time of 7 days is significantly higher than that for 28 days.

In order to clarify the improvement of strength by charcoal mixing, an unconfined compression test was also performed on the cement-treated soil with Toyoura sand. The test results are shown in Fig. 7. There is no increase in the strength of the cement-treated soils mixed with sand, because the water absorption capability of sand is very low as compared with that of charcoal. Another test result confirmed that there is no increase in the strength of the treated soils mixed with humid charcoal. From these test results, it is concluded that the apparent water-cement ratio in the treated soil decreases due to the absorption by charcoal, thereby improving its strength.

Figure 8 shows the total mass of hexavalent chromium Cr(IV) released during the leaching test on the treated soils mixed with charcoal. The concentration of Cr(IV) decreases with an

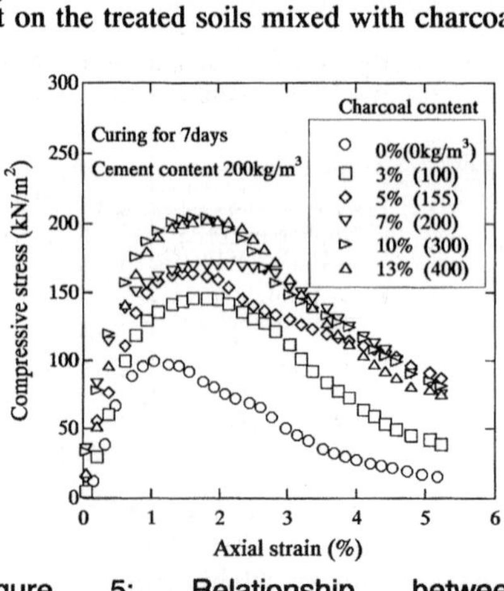

Figure 5: Relationship between compressive stress and axial strain of the charcoal mixed cement-treated soils

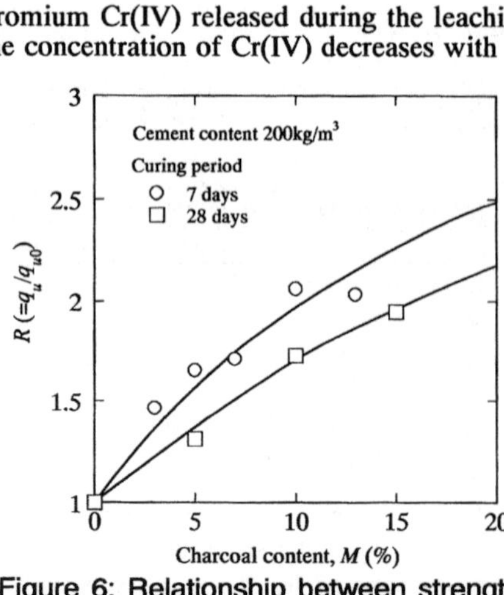

Figure 6: Relationship between strength ratio and charcoal content for different curing periods

increase in the charcoal content, and its value is below the value set by the environmental quality standards (0.05 mg/l). This suggests that the charcoal can possibly adsorb Cr(VI). Thus, the amount of Cr(IV) released from the cement-treated soil can be reduced by mixing charcoal.

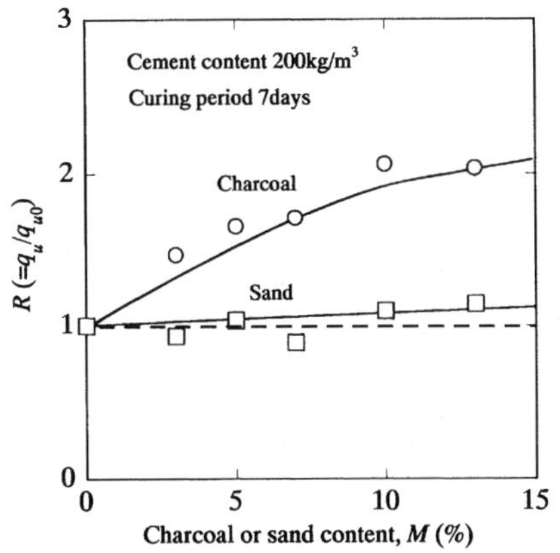

Figure 6: Relationship between the strength ratio and charcoal content after mixing charcoal or sand

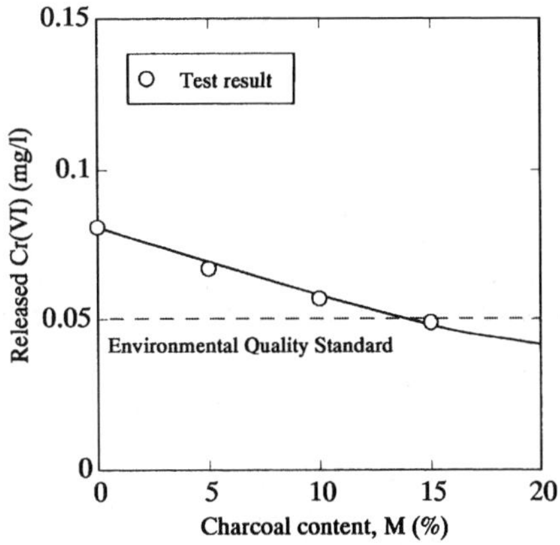

Figure 8: Relationship between the total mass of released hexavalent chromium and charcoal content

Improvement in Cement-treated Soil by Mixing Various Types of Carbides

Samples and Test Method

For investigating the improvement in the strength of cement-treated soil by mixing various kinds of carbides, two types of mown grass carbides (molded or uncompressed) and two types of fowl droppings carbides (pellet or powder) are used as mixing materials. Cement-treated soils mixed with these carbides were prepared using the abovementioned method. An unconfined compression test was then performed on the specimens.

Test results

Figure 9 (a) and (b) show the relationship between the unconfined compressive strength and carbide content for the curing periods of 7 and 28 days, respectively. The largest value of the strength can be observed for the treated soil with the mown grass carbide (uncompressed)

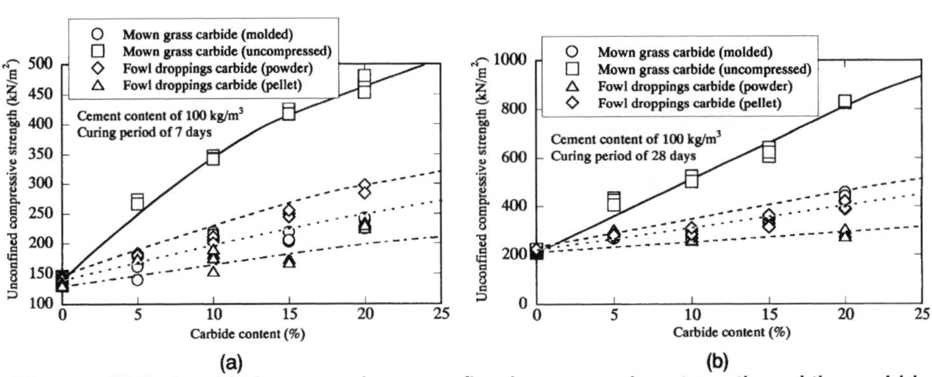

Figure 9: Relationship between the unconfined compressive strength and the carbide content

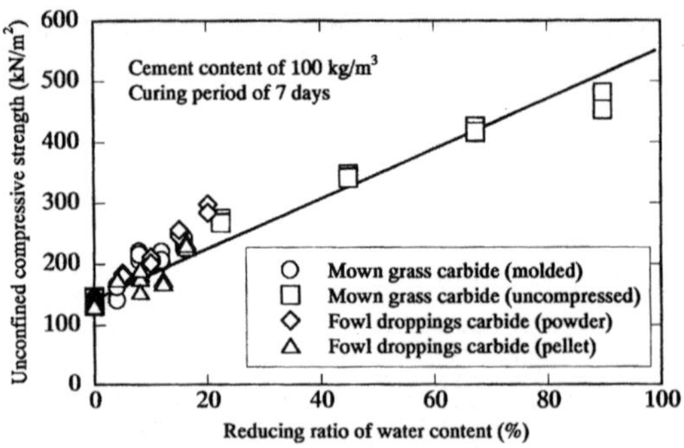

(b)

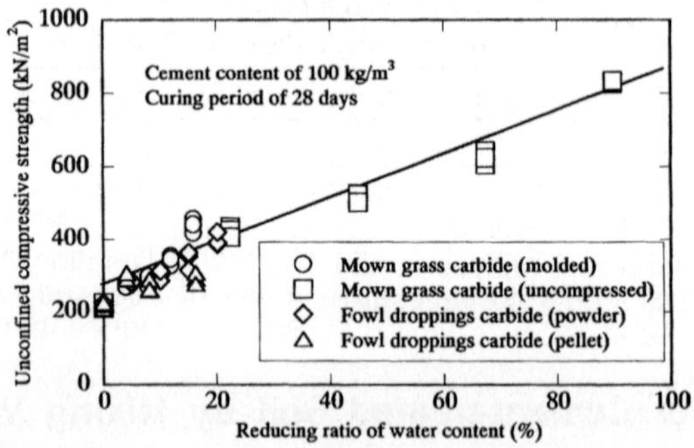

(a)

Figure 10: Relationship between the unconfined compressive strength and reducing ratio of the water content

having an identical carbide content that is independent of the curing period. It is considered that the large improvement in the strength of the treated soil by mixing it with the mown grass carbide can be accounted by its high water absorption capacity.

As shown in figure 9, it is seen that an improvement in the strength of the treated soil is closely connected to the water absorption capability of the carbide. Figure 10 (a) and (b) show the relationship between the unconfined compressive strength and reducing ratio of the water content for the curing periods of 7 and 28 days, respectively. The reducing ratio of the water content is calculated by Eq. (2). The strength of the treated soils with various types of carbides has a linear relation with the reducing ratio of the water content. Thus, the effect of an improvement in the strength of the treated soils by mixing the carbide can be estimated from the reducing ratio of the water content.

Improvement in Trafficability of Volcanic Cohesive Soil by Mixing Water Absorptive Materials

Samples and Test Method

Volcanic cohesive soil (loam) was used; this is a surplus soil at construction sites in the Oita prefecture, Kyushu, Japan. The physical properties of the volcanic cohesive soil are shown in Table 3. The cone penetration test was performed because it exhibits insufficient trafficability. Based on the standards of the Japanese Geotechnical Society, the specimens were prepared by the compaction of stabilized soil. The conditions for the cone penetration test are listed in Table 4. Charcoal, mown grass carbide, used paper, and Toyoura sand were used as mixing materials. These materials were mixed with a soil sample and the resulting mixture was cured for a day. The mixing content is represented in kg/m^3, where kg is the unit of mass of the mixing material and m^3 is calculated from a wet unit weight of in-situ original soil.

Table 3: Physical properties of volcanic cohesive soil

Soil	Gradation (%)			ρ_s (g/cm³)	W_L (%)	W_P (%)	I_P	I_L	W_n (%)	ρ_t (g/cm³)
	Coarse	Silt	Clay							
Volcanic cohesive soil (Loam)	8	51	41	2.81	119.4	60.2	59.2	0.88	112.5	1.33

Table 4: Conditions for the cone penetration test

Soil sample	Volcanic cohesive soil (less than 4.75mm)
Mixing materials	Charcoal, Mown grass carbides, Used paper or Toyoura sand
Mold	Height of 10cm, Volume of 1000cm³
Rammer	Weight of 2.5kg, Falling height of 30cm
Curing	Mixing and compaction the day before, Cone test the next day
Cone penetrometer	Point angle of 30°, Base cross section of 3.24cm²
Penetration speed	1cm/s
Cone index	Measurment of resistance force at penetration of 5, 7.5 and 10cm, and calculation of the average value devided by the cross section

Test results

The relationship between the cone index and content of mixing material in the volcanic cohesive soil mixed with various kinds of water absorptive materials is shown in Fig. 11. With the exception of sand that has a low water absorption capability, the cone index of the volcanic cohesive soil increases with the mixing material content; this improve trafficability. At the road construction site, a swamp bulldozer was used as a surface compaction machine, and the target value of the cone index in the laboratory mixing test was $q_c = 400$ kN/m² (after considering safety factor). The required content of the mixing material is 50 kg/m³ for the used paper and mown grass carbide (uncompressed), 100 kg/m³ for the charcoal, and 150 kg/m³ for the mown grass carbide (molded). This indicates that the improvement in trafficability is more significant for a mixing material with high water absorption capability.

Figure 12 shows the relationship between the cone index and reducing ratio of the water content of the volcanic cohesive soil. The reducing ratio of the water content is calculated by Eq. (2) as indicated in the previous chapter. The cone index and reducing ratio of the water content has a linear relation that is independent of the type of mixing material. Therefore, the

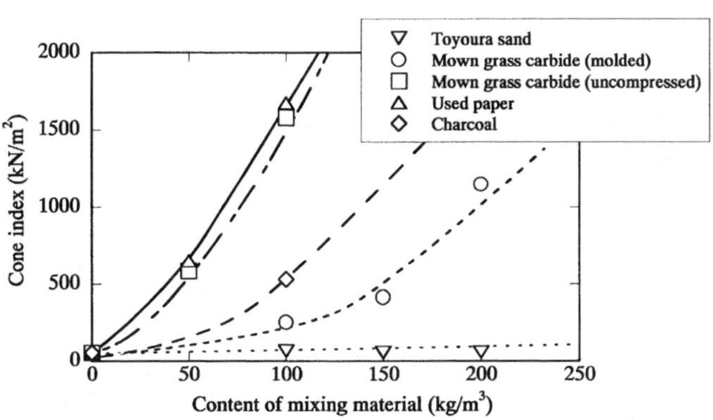

Figure 11: Relationship between the cone index and content of mixing material

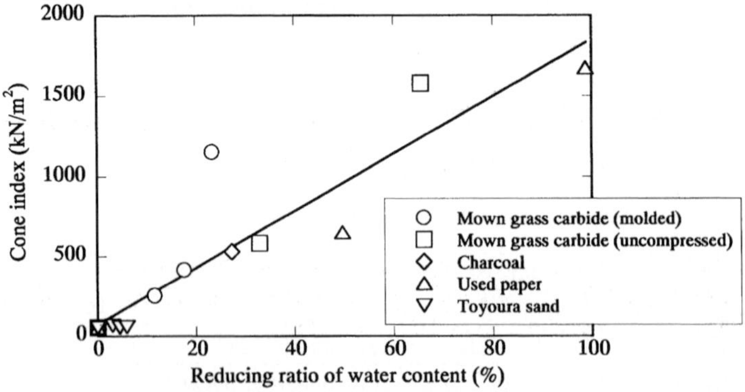

Figure 12: Relationship between the cone index and reducing ratio of the water content

cone index of the cohesive soil after mixing the water absorptive material can be estimated from the reducing ratio of the water content.

As indicated in Figures 10 and 12, the improvement in the strength and trafficability of the cohesive soil by mixing water absorptive materials such as carbides can be evaluated by introducing a new parameter for the reducing ratio of the water content.

Conclusions

In this study, the improvement in the strength of cement-treated soil and trafficability of volcanic cohesive soils by mixing water absorptive materials such as charcoal, carbides and used paper are investigated by performing unconfined compression and cone penetration tests. The property of charcoal to adsorb the hexavalent chromium is also confirmed. The main conclusions are summarized as follows:

1) It is observed that the unconfined compressive strength of cement-treated soil increases by mixing it with charcoal or carbide that exhibits a high water absorption capability. Furthermore, the amount of Cr(IV) released from the cement-treated soil can be reduced by mixing the charcoal.
2) The cone index of the volcanic cohesive soil (loam) that exhibits insufficient trafficability can be increased by mixing the water absorptive materials such as carbides or used paper.
3) The improvement in the strength of cement-treated soil and trafficability of volcanic cohesive soil by mixing water absorptive materials can be estimated from the reducing ratio of the water content.

Acknowledgements

The authors would like to thank Mr. Keiichirou Kuroki of Kyushu University for his cooperation in carrying out the laboratory tests as well as Kurimoto Ltd. for offering the carbides used in the test.

References

Kuroki, K., Ochiai, H., Yasufuku, N., Omine, K., Kobayashi, T. and Shiozaki, M. (2005): Improvement Effect of Trafficability of Volcanic Cohesive Soils by Mixing Carbide of Mown Grass, Proc. of 6th Geo-environmental Engineering Symposium, JGS, 91-96 (in Japanese).

Omine, K., Ochiai, H. and Yasufuku, N. (2005): Development of Geomaterials with Various Immobilisation Treatments for Heavy Metals and Evaluation of Environmental Impact, International Conference on Stabilisation/Solidification Treatment and Remediation, Balkema, 335-345.

Recycled and alternative materials in drainage systems of waste disposal areas

Ennio Marques Palmeira, University of Brasília, Brasília, Brazil
Antônio Rafael Leite Silva, State University of Maranhão, São Luis, Brazil
Haroldo Paranhos, Prodesivo S.A, Brasília, Brazil

Abstract

Drainage systems may involve the exploitation and movement of large amount of natural materials, which may have important impacts on the environment. On the other hand, a considerable volume of different waste items that might be better used are disposed in landfills and dumps on a daily basis, reducing the space available for the disposal of more hazardous materials. This work deals with the evaluation of the potentials of some these materials (PET bottles and tire shreds) found in the waste as well as processed rubble aggregate as drainage materials associated with geotextile filters for geotechnical and environmental protection works. The results obtained from large scale tests allowed the evaluation of relevant properties of these materials and showed that they may substitute natural sands and gravels in drainage systems, particularly those of waste disposal areas.

Introduction

Recycling of materials can save natural resources and minimise hazards to the environment being a common practice in several areas in developed countries, particularly for rubber and plastic materials. When disposed in a landfill these materials occupy space that might be more efficiently used to store waste and favour the occurrence of fires and proliferation of diseases. However, materials such as plastic waste, rubble and tires can be reused in a more appropriate and environmentally friendly way. Some of these uses are as drainage layers in substitution to natural gravels and sands.

In spite of the low cost of some recycled materials in comparison to natural ones, to be properly used as a construction materials in routine works they have to attend technical and economical requirements. Some of these requirements are: durability, not to be hazardous to the environment and to have appropriate mechanical properties. Because recycled materials are still little used as routine construction works there are few standardise design methodologies or testing procedures for such materials. However, the use of recycled materials in civil and environmental engineering will certainly continue to increase markedly in the coming years, demanding the development of design methods, codes and testing methodologies to allow for better utilizations of these materials.

Some recycled materials that can be used to substitute soils in drainage systems are PET bottles, tires (shreds or wholly) and processed rubble. PET bottles can be compressed resulting in a coarse granular like material. The same applies to whole tires or tire shreds.

Rubble, if processed accordingly, can result in a highly permeable aggregate. Alternative drainage materials can be combined to geotextiles, with the latter functioning as filters.

This paper describes the use of some of these materials in drainage systems of geotechnical works and waste disposal areas in combination with geotextile filters.

Materials and testing methodology

Materials

The materials tested in this work were compressed PET bottles, tire shreds and aggregate from processed rubble. Table 1 summarises the main characteristics of the materials tested and Figures 1 (a) to (c) show photographic views of these materials. Both the PET bottles and the tires were collected at the Jockey Club dump, which receives most of the domestic waste produced by the city of Brasilia, Brazil. Works are being planned for the dump rehabilitation and the potentials of the use of low cost alternative materials are being considered.

Table 1. Characteristics of the materials tested.

Property	Material		
	PET bottles	Tire shreds	Rubble aggregate
Shape	round	plate like	gravel like
D_{50} (mm)	95	37	21
D_{90} (mm)	95	45	35
CU	1.0	1.5	12.5
Density (kg/m^3)	92	367	1113
Void ratio	2.11	2.19	0.95

Notes: (1) D_{50} and D_{90} = particle diameters for which 50% and 90% of the remaining particles are smaller than those values, respectively;
(2) CU = coefficient of uniformity (D_{60}/D_{10})

(a) PET bottles. (b) Rubble aggregate. (c) Tire shreds.

Figure 1. Photographic views of the materials tested.

The compression of a PET bottle alters its original shape to something similar to a coarse aggregate. Initially the bottles were bathed in water at 100 °C to make then softer. After that a light compression was enough to produce its final shape (Fig. 1a). The tire shreds are produced using special mechanical processes (cracker mills and granulators, for instance). The aggregates from rubble were processed by crushing and sieving to attend specific grain size dimensions, similar to those of natural gravel, for instance. Elements of this aggregate consisted of concrete grains. It is estimated that the city of Brasilia alone produces approximately 85000 t/month (Zordan 2002) of rubble that if appropriately used might yield in significant cost savings in civil and environmental protection projects.

Tests performed

Hydraulic conductivity and compressibility tests were carried out to evaluate the performance of the alternative construction materials studied. Because of the dimensions of the materials investigated, large scale testing devices were developed, as shown in Figures 2 and 3. The large scale permeameter comprised a rigid cylindrical cell were a sample 1.2m diameter and 1.6m high could be accommodated (Fig. 2). The tests were performed under constant hydraulic head conditions and under a total hydraulic gradient of 0.2. For the compressibility tests the samples were prepared in a large and rigid cylindrical cell (Fig. 3). The sample in this test was 0.7m diameter and 0.4m high. A rigid metal platen transferred the vertical load from a hydraulic cylinder to the sample. A load cell and a dial gauge measured vertical loads and platen vertical displacements for each loading stage during the tests. Tests with normal stresses up to 400 kPa on the sample top were performed.

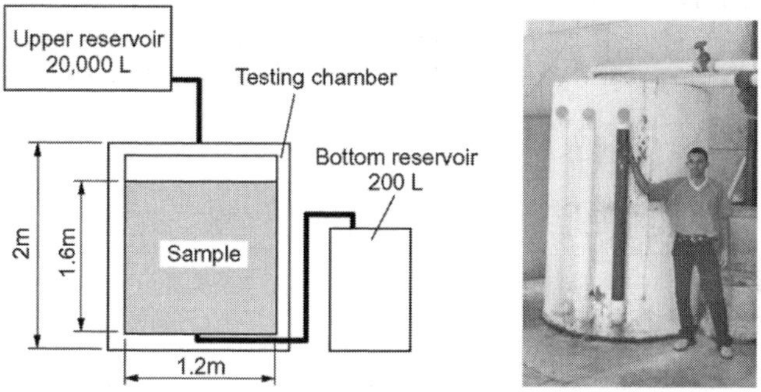

Figure 2. Large scale permeameter for hydraulic conductivity tests.

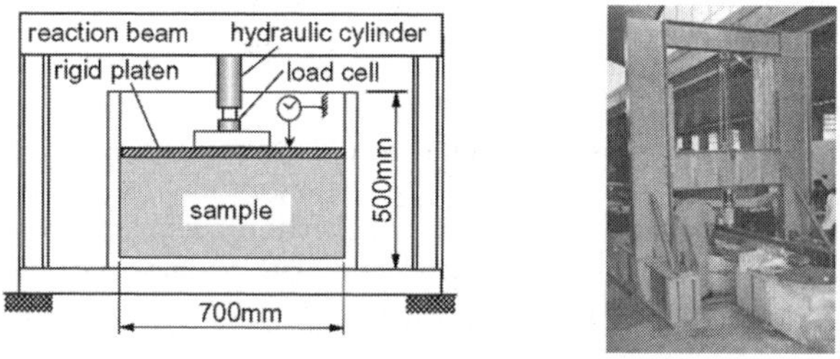

Figure 3. Compressibility tests cell.

The behaviour of the drainage systems made of alternative materials incorporating geotextile filters was investigated by using large rigid containers with domestic waste overlying those systems. Two containers (C1 and C2) were manufactured for this purpose and their general characteristics are schematically presented in Figure 4. The dimensions of the containers are 3.5m (length) x 2.5 (width) x 1.0 m (height). Along the largest side of the container there is a lateral trench that can be filled with the drainage material and which conducts the leachate to reservoirs adjacent to the containers. In the middle of the same side a 1.4m high glass window allows views of the drainage system during the tests. A set of perforated pipes at the container's top allowed water to be uniformly spread on the waste mass surface, simulating precipitation under controlled conditions. In container C1 the drainage system consisted only of the drainage trench filled with the rubble aggregate and a layer of geotextile covering the entire plan area of the container (Fig. 4a). The drainage system in container C2 consisted of a

0.1m thick layer of tire shreds underlying a geotextile filter (Fig. 4b). In this case the lateral drainage trench was also filled with tire shreds. The geotextile filter used in both containers was a needle-punched non woven geotextile, made of polyester, 2.3mm thick and with a mass per unit area of 200g/m^2.

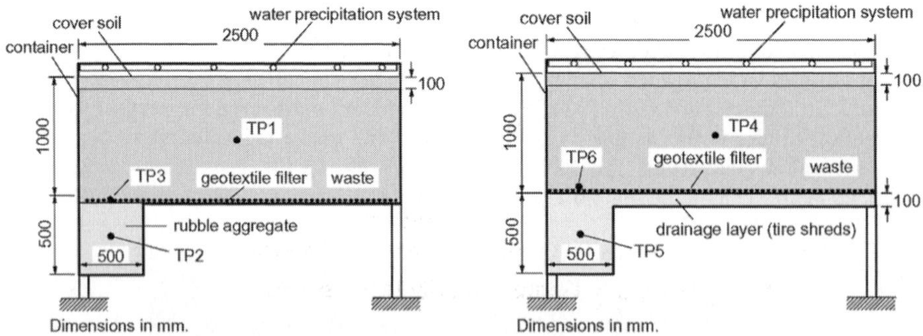

(a) Container C1. (b) Container C2.

Figure 4. Waste containers.

The domestic waste employed in the containers had the following composition: organic matter (44%), paper (16%), cardboard (10%), plastic (16%), metal (4%) and others (10%). The initial waste moisture content was equal to 58%. The total mass of waste to be disposed in the containers was initially mixed for homogenisation with each container having received approximately 2.5 tons of waste with an initial height of 0.9m. The waste was disposed in a loose state with an average mass per unit weight of 4 kN/m^3. A 0.1m thick clayey cover soil was placed on top of the waste layer. Seventy five days after waste disposal in the containers water started to be spread on the waste mass to increase its moisture content and favour biological degradation. During wet seasons water intakes occurred at a rate of 10L/week whereas during dry season this rate was equal to 2.5L/week. Water was added to the waste mass in a single moistening stage each week. Instrumentation installed in the containers allowed the measurement of waste settlement, temperatures (temperature transducers) in the middle of the waste mass, below the geotextile filter and inside the drainage trench (TP1 to TP6 in Figs. 4a and b) as well as the measurement of leachate flow rate variations with time. Leachate samples were collected with time for chemical analysis such as pH, nitrate content, ammonium content, chemical oxygen demand, sulphate content and solids contents.

Additional information on materials and testing methodology can be found in Paranhos (2002) and Silva (2004).

Results obtained

Figure 5 shows results of compressibility tests on the materials studied. As expected, the loose rubble aggregate was the least compressible material even under an initial loose state, followed by the compressed PET bottles and the tire shreds. The rubble aggregate was also the material that presented the largest plastic strain after unloading, whereas the tire shreds was the one with lowest relative plastic strains after testing. The average oedometric modulus values within the normal stress range between 0 and 100 kPa were equal to 1090 kPa for the rubble aggregate, 333 kPa for the compressed PET bottles and 235 kPa for the tire shreds for the initial conditions of each sample presented in Table 1. These results show that drains made of compressed PET bottles or tire shreds are significantly compressible and this aspect may have to be taken into account depending on the characteristics of the project. The low oedometric modulus for the rubble aggregate is a consequence of its loose state and breakage of particles. This also limits the utilization of this material to situations were particle breakage is not relevant or to applications under low stress levels.

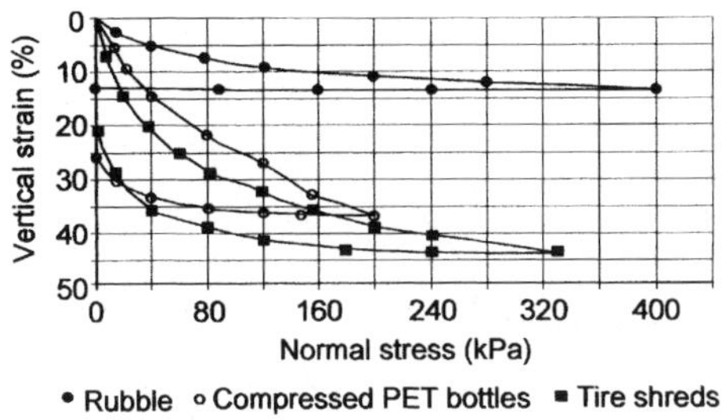

Figure 5. Compressibility tests results.

Large scale hydraulic conductivity tests yielded the values presented in Table 3. The results obtained showed high values of hydraulic conductivity, particularly for the compressed PET bottles, that was 3.3 and 5 times more permeable than the tire shreds and the rubble aggregate, respectively.

Table 3. Values of coefficients of hydraulic conductivity.

Material	Hydraulic conductivity (cm/s)
Rubble aggregate	10
Tire shreds	15
Compressed PET bottles	50

Regarding the behaviour of the waste in the containers, Figure 6 presents the cumulative volumes of leachate released and precipitation versus time in these tests. During the first stages of the test when no water was added liberation of leachate was only observed for the system with the tire shreds (container C2). The lack of leachate liberation from container C1 can be in part due to relatively low transmissivity of the geotextile layer at its bottom, which delayed the flow of leachate towards the drainage trench. On the other hand, container C1 liberated leachate immediately with the cumulative volume of leachate tending to stabilisation 70 days after waste disposal. After precipitation has started both drainage systems responded immediately with rates of leachate liberation similar to the rate of water intake.

The variations of waste temperature and temperatures in the drainage trenches of containers C1 and C2 with time are presented in Figures 7(a) and (b). Temperatures in the waste mass just after waste disposal in containers C1 and C2 were equal to 42 °C and 43 °C, respectively, and considerably above the environmental temperature (25 °C, T_{env}, if Fig. 7), as show in Figure 7(a). During most of the observation period the temperature of the waste in container C2 was greater than that in container C1. After 390 days the temperatures in the waste of both containers was similar to the environment temperature. Temperature in the drainage trenches in both containers also started with values significantly above the environment temperature, ranging from 32 °C to 37 °C, depending on the container considered. After approximately 200 days the temperatures in the drainage trenches were close to the environment temperature. During most of the first 200 days of observation the temperature in the tire shreds trench was slightly greater than that in the rubble aggregate trench. This may have been caused by the faster release of leachate from container C2.

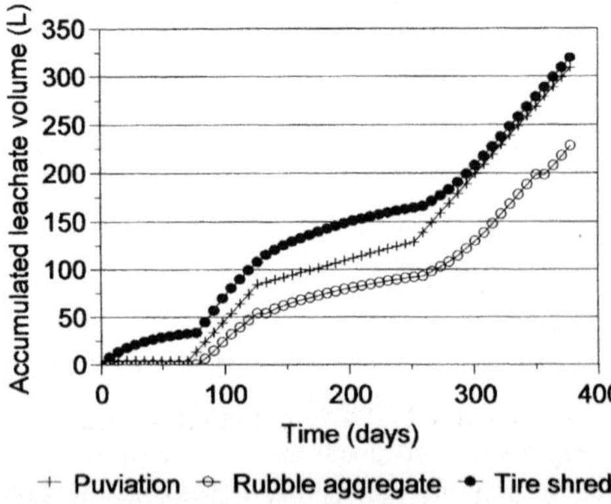

Figure 6. Cumulative leachate liberation with time.

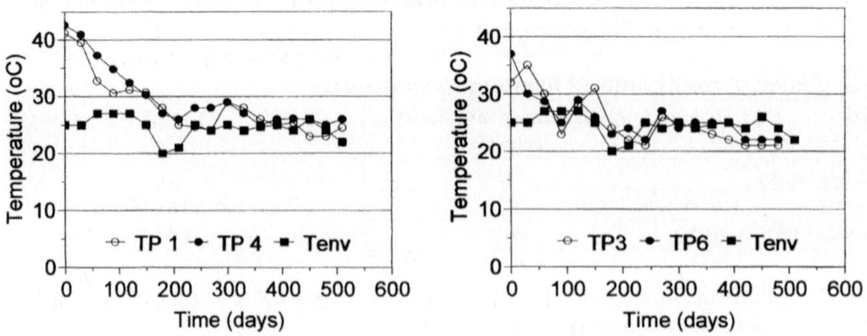

(a) Waste mass.

(b) Drainage trench.

Figure 7. Temperature measurements versus time.

Figures 8(a) and (b) shows the variations of leachate chemical oxygen demand (COD) and pH with time. This figure also shows the rate precipitation during the testing period. The COD values obtained for the leachate from C1 was considerably higher than those obtained for the leachate from C2. This difference can be a result of several factors, such as interaction between drainage material and leachate, better filtering action of the aggregate and the delay in the liberation of the leachate by the aggregate drain in comparison to the tire shreds drain. Silva (2004) also observed higher COD values in the leachate from a large test waste cell with tire shreds drain than those from another test cell with a drainage system consisting of natural gravel. An initial pH value of 8.5 of the leachate from C1 can be observed, which is likely to be a consequence of the interaction of the leachate and the concrete particles of the drainage system of this container. A pH value of 8.7 was observed in aggregate particles submerged in clean water. As time passed, the values of pH dropped ending at a value close to 7 (neutral condition) in C1. This continuous reduction in pH can be a result of the aggregate particles having been covered by a film of leachate which minimised the direct interaction between concrete and leachate. pH values started within the acid range (between 5 and 6) in the early stages of the test in container C2 and also ended close to the neutral value.

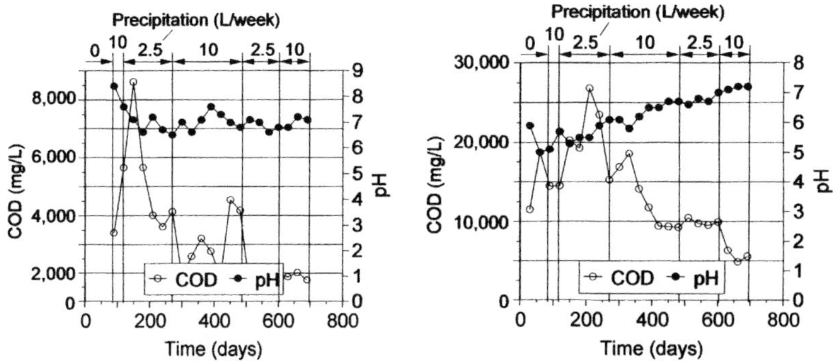

(a) Container C1 (b) Container C2

Figure 8. COD and pH versus time.

Figure 9 presents the results of total, suspended and dissolved solids obtained in the leachates released by containers C1 and C2 with time. The amount of total solids in the leachate from C1 showed a marked decrease with time, associated with the beginning of precipitation. On the other hand, the amount of total solids in the leachate from C2 oscillated between 13000 mg/L and 20000 mg/L during the observation period, with little influence of precipitation. These results also suggest the better filtering action of the drainage system of container C1 in comparison to C2. Figure 10 shows photographs of the leachate released by containers C1 and C2 and corroborate with the conclusions that the leachate from C1 had less organic matter and solids than that from C2.

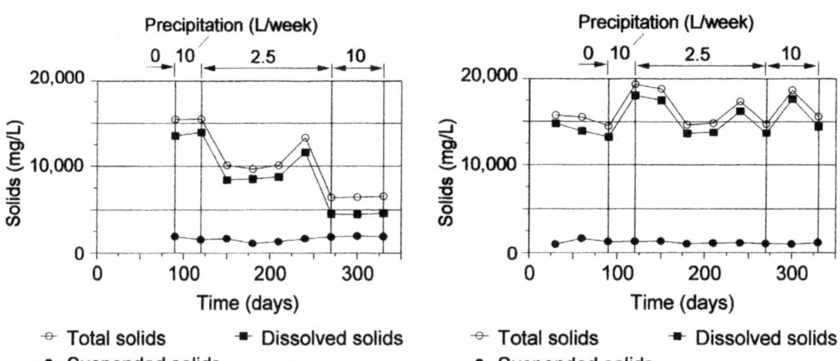

(a) Container C1 (b) Container C2

Figure 9. Solids in the leachates.

The promising results obtained in the test programme reported in this work have already led to some actions regarding the use of tires, tire shreds and compressed PET bottles as drainage systems in geotechnical works and in landfills. Figure 11(a) shows a horizontal drain consisting of tires enveloped by a geotextile filter in the Jockey Club dump in Brasilia, Brazil. Figure 11(b) shows a drain consisting of compressed PET bottles enveloped by a geotextile layer.

Figure 10. Leachate samples from containers C1 (left) and C2 (right).

(a) Horizontal drain consisting of tires enveloped by a geotextile

(b) "Sausage" of compressed PET bottles in a geotextile cover.

Figure 11. Applications of alternative materials and geotextiles as drainage systems.

Conclusions

This paper presented different types of tests to evaluate the potential use of alternative construction materials as drainage systems of waste disposal areas. Large scale hydraulic conductivity tests showed that the rubble aggregate, compressed PET bottles and tire shreds tested highly permeable, compatible with the usual requirements for drainage materials of geotechnical and geoenvironmental works. Loading tests evidenced the large compressibilities of the compressed PET bottles and of the tire shreds in comparison to the rubble aggregate. Thus, the former materials should not be used in applications were significant settlements are not allowed, which is not the case of usual waste disposal facilities.

The tests in the large waste containers showed that the type of material used in the drainage system affected the properties of the leachate released and that the geotextile filters used functioned well. For the duration of the experiments, the drainage materials and the geotextile filter presented a satisfactory performance. Some visually small level of degradation was observed in the rubble aggregate grains. Further investigations are required to better assess the long term behaviour of these materials under hazardous working conditions.

References

Paranhos, H. (2002). Alternative low cost drainage systems for geotechnical and geoenvironmental works. Research Report-RHAE/CNPq/UnB Programme, Graduate Programme of Geotechnics, University of Brasilia, Brasilia, Brazil, 79 p. (in Portuguese).

Silva, A.R.L. (2004). A study on the behaviour of drainage systems of waste disposal areas under different scales. DSc. Thesis, University of Brasilia, Brazil, 329 p. (in Portuguese)

Zordan, S.E. (2002). Rubble of the civil construction industry. In www.reciclagem.pcc.usp.br /entulho_ind_ccivil.htm (in Portuguese).

Preliminary static and seismic slope stability analyses for a Portuguese municipal solid waste landfill

F. Pardo de Santayana, LNEC, Lisbon, Portugal
A. J. Roque, LNEC, Lisbon, Portugal

Abstract

Gaeiras Landfill, in central Portugal, occupies an area of about 7 ha and received approximately 500 000 m^3 of municipal solid wastes from 1984 to 2001. Landfill operation finished in 2002. The landfill presents some problems related to excessive leachate volumes and erosion and slippage of the cover system. Besides, it is located at a short distance upstream from the perimeter of a future reservoir. In 2003, a preliminary evaluation of the landfill global slope stability was made, considering the risks of a waste mass slide as a consequence of the filling of the reservoir. A description of the works and calculations carried out is included in the paper. This case history arouses the necessity of establishing appropriate procedures to analyse the static and seismic stability of landfills in Portugal.

Introduction

Gaeiras Landfill occupies an area of 7 ha and contains about 500 000 m^3 of municipal solid wastes (MSW) deposited from 1984 to 2001. Landfill operation finished in 2002. Soon after closure some problems appeared related to an excessive production of leachate and a surface slippage in the South slope. The landfill closure design report only included a simple stability analysis, admitting very simplified landfill geometry and calculation procedures. Because a dam was being constructed in the area, with the reservoir perimeter located a few ten meters apart downstream from the landfill, in 2003 the Laboratório Nacional de Engenharia Civil (LNEC) was requested by the owner for carrying out a preliminary and urgent evaluation of the global slope stability in relation to the risk of a slide involving the mass of wastes deposited in the landfill. Works carried out by the LNEC included compilation and analysis of existing information about the landfill, site inspections, a first evaluation of the main geotechnical problems shown by the landfill and preliminary static and seismic stability analyses for the landfill. The study has disclosed some shortages in the way design stability analyses for landfills are carried out in Portugal.

Brief description of the landfill

Different documentation pieces about Gaeiras Landfill were analysed, including the following: the 1982's landfill original design; a preliminary study of the soil contamination around the landfill made in 2000; the 2001's landfill closure design; a geological and hydro-geological study made to determine the causes of the excessive volumes of landfill leachate, made in 2002; design documentation of the neighbouring dam, under construction; and a topographical survey map of the landfill carried out in 2003. As there was no possibility, within the limits of this study, to drill bore-holes or perform in situ tests, the geometrical configuration of the landfill, specifically the thickness of waste, was determined on the basis of the above-mentioned documents, which permitted also to locate excavation and filling zones and thicknesses under the landfill base. The base of the landfill, that occupies a 260 m long and 330 m wide square area, was divided in four cells, according to the available information, and was not provided with a bottom lining system, but only with a net of collection trenches to conduct the leachate out of the area.

Closure works carried out in 2002 left the top of the landfill as a relatively flat platform (minimum slope of 2%), between elevations 65 m, at the South side, and 70 m, at the North side, with up to 1V:3H side slopes (Fig. 1). Only the South slope is provided with a mid-height bench. The maximum thickness of waste is about 20 m. The cover system includes the following components, from top to bottom: a 0.3 m thick topsoil layer; a 0.7 m thick soil cover layer; a 9.6 mm thick geocomposite drain, with a weight of 500 g/m^2; a geosynthetic clay layer (GCL); and an approximately 0.15 m thick soil foundation layer placed directly on the regraded waste mass. A new leachate collection system was constructed during the closure works consisting of a 2 m deep collection trench around the perimeter of the landfill (except at the North side) and two leachate retention tanks at the South limit of the landfill.

Visual inspection to the landfill showed some problems at its surface: a superficial 12 m wide slippage at the central section of the South slope, between the mid-height bench and the toe, mobilizing the cover soil layers above the geocomposite drain; surface erosion signs, particularly at the lower half of the South slope; several leachate exfiltration points, also at the South slope; leachate leaks at one of the collection tanks located at the South limit of the landfill; erosion gully at the earth fill slope situated downwards from the South slope.

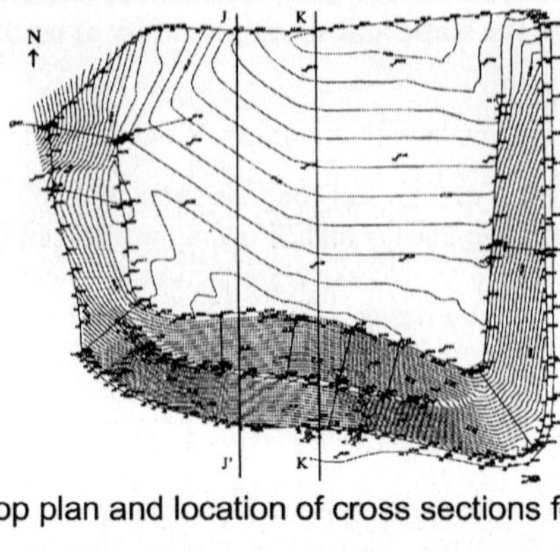

Figure 1. Landfill top plan and location of cross sections for stability analyses

Geological and hydrogeological characteristics of the site

Gaeiras landfill was constructed on top of Jurassic formations consisting of a highly heterogeneous siltstone-clay complex. Geomorphologically, the landfill is located on a relatively gentle slope at the right bank of Arnoia River, between elevations 45 and 75 m. The landfill construction design admitted that there was no relevant aquifer at the site. The elevation of the base of the different cells was fixed in order to place the landfill bottom over clayey formations as much as possible. No bottom lining system was provided, and only a net of leachate drainage trenches was constructed, connected to a collection tank discharging directly to the closest watercourse.

The 2001's design report for final closure admitted that after placement of the cover system, the volume of leachate would diminish substantially in time. The landfill was supposed to be located on top of a low permeability hydrogeological formation. Leachate exfiltration points were observed along the landfill lateral slope toes. Therefore, the above-mentioned 2 m depth perimeter drainage trench was constructed to collect the leachate and conduct it to a 50 m^3 storage tank, anticipating a production of 30 m^3/day. However, since a higher production of leachate was observed during the works, two collection tanks, with a total capacity of 120 m^3, were actually built to avoid discharges to the watercourse. Even so, after the works, the installed capacity of the tanks showed to be insufficient for the real leachate production. The level of leachate inside the landfill showed to be also very high (at 3 to 5 m of depth below the landfill top surface).

Later hydrogeological studies concluded that one of the most significant contributions to the excessive leachate production was groundwater infiltration through the bottom of the landfill. The landfill would be partially located on top of the edge of a small aquifer draining to the Arnoia River. The soil profile at the site would consist of, from top to bottom: a 0 to 15 m thick sandy layer (little consolidated white to yellow sandstone with silt), extending below the North-western half of the landfill base; an 8 to 17 m thick clayey layer (grey to reddish sandy-silty clays); an about 5 m thick consolidated micaceous sandstone layer, oriented NW-SE, 5° SE, extending over the whole area below the landfill; and a basal clayey complex (brown marls). Comparison between water levels at piezometers and other water measurement points around the landfill in different dates indicates that the variation of the piezometric levels of the aquifer is low. Differences of just 1 m, between 1982 and 2003, and 0.8 m, between 2002 and 2003, were registered. In view of this weak variability, and considering that the leachate mound is located well above the bottom of the landfill cells, the influence of the groundwater infiltration on the production of leachate must be very significant.

Stability analysis

Global static slope stability

The design report for closure of the landfill only included a simple static slope stability analysis. A very simplified landfill geometry was used, considering flat horizontal base and top surfaces, 19 m of landfill height, and 1V:3H lateral slopes. Waste parameters used were: unit weight, 14 kN/m^3; cohesion, 10 kPa; friction angle, 25°. A factor of safety of 1.75 was obtained. Neither seismic analyses nor specific slope stability analyses for the cover system were made.

New static global slope stability analyses were considered, therefore, necessary, and were carried out for the South slope of the landfill, as it is the highest and is oriented towards the Arnoia River and its future reservoir. Bishop's simplified method of slices was chosen for computer calculations, looking for failure mechanisms intersecting the waste mass. Two cross-sections were selected for analysis, JJ' and KK', separated 50 m from each other and oriented in the North to South direction (Fig. 1). Below the South slope, cross-section JJ' intersects natural excavated foundation soil, whereas cross-section KK' intersects an artificial fill that occupies a pre-existing depression created by a small affluent of the Arnoia River.

As no information on the geotechnical properties of the materials, or time for conducting site-specific testing, was available, the mechanical properties (specifically, unit weight, γ, and shear strength parameters: cohesion, c', and friction angle, φ') were estimated taking into account typical values for the different materials. Table 1 shows the materials considered and the values adopted. A sensitivity study was carried out to evaluate the influence of the waste mechanical parameters on the factors of safety. As waste unit weight values found in literature are normally of the order of 10 kN/m^3, above the leachate mound, and slightly higher in the saturated zone, values between 10 e 13 kN/m^3 were used in the analyses. Waste shear strength parameter values suggested by different authors differ significantly, not being uncommon, accepting the Mohr-Coulomb failure criterion, to propose curve or bilinear rupture lines. The results of the studies developed in the University of Cantabria, Spain (Sánchez Alciturri, 2000) suggest values, in the c-ϕ plane, located inside a quadrilateral formed by the following points: (0 kPa; 22°), (10 kPa; 16°), (20 kPa; 20°) and (0 kPa; 26°). For the present study, cohesion values between 10 and 20 kPa, and friction angles between 20 and 25° were considered.

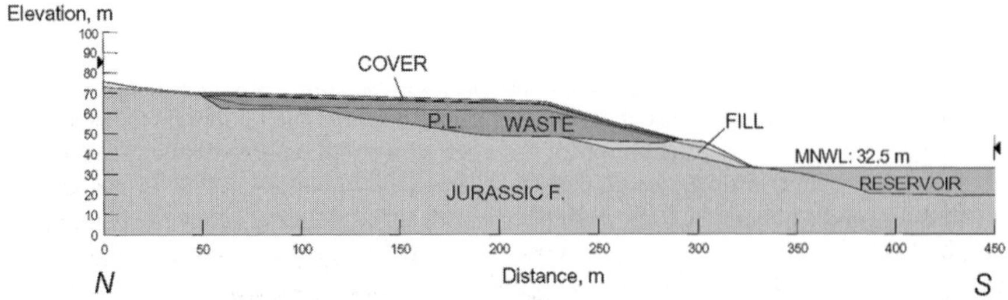

Figure 2. Cross-section KK'

Table 1: Mechanical properties adopted

Material	γ (kN/m³)	c' (kPa)	ϕ' (°)
Cover	18	0	25
Waste	10 to 13	10 to 20	20 to 25
Fill	20	10	30
Foundation	22	50	35

Pore water pressures were taken into account by defining the location of the water table and leachate mound. Three hypotheses were considered as for the location of the piezometric line in the cross-sections: *Case I)* Piezometric line at upper position: this seems to be the present situation, with the leachate mound very close (3 to 5 m) to the landfill surface; *Case II)* Piezometric line at lower position: For cross-section KK', analyses were also made considering the effect of lowering the leachate mound down to elevation 56 m close to the South slope by pumping from the gas wells at that zone; *Case III)* Piezometric line at upper position and intercepting the South slope surface: As an exfiltration point was observed about 5 m meters above the South slope toe, near the landfill South-west corner, another hypothesis was considered for cross-section JJ', by making the piezometric line to intercept the South slope face at the mid-height bench position.

The factors of safety obtained for the different cases performed on cross-sections KK' and JJ' are shown in Figs. 3 and 4, respectively. For cross-section KK', Fig. 3 shows the influence of the mechanical properties of waste and the piezometric line position (cases I and II), considering failure surfaces passing through the volume of waste (when the factor of safety is higher than 1.43, the critical failure surface is determined by the fill located downstream from the landfill, and does not intercept the waste mass). The influence of the waste unit weight is moderate, since the factor of safety only increases 0.1 between 10 kN/m^3 and 13 kN/m^3. On the contrary, the influence of cohesion and friction angle is significant. Reasonable values for waste parameters (γ, 12 kN/m^3; c', 10 kPa; ϕ', 25°) result in factors of safety higher than 1.5 (piezometric line in position I). With the piezometric line in position II, factors of safety are higher than 1.5, even if conservative low values are adopted for the waste properties (γ, 10 kN/m^3; c', 10 kPa; φ', 20°).

Analyses were performed for cross-section JJ' (Fig. 4) considering the piezometric line in positions I and III, waste unit weight values of 10 kN/m^3 and 12 kN/m^3, waste cohesion of 10 kPa and waste friction angles of 20°, 22° and 25°. Factors of safety lower than 1.5 were only obtained in case of the piezometric line in position III and waste friction angles of 20° or 22°. With a friction angle of 25° for the waste, the factors of safety were always higher than 1.5. In case of the piezometric line in position I, all the factors of safety obtained were also higher than 1.5.

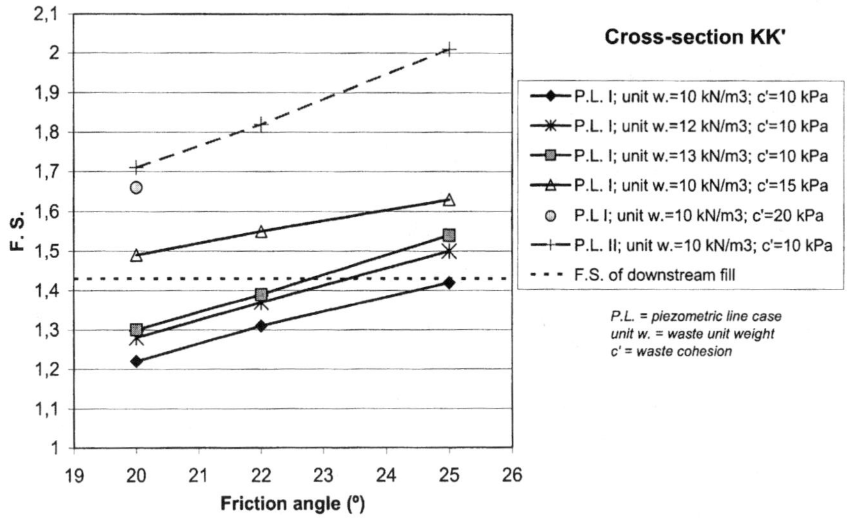

Figure 3. Factors of safety for cross-section KK', as a function of waste mechanical properties and position of piezometric line

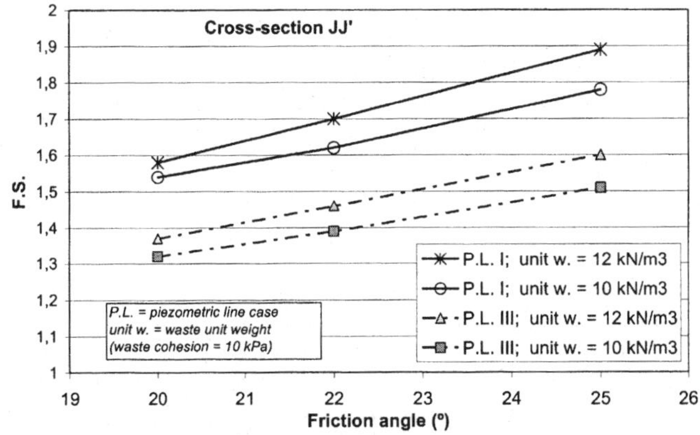

Figure 4. Factors of safety for cross-section JJ', as a function of the mechanical properties of waste and the position of piezometric line.

About the effect of the future reservoir on the global stability of the landfill (Fig. 2), the elevations of the maximum normal water level (32.5 m) and maximum flood water level (33.6 m) do not reach the lowest position of the artificial fill located below the toe of the landfill South slope. Taking into account the high position of the piezometric line considered in the stability analyses, the direct effect of the reservoir on the global stability of the landfill was considered small. Therefore, no specific analyses were performed in this phase for evaluating the landfill stability in the cases of full filling or fast draw-down of the reservoir.

Global seismic stability

A simplified approach, based on the pseudo-dynamic Makdisi-Seed (1978) method, was used to evaluate the seismic global slope stability of the landfill. It is a decoupled method that uses the sliding block concept for calculating permanent deformations, but is based on an evaluation of the dynamic response of the embankment rather than on rigid body behavior. It assumes that failure occurs on a well-defined slip surface and that the material behaves elastically at stress levels below failure but develops a perfectly plastic behavior above yield. The method was developed for earth dams, but was chosen because it is broadly used in Portugal. Limitations of its use for landfills were taken into consideration and some adjustments were introduced concerning the especial geometry of the landfill. It is considered that, as a seismic performance index, it may give a reasonable estimate of the landfill seismically induced permanent displacements. By applying a simplified methodology, the maximum acceleration at a slip surface (a_{max}) is evaluated as a function of the maximum crest acceleration ($\ddot{u}_{max}$), and the permanent deformation of the sliding mass is calculated as a function of the ratio between the yield acceleration (determined in a pseudo-static analysis) and the maximum acceleration at the sliding surface (k_y/k_{max}). The maximum crest acceleration is evaluated by an equivalent linear iterative procedure, calculating the variation of shear modulus (G) and damping ratio (D) with the equivalent cyclic shear strain amplitude (considered to be 65% of the maximum average shear strain) for the entire section. Shear modulus and damping ratio reduction curves proposed by Kavazanjian and Matasovic (1995) for MSW were used. The maximum crest acceleration was determined by:

$$\ddot{u}_{max} = \sqrt{(\ddot{u}_{1max})^2 + (\ddot{u}_{2max})^2 + (\ddot{u}_{3max})^2}$$

were: $\ddot{u}_{1max} = \varphi_1(0)\cdot Sa_1$; $\ddot{u}_{2max} = \varphi_2(0)\cdot Sa_2$; $\ddot{u}_{3max} = \varphi_2(0)\cdot Sa_3$. Coefficients $\varphi_1(0)$, $\varphi_2(0)$ e $\varphi_3(0)$ are the corresponding values, at the crest, of the mode participation factors for the first three modes, and Sa_1, Sa_2 e Sa_3 are the spectral accelerations determined from the power spectral densities defined by the Portuguese regulation (RSAEEP, 1988) for the respective natural frequencies (ω_1, ω_2, ω_3) of the embankment: $\omega_1 = \beta_1 V_s/h$; $\omega_2 = \beta_2 V_s/h$; $\omega_3 = \beta_3 V_s/h$. V_s and h are the mean shear wave velocity and the height of the fill, respectively. Coefficients $\varphi_n(0)$ and β_n depend on the geometry of the embankment. Values used for these coefficients in the case of earth dams (narrow crest) correspond to a triangular profile. As the landfill crest, however, is relatively flat and wide, another geometric profile, closer to the landfill case, a triangular section truncated at mid-height, was also considered. Table 2 shows the values of $\varphi_n(0)$ e β_n, obtained from Ambraseys (1960).

Table 2: Frequency coefficients, β_n, and mode participation factors $\varphi_n(0)$

Profile	$\varphi_1(0)$	$\varphi_2(0)$	$\varphi_3(0)$	β_1	β_2	β_3
Triangular (T)	1,60	1,06	0,86	2,40	5,52	8,62
Truncated triangle (TT)	1,40	0,60	0,45	1,79	4,80	7,91

The equivalent cyclic shear strain amplitude was determined at each iteration by: $(\gamma_{med})_{eq} = 0{,}65 \cdot 0{,}3\,(h/V_s^2)\,Sa_1$. Dissipation of energy through the foundation was taken into account in a simplified way by means of an equivalent damping ratio: $D_{eq} = (2\,\gamma_t\,V_s)/(\pi\,\gamma_{tf}\,V_{sf})$, where γ_t and γ_{tf} are the average unit weights, and V_s and V_{sf} are the average shear wave velocities, of landfill and foundation, respectively.

Portuguese seismic regulation (RSAEEP, 1988) can be followed to determine ground motion parameters. Thus, the response spectra included in the regulation were used as representative of the country. Gaeiras Landfill is located in seismic zone B, to which a seismicity coefficient, α, of 0.7 corresponds. Foundation soil can be considered of type I: "stiff soils and rocks". Based on bibliographical data, a value of 220 m/s was adopted for the shear wave velocity in the waste.

For the foundation, a value of 1500 m/s was considered, after analyzing the results of refraction seismic profiles carried out on lithologically similar soils for the dam project at the Arnoia River. A seismic motion equal to the one corresponding to the design base earthquake for the dam, with a_{max} equal to 0.13g (return period of 1000 years) was adopted. As in the case of the dam, the design base earthquake was considered to be of class 1 (magnitude 7.5; short distance from the source; predominat period, T_p, equal to 0.35 seconds).

Cross-section KK', considered to be the most unfavourable, was selected for the analysis. As to the mechanical properties of the waste and the position of the piezometric line, cases 7 and 2R, shown in Table 3, were chosen. The seismic coefficients obtained in the pseudo-static analyses are also shown in Table 3. Table 4 shows the permanent deformations determined for the different cases evaluated. Seismic motions were increased by a factor of safety of 1.5.

Table 3: Seismic coefficients obtained in pseudo-static analyses for cross-section KK'

Case	Piezometric line position	MSW properties			Seismic coefficient	Factor of safety
		γ (kN/m^3)	ϕ' (°)	c' (kPa)		
7-S	I	12	22	10	0.075	1.04
2R-S	II	10	22	10	0.18	1.03

Table 4: Seismic deformation determined by the Makdisi-Seed method

Cross-section geometry (*)	Case	Piezometric line	$\ddot{u}_{max}$ (cm/s^2)	a_{max} (cm/s^2)	a_{crit} (cm/s^2)	Deformation (cm)
T	7-S	I	739.8	369.9	73.6	38.9
T	2R-S	II	735.3	376.6	176.6	6.8
TT	7-S	I	472.2	236.1	73.6	16.5
TT	2R-S	II	481.6	240.8	176.6	0.9

(*) *T= triangle; TT= triangle truncated at mid height*

Some comments on the stability of the cover system

The design report for closure of the landfill specified a uniform inclination of 1V:3H (18.4°) for the South slope, but no specific slope stability analyses were included for the cover, and no information was found about shear strength characterization of the different interfaces. In addition, the geocomposite drain/GCL interface solution raises some doubts about its efficiency, as the drainage layer is in contact with the bentonite of the GCL. Possible friction angles values for this interface might not be very different from the slope inclination, though no signs of slippage were observed concerning it. Another relevant interface for the cover stability is the one between the cover soils and the geocomposite drain. Friction angle values available in literature about this kind of interface are quiet variable, depending of the nature of the soil and the type of geocomposite. To determine adequate shear strength parameters for analyses, specific laboratory direct shear tests for these interfaces, taking into account the particular conditions of the work, are considered necessary. Anyhow, a reformulation of the landfill cover system was recommended, as the main cause of the damages observed at the South slope was attributed to a deficient performance of the cover draining system. Signs of gully erosion processes (down to the geocomposite drain in depth) were observed over practically all the South slope, though they were more intense in the lower half. The slippage observed at the central section of the South slope probably started as an erosion process, due to the concentration of runoff water at that point. Although six drainage ditches in the direction of the maximum inclination were designed for the South slope, only one was actually built between the bench and the toe. Except at this section, no other signs of cover slippage were observed at the landfill slopes.

Conclusions

A preliminary analysis of the global stability of Gaeiras Landfill is described in the paper. The study, carried out with limitations on time and information, was, nevertheless, useful to arouse the issue of the current practice of landfill stability evaluation in Portugal.

The results of a sensitivity study performed in relation to the possible geotechnical properties of the waste indicate that Gaeiras Landfill presents sufficient global static slope stability, even with the leachate mound located close to the landfill surface. The effect of the future reservoir on the landfill slope stability can be consider insignificant, provided good geotechnical conditions be confirmed for the natural slope located downstream from the landfill. For evaluating the global seismic slope stability of the landfill, an approach based on the Maksidi-Seed pseudo-dynamic method was carried out, though the results must be regarded with caution, as it is a simplified method developed for earth dams. It was observed that the shape of the embankment exerts a considerable influence on the results. A truncated triangular cross-section gives smaller permanent deformations than the typical triangular shape used for earth dams. The influence of the position of the piezometric line is also significant. It can be admitted that the effects of seismic permanent deformations of the order of the ones obtained in the analysis would not be grave, although repair works would be necessary, specifically for the cover, the drainage and the gas collection systems. The cover system slope stability was not evaluated in this study, as it is necessary to determine reliable shear strength parameters for the different interfaces; however, no general signs of slope instability of the cover system were observed in the landfill. The slippage occurred at the South slope was probably caused by inefficiency of the surface drainage system.

This case history reveals the necessity to establish, in Portugal, appropriate procedures to evaluate the stability of landfills under static and seismic conditions, usable for design and that take into account the special characteristics of this kind of structures. It would be especially important to develop easily usable methods to estimate the seismic stability of landfills. The suitability of the Makdisi-Seed approach to landfill seismic stability analysis, considering the seismicity of the country and the broad experience acquired in the use of this method for earth dams in Portugal, was one of the issues aroused by this work which needs further research. Finally, this case shows the importance, not only to avoid instability problems, but also to ensure a satisfactory performance of the structure, of two basic design steps: an appropriate hydrogeological study of the site, and an adequate design for the cover system, with especial attention to potential erosion problems, to avoid excessive repair and maintenance costs.

Acknowledgements

The authors thank the Landfill's Owner, Resioeste, S. A. for the permission to publish this paper.

References

Ambraseys, N. N. (1960). *On the shear response of a two-dimensional truncated wedge subjected to na arbitrary disturbance.* Bull. of the Seism. Soc. of America. V. 50, No. 1, p. 45-56.

Kavazanjian, E. E and Matasovic, N. (1995). *Seismic analysis of solid waste landfills.* Proc. Geoenvironmental 2000, ASCE Specialty Conf. New Orleans, EUA, February.

Makdisi, F. I., Seed, H. B. (1978). *Simplified Procedure for Estimating Dam and Embankment Earthquake-Induced Deformations.* J. Geot. Eng. Div., ASCE Vol. 104, No GT7, pp. 849-867.

RSAEEP (1988), *Regulamento de segurança e acções para estruturas de edifícios e pontes;* Decreto-Lei nº 357/85 de 2 de Setembro. Imprensa Nacional, E.P., Lisbon (in Portuguese).

Sánchez Alciturri, J. M. (2000). *Comportamiento geotécnico de vertederos de residuos urbanos.* Seminar on geotechnical aspects of landfill design and construction; SPG, LNEC, Lisbon, (in Spanish).

Behaviour of coal ash as structural fill material

Dr. G. Ramasamy, Prof. Civil Engg., Indian Institute of Technology, Roorkee, India
Dr. S. S. Pusadkar, Lecturer Civil Engg., Govt. College of Engg., Amravati, India

Abstract: Compaction studies carried out show that the use of fresh / remoulded sample for each compaction and preconditioning period affect the maximum dry density (MDD) value. A representative MDD value of coal ash may be obtained as the average of at least three replicate compaction test results. Oedometer collapse studies carried out show that compacted fly ash is susceptible to significant collapse under different conditions of stress, moisture content and relative compaction while compacted bottom ash and a mixture of fly ash (FA) and bottom ash (BA) at equal proportion (FA50+BA50) show negligible collapse under similar conditions. Load tests on small plate (150mm diameter) were carried out on compacted ash in a circular tank. The results show that ash beds under as-compacted condition exhibit very high bearing capacity due to the effect of capillarity and compaction induced pre-consolidation stress. Bottom ash and FA50+BA50 exhibit much better performance than fly ash under submerged condition of the test bed.

Introduction

In dry mode of collection system, fly ash and bottom ash are separately collected and stored. The disposal of these coal ashes has become a serious environmental problem. Utilization of coal ash in construction of road embankment or filling of low-lying areas is a promising disposal alternative. In all these constructions, coal ash is compacted at site at desired water content. These compacted ash fills, like compacted soil fills, may undergo collapse on inundation. Literature shows that nearly all types of compacted soils are subjected to collapse under certain conditions (Lawton *et al.*, 1989,1992, Brandon *et al.* 1990, Pereira and Fredlund, 2000). Hence studies on various factors affecting collapse of compacted coal ash have been carried out.

Literature suggests that while determining maximum dry density and optimum moisture content (OMC) by standard or modified Proctor test, various investigators have used different methods of preparing the sample. Leonards *et al.* (1982) reported that the value of MDD and OMC of the Proctor test differed significantly when (i) the same sample is remoulded and reused and (ii) fresh samples are used for each test point. The value of MDD is lower and OMC is higher for fresh samples than for remoulded samples. Seals *et al.* (1972) used fresh sample for each individual compaction test on bottom ash.

The performance of structural fill depends upon the compaction or densification of the fill. It is believed that proper compaction is critical to the performance of a conventional soil embankment or structural fill and eventually it is true for coal ash fills also. The MDD and OMC obtained by Proctor test become the benchmark parameter for determination of quality of compaction. The density of fill is of primary importance since it is the major determinant of its strength and compressibility. (Grey and Lin, 1972, Indraratna *et al.* 1991). While standard procedures are available for the determination of MDD and OMC of soils (BS 1377 part 4), there is a need to evaluate the applicability of the same when dealing with coal ash. Therefore, a detailed experimental study has been carried on coal ash and a method has been suggested to determine MDD and OMC of coal ash.

Coal ash fills are often used as construction sites, and the bearing capacity and settlement are the required input for the design of foundations resting on such fills. Accordingly, load-settlement behaviour of compacted coal ash beds has been studied under different conditions of test bed such as, as-compacted, aged and submerged to bring out the effect of these factors on bearing capacity and settlement of foundations resting on coal ash fills.

Test material

Fly ash (FA) and bottom ash (BA), collected from a thermal power plant at Dadri, 50 km away from New Delhi, and a mixture of fly ash and bottom ash in equal proportion (FA50+BA50) are used as test material for studies. The physical properties of these ashes are presented in Table 1.

It is noted that bottom ash, a material consisting of sand size particles, exhibits a relatively high value of OMC. This may be due to the fact that the bottom ash particles are hollow or cenospheres and can hold a considerable quantity of water internally.

Table 1: Physical properties of coal ash from Dadri.

Sr. No.	Properties	Fly ash (FA)	Bottom ash (BA)	FA50+BA50
1	Particle size			
	Sand size (%)	2	82	38
	Silt size (%)	95	18	62
	Clay size (%)	3	0	0
	Coefficient of Uniformity (C_U)	2.8	6.5	4.67
	Coefficient of Curvature (C_C)	0.90	1.38	1.04
2	Sp. Gravity	2.06	1.86	1.93
3	Standard Proctor			
	a) MDD (kN/m^3)	13.44	8.89	11.87
	b) OMC (%)	19.8	42.00	26.75

Compaction properties

Standard Proctor tests (BS 1377 Part 4) were carried out on fly ash and bottom ash. The factors considered were (i) use of fresh or remoulded sample for each compaction, (ii) variation in preconditioning period and (iii) repeatability of test results.

Fresh / Remoulded samples

Compaction tests on fly ash and bottom ash were conducted with fresh sample (fresh sample for compaction at every moisture content) and remoulded sample (same sample repeated for compaction at every moisture content). The results of these tests are presented in Table 2. The results show that the MDD as obtained using fresh sample is less than that of remoulded sample while OMC is more or less the same in both the cases. This may be due to crushing of particles due to repeated compaction. Use of fresh samples better simulates the actual construction process and therefore, the same may be adopted for carrying out the compaction test.

Table 2: Effect of fresh and remoulded samples on MDD and OMC.

Sr. No.	Coal Ash	Fresh sample		Remoulded sample	
		MDD (kN/m^3)	OMC %	MDD (kN/m^3)	OMC %
1	Fly ash (FA)	12.88	20.88	13.51	18.00
2	Bottom ash (BA)	8.84	47.17	9.53	45.25

Preconditioning period

The results of compaction test conducted adopting 0hr, 1hr and 16 hr as preconditioning period are shown in Table 3. It can be seen that OMC of fly ash and bottom ash is marginally affected whereas MDD is significantly affected by preconditioning period. This may be due to free lime content (1%) in these ashes. As moistening of ash is generally done partly before transportation and partly at the site of construction immediately before compaction, a period equivalent to the expected time interval between wetting and compaction in the field may be adopted as preconditioning period during laboratory compaction test.

Table 3: Effect of preconditioning period on MDD and OMC

Sr. No.	Preconditioning period in hrs.	Fly ash (FA)		Bottom ash (BA)	
		MDD (kN/m^3)	OMC %	MDD (kN/m^3)	OMC %
1	0	12.88	20.20	8.81	43.9
2	1	13.44	19.80	8.89	42.0
3	16	12.65	21.70	8.74	44.5

Repeatability of test results

When a Proctor test was repeated on a given ash sample, the value of MDD obtained for each trial differed significantly as shown in Fig. 1. The difference in MDD between the maximum and minimum value was observed to be about 5% of the minimum value for the same sample tested. Therefore, compaction quality specification based on just one test may lead to poor quality fill. In order to keep proper quality control, the value of MDD to be reported shall be the average of at least three compaction tests.

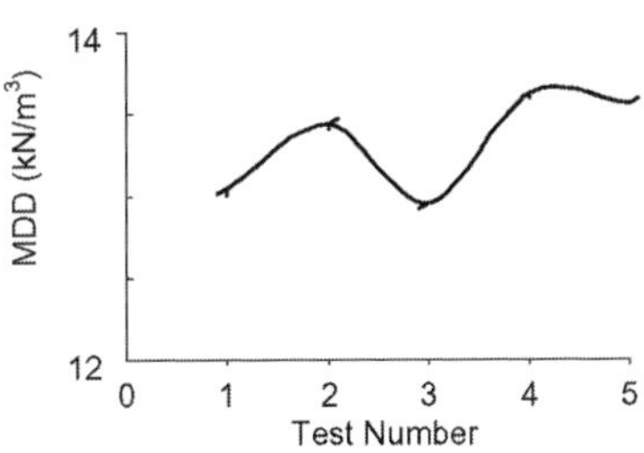

Figure 1: Repeatability of MDD in Standard Proctor test for fly ash

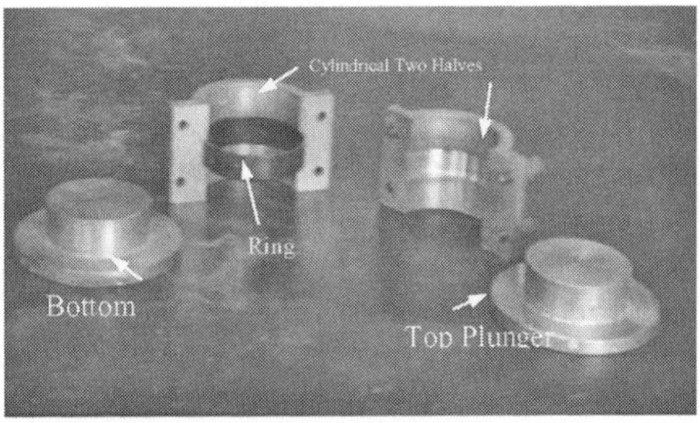

Figure 2: Photographic view of special mould

Collapse potential

Single oedometer collapse tests (ASTM: D5333-1992) were carried out keeping relative compaction (85, 90, 95, 100 % of Proctor MDD), compaction moisture content (OMC, OMC-5%, OMC-10%, OMC-15%) and inundation stress (25, 50,100, 200, 400 and 800 kPa) as variable parameters.

Preparation of test specimen

It was not possible to prepare test specimen of given relative compaction with varying moisture content by Proctor compaction method. Therefore, static compaction method was used for the preparation of samples of varying relative compaction and moisture content. Accordingly, a special mould was fabricated (Fig. 2) for the preparation of oedometer test sample.

Collapse potential of different types of coal Ash

The collapse potential of FA, BA and FA50+BA50 was determined for specimens prepared at the respective values of 95% relative compaction and optimum moisture content. These results are presented in Fig. 3. It was observed that the collapse potential for fly ash is substantially higher than bottom ash and FA50+BA50. This indicates that the fly ash is susceptible to collapse on inundation while BA and FA50+BA50 are not susceptible. Fly ash probably experiences higher suction pressure than BA or FA50+BA50 due to the following reasons,

i) Fly ash is finer.
ii) The OMC value of fly ash is lower.

The loss in suction pressure on inundation is high in fly ash resulting in higher collapse than observed in BA and FA50+BA50. Hence, further parametric studies were carried out only on fly ash.

Effect of inundation stress and compaction moisture content

The collapse potential of fly ash compacted to 95% relative compaction and at various moisture content values are presented in Fig. 4. These results reveal that:

i) The collapse potential increases with increase in inundation stress at all values of compaction moisture content. This is in agreement with the observation reported for silts by El-Sohby et al., 1987 and Basma and Tuncer, 1992 (Fly ash is a silt size material).
ii) The collapse potential, at a given relative compaction, increases with decrease in compaction moisture content.

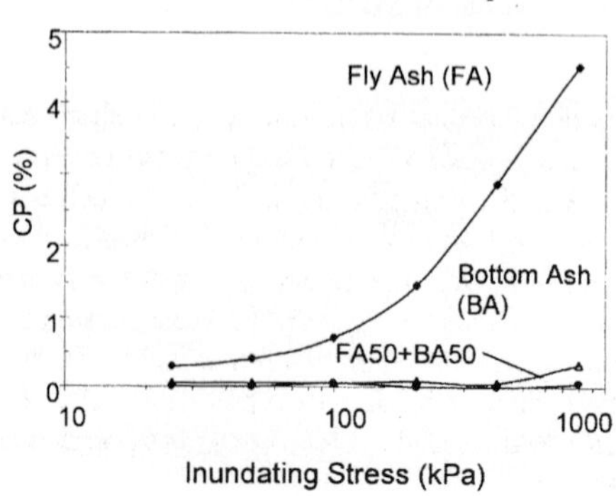

Figure 3: Collapse potential of different ashes

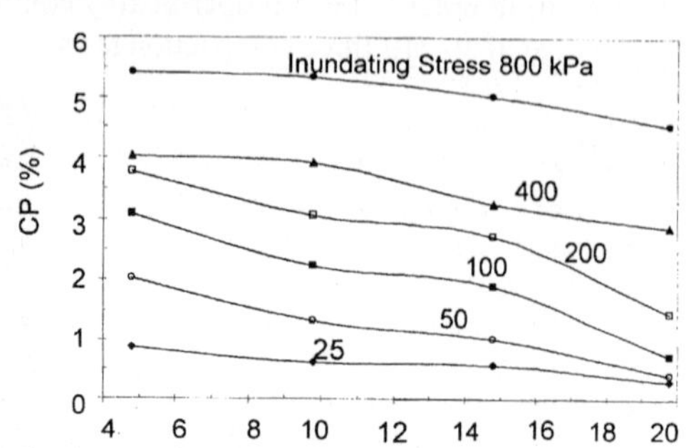

Figure 4: Collapse potential of fly ash

Effect of relative compaction

Figure 5 shows that as the relative compaction increases, the collapse potential decreases. This is due to increase in densification of the structure and reduction in the void ratio. Fills having collapse potential of 1 to 5% are identified as susceptible to causing moderate trouble to structures placed on it. It can be seen that even at a relative compaction of 95%, the value usually specified for acceptable compacted fills (Peck et al., 1974), the collapse potential becomes more than 1%.

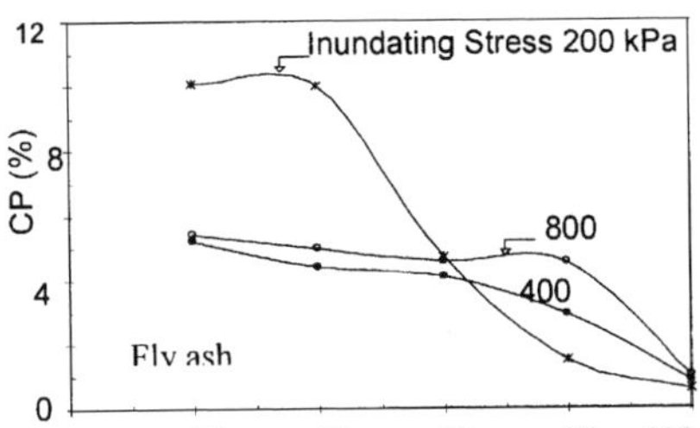

Figure 5: Influence of relative compaction on collapse potential

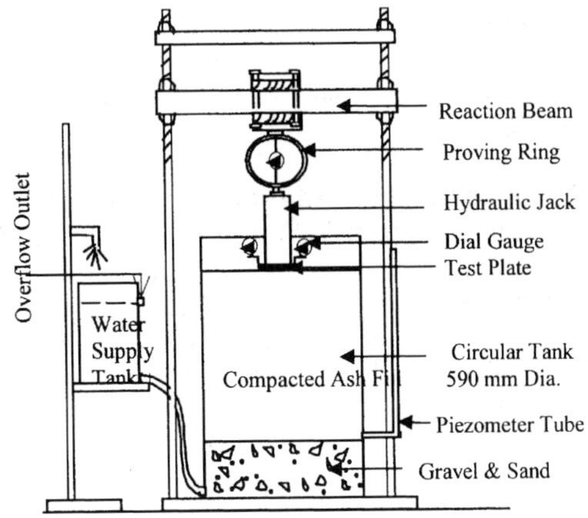

Figure 6: General arrangement of plate load test set-up

Load settlement behaviour

Laboratory plate load tests were conducted in a test set-up shown in Fig. 6 on 15 cm diameter circular plate on compacted ash beds. The test beds were compacted applying compaction energy equivalent to standard Proctor compaction. The tests were carried out under different conditions of test bed such as, as compacted, aged, submerged and at various relative compaction (Kaushik 2000). Results of these load tests are compared to bring out the effect of these parameters on the bearing capacity and settlement of foundations resting on compacted ash structural fills.

Effect of type of ash

Results of load tests on fly ash and bottom ash in as compacted condition (Fig. 7) show that the behaviour of compacted bottom ash is better than fly ash. Of the total coal ash produced in thermal power plants, bottom ash is about 20% and fly ash is about 80%. For effective disposal of coal ash, use of a mixture of fly ash and bottom ash in geotechnical applications may prove to be a better alternative. Therefore, load tests were conducted on various mixtures of bottom ash and fly ash and the results show that a mixture of fly ash (FA) and bottom ash (BA) in equal proportion (FA50+BA50) perform even better than bottom ash (Fig.7). It is also found that the performance of compacted coal ash of all types (FA, BA and FA50+BA50) in as-compacted condition is superior to the standard load-settlement behaviour (Peck *et al.* 1974) of 30 cm test plate on dense sand.

Effect of aging

Coal ash contains free lime which provides pozzolanic strength. Load tests were carried out on test beds aged for periods of 0, 28, 56 and 90 days. The results obtained for bottom are shown in Fig. 8. The results show that there is a significant increase in load carrying capacity due to aging and the full aged strength is obtained in a period of 28 days. Similar results were observed in fly ash and FA50+BA50.

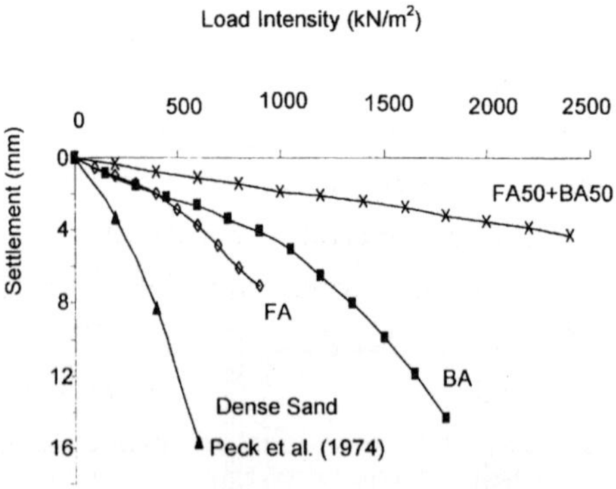

Figure 7: Load settlement behaviour of different coal ashes

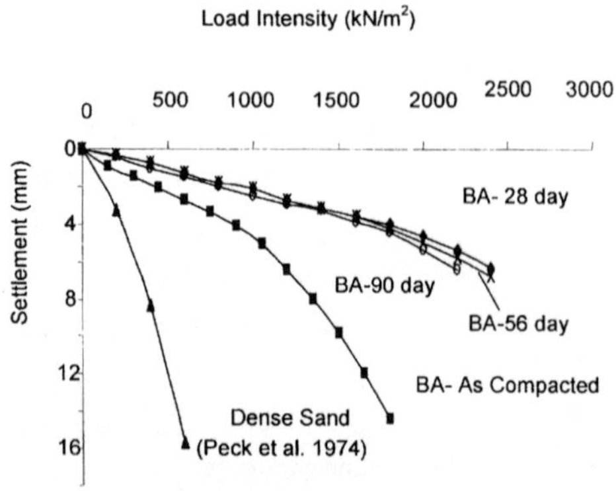

Figure 8: Effect of aging on load settlement behaviour of bottom ash

Effect of submergence

The results of load tests conducted under the submerged condition of the beds are shown in Fig. 9. It is found that the load carrying capacity is significantly reduced due to submergence (comparison of Fig. 7 and 9). However, the performance of BA and FA50+BA50 is found to be comparable or even better than that of a dense sand bed whereas, FA performed very poorly (Fig. 9).

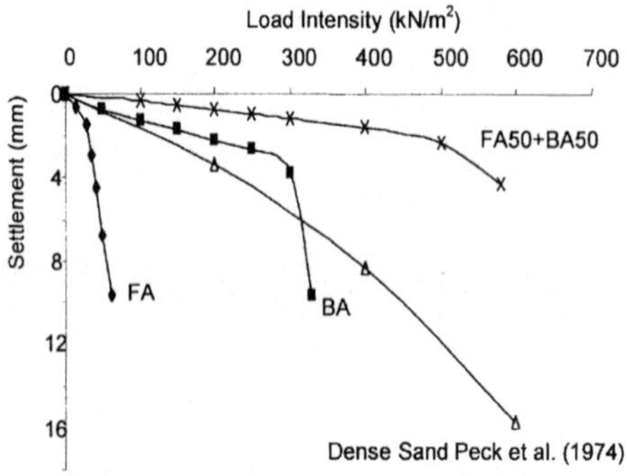

Figure 9: Load settlement behavior of coal ashes under submerged condition

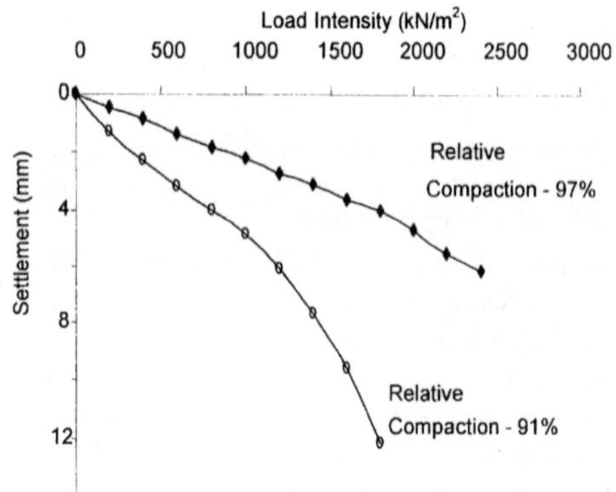

Figure 10: Effect of relative compaction on load settlement behavior of bottom ash

Influence of relative compaction

The influence of relative compaction on the performance of compacted test bed of bottom ash is shown in Fig. 10. The results show a significant reduction in load carrying capacity when the relative compaction reduces from 97% to 91%. Therefore, a more stringent specification than the usually adopted 95% relative compaction is recommended for adoption in the case of coal ash fills.

Conclusions

Based on the above studies the following conclusions have been drawn:

- Proctor test on coal ash may be standardized with the following specification:
  - (i) Use of the fresh sample for compaction at every water content.
  - (ii) A period equivalent to the expected time interval between wetting and compaction of ash in the field may be adopted as preconditioning period in laboratory compaction test for all types of ash.
  - (iii) Maximum dry unit weight (MDD) and optimum moisture content (OMC) values for coal ash may be obtained as the average of values obtained from at least three replicate compaction tests.
- Compacted fly ash (FA) is observed to be susceptible to collapse while bottom ash (BA) and mixture of fly ash and bottom ash of equal proportion (FA50+BA50) show negligible collapse on inundation.
- The ultimate load carrying capacity of the compacted coal ash increases with aging and 28 days are sufficient to gain full strength.
- Bottom ash and combination of fly ash and bottom ash in equal proportion (FA50+BA50) perform well under both as-compacted and submerged conditions. However, fly ash performs poorly under submerged condition.
- Bottom ash and FA50+BA50 are found to be competent structural fill materials for use in geotechnical applications.

References

ASTM: D5333-1992: Standard Test Method for Measurement of Collapse Potential of Soils, Annual Book of ASTM Standards, Vol. 04.08 (2002) pp. 1197-1199.

British Standard Institution, 1990, Methods of test for soils for civil engineering purposes, BS 1377 – 1990: Part 4, Proctor standard compaction test.

Basma A. A. and Tuncer E. R. (1992): "Evaluation and Control of Collapsible Soils", J. of Geotechnical Engineering ASCE, Vol. 118, No.10, October1992, pp. 1491-1504.

Brandon T. L., Duncan M. J. and Gardner W. S. (1990): "Hydrocompression Settlement of Deep Fills", J. of Geotechnical Engineering ASCE, Vol. 116, No.10, Oct.1990, pp. 1536-1548.

El-Sohby M. A., Rabbaa S. A., El-Khoraibi M. C. and El-Saadany M. M. (1987): "Influence of Soil Constituents on Collapsible Soils", 9th Regional Conference for Africa on Soil Mechanics and Foundation Engineering, Lagos, September, 1987, pp. 147-153.

Gray D. H. and Lin Y. (1972): " Engineering Properties of Compacted Fly Ash", J. of Soil Mechanics and Foundation Division, ASCE, Vol.98, No. SM4, April 1972, pp. 361-380.

Indraratna B., Nutalaya P., Kpp K. S. and Kuganenthira (1991): "Engineering Behaviour of a Low Carbon Pozzolanic Fly Ash and its Potential as Construction Fill" Canadian Geotechnical Journal, 28, pp. 542-555.

Kaushik N. P., (2000): "Performance of Compacted Ash and Sand fills Under Load." Ph.D. thesis, Department of Civil Engineering, University of Roorkee, Roorkee.

Lawton E. C., Fragaszy R. J. and Hardcastle J. H. (1989) " Collapse of Compacted Clayey Sand", J. of Geotechnical Engineering, ASCE, Vol.115, No.9, Sept. 1989, pp. 1252-1267.

Lawton E. C., Fragaszy R. J. and Hetherington M. D. (1992): " Review of Wetting Induced Collapse in Compacted Soil", J. of Geotechnical Engineering, ASCE, Vol.118, No.9, Sept. 1992, pp. 1376-1394.

Leonards G. A. and Bailey B. (1982): "Pulverized Coal Ash as Structural Fill", J. of Geotechnical Engineering, ASCE, Vol.108, No.GT4, April 1982, pp. 517-537.

Peck R.B., Hanson W. E. and Thornburn T.H. (1974): "Foundation Engineering" 2nd Edition, John Wiley and Sons, Inc.

Pereira J. H. and Fredlund D. G. (2000): "Volume Change Behavior of Collapsible Compacted Gneiss Soil", J. Geotechnical and Geoenvironmental Engineering, ASCE, Vol. 126, No.10, Oct. 2000, pp. 907-916.

Seals R .K., Moulton L. K., Roth B. E. (1972), " Bottom Ash: An Engineering Material", J. of Soil Mechanics and Foundation Division, ASCE, Vol. 98, No. SM4, April 1972, pp. 311-325.

Possibility of Using the Drinking Water Sludge as Geotechnical Material

António José Roque, Laboratório Nacional de Engenharia Civil (LNEC), Lisbon, Portugal
Marta Carvalho, AdP – Serviços Ambientais, S.A., Lisbon, Portugal

Abstract
During the drinking water treatment processes, a chemical sludge is produced in large quantities. In the European Community List of Waste, that sludge is classified by the code number 19 09 02 "Sludges from water clarification". In the Portuguese Water Group (AdP Group) there are several facilities that carry out water treatment, sludge production being estimated in about 20 x 10^3 tons/year. Therefore, sludge from Portuguese Water Treatment Plants represents a considerable annual quantity of waste, of which the management is a concern for the AdP Group. In this study the geoenvironmental and geotechnical characteristics of a Portuguese sludge are presented and the technical possibility of its application to geotechnical works is evaluated.

Introduction
The treatment of drinking water usually produces a chemical sludge, which is mainly formed by the substances existing in raw water and chemical products used in its treatment. In the European Community List of Waste, that type of sludge is classified by the code number 190902 "Sludges from water clarification".

Currently, the prevention of sludge production in Portugal is unlikely to be applicable. In fact, the compliance with the legislation in force and the fulfilment of the objectives defined in the Regional Development Plan for 2000-2006, have led, in the last few years, to a significant increase in: a) the population served by Water Treatment Plants (WTPs); and b) the sludge production by the facilities of the Portuguese Water Group (AdP Group). Presently, these facilities produce more than 20 x 10^3 tons of sludge per year. It is in this context that RECICLAMAS, a company created in 2001 by the AdP Group, has requested LNEC to evaluate the possibility of using the drinking water sludge for geotechnical works. For this investigation, geoenvironmental and geotechnical laboratory tests were performed on the sludge, like chemical analysis on the sludge and the leachate, and index, permeability, compaction, consolidation and triaxial tests.

Geoenvironmental and Geotechnical Characteristics of the Sludge

Sampling and Preparation of the Samples
Two types of samples were collected: one from the sludge removed from the belt filter press, which ensures a significant dehydration of the sludge; and another from the sludge dumped in a dump truck by a conveyor that transported the material from the belt filter. The sludge

sampled in the belt filter was collected in small plastic bags and the one from the dump truck in 50 kg plastic-nylon bags.

The sludge collected in the belt filter was not subject to any type of preparation prior to the determination of its natural water content and to the sending of a test sample to the laboratory that carried out the chemical analyses. The sludge collected in the dump truck was spread, in laboratory, on a plastic to dry. Considering that in these conditions the drying occurred very slowly, the sample was placed into trays and dried in a drying oven at a 50-60 ºC temperature. After drying, the sample was broken up and sieved in a 4.75 mm mesh sieve, returning subsequently to the drying oven (50-60 ºC temperature) for homogenisation of its water content. Three test samples were collected from this sample to determine the residual water content (3-4%). The oven-dried material was stored in plastic bags until the execution of the programmed tests.

Geoenvironmental Characteristics

The legislation in force imposes the execution of chemical analyses on both the sludge and the eluate so as to know, respectively, their composition and leachability. The contents obtained for the sludge and the eluate are compared in Table 1 and Table 2, respectively, with the maximum admissible values established in the legislation in force for inert, non-hazardous and hazardous waste. The only parameter determined on the sludge with a higher value than the maximum admissible was the Loss at 550ºC – Loss at 105ºC. Nevertheless, since this waste is not susceptible to leaven, the legislation allows the established Loss at 550ºC – Loss at 105ºC value to be exceeded. The results obtained on the eluate demonstrate that fifteen over nineteen parameters presented lower values, sometimes significantly lower, than the maximum admissible values required for inert waste.

Table 1: Chemical composition of the drinking water sludge

Parameter	Unit	Sludge composit.(1)	Waste category (Decree-Law n.º 152/2002)		
			Inert	Non-hazardous	Hazardous
Ignition loss at 105 ºC	%	3.0	65	65(6)	65(6)
Ignition loss at 550 ºC - 105 ºC	%	10.0	5(2)	15(7)	15(7)
Ignition point	ºC	> 70.0	55(3)	55(3)	55(3)
Lipophilic substances	%	0.4	0.5	4	10
Halogenated vol. org. comp.	%	< 0.01	0.05(4)	0.1	1
Non-halogenated vol. org. comp.	%	< 0.01	0.15(5)	0.3	3
Arsenic	mg/kg	119.0	250	2000	—
Cadmium	mg/kg	< 2.0	50	1000	—
Lead	mg/kg	< 20.0	2000	5%	—
Copper	mg/kg	49.0	6000	6%	—
Chromium	mg/kg	< 10.0	3000	5%	—
Mercury	mg/kg	< 0.1	25	250	—
Nickel	mg/kg	20.0	2000	5%	—
Zinc	mg/kg	142.0	8000	7.5%	—

(1)Reported to dry matter. (2)This value may be exceeded whenever the waste is not susceptible to leaven. (3)Minimum values. (4)No parameter can exceed 100 mg/kg individually. (5)No parameter can exceed 300 mg/kg individually. (6)Waste disposal can not admit, monthly, more than 10% of waste that present a higher value. (7)For waste disposals specially designed for admitting organic or non-leaving waste, this value can be exceeded. The symbol < (less than) represents the limit of quantification of the method used.

Table 2: Chemical composition of the drinking water sludge eluate

Parameter	Unit	Eluate composit.	Waste category (Decree-Law n.º 152/2002)		
			Inert	Non-hazardous	Hazardous
pH	(1)	12.6	5.5<x<12	4<x<13	4<x<13
Electrical conductivity	mS/cm	6.3	6<y<50	—	100
Ammonium	mg N/l	< 3.0	5	200	1000
Adsorb. Organic Halogens	mg Cl/l	0.52	0.3	1.5	3
Arsenic	mg/l	0.022	0.1	0.5	1
Cadmium	mg/l	< 0.03	0.1	0.2	0.5
Lead	mg/l	< 0.2	0.5	1	2
Cyanide	mg/l	< 0.01	0.1	0.5	1
Chloride	mg/l	28.4	500	5000	10000
Copper	mg/l	< 0.23	2	5	10
Carbon Org. Total	mg C/l	100.0	40	100(2)	200(2)
Chromium VI	mg/l	< 0.1	0.1	0.1	0.5
Total chromium	mg/l	< 0.1	0.5	2	5
Phenol	mg/l	< 0.025	1	10	50
Fluoride	mg/l	8.0	5	25	50
Mercury	mg/l	< 0.0002	0.02	0.05	0.1
Nickel	mg/l	< 0.2	0.5	1	2
Nitrate	mg/l	—	3	10	30
Sulphate	mg/l	< 2.0	500	1500	5000
Zinc	mg/l	0.02	2	5	10

(1)Sorensen scale. (2)For waste disposals specially designed for admitting organic waste, this value can be excedeed. The value can also be exceeded whenever the waste is not susceptible to leaven. The symbol < (less than) represents the limit of quantification of the method used.

Mention must also be made of the fact that: a) the organic matter content (reported to dry matter) obtained for the sludge is much higher than the one existing in most natural soils; and b) the aluminium content (6,7%) obtained for the sludge is much higher than the one of the other heavy metals. It derives from the use of aluminium sulphate as coagulant/flocculant in the water treatment process. The calcium content (37,4%) is the highest, and is due to the treatment of water with 30 g/m^3 of calcium hydroxide.

Geotechnical Characteristics

Physical properties

Comparatively with natural soils, the water content of the sludge is much higher, and the specific gravity of solids is lower (Table 3). As regards grain-size distribution, considering the grain-size ranges used in Soil Mechanics to describe natural soils, the sludge composition (in weight) is about: 21% clay-size particles, 65% silt-size particles and 14% sand-size particles.

The consistency limits obtained for the sludge were very high (Table 3), which indicates that it is a water sensitive material. In accordance with the Unified Classification of Soils (ASTM D 2487), it is a material of the type "MH" ("elastic silt"). Because of these high plasticity limits, the crystalline compounds of the sludge were studied. The following were identified: calcite ($CaCO_3$), the compound in higher proportion; hydrated calcium aluminates ($4CaO \cdot Al_2O_3 \cdot 13H_2O$ and $3CaO \cdot Al_2O_3 \cdot CaCO_3 \cdot 11H_2O$); traces of calcium hydroxide ($Ca(OH)_2$); quartz; hydrated aluminium sulphate ($Al_2(SO_4)_3 \cdot 16H_2O$); magnesite ($MgCO_3$);

and, probably, gypsum ($CaSO_4 \cdot 2H_2O$). The presence of the hydrated calcium aluminates and the hydrated aluminium sulphate, which contain several crystallisation water molecules, can explain the high capacity of water retention of the sludge, taking into account that the bonding forces are weak.

The values obtained for the maximum dry unit weight γ_{dmax}, and optimum water contents w_{op} (Table 3) are, respectively, much lower and much higher than the typical values for natural soils. The relative location of the standard and the modified Proctor compaction curves for this material, presented in Figure 1, are, nevertheless, in compliance with the behaviour observed in natural soils when the compaction effort is modified.

Table 3: Physical properties of the drinking water sludge

Parameter	Unit	Value
Water content, w	%	478.2
Specific gravity of solids, G_s	—	2.27
Grain-size distribution:		
o sand fraction (0.06 - 2.00 mm)	%	14
o silt fraction (0.002 - 0.06 mm)	%	65
o clay fraction (< 0.002 mm)	%	21
Consistency limits:		
o liquid limit, w_L	%	107.5
o plastic limit, w_P	%	81.7
o shrinkage limit, w_S	%	71.8
Plasticity index, I_P	%	25.8
Compaction:		
o standard Proctor		
◦ max. dry unit weight, γ_{dmax}	kN/m^3	6.27
◦ opt. water content, w_{opn}	%	84
o modified Proctor		
◦ max. dry unit weight, γ_{dmax}	kN/m^3	7.5
◦ opt. water content, w_{opm}	%	71

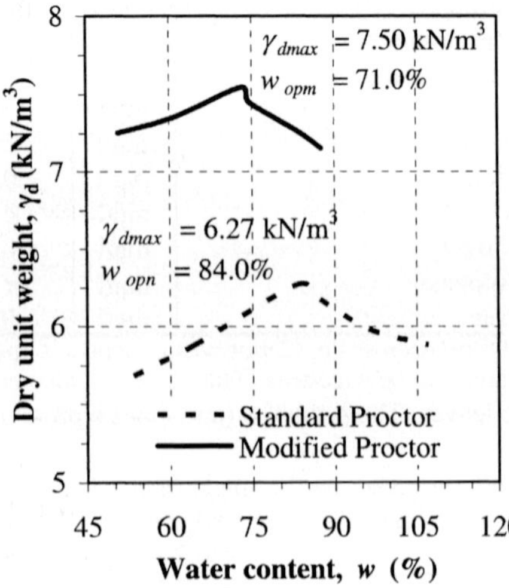

Figure 1: Standard and modified Proctor curves for drinking water sludge.

Hydraulic properties

To study the hydraulic conductivity k of the sludge, rigid-wall permeameters were used. Specimens were compacted directly in a stainless steel ring, with an inner diameter of 101.6 mm and a height of 40.0 mm. Demineralised water (pH $\cong$ 6.0 and electrical conductivity EC $\cong$ 5 μS/cm) and a young leachate produced in a municipal solid waste landfill (pH = 7.85 and EC = 21.9 mS/cm) were used in the permeability tests.

A different behaviour, from the one shown in Figure 1, was observed at the dry side of the curve, for the modified Proctor compaction energy (Figure 2). Because of that, compaction of P6 point was repeated, but, as shown in Figure 2, the values corresponding to P6 were confirmed. No clear reasons were found to explain the different results at the dry side of the optimum.

The k value of the specimens compacted according to the standard and modified Proctor tests, determined with demineralised water, was about 1 x 10^{-7} m/s and 1.5 x 10^{-8} m/s, respectively (Table 4). Unlike normal observations in natural fine soils (Roque, 2001), there was no significant variation in k value of the sludge with the increase in the water content. The k value of specimen P13, tested after 28 days of curing, being of the same magnitude as the one obtained with other specimens compacted by the modified Proctor test, suggests that the curing time did not influence the sludge permeability to demineralised water. Nevertheless, this assumption is not conclusive, taking into account the limited number of available results. The values obtained for k value of specimens P7 and P11 indicate that the sludge is less permeable to the leachate than to demineralised water. In specimen P7, after replacing the demineralised water by leachate, k decreased, continuously and gradually, from 1.7 x 10^{-8} m/s until 9.3 x 10^{-10} m/s, i.e., about 18 times. By comparing k value to the leachate of specimen P11 with k value to demineralised water of specimen P8, both having similar w and γ_d values, it can be observed that the value of k obtained in specimen P11 was about 90 times less than the one obtained in specimen P8. As regards specimen P12,w_{opm}-2%, tested after 7 days of curing, no effluent solution was recorded during the execution of the permeability test, i.e., during 112 days. Therefore, the study does not make it possible to assess the influence of the curing time on the sludge permeability to the leachate.

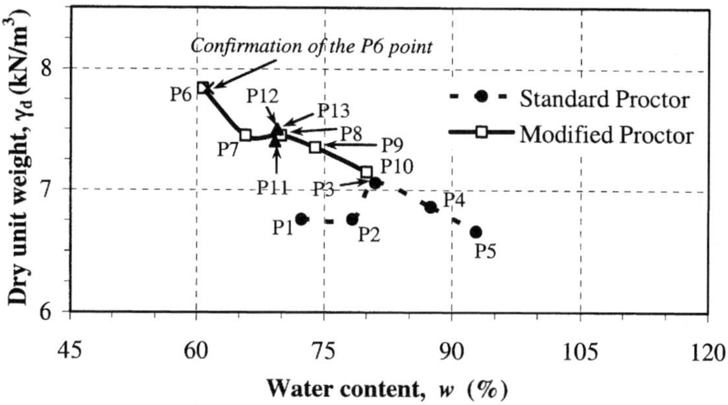

Figure 2: Sludge specimens used in the permeability tests.

Table 4: Hydraulic conductivity of the drinking water sludge

Specimen compacted by standard Proctor	P1 w_{opn}-12%	P2 w_{opn}-6%	P3 w_{opn}-3%	P4 w_{opn}+3%	P5 w_{opn}+9%
k_{Dw} at 20 ºC, m/s	9.0x10^{-8}	1.2x10^{-7}	2.3x10^{-7}	2.7x10^{-8}	1.8x10^{-7}
Specimen compacted by modified Proctor	P6 w_{opm}-10%	P7(1) w_{opm}-5%	P8 w_{opm}-1%	P9 w_{opm}+3%	P10 w_{opm}+9%
k_{Dw} at 20 ºC, m/s	1.6x10^{-8}	1.7x10^{-8}	1.1x10^{-8}	1.5x10^{-8}	2.0x10^{-9}
k_L at 20 ºC, m/s	—	9.3x10^{-10}	—	—	—
Specimen compacted by modified Proctor	P11 w_{opm}-2%	P12(2) w_{opm}-2%	P13(3) w_{opm}-2%		
k_{Dw} at 20 ºC, m/s	—	—	1.5x10^{-8}		
k_L at 20 ºC, m/s	1.2x10^{-10}	(4)	—		

k - Hydraulic conductivity. Dw - Demineralised water. L - Leachate. (1)Sludge specimen permeated with the acid leachate after percolation with demineralised water. (2)Sludge specimen with 7 days of curing. (3)Sludge specimen with 28 days of curing. (4)The effluent solution was not recorded during the 112 days of the permeability test.

Mechanical properties

The study of the mechanical strength of the sludge was carried out by consolidated undrained triaxial compression tests (CU tests) with pore pressure measurements on three specimens (101.6 mm, diameter, and 177.0 mm, height). The dry unit weight and the water content of these specimens were about the maximum dry unit weight and the optimum water content of the modified Proctor.

The maximum confining pressure applied to one of the specimens was 320 kPa. On the other two specimens, the confining pressures used were 160 and 80 kPa. The tests were stopped at an axial strain of 12%. Figure 3 illustrates the behaviour of the three specimens tested, in terms of the deviator stress versus the axial strain.

The values of the shear strength parameters, in terms of effective stresses, adopting as failure criterion the inversion of pore water pressure, were as follows:

$$c' = 77 \text{ kPa},$$
$$\phi' = 44^\circ.$$

In order to study the sludge compressibility characteristics, one-dimensional consolidation tests were performed. The void ratio versus logarithm of consolidation pressure relation of the specimen compacted with the modified Proctor compaction is shown in Figure 4. The following compressibility characteristics for the sludge, corresponding to a stress range from 138.2 kPa to 274.4 kPa, were determined:

- specimen compacted with a standard Proctor compaction energy and 84% water content
  - coefficient of compressibility, $a_v = 2.44 \times 10^{-4}$ kPa^{-1},
  - coefficient of consolidation, $c_v = 4.30 \times 10^{-8}$ m^2/s;
- specimen compacted with a modified Proctor compaction energy and 71% water content
  - coefficient of compressibility, $a_v = 1.83 \times 10^{-4}$ kPa^{-1},
  - coefficient of consolidation, $c_v = 1.80 \times 10^{-8}$ m^2/s.

The time values necessary to complete 90% of the consolidation, $t_{90;v}$, were 28.6 minutes, for the specimen compacted by standard Proctor, and 67.2 minutes for the specimen compacted by modified Proctor.

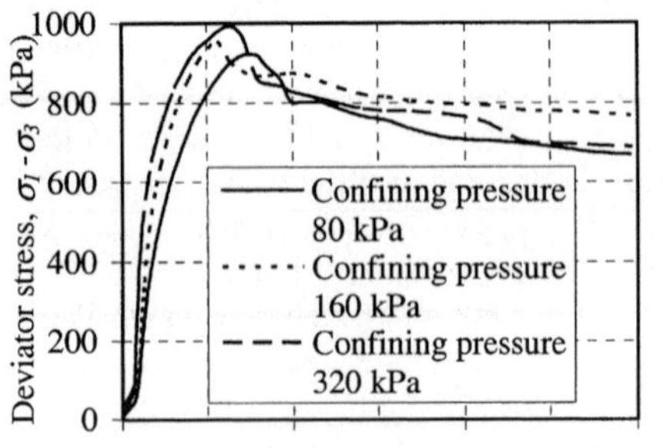

Figure 3: Deviator stress versus axial strain of the three specimens compacted with a modified Proctor compaction energy.

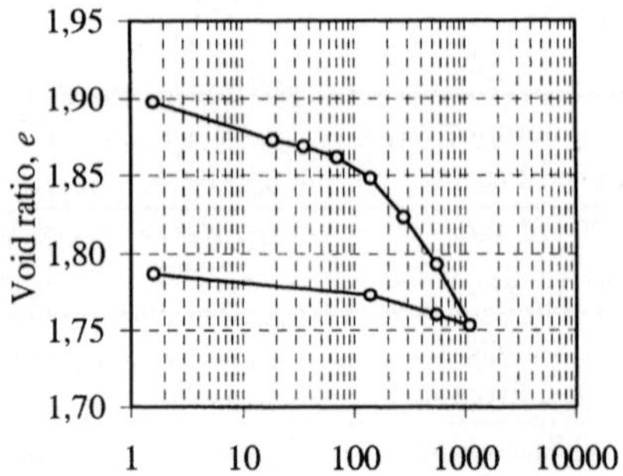

Figure 4: Void ratio versus consolidation pressure of the specimen compacted with a modified Proctor compaction energy.

Aplicability of Drinking Water Sludge to Geotechnical Works

By comparing the geoenvironmental characteristics of the sludge with the values defined in the Portuguese legislation for the different parameters, it can be observed that this waste cannot be included in the inert category. Nonetheless, its application to geotechnical works is not considered to represent a hazard for the environment or the public health, for the following reasons: a) comparatively with the values established in the Portuguese legislation for inert waste class, for most parameters, the sludge fulfils the specified values; b) the hydraulic conductivity of the layers constructed with sludge in geotechnical works will be low to very low, as shown by the values obtained; c) the sludge volume in each geotechnical work and the contact surface between the sludge and the foundation will, generally, be small; d) the direct discharge of leachates on aquifers or water lines is unlikely to occur; e) the main mechanism of lixiviation and transport of pollutants present in the sludge, i.e., rainfall water, is rather discontinuous. In one of the few available studies on the possibility of using the drinking water sludge as geotechnical material, Raghu et al. (1987) also conducted leaching tests on a drinking water sludge obtained from the Jersey City Water Treatment Plant. They reported that the concentrations of heavy metals and organic matter were too low to create any pollution problems.

In view of the values obtained for the hydraulic conductivity of the sludge and of the minimum specified value in the Portuguese legislation for this parameter on liner systems of landfills, the sludge compacted with the energy of the modified Proctor test will be appropriate for landfills of inert, non-hazardous and hazardous waste. Compacted with the standard Proctor test energy, the material will be suitable for landfill liners for inert waste. Considering that the present legal framework does not establish the minimum requirements for the hydraulic conductivity of the compacted soil layers for cover systems of non-hazardous and hazardous waste landfills, the use of the sludge in this application seems also possible. In the study carried out by Raghu et al. (1987), consolidation tests on sludge compacted to 90 percent of maximum modified Proctor dry density were performed. From these tests, the hydraulic conductivity was determined to be about 10^{-9} m/s. These results confirm the impervious nature of the drinking water sludge.

Two important geotechnical properties of the sludge to allow its use as an embankment material are the compressibility and the shear strength. In view of the high shear strength, the low compressibility and also the low dry unit weight (less than half of the dry unit weight generally observed in natural soils) of the sludge studied, the use of this type of waste can be particularly advantageous, for instance, in the landscape recuperation of waste landfills and waste dumps. It is obvious that the use of a low density material to reduce the loading on the underlying waste is relevant to carry out the earthworks that are usually necessary to model the surface of waste landfills and abandoned dumps. Raghu et al. (1987) also investigated the compressibility of the sludge obtained from the Jersey City Water Treatment Plant. They reported that no compressibility behaviour of the material was observed. Wang et al. (1992) evaluated the possibility of using a sludge obtained directly from a Water Treatment Plant in Chesapeake, Virginia for embankment construction. In this study, they used additives (a slaked lime, a fly ash, and a nature soil) to improve sludge properties. The results of the study indicated that admixing stabilization improved the compressibility and shear strength properties of the sludge tremendously. Nevertheless, for the range of conditions investigated, the treated sludge is still too compressible and without sufficient strength for use as an embankment material. They concluded that compressibility and shear strength can be improved by either dewatering further or by raising the additive content or by both.

Considering the results obtained, the technical specifications in force for road and railway works exclude beforehand the use of drinking water sludge in this type of works. Nevertheless, reference must be made to the fact that these specifications were developed for conventional soils, of which the origin and geo-historical evolution are very different from the conditions that led to the production of drinking water sludge. Therefore, it is considered that regardless of the need to carry out an appropriate laboratory characterisation of the materials, it would be necessary to construct experimental sections and to control those sections via specifications based on performance.

Conclusions

The results obtained in the experimental programme developed in laboratory made it possible to draw the main conclusions as follows:

a) the sludge from the Water Treatment Plant (WTP) studied can be compared to a natural soil, even though with some particularities related with the fact of it being a waste;
b) the application of the sludge to the construction of geotechnical works is not to be a significant pollution source for the environment and the public health, not only because its leacheability will be reduced and will occur in a discontinuous way, but also due to the dilution of the eluate that is to reach the water resources. Furthermore, since there is no external feeding, the leacheability of the pollutant mass will tend to depletion;
c) the compacted sludge presents a high shear strength and a low compressibility. Therefore, under proper design conditions to be specified in each case, no geotechnical problems related with its failure or with excessive settlement, either differential or total, are likely to occur;
d) the dry unit weight of sludge, being approximately half of the normally one observed in natural soils, indicates that its use in the construction of embankments on soft soils, or in the modelling of final covers of waste landfills and abandoned dumps, will be advantageous; and
e) from a geotechnical point of view, the use of drinking water sludge in geotechnical works (waste landfills: daily cover of waste, liner and cover systems; road and railway infrastructures; embankments, light embankments over soft soils; modelling of the final cover of waste landfills and abandoned dumps after closure; among others) seems possible and, in certain cases, advantageous.

References

Raghu, D., Hsieh H.-N., Neilan, T., and Yih, C.-T. (1987). Water treatment plant sludge as landfill liner. In: *Geotechnical Practice for Waste Disposal'87, Geotechnical Special Publication n.º 13*. Edited by R.D. Woods, New York: ASCE, pp. 744-758.

Roque, A.J. (2001). *Advective and diffusive transport of inorganic pollutants through compacted clay liners existing in the landfills. Application to Portuguese fine-grained soils*. Ph.D. Dissertation, INSA Lyon (France), 697 p. (in French).

Roque, A.J. (2002). *Valorisation of drinking water sludge. Geotechnical and geoenvironmental characterisation of the Asseiceira (Tomar) drinking water sludge*. Lisbon (Portugal): LNEC, Report (1st) n. 162/02 – NP, 60 p. (in Portuguese)

Roque, A.J. (2003). *Valorisation of drinking water sludge. Geotechnical and geoenvironmental characterisation of the Asseiceira (Tomar) drinking water sludge*. Lisbon (Portugal): LNEC, Report (final) n. 111/03 – NGE, 102 p. (in Portuguese)

Wang, M.C., Hull, J.Q., and Jao, M. (1992). Stabilization of water treatment plant sludge for possible use as embankment material. In: *Transportation Research Board (TRB), Record n.º 1345*. Washington, DC: National Research Council, pp. 36-43.

Elution Reduction of Hazardous Substances by Solidification of Coal Ash

Atsuko Sato, Civil Engineering Rsrch. Institute of Hokkaido, Sapporo, Japan
Satoshi Nishimoto, Civil Engineering Rsrch. Institute of Hokkaido, Sapporo, Japan
Teruyuki Suzuki, Kitami Institute of Technology, Kitami, Japan

Abstract
In response to calls for a recycling-oriented society, technology has been developed for effectively utilizing coal ash generated by coal-fired power plants as a material for civil engineering works. However, leaching of hazardous substances from some ash exceeds the Japanese standards for environmental soil quality. This means that some types of coal ash cannot be used as a soil material as they are. Examinations addressed whether it is possible to reduce leaching by solidifying coal ash. In the examination, the leaching of hazardous substances was measured for coal ash solidified after mixing with water and blast furnace cement B and for coal ash solidified after mixing only with water. Also, the change in leaching of hazardous substances over time was measured. The examination found that there are cases in which the leaching of hazardous substances was reduced to within the environmental standard, and that some types of coal ash meet the environmental standard and can be used as soil material after solidification. It was also found that the leaching rate of some hazardous substances changes over time.

Introduction
The characteristics of coal ash produced by coal-fired power plants depend on the characteristics of the source coal, the type of boiler furnace and the combustion temperature. It has been verified that coal ash is useful for application to embankments and for improving soils that are unsuitable for construction[1)].

Japanese society has become increasingly concerned over hazardous substances in the environment[2)]. There are cases where hazardous substances leach from some ash. In using ash as a construction material, measures should be taken to prevent the leaching of hazardous substances.

To prevent leaching, hazardous substances are rendered insoluble by the addition of chemicals or a solidification agent[3)], or they are encapsulated in the surrounding material[4)]. This study focuses on leaching prevention in which coal ash is solidified with blast furnace cement B. The leaching of hazardous substances was measured in two types of coal ash samples: one that was mixed with blast furnace cement B and water and then compacted[5)], and another that was mixed with blast furnace cement B and water and then granulated[6)]. Tests found that solidification and granulation reduce the leaching of hazardous substances. To clarify the long-term effect of leaching reduction in coal ash solidified or granulated using blast furnace cement B, measurements were done for the leaching of hazardous substances after one year. Also, measurements were done for the leaching of hazardous substances from embankments three to four years after construction. This paper reports the results of those measurements.

Coal Ash Tested

Characteristics and properties of coal ash as soil material

Table 1 shows the properties of the coal ash samples used in the laboratory tests. The ash was coal ash from five boiler furnaces: A, B, C, D, and E. The particle density of the tested coal ashes A, B and D is lower than that of generally used construction soil materials (2.5 to 2.7 g/cm^3)[7]. Water content of specimens is almost 0%, except for ash C, which has high water content because water was added to prevent scattering during transport. Ash C is coarse grained with sand fraction greater than 50% and a high distribution of fine-grained material. The consistency limit of all ash is NP, and the fine-grained material (75 μm or less) content is 80%. Therefore, it is necessary to prevent scattering when it is used as a soil material. The coefficient of uniformity of all coal ash is small, which indicates that all five of the coal ashes consist of fine, uniformly granular material. Because of this, the coal ash is prone to liquefaction by earthquake vibration. The addition of a solidifying agent is a liquefaction countermeasure. Planting on structures constructed using coal ash requires the addition of soil dressing, because some of the ash B samples are acidic and most of the other samples are strongly alkaline. Ash E has a calcium oxide content of 10%, because limestone is used in boiler furnace E to remove the sulfur that is generated in combustion. Because of calcium oxide inclusion, ash E solidifies when mixed with water.

Table 1: Physico-chemical properties of coal ash

Boiler		A	B	C	D	E
Soil particle density (g/cm^3)		2.120-2.334	2.238-2.501	2.361-2.522	2.167	2.522-2.690
Notional water content (%)		0.05-0.30	0.02-0.30	43.68-58.18	0.06	0
Grain size distrib.	> 2 mm (%)	0	0	0	0	0
	2 mm - 75 μm (%)	7.8-19.0	5.5-7.0	54.6-71.8	9.5	4.5-8.1
	75 μm < (%)	81.0-92.2	93.0-94.5	28.2-45.4	90.5	91.9-95.5
Consistency limit (%)		N.P.	N.P.	N.P.	N.P.	N.P.
Ground material classification		ML	ML	ML	ML	ML
Ignition loss (%)		8.2-17.3	1.0-3.7	5.2-16.5	16	9.3-18.8
pH		12.0-13.0	3.7-13.0	11.6-13.0	12	11.5-12.6
Maximum dry density (t/m^3)		0.923-1.085	1.020-1.592	0.839-1.013	0.956	1.038-1.005
Optimum moist. content (%)		36.1-43.0	17.0-31.2	47.8-61.1	48.1	45.5-46.0

Environmental concerns and hazardous substances

The Japanese government has been formulating various environmental soil quality standards to protect land from contamination[8]. One such example is the standard for soil reuse. Soil removed from sites where contamination may have occurred, including former factory sites, must meet this environmental standard when it is reused.

In civil engineering projects, soil materials improved by mixing with cement or a cement-based solidification agent must meet standards for hexavalent chromium leaching[9]. While the coal ash itself need not legally comply with the environmental soil quality standards, these standards are often applied to coal ash in consideration of its effects on the environment when it is applied to the ground as soil material.

The authors examined coal ashes A through E based on Japan's environmental soil quality standards. The standards require measurement of 27 items. After examining the raw material and the generation process for coal ash, the authors chose the 8 items of these contained in coal ash for examination of leaching. The authors measured cadmium (Cd), lead (Pb), hexavalent chromium (Cr (VI)), arsenic (As), total mercury (total Hg), selenium (Se), fluorine (F), and boron (B). The measurement results are shown in Table 2. For all types of coal ash measured, the cadmium and lead values fell within the leaching standards. However, the Cr (VI), As, Se, F, and B values exceeded the leaching standards in some coal ashes. It is required to take leaching-reduction measures to render these types of coal ash usable as soil materials.

Table 2: Leaching of hazardous substances: standards and measured values

Substance	Standard (mg/*l*)	Ash A	Ash B	Ash C	Ash D	Ash E
Cd	0.01	<0.001	<0.001	<0.001	<0.001	<0.001
Pb	0.01	<0.005	<0.01	<0.001	<0.001	<0.002
Cr (VI)	0.05	0.021-0.05	0.017-0.07	<0.07	0.011-0.014	0.018-0.004
As	0.01	<0.005	0.002-0.03	<0.02	<0.001	-0.004
total Hg	0.0005	-	-	<0.0005	<0.0005	<0.0005
Se	0.01	<0.005	<0.33	<0.001	0.122-0.181	<0.001
F	0.8	<0.68	1.2-3.4	1.1-2.6	<1.2	-
B	1	0.3-2	1.8-2.8	1.0-1.6	2.6-4.78	-

Leaching Test

Measurements of 8 items of the environmental soil quality standards showed that some types of coal ash release hazardous substances in excess of the standard. To reduce the leaching of these substances, the authors have generally used two methods: mixing of coal ash with water and a cement-based solidification agent followed by compaction; and, mixing of coal ash with water and a cement-based solidification agent followed by granulation. In Japan, a ministry notice states that the leaching of hexavalent chromium from solidified poor-quality soil is most effectively reduced by using blast furnace cement B. This paper examines how the leaching of hazardous substances can be reduced by using blast furnace cement B in two solidification methods.

Laboratory test

(a) Preparation of compacted samples

First, coal ash was mixed with a prescribed amount of water and agitated in a Hobart mixer. Then blast furnace cement B was added, agitation was performed again, and the mix was compacted. Samples produced this way were cured indoors at 20℃ for seven days.

(b) Preparation of granulated samples

Coal ash was mixed with a prescribed amount of water and blast furnace cement B and granulated in a rotating machine. The target mean grain diameter was 3 mm. Because the granulated ash had not solidified enough, the granulated samples were put into plastic bags and cured indoors for about one month before the leaching test.

(c) Measurements of 8 items in the soil quality standard

A solvent solution (500 ml) of purified water and hydrochloric acid (pH 5.8 – 6.3) was prepared. The samples prepared in (a) and (b) above were crushed to a grain diameter of less than 2 mm and put into this solution, with the sample accounting for 10% by weight. Mechanical shaking was conducted (200 rpm, shaking arc: 4 – 5 cm) at 20℃ for six hours under one atmospheric pressure. The liquid after shaking was left to rest for 10 – 30 minutes before being centrifuged for 20 minutes at 3000 rpm. The supernatant was then filtered using a membrane filter with a pore size of 0.45 μm.

Measurements were made of cadmium (Cd), lead (Pb), hexavalent chromium (Cr (VI)), arsenic (As), total mercury (total Hg), selenium (Se), fluorine (F), and boron (B) in the filtered liquid.

Change in leaching of hazardous substances from solidified coal ash

The authors examined the leaching of hazardous substances from solidified cola ash over time. Leaching of the 8 substances in (c) was measured after 7 days and after 1 year, for compacted samples (coal ash that had been mixed with water and a cement-based solidification agent followed by compaction) and for granulated samples (coal ash that had been mixed with water and a cement-based solidification agent followed by granulation). Leaching after 1 year was not measured for ash C. Also, leaching of the 8 substances was measured for samples from embankments constructed 3 to 4 years before the authors' experiment done in 2004 from solidified coal ash and granulated coal ash. Figure 1 shows a cross-section of the constructed embankments. To obtain samples that had not been exposed to the atmosphere, samples were taken at the depth of 5 cm in the embankments. By using the method described in (c), measurements were made of the leaching of Cd, Pb, Cr (VI), As, total Hg, Se, F, and B.

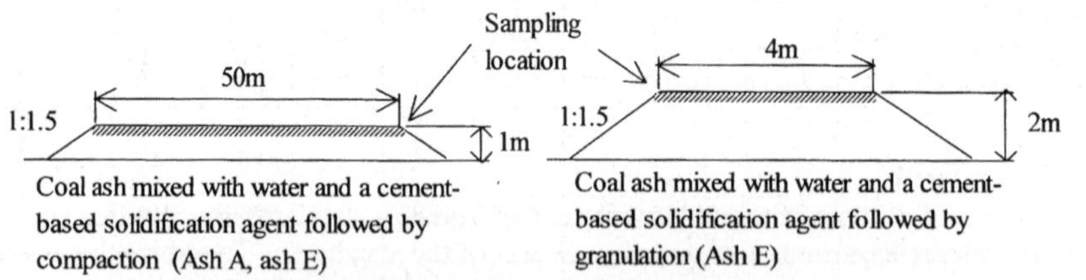

Figure 1: Cross-section of constructed embankment

Tested coal ash and addition of solidification agent

Table 3 shows the types, water content, and use amounts of solidification agents. As shown in the table, 5 types of coal ash samples were used. The leaching of hazardous substances was measured for all 5 types. Measurement in granulated state was done for ashes D and E. Blast furnace cement B was used to solidify or granulate coal ash, but cement was not used for some of the samples of ash from boiler furnace E (ash E), which was solidified by adding only water. In test construction using ash A, to examine the change of leaching of hazardous substances over time, blast furnace cement B was not added. Instead, the water content was adjusted to that which enables roller compaction.

Table 3: Tested coal ash samples

		Name of ash	A	B	C	D	E
Laboratory samples	Compacted, w/ water	Water content (%)	20	20		20,34,48	w_{opt}
		Addition of solidfic. agent (%)	5	5	2,6,10,20	5	0
	Granulated	Water content (%)				20~40	
		Addition of solidfic. agent (%)				0~25	0,2.5,5,7.5
Field samples (embankment)	Compacted, w/ water	Water content (%)	15.7				38
		Addition of solidfic. agent (%)	0				0
	Granulated	Water content (%)					40.9*
		Addition of solidfic. agent (%)					0

*optimal treatment

Test results

Mixing with water and a solidification agent followed by compaction

Table 4 shows the leaching test results for two samples each of ashes A, B, C, and E, one

Table 4: Leaching before and after solidification (mg/l)

Coal ash	Standard value less than mg/l	A-1			A-2		B-1			B-2	
Solidification agent %		Untreated	20		Untreated	20	Untreated	20		Untreated	20
Elapsed time			7 days	1 years		7 days		7 days	1 years		1 years
Cd	0.01	-	-	<0.001	-	-	-	-	<0.001	<0.001	<0.001
Pb	0.01	-	-	<0.001	-	-	-	-	<0.001	<0.01	<0.001
Cr(VI)	0.05	0.05	0.023	0.06	0.05	0.057	0.07	0.03	0.14	0.04	0.14
As	0.01	0.002	0.001	<0.001	0.005	0.006	0.002	0.001	<0.001	0.01	0.002
total Hg	0.0005	-	-	<0.0005	-	-	-	-	<0.005	-	<0.0005
Se	0.01	0.001	0.0009	<0.001	0.002	0.001	0.067	0.003	<0.001	<0.002	<0.001
F	0.8	0	0	0.3	0	0	1.09	0.49	0.3	1.3	0.4
B	1	0.47	0.11	0.18	0.92	0.4	2.4	0.13	0.28	2.5	0.22
pH	-	12.1	12.1	8	-	-	12.1	12.3	9	10.8	7

Coal ash	C-1		C-2		D-1		E-1			E-2		
Solidification agent %	Untreated	20	Untreated	20	Untreated	5	Untreated	0		Untreated	0	
Elapsed time		7 days		7 days		7 days		7 days	1 years		7 days	1 years
Cd	<0.001	<0.001	<0.001	<0.001	-	-	<0.001	-	<0.001	<0.001	-	<0.001
Pb	<0.001	<0.001	<0.001	<0.001	-	-	0.001	-	<0.001	0.002	-	<0.001
Cr(VI)	0.07	0.03	<0.01	0.06	0.014	0.012	0.047	0.013	0.01	0.084	0.016	0.01
As	<0.001	0.008	0.018	0.008	-	-	<0.001	-	0.001	<0.001	-	<0.001
total Hg	<0.0005	<0.0005	<0.0005	<0.0005	-	-	<0.0005	-	<0.0005	<0.0005	-	<0.0005
Se	<0.001	<0.001	<0.001	<0.001	0.161	0.008	<0.001	-	<0.001	<0.001	-	<0.001
F	1.4	0.6	1.9	0.8	1.2	<0.01	-	-	0.3	-	-	0.4
B	1.2	0.09	1.5	0.1	1	0.22	-	-	0.1	-	-	0.05
pH	-	-	-	-	12	-	12	11	8.5	12	11	8.5

sample of ash D, before and after hydration.
Cd, Pb and total Hg were not measured for some samples of coal ash before solidification. However, leaching of those substances after 1 year is less than the measurable level. It can be said that leaching of those substances does not occur over time.

Values for hexavalent chromium in ash samples of A-1, A-2, B-1, and C-1 before solidification exceed the standards. Some of the values after 7 days and 1 year also exceed the standards. There is not much leaching reduction by solidification in these ash samples. But values for hexavalent chromium in ash samples E-1 and E-2 after solidification are less than values before solidification. The values after 1 year remain smaller. Solidification is effective in reducing leaching of hexavalent chromium.

Values for arsenic, selenium and boron exceed the standards before solidification, but after solidification, they fall within the standards for measurements taken after 7 days and 1 year. Solidification is effective in reducing the leaching of those substances.

The leaching of fluorine is reduced by solidification. However, leaching from ash C-2 is great before solidification, and it exceeds the standards even after solidification.

Values for hexavalent chromium, fluorine, and boron in ash C exceed the standards. Figure 2 shows the leaching at different percentages of blast furnace cement B addition for those three substances. Leaching of hexavalent chromium is greater when the mixing ratio of blast furnace cement B is increased, and, in some samples, it exceeds the standards. Boron dropped to within the standard at 10% addition. Fluorine drops to within the standard at 20% addition.

Water content greatly influences the solidification of blast furnace cement B. The leaching of hazardous substances was examined for ash D for different quantities of water added, at

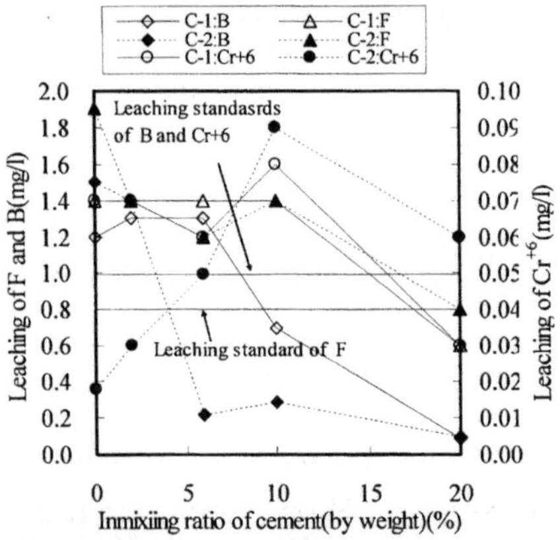

Figure 2: Leaching at different % of blast furnace cement B addition: ash C

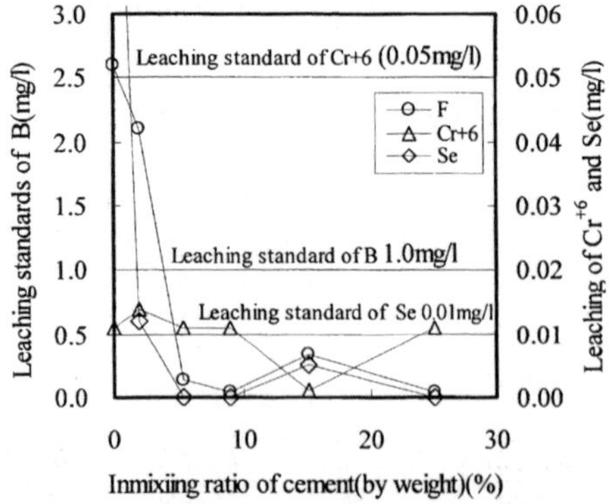

Figure 3: Relationship between leaching of selenium, boron, and hexavalent chromium, and the ratio of blast furnace cement B (by weight): ash D

Table 5. Leaching of selenium, fluorine and boron at different water content when solidified by mixing with blast furnace cement B (mixing ratio of 5 % by weight): ash D

Eluted substance	Standard value (mg/l)	Water content(%)			
		0	20	34	48
Selenium	0.01	0.161	0.008	0.01	0.011
Fluorine	0.8	1.2	<0.01	<0.01	<0.01
Boron	1	4.78	0.22	0.22	0.16

5 % constant mixing ratio of blast furnace cement B. Water content was varied at three levels: the minimum required for compaction, the optimum water content, and a level halfway between these two. Table 5 shows the leaching of selenium, fluorine and boron at different water contents when solidified by mixing with blast furnace cement B. The values for these substances did not satisfy the standards before solidification. However, all the values satisfy the standard after solidification by mixing with blast furnace cement B. The amount of water added does not influence the leaching of those substances.

Thus addition of blast furnace cement B was effective in restraining the leaching of fluorine and boron of every type of coal ash that was mixed with water and the blast furnace cement B and compacted. However, it should be noted that this solidification method does not reduce the leaching of hexavalent chromium for some types of coal ash.

Improvement by granulation

Leaching of selenium, fluorine and boron from some samples of ash D before solidification exceeded the standard. The leaching of these three substances and hexavalent chromium from granulated coal ash D was measured. Figure 3 shows the relationship between, leaching of selenium, boron, and hexavalent chromium, and the ratio of added blast furnace cement B (by weight).

Fluorine did not leach out of this sample. Granulation helps to restrain the leaching of selenium and fluorine. Table 6 shows the leaching of hazardous substances from coal ash compacted after mixing with water and 5% addition of blast furnace cement B, and from coal ash granulated after mixing with water and 5% addition of blast furnace cement B.

Table 6. Leaching of hazardous substances from coal ash compacted, and from coal ash granulated after mixing w/ water and blast furnace cement B (5% mix. ratio by weight)

Eluted substance	Standard value (mg/l)	mixing ratio of 5 % by weight: Coal ash mixed with water	Granulated
Selenium	0.01	0.008	<0.001
Fluorine	0.8	<0.01	<0.01
Boron	1	0.22	0.14

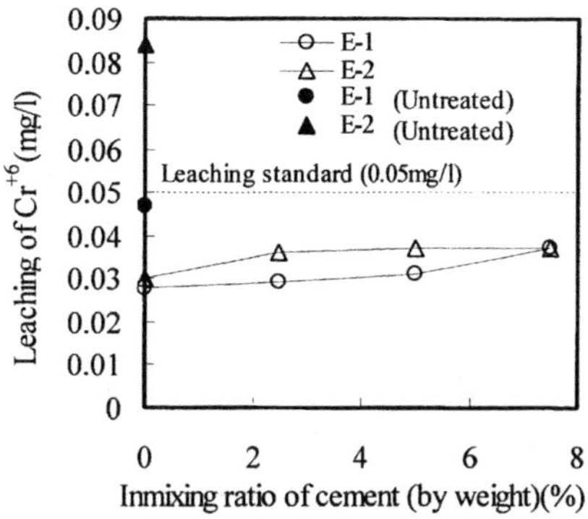

Figure 4: Hexavalent chromium leaching and the ratio of blast furnace cement B (by weight): ash E

Leaching of selenium, fluorine and boron show almost same values for these two types of coal ash. This suggests that the leaching reduction afforded by compaction equals that afforded by granulation.

Figure 4 shows the relationship between leaching of hexavalent chromium from granulated coal ash E and the ratio of added blast furnace cement B (by weight). The volume of hexavalent chromium released from coal ash E, which was granulated without mixing with the solidification agent, is within the environmental quality standard. The mixing ratio of blast furnace cement B does not influence the leaching of hexavalent chromium.

Leaching of hazardous substances from embankments built with coal ash

The leaching of hazardous substances from the embankments built with coal ash is shown in Table 7. At embankment No. 1, leaching of arsenic exceeds the environmental quality standard. Arsenic leaching was not measured when embankment No. 1 was constructed, because the environmental soil quality standards at the time of construction of embankment No. 1 did not specify testing for arsenic in embankment construction. However, leaching should be carefully monitored during embankment construction. At embankment No. 2, the leaching of fluorine exceeds the environmental quality standard 4 years after construction, even though it had met the standard 3 years after construction. At embankments No. 3 and

Table 7: Leaching of hazardous substances from embankments built w/ coal ash

Embankment No.	Standard value less than mg/l	1		2			3			4	
Coal Ash		A-3 hydrated and compacted		E-3 hydrated and compacted			E-4 granulated				
Elapsed time(years)		3	4	Untreated	3	4	Construction	3	4	2	3
Cd	0.01	<0.001	<0.001	<0.001	<0.001	<0.001	0	<0.001	<0.001	<0.001	<0.001
Pb	0.01	<0.001	<0.001	<0.001	<0.001	<0.001	0	<0.001	<0.001	<0.001	<0.001
Cr(VI)	0.05	<0.01	<0.01	0.018	0.02	<0.01	0.022	<0.01	<0.01	<0.01	<0.01
As	0.01	0.31	0.17	0.004	0.007	0.002	0	<0.001	<0.001	<0.001	<0.001
total Hg	0.0005	<0.0005	<0.0005	<0.0005	<0.0005	<0.0005	0	<0.0005	<0.0005	<0.0005	<0.0005
Se	0.01	0.005	0.006	<0.001	<0.001	0.005	-	<0.001	0.002	<0.001	0.002
F	0.8	0.6	0.7	-	0.2	1.7	0.1	0.7	0.7	0.9	0.6
B	1	0.55	0.74	-	0.58	0.5	0.31	0.48	0.46	0.74	0.65
pH	-	7.7	-	12.6	8	-	-	7.5	-	7.6	-

No. 4, leaching of fluorine and boron is greater than that measured immediately after construction. Fluorine leaching at embankment No. 4 exceeds the standard at 2 years after construction, but it falls within the standard at 3 years after construction. This requires investigation. Leaching of fluorine and boron may increase over time, and it may not be sufficient to render coal ash insoluble only by the addition of blast furnace cement B. Follow-up study needs to be made and leaching countermeasures need to be examined. In contrast, the leaching of hexavalent chromium remains small over time. At least with regard to hexavalent chromium, granulation effectively restrains the leaching. The pH of all the coal ash samples is about 12, but it became pH8 at 2 to 3 years after construction.

Conclusions

This study verified that hazardous substance leach from some types of coal ash in excess of the environmental soil quality standards and that addition of blast furnace cement B to coal ash helps to restrain such leaching. The findings are summarized here:

a. The leaching of hazardous substances from coal ash depends on the type of boiler furnace that produces the ash. Mixing with the same amount of solidification agent is not equally effective in restraining leaching for various types of coal ash.

b. Solidification of coal ash helps restrain leaching of fluorine, boron and selenium in the short term.

c. Some hazardous substances leach from coal ash over time, even when the ash is solidified.

Coal ash can be effectively utilized for building embankments. It has already been applied to many embankments in Hokkaido, but there are few cases in which environmental data have been taken to understand changes in the leaching of hazardous substances over time. In future study, the authors will focus on the role of coal ash chemical composition in restraining the leaching of hazardous substances.

References

1)Fumio Yamasawa, Jun'ichi Nishikawa, Atsuko Sato: Use of Coal Ash as a Building Material, Monthly Report of Civil Engineering Research Institute,, No. 563, pp 4-14, 2000.

2)Japan Environmental Technology Association, Coal Ash Handbook, October 1995.

3)Hideo Takahashi, Masato Niikado, Eiji Kaida, Seiji Hasegawa: Fundamental Research on Solidification of Industrial Wastes with Heavy Metals by Cement, Cement Technology Annual Report XXVII, Japan Cement Association, December 1973.

4)Independent Administrative Institute Public Works Research Institute, Manual of Measures against Soil Contamination at Construction Sites (tentative), PWRI Ref. No. 3903. July 2003.

5)Masahiro Matsuda, Satoshi Nishimoto, Atsuko Sato: Research on Use of Coal Ash Mixed with Water, Technical Report Vol. 44, The Japanese Geotechnical Society Hokkaido Branch, February 2004.

6)Kimihito Amano, Jun'ichi Nishikawa, Atsuko Sato, Masayuki Nishijima: Construction of Embankment Using Granulated Coal Ash, The 38th Geotechnical Engineering Research Conference, July 2003.

7)*Procedures for Testing Soils,* The Japanese Geotechnical Society, March 2000.

8)Notification No. 46 of the Ministry of the Environment, Environmental Quality Standards for Soil Contamination, Ministry of the Environment, Japan, August 2001.

9)Examination Procedure for Hexavalent Chromium Leaching from Soil Solidified with Cement or Cementitious Solidification Agent (Draft), Ministry of Land, Infrastructure and Transport, April 2001.

Application to Liquefied soil stabilization material of industry wastes

Hideaki SHIBATA Department of Civil and Environmental Engineering Kokushikan University TOKYO JAPAN
Masatomo TANAKA Department of Civil and Environmental Engineering Kokushikan University TOKYO JAPAN
Isamu ONO Department of Civil and Environmental Engineering Kokushikan University TOKYO JAPAN
Hideki SHIMIZU Engineer Maeda Corporation TOKYO JAPAN
Jun TSUKAHARA Engineer Chuo Kaihatsu Corporation OSAKA JAPAN

Abstract

Today, mass production and consumption to meet people's demand for efficiency, comfort and convenience in their daily lives result in the production of wide varieties of wastes. The treatment and disposal of wastes have been posing serious social problems. In urban areas in particular, final waste landfill sites have only a limited capacity. Reducing or recycling the wastes that are carried to the sites requires urgent action. Attempts have been made to recycle industrial wastes into soil materials because the approach can handle large quantities of wastes. This study examined the possibility of using liquefied soil stabilization materials that are made from various industrial wastes as soil materials. As a result, it was found that coal ash and volcanic ash are sufficiently usable as aggregate of liquefied soil stabilization material. Volcanic ash in particular is highly flowable even where it is subjected to high-density mixing and proved to be an excellent material for producing liquefied soil stabilization material with designated strength by adding only a small quantity of solidifiers.

Introduction

Today, mass production and consumption to meet people's demand for efficiency, comfort and convenience in their daily lives result in the production of wide varieties of wastes. The treatment and disposal of wastes have been posing serious social problems. In urban areas in particular, final waste landfill sites have only a limited capacity. Reducing or recycling the wastes that are carried to the sites requires urgent action. Attempts have been made to recycle industrial wastes into soil materials because the approach can handle large quantities of wastes. This study examined the possibility of using liquefied soil stabilization materials that are made from various industrial wastes as soil materials. Liquefied soil stabilization materials refer to soil materials that are made artificially by liquefying soils using slurry or water and stabilizing them with solidifier. The soil liquefying technology has been developed to use liquefied soil as a highly self-compacting backfill material in urban underground space crowded with lifeline utility facilities and structures of complicated shapes such as subway stations.

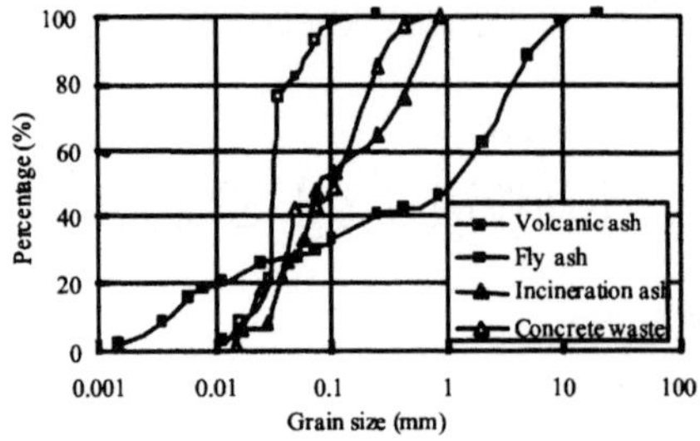

Figure 1 Grain size disstribution curve

Table 1 Engineering properties of Industrial wastes

samples	Remarks						
Concrete waste	Gs=2.53	Uc=7.36	Np	ϕ=37.5deg	$\rho_{d\,max}$=1.51 g/cm^3	ω_{opt}=24%	Sr=73%
Incineration ash	Gs=2.58	Uc=6.00	Np	ϕ=29.5deg	$\rho_{d\,max}$=1.48 g/cm^3	ω_{opt}=25%	Sr=85%
Fly ash	Gs=2.23	Uc=1.84	Np	ϕ=28.0deg	$\rho_{d\,max}$=1.33 g/cm^3	ω_{opt}=23%	Sr=74%
Volcanic ash	Gs=2.86	Uc=7.36	Ip=9.04	ϕ=36.0deg	$\rho_{d\,max}$=1.91 g/cm^3	ω_{opt}=14%	Sr=50%

Engineering properties of wastes

In this study, garbage incineration ash, coal ash and volcanic ash were used as waste samples. The volcanic ash was produced at the time of eruption on the Miyake Island. For comparison with ordinary soil materials, the physical and mechanical properties of these wastes were examined. As parameters of physical properties of the wastes, grain size distribution, density and plasticity index were investigated. Tests were conducted to identify grain size, density and consistency limit according to JIS A 1204, 1202 and 1205 standards, respectively. For the volcanic ash, components were analyzed and analysis was also made through X-ray diffraction. Then, compaction test was conducted and unconfined compressive strength was examined for the wastes according to JIS A 1210 and 1216 standards, respectively to investigate their mechanical properties. The test results are listed in Table 1. The grain size distributions of respective wastes are shown in Figure 1. Table 1 shows that incineration ash and coal ash had a specific gravity of around 2.1, lower than the specific gravity of ordinary soil, 2.6 to 2.7, and a relatively low maximum dry density of 1.33 to 1.445 g/cm3 determined in a standard compaction test. The volcanic ash had a higher specific gravity of 2.864 and a higher maximum dry density of 1.932 g/cm^3, and thus was highly compacted. Figure 1 shows that consistency tests were possible for the volcanic ash although it was a sandy material. As a result of analyses, the volcanic ash was found to contain approximately 10% of CaO and have high contents of SiO2 and Al2O3. Strength therefore could develop easily in the waste. The volcanic ash was also found to be composed mainly of andesite. It was also confirmed that the volcanic ash sample was fully freestanding without using any solidifiers unlike the other wastes, and that the volcanic ash had a strength of qu = 128 to 172 kN/m^2.

Requirements for liquefied soil stabilization material

Quality standards

The general quality requirements for liquefied soil stabilization materials vary according to the government agency concerned[1)2)3)]. Flow, bleeding and post-compaction strength (unconfined compressive strength) are generally adopted as parameters. The requirements specified by the Tokyo metropolitan government, Ministry of Land, Infrastructure and Transport and Japan Highway Public Corporation are listed below.

(1) Tokyo metropolitan government's guidelines[1)]

The Tokyo metropolitan government has specified requirements both for rapid- and slow-hardening materials. In this study, those for slow-hardening materials were used.

Flow: 180 to 300 mm

Percentage of bleeding: Less than 1%

Unconfined compressive strength:

At seven days: More than 127.5 kN/m^2

At 28 days: Less than 549.4 kN/m^2

(2) Ministry of Land, Infrastructure and Transport[2)]

Flow: More than 160 mm (pumping is possible)

Percentage of bleeding: 1% to less than 3% (specified for each use)

Unconfined compressive strength:

At 28 days: Specified for the site where the material is used

(3) Japan Highway Public Corporation[3)]

The Japan Highway Public Corporation has not yet specified requirements. Guidelines are listed below.

Flow: 140 to 200 mm (pumping is possible)

Unconfined compressive strength: 300 to 1000 kN/m^2

Wet density: 1.39 to 1.88 g/cm^3

Mixing test

In this study, the following three tests were conducted to evaluate materials with a view to verifying the applicability of industrial wastes as liquefied soil stabilization materials.

Flow test:

Flow was measured immediately following mixing according to JHS A313 (8 cm x 8 cm cylinder).

Bleeding test:

Bleeding was measured three and 24 hours after mixing according to JSCE-F-522.

Unconfined compressive strength test:

Test was conducted on three samples at seven and 28 days according to JIS A1108.

Test results

Test results for respective wastes are described below.

Incineration ash and concrete wastes

No slurry could be produced from incineration ash or concrete wastes probably because these wastes contained few fine particles of less than 0.075 mm and thus were difficult to mix with water.

Coal ash

Figures 2 through 4 indicate the relationship between C/e, one of the indicators of mixed state of liquefied soil stabilization material, and three common quality parameters, flow, bleeding and post-compaction strength (unconfined compressive strength). C/e is the ratio between solidifier (C) and void (e) that represents both the degree of cementation by the solidifier and the structural strength of soil particle. Some works reported a good correlation between C/e and the quality of liquefied soil stabilization materials[4)5)6)]. Figure 2 shows that the coal ash is highly flowable because flow ranges from 160 mm to 220 mm, the level at which pumping is possible, as long as C/e is 20% to 60%. Figure

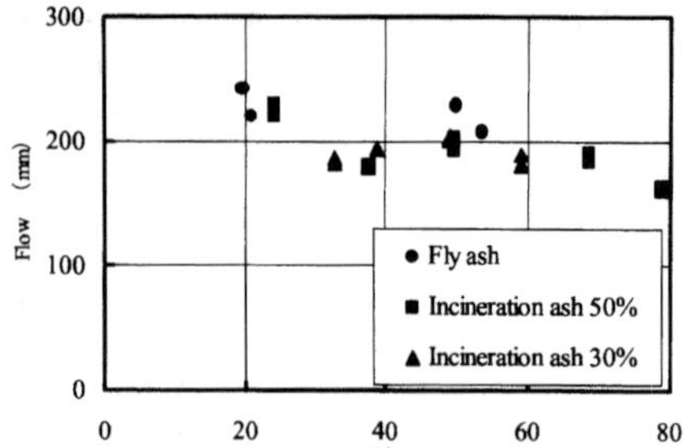

Figure 2 Relationship between c/e and Flow

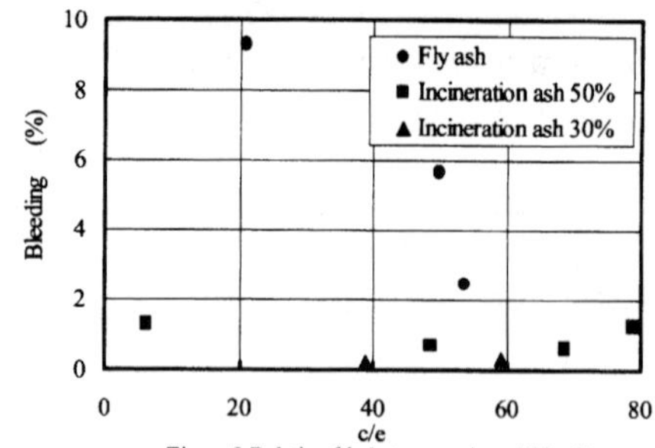

Figure 3 Relationship between c/e and Bleeding

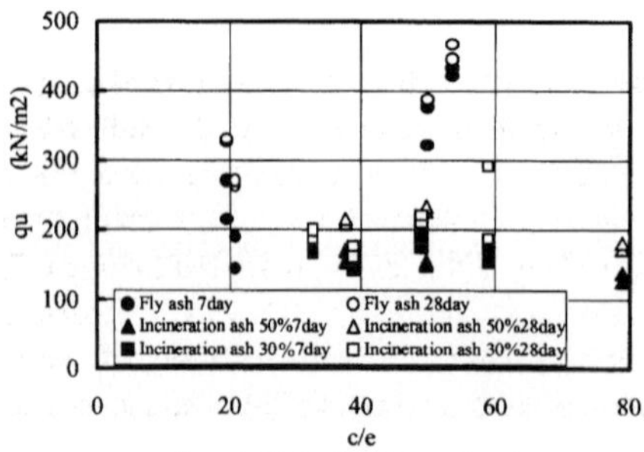

Figure 4 Relationship between c/e and qu

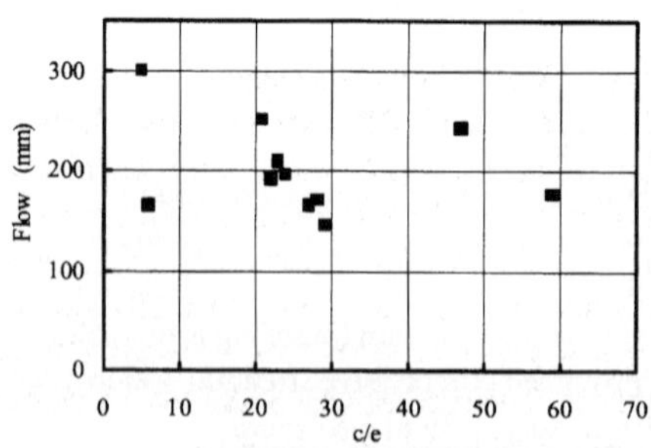

Figure 5 Relationship between c/e and Flow

3 shows that bleeding is high at 2.5% at C/e of 52%. Coal ash is therefore easily subjected to material separation as it is. Figure 4 shows that qu increases with C/e. qu is in the range between 127 kN/m^2 and 549 kN/m^2 specified by the Tokyo metropolitan government where C/e varies from 20% to 60%. Coal ash was thus found to be able to retain strength after hardening.

Composite material

A study was made of the possibility of using a composite of incineration and coal ash, either of which could not be used individually to produce slurry, as a liquefied soil stabilization material. Mix proportion of coal and incineration ash was 1:1 (referred to as composite material 1) or 7:3 (composite material 2). Test results according to the mix proportion are shown in Figures 2 through 4. Figure 2 shows that both of the composite materials had a flow of more than 160 mm at the range of C/e shown in the figure. Thus, they were found to be highly flowable. Figure 3 shows that both had a low bleeding percentage of less than 2%. Separation hardly occurs in these materials. Figure 4 confirms that both materials have post-consolidation strength qu of more than 127 kN/m^2 at seven days and less than 250 kN/m^2 at 28 days, meeting the requirements of the Tokyo metropolitan government. Composite materials, however, had a lower strength than the material made solely of coal ash, and were found to be difficult to apply to structures requiring high strength. Mixing coal ash with other wastes was found to enable wider use of composites.

Volcanic ash

Mixing tests were conducted in eleven cases. Tests were carried out in five cases of varying unit volumes of cement. High- or low-density mixing was adopted in each case. Figures 5 through 7 indicate the relationship between C/e, one of the indicators of mixed state of liquefied soil stabilization material, and one of three common quality parameters, flow, bleeding or post-compaction strength (unconfined compressive strength). The C/e-flow relationship in Figure 5 shows that mean flow is

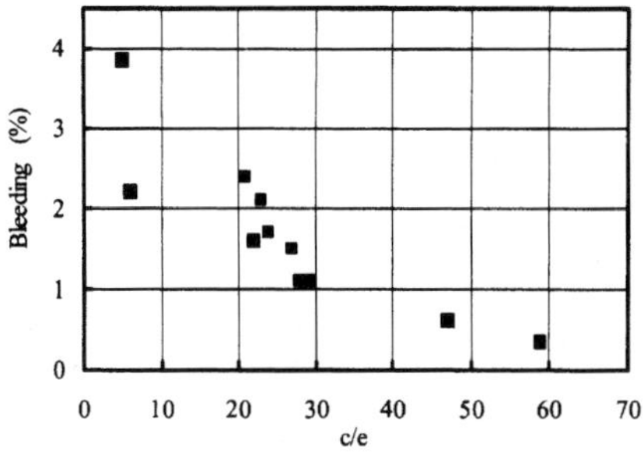

Figure 6 Relationship betweenn c/e and Bleeding

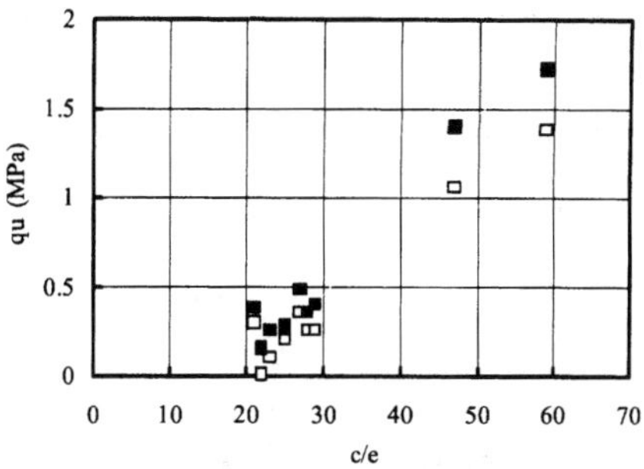

Figure 7 Relasionship between c/e and qu

never below 160 mm, the marginal flow at which sufficient self compaction can be secured, as long as C/e is less than 60% although flow fluctuates slightly. Figure 6 indicates a good C/e-bleeding correlation. It seems therefore necessary to set C/e at more than 30% to control bleeding below 1%, a general target. The knowledge obtained from Figures 5 and 6 suggests that C/e should be set at 30 to 60% to meet the requirements with respect to flow and bleeding. If the finding is applied to Figure 7, unconfined compressive strength at 28 days can be set at 0.5 to 1.75 MPa (= 103 kN/m^2). To obtain strength according to the use, adjusting C/e, or the quantity of solidifier and the density of slurry, suffices. The mixing tests produced the following information about the use of volcanic ash as liquefied soil stabilization material.

(1) Strength develops easily. This is because the composition of volcanic ash helps promote cement hardening.

(2) High-density mixing does not greatly deteriorate flowability. This is attributable to the effect of high soil particle density.

(3) Producing low-density liquefied material with a small quantity of cement (e.g. for backfill at locations that are expected to be re-excavated) is likely to increase bleeding.

As a result of verification of the applicability of volcanic ash as liquefied soil stabilization material, it was found to be nearly ideal material because it is highly flowable despite high density and provide designated strength with the addition of a small quantity of solidifiers. The only disadvantage is the need of measures to control bleeding where soil materials of poor mixture with low strength are required.

Conclusions

1.No slurry can be produced from incineration ash or concrete wastes. These materials are therefore unusable as aggregate of liquefied soil stabilization material.

2.Coal ash is easily subjected to material separation but is able to retain sufficient strength after hardening. It is therefore usable as aggregate of liquefied soil stabilization material.

3.In volcanic ash, strength develops easily and high-density mixing does not greatly deteriorate flowability. Volcanic ash is therefore considered to be an excellent aggregate of liquefied soil stabilization material.

Acknowledgment

The authors would acknowledge with gratitude that the Japan Iron and Steel Federation granted subsidies for this study in fiscal year 2004-2005.

References

1) Ministry of Land, Infrastructure and Transport: Technical manual concerning the use of liquefied soil stabilization materials, pp. 1-72, December 1997.

2) Matsumura, M., Abe, T., Sugimoto, T., Hiroshima, M. and Doiuchi, H.: Recycling of construction sludge to liquefied soil stabilization material, 1998 annual report of the Institute of Civil Engineering of the Tokyo metropolitan government, pp. 73-84, 1998.

3) Hokkaido Branch of the Japanese Geotechnical Society: Proceedings of symposium on the engineering property evaluation of volcanic ash soils, pp. 219-244, September 2002.

4) Kuno, G. (Ed.): Soil liquefying method - Technologies for recycling construction muck and sludge, Gihodo Shuppan Co., Ltd., 1997.

5) Shibata, H., et al.: A proposal for rehabilitation from the volcanic disaster on the Miyake Island - Use of volcanic ash for liquefied soil stabilization materials -, Soils and Foundations, 51-9, 2003.

6) Kuno, G., et al.: A few considerations on unconfined compressive strength of liquefied soil stabilization materials, Proceedings of 33rd conference of Japanese Geotechnical Society, 1998

Landfill Site Selection Using GIS

R.K. Srivastava, Professor, Department of Civil Engineering, MNNIT Allahabad, India -211 004,
Mahesh Reddy, P.G. Student, Department of Civil Engineering, MNNIT, Allahabad, India- 211004
P. Bala Ramudu, Research Scholar, Department of Civil Engineering, MNNIT, Allahabad, India, 211 004
R.P. Tiwari, Reader, Department of Civil Engineering, MNNIT Allahabad, India -211 004,

Abstract
In under developed and developing countries solid waste disposal was never looked after significantly and waste disposal kept on without much concern about any ill effects on land, air, groundwater or community. Nowadays the growing environmental and ecological concerns with the increasing population and reducing land availability for infrastructural development critically, the requirement of engineered landfill has become essential component of development process and also site selection regulations have become more rigid. The proper siting of a landfill takes into account all factors of the surrounding landscape and environment and assures that the landfill is in compliance with the engineering and environmental stipulations. The applications of GIS in waste management facilities are highlighted and a conceptual GIS model is developed for locating best/suitable sites for landfills for the city of Allahabad, India.

Introduction
In today's society, finding a location to site desirable facilities for disposal of solid waste is becoming a significant problem in the planning sector. The siting of landfills is becoming more of an issue due to "not in my backyard (NIMBY)" and "not in anyone's backyard (NIABY)" concerns from the public. It is also important because of the imperative nature of landfills, expanding population and their expanding volume of garbage. Factors associated are of spatial nature and a huge database is needed to be managed and analysed. Geographical Information System (GIS) is an ideal tool for such spatio-temporal problems owing to its versatility in handling large sets of data. It is a powerful set of tools for collecting, storing, retrieving at will, displaying and transforming spatial data. Many barriers to siting landfills have existed in the past and many of them are still a problem today. The greatest barrier that facility site planners face is the spatial problem itself; finding a location for the site that satisfies everyone. The main challenge is in dealing with outraged members of the community who do not want a certain facility located in their neighborhood. The second challenge is economic influence. It is difficult to find suitable land to develop a landfill that is reasonably close to the garbage source. Economics can become an even larger issue when the only option is to transport the garbage to far of locations. The planner has the challenge of determining which site can best serve those within certain boundaries.

In the city of Allahabad (India), open dumping landfilling is normally practised for disposal of solid wastes. Large quantities of waste are just deposited in low lying areas of the city or on city boundaries. There is urgent need to select a site and construct a engineered landfill. Hence a study has been undertaken to select site for such a landfill using GIS platform. In this case the collective information and overall view of the possible sites proved beneficial in deciding priorities for disposal with respect to the waste characteristics.

In landfill site location, large amounts of data have been analysed, such as data of soils, landuse/landcover. Slope and bedrock information has been utilised in specifications of landfill. Formulation of the multi-criteria framework has been done from the engineering (geological, geotechnical, geoenvironmental), ecological and finally economic criteria at stages. GIS is an immensely useful tool in waste management as it can help in management of the removal of wastes from populated areas and proper disposal of the waste. This tool has been used to identify the appropriate site locations of landfills, through spatial analysis.

Methodology

Landfill Location Criteria

The engineering, ecologic and environmental criteria for proper siting of landfills, is commonly prescribed by the local governments and environmental groups. One such is criteria for municipal waste landfills by Central Pollution Control Board, New Delhi for Indian conditions, is presented in Table 1.

Table 1. Location criteria CPHEEO (2000)

Features	Criteria
Lake/Pond	No landfill within 200m of Local surface water bodies
River	No landfill within 100m of river, stream
Flood plain	No landfill within 100yr flood plain
Highway	No landfill within 500m
Habitation	No landfill within 500m
Public parks	No landfill within 500m
Airports	No landfill within 1500m
Water supply well	No landfill within the radius of 500m
Power lines (electrical lines)	No landfill within 700m
Ground water level	Not less then 2m below the base of Landfill
Canals	No landfill within 200m

Ranking System

In any multi – criterion framework, suitability rating should be measured based on which the viability of alternate decisions could be compared. Suitability models attempt to identify locations that are best suited for a landfill site. Based on the extensive study a set of 8 parameters for municipal waste sites have been considered for ranking. Parameters are expressed on a common scale of 1 to 9, which is also known as sensitivity index. The value of sensitivity index multiplied by the corresponding weightage, results in score for each of the parameter. Individual score are added for total score for the site. The site which is having highest suitability score is taken as best site. (Kukesh and Babu, 2002). Table 2 presents parameters for MSW disposal sites and corresponding weightages. Table 3 presents ranking of sites as overall score.

Table 2 Suitability Parameters and weightage

Parameter	weightage
Slope of ground	60
Water table	100
Fluctuation of water table	80
Drainage (distance from rivers)	70
Soil type	100
Permeability	90
Distance from roads	40
Distance from waste producing units	50

Table 3 Ranking of Sites

Class	Overall score sites	Designation
I	1000-600	VERY GOOD
II	600-400	GOOD
III	400-200	FAIR
IV	200-000	POOR

Data Base Creation

The various steps followed in the creation of GIS based integrated database in the present work are as follows:

- Data Acquisition
- Scanning of maps
- Georeferencing
- Digitizing

Data Acquisition

- Topographic map 63G at 1:250,000
- Topographic map 63G-15 at 1:50000
- Topographic map of Allahabad city at 1:20000
- Land use plan of Allahabad at 1:50000
- Data of waste generated from each ward of Allahabad
- Bore hole data which provides details of the type of soil, depth of the water table, fluctuation of water table and permeability.
- Tube well data

Scanning the Maps

The data obtained in the data collection stage was in the form of paper maps and plans. This data is to be converted in to digital format so that it can be understood by GIS software. For this, the map 63G, 63G-15, Landuse Plan and map of Allahabad city were scanned at 300dpi.

Georeferncing

Assigning coordinate from known reference system, such as latitude/longitude, UTM, state planar map, to the page coordinate of a raster (image) or a planar map is known as georeferencing. Georeferencing raster data allows it to be viewed, queried, and analyzed with other geographic data. The topographic maps 63G, 63G-15, Allahabad city map is georeferenced in ArcGIS 9.0 using the Georeferencing toolbar. The control points chosen for the present work are the intersections of latitudes and longitudes in the topographical maps provided by Survey of India. Then the map is rectified to form a new georeferenced raster dataset. The control points are saved in the form of a text file for further reference.

The landuse map is rectified by image to image registration. For this, the georeferenced 63G and 63G-15 are map used. The important points which can be easily identified in both the raster datasets like the intersections of roads, railway level crossing, buildings and boundaries, *etc.* were identified. The X, Y input for these points in landuse is obtained from the georeferenced 63G and 63g-15 maps. For present work, geographic coordinate system is Everest modified projection and the assigned projection coordinate system is Polyconic.

Digitizing

Digitizing is the process of converting features on a paper map into digital format. To digitize a map, digitizing tablet (also know as a digitizer) connected to computer to trace over the features of interest are used. The x, y coordinates of these features are automatically recorded and stored as spatial data. Digitizing with a digitizing tablet offers another way, besides screen digitizing "freehand", to create and edit spatial data. One can convert features from almost any paper map into digital features. One can use a digitizer in conjunction with the editing tools in ArcMap to create new features or edit existing features on a digital map. For present work digitization has been done by free hand. Following features have been digitized.

- Lake/pond as polygon
- Rivers as line
- Flood plain as polygon
- Highways as line
- Habitation as polygon
- Airport as polygon
- Water supply well as point
- Power line (electrical line) as line
- Canal as line
- Parks as polygon
- City boundary as polygon

Buffer Analysis:

Buffering create zones of interest around an entity or feature. Buffer zones may be of fixed or varying width and according to necessity. In the present work buffer zones created around the features are of fixed width (Table 4).

Table 4 Features and Buffer Distances

Features	Buffer distance
Lake/Pond	200m
River	100m
Highway	500m
Habitation	500m
Public parks	500m
Airports	1500m
Water supply well	500m
Power lines (electrical lines)	700m

Overlay Analysis

Overlay analysis involves joining two existing sets of features into a single set of features and to identify spatial relationships between the input features. There are three types of vector overlay operations. For the present work union and erase are used. Union computes a geometric intersection of the input feature. All features are to be written to be output feature class with the attribute from the input features which it overlaps. Figure 1 presents a schematic diagram of union operation.

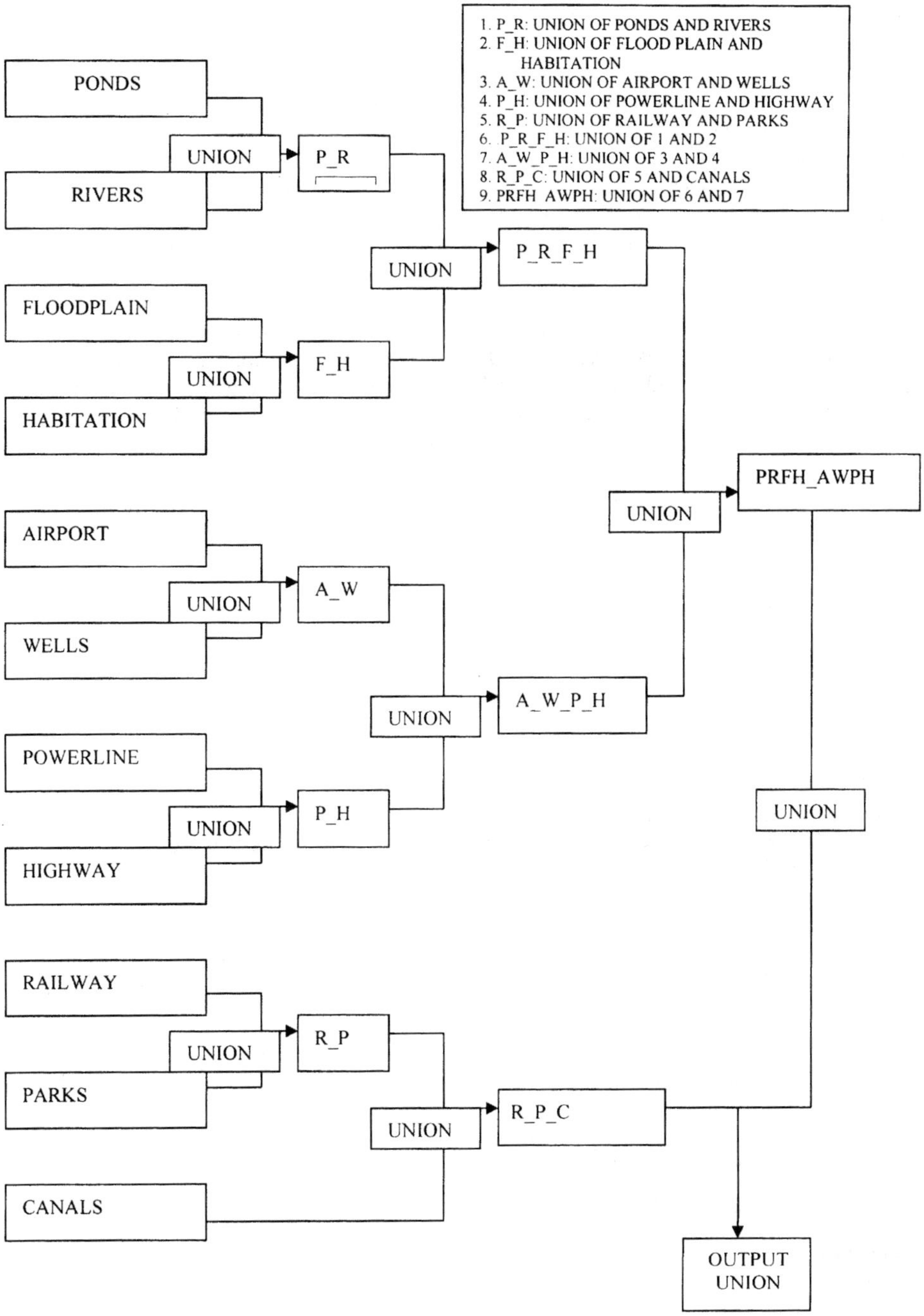

Fig.1 Union Operation

Final Sites

Boundary of Allahabad which is digitized as polygon is input feature and erase feature as final union output saved as output_erase.shp. Output_erase.shp is a feature class having single polygon feature. This output_erase.shp feature class was exported into coverage file. By this operation all features which are having closed boundary breaks into individual polygons and output is saved as output_erase.cov. Area of each polygon is calculated by using tool "calculate area" for output_erase.cov and output is saved as output _erase_area.cov. Again coverage file (erase_area.cov) is exported into shapefile and saved as erase_ area.shp, because one cannot perform manipulation on coverage file in ArcMap. Areas of features which is less than 60 hectares and one dimension of features are very small than the other dimension are eliminated. The polygon features which are remaining in file are potential sites for landfill which is saved as final sites.shp. The total number of sites whose area is more than 60 hectares and satisfying landfill selection criteria as per CPHEEO (2000) are 16. the minimum landfill area of 60 hectares has been chosen on the basis of amount of MSW produced in Allahabad city and for design life of landfill as 30 years.

Ranking of sites

Site ranking is carried out by adding suitability map and potential sites shape file to Arcmap. Overlaying potential sites on suitability map graphical, the site which fall in higher suitability area is ranked as 1 and which have lower suitability as 4 (Fig.4). Table 5 and Figure 4 present various potential MSW landfill site ranked from 1 to 4 and from very good to poor.

Conclusions

1. GIS is ideal tool for handling, modeling and analysis of landfill site selection data.
2. GIS based landfill site selection screening methodology has been developed.
3. Site suitability model has been developed for Allahabad city.
4. Total number of sites in Allahabad are 16 of which 2 are of ranked as very good, 5 are of ranked good, 6 are of ranked as fair and 3 are of ranked as 4 poor.
5. GIS model can only be effective with availability of accurate data sets.
6. As compared with conventional method, GIS based model is more efficient, time effective and accurate.
7. Database which is created for landfill site selection can be used for calculation of shortest path, optimum path, least cost path, cost weight path and finding nearest facility for disposal of waste.

References

1. Kukesh M.D and Babu S.G.L (2002) GIS Application in Locating Landfill Sites. *Proceedings IGC,* Allahabad, India, Vol.1, pp. 153-156
2. CPHEEO (2000), Manual on Municipal Solid Waste Management, Central Public Health and Environmental Engineering Organisation, Ministry of Urban Development, New Delhi.
3. Rao K.S. (1997) Site Selection for Landfills and Characterization of Landfill Site, Waste Disposal in Engineered Landfills, Narosa Publishing House, pp.56-87.

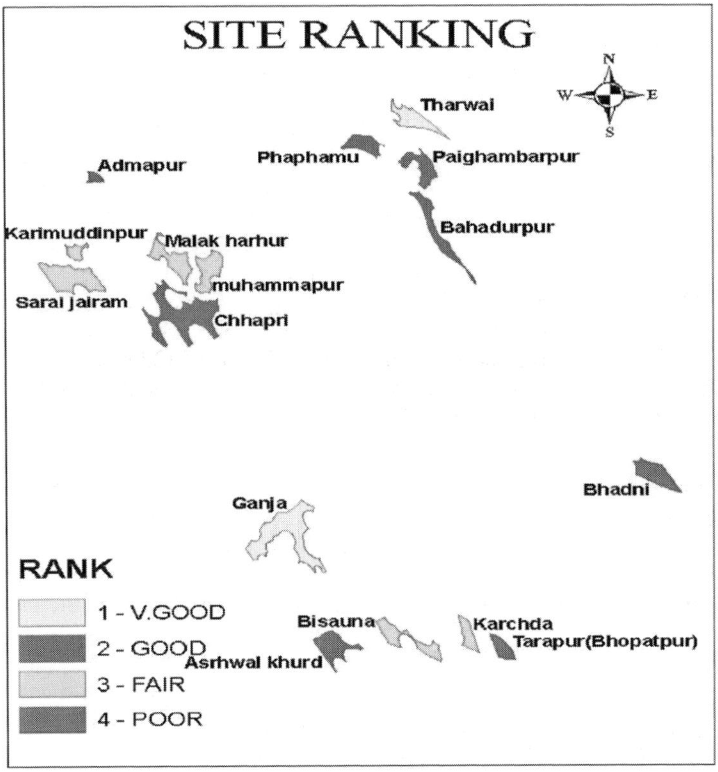

Fig.4 Ranking of Sites

Table 5 Ranking of Sites

S.no	Name of the site	Rank	Level of site
1	Ganja	1	Very Good
2	Tharwai	1	Very Good
3	Ashrwad Khurd	2	Good
4	Bahadurpur	2	Good
5	Chhapri	2	Good
6	Paighambarpur	2	Good
7	Phaphamu	2	Good
8	Bisauna	3	Fair
9	Karchda	3	Fair
10	Karimuddinpur	3	Fair
11	Malak Harhur	3	Fair
12	Muhammapur	3	Fair
13	Sarai Jairam	3	Fair
14	Admapur	4	Poor
15	Bhadni	4	Poor
16	Tarapur	4	Poor

Evaluation method for consistency and its application for estimating strength properties of coal fly ash

Daisuke SUETSUGU, Saga University, Saga, Japan (Formerly, National Defense Academy)
Yoshihisa MIYATA, National Defense Academy, Yokosuka, Japan

Abstract

This paper concerns geotechnical utilization of coal fly ash. A new consistency concept of coal fly ash is proposed to correspond to variety of its characteristics. And it is shown that strength properties are able to be evaluated by new consistency concept. In this study, six kinds of coal fly ash are applied to define consistency. Triaxial compression tests are carried out to investigate strength properties and to consider applicability of consistency concept for evaluating them. The summary of the results is shown as follows: (1) Coal fly ash-water mixtures are able to be classified into the flocked state, the viscoid state and the fluid state. The viscoid limit is defined as the boundary between the flocked and the viscoid state, and the fluid limit is defined as the boundary between the viscoid and the fluid state. (2) The optimum mixture ratio that shear strength appears the highest is founded. (3) The optimum mixture ratio is evaluated by a new index based on proposed consistency limits.

Introduction

Recently, the amount of coal fly ash discharged from thermal power plant tends to increase in Japan. It is necessary to still more utilize coal fly ashes. Coal fly ash is regarded as an high quality earth material. Because coal fly ashes have self-hardening characteristic and its weight is lighter than typical soils. Using as an earth material is an ideal utilizing method because of being necessary large amount for filling and reclaiming.

It is very useful to use coal fly ashes with slurry condition (Horiuchi et al, 1992, 1996). This method is able to fill into narrow spaces or under water in high density. The characteristic of coal fly ashes varies with combustion condition or the coal characteristic. So, the characteristic as an earth material also varies. In the case of using coal fly ash with slurry condition, it is important to determine the amount of mixing water.

In the present paper, we propose a new consistency concept of coal fly ash to correspond to variety of it and to use it efficiently as an earth material, observing conditions of coal fly ashes mixed with water. Next, we investigate strength properties of coal fly ashes from a series of triaxial test results and show that strength properties can be evaluated by consistency limits which are defined in this study.

Consistency for coal fly ash

Samples

In this study, six kinds of coal fly ash were used, which are produced from each six coals

Table 1: Physical properties of coal fly ashes.

Sample	Physical properties					
	Particle density (g/cm³)	Mean diameter (μm)	Silt fraction (%)	Clay fraction (%)	Plasticity index	Ig loss (%)
RS	2.155	10.6	83.9	15.0	NP	0.3
YP	2.382	31.2	87.3	4.5	NP	0.5
DD	2.333	21.5	79.9	9.5	NP	2.3
BT	2.264	31.7	89.4	4.2	NP	0.1
WW	2.174	26.5	91.1	2.5	NP	4.3
BS	2.128	26.1	87.9	3.6	NP	3.8

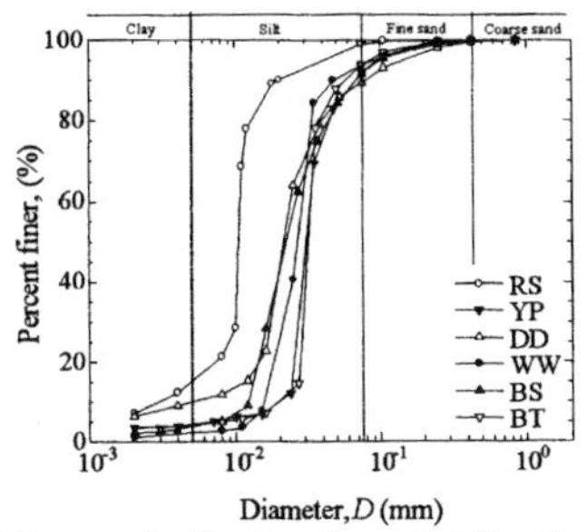

Figure 1: Grain size distribution curve

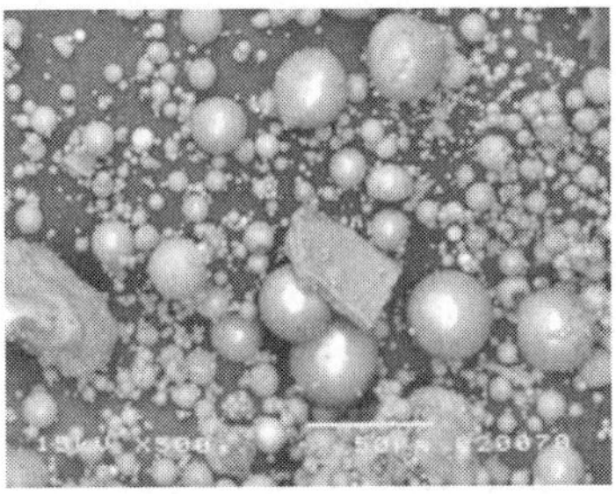

Figure2: Particle shape

Table 2: Chemical composition of coal fly ashes.

Samples	Chemical composition					Total (%)
	SiO_2 (%)	Al_2O_3 (%)	CaO (%)	MgO (%)	Fe_2O_3 (%)	
RS	47.6	30.1	7.8	4.0	3.0	92.5
YP	58.5	14.2	11.4	3.4	5.8	93.3
DD	56.7	22.9	3.0	-	9.2	91.8
BT	45.0	26.4	4.6	4.1	11.2	91.3
WW	71.8	23.6	0.3	-	2.1	97.8
BS	61.0	29.4	0.7	-	1.8	92.9

mined at different places. These are collected by the electric dust catcher at several thermal power plants in Japan. The coal fly ash sets when contacts with water. But coal fly ashes used in this study has maintained the condition when being discharged and never contacted with water. Physical properties of coal fly ashes are shown in Table 1, and grain size distribution curves are shown Figure 1. Particle density of coal fly ashes are 2.1-2.4g/cm^3. It is lighter than that of typical soils. Uniformity coefficients U_c are 1-4 and mean diameters are 10 to 30μm. The main texture of all coal fly ashes consists of a silt fracture. Coal fly ashes used in this study are classified as silt according to the conventional soil classification. The shape of particle, as in Figure 2 taken by SEM, is round. Chemical components analyzed by the fluorescent X-ray are shown in Table 2. Main components are SiO_2 and Al_2O_3. There components account for about seventy to ninety percentages of the whole component. CaO which influences the self-hardening characteristic is detected with less than one to over ten percentages. It is founded that characteristics of each coal fly ashes used in this study are different.

Examination for the consistency

To define consistency of the coal fly ash, conditions for six kinds of coal fly ashes when water is mixed were observed with increasing gradually mixing water to that water bleeds from them. Consistency is often used as a term of deformation resistance degree with amount of water content for plastic fine graded soil in geotechnical engineering. Generally, consistency is defined as a term of hard and fluid of the body. In this study, consistency is

Figure 3: Change of condition with adding water

Table 3: Observation results

Flocked	○	○	×	×
Viscoid	×	○	○	○
Fluid	×	×	×	○
Condition	NF	F-1	F-2	F-3

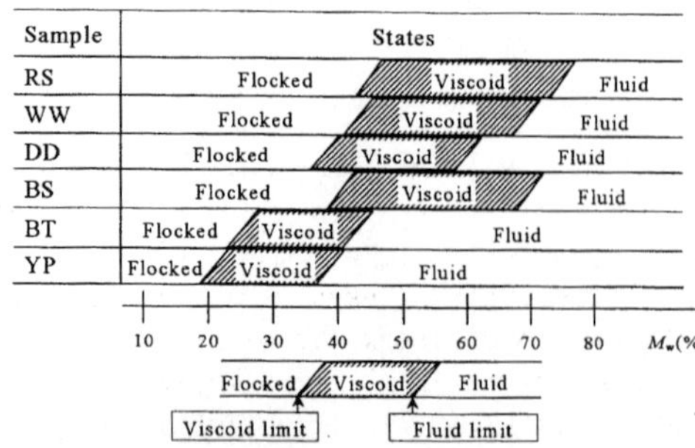

Figure 4: Consistency of coal fly ashes

defined as a term of solidity and deformation resistance degree with changing amount of water contents. Consistency was observed by a below method to classify states of coal fly ashes mixed with water. Mixture ratio M_w is defined as the ratio of amount of mixing water m_w to dry mass of a coal fly ash m_s. M_w is expressed as follows.

$$M_w = \frac{m_w}{m_s} \times 100 \quad (1)$$

In order to product coal fly ash – water mixtures homogeneously, water was added into two steps and then mixed well by a mixer. A bit of the mixture was collected to measure mixture ratio and then remaining the mixture was poured in a tray. States of the mixture poured in a tray were observed. Above operations were carried out from completely dry condition with increasing water gradually. Change of condition from dry condition to bleeding condition is shown in Figure 3. From a dry condition such as powder as adding water, coal fly ashes became to be what flocks conspicuously appears as Figure 3(a). Continuously, there became to be viscoid though flocks thinly appeared as Figure 3(b) and then flocks disappeared as Figure 3(c). Finally, there became to be a condition that excess water clearly appeared as Figure 3(d). If above observation results are summarized based on characteristic states, there are shown as Table 3. On this consideration, it is possible that states of coal fly ash are classified into four categories. NF, F-1, F-2 and F-3 in the table correspond to Figure 3(a), 3(b), 3(c) and 3(d) respectively. But it is difficult to quantitatively evaluate flocks appearance such as difference between F-1 and F-2 in Table 3. If concerning two characteristics, viscoid and fluid, states of the coal fly ash are able to be classified in three states as the flocked state, the viscoid state and the fluid state. The visciod state on above classification is defined as what combined F-1 with F-2 in Table 3. Observation results of all coal fly ashes based on this classification are shown in Figure 4. Mixture ratios on limit states is described in about 3% error with observation accuracy considering. It is found that mixture ratios on limit states of each coal fly ashes are different. We define the fluid limit M_{wf} as a mixture ratio of bound between the flocked state and the viscoid state, and the fluid limit M_{wg} as a mixture ratio of bound between the viscoid state and the fluid state. In the next chapter, we discuss applications of two limit indexes for evaluating strength properties.

Relationship between strength properties and consistency

Experiment

Strength properties of coal fly ashes are examined by triaxial compression tests, in order to discuss application of consistency limits defined in this study. Three kinds of coal fly ashes were used for a series of triaxial tests. Specimens applied for triaxial tests were prepared as follows. A coal fly ash-water mixture fixed in a mixture ratio was poured into the mould as in Figure 5. The mould has the size of 200mm in diameter and 400mm in height, and four small moulds whose size are 50mm in diameter and 100mm in diameter inside. After pouring the mixture, inside air of the mixture was removed with applying vibration on the side of the mould, and covered the mixture with a high polymer film to prevent water from evaporating and then cured during seven days at room temperature of 20 degrees Celsius. After seven days, inside water of the mixture was drained from the bottom of the mould, and specimen applied for triaxial tests were taken from small moulds. In this study, coal fly ashes contacted with water for only one week. We consider that effects of solidification by pozzolanic reaction on strength properties can be ignored because the reaction needs very long time with contacting water. Triaxial tests were carried out with undrained condition and four consolidation pressures.

Test results

Relationships between deviator stress q and axial strain ε_a and relationships between excess pore water pressure Δu and axial strain ε_a of each mixture ratios of RS-ash are shown in Figure 6 in a case of isotropic consolidation pressure p_0'=98kPa. In case of a small M_w specimen, q increased with Δu increasing just after starting shear, and Δu turned into decrease when ε_a reaches one value, q increased and showed a peak with Δu decreasing. After this, q decreased immediately with Δu converging to constant value. In case of a large M_w specimen, q increased with Δu increasing just after starting shear, and q tended to converge to a constant value though Δu turned into decrease. Decrease degree of Δu after turning decreased much smaller than that of small M_w specimens, and q did not show the peak but increased monotonously. Stress-strain relationship behaviors with amount of water changing are same even if the kind of coal fly ash varies. Internal friction angle $\phi'$and cohesion c' obtained from a series of triaxial test results are shown in Figure 7. Sedimented sample results are both described in Figure 10 to compare with strength properties. As M_w became lager, ϕ' also became larger. ϕ' of the maximum M_w specimen nearly equaled that of sedimented specimen. c' became smaller with M_w increasing and finally became zero. Shear strength of a small M_w specimen is larger than that of a large M_w specimen because of dilating based on stress-strain relationships as in Figure 6.

Maximum value of q is defined as shear strength q_f, relationships between q_f and M_w are shown in Figure 8. It is cleared that maximum of q_f appears regardless of the kind of coal fly ash. M_w when the maximum appears is different from each coal fly ashes. q_f- M_w relationships on various consolidation pressures are also same as this. Here, we define this M_w as the optimum mixture ratio M_{wopt}. From these test results, it is founded that strength properties of coal fly ashes varies with the kind and amount of mixing water, and the maximum of q_f appears regardless of the kind of the coal fly ash.

Relationships between strength properties and the consistency indexes

A relationship between shear strength q_f and mixture ratio M_w rearranged using consistency limits are shown in Figure 9. It is cleared that maximum q_f appears on the viscoid state. This result is invariable regardless of the kind of the coal fly ash and consolidation pressure. Each coal fly ashes have the own optimum mixture ratio respectively as in Figure 8. The relative

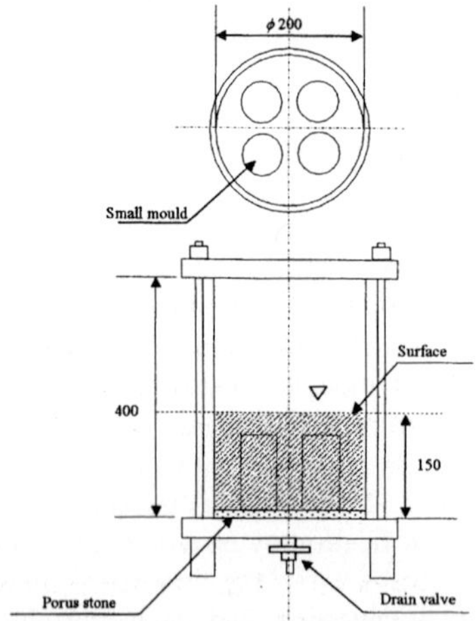

Figure 5: Schematic diagram of a specimen preparation mould

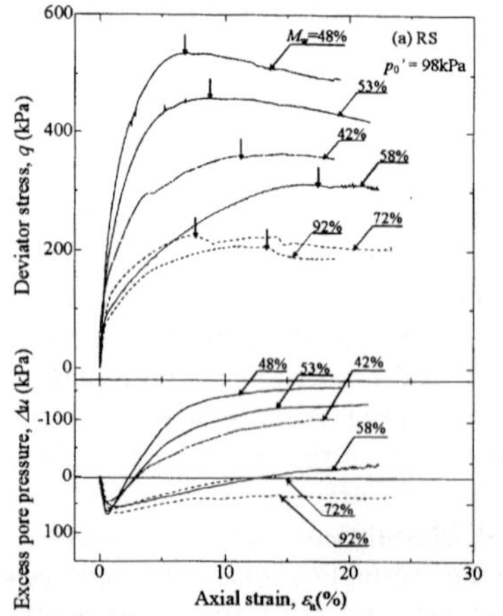

Figure 6: Stress-strain relationships of RS-ash

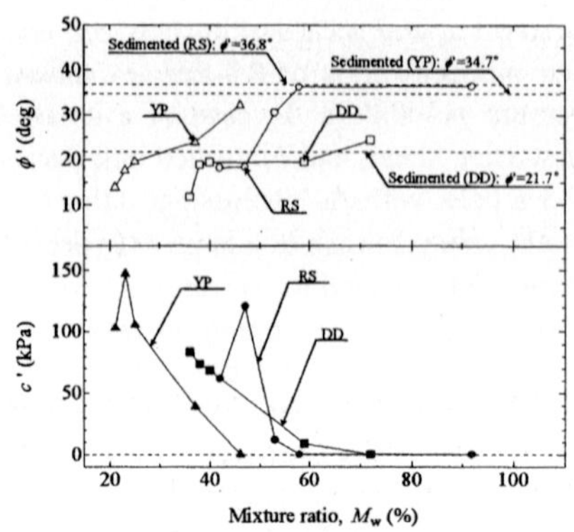

Figure 7: Relationships between ϕ',c' and M_w

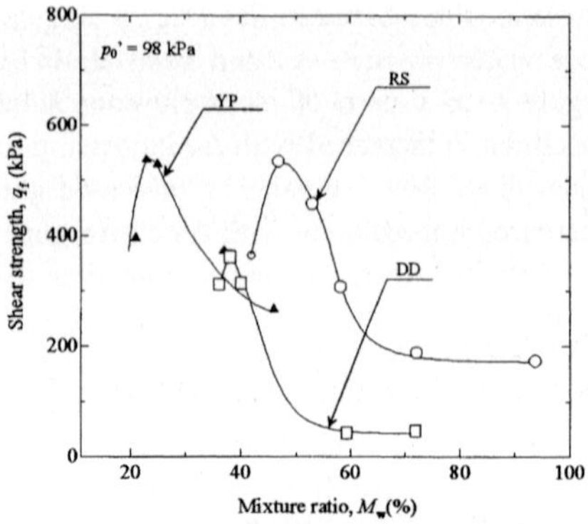

Figure 8: Relationships between q_f and M_w

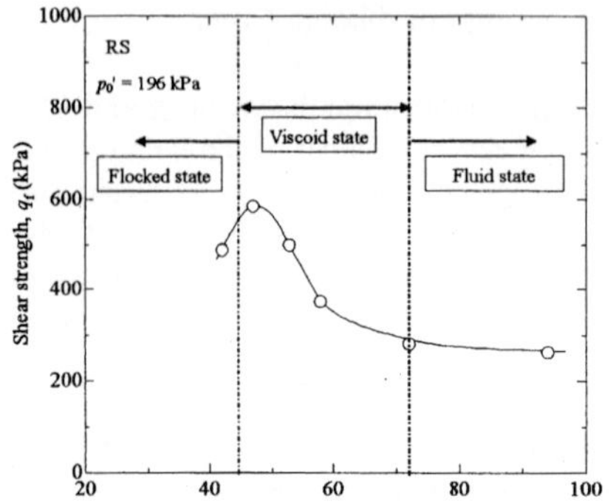

Figure 9: Shear strength on each states

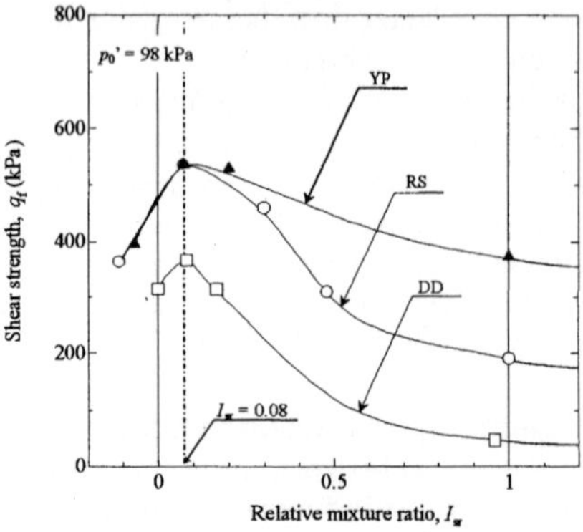

Figure 10: Relationships between q_f and I_{sr}

mixture ratio I_{sr} was introduced to evaluate strength properties on the viscoid state and expressed as follows.

$$I_{sr} = \frac{m_w - m_{wg}}{m_{wf} - m_{wg}} = \frac{m_w - m_{wg}}{I_s} \quad (2)$$

Here, I_s is the width of mixture ratio on appearing the viscoid state. I_{sr} gives viscoid limit if I_{sr} =0, and I_{sr} gives the fluid limit if I_{sr} =1. Relationships between shear strength q_f and relative mixture ratio I_{sr} are shown in Figure 10. Shapes of q_f - I_{sr} relationships are similar to each other. And I_{sr} values where q_f appears maximum value coincident regardless of the kind of coal fly ashes. Maximum shear strength of coal fly ashes used in this study appears on I_{sr} =0.08.

Conclusion

In the present paper, we proposed a new consistency concept of coal fly ash to correspond to variety of it and to use it efficiently as an earth material. Additionally, we investigated strength properties of coal fly ashes and showed that strength properties ware able to be evaluated by consistency limits which were defined in this study. The summary of the results is shown below.

(1) The coal fly ash-water mixture is able to be classified into the flocked state, the viscoid state and the fluid state. The viscoid limit is defined as the boundary between the flocked and the viscoid state, and the fluid limit is defined as the boundary between the viscoid and the fluid state.
(2) The optimum mixture ratio that shear strength appears the highest is founded.
(3) The optimum mixture ratio is evaluated by a new index based on proposed consistency limits.

References

Horiuch,S. Taketsuka, M., Odawara, T. and Kawasaki, H. (1992), Fly-Ash Slurry Island I: Theorical and experimental Investigation, *J. Merial Eng. Div., ASCE,* 4(2), pp.117-133.

Horiuchi, S., et al. (1996). Coal fly ash slurry for effective under water disposal. *Soils and Foundations*. 35(1), pp.1-10.

Heat and mechanical properties of waste glass aggregate as a base course material for taking measures of heat island

Hisashi Suzuki, The University of Tokushima, Japan
Yuji Itasaka, The University of Tokushima, Japan
Katsushi Makino, Makino Doboku Co. Ltd., Japan

Abstract

Glass bottles which should be disposed are 3.3 millions ton a year in Japan. The 70% of the bottles are brown and transparent glass bottles reused as a drinking bottle, however, the rest 30% are colored glass bottles which are abolished to reclaimed land. In consideration of the present state, it is the social mission that we immediately find out the valid recycling method of colored glass bottles. In this study, not only CBR-values, but also the permeability, water-holding capacity and heat properties in a ground are obtained from the laboratory tests as environmental factors. We also performed the field tests to investigate the heat environment of pavement yards, and found out that waste glass aggregate is very desirable as a base course material to prevent from heat island phenomena.

INTRODUCTION

In recent years, the waste materials produced by "large production and consumption" are increasing rapidly, for example, the amount of municipal waste is 0.5 million ton, and the amount of industrial waste is 4.0 millions ton a year at 2000 in Japan. The lack of waste disposal is a serious trouble, that is, the rest years of the final disposal are 8.5 years for municipal waste and 3.0 years for industrial waste. Therefore, Japanese Government is promoting the recycling of industrial wastes. Generally, the 3R are to reduce disposal, to reuse manufactured goods and to recycle products. This trial is the very important subject for the maintenance of 21st adequate environment and the development of Japanese continuous economy.

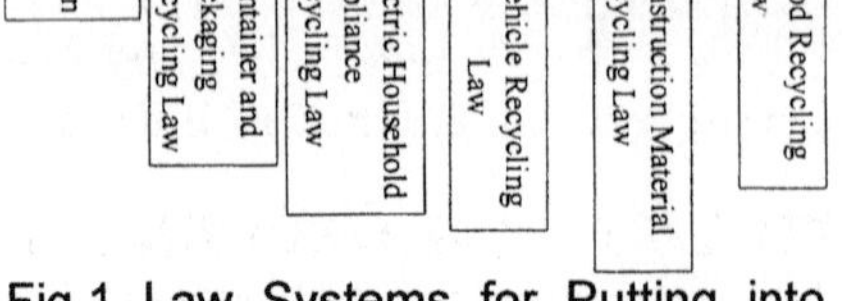

Fig.1 Law Systems for Putting into Practice the Society Circulating the Industrial Waste

Fig.1 shows the law system for putting into practice the society circulating the industrial waste. Under such as the outline of the law, Japanese Government aims at reducing the final disposal, which becomes the half of the 1996's amount until 2010. The Packaging Recycling law was already enacted to build up the new recycling and circulating society at June in 1995.

Waste glass aggregate dealt with this study were born from such leaderships of administration side. However, the uses are limited to a water purification construction of rivers, man-made fishing reef, tree planting at housetop and veranda which are small constructions. There are only 2 or 3 samples to apply waste glass to ground materials. In this study, we try to apply waste glass aggregate to a base course materials as an effective utilization of geotechnical engineering.

The reason why waste glass aggregate has not been applied to a base course material is not to expect an enough bearing capacity because of the particle breakage. This difficulty can be made clear from the modified CBR-test. On the contrary, we can expect as a base course material to take measures heat island for the small heat capacity caused by the very light specific gravity. The measures of heat island, namely, the utilizations of the heat properties of waste glass aggregate are investigated deeply in this study.

Heat island is rapidly advancing in Japanese big cities whose increases of air temperature are 2℃ at Tokyo, 1.5℃ at Osaka etc., on the other hand, the average increase of air temperature of whole globe is 0.6℃ for the recent 100 years. The numbers of sultry nights are 40-50/year at Tokyo, 30-40/year at Osaka and Nagoya. These sultry nights which we can not sleep increase rapidly owing to heat island phenomena.

The cause of heat island which directly connected with civil engineering depends on heat properties of asphalt, concrete, tile and brick. Considering building materials from view point of effecting on heat island, the accumulated heat by sunshine is large using building materials which have large heat capacity, therefore, it is difficult that surface temperatures of building falls in night. On the other hand, typical Japanese roofs made by tiles have rather high temperature by sunshine. However, they become low temperatures in night time by the radiation of heat without an effect of accumulated heat by sunshine. The point which we should pay attention is to use the materials having a small heat capacity, not to determine by a temperature property of a material in daytime.

In this study, the mechanical properties of waste glass aggregate as a base course material are examined by the modified CBR-test. The permeability for a permeable pavement and the water-holding for a tree planting are investigated by the laboratory tests as the environmental factors. The laboratory tests are carried out for obtaining the basic heat properties of the recycled ground materials and the full-scale pavements are constructed to investigate the actual heat properties. The materials used in this study are not only waste glass aggregate, but also the RC (Recycled Concrete) -material which is ordinary used at a pavement construction. The RC-material is also a recycled material made of concrete structures.

PHYSICAL AND MECHANICAL PROPERTIES OF SAMPLES

The samples are RC-material used in ordinary road construction and waste glass aggregate produced by calcinations processing of the mixtures of coloured glass bottles powders and shell powders. Table.1 shows the physical properties of the samples. The dry specific gravity of waste glass aggregate has very small value which is equal to 0.855. The aggregate includes many apertures and has large water absorption equal to 36.7%.

Table-1 Physical Properties of Samples

conditions / material	specific gravity in saturated surface dry	specific gravity in absolute dry condition	water absorption(%)
RC-material	2.395	2.258	6.07
waste glass	1.168	0.855	36.7

As waste glass aggregate has considerably large continuous apertures, they can be expected as a ground material preparing for high permeability and high water-holding capacity. Fig.2 shows the relationship between the coefficients of permeability and the mixture rates (= volume of waste glass/total volume×100%). As shown this figure, the higher mixture rate is, the larger coefficient of permeability is. Comparing 0%(mixture rate) with 100%, waste glass has about double higher permeability than RC-material. Fig.3 shows the relationship between the mixture rates and water-holding contents. A water-holding content is to be expressed by a water content which becomes residual volumetric content (D.G. Fredlund and Rahardjo, 1993). The higher mixture rate is, the higher water-holding content is. Comparing 0% (mixture rate) with 100%, waste glass has about seven times higher retention of water against RC-material.

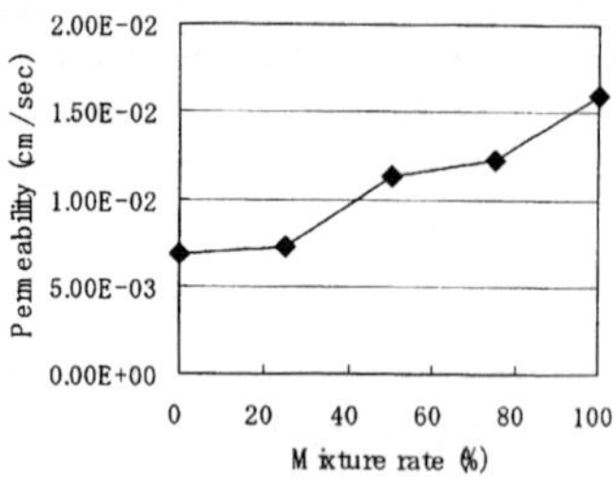

Fig.2 Relationship between Mixture Rate and Permeability

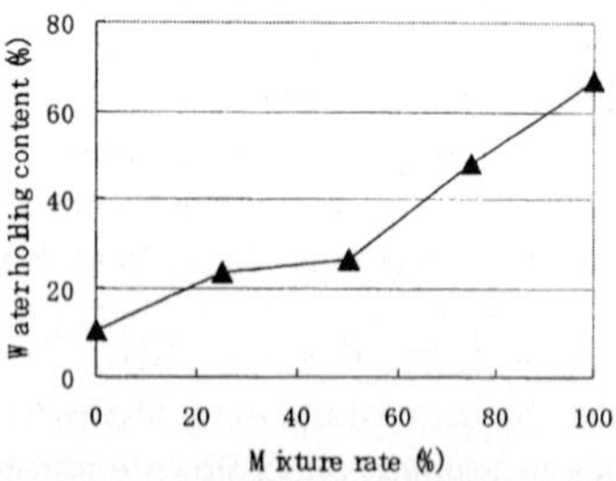

Fig.3 Relationship between Mixture Rate and Water-holding Content

The quality of the material as a base course is determined by the CBR-values obtained from the modified CBR-test. The mixture rates of the samples are 50% and 100% and the particle breakage is investigated by the sieve analysis after the modified CBR-test whose numbers of compaction are 17, 42 and 92 times.

Fig.4 shows the results of the modified CBR-test using the samples of 50% and 100% mixtures. From this figure, the CBR-values and the dry densities of both samples become large with the number of compaction times. The CBR-values of 100% mixture sample are more than 15% even in the 17 times. Similarly, the 50% mixture samples are also more than 10%. Both materials can be applied to a base course of a sidewalk, a bicycle road and a lower layer of simplified pavement. The CBR-values of 100% mixture are larger than those of 50% mixture.

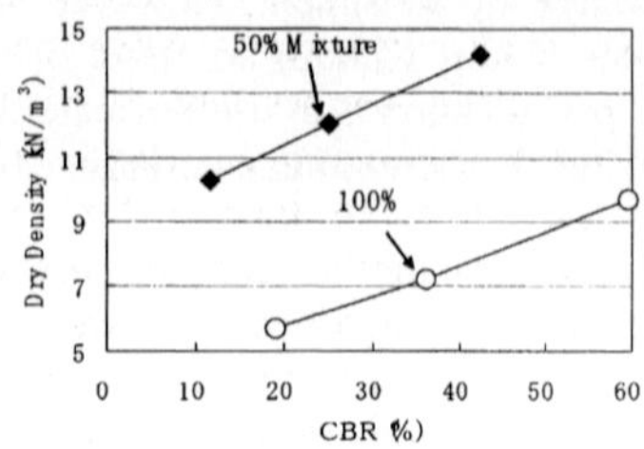

Fig.4 Comparison of Results of the Modified CBR Tests

EXPERIMENTS FOR INVESTIGATING HEAT PROPERTIES

Outline of Experimental Apparatus

Fig.5 shows the apparatus for measuring the temperatures of the soil samples under the constant quantity of heat. The samples are compacted into the mold used in the standard compaction test. The heat is supplied by the heater (100V, 375W). As shown

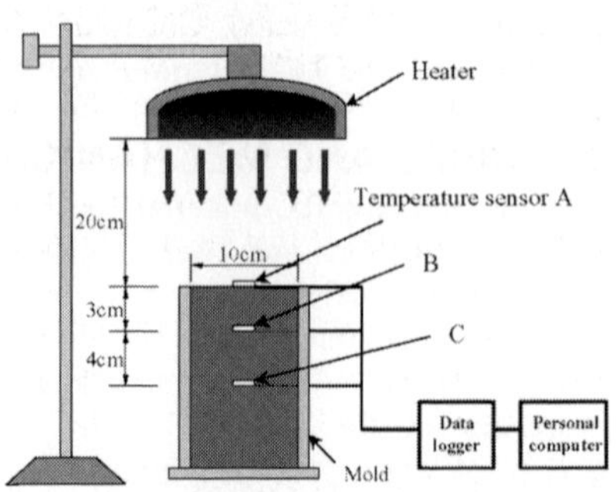

Fig.5 Laboratory Test for evaluating Heat Properties of Ground Materials

this figure, the three temperature sensors are installed the locations which are the surface of the sample, 3cm and 7cm depths from the surface respectively. These temperatures are automatically measured and recorded at 30 seconds intervals.

Procedure of Experiments

We explain the experiments for measuring temperatures of the ground materials. These samples are waste glass aggregate, 50% mixture and RC-material under air dry condition. The experiments have 1 hour heating and then 1 hour radiation of heat. In the time ~ temperature curves obtained from the laboratory experiments of the heat properties, the temperatures rapidly rise with time in the first stage and then gradually reach the equilibrium conditions. The heating for all specimens stop at 1 hour after to keep the constant quantity of heat. The heat capacities of the samples are estimated from these experiments.

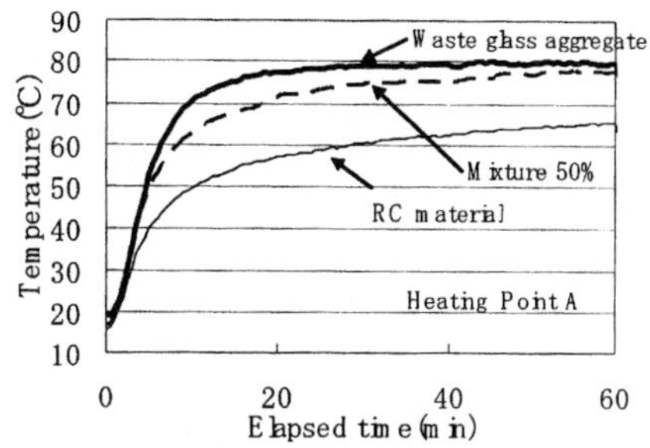

Fig.6 Relationships between Time and Temperature in Heating

Experimental results and considerations

Fig.6 shows the relationships between time and temperature of waste glass aggregate, 50% mixture and RC-material at point A in Fig.5. As shown in this figure, the temperatures at 60 minutes after become high with the increase of the mixture rate. The velocity of rising temperature of waste glass aggregate is most rapid and the velocity of rising temperature of RC-material is the slowest in heating. Fig.7 shows the relationships between time and temperature at point A in the radiation of heat. The velocity of falling temperature of waste glass aggregate is most rapid and the velocity of falling temperature of RC-material is the slowest. These tendencies mean that the heat capacities are gradually small with the increase of mixture rate of waste glass aggregate.

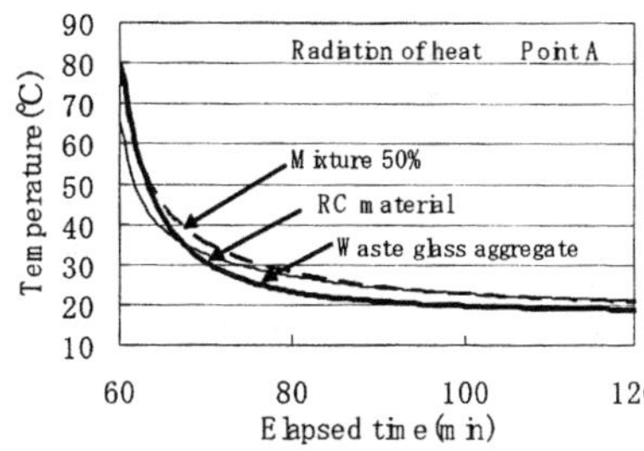

Fig.7 Relationships between Time and Temperature in Radiation of Heat

FIELD TESTS FOR OBTAINING HEAT PROPERTIES

System for measuring underground temperatures

Fig.8 shows the outline of the system for measuring underground temperatures. There are the twelve sensors, and they are connected to the body of recorder by way of the relaying basic plate. The recorded data are compiled to the smart media installed in this recorder. These data can be taken up to the personal computer, whenever it is necessary.

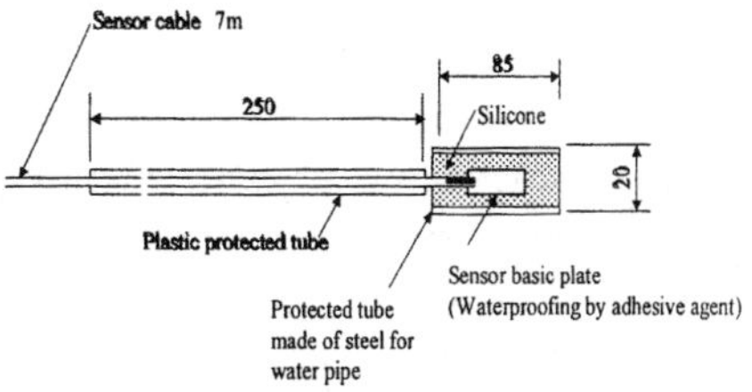

Fig.8 Outline of Underground Thermometer

Situation of installation of temperature sensors

The underground temperature sensors mentioned above were laid according to the actual sidewalk construction. There are 3 yards, first base course is made of RC-material, second is

50% mixture of waste glass and RC-material (volumetric ratio) and third is waste glass aggregate. The ground plans and section of these yards are shown in Fig.9. The sidewalks of the field test consist of the pavement whose thickness is 30mm from the ground surface, the base course whose thickness is 100mm, and then the subgrade. Especially, the surface layer is constructed by the fiber resin, and the heat decreasing effects of these layers are measured. The four underground temperature sensors are installed in each yard and they are located at ①ground surface, ②boundary between surface layer and base course, ③center of base course, ④ subgrade from the ground surface. We give the names of all sensors to simplify the explanations. Namely, the temperature sensors in the yard of RC-material are called as RC-1～RC-4, similarly, the sensors in the yard of 50% mixture are called as mixture-1～mixture-4, and the sensors in the yard of waste glass aggregate are called as glass-1～glass-4. However, the data of the mixture-4 are not shown, because the sensor can not measure the temperatures to mistake the installation of it. Of cause, all yard have not the shade, and have the same sunshine conditions.

Weather conditions

Changes of underground temperature are mainly dependent on the weather conditions. Before giving the results of field tests, we explain the weather conditions in Tokushima city of August 2004 when is the subject in this study. These data are obtained from the HP of the Meteorological Agency in Japan.

Accordingly, we simply write air temperature, precipitation and the natural phenomena at Tokushima in summer 2004. This summer was abnormal weather in Japan. Especially, in Tokushima not only hot weather, but also the big disasters of outflow of soils occurred due to abnormal precipitation caused by the typhoons. Fig.10 shows the precipitation, the maximum of air temperature and the minimum of air temperature of a day in August 2004. The maximum of air temperatures sometime reached 35℃, therefore, this summer was unusually a heat wave. The minimum of air temperatures became near 25 ℃ whose temperature is rather high, therefore, we realized

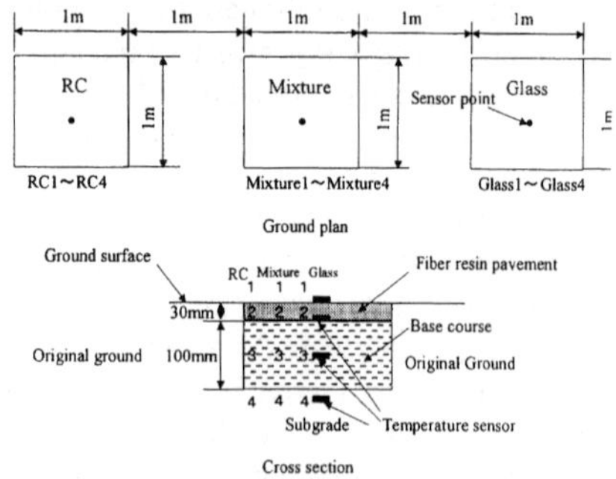

Fig.9 Outline of Field Test

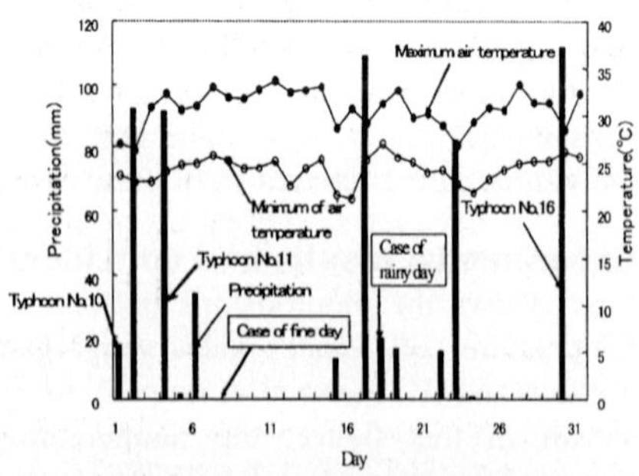

Fig.10 Weather Conditions (August 2004; Tokushima City)

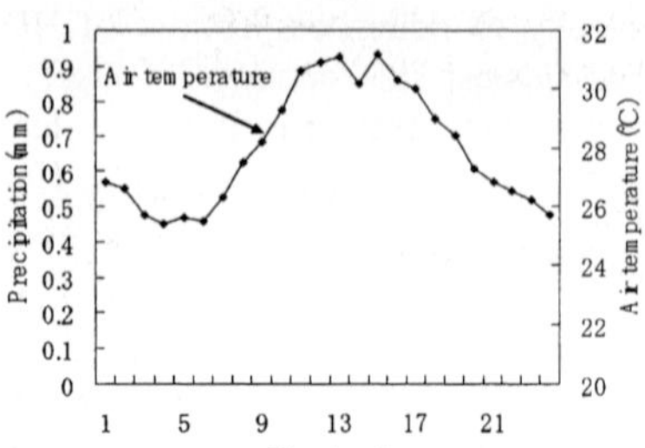

Fig.11 Weather Conditions (8th August.2004: Tokushima City)

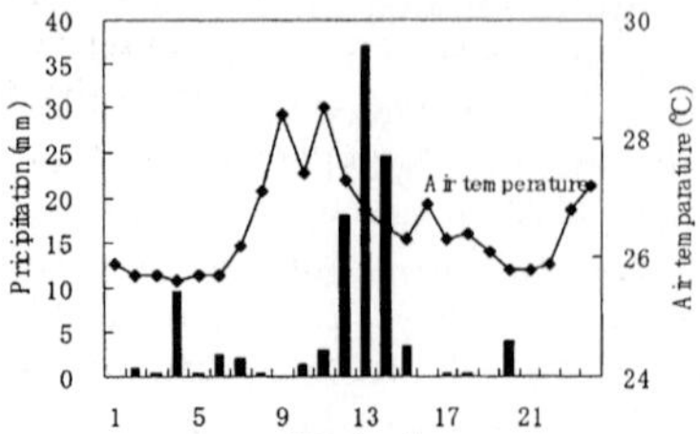

Fig.12 Weather Conditions (17th August 2004: Tokushima City)

that sultry nights continued for the long times. The Typhoons of No.10, 11, 16 bring the record-breaking heavy rain to Tokushima. Some of daily rainfalls reached 100mm and the precipitation became the cause of outflow of soils.

The heat properties of waste glass depend on the water conditions of the aggregate. Mentioned above, we expect the continuous effect of cooling by the heat of vaporization because of long time water-holding capacity. We discuss showing the results of measurements 8th August 2004 when continued for beyond two fine weather days and 17th August 2004 when was immediately the heavy rainfall reaching to 100mm a day. Fig.11 and Fig.12 show the weather conditions of these days. The former is the fine weather without rainfall and the latter is the rainy weather with intermittent precipitations. In latter case, we can expect the effects of cooling of ground caused by the heat of vaporization.

The measurement results and discussions

Fine weather

As the fine weather continued beyond 2 days in 8th August 2004, there was perfectly no water in the fiber pavement and the base course. Accordingly, we can not expect the effects of decreasing temperatures by the heat of vaporization mentioned above.

Fig.13 shows the temperatures within the pavements obtained from the field tests arranged at (a) ground surface, (b) boundary between surface layer and base course, (c) center of base course, (d) subgrade. Each figure also shows the air temperatures of Fig.11 for comparing with the underground temperatures. At first, the effects of temperature of waste glass aggregate are found out that waste glass is higher than the temperatures of other materials, especially, in 10 ～17 o' clock when are strong sunshine hours. This tendency is remarkable in (b), (c) of Fig.13. The differences between waste glass aggregate and other materials are not so remarkable. Especially, in (d) subgrade the temperatures of waste glass are lower than those of RC-material at every time. The temperatures of waste glass aggregate become lower 2～5℃ than those of other materials in 0～5 and 19～23 o' clock without sunshine. The hours when waste glass

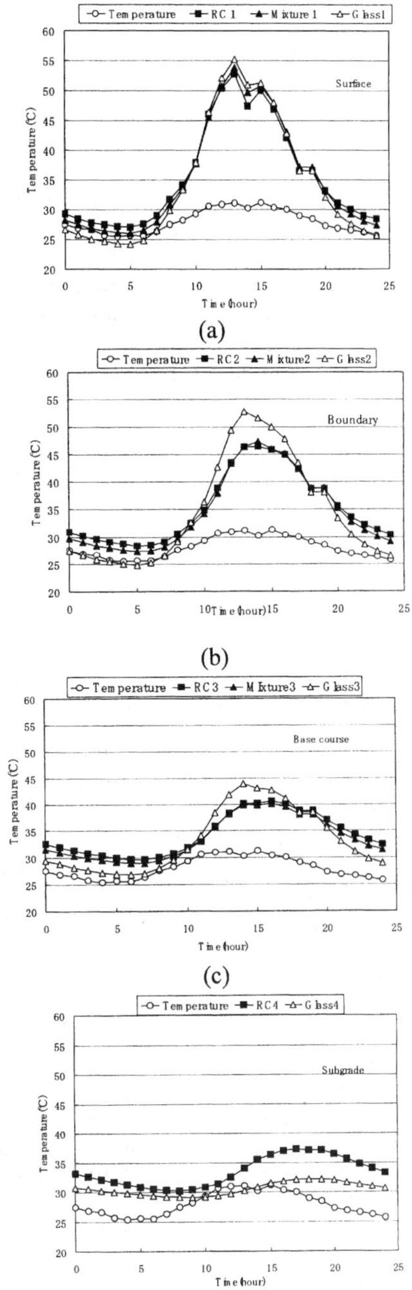

(d)

Fig.13 Comparison of Temperatures of Pavement Ground (Case of Fine Weather 8/8 2004:Tokushima City) material becomes high temperature is only 1/3 of a day and in the rest 2/3 the temperatures of waste glass are lower than those of other materials. The temperatures of waste glass aggregate are sensitively effected by air temperature. Therefore, waste glass aggregate has a small heat capacity. This property is also described in the results of the laboratory tests. If a sidewalk is constructed by this material, we can sleep comfortably without a sultry night.

Comparing with the properties of temperatures at 3 yards which are made of RC-material, waste glass aggregate and 50% mixture, the temperature property of 50% mixture is similar to that of RC-material. The 50% mixture has an effect of decreasing temperature by sunshine. However, its temperature becomes higher than those of waste glass aggregate in nighttime, namely, it is a material having a large heat capacity. In the fine day, waste glass aggregate has high temperatures in daytime, but it has an effect of decreasing temperature in night, namely, it has a small heat capacity. Therefore, waste glass aggregate can become a base course material for taking measures of heat island. We can construct a sidewalk which has a high permeability and high water-holding considering heat environment, if we use waste glass aggregate as a base course material. The material is also satisfied with the properties of mechanical, namely, the regular CBR-values. Waste glass aggregate is not used as a ground material. However, the result that waste glass aggregate can be applied to a base course material for taking measures of heat island is very useful to recycled products.

Next, it is examined that the effects of decreasing temperature under the fiber resin pavements constructed in all yards. Although the figures which can directly compare the temperatures are not shown in this paper, they can be examined by comparing RC1 with RC2, mixture 1 with mixture 2, and waste glass 1 with waste glass2. The effects of decreasing temperature is large in very strong sunshine, they have comparatively high temperatures which are about 5℃ in every cases. The decreasing temperatures at pavement surface

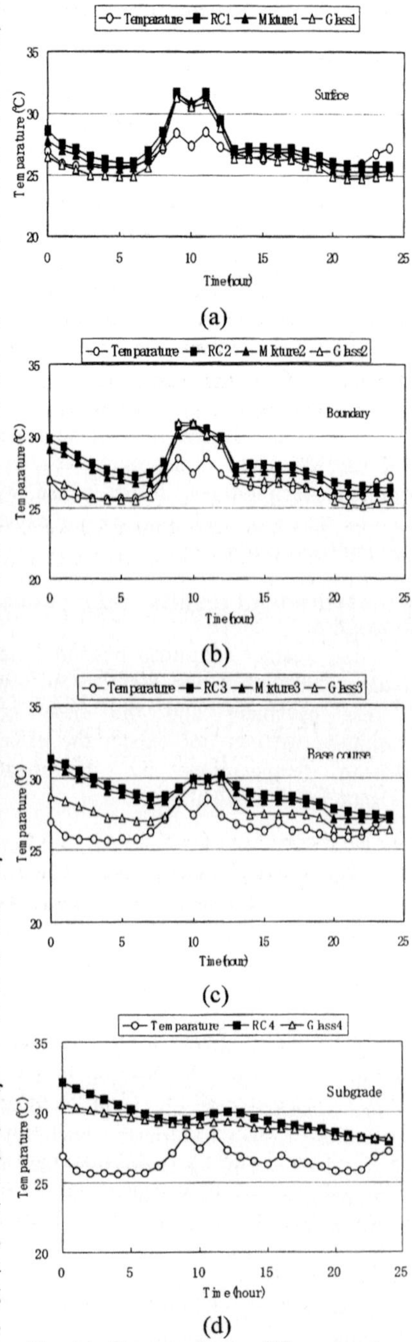

Fig.14 Comparison of Temperatures of Pavement Ground (Case of Rainy Weather 17/8 2004: Tokushima City)

have important role because of decreasing to base course and subgrade.

Rainy weather

17th August 2004 is not the perfect rainy day, which has discontinuous rain and sunshine as shown in Fig.12. In this case, waste glass aggregate sufficiently contains water, we expect the decreasing temperature by heat of vaporization.

Fig.14 is arranged the results which are similar to Fig.13. Underground temperatures are totally lower than 32℃ and they become rather lower compared with the fine weather. The underground temperatures of sidewalk used waste glass aggregate become lowest at all figures. Then the secondary temperatures are 50% mixture and the highest temperatures are RC-material. The effects of decreasing temperature caused by rainfall are conducting from ground surface to the lower part. Accordingly, the temperatures at base course and subgrade become about 5℃ higher than those at ground surface. Anyway, it is worth that waste glass has the lowest temperature caused by its small heat capacity in this case.

CONCLUSIONS

In this study, the application of waste glass aggregate which is a recycled product of colored glass bottles to a base course material was examined from not only the mechanical properties but also the considering of heat environment. The following conclusions were obtained.

(1) The values of the modified CBR-test for waste glass aggregate are 15～60% with the number of times of compaction. Therefore this material can be applied to not only a sidewalk and a bicycle road but also a lower layer of general road as a base cause material. We obtained the coefficients of permeability and water-holding capacity with the mixture rates of waste glass. These values became large with the increase of the mixture rates, therefore, we showed that waste glass was more useful than RC-material to construct the permeable and water-retentive sidewalks.

(2) The laboratory tests for obtaining the basic heat properties of the ground materials were performed for waste glass aggregate, 50% mixture and RC-material. As a result, the velocity of rising temperature of samples in heating becomes high with the increase of mixture rates. On the other hand, the velocity of falling temperature of waste glass aggregate becomes high in the radiation of heat. Waste glass aggregate has the sensitive property for a change of temperature. Therefore, it has a small heat capacity.

(3) We carried out the field tests to construct a sidewalk considering heat environment. As a result, waste glass aggregate has high temperature for several hours in strong sunshine, but it has lower temperatures of 2～5℃ in the other hours. This material has a small heat capacity, that is, the small accumulated heat. This property becomes weak with the increase of mixture rate of RC-material, therefore, RC-material has a large heat capacity. To bear a sultry night is more difficult for us than to bear a high air temperature by sunshine in a daytime. The heat properties of waste glass aggregate are desirable to prevent from heat island.

REFERENCES

D.G Fredlund and H.Rahardjo : Soil Mechanics for Unsaturated Soils, John Wiley & Sons, Inc., pp.112~113, 1993.

Jumikis, A.R : Thermal Geotechnics, Rutgers University Press, New Brunswick, New Jersey, 1977

The Strength Characteristics of Muddy Soils Mixed with Granulated Coal Fly Ash

Seishi Tomohisa, Professor, Akashi National College of Technology, Akashi, JAPAN
Kohei Sawa, Professor, Akashi National College of Technology, Akashi, JAPAN
Manabu Saito, Advanced Course, Akashi National College of Tech. , Akashi, JAPAN
Kiyohiro Michishita, Researcher, Nippon Steel Co. Ltd. Himeji, JAPAN
Tetsuo Ioku , Researcher, Hirokogiken Co. Ltd. Himeji, JAPAN
Hiroo Tanaka, Researcher, Kotoh Co. Ltd. Osaka, JAPAN

Abstract

Recently, the generation of un-reusable soils from various construction sites has been increasing. It is difficult to reuse them as construction materials owing to the high water content and low strength, and the disposal sites of the soils have been decreasing. Therefore, it is very important to reuse the muddy soils. The purpose of this study is to clarify the effectiveness as hardening additive of granulated coal fly ash which is formed by mixing coal fly ash with cement. The granulated coal fly ash has as merits the improvement of handling and the reduction of dust. As a result, the strength of the treated soil increases because of the formation of the granular structure and the decrease in water content of the treated soil due to the mixing of the granulated coal fly ash.

Introduction

In 2002, the annual amount of total construction wastes in Japan was about 83 million tons and the waste sludge was 10 million tons. The percentage of reuse of waste asphalt-concrete was 99%, whereas that of the waste sludge was 69% (Ministry of LIT, 2002). On the other hand, surplus soil does not include the category of wastes, whereas the percentage of reuse was only 65% (Table 1).

Most of the waste sludge and surplus soil (hereafter, called muddy soils) is difficult to be reused as construction materials owing to the high water content and low strength, and the disposal sites for wastes have been decreasing. Generally, these muddy soils are stabilized by cement or lime and reused as construction materials.

Table 1. The account of reuse for wastes (Actual results and targeted values) Unit:%

Year	2002	2005*	2010*
All of Construction Wastes	92	88	91
Waste Asphalt-concrete	99	98	98
Waste Concrete	98	96	96
Waste Sludge	69	90	95
Mixed By-products	30	25	50
Waste Timber	89	90	95
Surplus Soil	65	75	90

* : targeted value of Ministry of Land, Infrastructure, and Transport

Nevertheless, these methods have the following demerits (Sawa et al., 2003).
(1) When muddy soils contain much organic matter, they need many hardening additives and high cost.
(2) The treated soils become high in alkali.
(3) Excavation of treated soils is difficult because their strength is too high after long aging.

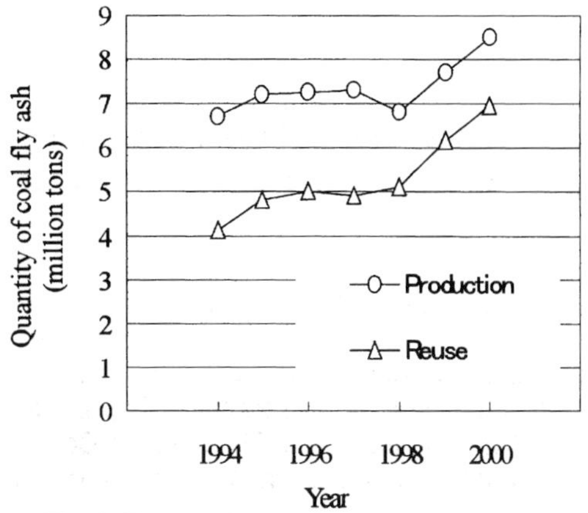

Fig.1. Production and reuse of coal fly ash

The coal fly ash of industrial waste also has seen the same problem. A number of thermal power stations have been constructed in order to diversify energy sources. Consequently, the product of coal fly ash has been increasing every year. Fig. 1 shows the quantity of production and reuse of coal fly ash in recent years in Japan (JSCE, 2003). The coal fly ash is reused for material in cement, and with dams and bridges as construction resources, nevertheless, when it is mixed with the soft soil on the ground improvement, a lot of coal fly ash is scattered in the air and makes a dust (Sawa et al., 1996), The dust is a serious problem for human health. Furthermore, about 1.5 million tons of coal fly ash have been thrown in disposal sites every year. It is considered that an immense amount of these muddy soils and coal fly ash will be produced continually from now on. Therefore, it is very important to reuse them from the viewpoint of natural resource preservation and environmental impact mitigation (Kamon, 2000) (Katsumi, 2000).

The purpose of this study is to reuse effectively the muddy soils mixed with granulated coal fly ash and to clarify the possibility of application of granulated coal fly ash as a hardening additive on soil stabilization. The improvement effects of the granulated coal fly ash are evaluated by the California Bearing Ratio (hereafter, called CBR) test.

Experimental methods

Samples

The sample used in this study is a muddy soil which is dredged from a river in Tokyo Prefecture, Japan. The water content of the muddy soil after dredge is reduced to about 60%. The properties of the muddy soil are shown in Table 2. The CBR of the muddy soil has a very low value of 0.01% which is difficult to reuse as a construction material. In the hardening treatment test, the water content of the muddy soil is adjusted to 40, 45, 50 and 60 %

Table 2. Properties of muddy soil

Density of soil grain	g/cm^3	2.66
Liquid limit w_L	%	62.6
Plasticity index I_P		33.0
Natural water content	%	60.8
CBR	%	0.01
Grain size distribution		
Sand (75 μ m~2mm)	%	6.2
Silt (5~75 μ m)	%	33.5
Clay (<5 μ m)	%	60.3

Table 3. Properties of GCF

Density of GCF grains	g/cm^3	1.86
Natural water content	%	18.9
Water absorption	%	44.1

respectively.

Additive

The hardening additive used for the muddy soil is the granulated coal fly ash (hereafter, called GCF). The GCF is produced by mixing fluidized bed combustion coal fly ash with normal Portland cement 5% and water 28%. Thereafter the granular grains are formed with a maximum diameter of 20mm. The GCF has some merits in working effectively for the reduction of the amount of flying dust and to facilitate the handling at work and during transportation. The properties of GCF are shown in Table 3. The water absorption of GCF is about four times as much as that of aggregate.

Combinations and testing methods of treated muddy soil

In this study, the examination of the treated muddy soil is performed following three methods, namely, the (1) strength test of the treated soil, (2) water absorption test and (3) scanning micro scope (SEM) observation of the GCF. The flow chart of the experimental procedure (strength test and absorption test) is shown in Fig. 2.

Strength test

The water content of the muddy soils is adjusted to 40%, 45%, 50% and 60%. In order to investigate the strengthening effect of the reduction of water content due to the absorption of the GCF in the treated soil, two conditions (dried and saturated) of GCF are used in this study. The dried GCF is able to absorb water in the soil. On the other hand, the saturated GCF has the property of no absorption. The GCF is added in dosages of 10, 20, 30, 40 and 50% by wet weight of the muddy soil. The GCF of three kinds of grain size distribution are used, namely grain size of over 5 mm and under 5 mm, and all grain sizes (0~20mm).

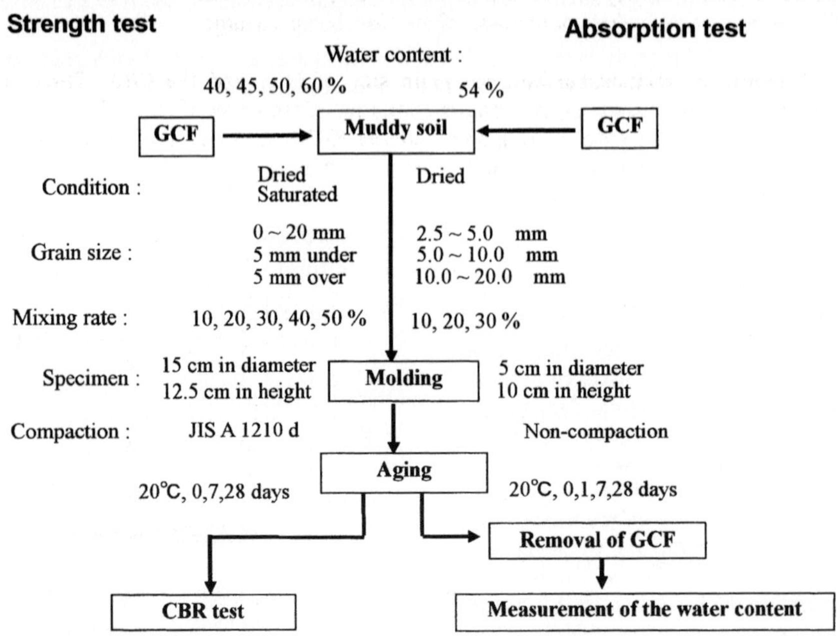

Fig. 2 Flow chart of experimental procedure

The cylindrical specimens (15cm in diameter and 12.5cm in height) are molded in accordance with the testing method for stabilized soil with compaction which is specified as the technical standard by the Japanese Industrial Standards Committee (JIS A 1210 (d)). Each specimen is sealed with resin film and is aged at a constant condition of 20℃ for 0, 7 and 28 days. A CBR test is performed on the specimens after aging.

Absorption test

This is the test to measure the absorption ability of GCF. The water content of the muddy soil is adjusted to 54%. The GCF is prepared in the three kinds of grain size, with the distribution of 2.5~5.0mm, 5.0~10mm and 10~20mm. The GCF is added in dosages of 10, 20 and 30% by wet weight of the muddy soil, respectively. Each specimen (5cm in diameter and 10cm in height) is sealed with resin film and is aged at a constant condition of 20℃ for 0, 1, 7 and 28 days. The GCF is removed from the treated soil, and then the amount of the water content of soil substance is measured.

SEM observation

The scanning electron microscopy observation is performed in order to confirm the influence of the GCF in absorbing water. A cross section with depth of 1-3mm from the GCF grain surface after 0, 7, 28 and 120 days of aging is observed.

Results and discussions

Strength characteristics

Fig. 3 shows the relationship between the mixing rate of GCF and the CBR of treated soil. In the case of high water content of the muddy soil (60%), an increase in the CBR owing to the mixing of GCF is not found. On the other hand, the improving effect of treated soil is remarkable for the muddy soil of low water content (50%). Especially, when the mixing rate of GCF is over 40%, the rate of increase of the CBR becomes large.

Fig. 4 shows the relationship between grain size of GCF and the CBR. There is not a difference in the improving effect with a mixing rate of 10% even if the grain size of GCF is different. However, when the mixing rate increases to 20% or more, the increase of the CBR depends on the grain size of GCF. The improving effect is largest for grain size of under 5 mm, then 0~20mm, and least for over 5 mm.

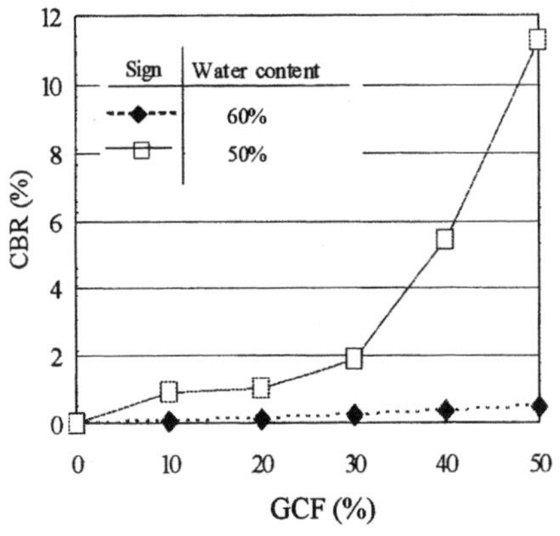

Fig. 3. Mixing rate of GCF and CBR

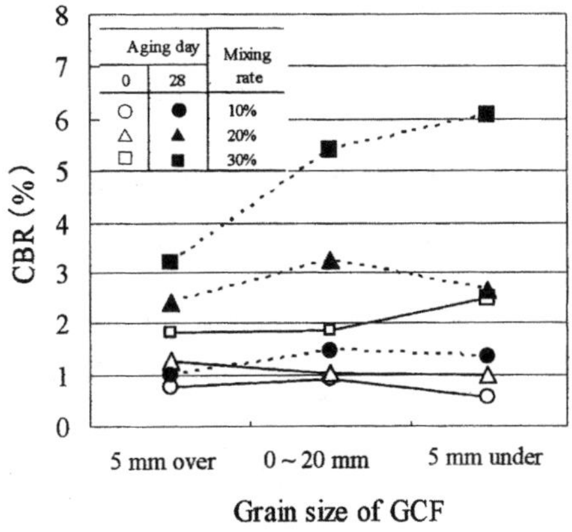

Fig. 4. Grain size of GCF and CBR

Fig. 5 shows the relationship between aging days and CBR. Even with any grain size of GCF, the strength of the treated soil increases with a long period of aging. The smaller diameter the grain of GCF contains, the larger the CBR of the treated soil becomes. Especially, in the case of the size under 5 mm, the effect of the increase in strength is large.

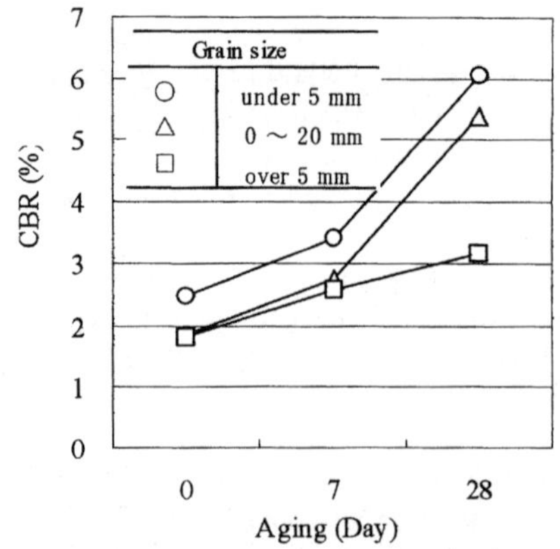

Fig. 5. Aging days and CBR (Mixing rate : 30%)

Water absorption characteristics

The authors have reported that the water content of soil substance in the treated soil has a large effect on the strength (Tomohisa et al., 2002). Therefore, in this study, the absorption ability of GCF is measured.

Fig. 6 shows the relationship between aging days and water content. It is clear that the decrease in water content in the treated soil appears immediately after the mixing with GCF, and continues for a long aging term. The fine (2.5-5.0mm) GCF decreases in water content more than the coarse (10.0-20.0mm) GCF, because of the large effect of water absorption ability with the fine GCF, which has a wider grain surface area than the coarse GCF.

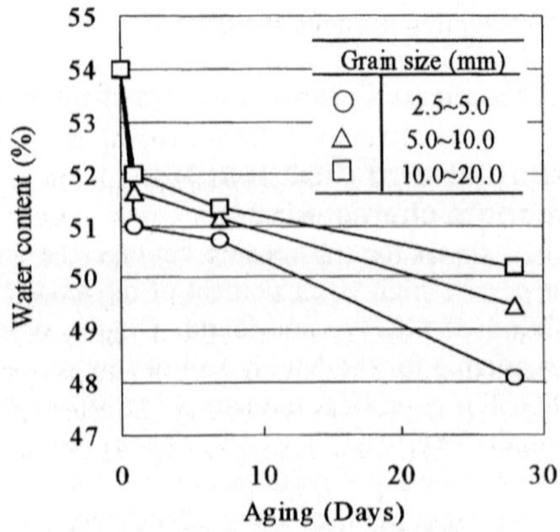

Fig. 6. Aging days and water content (Mixing rate : 10%)

Fig. 7 shows the relationship between aging days and the CBR. The solid marks show the dried GCF which has more water absorption ability, and the hollow marks show the saturated GCF which has less water absorption ability. There is a little increase in the CBR as the aging time increases for the saturated GCF. On the other hand, the treated soil with the dried GCF has a large increase of CBR through 28 days aging. The proportion of an increase in the CBR for the dried GCF is from 3 to 6 times higher in comparison with the saturated GCF.

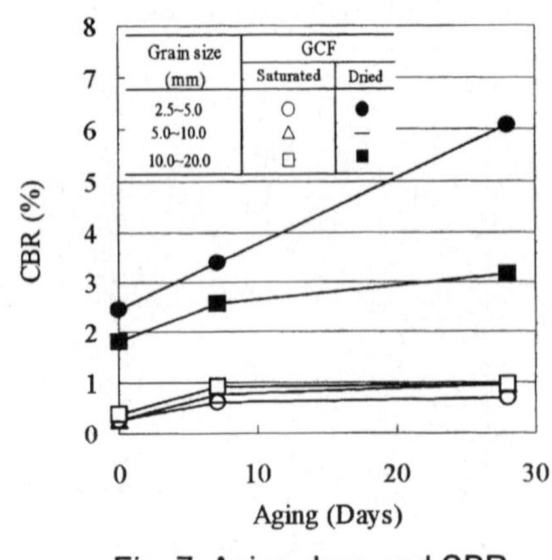

Fig. 7. Aging days and CBR (Mixing rate : 30%)

Fig. 8 shows the relationship between the water content of the soil substance and the CBR. When the water content of the soil substance in the treated soil is the same, the CBR of the treated soil for the GCF of over 5 mm is larger than that of under 5 mm. It is thought that the difference in CBR is due to the biting effect of coarse grain in the formation of the granular structure. Consequently, it is necessary to mix a lot of GCF of over 5 mm compared with under 5 mm, so the larger CBR is obtained. Furthermore, the lower the water content of the soil substance, the larger the CBR of the treated soil.

From these results, it is clear that the CBR of treated soil relates to the mixing rate, the grain size distribution of GCF, and the water content of the soil substance. Therefore, the results can be applied to the design of improvement of muddy soils.

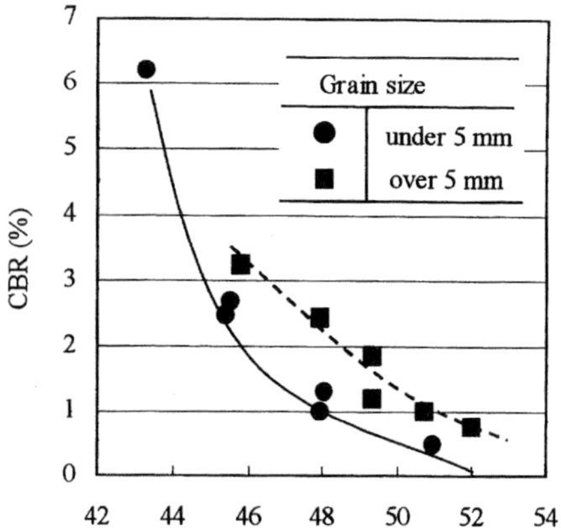

Fig. 8. CBR and water content

SEM observations

Fig. 9~12 show the SEM photographs of GCF.

Fig. 9 is the GCF before mixing with the muddy soil. On the surface of all grains, a large void and a number of minute reaction products such as thorn-shape calcium silicate hydrate (abbreviated as CSH), $Ca(OH)_2$ and so on, are observed. Fig. 10 is the GCF in the treated soil after aging for a week. Many individual grains combining with each other as reaction products. Fig. 11 is the GCF in the treated soil after aging for four weeks. The crystal of needle-shape ettringite and block-shape gypsum grows greatly with a longer period of aging. Furthermore, according to Fig. 12 after aging for four months, on the surface of the GCF, a number of reaction products such as ettringite, CSH, $Ca(OH)_2$ $CaCO_3$ and so on, are observed.

Fig. 9. SEM image of the GCF before mixing

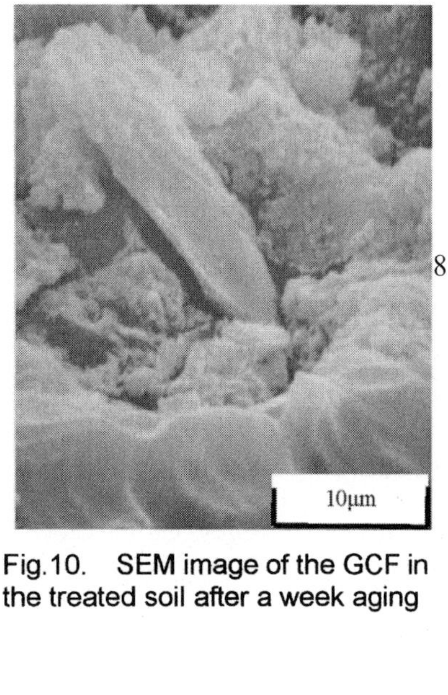

Fig.10. SEM image of the GCF in the treated soil after a week aging

(a) Gypsum

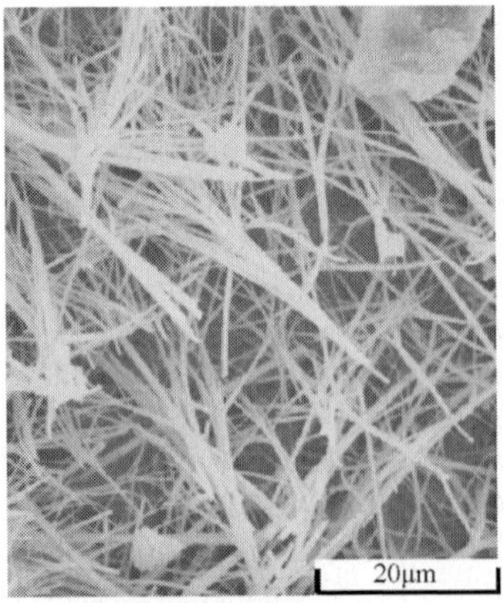

(b) Eettrigite

Fig.11. SEM image of the GCF in the treated soil after four weeks aging

Therefore, it is confirmed that the hydration have been continuing for a long term owing to the water absorption in the treated soil.

Fig.12. SEM images of the GCF in the treated soil after four months aging

Conclusions

In this paper, the possibility of reusing muddy soils of low strength and generated from various construction sites as construction materials by adding granulated coal fly ash was examined.

The results obtained are summarized as follows:

(1) The greater mixing rate of the granulated coal fly ash improves the strength of muddy soils. In addition, the increase in strength is remarkable in mixing rate of 40 % or more.
(2) As the water content of muddy soil has great influence on the strength of treated soil, the preprocessing of the muddy soils is very important.
(3) The increase in strength of treated soil is caused by the decrease of water content and the biting effect of coarse grain in the formation of granular structure in the soil.
(4) The finer the grain of the granulated coal fly ash, the higher the CBR of the treated soils. The reason is that the granulated coal fly ash of finer grain has a wide surface area and high absorption ability. On the other hand, the granulated coal fly ash of coarse grain contributes to the strength appearance due to the biting effect of the coarse grain.
(5) The effect of the water absorption of the granulated coal fly ash on strength development continues during the long aging period. Especially, it is remarkable within a week after

mixing.

(6) With the relation of the grain size distribution and the mixing rate of the granulated coal fly ash, and the water content of muddy soil, it is possible to estimate the CBR of the treated soil. Therefore, the results can be applied to the design of improvement of muddy soils.

(7) It is confirmed that the hydration have been continuing for a long term owing to the water absorption in the treated soil.

References

Ministry of Land Infrastructure and Transport (2002): Research Report of Construction By-product (in Japanese).

Sawa, K. et al. (2000): Ground Improvement and its Environmental Impacts, II: Influence of the Ground Improvement to Environment and it's Measures, *Journal of the Society of Materials Science,* Vol.49, pp. 1058-1063 (in Japanese).

Committee of Civil Engineering for Energy Equipment (JSCE), (2003): On the Utilization Technology of Coal Ash (in Japanese).

Sawa, K. et al. (1996): Dustless Hardening Material, *Proc. of the 2nd International Congress on Environmental Geotechnics,* pp. 1191-1196.

Kamon, M. (2000): Ground Improvement and Geo-environmental Problems, *Journal of the Society of Materials Science,* Vol.49, pp. 948-954 (in Japanese).

Katsumi, T. (2000): Ground Improvement and its Environmental Impacts, III: Ground Improvement Techniques for Geotechnical Waste Utilization, *Journal of the Society of Materials Science,* Vol.49, pp. 1160-1166 (in Japanese).

Tomohisa, S. et al. (2002): Utilization of Muddy Soils from Construction Sites by Mechanical Stabilization, *Proceedings of the Spring Symposium of the Iron and Steel Institute of Japan,* pp.32-39 (in Japanese).

Utilization of Irrigation Pond Muddy Soil to Embankment and Replanting Material

Minoru Yamanaka, Kagawa University, Takamatsu, Japan
Shuichi Hasegawa, Kagawa University, Takamatsu, Japan
Naoto Tomida, Howa Development Co, Sakaide, Japan

Abstract

Muddy soil of irrigation pond is difficult to use because it is ultra soft clay and has very high water content. But its use was ensured in this research. Basically, two methods of use as embankment and replanting materials were explored in this research. Firstly, possibility of the pond muddy soil as the embankment material was investigated. For this purpose, samples were mixed with cement and decomposed granite sand. The unconfined compressive strength test and permeability test were carried out to obtain strength and permeability of the sample. By mixing decomposed granite sand and pond muddy soil, high hardening strength was obtained. However, permeability increased with increasing of amount of decomposed granite sand, it was not possible to be applied as impervious zone material in embankment. But samples were considered to use efficiently in other zone of pond embankment. Likewise, in second test as replanting material, growth of the turf grass was analyzed in pond muddy soil. Plant height of turf grass was measured for 3 months from September to November 2004, it was noticed that mostly turf grass were well grown-up.

Introduction

In Japan, many irrigation ponds have the serious problem of deterioration. These ponds need to care not only for water leakage problem but also for probable damage in earthquakes and heavy rain. Presently, it came to know that about two thousands numbers of deteriorated irrigation ponds are necessary to reinforced urgently. There are mainly two problems for reinforcement of deteriorated ponds, namely, 1) it is difficult to ensure embankment material for reinforcement and improvement around the pond, 2) it is difficult to transport and disposal of pond muddy soil because it is ultra soft soil having very high water content.

Generally, it is disposed in present state after stabilization by cement solidification and there are some environmental issues from this waste. There are some researches for utilization of pond mud as soil dressing on farm land. Likewise, there are some earlier researches on pond muddy soil and it was found that pond mud is strongly acidic and has considerable amount of organic substance. So, its use as gardening soil was also considered (Tamaki et al., 1957, Maekawa et al. 1969). Similarly, for improvement of water environment artificial swamp was produced by using the dredging soil (Hosomi et al., 1997, Sakurai et al., 2003). Likewise, there is some research in which utilization of pond mud for embankment material is suggested (Fukushima et al., 2004). In the filed of use of pond mud as replanting materials, there are some research in which slope development by pond mud are suggested before plantation on slope. Moreover, research on use of mud to prepare water retention block was also carried out (Okuda et al., 1999, Yanagisawa et al., 2004).

The authors have also been studying basic physicochemical properties of irrigation pond mud regularly for the purpose of utilization of pond muddy soils (Tomida et al., 2004, Yamanaka et al., 2005). Thus, in this research also, for effective utilization of pond mud, two types of utilization method namely, as embankment material and as replanting material are discussed.

5th ICEG Environmental Geotechnics, Thomas Telford, London 2006.

Utilization to embankment material

Soil samples

The pond muddy soil used in this study was sampled at the Ooike pond in Kagawa Prefecture, south-west of Japan. There are more than 14,000 irrigation ponds in Kagawa Pref. and most of the ponds are very close to urban area. The Ooike is located in the rural area in suburb. The reservoir volume of Ooike is 153 thousands m^3, and the embankment height is 7.3 m, and the water surface area is 4.00 ha. Thus, the Ooike pond is considered as a medium-scale irrigation pond in Japan.

Figure 1: Sampling of the irrigation pond muddy soil at Ooike in Kagawa Pref.

Figure 1 shows the sampling at the Ooike in February, 2004. The water was removed 4 months ago for renovation work. The surface of the field was a little hardening because of drying. But the layer under the surface was so weak like swamp.

Figure 2 shows grain size distribution of pond muddy soil of Ooike pond. The average particle size D_{50} is 0.016mm, and fine-grain fraction content rate F_c is 88.9%. The plasticity index I_p is 40.4%, and it is classified as clay of high liquid limit with little fine-sand. Regarding other physical property, water content is 234%, pH is 6.0, and ignition loss is 13.9%.

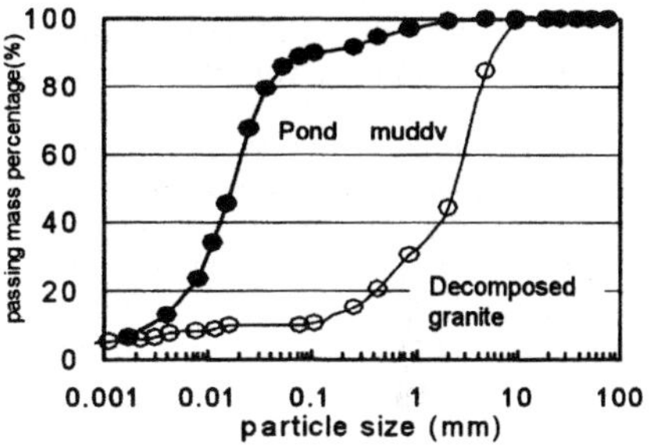

Figure 2: Grain size distribution of soil

Therefore it is found that hardening strength of pond muddy soil only is insufficient, so that the decomposed granite soil was also mixed. The decomposed granite soil is main soil type of Kagawa Pref. Its average particle size D_{50} is 2.1mm, and fine-grain fraction content rate F_c is 10.1%.

Specimen preparation

The specimen was made from the mixing of the pond muddy soil, decomposed granite soil and cement. There are three mixture ratios of pond muddy soil and decomposed granite soil, namely, 1:3, 1:1 and 3:1 in terms of mass ratio. Similarly, cement was added as 5%, 7% and 10% of mixture weight accordingly.

The curing condition kept constant at temperature of 20 degrees Celsius and at humidity over 90% for 7 days.

The day before testing, the specimen was submerged into water in order to raise the saturation ratio.

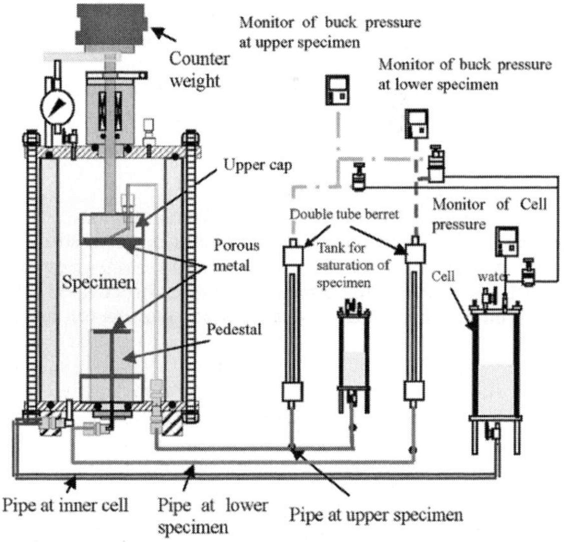

Figure 3: Triaxial stress type falling head permeability test apparatus

Test method

Unconfined compressive strength test and falling head permeability test were performed for the specimen after 7 days of curing.

Figure 3 shows the schematic diagram of the triaxial stress type falling head permeability test apparatus. This equipment is able to perform the permeability test under triaxial stress condition after consolidation process. It is possible that this equipment clarifies change of permeability by confining pressure. The permeability test was performed under head of water pressure of 20kN/m^2 between lower and upper specimen after consolidation process at 200kN/m^2 for 24 hours.

Results and discussions

Unconfined compressive strength

Figure 4 shows the relationship between mix ratio of cement and unconfined compressive strength q_u for curing of 7 days. The strength q_u increases with the increasing of cement content. On the mix ratio of the decomposed granite soil versus the pond muddy soil of 1:1 and 1:3, the strengths values are still low. It seems that these strengths are related with the amount of cement in some extent. However, it is felt that strength largely depends on mix ratio of decomposed granite soil and pond muddy soil with cement and when mix ratio of them is 3:1, value of strength is high although cement weight percentage is less.

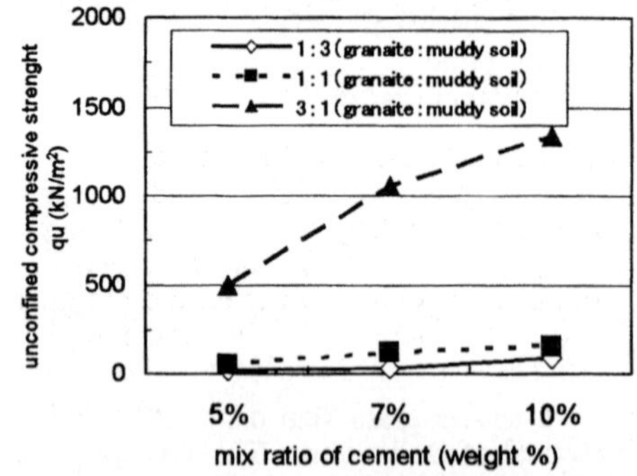

Figure 4: Change of hardening strength

Thus, pond muddy soil gives larger strength value in increasing amount of coarse grain decomposed granite soil. Generally, value of q_u should be over 200kN/m^2 for utilization as an embankment material. Therefore, it satisfy the condition of strength as an embankment material because the q_u is over 200kN/m^2 at more than 5% mix ratio of cement high amount of decomposed granite soil and suitable ratio is 3:1.

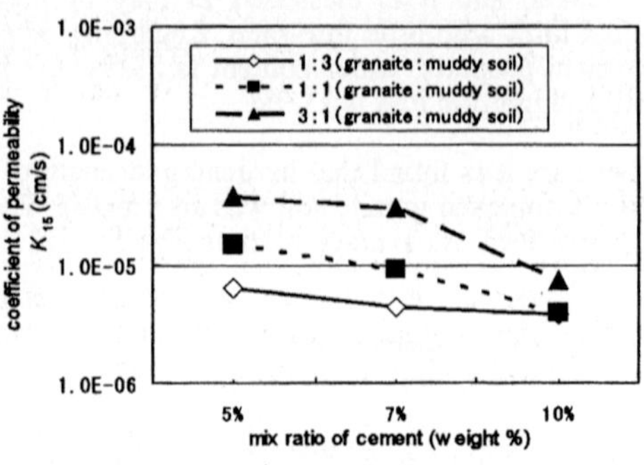

Figure 5: Change of permeability

Coefficient of permeability

Figure 5 shows the relationship between mix ratio of cement and coefficient of permeability k_{15} at 7 days of curing. The coefficient of permeability decreases with the increasing of mix ratio of cement, and increases with the increasing ratio of decomposed granite soil. It is clearly seen that permeability of sample is increased by adding decomposed granite soil to the pond muddy soil, because decomposed granite soil has permeability about 1×10^{-3}cm/s.

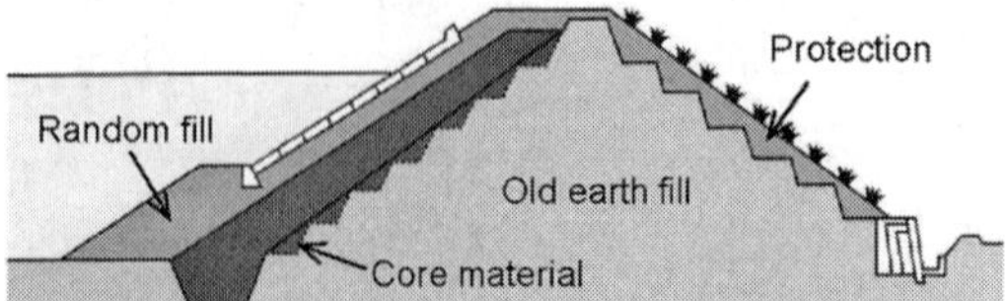

Figure 6: Cross-section of embankment renovation

Table 1: Range of coefficient of permeability as an embankment soil

Classification of embankment soil	Coefficient of permeability (cm/s)
Core material	under 1×10^{-5}
Random fill	$1\times10^{-5}\sim1\times10^{-4}$
Protection fill	$1\times10^{-4}\sim1\times10^{-2}$

Figure 6 shows the cross-section of embankment renovation. Most of the existed embankments generally have homogeneous fill by a single type of soil. In renovating and reinforcing works, it is a general practice to replace the weakened surface soil of the upstream by an impervious earth material called core material which is mechanically more resistant. The utilization method of pond muddy soil proposed in this study suggests applying pond mud in any of core materials such as random fill or protection fill of pond embankment.

Table 1 lists the range of coefficient of permeability required for an embankment soil. The applicable mixture condition for the core material in the case of decomposed granite soil vs. the pond muddy soil is 1:3 in 7% to 10% of cement mix ratio. It is also considered that the coefficient of permeability of the other mixture is applicable to use as random fill.

Utilization to replanting basis

Pond muddy soil burnt product

Authors examined how pond muddy soil can be used directly as a replanting material with the expectation of high organic content. However, vegetables (radish, carrot and lettuce) and turf grass did not grow because of lack of soil aeration in the replanting basis which was improved by pond muddy soil. So, to improve soil aeration, the pond muddy soil was burnt at high temperature.

Figure 7: Pond muddy soil burnt product (φ2-5 mm)

Figure 7 shows the burnt pond muddy soil product which has particle diameter of 2-5mm. The product was burnt at temperature of about 700 centigrade for 2 hours, and was crushed small. The color tone changed from the dark gray to the dark-brown and it became a light porous material.

Figure 8 shows the *pF*-soil moisture curves measured by the centrifuge method of the burnt product and some replanting materials. For the comparison, curves of Kanuma soil and Akadama soil and decomposed granite soil (Masa soil) are also added as a general replanting basis material. It is noticed that water retentivity is high for the burnt product in which particle size is small and it proves that water content volume of the burnt product is smaller than Kanuma soil and Akadama soil, and is bigger than the Masa soil,

Figure 9 shows the three phase of soil in the condition of moisture holding capacity after 24 hours (*pF*=1.8). Such condition is generally suitable for a growth of root system of plant because solid phase vs. liquid phase vs. vapor phase of soil is 40:30:30. The three-phase division of burnt product is also approximately near to the standard value (40:30:30).

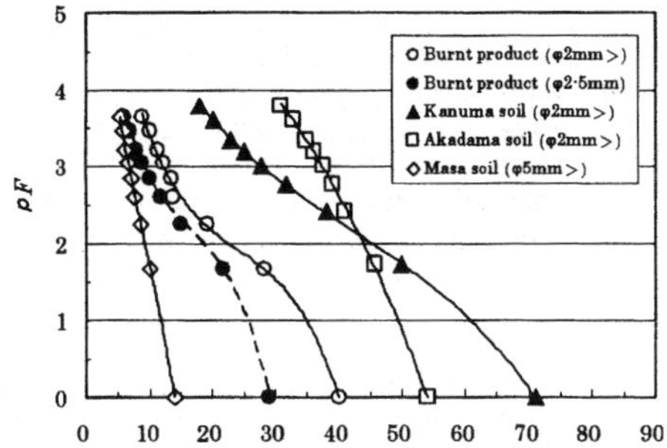

Figure 8: pF-soil moisture curves of the burnt product and some replanting materials.

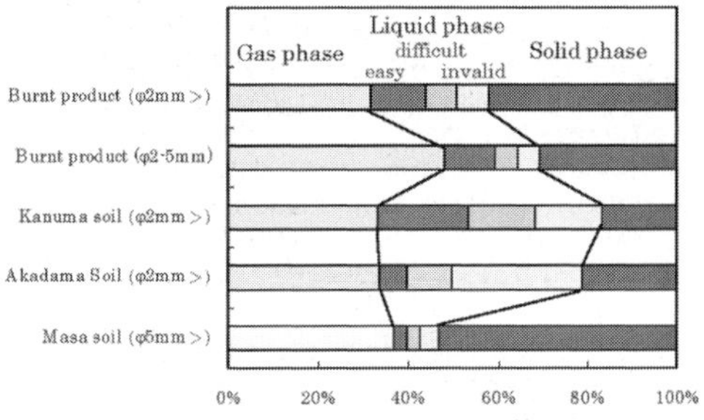

Figure 9: Comparison of three phase division

For the liquid phase, it can be divided into easy use water (water used by plant) and difficulty use water and invalid use water and it is noticed that percentage of easy use water of the burnt product is larger than other replanting materials. Therefore, from water retentivity and water absorbency point of view, the burnt product is excellent for plant growth.

Table 2 gives the physicochemical properties of the burnt product and other replanting basis. The superiority of burnt product as a light weight soil is comparatively low because it has dry density about 0.60g/cm^3 and it is more than other soil (Table 2). The average (50%) pore size of the burnt product is 1.68μm. Although pore size of burnt product is bigger than Akadama soil, it is smaller than Kanuma soil and Masa soil.

From the value of pH and cation exchange capacity (CEC), the burnt product can be ascertained as suitable material for growth of plant. Total nitrogen (T-N) of the burnt product is 0.02%, and this value is slightly below than limit normal value 0.03%. Even in limit normal value (0.03%), plant could not grow easily (Matsui, 1981). Therefore, it was able to be confirmed that the physicochemical property of the burnt product is excellent among one of the general replanting basis.

The capability test was also carried out using seeding of Bermuda-grass, and applicability of the burnt product was examined. Details of this test are given below.

Table 2: Physicochemical properties of the burnt product and some replanting basis

		Burnt product	Kanuma soil	Akadama soil	Masa soil
Water content (%)		0.3	49.9	49.5	6.6
Wet density (g/cm^3)		0.60	0.35	0.56	1.02
Dry density (g/cm^3)		0.60	0.23	0.37	0.96
Particle density (g/cm^3)		2.60	2.31	2.70	2.69
Pore size	90%(μm)	0.38	0.46	0.04	0.10
	50%(μm)	1.68	4.04	0.40	5.58
	10%(μm)	17.77	10.62	24.52	27.34
pH(H_2O)		6.2	6.2	6.4	7.0
EC (mS/cm)		0.22	1.14	0.03	0.55
T-P (%)		0.04	0.06	0.05	0.05
T-N (%)		0.02	0.10	0.20	0.04
K_2O (%)		0.25	0.01	0.04	0.25
C/N		3.5	2.5	7.3	1.5
CEC (me/100g)		25	160	310	67
Li (%)		0.4	11.6	13.1	1.6

Test method

Growth test of Bermuda-grass was carried out in burnt product by growing grass in the Wagner pot (1/5000a) for the period of 3 months from September to November in 2004. It is difficult to say that the period of this test was optimum, because Bermuda-grass is warm-season type. However, the effectiveness of the sample material as replanting basis was sufficiently analyzed. Five kinds of replanting basis of the burnt product (φ2mm > and 2-5mm), Akadama soil, Kanuma soil and Masa soil were used in this test. The gravel sand (φ5mm) was put into the lower layer of Wagner pot as drainage layer. Seeded pots were set at the roof of Kagawa University (see Fig. 10).

Various parameters were measured per a week. Basically, plant height, soil temperature, soil hardness, water content ratio and water content by volume, temperature and flux of solar radiation were measured. During the experiment, fertilizer was not added and water was equally sprinkled in all pots.

Figure 10: Situation of pot test

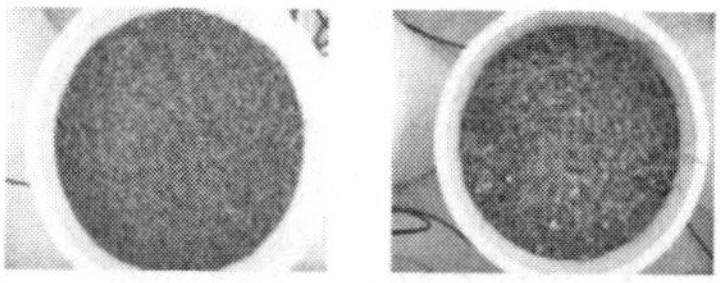

(a) Burnt product (φ2 mm>) (b) Burnt product (φ2-5mm)

(c) Kanuma soil (d) Akadama soil (e) Masa soil

Figure 11: Growth of Bermuda-grass for 7 weeks after seeding.

Results and discussions

Figure 11 shows the growth of Bermuda-grass after 7 weeks of the seeding. The burnt product (φ2mm >) has many grass (see Fig 11a). Likewise, plant height is increased well in burnt product of size φ2-5 mm (Fig.11b).

Figure 12 shows plant height changing patterns. In all replanting basis, post-third or fourth budding in seed was tested to examine whether there is large difference in the germination or not. The germination of grass has started from 3 or 4 days after seeding. It is found that the plant height of burnt product (φ2-5mm) is longest among other replanting basis. Similarly, it is also noticed plant height of the burnt product of size φ2mm> could not increase, because the vapor phase could not developed sufficiently.

Figure 13 shows change of water content by volume. It is noticed that water content volume increased in each replanting basis as per increment in precipitation. Similarly, it is noticed that burnt product (φ2mm>), Kanuma soil and Akadama soil have high water content volume.

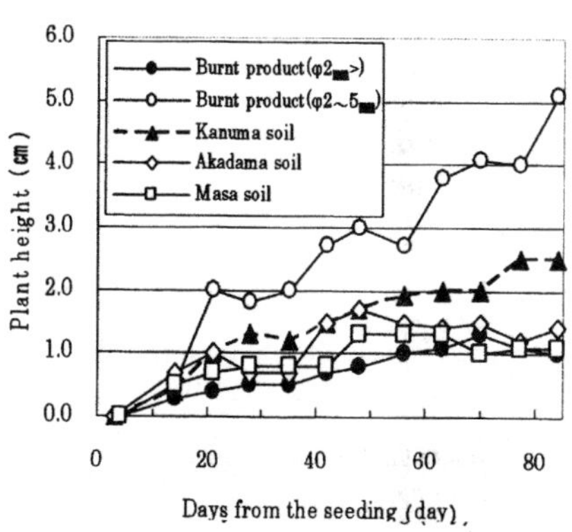

Figure 12: Growth of Bermuda-grass height

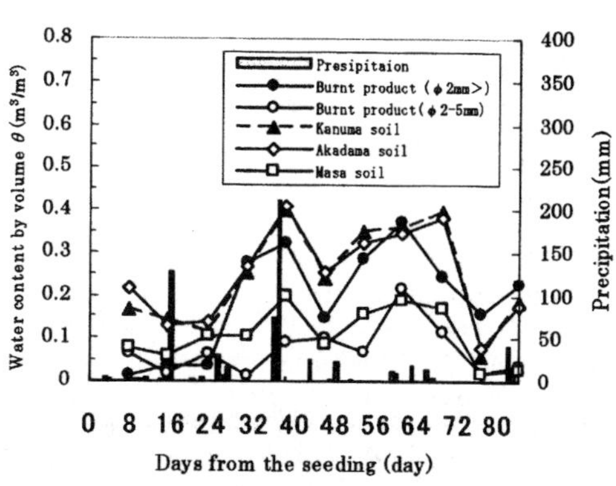

Figure 13: Change of water content by volume

Figure 14 shows change of soil hardness of replanting basis. Generally, growth of root system is retarded when soil hardness of replanting basis is over 20mm (Ministry of Construction City Bureau, 1999). In this research, soil hardness of burnt product (φ2mm>) is found to be high and soil hardness of them has sometimes exceeded mm. The soil hardness of the burnt product (φ2mm>) is highest because growth of grass was retarded in burnt product (φ2mm>).

Soil hardness of other replanting basis is approximately 10mm or less and they are used as replanting basis because they do not inhibit the root growth. Similarly, plant height in the burnt product of size φ2-5mm increased considerably because it was strongly influenced by low value of soil hardness and large value of gas phase.

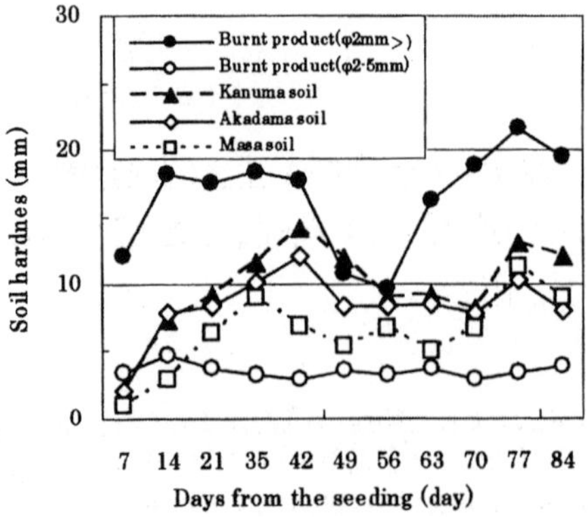

Figure 14: Change of replanting soil hardness

Conclusion

From this study, following conclusions are made.

1) For use of pond muddy soil as embankment material, it is found that it can be changed into suitable embankment materials by mixing the granite sand and cement in certain ratio.
2) For use as replanting bassis, pond muddy soil need to burn at high temperature in order to improve soil aeration. Performance of such burnt muddy soil was tested with the Bermuda-grass in Wagener pot and it is found that they were well grown than another general replanting basis.
3) Likewise, although the burnt product is a good replanting material, the burning treatment is costly. Therefore, efficient burning method in which the treatment cost can be lowered need to be examined from ceramics engineering point of view. It is noticed that cost reduction in waste treatment system of irrigation pond is possible by introducing these proposed utilization methods.

References

Fukushima,S., Tani,S. and Kitajima,A.,2004: Case study on reinforcement of steep and high embankment by cement-stabilized muddy soil in old irrigation dam. *J. of Japan Society of Civil Engineers*, 771/3-68, 289-308.

Hosomi,M.,Yoshigae,T.,Kashiuchi,T. and Sudo,R.,1997:Experimental study on development of wet-land system using dredging sludge. *J. of Water and Waste*, 39(7), 20-26.

Maekawa,T., Wakita,T. and Iida,Y., 1969: Studies on the phase of irrigational reservoirs -on the manorial value of the water-. *Technical bulletin of Faculty of Agriculture, Kagawa Univ.*,9(2),73-78.

Matsui,T., 1981: Point and problem of soil conservation in landscape architecture. *J. Japanese Institute of landscape Architecture*, 44(1),22-23

Ministry of Construction City Bureau (ed.),1999: *Technical maintenance manual for planting basis (draft)*, Japan greenery research and development center.

Okuda,K., Fukagawa,R.,1999: Fundamental study on slope replanting basis material which mixed lake bottom soil. *Proc. 54th annual conference of JSCE*, 54, 442-443

Tamaki,T., Hoshikawa,G.,1967:Syudy on pond mud- Degree of basis saturation and amount of water soluble sulfates. *Technical bulletin of Faculty of Agriculture, Kagawa University*, 8(2), 189-193.

Tomida,N., Yamane,K., Yamanaka,M., and Hasegawa,S.,2004: Fundamental physical property of pond bottom sludge burnt product. *Proc. annual conf. of JGS, Shikoku Branch*, 13-14.

Yamanaka,M., Hasegawa,S. and Morishita,K., 2005: Physico-chemical property of Pond sediment in Kagawa , *Proc. annual conference of the Japanese society of irrigation, drainage and reclamation engineering*, , 906-907.

Yanagisawa,H.,Yamadera,Y. and Miyazaki,T., 2004: Experimental study on development of nurture block using deposition mud. *J. of Japanese Society of Revegetation Technology*,30(1),289-292.

Watanabe,K.,Sugiyama,Y.,Inoha,H. and Ueda,M., 2004: The example of reclamation of the Artificial Tidal Flat which utilized the dredge ground as Local Resources. *J. of JSIDRE*,72(8),685-688.

Technical and Environmental Viability Study of the Use of Coal Ash as Covering Material for Mining Rejects

Mariluce de Oliveira Ubaldo, CETEM/MCT, Rio de Janeiro, RJ, Brasil.
Maria Claudia Barbosa,COPPE/UFRJ, Rio de Janeiro, RJ, Brasil,
Laura de Simone Borma, Federal University of Tocantins, Palmas, TO, Brazil.

Abstract: **This work studies the use of ashes arising from the burning of mineral coal as an alternative material to compose a dry coverage system to cover piles of rejects that produce acidity. This coverage system is based on the principle of capillary barriers, which consists of the overlaying of a fine-textured material over a granular material. The proposal is to verify the viability of the bottom ash and the fly ash as a layer of granular material. The ashes evaluated in the present study are bottom ash and the fly ash that were produced at the Jorge Lacerda Thermoelectric Plant. The results showed that the bottom ash proved to be viable for the proposal in the study.**

Introduction

The acid mine drainage (AMD) is one of the most significant environmental problems confronted by the mining-metallurgy industry. This takes place naturally in environments that expose Fe sulfates to oxidation; nevertheless, it also takes place as environmental impact mainly in the mining activities (BORMA, 2003). In the last few years, this problem has become a very serious question, requiring studies aimed at finding solutions to minimize the environment damage caused.

To minimize the production of acidity in mining residues piles, the alternative usually employed has been the so-called dry coverage. Dry coverage consists of covering the area of the deposit with layers of soil with the purpose of minimizing the entrance of oxygen and water inside the reactive matter, thus minimizing the process producing acidity. Its performance strongly depends on the materials employed and the climatic conditions. One type of dry coverage that has been suggested by humid climate is that employing the principle of the capillary barrier. The capillary barrier prevents the passage of water to the mining rejects and at the same time is capable of maintaining a high degree of inner saturation, even during the dry period, preventing the passage of oxygen to the rejects.

Another problem of Brazilian coal mining is the excess of ash produced during the burning. According to BORMA et al. (2003), the resulting ash contents of burning the coal are about 50%. Due to the large volume of ashes produced in the mines in the states of Santa Catarina and Rio Grande do Sul, research has been conducted with respect to its potential to be employed in the recovery of the areas degraded by mining.

With the purpose of employing a residue that has been disposed in an inappropriate way, the present research presents a viability study of the employment of the ash as an alternative material in the covering of deposits of coal rejects. Due to the behavior of granular matter, the ash would be employed to substitute sand in the capillary barrier, providing the needed contrast of permeability between layers.

Background

Acid mine drainage (AMD) takes place when residues from mines rich in metallic sulfates oxidize on exposure to water and oxygen. The production of acidic drainage results in the contamination of surface and ground waters, making the use of these waters for recreational, agricultural and consumption purposes unviable. The acidic drainage can take place in mines involving the commercial exploitation of minerals, such as copper, nickel, uranium, coal and gold. One of the largest problems in Brazil in the production of acidity occurs in the coal mining in the southern part of the country, especially in states of Rio Grande do Sul and Santa Catarina.

Covers with function of capillary barriers are considered one of the most effective ways to control the acid mine drainage (BUSSIÈRE, et al, 2004; NICHOLSON, et al., 1989; YANFUL, et al., 1993). The effect of a capillary barrier is obtained when a fine material is placed over a granular material. The two materials have different hydraulic properties mainly because of their different particle size distribution. During the process of desaturation, the fine material, on presenting smaller intersticial pores, shows a greater capacity to retain water than the granular material. In the layer of granular material, the presence of air after desaturation reduces the inter-connectivity of the pores filled by water, thus reducing the effective hydraulic conductivity. The lower the unsaturated hydraulic conductivity of the granular material, the less will be the drainage flow of vertical water from the layer of fine material and, consequently, the fine material remains with a high water content, thus reducing the oxygen flow from the atmosphere into the pile of mine waste.

Generally, the effect of a capillary barrier is increased with the increase in the contrast of the unsaturated hydraulic conductivity between the layers of fine and coarse material (SHACKELFORD, 1997).

The coal combustion process gives rise to residues generically called ashes, which can be of several types: slag, bottom ash and fly ash. The large quantity of ash produced in Brazil and its alkaline pH (between 8 and 10) led to a series a researches focusing on its alternative employment in mining itself, mainly as neutralizing agent.

SOUZA (2001) conducted a study on the use of ashes, aluminum-silicates and hydrated lime in the composition of geochemical barriers. According to the author, the use of ashes in the cover layer did not show any effect of a geochemical barrier. However, when the ash mixed with aluminum-silicates and lime was employed, the formation of a layer of low water permeability (about 10^{-7} to 10^{-8} cm/s) was observed.

BORMA (1999) conducted studies utilizing ash that came from the President Médici Thermoelectric Plant in Rio Grande do Sul and MENDONÇA et al (2002) employed ash coming from the Thermoelectric Plant Lacerda in Santa Catarina. In both studies buffer capacity tests were conducted. The buffer capacity test is described in YONG et al (1990), and the test consists of adding increasing amounts of acid in soil/water suspension samples placed in a centrifuge bottle and verifying if there has been a change in pH after a certain time of stabilization. The results obtained by the two authors are reproduced in figure 1.

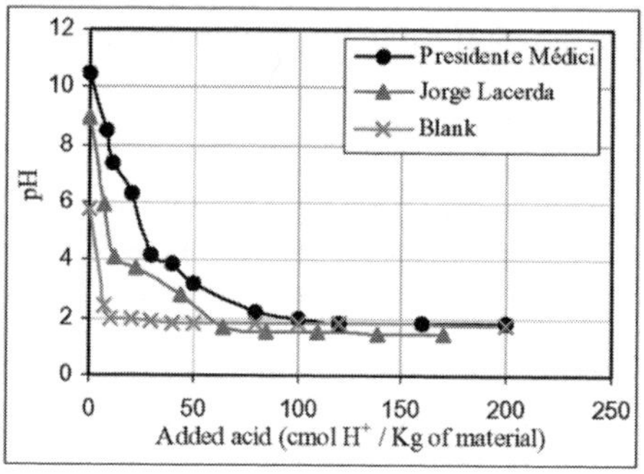

Figure 1 – Buffer Capacity Test data after BORMA (1999) and MENDONÇA et al (2002).

As it can be observed in figure 1, the Brazilian ash produced from coal combustion, despite presenting alkaline pH, does not show a high buffer capacity, thus not having, in this matter, the potential for neutralizing acidity.

The present work shows results of laboratory tests conducted in the ashes focusing on their potential to be employed in the rehabilitation of deposits producing acidity, as part of a cover system of waste deposits with the purpose of minimizing the entrance of oxygen and water into the deposit, blocking the process of AMD generation. The results are part of the research conducted by UBALDO (2005).

Methodology

Materials

This paper deals with two types of ashes: bottom ash and fly ash. The ashes come from the burning of mineral coal at the Jorge Lacerda Thermoelectric Plant in Tubarão-SC. The classification, according to the Soil Unified Classification System (SUCS), and the physical characterization of the ashes are presented in table 1.

Table 1 – Physical Characterization and Classification.

Sample	*Bottom ash*	*Fly ash*
Clay (%)	04	08
Silt (%)	36	60
Sand (%)	58	32
Gravel (%)	02	-
SUCS	SM	ML
Gravity density (G)	2.043	2.090

The tests to determine the Atteberg limits showed that both bottom ash and fly ash are non-plastic materials. CHIES et al., (2003) observed the same behavior for all the bottom ashes studied, coming from different Brazilian Thermoelectric Plants.

Methods

The research consisted of conducting laboratory experiments for technical viability study of the ashes as the coarse layer material in the waste pile cover system. The following tests were performed: Standard Proctor compaction curve, water retention characteristic curve, saturated hydraulic conductivity, chemical composition, Brazilian Standards lixiviation and solubilization tests for waste classification, and mineralogical characterization. Testing procedures used in the experiments are presented in table 2.

Table 2 – Tests procedures

Test	*Standard*
Grain size analysis	NBR 7181/84
Gravity density	DNER-ME 093/94
Liquid limit	NBR 6459/84
Plastic limit	NBR 7180/84
Saturated hydraulic conductivity	NBR 13293
Compaction curve	NBR 7182/86
Filter paper	ASTM D 5298/03
Classification of residues	NBR 10004/2004
Lixiviation	NBR 10005/2004
Solubilization	NBR 10006/2004

Experimental Results

Compaction Test

The compaction test was conducted at the Normal Proctor energy. The compaction curves are shown in figures 2 and 3 for the bottom and fly ash samples respectively, as well as the values of maximum dry density ($\gamma_{dmáx}$) and optimum water content (W_{otm}) obtained.

The ashes did not show a good response to the compaction process, and the optimum condition is difficult to be established from the compaction curves.

The same behavior was observed in other studies, for example, NISHIKAMA et al. (2002) and HEINECK (2000). However, the results are useful as criteria for molding the test samples.

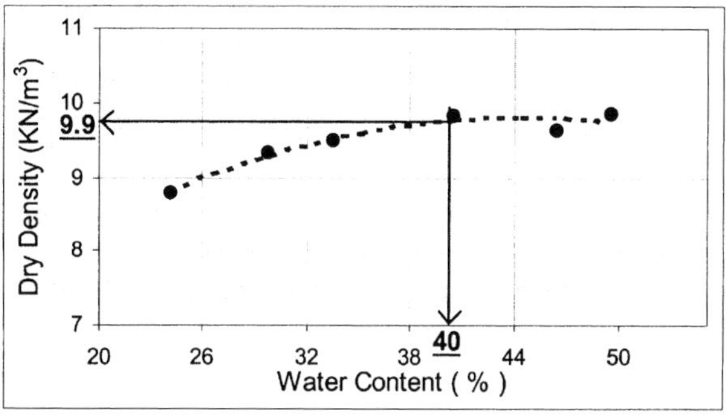

Figure 2 – Bottom Ash Normal Proctor Compaction Curve.

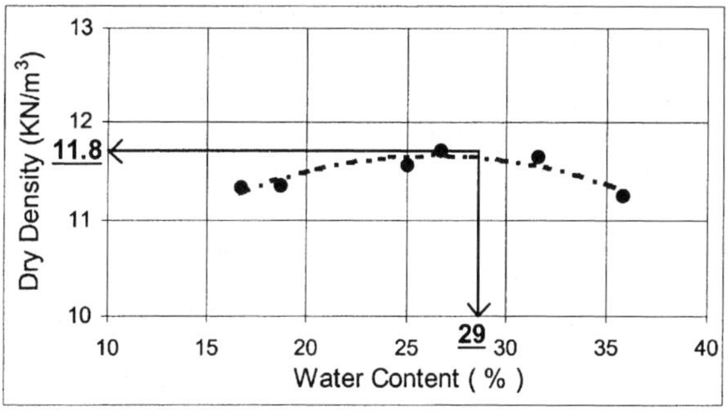

Figure 3 – Fly Ash Normal Proctor Compaction Curve.

Saturated Hydraulic Conductivity

The saturated hydraulic conductivity was determined in falling head permeability tests. The test samples were molded by static compaction procedure at optimum water content and maximum dry density of the corresponding Proctor compaction curve. The measured saturated hydraulic conductivity values are showed in table 3.

Table 3- Saturated Hydraulic Conductivity Data.

Samples	$K_{20°Csat}$ *(m/s)*
Bottom ash	2.45×10^{-6}
Fly ash	6.00×10^{-7}

Water Retention Characteristic Curve

The water retention characteristic curves were obtained by the filter paper method and are shown in figure 4. The samples were molded in the same way as described for the permeability test, at the same void ratio.

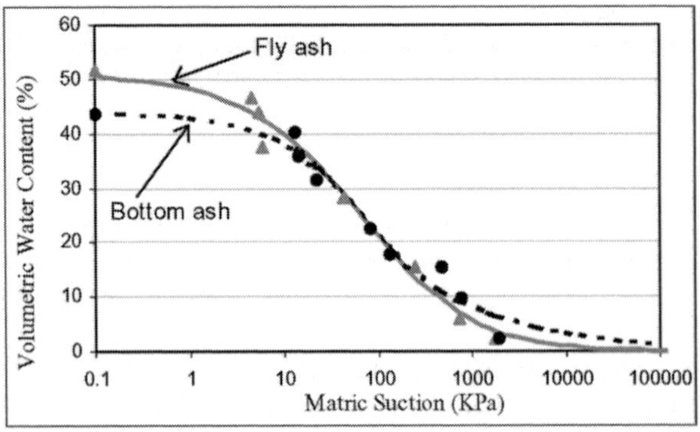

Figure 4 – Water Retention Characteristic Curves of Bottom and Fly Ashes.

Chemical Characterization

The chemical characterization of the ash samples was performed by X-ray fluorescence – a semi-quantitative analysis, and the results are summarized in table 4. Both ashes are composed basically of SiO_2, Al_2O_3 and Fe_2O_3.

Table 4 – Chemical analysis of the samples

Elements	*Bottom ash (%)*	*Fly ash (%)*
SiO_2	57.9	57.1
Al_2O_3	27.2	28.7
Fe_2O_3	5.5	4.4
Cl^-	1.7	0.09
CaO	1.4	2.0
K_2O	2.5	2.6
MgO	0.64	0.72
TiO_2	1.1	1.3
ZrO_2	0.12	0.13
SO_3	< 200ppm	1.0
S	< 200ppm	0.4

Ashes Classification as Residues

According to the NBR 10004/2004 Brazilian Standards classification of residues, both the bottom and the fly ashes are classified as Residue Class II A -. Non-inert. In the bottom ash test only Aluminum was found above maximum tolerable limits, while in the fly ash test the elements "Al", "Cr" and "SO_4" exceeded the limits defined in the standard.

Mineralogical Composition

According to the X–ray diffraction analysis, the dominant mineralogical constituents of both bottom and fly ashes are: quartz, kaolinite, muscovite, gibbsite and pyrite.

Discussion and Conclusions

Related to the results of the experiments detailed, one can conclude that: the ashes present material granular characteristics and according to the definitions of NBR 10004/2004 for solid residues, both were classified as Residue Class II A – Non-Inert.

The bottom ash is more feasible than the fly ash as an alternative material in a capillary barrier used as a cover layer for deposits of coal mine tailings, because bottom ash the saturated hydraulic conductivity coefficient was greater than that of fly ash, thereby permitting a greater contrast of hydraulic conductivity between the granular material and the clay material that makes up the capillary barrier. Besides, the Jorge Lacerda thermo-electric power plant (SC) has a greater quantity of bottom ash available than fly ash, since the fly ash produced is sold to cement factories and despite the fact that both ashes are classified as non-inert material, bottom ash had only one element above the value permitted by the norm, while fly ash had three.

References

-ABNT/NBR 10005:2004, “Procedimento para obtenção de extrato de lixiviado de resíduos sólidos” (in Portuguese).
-ABNT/NBR 10006:2004, “Procedimento para obtenção de extrato solubilizado de resíduos sólidos” (in Portuguese).
-ABNT/NBR 10004:2004, “Resíduos sólidos - Classificação” (in Portuguese).
-ABNT/NBR 7181:1984, “Solo - Análise granulométrica” (in Portuguese).
-ABNT/NBR 13292:1995. “Solo - Determinação do Coeficiente de Permeabilidade de solos granulares e de carga constante” (in Portuguese).
-ABNT/NBR 13293:2000. “Solo - Determinação do Coeficiente de Permeabilidade de solos argilosos e de carga variável” (in Portuguese).
-ABNT/NBR 6459:1984, “Solo - Determinação do Limite de Liquidez” (in Portuguese).
-ABNT/NBR 7180:1984, “Solo – Determinação do Limite de Plasticidade” (in Portuguese).
-ABNT/NBR 7182:1986, “Solo – Ensaio de compactação” (in Portuguese).
-ASTMD 5298 - 03, “Standard test method for measurement of soil potential (suction) using filter paper”.
-Borma, L.S., Mendonça, R. G., Souza, V. P., Soares, P. S. M. (2003), “Utilização de cinza no contexto da reabilitação de áreas de mineração de carvão”. V REGEO – Congresso Brasileiro de Geotecnia Ambiental, Porto Alegre, RS (in Portuguese).
-Borma, L. S. (2003), “Ensaios para determinação da velocidade intrínseca de oxidação em resíduos geradores de acidez”. V REGEO – Congresso Brasileiro de Geotecnia Ambiental, Porto Alegre, RS (in Portuguese).

-Borma , L. S. (1999), "Determinação da velocidade de oxidação em estéreis de mineração e em sedimentos dragados ricos em sulfetos de Fe". Relatório de Atividades Recém Doutor – CETEM/MCT – Centro de Tecnologia Mineral, Rio de Janeiro, 22p (in Portuguese).

-Bussière, B.; Benzaazoua, M.; Aubertin, M.; Mbonimpa, M. (2004), "A laboratory study of covers made of low-sulphide tailings to prevent acid mine drainage". Environmental Geology 45:609-622.

-Chies, F., Silva, N. I. W., Zwonok, O. (2003), "Desenvolvimento de blocos e tijolos a partir de cinzas de fundo de carvão - CEPECAL".In: Utilização de Resíduos na Construção Habitacional - Coletânea Habitare - Volume 4 – Capítulo 7 - Porto Alegre-RS, Editora ANTC (in Portuguese).

-DNER – ME 093 (1994) "Solos – Determinação da densidade real dos grãos" 4 p (in Portuguese).

-Heineck, K. S. (2000), "Estudo do comportamento hidráulico e mecânico de materiais geotécnicos para barreiras horizontais impermeáveis". Tese de doutorado, Universidade Federal do Rio Grande do Sul - Porto Alegre, RS (in Portuguese).

-Mendonça, R. M. G., Borma, L. S., Soares, P. S. M. (2002), "Characterization of residues from coal processing". Proceedings of the 4ICEG - Fourth International Congress on Environmental Geotechnics, Rio de Janeiro, Brazil, pp.377-382.

-Nicholson, R. V., Gillham, R.W., Cherry, J. A., Reardon, E. J. (1989), "Reduction of acid generation in mine tailings through the use of moisture-retaining cover layers as oxygen barriers". Canadian Geotechnical Journal, 26: 1-8.

-Nishikawa, J., Sato, A., Amano, K. (2002), "Utilization of coal ash as filling materials". 4º ICEG - Rio de Janeiro, Brazil, 615-618.

-Shackelford, C. D. (1997), "Modeling and analysis in environmental geotechnics: An overview of practical applications". Second International Congress on Environmental Geotechnics; V 3, pp. 1375 - 1403, Osaka, Japan.

-Souza, V. P. (2001), "Formação de barreira geoquímica para abatimento de drenagens ácidas de pilhas de estéril piritoso". Tese de Doutorado, Universidade de São Paulo - SP. Março -142p (in Portuguese).

-Ubaldo, M. O. (2005), "Uso de cinzas de carvão na composição de uma cobertura de rejeitos de mineração". M.Sc. thesis, COPPE/UFRJ, may (in Portuguese).

-Yanful, E. K., Riley, M.D., Woyshener, M. R., Duncan, J. (1993), "Construction and monitoring of a composite soil cover on an experimental waste-rock pile near Newcastle, New Brunswick, Canada". Canadian Geotechnical Journal, Vol. 30, pp. 588-599.

-Yong, R. N., Wakentin, B. P., Padhungchewit, Y., *Et Al.* (1990), "Buffer capacity and lead retention in some clays materials water". Air and Soil Pollution, 53, p.53-67.

OTHER

Modelling the Behaviour of Freezing and Thawing Soil Slopes

P.J. Cleall, Geoenvironmental Research Centre, Cardiff School of Engineering, Cardiff University, Cardiff, UK
H.R. Thomas, Geoenvironmental Research Centre, Cardiff School of Engineering, Cardiff University, Cardiff, UK
M.C. Glendinning, Geoenvironmental Research Centre, Cardiff School of Engineering, Cardiff University, Cardiff, UK

Abstract
Mass movement processes on soil slopes where the ground thermal regime is dominated by seasonal freezing and thawing is often characterised by widespread periglacial solifluction (gelifluction) and shallow landslides. Observations in permafrost regions show that when ice-rich frozen slopes thaw, high moisture contents, raised pore water pressures and significant loss of strength results causing a detrimental effect to the stability of the slope through an increase in slope mass and reduction in effective stress. A Thermo-Hydro-Mechanical (THM) coupled theoretical formulation that incorporates an elasto-plastic Camclay stress-strain constitutive relationship is presented. Furthermore a numerical solution of this formulation and results of its application to a simple soil column during an annual cycle of freezing and thawing are presented.

Introduction
The effects of climate change and global warming are leading to increases in average temperatures. In periglacial environments these rises in temperature have the potential to alter the stability of slopes, potentially resulting in increased pre-failure strains and possibly failure. It is of considerable importance to be able to quantify the change in solifluction rates due to such climatic changes in these environments. With increasing temperatures slopes are experiencing an increase to the active layer depth, resulting in parts of the slope that would have normally been frozen starting to thaw and move. The development of models capable of representing these processes will lead to an improved ability to assess the effects of climate change on such slopes.

Freeze–thaw action induces downslope displacement of soils in cold, non-glacial environments, where vegetation is lacking or sparse (Ballantyne and Harris 1994). This process, broadly referred to as solifluction, operates slowly, in general at a rate of at most 1 m per year. Its widespread distribution on mountain slopes means that solifluction contributes greatly to the evolution of mountain landscapes. Moreover, landforms and subsurface structures resulting from solifluction are strongly dependent on climatic conditions (Matsuoka 2001). Rates and processes of solifluction depend on climate, hydrology, geology and topography. Prediction of landscape evolution in periglacial mountains requires quantitative relationships between the rate of solifluction and these variables (Kirkby 1995).

5th ICEG Environmental Geotechnics, Thomas Telford, London 2006.

Extensive research has been carried out in the field to monitor and evaluate the conditions which cause the gradual mass wasting of slopes. For this purpose, field measurements have been undertaken in a variety of geographical situations ranging from polar hill slopes to tropical high mountains and, as a result, a large number of data-sets on solifluction rates and associated parameters have been obtained (Matsuoka 2001). Laboratory simulations using full scale and centrifuge modelling have further contributed to the understanding of the mechanisms that are involved in the solifluction processes (Harris and Davies 2000; Harris et al. 1997; Harris et al. 2003; Harris et al. 1993). The term 'solifluction' has not yet been defined unequivocally (Matsuoka 2001). The original meaning was defined as the slow downslope movement of saturated soil occurring in cold regions. It has been subsequently shown that it is not a necessary requirement for the soil to be saturated for slow mass movements to occur. It is generally accepted today that the term solifluction represents slow mass wasting associated with freeze–thaw action, whilst saturated soil movement associated with ground thawing is designated as gelifluction (Ballantyne and Harris 1994). Solifluction precludes rapid slope failures that reflect slide or flow over a shear plane, such as active-layer detachment slides and skinflows (Ballantyne and Harris 1994). Solifluction is most prevalent in sandy to silty soils having low liquid limits and plasticity indices (Matsuoka 2001).

Despite extensive field studies, progress in understanding gelifluction processes has been limited. Controlled laboratory simulation experiments offer an alternative and potentially extremely effective approach (Harris and Davies 2000). Results from two-dimensional vectors of soil surface movement with evidence from excavated displacement columns suggest that gelifluction occurred only during the thaw consolidation of the upper parts of the soil profile: thawing of the deeper layers caused thaw consolidation but little downslope displacement (Harris and Davies 2000). It was also observed that the cryogenic processes caused a decrease in void ratio and moisture content with depth with an increase in the shear strength within the continuous soil matrix that separates ice lenses. Since self-weight stress levels are low, thawing leads to significant shear strain only in the softer, wetter near-surface soil layers (Harris and Davies 2000). For slopes of low angles and low overburden stress, the rate of thaw induced shear strain is highly sensitive to changes in soil shear strength and viscosity. The potential for gelifluction is therefore likely to decrease with depth.

In order to formulate constitutive equations, the timings and mechanisms of soil displacements during gelifluction have only rarely been measured in the field (Matsuoka et al. 1997). Due to the complexity of field conditions, determining good quality data to use in the formulation of constitutive equations is difficult and efforts are often directed towards laboratory based studies. In the laboratory the material properties and conditions can be controlled and offer the most effective approach to process understanding (Harris et al. 1996; Matsuoka 2001). Two scaled physical modelling experiments at elevated gravity were conducted in the geotechnical centrifuge (Harris et al. 2003). Their investigation was to test the hypothesis that gelifluction may be modelled by assuming soils flow as viscosity controlled fluids. Anderson in 1906 originally defined solifluction as 'slow flowing from higher to lower ground of masses of waste saturated with water' (Harris et al. 2003) and several authors have used 'viscous flow' to described gelifluction. It was concluded that gelifluction is not a time-dependent viscosity-controlled flow phenomenon, but rather elasto-plastic in nature (Harris et al. 2003). A flow law was then proposed based on the 'Cam Clay' constitutive model for soil as a first approximation gelifluction model.

Numerical modelling of freezing and the associated frost heave that accompanies freezing has attracted attention since the 1970s with many studies and models proposed. These models have nearly all been based on coupled heat and moisture transfer (Harlan, 1973; Taylor and Luthin, 1978) or conceptual models such as the segregation potential (Konrad and Morgenstern 1981; Nixon 1991). The major limitation to these models was the inability to consider the effect of external loading on the freezing process and assumed that there was no applied loading and the ice was under atmospheric pressure (the segregation potential was later modified to account for external loads).

One of the first models to incorporate applied loading was by Hopke (1980) with further advancements in understanding presented by Gilpin (1980) allowing the prediction of ice lens location and the introduction of the rigid ice model (O'Neill and Miller, 1985) considered the application of external loading. These three models were still limited as the applied loading was only considered as a factor influencing the frost heave, while the stress field in the soil due to loading, volumetric expansion and creep is not considered and calculated in these models (Shen and Ladanyi, 1987). Traditionally, frost heave modelling has been one-dimensional however the theory is being extended to include continuum models with three-dimensions (e.g. Talamucci, 1997; Selvadurai et al, 1999; Mikkola and Hartikainen, 2001).

Modelling of frozen soils requires knowledge of the complex thermo-hydro-mechanical processes that occur as a result of freezing and thawing in soil in order to provide a model that can be applied to practical applications. The adaptability and reliability of these developments, for practical application to problems such as frost heave induced mechanics of civil engineering structures is as yet unproven and currently, there appears to be no satisfactory model which can accommodate these aspects in a comprehensive fashion (Selvadurai et al, 1999). Numerical modelling of such complicated processes nearly always results in simplifying assumptions and under highly idealised conditions.

This paper presents a model of the thermo-hydro-mechanical behaviour of freezing and thawing soil. In the following sections, the theoretical and numerical formulations are introduced first. The model assumptions are also presented. Finally, some initial results from the model are presented. Work is continuing on the development of the model and as such the results are preliminary.

Theoretical Formulation

A coupled thermo-hydro-mechanical model is presented here based on the concepts of mass balance and stress equilibrium. In the development of this model a number of assumptions have been made. These are listed below:

- Moisture transport in both the frozen and unfrozen zones occurs in the liquid phase only. Air and Vapour phases have negligible effects on the moisture transfer.
- The water flux in the unfrozen soil and frozen fringe is given by Darcy's law.
- Salt exclusion is negligible.
- The porous matrix in the unfrozen soil and frozen fringe is deformable. Frozen soil is non-deformable.
- Pore ice in the frozen fringe is at rest with respect to the porous matrix with no capillary interfacial effects between ice and water in the frozen fringe.
- Freezing point depression of ice in the soil under loading is negligible.
- Unfrozen and partially frozen soil are non-linear elasto-plastic mediums.

Moisture transfer

Utilising the law of conservation of mass with Darcy's law and the hydraulic head gradient for steady or unsteady flow in saturated soil during freezing gives the moisture transport equation as

$$[(1-S_i)\rho_l + S_i\rho_i]\mathbf{m}^T\mathbf{P}\frac{\partial \mathbf{u}}{\partial t} + n(\rho_i - \rho_l)\frac{\partial S_i}{\partial T}\frac{\partial T}{\partial t} = \nabla\left[\frac{\rho_l k_l}{\gamma_l}(1-S_i)\nabla u_l\right] + \rho_l(1-S_i)\nabla k_l \nabla z \quad (1)$$

where ρ_l, ρ_i are the densities of liquid water and ice, k_l the hydraulic conductivity, γ_l the unit weight of water, n the porosity, S_i the degree of ice saturation i.e. the volume of the pore space that is ice, t is the time, **P** and **m** the strain matrix and differential operator, T the temperature, **u** displacement, u_l pressure, ∇ the laplace operator and z the vertical axis.

Heat transfer

From the conservation of energy the heat transport equation can be written as:

$$[((1-S_i)\rho_l c_l + S_i\rho_i c_i)(T-T_r) - L\rho_i S_i]\mathbf{m}^T\mathbf{P}\frac{\partial \mathbf{u}}{\partial t} + \left[H_c + (n(T-T_r)(\rho_i c_i - \rho_l c_l) - L\rho_i n)\frac{\partial S_i}{\partial T}\right]\frac{\partial T}{\partial t}$$

$$= \nabla\left[\left[\lambda_T + c_l\rho_l k_l(1-S_i)\left(\frac{u_l}{\gamma_l} + z\right)\right]\nabla T\right] + \nabla\left[\frac{c_l\rho_l k_l}{\gamma_l}(T-T_r)(1-S_i)\nabla u_l\right]$$

$$+ c_l\rho_l(T-T_r)(1-S_i)\nabla(k_l\nabla z) \quad (2)$$

where c_l, c_i is the specific heat capacity of liquid and ice, L is the Latent heat of fusion, H_c the heat capacity, and λ_T the thermal conductivity.

Stress – strain behaviour

The governing equation for deformation is derived from considerations of stress equilibrium. The model used in this study is based on a modified Camclay model (Wood 1990). During the freezing process the phase change of moisture to ice causes the volumetric change in freezing soil. This freezing strain is given by:

$$d\varepsilon_f^e = \left(\frac{\sigma_l - \sigma_i}{\sigma_i}\right) n dS_i \quad (3)$$

The stress-strain relationship given can now be written in terms of the individual components of plastic strain increments due to changes in stress;

$$d\sigma' = \mathbf{Dep}(d\varepsilon - d\varepsilon_p^e - d\varepsilon_f^e - d\varepsilon_p^p) \quad (4)$$

where $d\sigma'$ is the stress increment vector, **Dep** the elasto-plastic stress-strain matrix, $d\varepsilon$ the strain increment vector, and $d\varepsilon_p^e$, $d\varepsilon_f^e$, $d\varepsilon_p^p$ the elastic strain increment, freezing strain increment and plastic strain increment respectively.

Numerical Formulation

The governing differential equations are solved spatially using a finite element technique and temporally using a finite difference technique. In this paper the Galerkin weighted residual approach is adopted and the domain is discretised using eight noded isoparametric axisymmetric elements. This approach has been successfully applied to examine transient coupled problems and has been found to be effective and robust (Zienkiewicz and Taylor, 1989; Thomas et al., 1998). The coupled processes are solved simultaneously within a time step and a converged solution is assumed to have been achieved when the difference between successive iterations falls below a certain tolerance.

Numerical Simulation

A two-dimensional column of a hypothetical soil has been considered to examine the proposed freezing and thawing model. The dimensions of the soil column are 1000mm x 200mm, the finite element mesh divided into 80 elements high by 1 element wide. The sample is restrained in the vertical direction at the base and on one side restrained in the horizontal direction. The pore water pressure is fixed constant throughout the domain. A sinusoidal temperature profile is applied to the surface and base of the column as this is typical of the climate in periglacial regions, with a starting and maximum value of 278K and minimum of 270K. One complete cycle was applied over a period of one year, with an initial timestep of 600s increasing to 6000s. Both the mesh and timestep used have been checked for numerical convergence.

It should be noted that the problem is a challenging one due to the moving boundary problem related to the frozen fringe and the rapid changes that occur in a spatially small area. Modelling this problem within a regular finite element mesh requires care to avoid numerical errors from too coarse a mesh or too long a time step.

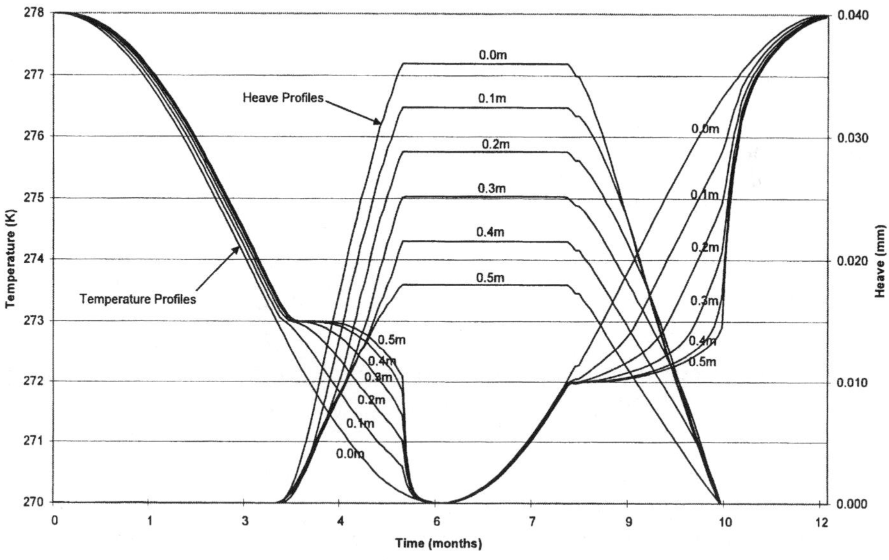

Figure 1. Frost heave and temperature variation with depth.

The heat capacity is defined by

$$H_c = (1 - S_i)n\rho_l c_l + nS_i\rho_i c_i + (1 - n)\rho_s c_s \tag{5}$$

where c is the specific heat capacity, ρ is the density, n is the porosity and S_i is the volumetric ice content. The subscripts s, l, and i refer to the solid, liquid and ice phases.

The thermal conductivity of the soil is defined by

$$\lambda_T = \lambda_s(S_i) = (1 - S_i)n\lambda_l + nS_i\lambda_i + (1 - n)\lambda_s \tag{6}$$

where λ_s, λ_l, and λ_i are the thermal conductivity of solid, liquid and ice respectively.

The volumetric ice content is used to model the hydraulic conductivity as a temperature dependant property.

$$k = k \cdot S_i \text{ where } S_i = f(T) \tag{7}$$

The frozen fringe is the region between the frozen and unfrozen soil where a mixture of the two phases co-exist together and the release or absorption of latent heat of fusion occurs. In the simulation the frozen fringe is defined as the region of soil where the temperature is within the range 272-273K. This range is higher than would be encountered in the field under low overburden conditions but serves to explore the performance of the numerical model. The parameters used in the simulation are shown in Table 1 and Table 2.

Property	Value	Property	Value
ρ_s	1800 kg m^{-3}	c_s	1000 $\text{J kg}^{-1} K^{-1}$
ρ_l	1000 kg m^{-3}	c_l	4180 $\text{J kg}^{-1} K^{-1}$
ρ_i	917 kg m^{-3}	c_i	2050 $\text{J kg}^{-1} K^{-1}$
λ_s	1 $\text{Wm}^{-1}K^{-1}$	k_l	$1\text{x}10^{-5}\ m s^{-1}$
λ_l	0.58 $\text{Wm}^{-1}K^{-1}$	n	0.4
λ_i	2.22 $\text{Wm}^{-1}K^{-1}$	L	$334\text{x}10^3$ J kg^{-1}

Table 1 – Model Flow Properties

Property	Value	Property	Value
λ	0.25	k	0.05
M	0.95	P_0'	$1\text{x}10^4$ MPa
G	$2.5\text{x}10^3$ kPa		

Table 2 – Model Deformation Properties

Figure 1 shows the temperature profile and the location of the frozen fringe can clearly been seen. Upon freezing the sinusoidal temperature profile changes rapidly at the 273K isotherm as the latent heat of fusion is released and the pore water changes phase to ice. This phenomenon is known as the 'zero curtain' where the temperature remains nearly constant at 273K between the unfrozen and frozen regions while the latent heat energy is being released. Once the latent heat has been released the temperature profiles abruptly changes due to ice having a lower specific heat capacity and much higher thermal conductivity. The reverse is shown as the sample thaws and the profile change is visible at the segregation freezing temperature (272K).

Heaving occurs due to the volumetric expansion of water in the pores as it changes phase to ice. Thaw settlement is this process in reverse. At this stage of development there is no mechanism for the migration of moisture to the freezing front so the calculated heave is solely due to the expansion of the fluid in the pores. Figure 1 also shows the heaving profile during the simulation and it can be seen that heaving only occurs while the temperature falls within the frozen fringe and latent heat is being released or absorbed. The magnitude of heave at all the points agrees with the quantity of heave expected when the volumetric expansion due to phase change is considered. As would be expected upon thawing the sample returns to its initial position.

Conclusion

In this study initial results of a thermo-hydro-mechanical numerical model have been presented to examine the coupled problem of freezing and thawing in porous media. While it is acknowledged that the model, at this stage, does not incorporate cryogenic effects such as water migration to the freezing front and ice segregation that are exhibited in the field the model does exhibit the 'zero curtain effect' as a result of the latent heat and the large energies involved with the change of phase from water to ice and vice versa. Numerically the problem is a challenging one due to both the frozen fringe and the rapid changes that occur in a spatially small area.

This paper presents the initial results from the model and it is acknowledged that the model is in early stages of development and the inclusion of additional cryogenic processes is required. Work is continuing on the development of the model to incorporate these processes.

Acknowledgements

The authors would like to thank Professor Charles Harris, Cardiff University, for the many informal discussions on the subject area and to gratefully acknowledge Engineering and Physical Sciences Research Council (EPSRC) support for the studentship of the third author.

References

Ballantyne, C.K and Harris, C. The Periglaciation of Great Britain. *Cambridge University Press,* 1994.

Gilpin, R. R. 1980. A model for the prediction of ice lensing and frost heave in soils. *Water Resources Research* 16(5), pp. 918-930.

Harlan, R. L. 1973. Analysis of coupled heat-fluid transport in partially frozen soil. *Water Resources Research* 9(5), pp. 1314-1323.

Harris, C. and Davies, M. C. R. 2000. Gelifluction: Observations from Large-Scale Laboratory Simulations. *Arctic Antarctic and Alpine Research* 32(2), pp. 202-207.

Harris, C. Davies, M.C.R. and Coutard J-P. 1997. Rates and Processes of Periglacial Solifluction: An Experimental Approach. *Earth Surface Processes and Landforms* 22(9), pp. 849-868.

Harris, C. Davies, M.C.R. and Rea, B.R. 2003. Gelifluction: Viscous Flow or Plastic Creep? *Earth Surface Processes and Landforms* 28(12), pp. 1289-1302.

Harris, C. Gallop, M. and Coutard, J-P. 1993. Physical Modelling of Gelifluction and Frost Creep: Some Results of a Large-scale Laboratory Experiment. *Earth Surface Processes and Landforms* 18(5), pp. 383.

Harris, C. Davies, M.C.R. and Coutard J-P. 1996. An Experimental Design for Laboratory Simulation of Periglacial Solifluction Processes. *Earth Surface Processes and Landforms* 21(1), pp. 67-76.

Hopke, S. W. 1980. A model for frost heave including overburden. *Cold Regions Science and Technology* 3(2-3), pp. 111-127.

Kirkby, M. J. 1995. A model for variations in gelifluction rates with temperature and topography: implications for global change. *Geografiska Annaler Series A* 77(4), pp. 269-278.

Konrad, J. M. and Morgenstern, N. R. 1981. The segregation potential of a freezing soil. *Canadian Geotechnical Journal* 18, pp. 482-491.

Matsuoka, N. 2001. Solifluction rates, processes and landforms: a global review. *Earth Science Reviews* 55(1-2), pp. 107-134.

Matsuoka, N. Hirakawa, K. Watanabe, T. Moriwaki, K. 1997. Monitoring of Periglacial Slope Processes in the Swiss Alps: the First Two Years of Frost Shattering, Heave and Creep. *Permafrost and Periglacial Processes* 8(2), pp. 155-178.

Mikkola, M. and Hartikainen, J. 2001. Mathematical model of soil freezing and its numerical implementation. *International journal for numerical methods in engineering.* 52: pp 543-557

Nixon, J. F. 1991. Discrete ice lens theory for frost heave in soils. *Canadian Geotechnical Journal* 28, pp. 843-859.

O'Neill, K. and Miller, R. D. 1985. Exploration of a rigid ice model of frost heave. *Water Resources Research* 21(3), pp. 281-296.

Selvadurai, A.P.S., Hu, J. and Konuk, I. 1999. Computational modelling of frost heave induced soil-pipeline interaction. *Cold regions science and technology*. 29: pp 215-228.

Shen, M. and Ladanyi, B. 1987. Modelling of coupled heat, moisture and stress field in freezing soil. *Cold regions science and technology*, 14: pp 237-246

Talamucci, F. 1997. Some recent mathematical results on the problem of soil freezing In: Knutsson, S (Ed) 1997. *Ground freezing and frost action in geotechnical engineering Lulea; Sweden: Balkema*, pp 179-186.

Taylor, G.S. and J.N. Luthin, 1978. A model for coupled heat and moisture transfer during soil freezing. *Canadian Geotechnical Journal*, 15: pp 548-555.

Thomas, H.R., He, Y., and Onofrei, C., 1998. An examination of the validation of a model of the hydro/thermo/mechanical behaviour of engineered clay barriers, *International Journal of Numerical and Analytical Methods in Geomechanics*, 22, pp 49-71.

Wood, D. M. 1990. Soil behaviour and critical state soil mechanics. *Cambridge University Press.*

Zienkiewicz, O.C., and Taylor, R.L., 1989. The finite element method, McGraw Hill, 4th edition.

Alluvial substratum influence on the main stream (thalweg) orientation and the stability of regulation structures based on the Middle Vistula River reaches

Tomasz Falkowski Department of Geotechnical Engineering, Warsaw Agricultural University, Warsaw, Poland

Abstract
Investigations carried out in the Middle Vistula River Valley reveal the occurrence of erosion-resistant culminations in the channel zone. The culminations are built of Mesozoic marls, gaizes and limestones, Tertiary cohesive deposits, as well as different glaciogenic deposits. Series of Cainozoic sediments are often disturbed glaciotectonically. The surface of the alluvia substratum is covered by residual lags, composed of gravels and boulders.
Echo-soundings carried out at different stages of the channel water show a constant trend of the main stream concentration as well as directions of flood flow streams, forced by culminations of the alluvia substratum morphology. This phenomenon is the reason of periodical break down of the regulation structures (bank protection, groynes).
Occurrence of culminations in the alluvial substratum is the reason of variability of contemporary alluvial facies.

Introduction
Studies carried out in the Middle Vistula River valley between Annopol and Modlin (Fig. 1) have indicated the occurrence of culminations of the alluvial substratum built of deposits more resistant to erosion in its main stream. In the southern part of the analysed reach, up to the vicinity of Puławy, the culminations are composed of firm Mesozoic and Palaeogene rocks – limestones, opokas and gaizes. Below Puławy, the alluvial substratum comprises Palaeogene and Neogene sands and clays (mainly Pliocene clays), as well as various glacial deposits. Cainozoic sediments are often glaciotectonically disturbed. The culminations are commonly covered with residual lag built of gravels and pebbles.

Culminations of the alluvial substratum, which are resistant to weathering, play the role of local erosion bases. They are also crucial for the orientation and morphology of the thalweg.

Aim of studies
The studies carried out in selected reaches of the Middle Vistula River valley were focused on:

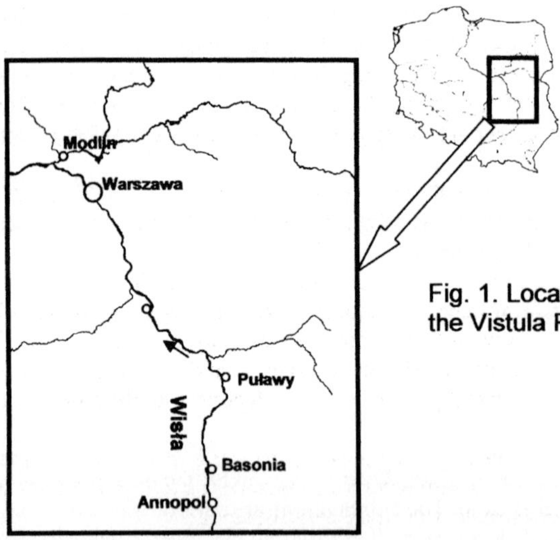

Fig. 1. Location of the analysed reach of the Vistula River.

- determining the relation between the morphology of the alluvial substratum and the orientation of the thalweg during medium and high stages, as well as the orientation of the flood water streams before regulation and construction of flood embankments;
- in zones with culminations of the alluvial substratum – recognising morphological features of the terrace surface and determining structural and textural features of contemporary deposits, which can facilitate the geological-engineering characteristics of the valley bottom environment; followed by formulating prospects on the stability of hydrotechnical structures.

In hydraulic engineering, the main aim in designing the course of a regulated channel is the choice of such parameters, i.e. width, depth, radius and character of curves, so as to assure the stability of erosion and accumulation processes and a required output of the channel for water and debris. Presently, most procedures applied are based on the relation between the geometry of the channel and the mean discharge in the river. Dębski (1968) and Przedwojski, Błażejewski and Pilarczyk (1996) present a review of the applied rules in designing the channel course.

The rule of balanced development applied in river engineering requires taking into account the natural elements stabilising the channel morphology in channel regulation projects or in the management of river valley bottoms. This should increase the efficiency of the regulation works as well as the durability of the regulation structures. Moreover, such actions should restrict human influence on the natural environment.

Zones, the presence of which due to their geological structure is crucial for the stability of the channel, can be found based on the geomorphological analysis of the terrace surface (Falkowski 1990). The role of geomorphological analyses of the fluvial environment in river engineering has been pointed out by Gilvear (1999), Florsheim and Mount (2003) and Eaton and Millar (2004).

Methodology of studies

The studies have been carried out along a 250 km reach of the Vistula River between Annopol and Modlin (Fig. 1), and included:

1. Geomorphological analysis of the terrace surface; its main aim was discovering traces of erosion by flood waters (relief channels or transformed series of oxbow lakes) on the

surface of the Holocene terrace. These traces represent zones with culminations of the resistant alluvial substratum, formed as a result of the impossibility to increase the cross-section of the active channel by its deepening during flooding (escarpment resistance).

Erosion by the stream of flood waters can be accompanied by deposition of the captured sediments. Therefore, alluvial fans should occur at the mouths of the relief channels below the escarpment.

The formation of relief channels and alluvial fans on surfaces of channels can also be linked with ice-jams. The presence of culminations of the resistant alluvial substratum does not allow the flow of the dammed water under the ice-jam.

2. Analysis of archival data referring to the occurrence of ice-jams in the analysed reach.
3. Drillings and soundings have been carried out in the channel of 18 zones selected after data analysis, each averagely 5-7 km long; samples of the alluvia and the alluvial substratum have been also collected for laboratory studies. These were focused on determining the physical-chemical properties of the deposits and the stratigraphic position of the beds.
4. Sedimentary structures in contemporary deposits exposed in the channel margins have been registered and analysed. These sediments were deposited by the river during high stages, often before the construction of flood embankments.
5. Echo-sounding of the channel at different water stages has been carried out in 18 selected reaches. An integrated device including an echo-sounder and a GPS receiver (DGPS) has been used in the studies, which enabled the measurement and registration of the coordinates of the measuring point and the depth of the channel at a selected time interval (measurements were taken every 3 seconds).
6. Archival data of the hydrogeological survey (RZGW), the administrator of each reach, referring to the operation failure of the regulation structures and the occurrence of zones with intense erosion processes in the channel have also been analysed.

Results of the study were presented in a GIS database.

Results

In the particular reaches, selected after analysis of archival data, the presence of culminations of the alluvial substratum has been observed. The top of resistant soils has been observed in these zones at depths between 5 and 10 m below mean water level, averagely at 7 m below mean water level. Outside the culmination zones the top of the alluvial substratum occurs at depths exceeding 15 m below mean water level. Regardless their lithology, the surface of the culminations is covered by residual lag. Above the lag, deposits of the flood facies are poorly dense. This phenomenon points to the exposure of the embankment surface during high stages.

Type of culminations

Two general types of culminations can be distinguished in the morphology of the alluvial substratum. They include troughs concentrating the stream of flood waters and bars causing its deflection. Two examples showing the morphology and lithology of the alluvial substratum culminations, as well as their influence on the channel deposits are presented below. The examples are from two reaches of the Vistula valley, within which the alluvial substratum is composed of rocky and non-rocky soils.

Reach in the vicinity of Basonia

In the vicinity of Basonia (Fig. 2), the rocky substratum of the contemporary alluvia forms a trough, concentrating the stream of high water.

In this locality, the Vistula valley is ca. 2 km wide. The plateau rising at 50 m above the valley bottom is built of solid rocks – Upper Campanian marls, opokas and gaizes, locally covered with glacial or aeolian deposits. At the mouths of valleys cutting the plateau margin and below the slopes occur zones of deluvial deposits and alluvial fans.

Within the valley bottom three structural elements can be distinguished (Fig. 2):

• channel zone of the contemporary Vistula River built of braided river deposits, along with flow zones of flood waters below the plateau;

• terrace zone of the meandering river, transformed by water of contemporary floodings (of the braided river);

• alluvial fan zone (10), influencing the orientation of the erosion trough of the flood flow (Fig. 2).

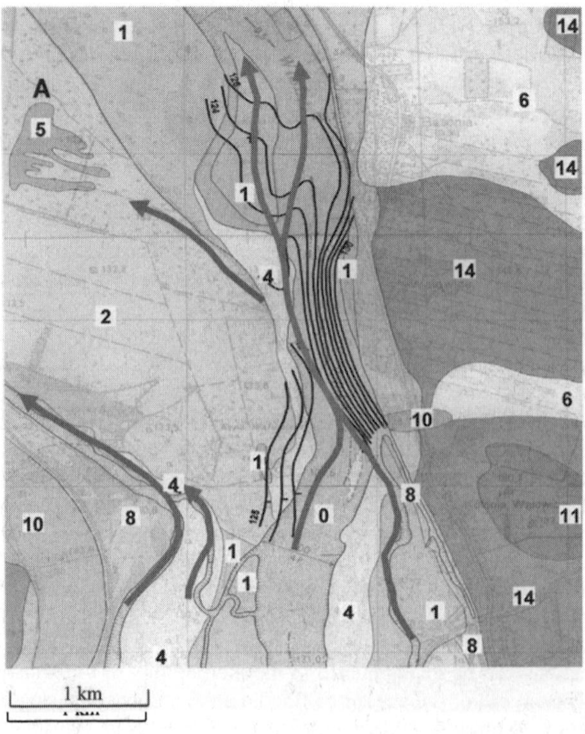

0 – river channel,
1 – silty-sandy muds of the contemporary river,
2 – silty-sandy muds on clayey muds of the meandering river,
3 – clayey muds of the meandering river,
4 – sands of the channel facies of the contemporary river,
5 – humus sands, alluvia and peats of the ox-bow facies
6 – aeolian sands,
7 – 'brown' silty-sandy muds of the braided river,
8 – other humus soils,
10 – slope deposits,
11 – glacial deposits,
14 – Upper Cretaceous opokas, gaizes and marls
arrows point the position of the main channel and the direction of flood waters flow;
A – alluvial fan of the flood waters.

Fig. 2. Isolines of the top of the alluvial basement and lithology of the surface deposits of a fragment of the Vistula reach in the vicinity of Basonia (Fig. 1) (fragment of GIS database);

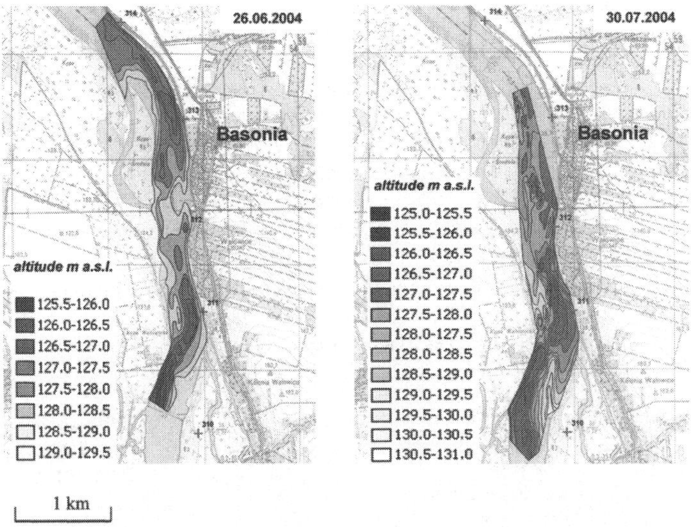

Fig. 3. Echo-sounding results of the Vistula channel in the vicinity of Basonia (fragment of GIS database).

In the analysed reach, the alluvial substratum comprises Upper Campanian deposits, mainly opokas and gaizes, covered by glacial till. Locally, the presence of residual lag could be observed on its surface, built of the substratum rocks and admixtures of other material from this part of the Vistula River catchment area.

The substratum surface forms here an asymmetric trough, shallowing northwards along the river course. This pattern of the main channel is confirmed by echo-soundings carried out at different water stages (Fig. 3). The flow directions of the main flood waters on the surface of the flood terrace can be reconstructed based on geomorphological analysis of its surface. These waters pass round an alluvial fan formed of deposits flowing from the plateau (10 on Fig. 2), flowing further between the plateau and a fragment of the terrace of the braided Vistula River. The surface of this element of the river valley bottom (2) is covered by an over 3.5 m thick complex of flood deposits, mainly silty clays, resistant to washing out. This stabilized fragment of the terrace surface is washed from the north-east by streams of flood water flowing from the culmination zone of the alluvial substratum towards the north-west. Numerous troughs on the terrace surface and an alluvial fan at their mouths are considered as traces of such flows (A, Fig. 2).

Reach in the vicinity of Magnuszew

In the vicinity of Magnuszew (Fig. 1), the alluvial substratum is composed of coarse-grained, alluvial pre-Pleistocene deposits – sands, gravels and sand-gravels with inserts of silty clays and silts. The deposit is extremely dense. Residual lags occur on its surface. The morphology of this culmination of the alluvial substratum causes blocking of high waters flow and their deflection towards the flood terrace westwards of the channel. Besides traces of erosion caused by different generations of flood waters, the terrace surface is covered by a vast zone of alluvial fans (A), formed due to decrease of flow rate of these waters beyond the culmination zone (Fig. 4). The studied reach of the Vistula River is also often ice-jammed (Falkowski and Popek 2000).

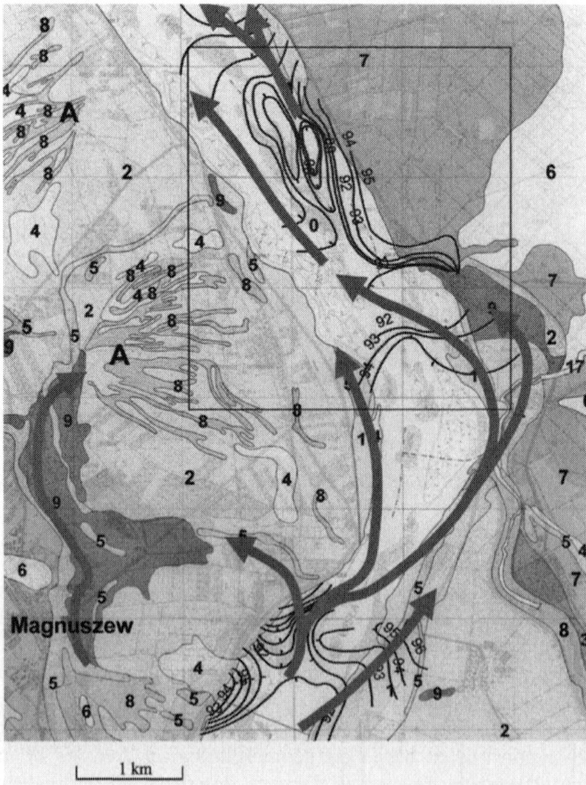

Fig. 4. Isolines of the top of the alluvial basement and lithology of surface deposits for fragment of the Vistula River reach in the vicinity of Magnuszew (location Fig. 1) (fragment of GIS database); explanations on Fig. 2; the rectangle marks the fragment shown on Fig. 5.

Sedimentary structures of the deposits

The influence of substratum morphology on the orientation of the main channel is marked also in the structural and textural features of contemporary alluvial deposits. In the vicinity of Magnuszew, in the zone where the stream of flood waters concentrates, the deposited sediments bear sedimentary structures typical of the channel environment – cross-stratified sands and sands with gravel (lithofacies Sp, SGp, St and SGt of Miall 1996). In zones of culminations of the alluvial substratum, where the resistances of flood water flow are higher, the resulting deposits bear sedimentary structures typical of the flood plain environment – sands, silts and clays (S, F) with massive, flaser and wavy structures (m, f, w) (Fig. 5). The variability of sedimentary structures can be used in forecasting the stability of hydrotechnical structures. It has been observed that in zones where the regulation structures lie beyond the zone of flood water concentration the sediments, the deposition of which is caused by

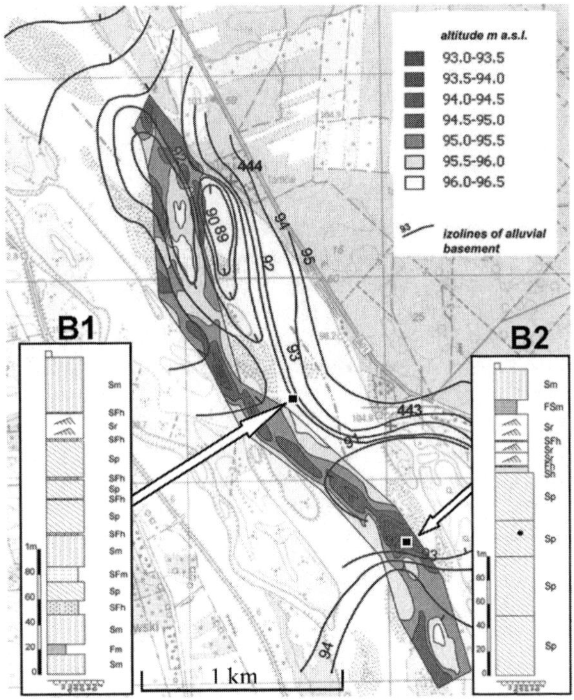

Fig. 5. Sedimentary structures of the contemporary Vistula deposits (exposures B1 and B2) in relation to the morphology of the alluvial basement and echo-sounding results.

the reaction of hydrotechnical structures, bear features characteristic of the flood plain environment. In zones where the flood waters concentrate and where failures of regulation structures often occur, sediments with structures typical of the channel environment are deposited during high stages.

The analysis of the channel morphology, based on aerial photographs from between 1990 and 1999, as well as the comparison of the channel morphology on numerous editions of topographic maps has shown that in zones of culminations of the resistant alluvial substratum occur stable depositional forms, which to a very low degree undergo transformation by flood waters.

Stability of the longitudinal channel cross-section

Culminations of the resistant alluvial substratum play also the role of local erosion bases. On longitudinal cross-sections, assembled from several series of echo-sounding measurements, these zones are marked by the stable location of its surface.

Artificial lowering of the embankment level, e.g. during regulation works, might cause reversed erosion and correction of the channel bottom (high water) to a new level of the local erosion base.

This phenomenon is presently also observed in the channel zone of Vistula River in Warsaw. It might be dangerous for the regulation structures, bridges and river water intakes.

Conclusions

1. The morphology of the alluvial substratum in zones of its culminations influences the orientation of the thalweg, as well as the flows of flood waters.
2. The durability of regulation structures in zones of alluvial substratum culminations depends to a large degree on the trend of the channel to form its main current concordantly with the substratum morphology.
3. The phenomenon of the main channel adaptation to the substratum morphology is generally linked with the activity of a braided river, characterised by large differences between high and low stages. The influence of the resistant substratum is the largest during high stages, when the depth of alluvial reworking reaches the top of the substratum, and the short duration of such stages does not allow the correction of its morphology.

It can be assumed that further increase of differences between the extreme flows presently observed in rivers of the Polish Lowlands (Ozga-Zielińska 1997) will fix and deepen this trend.

Acknowledgements

The presented studies have been carried out in the frame of the topic 'Relation of the dynamics of selected channel processes with the variable morphology and lithology of the alluvial substratum based on examples from the Vistula valley between Annopol and Modlin' (KBN Grant no. T07G 020 21).

References

Dębski K., 1978: River regulation, PWN, Warszawa.

Eaton B. C., Millar R. G., 2004: Optimal alluvial channel width under a bank stability constrain; Geomorphology 62, ELSEVIER, 35-45.

Falkowski E., 1990: Morphogenetic classification of river valleys developing in formerly glaciated areas for needs of mathematical and physical modeling in hydro technical projects. Geographia Polonica vol. 58, 55-67.

Falkowski T., Popek Z. 2000: Zones of ice-jams formation on the Middle Vistula River reach in relation to variable of river valley morphology; Annals of WAU, Land Reclamation No 30, 77-90.

Florsheim J. L., Mount J. F., 2003: Changes in lowland floodplain sedimentation processes: pre-disturbance to post-rehabilitation, Cosumnes River CA, Geomorphology 56, ELSEVIER, 305-323.

Gilvear D. J., 1999: Fluvial geomorphology and river engineering: future roles utilizing a fluvial hydrosystems framework; Geomorphology 31, ELSEVIER, 229-245.

Miall A. D., 1996: The geology of fluvial deposits, sedimentary facies, basin analysis and petroleum geology; Springer (Berlin, Heidelberg, New York), 1-582.

Ozga-Zielińska M., 1997: On the need to determine the maximal reliable floodings resulting from maximal reliable rainfall for Polish rivers. Forum naukowo-techniczne – POWÓDŹ 1997, IMGW Warszawa, tom 2, 1-10.

Przedwojski B., Błażejewski R., Pilarczyk K.W., 1995: River training techniques, fundamentals, design and applications; A.A. Balkema, pp 625.

Kinematic and Block Theory Applications to Rock Slope Stability Analysis at Fraser's Hill, Pahang, Malaysia

Haswanto, Department of Mining, University of Trisakti, Jakarta, Indonesia.
E-mail : haswanto_wa@yahoo.com
Ghani Rafek A., Department of Geology, UKM, Bangi, Malaysia.

Abstract

The granite rock mass at cut slope located at the start of the entry road to Fraser's Hill in Pahang, Malaysia is characterized by a number of major and minor discontinuities. The persistence of most of these major discontinuities is expected to be between 15 m and 50 m. These discontinuities have been mapped at the mentioned site within and area covering 50 m x 400 m. The major discontinuities were used to perform a kinematic and block theory analysis of the rock slope stability. The kinematic analysis has been used in order to find the stability of the granite rock for cutting the rock slope at Fraser's Hill. The block theory has been applied to identify the different block types found at the surface of researched area and to estimate the maximum safety angle of the rock slope at the researched area. The orientations of the major discontinuities that occur in the researched area have been considered in this analysis. Which the data have been taken. The data was analyzed by stereo-plot and then analyzed using kinematic to determine the maximum safe slope angle (MSSA) for the rock slope and block theory to determine the key blocks types (Type I) and potential key blocks types (Type II) of the rock slope. The main purpose of this study is to determine a safe angle for the rock slope at entry road cut slope Fraser's Hill.

Key-word : *Kinematic, Block Theory Analysis, Granite, Rock Slope Stability, Fraser's Hill.*

Introduction

The Fraser's Hill is the tourism region at Pahang, Malaysia. The granitic rock mass which exist in the Fraser's Hill region contains a number of major discontinuities, and number of set of minor discontinuities. Therefore, the rock engineering problems of high steep rock slope in the Fraser's Hill are very complicated. In the natural state, These rock slope are in a stable condition. When the rock-mass is excavated, for example for road, some blocks will develop sliding according to structural planes, and then a chain reaction may occur, which may lead the collapse of whole slope. The block theory is a new method to analyze rock mass stability with graph theory, set theory, geometry and vector algebra. The reason for the rock mass collapse is due to structural blocks sliding over each other at discontinuities.

The Block theory is used to identify different block types that exist at entry road cut slope Fraser's Hill faces and the main purpose of this study in to determine a safe angle for the rock slope at this location at entry road cut slope Fraser's Hill. Then analyzed using kinematic to determine the maximum safe slope angle for the rock slope.

Geology

Comprehensive geological research for the Fraser's Hill region has been in progress since 1951 and is still continuing. Part of the main range granite forms about 130 square miles of mountainous country, occupying the northeastern portion of the area shown us Fraser's Hill and extending eastwards into the Raub and Bentong districts of Pahang, Malaysia. The granite was probably introduced into an anticlinal complex of folds, the eroded crest of which is now partly exposed by removal of some of the sedimentary rocks. In the area under survey the granite is considered to be of post-Triassic age. Phorphyritic granite, generally coarse-grain and rich in biotite, is the commonest variety of granite in Main Range granite. (see Roe, 1951)

Figure 1. Fraser's Hill region for this studied

Major Discontinuities in The Fraser's Hill Region

These discontinuities have been mapped. A total fivety nine discontinuities have been at entry road cut slope Fraser's Hill area 50 x 400 m. Figure 2 and 3 shows the pole distribution of these discontinuities. High orientation variability of the major discontinuities is well depicted by the polar diagram. Predominant strike direction of the major discontinuities is along SW and NE directions. The persistence of most of these major discontinuities is expected to be between 15 m and 50 m. Most of them have steep dip angles varying between 60^0-90^0. The results of analysis stereographic projection (Ibrahim, and Ibrahim., 1983) for major discontinuities in dip-direction and dip can shown in Table 1. Figure 2 and 3. For the kinematic and block theory analysis, a somewhat conservative friction angle of 30^0 was assumed for the joint plane surface, base on field observations of surface roughness.

Table 1. Major Joints at entry road cut slope Fraser's Hill.

No	Joint	Dip-Direction	Dip
1	J_1	301^0	87^0
2	J_2	38^0	83^0
3	J_3	200^0	78^0
4	J_4	60^0	84^0

Note : $J_{1...n}$ = Joint set 1...n

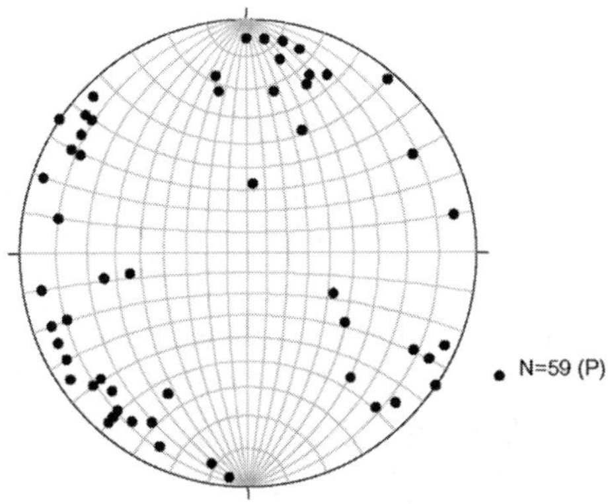

Figure.2. Orientasi distribution of major discontinuity poles at entry road cut slope Fraser's Hill

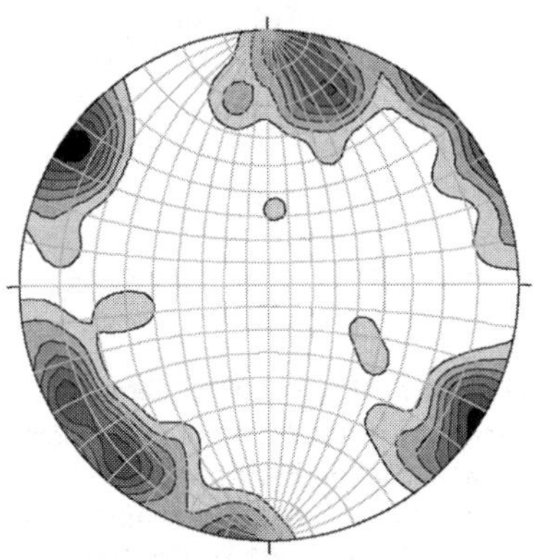

Figure. 3. Density stereogram of major discontinuity poles at entry road cut slope Fraser's Hill.

Kinematic Analyses

"Kinematic" refers to the motion of bodies without reference to the forces that cause them to move (Goodman, 1989). Many rock cuts are stable on steep slope even thought they contain steeply inclined planes of weakness with exceedingly low strength; this happens when there is no freedom for a block to move along the weak surface because other ledges of intact rock are in the way. Kinematic analyses are very useful to investigate possible failure modes of rock masses which contain discontinuities (Jeongi-gi Um and Kulatilake, 2001).

Failure involving movement of rock blocks on discontinuities combine one or more of the three basic modes: plane sliding, wedge sliding and toppling.

For the rock slope investigated at the entry road to Fraser's Hill, kinematic analyses was performed to estimate the maximum safe slope angles with respect to the aforementioned three basic failure modes. The basic concepts related to estimation of maximum safe slope angles for the three basic modes of failure are discussed by Goodman (1989).

Kinematic analysis will demonstrate that slopes of different strike have vastly different requirement for safety. By considering individually each of the element $D_{(1..4)}$, $I_{(1..3;\ 2...4)}$, $T_{(1...4)}$, maximum safe cut slopes were determined every 30^0 around at entry road slope Fraser's Hill. The lowest value of α for each of the mode of failure is retained as the maximum safe value for design. Table 2. show the slope values and the critical modes of failure all the way 30^0 around at entry road slope Fraser's Hill.

The Kinematic analysis of sliding consists of finding the dip of cut with given strike and containing I and D. And the shown these kinematic analysis in Table 2. For the strike for cut of slope at entry road Fraser's Hill is N80E and dip 75^0, and show the slope value and the critical mode are 54^0 and I_{23}. This analysis shows the slope is not safe.

This analysis shows the slope is not safe on the strike N, N30E, N60E, E, N30W, for the design slope $\alpha=75^0$.

Block Theory Analysis

The principal idea behind block theory analysis (Goodman and Shi, 1985; Goodman, 1995) is that it allows many different combinations of discontinuities to be passed over and to directly identify and consider critical rock block known as "key blocks". Types of blocks can be divided into infinite and finite blocks (see figure 4.). An infinite block (type V), as shown in Figure 4(a) is not dangerous as long as it is incapable of internal cracking. Finite blocks can be categorized into non-removable and removable blocks. Figure 4(b) is an example of type IV non-removable tapered blocks. It is finite, but it cannot come out to free space because of its tapered shape (Jeongi-gi Um and Kulatilake, 2001).

Finite and removable blocks can be separated into three categories, namely type III, type II, and type I, and identification of these blocks plays an important role in rock slope design. As shown in Figure 4(c), a type III block is stable without friction under gravity alone. A type II block which is in Figure 4(d) can be remained stable as long as the sliding force on the block is less than its frictional resistance. Under only gravitational loading, type II blocks are stable. Type II blocks are also called potential key blocks.

Finally, a key block which is denoted by type I shown in Figure 4(e) can slide into free space under gravitational loading without any external force unless a proper support system is provided. Therefore, the identification of key blocks is one of the most important parts in a rock slope stability analysis.

Table. 2. Result of the kinematic analysis for maximum safe slope angles for different failure modes for strike for cut of slope 30^0 around at entry road cut slope Fraser's Hill

No	Strike deg	Di\p deg	D_1 deg	D_2 deg	D_3 deg	D_4 deg	I_{12} deg	I_{13} deg	I_{14} deg	I_{23} deg	I_{24} deg	I_{34} deg	T_1 deg	T_2 deg	T_3 deg	T_4 deg	Mode	MSSA
1	N	E	90	85	90	86	88	90	89	**47**	90	73	90	90	90	90	I_{23}	47
2	N30E	SE	90	90	88	87	90	90	90	**43**	90	66	90	90	90	90	I_{23}	43
3	N60E	SE	90	90	83	90	90	87	90	**46**	90	65	90	90	90	90	I_{23}	46
4	E	S	90	90	79	90	90	81	90	**62**	90	81	90	90	90	90	I_{23}	62
5	S60E	SW	90	90	**78**	90	90	**78**	90	90	90	82	90	90	90	90	I_{13},D_3	78
6	S30E	SW	89	90	81	90	90	**77**	90	90	90	90	90	90	90	90	I_{13}	77
7	S	W	87	90	85	90	90	**80**	90	90	89	90	90	90	90	90	I_{13}	80
8	S30W	NW	87	90	90	90	87	87	87	90	**83**	90	90	90	90	90	I_{24}	83
9	S60W	NW	87	87	90	90	82	90	83	90	**80**	90	90	90	90	90	I_{24}	80
10	W	N	88	84	90	88	81	90	82	90	**79**	90	90	90	90	90	I_{24}	79
11	N60W	NE	89	81	90	85	**80**	90	81	90	**80**	90	90	90	90	90	I_{24},I_{12}	80
12	N30W	NE	90	82	90	84	82	90	84	**62**	84	84	90	90	90	90	I_{23}	62
13	**N80E** [(1)]	**75SE**	90	90	80	90	90	82	90	**54**	90	68	90	90	90	90	I_{23}	54

Note:
[(1)] is Strike of cut slope at entry road Fraser's Hill.
D = Plane Sliding I = Wedge Sliding; T = Toppling; α= Slope angle; MSSA = Maximum Safe Slope Angle

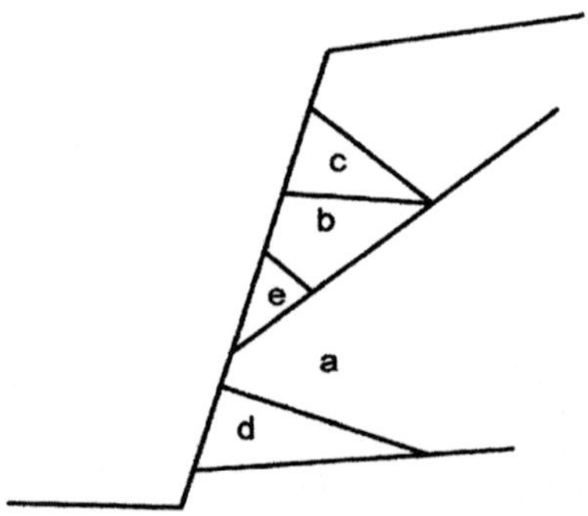

Figure 4. Blocks in a surface cut. (a) infinite; (b) tapered; (c) stable; (d) potential key block; (e) key block.

Results of Block Theory Analysis

As mentioned before block theory analysis was performed for one segment using the major discontinuities shown in table 1, that at least intersect investigated slope at the entry road to Fraser's Hill. Maximum safe slope angles (MSSA) were estimated with respect to possible single plane sliding, double plane sliding and free falling. The maximum safe slope angle (MSSA) corresponding to a possible mode of failure can be considered as the maximum dip of a cut slope which will not allow formation of a type I block under the mode considered. Analysis was done separately at entry road cut slope Fraser's Hill face for the strike is N80°E and dip is 75°.

First, removable blocks (types I, II and III) were identified for these segment. Then the mode analysis was performed to separate type III blocks from types I and II blocks. Out of all the segments analyzed, only two segments produced type I or type II blocks. Finally, type I blocks were separated from type II blocks for each segment analyzed.

Figure 5. shows the identification of removable blocks for the section at entry road cut slope Fraser's Hill using the stereographic projection. The orientations of the discontinuities and the excavation surface for the sections are listed in Table 3. When the dip of the excavation surface is between 0° and 50°, the Joint Piramids (JPs) 0100, provide removable blocks. The JPs 1000, provide removable blocks when the dip of the excavation surface is between 50° and 70°. For the dip of the excavation between 70° and 90°, the JPs 1000, and 1100 provide removable blocks (key block).

Table 3 shows the maximum safe slope angle (MSSA) for all identified types I and II blocks. This table also shows the maximum safe slope angle obtained for these segments that produced either a type I or type II block. It is important to note that when the cut slope angle is less than MSSA, both types I and II blocks become infinite (type V) blocks. To calculate the required support, limit equilibrium analysis can be used incorporating the selected slope value together with external forces.

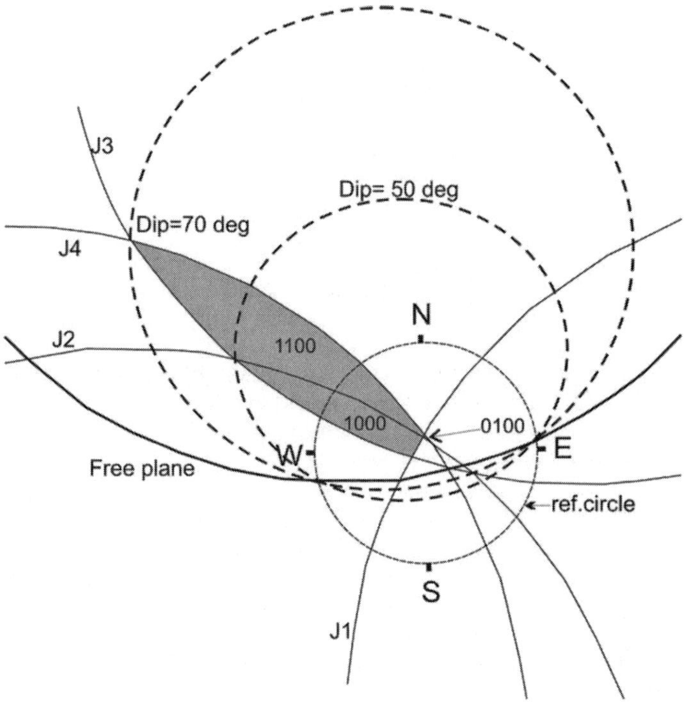

Figure 5. Identification of removable blocks for the Fraser's Hill using stereographic projection.

Table 3. Results of block theory analysis for strike N80^{0}E/dip 75^0

No	Dip of cut slope (deg)	Strike of cut slope	key block (type I or II)	Removal Block.	MSSA
1	0^0 - 50^0	N80^{0}E		0100	50^0
2	50^0 - 70^0	N80^{0}E	1000 (I)	0100	70^0
3	70^0 - 90^0	N80^{0}E	1000 (I); 1100 (II)		

Block theory analysis results three JPs. However, the results with respect to the sliding modes corresponding to type I blocks, sliding modes corresponding to type II blocks and MSSA would be exactly the same as that for the analysis using all four discontinuities together at one time. The only difference would be the individual blocks belonging to type I or type II. For types I or II blocks, to have a design slope angle greater than MSSA, it is necessary to perform a limit equilibrium analysis incorporating all the forces to estimate the required external rock support system.

Comparison Between the Results of Kinematic and Block Theory Analyses

Maximum safe slope angles (MSSA) for type I blocks almost correspond to the same sliding modes found in case of the kinematic analysis (compare Tables 2 and 3). Both methods show that the lowest maximum safe slope angle for type I blocks for the region investigated is 54° for kinematic analysis (I_{23}) and 50^0 for block theory (intersection J_2 with J_3). Type II blocks and MSSA is 70^0 (intersection J_3 with J_4) for the Block Theory and comparison between the results with MSSA is 68^0 for kinematic analysis (I_{34}).

MSSA can be 90°. According to the procedure used in the kinematic analysis, the dip vectors and lines of intersections having a plunge angle less than the friction angle produce maximum safe slope angles of 90°.

These blocks under kinematic analysis produce maximum safe slope angles less than 90° or exactly 43^0 to 83^0. It is important to note that kinematic analysis produce maximum safe slope angles less than 90° for the strike slopes every 30^0 around at entry road cut slope Fraser's Hill.

The primary advantage of block theory over traditional kinematic analyses is that it gives the ability to identify the keyblocks that require immediate attention. It separates the most important and dangerous blocks from the less critical ones.

Conclusions

Kinematic analysis produce maximum safe slope angles exactly 43^0 to 83^0. It is important to note that kinematic analysis produce maximum safe slope angles less than 90° for the strike slopes every 30^0 around at entry road cut slope Fraser's Hill.

For the block theory analyses, the result showed that the maximum safe slope angle (MSSA) is 50^0 for the type I (keyblock) and that the MSSA is 70^0 for the type II (potential keyblock) for the investigated slope.

It is noted that for the cut slope at the entry road to Fraser's Hill, the average cut slope angle is 75^0 within fresh to slightly wethered granite therefore this analysis shows the need for installation of a proper support system in order to maintain the long term stability of this rock slope.

References

Goodman R.E., dan Shi Gen-Hua., 1985, Block Theory and its application to rock engineering, Englewood Cliffs, NJ:Prantice-Hall.

Goodman R.E., 1989, Introduction to rock mechanics, 2nd ed. New York, Willey.

Goodman R.E., 1995, Block Theory and its application, Geotechnique, 45(3), 383-423.

Ibrahim Komoo, dan Ibrahim Abdullah., 1983, Ketakselanjaran dan kaedah pengukuran di lapangan, Sains Malaysiana, 12 (2), 119-140.

Jeongi-gi Um, Kulatilake P.H.S.W., 2001, Kinematic and block theory analyses for shiplock slope of the three gorges dam site in china, Geotechnical and Geological Eng. 19: 21-42.

Roe F.W., 1951, "The geology and mineral resources of the Fraser's Hill area Selangor, Perak, Pahang, Federation of Malaya, with an account of the mineral resources", Memoir No.5. Geology Survey Departement, Federation of Malaya.

Authors index